# 55th International Conference on Experimental Stress Analysis (EAN 2017)



Novy Smokovec, Slovakia
30 May – 1 June 2017

**Editors:**

**František Trebuňa**
**Jozef Bocko**
**Peter Frankovský**
**Róbert Huňady**
**Ján Kostka**
**Tomáš Kula**

ISBN: 978-1-5108-4224-3

**Printed from e-media with permission by:**

Curran Associates, Inc.
57 Morehouse Lane
Red Hook, NY 12571

**Some format issues inherent in the e-media version may also appear in this print version.**



Printed by Curran Associates, Inc. (2017)

For permission requests, please contact the Czech Society for Mechanics
at the address below.

Czech Society for Mechanics
Dolejskova 5
Praha 8
182 00
Czech Republic

Phone: +420 266 053 045
Fax: +420 286 587 784

csm@csm.cz

**Additional copies of this publication are available from:**

Curran Associates, Inc.
57 Morehouse Lane
Red Hook, NY 12571 USA
Phone: 845-758-0400
Fax: 845-758-2633
Email: curran@proceedings.com
Web: www.proceedings.com

EAN 2017
55th conference on experimental stress analysis

# Table of contents

## SECTION 1: Development of experimental methods in mechanics and biomechanics

## SECTION 2: New methods and applications of deformation and stress analysis in mechanical, civil and other structures

## SECTION 3: Experiment as a tool for the verification of analytical and numerical methods

## SECTION 4: Experimental research and prediction of strength, life and operational reliability of structures and devices

## SECTION 5: Monitoring of operational loadings and operational states of structures and devices

## SECTION 6: Modelling and design of mechatronic systems

# EXPERIMENTAL STRESS ANALYSIS 2017

## 55th International Scientific Conference

**May 30th - June 1st, 2017, Nový Smokovec, Slovakia**

EAN 2017
55th conference on experimental stress analysis

## Conference Proceedings

**(Full Text of Papers)**

**Edited by**

František Trebuňa, Jozef Bocko, Peter Frankovský
Róbert Huňady, Ján Kostka, Tomáš Kula

**Technical University of Košice – Faculty of Mechanical Engineering**

**Czech Society for Mechanics**

The editors did their best to ensure the printing quality of the proceedings. However, some manuscripts did not comply with the proceedings instructions. Therefore, the editors cannot accept responsibility for the final appearance of these manuscripts.

Peer Reviewed Proceedings

Title: Experimental Stress Analysis 2017
Conference Proceedings

Reviewers: František Trebuňa, František Šimčák, Jozef Bocko, Alexander Gmiterko, Michal Kelemen, Peter Trebuňa, Marián Buršák, Piotr Kuryło, Władysław Papacz, Peter Frankovský, Róbert Huňady, Miroslav Pástor, Ingrid Delyová, Ivan Virgala, Peter Sivák, Daniel Kottfer

Editors: František Trebuňa, Jozef Bocko, Peter Frankovský
Róbert Huňady, Ján Kostka, Tomáš Kula

Published by: Technical University of Košice,
Faculty of Mechanical Engineering

The complete set of EAN 2017 Conference Proceedings consist of two parts:

- Book of Extended Abstracts
- CD-ROM (Full Text of Papers)

# Program Committee

**Chairman**

- Dr.h.c. mult. prof. Ing. František Trebuňa, CSc. (Technical University of Košice, Slovakia)

**Members**

- prof. Ing. Jozef Bocko, CSc. (Technical University of Košice, Slovakia)
- prof. Ing. František Šimčák, CSc. (Technical University of Košice, Slovakia)
- prof. Ing. Štefan Segľa, CSc. (Technical University of Košice, Slovakia)
- prof. Ing. Michal Kelemen, PhD. (Technical University of Košice, Slovakia)
- prof. Ing. Alexander Gmiterko, PhD. (Technical University of Košice, Slovakia)
- prof. Ing. Peter Trebuňa, PhD. (Technical University of Košice, Slovakia)
- doc. Ing. Peter Frankovský, PhD. (Technical University of Košice, Slovakia)
- doc. Ing. Róbert Huňady, PhD. (Technical University of Košice, Slovakia)
- doc. Ing. Miroslav Pástor, PhD. (Technical University of Košice, Slovakia)
- doc. Ing. Daniel Kottfer, PhD. (Technical University of Košice, Slovakia)
- prof. Dr. Ing. Milan Sága (University of Žilina, Slovakia)
- prof. Ing. Milan Žmindák, CSc. (University of Žilina, Slovakia)
- prof. Ing. Jozef Melcer, DrSc. (University of Žilina, Slovakia)
- Ing. Michal Gregor, PhD. (University of Žilina, Slovakia)
- prof. Ing. Ján Vavro, PhD. (Alexander Dubček University of Trenčín, Slovakia)
- Dr.h.c. prof. Ing. Pavol Božek, PhD. (Slovak University of Technology, Trnava, Slovakia)
- prof. Ing. Ľubomír Šooš, PhD. (Slovak University of Technology in Bratislava, Slovakia)
- prof. Ing. Pavel Élesztös, PhD. (Slovak University of Technology in Bratislava, Slovakia)
- prof. Ing. Justín Murín, DrSc. (Slovak University of Technology in Bratislava, Slovakia)
- prof. Ing. Marián Peciar, PhD. (Slovak University of Technology in Bratislava, Slovakia)
- doc. Ing. František Duchoň, PhD. (Slovak University of Technology in Bratislava, Slovakia)
- prof. Ing. Stanislav Holý, CSc. (Czech Technical University in Prague, Czech Republic)
- prof. Ing. Milan Růžička, CSc. (Czech Technical University in Prague, Czech Republic)
- prof. Ing. Nikolaj Ganev, CSc. (Czech Technical University in Prague, Czech Republic)
- prof. Ing. Josef Jíra, CSc. (Czech Technical University in Prague, Czech Republic)
- Ing. Karel Doubrava, PhD. (Czech Technical University in Prague, Czech Republic)
- Ing. Pavel Padevět, PhD. (Czech Technical University in Prague, Czech Republic)
- prof. Ing. Vladislav Laš, CSc. (University of West Bohemia, Czech Republic)
- prof. Ing. František Plánička, CSc. (University of West Bohemia, Czech Republic)
- prof. RNDr. Miroslav Hrabovský, DrSc. (SLO UP & FZÚ AV ČR, Olomouc, Czech Republic)

# Organizing Committee

**Chairman**



**Members**

**Technical University of Košice**

**Faculty of Mechanical Engineering**

www.sjf.tuke.sk

# Department of Applied Mechanics and Mechanical Engineering

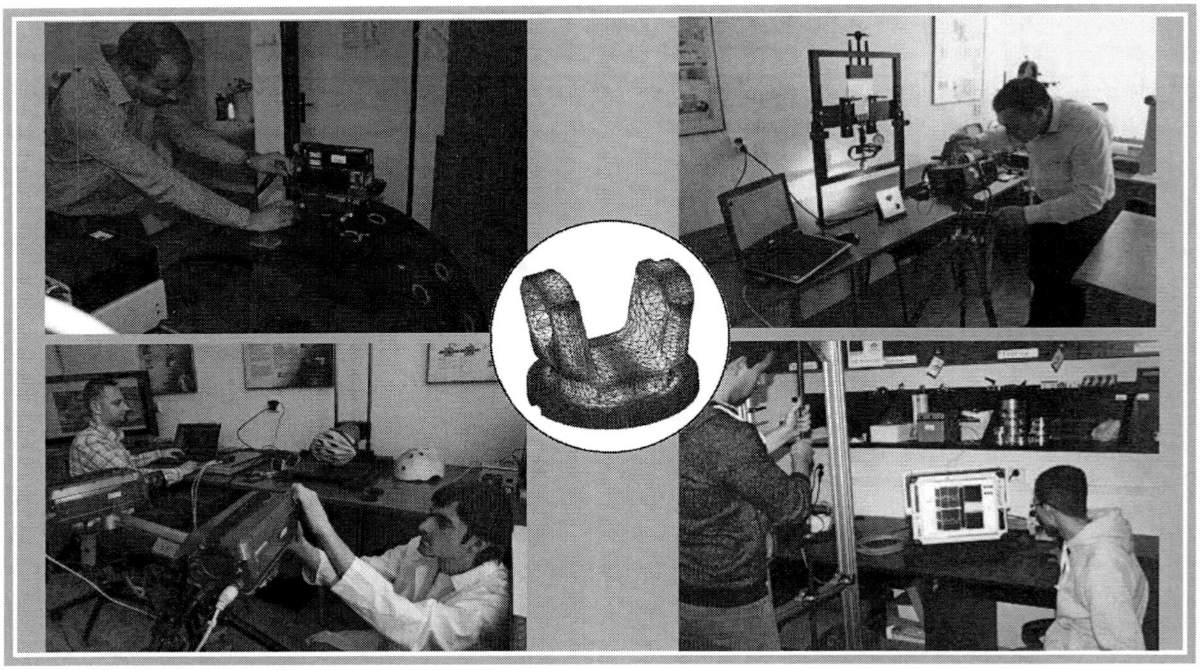

www.sjf.tuke.sk/kamasi/

## RESEARCH AT THE DEPARTMENT

**Area of research:**

- ✓ Development of non - traditional experimental methods for mechanical and mechatronic systems
- ✓ Development of computer methods and algorithms for numerical simulation and optimization of systems
- ✓ Methods of experimental and numerical modelling of mechanical systems
- ✓ Using of experimental methods for development of methodology for identification and prediction of failures in supporting elements of mechanical systems
- ✓ Stochastic processes and limit states of mechanical systems
- ✓ Development, verification and using of modern experimental methods of mechanics
- ✓ Measurement of residual stresses by the hole-drilling and Ring-Core methods
- ✓ Modal analysis of mechanical systems
- ✓ Measurement of deformation and stress states by the methods of photoelasticity
- ✓ Fatigue analysis and determination of life-span of mechanical systems
- ✓ Research of modules for intelligent robotic systems

# SECTION 1

# Development of experimental methods in mechanics and biomechanics

# CALIBRATION OF MODEL FOR LAMINATED GLASS POLYMER INTERLAYER BASED ON RHEOMETER DATA

Jaroslav Schmidt[1, a], Tomáš Janda[2,b] and Michal Šejnoha[3,c]

[1]Czech Technical University in Prague, Faculty of Civil Engineering,

Thákurova 7, 166 29 Praha 6, Czech Republic

[a]jarasit@gmail.com, [b]tomas.janda@fsv.cvut.cz, [c]sejnom@fsv.cvut.cz

**Keywords:** Laminated glass, rheometer, generalized Maxwell-chain model, calibration, fitting.

## Introduction

The application of glass as a structural element has shown a continuous rise recently. Owing to the introduction of laminated glass, its use expanded into building construction. Examples include roof and floor systems, columns, staircases etc. For this reason, it is necessary to apply design methods and material models for the prediction of the laminated glass behavior. In the elastic range glass is considered as a homogenous purely elastic material. The polymer interlayer has a more complex, time and temperature depend behavior and the experimentally obtained properties have crucial influence on the material model performance. The behavior of the plastic interlayer is assumed to follow the principles of viscoelasticity and is described by the generalized Maxwell chain model. Therefore, the material is characterized by a set of stiffness moduli $G_i$ and $G_\infty$ and viscous elements $\eta_i$, see Fig. 1.

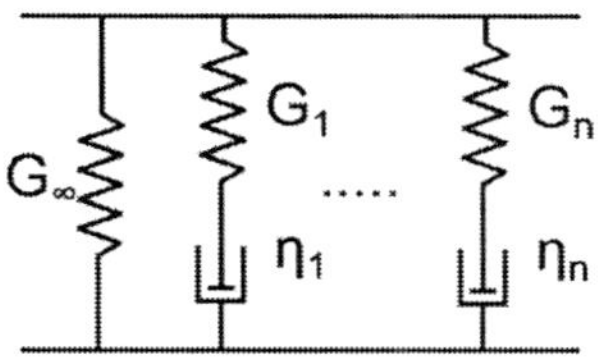


*Fig. 1: Generalized Maxwell chain model of the glass interlayer*

The temperature-depend behavior is described by the shift factor $a(T)$ expressed using the William, Landel and Ferry (WLF) equation [1]:

$$\log a(T) = \frac{-C_1(T - T_R)}{(C_2 + T - T_R)} \quad (1)$$

This relation introduces two additional parameters $C_1$ and $C_2$ into the model. All necessary details describing the calibration of parameters $G_i$, $G_\infty$, $C_1$ and $C_2$ for the chosen characteristic times of the Maxwell cells are provided below.

The advantage of the generalized Maxwell chain model reside in the description of the material response being straightforward both in the time and frequency domain. This property is exploited in the calibration procedure where the model parameters are fitted to the rheometer measurements in which the samples are exposed to the harmonic shear stress.

**Experiments**

The experiments were carried out in footstep of [2] in the dynamic shear rheometer HAAKE MARS, which operates on the plate-plate shear principle, see Figure 2. This device is mainly used for measurement of asphalt properties, therefore evaluating software calculate storage and loss moduli assuming that rotary displacement is linearly distributed over the height of the sample. This does not hold for laminated glass setup. Glass has generally higher stiffness than polymer interlayer and can be assumed perfectly rigid when evaluating the rheometer measurements. This means, that whole rotary displacement takes place in the interlayer. For this reason, the resulting moduli evaluated directly in the rheometer software must be transformed by equation

$$G_f^* = \frac{R_a^4}{R_s^4} \cdot \frac{h_f}{h_s} \cdot G_s^*, \tag{2}$$

which is based on the ordinary theory of elasticity and where $G_s^*$ is modulus from the software and $G_f^*$ is transformed moduli of the polymer. Other parameters are specified in Figure 3. Equation (2) also includes effect of different radius of the sample $R_s$ and radius of the adapter $R_a$.

*Fig. 2: Specimen glued to rheometer plates*

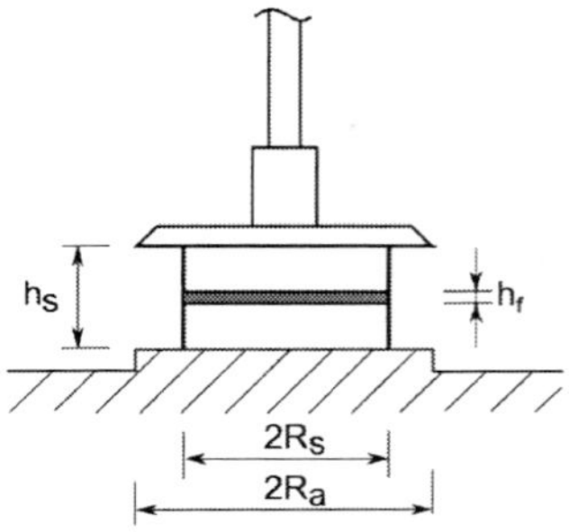


*Fig. 3: Rheometer setup for laminated glass*

The cylindrical samples drilled out of the laminated glass had the diameter of 20 mm and the thickness of 5+0.76+5 mm. For this specimens, height of glass layers was nearly equal, but value of interlayer height had a higher variability and moreover this height was variable with change of temperature during measurement. This fact is included in the calculation by equation (2). As for the interlayer, attention is limited to the application of EVA (ethylene-vinyl acetate) polymer ply. The samples were tested in pure shear dynamic excitation. The measurements were performed on 7 samples each assuming the temperature sweep of 10°C, 20°C, 30°C, 40°C, 50°C, 60°C and the frequency sweep of 0.001Hz, 0.01Hz, 0.05Hz, 0.1Hz, 0.5Hz, 1Hz, 5Hz, 10Hz, 20Hz, 30Hz, 40Hz, 50Hz for each sample. Frequencies in the range of 50Hz to 100Hz were also examined, the results however showed an inapplicably high degree of volatility. This is obvious from the sample graph displaying the variation of the storage modulus in Figure 3. Graph of loss modulus of the same sample is shown in Figure 5. Each sample was tested the same measurement scenario. First day specimen was glued to rheometer by high stiff epoxide glue to prevent any relative rotation between the sample and the adapter. The storage and loss moduli for each temperature over whole frequency domain was measured the following day, which took about 9 hours including the warming periods. The third and the fourth day this scenario was repeated without the measurements at frequency of 0.001Hz. The data for temperature 20°C for these three cycles are shown in Figure 6 and indicate then $G^*$ is not only time and temperature depend, but depends also on load history.

Other methods of experimental measurement of laminated glass interlayer properties is summarized in [4].

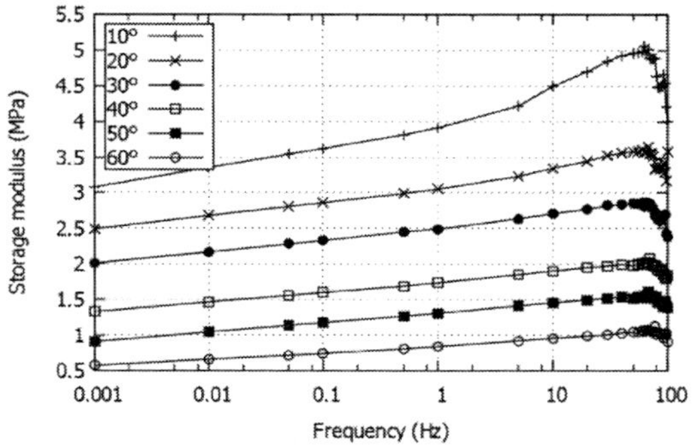


*Fig. 4: Storage modulus*

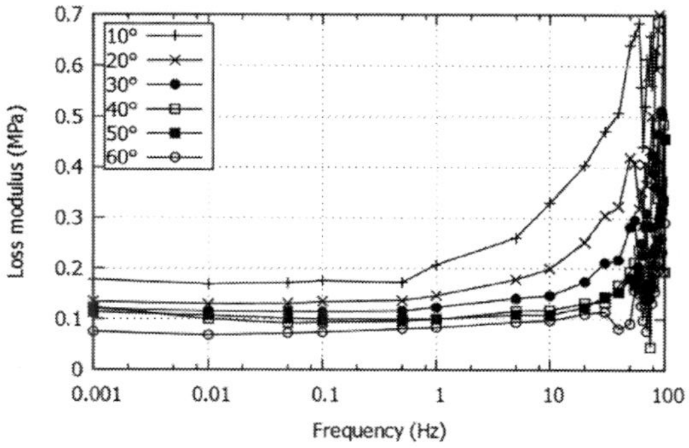


*Fig. 5: Loss modulus*

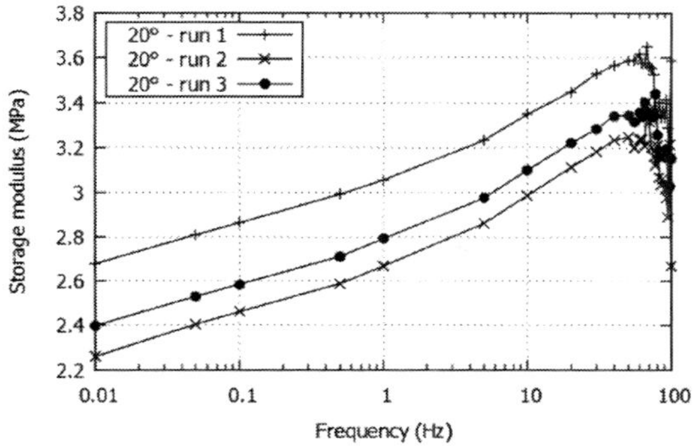


*Fig. 6: Storage modulus for temperature 20°C during three consecutive cycles*

## Calibration

The rheometer is able to measure discrete values of the storage and loss moduli for a specified set of frequencies and temperatures. The goal of the calibration procedure is to describe the components of the complex shear modulus $G^* = G' + iG''$ by two continuously curve in time and frequency domain. The generalized Maxwell-chain model is used here which fully describes these two curves by a series of the real shear moduli of the Maxwell cells $G_i$ and $G_\infty$ accompanied by the shift factor parameters $C_1$ and $C_2$. The calibration of the Maxwell model without temperature dependency is performed first since it lead to linear regression problem.

Storage and loss moduli for one temperature and $n$ Maxwell cells have following forms

$$G'(\omega) = G_\infty + \sum_{i=1}^{n} G_i \frac{\omega^2\tau_i^2}{\omega^2\tau_i^2 + 1},$$

$$G''(\omega) = \sum_{i=1}^{n} G_i \frac{\omega\tau_i}{\omega^2\tau_i^2 + 1},$$

in which unknowns are the set of shear moduli $G_i$ and $G_\infty$ and relaxation times $\tau_i$. Relaxation times are usually chosen so they cover the desired range of frequencies and are treated as constants during calibration. With this assumption, storage and loss modulus are linear in parameters $G_i$. For fitting is used common least square method. It is minimizing following square of residues

$$F(G_i, G_\infty) = r^2 = \sum_{j=1}^{m} \left(G'(\omega_j) - \overline{G}_j'\right)^2 + \sum_{j=1}^{m} \left(G''(\omega_j) - \overline{G}_j''\right)^2,$$

where $m$ is number of measured discrete points and $\overline{G}_j'$, $\overline{G}_j''$ is set of measurement. Minimizing conditions

$$\frac{\partial F}{\partial G_i} = 0, \frac{\partial F}{\partial G_\infty} = 0$$

leads to a system of linear equations. Implicit solution for unknown parameters was shown in [3].

For fitting a master curve is necessary to include temperature effect to the calibration. Certain temperature is chosen as reference (for this paper it is $T_r = 20°C$) and other curves are horizontal shifted in logarithm scale by WLF shift factor (recall eq. (1)). This bring nonlinearity to calibration and an iteration algorithm must be employed. The Gauss-Newton algorithm, developed by combining the Newton numerical method and the least square method, appears as a suitable method for this task. This method iteratively minimizes a nonlinear sum of squares of differences between the measured and computed moduli for given frequencies and temperatures. In particular, the function to be minimized attains the following form

$$F(G_i, G_\infty, C_1, C_2) = \sum_{j=1}^{m} \left(G'\left(\omega_j \cdot a(T_j)\right) - \bar{G}_j'\right)^2 + \sum_{j=1}^{m} \left(G''\left(\omega_j \cdot a(T_j)\right) - \bar{G}_j''\right)^2, \qquad (3)$$

where again $\bar{G}'$ and $\bar{G}''$ are the measured values. Parameters $G_i$ appear in the function of storage modulus $G'(\omega)$ and loss modulus $G''(\omega)$ in a linear form whereas parameters $C_1$ and $C_2$ appear in the WLF equation in a nonlinear form. In Gauss-Newton method it is favorable to rewrite problem in matrix form. First let us arrange fitting parameters to vector $\boldsymbol{\beta}$ with length $n + 3$, thus

$$\boldsymbol{\beta} = \{G_1, \dots, G_n, G_\infty, C_1, C_2\}^{\mathrm{T}}.$$

The residues are stored in vector $\boldsymbol{r}$ with length $2m$, having the form

$$\boldsymbol{r}(\boldsymbol{\beta}) = \left\{G'\left(\omega_j \cdot a(T_j)\right) - \bar{G}_j',\ G''\left(\omega_j \cdot a(T_j)\right) - \bar{G}_j''\right\}^{\mathrm{T}}.$$

With this notation, function to be minimized have new vector form

$$F(\boldsymbol{\beta}) = \boldsymbol{r} \cdot \boldsymbol{r}^{\mathrm{T}}$$

and minimizing condition is

$$\frac{\partial F}{\partial \boldsymbol{\beta}} = \mathbf{0}.$$

With vector notation, one iteration step in Gauss-Newton iterative method can be written as

$$\boldsymbol{\beta}^{(s+1)} = \boldsymbol{\beta}^{(s)} - (\mathbf{J}_s^{\mathrm{T}} \mathbf{J}_s)^{-1} \mathbf{J}_s^{\mathrm{T}} \cdot \boldsymbol{r}(\boldsymbol{\beta}^{(s)}),$$

where $\boldsymbol{\beta}^{(k)}$ is the parameter vector in $k$-th iteration and $\mathbf{J}_s$ is the Jacobian matrix, which is evaluated at every step by the expression

$$\mathbf{J}_s = \frac{\partial}{\partial \boldsymbol{\beta}^{(s)}} \cdot \left(\boldsymbol{r}(\boldsymbol{\beta}^{(s)})\right)^{\mathrm{T}}.$$

Form of Jacobian matrix in general form expressed analytically is placed in appendix.

Another possible approach is divide 2D state space of parameters $C_1$ and $C_2$ to discrete mesh. This state space is potentially infinite, so reasonable boundaries for these parameters must be predefined. In every node characterized by a pair $C_{1,i}$ and $C_{2,i}$ the optimization problem becomes linear and the linear least square method can be directly used. The nonlinear optimization problem then reduces to search of $C_1$ and $C_2$ only and can be managed by directly exploring a 2D parameter space. It seems to be difficult set ideal boundary to find appropriate pair of parameters, but fortunately objective function (3) remains almost constant in direction of constant ration of $C_1$ and $C_2$. Thus it is sufficient to find quasi minimum in adequately selected mesh space.

*Tab. 1: Fitted parameters*

| Parameter | Value | Parameter | Value |
|---|---|---|---|
| $G_1$ | 1377617.035 | $G_{12}$ | 114046.6687 |
| $G_2$ | 348848.0308 | $G_{13}$ | 269952.2981 |
| $G_3$ | 365819.0478 | $G_{14}$ | 156390.66 |
| $G_4$ | 222148.234 | $G_{15}$ | 184947.1187 |
| $G_5$ | 352833.5303 | $G_{16}$ | 194636.113 |
| $G_6$ | 301196.1553 | $G_{17}$ | 219579.655 |
| $G_7$ | 230536.3582 | $G_{18}$ | 107275.2708 |
| $G_8$ | 210105.2379 | $G_{19}$ | 242276.0251 |
| $G_9$ | 381157.4811 | $G_\infty$ | 901388.198110 |
| $G_{10}$ | 169933.1055 | $C_1$ | 272.830000 |
| $G_{11}$ | 231241.6111 | $C_2$ | 706.590000 |

These methods were used to calibrate the generalized Maxwell model and the WLF equation parameters given the experimentally obtained data from one specimen. Measurement was performed according to the scenario described in section Experiments. Output of this experiment are three sets of data, each for individual loading cycle. The calibration for first loading cycle, i.e. the first day measurement, is provided below. The reference temperature of 20°C, number of Maxwell cells $n = 19$ and the relaxation times $\tau_i = \{0.001, 0.01, 0.1, \dots, 1e14, 1e15\}$ were considered. The obtained values of parameters are shown in tab. 1 and the fitted storage modulus is illustrated in Fig. 7, showing the computed curve as a solid line and the original measured data by crosses. The measured data above 50Hz (decreasing branch of curve) are also shown in the figure, but they were not used during the calibration. The same results are shown in Figure (8) in form of a master curve. Crosses, which lies on solid curve, represent first loading cycle while the points under solid line are second loading cycle. It follows from this graph that the master curve can precisely represent data for specific loading cycle, but cannot include difference between loading cycles. If calibration is done for data from both loading cycles, the optimal master curve is slightly more compliant. The choice about which loading cycles should be used for the calibration depends mainly on the domain of application of the model.

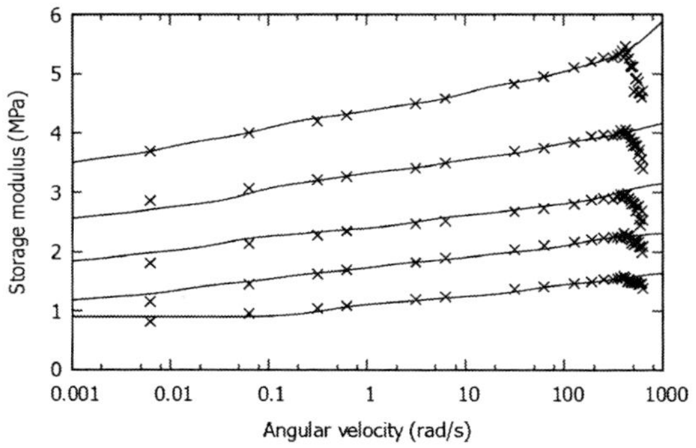

*Fig. 7: Fitted storage modulus*

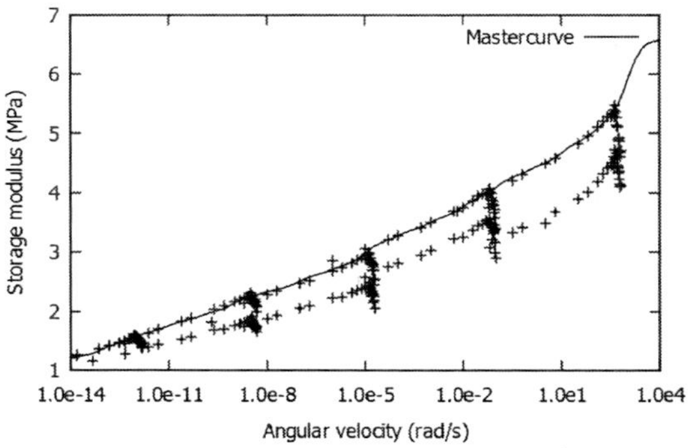

*Fig. 8: Fitted mastercurve for 1st run*

## Conclusions

The text briefly described the methodology for measuring viscous properties of laminated glass interlayer in rheometer apparatus. The paper also summarized the post processing of the measured data considering the layered configuration of the glass specimen. The data show that the measurements done in one cycle over all selected temperatures give relatively consistent results but repeating the measurement cycle on the same sample after 24 hours give less stiff response suggesting that the polymer interlayer has a complex load history dependence. The calibration of the viscoelastic generalized Maxwell chain model based on the measured data was presented. The linear problem of calibrating the model to certain fixed temperature curve becomes nonlinear when the measurements at more temperatures is considered and the calibration procedure extends to shift factor parameters as well. This is caused by nonlinearity of the William-Landel-Ferry equation. Two calibration strategies are described in the paper solving this nonlinear problem and the results of calibration of specific specimen with EVA polymer interlayer are provided. It is evident from the measured data that the shear modulus obtained at certain frequency and temperature generally decreases with rising number of measurements. This makes the calibration more difficult because calibrated parameters depend on which data is selected. In other words, the model can be fitted to one selected measurement cycle quite precisely but when all measurement cycles are utilized, the polymer interlayer behavior can be characterized with a single master curve only with certain degree of precision.

## Acknowledgement

The financial support provided by the GAČR project No. GA16-14770S and project SGS17/043/OHK1/1T/11 is gratefully acknowledged.

## References

[1] Malcolm L. Williams, Robert F. Landel, and John D. Ferry, The Temperature Dependence of Relaxation Mechanisms in Amorphous Polymers and Other Glass-forming Liquids, Journal of the American Chemical Society, vol. 77 (1955) 3701-3707

[2] L. Andreozzi et al., Dynamic torsion tests to characterize the thermo-viscoelastic properties of polymeric interlayers for laminated glass, Construction and Building Materials, vol. 65 (2014) 1-13

[3] T. Janda, A. Zemanová, J. Zeman, and M. Šejnoha, Finite element models for laminated glass units with viscoelastic interlayer for dynamic analysis, High Performance and Optimum Design of Structures and Materials II (2016)

[4] Zulli, F., Andreozzi, L., Giovanna, R. and Fagone, M., Test Methods for the Determination of Interlayer Properties in Laminated Glass, Journal of Materials in Civil Engineering, 2016

**Appendix**

Jacobian matrix have following form (matrix have size $(n+3) \times 2m$):

$$\begin{bmatrix} \frac{\partial r_1}{\partial G_1} & \cdots & \frac{\partial r_{2m}}{\partial G_1} \\ \vdots & \cdots & \vdots \\ \frac{\partial r_1}{\partial G_n} & \cdots & \frac{\partial r_{2m}}{\partial G_n} \\ \frac{\partial r_1}{\partial G_\infty} & \cdots & \frac{\partial r_{2m}}{\partial G_\infty} \\ \frac{\partial r_1}{\partial C_1} & \cdots & \frac{\partial r_{2m}}{\partial C_1} \\ \frac{\partial r_1}{\partial C_2} & \cdots & \frac{\partial r_{2m}}{\partial C_2} \end{bmatrix}$$

where for $j \in < 1; m >$

$$\frac{\partial r_j}{\partial G_i} = \frac{(a\omega_j)^2\tau_i^2}{(a\omega_j)^2\tau_i^2 + 1}$$

$$\frac{\partial r_j}{\partial G_\infty} = 1$$

$$\frac{\partial r_j}{\partial C_1} = \sum_{i=1}^{n} G_i \frac{2a\omega_j^2\tau_i^2 \cdot \frac{\partial a}{\partial C_1}}{\left(a^2\omega_j^2\tau_i^2 + 1\right)^2}$$

$$\frac{\partial r_j}{\partial C_2} = \sum_{i=1}^{n} G_i \frac{2a\omega_j^2\tau_i^2 \cdot \frac{\partial a}{\partial C_2}}{\left(a^2\omega_j^2\tau_i^2 + 1\right)^2}$$

where for $j \in < m+1; 2m >$

$$\frac{\partial r_j}{\partial G_i} = \frac{a\omega_j\tau_i}{(a\omega_j)^2\tau_i^2 + 1}$$

$$\frac{\partial r_j}{\partial G_\infty} = 0$$

$$\frac{\partial r_j}{\partial C_1} = \sum_{i=1}^{n} G_i \frac{\omega_j\tau_i \cdot \frac{\partial a}{\partial C_1} \cdot \left(1 - a^2\omega_j^2\tau_i^2\right)}{\left(a^2\omega_j^2\tau_i^2 + 1\right)^2}$$

$$\frac{\partial r_j}{\partial C_2} = \sum_{i=1}^{n} G_i \frac{\omega_j\tau_i \cdot \frac{\partial a}{\partial C_2} \cdot \left(1 - a^2\omega_j^2\tau_i^2\right)}{\left(a^2\omega_j^2\tau_i^2 + 1\right)^2}$$

and where

$$\frac{\partial a}{\partial C_1} = \exp\left(\frac{-C_1(T - T_R)}{(C_2 + T - T_R)}\right) \cdot \frac{T_R - T}{C_2 + T - T_R}$$

$$\frac{\partial a}{\partial C_2} = \exp\left(\frac{-C_1(T - T_R)}{(C_2 + T - T_R)}\right) \cdot \frac{C_1(T - T_R)}{(C_2 + T - T_R)^2}$$

EAN 2017
55th conference on experimental stress analysis

# The issue of recommended hole diameters in ASTM E837-13a

Dávid Halabuk[1, a], Tomáš Návrat[1,b] and Miloš Vlk[1,c]

[1] Faculty of Mechanical Engineering, Brno University of Technology, Technická 2896/2, 616 69 Brno, Czech Republic

[a]david.halabuk@vutbr.cz, [b]navrat@fme.vutbr.cz, [c]vlk@fme.vutbr.cz

**Keywords:** residual stress, hole-drilling method, hole diameters, ASTM E837-13a, EVAL-7

**Abstract.** The paper focuses on assessment of method for non-uniform residual stress calculation published in ASTM E837-13a for case of various hole diameters. The method is an integral method which uses calibration coefficients calculated by finite element method for strain gauge rosette 1/16" and hole diameter of 2 mm. In case of using different hole diameter these calibration constants have to be recalculated using simple formula. The aim of the study was to verify this method of recalculation. Hole-drilling experiment was simulated by using finite element method and residual stresses were evaluated from obtained strains by EVAL-7 software. Simulations were done for 6 different hole diameters. From obtained results it was shown that relative error increases with varying diameter due to recalculation of calibration coefficients. This error can reach values which significantly devalue results of measured residual stresses.

## Introduction

Residual stress included in material can be significant factor during loading, because it decreases or increases total stress in a component. One of the most popular and widely used technique for measurement of residual stresses is the hole-drilling method [1]. This method is based on drilling a small hole to a measured component and this causes redistribution of residual stresses around it. After that, relaxed strains are measured by strain gauge rosette and residual stresses are evaluated from these measured strains. Accuracy of evaluated residual stresses depends on many factors, e.g. eccentricity of drilled hole, plasticity effect, process of hole drilling, accuracy of measurement, etc. [1]. One of the significant factors is also a diameter of a drilled hole. The choice of an appropriate diameter of a drilled hole is important because a larger hole diameter leads to bigger strain relaxation, what decreases inaccuracy in calculation of residual stress values, but the use of large diameter can also cause damage of strain gauge rosette [1].

## Methods

**ASTM E837-13a standard.** Residual stresses can be evaluated by EVAL-7 software developed by SINT Technology. This software contains few different methods for evaluating residual stresses, but we focused on ASTM E837-13 Non-Uniform calculation method, especially on its extended version, which contains wide range of usable strain gauge rosettes [2]. ASTM E837-13 Non-Uniform calculation method is integral method which means that in every j-th depth increment relaxed strains are measured by threes gauge grids A, B, C and expressed as

$$p_j = \frac{\varepsilon_A + \varepsilon_C}{2} = \frac{1+\mu}{E}\sum_{k=1}^{j} \bar{a}_{jk} P_k \quad (1)$$

$$q_j = \frac{\varepsilon_A - \varepsilon_C}{2} = \frac{1}{E}\sum_{k=1}^{j} \bar{b}_{jk} Q_k \quad (2)$$

$$t_j = \frac{\varepsilon_A + \varepsilon_C - 2\varepsilon_B}{2} = \frac{1}{E}\sum_{k=1}^{j} \bar{b}_{jk} T_k \quad (3)$$

where $p_j, q_j, t_j$ are measured strain relaxations after the j-th hole depth increment, $\varepsilon_A, \varepsilon_B, \varepsilon_C$ are strain relaxations measured by gauge grids A, B, C, $\mu$ is Poisson's ratio, E is elastic modulus, $\bar{a}_{jk}, \bar{b}_{jk}$ are strain relaxations due to a unit stress within increment k of a hole j increments deep, $P_k, Q_k, T_k$ are equivalent uniform stresses within the k-th hole depth increment [3,4]. Residual stresses within the k-th hole depth increment are

$$P_k = \frac{\sigma_{xk} + \sigma_{yk}}{2} \quad (4)$$

$$Q_k = \frac{\sigma_{xk} - \sigma_{yk}}{2} \quad (5)$$

$$T_k = \tau_{xyk} \quad (6)$$

where $\sigma_{xk}$, $\sigma_{yk}$ are normal stresses in x, y directions and $\tau_{xyk}$ is shear stress in xy direction within the k-th hole depth increment [3,4]. Value of principal stresses and their orientation can be found by solving system of Eq. 1 - 6, where $\bar{a}_{jk}, \bar{b}_{jk}$ are matrixes of calibration coefficients obtained by simulation of hole-drilling using finite element method [3,4].

ASTM E837-13a standard presents two types of residual stresses: uniform, where stresses are uniform in depth direction and non-uniform, where stresses vary with depth direction [3]. In practice, there is usually not relevant information about type of residual stresses in depth direction and it is more common, due to technological operations, that residual stresses are non-uniform. ASTM E837-13a standard also presents three types of strain gauge rosettes, which are shown in Fig. 1 [3]. The most commonly used strain gauge rosette is type A and its basic dimensions are shown in Table 1, where D is a diameter of a gauge circle, L is length of a gauge grid, W is width of a gauge grid and $D_1$ is a diameter of a circle on which gauge grids inner edges lie.

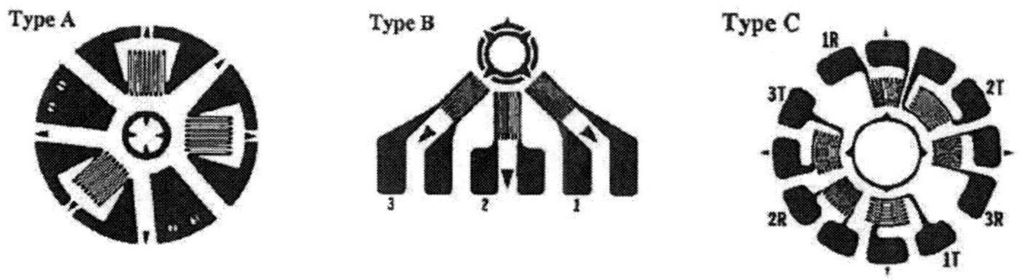


Fig. 1 Strain gauge rosette for hole drilling published in ASTM E837-13a standard [3]

Table 1 Basic dimensions of strain gauge rosettes type A [3]

| | D [mm] | L [mm] | W [mm] | $D_1$ [mm] | $D_1/D$ [mm/mm] |
|---|---|---|---|---|---|
| 1/32" | 2.57 | 0.79 | 0.79 | 1.77 | 0.691 |
| 1/16" | 5.13 | 1.57 | 1.57 | 3.56 | 0.694 |
| 1/8" | 10.26 | 3.18 | 3.18 | 7.08 | 0.690 |

ASTM E837-13a standard publishes only recommended minimal and maximal hole diameters, which differ for uniform and non-uniform residual stresses [3]. It is not clear if these recommended limits are also related to another methods evaluating non-uniform residual stresses and how these limits were obtained. Values of minimal ($D_{0,min}$) and maximal ($D_{0,max}$) hole diameters for strain gauge rosette type A are shown in Table 2.

Table 2 Minimal and maximal hole diameters for strain gauge rosette type A for uniform and non-uniform residual stresses [3]

| | uniform | | non-uniform | |
|---|---|---|---|---|
| | $D_{0,min}$ [mm] | $D_{0,max}$ [mm] | $D_{0,min}$ [mm] | $D_{0,max}$ [mm] |
| 1/32" | 0.61 | 1.01 | 0.93 | 1.00 |
| 1/16" | 1.52 | 2.54 | 1.88 | 2.12 |
| 1/8" | 3.35 | 5.59 | 3.75 | 4.25 |

We can generalize values of recommended hole diameters also for another similar strain gauge rosette type with linear regression as
-uniform residual stresses

$$D_{0,min} = 0.356 \cdot D - 0.307 \tag{7}$$

$$D_{0,max} = 0.595 \cdot D - 0.518 \tag{8}$$

-non-uniform residual stresses

$$D_{0,min} = 0.366 \cdot D - 0.007 \tag{9}$$

$$D_{0,max} = 0.422 \cdot D - 0.067 \tag{10}$$

Matrixes of calibration coefficients published in ASTM E837-13a standard were calculated by finite element method only for 1/16'' size type A strain gauge rosette and hole diameter of 2 mm [3]. Published calibration coefficients depend on hole depth and stress depth, which have to be multiplied by 0.5 for case of 1/32'' rosette and by 2 for case of 1/8'' rosette [3]. In case of another hole diamter, ASTM E837-13a standard provides a recalculation formula

$$(\bar{a}_{jk})_{D_0} = (\bar{a}_{jk})_P \left(\frac{D_0}{2}\right)^2 \tag{11}$$

$$(\bar{b}_{jk})_{D_0} = (\bar{b}_{jk})_P \left(\frac{D_0}{2}\right)^2 \tag{12}$$

where $(\bar{a}_{jk})_{D_0}$ and $(\bar{b}_{jk})_{D_0}$ are calculated calibration coefficient, $(\bar{a}_{jk})_P$ and $(\bar{b}_{jk})_P$ are calibration coefficients for a drilled hole of 2 mm, and $D_0$ is a diameter of a drilled hole [3].

**Simulation.** Hole-drilling experiment was simulated using the finite element method. By this process, we could avoid adding additional errors due to uncertainty of material properties, technology-related factors (the shape and blunt of the tool, additional stresses created by drilling, etc.) and process of measurement. Parametric model, which allowed easy change of important parameters (geometry, material properties, boundary conditions), was created in ANSYS software. The model is based on the "hole after residual stress" procedure which prescribes residual stress to the solid and relaxed strains are measured after removal of material from the hole. The hole-drilling process was simulated by progressive removal of individual layers of elements. The total number of removed layers and therefore increments of depth was 20.

The model used in calculation consists of thick solid body with uniform uniaxial tension in depth direction. Due to the symmetry of the model, the simulated model consists only of one quarter of the whole geometry. Fig. 2 depicts the geometry with boundary conditions (left) and the finite element mesh around a drilled hole (right) used in simulations.

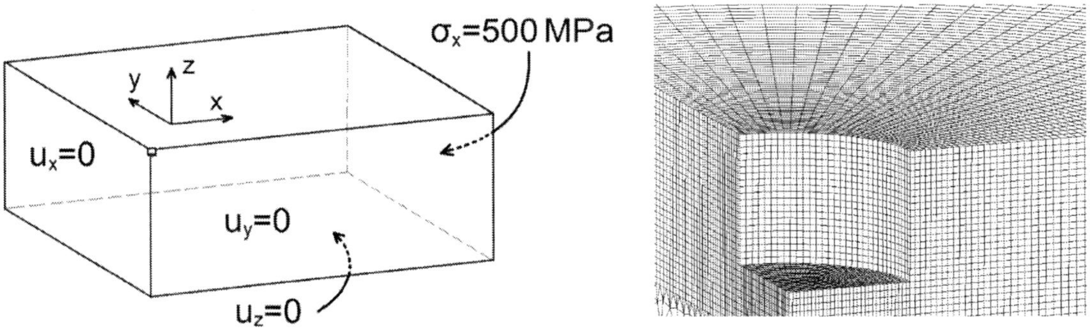


Fig.2 Model of geometry with boundary conditions (left) and the finite element mesh around the drilled hole (right)

For measurement of relaxed strains, a strain gauge rosette 1-RY61-1,5/120S from company HBM was considered. This rosette has diameter of gauge circle D = 5.1 mm, width of gauge grid W = 0.7 mm and length of gauge grid L = 1.5 mm [5]. The relaxed strains were obtained from simulation by averaging nodal strains across the virtual strain gauge grid surface at progressive removal of hole layers. Three strain gauge grids were used; one in principal direction and remaining two rotated by 45 and 90 degrees from principal direction of residual stress. Computational simulations were done for six different hole diameters: 1.6, 1.7, 1.8, 1.9, 2.0 and 2.1 mm. Depth of all holes was 1 mm. Recommended hole diameters for the considered rosette calculated by Eq. 7 - 10 are: $D_{0,min}$ = 1.51 mm, $D_{0,max}$ = 2.52 mm for uniform stresses and $D_{0,min}$ = 1.86 mm, $D_{0,max}$ = 2.09 mm for non-uniform stresses. For study reasons there is a wider range of hole diameters in our simulations than it is recommended for non-uniform residual stress by ASTM E837-13a standard.

Eccentricity of drilled hole and plastic deformations were not considered in simulation since evaluation of results depended mainly on hole diameters. The bilinear model of material with kinematic hardening was used. Properties of the material were the following: elastic modulus $E = 210$ GPa, tangent modulus $E_t = 0$ GPa, Poisson's ratio $\mu = 0.3$ and yield stress $\sigma_Y = 2100$ MPa. Yield stress was considered so high in order to prevent creation of plastic deformations.

Strains obtained from computational simulations were used as input for EVAL 7 software and ASTM E837-13 Extended Non-Uniform calculation method was used for evaluation of residual stresses. A non-uniform method was used because, in practice, it is common that the residual stress distribution is not known and therefore non-uniform residual stresses in depth direction are assumed. EVAL-7 software contains several calculation methods, so strains obtained from simulations were also evaluated by Integral, HDM and ASTM E837-13 Extended Uniform calculation method [2].

Absolute error of equivalent residual stress was calculated as

$$\Delta = \widehat{\sigma}_{eq} - \sigma_{eq} \tag{13}$$

where $\widehat{\sigma}_{eq}$ is equivalent evaluated residual stress and $\sigma_{eq}$ is equivalent simulated residual stress. Equivalent stress was evaluated according to HMH theory. Relative errors were calculated using following equation

$$\delta = \frac{\Delta}{\sigma_{eq}} \cdot 100 = \frac{\widehat{\sigma}_{eq} - \sigma_{eq}}{\sigma_{eq}} \cdot 100\ [\%] \tag{14}$$

**Results and discussion**

Dependences of relative errors on stress depth, evaluated by ASTM E837-13 Extended Non-Uniform calculation method, for different hole diameters are shown in Fig. 3.

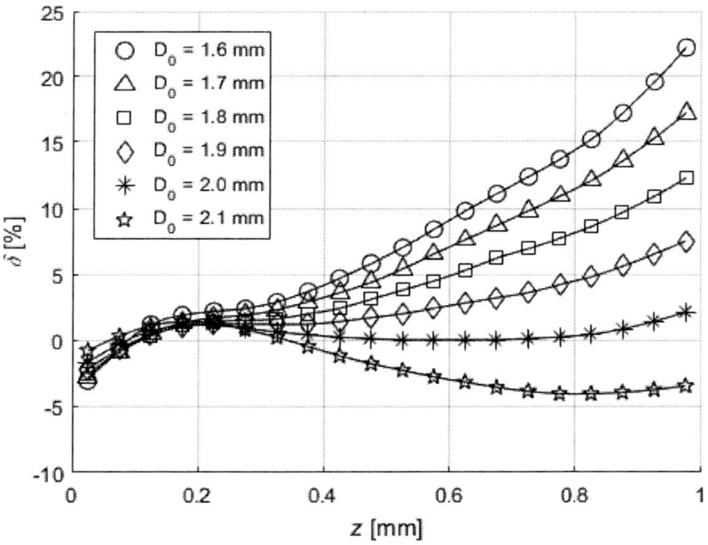


Fig. 3 Relative errors of residual stresses evaluated by ASTM E837-13 Extended Non-Uniform calculation method

Relative errors of evaluated residual stresses in case of drilled hole diameter of 2 mm are small with maximum value 2.1 %. For different hole diameters, relative error increase with increase or decrease in diameter size compared to hole diameter of 2 mm.
These errors, which are a consequence of recalculation process, can reach values which significantly devalue results. ASTM E837-13a standard published range of recommended hole diameters, however from the obtained results, due to high error that occurs for diameters out of this range, these hole diameters should be rather presented as acceptable, not recommended. The choice of range of recommended hole diameters for this method relates to preservation of relative errors in acceptable values.
Relative errors evaluated by ASTM E837-13 Extended Uniform calculation method were very small and for hole diameters of 1.6 - 2.1 mm varied from -0.04 to 0.53 %. Recommended hole diameters for this method are given by requirements for sufficient relaxed strains and for exclusion of strain gauge rosette damage.
Residual stresses were also evaluated by Integral and HDM method. Dependence of relative errors on stress depth evaluated by these methods for different hole diameters are shown in Fig. 4 and Fig. 5.

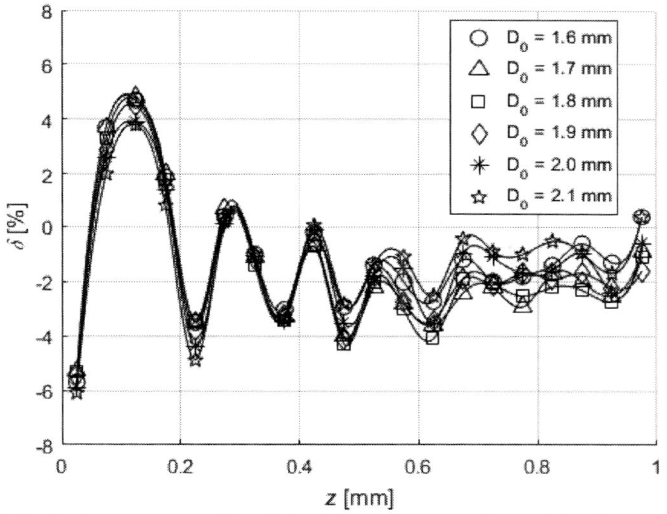


Fig. 4 Relative errors of residual stresses evaluated by Integral method

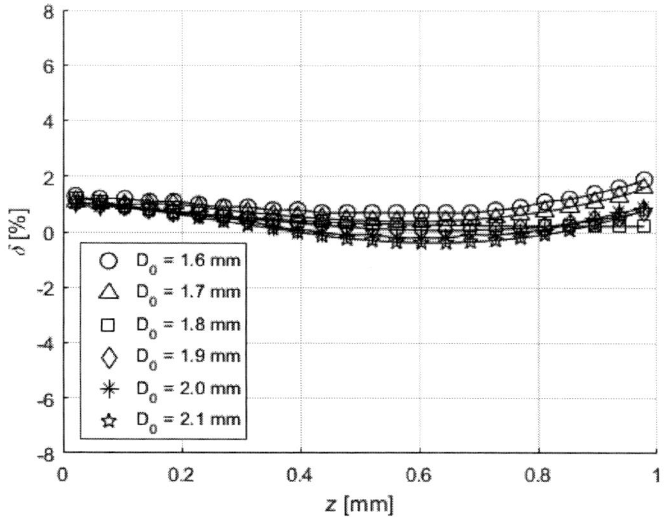


Fig. 5 Relative errors of residual stresses evaluated by HDM method

For both methods, Integral and HDM, errors don't significantly vary with the change of hole diameter and total errors are small. The range of usable hole diameters for use of these two non-uniform calculation methods is wider than it is published for non-uniform residual stresses by ASTM E837-13a standard.
Because of small errors that occur when evaluating by Integral, HDM and ASTM E837-13 Extended Uniform calculation method, we assume that recommended hole diameters don't relate to type of residual stresses in depth direction but to a method which is used for evaluation.

## Conclusion

Recalculation of calibration coefficients, from their original values for a hole diameter of 2 mm, using formula described in ASTM E837-13a, introduce additional errors into results. These errors can reach values which significantly devalue results of measured residual stresses. It has been also found that the range of recommended hole diameters for non-uniform residual stresses published in ASTM E837-13a is not generally valid for all non-uniform methods, but only for ASTM E837-13a non-uniform calculation method because of recalculation process. Another non-uniform methods (Integral, HDM), included in EVAL 7, have wider range of usable hole diameters than it is published in ASTM E837-13a for non-uniform residual stresses.

## Acknowledgment

This work is an output of project NETME CENTRE PLUS (LO1202) created with financial support from the Ministry of Education, Youth and Sports under the „National Sustainability Programme I".

## References

[1] Schajer, G.S., Practical Residual Stress Measurement Methods, 1st ed. West Sussex: Wiley, 2013. ISBN 978-1-118-34237-4

[2] SINT Technology EVAL – Software for Measuring Residual Stress by the Hole Drilling Method EVAL Back Calculation Software Manual, 2014.

[3] ASTM Designation E-837-13a Standard Test Method for Determining Residual Stresses by the Hole-Drilling Strain Gauge Method.

[4] Schajer, G. Measurement of Non-Uniform Residual Stresses Using the Hole-Drilling Method. Part I—Stress Calculation Procedures. Journal of Engineering Materials and Technology, 110(4), p. 338-343, 1988.

[5] HBM, Strain Gauges and Accessories, (company cataloge of products).

# Determination of Fracture Toughness of Human Enamel by Nanoindentation Based on Energy Release Method

HÁJKOVÁ Petra[1,a], JÍRA Aleš[1,b]

[1]Czech Technical University in Prague, Faculty of Civil Engineering, Department of Mechanics, Thákurova 7, 166 29, Prague, Czech Republic

[a]petra.hajkova@fsv.cvut.cz, [b]jira@fsv.cvut.cz

**Keywords:** Enamel; fracture toughness; cracks; nanoindentation; energy dissipation.

**Abstract.** Enamel is a brittle material. The main property showing its fracture resistance is fracture toughness. The aim of this study is to determine parameters necessary to estimate fracture toughness by the nanoindentation method. The method of fracture toughness evaluation based on dissipation of energy is described. In the first section of study, changes in values of hardness ($H_{IT}$) and reduced modulus ($E_r$) are observed with regard to crack initiation and growth. The downward trend of $H_{IT}$ was found with the increasing peak load. The next section is focused on the loading rate impact on crack formation. The results supported the assumption that higher rates of loading induce greater crack formation.

## Introduction

Enamel, like dentin or pulp, is an integral part of human teeth. Each of these parts has a specific function with corresponding material properties. The enamel protects the internal part of the tooth and is strained the most during mastication. This is the reason why it has the highest hardness among human tissues. Unfortunately, the high hardness is usually related to brittleness. The main property showing fracture resistance is fracture toughness.

The problem of enamel fracture toughness determination is that small volume of specimens limits the application of conventional methods. Consequently, the indentation method started to be used. In addition to fracture toughness determination, the method enables evaluation of other properties of enamel such as hardness or Young´s modulus.

The most frequently used method of fracture toughness determination based on indentation is Vickers indentation fracture (VIF) test. VIF enables direct determination of static fracture toughness ($K_{IC}$) from applied load and crack length originated in indentation. VIF was first introduced by Palmqvist in the 1950s [1], but it was advocated in 1976 by Evans and Charles [2]. Although it was supposed to evaluate fracture parameters of hard metals and ceramics, the method spread fast to the field of biomaterials. The main reason is the possibility of usage of small specimens and a survey of material properties at the micro level. However, in the last years the method has been criticised due to unreliable results. Authors [3, 4] have pointed out a large number of $K_{IC}$ equations which rarely provide the same results. Even so, most papers [5, 6] evaluate fracture toughness of enamel by VIF. Rarely, authors [7] apply other methods as single edge notched bending (SENB) specimen. The unreliability of VIF led to research of other methods for fracture toughness determination. Kruzic et al. [8] compare application of different methods (VIF, CCIF, VCOD and IIF) aimed at fracture toughness measurement of biomaterials. Hrbek, Malzbender et al. [9-11] deal with the possibility of determining fracture properties and erosion behaviour by a method based on dissipation of energy during indentation.

The aim of this study is to determine parameters necessary to estimate fracture toughness of enamel by the method of nanoindentation. The nanoindentation method based on dissipation of energy would reduce some undesirable aspects which are found to be wrong in VIF application. Additional advantage is the use of nanoindentation instead of microindentation. It allows to load the specimen by smaller maximum forces, so as to examine each part and trend in the enamel more precisely. In this paper, changes in values of hardness ($H_{IT}$) and reduced modulus ($E_r$) are observed with regard to crack initiation and growth. The next goal is an assessment of loading/unloading rate impact on extent of damage during indentation with the same peak load.

## Materials and Methods

**Preparation of Specimens.** A sample was prepared from a molar which was extracted due to orthodontic reasons. The clean, dry molar was embedded in Dentacryl. A 15 mm thick section was sliced perpendicularly to the longitudinal axis of the tooth by a water-cooled saw (ATM, Germany) equipped with a diamond blade. A required depth in enamel was reached by grinding by abrasive paper (grit size of 320). Then, the sample was polished by finer abrasive papers (grit size of 1000, 2500) and a diamond paste (0,25 μm).

**Nanoindentation.** The indentation was performed on the indenter CSM Instruments Nano Hardness Tester (Anton Paar, Australia) with a wedge tip. The methodology used for determination of hardness and reduced modulus was described by Oliver and Pharr [12]. Five indentation matrices were performed for observing changes in values of $H_{IT}$ and $E_r$ with regard to crack initiation and growth. Each matrix consisted of 16 indents. Force controlled tests with a different peak load were carried out for each cycle. The maximum forces were 10, 20, 40, 80 and 150 mN. The parameters of each cycle: loading 120 mN/min, pause 10.0 s, unloading 120 mN/min.

The next four matrices were performed for the comparison of loading/unloading impact on extent of damage. Three matrices were located in enamel, the last one in dentin. Each matrix consisted of 25 indents. Force controlled tests with a maximum force of 150 mN were carried out for each cycle. The loading and unloading rate was different for each matrix – 60, 250, 450 mN/min for enamel and 450 mN/min for dentin.

The axial distance of the indents was selected with regard to the contact area size of the indents. The position of indents towards the dentino-enamel junction (DEJ) and occlusal surface differs a lot due to different sizes of the matrices.

**Energy Release Method.** Energy release method is an alternative method for fracture toughness evaluation. It enables calculation of $K_{IC}$ using the energy dissipated during indentation. The total work of indentation ($W_{tot}$) can be easily obtained by integration of the loading curve. Consequently, the total work of indentation can be divided into reversible ($W_{rev}$) and irreversible work ($W_{irr}$) (Eq. 1) or more different energy dissipation mechanisms (Eq. 2) including work of fracture ($W_{fracture}$).

$$W_{tot} = W_{rev} + W_{irr} \quad (1)$$

$$W_{tot} = W_{plastic} + W_{elastic} + W_{fracture} + W_{creep} + W_{other} \quad (2)$$

Reversible and irreversible work can be as like as total work obtained from the load-displacement curve. Fig. 1 shows that $W_{rev}$ is the area under unloading curve, whereas $W_{irr}$ is the area enclosed between the loading and unloading curve.

The load-displacement curve as shown in Fig. 1 can also indicate crack formation. In this case, errors (pop-in) in the loading curve appear, displacement grows suddenly ($\delta$) and the loading curve is shifted to the right.

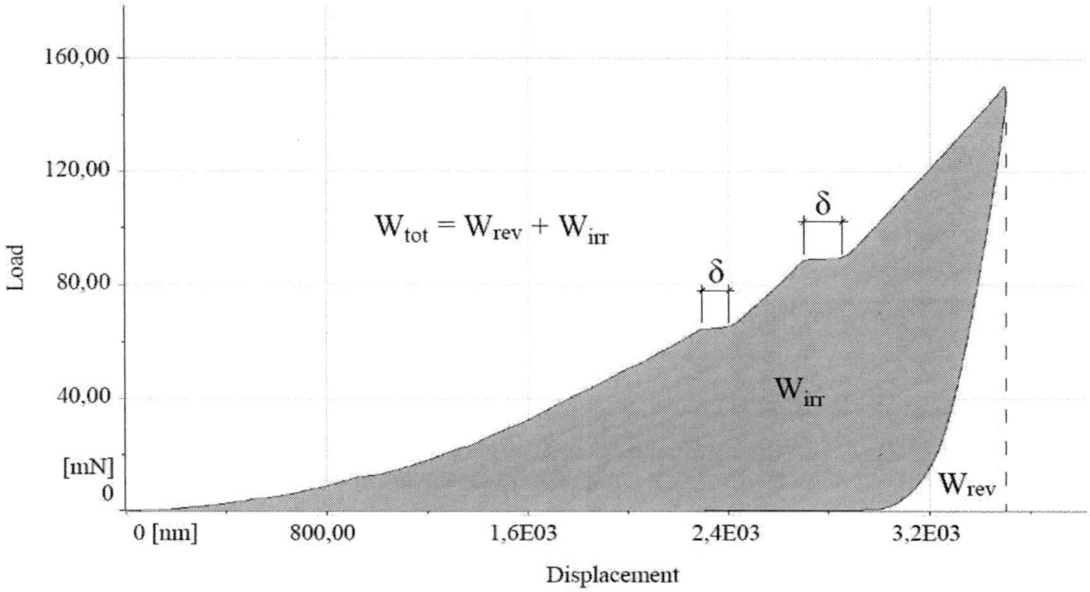


Fig. 1: Load-displacement curve showing total work ($W_{tot}$) divided into reversible ($W_{rev}$) and irreversible ($W_{irr}$) work. Shift of loading curve (δ) is caused by crack formation.

**Results**

**Property Changes Related to Cracks.** Changes of hardness ($H_{IT}$) and reduced modulus ($E_r$) were observed with regard to crack formation. Five indentation matrices with 16 indents were performed by force controlled test with a different peak load for each matrix. Table 1 shows mean values of $H_{IT}$ and $E_r$. There were significant changes of properties. It was observed that the hardness was the greatest for the minimum loading force of 10 mN ~ 3791 MPa and decreased up to ~ 3246 MPa for the maximum loading force of 150 mN. Reduced modulus showed the opposite trend. Fig. 2 shows variation of $H_{IT}$ and $E_r$ with the indentation peak load.

Table 1: Mean values of observed properties ($H_{IT}$, $E_r$) and ratio of irreversible ($W_{irr}$) to total ($W_{tot}$) obtained by force controlled tests with the different peak load.

| Maximum force [mN] | 10 | 20 | 40 | 80 | 150 |
|---|---|---|---|---|---|
| $H_{IT}$ [Mpa] | 3790.40 | 3707.57 | 3498.63 | 3275.21 | 3246.29 |
| $E_r$ [Gpa] | 79.65 | 80.33 | 80.95 | 81.55 | 85.66 |
| $W_{irr}/W_{tot}$ [-] | 0.874 | 0.880 | 0.884 | 0.887 | 0.889 |

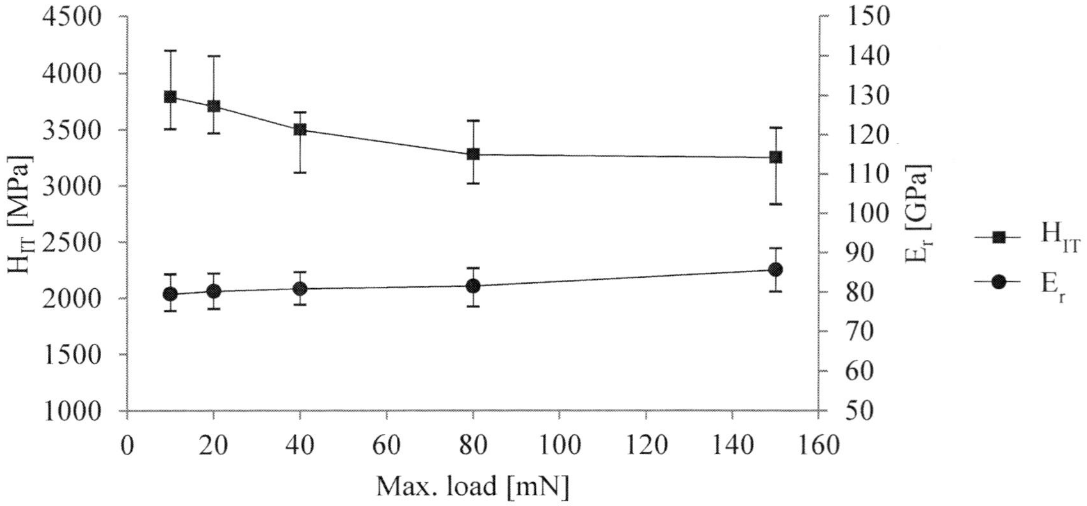


Fig. 2: Variation of hardness ($H_{IT}$) and reduced modulus ($E_r$) connected to the maximum loading force.

Compared with our last study [13], enamel showed significant changes of $H_{IT}$ whereas hardness of dentin seems to be almost constant with the increasing maximum loading. Property changes in enamel respond to pop-in events observed in load-displacement curves and increasing ratio of $W_{irr}$ to $W_{tot}$ which indicates crack initiation and growth. Dentin which is much less brittle material does not show these characters.

**Loading Rate Impact on Crack Formation.** Three indentation matrices with 25 indents were performed to assess the loading/unloading rate impact on crack formation. The fourth matrix was located in dentin to obtain property values for illustrative comparison of dentin and enamel. Just as in the first parts of study, hardness ($H_{IT}$) and reduced modulus ($E_r$) were observed. In addition, Table 2 shows the mean reversible ($W_{rev}$), irreversible ($W_{irr}$) and total ($W_{tot}$) work of indentation and ratio of $W_{irr}$ to $W_{tot}$. The changes of $H_{IT}$ and $E_r$ were not as significant as in the test of property changes which showed a difference between formed cracks as a result of greatly varying peak loads. But there was a decline in $H_{IT}$ from ~ 3287 MPa at the loading rate of 60 mN/min to ~ 3202 MPa at the loading rate of 450 mN/min. In contrast to the first section of the study, values of $E_r$ decreased.

Table 2: Mean values of observed properties ($H_{IT}$, $E_r$), work of indentation and ratio of irreversible ($W_{irr}$) to total ($W_{tot}$) work obtained by force controlled tests with different loading/unloading rate.

| Loading rate [mN/min] | | Enamel | | Dentin |
|---|---|---|---|---|
| | 60 | 250 | 450 | 450 |
| $H_{IT}$ [Mpa] | 3286.60 | 3309.95 | 3201.79 | 1153.21 |
| $E_r$ [Gpa] | 85.33 | 86.16 | 82.22 | 24.99 |
| $W_{rev}$ [pJ] | 18613 | 16741 | 15634 | 27773 |
| $W_{irr}$ [pJ] | 167390 | 168153 | 181320 | 310837 |
| $W_{tot}$ [pJ] | 186003 | 184894 | 196955 | 338610 |
| $W_{irr}/W_{tot}$ [-] | 0.899 | 0.909 | 0.920 | 0.918 |

The upward trend of the $W_{irr}/W_{tot}$ ratio connected with increasing loading/unloading rate supports an assumption that higher rates induce greater crack formation. The extent of damage can be observed in Fig. 3 as well.

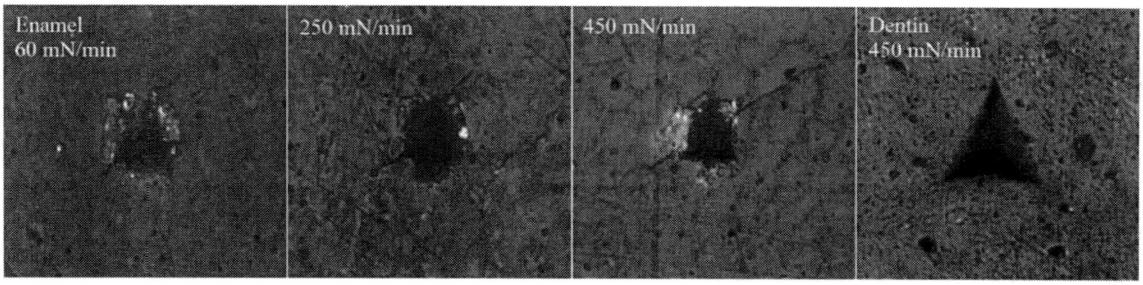


Fig. 3: Indents performed by force controlled test with the maximum force of 150 mN. Loading/unloading rate of indentation - 60, 250 and 450 mN/min. There are observable cracks in enamel. In contrast, the fourth indent was located in dentin where no cracks were recognized.

## Conclusions

The parameters for fracture toughness determination were obtained. Although the energy release method is based on constant material properties, it was suggested that fracture events affect hardness ($H_{IT}$) and reduced modulus ($E_r$) in the first section of the study. It was compared to the study of dentin which did not show any significant changes. The upward trend of the $W_{irr}/W_{tot}$ ratio connected with the increasing peak load indicated the crack initiation and growth. It was supported by load-displacement curve errors (pop-in) and investigation of indents under a microscope. In the next study, it is necessary to focus more on the relationship between $H_{IT}/E_r$ and $W_{irr}/W_{tot}$ which appear often in fracture toughness equations.

The loading rate impact on crack formation was observed in the second section of the study. The results showed a greater $W_{irr}/W_{tot}$ ratio when the enamel was loaded at higher loading rates. It supported the assumption that higher rates induce greater crack formation. There was a decrease of $H_{IT} \sim 2.6\%$ and a decrease of $E_r \sim 3.7\%$ by performing indents at the highest rate of 450 mN/min and the lowest rate of 60 mN/min.

This study precedes the fracture toughness evaluation. The problem of work of fracture ($W_{fracture}$) extraction will be crucial. It is also necessary to suppress other impacts on crack formation - creep, test conditions etc. as much as possible. The next issue will be the identification of fracture geometry using computed tomography scanning (CT scan).

## Acknowledgement

The financial support by the Faculty of Civil Engineering, Czech Technical University in Prague (SGS project No. SGS17/168/OHK1/3T/11) is gratefully acknowledged.

## References

[1] S. Palmqvist, Method att Bestamma Segheten hos Sproda Material, Sarskilt Hardmetaller, Jernkontorets Ann. 141 (1957) 300-307.

[2] A. G. Evans, E. A. Charles, Fracture Toughness Determinations by Indentation, J. Am. Ceram. Soc. 59 (1976) 371-372.

[3] G. D. Quinn, R. C. Bradt, On the Vickers Indentation Fracture Toughness Test, J. Am. Ceram. Soc. 90 (2007) 673-680.

[4] G. D. Quinn, Fracture Toughness of Ceramics by the Vickers Indentation Crack Length Method: A Critical Review, Ceram. Eng. Sci. Proc. 27 (2006) 45-62.

[5] Q. Zheng et al., Spatial Distribution of the Human Enamel Fracture Toughness with Aging, J. Mech. Behav. Biomed. Mater. 26 (2013) 148-154.

[6] S. Hayashi-Sakai, J. Sakai, M. Sakamoto, H. Endo, Determination of Fracture Toughness of Human Permanent and Primary Enamel Using an Indentation Microfracture Method, J. Mater. Sci.: Mater. Med. 23 (2012) 2047-2054.

[7] Bechtle S. et al., The Fracture Behaviour of Dental Enamel, Biomaterials 31 (2010) 375-384.

[8] J. J. Kruzic et al., Indentation Technique for Evaluating the Fracture Toughness of Biomaterials and Hard Tissues, J. Mech. Behav. Biomed. Mater. 2 (2009) 384-395.

[9] J. Němeček, V. Hrbek, Nanoindentation Assessed Fracture Toughness of Cement Paste, Defect and Diffusion Forum 368 (2016) 186-189.

[10] J. Malzbender, G. de With, Energy Dissipation, Fracture Toughness and the Indentation Load-displacement Curve of Coated Materials, Surf. Coat. Technol. 135 (2000) 60-68.

[11] S. J. Bull, Using Work of Indentation to Predict Erosion Behavior in Bulk Materials and Coatings, J. Phys. D: Appl.Phys. 39 (2006) 1626-1634.

[12] W. C. Oliver, G. M. Pharr, An Improved Technique for Determining Hardness and Elastic Modulus Using Load and Displacement Sensing Indentation Experiments, J. Mater. Res. 7 (1992) 1564-1583.

[13] P. Hájková, A. Jíra, Micromechanical Analysis of Complex Structures by Nanoindentation, Key Eng. Mater. 731 (2017) 60-65.

EAN 2017
55th conference on experimental stress analysis

# Numerical Model of a 3D-Printed Titanium Structure For Use in Implant Dentistry

ŘEHOUNEK Luboš[1, a], JÍRA Aleš[1,b] and DENK František[1,c]

[1] Department of Mechanics, Czech Technical University in Prague, Thákurova 7; 166 29, Prague; Czech Republic

[a] lubos.rehounek@fsv.cvut.cz, [b] jira@fsv.cvut.cz, [c] frantisek.denk@fsv.cvut.cz

**Keywords:** Numerical Model, 3D Printing, Titanium, Biomechanics, Trabecular.

**Abstract.** The aim of this work is to describe the mechanical behavior of the novel trabecular structure. It is a porous structure that comprises of beams of equal length embedded into a 3D matrix. 3D printing was used to create test specimens. To determine the values of Young's modulus E and reduced modulus $E_r$, we performed nanoindentation and tensile and compression tests. The tensile test was then used as a baseline for the development of a numerical model. The next effort will be to apply the model onto a real implant that was invented and patented in Czech Republic.

## Introduction

The trabecular structure has only recently been recognized as a viable option for dental restorations and other prosthetics. There are no known documented material tests yet, and since the parameters of every structure can be different, the outcomes may significantly vary. The stress-strain diagrams of this specific structure were provided by uniaxial tests. These tests were carried out on specifically designed implant specimens shown in Fig.1. Traditionally machined implants have a long history of success [1, 2], but they are still prone to failure in the early stages of the bonding process, because the surface properties play an important role in osseointegration [3]. The trabecular structure can potentially have many benefits, such as increased osseointegration, reduced stress-shielding and lower modulus [4]. This reduction is a significant benefit because it provides a smoother transition region in between the faces of the dental materials. The trabecular shell of an implant (Fig. 7) also provides an environment for bone ingrowth, as opposed to conventional implants, which only allow for bone ongrowth. The specimens were 3D-printed using the M2 Cusing Concept Laser machine and a specialized Rematitan Cl metal powder in cooperation with ProSpon spol. s. r. o.

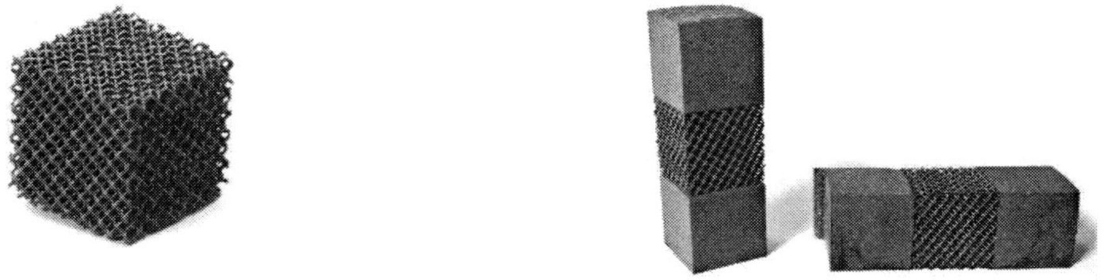

a) compression test specimen

b) tensile test specimen

Fig. 1: Trabecular structure specimens for global mechanical tests and nanoindentation.

## Trabecular Structure

Since the structure is porous, its geometry allows for bone ingrowth and creates a scaffold for the bone cells to grow into [5, 6]. With conventional implants, bone cells are only able to grow onto the surface of the implant. With the trabecular structure, the cells can grow inside and create an interconnected material comprising of both bone and metal bonded together (Fig. 2).

**Osseointegration.** The trabecular structure can potentially provide better osseogration because its morphology resembles a material with high surface roughness (trabecular portion, Fig. 2 left). This assumption is, however, still to be proven as there is not enough evidence and history to support this claim. Surface rougness has been proven to play an important role in the bonding process [7]. The greater the surface area of the implant, the greater the interlocking between the implant surface and bone.

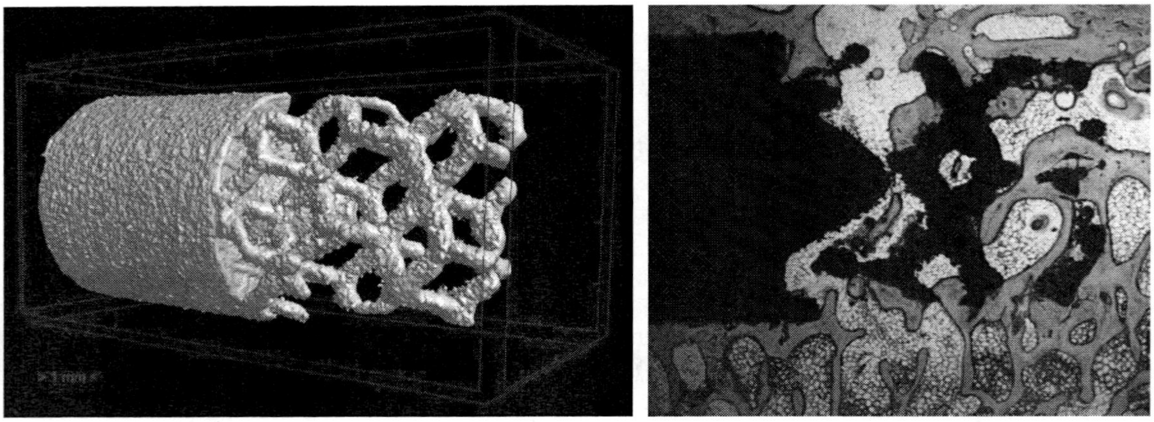

Fig. 2: Scaled micro-CT scan of the implant specimen with a trabecular end portion (left) and a micrograph of a longitudinal section of an extracted implant (right). White and light yellow color represents the beams of newly formed bone, orange represents fibrous tissue and black color represents the trabecular implant. These figures were obtained from in-vivo tests of the trabecular implants.

**Young's Modulus.** It is very important to distinguish between the values of Young's modulus and reduced modulus. The reduced modulus of the material remains unchanged, but the Young's modulus is reduced as the cross-section area of the implant is also reduced (Fig. 3). One of the causes of implant failure during the early stages of osseointegration is its loosening. This reduction eliminates the concentration of stresses in the intraosseous implant stem and is beneficial to the healing process.

**Stress-Shielding.** Stress-shielding is an unwanted factor that represents uneven distribution of stress between the implant, peri-implant area and bone [8]. The values of Young's modulus of human bone are significantly lower (approximately 20-30 GPa [9]) than those of the implant (90-110 GPa for conventional titanium implants [4]), the bone is left without sufficient stimulus and the stresses are transferred into the implant.

When the bone material is not subjected to sufficient loads, it becomes overly porous and is no longer able to hold the implant in position, making it eventually slip out, resulting in failure of the implant. This is true due to the fact that bone is always adapting to the conditions it is placed under. This process is described by the Wolff's law [10].

## Mechanical Tests

**Nanoindentation.** In order to determine the values of properties on the micro level, we performed nanoindentation tests considering reduced modulus of elasticity, hardness and contact depth. The nanoindentation tests were performed using the Oliver & Pharr method and their main purpose was to compare the conventional Ti-6Al-4V machined material to our 3D-printed specimens. The micromechanical analysis was performed using the CSM Instruments nanoindenter in the mode of directed force and repeated loading. The load program was set with consideration of eliminating surface tension and shear stiffness in the atomic material structure. The values of reduced modulus $E_r$ were in the range of 118-131 GPa.

**Uniaxial Tests.** For the purpose of global mechanical tests, we used the 3D Dode-Thick [MSG] structures with dimensions of 14×14×14 mm (a cube for the compression test) and a 14×14×42 mm (a block for the tensile test). A total of 9 specimens were tested beyond the point of failure on the MTS Alliance RT-30 machine. The results of the tests are shown in Tab. 1.

Tab. 1: Values of Young's modulus obtained by global mechanical analysis.

| | Young's modulus E [MPa] | | | | | | | | | |
|---|---|---|---|---|---|---|---|---|---|---|
| Specimen | 1 | 2 | 3 | 4 | 5 | 6 | 7 | 8 | 9 | Mean |
| Tens. test | 964.7 | 975.9 | 982.2 | - | - | - | - | - | - | **974.3** |
| Comp. test | - | - | - | 1114.2 | 1080.6 | 947.2 | 818.8 | 999.6 | 803.8 | **960.7** |

By incorporating the trabecular structure, we were able to reduce the modulus more than 100 times as compared to traditionally machined implants [4]. This reduction is connected with the reduction of the cross-section area (Fig. 3). This outcome is purely experimental as the structure does not have any documented material tests and its morphology is individual. However, we believe that the values of modulus do not necessarily need to exactly match human bone as more aspects come into consideration. These aspects are such as magnitude of pores, their geometry or character of bone ingrowth in relation to the geometry of the structure.

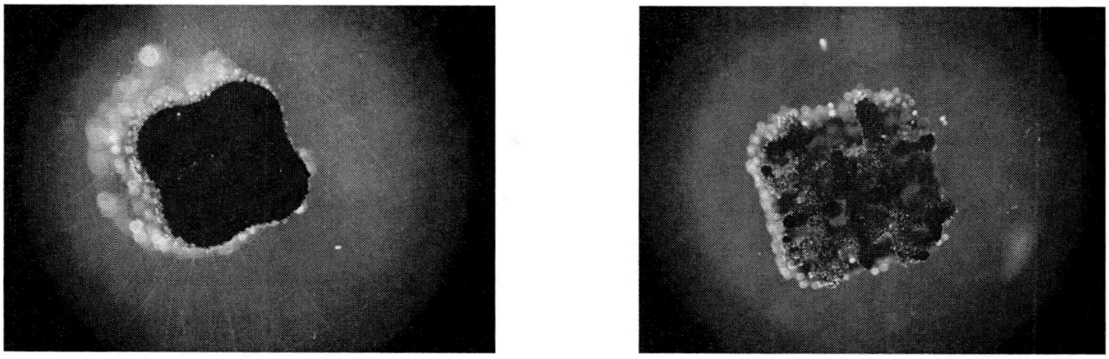

a) homogeneous (full) cross-section of an implant test specimen

b) trabecular cross-section of an implant test specimen

Fig. 3: Micrographs of full and trabecular cross-sections of an implant test specimen. Note the difference between the cross-section area, which is reduced approximately by 90%.

## Numerical Model

**Geometry.** The geometry of the model was created in order to minimize the computational requirements. Since the original geometry contained too many beams and its use would be complicated, we decided to create a new geometry that would require fewer elements. Comparison of the two geometries is shown in Fig. 4 and Fig. 5. This new geometry shown in Fig. 5 used only three basic structure elements per an edge of the imaginatory circumscribing cube. This reduction has proven to be beneficial as the computation times decreased.

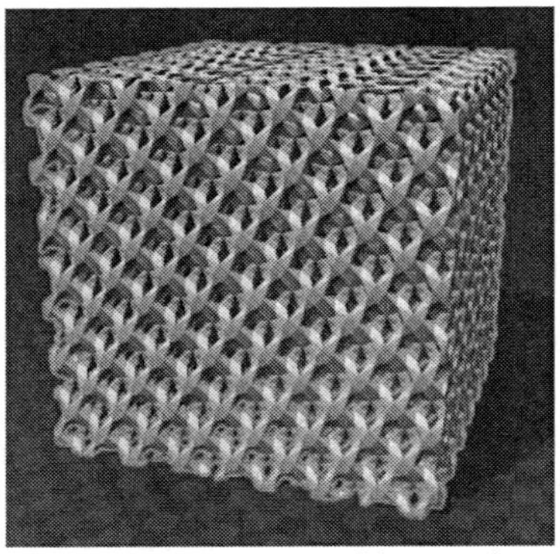

Fig. 4.: The former STL file used to create the test specimens.

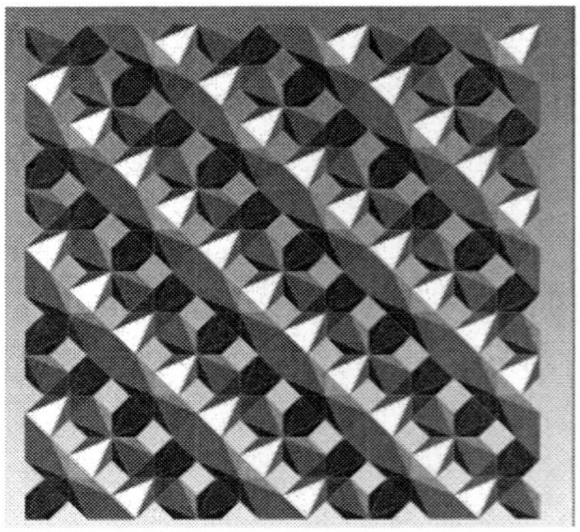

Fig. 5: The new 3D geometry created in Ansys Workbench Design Modeler.

**Curve-fitting.** There is only one time-dependent load (the displacement recorded by the machine), so the analysis can be performed in one computation step. We were trying to curve-fit the stress-strain diagram of the tensile test of specimen number 3 (Fig. 6 black). The convergence of the solution was attained by the Newton-Raphson method. We used the Ti-6Al-4V material model with a bilinear stress-strain diagram with isotropic hardening.

In order to create the virtual material's properties, we experimentally manipulated the values of mechanical properties in the model. The final values which have been fine-tuned during the curve-fitting process were saved and recorded for further development of the model. The final curve is shown in Fig. 6.

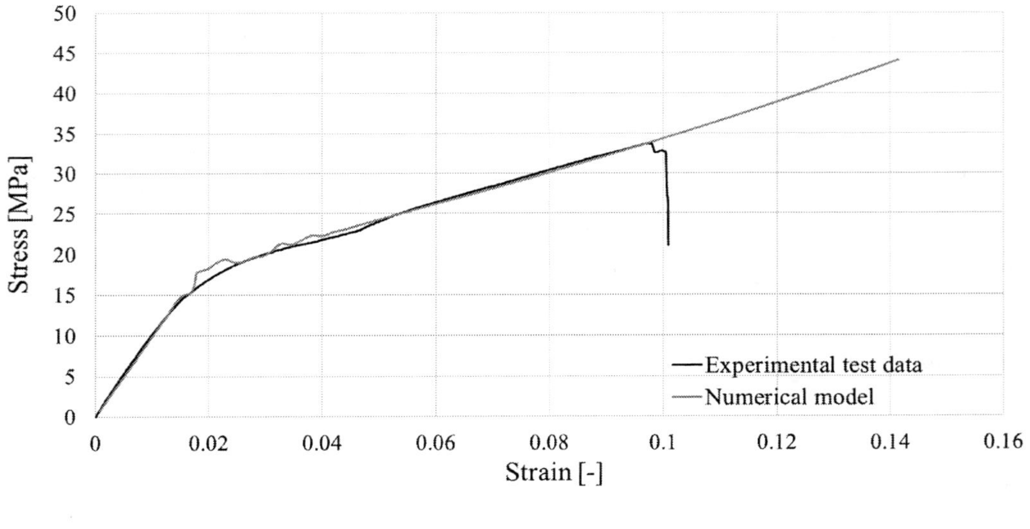

Fig. 61: The final achieved curve of the numerical model (green) compared to the tensile test stress-strain diagram of specimen number 3 (black). The model did not achieve failure beyond the range of ultimate strength, which is an acknowledged fact and will be addressed in our future efforts.

**Future Prospects of the Model.** The first feature of the model that needs to be addressed is the unattained failure. This is due to the fact that the solver cannot, by default, provide an accurate solution beyond the range of ultimate strength for this material model. We expect that we will be able to solve this shortcoming by adding additional algorithms to the solver.

The next major step is to apply this numerical model on a real implant. This implant has been invented and patented in Czech Republic and is shown in Fig. 7 (middle). On the basis of the patent application PV 2014-795, the patent n.306457 has been granted for the „four clover" implant variant by authors F. Denk Jr., A. Jíra and F. Denk Sr [11]. The methodology will probably comprise of applying the numerically obtained properties of the virtual material onto a homogeneous structure that will substitute the trabecular outer shell of the implant (Fig. 7 right).

## Conclusions

The main benefits that the trabecular structure can introduce are bone ingrowth, reduced modulus and reduction of stress-shielding. We have been able to successfully determine the values of the trabecular's structure modulus (Tab. 1), which is over a 100 times lower than the modulus of conventionally machined homogeneous implants [4], but other assumptions are still to be proven true with time and more experiments. This reduction is beneficial as the mechanical properties of the implant are more similar to human bone. The curve-fitting process was completed successfully with the exception of attaining failure beyond the range of ultimate strength, which is going to be addressed in our future work. The trabecular structure, unlike conventional homogeneous structures, allows for both bone ongrowth and ingrowth.

With the curve-fitted material properties available, we will be able to determine the mechanical behavior of an implant comprising of a homogeneous stem and a trabecular outer shell (Fig. 7). This conjunction will be the final purpose of the model. The STL files needed to create the model and merge the two materials together are already available to us (Fig. 7 left). However, since the full analysis of the trabecular structure would be very demanding in computation times, it is expected that its mechanical properties will be applied onto a homogeneous cross-section (Fig. 7. right).

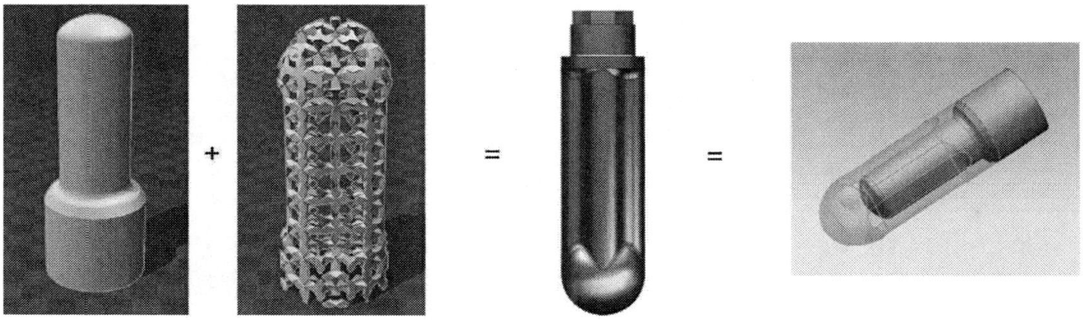

Fig. 7: The "four leaf clover" implant variant, comprising of a cylindrical body and four rounded grooves encompassed in a hemisphere (in the middle) and the two potential approaches towards modelling the structure. Two STL files for direct characterization (left) and a model using the homogenized structure (right).

**Acknowledgement**
The financial support by the Faculty of Civil Engineering, Czech Technical University in Prague (SGS project No. SGS17/168/OHK1/3T/11) is gratefully acknowledged.

**References**

[1] T. Hasegawa et al., Survival of Brånemark System Mk III implants and analysis of risk factors associated with implant failure, Int. J. Oral Maxillofac. Surg., 46 (2016) 267–273.

[2] F. De Angelis, P. Papi, F. Mencio, D. Rosella, S. Di Carlo, and G. Pompa, Implant survival and success rates in patients with risk factors: results from a long-term retrospective study with a 10 to 18 years follow-up, Eur. Rev. Med. Pharmacol. Sci., 21 (2017) 433–437.

[3] L. Le Guéhennec, A. Soueidan, P. Layrolle, and Y. Amouriq, Surface treatments of titanium dental implants for rapid osseointegration, Dent. Mater., 23 (2007) 844–854.

[4] M. Niinomi, Mechanical properties of biomedical titanium alloys, Mater. Sci. Eng. A, 243 (1998) 231–236.

[5] F. E. Wiria, J. Y. M. Shyan, P. N. Lim, F. G. C. Wen, J. F. Yeo, and T. Cao, Printing of Titanium implant prototype, Mater. Des., 31 (2010) 101–105.

[6] A. El-Hajje et al., Physical and mechanical characterisation of 3D-printed porous titanium for biomedical applications, J. Mater. Sci. Mater. Med., 25 (2014) 2471–2480.

[7] K. Gotfredsen, T. Berglundh, and J. Lindhe, Anchorage of titanium implants with different surface characteristics: an experimental study in rabbits, Clin. Implant Dent. Relat. Res., 2 (2000) 120–128.

[8] M. I. Z. Ridzwan, S. Shuib, A. Y. Hassan, A. A. Shokri, and M. N. Mohammad Ibrahim, Problem of stress shielding and improvement to the hip implant designs: A review, Journal of Medical Sciences, 7 (2007) 460–467.

[9] J. Y. Rho, R. B. Ashman, and C. H. Turner, Young's modulus of trabecular and cortical bone material: Ultrasonic and microtensile measurements, J. Biomech., 26 (1993) 111–119.

[10] S. C. Cowin and D. H. Hegedus, Bone remodeling II: theory of adaptive elasticity, J. Elast., 6 (1976) 313–326.

[11] F. Denk Sr., F. Denk Jr. and A. Jíra, CZ Patent 306,457. (2017)

# Drop Test of Foams used in Motorbike Protectors

ŠOLTÉS Lukáš [1, a], BOŃKOWSKI Tomasz [1, b], KOTTNER Radek [1, c] and HYNČÍK Luděk [1, d]

[1] NTIS - New Technologies for the Information Society, Faculty of Applied Sciences, University of West Bohemia, Technicka 8, 301 00 Pilsen, Czech Republic
[a]soltesl@ntis.zcu.cz, [b]tomasz@ntis.zcu.cz, [c]kottner@kme.zcu.cz, [d]hyncik@ntc.zcu.cz

**Keywords:** drop test, foam materials, protectors, acceleration, motorcycle rider

**Abstract.** This article is focused on the experimental drop testing of foam materials which are used in motorbike protectors. The EN 1621-1 standard, which describes the test methodology of motorcycle protectors, determines only one type of the drop test. Every protector is tested only once [4], however, there is a high probability of multiple hits to the same spot during motor rider fall. Every sample of the motorcycle protector was tested ten times with 60 seconds pause, using methods described in this article. Three impact velocities 1 $m \cdot s^{-1}$, 2 $m \cdot s^{-1}$ and 3 $m \cdot s^{-1}$ were applied. The experiment showed that, in many cases of the impact, protectors are not destroyed.

## Introduction

In the year 2014, there were 4570 motorcycle accidents in the European Union countries. A large amount of them were fatal [1, 2]. Many motorcycle riders consider the motorcycle protective clothing only as a protection against the bad weather conditions [3]. In our experiment, two types of foam materials were compared. First type of samples was cut out from the motorcycle protectors SAS-TEC SC-1/42 (Fig. 1, black color), second type of samples was cut out from motorcycle protectors SAS-TEC SLC 2 (Fig. 1, gray color). The size of black samples was 19.5 mm x 20.2 mm x 10.8 mm. The target area was 395.7 $mm^2$. The size of gray samples was 19.2 mm x 20.8 mm x 13 mm. The target area was 40 $mm^2$.

Fig. 1: Experimental samples.

Both types of protectors were classified according EN 1621-1 standard as class 2. In the protector verification test described by EN 1621-1 standard, the impactor falls with the weight of 5 kg from 1m height. The maximum ground reaction force must be equal or less than

20 kN [4]. In this article unstandardized test for motorcycle protectors was applied. Compared to the standard test, every sample was tested ten times. While the ground reaction forces were not compared but only acceleration and compression during impact was measured. A lot of motorbike protector manufacturers recommend change of all protectors after a motorbike fall [5].

## Experimental methodology

The experiment was performed on the device developed by our institution. This machine was designed to simulate free fall and its functionality was verified by leaser sensors. Mass of impactor used in the experiment was 2.4 kg. Dimensions of rectangular shaped impactor were 30 mm x 30 mm. By application equation

$$y = \frac{v^2}{2g} \tag{1}$$

three impact heights (51 mm, 204.1 mm and 459.2 mm) were calculated from which the impactor was released. These three heights represented impact speeds 1 $m \cdot s^{-1}$, 2 $m \cdot s^{-1}$ and 3 $m \cdot s^{-1}$. The measuring apparatus is showed on figure 1. Measuring chain was composed from piezo-ceramic accelerometer KISTLER 8702B500 with range ±500 g and optical distance sensor optoNCDT2300-500 with rage ±50 mm. The first of sensors was connected to measuring card NI 9215 by coaxial cable, the second sensor was also connected by coaxial cable to measuring card NI 3234. Both cards were connected to NI eDAQ-9178 bus. The sampling frequency was set to 25.4 kHz. Data acquisition was performed in Nation Instruments Signal Express environment and post processed in MATLAB software.

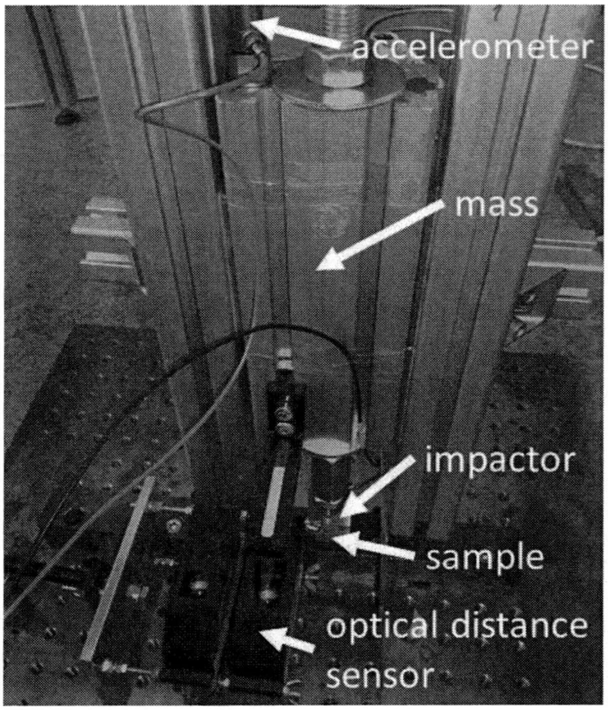


Fig. 2: Measuring apparatus.

### Experimental results

Fig. 3 shows the maximal values of acceleration during the impact. In the first impact velocity 1 $m \cdot s^{-1}$, the difference in the acceleration between black and gray samples was very small 0.47 g ± 1.34 g (Fig. 3, the bar markers). The differences were increasing with rising impact velocity. At impact speed 2 $m \cdot s^{-1}$, the difference between the black and gray samples was 13.25 g ± 1.39 g (Fig. 3, the circular markers). The maximal difference 28.46 g ± 3.98 g was detected at the impact speed 3 $m \cdot s^{-1}$ (Fig. 3, the square markers). The minimum acceleration 19.45 g was measured on gray sample G1 in first impact at impact speed 1 $m \cdot s^{-1}$. Also, the maximum acceleration 184.7 g was recorded on gray sample G3 during the tenth impact at speed 3 $m \cdot s^{-1}$.

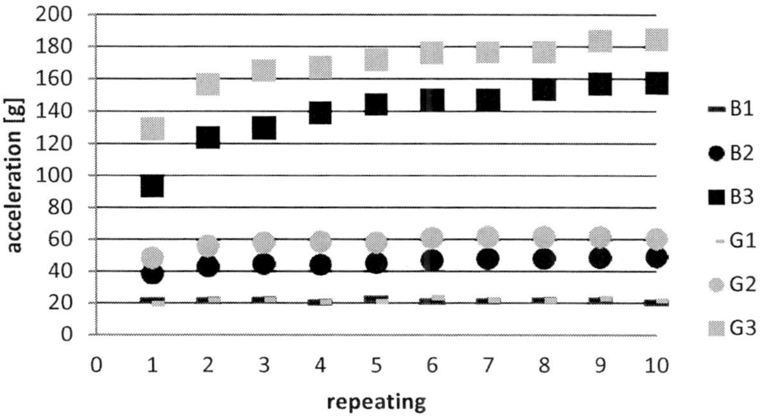


Fig. 3: Results of accelerations from drop tests with different speed of impact.

The maximal values of the compression during the impact was in the Fig. 4 showed. At the first impact velocity 1 $m \cdot s^{-1}$, the difference of the compression between black and gray samples was 0.26 mm ± 0.14 mm (Fig. 4, the bar markers). At impact speed 2 $m \cdot s^{-1}$, the difference between the black and gray samples was 0.79 mm ± 0.11 mm (Fig. 4, the circular markers). The maximal difference 1.63 mm ± 0.51 mm was detected at the impact speed 3 $m \cdot s^{-1}$ (Fig. 4, the square markers). In contrast with accelerations, the minimum compression 1.51 mm was measured on the black sample B1 at impact speed 1 $m \cdot s^{-1}$ in the sixth and the eighth falls. The maximal compression 6.55 mm was measured on the gray sample G3. This value was measured during the fifth drop at impact speed 3 $m \cdot s^{-1}$.

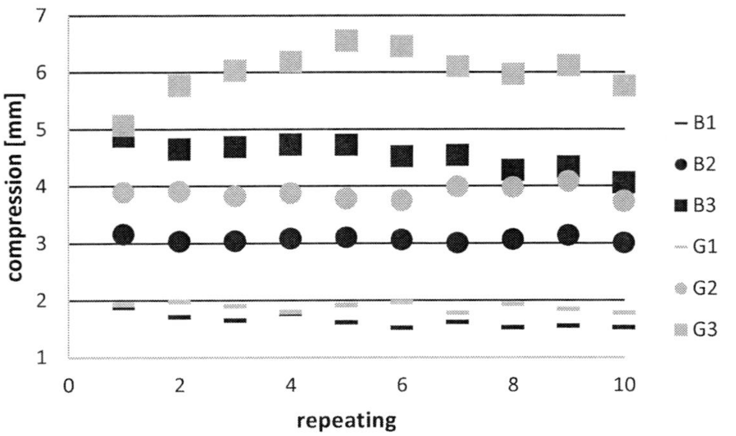


Fig. 4: Results of compression from drop tests with different speed of impact.

## Conclusions

All of experiments showed that protectors were not at first impact destroyed. During slow speed impact were increasing the acceleration and deformation of both samples on minimum. The effect of the increasing acceleration was on the maximum impact speed 3 $m \cdot s^{-1}$ of both samples (B3, G3) measured. Unusual date of the G3 sample was found. This foam sample show significantly increasing acceleration but the deformation during the maximal impact speed 3 $m \cdot s^{-1}$ was decreased.

## Acknowledgment

The research leading to these results has received funding from the People Programme (Marie Curie Actions) of the European Union's Seventh Framework Programme FP7/2007-2013/ under REA grant agreement no 608092. This work was supported by the internal grant project SGS-2016-059 and by the project LO1506 of the Czech Ministry of Education, Youth and Sports under the program NPU I. The authors would like to thank PSí Hubík for providing foam samples.

## References

[1] Information on: http://ec.europa.eu/transport/road_safety/sites/roadsafety/files/pdf/statistics/dacota/bfs2016_motomoped.pdf

[2] Information on: http://ec.europa.eu/transport/road_safety/specialist/statistics_en

[3] L. de Rome, R. Ivers, M. Fitzharris, W. Du, N. Haworth, S. Heritier, D. Richardson, Motorcycle protective clothing: Protection from injury or just the weather?, Accident Analysis and Prevention, Volume 43, Issue 6, Pages 1983-1990, (2011)

[4] Motorcyclists' protective clothing against mechanical impact - Part 1: Motorcyclists' limb joint impact protectors - Requirements and test methods, EN 1621-1

[5] Information on: http://www.motojomax.cz

EAN 2017
55th conference on experimental stress analysis

# The Proposal of Methodology for Finding of Proper Smoothing Parameters for Stress/Strain Analysis Performed by Digital Image Correlation

HAGARA Martin[1,a], PÁSTOR Miroslav[1,b] and ČARÁK Peter[1,c]

[1]Technical University of Košice, Faculty of Mechanical Engineering, Department of Applied Mechanics and Mechanical Engineering, Letná 9, 040 01 Košice, Slovakia

[a]martin.hagara@tuke.sk, [b]miroslav.pastor@tuke.sk, [c]peter.carak@tuke.sk

**Keywords:** Digital Image Correlation, Stress/Strain Analysis, Smoothing Parameters, Photoelasticity, Finite Element Analysis.

**Abstract.** The paper describes the proposal of methodology for finding the most proper smoothing parameters, used by evaluation of full-field stress/strain analysis performed through device working on a principle of digital image correlation method. The proposal for finding of mentioned parameters follows from comparison of the experimentally obtained principal stress difference, whereby the authors did not consider the overlapping of the facets, with the results acquired numerically using finite element method. The correctness of the used numerical model was verified by an experimental analysis performed by transmission photoelasticity method.

## Introduction

Digital image correlation is an experimental full-field method based on comparison of digital images obtained during the process of investigated object loading, called correlation. The images are compared along small picture elements, called facets, with size (in square pixels) defined by the user. Typical facet size used is from 15x15 up to 30x30 pixels, whereby using smaller facets the results are influenced by an increasing systematical error [1, 2]. To ensure the comparison of identical image parts it is necessary to create black and white speckle pattern on the object surface, and then each of the facets becomes unique with specific content of black and white color.

The correlation can be carried out using one or more cameras. By the use of one camera, the measurement is 2-D and then it is needful to ensure the parallelism between camera's image plane and object surface (Fig. 1a) during the whole process of measurement. In the case if two (or more) cameras are used, mentioned condition does not exist and the output of the experiment is in the form of 3-D displacements (Fig. 1b).

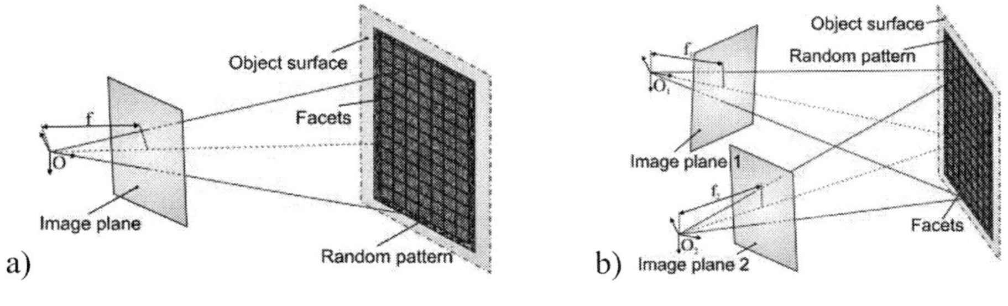


Fig. 1 Basic principle of: a) 2-D digital image correlation, b) 3-D digital image correlation.

The results of the analyses, performed by correlation system Q-400 Dantec Dynamics, are significantly affected by the use of smoothing filters involved in Istra4D software, delivered as a control unit of correlation systems Dantec Dynamics. Istra4D v.4.3.0., which the authors of this contribution work with, evaluate only displacement and strain fields. It comprises two types of smoothing filters – the first one is based on the adaptive spline polynomial algorithm (ACSP) and the second one is denoted as smoothing spline, where Grid reduction factor (GRF) and Smoothing factor (SF) are set for contour and displacement as well [3]. Fig. 2 shows a typical change of the Guest stress field distribution acquired on the rectangle shaped specimen (200x30mm) containing a small hole (diameter of 3 mm) created on the intersection of specimen axis of symmetry, loaded by uniaxial tension loading force using various levels of smoothing filters.

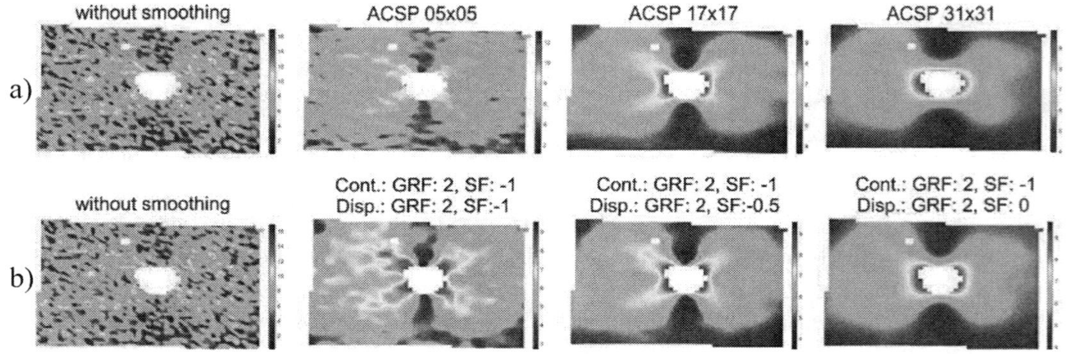


Fig. 2 The influence of: a) ACSP, b) Smoothing spline on the distribution of Guest stress fields acquired by the loading of rectangle shaped specimen with a hole through tension force.

According to information available to authors, there is no instruction how to set the smoothing parameters to get the proper results, which correspond to the real stress/strain state of the structure at most. For that reason, they decided to realize such analyses, which could lead to find the proper values of ACSP smoothing used for evaluation of measurements on the specimen with common size.

### Methodology for finding of proper values of ACSP smoothing used in Istra4D

The analyses was realize on the specimen with a shape and dimensions, adjustment to the realization of measurements in laboratory conditions, depicted in Fig. 3.

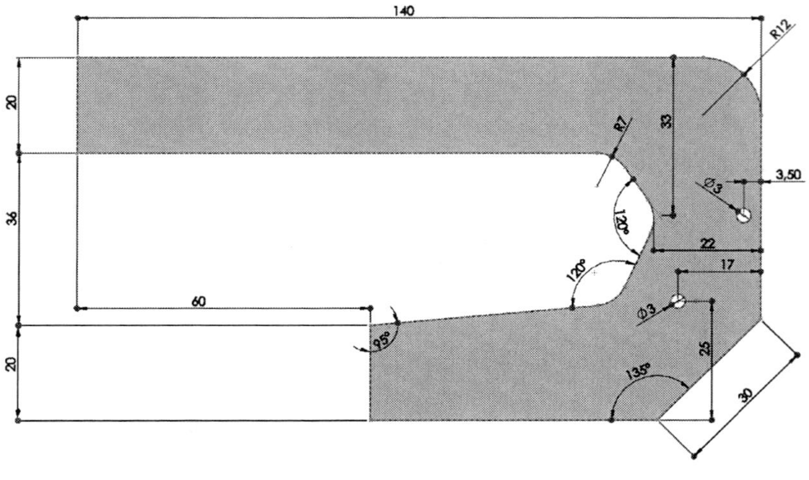


Fig. 3 Shape and dimensions of the specimen.

Such shape of the specimen was chosen to ensure its in-plane bending behavior (without out-of-plane deformation) through the experimental analysis and several stress concentrators served for creation of necessary stress gradient.

The PSM-1 material, used in transmission photoelasticity method and which the specimen was made from, has a thickness of 3.05 mm. Such material has the following mechanical properties: Young modulus E = 2500 MPa and Poisson ration $\mu = 0.38$.

The proposal of methodology leading for finding proper smoothing levels in Istra4D software consists in a comparison of strains and stresses acquired in several chosen points by digital image correlation and finite element method. The numerical model with boundary conditions, used in numerical analyses, is depicted in Fig. 4. The size of the finite element used for calculation was set to 0.3 mm. The mesh was refined in the curved parts of specimen. The study was solved as a static one with consideration of plane model with thickness of 3.05 mm and plane stress state. The loading force, act on the perpendicular edge of specimen lower part, was set to 20 N.

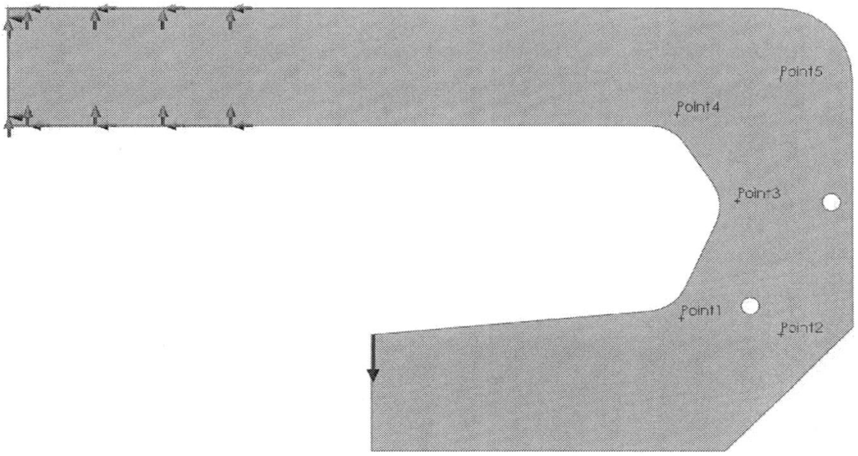


Fig. 4 Boundary conditions used by the solving of numerical analyses and five chosen points which the difference of principal stresses was analyzed in.

The obtained results has to be compared to be sure that the numerical model is correct. The comparison was realized experimentally using Model 060 polariscope working on the principle of transmission photoleasticity [4]. On the specimen surface five points was labeled, which the difference of principal stress was determined in and compared with the results obtained numerically. The pictures, present in Fig. 5, show the shift of fringes caused by the use of compensator Model 067. The obtained results are shown in Table. 1.

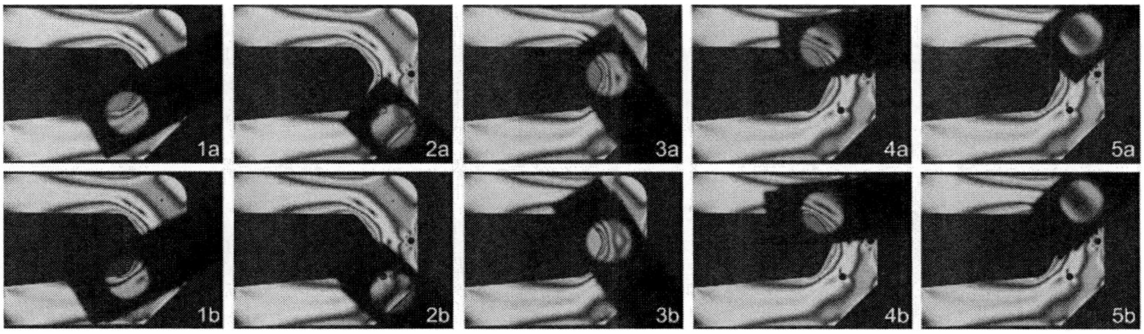


Fig. 5 The use of compensator for determination of fringe order in five chosen points: a) state before its using, b) state after its using.

Table 1. Principal stress difference ($\sigma_1$ - $\sigma_2$) obtained by FEA and transmission photoelasticity

| Finite element analysis | Point 1 | Point 2 | Point 3 | Point 4 | Point 5 |
|---|---|---|---|---|---|
| Principal Stress 1 [MPa] | 1.564 | 0.321 | 1.936 | 1.599 | 0.566 |
| Principal Stress 2 [MPa] | 0.617 | -0.731 | 0.495 | 0.497 | -0.613 |
| $\sigma_1$ - $\sigma_2$ [MPa] | 0.947 | 1.052 | 1.441 | 1.102 | 1.179 |

| Transmission photoelasticity | Point 1 | Point 2 | Point 3 | Point 4 | Point 5 |
|---|---|---|---|---|---|
| Fringe order | 0.833 | 0.938 | 1.250 | 0.979 | 1.042 |
| $\sigma_1$ - $\sigma_2$ [MPa] | 0.972 | 1.094 | 1.458 | 1.142 | 1.215 |

As can be seen the values of principal stress difference acquired numerically and experimentally via transmission photoelasticity are nearly the same (the data correspond to the comparison of results acquired numerically and experimentally). For that reason, it can be stated that the numerical model should be correct and the obtained results can be used for comparison with the results obtained by digital image correlation.

On the specimen surface, a vinyl foil with printed random speckle pattern was attached. The specimen was constrained and loaded in the same way as by experimental analysis performed by transmission photoelasticity. The fields of view from two cameras with 5 Mpx resolution, creating the correlation system Q-400 Dantec Dynamics, can be seen in Fig. 6. By the capturing of images, the whole resolution of the cameras (2452 x 2056 px) was used. For the evaluation of the measurement using digital image correlation, the facet size was changing from the value 15x15 px up to 30x30 px with increment 1. Moreover, the evaluation was realized in such a way, that the facets did not overlap themselves.

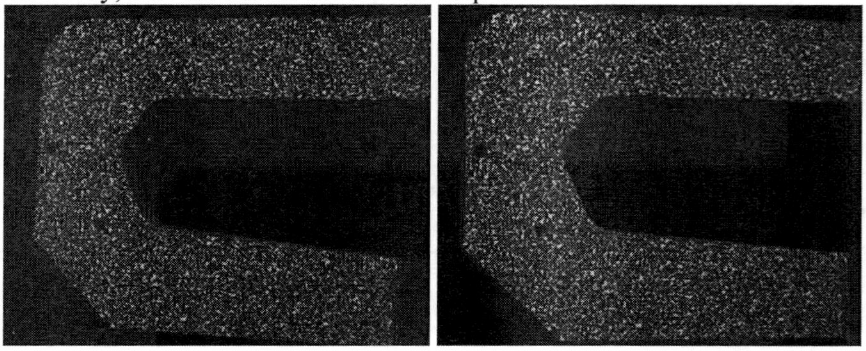

Fig. 6 Fields of view captured by two cameras of correlation system Q-400 Dantec Dynamics.

As the Istra4D allows only visualization of displacement and strain fields, it was necessary to export these data from the numerical analyses. The normal strains, calculated via FEA in analyzed points, are present in Table 2.

Table 2. The values of tangential strains obtained in analyzed points via FEA

| Finite element analysis | Point 1 | Point 2 | Point 3 | Point 4 | Point 5 |
|---|---|---|---|---|---|
| Normal strain $\varepsilon_x$ [-] | 0.000363 | 0.000087 | -0.000096 | 0.000513 | 0.0000075 |
| Normal strain $\varepsilon_y$ [-] | 0.000178 | -0.000189 | 0.000699 | 0.0000067 | -0.000019 |

Smoothing in Istra4D is performed on contour and displacements particularly, whereby for smoothing of strain fields it is necessary to filter both quantities simultaneously, otherwise the strain fields will be calculated just from the facet deformation (convenient rather for large deformation). Using local regression (ACSP), the strains are computed not only from facet deformation, but also from deformation gradient. The higher the level of smoothing, the higher the influence of deformation gradient on the quality of obtained results. Practically only the deformation gradient is used for the computation of strains using ACSP smoothing level of 7x7 and higher. Higher filter makes the results more smoothed and should decrease the standard deviation of the obtained results, however, it also decreases the spatial resolution of the image. There are fifteen levels of ACSP in Istra4D, starting with 3x3 and ending with 31x31 of increment two. The second type of smoothing used in Istra4D is Smoothing spline, a global filter used commonly in cases when homogenous results on the whole object surface are expected.

For these reasons, for various facet sizes used for evaluation of measurement ACSP smoothing filter was determined, in order to find the results which correspond with numerical model at most. From the evaluated data, the dependence depicted in Fig. 7 was achieved.

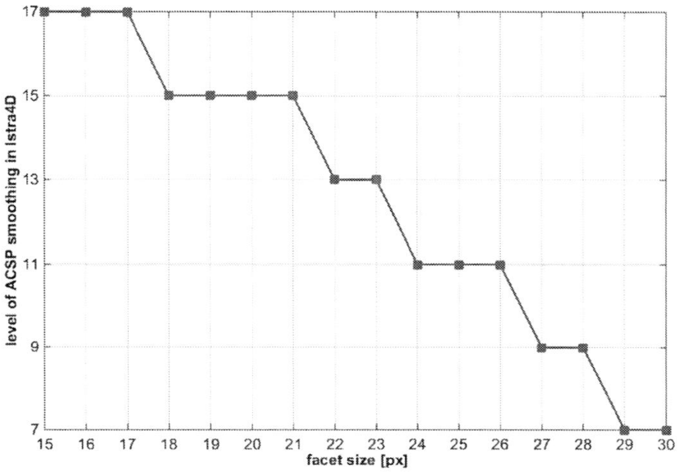


Fig. 7 Dependence of the ACSP smoothing level on the facet size used for evaluation of the measurement.

Principal stress differences ($\sigma_1$ - $\sigma_2$) acquired by DIC were compared to the results acquired in five chosen points by numerical analysis. The relative differences between both methods are depicted in Fig. 8-10 particularly for the facet sizes set to from 15x15 px up to 19x19 px, from 20x20 up to 24x24 px as well as from 25x25 px up to 30x30 px.

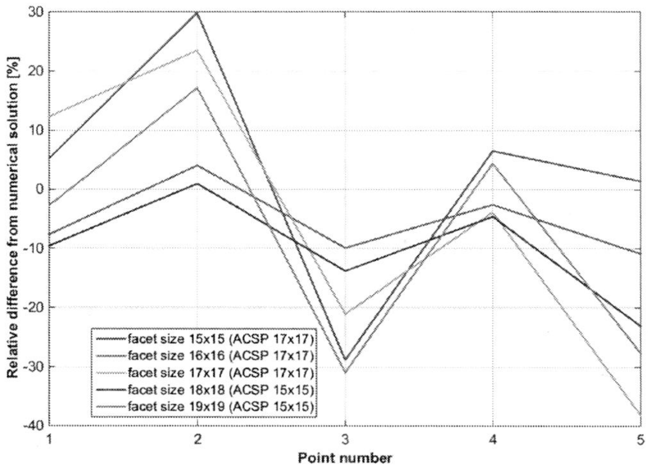


Fig. 8 The relative difference of ($\sigma_1$ - $\sigma_2$) acquired by DIC by facet sizes from 15x15 px up to 19x19 px with corresponding ACSP from the results obtained in chosen points numerically.

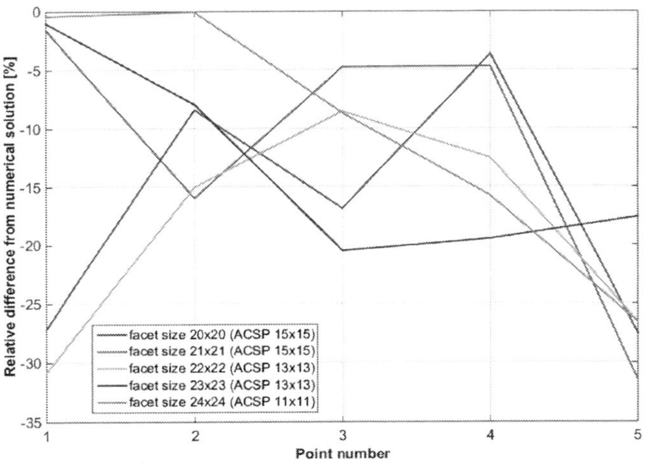


Fig. 9 The relative difference of ($\sigma_1$ - $\sigma_2$) acquired by DIC by facet sizes from 20x20 px up to 24x24 px with corresponding ACSP from the results obtained in chosen points numerically.

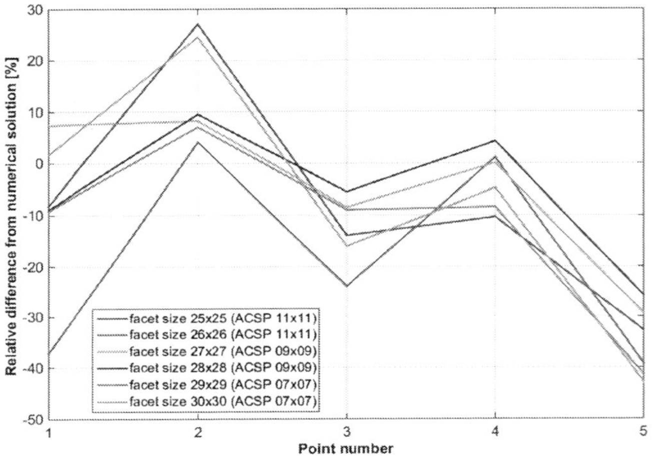


Fig. 10 The relative difference of ($\sigma_1$ - $\sigma_2$) acquired by DIC by facet sizes from 25x25 px up to 30x30 px with corresponding ACSP from the results obtained in chosen points numerically.

From all the evaluated data, the best correspondence between the numerical and DIC analyses acquired in analyzed points was achieved by the facet size of 16x16 px and level of ACSP set to 17x17 as well as facet size 27x27 px and level of ACSP 9x9. The obtained results in a form of normal strain $\varepsilon_x$, $\varepsilon_y$ and shear strain $\gamma_{xy}$ acquired by the mentioned settings of facet size and ACSP smoothing can be seen in Fig. 11-12.

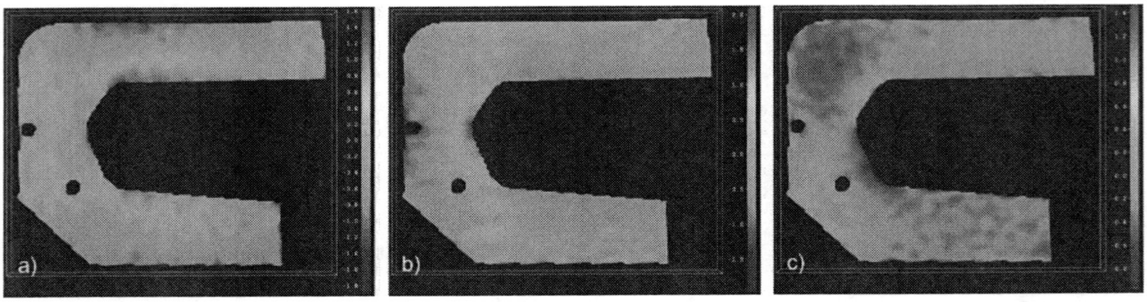


Fig. 11 The fields of: a) normal strain $\varepsilon_x$, b) normal strain $\varepsilon_y$, c) shear strain $\gamma_{xy}$ acquired in Istra4D by the facet size of 16x16 px and level of ACSP set to 17x17

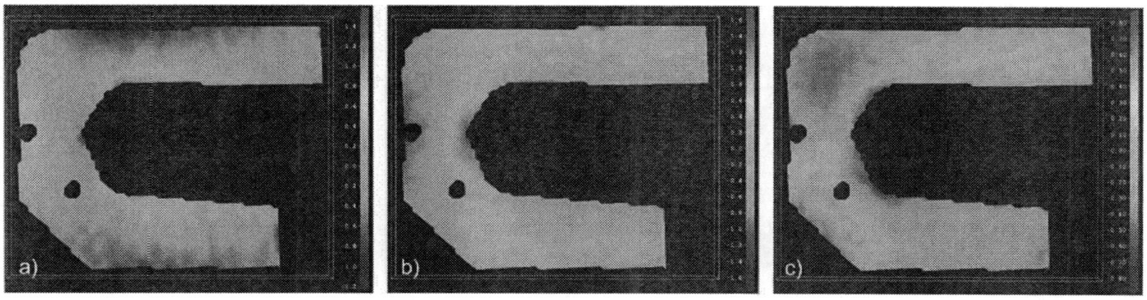


Fig. 12 The fields of: a) normal strain $\varepsilon_x$, b) normal strain $\varepsilon_y$, c) shear strain $\gamma_{xy}$ acquired in Istra4D by the facet size of 27x27 px and level of ACSP set to 9x9

In Table. 3 the results of principal stress difference ($\sigma_1$ - $\sigma_2$) obtained in analyzed points via digital image correlation system Q-400 Dantec Dynamics can be found. The first row describes the results obtained by the facet size set to 16x16 px and the second one describes the results obtained by the facet size set to 27x27 px.

Table 3. Principal stress difference ($\sigma_1$ - $\sigma_2$) obtained by digital image correlation

| Digital image correlation | Point 1 | Point 2 | Point 3 | Point 4 | Point 5 |
|---|---|---|---|---|---|
| **($\sigma_1$ - $\sigma_2$) for facet size of 16x16 px [MPa]** | 1.019 | 1.009 | 1.298 | 1.131 | 1.307 |
| **($\sigma_1$ - $\sigma_2$) for facet size of 27x27 px [MPa]** | 0.877 | 0.965 | 1.318 | 1.102 | 1.524 |

Comparing the results of the both evaluations it can be stated, that the differences of DIC analyses from the results of numerical analyses are approximately the same except the fifth point, in which the higher difference was achieved by the use of higher facet size. This phenomenon can be caused by the technical possibilities of the correlation system Q-400 Dantec Dynamics, which is assigned to measure with convenient accuracy deformation of $1e^{-4}$ and higher level. Due to the facet size used for the evaluation of the measurement, it is clear that the evaluated contour does not cover the real shape of the specimen up to the edges, especially in the second case where the spatial resolution of measurement is decreasing with higher facet size. For that reason, that the evaluated contour obtained by the use of facet size of 16x16 px corresponds better with the real surface of the specimen, the maximal values of strains reach higher levels.

## Conclusions

The authors introduce a methodology, which can be used for the "calibration" of the ACSP smoothing filter used in Istra4D (software delivered with correlation systems Dantec Dynamics) according to the facet size, which has to be set for the evaluation of the measurement. In the paper, there is a comparison of the results obtained by digital image correlation, transmission photoelasticity and numerical analyses. The authors propose to perform several other measurements, where the effect of higher strain level, size of the specimen cover the spatial resolution of the cameras as well as the overlapping of the facets concerning the level of smoothing, used by the evaluation, will be study.


## Acknowledgement

This contribution was created with a support of project *VEGA 1/0751/16* and *APVV 15-0435.*

## References

[1] T. Becker, K. Splitthof, T. Siebert, P. Kletting, Error Estimations of 3D Digital Image Correlation Measurements, Technical Note: T-Q-400-Accuracy-3DCORR-003-EN, Dantec Dynamics, 2006.

[2] F. Trebuňa, R. Huňady, M. Hagara, Experimentálne metódy mechaniky – Digitálna obrazová korelácia, Edícia vedeckej a odbornej literatúry, Košice, 2015.

[3] Istra4D v.4.3.0. software manual: Q-400 System

[4] F. Trebuňa, F. Šimčák, Príručka experimentálnej mechaniky, Emilena, Košice, 2012.

EAN 2017
55th conference on experimental stress analysis

# Sensing damage in CFRP composite beam by electrical resistance measurement

Nikola Schmidová[1, a], Milan Růžička[1,b] and Milan Dvořák[1,c]

[1]Czech Technical University in Prague, Faculty of Mechanical Engineering, Technická 4, 166 07 Prague 6, Czech Republic

[a]nikola.schmidova@fs.cvut.cz, [b]milan.ruzicka@fs.cvut.cz, [c]milan.dvorak@fs.cvut.cz

**Keywords:** composite, CFRP, electrical resistance measurement, self sensing, structural health monitoring.

## Abstract

This paper presents a study on electrical resistance (ER) measurement on unidirectional CFRP composite beam during cyclic flexural loading. ER measurement was conducted in the transverse direction to the fiber direction. Three measuring configuration, each in two copies were measured during shutdowns of loading. One measuring point was monitored by ER measurement also during cyclic loading. Artificial damage was created in a form of a cut in the middle of a specimen so that the cross section of the specimen was reduced. According to performed measurements further tests on possibility of damage detection during cyclic flexural loading by measuring ER in transversal direction must be performed. Possibility of fiber crack detection by presented type of measurement was confirmed.

## Introduction

Carbon Fiber Reinforced Polymer (CFRP) composites are nowadays widely used for structural parts in aircraft and automotive industry, where high level of safety has to be ensured.

During the service composites are frequently exhibited to the dynamic loading. Damage detection can be conducted using NDT methods such as ultrasound or X-ray inspection, which is expensive and time-demanding. New approaches, which would enable also in-situ monitoring are being seeked. One of the possible method of Structure Health Monitoring (SHM) is electrical resistance measurement.

This method is based on a fact, that carbon fibers are electrically conductive but the matrix used for manufacturing CFRP is not. Carbon fibers in CFRP composite touch each other so that the material is conductive in all three directions. The conductivity changes depending on deformation and damage of the material. In previous work [1, 2] it was confirmed that it is possible to monitor damage during cyclic tension loading by electrical resistance measurement in the longitudinal and through-thickness direction. Damage was also monitored by this method during cyclic bending in both above mentioned directions.

So far no measurements on CFRP composite during cyclic bending were conducted in transversal direction (Fig. 1). Change in electrical resistance in transversal resistance might be also valuable indicator of damage in beams and other parts, where only some surfaces are accessible. This paper presents first set of measured data.

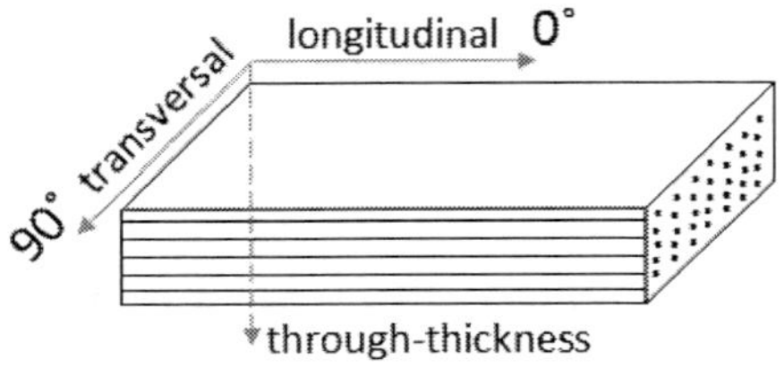


*Fig. 1 Possible direction of resistance measurement on CFRP.*

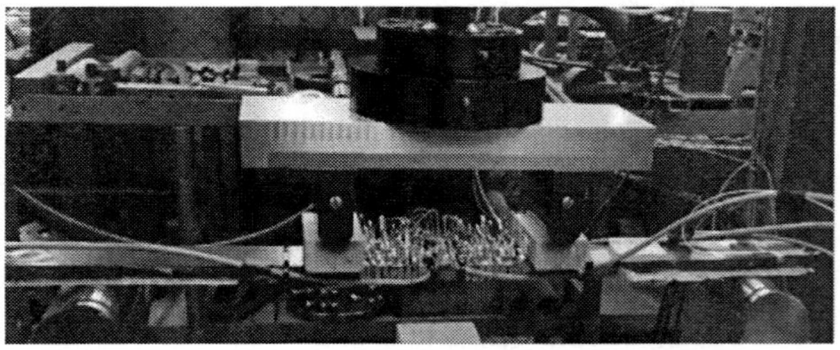

*Fig. 2 Specimen during loading.*

## Experimental Methods

**Materials.** The specimen used for the experiment was prepared from the unidirectional carbon/epoxy prepreg sheets. The composite was processed by prepreg moulding in an autoclave (90 min at 125°C and -80 kPa). Dimensions of the specimen are 800x60x13 mm. Electrical contacts were made using conductive epoxy resin and thin copper strips. Before preparing contacts specimen was sanded and cleaned with degreasing agent.

**Measurement configuration.** Configuration of the four-point bending tests is shown in Fig. 2 and Fig 3. A hydraulic testing system IST PL40N was used for four-point bending test of the specimen. Three types of electrode configurations were tested. Configuration of measuring points is shown in **Chyba! Nenašiel sa žiaden zdroj odkazov.**. To perform more accurate measurement, four-wire series comparison measuring method is used. Each measuring configuration (A, B, C) consists of two voltage contacts, which are between current contacts manufactured on the opposite sites of the beam. Each measuring configuration was prepared in two copies, so there were six measuring points. Measuring point was prepared in a such a way, that on one side of the specimen there was the voltage contact positioned closest to the top edge (depicted on the Fig. 4) and on the opposite side there was the voltage contact positioned closest to the bottom edge of the specimen. HP E3631A Current source was used for supply of 80 mA. Agilent 34461A device was used for voltage measurement.

**Cyclic loading.** Cyclic sinusoidal loading was applied to the specimen with the mean value of 2 kN and amplitude of 1 kN. This corresponds to the nominal value of the maximum stress on the sample surface 178 MPa. Before loading and during stops (shutdowns) between cyclic loading electrical resistance was measured at each measuring point under loading of 1, 2 and 3 kN. Besides this, electrical resistance at measuring point C2 was measured also during the cyclic loading. The temperature of the specimen was also measured during the loading using Pt-1000 sensor mounted to the top surface of the specimen.

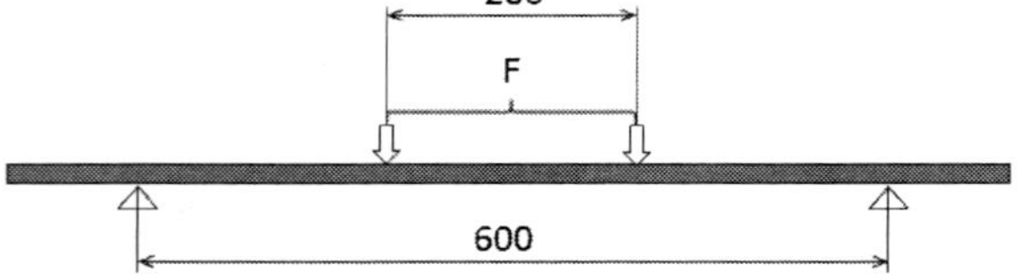


*Fig. 3 Configuration during loading of the specimen.*

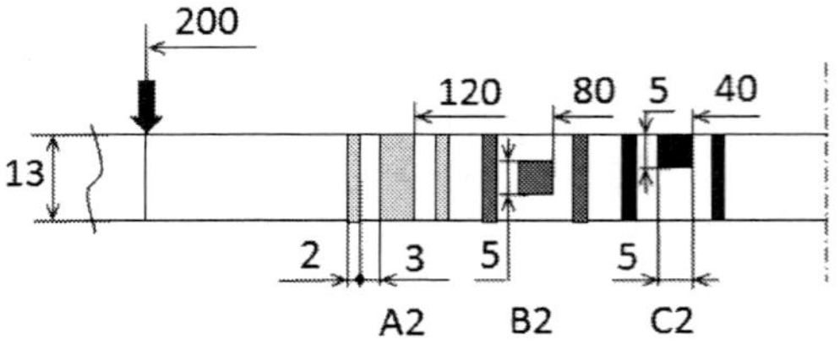


*Fig. 4 Configuration of measuring points.*

## Results and discussion

**Measuring during shutdowns.** A change in relative resistance measured during stops between loading blocks for all measuring points is shown in picture Fig. 5**Chyba! Nenašiel sa žiaden zdroj odkazov.**. Relative resistance change is defined according to equation Eq.1.

$$\frac{\Delta R}{R_0} = \frac{R - R_0}{R_0} \quad (1)$$

In this formula $R_0$ is electrical resistance measured at load level 1 kN. It can be seen that measuring points C1 and C2 exhibit the biggest relative resistance change during loading. Because of this, measuring point C2 was chosen for measuring also during cyclic loading. When we compare measured data at the same measuring configurations, we can see that measuring points on the opposite halves of the beam have inverse dependence on loading. It is obvious that trend is similar for all measuring configurations on one side (A1, B1, C1 vs. A2, B2, C2). So we suppose that it is not caused by the method of preparing electrical contacts or measuring configuration. On the base of observed behavior of the beam there is an assumption, that it is caused by imperfection in the specimen itself. This hypothesis needs to be verified by further measurement.

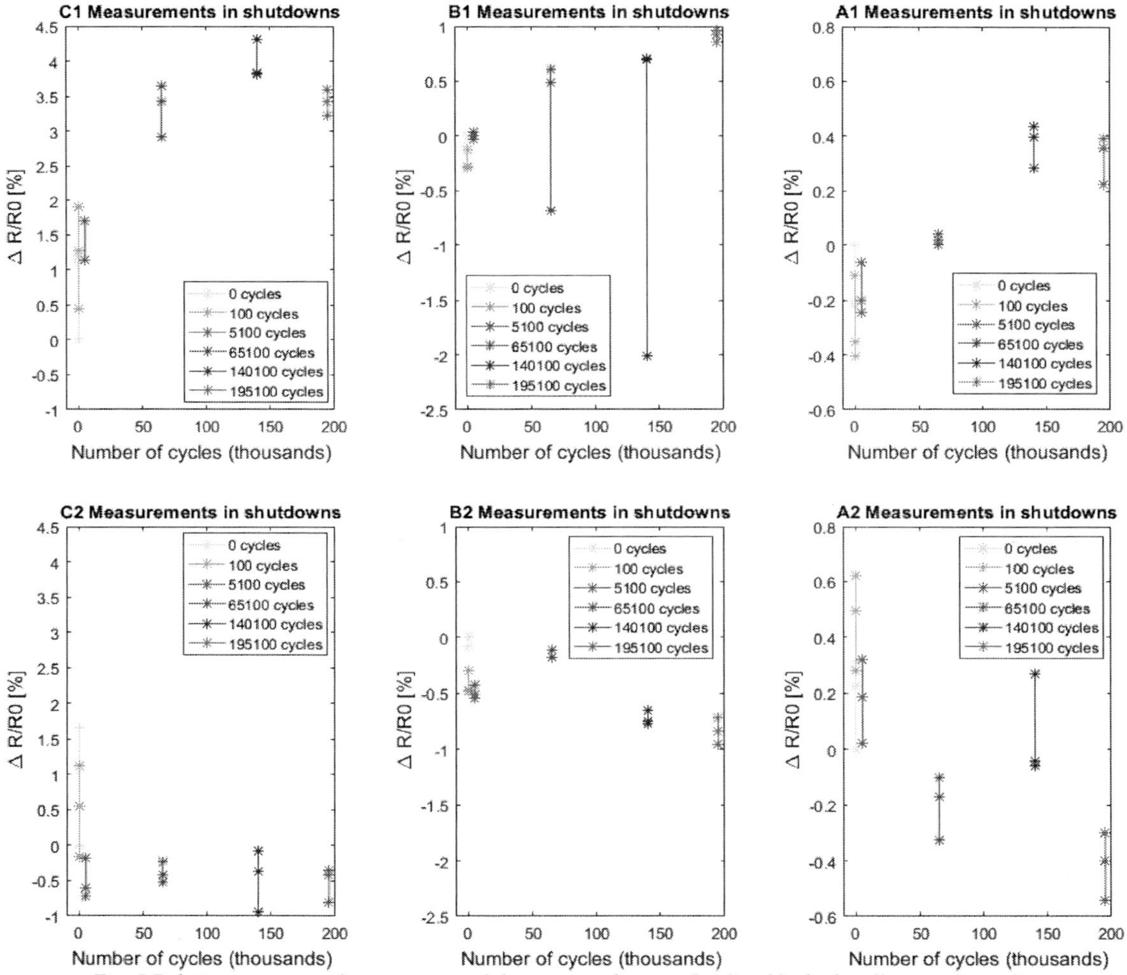


*Fig. 5 Relative resistance change measured during stops between loading blocks for all measuring points.*

After 195100 cycles artificial damage was created in a form of a cut in the middle of a specimen so that the cross section was reduced to 57x13 mm. So the depth of the cut was 3 mm and the width of the cut was 2 mm. After this weakening of the beam, measured electrical resistance significantly increased by more than 20 % in the measured point C2. Change in electrical resistance was observed in all measured points (Fig. 6). From this it can be drawn, that interruption of carbon fibers can be detected also by measurement in transversal direction.

**Cyclic loading.** Measured data at measuring point C2 are shown in figure Fig. 7. During first 2000 cycles we could observe significant change in measured electrical resistance, afterwards slight decreasing of electrical resistance can be observed.

In figure Fig. 7 temperature measured during cyclic loading is also depicted. On measured data during loading cycles 5100-15100 and 17820-195600 cycles it can be seen that at the same temperature of the specimen electrical resistance decreased. So the influence of the temperature on the measured resistance change doesn't need to be considered.

In figure Fig. 9 displacement under the load site is depicted. A small change in the value of displacement can be observed after the lowering of the specimen's cross section at 195100 cycles.

During the following loading cycles we can observe that measured resistance is decreasing, but it is not approaching its previous values, see data in figure Fig. 8.

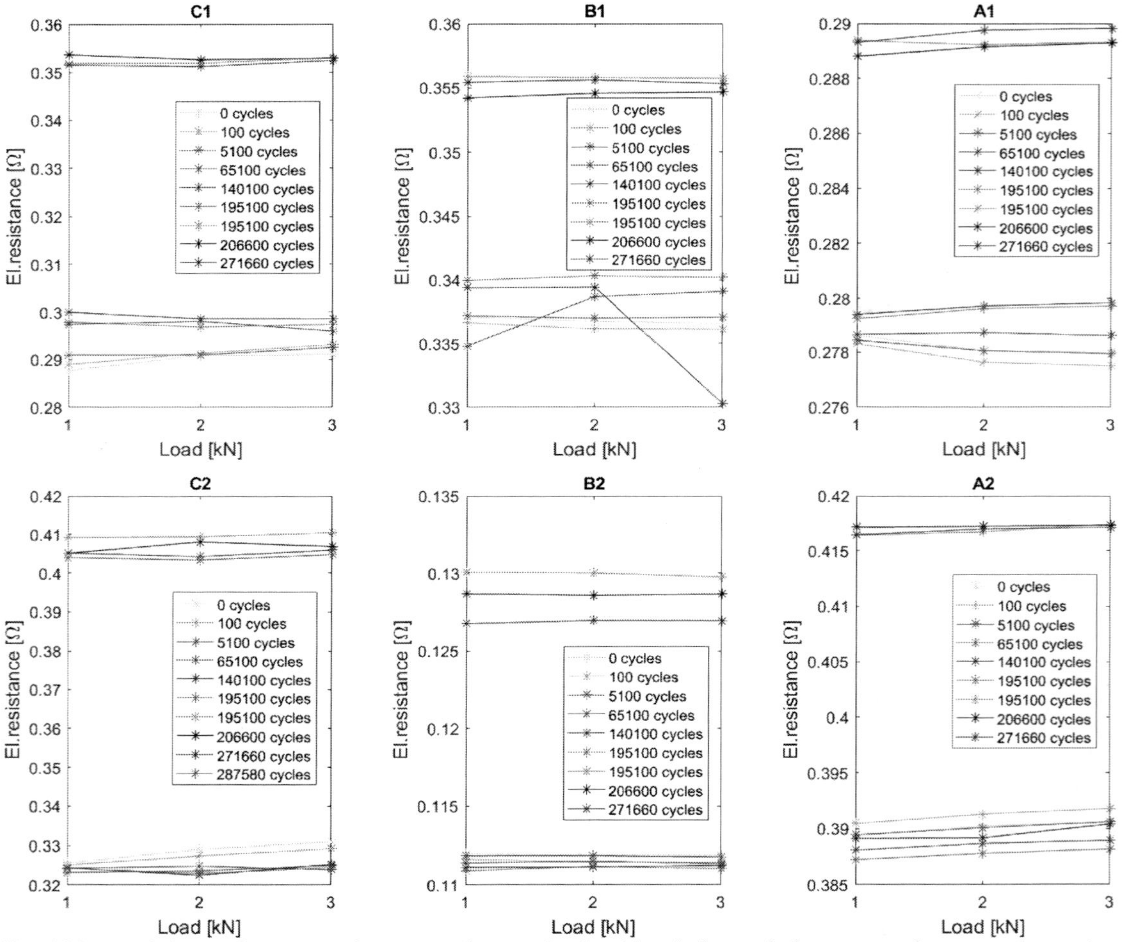


*Fig. 6 Measured electrical resistance during stops between loading blogs before and after creating damage on the specimen.*

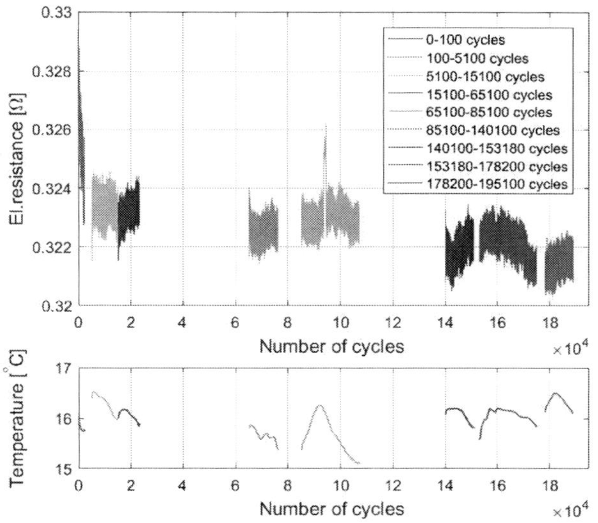


*Fig. 7 Electrical resistance measured during cyclic loading at measuring point C2 and measured temperature of the specimen.*

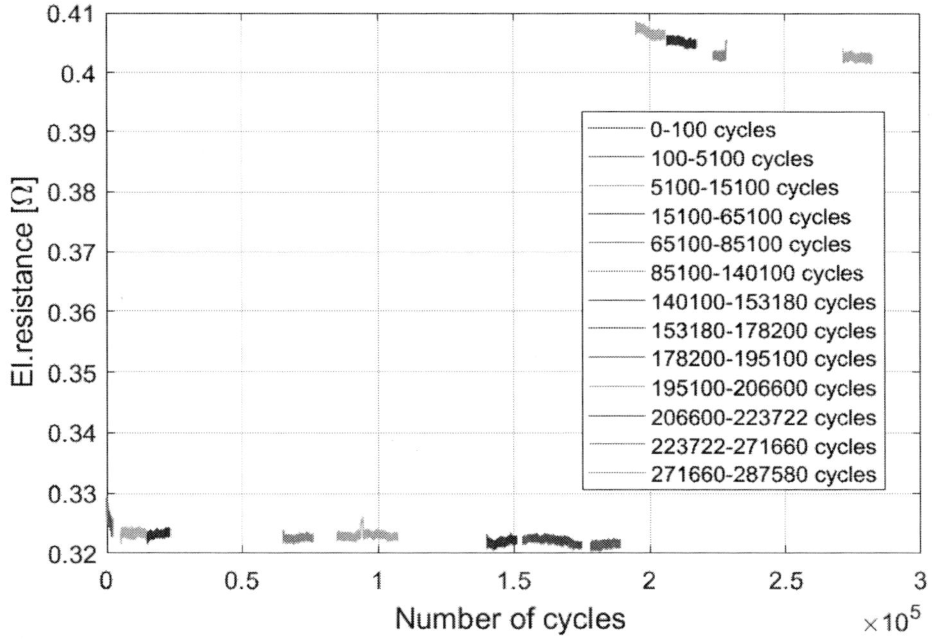


*Fig. 8 Electrical resistance measured during cyclic loading at measuring point C2 before and after creating damage on the specimen.*

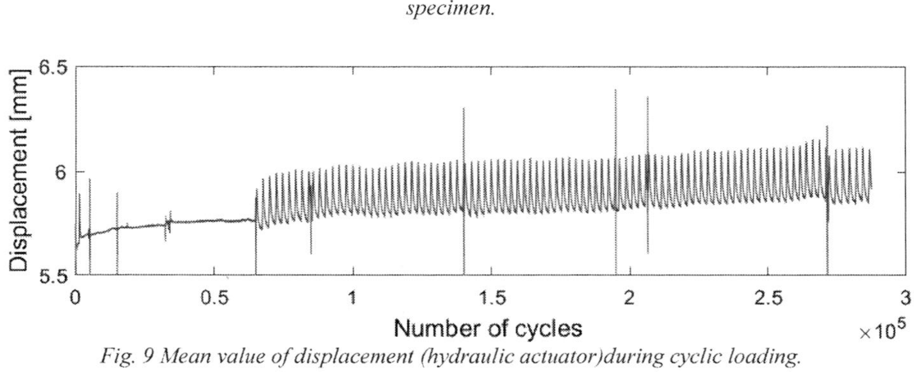


*Fig. 9 Mean value of displacement (hydraulic actuator)during cyclic loading.*

## Conclusions

The performed study suggests that the best measuring configuration for transversal electrical resistance measurement is type C, because the average relative change in measured resistance before and after loading was the most significant. Regarding possibility of damage detection during cyclic flexural loading, so far no definite conclusions can be drawn. Experimental data measured during shutdowns at the same measuring configurations exhibit the opposite trend. On the base of observed behavior of the beam there is an assumption, that it is caused by imperfection in the specimen itself. By measuring electrical resistance during cyclic loading decreasing of electrical resistance especially at the beginning of the loading was observed. Artificial damage confirmed possibility to detect interrupting of carbon fibers in unidirectional composite also by measuring in transversal direction.

## Acknowledgement

This work has been supported by Grant Agency of the Czech Technical University in Prague, grant No. SGS15/188/OHK2/3T/12

## References

[1] Wang, Shoukai a Chung, D. D.L. Self-sensing of flexural strain and damage in carbon fiber polymer-matrix composite by electrical resistance measurement. Carbon. Vol. 44 (2006) 2739-2751.

[2] X. Wang and D.D.L. Chung, Sensing Delamination in a Carbon Fiber Polymer-Matrix Composite During Fatigue by Electrical Resistance Measurement, Polymer Composites. Vol. 18, No. 6 (1997) 692-700.

[3] Todoroki, Akira, et al. Electrical Resistance Change of Carbon/Epoxy Composite Laminates under Cyclic Loading under Damage Initiation Limit. Open Journal of Composite Materials. 4, 2014, p. 22-31.

[4] Louis, M., Joshi, S.P. a Brockmann, W. An experimental investigation of through-thickness electrical resistivity of CFRP laminates. Composites Science and technology. 2001, p. 911-919.

EAN 2017
55th conference on experimental stress analysis

# Comparation of Mechanical Responses of Dental Implants Manufactured by 3D Printing and Machining

JÍRA Aleš[1, a], DENK František[1,b] and DENK František[2,c]

[1]Czech Technical University in Prague, Faculty of Civil Engineering, Department of Mechanics, Thákurova 7, 166 29 Prague, Czech Republic

[2]Czech Technical University in Prague, Faculty of Mechanical Engineering, Department of Materials Engineering, Karlovo nám. 13, 121 35 Prague, Czech Republic

[a]jira@fsv.cvut.cz, [b]frantisek.denk@fsv.cvut.cz, [c]fdenk@seznam.cz

**Keywords:** dental implant; titanium alloy; mechanical test; 3D printing; machining

**Abstract.** This paper deals with the basic comparison of dental implant stems behavior during static loading tests. The implants used for the tests were manufactured by the mechanical machining method with use of the standard alloy Ti6Al4V and experimental alloy Ti35Nb6Ta. Another implant used for loading tests was produced by the method of 3D printing from the standard alloy Ti6Al4V. The shape solutions of used implants are based on patent files no. CZ 306456 and CZ 306457. The method of loading test is based on valid legislative, which is necessary for product launch on the market and it is managed by ČSN EN ISO 140801 (Dentistry – Implants – Dynamic loading test for endosseous dental implants).

## Introduction

The stems of dental implants are conventionally manufactured by machining the titanium alloy Ti6Al4V. To ensure a good stability of the implant in the bone, a very good level of material biocompatibility is required [1]. The Ti6Al4V alloy has a long recorded history of success, but its use is usually coupled with applying a special coating (i.e.: plasmatic Ti coating, acid etching, blasting) and limiting the complexity of the intraosseous implant geometry. The overall stiffness of the implant then substantially exceeds the stiffness of the former tooth and the surrounding bone tissue [2]. This paper is dedicated to the comparison of mechanical load capacities of four different dental implant stems. For the purpose of mechanical tests, we used two different implant stem geometries, two different materials and two different manufacturing methods. The methods and materials were chosen with consideration of utilizing more complex geometries (3D printing), high biocompatibility (Ti35Nb6Ta) and lower material stiffness. The main goal of this research was to determine the ultimate bearing capacity of various implants according to the ČSN EN ISO 14801 (Dentistry – Implants – Dynamic loading test for endosseous dental implants) [3] and their comparison.

## Methods

Two basic intraosseous implant geometries were used in the experiment - the „ribbed“ [patented CZ patent n. 306456] and the „four leaf clover“ [patented CZ patent n. 306457] variants (Fig. 1). A total of 24 specimens were created and divided into four groups (Table 1).

Fig. 2: Computer 3D models: left – "ribbed" implant, right – "four-leaf clover" implant.

The use of alternative material with beta-structure Ti35Nb6Ta and the optimization of the intraosseous part of the implant is based on the need to lower the total stiffness of the implant stem. High stiffness of the stem causes stress-shielding and thus an unwanted distribution (lowering) of stress in the surrounding bone tissue.

Table 1.: Implant specimens specificatitons

| Implant type | Geometry | Material | Manu. method | Diameter [mm] |
|---|---|---|---|---|
| Variant I | ribbed | Ti6Al4V | machined | Ø3.8-10 |
| Variant II | ribbed | Ti35Nb6Ta | machined | Ø3.8-10 |
| Variant III | four leaf clover | Ti35Nb6Ta | machined | Ø3.8-10 |
| Variant IV | four leaf clover | Ti6Al4V | 3D printed | Ø3.8-10 |

Accurate comparison of used materials behavior was based on comparative measurement of micromechanical properties (reduced elastic modulus and hardness) analyzed by the nanoindentation method [4]. Material samples were fixed in epoxy resin, cut, polished and measured by the CSM Instruments device. Reduced modulus $E_r$ values are around ~ 126 GPa in case of the Ti6Al4V material, around ~ 122 GPa in case of the Ti6Al4V material manufactured by the 3D printing method and around ~ 82 GPa in case of the Ti35Nb6Ta material. Hardness $H_{it}$ values are around ~ 4.58 GPa in case of the Ti6Al4V material, around ~ 5.21 GPa in case of the Ti6Al4V material manufactured by the 3D printing method and around ~ 2.98 GPa in case of the Ti35Nb6Ta material [5].

Other components have to be produced for mechanical loading text realization, for example grouting plate, loading adapters with cylindrical surface, contact caps and connecting screws (Fig. 2). The grouting plate meets the requirements of the norm and allows repeated precision grouting of the implant's intraosseous part by the two-component methyl methacrylate resin (Dentacryl). Other used components replace the extension part of the implant (abutment with crown) and ensure the force transmission to the implant body.

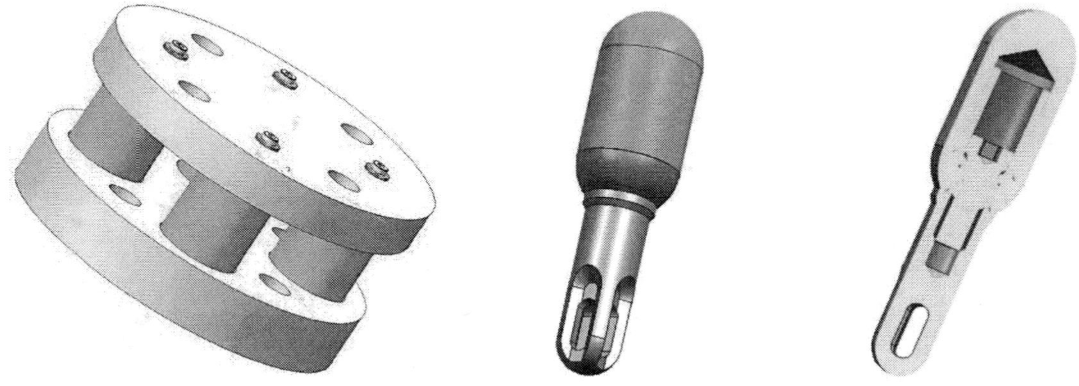

A B

Fig. 2: Computer 3D models: A – grouting plate for samples preparation, B – ribbed implant with loading adapter.

For the purpose of testing, we used the MTS Mini Bionix 8_58.02 system and the methodology according to the ČSN EN ISO 14801 standard. The mechanical tests were carried out by anchoring the test specimens in a special mount, fitting on a special cover and loading. Static loading was done at a constant speed of 4.0 mm/min. The air temperature was 20±5°C.

**Results**

For the 4 tested groups of newly designed implants, we determined the values of ultimate bearing capacities for static load $F_{ST}$ [Table 2]. During the next phase of measurement, we will determine the limit fatigue $F_{FL}$ for $5 \times 10^6$ cycles.

Table 2. Ultimate bearing capacities for static load of tested implants

| | Variant I | Variant II | Variant III | Variant IV |
|---|---|---|---|---|
| Ult. bearing capacity $F_{ST}$ [N] | 586.0 | 446.0 | 335.0 | 484.4 |
| Std. dev. [N] | 35.0 | 17.0 | 8.0 | 4.5 |
| Expanded uncertainty [N] | 69.0 | 33.0 | 17.0 | 9.3 |

Different type of damage was detected for individual implant variants after the loading tests (Fig. 3). Bearing capacity was achieved in case of variant I and II by screw collapse at the connection between the implant and the abutment. The intraosseal part of the implant was not damaged by fracture. In case of variant III and IV the fracture damage was detected at the interface of the implant and the grouting resin in accordance with the assumption of lower stiffness of the implant stem. The most frequent response of variant III during the loading test is the creation of a great plastic deformation with following non-increasing loading force. The most frequent collapse of stem variant IV was caused by fracture damage inside the body of the intraosseal part of the implant.

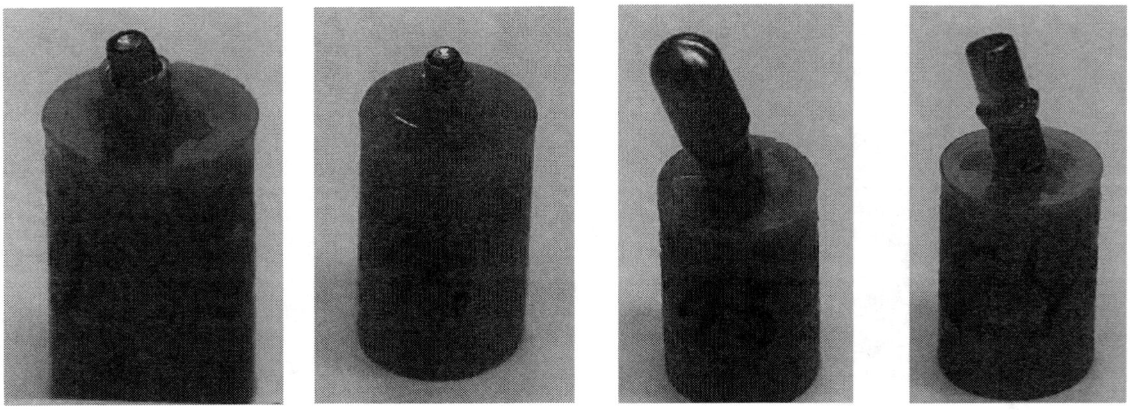

Variant I Variant II Variant III Variant IV

Fig. 3: Example of collapse in case of individual implant types after the loading tests.

## Conclusions

The experimental tests have shown a great difference in the ultimate bearing capacities of different implant variants. As predicted, the highest attained load capacity was recorded in the conventionally machined Ti6Al4V specimen. By introducing the beta-structured titanium alloy, a great reduction in stiffness was observed. The „four leaf clover" geometry scores a lower value of load capacity as well. The reduction of overall stiffness of the implant is beneficial in regard to the elimination of stress-shielding in the intraosseous implant stem.

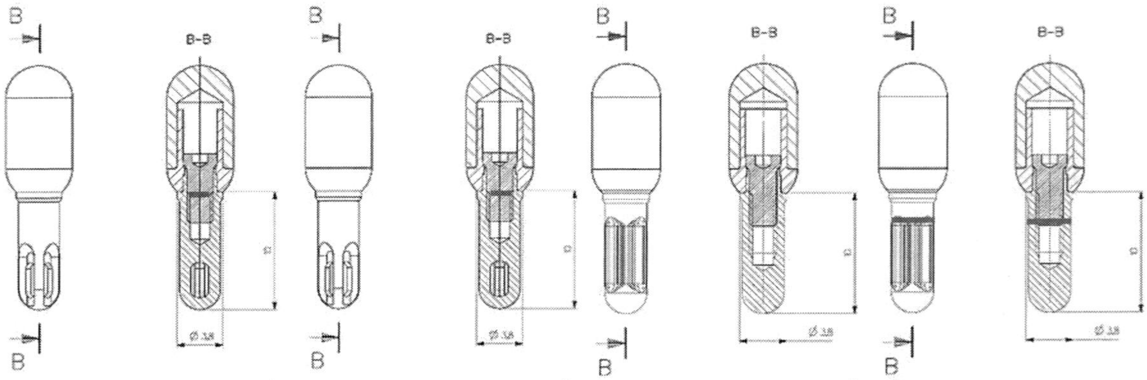


Fig. 4: Schematic illustration of common fracture localities (red line) in case of tested implants after the loading tests. From left side: variant I (screw damage), variant II (screw damage), variant III (plastic deformation without fracture), variant IV (implant damage at the anchoring interface)

## Acknowledgement
The financial support by the Technology Agency of the Czech Republic (TAČR project no. TA03010886) and Faculty of Civil Engineering, Czech Technical University in Prague (SGS project No. SGS17/168/OHK1/3T/11) is gratefully acknowledged.

## References

[1] P.I. Branemark, Osseointegration and its experimental background, J Prosth Dentistry, 50 (1983) 399–410.

[2] A. Jíra, M. Petrtýl, Elastic Properties of the Peripheral Surface of Implants, Human Bimomechanics, Czech Republic (2010) 139-143.

[3] ISO 14801. Dentistry -- Implants -- Dynamic loading test for endosseous dental implants. 2016. 16 p.

[4] W. Oliver, G. Pharr, An improved technique for determining hardness and elastic modulus using load and displacement sensing indentation experiments, J. Mat. Res. 7 (1992) 1564-1583.

[5] L. Řehounek, F. Denk, A. Jíra, Trabecular structures of dental implants formed by 3D printing and their mechanical properties, 54th International Conference on Experimental Stress Analysis, Czech Republic (2016)

EAN 2017
55th conference on experimental stress analysis

# Problematics of Twin-disc Wear Resistance Tests for Selected Wheel Rim Materials

Jiří ŠMACH[1,a], Michal ŠOFER[1,b], Lenka ČEPOVÁ[2,c] and Radim HALAMA[1,3,d]

[1] Department of Applied Mechanics, Faculty of Mechanical Engineering, VŠB-Technical University of Ostrava, 17. listopadu 15/2127, 708 33 Ostrava-Poruba, Czech Republic

[2] Department of Machining, Assembly and Engineering Metrology, Faculty of Mechanical Engineering, VŠB-Technical University of Ostrava, 17. listopadu 15/2127, 708 33 Ostrava-Poruba, Czech Republic

[3] IT4Innovations, VŠB-Technical University of Ostrava, 17. listopadu 2172/15, 708 33 Ostrava-Poruba, Czech Republic

[a]jiri.smach@vsb.cz, [b]michal.sofer@vsb.cz, [c]lenka.cepova@vsb.cz, [d]radim.halama@vsb.cz

**Keywords:** twin-disc test, wear resistance.

**Abstract.** Presented paper deals with problematics of methodology of twin-disc wear resistance tests for various types of commercially available materials, which are used for production process of rail wheel rims. Currently, there doesn´t exist an integrated document, which will deal with problematics of loading force correction according to the actual width of the contact path. The aim of this paper is to present findings related to the effect of correction of loading force according to the actual width of the contact path in the transversal direction on the weight loss, resulting contact path width and sample diameter. Proposed paper presents two techniques of measurement the contact path width, which is further used for computation of the real maximum contact pressure.

## Introduction

Thanks to progressively higher traveling speeds with increasing axle loads takes the railway industry ever greater demands on the materials, used for production of wheel rims as well as rails. Currently, the more attention is paid on various tests [1,2,3], which will as best as possible simulate real conditions. Twin-disc test is nowadays widely used as a standard experiment for qualitative analysis in terms of obtaining the information related to wear resistance as well as resistance to contact fatigue [4,5,6]. In most cases, the authors do not include the effect of change of contact path width on the number of realized cycles [7,8]. This phenomenon can, however, significantly influence the final results in terms of the wear rate or the profile of accumulated plastic shear deformation below contact surface. Proposed paper deals with the problematics of the variation of contact pressure as a consequence of the overall sample deformation in case of twin-disc test experiments. The authors also focused on the effect of the change of the contact path width with increasing number of cycles. The results point to the fact, that to each couple of samples (wheel rim/rail material) needs to be proposed a specific test procedure, which will take into account the specimen's geometry, testing conditions and as well as mechanical properties of tested materials in itself.

## Experimental procedure

The experiments have been realized on twin-disc experimental rig (Fig. 1), manufactured by INOVA company. Authors focused on two wheel rim/rail material sets of specimen. Set 1 consists from Class C (AAR M107 Standard) wheel rim sample and Class C rail sample hardened up to 350 HB. Set 2 consists from ER7 (EN 13262) wheel rim sample and the same rail sample as used in Set 1. The initial diameter of wheel rim specimen was for both sets equal to 70 millimeters with initial contact path width of 10 millimeters. The initial diameter of the rail samples was equal to 220 millimeters. The width of the rail samples was equal to 30 millimeters (See Figs. 2 and 3). Both sets carried out 100000 cycles. The slip and maximum Hertz contact pressure were equal to 2% and 1200 MPa respectively.

*Fig. 1: INOVA twin-disc experimental rig*

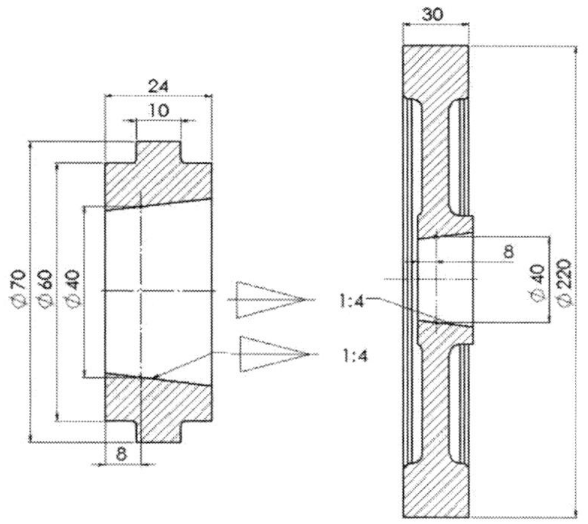


*Fig. 2: Geometry of wheel and rail specimen*

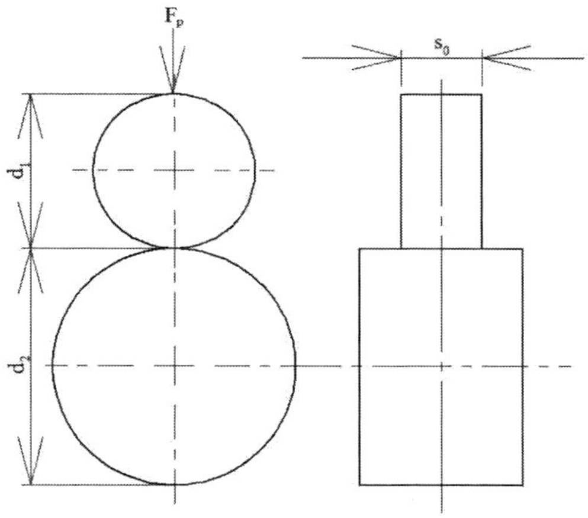


*Fig. 3: Dimensions for calculation of maximum Hertz contact pressure*

According to the known value of contact pressure, material properties of the specimen, which are in contact and their geometries, it is then possible to calculate the initial loading force from following expression [9]:

$$F_p = 0{,}5 \cdot \left(\frac{p_0}{0{,}59}\right)^2 \cdot \frac{d_1 \cdot d_2}{d_1 + d_2} \cdot \left(\frac{1}{E_1} + \frac{1}{E_2}\right) \cdot s_0, \tag{1}$$

where $F_p$ [N] is loading force, $p_0$ [MPa] is maximum Hertz contact pressure in terms of line contact, $d_{1,2}$ [mm] are diameters of the specimens, $E_{1,2}$ [MPa] represents Young modulus and $s_0$ [mm] is initial contact path width. For further evaluation of contact pressure, it is necessary from eq. 1 to derive the maximum contact pressure as the function of loading force, material and geometric properties:

$$p_0 = 0{,}59\sqrt{\frac{2 \cdot F_p}{s_0 \cdot \left(\frac{1}{E_1} + \frac{1}{E_2}\right) \cdot \frac{d_1 \cdot d_2}{d_1 + d_2}}} \qquad (2)$$

Every 25000 cycles, both specimens were subjected to measurement of the transversal profile in order to obtain more detailed information related to contact path width. The measurements were realized with use of Wenzel LH65 CNC X3M premium 3D coordinate measuring machine. Additionally, authors have also used a red contrast penetrant in order to estimate the contact path width during the test itself, when the specimen were subjected to predefined loading force. The method is based on the application of the contrast penetrant, used for NDT penetration testing, on the contact path of loaded specimen under low speed rotation.
Resulting measured transversal profiles of wheel and rail specimen were then used for estimation of contact path width after 25000, 50000, 75000 and 100000 cycles and as well as for further calculation of real value of maximum contact pressure based on preset loading force.

## Results

Figure 4 shows transverse profiles of wheel and rail specimen for both sets after 25000, 50000, 75000 and 100000 cycles, which have been measured with use of Wenzel LH65 3D coordinate measuring machine. According to the similarity of both profiles can be estimated the width of the contact path, which is listed, together with the results determined by use of contrast penetrant, in table 1. As can be seen, the conformity between the wheel and rail profile is higher in case of Set 1, where the wheel as well as rail specimen is manufactured from Class C material with additional hardening of the rail specimen up to 350 HB. The higher hardness of the Class C wheel specimen as well as its different mechanical properties causes the higher rates of plastization of the contact path´s surroundings compared to ER7 wheel specimen (Set 2).

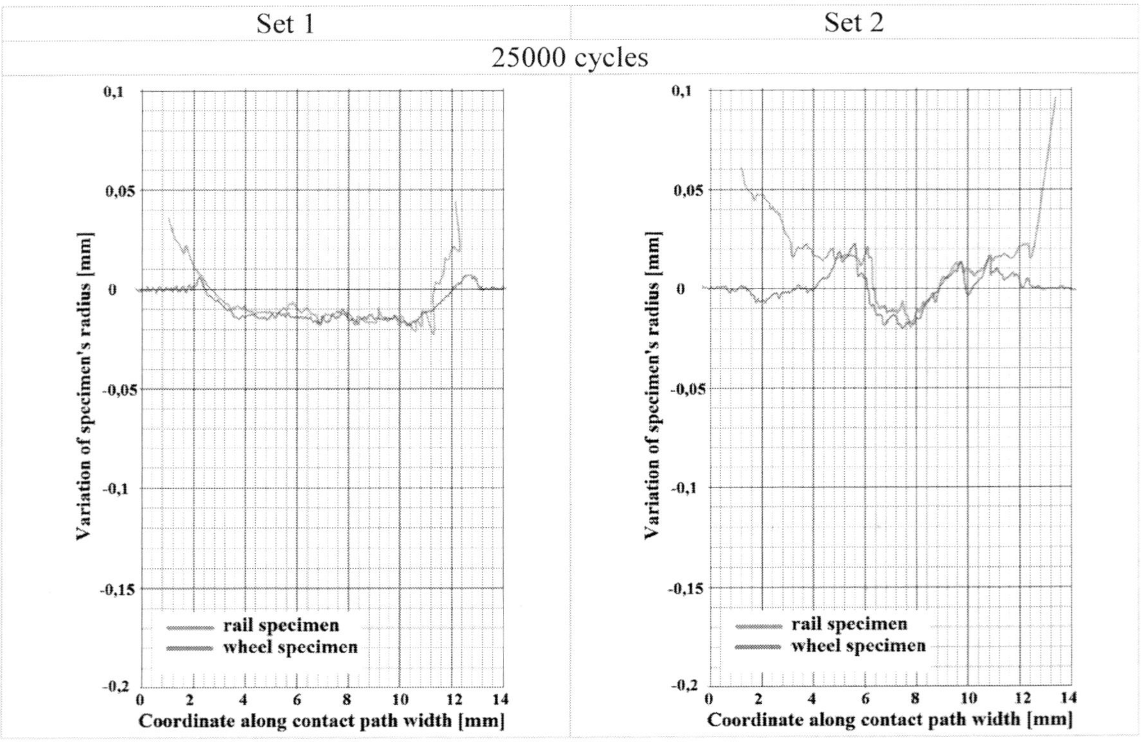

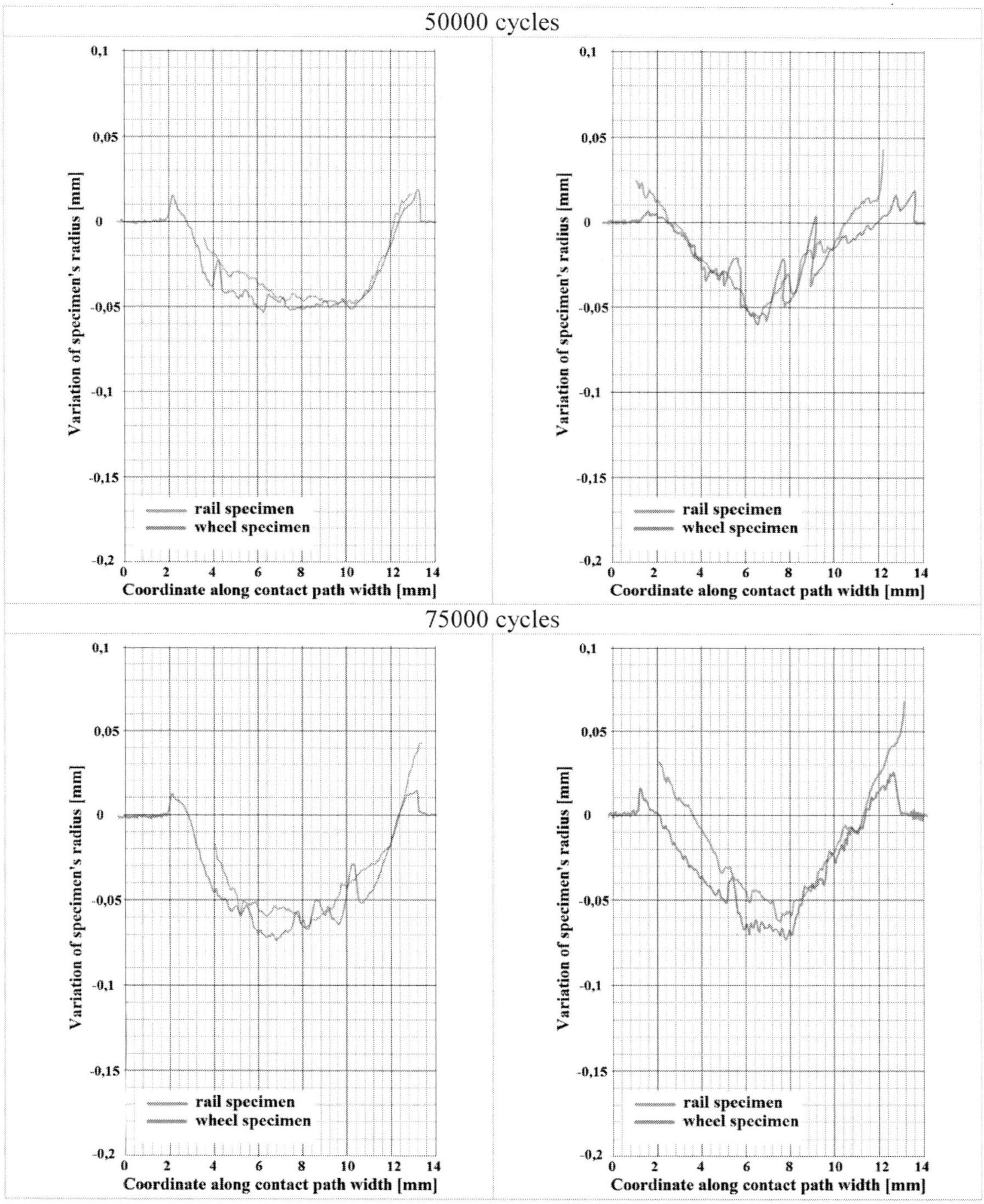
50000 cycles
75000 cycles
Variation of specimen's radius [mm]
Coordinate along contact path width [mm]
rail specimen
wheel specimen
0,1
0,05
0
-0,05
-0,1
-0,15
-0,2
0 2 4 6 8 10 12 14

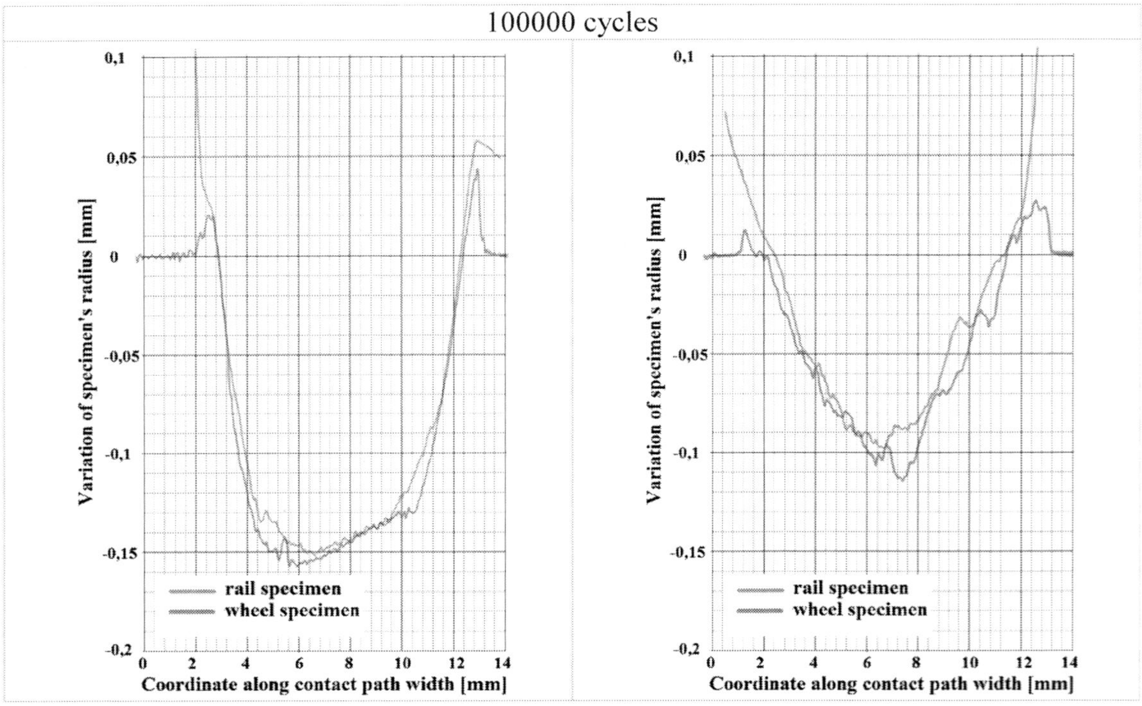


*Fig. 4: Transverse profiles for Set 1 and Set 2 after 25000, 50000, 75000 and 100000 cycles obtained by use of Wenzel LH65 3D coordinate measuring machine*

On the other hand, the lower hardness of the ER7 wheel specimen together with the hardened Class C rail specimen results in the lower plastization rates in both, horizontal as well as vertical, directions across the transverse profile (see Fig. 4 – Set 2).

Table 1 contains data of contact path width, which have been measured with use of 3D coordinate measuring machine (CMM) and contrast penetrant method (CP). In most cases, the CP method exhibit higher value of contact path width compared to CMM method. This finding is mainly caused by the measurement method in itself. In case of CP method, the specimen are loaded and are in contact, while in case of CMM method, the specimen are during their measurement unloaded. Overally, the ER7 wheel specimen exhibit during its test slightly higher values of contact path width compared to Class C wheel specimen. This fact is mainly caused by the lower values of specimen´s surface hardness.

*Tab. 1: Contact path width measurement with use of 3D coordinate measuring machine (CMM) and contrast penetrant (CP)*

| Number of cycles | Class C/Class C | | ER7/Class C | |
|---|---|---|---|---|
| | CP | CMM | CP | CMM |
| 0 | 9,97 mm | - | 10,00 mm | - |
| 25.000 | 8,45 mm | 8,13 mm | 9,06 mm | 8,12 mm |
| 50.000 | 9,12 mm | 8,52 mm | 9,49 mm | 9,03 mm |
| 75.000 | 8,74 mm | 7,99 mm | 10,03 mm | 9,80 mm |
| 100.000 | 10,1 mm | 9,76 mm | 9,15 mm | 9,67 mm |

Another subject of discussion is the real value of maximum Hertz contact pressure. In the beginning of each test, the loading force has been set in order to induce maximum Hertz contact pressure of 1200 MPa for 10 millimeters wide specimen. The value of the loading force hasn´t been changed during the entire test in order to evaluate the real value of the maximum Hertz contact pressure and compare it to initially preset value of the contact pressure. Table 2 and Fig. 5 summarize the values and trends of calculated maximum Hertz contact pressure (Eq. 1) according to measured contact path width by CP and CMM method. For 25000 cycles, the value of maximum Hertz contact pressure is in case of both sets increasing up to 1330 MPa. Between 25000 and 100000 cycles, the trend is for both sets almost the same. The value of the maximum contact pressure is progressively decreasing or remains almost stable (Set 1) with the final decrease within the interval of 1200÷1250 MPa.

*Tab. 2: Calculated values of maximum Hertz contact pressure according to the values of measured contact path width*

| Number of cycles | Class C [MPa] | | ER7 [MPa] | |
|---|---|---|---|---|
| | CP | CMM | CP | CMM |
| 0 | 1200 | 1200 | 1200 | 1200 |
| 25.000 | 1304 | 1329 | 1267 | 1332 |
| 50.000 | 1257 | 1298 | 1239 | 1263 |
| 75.000 | 1286 | 1341 | 1198 | 1212 |
| 100.000 | 1191 | 1213 | 1255 | 1220 |

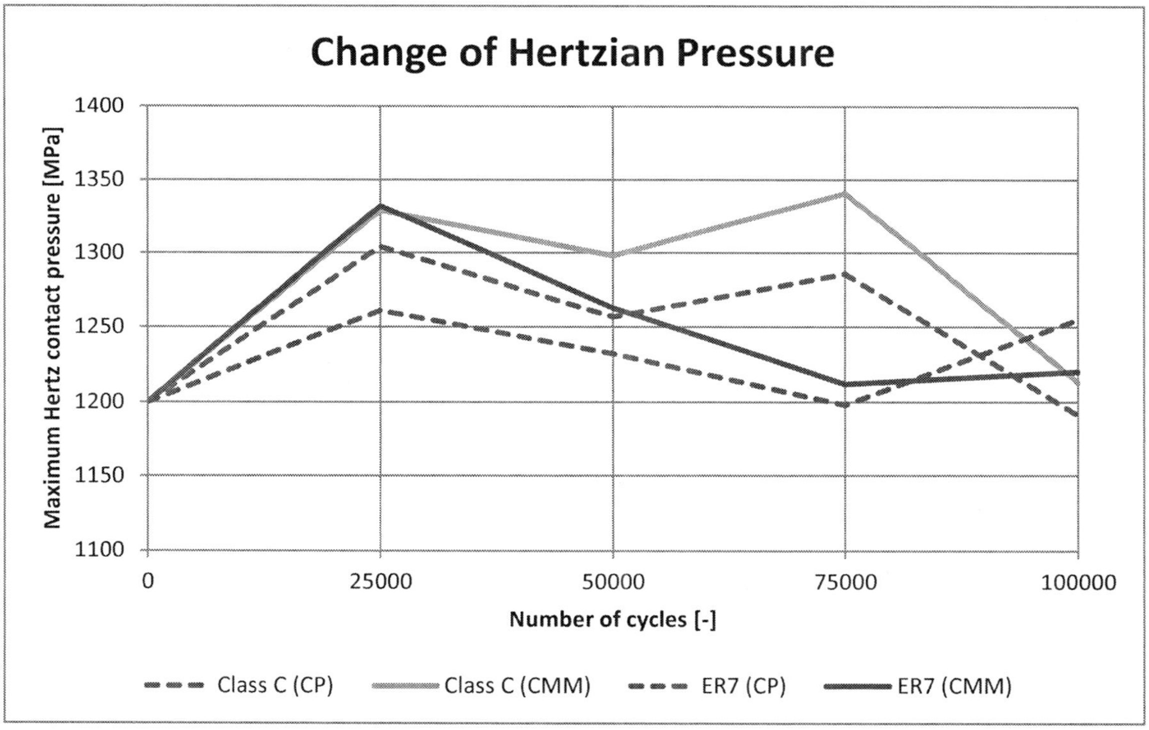


*Fig. 5: Computed maximum Hertz contact pressure*

## Conclusion

Presented results and findings point to the need to deal with precise measurement of contact path width and its evolution during the entire wear resistance test in more detail. The attention must be also paid to the specific combination of wheel/rail material, their material properties and, last but not least, to the desired value of maximum Hertz contact pressure including the number of simulated cycles or the slip ratio value. Proposed paper demonstrates two different methods of measurement the contact path width, the measurement by use of 3D coordinate measuring machine, which is capable to deliver precise results. However, the main drawback of this method is time-consuming measurement with additional complex data processing process. On the other hand, the incorporation of contrast penetrant method for contact path width measurement in case of loaded specimen seems to be a universal and inexpensive way how to relatively quickly obtain reasonable results, which will improve the informative character of the wear resistance test in itself.

**Acknowledgement**

This work was supported by Specific Research (SP2017/136) and by The Ministry of Education, Youth and Sports from the National Programme of Sustainability II (LQ1602). The support of GHH-BONATRANS Company is also acknowledged.

**References**

[1] W.R. Tyfour, J.H. Beynon, A. Kapoor, The steady state wear behavior of pearlitic rail steel under dry rolling–sliding contact condition. Wear 180 (1995) 79–89.

[2] A. Ramalho, M. Esteves, P. Marta, Friction and wear behavior of rolling-sliding steel contacts. Wear 302 (2013) 1468–1480.

[3] S. Jung-Won, K. J. Hyun-Kyu, Seok-Jin, L. Dong-Hyeong, Rolling contact fatigue and wear of two different rail steels under rolling-sliding contact. Fatique 83 (2016) 184–194.

[4] W.R. Tyfour, J. H. Beynon, A. Kapoor, Deterion of rolling contact fatigue life of pearlitic rail steel due to dry-wet rolling-sliding line contact. Wear 197 (1996) 255–265.

[5] S. Maya-Johnson, J.F. Santa, A. Toro, Dry and lubricated wear of rail steel under rolling contact fatigue – wear mechanism and crack growth. Wear 380-381 (2017) 240–250.

[6] F. Bucher, A.I. Dimitriev, M. Ertz, K. Knothe, V.L. Popov, S.G. Psakhie, E.V. Shilko, Multiscale simulation of dry friction in wheel/rail contact. Wear 261 (2006) 874–884.

[7] S. Jung-Won, J. Hyun-Kyu, K. Seok-Jin, L. Dong-Hyeong, Analysis of contact fatique crack growth using twin-disc tests and numerical evaluations. Fatique 55 (2013) 54–63.

[8] S.R. Lewis, R. Lewis, G. Evans, L.E. Buckley-Johnstone, Assessment of railway curve lubricant performance using a twin-disc tester. Wear 314 (2014) 205–212.

[9] M. Šofer, Description of surface degradation of construction components due to repeated contact loading, Dissertation Thesis, Ostrava, 2012, VŠB-TUO, Faculty of Mechanical Engineering, Department of Mechanics of Materials. (In Czech)

# Identification of impact force on composite textile plate for impacts with various energies

Jan Bartošek[1, a], Vladislav Laš[1], Tomáš Mandys[1] and Robert Zemčík[1]

[1] NTIS – New Technologies for Information Society, Faculty of Applied Sciences, University of West Bohemia, Univerzitní 8, 306 14, Plzeň, Czech Republic

[a]bartose4@ntis.zcu.cz

**Keywords:** impact force identification, inverse problem, drop weight test, composite material, delamination, damage, structural health monitoring

**Abstract.** Identification of impact force from measured response of composite plate is performed in this paper for various impact energies. Method based on transfer functions in time domain is used. Identification was performed on glass-epoxy textile laminated plate, which was simply supported and impacted by drop weight with various impact velocities up to occurrence of significant damage of the plate. Identification results and possibility to identify impact force on damage structure is discussed.

## Introduction

Identification of impact force is problem of estimation of unknown force from signals measured by sensors placed on construction. Different quantities, which are not directly related to impact force, can be measured by sensors. Mostly piezoelectric sensors or accelerometers are used to measure strains or acceleration of construction. This approach is also called indirect measurement and can be used as a part of structural health monitoring (SHM) system of composite structure. Low-velocity impact events on composite structure can induce delamination of individual plies which can be inside of the material and hidden for classical visual inspections. SHM can replace time and cost expensive methods that would be needed for finding of such damage in construction.

Several methods were used to identify impact forces but mostly in laboratory conditions and for limited amount of loading [5]. Additionally, impact force is often identified virtually by simulated experiment, when numerical model needs to be calibrated [3] and more complex problems like sensor distribution can be solved [4]. Nearly industrial problem of identification of impact force on pressure vessel was solved by Atobe et al. [1].

## Identification method

The identification method used in the paper is based on the deconvolution of signals in time domain. The assumptions are validity of principle of superposition, impact loading limited to one place and measurement in $K$ discrete time steps. Then the response of sensor $s$ can be expressed in matrix form

$$\mathbf{u}^s = \mathbf{G}^s \cdot \mathbf{f}, \tag{1}$$

where $\mathbf{u}^s = \left[u_1, u_2, \ldots, u_K\right]^T$ is vector of response, $\mathbf{f} = \left[f_1, f_2, \ldots, f_K\right]^T$ is vector of applied forces and $\mathbf{G}^s$ is matrix composed of time shifted impus resoponse vetctors of sensor $s$.

This matrix must be determined before the identification process. The impulse response vectros can be estimated directly from numerical model or indirectly from measurement by

$$\mathbf{F}\cdot\mathbf{g}^s=\mathbf{u}^s, \tag{2}$$

where $\mathbf{F}$ is matrix simlarly composed as matrix $\mathbf{G}^s$ but from vectors of impact forces, $\mathbf{u}^s$ is vector of coresponding responses and $\mathbf{g}^s$ is unknown impuse response. The Eq. (2) can be rewriten for $N$ measuremetns in form

$$\begin{bmatrix}\mathbf{F}_1^s\\ \mathbf{F}_2^s\\ \vdots\\ \mathbf{F}_N^s\end{bmatrix}\cdot\mathbf{g}^s=\begin{bmatrix}\mathbf{u}_1^s\\ \mathbf{u}_2^s\\ \vdots\\ \mathbf{u}_N^s\end{bmatrix}. \tag{3}$$

This equation was solved in the paper by the least square method and estimation of impulse response $\hat{\mathbf{g}}^s$ was obtained.

The impact force history estimation $\hat{\mathbf{f}}$ is set as a solution of Eq. (1). For the measurement using $S$ sensors the equation can be rewritten in form

$$\begin{bmatrix}\mathbf{G}^1\\ \mathbf{G}^2\\ \vdots\\ \mathbf{G}^S\end{bmatrix}\cdot\mathbf{f}=\begin{bmatrix}\mathbf{u}^1\\ \mathbf{u}^2\\ \vdots\\ \mathbf{u}^S\end{bmatrix}, \tag{4}$$

which was solved in the paper by Tikhonov regularization method with the non-negative constraint representing positive values of contact forces [2].

The correctness of the inverse solution can be characterized by Euclidean norm of residuum vector which is comparison of measured and calculated response

$$r=\left\|\mathbf{u}-\hat{\mathbf{u}}\right\|. \tag{5}$$

Response $\hat{\mathbf{u}}$ calculated from the inverse solution can be expressed from Eq. (1) for estimated impulse responses

$$\hat{\mathbf{u}}^s=\mathbf{G}^s\cdot\hat{\mathbf{f}}, \tag{6}$$

or from Eq. (2) for estimated impact forces

$$\hat{\mathbf{u}}^s=\mathbf{F}\cdot\hat{\mathbf{g}}^s. \tag{7}$$

Impact force identification error was defined as weighted comparison of identified and measured impact force

$$e_{\mathbf{f}}=\frac{\sqrt{\frac{1}{I}\sum_{i=1}^{I}\left(f_i-\hat{f}_i\right)^2}}{\max(\mathbf{f})}\times 100\ [\%],\qquad \forall i: f_i>\frac{\max(\mathbf{f})}{2}\vee\hat{f}_i>\frac{\max(\mathbf{f})}{2}, \tag{8}$$

where $I$ is total number of compared elements. The constraint for choice of individual elements for comparison was considered to suppress errors from small forces which have low importance.

## Experiment

The impacts on composite plate were performed experimentally on the impact device which consists of impactor with total mass of 2.336 kg and aluminum support frame. Hemispherical tip of the impactor with radius of 15 mm was equipped with force sensor *Kistler 9712B5000*. The impacts were performed on square composite plate made from three layers of glass fabric with plain weave (816 g/m$^2$) and epoxy resin *Epicote MGS LR 385*. The plate was simply supported around its perimeter by steel support and it was impacted in the center. Displacements of the plate were measured by three laser contactless displacement sensors (two pcs. of *OptoNCDT 2300-50* and one pc. of *OptoNCDT 2200-50*). Data acquisition of signals from lasers and force sensor were performed by *NI cDAQ-9178*, with modules NI 9215 and NI 9234, with sampling frequency of 10 kHz. Total length of recorded data was 50 ms which corresponds to 500 samples. The experimental setup is shown in *Fig. 1* and the important dimensions are in *Table 1*.

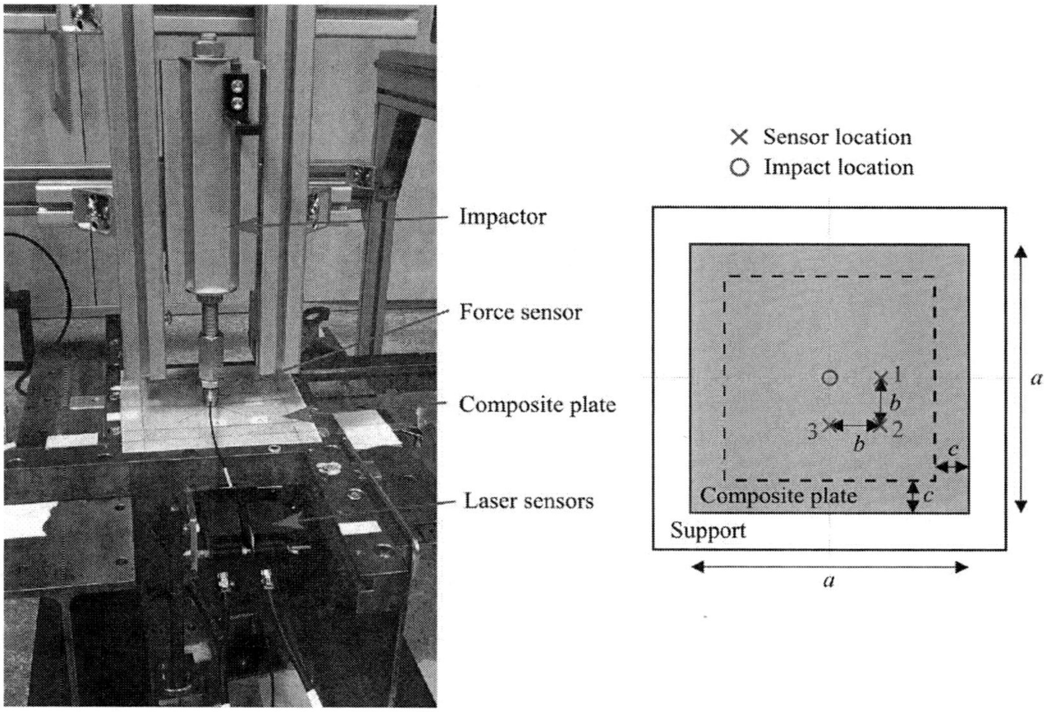


Fig. 1 Photograph (left) and schema (right) of the experimental setup

Table 1 Important dimensions of the experiment

| Dimensions | $a$ | $b$ | $c$ | $t$ |
|---|---|---|---|---|
| [mm] | 190 | 40 | 20 | 0.9 |

The impact velocities were continuously increased from initial impact velocity $v = 1$ m/s to the final velocity $v = 4$ m/s with constant step of 1 m/s. First damage of matrix, limited to the area under the tip of impactor, occurred for the impact velocity $v = 2$ m/s. Two measurements were performed for that impact velocity to verify influence of the induced damage on the sensors responses. The responses for both measurements was almost identical (see Figure 3). Damage evolution from the plate center in the diagonal directions was observed for impact with velocity $v = 4$ m/s. Overview of measured data is shown in *Fig. 3*.

Fig. 2 Composite plate after impact with impact velocity 4 m/s

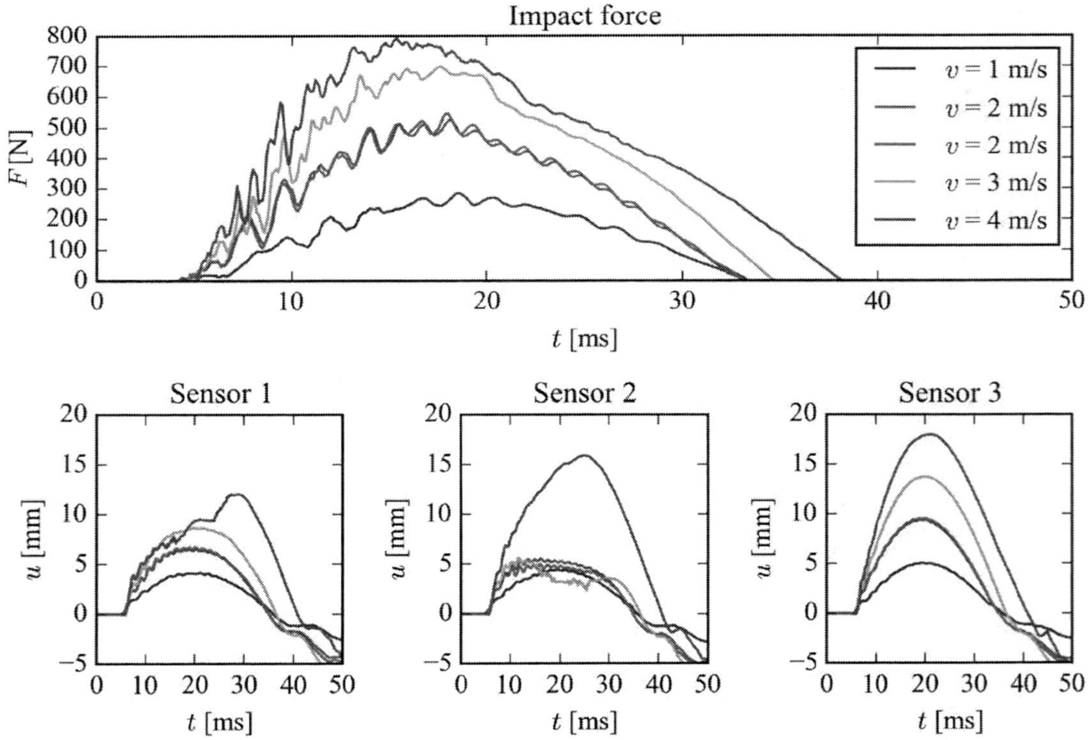


Fig. 3 Measured data

**Identification of impact force**

The measured data were used for the identification process. The impulse responses and corresponding residuals were determined from the first two measurements, where measured impact force and sensor responses were used. From the values of residual (Eq. 5) shown in *Table* is clear that the worst value is for sensor 2, where the estimated impulse response does not fit both measurements, which can be caused by non-linear response of the sensor. Therefore, sensor 2 was not used for the impact force identification.

Table 2 Residuum for impuse response idnetification for individual sensors

| Sensor | 1 | 2 | 3 |
|---|---|---|---|
| Residuum $r$ [mm] | 9.17 | 16.15 | 4.79 |

Remaining measurements were used for the impact force identification, where measured sensor responses and determined impulse responses were used. Comparison of identified and measured impact forces are shown in *Fig. 4* and calculated identification error is shown in *Table* .

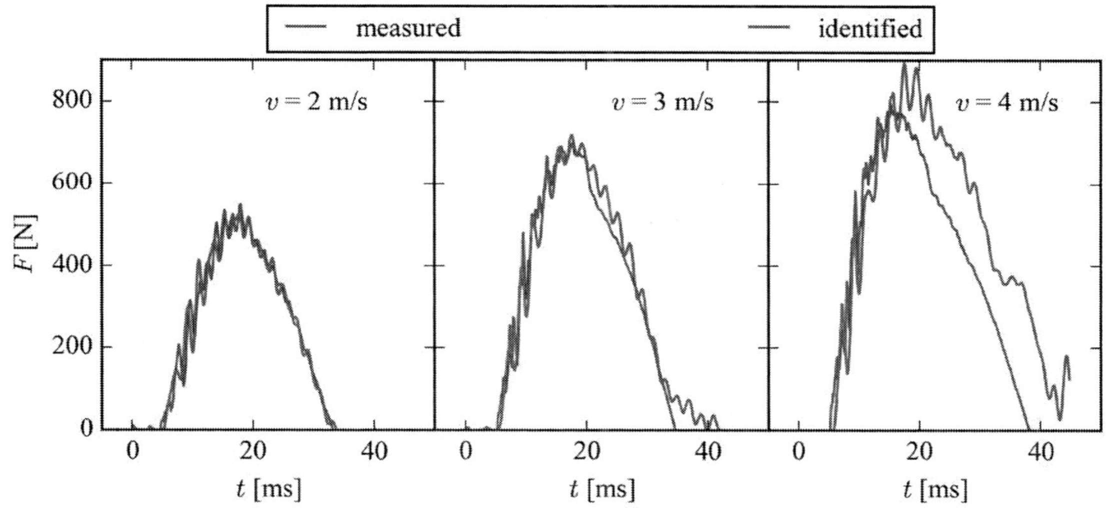


Fig. 4 Comparison of measured and identified impact forces.

Table 3 Identification error

| Impact velocity $v$ [m/s] | 2 | 3 | 4 |
|---|---|---|---|
| Identification error [%] | 5.93 | 6.53 | 17.86 |

From the results it can be concluded, that impact force was identified with good agreement to measured one for tests with the damage localized under the impact area ($v$ = 2 m/s and $v$ = 3 m/s). For impact with the highest impact velocity, the correspondence is acceptable only up to impact force peak value. Then, the identified force is overestimated due to high responses of sensors caused by loss of stiffness in the significantly damaged areas.

Identification procedure can be therefore used in SHM system for estimation of applied loading up to certain level, when the significant damage of constriction is expected. Exceed of that level can be trigger for the monitoring system to raise warming message and to activate other procedures to characterize the damage extent.

Determination of damage extent directly form the identified impact force is problematic. Numerical model can be used to estimate delamination area for impacts when the identified force correspond well with the applied one. For impacts with higher energies, when the damage significantly influence linearity of sensor responses and therefore the identified force does not correspond to applied one, other methods of SHM can be utilized like guided Lamb waves [6].

## Conclusions

Impact force identification procedure was applied on the rectangular composite plate which was experimentally impacted by drop weight mass with various impact velocities. Suitability of displacement sensors was characterized and one of the sensors was excluded from the identification process. Identified impact forces are in good agreement to measured one for two tests where no significant damage occurred. For the last test with the highest impact velocity, the identified force correspond only to the peak value of measured impact force and then is overestimated.

From the results it can be concluded that the identification procedure can be used for decision of exceeding of critical loading on construction. On the other hand, estimation of damage extent from identified force is problematic because the impact force is not correctly identified after occurrence of significant damage.

## References


[1] ATOBE, S., SUGIMOTO, S., HU, N., FUKUNAGA, H. Impact damage monitoring of FRP pressure vessels based on impact force identification, *Advanced Composite Materials,* 2014, vol. 23(5–6), pp. 491–505.

[2] BARTOŠEK, J. Identification and localization of impact force on composite structure (in Czech), *Doctoral dissertation,* University of West Bohemia in Pilsen, 2016.

[3] BARTOŠEK, J., KROUPA, T., SMOLÍK, L. Numerical model of composite airfoil segment with piezoelectric sensors. *Materiali in Tehnologije*, 2014, vol. 48(6), pp. 841–845.

[4] BARTOŠEK, J., KROUPA, T., ZEMČÍK, R., LAŠ, V. Optimal sensor distribution for impact loading identification. *In 11th European Conference on Non-Destructive Testing.* Brno: University of Technology, 2014. pp. 1–6.

[5] ZEMČÍK, R., LAŠ, V., KROUPA, T., BARTOŠEK, J.: Reconstruction of Impact on Textile Composite Plate Using Piezoelectric Sensors; *Proceedings of the, Ninth International Workshop on Structural Health Monitoring*, September 10–12, 2013, Stanford University, Stanford, vol. 1, pp. 393–400, ISBN: 978-1-60595-115-7.

[6] ZHONGQING S., LIN Y., YE L., Guided Lamb waves for identification of damage in composite structures: A review, *Journal of Sound and Vibration*, 2006, vol. 295(3–5), pp. 753–780, ISSN 0022-460X.


## Acknowledgment


This publication was supported by the project LO1506 of the Czech Ministry of Education, Youth and Sports and by project SGS-2016-038.

EAN 2017
55th conference on experimental stress analysis

# Motion Analyses of Human Body on a New Exercise Trainer: Pilot Study

Zbyšek Nový[a], Miroslav Urbánek[1,b] and Lukáš Čapek[2,c]

[1]Comtes FHT, a.s., Průmyslová 995, Dobřany, Czech Republic

[2]Technical University of Liberec, Studentská 2, Liberec, Czech Republic

[a]znovy@comtesfht.cz, [b]murbanek@comtesfht.cz, [c]lukas.capek@tul.cz

## Introduction

The history of exercise equipment goes back to early man´s need physical strength and speed. During the centuries, human being became more sedentary and beside mentioned need the rehabilitation role became also important. If we focus only to cardiovascular equipment, there are three basic equipment: bicycle trainer, treadmill and elliptical trainer. The choice of above mentioned training devices is influenced by the therapeutic goals and by the functional abilities of the patient. A lot of effort was put into a biomechanical evaluation of these trainers.

Cycling can augment rehabilitation effect and can control lower extremities stress [1]. The load of lower extremities muscles depends on various parameters including seat height, pedal locations and others. By omitting these factors several injuries can occurs. This include knee pain, patellar quadriceps tendinitis, iliotibial band syndrome, hip pain, medial tibial stress syndrome, stress fracture, compartment syndrome, numbness of the foot, and metatarsalgia [2]. Injury is usually caused by a combination of inadequate preparation, inappropriate equipment, poor technique, and overuse. Another dark side of cycling is loading of spine column. Patients with back problems should not cycle shortly after surgery as a precaution [3]. It means that this type of trainer cannot be used for these patients. Elliptical training appears to be a safe and feasible training alternative. Jackson et al. demonstrated that patient's rehabilitation at elliptical trainer resulted in no improvements in walking speed; however, participants did demonstrate variable improvements in endurance, balance, and functional mobility [4]. Wu-Lu et al pointed out, when using elliptical trainer, that users` joint function and muscle strength, especially at the knee must be monitored, to avoid injuries [5].

It seems, from the literature, that none of the training machine has a capability for complex and save training after rehabilitation. The aim of this article is to introduce a new exercise machine that would be more complex for various exercises.

## Design of the trainer

The principles of today's machines for aerobic-cardio exercise are derived from those of the bicycle. This new exercise machine uses a rocking platform whose range of rotary motion is ±10°. The rocking platform is V-shaped with an angle of 160°, where each of its halves has a size of 600×500 mm. The center of the rocking machine is pivotably attached to a fixed frame of the machine. The exercise machine includes a mechanism which provides defined resistance to the practitioner's movement, and thus determines the levels of load during the exercise. In the version of the machine presented here, this mechanism comprises a crank mechanism with a flywheel. The machine provides resistance through an electromagnetic brake which is installed on the flywheel and controlled by a transformer. Thanks to the design of the training machine, the practitioner can perform various kinds of exercise engaging large groups of muscles of the entire body or only selected parts in a highly comprehensive manner.

The design study of the training machine (fig.1) comprises all structural parts and user settings of the current version of the machine. Besides various kinds of stance (with feet together, legs astride sideways or forward, etc.) on the machine, the practitioner may also use fixed or movable grips which have a major effect on exercise control. Both grips are vertically adjustable and are provided with a training computer which displays and records the speed, time, the practitioner's performance, and other data. The surface of the rocking platform has anti-slip finish.

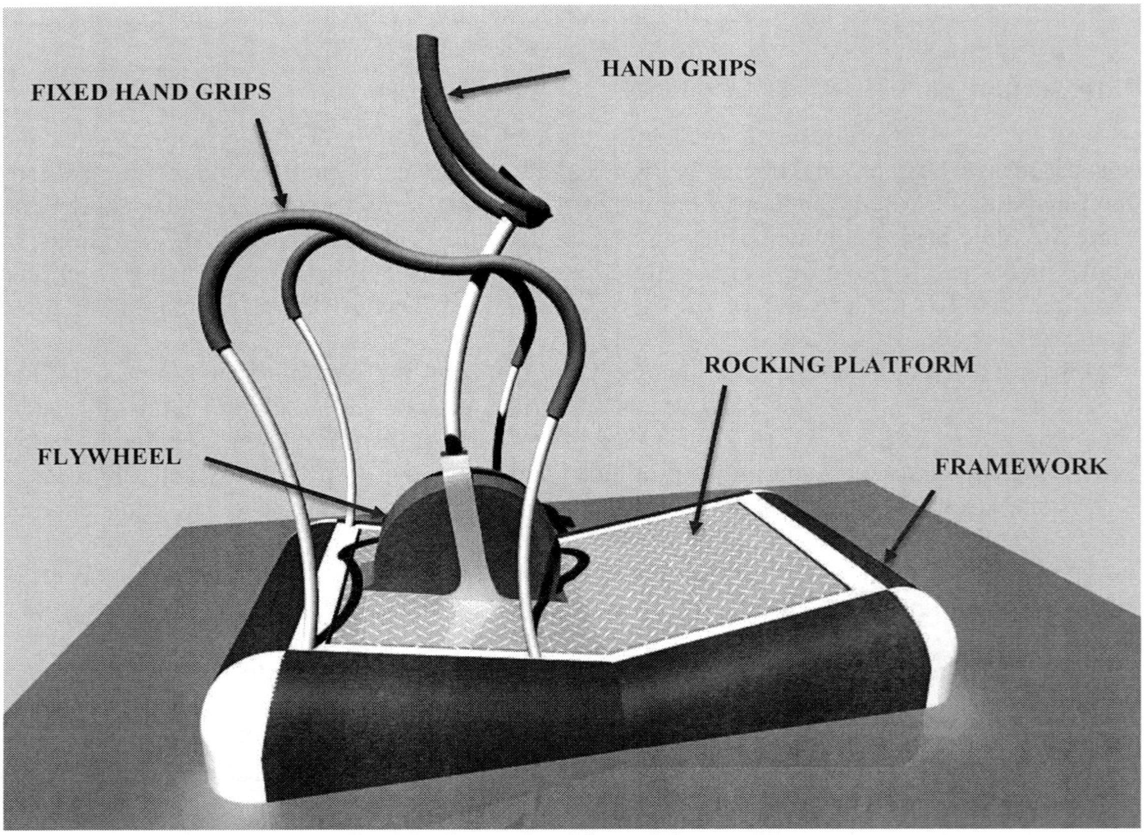


**Fig. 1 Design study of the training machine**

This machine is of simplified construction, comprising only load-bearing and functional parts, which means that the prototype lacks safety and design covers and specially-shaped grips. By contrast, a great emphasis was laid on verifying its function and on the crank mechanism (fig. 2b) with a flywheel (4), the parts which, together with the electromagnetic brake (5), make up the machine's source of resistance to be overcome by the practitioner.

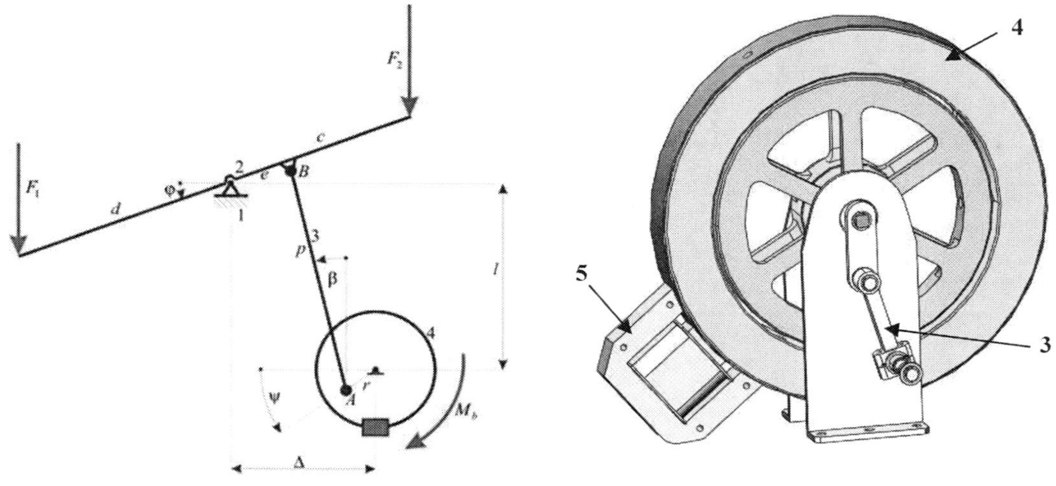


**Fig. 2 Four-bar linkage a) calculation diagram b) design solution**

The use of the Newton's method for finding kinematic dependences (fig. 2a) helped solve the movement of the platform (2) and crank (3) on the flywheel (4). Then, equations of motion for the four-bar linkage were constructed using Lagrange equations of the second kind. The generalized force on the right side of the equation can be obtained by comparing its virtual work with the virtual works of all generalized forces that act on the system as it is given a virtual motion. The equation which is constructed in this manner is a non-linear differential equation of the second order to which the analytical method cannot be applied, and therefore a procedure for obtaining the numerical solution is outlined here. It is based on a transformation of this second-order differential equation to two first-order differential equations. The problem can then be solved using one of standard numerical methods, such as the Runge-Kutta one. In MATLAB environment, this method can be implemented using instructions such as ode23 or ode45.

$$\dot{x}_1 = \frac{cos\varphi(x_2)\{F_1 d h(p_1 x_1) p_1 + F_2 c[h(p_1 x_1) - 1] p_1\} - M_b - I^* p_1 p_1' x_1^2 - I_{kor} p_2 p_2' x_1^2 - b s^2 p_1^2 x_1}{I^* p_1^2 + I_4 + m_B r^2 + I_{kor} p_2^2}.$$

$$\dot{x}_2 = x_1. \qquad (1)$$

To obtain the numerical solution, mass and geometric parameters of the mechanism shown in Fig. 1.1 were substituted into the equation. Based on the calculation, some dimensions or mass parameters can be adjusted. An example of the time course of the displacement, angular speed and acceleration of the flywheel in dependence on the rocking platform is shown in Fig. 3.

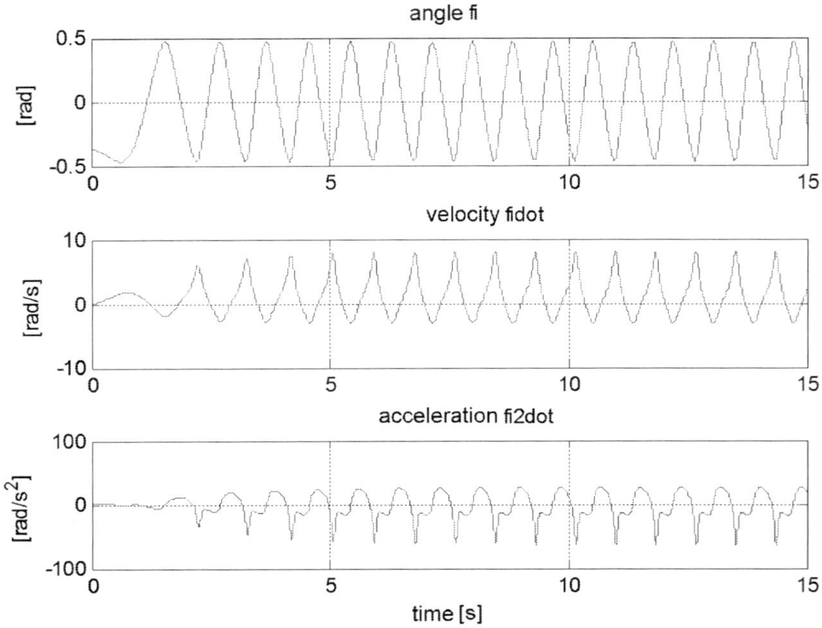


**Fig. 3 Time course of flywheel parameters**

## Motion analyses

The aim of the second step of this study was to analyze motion of human on this trainer. Physically healthy male subject (37 years, 82 kg) volunteered for this investigation. The resistant load was set to 110 W for all cases. 3D motion capture was provided by the optical system Pontos M4, 15 Hz speed cameras (GOM GmbH, Germany). The exercises were done in two ways:

a) Legs are positioned at the opposite side of V shaped plane and the movement is dominantly done by arm movement on handlebars.
b) Left leg is positioned at one side of the V shaped plane and the right leg on the opposite one. The arms are holding the handlebars.

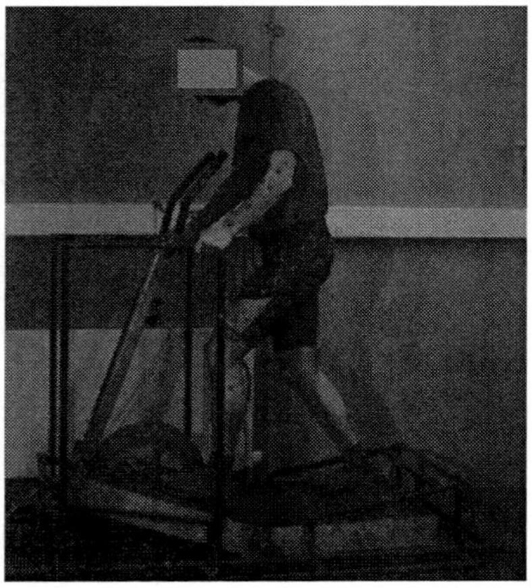

**Fig. 4 Possible type of exercising a) (left), b) (right)**

The region of arm, leg and handlebars were covered by markers.

## Results

Displacement, velocity and acceleration of selected region were treated for every type of exercising. The typical gained functions are on fig. 5.

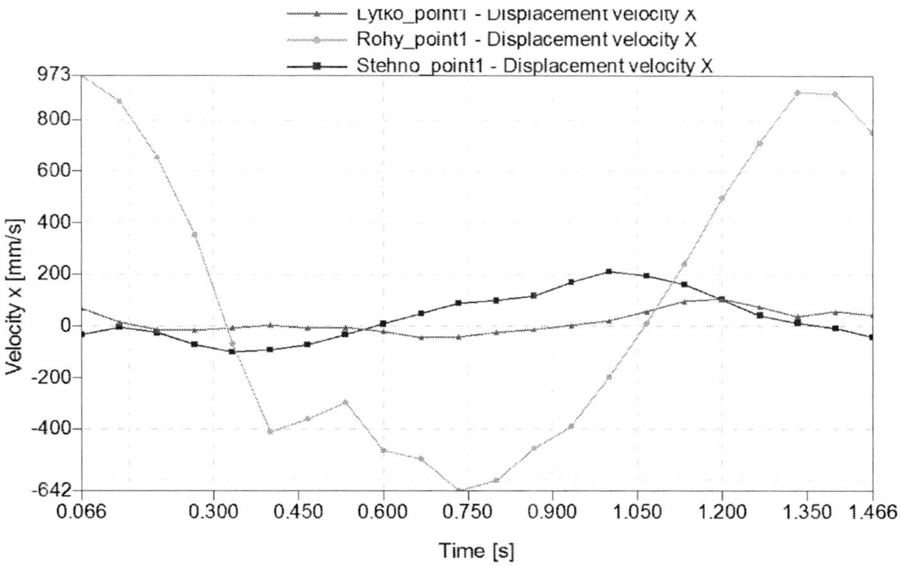


**Fig. 5 Velocity in axis X for case b. Calf (triangle), handlebars (circle), thigh (rectangular)**

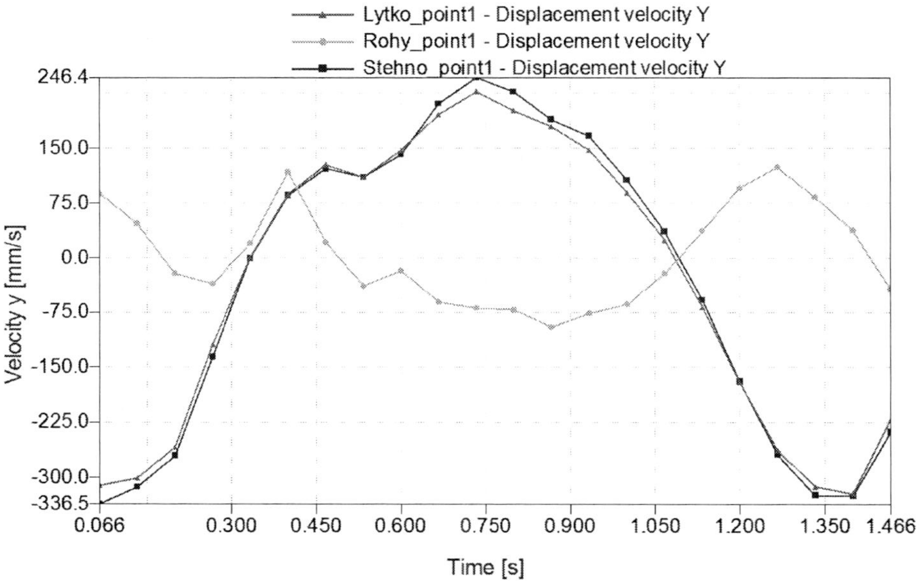


**Fig. 6 Velocity in axis Y for case b. Calf (triangle), handlebars (circle), thigh (rectangular)**

The maximal velocity in X direction reaches 904 mm/s in case a) and 1472 mm/s in case b).

| **Velocity x [mm/s]** | Exercise a | Exercise b |
|---|---|---|
| Handlebar | 904 | 1472 |
| Thigh | 201 | 283 |
| **Acceleration x [mm/s2]** | | |
| Handlebar | 3842 | 12019 |
| Thigh | 811 | 2851 |

**Tab.1: Maximal measured velocity and acceleration in x directions**

### Conclusions

There are many articles aimed on comparison of different type of trainer. Usually following conclusion is made: treadmill walking most closely resembled overground walking. Cycling showed the largest difference from overground walking, with elliptical closer but still a significant distance from all three. Cycling showed greater hip reciprocation. Cycling and elliptical training showed stronger intralimb synergism at the hip and knee than the other two. Based on kinematics, results suggest that elliptical training may have greater transfer to overground walking than cycling and cycling may be more useful for enhancing reciprocal coordination [6]. In this study a new type of trainer based on a completely new philosophy was introduced. Based on pilot experiments it is assumed that it would have more complex use in the field of rehabilitation.

### Acknowledgment

This study was created by project Development of West-Bohemian Centre of Materials and Metallurgy No.: LO1412, financed by the MEYS of the Czech Republic.

### References

[1] Wozniak Timmer CA. Cycling biomechanics: a literature review. J Orthop Sports Phys Ther. 14 (1991) 106-13.

[2] Wanich T, Hodgkins C, Columbier JA, Muraski E, Kennedy JG. Cycling injuries of the lower extremity. J Am Acad Orthop Surg. 15 (2007) 748-56.

[3] Rohlmann A., Zander T., Graichen F., Schmidt H., Bergmann G. Spinal Loads during Cycling on an Ergometer. PLoS One. 9 (2014) e95497.

[4] Jackson K., Merriman H., Campbell J. Use of an elliptical machine for improving functional walking capacity in individuals with chronic stroke: a case series. J Neurol Phys Ther. 34(2010)168-74.

[5] Lu TW, Chien HL, Chen HL. Joint loading in the lower extremities during elliptical exercise. Med Sci Sports Exerc. 39(2007)1651-8.

[6] Damiano DL, Norman T, Stanley CJ, Park HS. Comparison of elliptical training, stationary cycling, treadmill walking and overground walking. Gait Posture. 34(2011):260-264.

[7] Dupal J.: Dynamická analýza a optimalizace fyzioterapeutického trenažéru Wobble, výzkumná zpráva číslo: NTIS-VP3-0003/2013 (2013)

EAN 2017
55th conference on experimental stress analysis

# The optimization of the materials properties for the freight tyres in dependence on defect propagation

Ján Vavro[1, a], Ján Vavro jr.[1,b], Petra Kováčiková[1,c], Jakub Híreš[1,d]

[1]Alexander Dubček University of Trenčín, Faculty of Industrial Technologies in Púchov

[a]jan.vavro@fpt.tnuni.sk, [b]jan.vavro.jr@fpt.tnuni.sk, [c]petra.kovacikova@fpt.tnuni.sk, [d]jakub.hires@fpt.tnuni.sk

**Keywords:** measurement, pressure, dynamic loading, non-destructive method, holography

**Abstract.** The paper deals with the optimisation of the materials properties for the freight tyres in dependence on defect propagation at the dynamic loading. The given investigation was mainly focused on cracks and defects in tyre casing, which is commonly used for passengers cars. The mentioned measurement or investigation was performed under the strictly specified conditions. The given measurement of crack propagation in car tyre casing was based on changes of speed, changes of loading as well as changes of service time interval. It is important to point out that the pressure in the tyre was predetermined to be the constant value. The experimental procedures were carried out by help of special drum testing machine with designation: IGTT 25kN – C and with the 1707mm drum diameter. The mentioned measurement was performed for the area of tyre casing shoulder and it was in the site between the first and the second buffer lining. Using the ITT- 1 program relating to non-destructive testing analyser, the CCD camera for testing procedure of tyre casing was used for the camera images which were taken from the internal side of the tyre casing. The inspection of the tyre casing was performed for the specified points or segments around the periphery of he tyre casing. The measurement was based on simple cycle. The analyser testing led to creation of the resulting protocol based on ITTView program and on the basis of this program, it was possible to display or print this protocol in the form of the phase image mode or video mode. The crack propagation was evaluated on the basis of tyre casing loading as well as kilometres which were overrun during the car service. Shearograpy is a method which can be used for specification of tyre casing cracks under the laboratory conditions and it can be also useful during the quality inspection of the new made products. The obtained knowledge relating to crack initiation and propagation in tires can be used for easier solutions referring to different areas and sites of the tyre casing and by this way, it is possible to determine which one of variants is the most suitable for utilisation in the production.

## Introduction

The effort to improve the road-traffic safety leads to systematic enhancement of tyres. All producers endeavour to manufacture such tyres which would be reliable, safety, easily mountable as well as they would have long-term lifetime which is connected with wear-resistance. Moreover, the given tyres should be available from the aspect of cost and they should not have any negative influence on living environment. Enormous attention is also paid to occurrence of various material defects and failures, such as separations, air pockets, bulge and so on. Mutual adhesion between particular materials of which the tyre is made is also seriously investigated [1, 2]. Detection of defects, failures, separations in a tyre casing can be done by several methods. Destructive as well as non-destructive methods can be used for investigation and they help us to find defects occurring in new, re-treaded as well as old and worn-down tyres. Real disadvantage in relation to destructive methods is closely connected with the fact that the given tyre casings which are tested by destructive method are not able to be used again and again for testing, because destructive testing means destruction or rupture of tested object and in our case it is the tyre casing. It means that destructive methods are difficult from these aspects:

- from the economic aspect, because destructive testing means usage of high number of tested tyres,
- from the aspect of preservation of specific homogeneity for sample.

Therefore, there is the tendency to use non-destructive methods of testing in relation to tyre casings and the main contributions and advantages are:

- analysis of tyre casing without its damage or rupture,
- possibility to control the product and its following utilization for other purposes,
- the given method can be used for the process of tyre development as well as during the changes relating to manufacturing, during the control of quality of tyre casings and moreover, the mentioned method of testing can be also used for fatigue tests, speed tests and lifetime tests
- non-destructive analysis can be done again and again in relation to the same testing sample or product
- according to the type of used non-destructive method, it is possible to detect product defects in the course of a few minutes.

**Structure of the tyre casing**

It is necessary to have knowledge relating to structure of the tyre casing because only this is the correct way for specification of areas where the separations are initiated and moreover, it is also much easier to specify the distribution of the mentioned separations. The structure of tyre casing [3] consists of:

1. *Tread* – it is important part of tyre casing, because it is in the close or direct contact with the road surface. It is made of blend which has good adhesive properties and high wear resistance;
2. *Buffer linings* – it absorbs circumferential stresses as well as side forces and it absorbs also impacts which occur during the contact with road. It consists of individual layers of rubberized cord while the given layers are laid in a criss-cross formation;
3. *Cord carcass of casing* – it is the basic and supporting part which comprise of one or several linings from rubberized cord and they are laid around the bead plies;
4. *Filling linings* – they are the shaped rubber profiles and they are used for better and smoother mutual junction of individual construction parts relating to tyre casing;
5. *Bead bundle* – it reduces deformation in the area where there is the end of cord carcass and protection of bead;
6. *Sidewall* – it protect side part against damage and weather conditions. It is made of blend which is resistant to flex cracking and cracking in common;
7. *Inner skim rubber* – it is rubber lining which can be found on the internal side of tyre casing. It is used for protection of cord carcass and in the case of tubeless tyre casings, it avoid diffusion of air into the cord carcass of casing;
8. *Bead reinforcement* – it can be fabric as well as steel;
9. *Bead plies* – they consist of steel wires or strips which have high strength. They provide smooth and safe anchoring of cord carcass linings as well as attachment of tyre casing to wheel rim.

Fig. 1 shows the given parts mentioned hereinbefore.

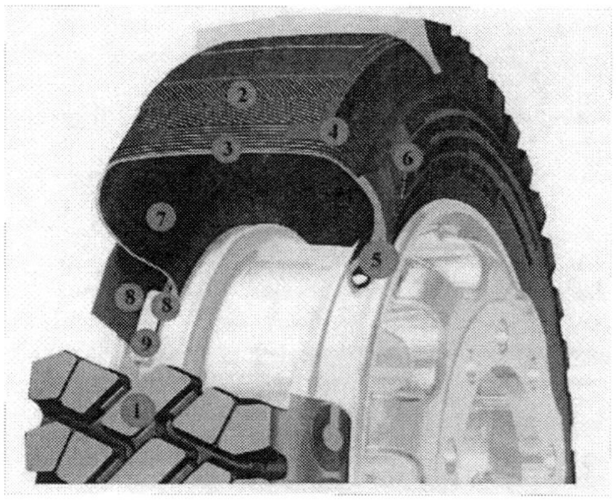

Fig. 1. Structure of tyre casing: 1. tread; 2. buffer linings; 3. cord carcass; 4. Filling linings; 5. bead bundle; 6. Sidewall; 7. Inner skim rubber; 8. bead reinforcement; 9. bead plies.

**Process of measurement with non-destructive analyzer**

Non-destructive analyser enables us to recognize tyre structure defects quickly and easily – it is connected with closed separations (Fig. 2), the extension of which we will observe by the dynamic test. Typical arrangement of the tyre control with help of the simple cycle for tyre impurity finding is in the Fig. 3.

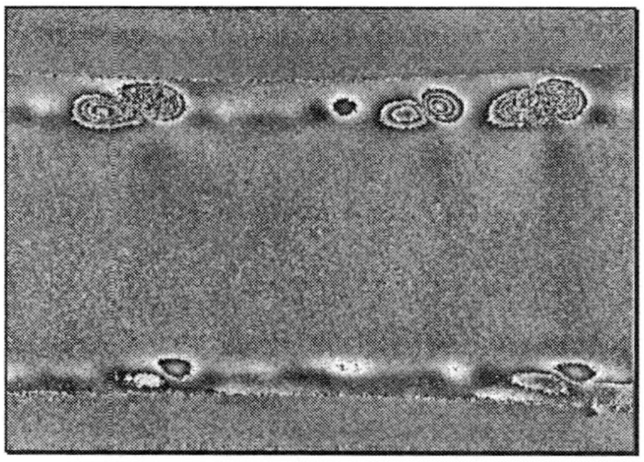

Fig. 2. Defect displayed on monitor.

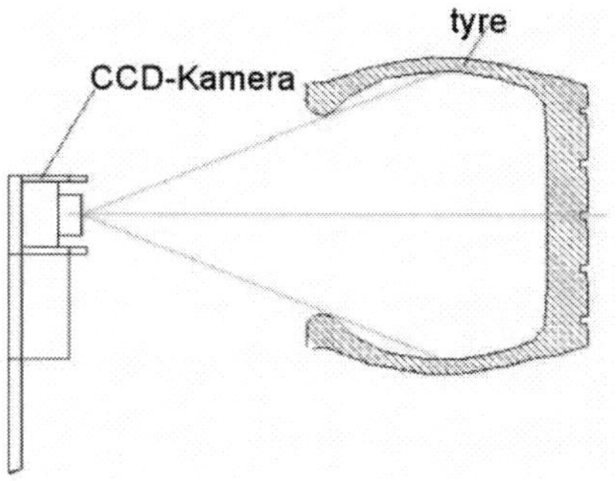


Fig. 3. Testing assembly by the simple cycle.

For observation of extension of the tyre separations, the specific tyre was selected and as it can be seen in Fig. 4, the separations in the shoulder area as well as crown area were found by the help of the ITT-1 testing analyser. After 20 hours of the dynamic test, there were only small changes in separations extension, therefore we introduce the display only after 200 hours of the dynamic test (Fig. 5) [4–6].

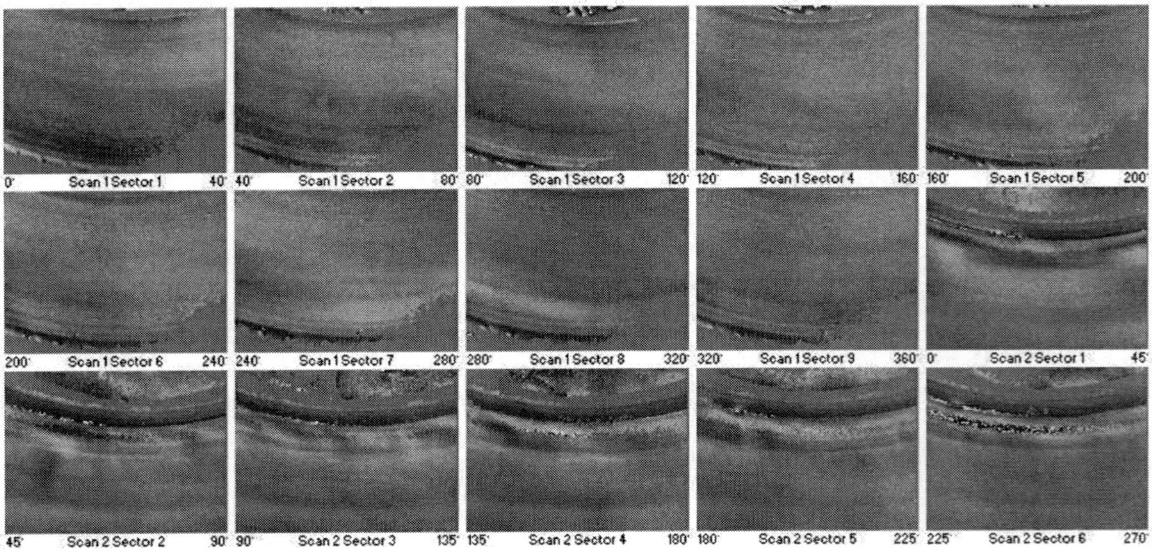


Fig. 4. First measurement before dynamic test.

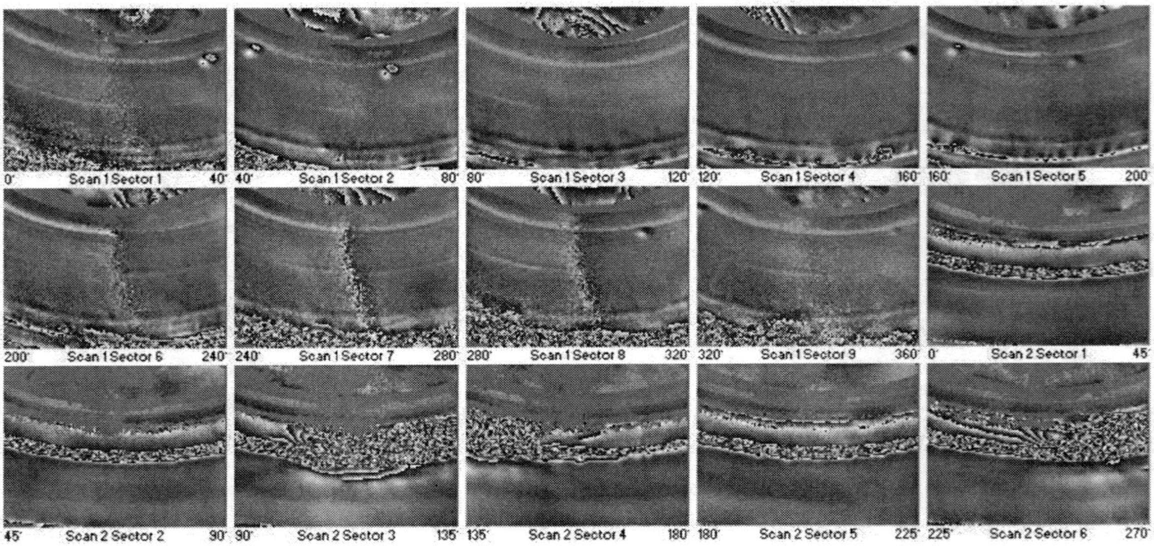


Fig. 5. 12th measurement after 200 hours of the dynamic test.

## Summary

On the basis of individual tyre figures it can be concluded that defects and separations on the tyre periphery are propagated and they join after dynamic loading. In Fig. 5, separations created almost around the whole periphery can be observed. Separations in the crown of the tyre casing were not detected after 10 000 km of running. After this detection by means of the ITT-1 machine, we cut the tyre and studied it for specification of separations which were seen in the individual scans. The occurrence of separations was confirmed, while small separations were detected for both shoulders in the end area in relation to the buffer lining and these separations were propagated around the whole tyre periphery – see Fig. 6 and 7.

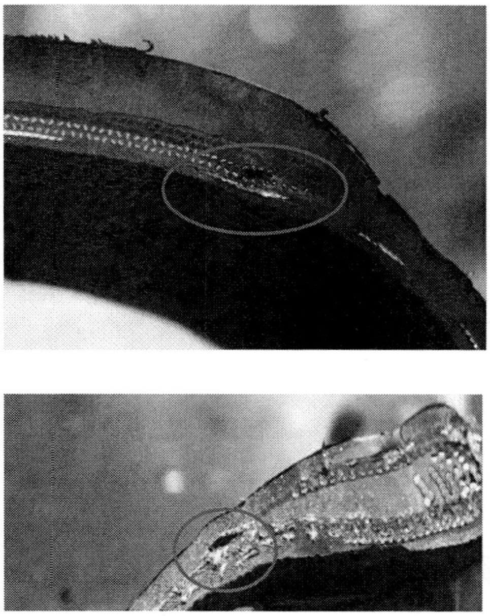

Fig. 6. Tyre section after 100 hours of dynamic testing

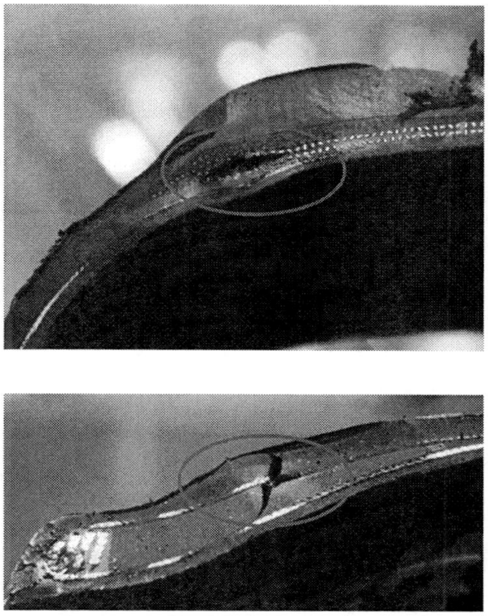

Fig. 7. Tyre section after 200 hours of dynamic testing.

Taking individual variants of tyres into account, the mutual comparison of tyres, which were made of different rubber blends, was carried out, while the dependence on the defect area (measured in $mm^2$) represented the crucial factor in relation to comparison. The given comparison can be seen in Fig. 8.

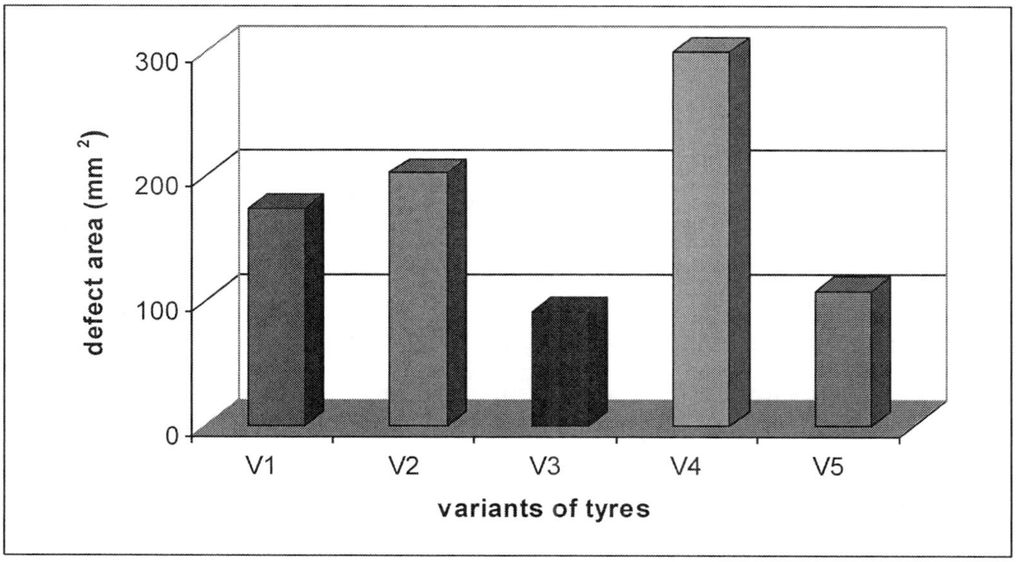


Fig. 8. Mutual comparison of tyres made of different rubber blends.

The given comparison of defect propagations in tire casing in dependence on increasing speed during the fatigue test can be seen in Fig. 9.

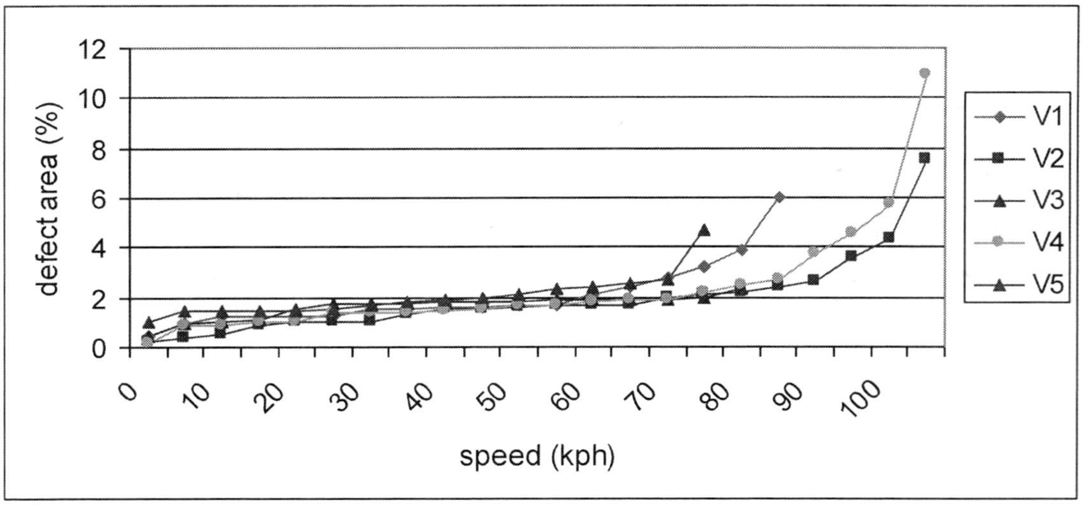


Fig. 9. The mutual comparison of defect propagations in tire casing in dependence on increasing speed during the fatigue test


## Acknowledgement

The work presented in this paper was supported by VEGA grant No. 1/0649/17, VEGA grant No. 1/0589/17 and KEGA No. 007TnUAD-4/2017.


## References


[1] I. Kováč, J. Krmela, D. Bakošová, Parametrizing of Material Input for Modal Analyses of FEA Tire Models, Metallurgical Journal LXIV(7) (2011) 73–78.

[2] D. Ondrušová, M. Pajtášová, Rubber components and their influence on rubber properties and enviromental aspects of production. 1st Ed. Spolok Slovákov v Poľsku, Krakow, 2011, 166 p., ISBN 978-83-7490-385-1.

[3] J. Marcín, P. Zítek, Rubber productions I – Tyres, SNTL, Prague, 1985. (in Czech)

[4] J. Vavro, M. Kopecký, P. Koštial, H. Hajská, M. Fandáková, The Experimental Measurement of the Tyre Impurities Circulation by the Dynamic Loading and Application of the Experiment Evaluation by the Method of the Active Factor Experiment, Machine Dynamics Problems, Warsaw University of Technology, 2004.

[5] J. Krmela, V. Krmelová, L. Beneš, Experiment of tire-crown for computational modeling of tire, Scientific papers of the University of Pardubice: Series B 19 (2014) 81–88.

[6] J. Krmela, Systems Approach to the Computational Modeling of Tyres, Monograph, 2008; 102 p., ISBN 978-80-7399-365-8.

# Determination of group velocities of Lamb waves in unidirectional carbon-epoxy plate

Zuzana Lašová[1,a], Robert Zemčík[1,b]

[1]Faculty of Applied Sciences, University of West Bohemia in Pilsen, Czech Republic

[a]zlasova@kme.zcu.cz, [b]zemcik@kme.zcu.cz

**Keywords:** Lamb waves, group velocity, wavelet transform, explicit solver.

**Abstract.** The group velocity of zero-order symmetric Lamb wave mode ($S_0$) in unidirectional composite plate is strongly dependent on the direction of propagation. The group velocities in chosen directions were determined experimentally using piezoelectric patches. The collected data were processed by wavelet transform to find a time-of-arrival of wave packages relevant to the $S_0$ mode. Group velocities in more directions of propagation were calculated using finite element analysis. The numerical results show good agreement with measured velocities.

## Introduction

Lamb waves are a kind of elastic waves propagating in plates and hollow cylinders. They have attracted attention of many researchers due to their potential use in detection of structural flaws known as structural health monitoring. This is especially useful in case of layered composite structures suffering by internal hidden flaws such as delamination [3, 4]. Structural health monitoring is based on extraction of information from time signals collected by a network of sensors applied on the structure. In each time record, forward and reflected wave packages need to be identified precisely to identify those reflected by the damage.

The group velocity can be understood as a speed of transfer of energy carried by wave packages. It is generally defined by

$$c_g = \frac{\partial \omega}{\partial k}, \tag{1}$$

where $\omega$ is angular frequency and $k$ is wavenumber. With the knowledge of group velocities it is possible to distinguish the wave packets associated with unique wave modes. There is theoretically infinite number of Lamb wave modes, but they can be divided to two families: symmetric ($S_n$) and anti-symmetric ($A_n$) modes in respect to the middle-plane. In low frequency range only two fundamental modes, $S_0$ and $A_0$, exist.

The group velocity of symmetric zero mode ($S_0$) in transversally isotropic composite plate is strongly dependent on angle of wave propagation in respect to direction of fibers. Rhee et al. [2] used ultrasonic pulser-receiver to find a group velocity curve in respect to angle of propagation in unidirectional composite plate. Jeong [1] used a lead break as a source of acoustic emission and two piezoelectric sensors to determine group velocities in quasi-isotropic laminate plate.

In this work, three piezoelectric transducers were used for experimental determination of group velocities in directions of propagation 0°, 25°, 45° and 90° in respect to the fiber direction. It is not always possible to localize time of arrival of Lamb wave modes in temporal domain, therefore methods of time-frequency analysis have to be applied. For the processing of recorded time signals wavelet transform (WT) is preferred by most of authors [4, 5].

### Wavelet transform

The time signals obtained by measurement or by finite element analysis were processed by wavelet transform. This method is advantageous mainly because the time signals are non-stationary and have character of harmonic waves. Continuous wavelet transform (CWT) of arbitrary function $f(t)$ is defined as [1]

$$CWT_f(\tau, s) = \int_{-\infty}^{\infty} f(t)\psi_{\tau,s}^{*}\mathrm{d}t. \tag{2}$$

The CWT formally resembles Fourier Transform, but the difference is in the basis function $\psi_{\tau,s}$ which is derived from mother wavelet function $\psi$ by scale $s$ and time shift $\tau$:

$$\psi_{\tau,s} = \frac{1}{\sqrt{s}}\psi\left(\frac{t-\tau}{s}\right). \tag{3}$$

Variable scale and time shift provide better resolution both in frequency and time, which is crucial in determination of time of arrival of wave packets. There are some options for the choice of mother wavelet function, in this work it was a complex Morlet wavelet:

$$\psi = e^{i2\pi ft}e^{-\frac{t^2}{2s}}. \tag{4}$$

### Measurement of group velocities

The group velocity was measured in unidirectional carbon-epoxy plate having dimensions 770×770×1.8 mm hanged on a frame by elastic strings and secured in vertical position by a clip (Fig. 1).

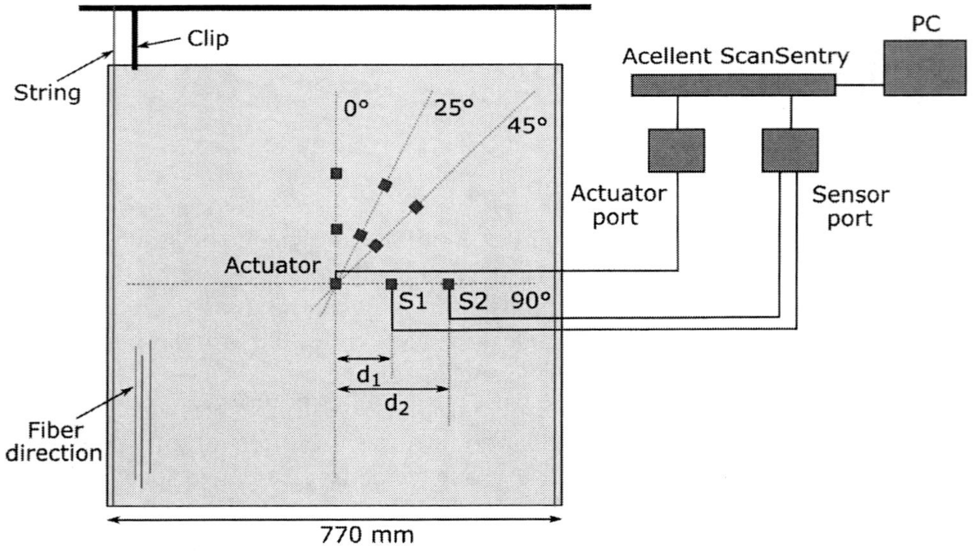


Fig.1 Measurement set-up

The transducers were piezoelectric patches DuraAct P-876.SP1. These transducers consist of piezoceramic layer (10×10×0.5 mm) and a protective plastic foil (15×15×1 mm). The patches were attached to the surface of the plate by double-sided tape. The actuator was placed in the center of the plate and the pair of sensors was sequentially collocated in directions 0° (direction of fibers), 25°, 45° and 90° in constant distances $d_1 = 100$ and $d_2 = 200$ mm from the actuator.

The elastic waves in the plate were excited and measured by Acellent ScanSentry, which is special system for the structural health monitoring. This device serves both as pulse generator (with built-in amplifier) and one-channel data acquisition system. The exciting pulse was five-cycle sine wave burst with central frequency of 200 kHz modulated by Hann window (Fig. 2).

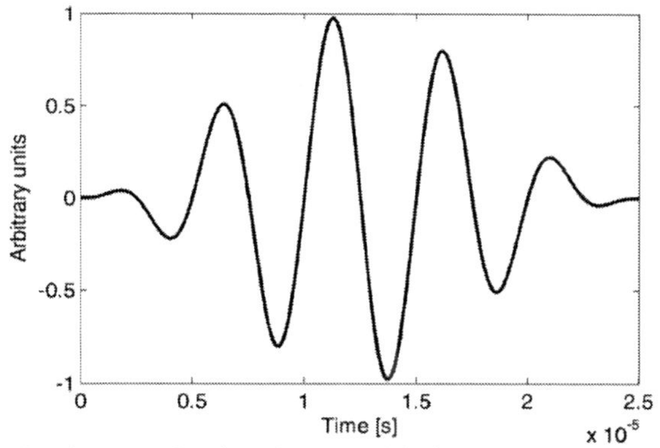


Fig. 2 Five-cycle sine burst with frequency 200 kHz

The time signals from pairs of sensors (see Fig. 3 a) and b)) were processed by wavelet transform. The time history of frequency component 200 kHz (Fig. 3 c)) was obtained as a relevant row of the scalogram (matrix of real parts of WT coefficients). The peaks in this graphs correspond to the passages of wave packages. $S_0$ has the highest group velocity, therefore it can be identified as the first significant peak [5].

As we know the mutual distance of the sensors, the group velocity was determined from the difference of times of arrival of corresponding wave packages. Calculated group velocities of S0 mode in investigated directions of propagation are presented in Table 1.

Tab. 1 Experimentally determined group velocities for frequency 200 kHz.

| Angle of propagation [°] | 0 | 25 | 45 | 90 |
|---|---|---|---|---|
| Group velocity [ms$^{-1}$] | 8791 | 4600 | 2860 | 2272 |

**Finite element model of the plate**

The finite element model was a one half of the plate of same thickness 1.8 mm as measured plate and dimensions 770×375 mm with directions of fibers parallel to $x$-axis. The reason for using this model was that the group velocity in the longitudinal direction is expected to be up to 4 times higher than the group velocity in transversal direction. Otherwise the wave packets in longitudinal direction reflected from the edges may reach the measured nodes before transversal wave packets.

The model was created using linear hexahedral elements with 2 elements per thickness. Material parameters were partly measured and taken from literature. Density $\rho$ was determined experimentally. Tensile moduli $E_{11}$, $E_{22}$ and Poisson's ratio $\nu_{12}$ were measured by tensile tests of specimens 0° and 90° cut from the plate. The shear moduli $G_{12}$ and $G_{23}$ were taken from [5]. All material data are presented in Tab. 2. The propagation of Lamb waves was modelled in Abaqus v.6-14 explicit solver.

It is not possible to include piezoelectric behavior in the explicit solver, therefore Lamb waves were excited mechanically. The time-variable displacements in-plane $u_x$ and $u_y$ identical with the actuating pulse in measurement (five-cycle sine burst with frequency 200 kHz) were applied to the nodes in the central location of the actuator.

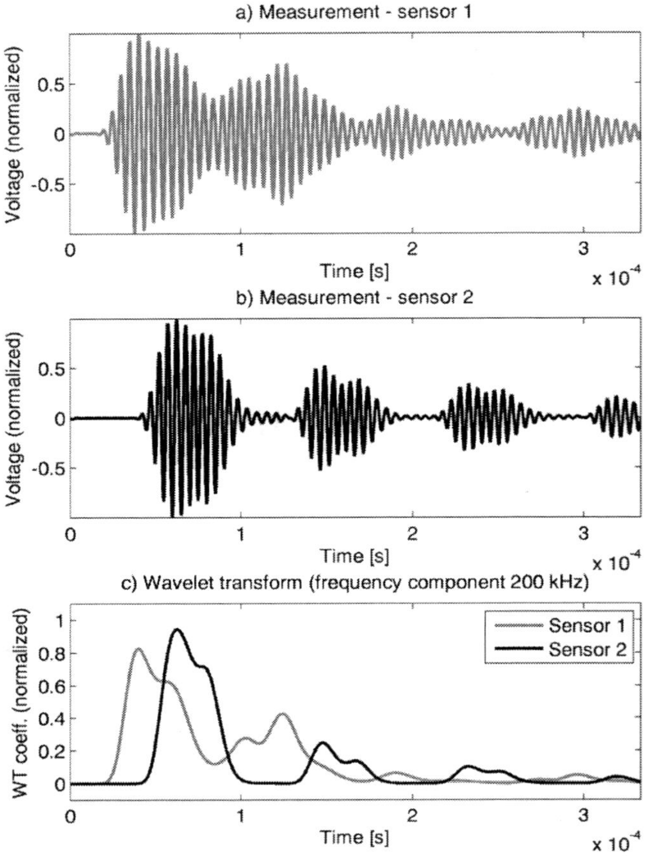


Fig. 3 Time records from a pair of sensors (a, b) and time history of frequency component 200 kHz (c).

Tab. 2 Material parameters of composite plate.

| $\rho$ [kg m$^{-3}$] | $E_{11}$ [GPa] | $E_{22}$ [GPa] | $\nu_{12}$ | $G_{12}$ [GPa] | $G_{23}$ [GPa] |
|---|---|---|---|---|---|
| 1560 | 122.0 | 8.0 | 0.308 | 5.97 | 3.75 |

The time histories of out-of-plane displacement $u_z$ were collected from pairs of nodes in directions of propagations in range [0, 90]° with step 5°. The nodes closest to required angles and distances $d_1$ = 10 mm and $d_2$ = 20 mm from actuator were used as sensors.

Results were processed same way as experimental data. The calculated group velocities are presented and compared with the experimental results in Fig. 4. It is apparent the approximated dependency curve approaches an ellipse, which corresponds with the group velocity wave curve in unidirectional composite determined by Rhee et al. [2]. The variations of group velocities in angles 70° and 75° are caused by an interference of $S_0$ mode with shear wave mode, as can be seen in Fig. 5.

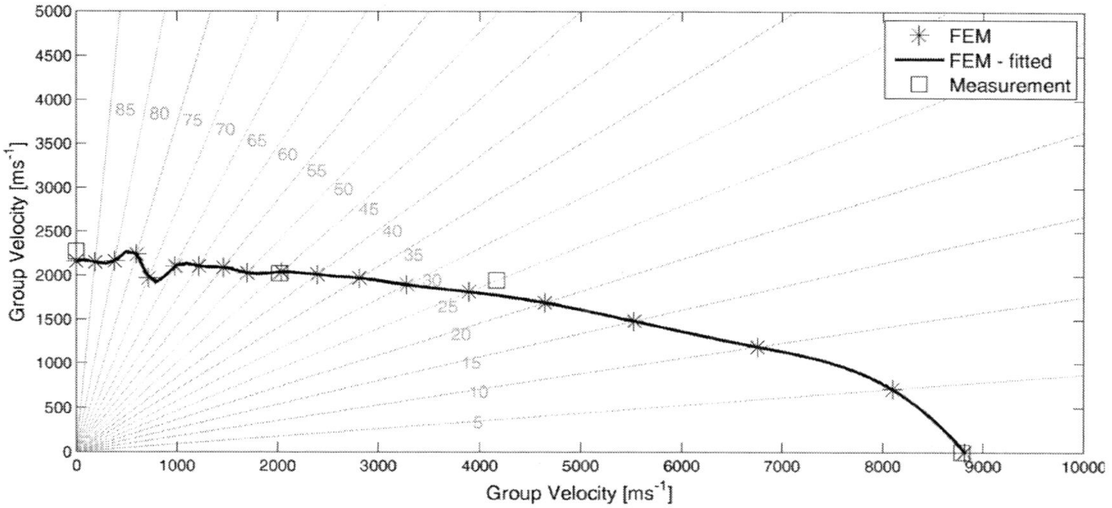


Fig. 4 Angular dependency of group velocity of $S_0$ mode (frequency 200 kHz).

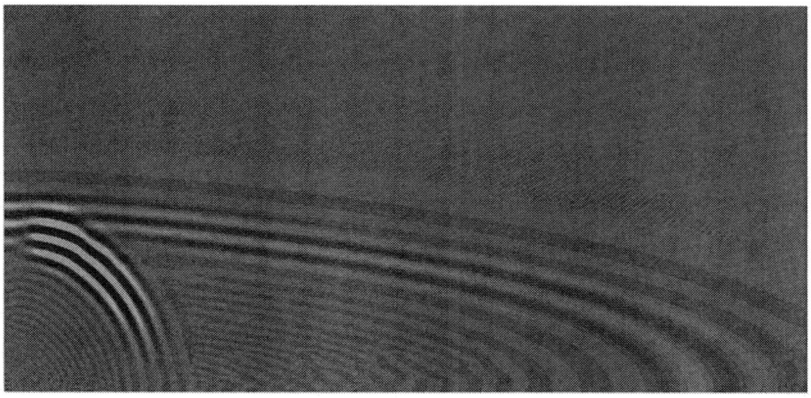

Fig. 5 Out-of-plane displacement of the plate in time $t$ =180 ms from numerical analysis.

## Conclusions

Dependency of group velocity on direction of propagation of Lamb waves in unidirectional carbon-epoxy plate was determined using finite element model. The values in directions 0°, 25°, 45° and 90° were compared to the experimental results. The difference between numerical and measured velocities does not exceed 7%. However, group velocities in more directions have to be measured to verify the numerical model.

## Acknowledgements

This publication was supported by project LO1506 of the Czech Ministry of Education, Youth and Sports and by project SGS-2016-083.

## References

[1] H. Jeong, Analysis of plate wave propagation in anisotropic laminates using a wavelet transform. NDT & E International, 34(3) (2001) 185-190.

[2] S. H. Rhee, J. K. Lee, J. J. Lee, The group velocity variation of Lamb wave in fiber reinforced composite plate, Ultrasonics 47 (1) (2007) 55-63.

[3] S. C. Rosalie, M. Vaughan, A. Bremner, W. L. Chiu, Variation in the group velocity of Lamb waves as a tool for the detection of delamination in GLARE aluminium plate-like structures. Composite Structures, 66(1) (2004) 77-86.

[4] H. Sohn, G. Park, J. R. Wait, N. P. Limback, C. R. Farrar, Wavelet-based active sensing for delamination detection in composite structures. Smart Materials and Structures, 13(1) (2003) 153-160.

[5] L. Wang, F. G. Yuan, Group velocity and characteristic wave curves of Lamb waves in composites: Modeling and experiments, Composites science and technology 67(7) (2007) 1370-1384.

EAN 2017
55th conference on experimental stress analysis

# Determining modal parameters of mechanical system by using enhanced frequency response function

PAVELKA Peter[1, a] and HUŇADY Róbert[1,b]

[1]Department of Applied Mechanics and Mechanical Engineering, Faculty of Mechanical Engineering, Letná 9, 042 00 Košice, Slovakia

[a]peter.pavelka@tuke.sk, [b]robert.hunady@tuke.sk

**Keywords:** modal parameters, complex mode indicator function (CMIF), enhanced frequency response function (EFRF), rational fraction polynomial (RFP)

**Abstract.** The paper deals with determining the modal parameters of mechanical system by using Enhanced Frequency Response Function which is obtained from Complex Mode Indicator Function. The aim of the paper is to present the advantages of this approach for a case when the FRF matrix consists of many degrees of freedom. The analyzed object is a system with four degrees of freedom with viscous damping. A model of the system and all calculations are realized in Matlab. The modal parameters are determined by two different ways. In the first case, the modal parameters are determined for a spatial model by eigenvalue calculation. In the second case, the modal parameters are estimated in the frequency domain using an iterative Rational Fraction Polynomial method from enhanced frequency response functions. At the end, the computational efficiency of the proposed approach is studied.

## Introduction

Modal analysis is a part of mechanics. It serves to analyze and to determine dynamic properties of structure. These features include natural frequency, mode shape and damping. Analyzed system can be described by a spatial model (mass matrix $[M]$, damping matrix $[B]$ and stiffness matrix $[K]$), modal model (spectral matrix $\left[\lambda^2\right]$ and modal matrix $[\Phi]$) and response model (frequency response function matrix $\left[H(\omega)\right]$) [1, 2]. Modal analysis can be divided into the following approaches: a) experimental modal analysis, where frequency response functions are obtained experimentally; b) numerical or analytical modal analysis, where frequency response functions are computed from spatial model [3].

## Analytical solution

Let us consider four degree of freedom system (Fig. 1) with vicious damping. Its mathematical model is given by system's matrices:

$$[M]=\begin{bmatrix} 2 & 0 & 0 & 0 \\ 0 & 3 & 0 & 0 \\ 0 & 0 & 4 & 0 \\ 0 & 0 & 0 & 5 \end{bmatrix},\ [K]=\begin{bmatrix} 3.10^5 & -2.10^5 & 0 & 0 \\ -2.10^5 & 5.10^5 & -3.10^5 & 0 \\ 0 & -3.10^5 & 5.10^5 & -2.10^5 \\ 0 & 0 & -2.10^5 & 3.10^5 \end{bmatrix},\ [B]=\begin{bmatrix} 30 & -20 & 0 & 0 \\ -20 & 50 & -30 & 0 \\ 0 & -30 & 50 & -20 \\ 0 & 0 & -20 & 30 \end{bmatrix}. \quad (1)$$

where $[M],[B],[K]$ are the mass, damping and stiffness matrices.

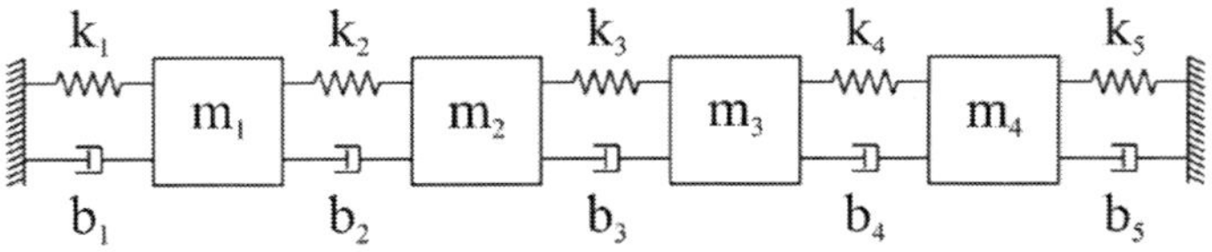


Fig. 1 Four degrees of freedom system.

The frequency response functions (FRFs) of the system was calculated analytically and are shown in Fig. 2a. All together represent the FRF matrix $[H(\omega)]$ that is used in the next calculations. Modal parameters were determined by eigenvalue solution and are listed in Table 1. Mode shapes are shown in Fig. 2b.

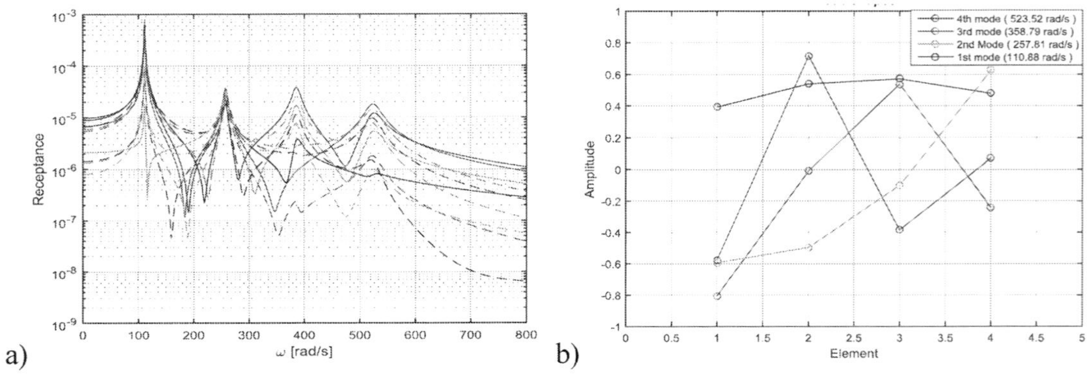


Fig. 2 a) FRFs of the system, b) Mode shapes of vibration.

Table 1 Modal parameters of the system determined analytically.

| **Mode** | **1.** | **2.** | **3.** | **4.** |
|---|---|---|---|---|
| **Frequency [Hz]** | 17.65 | 41.03 | 61.40 | 83.32 |
| **Damping ratio [%]** | 0.55 | 1.29 | 1.93 | 2.62 |

## Modal parameter estimation

The complex mode indicator function (CMIF) is based on singular value decomposition of FRF matrix and can be used to initial estimate of natural frequencies of the system. Singular value decomposition can be represented by equation:

$$[H(\omega)]=[U(\omega)][\Sigma(\omega)][V(\omega)]^H, \quad (2)$$

where $[\Sigma(\omega)]$ is a diagonal matrix of singular values, $[U(\omega)]$ is a left singular matrix and $[V(\omega)]$ is a right singular matrix. Mode shapes are represented by the left singular vectors of the matrix $[U(\omega)]$. The right singular vectors in matrix $[V(\omega)]$ represent the corresponding modal participation factors. CMIF is equal to the square of the singular value magnitude: [6]

$$\mathrm{CMIF}_k(\omega)=\Sigma_k^2(\omega). \quad (3)$$

Peaks detected in the CMIF plot indicate the existence of modes, and the located frequencies give the corresponding damped natural frequencies. The CMIF method is able to distinguish closed and also coupled modes.

The CMIF plot of the example problem is shown in Fig. 3. Four peaks can be seen there and every one of them indicates a mode. Peak picking method is used to initial estimate of natural frequencies. Initial estimate is important to create enhanced frequency response function (EFRF) for each mode.

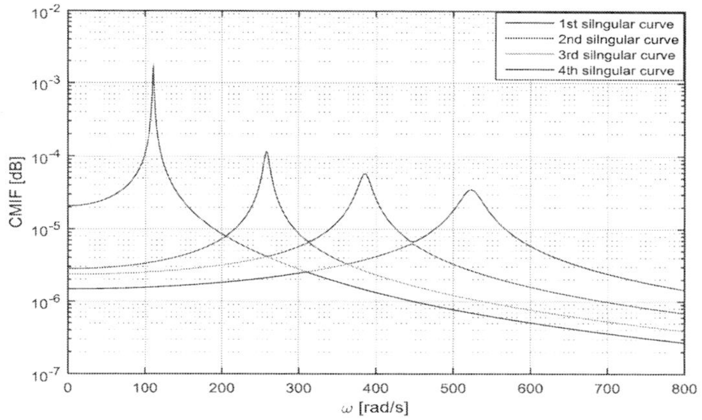


Fig. 3 CMIF plot.

The EFRF method is used to identify natural frequencies, and scaling of an equivalent single DOF characteristic that is associated with each peak in the CMIF [7, 8]. The EFRF is based on the concept of physical to modal coordinate transformation and can be defined as a weighted average of all of the measured FRFs, where the left and right singular vectors used as discrete modal filters. This is the way how the modes are isolated. Enhanced frequency response function is defined as follow: [4, 9]

$$eFRF(\omega)_r = \{U(\omega_r)\}^H [H(\omega)] \{V(\omega_r)\}, \tag{4}$$

where $eFRF(\omega)_r$ is an enhanced frequency response function of $r$-th mode. The EFRFs obtained for each mode of the analyzed system are shown in Fig 4.

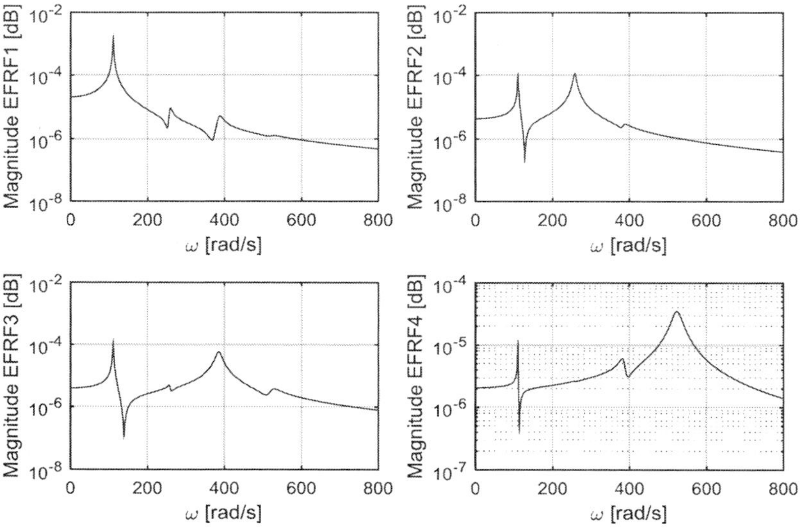


Fig. 4 EFRFs of the system.

As the next step, the rational fraction polynomial was applied to these functions to estimate modal parameters. The rational fraction polynomial is an iterative method to estimation of modal parameters from function, which describes dynamic behavior of mechanical system. This method says that frequency response function can be written as a ratio of two polynomials as follow [5, 10]

$$H(\omega) = \frac{\sum_{k=1}^{m} a_k \cdot (\omega)^k}{\sum_{k=1}^{n} b_k \cdot (\omega)^k}, \tag{5}$$

where $a_k, b_k$ are sought unknown polynomial coefficients. These coefficients are not the modal properties direct. Unknown coefficients are achieved by minimizing the error function $e_i$ which is defined by equation [9, 10]

$$e_i = \sum_{k=0}^{m} a_k (\omega_i)^k - h_i \left[ \sum_{k=0}^{n} b_k (\omega_i)^k + (\omega_i)^n \right], \tag{6}$$

where $h_i$ represents FRF data at frequency $\omega_i$. Minimizing of the error function is performed by using least-squares technique. Eq. (6) can be also written for EFRF. This estimation method was applied to each of EFRFs. Modal parameters obtain by this procedure are listed in Table 2. The comparison of the results in Table 1 and Table 2 shows the good agreement between the analytical solution and the estimation process.

Table 2 Modal parameters estimated from EFRFs.

| **Mode** | **1.** | **2.** | **3.** | **4.** |
|---|---|---|---|---|
| **Frequency [Hz]** | 17.680 | 39.372 | 59.763 | 81.771 |
| **Damping ratio [%]** | 0.61 | 1.14 | 1.90 | 2.81 |

**Computational efficiency of the proposed approach**

The standard estimation procedure is based on the extraction of modal parameters directly from FRF matrix. This is effective only if FRF matrix is not too large, otherwise there may be problems with the time-consuming computations and lack of memory. In order to avoid these problems, the modal parameters can be estimated from Enhanced Frequency Response Functions derived from FRF matrix. This allows to include all measurement data while retaining the maximum information about the dynamic behavior of the structure. For the purpose of comparing the computational efficiency of both approaches, a special program was created in Matlab. The program generated FRF matrices with a different number of measurement DOFs and measured computational time of the estimation process. The length of discrete frequency spectrum was the same in all cases (500 spectral lines). In this study, RFP method was used as the estimation algorithm. The results of the comparison are shown in Table 3. The results show the significant differences in the duration of the computation and in the memory usage, particularly for the large FRF matrix. The modal parameter estimation from EFRFs is much more effective. The parameters are determined for a relatively short time with minimum memory usage even in FRF matrix with tens of thousands of measurement DOFs. If the modal parameters are estimated directly from FRF matrix, the computation time and the memory usage increase significantly with an increasing number of measurement DOFs. As can be seen, when FRF matrix had 5929 DOFs, the computation failed after 3 h and

49 min due to lack of memory. The proposed approach was advantageously implemented in software application DICMAN 3D [9] that is used to measure modal parameters by high-speed digital image correlation method.

Table 3 Comparison of computational time and memory usage between the standard and proposed estimation procedure.

| Number of DOF of the system | Modal parameters estimated directly from FRF matrix | | Modal parameter estimated from EFRFs | |
|---|---|---|---|---|
| | Computational time [s] | Memory used by Matlab [%] | Computational time [s] | Memory used by Matlab [%] |
| **256** | 6.525 | 15.2 | 6.951 | 4.5 |
| **1024** | 53.96 | 39.6 | 11.89 | 4.6 |
| **2116** | 165.4 | 94.0 | 16.52 | 4.7 |
| **4096** | 1906 | 94.0 | 27.65 | 5.0 |
| **5929** | Out of memory | - | 34.03 | 5.1 |
| **22500** | - | - | 112.2 | 6.2 |
| **40000** | - | - | 209.8 | 7.0 |

## Conclusions

The aim of the paper was to consider the accuracy and efficiency of modal parameter estimation process when modal parameters are extracted from enhanced frequency response functions. These functions represent responses of separated modes in modal space and they are described as the response of an equivalent SDOF system. By this way, the data contained in the FRF matrix are reduced to a few functions. It is advantageous when the matrix is relatively large. The main advantage of the proposed approach lies in a significant reduction in memory requirements and computational time while keeping a high accuracy of modal parameters.


## Acknowledgement

The authors would like to express their gratitude to Scientific Grant Agency VEGA MŠ SR for the support of this work under the project no. 1/0751/16 and 1/0393/14.


## References


[1] J. He, Z.-F. Fu, Modal analysis, Butterworth-Heinemann, Great Britain, 2001.

[2] F. Trebuňa, F. Šimčák, R. Huňady, Vibration and modal analysis of mechanical systems, (In Slovak), Typopress, Košice, 2012.

[3] A. Bilošová.: Experimental modal analysis, (In Czech), VŠB TU Ostrava, 2011.

[4] J. Bocko, Š. Segľa, R. Huňady, Kmitanie mechanických sústav, Typopress, Košice, 2016.

[5] J. D. Ewins, Modal testing: Theory, Pracice and Application, second ed., Research Studies Press, Baldock, 2000.

[6] C.Y. Shih, Y.G. Tsuei, R.J. Allemang, D.L. Brown, Complex mode indication function and its application to spatial domain parameter estimation, *Mechanical Systems and Signal Processing* 2 (4) (1988) 367-377.

[7] A.W. Phillips, R.J. Allemang, The complex mode indicator function (CMIF) as a parameter estimation method, *Proceedings of the 16th International Modal Analysis Conference (IMAC),* Santa Barbara, California, 1998, p. 705-710.

[8] R.J. Allemang, D.L. Brown, A complete review of the complex mode indicator function (CMIF) with applications, *Proceedings of International Conference on Noise and Vibration Engineering (ISMA)*, Katholieke Universiteit Leuven, Belgium, 2006.

[9] R. Huňady, M. Hagara, A new procedure of modal parameter estimation for high-speed digital image correlation, Mechanical Systems and Signal Proscessing. 93 (2017) 66-79.

[10] M. H. Richardson, D. L. Formenti, Parameter estimation from frequency response measurements using rational fraction polynomials, Proceedings of the 1st International Modal Analysis Conference (IMAC), Orlando, Florida, 1982.

# Influence of Penetrated Roving Surface Treatment on the Crack Width Using Pull-out Test

Tomáš Vlach[1,2,a], Lenka Laiblová[1,2,b], Jakub Řepka[1,2,c], Michal Ženíšek[1,2,d] and Petr Hájek[1,2,e]

[1]University Centre for Energy Efficient Buildings, CTU in Prague, Trinecka 1024, 273 43 Bustehrad, Czech Republic

[2]Faculty of Civil Engineering, CTU in Prague, Thakurova 7, 166 29 Prague, Czech Republic

[a]tomas.vlach@cvut.cz, [b]lenka.laiblova@cvut.cz, [c]jakub.repka@cvut.cz, [d]michal.zenisek@cvut.cz, [e]petr.hajek@cvut.cz

**Keywords:** Textile reinforced concrete, pull-out test, bond behavior, crack width, surface treatments.

**Abstract.** This paper presents the possibility and suitability of chosen pull-out test method in the case of surface treatment using fine grain technical silica sand. Surface treatments in general significantly improve bond between composite reinforcement and cement matrix and thereby the effectiveness of roving single fibers. The specimen according to used testing method has a small thickness 6 mm. The advantage is that concrete cover less than 3 mm corresponds to concrete cover in real structures. But this cover is double side around one layer of reinforcement and in structures this situation does not occur. The disadvantage is that by the bond parameters improving specimens were damaged in their concrete part much before reaching the maximum force required for the reinforcement break.

## Introduction

With the development of high performance concretes (HPC) [1] come also demands for some new noncorrosive materials as reinforcement in extremely subtle structures with similar or better mechanical parameters compared to the conventional steel reinforcement. For this reason technical textiles began to be more and more popular. They use most commonly carbon, AR-Glass and basalt roving penetrated with polymer matrix. It is a relatively new composite material which is different from the traditional steel reinforcement also in the case of interaction with the cement matrix. In the past many numerical studies were created with influence of cohesion [2]–[4]. Aim of this presented study is to show the difference in results of pull-out test with and without surface treatments of penetrated roving using selected testing method. It was also investigated the effect of filler added into the epoxy resin matrix on bond behavior and controlled crack width.

## Testing methodology and set-up

The testing methodology was inspired by [5] with unsymmetrical anchoring length. On one side of the specimen a penetrated roving pull-out was secured using the short length 20 mm. On the opposite side was penetrated roving anchored along the remaining length of the HPC specimen. Specimen dimensions and also testing set-up are clearly presented in Fig. 1. The HPC plate with dimensions of 60 x 278 mm has a thickness of only 6 mm. The only one penetrated roving using epoxy resin was embedded in HPC specimen matrix in its axis. HPC recipe and its basic mechanical parameters, mechanical parameters of used AR-Glass roving and parameters of used epoxy resin matrix are listed in [6], [7]. It was used the same material in order to comparison of obtained results with previous research. Surface treatment was done using technical silica sand with grain size 0.1 – 0.6 mm. As filler into epoxy matrix Stellmittel T (Extender F) from the Sika Company was used. Four sets of specimens including five pieces for each group were created. The first group was penetrated roving by epoxy matrix without surface treatment (labeled "Pure 0 %"), the second group was penetrated roving with surface treatment using technical silica sand ("Pure + sand 0 %") and then this principle was repeated with Sika filler added into the epoxy resin matrix during the penetration process. Thus penetrated roving by epoxy matrix with filler without surface treatment ("Pure 2 %") and with surface treatment ("Pure + sand 2 %"). The amount of filler was selected based on the results of previous experiments – tensile test of penetrated single roving with different types of fillers and with different concentrations. Tensile test is not the subject of this paper.

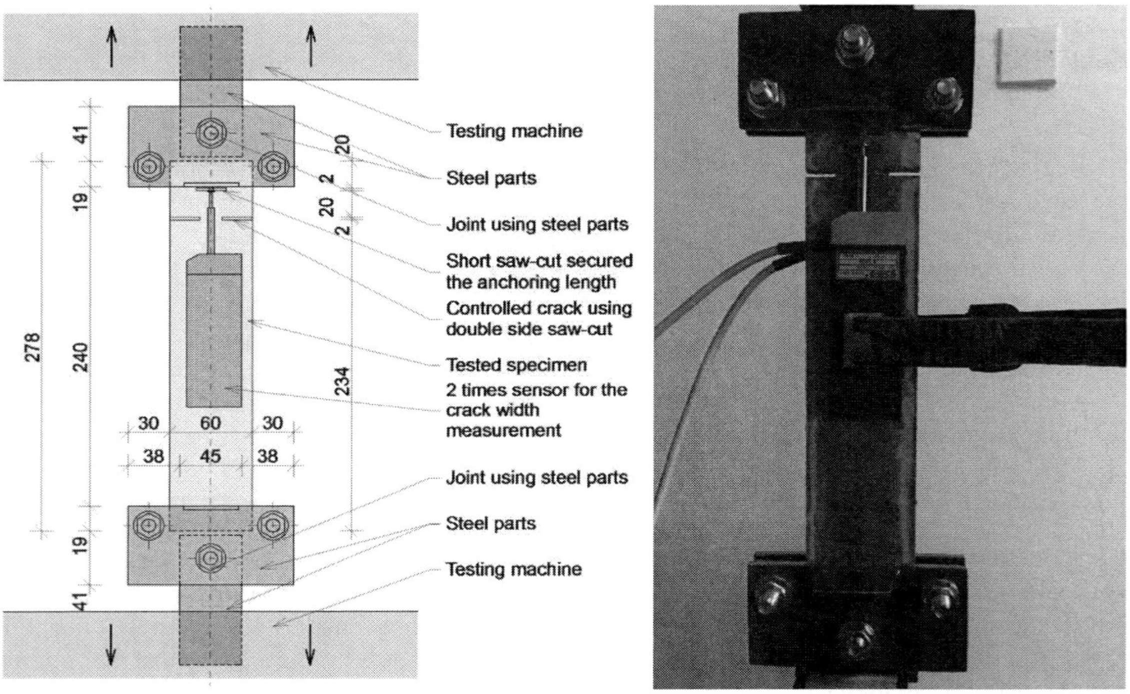


Fig. 1: Specimen dimensions and testing set-up presented at the colored scheme with descriptions and picture just before the start of measurement.

The anchoring length approximately 20 mm is secured using transverse very short saw-cut through the penetrated roving just below the steel part for the installation of the sample into the test machine and controlled crack at the required distance from the upper saw-cut. Double-side saw-cut was used for the predetermined breaking point – controlled crack and this point is also the border of unsymmetrical anchoring length. The length approximately 230 mm in the bottom part is certainly sufficient for the anchoring securing.

During the test procedure for determination of bod behavior were measured time, force, crosshead displacement and development of the crack width using two external potentiometers to obtain an accurate values in the axis of specimen. This experiment was also inspired by similar testing procedure according to ACI 440.3R-04 standard about the test methods for fiber-reinforced polymers (FRP) for reinforcing or strengthening concrete structures. The speed of loading was 2.0 mm/min according to prescribed tensile stress increasing 2.0 MPa/s in mentioned standard. Results are presented in Figure 2 in the form of two graphs. Stress in the first graph was calculated to the cross-sectional area of roving according to technical data sheet without the influence of epoxy matrix. Therefore the maximum theoretical tensile strength is 1.700 MPa but for no specimen was damaged the reinforcement as shown in Fig. 2 and Fig. 3. The HPC was always damaged due to its small cross-sectional area and transverse stresses initiation due to the location of the saw-cuts.

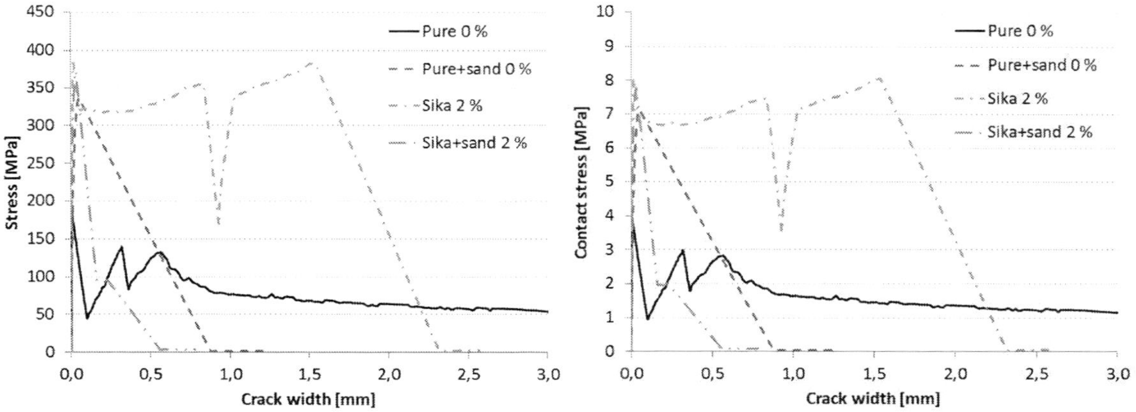


Fig. 2: Tensile stress in composite reinforcement and contact stress between composite reinforcement and cement matrix depending on the crack width development.

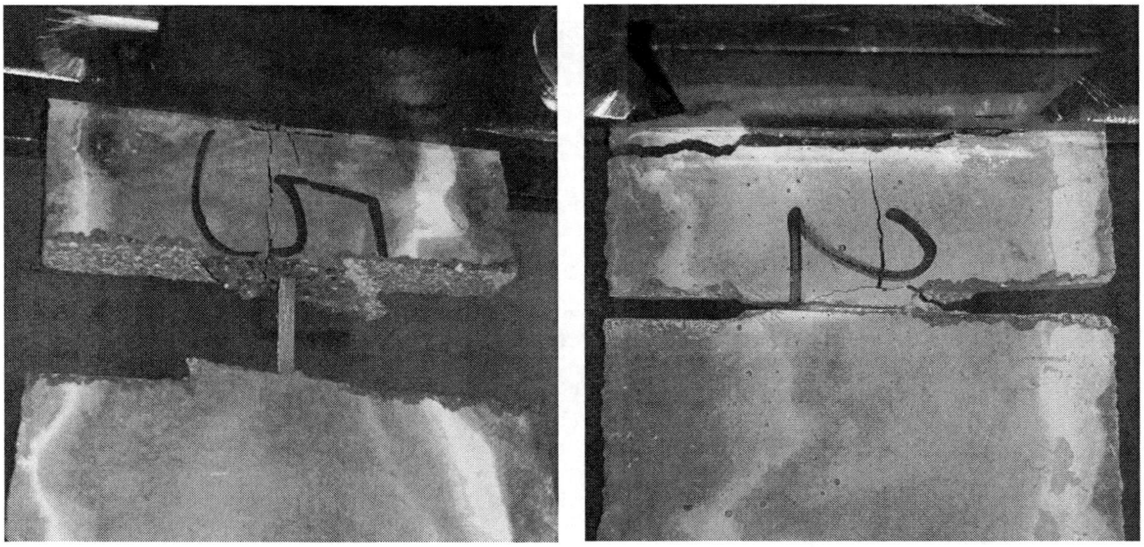

Fig. 3: The difference between specimens with and without surface treatments, pictures were made immediately after the test was interrupted due to the sample breaking.

In Figure 3 is presented the difference between specimens with and without surface treatments. The left part of Figure 3 for the sample without surface treatment shows gradual pulling-out of reinforcement from the HPC sample. Area in the contact with concrete is reduced during pulling-out and therefore there is a gradual decrease of force. Or as in the case presented in Figure 3 in the left is damaged the concrete specimen due to the irregular cross-section shape and concentration of radial stress around this point. This results in the longitudinal crack formation in the place of composite reinforcement. The right part of Figure 3 presented the specimen with surface treatment. It is seen the localization of controlled crack. In a short time after its formation and minimal crack opening occurred violation of concrete in longitudinal direction of reinforcement without expressive slipping of composite reinforcement and without violation of composite reinforcement.

## Conclusions

It is known filler addition and surface treatment should have positive influence on the mechanical properties. Graphs in Fig. 2 present the behavior just after the controlled crack initiation. In the case of pure specimen without any treatment there was a gradual pulling out with low bond. Specimen with filler had similar behavior with much better bond behavior until the damage of HPC. Both groups with the surface treatments have similar great results. There was almost no crack opening after the crack initiation. In the moment of critical stress in HPC part specimen were damaged and the stresses dropped to zero. This effect was enhanced by the shape of specimen with its short saw-cut because it led to the transverse tensile stresses initiation. Based on the results this method of the pull-out test should be modified for penetrated roving with surface treatments or filler addition into the epoxy resin matrix. Bigger cross-sectional area of concrete outside of controlled crack is required or concrete for the pull out test should be reinforced. Also short saw-cut for securing of anchoring length should be as short as possible or made by driller because of the transverse tensile stresses initiation which reduce the resistance of the concrete profile.

## Acknowledgement

Authors would like to thank for the financial support of CZ.01.1.02/0.0/0.0/15_019/0004908 - Advanced Concrete Elements with Woven Reinforcement project. This article also has been supported by Student Grant Competition SGS16/131/OHK1/2T/11 – Experimental Facade Panels from UHPC as a Base for LED Display.

## References

[1] P. Reiterman, M. Jogl, V. Baumelt, and J. Seifrt, Development and Mix Design of HPC and UHPFRC, Adv. Mater. Res., 982 (2014).

[2] J. Hartig, U. Häußler-Combe, and K. Schicktanz, Influence of bond properties on the tensile behaviour of Textile Reinforced Concrete, Cem. Concr. Compos., 30 (2008) 898–906.

[3] R. Chudoba, M. Vořechovský, and M. Konrad, Stochastic modeling of multi-filament yarns. I. Random properties within the cross-section and size effect, Int. J. Solids Struct., 43 (2006) 413–434.

[4] T. Vlach, M. Novotná, C. Fiala, L. Laiblová, and P. Hájek, Cohesion of Composite Reinforcement Produced from Rovings with High Performance Concrete, Appl. Mech. Mater., 732 (2015) 397–402.

[5] E. Lorenz and R. Ortlepp, Bond Behavior of Textile Reinforcements - Development of a Pull-Out Test and Modeling of the Respective Bond versus Slip Relation, High Performance Fiber Reinforced Cement Composites 6 (2012) 479–486.

[6] T. Vlach, A. Chira, L. Laiblová, C. Fiala, M. Novotná, and P. Hájek, Numerical Simulation of Cohesion Influence of Textile Reinforcement on Bending Performance of Plates Prepared from High Performance Concrete (HPC), Adv. Mater. Res., 1106 (2015).

[7] T. Vlach et al., Comparison of Different Methods for Determination of Modulus of Elasticity of Composite Reinforcement Produced from Roving, Adv. Material Research, 1054 (2014).

EAN 2017
55th conference on experimental stress analysis

# The reliability analyses of gear and centrifugal pumps at different rotational speeds

QAZIZADA Mohammad Emal, PIVARČIOVÁ Elena

Technical University in Zvolen, Faculty of Environmental and Manufacturing Technology, Department of Machinery Control and Automation Technology, Masarykova 24, 960 53 Zvolen, Slovakia

m.emalqazizada@yahoo.com, pivarciova@tuzvo.sk

**Keywords**: reliability, efficiency, characteristics, gear pumps, centrifugal pumps, VSD

**Abstract.** This paper concentrates the reliability analyses of gear and centrifugal pumps at different rotational speeds and flow rates. The gear and centrifugal pumps are often controlled by adjusting their rotational speed, which affects the resulting flow rate and output pressure of the pumped fluid. Initially, focused to analyze the behavior of individual characteristic curves of the pump, to find how can be applied for the determination of the most efficient frequency of rotation a pump, then the determination of energy efficiency– and reliability–based limits for the recommendable operating region of a variable speed–driven (VSD) centrifugal and gear pumps are discussed.

## Introduction

The study use of reliability based criteria in the determination by characteristics of centrifugal and gear pumps, to measure the basic representative functional curves, to evaluate the behavior of reliability, the head, and flow rate of fluid transported at different frequencies of rotation. From a reliability point of view gear and centrifugal pumps have received considerable attention in recent years [1]. The centrifugal pumps are one of the most important components in any industry which have to deal with fluids. The reliability and maintainability of centrifugal pump systems have in the overall device availability plays a very important role of a suitable maintenance strategy [2]. The type of gear pump is used for transferring and metering high viscosity fluids and power transfer in industrial processes usually at high pressure rates [3]. The gear pumps are simple and robust devices that can work at a wide range of pressures and rotational speeds, providing at the same time a high reliability. Their main applications can be found as hydraulic pumps in machine tools, also as oil pumps in engines or in fluid power transfer units [4]. The centrifugal and gear pumps operate on completely different principals [5]. The basic difference between gear and centrifugal pumps are evident in the pump's response to a system's head-flow rate curves. Several pump applications are presented to illustrate the selection process needed to insure pumps reliability. The performance overlap region, where both pump types should be considered [6]. Many authors were completed their research in the area of reliability analysis of centrifugal pump. Ferman [7], provided independent, objective engineering consulting services on new and existing pumping equipment reliability and systems. Frith and Scott studied comparison of an external gear pump wear model with test data. This research was about measuring the actual material lost due to wear is impossible in a practical sense [8].

## Material and Apparatus

Following experiments is proceeded in Edibon equipment. Equipment have a four pumps. This equipment designed to determine the operating physical appearance of several types of pumps, such as centrifugal pump, axial flow pump, gear pump and Peripheral pump. The scheme of four pumps is illustrated in Fig. 1 [9]. Users are generally much more familiar with one type than the other [5].

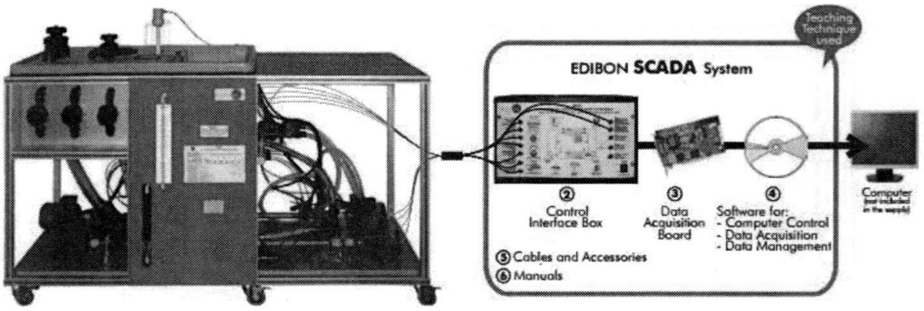


Fig. 1 Structure of a pumping system [9]

Pumping system equipped with measurement sensors that are used for control and monitoring purposes. The volumetric tank has been designed to accommodate low or high flows and it includes a level sensor (capacitive, length: 300 [mm]) to determine the water flow and the water level in the volumetric tank, an indicating transparent tube, and a graduated scale. Admission pressure sensor and discharge pressure sensor for each pump. Reading of speed (rpm) and torque [Nm] of the pump. Sensors connectors in the interface have different pines numbers (from 2–16), to avoid connection errors.

## Theoretical overview and applied formulas for calculation of pumps characteristics

The same equations applied to calculate gear and centrifugal pumps characteristics.

Net head $H$ is the height of the fluid column in the open pipe after the pump [10].

$$H = \frac{P_2 - P_1}{\rho g} \tag{1}$$

Where $P_1$, $P_2$ shows the pressure drop of the pump system. Or specific work $\varepsilon_w$ done by the pump is:

$$\varepsilon_w = Hg = \frac{P_2 - P_1}{\rho} \tag{2}$$

The power given to the fluid $P_w$ is denominated hydraulic power or output power and calculated [11]:

$$P_w = \rho g Q H \tag{3}$$

The mechanical power $P_f$ given to the pump by the activator motor is denominated control power or input power and can be calculated as [12]:

$$P_f = \omega T = \frac{2\pi}{60} nT \tag{4}$$

$$P_f = \frac{2\pi}{60} nT \tag{5}$$

The $\omega$ is the angular axis speed in [rad $sec^{-1}$], $n$ frequency of rotation in (rpm. revolutions per minute) and $T$ the torque in the axis in [Nm], to obtain the power in [W] [13].

Pump efficiency $\eta$ is defined as the ratio of the power imparted on the fluid by the pump in relation to the power supplied to drive the pump [10].

$$\eta = \frac{P_w}{P_f} = \frac{\rho g Q H}{\frac{2\pi}{60} nT} \tag{6}$$

NPSH has two parts: NPSH required ($NPSH_R$) and NPSH available ($NPSH_A$). ($NPSH_R$) is a function of the pump and it is defined as:

$$\mathrm{NPSH}_R = \left(\frac{P_1}{\rho g} + \frac{w_1^2}{2g} - \frac{P^*}{\rho g}\right) \tag{7}$$

Where $P_1$ and $w_1$ are the pressure and velocity at the inlet of the pump and $P^*$ is the transported liquid's steam pressure [14]. Velocity is calculated as:

$$w = \frac{Q}{A} \tag{8}$$

Where $A$ [$m^2$] is the circular surface area of the impeller which is calculated as:

$$A = \frac{\pi}{4} D^2 \tag{9}$$

Where $D$ [m] is impeller diameter [15].

## Calculation of gear and centrifugal pumps characteristics

The centrifugal and gear pumps were driven at ten different relative flow rates and at a rotational speed ranging 3000 and 450 rpm individually. To calculate the experimentation data at the begging of measuring new data we found $t_{av}$, for obtaining of average temperature, we measured water temperature $t_1 = 21$ [°C] at the beginning and $t_2 = 23$ [°C] at the end of the experiment, then $t_{av} = 22$ [°C]. According to average temperature acceleration of gravity and density of water are chosen from physical properties' table of water 9,81 [m $s^{-1}$] 997,9 [kg $m^{-3}$] respectively. Experiments recorded values with different frequency for centrifugal pump set in Table 1.

Table. 1 Calculated characteristics values of centrifugal pump at 3000 rpm

| Number | $Q$ [l min$^{-1}$] | $H$ [m] | $P_w$ [W] | $P_f$ [W] | $\eta$ [%] |
|---|---|---|---|---|---|
| 1 | 18,12 | 8,88 | 26,2 | 355 | 9,4 |
| 2 | 17,7 | 8,99 | 25,9 | 311,0 | 8,3 |
| 3 | 16,38 | 9,09 | 24,2 | 301,5 | 8,0 |
| 4 | 15,34 | 9,40 | 23,5 | 292,1 | 8,0 |
| 5 | 13,61 | 9,29 | 20,6 | 282,7 | 7,3 |
| 6 | 12,16 | 9,29 | 18,4 | 273,3 | 6,7 |
| 7 | 10,94 | 9,29 | 16,5 | 263,8 | 6,2 |
| 8 | 8,71 | 9,50 | 13,5 | 254,4 | 5,3 |
| 9 | 6,77 | 9,80 | 10,8 | 241,9 | 4,4 |
| 10 | 4,55 | 9,70 | 7,20 | 226,1 | 3,1 |

Intersection curves of efficiency with pump curve Fig. 2 the efficiency is always below 100 % since the supplied power is always larger than the hydraulic power due to losses in controller, motor, and pump components.

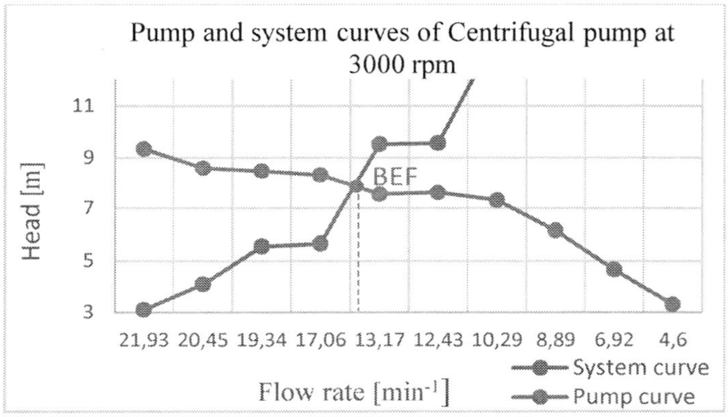


Fig. 2 Pump and system curves of centrifugal pump

Laboratory measurements were carried out for a gear laboratory pump too, the pump was driven at ten different relative flow rates and at a rotational speed ranging 450 rpm by adjusting its operation with the frequency converter and control valves. The measurement data was stored which has shown in Table 2.

System curve with pump curve see Fig. 3 which indicate the intersection point of pump curve and system curve of pump, at constant speed 450 rpm revolutions are plotted. The intersection of pump curve with system curve shows operating point [12].

Table. 2 Measured characteristics values of gear pump at 450 rpm

| Number | $Q$ [l min$^{-1}$] | $T$ [Nm] | $H$ [m] | $P_W$ [W] | $P_f$ [W] | μ [%] |
|---|---|---|---|---|---|---|
| 1 | 12.67 | 3.99 | 1.67 | 3.46 | 183.02 | 1.84 |
| 2 | 12.48 | 4.31 | 1.64 | 3.34 | 202.95 | 1.65 |
| 3 | 12.69 | 3.93 | 1.96 | 4.06 | 185.26 | 2.19 |
| 4 | 11.48 | 5.61 | 3.81 | 16.53 | 264.60 | 6.25 |
| 5 | 10.83 | 5.74 | 3.74 | 15.55 | 270.77 | 5.74 |
| 6 | 9.39 | 6.35 | 12.14 | 18.64 | 299.20 | 6.23 |
| 7 | 9.24 | 6.41 | 12.12 | 18.31 | 302.13 | 6.06 |
| 3 | 7.99 | 9.00 | 22.19 | 23.99 | 424.23 | 6.83 |
| 9 | 6.18 | 10.21 | 23.20 | 23.52 | 481.22 | 6.93 |
| 10 | 5.26 | 11.72 | 33.43 | 28.75 | 552.37 | 7.20 |

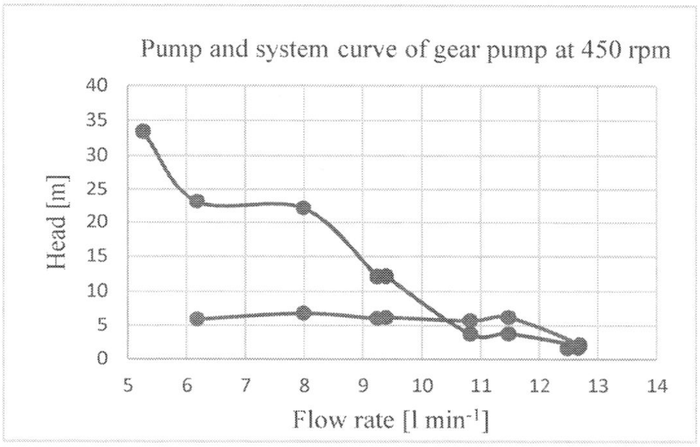


Fig. 3 Pump and system curve of gear pump

## Results

The experimental data from comparison of centrifugal and gear pumps showed that these pumps can operate at different rotational speeds, various heads, and flow rates without any mechanical problem [16]. Pumping system equipped with measurement sensors that are used for control and monitoring purposes. The results are shown in (Table 1) and (Table 2) the flow rate, Head, Powers, efficiency and torque are measured in different speeds of rotation from (3000, 450 rpm) for centrifugal and gear pump respectively, and the result are calculated from equations (1-3-5 and 6). The $H$, $P_{w}$, $P_f$ and $\eta$ for centrifugal pump at 3000 rpm are calculated in ten steps of different flow rate see (Table 1). According to these data the Fig. 2 is plotted that shows the pump runs reliable at its operation region, the figure shows the pump curve at 3000 rpm represented that in the value of $H$, is 9,7 [m] the flow rate $Q$ is 4,55 [l min$^{-1}$], decreasing of $H$ to 8,8 [m] cased to increase the flow rate to 18,12 [l min$^{-1}$]. For gear pump the head, hydraulic power, mechanical power, efficiency, and net position suction head at 450 rpm are measures too in ten different flow rate see (Table 2). Based on these data the Fig. 3 is shows that the gear pump runs reliable at its operation region too, the figure shows the pump curve at 450 rpm represented that in the value of $H$, is 33,43 [m] the flow rate $Q$ is 5,26 [l min$^{-1}$], decreasing of $H$ to 1,67 [m] cased to increase the flow rate to 12,67 [l $min^{-1}$]. The $NPSH_R$, velocity, and surface area for centrifugal pump, by using equation (7, 8 and 9) are calculated 11 [m], 0,47 [m $s^{-1}$] and 0,0007 [$m^2$] respectively. The measured data results of gear and centrifugal pumps indicated that often gear pump is used for producing high pressure then centrifugal pump.

## Conclusion

From a reliability point of view centrifugal and gear pumps have received considerable attention in recent years. Therefore it is necessary to determine the effective of reliability. The reliability and maintainability of centrifugal and gear pumps systems have in the overall device availability plays a very important role of a suitable maintenance strategy [17].

Gear pump constructions have been used and developed for over 400 years. Johannes Kepler created the first gear unit in 1604. The primary application of the patented solution at that time was to pump water in dehydrated mines [19].

This article relates to the problem of reliability analyses of gear and centrifugal pumps at different rotational speeds and flow rates [18].

The gear pumps are capable of moving a wide range of fluids. Entrained gasses, solids, low viscosity to high viscosity, and low net positive inlet pressure available can all be designed. The high mechanical efficiency offers energy savings. Peripheral pump can be used for pressure increasing, fluid transfer and distribution. They are suitable for flooded suction applications. In Table 2 from characteristic calculation of gear pump achieved that in 450 rpm in the low flow rate 5,26 [l $min^{-1}$] can produced a high head 33,43 [m]. However from characteristics calculation of the centrifugal pump which is run in 3000 rpm see Table 1, in low flow rate 4,55 [l $min^{-1}$], produced the head of 9,7 [m]. Furthermore in Table 1 and Table 2 the consumption of hydraulic power and mechanical power in gear or positive displacement pump is from low flow rate to high gradually increasing too much, but in centrifugal pump from low flow rate to high flow rate the consumption of hydraulic power and mechanical power gradually increasing is not reached to higher values of consumption as gear pump.

## Acknowledgments

This paper was prepared within the work on a research project KEGA MŠ SR 003TU Z-4/2016: Research and education laboratory for robotics.

## References

[1] R. B. Erickson, A. Budris, E. P. Sabini, Pump Reliability – Correct Hydraulic Selection Minimizes Unscheduled Maintenance. PumpLines, 2001, pp. 1–3

[2] D. Singh, A. Suhane, Study of Centrifugal Pump Using Failure Mode Effect and Critical Analysis Based on Fuzzy Cost Estimation: A Case Study. International Journal of Science and Research (IJSR), 2013, ISSN (Online): 2319-7064

[3] N. Erturk, A. Vernet, R. Castilla, P. J. Gamez-Montero, J. A. Ferre, Experimental analysis of the flow dynamics in the suction chamber of an external gear pump. Tarragona, Spain. International Journal of Mechanical Sciences, 2010

[4] E, Mucchi, G. Dalpiaz, A. Fernandez Del Rincon, Elastodynamic analysis of a gear pump. Part I: Pressure distribution and gear eccentricity. Ferrara, Italy. Mechanical Systems and Signal Processing 24, 2010, 2160–2179

[5] J. E. Purcell, J. A. Silvaggio, A comparison of positive displacement and centrifugal pump applications. Trenton, New Jersey, 2012, pp. 99–104

[6] D. B. Parker, Positive displacement pumps-performance, and application. Texas. Proceedings of 11th International Pump Users Symposium, 1992

[7] R. Ferman, The user's role in pump reliability, 2015. Available in: http://info.empoweringpumps.com/pump–reliability

[8] R. H. Frith, W. Scott, Comparison of an external gear pump wear model with test data. Australia. Elsevier wear 196, 1996, 64–71

[9] P. Edibon, Technical Teaching uipment. Computer Controlled Multi pump Testing Bench. with SCADA, 2014

[10] P. Blišt'an, H. Pacaiova, Modelling environmental influence on the pipelines integrity. 11th International Multidisciplinary Scientific Geo conference and EXPO, Varna; Bulgaria; 2011, Vol. 2, pp. 645–652, Code 101584

[11] I. Chalghoum, Transient behaviour of a centrifugal pump during starting period. Tunisia 2016. Available in: http://ac.els–cdn.com/S0003682X16300238/1–s2.0–S0003682X16300238–main. pdf?_tid=0ce4b116–1824–11e6–99c8–00000aab0f01&acdnat=1463045559_ba8aef040e0a 29877c1ea1fe568ac0d8,

[12] E. Larralde, R. Ocampo, Centrifugal pump selection process, 2010, pp. 24–28. Available in: http://ac.els-cdn.com/S0262176210700298/1-s2.0-S0262176210700298-main.pdf?_tid=bb4f322c -1823-11e6-b16b-00000aab0f02&acdnat=1463045422_feccae9ea7c5efa40934b952b18f95de

[13] J. Haidary, MANUAL. Chemical engineering laboratory. Kabul University Publisher, 2013

[14] B. A. Sentyakov, V. M. Sviatskii, K. B. Sentyakov, Calculation of the average velocity and modelling the air flow in the working area of the blow head, Automation and modern technologies, 2013, №6, Moscow, pp. 20–24. ISSN 1585–1558

[15] D. Wang, Application of the two–phase three–component computational model to predict activating flow in a centrifugal pump and its validation. Xuefu Road, Zhenjiang China. 2016. Available in: http://www.sciencedirect.com/science/ article/pii/S0045793016300834

[16] M. E. Qazizada, V. Sviatskii,V., P. Božek, Analysis performance characteristics of centrifugal pumps, MM science journal. 2016, No. October, p. 1151–1159. ISSN 1803-1269

[17] P. Božek, Reliability control systems – methodology guide (in slovak). SP Synergia, Trnava, 2008, ISBN 978-80-89291-14-4

[18] P. Frankovský, O. Ostertag, E. Ostertagová, F. Trebuňa, J. Kostka and M. Výrostek (2017). Experimental analysis of stress fields of rotating structural elements by means of reflection photoelasticity. Applied Optics, 56(11), 3064-3070.

[19] A. Deptuła, M. A. Partyka, Inductive Decision Tree Analysis of the Validity Rank of Construction Parameters of Innovative Gear Pump after Tooth Root Undercutting. International Journal of Applied Mechanics and Engineering. 2017; 22(1): pp. 25–34 DOI 10.1515/ijame-2017-0002

# SECTION 2

## New methods and applications of deformation and stress analysis in mechanical, civil and other structures

# Evaluation of the Influence of Corrosion Fatigue on Steel Alloy for Steam Turbines

M. J. CERNY[1,a]

[1] Czech Technical University in Prague, Klokner Institute, Solinova 7, 166 08 Prague 6, Czech Republic
[a] cerny@hppro.klok.cvut.cz

## Introduction

To evaluate the influence of corrosion pits on fatigue [1], linear elastic fracture mechanics (LEFM) is used. It has been observed, that most pits have roughly a semielliptical shape with the width at the surface, 2c, and the pit depth. Investigations show that the pits can be treated as semi-circular surface cracks with a geometry described above.
Fatigue failure in steam turbine blades is a critical issue since high frequency fatigue loading is induced by unsteady steam flow. The cyclic stresses are low compared to centrifugal forces due to the rotation which leads to high stress ratios.
In the low pressure part of engines, fatigue crack initiation at corrosion pits is a frequently observed cause of damage. Especially in the last two blading stages – where transition from dry to wet steam takes place – pitting may occur when unexpected operating conditions, for example, due to leaking of the condenser are present.

## Modeling of corrosion fatigue damage

Since fatigue crack is expected to nucleate at the mouth of the pit (i.e. at the end of the c axis), the dimension c (half pit) will be used to represent the pit or crack size. With this approximation and that the pit size is small in terms of crack behavior, the stress intensity factor of pits $\Delta K$, FCGR diagram for long cracks and cracks that were initiated at corrosion pits (Fig.1) and finally a Kitagawa- Takahashi (K-T) diagram can be created.
The shown diagram is for 403/410 12% Cr martensitic steel and temperature 90°C. The K-T diagram correlates pit-to-crack fatigue data with the use of the approach El Haddad for short cracks (Fig.2). The data are shown for specific values of R (ratio of minimum to maximum stress for combined steady and cyclic stress). As a result, K-T diagrams can be used for assessment of the fatigue limit and life-time of corroded parts of steam turbines.

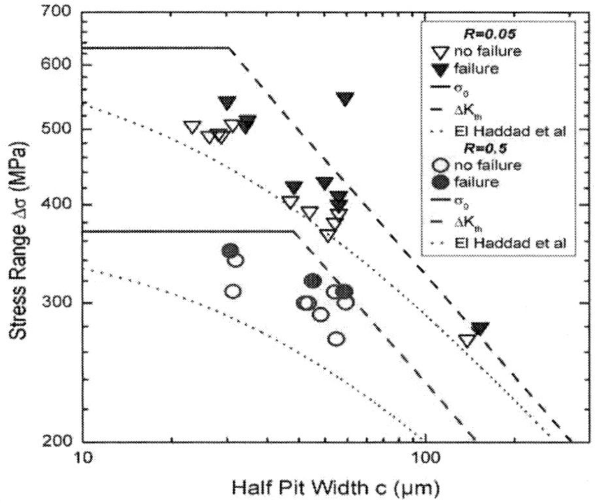


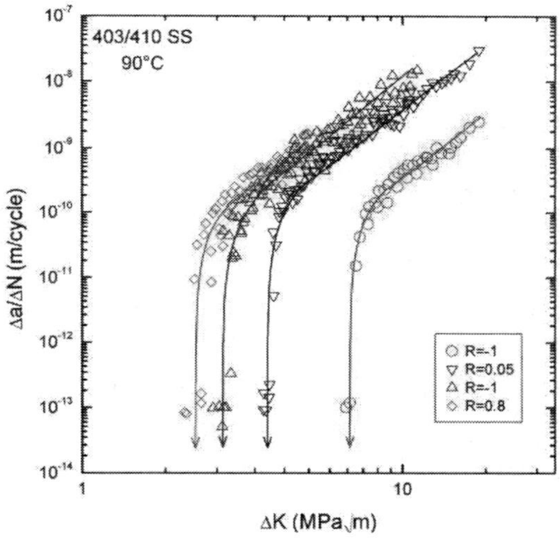


Fig. 1: Fatigue Crack Growth Rate (FCGR)

Fig.2 Kitagawa- Takahashi Diagram

An important parameter in the diagram is R, which is the ratio of minimum to maximum stress in a combined steady and cyclic stress environment, as shown in following relation

$$R \equiv \frac{\sigma_m - \sigma_a}{\sigma_m + \sigma_a}$$

where:

$\sigma_m$ is the steady (mean) stress, and
$\sigma_a$ is the cyclic (alternating) stress.

Further, stress intensity factor ΔK of the corrosion pits in the diagram can be calculated from the relationship:

$$\Delta K = \Delta\sigma\sqrt{\pi \cdot (c + c_0)} \cdot Y$$

where the geometry factor, Y , for semi-circular surface cracks is 0,65 and $c_0$ depends on the material according to empirical relationship, c is half pit width.

$$c_0 = \frac{1}{\pi} \cdot \left( \frac{\Delta K_{th,lc}}{Y \cdot \Delta\sigma_0} \right)^2,$$

where $\Delta K_{th,lc}$ ist he treshold stress intensity factor for long cracks.

The description of true pit geometry has been published by author earlier [2]. The max. and min. depths of pits and their densities have been found. In [3] are shown the mentioned characteristics and Table 1 comprises the parameters of corrosion evaluated for part of turbine packing ETU23 (alloy X22CrMoV12-1).

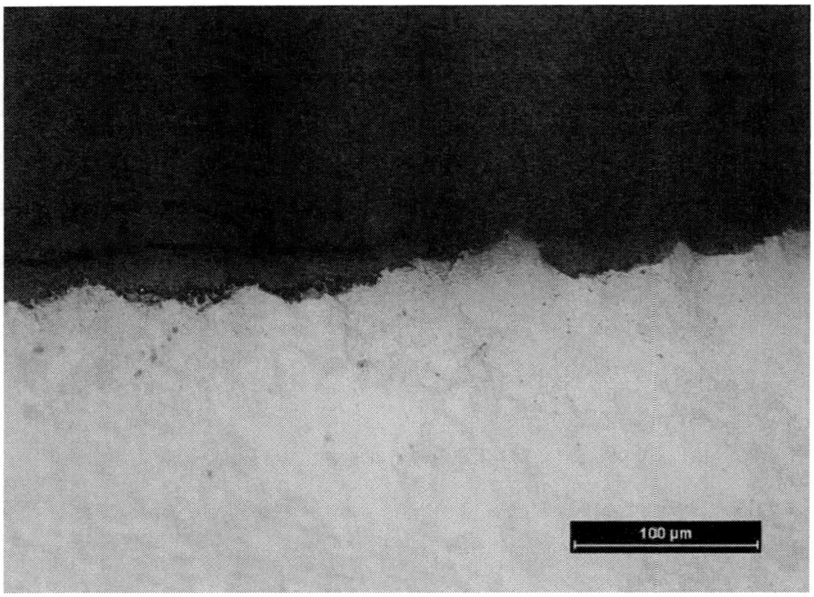


Fig. 3. Pitting corrosion after 6700 hr in 650 °C steam

Table 1. Parameters of pitting observed by microscopy Nikon Eclipse ME600 and image analysis by Nikon Elements

| | | | | | Depth [µm] | | | |
|---|---|---|---|---|---|---|---|---|
| **Area** | **No of Pits** | **Σ L i [mm]** | **Density** | **Oxide [%]** | **Max** | **Min** | **Mean** | **St.Deviation** |
| 1 | 40 | 4.1944 | 9.54 | 80.48 | 27.04 | 8.45 | 18.57 | 4.944 |
| 2 | 58 | 6.6833 | 8.68 | 68.26 | 29.54 | 7.87 | 20.71 | 5.381 |
| 3 | 17 | 3.3763 | 5.04 | 83.93 | 30.05 | 13.02 | 21.36 | 4.900 |
| 4 | 28 | 2.0299 | 13.79 | 63.79 | 31.33 | 11.13 | 20.66 | 5.028 |
| 5 | 18 | 2.4452 | 7.36 | 78.46 | 29.57 | 9.68 | 21.36 | 4.815 |
| 6 | | N/A | | | | | | |
| 7 | 39 | 2.8553 | 13.66 | 49.98 | 35.68 | 8.72 | 19.52 | 5.855 |
| 8 | 14 | 1.5375 | 9.11 | 80.08 | 31.88 | 18.06 | 24.11 | 3.682 |

## Acknowledgement

The presented work has been supported by grant No TE01020068.


## References


[1] Černý M.: Damage and Localized Corrosion in Steam Turbines, Conference EAN 2014, Proceedings (2014)

[2] Černý M.: Localized Corrosion in Alloy Steel of Steam Turbines, Applied Mechanics and Materials, vol.827, Trans Tech Publications (2016) doi:10.428/www.scientific.net/AMM.827.177

[3] Černý M.: Modeling of Corrosion Fatigue Crack Initiation in Steel Alloy for Steam Turbines, Conference EAN 2016, Proceedings (2016)

# Use of FBG Sensors for Delamination Growth Measurement under Mode I loading

Milan Dvořák[1, a], Nikola Schmidová[1, b], Martin Kadlec[2, c] and Milan Růžička[1, d]

[1]Czech Technical University in Prague, Faculty of Mechanical Engineering, Technická 4, 166 07 Prague 6, Czech Republic

[2]VZLU – Aerospace Research and Test Establishment, Beranových 130, Prague, Czech Republic

[a]milan.dvorak@fs.cvut.cz, [b]nikola.chmidova@fs.cvut.cz, [c]kadlec@vzlu.cz, [d]milan.ruzicka@fs.cvut.cz

**Keywords:** FBG; DCB; SHM; delamination; optical sensor; composite.

## Abstract

This paper summarizes the experience gained during the experimental research on sensitivity of optical Fiber Bragg Grating (FBG) sensors. The sensors were embedded inside composite specimens that were subjected to cyclic loading in mode I to achieve delamination growth. Data from the FBG sensors were compared to camera readings to evaluate the ability of the sensors to detect the damage in advance. Average delamination detection advance in loaded state was 6.7 mm and 8.9 mm for the first and the second sensor, respectively. The obtained results will be used to design proposal of the sensor network for an air intake.

## Introduction

The presented experimental work was focused on a sensitivity research of the FBG sensors. The technical problem to be solved is the impact or acoustic damage detection on an air intake carbon/epoxy composite structure. This structural part has no flight load bearing capacity, but it is loaded by acoustic pressure from the engine, resulting in presumed combination of mode I (opening) and mode II (shear) loading. FBG sensors are small enough to be embedded inside the prepreg composite structure. The embedding procedure was successfully tested, but there is a question of the sensitivity of such sensors in the given conditions. The first test of the damage detection capabilities of the FBG sensors was focused on the evaluation of the distance, from which can this sensor detect the delamination. Research was performed during the delamination growth testing in a mode I loading case using a double cantilever beam (DCB) specimens [1].

## Experimental Method

FBG sensors are based on a periodic variation in the refractive index of the optical fiber core. When a broadband light spectrum from an optic source is guided by the optical fiber into the FBG sensor, part of the light is reflected by the grating (in a form of narrow Gaussian peak) and the rest is transmitted (more details can be found in [2, 3]). For sensing purposes, the reflected peak is commonly measured using a spectrometer. Each Bragg grating is characterized by so-called Bragg wavelength, which is shifting according to deformation of grating. The elongation causes a positive shift, shortening causes a negative shift (see Fig. 1).

However, the FBG sensors that are embedded inside the composite structure are not only stretched. They are exposed to complex non-uniform stress field because of anisotropy and microbends which can lead to a distortion of the spectrum (see Fig. 2). These two effects (shift and distortion of the peak) were used for detection of near fatigue delamination inside the tested DCB specimen.

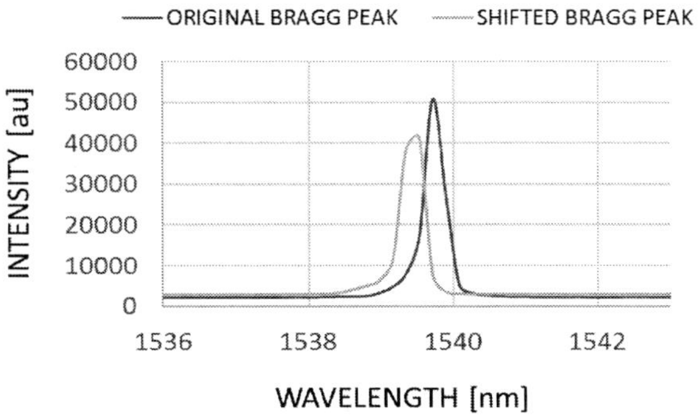


Fig. 1 Shifted peak (uniform strain) caused by pressure.

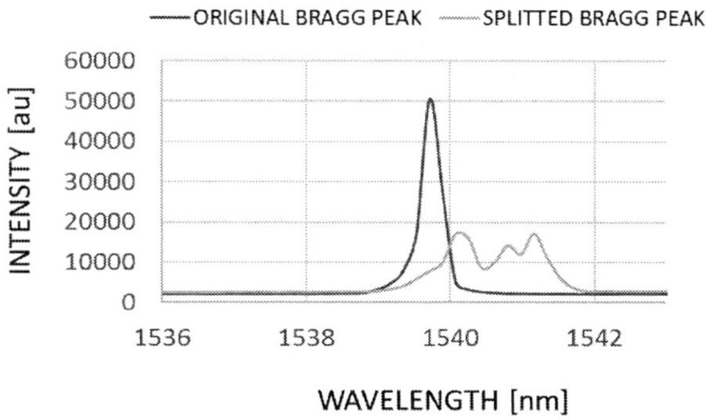


Fig. 2 Distortion of the peak caused by the complex non-uniform stress field.

So-called FBG chains were used for testing of five specimens (DCB01-05). The chains consisted of multiple FBG sensors on a single optical fiber with the following parameters: outer diameter of 0.195 mm, central wavelength of 1550 nm. Spacing between the two successive sensors was 25 mm. ORMOCER® primary coating was used.

## Experimental Specimens

Carbon fibre-reinforced composite Hexply AGP 193PW/8552S RC40 with the 16 layers lay-up of $[45/0]_{4S}$ was manufactured and five DCB specimens were extracted (dimensions: 250 x 25 x 5 mm). A non-adhesive plastic foil sheet was inserted to the plate edge during manufacturing to the neutral axis to initiate the delamination growth between the $8^{th}$ and $9^{th}$ layer. Optical fibers were embedded into the composite structure between the same layers (see Fig. 3). One end of the optical fiber was led out from the composite lay-up and it was fusion spliced to the protective cable with FC/APC connector.

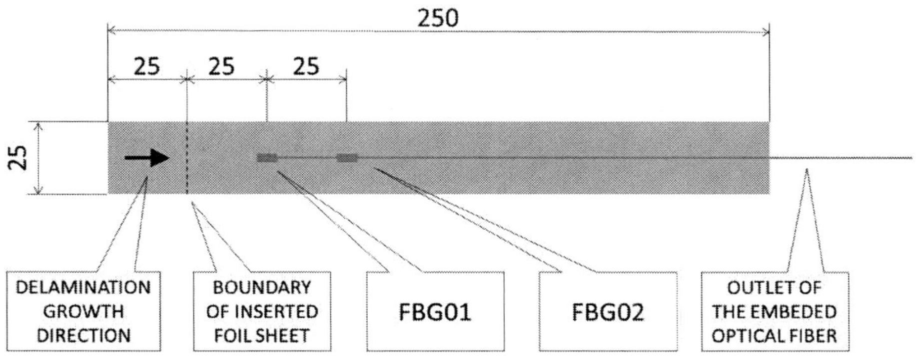


Fig. 3 Configuration of specimen and embedded FBG sensors (DCB01-05).

**Experimental Procedures**

**DCB Specimens Fatigue Testing**. The specimens were clamped using bonded aluminum blocks into INOVA ZUZ 50 loading machine, with a LTC-116-0.1 load cell, which has a capacity of 1 kN. The end of the specimen was supported by a flexible wire to maintain the specimen in the horizontal direction, and to reduce any additional forces that could originate from the specimen weight (Fig. 4).

The test was driven by displacement amplitude. The goal was to find loading conditions that will enable to test specimen within demanded time period with approximately constant crack growth rate (achieved by regularly increased displacement) and with one constant displacement cycling part included for the material properties evaluation.

The ASTM D6115 [4] is the only released standard considering fatigue delamination in CFRP and it focuses only on the crack growth onset (initiation from the insert). Therefore, the used fatigue crack growth test procedure was based on a draft standard for fatigue delamination crack growth that is being prepared for ASTM International. This procedure is well described in [5]. The displacement was zeroed at a value that corresponded with the force of 1 N. This step was performed because the displacement-controlled cycling could cause negative force values for very low displacements when the displacement would not be zeroed properly. The crack propagation during the following loading was monitored using video cameras. Sinusoidal loading with displacement control and a frequency of 5 Hz was applied. Load and displacement were recorded by the INOVA machine at a frequency of 100 Hz. The overall duration of all cyclic loading blocks was nearly 700 000 cycles. Typical fatigue crack (delamination) length propagation curve is pictured in Fig. 5.

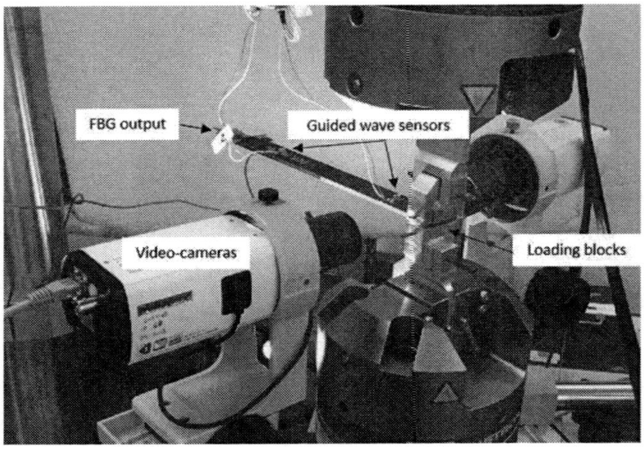


Fig. 4 Experimental set-up.

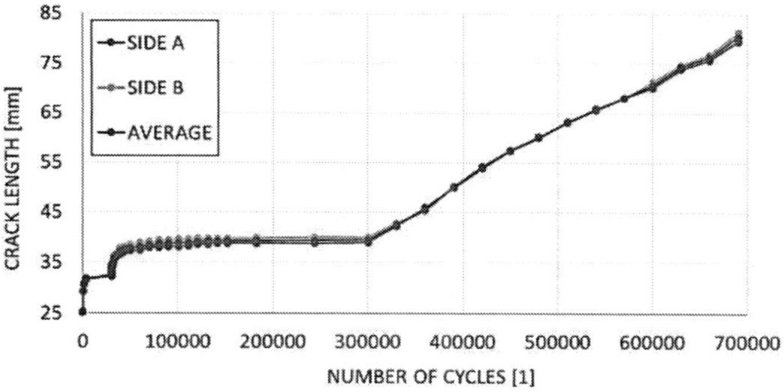


Fig. 5 Fatigue crack length propagation curve (specimen DCB01).

**FBG Sensors.** Safibra FBGuard 1550 DAQ device was used to capture signal from the FBG sensors. Bragg spectra during two static load cases (loaded / unloaded specimen) were recorded after the each cyclic loading block. Mechanical strain from FBG sensors versus time was also measured with data sampling frequency of 50 Hz.

## Experimental Results

Analysis of the Bragg spectra for all static load cases was performed. There are two characteristic peaks, one for each FBG sensor. Damage is detected by the shift of the Bragg wavelength (as shown in the Fig. 6 for the unloaded state) of each peak and distortion of its shape (original peak is splitted into multiple peaks of lower intensity, see Fig. 7). For the number of cycles, where FBG sensor has detected damage, the actual damage length (from camera signal analysis) was evaluated. Damage detection advance (FBG sensor vs. camera reading) was calculated as a difference between the position of the FBG sensor (50 or 75 mm, measured from the specimen's front edge) and the value obtained from cameras. Results are shown in the Table 1. It can be seen that in all cases the FBG sensor was able to detect damage in composite before it reaches sensor's position. Average damage detection advance in loaded state was 6.7 mm (first sensor) and 8.9 mm (second sensor, under the higher bending stress). These values are higher than in the unloaded state (5.9 and 6.9 mm), as expected.

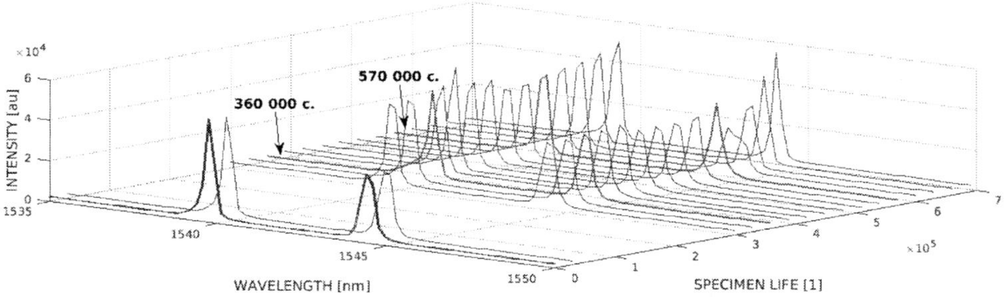


Fig. 6 Specimen DCB01 Bragg spectra reading, Unloaded state.

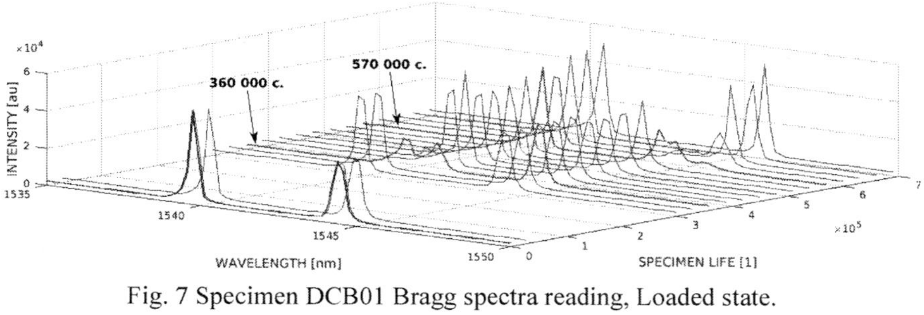


Fig. 7 Specimen DCB01 Bragg spectra reading, Loaded state.

Table 1 Summary of results for specimens DCB01-05.

| | distance from specimen's front edge [mm] | | 50 | 75 |
|---|---|---|---|---|
| DCB01 | damage detection advance (FBG sensor vs. camera reading) [mm] | loaded | 4.4 | 9.2 |
| | | unloaded | 4.4 | 6.9 |
| DCB02 | | loaded | 6.4 | 10.0 |
| | | unloaded | 6.4 | 8.1 |
| DCB03 | | loaded | 5.9 | 8.1 |
| | | unloaded | 5.9 | 5.3 |
| DCB04 | | loaded | 9.5 | 7.8 |
| | | unloaded | 9.5 | 7.8 |
| DCB05 | | loaded | 7.4 | 9.4 |
| | | unloaded | 3.1 | 6.4 |
| | average damage detection advance [mm] | loaded | 6.7 | 8.9 |
| | | unloaded | 5.9 | 6.9 |

## Conclusions

The conducted experiment demonstrated that embedded FBG sensors are able to detect delamination in composite specimen loaded in the mode I, both in loaded and unloaded state. Average detection advance of FBG sensors before the damage is visible reaches up to 6.9 mm in the unloaded state and 8.9 mm in the loaded state. The obtained results will be used to design proposal of the sensor network for an air intake. Density of the FBG sensors will depend on minimum required detectable damage size, in combination with evaluated sensitivities. However, it is obvious that sensor density will be significant and therefore it will be necessary to monitor only critical parts of the air intake. These will be found during the full-scale demonstrator test.


## Acknowledgement

The authors would like to thank the Technology Agency of the Czech Republic for supporting this research with project no. TE02000032.


## References


[1] AITM 1.0005. Determination of Interlaminar fracture toughness energy; Mode I; GIc test. Issue 2. Airbus Industrie, Blagnac (FR), 1994, 8 p.

[2] Measures R.M., Structural Monitoring with Fiber Optic Technology (Academic Press, San Diego, 2001). ISBN 0-12-487430-4.

[3] Růžička, M. - Dvořák, M. - Doubrava, K.: Strain measurement with the Fiber Bragg Grating optical sensors. In Proceedings of the 50th Annual Conference on Experimental Stress Analysis. Praha: Czech Technical University in Prague, 2012, p. 385-392. ISBN 978-80-01-05060-6.

[4] ASTM Standard D6115 – 97. Standard test method for mode I fatigue delamination growth onset of unidirectional Fiber-reinforced polymer matrix composites, Annual Book of ASTM Standards, ASTM International, West Conshohocken, PA. 2011, 7 p.

[5] Gretchen B. Murri: Evaluation of Delamination Growth Characterization Methods Under Mode I Fatigue Loading, ASC 27th Technical Conference, ASTM D30 Meeting, Arlington, TX, October 1-3, 2012.

EAN 2017
55th conference on experimental stress analysis

# Remark on repairing broken pipelines

Eliška Cézová, Jan Kanaval, František Starý

Czech Technical University in Prague, Faculty of Mechanical Engineering, Department of Designing and Machine Components, Czech Republic

eliska.cezova@fs.cvut.cz, jan.kanaval@fs.cvut.cz, frantisek.stary@fs.cvut.cz

## Introduction

Broken pipelines can be repaired in different ways. The standard procedure is to simply replace the damaged segment of the pipe. However, it is usually quite expensive solution requiring non-negligible human and financial resources. Instead, in certain cases, much cheaper and faster solution is available. It is based on the use of repair clamps, such as e. g. in the Figure 1. In the case this repair clamp is properly chosen and mounted, it can last for many years, almost as long as if the pipe segment has been completely exchanged.

**Fig. 1**. Example of a repair clamp and the tightening screwed joint - Laboratories of the Faculty of Mechanical Engineering CTU in Prague.

Unfortunately, this type of repair procedure also has certain limitations. The grip force in the repair clamp is ensured by a stainless steel screwed joint (screw and nut). The use of austenitic stainless steel for connecting parts can however cause some problems, due to higher friction coefficient causing the joint seizure. To improve the friction properties during the assembly, specific types of coatings or dry lubricants can be applied. Despite of using stainless steel parts, some kind of additional protective cover is necessary to avoid problems arising due to microbiological corrosion in the ground. In addition, prior to pipeline use, careful anti-leak tests have to be performed in site.

In order to find the optimal combination of connecting parts, their coatings-lubricants, a series of experiments was performed to analyze the torque required for tightening and loosening the screw connections. Two sizes of threads were considered together with various surface finishes of the nut thread. It was found that tested coatings significantly reduce the friction coefficient in the threaded joint, facilitate the assembly and also increase the corrosion resistance in the extremely aggressive corrosive environment (due to Sulphur, chlorine or microbiological effects). Some experiments were also performed in order to determine the friction coefficient in the threaded joints.

Such measurements of screwed joints are very expensive and time-consuming. Therefore the experiments were only performed on an economically reasonably sized sample group. The results of these tests are presented hereafter, accompanied by a short analysis of the structure of the corresponding measurements costs.

The tests have confirmed the justification of the replacement of coated nuts by uncoated ones. However, the price of the coated parts is about 30% higher than the standard, uncoated ones [5], [6], [7].

## Measurement methodology

To ensure the consistency of acquired data, specific measurement methodology had to be designed. This task was quite time consuming, as the appropriate measurement approach had to be chosen to guarantee the desired accuracy and completeness of the data. Preparation of the draft of the methodology took about 8 hours to an experienced professor. During this time it was necessary to choose the measuring apparatus, design suitable fixtures, measuring tools and instruments required to perform the whole experimental task. Experimental workplace and measuring stand are shown in the Figure 2.

During the experiment, the following parameters were *monitored*:

- Axial force in bolts.
- Applied torque.
- Type of the coating.
- Type of the lubricant.

Based on the measured data, the following additional quantities were *calculated*:

- Friction angle of the thread.
- Coefficient of friction in the thread.
- Friction angle for the tribological pair (combinations of materials, conditions of sliding friction and surfaces).
- Coefficient of friction for the tribological pair (the combination of material conditions of sliding friction and surfaces) [2], [3].

**Fig. 2.** Details of experimental measuring stand - Laboratories of the Faculty of Mechanical Engineering CTU in Prague.

Before the actual measurement started, the screwed joints had to be prepared at the test stand. These activities took roughly 90 minutes. The preparatory phase of the measurements included, among others, preparation of recording sheets (measurement tables), preparation of weights for given measurement, etc. To ensure the accuracy of the measurements, the torque sensor had to be calibrated in a time-frame of 5 minutes. Generally, one measurement usually took about 40 minutes. Since two sizes of matrices were tested (M12 and M16), the time for their exchange, taking approximately 5 minutes, has to be considered as well. In summary, the actual measurements lasted for 5 days, typically 8 hours per day. This time includes the preparatory phase, sensor calibration and instruments cleaning after the measurement.

### Overall costs breakdown

The overall costs include:

1. Consumption of material (laboratory gloves and overalls, cleaning material, DVDs for storing data, etc.)
2. Production services (clamping device production).
3. Nonproduction services (assembly, installation software and verification of the testing equipment).
4. Depreciation of tools and machines used.
5. Personal costs including the salaries, that can be further quantified according to the time requirements described above.
6. Taxes and fees.
7. Financial expenses.

### Economic issues in detail

- The selection of suitable machines and instruments needed for each measurement took about 8 hours of time to an assistant professor.
- Manufacturing and preparation of the test fixtures needed for the measurements took (on average) 8 hours of time to a skilled worker.
- Preparation of the experimental facility and its testing took approximately 1 hour, being done by three investigators.
- The actual measurement, i.e. data capturing, took about 16 hours, being performed by three investigators.
- Data preprocessing, i.e. sorting, editing and transformation for further analysis took (on average) 16 hours. This work has handled by three investigators.
- Data analysis, post processing and writing of the final report took about 40 hours of a joint work of three lecturers.

## Remarks on model for measurements

The basic set of measurements contained 16 nuts. They were evaluated in two sizes, M12 and M16. In the second part of measurements, it was possible to determine the damage of the coating of various nuts.

According to the theory, the torque should increase linearly depending on the axial force in the bolt, considering the constant friction coefficients in the thread (see equation (1))[3].

$$M_{KK} = M_K + M_{TH} = Q_o \cdot \frac{d_2}{2} \cdot tg(\gamma + \varphi') + Q_o \cdot \rho_{TH} \cdot f_{TH} \qquad (1)$$

In our case, the coefficient of friction in the thread changes with the load (see Figures 3-10). This is probably due to a variable quality of the surface of the thread, abrasion of the applied coating and a non-uniform speed of nuts tightening (manual assembly).

From the torque depending on the axial force, see the diagrams in Figure 3 and Figure 4, it is possible to compare the size of the friction coefficients of the screw connections with fixed coatings versus those where the lubricating grease was used. The best is the case in which the measured torque achieves its lowest value. Less energy is consumed to overcome the friction in the thread during the tightening. The assembly is easier and at the same time a higher axial pretension is achieved in the screw joint, thus the connecting clamp provides the maximum sealing capability [1], [2], [4].

## Conclusions

From the Table 1 it is evident that from the experiments performed on the M16 sized nuts, in case of using new as well as re-used (after one use) stainless nuts, the best is the performance of nuts with the Coating A. In the case of new nuts the Coating C (sliding paint) works quite well, and the lubricated stainless steel nuts (using good quality lubricating grease) are also an acceptable option.

This conclusion is also documented by the Fig. 3 and 4, showing the torque necessary for lifting and lowering the weights during individual experiments. The best choice requiring the lowest torque for M16 nuts is again Coating A. In the case of new nuts the Coating C (sliding paint) seems to be the best, but the new stainless steel nuts can equally be treated by a good quality lubricant (e.g. graphite grease).

| Variant<br>Nut A4-80 | Friction coefficient-theoretical | Friction coefficient – measured (M16) | Friction coefficient – measured (M12) |
|---|---|---|---|
| Coating A<br>zinc (corrosion-proof) | **0.09 - 0.14** | 0.097 | 0.204 |
| Coating A (re-used nut)<br>zinc (corrosion-proof) | | 0.188 | 0.240 |
| Coating B - based on PTFE (Polytetrafluorethylene) | **0.02 - 0.2** | 0.296 | 0.271 |
| Coating B (re-used nut)<br>based on PTFE (Polytetrafluorethylene) | | 0.290 | 0.312 |
| Coating C - dry sliding paint with molybden suplhide additive | **0.05 - 0.1** | 0.190 | 0.225 |
| Coating C (re-used) - dry sliding paint with molybden suplhide additive | | 0.389 | 0.445 |
| Stainless nut | | 0.401 | 0.553 |
| Stainless nut with graphite grease | | 0.250 | 0.202 |

**Tab. 1.** Summary of the experimentally determined initial values of the friction coefficient *f* for the tested and reference M16 and M12 nuts. [2]

It is apparent from the Table 1, that from the experiments performed on the M12 nuts, in the case of using new stainless nuts, the best choice seems to be the use of good quality lubricant (e.g. graphite grease), or alternatively, the Coating A can provide a good performance for both new and re-used nuts. For new nuts, the Coating C is also acceptable.

Similar conclusion can be drawn from the Fig 5 and 6, showing the torque necessary for lifting and lowering the weights during individual experiments. The best choice, requiring the lowest torque, for the case of M12 nuts is again the combination of stainless nut with suitable lubricant, followed by the nuts with Coating A as the second choice and nuts with Coating C at the third place.

The essential finding is the fact that in all cases, on the working surfaces of the threads of used nuts (i.e. after being used for one mounting of the repair fixture), there was observed marked abrasive wear of the applied coatings (as well as of the sliding paint), often accompanied by the deformation (plastization) of the working surface of the thread of the nut. Due to high values of contact pressures in the threaded surfaces of the bolts and nuts during their mutual relative motion in the course of the assembly (tightening) of the screwed joint, there appears an intensive abrasion of the coating (and paint). In all cases the whole coating (paint) layer was rubbed off from the used nuts.

Significant abrasion of the used nuts coating was also visible on the annulus contact surfaces of nuts. This effect was also observed during the disassembly of the test joints of repair clamps, where appeared "scurfs" of the coating rubbed off from the bearing surfaces of the nuts. Traces of the rubbed of coating were also found in the threaded surfaces of the stainless bolt after the disassembly of the joint.

As a conclusion, it's possible to say that the positive properties of the studied coatings can only improve the mounting behaviour of the screwed joints till the moment, when the coating layer is rubbed off, i.e. till the beginning of the mounting (tightening) process. During the final stage of the screwed joint tightening, the friction properties of the tribological pair will be significantly worse.

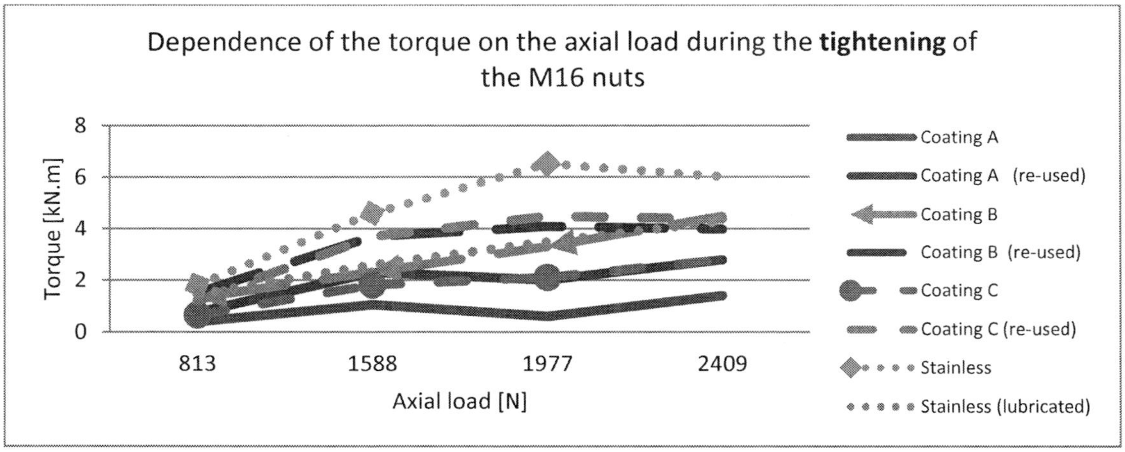


**Fig. 3.** Dependence of the torque on the axial load during the **tightening** of the M16 nuts.

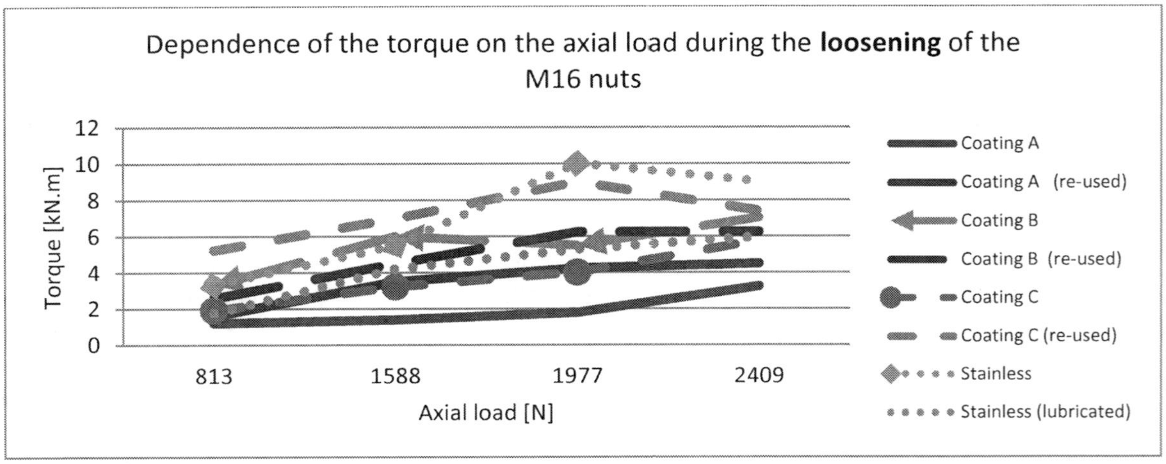


**Fig. 4.** Dependence of the torque on the axial load during the **loosening** of the M16 nuts.

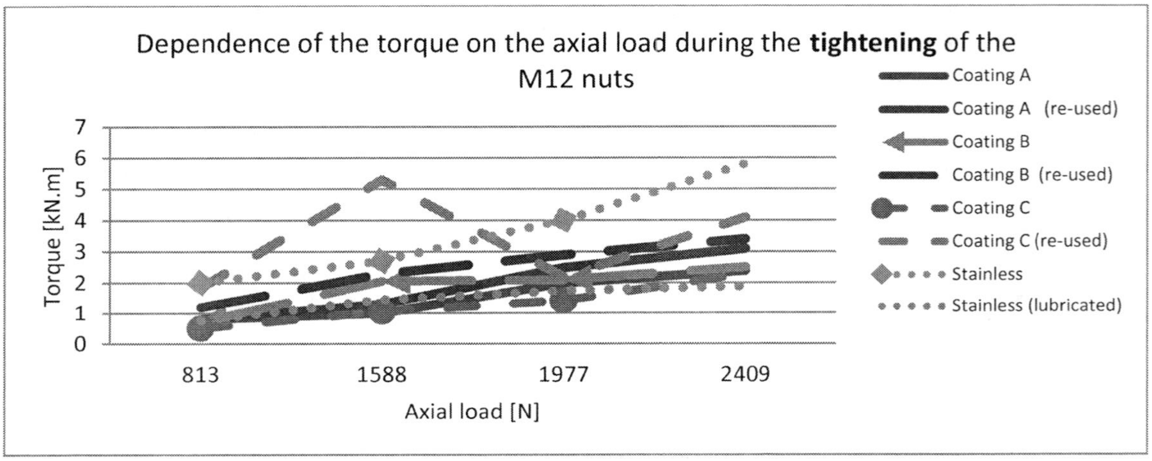


**Fig. 5.** Dependence of the torque on the axial load during the **tightening** of the M12 nuts.

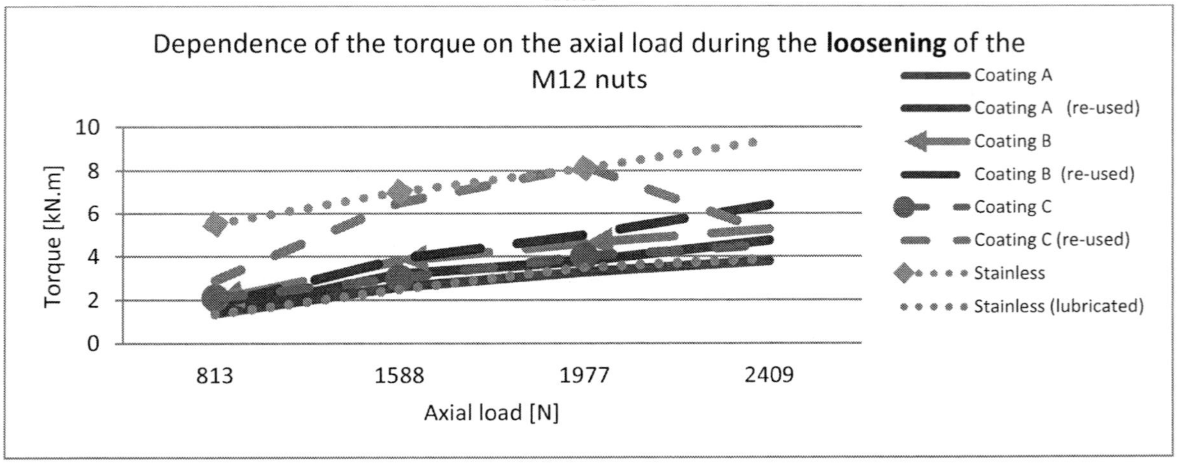


**Fig. 6.** Dependence of the torque on the axial load during the **loosening** of the M12 nuts.

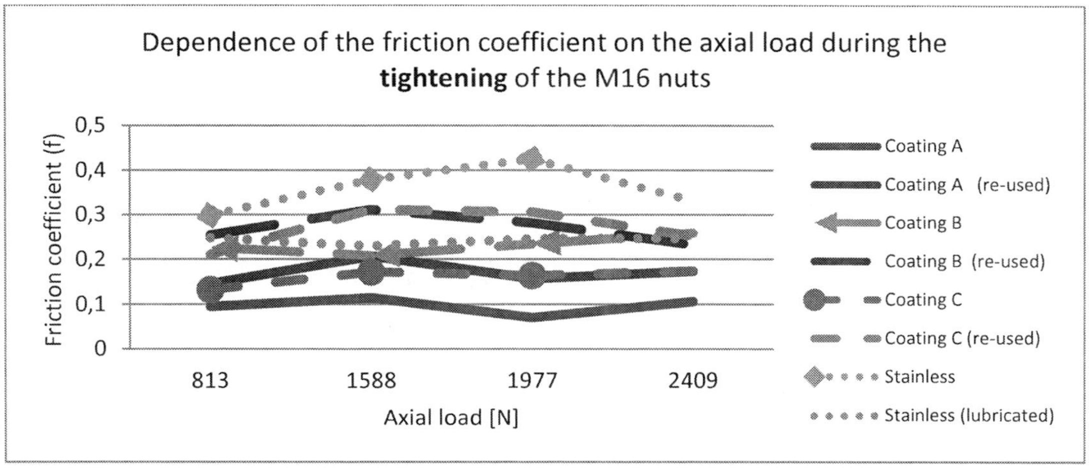


**Fig. 7.** Dependence of the friction coefficient on the axial load during the **tightening** of the M16 nuts.

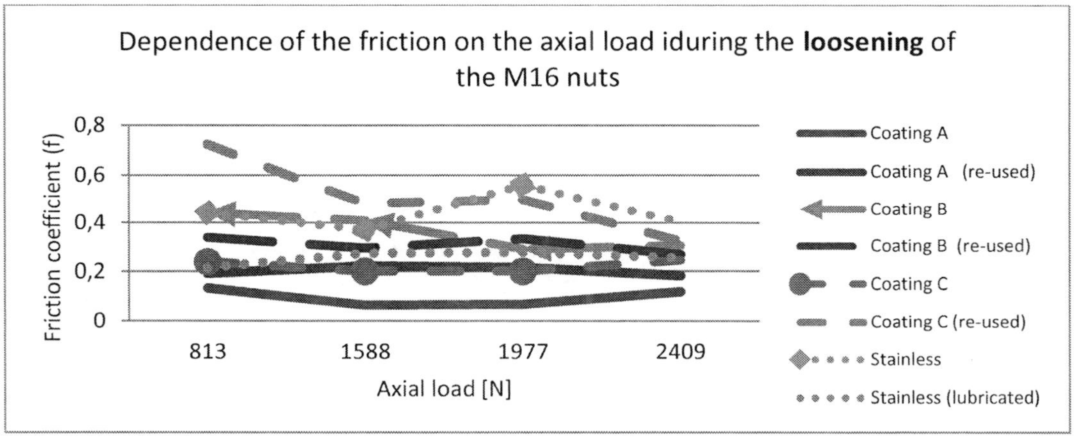


**Fig. 8.** Dependence of the friction coefficient on the axial load during the **loosening** of the M16 nuts.

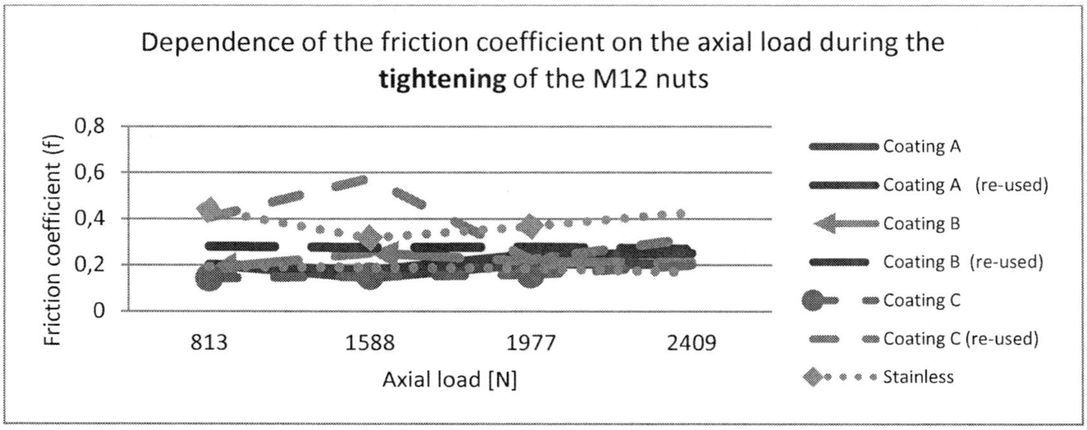


**Fig. 9.** Dependence of the friction coefficient on the axial force during the **tightening** of the M12 nuts.

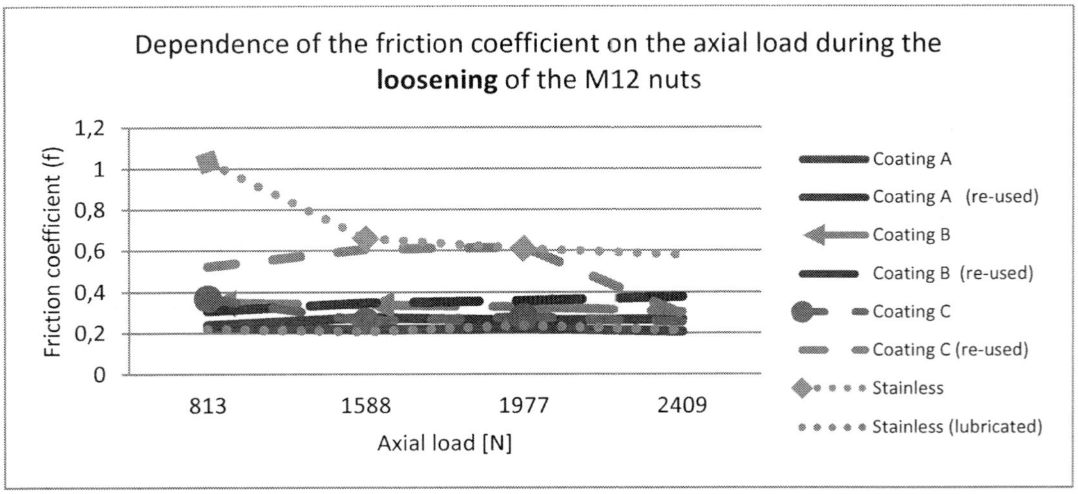


**Fig. 10.** Dependence of the friction coefficient on the axial load during the **loosening** of the M12 nuts.

## References


[1] J. Dejl: Design of machines and devices I. Connecting parts of machines. Design, calculation and construction. Ostrava: Montanex, a.s., 2000. ISBN 80-7225-018-3. (in Czech)

[2]J. Kanaval, F. Starý, E. Cézová: Screw connections complex analysis in the application of advanced coatings and lubricants. In: Book of proceedings of the 57th International Conference of Machine Design Departments. Plzeň: Západočeská universita, 2016, pp. 83-86. ISBN-978-80-261-0609-8.

[3]V. Švec: Parts and mechanisms of machines. Joints and connecting parts. Praha: ČVUT, (2008) (in Czech)

[4]ČSN EN ISO 4762:2004 Hexagon socket head cap screws (ČSN 02 1143, DIN 912)

[5]http:/www.fabory.cz, (2016-07-29).

[6]http:/www.fuchslubricants.com, (2016-07-29).

[7]http:/www.technicoat.cz/-data/files/Image/content/xylan, (2016-07-29)

# Grain Orientation Influence on Residual Stress Distribution of Ground Surface of Blade Roots Made of Inconel 713LC

ČAPEK Jiří[1,a], GANCARCZYK Kamil[2,b], PITRMUC Zdeněk[3,c], KOLAŘÍK Kamil[1,d], MÁLEK Jaroslav[4,e], BERÁNEK Libor[3,f], GANEV Nikolaj[1,g] and TROJAN Karel[1,h]

[1] Department of Solid State Engineering, Faculty of Nuclear Sciences and Physical Engineering, Czech Technical University in Prague, Trojanova 13, 120 00 Prague 2, Czech Republic

[2] Department of Materials Science, Rzeszow University of Technology, al. Powstańców Warszawy 12, 35-959 Rzeszów, Poland

[3] Department of Machining, Process Planning and Metrology, Faculty of Mechanical Engineering, Czech Technical University in Prague, Technická 4, 166 07 Prague 6, Czech Republic

[4] Department of Materials Engineering, Faculty of Mechanical Engineering, Czech Technical University in Prague, Karlovo náměstí 13, 121 35 Prague 2, Czech Republic

[a]jiri.capek@fjfi.cvut.cz, [b]kamilgancarczyk@prz.edu.pl, [c]zdenek.pitrmuc@fs.cvut.cz, [d]kamil.kolarik@fjfi.cvut.cz, [e]jardamalek@seznam.cz, [f]libor.beranek@fs.cvut.cz, [g]nikolaj.ganev@fjfi.cvut.cz, [h]karel.trojan@fjfi.cvut.cz

**Keywords:** Inconel 713LC, coarse-grain, grain orientation, omega-scan method, residual stresses.

**Abstract.** The aim of this contribution was to find whether there exists a relation between grains orientation in coarse-grained bulk material and macroscopic residual stress distribution after surface grinding. It could enable to predict that several coarse-grains are parallel to the primary (total) grinding force. Therefore, even smaller force can cause the slip (plastic deformation) and thereby smaller compressive residual stresses are determined in the very thin deformed surface layer. This effect can lead to degradation of some utility properties. In the case of a dynamically loaded nickel-based superalloy, it could lead to decrease of both the service life and the creep resistance.

## Introduction

Inconel is two-phase material. The microstructure contains the nickel matrix (γ phase) and $Ni_3Al$ precipitates (γ' phase). Nickel has a face centred cubic lattice (fcc) with regular structure, which is close-packing structure of atoms, therefore, the primary slip system is *<110>{111}*. The number of slip systems is 12, which is sufficient amount to plastic deformation and it is the reason of plasticity of austenite-type materials. Precipitates $Ni_3Al$ have ordered regular structure ($L1_2$), i.e. the primary slip system is *<110>{111}* too[1]. Because bonds A-B are stronger than A-A type, yield strength of $Ni_3Al$ phase is much higher in comparison with nickel matrix [2,3]. Moreover, these phases have coherent boundaries with

[1] In comparison with γ phase, $L1_2$ materials have anomalous yield strength. This anomalous originates from creation of four partial dislocations, two superintrisic stacking fault and one antiphase boundary [1].

nickel grains, i.e. the dislocation movement through γ' phase is almost stopped due to its structure [2]. Therefore, these materials are hard, and very high strength or temperature is necessary to use to their phase deformation [4,5]. Moreover, Inconel 713LC is a work-hardening and creep-resistant material. Above temperature 600–800 °C, the precipitates lose their coherency and the strengthening is decreasing.

According to articles [6,7], the other material properties are poor thermal diffusivity, high temperature strength, presence of hard carbide phases or production of tough chip. Therefore, nickel-based superalloys are often used in automotive and aerospace industry as rotor turbine blades [8]. However, very stringent requirements on surface integrity specifically, residual stresses (RS) and microstructure are required, especially on blade roots [9]. Because of susceptibility to mechanical reinforcement of ground surface, the compressive RS are created in the surface layer in the case of adequate cooling. Nevertheless, the areas with potential accumulation of unfavourable tensile RS of γ/γ' phases on a macroscale can be localized by X-ray diffraction (XRD) techniques. Since the precipitates (γ' phase) are too hard to be plastically deformed, the local strain is very heterogeneous on a microscale – in the matrix channels, i.e. γ phase [2].

The anisotropy of polycrystalline bulk material is generally more reduced in comparison with single crystal because of the distribution of crystal orientation. The coarse-grained surface layers affected by grinding are crumbled to a fine-grained structure. In the case of coarse-grained bulk material, there may be several grains whose slip directions are parallel to the total grinding force[2]. Therefore, even smaller force could cause the plastic deformation and thereby smaller compressive RS of surface fine-grained layers are determined [12]. Decreasing of favourable compressive RS can lead to reduction of the service life of dynamic loaded blades. For microstructure visibility, the electro-chemical etching of plastically deformed surface layer has to be done. Subsequently, XRD omega-scan (Ω-scan) method [13] or electron backscatter diffraction method (EBSD) [14] can be used for determination of the orientation of coarse-grains of bulk material from Inconel.

## Experiment

The tested samples (blades) were casted from Inconel 713LC which is a nickel based superalloy. The nominal chemical composition (in wt.%) was following: 12.0 Cr, 5.7 Al, 4.6 Mo, 2.0 Nb, 0.7 Ti, 0.2 Fe, 0.1 Zr, 0.08 Co, 0.05 C, 0.013 B and balance of Ni.

Surfaces of three turbine blade roots (sample 1–3) with dimension 52×15 $mm^2$ were face ground. For grinding conditions see Figs. 3a–5a. Three areas were analysed by X-ray diffraction for residual stress determination on the blade root, namely I, II, see Fig. 1.

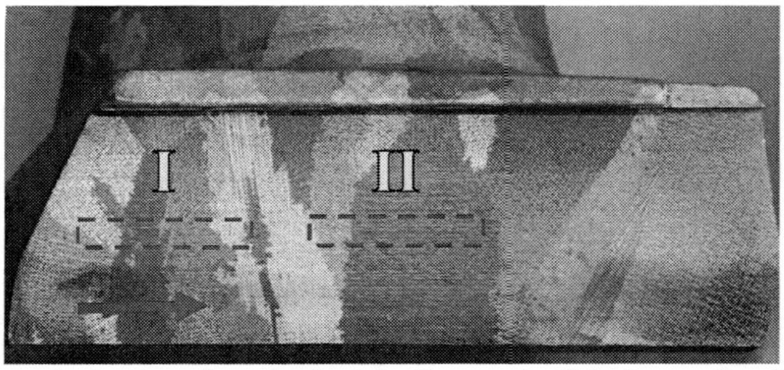


Figure 1: Experimental electro-chemical etched sample (blade root) with marked grinding direction and analysed areas I–II.

[2] Total grinding force is the sum of cutting, feed and passive forces whose magnitudes are given by grinding conditions, properties of grinding wheel and ground material: $\boldsymbol{F} = \sum_i \boldsymbol{F}_i$ [10,11].

*PROTO iXRD COMBO* diffractometer in ω-goniometer set-up was used to measure lattice deformations using manganese radiation. The diffraction lines *{311}* of γ/γ' phases were analysed for obtaining macroscopic residual stresses [2]. *Rachinger's* method and *Absolute peak* method were used for determination of the diffraction angles $2\theta^{311}$ of the diffraction lines $K\alpha_1$. For the residual stress determination, $\sin^2\psi$ method with X-ray elastic constants $\frac{1}{2}s_2 = 6.57\ \mathrm{TPa}^{-1}$, $s_1 = -1.56\ \mathrm{TPa}^{-1}$ calculated from components of stiffness tensor ($C_{11} = 234.6$ GPa; $C_{12} = 145.4$ GPa; $C_{44} = 126.2$ GPa) using Hill elastic model [15] were used. The cylinder collimator with diameter of 2 mm was inserted into the primary beam path. The blades were moving by ±5 mm, therefore, the irradiated area was approx. 12×2 mm², see Fig. 1. The average value of the effective penetration depth of Mn*Kα* radiation into the nickel-based material was approx. 4–5 μm [16].

The *ISO DEBYEFLEX 3003* equipment with chromium X-ray tube was used to back-reflection experiment for qualitative real structure determination. Using 30mm distance between radiation source and plate detector, the diffraction lines (Debye rings) *{220}* of *Kα* radiation and *{311}* of *Kβ* radiation of γ/γ' phases were obtained.

Evaluation of crystal orientation was based on measurements of α angle; i.e. the angle between crystallographic growth direction *<100>* and the main axis of the blade (perpendicular to grinding direction, see Fig. 1). The investigation was conducted with the use of X-ray diffraction method called omega-scan, utilizing prototype *OD-EFG* diffractometer invented and produced by *EFG* company (Berlin). The apparatus enables determination of the crystal orientation in the root and airfoil of turbine blade made of nickel-based superalloys. The size of the measuring spot was approx. 2 mm² and by this size; the orientation of whole surface of turbine blade can be mapped with an accuracy of ±0.01°. Because of the method limitation, the unmeasurable α angles were set to zero.

The microstructure of specimens was studied with a *JEOL JSM 7600F* microscope equipped with *Nordly's EBSD* detector. The results were processed by *HKL Channel 5* software equipment.

## Results and discussions

The microstructure of thin surface layers after grinding was fine-grained, see Fig. 2a. Analysing this surface, the inhomogeneous RS were determined on the ground surface, i.e. areas I–II, see Figs. 3a–5a. Therefore, the surfaces were electro-chemical etched for removing of surface layers affected by grinding (approx. 0.2 mm) to bulk material microstructure visibility, see Fig. 1, where the coarse-grained microstructure of bulk material was evident, see Fig. 2b.

Figure 2a: The Debye rings of ground surface – fine grained structure.

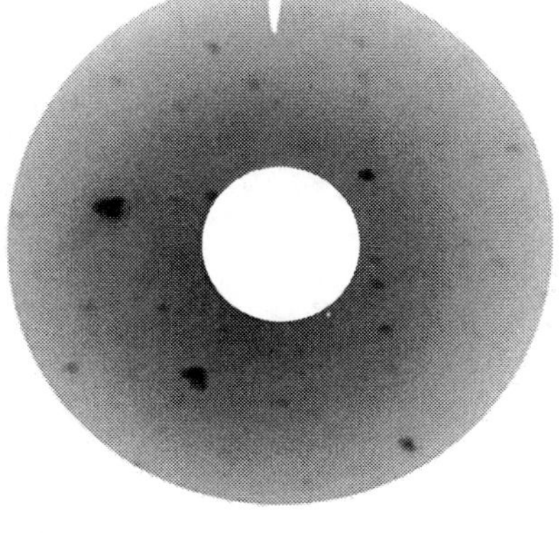

Figure 2b: The Debye rings of bulk material – coarse grained structure.

Although grinding usually leads to a creation of the tensile RS, the compressive RS were determined on the all surfaces of ground blades. Nevertheless, the tensile RS are predicted to the increasing of depth [11]. The compressive RS in the thin surface layers are caused by plastic deformation during grinding of the two-phase material which is prone to work-hardening. Moreover, the thermal conductivity of Inconel 713LC is approx. 4× smaller in comparison with classic carbon steel which leads to insufficient heat distribution into chip and workpiece.

The etched samples were measured using the omega-scan method for gaining grain orientation of bulk material, see Figs. 3c–5c. The omega-scan method is relatively fast method in comparison with the EBSD, see Figs. 3b–5b. EBSD was selected as the comparative and verification method. It is necessary to note that X-ray radiation has an order of magnitude greater penetration depth than electrons. However, for this size of grains, it is not in this case important parameter.

Because of down grinding, i.e. the direction of the total grinding force points to the material, smaller force is necessary to use for deformation of grains with orientation in most cases between 20° and 70° in relation to surface, for facilitation around 45°. Using omega-scan method, deviation α angle was analysed. Because of three slip direction *<110>* of slip planes *{111}*, two possible values of α angle should be fit to gain. From relation of crystallography follows that 0° or 45° of α angle [17] is in relative to the orientation of total grinding force of 45°. Similarly, using EBSD method, two possible values of grain orientation are necessary to have, namely *<100>* or *<111>*.

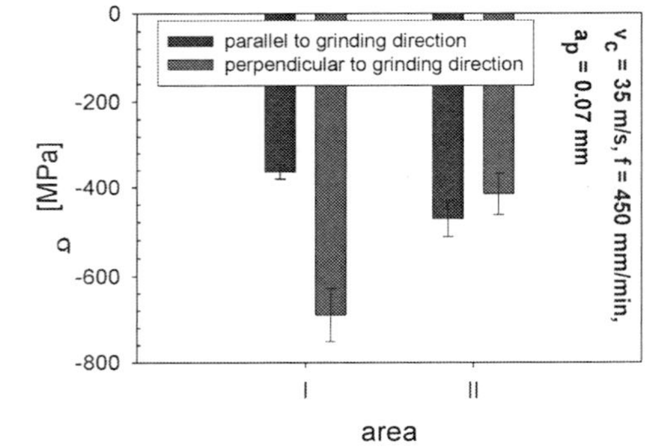


Figure 3a: RS of areas I–II on the surface of the sample 1.

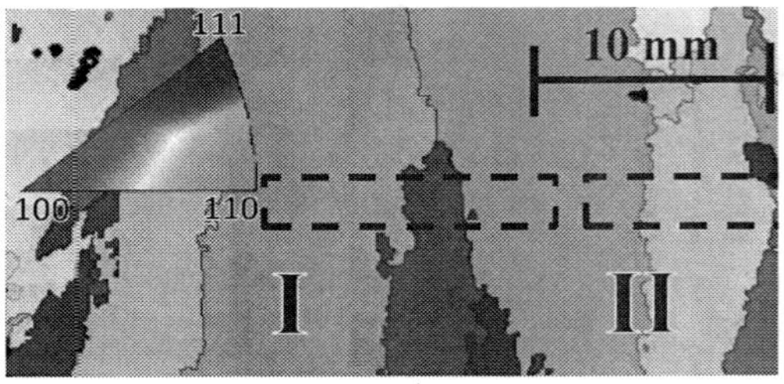


Figure 3b: EBSD orinetation map of the sample 1.

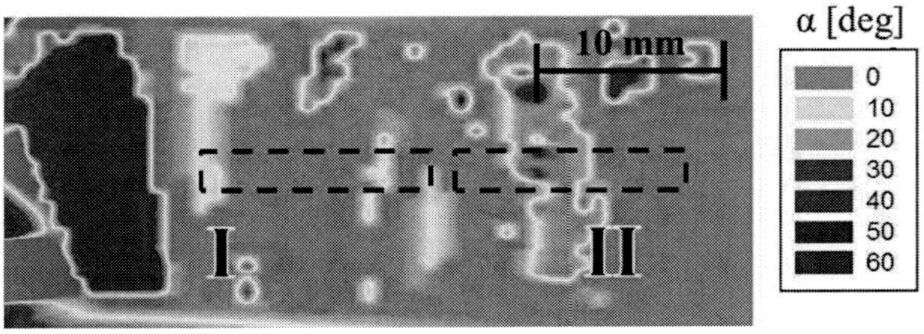


Figure 3c: Deviation angle α of omega-scan method of the sample 1.

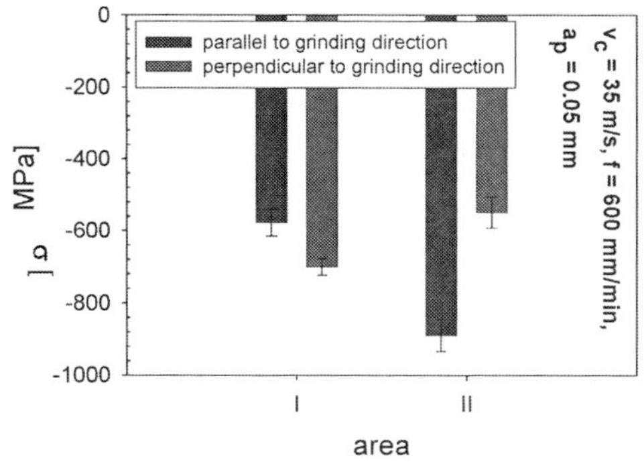


Figure 4a: RS of areas I–II on the surface of the sample 2.

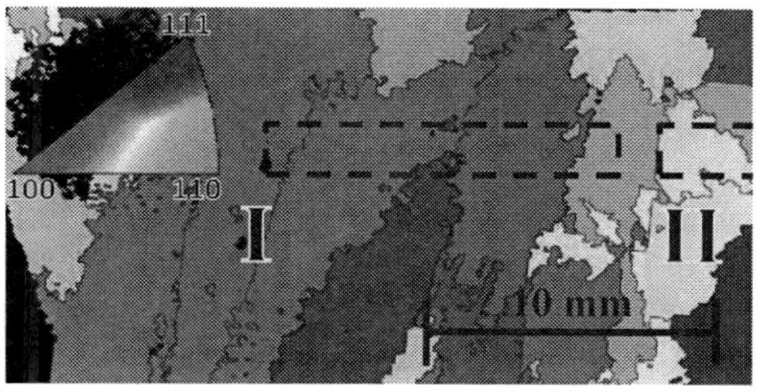


Figure 4b: EBSD orinetation map of the sample 2.

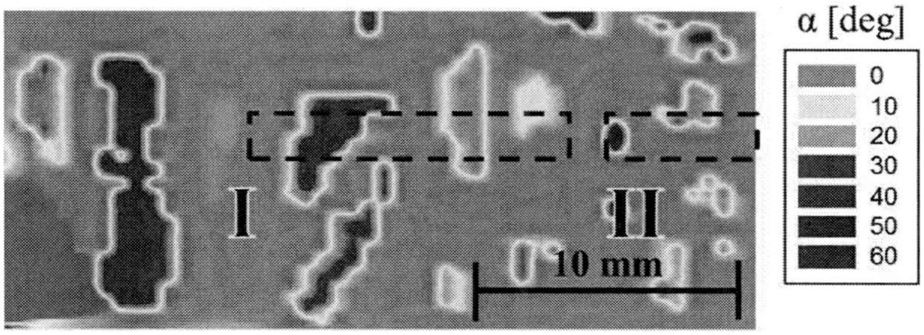


Figure 4c: Deviation angle α of omega-scan method of the sample 2.

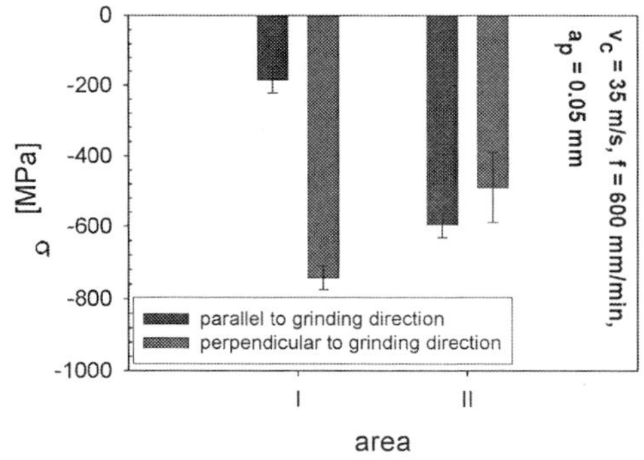


Figure 5a: RS of areas I–II on the surface of the sample 3.

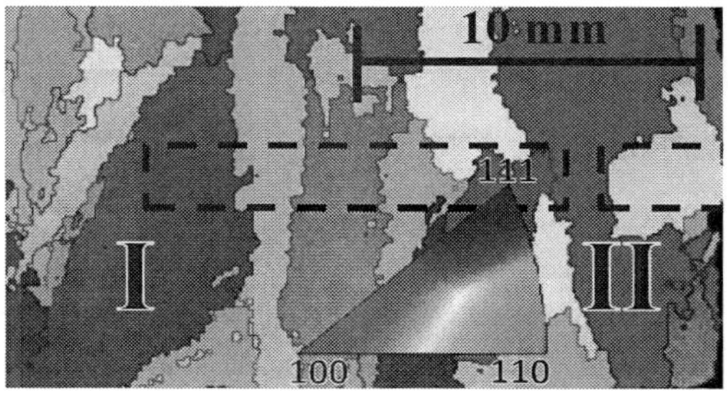


Figure 5b: EBSD orinetation map of the sample 3.

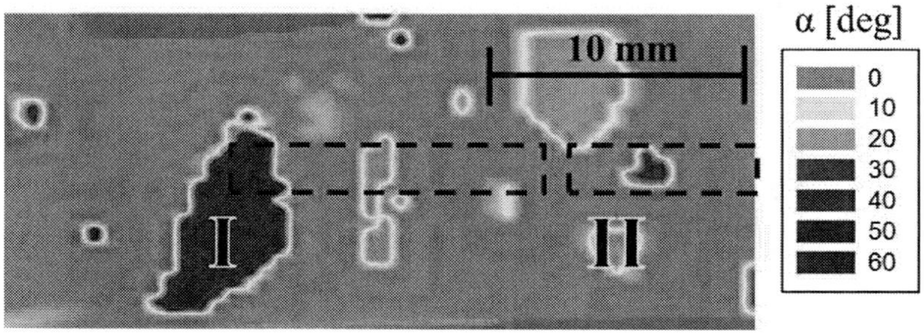


Figure 5c: Deviation angle α of omega-scan method of the sample 3.

In the case of areas I of sample 2 and 3, the compressive RS decreased in comparison with the area II in the grinding direction, see Figs. 4a and 5a. The decreasing should be caused by irradiated grains which have the deviations of α angle 40–50°, see Figs. 4c and 5c. The orientation of grains *{111}* or *{100}* is evident from EBSD, see Figs. 4b and 5b. These results are in accordance with expectation and before explained theory.

The RS are dependent on the grinding force in the direction perpendicular to grinding direction too. In general, the macroscopic RS on the ground surfaces should be compressive when cooled substantially during grinding and, moreover, the compressive RS in direction perpendicular to grinding are expected to be greater than in the grinding direction [9]. This statement could be not valid in the case of coarse-grained material. From Fig. 3c, it is evident that the orientation of irradiated grain of area II is approx. 20°. Therefore, the decreasing of the RS is significant too in the direction perpendicular to grinding direction, see Fig. 3a. The biggest decreasing of the RS values is for area II of sample. The influence on the total grinding force is not significant as in the case of α angle deviation 45°. From EBSD, the more general orientation of grains is evident, see Fig. 3b. On the other hand, incomplete area II was measured.

## Conclusions

The present study showed:

- The thin ground surface layers determined by back-reflection X-ray diffraction show the fine-grained crystallographic real structure due to plastic deformation (the diffraction line is relatively continuous, and has homogeneous intensity around its perimeter, without any sign of texture). This is the reason why the surface RS of the samples can be determined by using X-ray diffraction.
- After removing of a surface layer 0.2 mm in thickness, the discrete diffraction spots were obtained which proves the considerable coarse-grained material comprised of single-crystal domains. This result is in correlation with metallographic pattern.
- The compressive RS of all ground surfaces with effective thickness approx. 4–5 μm were determined. In most cases, higher compressive RS were determined in the direction perpendicular to grinding direction. Due to mechanical interaction between cutting tool and material, the surface layers have tendency to elongate.
- The correlation between the decreasing of RS values of and orientation of the irradiated grains was established for both methods. Despite higher resolution and lower penetration depth of electrons, EBSD method with omega-scan method shows similar results.
- The limitation of α angle measurement is a disadvantage of this method. On the other hand, relatively fast measurements and easier preparation of samples are advantages of omega-scan method.

## Acknowledgement

This work was supported by the governmental funding of Technological Agency of Czech Republic – project No. TA04010600 and the Grant Agency of the Czech Technical University in Prague, grant No. SGS16/245/OHK4/3T/14.

## References

[1] O.N. Mryasov, Y.N. Gornostyrev, M. Van Schilfgaarde, A.J. Freeman, Superdislocation core structure in $L1_2$ $Ni_3Al$, $Ni_3Ge$ and $Fe_3Ge$: Peierls–Nabarro analysis starting from ab-initio, Acta Mater. 50 (2002) 4545-4554.

[2] J. Li, R.P. Wahi, Investigation of $\gamma/\gamma'$ lattice mismatch in the polycrystalline nickel-base superalloy IN738LC: Influence of heat treatment and creep deformation, Acta Metall. Mater. 43 (1995) 507-517.

[3] P.C. Xia, J.J. Yu, X.F. Sun, H.R. Guan, Z.Q. Hu, Influence of $\gamma'$ precipitate morphology on the creep property of a directionally solidified nickel-base superalloy, Mater. Sci. Eng.: A. 476.1 (2008) 39-45.

[4] W. Österle, P.X. Li, Mechanical and thermal response of a nickel-base superalloy upon grinding with high removal rates, Mater. Sci. Eng.: A. 238.2 (1997) 357-366.

[5] K. Kolařík, Z. Pala, L. Beránek, J. Čapek, Z. Vyskočil, N. Ganev, Non-Destructive Inspection of Surface Integrity in Milled Turbine Blades of Inconel 738LC, Appl. Mech. Mater. 486 (2014) 9-15.

[6] E.O. Ezugwu, Z.M. Wang, A.R. Machado, The machinability of nickel-based alloys: a review, J. Mater. Process. Tech. 86.1–3 (1999) 1-16.

[7] F. Tehovnik, J. Burja, B. Podgornik, M. Godec, F. Vode, Microstructural Evolution of Inconel 625 During Hot Rolling, Mater. Tech. 49.5 (2015) 801-806.

[8] M. Petrenec, K. Obrtlík, J. Polák, T. Kruml, Fatigue behaviour of a cast nickel-based superalloy Inconel 792-5A at 700 °C, Mater. Tech. 40.5 (2006) 175-178.

[9] J. Čapek J. Kyncl K. Kolařík, L. Beránek, Z. Pitrmuc, J. Medřický, Z. Pala, Grinding of Inconel 713 Superalloy for Gas Turbines, Manuf. Technol. 16.1 (2016) 38-45.

[10] C. Xun, W.B. Rowe, Analysis and simulation of the grinding process. Part II: Mechanics of grinding, Int. J. Mach. Tool. Manu. 36.8 (1996) 883-896.

[11] K. Vasilko, Teória a prax trieskového obrábania, COFIN, Prešov, 2009.

[12] X. Zhang, T. Murakumo, Y. Koizumi, T. Kobayashi, H. Harada, Slip geometry of dislocations related to cutting of the $\gamma'$ phase in a new generation single-crystal superalloy, Acta Mater. 51.17 (2003) 5073-5081.

[13] K. Gancarczyk, R. Albrecht, C. Olesch, K. Kubiak, J. Sieniawski, Novel method for the measurement of crystalline perfection in single crystal turbine blades made of CMSX-4 superalloy, Mater. Eng. 208.6 (2015) 154-160.

[14] T. Garcin, J.H. Schmitt, M. Militzer, In-situ laser ultrasonic grain size measurement in superalloy INCONEL 718, J. Alloy. Compd. 670 (2016) 329-336.

[15] R. Hill, The elastic behaviour of a crystalline aggregate, P. Phys. Soc.: A. 65.5 (1952) 349.

[16] J. Čapek, Z. Pala, Auxiliary programs for diffraction experiments, Mater. Struct. 22 (2015) 78-81.

[17] I.L. Dillamore, Plasticity of Crystals with Special Reference to Metals, Phys. Bull. 20.3 (1969) 107.

EAN 2017
55th conference on experimental stress analysis

# The in Situ Application of a New Approach to Experimental Estimation of Tensile Forces in Prestressed Structural Elements by Method Based on the Magnetoelastic Principle

KLIER Tomáš[1,a], MÍČKA Tomáš[1,b], POLÁK Michal[2,c], PLACHÝ Tomáš[2,d], ŠIMLER Miloš[3,e] and SMETÁK Tomáš[3,f]

[1]Department of Diagnostics, Pontex spol. s r. o., Bezová 1658, 147 14 Prague 4, Czech Republic

[2]Department of Mechanics, Faculty of Civil Engineering, Czech Technical University in Prague, Thákurova 7, Praha 6, 166 29, Czech Republic

[3]Freyssinet CS, a. s., Zápy 267, 250 01 Brandýs nad Labem, Czech Republic

[a]klier@pontex.cz, [b]micka@pontex.cz, [c]polak@fsv.cvut.cz, [d]plachy@fsv.cvut.cz, [e]simler@freyssinet.cz, [f]smetak@freyssinet.cz

**Keywords:** tensile force, prestressed strand, prestressed cable, prestressed wire, magnetoelastic principle, magnetoelastic sensor.

**Abstract.** The important structural elements loaded by great tensile forces (prestressed reinforcements of concrete bridges for example) significantly affect the reliability of the whole structure. The force in such element fluctuates considerably in time. Accurate information about the current force value is very important for exact assessment of the structural reliability at any of the life cycle stages of the structure. The five experimental approaches are most often applied for determination of axial tensile forces in civil engineering practice. They are the direct measurement of the force by a pre-installed load cell, the approach founded on a strain measurement with strain gauges, the force determination in a flexible structural element based on the relation between the transverse force and the caused transverse displacement, the vibration frequency method and the magnetoelastic method. At first, the advantages and disadvantages of the above mentioned approaches have been discussed in the paper. Subsequently, the basic principles, results of the experimental verification and the in situ application of the new approach to experimental estimation of tensile forces by the method based on the magnetoelastic principle are described.

## Introduction

The significant structural elements of civil engineering structures loaded by large tensile forces (prestressed reinforcements of concrete bridges, stays of cable stayed bridges, suspension cables of suspended roofs, ground and rock anchors for example) significantly affect the reliability of the whole structure. The force in such element fluctuates considerably in time. Accurate information about the current force value is very important for exact assessment of the structural reliability at any of the life cycle stages of the entire structure.

The five experimental approaches are generally used for evaluation of axial tensile forces in civil engineering practice. They are the direct measurement of the force by a pre-installed load cell, the approach founded on a strain measurement with strain gauges, the force determination in a flexible structural element based on the relation between the transverse force and the caused transverse displacement [1, 2], the vibration frequency method [3, 4] and the magnetoelastic method [5, 6, 7]

The main objective of the research work described in the paper was a solution of the issue that could be named as "an accurate experimental estimation of the current value of the tensile force in a prestressed structural element on an existing concrete structure". The need to solve this problem is very frequent in praxis. Nevertheless, it is not often soluble by any other experimental or theoretical method.

## The Brief Description of the Experimental Approaches Generally Used in the Practice

In this chapter, the brief description and the discussion of advantages and disadvantages of the five above mentioned approaches are summarized that are generally used in the civil engineering practice for experimental estimation of axial tensile forces in significant structural elements.

**The Direct Measurement of the Force by a Pre-Installed Load Cell.** The professionally made force sensor is usually used by this experimental approach. The transducer has to be connected in series between the investigated structural element and its anchoring to the structure. The basic advantage of this approach is the possibility to calibrate properly the force sensor and so increase the accuracy of the force measurement. The two fundamental disadvantages are that the sensor has to be installed on the element before a start of activation of the element and that the experiment has to be terminated in case of a sensor failure. The potential sensor replacement is tied to deactivation and reactivation of the observed element. For the mentioned disadvantages, the approach is unusable for a task of the experimental estimation of the current value of the tensile force in a prestressed structural element on an existing concrete structure.

**The Approach Founded on a Strain Measurement with Strain Gauges.** This approach is unusable for the stated main research objective too. The measurement with strain gauges fixed on the element surface is realized comparatively. The initial reading has to be realized on unloaded investigated structural element if the total force is to be determined. A damage of the installed strain gauges or a failure of the used datalogger lead to termination of the experiment.

**The Force Determination in a Flexible Structural Element Based on the Relation between the Transverse Force and the Caused Transverse Displacement.** This experimental approach [1, 2] can be used for the precise determination of the total force in the structural element at any of its life cycle stages. However, the applicability of the approach is significantly limited only for the flexible elements with a relatively small cross section and with a relatively long free length. The measurable transverse displacement increase substantially the tensile force in the considerably short element. It means, the approach is unusable for the embedded prestressed reinforcement which is minimally uncovered only.

**The Vibration Frequency Method.** The total force in the investigated structural element at any of its life cycle stages can be determined by this experimental approach too [3, 4]. However, the method needs relatively long free vibrating length of the studied element and the evaluated force uncertainty is significantly dependent on the element bending stiffness, the vibrating length and the element boundary conditions. Especially, the boundary conditions cannot be sometimes reliably specified, particularly for very short elements. The uncertainty of an evaluated force in some embedded prestressed reinforcement, which is minimally uncovered, is evidently greater than for the new approach based on magnetoelastic principle described in the paper.

**The Magnetoelastic Method – Standard Approach.** Measuring equipment, that is based on the magnetoelastic principle and that is standardly used at present [5, 6, 7], evaluates measured data in a relatively simple way. The standard magnetoelastic sensors, that have hitherto been used in civil engineering practice, are composed from the primary and secondary coils only (compare with the Fig. 1). Based on the author's knowledge, the standard magnetoelastic sensors [5, 6] are situated on an unloaded observed element before its activation. The sensitivity of the used measuring equipment, which is usually composed from the data logger and the sensors, is specified during tensioning of the element using the information about the actual prestressed force from a tension hydraulic jack. The evaluation of the measured signal is based on observation of one selected parameter in the specific magnetic conditions around the sensor in which its sensibility was specified. There is a risk of an inaccurate determination of the results that is attached to the possible change of magnetoelastic characteristic of the standard magnetoelastic sensor caused by a variation of sensor magnetic surroundings in the time between its sensibility specification during the observed element activation and a carried out force measurement. The changeability in the sensor magnetic surroundings is common during building of a civil engineering structure. It could be caused by a concreting of steel-fibre concrete or a removal of a steel massive falsework from the sensor surroundings after concrete hardening or the simple removal of the tension hydraulic jack after tensioning for example. The standard approach is only useable for the observation of the comparative change of the tension force compared with the realized initial reading which has to be carried out in the unloaded state of the element.

## The Brief Description of the Physical Principle of the Magnetoelastic Method

The magnetoelastic method is based on an experimental estimation of the magnetic response of the tensile stressed structural element on an external magnetic field. It means, the method is applicable for elements made from ferromagnetic materials only, as the steel used for production of prestressed reinforcement for example.

The magnetic field intensity H and the magnetic flux density B are ones of the fundamental physical quantities describing the magnetic field arrangement. The magnetic flux density B compared to the magnetic field intensity H describes a force effect of the magnetism.

The relation between B and H, that is characterized by a curve called the hysteresis loop, depends on the quantity named the magnetic permeability of material $\mu$ that is influenced by particular materials subjected to the effect of the magnetic field and by momentary state of these materials. Certain materials or surroundings amplify the force effect of the magnetic field more or less, some others on the other hand waken it.

As was mentioned above, the permeability is given by the type of material exposed to the magnetic field influence, its features and current conditions. Usually, the value of permeability depends on the grade of the affected magnetic field, and not only on its actual value, but also on its history and rate of change, on the actual material temperature and on the current material stress. The ferromagnetic materials are the typical example of ones with this general dependence.

## The Basic Principle of the Magnetoelastic Method and the New Approach

As was already stated above, the magnetoelastic method is based on the magnetoelastic principle. It is known, the magnetic characteristics of the ferromagnetic materials are dependent on the level of their mechanical stress among other things. The practical effect is that the permeability of steel is the function of the tension stress or the force straining the investigated steel element and this matter of fact is applied by the magnetoelastic method for determination of its mechanical stress or tension force.

The principle of the newly designed arrangement of the magnetoelastic sensor consisting of the four basic parts is shown in Fig. 1. The first segment of the sensor is the controlled source of the variable magnetic field, as is the primary coin supplied by changing current. The second one is the appropriately structured system of Hall sensors for detailed investigation of the magnetic field intensity H in the investigated cross section of the observed prestressed structural element. The third one is the secondary coil scanning the magnetic flux density B in the investigated cross section. The fourth one is the careful sensor shielding made from steel that significantly reduces the influence of the sensor magnetic surroundings.

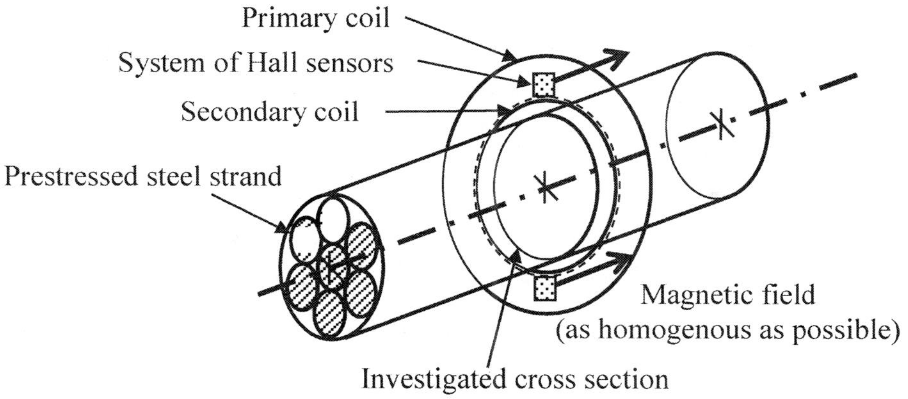


Fig. 1 The principle of the newly proposed arrangement of the magnetoelastic sensor

The Hall sensor is an electronic component that is applied for measurement of the magnetic field intensity H. The modern Hall sensors are small in size and they are satisfactory sensitive for accurate enough observation of the tangential component of the magnetic field close to the surface of the observed prestressed element. According to manufacturer's catalogs, the manufacturing tolerances of the sensitivity of the commonly available Hall sensors vary by up to about +/- 5%. Therefore, for a more precise measurement, the more accurate sensitivity has to be specified for each Hall sensor before its application by a special calibration device.

The magnetic flux density B in the observed element cross section is evaluated base on the analog integration of the voltage induced in the secondary coil.

## The Verification of the New Approach

Several laboratory experiments were carried out during the verification process of the new approach to experimental determination of tensile forces in prestressed structural elements based on the magnetoelastic principle. Various sensor configurations and variants for measurement characteristics were studied during the verification experiments. For example, an appropriate time behavior and time duration of the changing current signal supplying the primary coil, a dimensionless parameter P characterizing most suitably a measured hysteresis loop, that can be applied for determination of experiment final results, a suitable placement of the Hall sensors around the investigated structural element for the appropriate description of the magnetic field intensity H in the close proximity of the element, a consequence of the sensor shielding and an effect of the sensor magnetic surroundings were analyzed.

In November 2016, the verification process of the new approach was finalized by a laboratory experiment concentrated on the systematic study of variations in the magnetic behavior of selected standard prestressed elements in dependence on its immediate temperature and rate of the mechanical stress. The laboratory experiment was realized in the experimental centre of the Klokner institute (the Czech Technical University in Prague).

The magnetoelastic parameters of several standard prestressed elements were investigated in the course of the experiment. Four currently used elements (one prestressed strand

Lp15.7/1860 MPa and three prestressed bars 15/17 made by companies Dywidag and Mikusol) and three formerly applied prestressed elements, that were obtained by demolitions of old prestressed concrete structures, (three patented wires P4.5 of different age) were studied.

Fig. 2 The exterior view on the climatic chamber and the steel tensile testing machine (on left) and on the magnetoelastic sensor installed on the prestressed wire inside the chamber (on right).

In the course of the experiment, the investigated elements were placed in a climatic chamber (see Fig. 2) and loaded in a steel tensile testing machine. The magnetic features of the studied elements were investigated in four temperature steps namely, at -20° C, at ±0° C, at +20° C and at +40° C. The studied prestressed elements were loaded in a defined number of force steps for each temperature step. The used number of the force steps was nine or seven and it was chosen differently for each particular investigated element according to its design resistance.

For each element, each particular temperature and force step, the hysteresis loop was measured and determined. It was not effective to evaluate the measured hysteresis loops in their whole range because of the quantity and complexity of the obtained experimental results. In the course of the verification experiments, several various definitions of the dimensionless parameters P were defined and studied that represented in a simple way the character and shape of every measured hysteresis loop. The sensitivity to the variation of the mechanical stress and temperature was evaluated and critically assessed for each particular evaluated dimensionless parameters P. Base on this rating, the resultant dimensionless parameter P was selected that is defined as the difference between two values of the quantity B in exactly determined points of the hysteresis loop divided by the greater quantity B. It has been used for finally evaluation of all measured data.

The example of two evaluation curves resulting from the verification experiment for the patent wire labelled D1, that describes the dependence between the stress in the wire and the resultant dimensionless parameter P for two defined temperatures, is shown in Fig. 3.

The evaluated magnetoelastic characteristics for the particular wires are slightly different. But their differences are small and the values of the stress evaluated based on the average relation between the stress and resultant dimensionless parameter P deviate from the actual values about ±3% of the largest stress at maximum.

In the course of the verification experiments, the magnetoelastic characteristics of the prestressed strands were investigated only for one kind of this material which is used in the civil engineering practice at present. Though, several different test samples of this material were used during different verification experiments. The stress values were afterwards specified for almost all previously studied samples based on the final evaluation curve similar to the one shown in Fig. 3 that describes dependence between the stress and resultant dimensionless parameter P for the prestressed strands. The obtained results deviate from the actual stress values about ±2.0 % of the largest stress at maximum. The actual stresses in the studied strand samples were determined from the measured force values.

On the other hand, the large differences between the magnetoelastic characteristics of the prestressed bars made from various steel materials were identified. Moreover, it was found, the screw thread of the bar influences negatively the experiment. For obtaining the acceptably accurate results by the magnetoelastic method, the screw thread has to be removed from the bar surface in the proximity of the cross section, where the magnetoelastic sensor would be installed. And it is practically not feasible on an existing structure in a real situation.

The sensitivity of the dimensionless parameter P on the mechanical stress in the studied prestressed bars is considerably less than these ones evaluated for the prestressed strands or wires. This matter of fact is caused partially by the notably lower design resistance of the bar materials and also by their partly different chemical compositions.

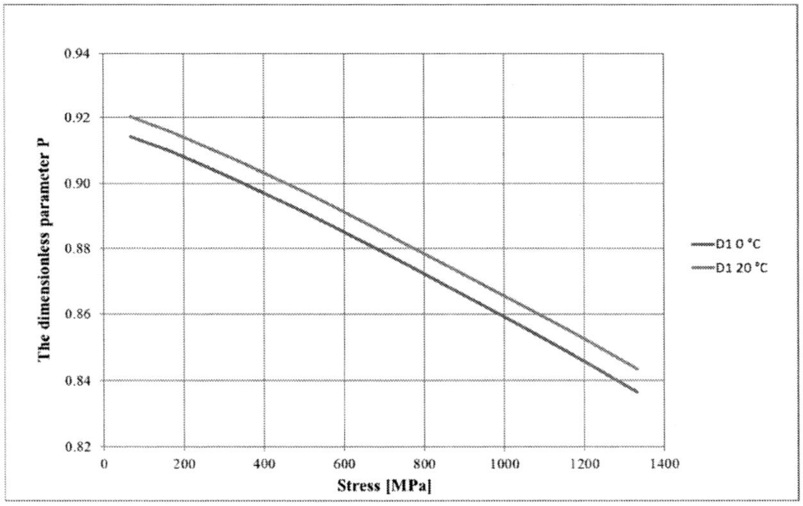


Fig. 3 The evaluation curve resulting from the verification experiment for the patent wire D1 and for the temperature steps ±0°C and +20°C, that describes the relation between the stress in the observed wire and the resultant dimensionless parameter P.

Another issue, that had to be solved during the evaluation of the experimental results, was the transformation of the magnetic field intensity H measured in the discrete specific positions by the system of the Hall sensors to a continuous field in the whole cross-section area. For the more accurate result evaluation, it is significant to know the value of the magnetic field intensity in the cross section sectors where the Hall sensor cannot be placed, as is the cross-section center of the strand for example.

The interpolation of measured values H for the significant points in the cross-section can be significantly improved by a numerical analysis. A suitable computing tool base on the finite element method (FEM) should be used for this activity. Rotationally symmetrical (see Fig. 4) and also general problems were solved during the new approach verification.

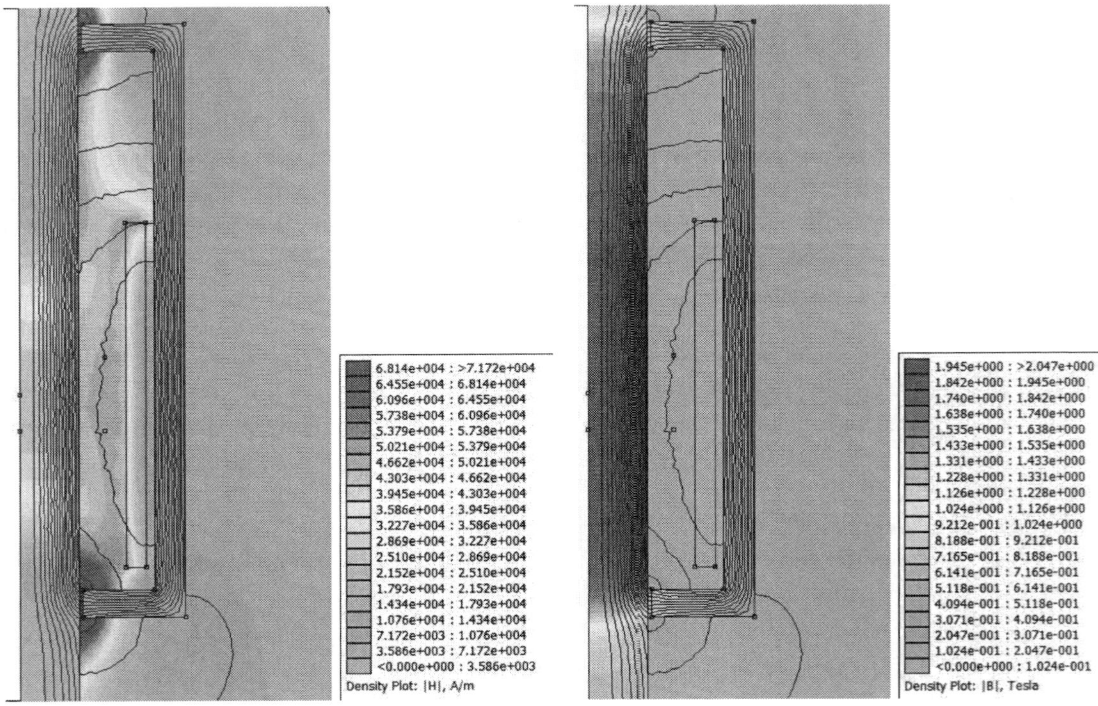


Fig. 4 The graphic visualization of the arrangement of the magnetic field calculated by the software FEMM [8] for the FEM model of the rotationally symmetrical magnetoelastic sensor (the axis of symmetry is in the left part of both images), the magnetic field intensity H (on the left) and magnetic flux density B (on the right).

## The Application of the New Approach on the Existing Structure

The applicability of the new approach described in the paper was realized by an implementation on a prestessed concrete structure, namely on the existing footbridge across Vltava River in Prague constructed in 1984 (see Fig. 5). The footbridge load bearing structure is built as the precast segmental construction and its static system is the stress ribbon bridge with three spans 85.5 m, 96.0 m and 67.5 m.

Fig. 5 The overview on the investigated footbridge (on the left) and on the footbridge section, where the magnetoelastic sensors were installed (on the right).

The substantial part of the footbridge deck was overflowed by two great floods that occurred in August 2002 and in June 2013. Subsequent diagnostics of the footbridge detected several serious problems of the load bearing structure. Especially, justified doubts about the real actual value of the prestress in the prestressed cables were the basic reason for the realization of the verification experiment on this specific footbridge. It was supposed, the strong local corrosion of several prestressed cables and the partial deterioration of the construction conditions of the load bearing structure both detected by diagnostics could significantly affect the cable prestressed forces.

The archived footbridge design documentation was used to obtain the basic knowledge about the load bearing structure geometry, about the used fundamental structural elements and about the installation procedure of the footbridge structure.

Two types of the steel cables were used in the course of the footbridge construction process. At first, the cables of type A were situated. They are the carrying ones of the stress ribbon bridge. Afterwards the installation of the footbridge concrete parts, fourteen cables of type B assembled from six strands (see Fig. 6 and Fig. 7) were activated. They are the prestressed ones preventing the dilatation of transversal gaps between the concrete segments of the footbridge deck.

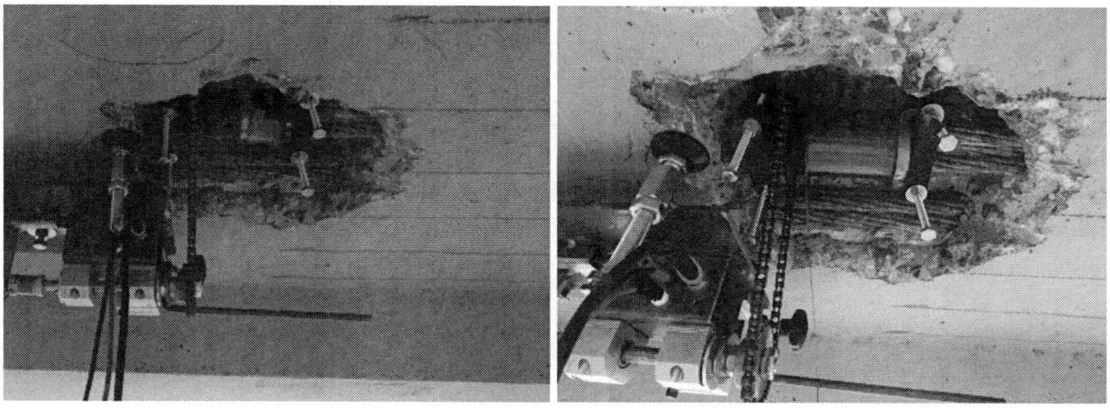

Fig. 6 The view on the created hollow, the studied cable, the coiler, the black plastic sensor body, the partially reeled secondary coil (on the left) and the partially reeled primary coil (on the right).

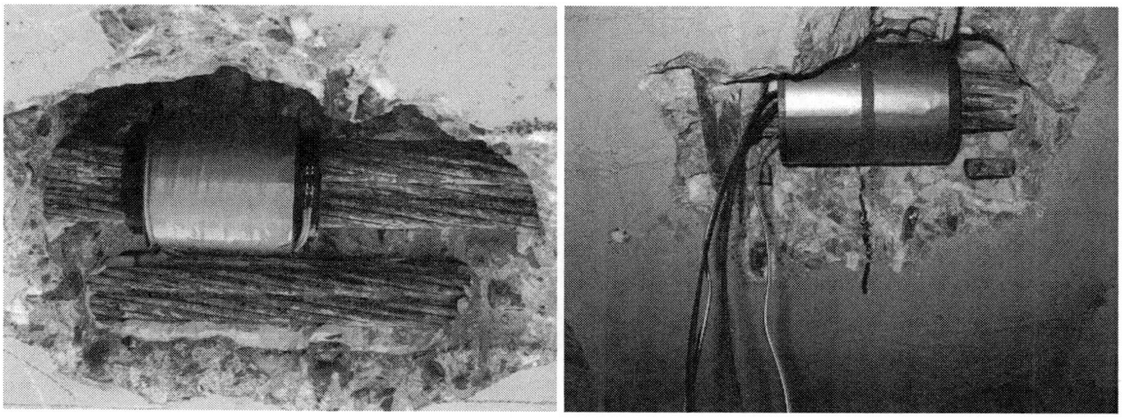

Fig. 7 The view on the completed magnetoelastic sensor (on the left) and on the sensor covered by the steel shielding (on the right).

In November 2016, the first magnetoelastic sensor designed for the assembly on the existing structure was installed on the prestressed cable of type B. In April 2017, the next three magnetoelastic sensors with the same parameters as the first one were made on another three cables of type B. The positions for the installation of all four sensors on the selected cables were chosen not to be significantly affected by cable corrosion in these places. All four observed cables were studied as the whole because of there was not gaps between the strand wide enough for the sensor installation on the individual strands.

At first, the adequately large hollows (see Fig. 6 and Fig. 7) were created carefully around the selected cables in which the sensors could be assembled. Secondly, the sensor bodies were assembled on the cable from two segments. Thirdly, the secondary coils were reeled on the sensor bodies by the help of the specially made coiler rotating with the plastic sensor body (see Fig. 6). Fourthly, the primary coils were made in the same way as the secondary ones (see Fig. 6). Next, the systems of the calibrated Hall sensors were placed inside the cables in small free hallows between strands and then their positions were specified accurately relative to both coils. One Hall sensor has always touched the surface of the middle strand. At the end, the thorough sensor steel shieldings were positioned (see Fig. 7) that were assembled from two segments again. Both experiments in situ were completed with the measurement of the hysteresis loops obtained as the primary experimental results. The results measured in April 2017 are still being processed.

The dimensionless parameter P was evaluated from the hysteresis loop measured in November 2016 and then the resulting value 825 MPa of the normal stress in the studied cable was obtained from the evaluation curve similar to the one depicted in Fig. 3.

The assessment of acceptability of the determined normal stress in the cables was realized based on comparison with expected stress in the prestressed cables of type B that was determined from the available information about the footbridge structure. The study of the static calculation of the footbridge showed that the normal stress in the cables of type B could be anticipated in the interval from 680 MPa to 1000 MPa. The large range of the stress interval is related to the prestress losses due to successive prestressing of the subtle footbridge structure by fourteen cables of type B. A protocol about the correct procedure of the successive prestressing of the prestressed cables in the course of the footbridge construction enabling a more exact assessment of the stress in the observed cables has not been found unfortunately. The experimentally determined resulting value of the normal stress in the first observed cable is practically located in the center of the anticipated interval taken from the static calculation. It means that the resulting value specified experimentally for the first investigated cable may be characterized as reliable.

## Conclusions

The paper introduces the concise description of the principle of the new approach to the experimental determination of tensile forces in prestressed structural elements of civil engineering structures and the basic validation of its results. The new approach is based on the magnetoelastic principle and was designed and developed especially for application on existing structures built from prestressed concrete.

The applicability of the method was confirmed by the experiments performed both in the laboratory and on existing prestressed concrete structures in situ. This method is especially suitable for a supervision of the stress in the steel prestressed elements assembled from wires or strands on the existing structures made from prestressed concrete on which another nondestructive method [1, 2, 3, 4] cannot be used as it is in most of actual cases in the civil engineering praxis.

On the other hand, the supervison of the presstressed bars using the new approach does not seem appropriate. These bars are usually used on civil engineering structures as temporary mounting prestressed elements. In these cases, another experimental tools can be easily used for the tensile forces monitoring e.g. a strain gauge measurement or a professionally made force transducer.

Based on the author's knowledge, the standard magnetoelastic sensors [5, 6] composed from the primary and secondary coils only, that have been used to the present days, are placed usually on an unloaded monitored element before its activation. Then the actual prestressed force from a tension hydraulic jack during tensioning of the element is used for the sensitivity specification of the used measuring equipment. It usually consists of the data logger and the sensors. The disadvantage of the standard approach is that it is only useable for the observation of the tension force compared with the initial reading which has to be realized in the unloaded state of the element. Based on the above mentioned knowledge, the total normal force in the prestressed structural elements on an existing concrete structure cannot be determined with enough precision by this approach.

The described new approach also provides an advantage for experiments carried out on newly built prestressed concrete structures compared to experiments performed by standard magnetoelastic sensors [5, 6, 7].

In the performed experiments, another advantage of the new approach was found. It is the possible significant elimination of risks of an inaccurate determination of the current stress in the prestressed elements monitored already in the moment of their activation. These risks are particularly attached to the possible change of magnetoelastic characteristics of the standard magnetoelastic sensor [5, 6, 7] caused by a change of sensor magnetic surroundings in the time between its calibration during the prestressed element activation and a performed stress measuring. The changeability in the sensor magnetic surroundings is usual during construction of a civil engineering structure. It could be caused by concreting of steel-fibre concrete, a removal of a steel massive falsework from the sensor surroundings after concrete hardening or the simple removal of the tension hydraulic jack after tensioning for example.

**Acknowledgement**

This paper has been supported by the Technology Agency of the Czech Republic project No. TA04030307.

**References**

[1] P. Fajman, M. Polák, Measurement of structural cable of membranes, in: M. Růžička, K. Doubrava, Z. Horák (Eds.), Proceedings of the $50^{th}$ Annual Conference on Experimental Stress Analysis, EAN 2012, Tábor, Czech Republic, 2012, pp. 61-64.

[2] P. Fajman, M. Polák, J. Máca, T. Plachý, The Experimental Observation of the Prestress Forces in the Structural Elements of a Tension Fabric Structure, Appl. Mech. Mater. 486, (2014) 189-194.

[3] M. Polák, T. Plachý, Determination of Forces in Roof Cables at Administrative Center Amazon Court, Procedia Engineer. 48, (2012) 578-582.

[4] M. Polák, T. Plachý, Experimental Evaluation of Tensile Forces in Short Steel Rods, Appl. Mech. Mater. 732, (2015) 333-336.

[5] M. Chandoga, P. Fabo, A. Jaroševič, Measurement of Forces in the Cable Stays of the Apollo Bridge, in: Proceedings of the $2^{nd}$ fib Congress, Naples, Italy, 2006, pp. 674-675.

[6] A. M. Sarmento, A. Lage, E. Caetano, J. Figueiras, Stress measurement and material defect detection in steel strands by magneto elastic effect. Comparison with other non-

destructive measurement techniques, in: Proceedings of the 6th International Conference on Bridge Maintenance, Safety and Management IABMAS 2012, Stresa, Lake Maggiore, Italy, 8 - 12 July, 2012, pp. 914-921.

[7] H. J. Wichmann, A. Holst, H. Budelmann, Magnetoelastic stress measurement and material defect detection in prestressed tendons using coil sensors, in: Proceedings of 7th International Symposium on Non-Destructive Testing in Civil Engineering NDTCE'09, Nantes, France, 30 June – 3 July, 2009.

[8] Information on http://www.femm.info/wiki/HomePage

# Residual Stress Distribution in Spherical Storage Tank Butt Weld

Ľuboš Mráz[1,a], Charles Hervoches[2,b], Miroslav Vrána[2,c], Pavol Mikula[2,d], Peter Baksa[1,e] and Ján Bošanský[1,f]

[1]Welding Research Institute – Industrial Inst., Račianska 71, 832 59 Bratislava, Slovak Rep.

[2]Nuclear Physics Institute ASCR, v.v.i. 250 68 Řež, Czech Rep.

[a]mrazl@mail.t-com.sk, [b]hervoches@ujf.cas.cz, [c]vrana@ujf.cas.cz, [d]mikula@ujf.cas.cz, [e]baksap@vuz.sk, [f]bosanskyj@vuz.sk

**Keywords:** Weld residual stresses, butt weld, normal, transversal and longitudinal stress components, neutron diffraction, stress corrosion cracking.

## Introduction

Residual stresses are typical phenomena associated with welding of any kind of structural material. Stresses are generally responsible for the deformation of welded structures during production and subsequently influence the behaviour of these structures during service. This factor directly influences the properties of real welds and sensitivity to the initiation of different type of service cracking such as reheat and stress corrosion cracking. It is generally known that the initiation of stress corrosion cracking is influenced by the combined effect of material quality, environment and level of stresses (internal – residual, external – service).
Spherical storage tanks are used for holding of anhydrous ammonia, LPG, NGL, gasoline, naphtha, butadiene, ethylene, hydrogen, oxygen, nitrogen, argon, LNG (liquefied natural gas), biogas, sewage gas and waste water.
There is a number of publications available in the related literature which report about the failures of this equipment [1]. The choice of a suitable structural material, modification of environment or lower stress level can help to avoid the failure. In this case the stresses play certainly an important role and the knowledge of their level can provide important information to designers. In order to clarify the role of stresses on individual failures it is important to identify their level. Subsequently, it opens possibilities to repair failed structure and to define the conditions for further service. The purpose of this contribution is to present the results of strain/stress distribution measurements in the vicinity of the butt weld from the spherical storage tank which was obtained by the neutron diffraction method.

## Experimental part

Residual stress distribution of the manual metal arc butt weld of the spherical storage tank has been experimentally analyzed. The storage tank of the diameter of 900 mm has been manufactured of C-Mn unalloyed normalized steel of the strength of approx. 480 MPa and of the thickness of 30 mm to store liquid anhydrous ammonia. Chemical composition of steel is summarized in Table 1 and mechanical properties in Table 2. Covered basic electrodes E-44.83 have been used for the butt weld preparation. Macrostructure of the butt weld is shown in Fig. 1.

Table 1. Chemical composition of C-Mn steel.

| Material | mass concentration [%] | | | | | |
|---|---|---|---|---|---|---|
| | C | Mn | Si | P | S | Mo |
| **C-Mn steel** | **0.224** | **0.49** | **0.24** | **0.024** | **0.043** | - |
| **11 418.1 according to STN 41 1418** | **max. 0,20** | **0,50 1,40** | **max. 0,35** | **max. 0,030** | **max. 0,025** | **max. 0,08** |

Table 2. Mechanical properties of C-Mn steel according to STN 411418.

| Material | $R_e$ [MPa] | $R_m$ [MPa] | A [%] | Z [%] |
|---|---|---|---|---|
| **C-Mn steel** | **259** | **421** | -* | -* |
| **11 418.1 according to STN 41 1418** | **Min. 255** | **410 530** | **Min. 26** | - |

* not estimated

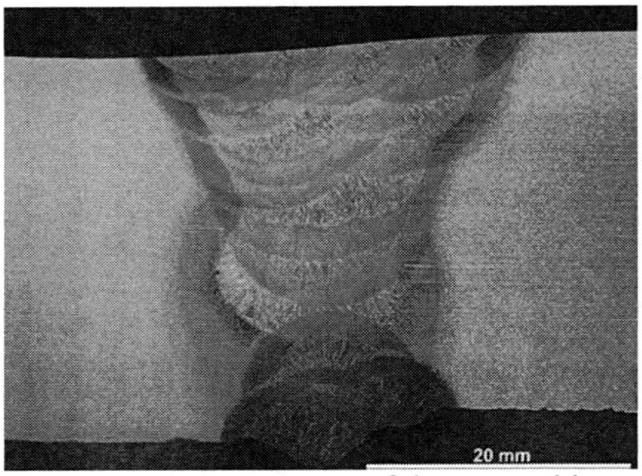


Fig. 1. Macrostructure of the test weld.

The specimen dimension of 150 x 200 x 15 mm$^3$ was cut off the circumferential weld of the storage tank for residual stress distribution measurements. The principle of the neutron diffraction method is quite simple. It consists in the precise determination of the $d_{hkl}$-spacing of particularly oriented crystal planes [2,3]. In neutron and X-ray diffraction the angular positions of the diffraction maxima are directly related to the values of the lattice spacing through the Bragg equation $2d_{hkl} \cdot \sin \theta_{hkl} = \lambda$ ($d_{hkl}$ - lattice spacing, $\theta_{hkl}$ - Bragg angle, $\lambda$ - the neutron wavelength) and thus offer a unique non-destructive technique for the investigation of stress fields. When a specimen is strained elastically, the lattice spacing changes. Then, when defining the strain $\varepsilon$ as $\varepsilon = \Delta d/d_{0,hkl}$ , it is related to a change in the lattice spacing, i.e. to a component parallel to the scattering vector $\boldsymbol{Q}$ perpendicular to the reflecting set of planes. Therefore, the knowledge of the $d_{0,hkl}$ value ($d_{0,hkl}$ is the lattice spacing of the strain-free material) is a crucial task [3]. Then, by the differentiation of the Bragg condition we arrive at $\varepsilon = -\cot \theta_{hkl} \cdot \Delta\theta_{hkl}$. The relation for the strain $\varepsilon$ indicates that it gives rise to a change in the scattering angle $2\theta_{hkl}$ resulting in an angular shift $\Delta(2\theta_{hkl})$ of the peak position for a particular reflecting plane illuminated by a fixed wavelength. In such a way, the shift in the Bragg angle (relative to that of the stress-free material) permits the determination of the average lattice macrostrain over the irradiated gauge volume (see Fig. 2). The conversion of strains to stresses is carried out by means of the relation

$$\sigma_x = \frac{E_{hkl}}{(1-2\nu_{hkl})(1+\nu_{hkl})}\left[(1-\nu_{hkl})\varepsilon_x^{hkl} + \nu_{hkl}(\varepsilon_y^{hkl} + \varepsilon_z^{hkl})\right] , \tag{1}$$

where $\varepsilon_{x,y,z}^{hkl}$ is the *x,y,z*-component of the lattice strain measured at the *hkl* crystal lattice planes, $E_{\mathrm{hkl}}$ and $\nu_{\mathrm{hkl}}$ are the diffraction elastic Young modulus and diffraction Poisson ratio, respectively. For the determination of the stress tensor in this case of the steel sample, three strain components should be determined as schematically shown in Fig. 3. Experimental measurements were carried out on the dedicated neutron strain scanner installed at the medium-power (10 MW) research reactor LVR-15 in Řež, Czech Republic [4,5].

The strain/stress instrument schematically shown in Fig. 2 is equipped with a focusing curved Si(111) monochromator and a high-resolution position-sensitive detector for fast recording of diffraction profiles. The monochromator provides the neutron beam with the wavelength of 0.232 nm. The measurement of the shift in the Bragg angle $\theta$ relative to that of the stress-free material $\theta_o$ permits determination of the average lattice macro-strain over the irradiated gauge volume. The strain scanning was carried when using diffraction on α-Fe(110) lattice planes at the scattering angle $2\cdot\theta_{110} = 70^{\circ}$. For the evaluation of the strains it was necessary to determine the angular position $2\cdot\theta_{0,\mathrm{hkl}}$ of the diffracted neutron beam for the strain free

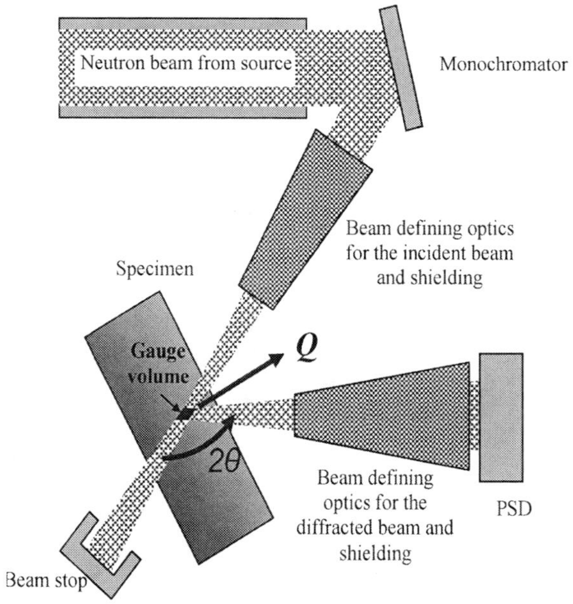


Fig. 2. Schematic illustration of a reactor source based diffractometer for strain measurement in parallel diffraction geometry.

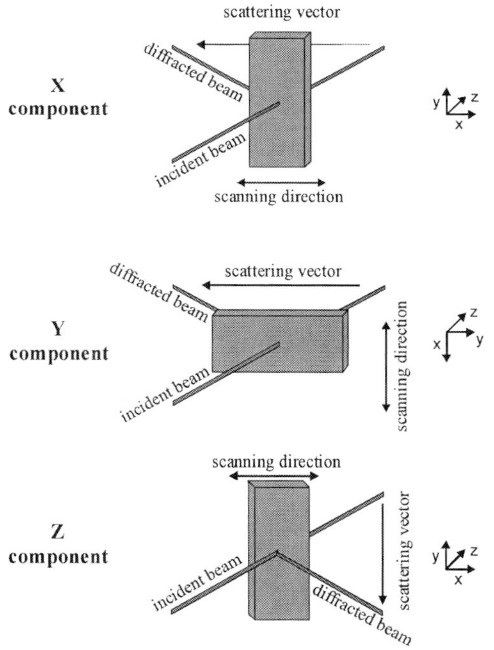


Fig. 3. Sketch of the sample setting for determination of three strain/stress components.

material. Normal, transversal and longitudinal strain components were estimated for three through thickness positions 3.5 mm, 7.5 mm and 11.5 mm below the weld root surface of the test specimen (see Fig. 4). Then, the results of residual stress estimation are summarized in Fig. 5, Fig. 6 and Fig. 7.

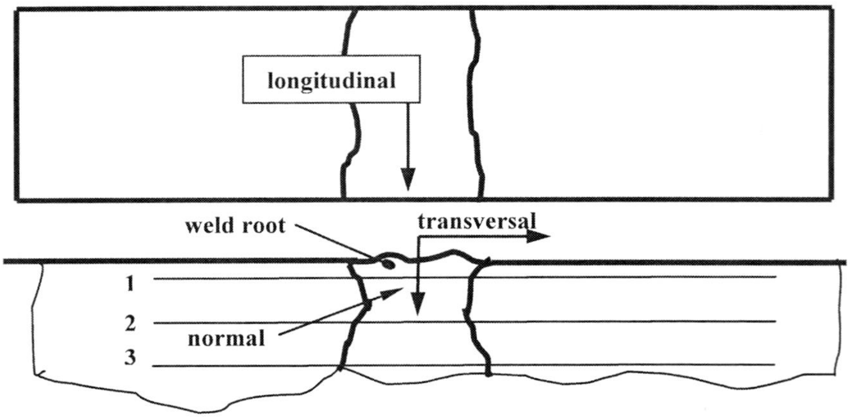


Fig. 4. Residual stress components

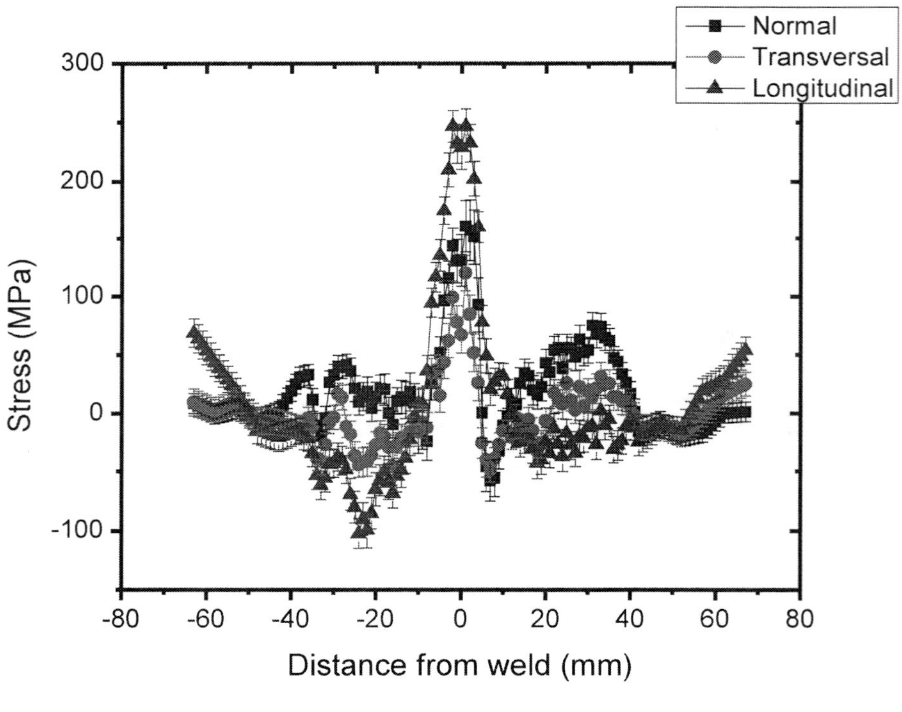


Fig. 5 Residual stress distribution 3.5 mm below the weld root surface (line 1 in Fig. 4).

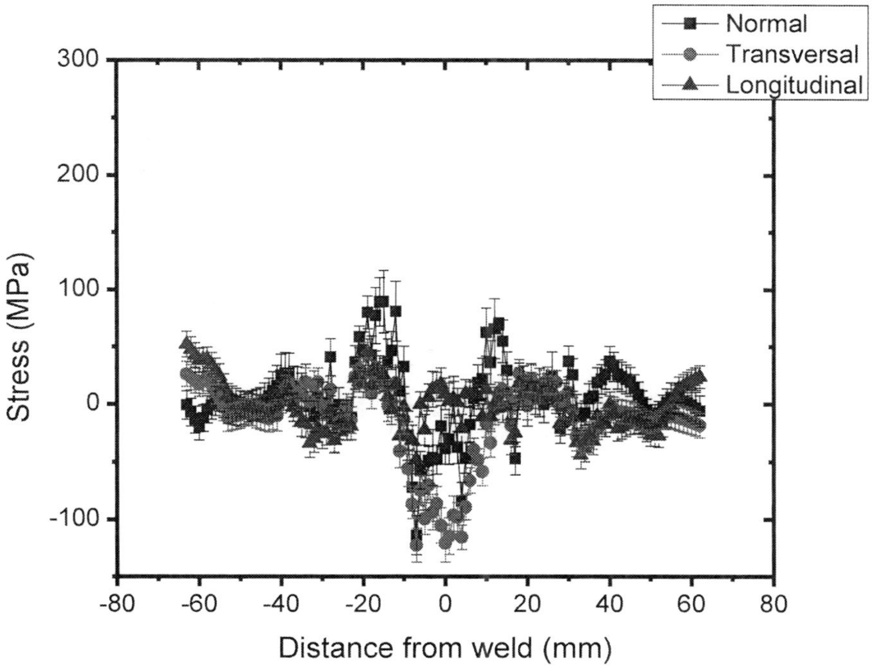


Fig. 6. Residual stress distribution 7.5 mm below the weld root surface (line 2 in Fig. 4).

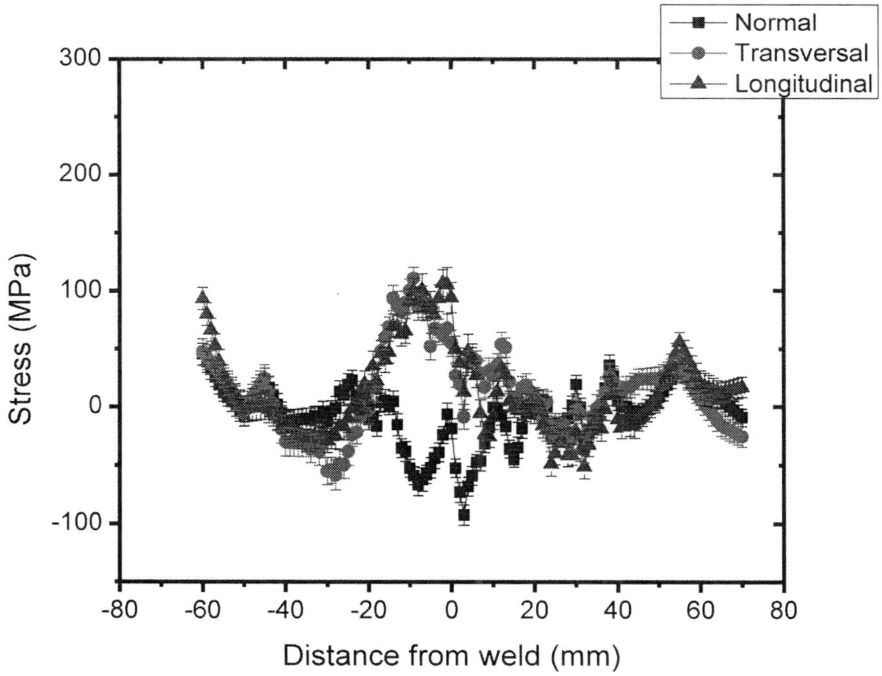


Fig. 7. Residual stress distribution 11.5 mm below the weld root surface (line 3 in Fig. 4).

## Discussion

The purpose of the residual stress measurements of the storage tank butt weld was to identify maximum stress level in the vicinity of the weld. Generally it has been recognized that maximum stresses are concentrated in the weld centre 3.5 mm below the surface and in the weld HAZ (heat affected zone) 7.5 mm below the surface. Maximum stresses are in the weld metal close to the weld surface. The level of 250 MPa has been estimated, which is practically the same as yield strength level of the parent material (259 MPa). Residual stresses 7.5 mm and 11.5 mm below the weld root surface are much lower (approx. 100 MPa). Compressive normal and transversal stresses have been identified especially 7.5 mm below the surface. According to the service of the storage tank the most critical seem to be stresses close to the weld root surface due to the fact that shallow stress corrosion cracks were observed at the surface of the weld root in the weld metal after long term exposure.

## Conclusions

Residual stress distribution in the vicinity of the weld taken out of spherical tank used for storage of anhydrous ammonia has been estimated using neutron diffraction. It has been recognized that the highest tensile stresses at the level of 250 MPa have been identified close to the weld root surface in the weld metal.


## Acknowledgement

The related measurements were carried out at the CANAM infrastructure of the NPI ASCR Rez. The residual stress investigations are in the Czech Republic supported by GACR project no. 14-36566G.


## References


[1] W. Atherton and J.W. Ash, Review of failures, causes & consequences In the bulk storage industry, http://lightningsafety.com/nlsi_lls/Causes-of-Failures-in-Bulk-Storage.pdf

[2] V. Stelmukh, L. Edwards, J.R. Santisteban, S. Ganguly and M.E. Fitzpatrick, Weld stress mapping using neutron and synchrotron x-ray diffraction, Materials Science Forum, 404-407 (2002) 599-604.

[3] A.D. Krawitz and R.A. Winholtz, Use of position-dependent stress-free standards for diffraction stress measurements, Materials Science and Engineering, A185 (1994) 123-130.

[4] P. Mikula, M. Vrána, P. Lukáš, J. Šaroun, P. Strunz, H.J. Ullrich and V. Wagner, Neutron diffractometer exploiting Bragg diffraction optics - A high resolution strain scanner, in: T. Ericsson, M. Odén and A. Andersson (Eds.), Proc. of ICRS-5, June 16-18, 1997, Linköping, Vol. 2, 1997, pp. 721-725.

[5] P. Mikula, M. Vrána, P. Lukáš, J. Šaroun and V. Wagner, High-resolution neutron powder diffractometry on samples of small dimensions, Materials Science Forum, 228-231 (1996) 269-274.

EAN 2017
55th conference on experimental stress analysis

# Strain Gauge Measurement on Clinched Joints Specimens

Pavel Malý[1, a], František Lopot[1,b], František Starý[1,c] and Jiří Sojka[1,d]

[1]CTU in Prague, Department of Designing and Machine Components, Technická 4, 166 07 Prague 6, Czech Republic

[a]pavel.maly@fs.cvut.cz, [b]frantisek.lopot@fs.cvut.cz, [c]frantisek.stary@fs.cvut.cz, [d]jiri.sojka@fs.cvut.cz

**Keywords:** Clinched Joint, Experimental Testing, Strain Gauge, Measurement, Finite Element Analysis

**Abstract.** This paper describes the preparation, implementation and execution of experimental measurement on clinched joint specimens using strain gauges. The real behavior of the clinched joint specimen during the specified loading was obtained. The results from the experiment were compared with the FE simulations. The designed testing method was further used for testing of more complex specimens.

## Introduction

Experimental measurements using different sensors are important in these days because they provide the information about the real conditions and behavior of various components or machines. This information is often one of the basis for subsequent optimizations or validation of computational simulations and analysis, [4], [5].

Clinching represents one of the modern methods for the joining of the sheet metal plates used in many branches of production. The joining method is based on the plastic deformation of the base material and is characterized with many advantages. The creation of the round clinched joint is shown in figure 1.

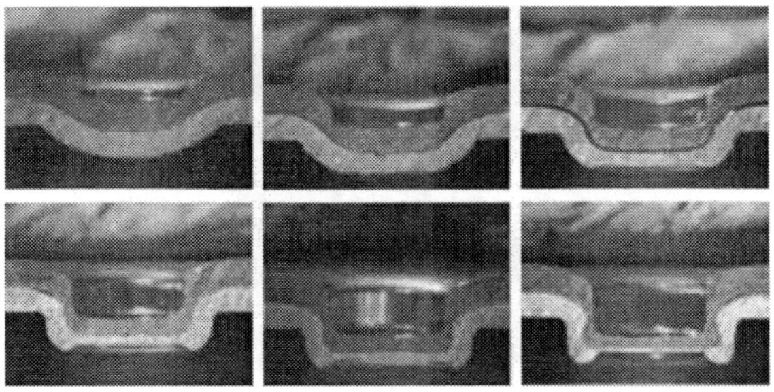

Fig. 1: Creation of the round clinched joint, [2].

## Method

The experimental testing of the specimens is described by the unidirectional nondestructive test when the uniaxial stress near the joint is measured using strain gauge during the loading, [1].

**Instrumentation and measurement protocol.** Special testing stand was designed and built for the realization of the measurement, the scheme is shown in figure 2. The loading is applied by a hydraulic cylinder (HM), velocity of loading is controlled by proportional valve (PV) and the size is measured by HBM force sensor. The controlled variable is force that smoothly increases from 70 N to the maximum value and decreases back to the minimum value.

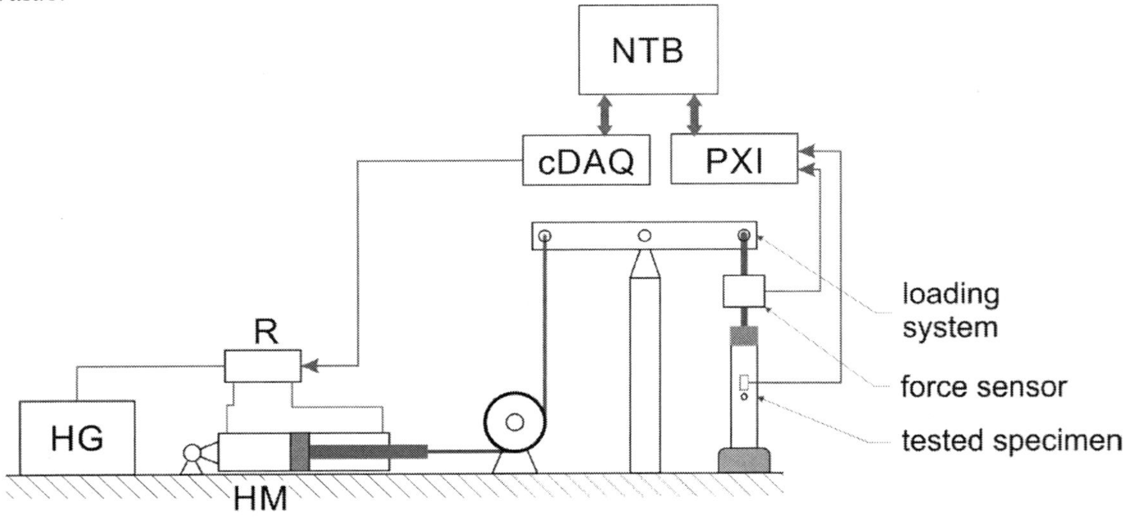


Fig. 2: Scheme of the testing stand

**Investigated set of specimens.** The specimen for the initial measurement and verification of the method was a connection of two overlapped sheet-metal plates from low carbon steel (1.0226) joint by the single clinched joint, [2]. The deformation in the place of interest was supposed to be uniaxial, therefore the HBM uniaxial wire strain gauge in quarter-bridge configuration was used for the measurement. Placement of the strain gauge on the specimen near the clinched joint is shown in figure 3.

Fig. 3: Strain gauge placement near the joint

**Data acquisition and processing.** The testing stand is controlled and the data are measured with National Instruments apparatus (cDAQ, PXI) and LabView software. Following data processing is carried out in Matlab. Example of measured data, i.e. loading force and output from the strain gauge converted to the stress, is shown in figure 4. It is obvious from the graph that the total testing sequence consists of three consecutive cycles.

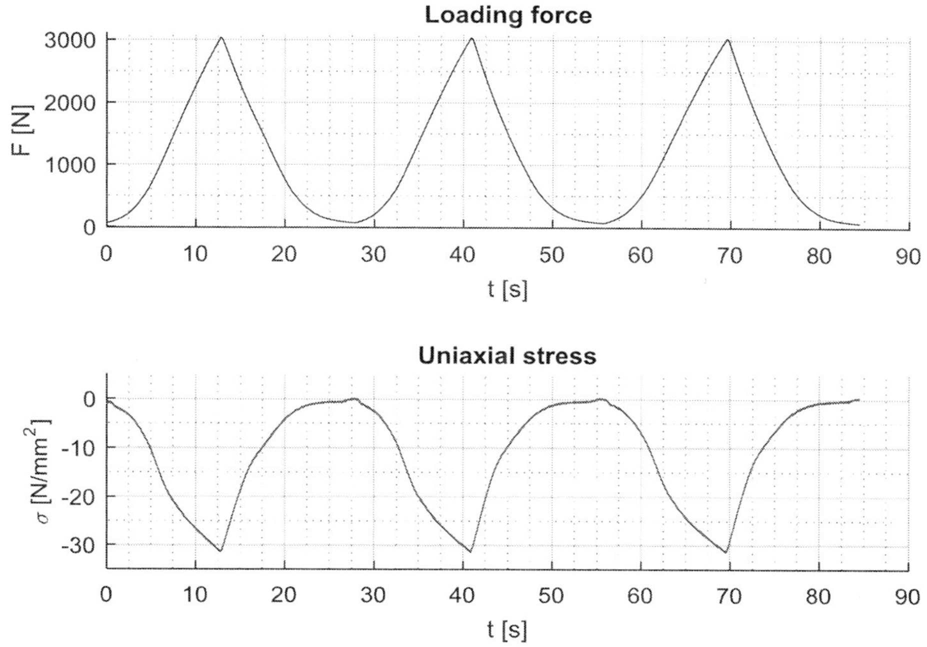


Fig. 4: Measured data

Processing of the measured data represents the identification of the loading force and the dependency between the measured data (loading force and uniaxial stress) as is shown in figure 5. Each cycle consists of loading and unloading part that can be described by the tangent in the linear part of the force course (defined velocity of loading and unloading). The velocities were chosen so that the loading and unloading of the specimen was quasi-static.

The dependency of the measured stress on the loading is plotted in the graph for evaluation of the experiment. The course of the stress depending on the force is different for loading and unloading. This fact can be described by the influence of the testing stand properties, or by the additional bending load of the specimen.

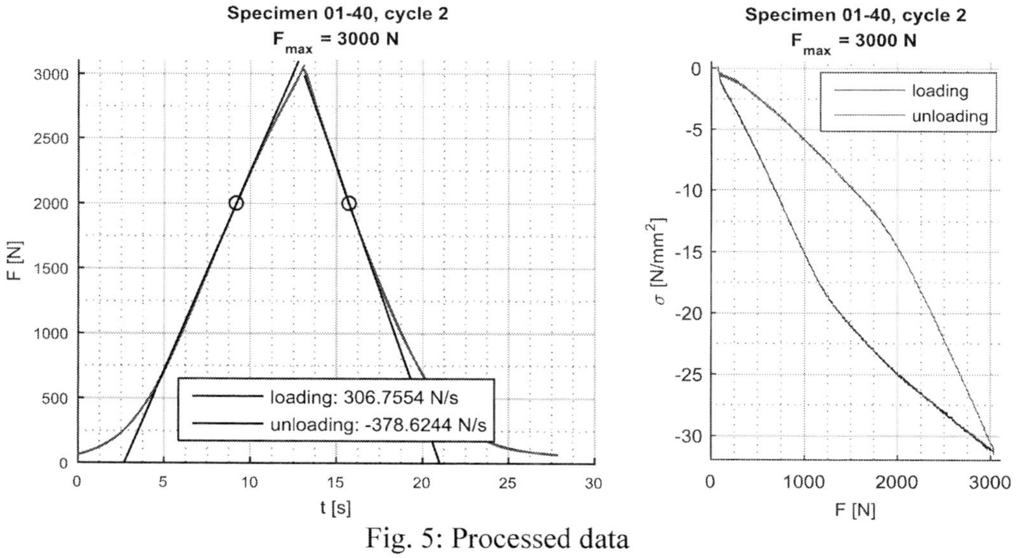


Fig. 5: Processed data

**FE simulations**

The simulations of clinched joint loading were performed in the engineering software Abaqus CAE using 3D model of the clinched joint, [3]. The uniaxial stress near the clinched joint in the place of the strain gauge (figure 6) was used for evaluation of the simulations and comparison with the experiment.

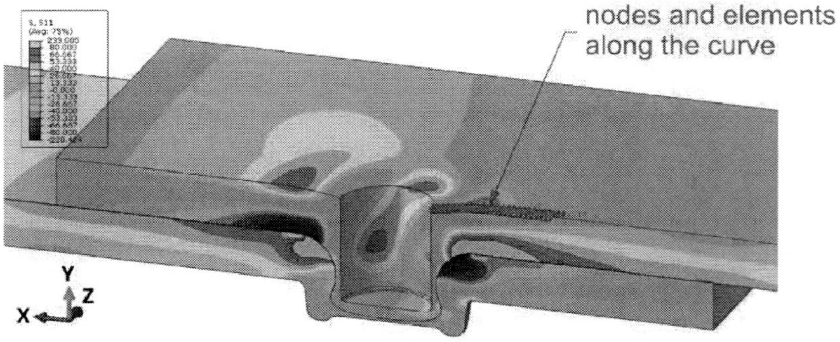


Fig. 6: FEA of the clinched joint model

**Results and conclusions**

The experimental method of nondestructive clinched joint loading with concurrent measurement of uniaxial stress provides the information about the behavior and properties of the joint. Moreover, the results of FE simulations were used for experiment assessment and mutual comparison. The uniaxial stress from FE simulations in the place of the strain gauge for the loading force of the 3000 N is in the range of -28 N/mm² to -32 N/mm² depending on the conditions of simulations. The measured stress is -29,5±2,6 N/mm² which implies that the experimental measurement and simulations correspond. Next step of the testing was the application of the method on the specimens with three joints in the row varying in the pitch.

**References**

[1] H. Huth, Influence of fastener flexibility on the prediction of load transfer and fatigue life for multiple-row joints. ASTM Special Technical Publication (1986) 221-250.

[2] Information on http://www.tox-cz.com

[3] P. Malý, F. Lopot, J. Sojka, FEM model and experimental measurement of clinched joint. IOP Conference Series: Materials Science and Engineering, volume 179, number 1 (2017)

[4] M. Dub, J. Kolář, F. Lopot, V. Dynybyl, O. Berka, Experimental analysis of stress state and motion of tram gearbox hinge, EAN 2014 - 52nd International Conference on Experimental Stress Analysis (2014).

[5] O. Berka, M. Dub, F. Lopot, V. Dynybyl, Experimental analysis of stress state and motion of tram gearbox hinge, EAN 2015 - 53rd Conference on Experimental Stress Analysis (2015).

EAN 2017
55th conference on experimental stress analysis

# Strain Response Determination in Notched Specimens under Multiaxial Cyclic Loading by DICM

Zbyněk Paška[1,a], Radim Halama[1,2,b], František Fojtík[1,2,c], and Petr Gajdoš[3,d]

[1]Department of Applied Mechanics, Faculty of Mechanical Engineering, VŠB – Technical University of Ostrava, 17. listopadu 2172/15, 708 33 Ostrava, Czech Republic

[2]Národní superpočítačové centrum IT4Innovations, VŠB – Technical University of Ostrava, Studentská 6231/1B, 708 33 Ostrava, Czech Republic

[3]Sobriety s.r.o., Loosova 579/10 638 00 Brno, Czech Republic

[a]zbynek.paska@vsb.cz, [b]radim.halama@vsb.cz, [c]frantisek.fojtik@vsb.cz, [d]petr.gajdos@sobriety.at

**Keywords:** Multiaxial cyclic loading, DICM, full-field strain analysis, fatigue, deformation response.

**Abstract.** Focus of the contribution is to determine strain tensor's components in root of notched specimens by DICM. Three types of samples were examined with three types of notch: U, V and "fillet". The behaviour of these samples during low-cycle fatigue testing were examined. Specimens were subjected to five different types of loading, which caused uniaxial and also biaxial state of stress. The results are deformation responses in the notch of the each sample, which will serve to validate numerical results.

## Introduction

Digital Image Correlation Method (DICM) belongs to the most progressive methods suitable for 3D full-field strain analysis of structural components [1,2]. In present time the DICM enables to find out deformations under uniaxial as well as multiaxial loading during fatigue testing [3]. Measurement of deformation, vibration or displacements take place in real time. The DICM is full field methodology of measurement of deformations or displacements. DIC is a non-contact optical method, which enable to determine primarily displacements even in 3D space. Application of DICM for strain determination inside the notch have clear advantages, because the usage of extensometer or strain gauges is excluded. Each specimen must be equipped by high contrast speckle pattern. Measuring set-up consists of two cameras with high-resolution, which are placed on tripod and controlled by appropriate software.

This contribution shows some interesting results of deformation response on cyclic loading in notch roots obtained by DICM. The samples were loaded on biaxial testing machine LabControl 100kN/1000Nm. Behaviour of specimens with three different types of notch was investigated: U-notch, V-notch and "fillet". Schemes of these notches are shown on Fig. 1. Specimens with these types of notch were exposed to cyclic loading in five modes: Tension-compression (path 1), torsion (path 2), tension-compression/torsion proportional (path 3), tension-compression/torsion non-proportional 90° out of phase loading (path 4) and 45° out of phase loading (path 5).

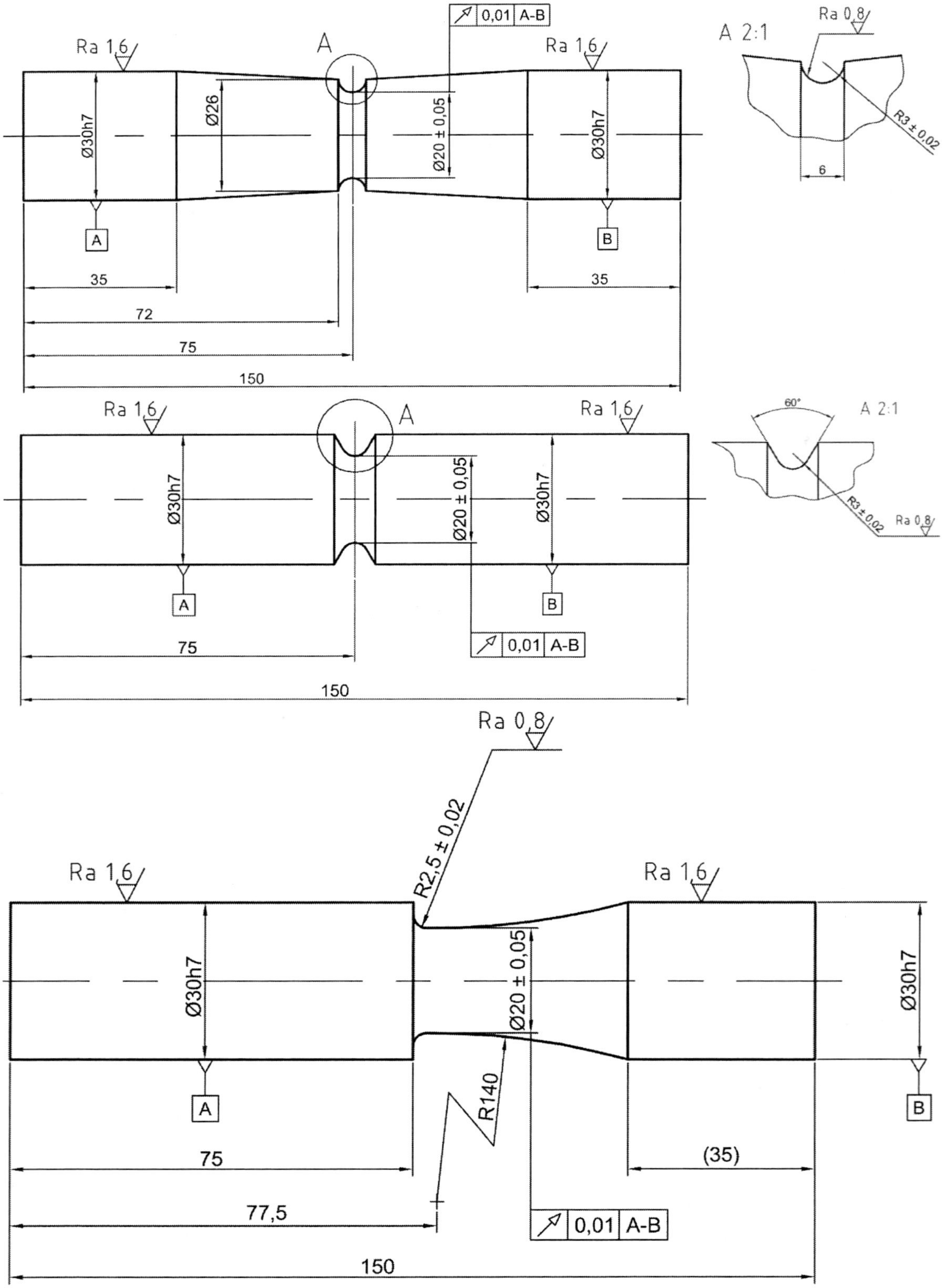


Fig. 1. Three types of notched specimens used for testing.

## Experimental Setup

Specimens were made from the aluminium alloy 2124-T851. The deformations in notch root cannot be investigated by strain gauges because of strain gradient, therefore the DIC technology was applied. Mercury RT® system (2x2.3Mpx@40Hz) provided by Sobriety company was used (Fig. 2b). The optical contrast coating was created on each specimen (generally known as pattern).

The samples were loaded up to failure on the universal testing machine LabControl 100kN/1000Nm (Fig. 2a.). The testing machine is able to exert uniaxial (tension-compression) and biaxial (tension-compression-torsion with and without phase shift) state of stress inside the specimen. Fatigue tests were done in low-cycle fatigue mode and all of them were done under load control. Overall 15 tests and DIC measurements were carried out, 5 tests for each specimen type under 5 different load setting.

a) b)

Fig. 2. Photographs of used equipment: a) Biaxial testing machine LabControl 100kN/1000Nm, b) Mercury RT® system.

Frequency of all five loading cases was 0.5 Hz. Deformation responses on cyclic load were investigated on all specimens. Main attention was concentrated on determination of longitudinal, transverse and shear strains in root of a notch. With the aid of DIC apparatus first hundred cycles of test was recorded for afterwards processing.

Processing of measured data was done in software Mercury RT® and the basic GUI layout can be seen on Fig. 3.

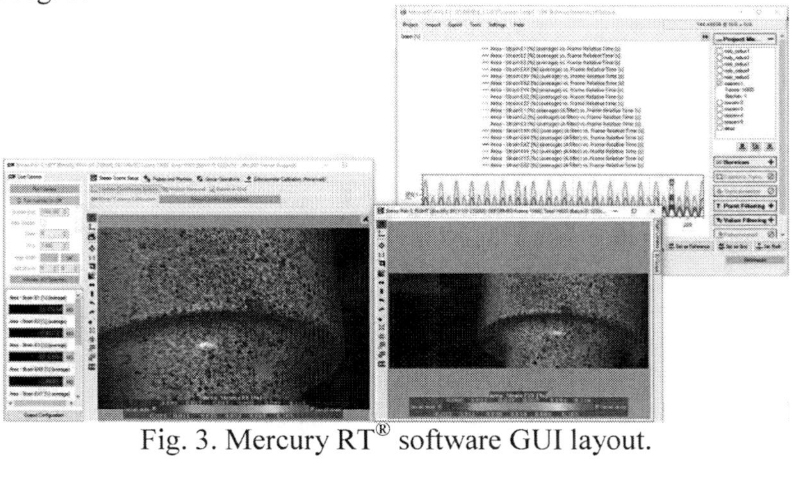

Fig. 3. Mercury RT® software GUI layout.

Software mentioned above enables, among other things, to configure the measuring cameras and to do their calibration. In the case of periodical loading the system is able to record and evaluate deformation response up to frequency of 40 Hz in full resolution of cameras. Into the area equipped by contrast coating an optical probe is subsequently placed (virtually in Mercury RT® software), which has to be aligned in both images (from two cameras). The testing machine was started together with the DIC measuring system. The data acquisition was done in real time and the calculation of desired outputs takes place in post-processing procedure.

### Theory of Digital Image Correlation Method

Digital image correlation method is modern non-contact technique for measuring contours as well as measuring displacements or deformations on surfaces of solid bodies. For 3D measuring two optical sensors are desirable (two high resolution cameras) in the same way as live organisms have two eyes. For 2D application one CCD camera is sufficient. DIC methodology is based on digital images processing in grayscale mode so the RGB format is not required. The DIC method is widely used for determination of Young modulus, Poisson's ratio or for capturing material elasto-plastic behaviour. Motion or displacement are derived from two images taken before and after the deformation occurs by minimizing the grayscale difference on these two images in a small area called subset. The correlation algorithm is based on minimizing the sum [4]

$$C = \sum[G_t(x_t, y_t) - G(x_t, y_t)]^2, \tag{1}$$

$$G_t(x_t, y_t) = g_0 + g_1\, G(x_t, y_t), \tag{2}$$

$$x_1 = a_0 + a_1\, x + a_2\, y + a_3\, xy, \tag{3}$$

$$y_1 = a_4 + a_5\, x + a_6\, y + a_7\, xy, \tag{4}$$

where $g_0, g_1$ are parameters of illumination, $(a_0, \ldots, a_7)$ are parameters of affine transformation and $(x, y)$ are coordinates of the point inside a subset.

### Results

All specimens were subjected to low-cycle fatigue testing in five modes and each test was force-controlled. Values of applied force or torque were recorded by testing machine. From these records it is possible to depict five loading paths (according to Fig. 4), where on the horizontal axis are values of axial force and on vertical axis are values of torque.

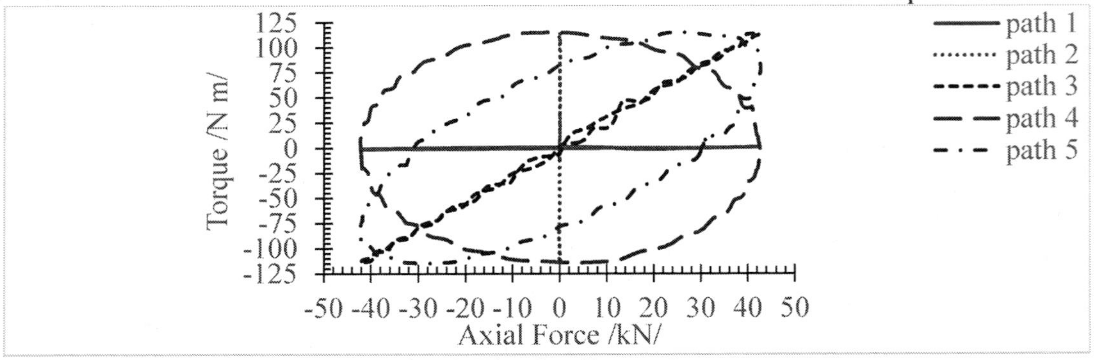


Fig. 4. Five loading paths.

In testing mode according to path 1 the specimen is subjected only to axial tension/compression load. Path 2 depicts loading by torque. Path 3 is combination of axial force and torque with no phase shift between them. Paths 4 and 5 are also combination with phase shift 45° and 90°. One cycle took two seconds for all testing modes. Our attention was aimed particularly on determination of longitudinal and shear strain in the tangential plane of the notch (Fig. 5a, Fig. 6a and Fig. 7a). Strain responses (near the hundredth cycle) are shown: For notch U in Fig. 5b, for notch V in Fig. 6b and for notch "fillet" in Fig. 7b.

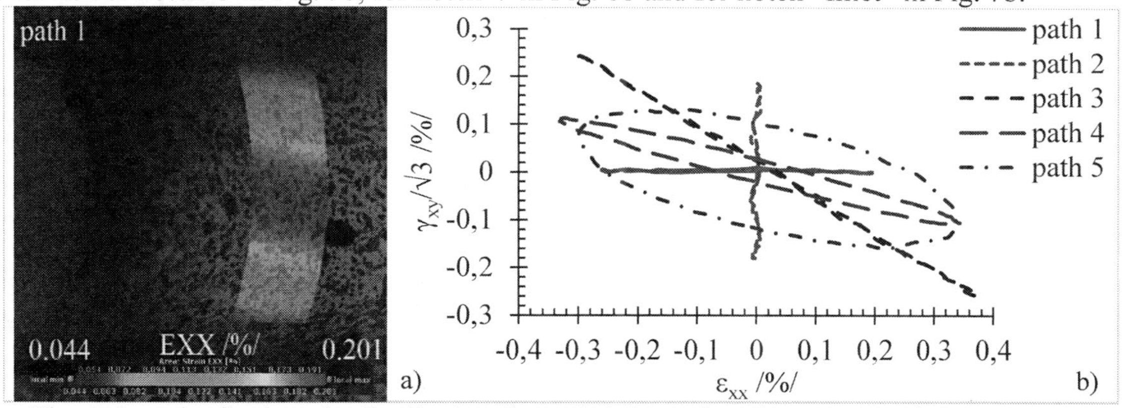


Fig. 5. Longitudinal strain distribution for path 1 (a) and strain responses in U-notch (b).

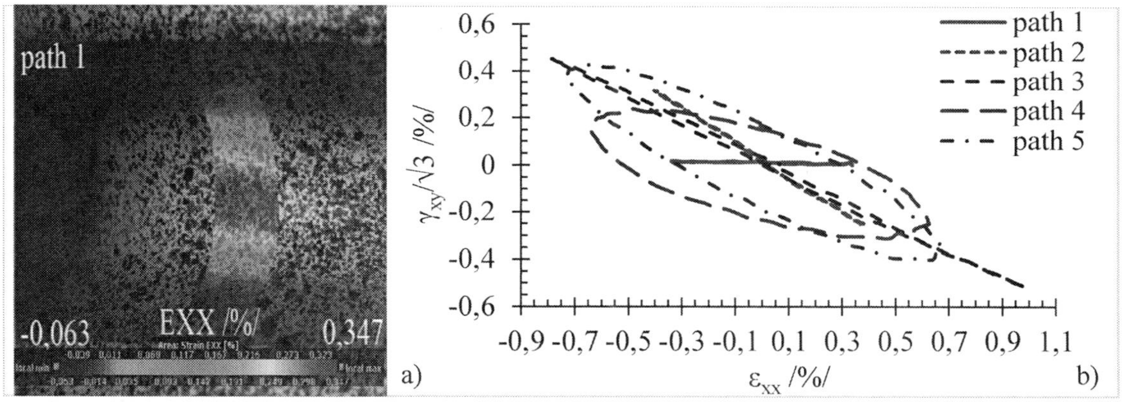


Fig. 6. Longitudinal strain distribution for path 1 (a) and strain responses in V-notch (b).

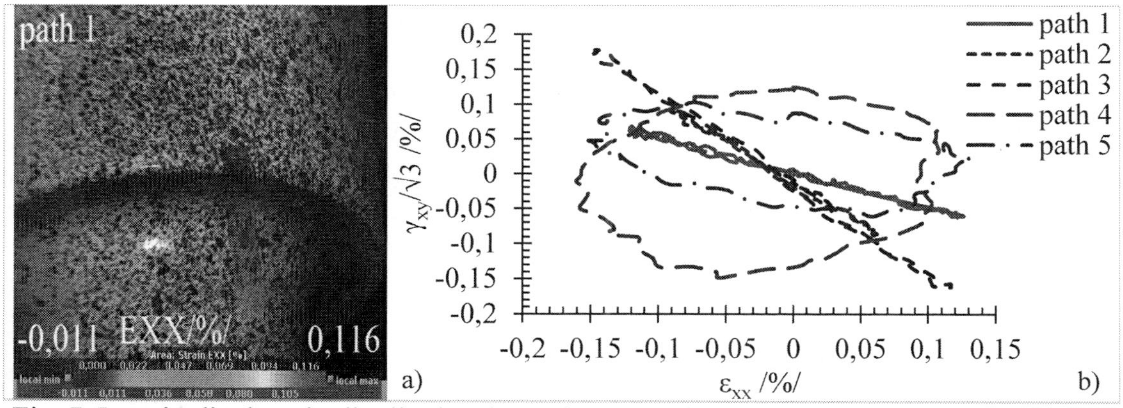


Fig. 7. Longitudinal strain distribution for path 1 (a) and strain responses in fillet-notch (b).

Now the question is how to compare the behaviour of different samples under the loading? The simplest way is to determine an equivalent strain amplitude according to formula

$$\varepsilon_a^{eq} = max\left[\sqrt{\varepsilon_{x,i}^2 + \left(\frac{\gamma_{xy,i}}{\sqrt{3}}\right)^2}\right]. \quad (5)$$

In more precise approach the value of $\varepsilon_a^{eq}$ is calculated as the radius of envelope of each strain path. The calculated equivalent strain amplitudes for each geometry of specimen are pictured on Table 1 and visualised on Fig. 8.

Table 1. Values of equivalent strain amplitude for each geometry.

| Specimen Geometry | $\varepsilon_a^{eq}$ [%] | | | | |
|---|---|---|---|---|---|
| | Path 1 | Path 2 | Path 3 | Path 4 | Path 5 |
| U-notch | 0.260734 | 0.186903 | 0.453987 | 0.359194 | 0.358940 |
| V-notch | 0.338717 | 0.509316 | 1.113471 | 0.674877 | 0.826682 |
| Fillet-notch | 0.139909 | 0.087370 | 0.177957 | 0.125172 | 0.089847 |

The highest values of equivalent strain occur in proportional combination loading. This behaviour is generally known. On the other hand the smallest values are presented in case of loading according to path 2 (torsional loading).

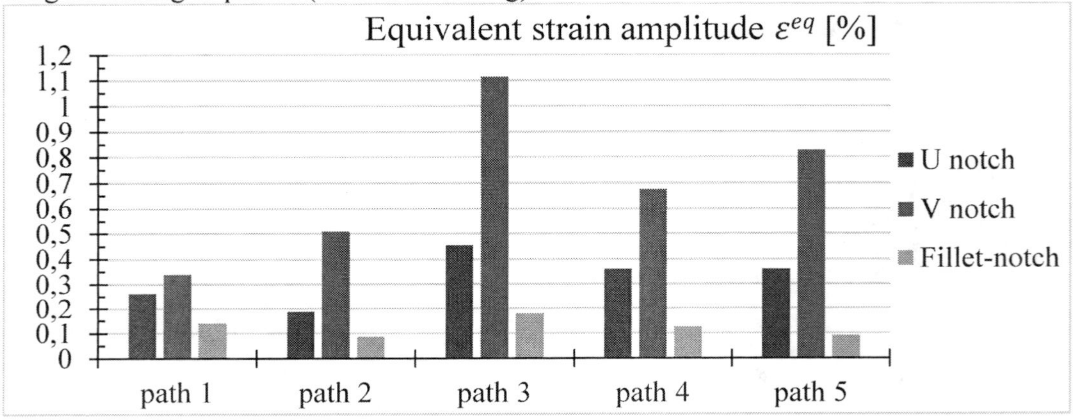


Fig. 8. Values of von Mises equivalent strain.

The second question is, if the measured values by DICM reflected reality. In order to prove that, some tests were performed on notch-free specimen and the deformations were measured by extensometer and by DICM. An acceptable correspondence was observed.

**Conclusions**

Objective of performed low-cycle fatigue tests was to determine complete strain tensor history in the root of notches by DICM as well as to investigate lifetime. Three types of specimen geometry were investigated: Specimen with notch U, V and fillet. Specimens were made from Aluminium Alloy 2124-T851. An application of strain gauges or extensometers in notch roots is problematic. The DIC methodology is good alternative here. DICM is contact less method and can stand to even large strains. Some basic characteristics of this modern optical method are stated above including a brief mathematical description. Specimens were subjected to the same five cyclic loading paths (Fig. 4), thus lifetimes can also be compared. The loading cases were without the mean value of stress, so there was no influence of mean stress on fatigue life.

Obtained data will be used for notch effect investigation in low cycle fatigue, for FEA verification and for evaluation of various Neuber-like methods.

## Acknowledgement

This work was supported by The Ministry of Education, Youth and Sports from the National Programme of Sustainability (NPU II) project „IT4Innovations excellence in science - LQ1602“, by the Czech Science Foundation (GA15-18274S) and by the Specific Research (SP2017/136). The support is gratefully acknowledged [5].


## References


[1] SUTTON, Michael Albert, Jean-José. ORTEU and Hubert W. SCHREIER. *Image correlation for shape, motion and deformation measurements: basic concepts, theory and applications.* New York, N.Y.: Springer, c2009. ISBN 978-0-387-78746-6.

[2] TREBUŇA, František, Róbert HUŇADY and Martin HAGARA. *Experimentálne metódy mechaniky - Digitálna obrazová korelácia.* Košice: Technická univerzita v Košiciach, 2016. ISBN 978-80-5532-346-6.

[3] HALAMA, R., M. FUSEK, F. FOJTÍK and J. BRUMEK, *Strain Measurement in Ratcheting Tests and Prediction by FEM,* in proc.: EAN (2009), Technical University of Liberec. ISBN 978-807372483-2.

[4] Dantec Dynamics: Measurement Principles of (DIC). *Dantec Dynamics A/S* [online]. Skovlunde: Dantec Dynamics A/S, c2017 [cit. 2017-04-06]. Dostupné z: https://www.dantecdynamics.com/measurement-principles-of-dic

[5] HALAMA, Radim, Martin FUSEK and Zdeněk PORUBA. Influence of mean stress and stress amplitude on uniaxial and biaxial ratcheting of ST52 steel and its prediction by the AbdelKarim–Ohno model. *International Journal of Fatigue.* 2016, **91**, 313-321. DOI: 10.1016/j.ijfatigue.2016.04.033. ISSN 01421123.

# Influence of the Body Wheel Shape on the Tooth Stiffness

Silvia Medvecká-Beňová[1, a], František Trebuňa[1,b], Peter Frankovský[1,c] and Daniel Kottfer[1,d]

[1]Faculty of Mechanical Engineering, Technical University of Košice, Košice, Slovakia

[a]silvia.medvecka@tuke.sk, [b]frantisek.trebuna@tuke.sk, [c]peter.frankovsky@tuke.sk, [d]daniel.kottfer@tuke.sk

**Keywords:** Spur gear, tooth deformation, tooth stiffness, finite element method.

## Introduction

At present it is still higher demands on machine parameters. This is manifested by the increasing load and increasing the speed of individual parts of machine [1, 3]. All while of increasing requirements for performance, accuracy, lifetime reliability of the machinery itself. It is necessary to consider the action of internal and external load and the dynamic processes in machines. Increasing performance and improving load machines with gear unit leads to increase the technical level of machines. This process is often at the cost of degradation of the environment. One of the factors that aggravate environmental is a noise. Periodic changes the stiffness of the tooth during meshing in gear drives mainly affects of the noise in the transfer.

The work is devoted to the analysis of influence of the body wheel shape on the tooth stiffness. The problem is solved for spur gears. As the basis for calculating the tooth stiffness are results of teeth deformation. The teeth deformation has be solution problem by finite element method.

## Tooth Stiffness of spur gears

Deformation of teeth is usually expressed quantitatively as a teeth stiffness during by gear mesh [4]. In general, stiffness defined as the ratio of load to deformation. The stiffness of the tooth is defined as the force per unit width, which is necessary for the deformation about 1μm [5, 9]. During the mesh cycle, the stiffness of the teeth is changes, it is mainly due to the change arm bend. Basically the mesh stiffness of a given gear pair depends on several parameters, particulartly the load and the rotational position of the gears. Provided we assume linear-elastic material properties, which is reasonable for metal gears, the contact stiffness varies with the load (as the contact zone gets larger and thus stiffer with higher contact forces) where as the other portions, like tooth bending a.o. are constant. But the stiffness definitely depends on the point of force application on the tooth flank and on the contact ratio (i.e. how many teeth are sharing the load at a given mesh position) and thus on the angular position of the gears. These variations are periodic.

As has been said, the teeth of the gear wheels are deformed due to load. This is the cause of some negative but also positive consequences. Therefore, the knowledge of the deformation properties of the teeth is very important [6, 8]. The theoretical determination of the teeth deformation is difficult due to the complex shape of the gear teeth. In recent years, the question of teeth deformation has been solved using modern methods of calculation, for example by a finite element method. The finite element method is used to determine the teeth deformation. We will focus on the value of the total deformation in the direction of action forces (Fig. 3). On the basis of the teeth deformation, the stiffness of the teeth is calculated. In general the teeth stiffness c is defined by equation (1):

$$c = \frac{w}{\delta} \tag{1}$$

where c – tooth stiffness [N/mm.μm],
w – load across the width of the teeth [N/mm],
$\delta$ – tooth deformation [μm].

The teeth stiffness values are the basis for assessing the suitability of design of the wheels in this paper.

## Construction of large-dimensional gear wheels

Achieving the smallest weight of the spur gear wheel is the designer's effort in designing the shape of a large gear wheel body. However, the reduction of the load-bearing cross-sections is usually limited by the requirement for sufficient stiffness of the bodies. In the case of gears it is a gear mesh stiffness.

Large-dimensional gear wheels can be made as monolithic one, the most commonly are a forged and cast gear wheels. Spur gears with a diameter of 200 to 500mm are produce of forged blanks most commonly. Larger spur gears with a diameter of over 500mm may be produce as a cast blanks. But striving to improve the design of the gears, aimed at reducing the dimensions and weight, has be led to the used of gears with thin-walled gear rim and with thin gear rim. Fig.1 shows an example of design a forged and cast body of spur gear wheel.

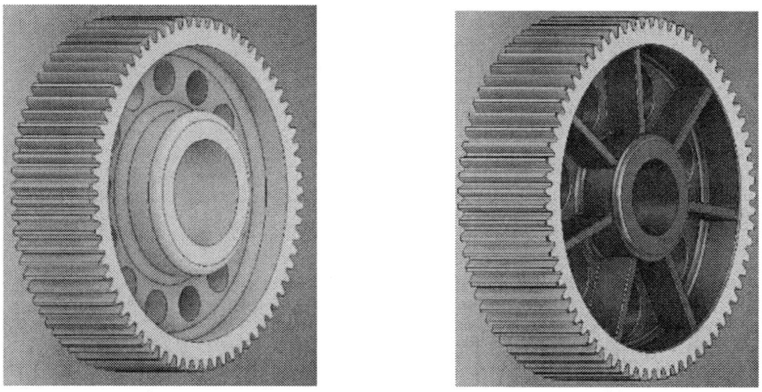

a) b)

Fig. 1 Design of body gear wheel of a) forged and b) cast blanks of the spur gears

The wheel rim thickness ($s_R$) and wheel web thickness ($b_S$) are important parameters which influence of weight of gear wheel. In the next section, the paper will focus on the influence of wheel rim and web parameters on the tooth stiffness of spur gears.

## Influence of the wheel rim thickness on the tooth stiffness

The impact of the rim thickness (value $s_R$ in Fig.2) on the deformation and stiffness of the tooth will be determined on the spur gear with a number of teeth of z=61 and a module $m_n$=4mm and a tooth width b=80mm.

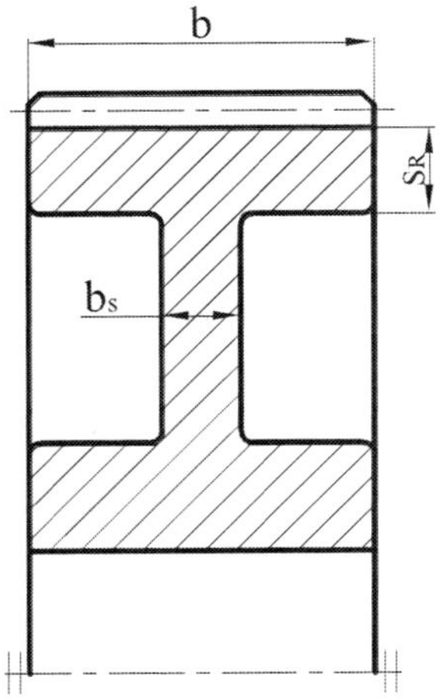


Fig. 2 Design of gear wheel

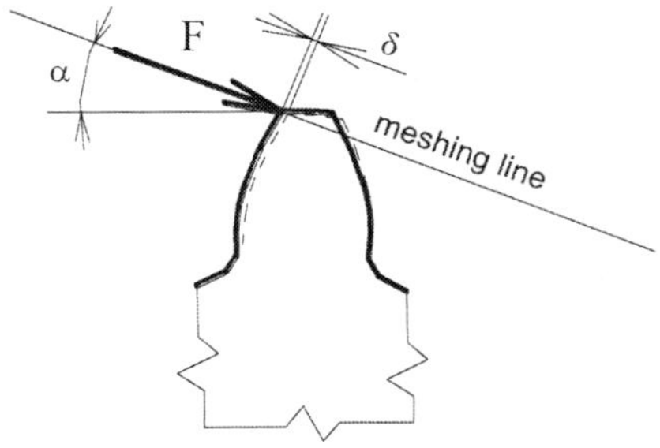


Fig. 3 Deformation of tooth

Will be examined the tooth stiffness under load F =5000 N, if the if the force is applied to the head of the tooth (the largest bend) according to Fig.3. The thickness of the wheel rim is changed from the value $s_R$=1.75$m_n$ (2.5$m_n$, 3.5$m_n$, 5$m_n$, 8$m_n$ and full wheel) to the full body of the wheel. These results are determined by the finite element method and are processed in the graph in Fig.4.

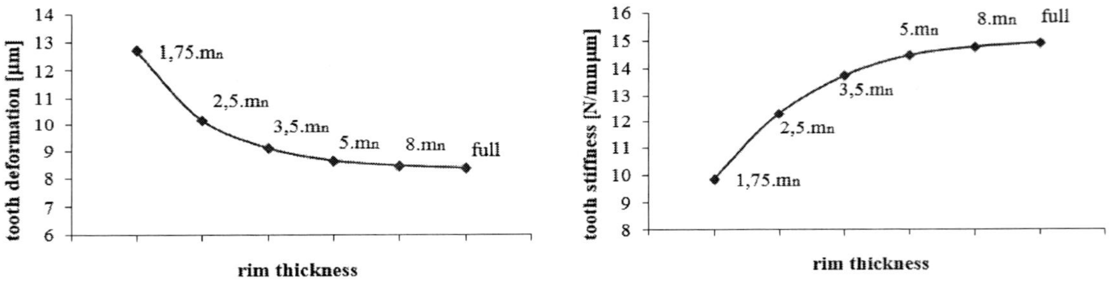


Fig. 4 Influence of wheel rim on the tooth deformation and stiffness, if z=61, $m_n$=4mm, b=80mm, F =5000 N

As the results show, decreasing the thickness of the rim wheel is increases the tooth deformation and the tooth stiffness is decreases. The minimum permissible thickness of the wheel rim according to [10] is a $s_R$ = 3.5$m_n$, where $m_n$ is a modul of the gear wheel. Thickness of wheel rim less than value 3.5 $m_n$ has more affects to the tooth deformation and stiffness. Thickness of wheel rim bigger than value 3.5 $m_n$ has smaller affects to the teeth deformation and stiffness, as show in the Fig.4.

**Influence of the wheel web thickness on the tooth stiffness.**

The impact of the rub thickness (value $b_S$ - Fig.2) on the tooth stiffness will be determined on the spur gear with a number of teeth of z=61 and a module $m_n$=4mm and a tooth width b=80mm, the value of wheel rim is $s_R$=22mm.The web is located at the center of the gear wheel width and its thickness will change from 10 mm to the 80 mm, when it is a full gear wheel without a web. The force is applied to the head of the tooth (the largest bend) according to Fig.3 and the value of load across the width of the tooth is a w=40N/mm.

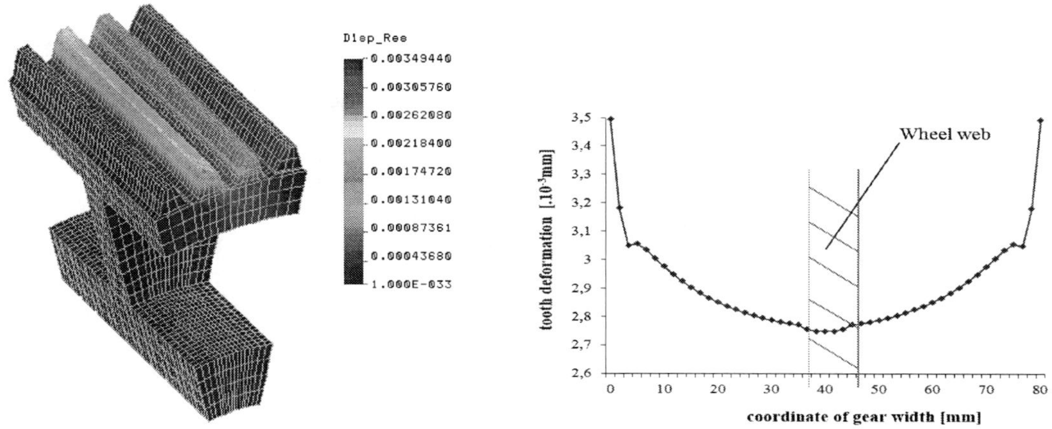


Fig. 5 Example of the tooth deformation on the spur gear with $z$=61, $m_n$=4mm, $b$=80mm, $w$=40N/mm, $b_S$=10mm solution by FEM

The results of the deformation distribution along the gear wheel width and solution by finite element method are shown in Fig. 5 and for the above mentioned gear wheel model. In this figure is the local influence of the tooth deformation with is distribution on the gear width. The lesser value on tooth deformation is in the locality with wheel web.

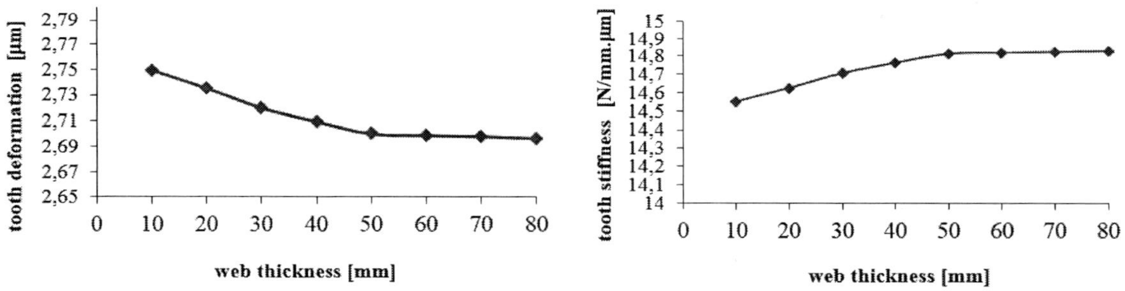


Fig. 6 Influence of wheel web on the tooth deformation and tooth stiffness, if $z$=61, $m_n$=4mm, $b$=80mm, $s_R$=22mm, $w$=40N/mm

The tooth deformation is reduced and the tooth stiffness is expanded, due to the increase in the thickness of the wheel web located at the center of the gearing width, it is show in Fig. 6. This change in tooth deformation and tooth stiffness is more pronounced to the first half of gearing width.

**Influence of the wheel web localization on the tooth stiffness**

The influence of the wheel web (value $b_S$ - Fig.2) localization on the tooth stiffness will be determined on the spur gear with a number of teeth of $z$=61 and a module $m_n$=4mm and a tooth width $b$=80mm, the value of wheel rim is $s_R$=22mm.The force is applied to the head of the tooth (the largest bend) according to Fig.3 and the value of load across the width of the teeth is a $w$=40N/mm. The wheel rub thickness is $b_S$=10mm and its locality varies according to Fig.7.

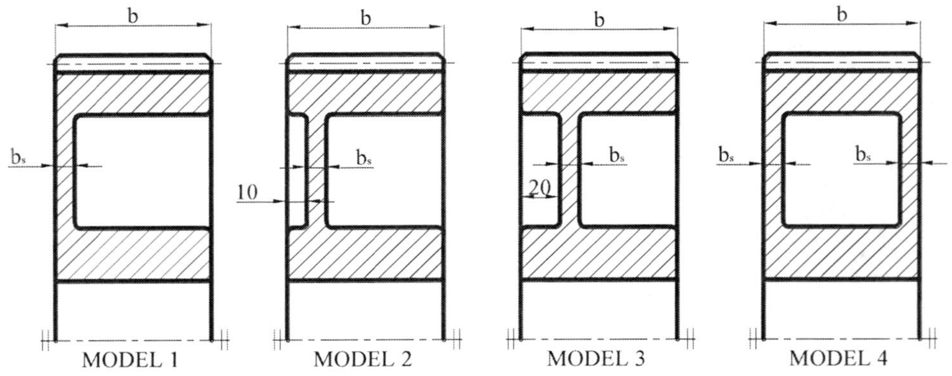


Fig. 7The locality of wheel web

Example of the tooth deformation on the spur gear for spur gear with wheel web localization as model 2 (Fig. 7), solution by finite element method is show in Fig.8. The results of the tooth deformation are solution by along the gear wheel width.

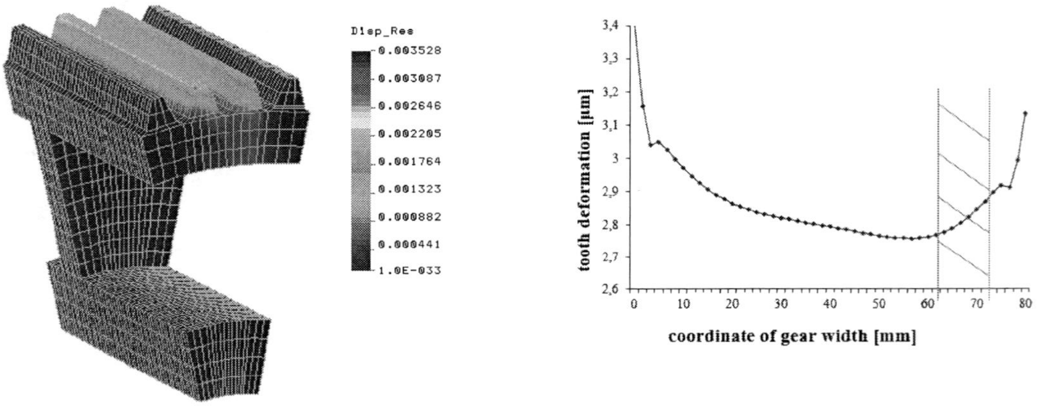


Fig. 8 Example of the tooth deformation on the spur gear with z=61, $m_n$=4mm, b=80mm, w=40N/mm, $b_S$=10mm solution by FEM, model 2

The results of the maximum tooth deformation examined at the loading point, if the force acts on the tip of the tooth (Fig. 3), for wheels web with different station locations, according to the models 01 to 04 (Fig.7), are shown in the Fig.9. The value of the tooth deformation in the place of loading we consider the second highest value.

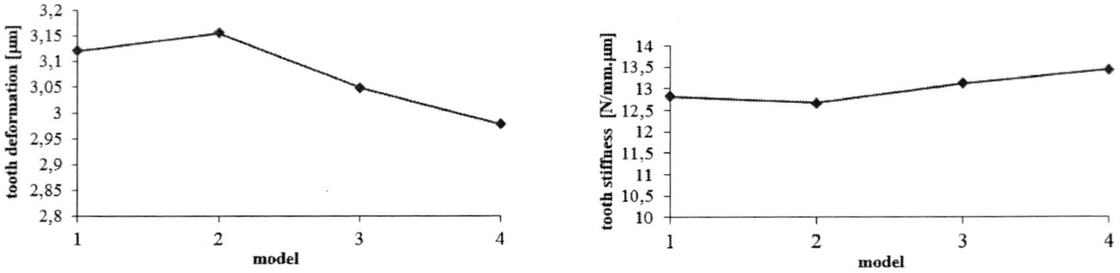


Fig. 9 Influence of wheel web localization on the tooth deformation and stiffness, if z=61, $m_n$=4mm, b=80mm, $s_R$=22mm, w=40N/mm, for model 1, 2, 3 and 4

The minimum tooth deformation, the maximum tooth stiffness, is for the gear whit two wheel web. These wheel webs are located at the edges of the wheel width. This is due to the supportive effect of wheel web. The wheel web placement in generally is affects the teeth deformation as well as the teeth stiffness over the width of the spur gear.

## Conclusions

Development of the modern machinery and means of production is characterized by steadily increasing performance factors in decreasing the weight of the device. It is a feature of the design for choice of the shape of the body larger gear. Lightening of body wheel gears has affects on the deformation and stiffness of the teeth. Teeth stiffness is decreases with the decrease in thickness of the wheel rim. Teeth stiffness is not constant even after the width of tooth. If the ends of mesh contact line identic of the edge of the teeth, the teeth stiffness is less on this locality. The teeth deformation is reduced and the teeth stiffness is expanded, due to the increase in the thickness of the wheel web located at the center of the gearing width. The localization of wheel web has a influence of the teeth stiffness.

**Acknowledgement**

The work has been supported by the VEGA 1/0872/16 and VEGA 1/0473/17.

## References

[1] M. Batsch, Surface strength of novikov covexo-concave gear, Scientific Journal of Silesian University of Technology. SeriesTransport. 90 (2016) 17-24.

[2] CH. G. Cooley, Ch. Liu, X. Dai, R. G. Parker, Gear tooth mesh stiffness: A comparison ofcalculation approaches, Mechanism and Machine Theory. 105 (2016) 540-553.

[3] P. Czech, Diagnosis of Industrial Gearboxes Condition By Vibration and Time-Frequency, Scale-Frequency, Frequency-Frequency Analysis. Metalurgija, Volume: 51, 4 (2012) 521-524.

[4] Z. Chen, Y. Shao, Mesh stiffness calculation of a spur gear pair with tooth profilemodification and tooth root crack, Mechanism and Machine Theory. 62 (2013) 63-74.

[5] A. Femandez del Rincon, F. Viadero, M. Iglesias, P. García, A. de-Juan, R. Sancibrian, A model for the study of meshing stiffness in spur gear transmissions, Mechanism and Machine Theory. 61 (2013), 30-58.

[6] P. HenrykMadej, P.Czech, Discrete Wavelet Transform and Probabilistic Neural Network in IC Engine Fault Diagnosis. Eksploatacja I niezawodnosc - maintenance and reliability. 4, (2010) 47-54.

[7] P.Kuryło, Rozkład naprężeń w warstwie wierzchniej napawanych odlewów żeliwnych Proizvodstvo, Tehnologiâ, Ekologiâ, 10 (2007), 573-584.

[8] S. Medvecká-Beňová, Influence of the face width and length of contact on teeth deformation and stiffness, Scientific Journal of Silesian University of Technology: Series Transport. Vol. 91 (2016), 99-106.

[9] S. Noga, T. Markowski, R. Bogacz, Method of determining thenormal modes of toothed gears with complex geometry, Scientific Journal of Silesian University of Technology. Series Transport. 89 (2015) 119-127.

[10] Standard STN 01 4686.

EAN 2017
55th conference on
experimental stress analysis

# Diagnostics brushless DC motors

Ivan Abramov[1, a], Pavol Božek[2,b] Andrej Abramov [1,c], Ella Sosnovich[1,d] and Yuri Nikitin[1,e]

[1]The "Mechatronic Systems" Department, Kalashnikov Izhevsk State Technical University, Izhevsk, Russian Federation

[2]Full Slovak University of Technology, Faculty of Materials Science and Technology, Institute of Production Technologies, J. Bottu 25, 917 24 Trnava, Slovakia

[a]ms@istu.ru, [b]pavol.bozek@stuba.sk, [d]ms@istu.ru,

[c]ellasosnovich@istu.ru, [e]nikitin@istu.ru

**Keywords:** model, brushless DC motor, control, diagnostics, fuzzy logic

**Abstract.** This paper presents the BLDC motors - one of the types of electric motors, which rapidly gaining popularity due to its good performance and the development of microprocessor-based control. New PWM switching strategy to minimize the torque ripples in BLDC motor which is based on sensored rotor position control is discussed in the paper. Analyzed defects and diagnostic parameters of BLDC motors. A mathematical model for diagnosing BLDC motors and its implementation in MatLab software product based on fuzzy logic are presented. Simulation results of BLDC motors are presented.

## Introduction

BLDC motors - one of the types of electric motors, which rapidly gaining popularity due to its good performance and the development of microprocessor-based control. New PWM switching strategy to minimize the torque ripples in BLDC motor which is based on sensored rotor position control is discussed in the paper [1]. Tuning methodology for the parameters of adaptive current and speed controllers in a permanent-magnet BLDC motor drive system is presented in paper [2]. Two Fault Detection and Diagnosis strategies for detecting Brushless DC Motor faults were considered involving wavelets and state estimation [3]. Bearing faults and stator winding faults, which are responsible for the majority of motor failures, are considered [3]. A novel method using windowed Fourier ridges is proposed in paper [4] for the detection of rotor faults in BLDC motors operating under continuous non-stationarity. The use of quadratic TFRs is presented as a solution for the diagnostics of rotor faults in brushless DC (BLDC) motors operating under constantly changing load and speed conditions [5]. Four time-frequency representations are considered short-time Fourier transform (STFT), Wigner-Ville distribution (WVD), Choi-Williams distribution (CWD), and the Zhao-Atlas marks distribution (ZAM) [5]. Three new algorithms for the detection BLDC motors faults are proposed that can track and detect rotor faults in non-stationary or transient current signals [6]. Park's vector method was used to extract the features and to isolate the BLDC motors faults from the current measured by sensors [7]. Proposed a model of a fault diagnosis expert system with high reliability to compare identical well-functioning BLDC motors [8].

The authors propose the integration of control and diagnostic systems. The economic efficiency of diagnostic systems due to an increase in reliability and quality, reduction of accidents, reduced rejects, decrease downtime of expensive equipment, reducing costs for maintenance and repair, increase service life. Currently, artificial intelligence technologies are widely used for control and diagnostics of electric motors and drivers [9-13].

## Theoretical part

*Simulation model of Brushless DC motors*

To investigate the **BLDC motor** control proposed a simulation model in a natural (phase) coordinate system in the software **complex "Modeling In Technical Devices"** (MITD) developed in Bauman Moscow State Technical University [10], shown in Fig. 1. MITD is an alternative software products such as SIMULINK, VisSim, MATRIXx etc

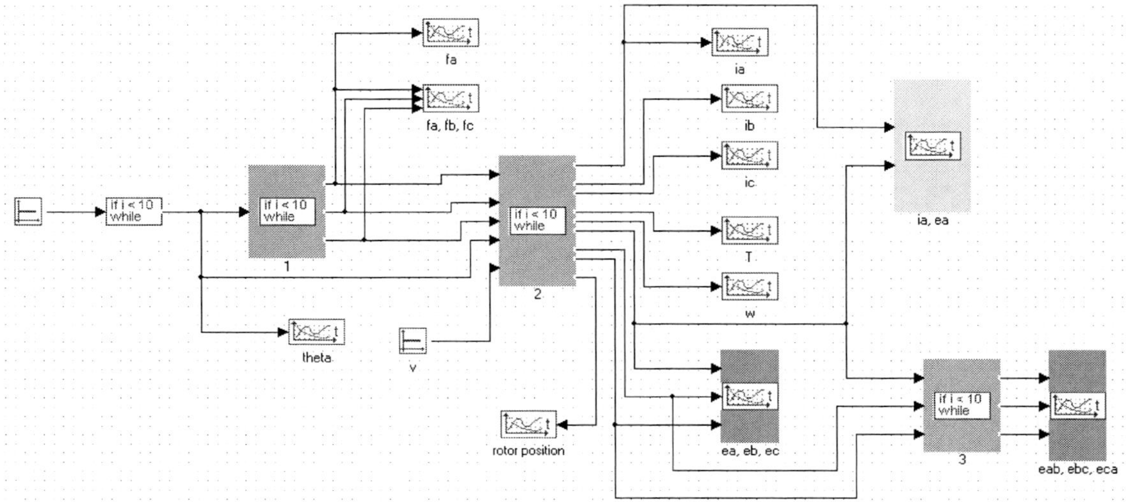


Fig. 1. Simulation model of BDSM

Input data of the simulation model BLDC motor are angle and rotor speed. Relation electrical and mechanical parameters of the angle are obtained. The resulting functions are shown in Fig. 2.

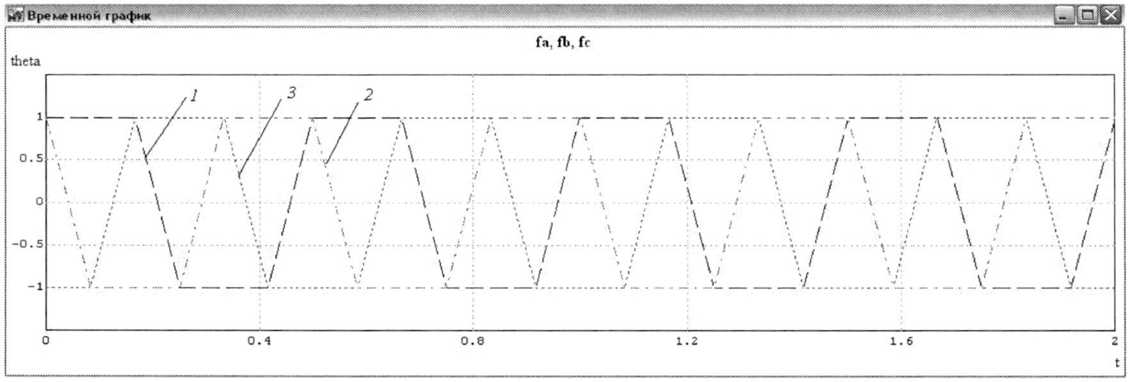


Fig. 2. Dependence functions $f_a(\theta)$, $f_b(\theta)$, $f_c(\theta)$ of the angle of rotor rotation, where 1 - function $f_a(\theta)$, 2 - function $f_b(\theta)$, 3 – function $f_c(\theta)$

Dependence the phase a current ia (on the graph function 1) and electromotive force ea (on the graph function 2) of the rotor rotation angle is shown in Fig. 3.

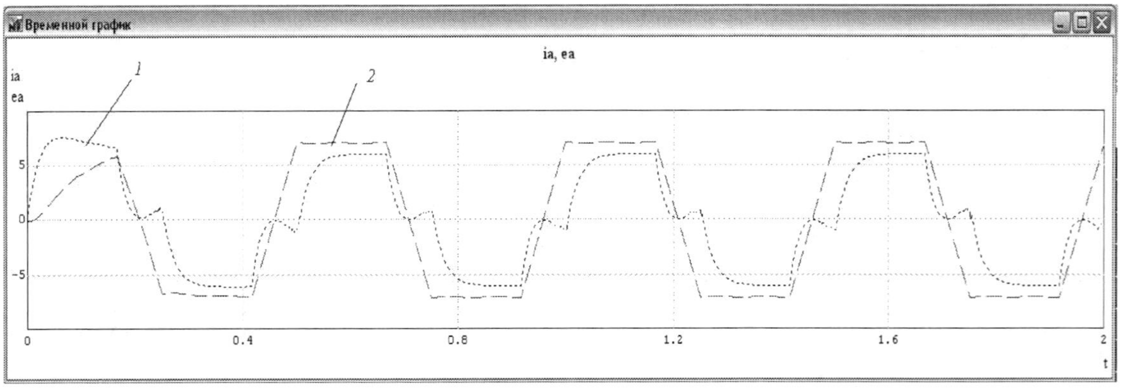


Fig. 3. Dependence the phase *a* current $i_a$ (on the graph function 1) and electromotive force $e_a$ (on the graph function 2) of the rotor rotation angle

Organization of the Text Simulation model allows you to control BLDC motor dependence of the phase currents and electromotive force for the perfect BLDC motor that allows you to use these results as a diagnostic parameter [11,12,13]. To improve the reliability of diagnosis is necessary to analyze defects and other diagnostic parameters of BLDC motor.

*BLDC motor defects and diagnostic parameters*

BLDC motor defects are divided into two classes: electrical and mechanical. BLDC motor defects are shown in Tab. 1.

Tab.1. BLDC motor defects

| Electrical defects: | Mechanical defects: |
|---|---|
| open conductors in the winding;<br>insulation between windings;<br>unacceptable reduction of insulation resistance due to its aging or excessive moisture;<br>disorders contacts and connections. | degradation processes in the bearings - the destruction of the separator, balls or rollers;<br>deterioration in heat transfer due to fouling and dusty coils;<br>deformation of the rotor shaft. |

Selected diagnostic parameters: current, vibration, temperature, which are shown in Tab. 2.

Tab. 2. BLDC motor diagnostic parameters

| Diagnostic parameters | Reason for change diagnostic parameter |
|---|---|
| Current | overload;<br>open or shorted winding;<br>changing the mains voltage. |
| Vibration | shaft misalignment;<br>bearing defects. |
| Temperature | overload;<br>winding circuit;<br>ambient temperature changes. |

The measurement results of the diagnostic parameters should be processed for a decision on the BLDC motor technical condition. Most appropriate mathematical tools to construct diagnostic model is fuzzy logic [14,15,16].

*Logical-linguistic diagnostic model of BLDC motor*

Logical-linguistic diagnostic model of BLDC motor based on fuzzy logic is represented by a system of equations:

$$\begin{cases} x(t) = F(x1(t), x2(t), x3(t)), \\ D(t) = G(x(t), t), \\ Z(t) = H(x(t), D(t), t), \end{cases} \quad (1)$$

where $x(t) = F(x1(t), x2(t), x3(t))$ – the equation of diagnostic parameters;

$x(t)$ – vector diagnostic parameters;

$x1(t), x2(t), x3(t)$ – a set of measurements of diagnostic parameters;

$D(t) = G(x(t), t)$ – calculation of trend vector equation diagnostic parameters;

$t$ – worn-out time;

$Z(t) = H(x(t), D(t), t)$ – evaluate the technical condition equation.

Logical-linguistic diagnostic model of BLDC motor is implemented by a software MatLab in Fuzzy Logic Toolbox. Fuzzy inference system technical condition assessment is implemented on a fuzzy knowledge base of Mamdani type with three input variables x, D, t. Scheme fuzzy inference system is shown in Fig. 4.

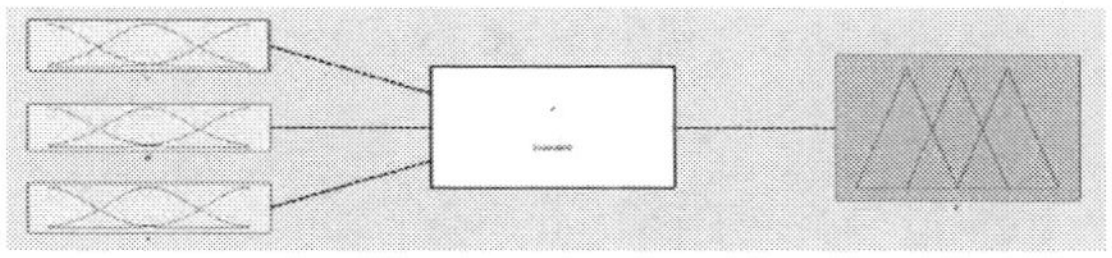

Fig. 4. Scheme fuzzy inference system technical condition assessment with three input variables x, D, t

As the term of membership functions of the linguistic variable M is selected Gaussian function, as it is quite simple, differentiable, given only 2 parameters, which reduces the computational complexity of the algorithm. As the membership functions of terms L, H linguistic variable chosen z, s functions.

Selected Mamdani fuzzy inference, as t-norm selected maximum deffuzification conducted by the method of the center of gravity, as it provides good accuracy and speed settings fuzzy knowledge base [17, 18, 19]. As adjustable parameters used weights rules coordinates highs membership functions M linguistic variable term. Example of terms L, M, H membership functions of input and output variables are shown in Fig. 5.

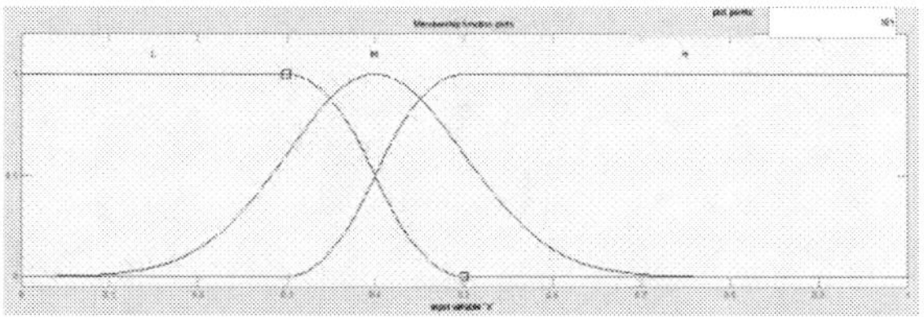

Fig. 5. Example of terms L, M, H membership functions of input and output variable

When using three linguistic variables with three terms by combining logical operations AND, OR, received 7 rules reflecting the dependence of the technical condition of the values of diagnostic parameters, the trend of diagnostic parameters and to develop the resource as shown below:

If (x is L) and (d is L) and (t is L) then (z is L)
If (x is M) and (d is L) and (t is L) then (z is M)
If (x is L) and (d isM) and (t is L) then (z is M)
If (x is L) and (d is L) and (t is M) then (z is M)
If (x is H) then (z is H)
If (d is H) then (z is H)
If (t is H) then (z is H)

Response surface fuzzy inference system evaluate the technical condition shown in Fig. 6. This drawing shows how the BLDC motor technical condition (output variable z) depending on the input variables x, D.

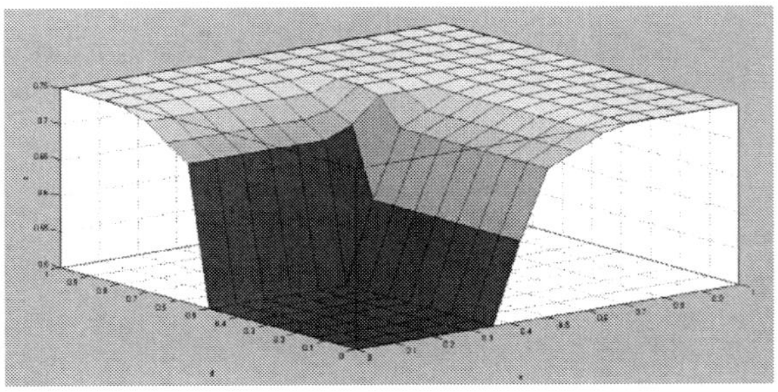

Fig. 6. Response surface fuzzy inference system technical condition assessment

Fig. 6 shows that for small values of integral diagnostic parameter x and low values of the diagnostic parameter trend the BLDC motor condition is serviceable. Examples of visualization of the rule base of fuzzy inference system for assessing the technical condition of the engine shown in Fig. 7, 8.

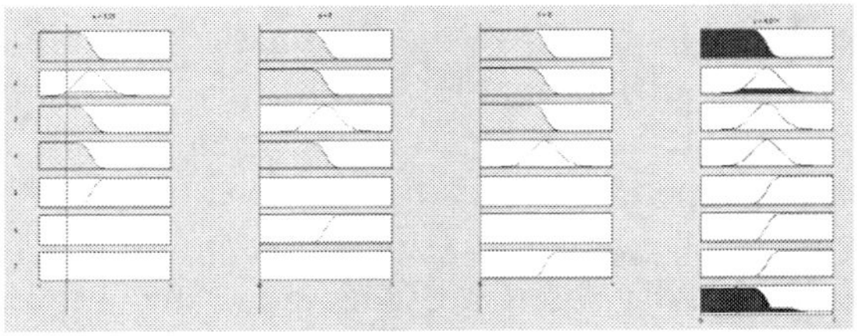

Fig. 7. Example visualization of the rule base of fuzzy inference system technical condition assessment BLDC motor (x = 0.21, D = 0, t = 0, Z = 0.271 - good technical condition of BLDC motor)

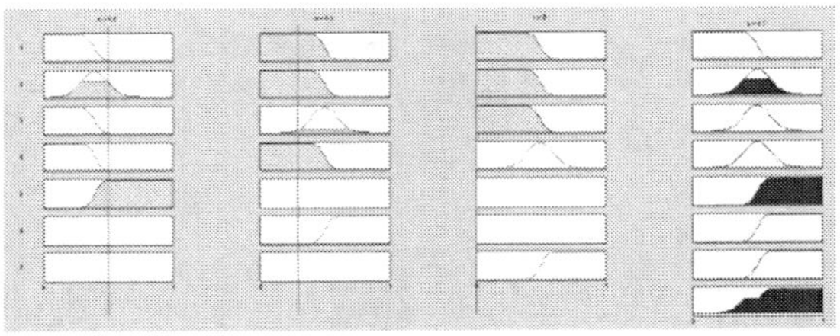

Fig. 8. Example visualization of the rule base of fuzzy inference system for assessing the technical condition of the engine (x = 0.5, D = 0.3, t = 0, Z = 0.7 - BLDC motor with defects)

*Simulation model of BLDC motor diagnostics system*

Model diagnostic system based on the mathematical apparatus of fuzzy logic, allows to determine the BLDC motor condition based diagnostic parameters (current, vibration, temperature). A simulation model of a BLDC motor diagnostics system designed by MITD. Fig. 9 shows the simulation model of diagnosis, determining BLDC motor condition.

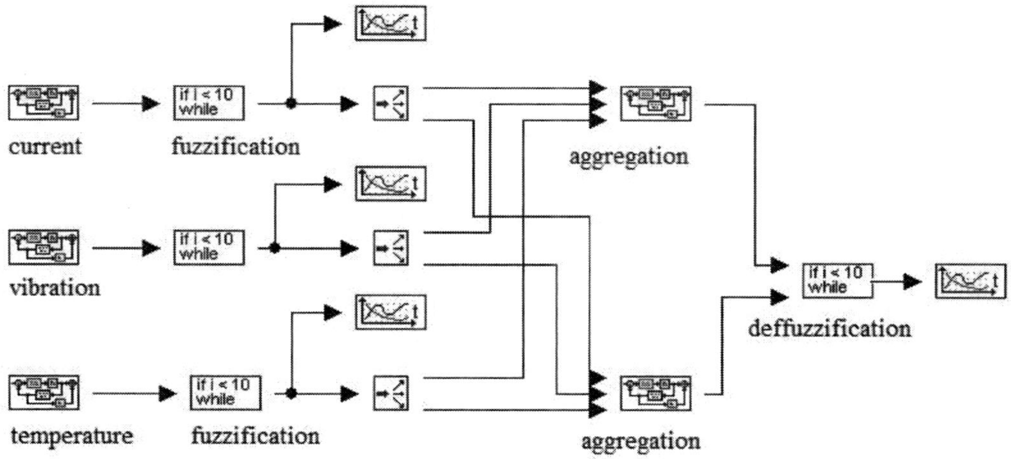


Fig. 9. Simulation model of BLDC motor diagnostic systems

## Experimental part

As an example, the simulation results show changes depending on the input signals (current, vibration, temperature) from time to time in relative units with the appearance of BLDC motor defects in Fig. 10. Fig. 11 shows an example of change in the output variable characterizing the degree of development of BLDC motor defects.

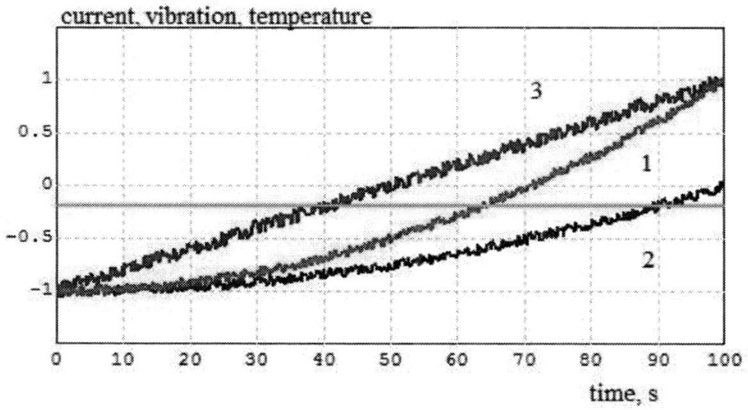


Fig. 10. Simulation results change depending on input signals (current, vibration, temperature) from time to time in relative units with the appearance of defects

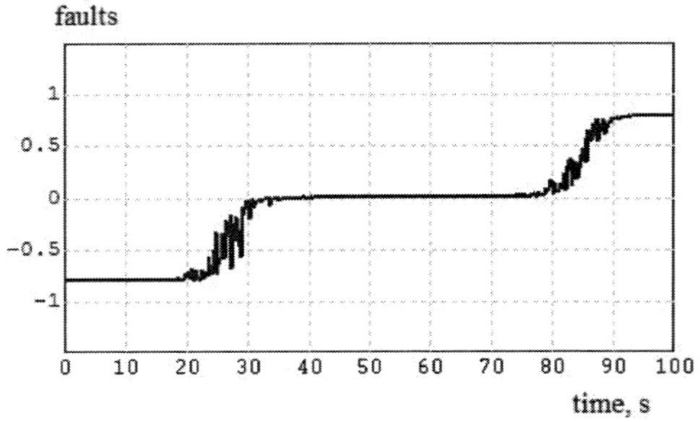


Figure 11 - Condition of the diagnosis depending on the diagnostic parameters in the process of BLDC motor degradation

**Discussion**

Upon receipt of the schedule can be concluded that the BLDC motor condition for range 0 ... 20 s. can be described as a workable development BLDC motor degradation processes occur in the interval 30 ... 80 s., BLDC motor alarm condition occurs in the interval 90 ... 100 s. By the nature of the graph can be judged on the dynamics of degradation processes - change of state of emergency comes to working-not once, but with the rise of increasing diagnostic parameters.

Thus, analyzing the diagnostic parameters, the trend and the operating time can foresee the emergence, to minimize the risk of accidents, timely schedule a maintenance. Experiments in which the diagnostic parameters were within the normal parameters and the output of the tolerance range. Experiments were carried out at the same supply voltage and load torque. A simulation model of the system control and diagnostics BLDC motor allows to investigate the influence of the diagnostic parameters on the performance.

**Conclusions**

In this paper we show that for the diagnosis BLDC motor can be effectively applied to the simulation model of the system diagnostic software developed by MITD, is based on the mathematical apparatus of fuzzy logic, which determines the BLDC motor condition based diagnostic parameters - current, vibration and temperature.

The BLDC motor simulation obtained depending phase currents, which can serve as the basis for selection of diagnostic parameters that are associated with BLDC motor physical models.

Analysis of diagnostic parameters was performed using the original software package that implements a fuzzy inference system MatLab. Feature of the system is the use of fuzzy inference integral parameter diagnostics combining information about diagnostic parameters and their derivatives.

Adapting this BLDC motor model accomplished by injecting a BLDC motor specific technical parameters.

Trend of change of the integral parameter allows to foresee the time to reach the limit of technical condition and through organizational and technical measures to minimize the risk of premature failure, plan maintenance and repair. With the development of BLDC motor degradation processes in the operation mechatronic system may change the control action on BLDC motor by reducing the current and move to more lenient modes of operation.

Computational experiments conducted confirmed the effectiveness of the proposed model of diagnosis and the prospect of its application in the diagnosis of mechatronic systems with real BLDC motor.


## Acknowledgements

The contribution is sponsored and elaborated within the running Project of VEGA MŠ SR No. 1/0367/15 titled: Research and development of a new system of the robot trajectory autonomous control and study was supported by Ministry of Education and Science of the Russian Federation, project "The development of intelligent control and diagnostics systems of mechatronic drives" under the State jobReferences


## References


[1] Wael, A. S. – Dahaman, I. – Khaleel, J. H.: PWM switching strategy for torque ripple minimization in BLDC motor, Journal of Electrical Engineering, Vol 62, 3 (2011), 141-146.

[2] Mohamed, A. A. – Ehab H. E. B. – Hisham M. S.: Adaptive deadbeat controllers for brushless DC drives using PSO and ANFIS techniques, Journal of Electrical Engineering, Vol 60, 1 (2009), 3-11.

[3] Zhang, W.: A Fault detection and diagnosis strategy for permanent magnet brushless DC motor (2013). Open Access Dissertations and Theses. Paper 7598, URL:http://digitalcommons.mcmaster.ca/opendissertations.

[4] Rajagopalan, S. – Habetler, T.G. – Harley, R.G. – Aller, J.M. – Restrepo, J.A.: Diagnosis of rotor faults in brushless DC (BLDC) motors operating under non-stationary conditions using windowed Fourier ridges, Industry Applications Conference, 2005. Fourtieth IAS Annual Meeting. Conference Record of the 2005 (Volume:1 ), 2-6 Oct. 2005, 26 – 33.

[5] Rajagopalan, S. – Restrepo, J.A. – Aller, J.M. – Habetler, T.G. – Harley, R.G.: Selecting time-frequency representations for detecting rotor faults in BLDC motors operating under rapidly varying operating conditions, Industrial Electronics Society, 2005. IECON 2005. 31st Annual Conference of IEEE, 6-10 Nov. 2005.

[6] Rajagopalan, S: Detection of rotor and load faults in brushless DC motors operating under stationary and nonstationary conditions, Georgia Institute of Technology, 2006, URL:https://smartech.gatech.edu/handle/1853/11524?show=full.

[7] Bae, H. – Kim, S. – Vachtsevanos, G.: Fault detection and diagnosis of winding short in BLDC motors based on fuzzy similarity, International Journal of Fuzzy Logic and Intelligent Systems, Vol. 9, No. 2, June 2009, 99–104.

[8] Baek, G. – Kim, Y. – Kim, S.: Fault diagnosis of identical brushless DC motors under patterns of state change, Fuzzy Systems, 2008. FUZZ-IEEE 2008. (IEEE World Congress on Computational Intelligence), 2083–2088.

[9] Nikitin, Yu.: Development of intellectual mechanotronic modules with diagnostic functions, Mechatronika 2009, Proceedings of 12th International Conference on Mechatronics. Trenčianske Teplice, Slovak Republic, 3–5.6.2009, 133–137.

[10] Nikitin, Yu. – Abramov, I.: Algorithms for mechatronic systems diagnosing, AIM 2010, Proceeding of 5th International Symposium. Advances in Mechatronics, Dec. 7-9, 2010, 60–62.

[11] Nikitin, Y., – Abramov, I.: Mechatronic modules diagnosis by use of fuzzy sets, Mechatronika 2011, Proceedings of 14th International Conference on Mechatronics, Trenčianske Teplice, Slovakia, 2011, 109-111.

[12] Stepanov, P., – Nikitin, Y.: Diagnostics of mechatronic systems on the basis of neural networks with high-performance data collection. Mechatronics 2013. 10th International Conference on Mechatronics 2013; Brno; Czech Republic; p 433-440.

[13] Kelemen, M., – Virgala, I., – Frankovský, P., – Kelemenová, T., – Miková, L.: Amplifying System for Actuator Displacement. International Journal of Applied Engineering Research, 11(15), 8402-8407.

[14] Kelemen, M., – Virgala, I., – Kelemenová, T., – Miková, Ľ., – Frankovský, P., – Lipták, T., – Lörinc, M.: Distance Measurement via Using of Ultrasonic Sensor. Journal of Automation and Control, 3(3), 71-74.

[15] Božek, P., - Pokorný, P.:Analysis and evaluation of differences dimensional products of production system. In Applied Mechanics and Materials. Vol. 611 (2014), pp. 339-345. ISSN 1660-9336.

[16] Virgala, I. - Frankovský, P. - Kenderová, M.: Friction Effect Analysis of a DC Motor. In: American Journal of Mechanical Engineering. Vol. 1, no. 1 (2013), p. 1-5, ISSN 2328-4110.

[17] Nikitin, Yu. R. – Abramov, I.: CNC machines diagnostics, Mechatronika 2010, Proceedings of 13th International Symposium on Mechatronics, June 2-4, 2010, 89–91.

[18] Nikitin, Yu. R. – Abramov, I.: Information Processes Models of Mechatronic Systems Diagnosis, University Review, 2011. – Vol. 5, No. 1, 12-16. URL:http://www.tnuni.sk/fileadmin/dokumenty/univerzita/casopisy/university_review/UR_2011_01.pdf

[19] Site of software package "Simulation in technical devices", URL:http://mvtu.power.bmstu.ru/.

EAN 2017
55th conference on experimental stress analysis

# The Thickness Influence of Fibre-reinforced Slabs on Extreme Load Resistance

Romana Lovichová[1,a], Jindřich Fornůsek[1,b], Michal Mára[1,c] and Přemysl Kheml[1,d]

[1]Experimental Centre, Faculty of Civil Engineering, Czech Technical University in Prague, Thákurova 7, Prague, 166 29, Czech Republic

[a]romana.lovichova@fsv.cvut.cz, [b]jindrich.fornusek@fsv.cvut.cz, [c]michal.mara@fsv.cvut.cz, [d]premysl.kheml@fsv.cvut.cz

**Keywords:** UHPFRC, Slab Thickness, Projectile Impact, Ballistic Protection, Penetration, Fibre orientation.

**Abstract.** This paper deals with slabs made of ultra-high-performance fibre-reinforced concrete (UHPFRC) and experimentally verifies the influence of their thickness on projectile impact resistance for different calibers and the different resulting impact energies. It describes the classification of different thickness of UHPFRC slabs according to available European standards and military standards into the respective degrees of ballistic protection. The effectiveness of ballistic protection of tested slabs are checked according to standardized testing procedures for a personal ballistic protection. The aim is to determine the necessary thickness of the slabs to stop bullets of a certain impact energy. Therefore, optimize the amount of material used with regard to the required ballistic protection.

## Introduction

Sandbags and sacks filled with other loose materials are in these days used as a main protection against shooting from handguns at military and police posts in areas of armed conflicts. Construction of sack walls, in many cases created in combination with trenches, is overall technically and physically complicated task which is often carried out in dangerous areas with a threat of shooting. Moreover, sacks are mostly made of jute because of its low price but they degrade relatively quickly under weather conditions and damaged bags spill out filing material, which reduce protection. Therefore, the aim is to speed up and simplify the construction process of ballistic protection systems.

The slabs of UHPFRC provide good ballistic protection and their relatively low weight allows their quick assembly and disassembly of protective constructions. Thereby the time, when people are in danger during a building up protection, is reduced. Moreover high quality products of UHPFRC enable a long service life even when they are exposed to heavy weather conditions.

## Material Description

Ultra-high-performance fiber-reinforced concrete containing fine-grained aggregate up to fraction 1 mm was used for our experiments. The UHPFRC with 1.5 % steel fiber content in mixture volume was chosen. The devised mix for UHPC (containing mixture of cement, microsilica, aggregate and plasticiser) was specially prepared and delivered in 12 kg bags.

The volume of fibers and water were weighed to suit the required resulting strength of the mixture and were added into the mixture during the mixing process.

The mixture was poured into the horizontal formwork with dimensions 300×400 mm and variable thickness *t* of 1, 2, 3, 4 and 5 cm. The samples were de-molded after 24 hours. The impenetrability of UHPFRC slabs was experimentally tested for respective slab thicknesses *t*.

Average results of obtained mechanical properties (such as flexural strength $f_b$ and compressive strength $f_c$) in 7 and 28 days after casting of mixture are shown in Table 1. The flexural strength was tested by three point bending test on prisms 40×40×160 mm. The compressive strength was tested on the one part of specimen left after the three point bending test, in a press with dimensions of contact area 40×40 mm. Amount of samples was set on twelve for both tests.

Table 1. Flexural and compressive strength of used mixture.

| | $f_b$ [MPa] | $f_c$ [MPa] |
|---|---|---|
| 7 days | 13.8 | 108.3 |
| 28 days | 14.2 | 122.0 |

## Experimental Program

Samples with a minimum age of 28 days were tested on the shooting range. Testing conditions were adjusted as close as possible according to a standard EN-1522 and to the shooting range options [1]. Each specimen was placed into the special mount with simulation of point supports by two screws in each corner. With regard to the safety of the shooter, some non-standardized distances of the shooter and target were also selected [2]. Projectiles were fired from the distance 6 meters to center of specimen for handguns ("short firearms") and 20 meters for rifles ("long firearms"). A shooting system is schematically shown below, see Fig. 1. The slabs were oriented such that the bullets hit the slabs in direction perpendicular to its own plane, thus against the smallest mass of material which should be the worst possible scenario.

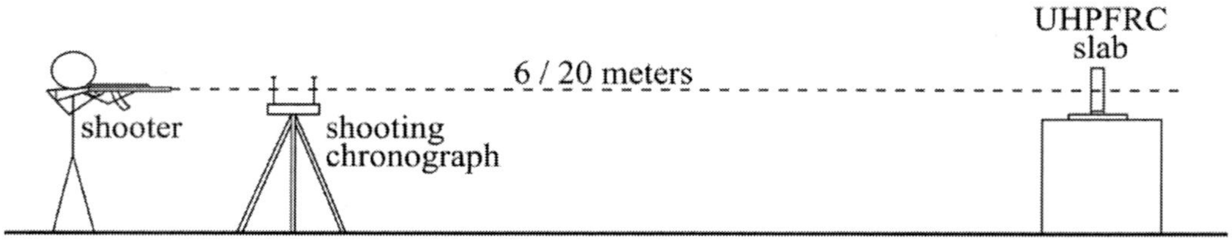


Fig. 1 Schema of the applied shooting system.

## Results

Low-thickness slabs have not been optimally created because fibres formed clumps in mixture that could change the thickness of the slabs in the range of ± 5 mm. For slabs with higher thickness, this problem did not occur that much and maximal variance of thickness was within ± 2 mm on the edges. Therefore, thin slabs (mainly 1 cm) are difficult to make in appropriate quality. For that reasons, slabs with thickness 1 cm fulfilled the supplementary assumption as the slabs failed to stop bullets of the smallest .22 caliber (5.6x15 mm).

Fig. 2 1-cm slab – .22 caliber on the left, 9 mm Luger caliber on the right side of the slab.

Fig. 3 1-cm slab – back side – crater after 9 mm Luger on the left and crater after .22 caliber on the right side oft he slab.

Another series of slabs with 2 cm in thickness easily stops the bullets of .22 caliber but 9 mm Luger caliber still penetrate the slabs through. The 9 mm Luger caliber from short or long firearms and even the 7.62x25mm Tokarev caliber are unable to penetrate the 3 cm thick slabs. These slabs stops first bullet of the .44 caliber but cannot withstand other following bullets of this caliber.

Fig. 4 2-cm slab – three small craters are from .22 caliber, one bullet hole in the middle is from 9 mm Luger caliber.

Fig. 5 2-cm slab – back side – nothing visible from .22 caliber and bullet hole from 9 mm Luger caliber.

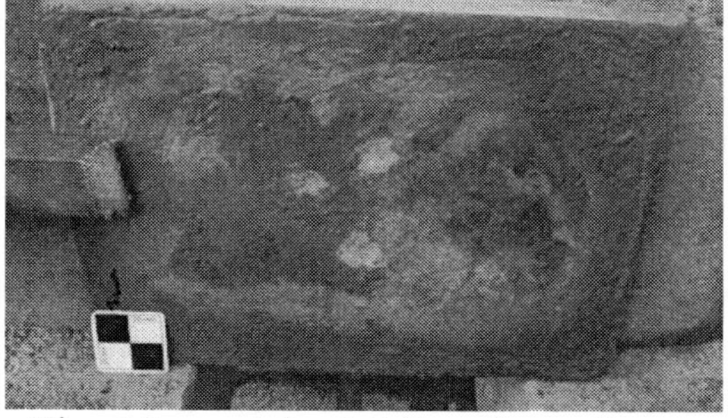

Fig. 6 3-cm slab – hit from 9 mm Luger caliber on the left and two hits from 62×25 mm Tokarev caliber on the right side.

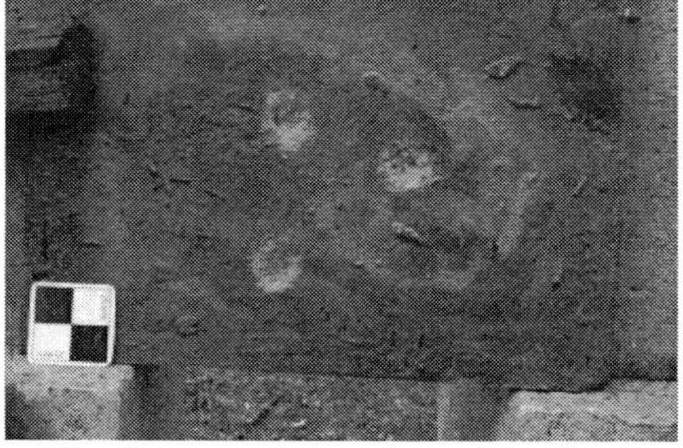

Fig. 7 3-cm slab – three hits from .44 Magnum caliber.

Slabs of 4 cm thickness stop with certainty bullets of .44 caliber, however all other bigger caliber (rifles) with much higher impact energy were able to penetrate the whole thickness of the slab. The last group of 5 cm thick slabs was tested on 7.62x39 mm caliber with soft or steel core and .223 Remington civil caliber. Bullets from 7.62x39 mm soft core and .223 calibers were stopped by the slab, but the back side of the slab was heavily damaged. The 7.62x39 steel core caliber was able to penetrate the slabs though.

Some samples of the hit slabs of all the tested thicknesses are shown on Fig. 2 – Fig. 11.

Fig. 8 4-cm slab – craters after .7,62x39 mm soft core caliber (two upper did not go through, lower did).

Fig. 9 4-cm slab – back side – craters after .7,62x39 mm soft core caliber (two upper did not go through, lower did).

Fig. 10 5-cm slab – craters after .7,62x39 mm steel core caliber (two upper did not go through, lower did).

Fig. 11 5-cm slab – back side – craters after .7,62x39 mm steel core caliber (two upper did not go through, lower did).

## Conclusions

Overall, the experiment can be considered as successful. The slabs made of UHPFRC provide good ballistic protection and their relatively low weight allows quick assembly and disassembly. Thereby, the people who build up the protection are exposed to risk for reduced time. It may be assumed that a police application does not need protection against military calibers, but there is more needed fast assembly and disassembly of any inhibitions. Therefore, slabs with lower thickness (3 – 5 cm) but sufficient protection against short guns can be used. While military applications can utilize the full potential of ballistic protection of thick (5 cm and more) UHPFRC slabs. Furthermore, high quality of the UHPFRC products allows a long service life even if they are exposed to strong weather conditions which allows various positions on the market for differently thick slabs according to various levels of ballistic protection.

Most of the slabs can be classified according to EN-1522 to the appropriate category of ballistic protection but even though the standard EN-1522 is widely recognized as sufficient, it is focused more on the handguns than on rifles, for which the difference in impact energy is multiple. Nevertheless, for civil usage is the standard EN-1522 sufficient and slabs with thickness of at least 5 cm can withstand most of the civil used guns [3].


### Acknowledgement

The authors gratefully acknowledge the financial support from the Ministry of Industry and Trade of the Czech Republic, the grant No. MPO FV10547.


### References


[1] European Comittee for Standardization, EN 1522, Windows, doors, shutters and blinds – Bullet resistance – Requirements and classification, 1998, Brussel.

[2] European Comittee for Standardization, EN 1523, Windows, doors, shutters and blinds – Bullet resistance – Test method, 1998, Brussel.

[3] KNEUBUEHL, Beat P. *Ballistic protection* [online]. Switzerland, 2002 [cit. 2017-04-16]. Dostupné z: http://www.vpam.eu/fileadmin/Wissenswertes/Ballistic-protection.pdf.

EAN 2017
55th conference on
experimental stress analysis

# Modal Analysis of Graphene Sheets Embedded in Elastic Medium Using Finite Element Method

Pavol Lengvarský[a] and Jozef Bocko[b]

Department of Applied Mechanics and Mechanical Engineering, Faculty of Mechanical Engineering, Technical University of Košice, Letná 9, 042 00 Košice, Slovakia

[a]pavol.lengvarsky@tuke.sk, [b]jozef.bocko@tuke.sk

**Keywords:** Modal analysis, graphene sheet, elastic medium, finite element method.

**Abstract.** The aim of this paper is to investigate natural frequencies of single-layered graphene sheets embedded in an elastic medium. The graphene sheets embedded in the elastic medium are modeled using the finite element method. The model of graphene sheet is represented by beam elements and mass elements; the elastic medium is modeled by spring elements. The stiffness of the spring elements varies and the modal analysis is performed for different magnitudes of stiffness. If the stiffness exceeds a certain limit value, the magnitudes of natural frequencies increased significantly and they are almost the same for all configurations and dimensions of the single-layered graphene sheets.

## Introduction

A plate with hexagonal pattern of carbon atoms connected by $sp^2$ hybridization is called graphene sheet. Graphene sheets are modern structures in nanoscale dimensions with remarkable physical, mechanical and electrical properties. They are the next step in the evolution of carbon materials. Due to their remarkable properties, graphene sheets have been extremely investigated from the beginning of their applications [1-4]. Since experimental investigation in nanoscale is very difficult, the modern numerical approaches and methods are used by different researchers.

In this paper, a modal analysis of the single-layered graphene sheets (SLGSs) embedded in elastic medium is investigated using the finite element method. The SLGSs with different chiralities are modeled by beam elements and the elastic medium is represented by spring elements. Finally, the modal analysis is performed and natural frequencies of the SLGSs are obtained. The effect of the elastic medium is compared with the natural frequencies of the SLGSs without the elastic medium.

## Single layer graphene sheets in continuum mechanics

The SLGSs are modeled using the finite element method. The interatomic interactions (covalent bonds – Fig. 1) are represented by the beam elements and the carbon atoms are represented by the joints (nodes) of the beam elements (Fig. 2). The input data of the beam elements are obtained from connection between molecular and continuum mechanics [1,2]. The following relations between the parameters in molecular mechanics, namely, $k_r$, $k_\theta$ and $k_\tau$, and the parameters in continuum mechanics, namely, EA, EI and GJ, are obtained using the following relations [1,2,5]:

$$k_r = \frac{EA}{L}, \tag{1}$$

$$k_\theta = \frac{EI}{L}, \tag{2}$$

$$k_t = \frac{GJ}{L}, \tag{3}$$

where $k_r$, $k_\theta$ and $k_\tau$ are respectively force constants (stiffness) of bond stretching, bond bending, and torsional resistance, and E, G, L, A, I and J are respectively Young's modulus, shear modulus, length of beam, cross-section area, quadratic moment of inertia, and polar moment of inertia [1,2,5]. The input data as diameter d = 0.1466 nm, length $L = a_{C\text{-}C} = 0.1421$ nm, Young's modulus E = 5.488 TPa and shear modulus G = 0.871 TPa are obtained for the circular beam section of the beam element.

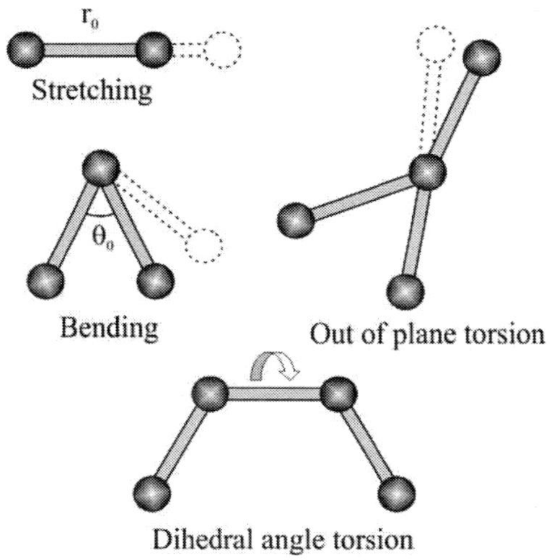


Fig. 1: Interatomic interactions in molecular mechanics.

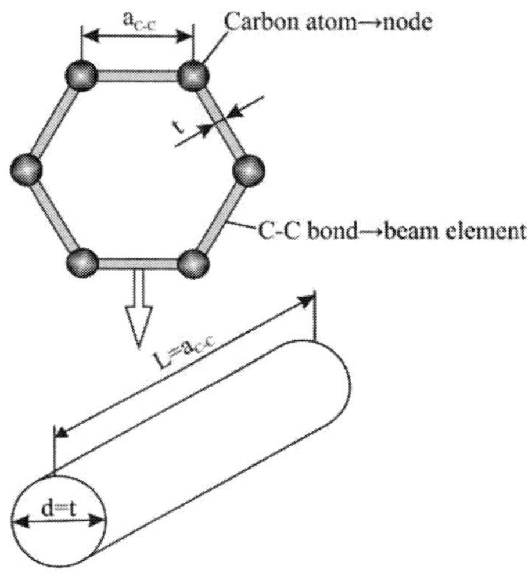


Fig. 2: Interatomic interactions modeled as beam elements.

**Analysis of the single layer graphene sheet**

The modal analysis of the SLGSs is performed for the three configurations called as armchair (n, n), chiral (n, m) and zigzag (n, 0), with lengths 3, 6, 9, 12 and 15 nanometers. Chirality varies from (10, 0) to (10, 10). With the change of chirality, the width of the SLGSs is changed too. All configurations of the SLGSs are modeled by the beam elements. Since the modal analysis of SLGSs is performed, at every node, the mass element is used with the mass of the carbon atom $m_C = 1.99 \times 10^{-26}$ kg. The density of the beam element is supposed to be zero. The SLGSs are embedded in the elastic medium with defined stiffness. The effect of the elastic medium is represented by the spring elements. They are connected to the one end with nodes of beam elements and on the second side they are fixed. The stiffness of the spring elements varies, but all springs have the same stiffness over individual analysis. In the Figs. 3-8 are seen the variations of natural frequencies of the graphene depending on changes of spring stiffness.

The natural frequencies of the SLGSs decrease with increasing dimensions of the graphene sheets. We can see small effect of the chirality on the values of natural frequencies. The natural frequencies for the small values of stiffness of the spring elements is almost the same. If the stiffness exceeds a certain limit value, the magnitudes of natural frequencies increased significantly and they maximal values are almost the same for all configurations and dimensions of the SLGSs.

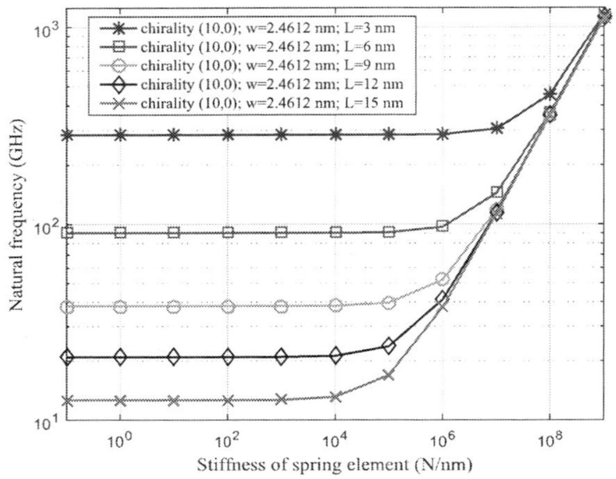


Fig. 3: Variation of natural frequencies of graphene sheets (chirality 10, 0) with variation of spring stiffness.

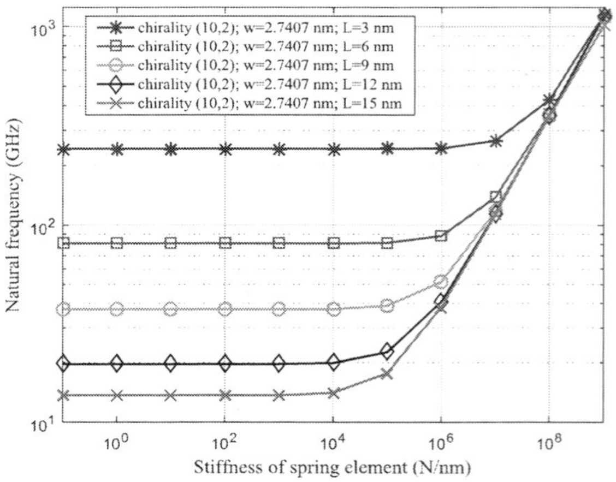


Fig. 4: Variation of natural frequencies of graphene sheets (chirality 10, 2) with variation of spring stiffness.

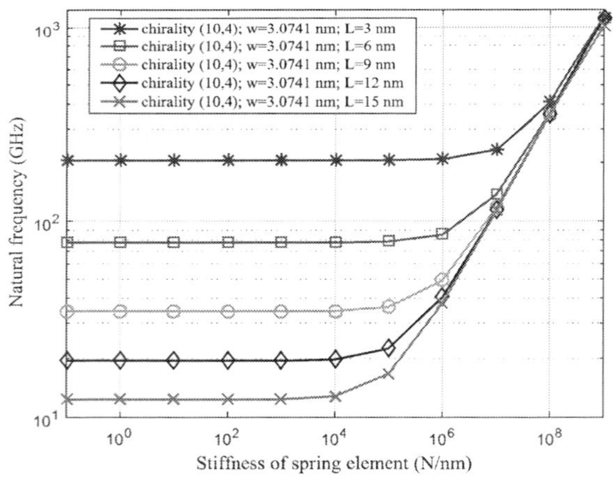


Fig. 5: Variation of natural frequencies of graphene sheets (chirality 10, 4) with variation of spring stiffness.

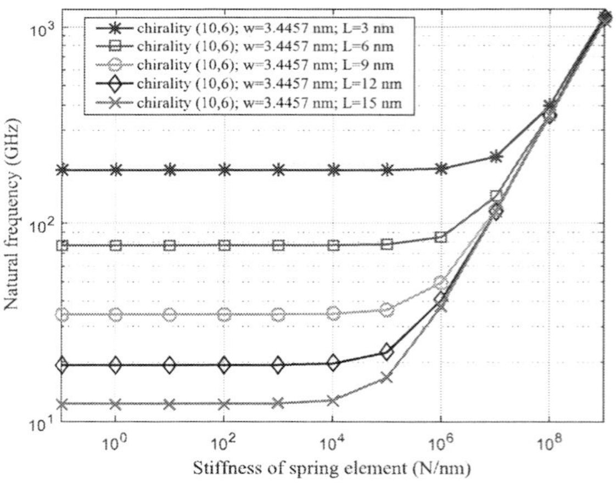


Fig. 6: Variation of natural frequencies of graphene sheets (chirality 10, 6) with variation of spring stiffness.

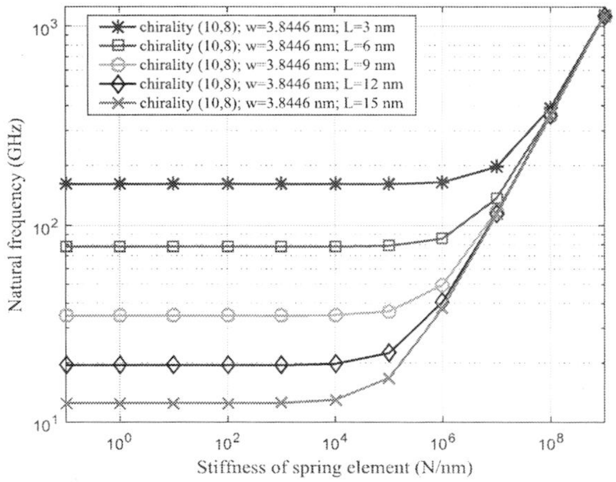


Fig. 7: Variation of natural frequencies of graphene sheets (chirality 10, 8) with variation of spring stiffness.

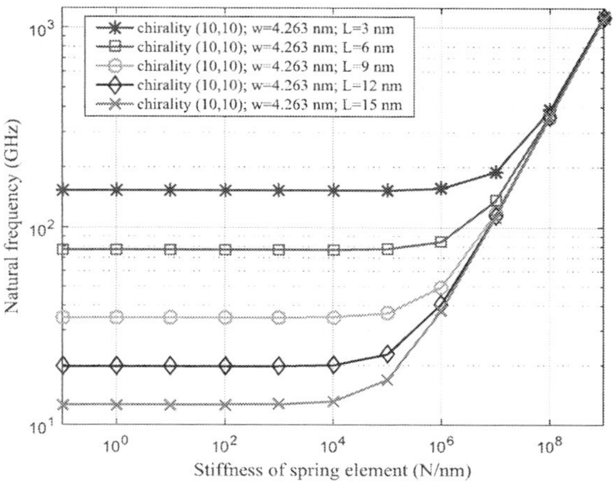


Fig. 8: Variation of natural frequencies of graphene sheets (chirality 10, 10) with variation of spring stiffness.

## Conclusions

The modal analysis of the SLGSs embedded in the elastic medium was performed using the finite element method. The graphene sheets were modeled using the beam elements and the elastic medium was modeled using the spring elements. The stiffness of the spring elements was changed and its influence to the natural frequencies was studied. The natural frequencies depend on the variation of stiffness but for small stiffnesses there is a negligible variation of frequencies. If the stiffness exceeds a certain limit value, the magnitudes of natural frequencies increased significantly and they are almost the same for all configurations and dimensions of the SLGSs.

## Acknowledgement

The authors would like to thank the Slovak Grant Agency VEGA for financial support for this study under grant VEGA no. 1/0731/16 - Development of Modern Numerical and Experimental Methods of Mechanical System Analysis.

## References

[1] C. Li, T. Chou, A structural mechanics approach for the analysis of carbon nanotubes, Int. J. Sol. Struc. 40 (2003) 2487-2499.

[2] K.I. Tserpes, P. Papanikos, Finite element modeling of single-walled carbon nanotubes, Compos. Part B 36 (2005) 468-477.

[3] R.C. Andrew, R.E. Mapasha, A.M. Ukpong, N. Chetty, Mechanical properties of graphene and boronitrene, Phys. Rev. B 85 (2012) 125428.

[4] C.P. Wu, W. Li, Free vibration analysis of embedded single-layered nanoplates and graphene sheets by using the multiple time scale method, Comput. Math. Appl. 73 (2017) 838-854.

[5] J. Bocko, P. Lengvarský, Buckling of single-walled carbon nanotubes with and without defects, J. Mech. Sci. Technol. 31 (2017) 1825-1833.

# Numerical Investigation of 3-D Strain Constraint in Lab Test Specimens

ŠEDEK Jakub[1,a]

[1]VZLÚ – Aerospace Research and Test Establishment, Czech Republic

[a]sedek@vzlu.cz

**Keywords:** Constraint, FEM, Lab test specimen

**Abstract.** The strain constraint investigation using finite element method is presented. Three-dimensional (3-D) models of standard test specimens (M(T), C(T) and SE(T)) are analysed from the point of view of out-of-plane constraint. The constraint factor $\alpha$ was defined according to Newman Jr. J. C. New relationship of constraint in dependence on the ratio of plastic zone size versus thickness is proposed. The global and local behaviors of constraint factors are presented. The local constraint is adopted via equivalent thickness in dependence on the through-thickness position on the crack front.

## Introduction

Crack growth in metal materials is significantly influenced by stress state in the vicinity of crack tip. Stress state changes due to finite dimensions of cracked element according to the distance from free surface, where plane-stress state is dominant. In contrast the plane-strain can prevail when moving inside the element in dependence on the distance from free surface. The stress state is being considered according to parameters which characterize strain constraint, i.e. strain restriction. The rate of constraint is generally determined based on the stress and stain fields.

Strain constraint can be assessed by analytical methods of fracture mechanics, but often it is necessary to utilize numerical methods. Flat specimens are generally analyzed by simplification to two-dimensional problem with assumption of plain-strain or plain-stress. However more complex view on the problem is suitable to analyze using three-dimensional (3-D) approach. In many cases the symmetry can be applied and the cost of problem will be simplified a lot; e.g. middle crack tension specimen (M(T)) analyzed by application of symmetry in three orthogonal planes. The results of the analysis are 3-D stress and strain fields, which serves determination of strain constraint.

## Analysis of Strain Constraint Behavior in Test Specimens

**Analysis basis.** In the plane of test specimen the crack from certain length is influenced by the finite specimen width. The effect is called in-plain constraint. On the contrary the specimen thickness influences the stress in the plane perpendicular to the plane of flat specimen and the effect is called out-of-plain constraint. In this work the latter case the out-of-plain constraint is analyzed.

The 3-D finite element elasto-plastic analysis of test specimens is carried out according to pioneer work of Newmann and Bigellow [1]. The elastic-perfectly plastic material is used in order to apply the results on modified Dugdale model of crack employed for crack growth prediction. 3-D finite element models of lab specimens, namely middle crack tension M(T), compact tension C(T) and side edge crack tension SE(T) specimens were created and analyzed in ABAQUS SW (for example of C(T) finite element model see Fig. 1). One-quarter models were created to reduce computational cost. 57 000 elements of C3D8R type with reduced integration scheme were used. The smallest element size in region of yield material was 0.03 mm.

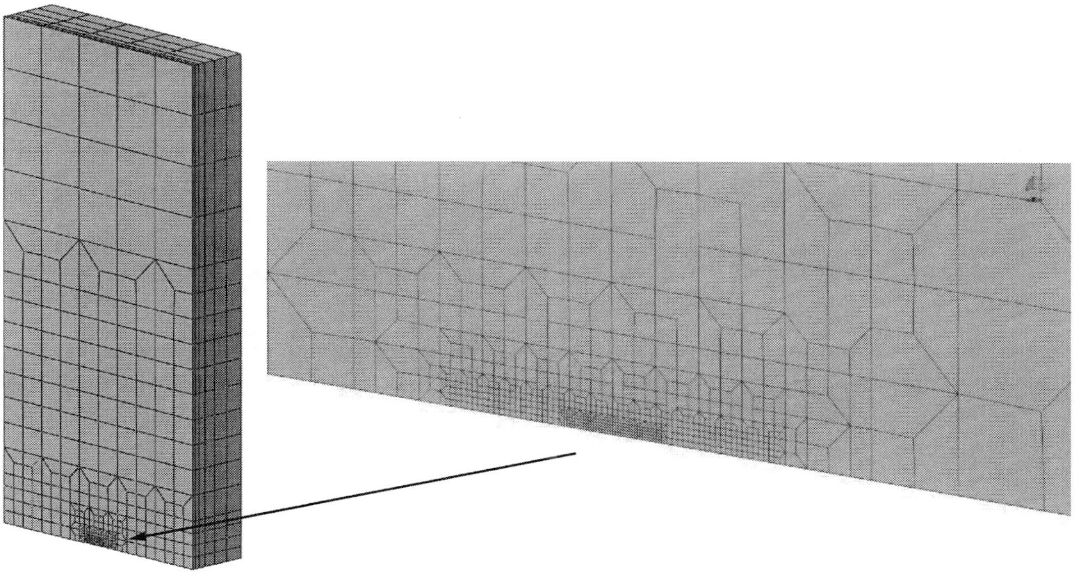

Fig. 1: One-quarter finite element model of M(T) lab test specimen.

Constraint factor $\alpha$ according to Newman [1] was determined through whole thickness and in several through-thickness layers individually. The layers were gradually thinner close to free surface. The definition of global constraint factor $\alpha_g$ is as follows:

$$\alpha_g = \frac{1}{A}\iint \alpha da \quad , \tag{1}$$

where $A$ denotes normal cross-section area through yielded material in the uncracked ligament and $\alpha$ is local constraint factor defined as the ratio of normal stress $\sigma_{yy}$ to the flow stress $\sigma_0$ along the crack front; Eq. 2.

$$\alpha = \frac{\sigma_{yy}}{\sigma_0} \tag{2}$$

Flow stress is defined as average of strength and yield stress; Eq. 3.

$$\sigma_0 = \frac{\sigma_{Rm} + \sigma_{yield}}{2} \tag{3}$$

When applying finite element analysis, the Eq.1 can be expressed as Eq. 4.

$$\alpha_g = \frac{1}{A_T} \sum_{i=1}^{N} \left( \frac{\sigma_{yy}}{\sigma_0} \right)_i A_i \tag{4}$$

$A_i$ is the area of yielded element $i$ in the plane $y = 0$, $\left( \frac{\sigma_{yy}}{\sigma_0} \right)_i$ is normalized normal stress in the element $i$ and $A_T$ is the area of all yielded $N$ elements in the crack plane.

Guo [2] presented the idea of equivalent thickness $B_{eq}$ in order to describe the constraint change through the thickness and expressed Newman's data in dependence of the rate of specimen thickness $B$ and plastic zone size. The equivalent thickness $B_{eq}$ is expressed in dependence on the through-thickness location according to Eq. 5, where $z$ denotes the distance from the center and $B$ is specimen thickness.

$$B_{eq} = 1 - (2z / B)^2 \tag{5}$$

**Analysis Results.** In present work, the constraint factor $\alpha$ was determined for several types of lab test specimens (M(T), C(T) and SE(T)) using finite element method. Constraint factor $\alpha$ was depicted in relation to the logarithm of the rate of specimen thickness $B$ and plastic zone size derived from FE analysis with elastic material ($B_{eq}/rp^{\,FEM\,lin}$); see Fig 2.

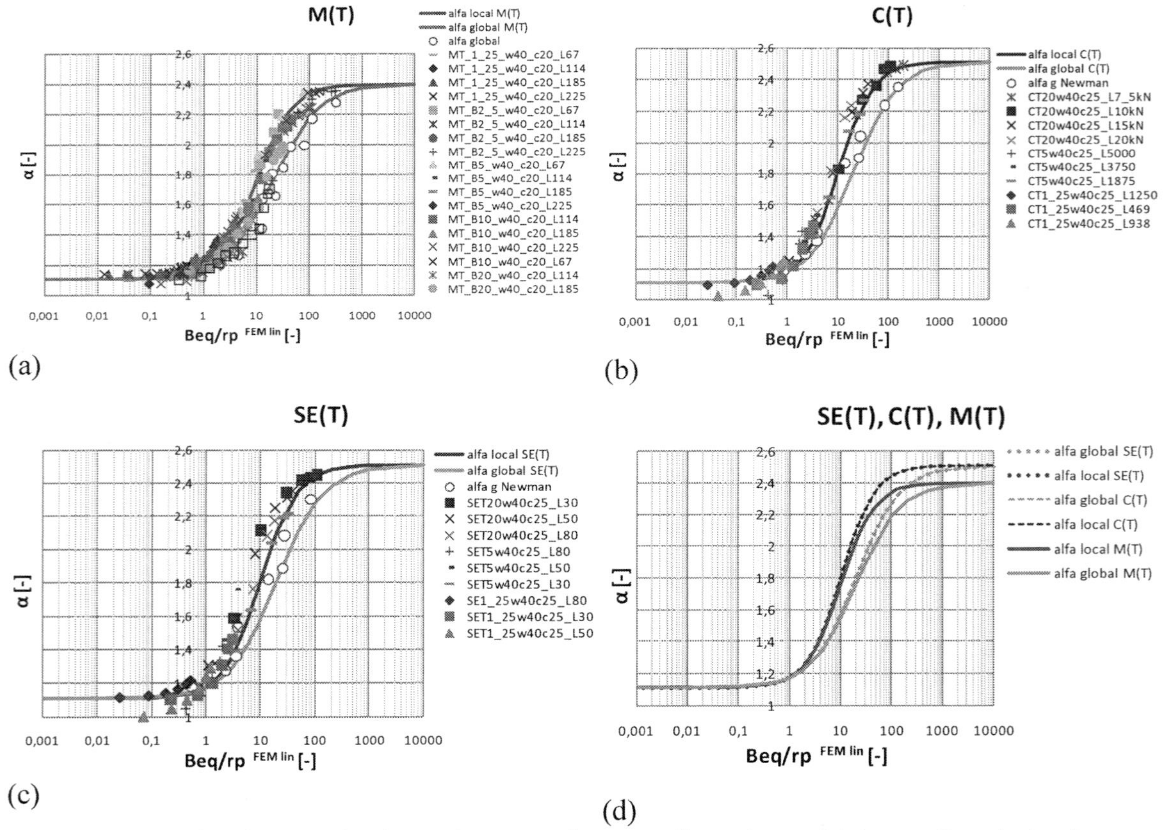


Fig. 2: Constraint factor $\alpha$ in dependence on the rate of specimen thickness $B$ and plastic zone size according to FE elastic analysis $r_p^{FEM\,lin}$; (a) M(T); (b) C(T); (c) SE(T); (d) Comparison of derived $\alpha$ curves.

Based on the results of FE analyses the new relationship of constraint factor $\alpha$ was derived. The global factor $\alpha_{global}$ for whole specimen and local factor $\alpha_{local}$ for equivalent thickness are expressed via Eq. 6.

$$\alpha = 1{,}11 + \frac{\frac{1}{1-2\upsilon} - C}{1 + 20\left(\frac{r_p}{B}\right)^{\beta}} \tag{6}$$

For $\alpha_{global}$: $\beta = 1$ and $B$ denotes specimen thickness and

for $\alpha_{local}$: $\beta = 1{,}3$ and $B$ is replaced with equivalent thickness $B_{eq}$ according to Eq. 5.

Parameter $C$ is expressed by relation $C = a + b v^{\gamma}$, as a function of Poisson's ratio $v$. The relation incorporates the change of maximum value according to specimen type with coefficients $a$, $b$ and $\gamma$ dependent on specimen type.

Maximum values of constraint factor $\alpha$ in dependence on Poisson's ratio $v$ are shown in Fig. 2.

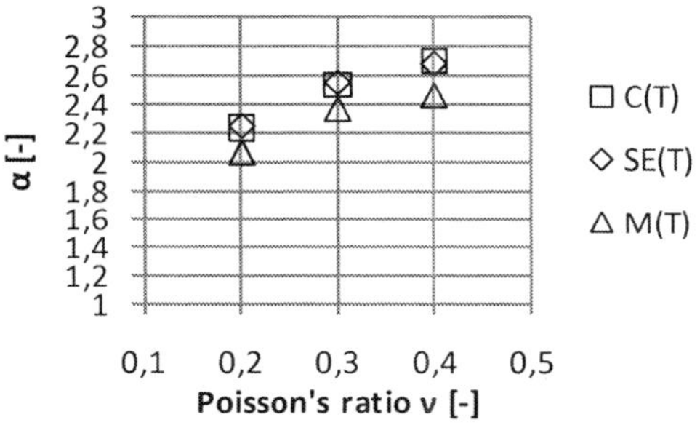


Fig. 2: Constraint factor $\alpha$ under plain-strain conditions in dependence on Poisson's ratio $v$; elasto-plastic FE analysis

## Conclusions

The constraint factor $\alpha$ according to Newman is analyzed in this work. New relationship in the form of sigmoid curve in dependence on the logarithm of the rate of specimen thickness $B$ and plastic zone size is presented. The maximum value of constraint factor $\alpha$ in plane-strain conditions differs a little according to specimen type, but the minimum value of 1,11 in plain-stress conditions for all analysed specimen types (M(T), C(T) and SE(T)) is identical.


## Acknowledgement

This work was realized under project ExANuM (IP3212) with the financial support for research institutions by foundation of Ministry of Industry and Trade of the Czech Republic.


## References


[1] Newman Jr. J. C., Bigelow C. A., Shivakumar K. N., Three-Dimensional Elastic-plastic Finite-Element Analyses of Constraint Variations in Cracked Bodies, NASA LRC, 1993, NTM 107704.

[2] Guo W., Three-Dimensional Analyses of Plastic Constraint for Through-thickness Cracked Bodies, Elsevier, 1999, Engineering Fracture Mechanics 62, pp. 383-407.

# Finite element analysis of EPS foams used in bicycle helmets

ČAJKA Martin[1, a], BOCKO Jozef[2,b] ŠARLOŠI Juraj[3,c]

[1]Technical university in Košice, Letná 9 042 00 Košice, Slovakia

[2]Technical university in Košice, Letná 9 042 00 Košice, Slovakia

[3]Technical university in Košice, Letná 9 042 00 Košice, Slovakia

[a]martin.cajka@tuke.sk, [b]jozef.bocko@tuke.sk, [c]juraj.sarlosi@tuke.sk

## EXTENDED ABSTRACT

### Introduction

Task of increasing protective properties of bicycle helmets is in the main scope of not only manufacturers of protective equipment, but also engineers and scientist all over the world. Recently, expanded polystyrene (EPS) foams became widely used as energy absorption material during impact, not only in bicycle helmets, but also in other protective equipment [1]. It is due to the fact of their unique material properties and low cost. In this paper, the parametrical study on EPS foams of different densities with and without polycarbonate (PC) outer shell is performed. Numerical simulations were performed using Abaqus software, where we conducted tests at different velocities and different impact angles. Output variables were translational velocities and accelerations and angular accelerations at the centre of gravity of head model, because these values are used in head injury criterions used for evaluation of possible head injuries that can occur as a result of such an impact. Comparison of peak values of accelerations, or velocities and duration of linear and angular accelerations at the centre of gravity of the head model was carried out. Based on the results some conclusions, concerning influence of density, presence of outer shell, impact angle, etc. were drown.

### Finite element model and simulation parameters

For all the numerical analysis, ABAQUS was used as a solver and also as modeller because of the simple geometry of all instances. Numerical tests were done to assess suitability of material model with comparison to real data obtained from experimental compression test, which was carried out according to standards. Abaqus/Implicit was used for compression test simulation, where we modelled block of EPS with dimensions 50x50x23 mm and densities of 40, 80 and 120 kg/m$^3$ using CRUSHABLE FOAM material model for isotropic material with volumetric hardening [2, 3]. This kind of material model must be used with linear elastic material model, where we have to define Young's modulus and Poisson's ratio. Poisson's ratio is for crushable foams considered to be equal to zero [4]. Young's modulus values we obtained from experimental data and they can be found in Table 1.

Table 1 Young's modulus values for EPS of different densities

| EPS density [kgm$^{-3}$] | 40 | 80 | 120 |
|---|---|---|---|
| Young's modulus | 11 | 44 | 48 |

Further input data for this model were compression yield stress ratio $k$ and hydrostatic yield stress ratio $k_t$, as we did not tested EPS in hydrostatic tension we used input data adopted by Mills **Chyba! Nenašiel sa žiaden zdroj odkazov.** , k=1.933 and $k_t$=1, which showed as reliable also for our material models. We also had to provide input for the hardening law by specifying value of yield stress in uniaxial compression as a function of the absolute value of the axial plastic strain. It was inserted in tabular way and those data were obtained from compression tests. For the dynamic analysis we had to take into consideration, that as a strain rate increase, EPS show increase in the yield stress. This increase is significant also in our case, so we defined strain-rate dependent material behaviour in Abaqus and we used tabular input of yield ratio, where rate-dependent behaviour is specified by giving a table of the ratio $r = \frac{\bar{\sigma}_c}{\sigma_c}$ as a function of the absolute value of the axial plastic strain, where $r$ is the uniaxial compression yield stress ratio, $\sigma_c$ is uniaxial compression yield stress at a given value of $\varepsilon^{pl}_{axial}$ for the experiment with the lowest strain rate [3]. Input data for strain-rate dependency where obtained from experimental results.

**Geometry of headform, anvil and EPS foam.** The aim was to set all the parameters for this numerical study in such a manner that it will be efficient for computation as we had to run numerous set of analysis, our FEA assembly was based on experimental assembly, which consist of three main parts: headform, EPS foam sample and anvil.

*Headform:* As a Hybrid III 50th Male Dummy head (used in experiments) is much stiffer than EPS foam we considered it as a rigid body, and in our case we have decided to model a simplified spherical model, using analytical rigid part, which is computationally more efficient than using discrete rigid part, but at cost of being not able to fully copy headform geometry of Hybrid II 50th Male Dummy head. We decided to use this simplification as it was parametrical study, where we were focused on influence of changing parameters of EPS foam on linear and angular acceleration. Radius of spherical model was 85.7 mm and weight was 4.5 kg. As a reference point for this rigid body we set the centre of mass.

*Anvil:* Rigid steel anvil was also modelled as analytical rigid part as its stiffness is also much higher than stiffness of EPS. On anvil we put a reference point where we restrained all degrees of freedom with implied boundary condition.

*EPS foam sample:* EPS was modelled directly in Abaqus. It was modelled using deformable solid part, where we used material model for crushable foams with aforementioned properties.

*Polycarbonate (PC) shell:* The outer shell is modelled as linear elastic material and is perfectly bonded to outer surface of EPS foam. Its thickness is 0.5 mm. Material properties of PC were based on the data found in literature so Young's modulus was 7250 MPa, Poisson's ratio was 0.3 and density was 1200 kg/m$^3$ [7].

**Meshing**. As analytical parts does not need to be meshed, only EPS foam and PC shell was meshed. For EPS foam linear C3D8R elements were used, with distortion control on (we used default value of 0.1), which does not allow elements to invert during large deformations. Element size was set to 4 mm, with 11 elements throughout the thickness of EPS foam. These values were set according to mesh sensitivity analysis. Shell was modelled with four node doubly curved thin shell, reduced integration, hourglass control, finite membrane strain model elements S4R. Element size was set to 3 mm. On the Fig. 1 are shown meshed EPS foam and meshed PC shell.

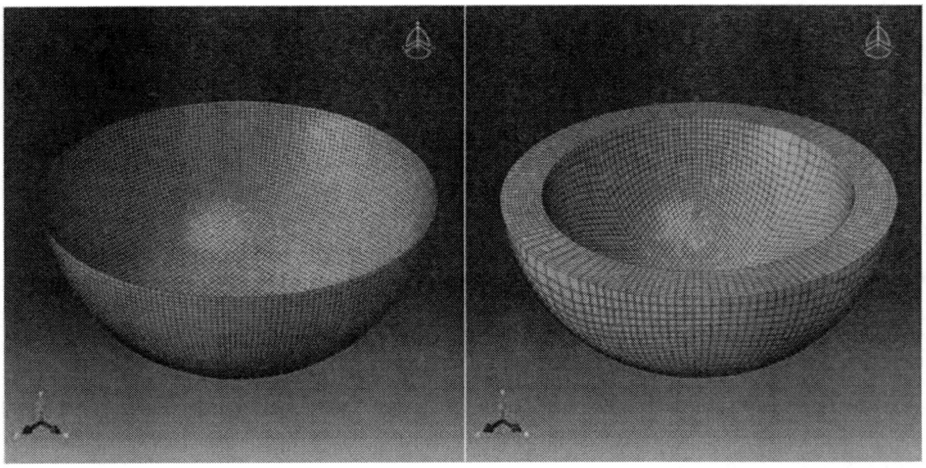

Fig. 1 Meshed PC shell (on the left) and meshed EPS foam (on the right)

**Contact conditions.** We used general contact which was satisfactory for our analysis. Friction in tangential direction was set to $f=0.2$ for PC to anvil contact and $f=0.37$ for EPS to anvil contact. In normal direction we used hard-contact option. EPS foam was connected to headform using coupling, where we restrained all the degrees of freedom (except rotations) for the surface which was in contact with headform.

**Simulation parameters.** To evaluate the energy absorption performance of the EPS foam with different densities of 40, 80 and 120 $kg/m^3$, impact simulations were performed with three impact positions: high energy (referred to 6500 mm/s), medium energy (referred to 5425 mm/s) and low energy (referred to 3962 mm/s). These impact positions were implied at the model in form of predefined field, that was defined on the whole model except the anvil and should simulate the state right before the impact. Whole time of simulation was 12 ms. For complete parametrical study we measured rotational and linear accelerations at the centre of mass of headform model as an output variables. In the Table 2 you can see all the parameters that were combined in this study [4, 5, 6].

Table 2 Numerical simulation parameters, that were involved in the study

<table>
<tr><th>Parameter</th><th colspan="5">Modification of parameter</th></tr>
<tr><td>EPS density [$kgm^{-3}$]</td><td colspan="2">40</td><td>80</td><td colspan="2">120</td></tr>
<tr><td>Outer PC shell</td><td colspan="2">yes</td><td colspan="3">no</td></tr>
<tr><td>Impact velocity [$ms^{-1}$]</td><td colspan="2">3.96</td><td>5.42</td><td colspan="2">6.5</td></tr>
<tr><td>Impact angle</td><td>90</td><td>75</td><td>60</td><td>45</td><td>30</td></tr>
</table>

On the Fig. 2, it is shown basic setup for impact simulations, where all parameters are in bold and the output variables are in italic. Anvil and simplified head model were created as rigid bodies as their stiffness is much higher than stiffness of EPS.

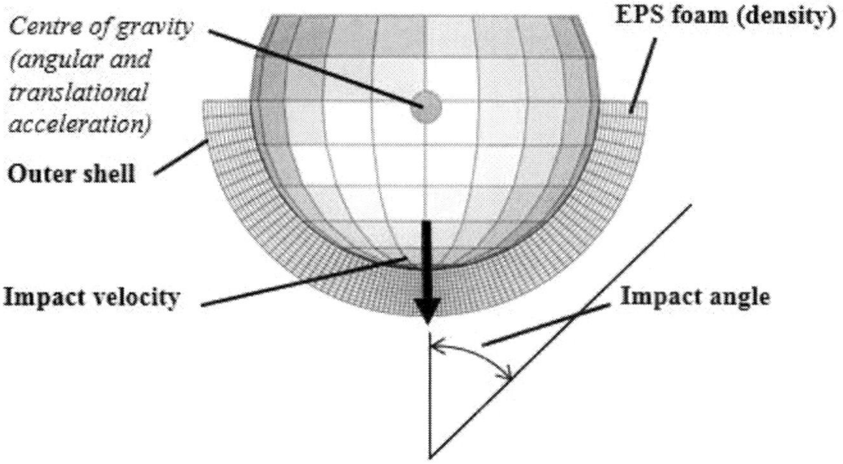


Fig. 2 Example of numerical model assembly with highlighted parameters of interest. Side view, EPS foam is meshed and in side section view.

On the Fig. 3 is shown assembly in Abaqus environment with predefined field of initial velocity of whole model and with boundary conditions implied on anvil. The anvil was rotated to match all the angles of impact that were of our particular interest.

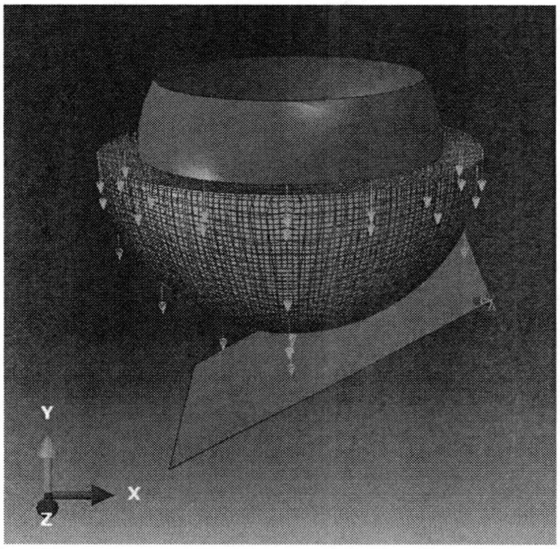


Fig. 3 Model in Abaqus environment with implied boundary conditions

## Results

For complete evaluation of helmet quality is necessary to use one of the head injury criterions, which relate linear or rotational accelerations, or linear and rotational velocities, or other quantities like von Mises stress, principal strains or strain rate (depending on used criterion) with possibility of severe head injuries like skull fracture, focal injuries or diffuse axonal injury. Nowadays most of those criterions use either linear or rotational accelerations for their input (HIC, GAMBIT, HIP or some other), so for our parametrical study with simplified model we decided that we will focus only on linear accelerations in X and Y direction and angular acceleration around Z axis as all the other accelerations for our study are negligible. More information about head injury criterion can be found in literature [8]. There are two main factors which are crucial for head injury criterions. It is peak value of acceleration and its duration. In Table 3, 4 and 5 are shown all the peak values of accelerations.

Table 3 Resluts of peak angular accleration around z axis in rads$^{-2}$

| | **Shell** | | | | **yes** | | | | | |
|---|---|---|---|---|---|---|---|---|---|---|
| | **EPS density** | 40 | | | 80 | | | 120 | | |
| | **Impact velocity** | L | M | H | L | M | H | L | M | H |
| Impact angle | 90 | 41.2 | 206.8 | 36.4 | 7.4 | 13.4 | 20.0 | 47.5 | 30.0 | 18.0 |
| | 75 | 4691.1 | 6748.9 | 8157.3 | 7167.2 | 9976.4 | 11765.2 | 9176.2 | 12309.3 | 14756.6 |
| | 60 | 6801.9 | 9126.3 | 10950.0 | 10432.1 | 14435.2 | 17925.9 | 13006.2 | 18291.3 | 22147.0 |
| | 45 | 5769.0 | 7869.0 | 8949.0 | 8988.0 | 12344.1 | 15210.7 | 11575.1 | 16207.0 | 19108.5 |
| | 30 | 3811.8 | 4932.6 | 5676.6 | 6371.5 | 8720.7 | 10157.6 | 8225.0 | 11294.9 | 13515.1 |
| | **Shell** | | | | **no** | | | | | |
| Impact angle | 90 | 27.9 | 38.2 | 89.3 | 53.0 | 39.9 | 54.1 | 86.0 | 107.5 | 92.8 |
| | 75 | 2361.5 | 3295.0 | 4256.5 | 4209.4 | 5612.5 | 6702.0 | 5922.6 | 7854.7 | 9255.2 |
| | 60 | 4546.9 | 6289.5 | 7878.6 | 7933.0 | 10697.3 | 12742.3 | 11032.2 | 14715.4 | 17537.0 |
| | 45 | 6182.0 | 8555.6 | 10448.6 | 10738.8 | 14468.3 | 17336.3 | 14538.6 | 19521.7 | 23273.2 |
| | 30 | 5425.37 | 7148.37 | 8293.41 | 9868.10 | 13120.6 | 15573.1 | 12915.1 | 17579.8 | 20727.1 |

Table 4 Resluts of peak linear accleration in x direction in ms$^{-2}$

| | **Shell** | | | | **yes** | | | | | |
|---|---|---|---|---|---|---|---|---|---|---|
| | **EPS density** | 40 | | | 80 | | | 120 | | |
| | **Impact velocity** | L | M | H | L | M | H | L | M | H |
| Impact angle | 90 | 1.3 | 3.3 | 0.9 | 0.2 | 0.4 | 0.7 | 0.9 | 0.7 | 0.8 |
| | 75 | 396.6 | 556.6 | 713.5 | 590.5 | 834.3 | 1014.5 | 768.9 | 1066.3 | 1291.0 |
| | 60 | 564.1 | 738.1 | 879.1 | 862.9 | 1176.4 | 1420.1 | 1056.8 | 1447.6 | 1742.1 |
| | 45 | 479.7 | 685.5 | 835.4 | 729.4 | 1021.3 | 1247.6 | 932.6 | 1313.3 | 1580.2 |
| | 30 | 424.7 | 582.8 | 715.5 | 676.2 | 933.5 | 1118.1 | 865.2 | 1202.4 | 1436.4 |
| | **Shell** | | | | **no** | | | | | |
| Impact angle | 90 | 0.9 | 1.3 | 2.3 | 1.4 | 1.9 | 2.3 | 3.2 | 2.5 | 3.1 |
| | 75 | 250.7 | 385.3 | 717.4 | 429.4 | 586.8 | 708.3 | 569.8 | 775.0 | 932.9 |
| | 60 | 411.5 | 592.8 | 829.6 | 723.0 | 981.3 | 1176.5 | 963.4 | 1300.5 | 1559.0 |
| | 45 | 417.2 | 566.7 | 703.0 | 755.6 | 1006.4 | 1181.3 | 1004.8 | 1349.1 | 1591.3 |
| | 30 | 307.8 | 425.7 | 513.3 | 540.9 | 739.1 | 889.3 | 711.9 | 982.4 | 1177.7 |

Table 5 Results of peak linear acceleration in y direction in ms$^{-2}$

| | **Shell** | | | | **yes** | | | | | |
|---|---|---|---|---|---|---|---|---|---|---|
| | **EPS density** | 40 | | | 80 | | | 120 | | |
| | **Impact velocity** | L | M | H | L | M | H | L | M | H |
| Impact angle | 90 | 1328.8 | 1988.3 | 2574.4 | 1930.9 | 2829.8 | 3476.9 | 2382.6 | 3388.3 | 4172.6 |
| | 75 | 1210.5 | 1791.6 | 2354.0 | 1750.6 | 2564.8 | 3147.7 | 2164.9 | 3080.4 | 3820.5 |
| | 60 | 1008.7 | 1459.2 | 1851.3 | 1483.3 | 2117.6 | 2617.3 | 1865.2 | 2632.7 | 3219.3 |
| | 45 | 706.0 | 1010.0 | 1215.5 | 1061.7 | 1493.9 | 1825.6 | 1350.8 | 1894.0 | 2294.3 |
| | 30 | 363.3 | 494.4 | 611.3 | 574.8 | 795.2 | 957.0 | 733.0 | 1024.6 | 1228.5 |
| | **Shell** | | | | **no** | | | | | |
| Impact angle | 90 | 1056.5 | 1722.3 | 3891.5 | 1705.3 | 2416.0 | 2972.2 | 2209.9 | 3116.7 | 3800.6 |
| | 75 | 989.1 | 1570.0 | 2757.5 | 1598.7 | 2269.9 | 2785.0 | 2074.1 | 2913.7 | 3568.1 |
| | 60 | 806.3 | 1219.7 | 1722.2 | 1324.4 | 1860.6 | 2295.5 | 1723.4 | 2402.7 | 2934.9 |
| | 45 | 594.0 | 856.7 | 1089.5 | 995.3 | 1391.7 | 1688.6 | 1301.1 | 1821.1 | 2187.8 |
| | 30 | 356.3 | 496.7 | 601.2 | 625.8 | 859.0 | 1034.9 | 818.9 | 1137.2 | 1361.8 |

After running complete set of sets few conclusions can be made based on these results. As we expected with increasing impact velocity and with increasing density of EPS foam both translational and linear accelerations are increasing. On the other hand peak duration with changing velocity is not varying too much, but with increasing density the peak duration is decreasing. This decrease is significant from EPS40 to EPS80, where peak duration is approximately 35 % shorter, whereas from EPS80 to EPS120 it is only around 12 %. There are few other interesting trends that are similar for EPS of all the densities and in this paper they are shown on the example of EPS80 on the following Fig.4. Based on this graphs we can see that for EPS without shell the most critical angle of impact is 45° on the other hand for EPS with shell critical angle of impact is 60° when combination of angular and linear acceleration has the highest values. Another important thing is that for impact angles 30° and

45°, EPS with shell has a lower values of peak accelerations that is because of lower friction coefficient between PC shell and steel anvil, then between EPS and anvil. Peak duration is affected by presence of shell in positive way (it means that peak duration is shorter) and with decreasing impact angle it decreases peak duration remarkably as it is shown on Fig. 5, where you have comparison of angular acceleration curves for EPS 80 at high impact velocity with and without shell.

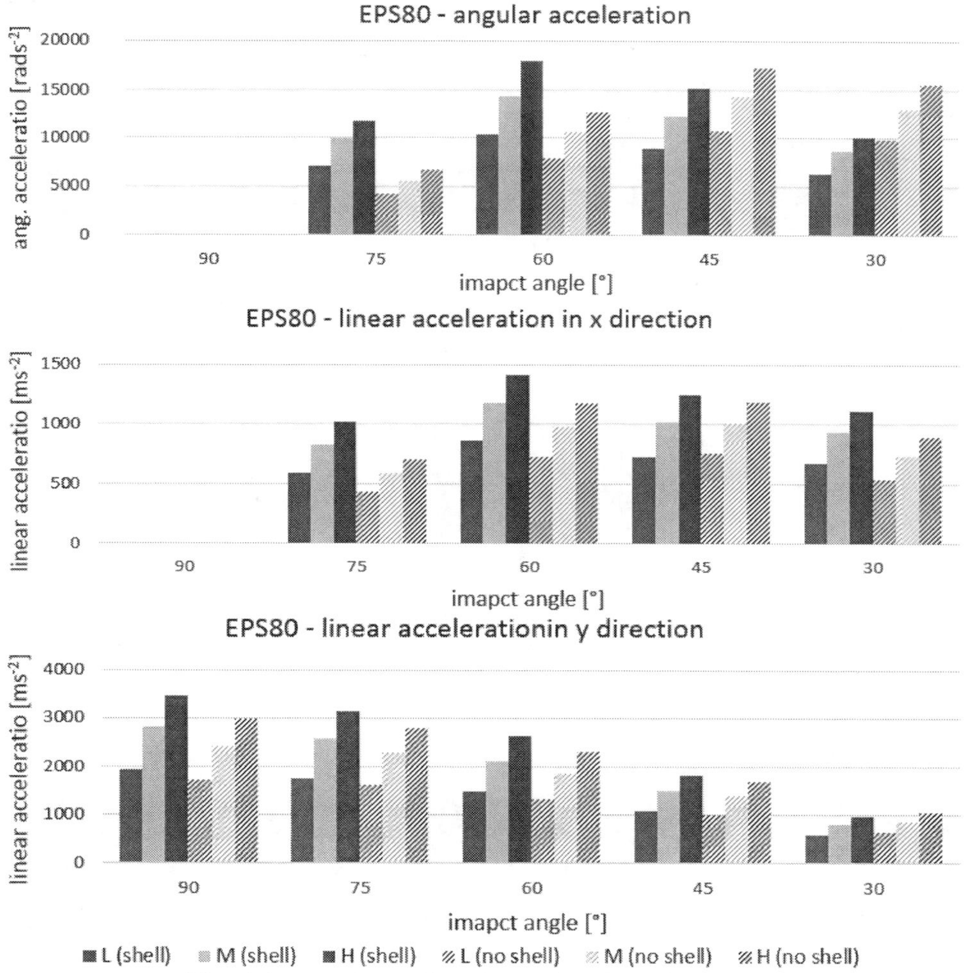


Fig. 4 Comparison of peak accelerations for EPS80

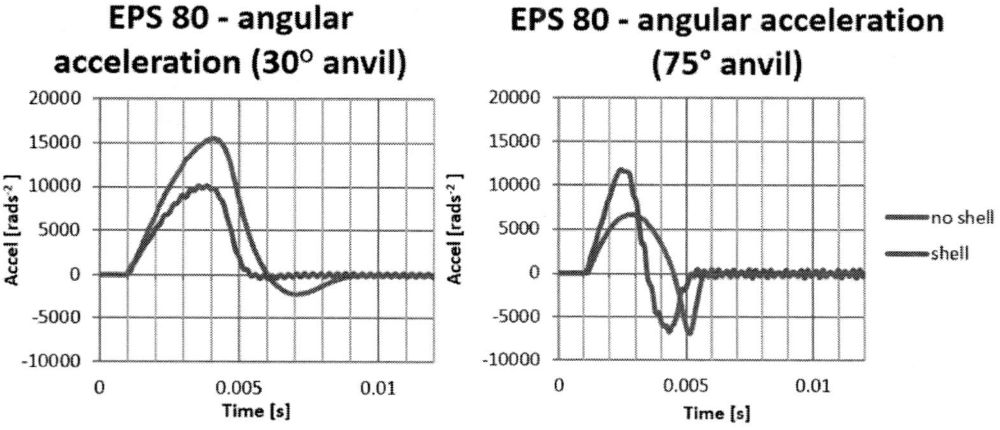


Fig. 5 Comparison of peak duration for EPS80 at high impact velocity

On the Fig. 6 and Fig. 7 is comparison of stress distribution at the cross-section of EPS80 and on the bottom part of EPS80, respectively shell in case with shell on EPS (Fig. 6) in the moment of peak value of accelerations.

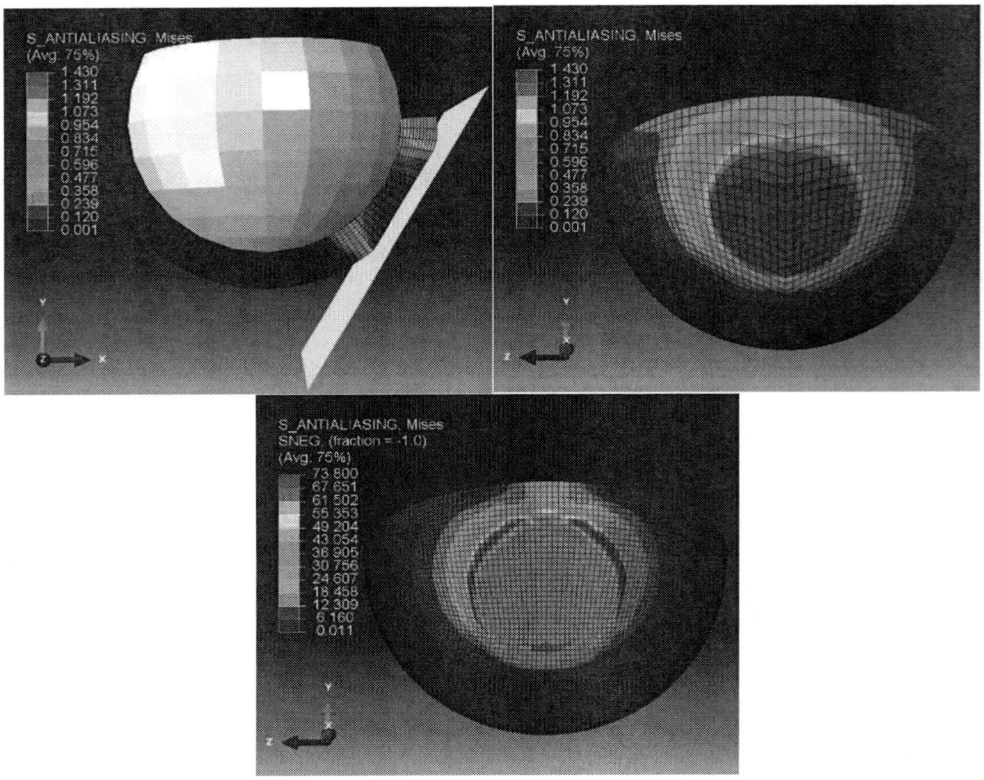


Fig. 6 Stress distribution for EPS 80 with shell. Cross-section (upper left picture), bottom view (upper right picture, bottom view of shell (lower picture)

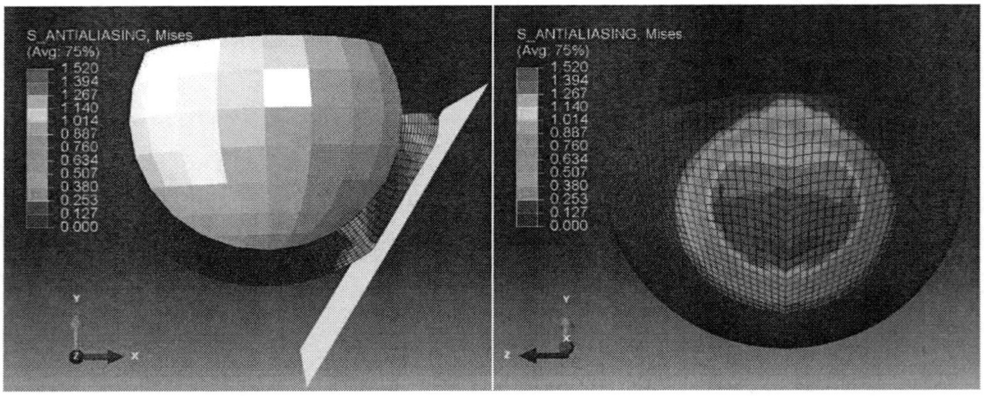


Fig. 7 Stress distribution for EPS 80 withou shell. Cross-section (left picture), bottom view (right picture)

## Conclusions

Based on this simulations we got global insight on how different conditions affect duration and peak value of angular and translational acceleration at centre of gravity of simplified head model. Based on this observations we can differ more significant from less significant parameters. This work should lead to deeper understanding of the relations among parameters and can lay foundation for further research on other parameters as friction between head and

EPS foam, influence of retention straps, different materials from which outer shell is made of, etc.

### Acknowledgement

The authors would like to express their gratitude to Scientific Grant Agency VEGA MŠ SR for the support of this work under the project no. VEGA 1/0731/16.

### References

[1] B. Depreitere et al., Bicycle-related head injury: a study of 86 cases. In: Accident Analysis and Prevention, Vol. 36, No. 4 (2004), p. 561–567.

[2] V.S. Deshpande, N.A. Fleck, Multi-axial yield behaviour of polymer foams. In: Acta Materialia, Vol. 49 (2001), p. 1859-1866.

[3] ABAQUS. Abaqus analysis user's guide 6.13. [online]. Vélizy-Villacoublay: Dassault Systemes SIMULIA Ltd, 2013. Information on: < http://129.97.46.200:2080/v6.13/books/usb/default.htm>

[4] N.J. Mills, A. Gilchrist, Finite-element analysis of bicycle helmet oblique impacts. In: Journal of Impact Engineering. Vol. 35, No. 9 (2008), p. 1087-1101.

[5] Q.H. Shah, A. Topa, Modelling large deformation and failure of expanded polystyrene crushable foam using LS-DYNA. In: Modelling and Simulation in Engineering. (2014), p. 1-7.

[6] K. Vanden Bosche, et al, Combination shear-compression testing of foam materials for their application in bicycle helmets or other complexly loaded structures. In: 18th International Conference on Composite Materials. Jeju, (2011).

[7] M. A. Forero Rueda, L. Cui, M.D. Gilchrist, Optimization of energy absorbing liner for equestrian helmets. Part I: Layered foam liner. In: Materials and Design. Vol. 30, (2009). p. 3405-3413

[8] W. Goldsmith, The state of head injury biomechanics: past, present and future (part 1). In: Critical reviews in Biomedical Engineering, Vol. 29, No. 5 (2001), p. 441–600

EAN 2017
55th conference on experimental stress analysis

# Some Important Experimental Details During Verification of Fatigue Durability of Railway Axles and Wheelsets

Ivo Černý[1,a] and Jan Kec[1,b]

[1]SVÚM a.s., Tovární 2053, 25088 Čelákovice, Czech Republic

[a]Ivo.Cerny@seznam.cz, [b]kec@svum.cz

**Keywords:** Railway axles, railway wheelsets, fatigue durability, fretting fatigue, press fit

**Abstract.** Railway wheelsets are crucial components of bogies from the viewpoint of service reliability and safety. They are compound of two subcomponents, wheels and axle, connected to each other by press fitting or shrink fitting. Wheelsets are components extremely dynamically loaded in service, by a very high number of cycles, usually more than $10^9$ during the life. That is why fatigue resistance of these components have to be adequately proved during the certification process, when such proves are strictly requested Two or three characteristics are defined in the standards, which have to be proved, namely (i) fatigue resistance of the smooth part of the axle body on its free surface, so called characteristics F1 and (ii) fatigue resistance under press fit (or shrink fit) - characteristics F3.

This contribution describes a new experimental approach to verification of press fitting quality and homogeneity by measurement of mutual circumferential displacements between the axle and wheel hub during the full scale fatigue test. Though the press fit of the tested wheelset met standard requirements for the press fitting diagram in terms of minimum and maximum forces, the mutual microscopic displacements under the press fit were not homogeneous for the whole circumference, the maximum value being almost twice as much higher than the minimum one. Another effect was a dependence of these displacements on number of cycles during rotating bending fatigue tests. The results are discussed considering circumferential dynamic stresses on the axle surface.

## Introduction

Railway wheelsets and axles are safety-critical components. Designing failsafe mechanisms is very difficult and the safety of the component is determined through a good understanding of the structural integrity and through effective management policies [1]. Railway wheelsets are compound of two subcomponents, wheels and axle, connected to each other by press fitting or shrink fitting. Wheelsets are components extremely dynamically loaded in service in terms of a very high number of cycles, usually more than $10^9$ during the life, corresponding to several tens of service years. In spite of that axle failures occur relatively infrequently [1], consequences are usually very serious, from financial damages to heavy casualties. An example is the railway accident of the high-speed train near Eschede almost twenty years ago, where more than 100 people were killed and even more injured. The disaster was followed by a very complicated court case, in which more than 13 experts from different countries participated in the evaluation of causes of the wheel fatigue failure. It was stated in the official verdict that no unambiguous causes were found, likely just a single incident or a technological material phenomenon [2]. This is one of numerous examples, why fatigue resistance of these components have to be adequately proved during the certification process, when such proves are strictly requested. A system of European and worldwide standards [e.g. 3, 4] have been elaborated and issued and is being systematically improved according to needs of the railway transport, its increasing demands on service safety and reliability in connection with increasing train speeds, both freight and passenger, with increasing demands on payload and other aspects [5]. Furthermore, fracture mechanics starts to be used to assist in the timing of crack inspection intervals. Nevertheless a small number of failures still occur [6]. Some current pressures, particularly the need to reduce unsprung mass at high speeds, continue to challenge axle designers.

Though strict standard criteria for the press or shrink fitting methods exist like overlapping dimensions of the axle seat diameter and wheel hole, dependence of press forces during the press fitting process, press fits still remain one of the critical areas where possible cracking occur. As an example, in [7] the authors investigated the effects of rotary bending and of the press fits at the wheel and gear on the stress intensity factor, fatigue crack growth and residual lifetime so as to provide essential input information for setting up inspection intervals. It was found that the effect of rotary bending, although existent, was rather moderate whereas the press fits had a large and detrimental effect.

Another problem is that unlike fatigue resistance of most of metallic materials without defects on free smooth surface increases with increasing static material strength, in case of fretting fatigue mechanism this need not be true. A comprehensive experimental programme, which included rotating bending tests of press fitted axle models reduced to the scale 1 : 5 showed that resistance to fretting fatigue under press fit of high strength railway axle steels is comparable or even worse than the standard EA1N normalized steel with the lowest strength. Endurance limit of the models of different steels was comparable and it was indicated that retardation and arrest of initiated microcracks or short fatigue cracks in compressive residual stress field are crucial mechanisms deciding, whether the main fatigue crack will be created and final failure occur [8].

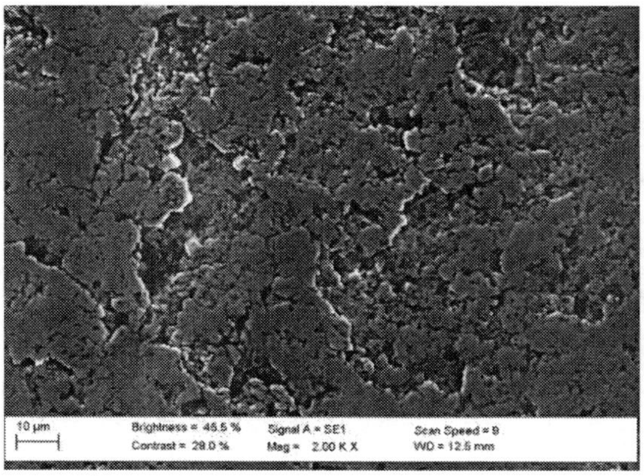

Fig. 1 Example of surface fretting fatigue damage of axle seat under press fit with microcracks and short fatigue cracks

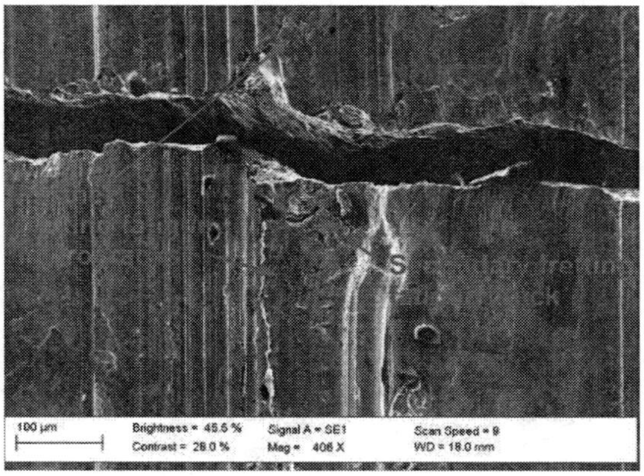

Fig. 2 Main fatigue crack under press fit with secondary crack and further fretting fatigue microcracks

Though fracture mechanics approaches to safety and reliability of railway axles start to be used particularly with the aim to specify necessary inspection intervals [9], standard requirements are still based on the safe life approach, i.e. no failure after unlimited number of cycles. Two or three characteristics are defined in the standards, which have to be proved, namely (i) fatigue resistance of the smooth part of the axle body on its free surface, so called characteristics F1 and (ii) fatigue resistance under press fit (or shrink fit) - characteristics F3.

Note that characteristics F2 does not concern solid axles, but resistance to fatigue damage of hollow axle internal surface. The fatigue resistance is usually proved by rotating bending full scale half wheelset.

Unlike the axle body, where fatigue resistance F1 is determined only by the material and surface conditions, the resistance under press fit, where different fatigue mechanisms occur, fretting fatigue being the most crucial, is affected by further factors, like quality of press fitting, which affect mutual displacements between axle seat surface and wheel hub. Too large displacements promote the fretting fatigue mechanisms and can results in premature cracking under press fit. In Figs. 1 and 2, there are characteristic examples of fretting fatigue damage, which occurred and was analysed during one of the recent full scale axle tests performed at SVUM laboratory. Fig. 1 shows system of surface fretting fatigue microcracs and pits at long initial stages of the process. The microcracks were eventually developed to the main fatigue crack fast growing to the final failure. The general conclusion in this case was that the axle seat had an insufficient value of the F3 characteristics, though it could not be said, whether the actual reason were insufficient material properties or just incorrect press fitting resulting in excessive mutual friction in the axle – hub interface.

This contribution describes a new experimental approach to verification of press fitting quality and homogeneity by measurement of mutual circumferential displacements between the axle and wheel hub during the full scale fatigue test. The results are presented and discussed considering possible effects on fretting fatigue conditions and damage.

## Experiments

Fig. 3 Total view of the axle prepared for fatigue test on the Sincotec machine

Fig. 4 LVDT extensometers measuring displacement between axle and wheel

The test piece was represented by a kind of full scale half wheelset, more exactly the axle press fitted into a substitution of actual wheel – a steel disc with the same dimensions in the area of the central hole, into which the axle was press fitted. The fatigue test was performed on the very sophisticated facility Sincotec GmbH of the last generation, characteristic by very exact circumferential load distribution. The test was prepared according to standard procedure, to ensure all the requested loading conditions. These steps contained detailed static measurement of the maximum stress concentration on the axle surface, evaluation of the stress concentration factor, calibration of the machine and independent measurement of circumferential dynamic stresses during the whole test to $10^7$ cycles. The total view of the test arrangement is in Fig. 3. The experimental procedure is described more in detail e.g. in [10].

Beyond the standard requirements, continuous measurement of mutual displacements between the axle and the disc simulating the wheel was performed and values recorded at three circumferential positions – for $120^{\circ}$ during the whole test. This measurement was performed using three LVDT extensometers Sangamo fixed to the axle using magnetic holders attached to the axle at the distance of approximately 230 mm from the hub edge – Fig. 4.

The rotating bending test frequency was approximately 25 Hz, given first of all by stiffness of the whole component and the requested load – stress amplitude 200 MPa at critical point of the maximum stress concentration. Dynamic values of the mutual displacement between the axle and hub, more exactly maximum and minimum peaks, were evaluated without interrupting the test, using separate high speed analog – digital converters Adicom PCA.

## Results and Discussion

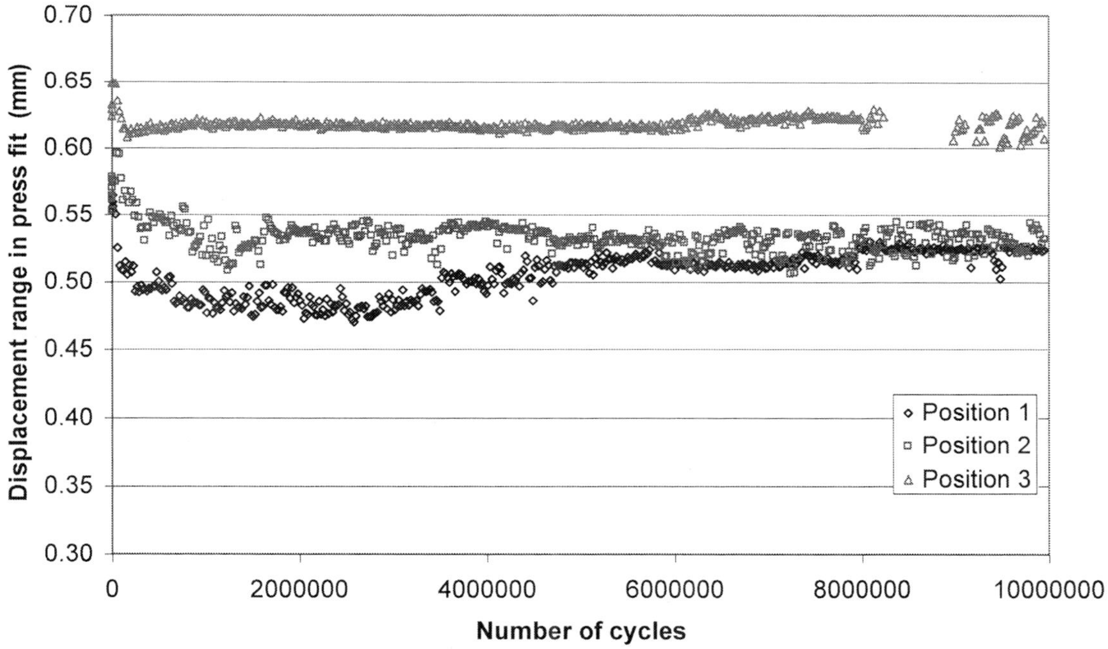


Fig. 5 Displacement range as measured by the extensometers at three circumferential points as dependence on number of fatigue load cycles

Continuous records of mutual displacement between the axle surface and the hub measured at three circumferential positions, as recorded by the LVDT extensometers, evaluated in terms of their range, difference between the maximum and minimum values as a dependence on number of load cycles is in Fig. 5.

It follows from Fig. 5 that differences between displacement values measured at different circumferential points were significant. At the beginning of the test, there were small differences between the positions 1 and 2, which, however, differed from the position 3, which was higher. The common feature of all three positions is quite significant drop of mutual displacement early after the test start. The values were saturated after 230000 cycles in case of position 3 and after more than one million cycles in case of positions 1 and 2. It indicates that some strengthening of the press fitting attachment occurs due to the repeated loading, which looks to be a common characteristic. It may be caused by an occurrence of first oxide debris by the fretting fatigue mechanism. This strengthening effect had its maximum between one and three million of cycles, then some release could be observed. Note that practically no long time strengthening effect of the mutual fixation between the axle and hub occurred at position 3 with the largest mutual displacement. It is an interesting phenomenon, because one would expect the highest oxide debris occurrence and increasing fixation particularly at points of maximum mutual displacement.

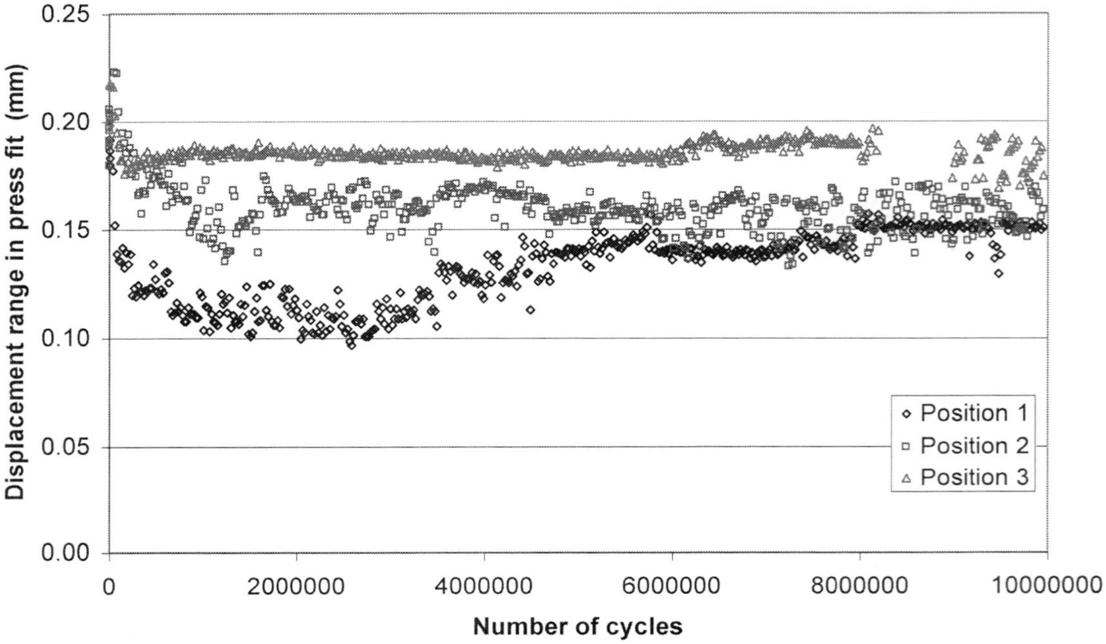


Fig. 6 Pure values of displacement between axle and hub during fatigue test after correction for elastic deformation of the part of the axle with magnetic holders

Displacement measurement shown in Fig. 5 contains, besides actual mutual displacement between the axle and hub, also the part of elastic deformation of the axle surface near the hub, at the distance of approximately 230 mm from the hub edge, where the centre of the magnetic holder was always attached. This elastic deformation was calculated and subtracted from the total displacement values shown in Fig. 5. This correction resulted in practically pure values of the mutual displacements between the axle and hub. Such corrected values are shown in Fig. 6.

Looking at Fig. 6, the features indicated already in Fig. 5 are even more distinct. The differences of displacements between positions 1 and 3 are very significant, values at position 3 being almost twice as much as those at position 1 during the first third of the test.

It should be pointed out that the differences between displacement at different measured points have nothing to do with symmetry of the component loading, which was checked by dynamic strain measurements at eight circumferential points. All the strain values were equal within the range of less than ± 1.5 %, which demonstrates that local mutual displacement was determined by local strength of the contact between the axle and hub surface, not by loading asymmetry.

The results of the displacement measurements show that even though the press fitting diagram of the axle into the wheel met the standard requirements and the pressing force was within the requested range, the homogeneity of the press fit in terms of constant dynamic displacements along the circumference during the fatigue loading was far from optimum conditions. Note that fretting fatigue mechanism and rate of the fretting fatigue damage are very sensitive on both displacement values and contact forces [11]. It would be therefore useful and interesting to investigate, whether there are some indications that the fretting fatigue damage process already started after the test interruption at $10^7$ cycles and if yes, whether the intensity of this process could be correlated with the circumferential position on the axle surface, i.e. with the values of mutual displacement between the axle and hub.

## Conclusions

Rotating bending fatigue test of full scale railway axle press fitted in the disc simulating actual wheel was performed according to European standards to verify sufficient fatigue durability either on the axle shank or under the press fit. Beyond the parameters monitored according to the standard requirements, shear contact displacements between the axle surface and the hub along the axle circumference were evaluated and recorded. The main results can be summarised as follows:

- In spite of that the pressure diagram – the diagram of pressure forces during pressing the axle into the hub fully satisfied the standard criteria, the mutual displacements during fatigue loading along the circumference were far from ideal homogeneity, the maximum values being almost twice as high as the minimum values. This inhomogeneity was not affected by loading itself, which was almost ideally homogeneous along the circumference.
- The inhomogeneity of the shear displacements may unfavourably affect the fretting fatigue process, which is strongly dependent on both mutual shear displacement and contact pressure forces at the axle and hub interface.

## Acknowledgements

Institutional support for long-time conceptual development of research organisation SVÚM a.s., used for this work, is strongly acknowledged.

## References

[1] S. Hillmansen, R.A. Smith, The management of fatigue crack growth in railway axles, P. I. Mech. Eng. F-J. Rai. 218, 4 (2004) 327-336.

[2] V. Esslinger, R. Kieselbach, R. Koller, B. Weisse, The railway accident of Eschede - technical background, Eng. Fail. Anal. 11, 4 (2004) 515-535.

[3] BS EN 13260:2009+A1:2010 Standard. Railway applications. Wheelsets and bogies. Wheelsets. Product requirements. British Standard Institution, April 2009, 40 p.

[4] BS EN 13261:2009+A1:2010. Railway applications. Wheelsets and bogies. Axles. Product requirements. British Standard Institution, April 2009, 60 p.

[5] S. Cervello, Fatigue properties of railway axles: New results of full-scale specimens from Euraxles project, Int. J. Fatigue 86 (2016) 2-12.

[6] R.A. Smith, S. Hillmansen, A brief historical overview of the fatigue of railway axles, P. I. Mech. Eng. F-J. Rai. 218, 4 (2004) 267-277.

[7] M. Madia, S. Beretta, U. Zerbst, An investigation on the influence of rotary bending and press fitting on stress intensity factors and fatigue crack growth in railway axles, Eng. Fract. Mech. 75, 8 (2008) 1906-1920.

[8] V. Linhart, I. Černý, An effect of strength of railway axle steels on fatigue resistance under press fit, Eng. Fract. Mech. 78, 5 (2011) 731-741.

[9] U. Zerbst et al., Safe life and damage tolerance aspects of railway axles—A review, Eng. Fract. Mech. 98 (2013) 214-271.

[10] I. Černý, M. Čipera, Static and dynamic experimental stress-strain analysis of axles loaded by rotation bending, EAN 2014 – 52nd Int. Conf. on Experimental Stress Analysis.

[11] Y. Mutoh, J.-Q. Xu, K. Kondoh, Observations and analysis of fretting fatigue crack initiation and propagation, ASTM STP 1425 (2002) 61-75.

# Analysis of Axially Compressed Cylindrical Shells Influenced by Geometrical Imperfections Using Finite Element Method

Tomáš Kula[a], Jozef Bocko[a], Ján Kostka[a] and Peter Frankovský[b]

Department of Applied Mechanics and Mechanical Engineering, Faculty of Mechanical Engineering, Technical University of Košice, Letná 9, 042 00 Košice, Slovakia

Department of Mechatronics, Faculty of Mechanical Engineering, Technical University of Košice, Letná 9, 042 00 Košice, Slovakia

[a]tomas.kula@tuke.sk, [a]jozef.bocko@tuke.sk, [a]jan.kostka@tuke.sk, [b]peter.frankovsky@tuke.sk

**Keywords:** Cylindrical shell, geometrical imperfection, critical load, FEM.

**Abstract.** This article deals with the influence of imperfection on the resulting stiffness of the cylindrical shell. In the first step, it is pointed out that, the imperfection that is oriented towards middle of the shell has negative effect, because it reduces the magnitude of the critical force. The second part of the article deals with the influence of the length to imperfection and how this geometric dimension fundamentally affects the magnitude of critical force. The last part examines the imperfection in relation to the three defined force positions.

## Introduction

Positive properties of shell elements are connected with their high strength with respect to their stiffness at minimum weight design. For this reasons they are used in designs, where steel structures have to have such basics properties, for example in planes, cranes, gas pipelines, etc. Design of thin-walled structures can be divided into two basic theories. Initial theory is considered to be linear theory, which considers the perfect design without imperfection. This approach can be used only for simple tasks. Nowadays, problem of nonlinear theory is developing for complicated technical tasks, where designers have to consider the imperfection.

However, this type of structure is considerably sensitive to different geometric variations, which in technical terms are called geometric imperfections. These imperfections have a major negative impact. Of their existence reduces the level of critical force that causes a loss of stability - buckling, as the authors indicated in [1-4]. The article deals with symmetrically cylindrical shell loaded by axial compressive load one end of which is fixed. The geometric imperfections are taken to be control parameters of our investigation with relation to the critical force magnitude. The computations are based on numerical calculations connected with linear buckling analysis.

## Basic knowledge and problem presentation

This paper is focused to the modelling of geometrical imperfections of axial loaded cylindrical shells. According to publication [5] there are five different methods that can be used for establishing relations between geometrical imperfections and knock down factors. The methods are:

- Linear buckling mode-shaped imperfection (LBMI),
- Single perturbation load imperfection (SPLI),
- Geometrical dimple imperfection (GDI),
- Axisymmetric imperfections (ASI),
- Mid-surface imperfection (MSI).

For our case, we will consider GDI method. The displacement field of the GDI is given as radial displacement and it is defined as a dimple cosine with wavelengths along the circumference ($a$) and the meridian ($b$):

$$\Delta r(\varphi,\zeta) = \frac{w_0}{4}\left[1-\cos\left(\frac{2\pi r}{a}\varphi\right)\right]\left[1-\cos\left(\frac{2\pi}{b}\zeta\right)\right], \tag{1}$$

where:

- $w_0 \rightarrow$ imperfection amplitude [m],
- $a \rightarrow$ wavelength along the circumference [m],
- $b \rightarrow$ wavelength along the meridian [m],
- $r \rightarrow$ radius [m].

For numerical approach we can consider linear bifurcation analysis. Linear bifurcation analysis is carried out to obtain the elastic critical buckling resistance of the perfect structure. In numerical software, the elastic critical buckling resistance is obtained through the eigenvalue problem given in (2), where the load for which the stiffness matrix becomes singular is sought:

$$(\lambda_i \mathbf{K})\bar{v}_i = 0, \tag{2}$$

where:

- $\lambda_i \rightarrow$ eigenvalue (elastic critical buckling resistance),
- $\mathbf{K} \rightarrow$ tangent stiffness matrix,
- $\bar{v}_i \rightarrow$ buckling mode shapes (eigenvectors) [6].

Let's introduce a structural element for which the analysis was accomplished. The constant dimensions are:

- $L$ (shell length) $\rightarrow$ $L = 1$ m,
- $D$ (shell diameter) $\rightarrow D = 0.2$ m.

For a better understanding of behaviour of the shell imperfections, the following variable parameters are chosen:

- $t_{min}$ (minimal wall thickness) $\rightarrow t_{min} = 0.5$ mm,
- $t_{max}$ (maximal wall thickness) $\rightarrow t_{max} = 10$ mm,
- $i_t$ (increase increment for wall thickness) $\rightarrow i_t = 0.5$ mm,
- $e_{min}$ (minimal imperfection diameter) $\rightarrow e_{min} = 0$ mm,
- $e_{max}$ (maximal imperfection diameter) $\rightarrow e_{max} = 10$ mm,
- $i_e$ (increase increment for imperfection diameter) $\rightarrow i_e = 0.5$ mm,

It is supposed that the structure is made of material with Young modulus $E = 2.1\text{e}5$ MPa and Poisson ratio $\mu = 0.3$.

In order to have the most accurate data, the parametric model was created in MATLAB. To simplify work, the MATLAB script was created such a way that it directly generates data files for ABAQUS software which subsequently carried out numerical simulations. This resulted to more than 700 values that were evaluated.

## Simulation of geometric imperfection for axisymmetric cylindrical shell

In this section the shell behaviour with relation to different magnitudes of geometric imperfections is described. The boundary conditions for FEM analyses are as follows. The shell is considered as a cantilever beam with one end fixed (removed all displacements and rotations) and the second one loaded by compression force of magnitude 1 N, fig. 1.

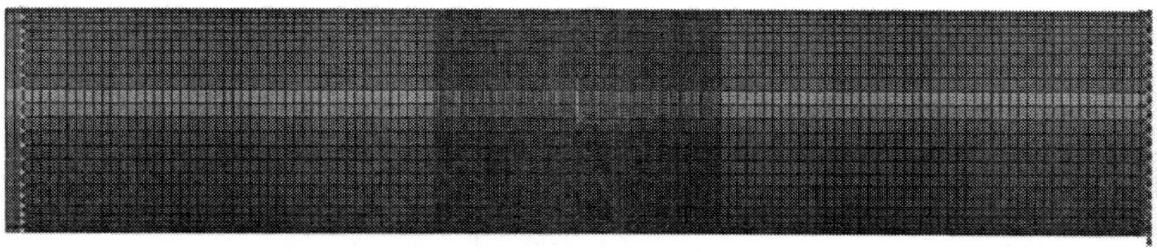

**Fig. 1 Defined boundary condition for FEM analyses**

At the beginning, we will change the shell diameter, so that it will rise to positive values and we will compare results with the state, when the shell diameter in location of imperfection decreases fig. 2. If the diameter of the shell at the location of imperfection decreases, decreases also critical force. From the numerical computations we see that the diameter decreasing leads to significant decreasing of the critical force $F_{cr,}$ as shown in fig. 3a and fig 3b. For the case, when the diameter of shell at the location of imperfection increases, we can meet with obvious phenomenon that the magnitudes of critical force are growing

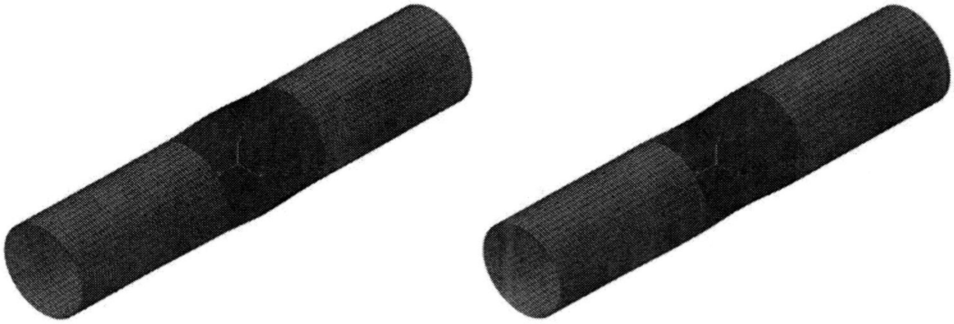

**Fig. 2 Created imperfection**

For further studies, we will only deal with cases, where the diameter at the location of imperfection decreases, because this leads to smaller critical forces.

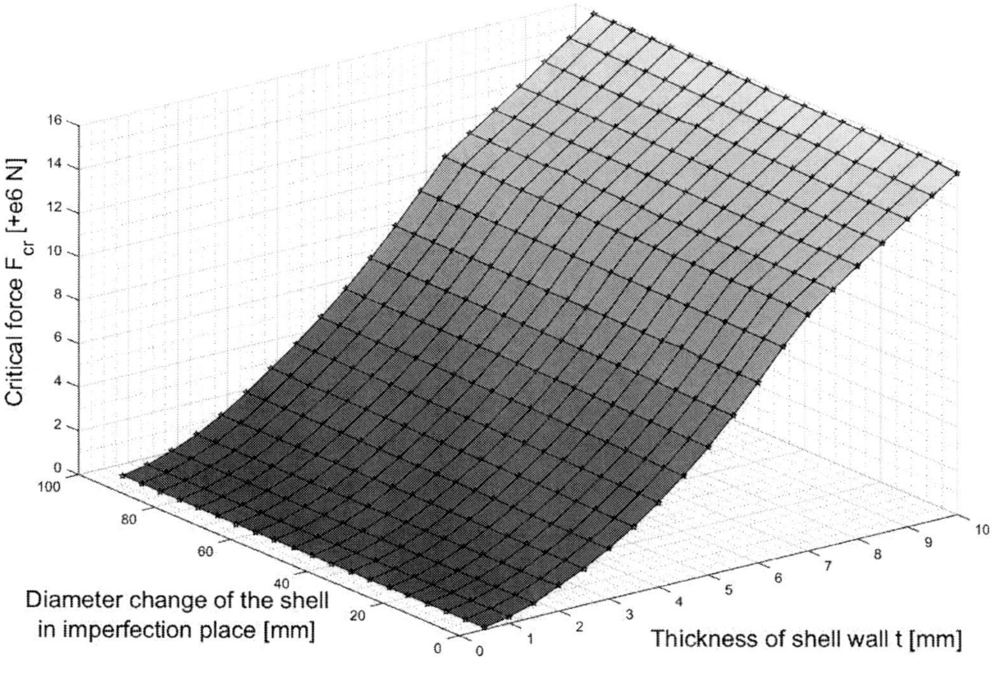


**Fig. 3a Field values depending on the size of shell diameter and critical force**

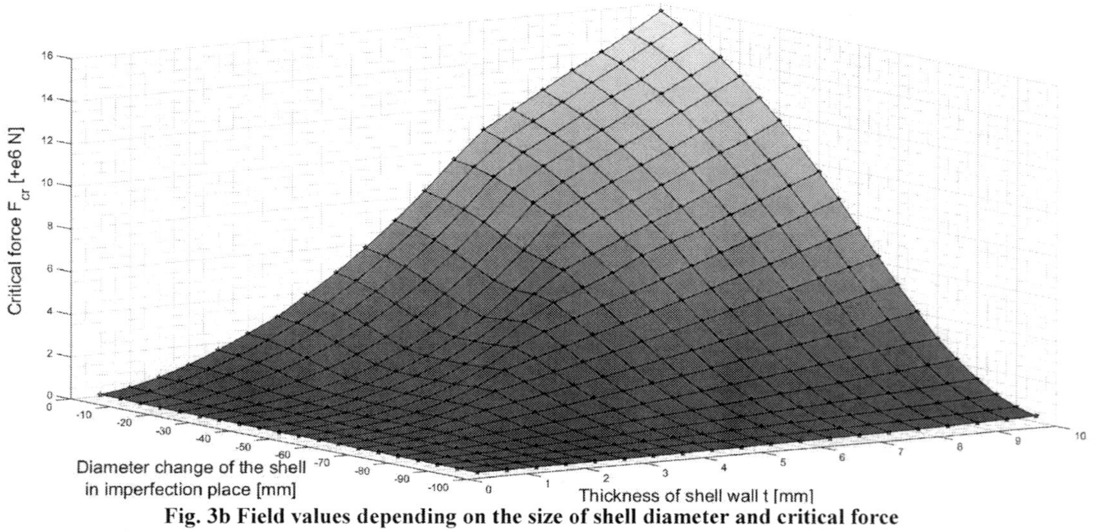


**Fig. 3b Field values depending on the size of shell diameter and critical force**

In the next step, we describe, how to length of imperfection affects the critical force $F_{\mathrm{cr}}$. There were three models prepared with defined imperfection length:

- $e_{l1} = 500$ cm,
- $e_{l2} = 400$ cm,
- $e_{l3} = 300$ cm.

These imperfections were placement in geometrical centre of shell length. Since we can assume shell behaviour, therefore, it is not necessary to count with 400 values as previous calculations. The changes are relates to increase increment for wall thickness. The previous value is $i_t = 0.5$ mm and current is $i_t = 1$ mm. And last change in this case, is reduce the diameter range of imperfection $e_d = 0.25$; 0.5; 0.75; 1; 1.25; 1.5 mm. These dimensions were chosen because, if they were produced during technological manufacturing of the structural element and if in manufacturing process is absence of quality checking, such these damaged elements could be shipped to the seller. In the fig. 4a, 4b and 4c, shows the individual fields of dependence on the length of the imperfection $e_l$.

In the last step, we will examine shell behaviour, if we change the position of imperfection. Firstly was again changed variable parameter, due to faster numerical calculation. The change occurs diameter range imperfection actual values are $e_d = 0.25$; 0.5; 0.75; 1 mm and imperfection length is constant $e_l = 250$ cm.

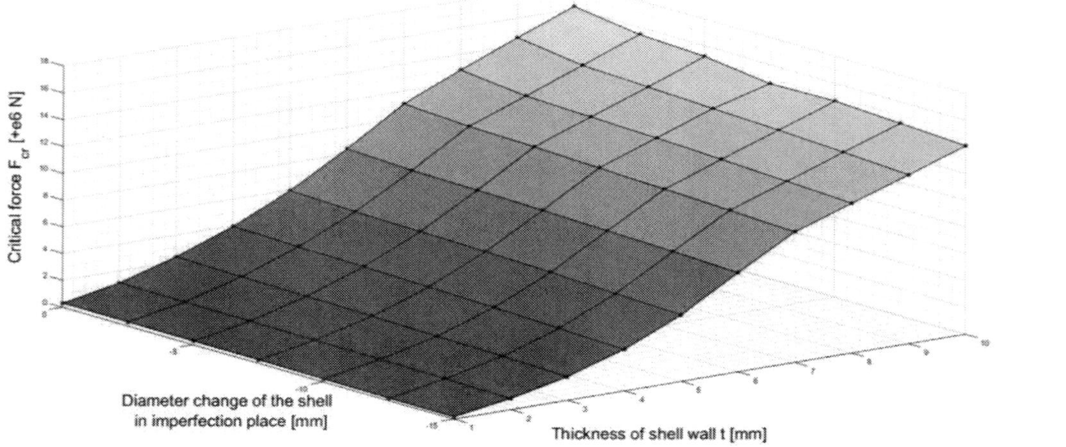


**Fig. 4a Field of critical force in dependence on the length of imperfection $e_{l1} = 500$ cm**

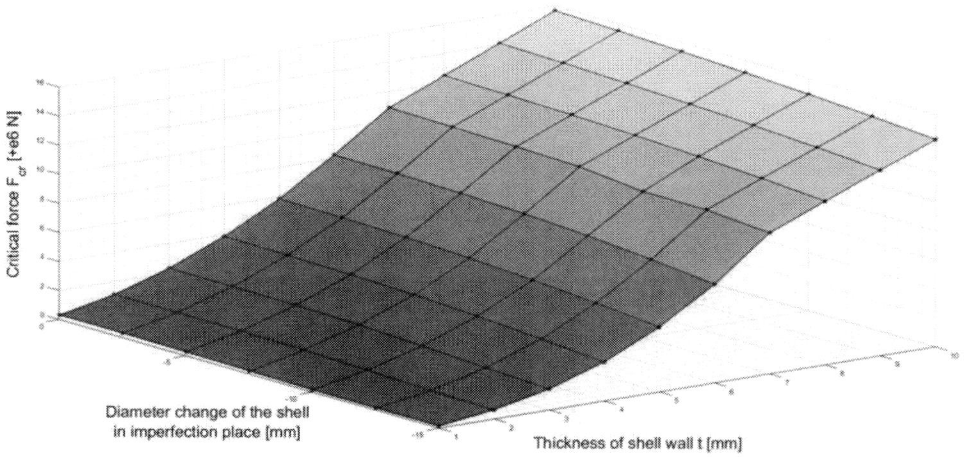


**Fig. 4b Field of critical force in dependence on the length of imperfection $e_{l1}$ = 400 cm**

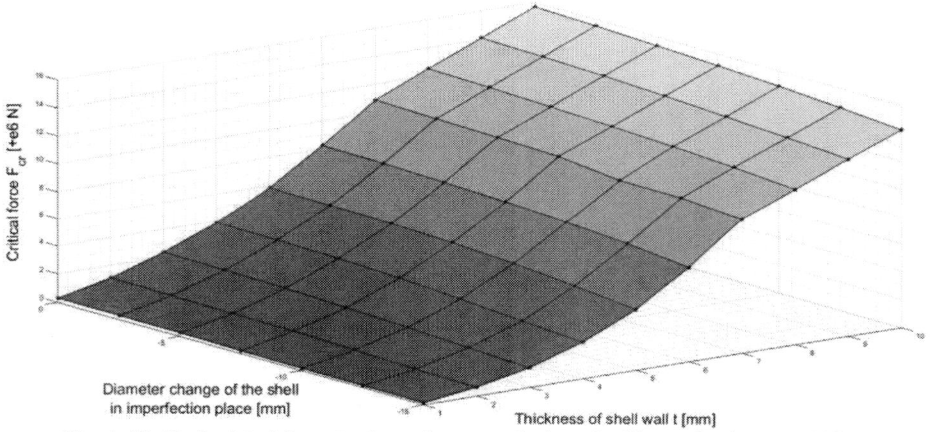


**Fig. 4c Field of critical force in dependence on the length of imperfection $e_{l1}$ = 300 cm**

First position was placed on the symmetrical centre of the shell length. Second imperfection position is at a distance from the free end - 25 cm and the last position is at a distance from the fixed end - 25 cm. In the fig. 5a, 5b and 5c shows the individual fields of dependence on the positon of the imperfection.

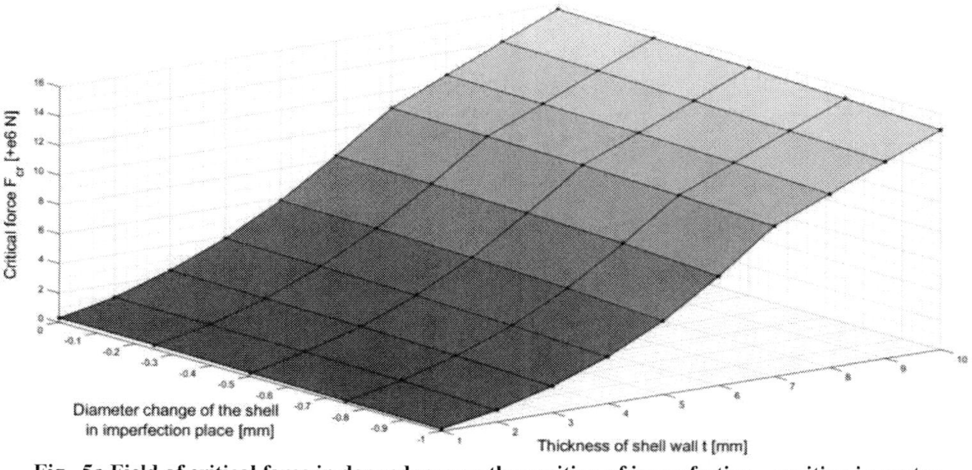


**Fig. 5a Field of critical force in dependence on the position of imperfection - position in centre**

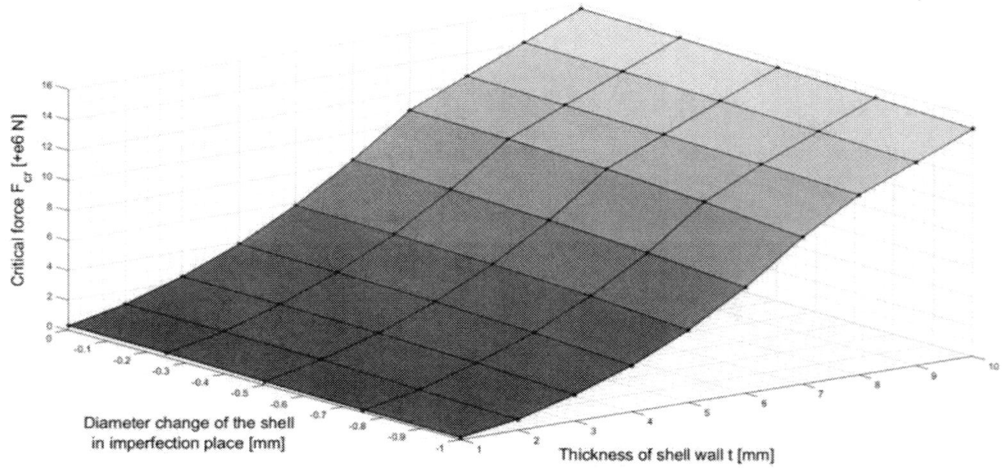


**Fig. 5b Field of critical force in dependence on the position of imperfection - upper position**

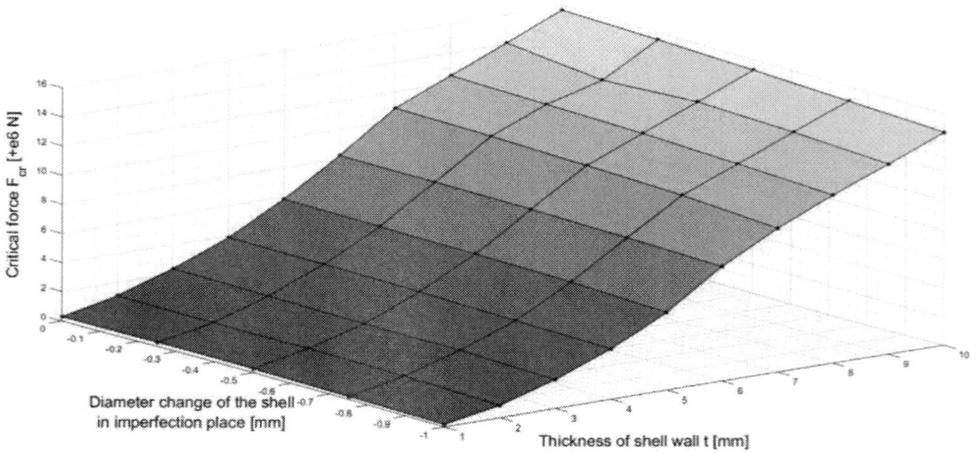


**Fig. 5c Field of critical force in dependence on the position of imperfection - bottom position**

## Results

This chapter deals with evaluation of the results obtained from simulations, which were described in previous chapter. For first case, we can evaluate the imperfections that cause the tapering of the shell casing has a negative impact for critical force $F_{cr}$. But, an imperfection which causes the shell casing extension have positive effect on critical force $F_{cr,}$ for a better understanding see fig. 6.

We can conclude that the positive contribution of imperfection is manifested for wall thicknesses greater than 3 mm.

For second case fig. 7, we can observe, that the imperfection length significantly negative affects value of critical force for smaller wall thicknesses. For wall thicknesses greater than 7 mm, the imperfection length of imperfection does not have such a major impact for critical force.

For last case fig. 8, we can observe, when the imperfection is located in the part of shell, where is defined fixture, again for smaller wall thicknesses cause a decrease value of critical force.

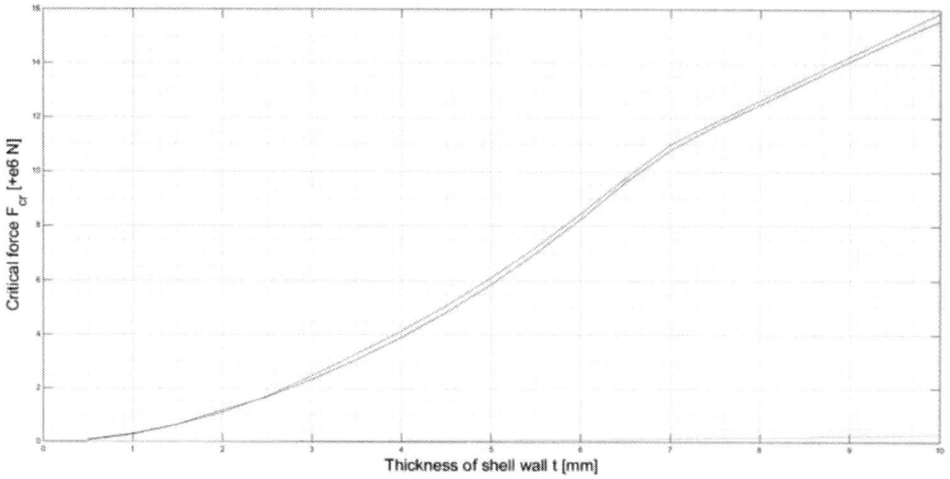


**Fig. 6** The course of critical force at defined maximum eccentricity $e_{max}$ = 95 mm, blue line – reference values without imperfection, red line – line with values + $e_{max}$ , yellow line – line with values - $e_{max}$

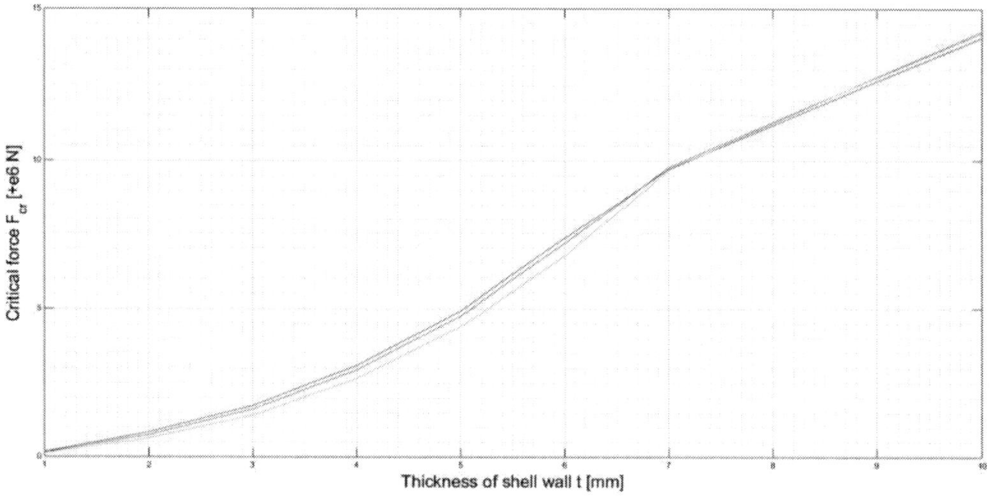


**Fig. 7** The course of critical force influenced by the imperfection length, $e_{max}$ = 1.5 mm, blue line → $e_{l1}$ = 500 cm , red line → $e_{l2}$ = 400 cm, yellow line → $e_{l3}$ = 300 cm

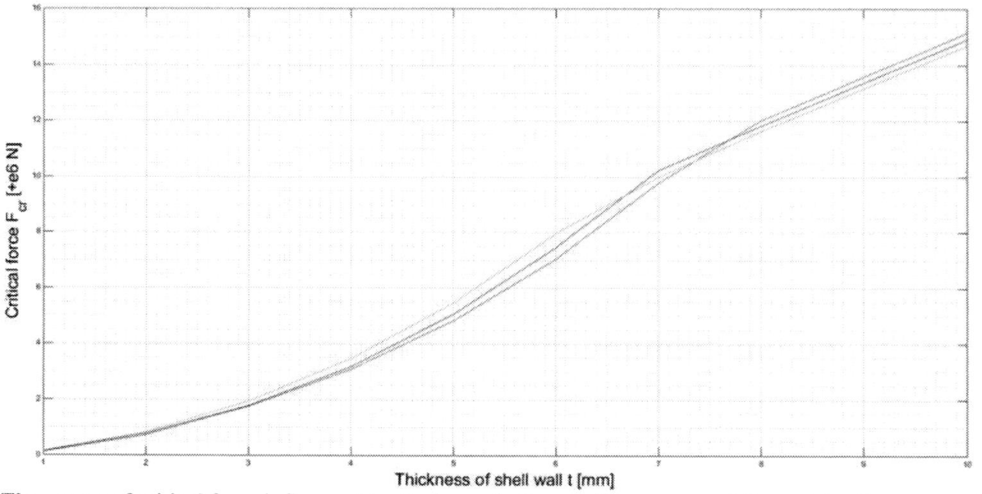


**Fig. 8** The course of critical force influenced by the imperfection position, emax = 1.5 mm, blue line → $e_{p1}$ = 0 cm , red line → $e_{p2}$ = + 25 cm, yellow line → $e_{p3}$ = - 25 cm

## Conclusions

The article deals with the influence of geometrical imperfections for case of Axially Compressed Cylindrical Shells. In general, the issue of imperfection is considered to be an undesirable phenomenon, which was also confirmed by numerical calculations, when the value of critical force was reduces by the influence of geometric imperfections.

## Acknowledgement

The authors would like to thank the Slovak Grant Agency VEGA for financial support for this study under grants VEGA no. 1/0731/16 and VEGA no. 1/0872/16.

## References

[1] Škaloud, M.: Návrhování pásů a stěn ocelových konstrukcí z hlediska stability. Praha: Academia - nakladatelství Československé akademie věd, 1988. 446 s.

[2] Ivanyi, M. Škaloud, M.: Stability problems of steel structures. Wien – New York: Springer – Verlag, 1992. 415 s. ISBN 3–211 82398-0.

[3] Donnell, L.H.: Beams, Plates and Shells. New York: Library of Cataloging in Publication Data, 1978. 453 s. ISBN 0-07-017593-4.

[4] Narayanan, R.: Shell Structures Stability and Strength. London: Elsevier Applied Science Publishers, 1985. 345s. ISBN 0-85334-343-8.

[5]Castro, Saullo and Zimmermann, Rolf and Arbelo, Mariano and Khakimova, Regina and Hilburger, Mark and Degenhardt, Richard (2014) *Geometric imperfections and lower-bound methods used to calculate knock-down factors for composite cylindrical shells.* Thin-Walled Structures, 74, pp. 118-132. Elsevier Ltd. DOI: 10.1016/j.tws.2013.08.011 ISSN 0263-8231.

[6] Simulia Corp. (2013). ABAQUS/CAE User's Manual. Dassault Systemes

EAN 2017
55th conference on experimental stress analysis

# Bending of Symmetrical and Nonsymmetrical Composite Sandwich Beam with Visco- elastic Core

M. J. CERNY[1,a], P. SLAPAK[1,a]

[1] Czech Technical University in Prague, Klokner Institute, Solinova 7, 166 08 Prague 6, Czech Republic
[a] cerny@hppro.klok.cvut.cz

**Keywords:** viscoelasticity; sandwich beam; symmetrical; nonsymmetrical; light core

## 1 Introduction

Foam-core sandwich panels and beams are increasingly used for load-bearing components in civil and mechanical engineering. Sandwich panels and beams offer a high stiffness per unit weight, but creep even at room temperature, what is limiting their use in structural applications. In this study, the viscoelastic bending of the sandwich beams will be given. The sandwich beams will be of symmetrical [1] and nonsymmetrical structure. The bearing layers are elastic, the internal core is linear viscoelastic. The model of deflection for a sandwich beam with laminate skins and a polymer core loaded by concentrated load is given. Theoretical results are compared with data from tests on sandwich beams with polymer based composite and rigid foam core.

## 2 Viscoelastic modeling of sandwich beam

### 2.1 Symmetrical sandwich beam

For the vertical displacement of the beam loaded by P at the middle applies

$$w\left(\frac{l}{2},t\right) = \frac{\overline{P}\,l^3}{48\,E.2rs_0^{\;2}} + \frac{\tilde{C}\,\overline{P}\,l}{8G_c\,s_0}$$

where $\overline{P} = \dfrac{P}{b}\,\Phi(t)$ and $\tilde{C}$ is a time integral operator, P is load and l,b beam length and width respectively, Φ(t) is a Heaviside function.

Sandwich beam consists of a core of thickness $2s_0$ and the skins with thickness $r$. Physical constants are Young modules of the skins $E$ and a constant for core $G_c$.

### 2.2 Nonsymmetrical sandwich beam

We consider geometric and physical nonsymmetrical structure of the beam.

Sandwich beam consists of a core of thickness $2s$ and the skins, from which the upper has thickness $r_h$ and lower $r_d$, respectively. Physical constants are Young modules $E_h$, $E_d$ and shear modulus of core $G_c$.

For a maximal vertical displacement under concentrated load P we get

$$w\left(\frac{l}{2},t\right) = \frac{1}{k\,\overline{s}}\,\frac{\overline{P}\,l^3}{24} + \frac{\tilde{C}\,P\,l}{4G_c\,\overline{s}}$$

where $\bar{s} = s_h + s_d = 2s_0$, when $s_h = s + \frac{r_h}{2}$ $s_d = s + \frac{r_d}{2}$

$k = E_h r_h (s_h - \delta) + E_d r_d (s_d + \delta)$, where $\delta = \dfrac{E_h r_h \left(s + \frac{r_h}{2}\right) - E_d r_d \left(s + \frac{r_d}{2}\right)}{E_h r_h + E_d r_d}$

Further $\bar{P} = \dfrac{P(t)}{b} = \dfrac{P}{b}\Phi(t)$, where $\Phi(t)$ is a Heaviside function, b is a beam width.

Operator $\tilde{C}$ is of the form $\tilde{C} = 1 + \lambda \; C^*(\alpha)$. Parameters $\lambda, \alpha$ are changed according to adopted mechanical model, e.g. for the Poynting-Thomson model (linear solid) and arrangement $G_c - G_x / K_2$ we get the following relationship:

$$\tilde{C}\,\Phi(t) = \Phi(t) + \frac{G_c}{K_2}\int_0^t e^{-\frac{G_x}{K_2}(t-\tau)}\Phi(\tau)\,d\tau = 1 + \frac{G_c}{G_x}\left(1 - e^{-\frac{G_x}{K_2}t}\right)$$

Relation $\tilde{C}\,\Phi(t)$ will be substituted into the above equation and we get similarly as in symmetrical case:

$$w\left(\frac{l}{2},t\right) = \frac{\bar{P}\,l^3}{48E.2rs_0{}^2} + \frac{\tilde{C}\,\bar{P}\,l}{8G_c\,s_0} = \frac{Pl}{8b\,s_0}\left(\frac{l^2}{12E\,r\,s_0} + 1 + \frac{G_c}{G_x}\,(1 - e^{-\frac{G_x}{K_2}t})\right)$$

We evaluate the influence of physical constants on the vertical displacement of sandwich beam. The mentioned integral equations depend on ratio $\dfrac{G}{K}$. In a relationship for $\tilde{C}\,\Phi(t)$

$$\tilde{C}\,\Phi(t) = 1 + 1 - e^{-\frac{G}{K}t}$$

Where the second part is $1 - e^{-\frac{G}{K}t}$, i.e. increment to value 1 . We take $\alpha = \dfrac{G}{K}$ stepwise 0,001, 0,003, 0,005, 0,007. The increments $\Delta = 1 - e^{-\frac{G}{K}t}$ pro $t = 10, 100, 1000$ h are given in Fig.1.

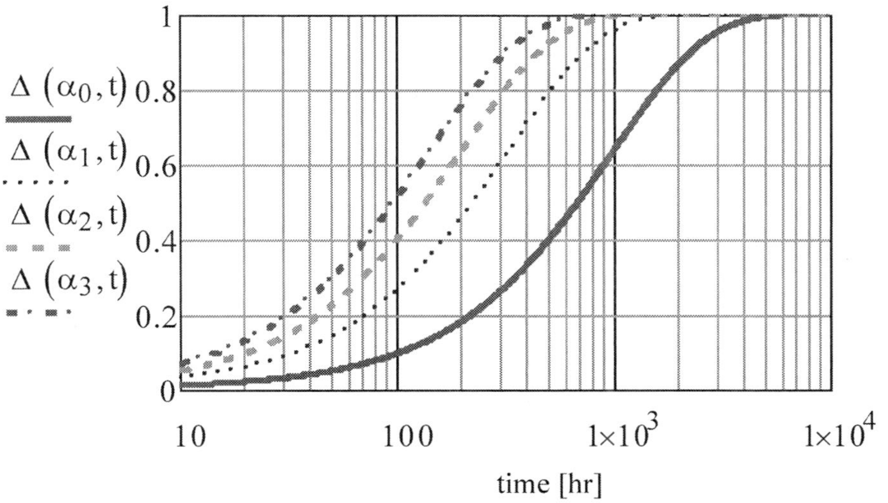


Fig.1 Influence of α= G/K on sandwich beam bending

### 2.3 Mechanical testing

Mechanical tests have been performed on a symmetrical sandwich beam with bearing layers from composite, reinforced by glass fibres and polyester matrix [2]. Further, the cube testing samples were loaded by constant load in a set-up developed and manufactured in Klokner Institute. Vertical displacements were measured by LVDTs and strains by strain gauges. Testing samples were loaded by 30 and 60 N and unloaded at least for 100 h. The tests show a good agreement with calculated values (Fig.2).

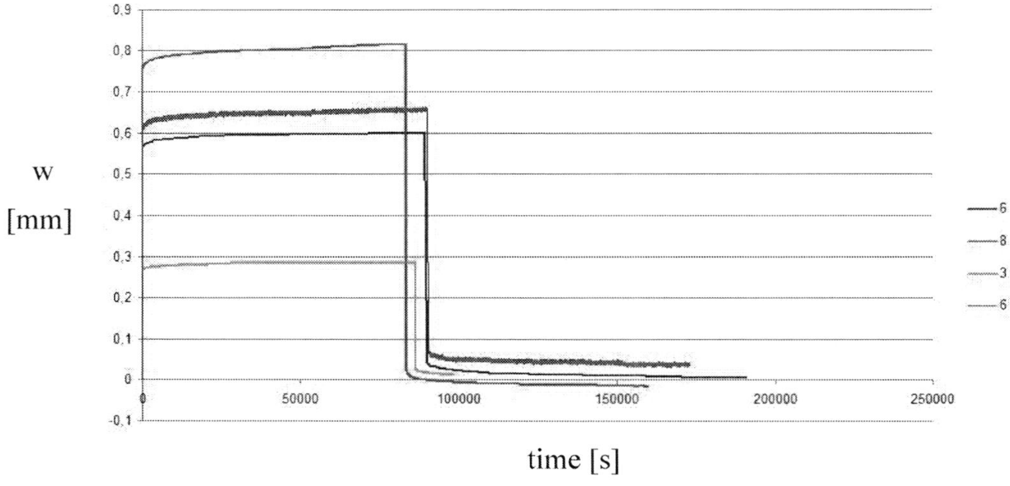


Fig.2 Experimental evaluating of creep for the symmetrical sandwich beam

**Acknowledgement**

The presented work has been supported by grant No UNMZ/73/0014/16.

**References**

[1] Cerny M.J., Slapak P., Behavior of Composite Sandwich Beam in Bending, Proceedings, ECCM15, Venice (2012)
[2] Cerny M.J., Slapak P., Visco-elastic Bending of Sandwich Beam with Foam Core, Proceedings, EAN2016, Srni (2016)
[3] Vinson J.R., 1999, Sandwich Structures of Isotropic and Composite Materials, Technomic.

EAN 2017
55th conference on experimental stress analysis

# An Alternative Material Model for Finite Strain $J_2$ Plasticity with Isotropic Hardening

Ladislav Écsi[1, a] and Pavel Élesztős[1,b]

[1]Institute of Applied Mechanics and Mechatronics, Faculty of Mech. Engrg., Slovak University of Technology in Bratislava, Námestie slobody 17, 812 31 Bratislava 1, Slovakia

[a]ladislav.ecsi@stuba.sk, [b]pavel.elesztos@stuba.sk

**Keywords:** finite-strain elastoplasticity, alternative $J_2$ material model, thermodynamically consistent formulation, finite element method

**Abstract.** In this paper an alternative $J_2$ material model with isotropic hardening for finite-strain elastoplasticity is presented. The model is based on a new non-linear continuum mechanical theory of finite deformations of elastoplastic media which allows us to describe the plastic flow in terms of various instances of the yield surface and corresponding stress measures in the body initial and current configurations. The approach also allows us to develop thermodynamically consistent material models in every respect. Consequently, the models not only do comply with the principles of material modelling, but also use constitutive equations, evolution equations and even 'normality rules' during return mapping which can be expressed in terms of power conjugate stress and strain measures or their objective rates. Therefore, such models and the results of the analyses employing them no longer depend on the description used in the model and the particularities of the material model formulation. Here we briefly present an improved version of our former material model capable of imitating ductile-to brittle failure mode transition at high strain rates in a ductile material and demonstrate the model in a numerical example.

## Introduction

Modelling of materials within the framework of finite-strain elastoplasticity represents a challenging task in computational mechanics. While plastic behaviour of structural materials within the framework of small-strain elastoplasticity is now well understood, due to the fact, that small-strain flow plasticity theories work well and their results are in agreement with experiments, the same cannot be said about finite-strain flow plasticity theories [1]. Although innumerable material models for finite-strain elastoplasticity have by now been proposed [2-10], the models in general lack universality, as their analysis results depend on the description used in the model and the particularities of the model formulation. The modelling method might simply need some developments in the non-linear continuum theory of finite deformations of elastoplastic media in order that the corresponding material models could be considered to be complete and thermodynamically consistent.

There are two modelling methods or techniques in contemporary phenomenological plasticity to model irreversible finite deformations in the material of a deformable body. The first is an ad hoc extension of infinitesimal flow plasticity theories into the area of finite deformations of elastic media to cover large displacements, but small strains of the deforming body. The related material models use an additive decomposition of a strain rate tensor into an elastic part and a plastic part and are based on a hypoelastic stress-strain relationship while utilizing the theory of nonlinear continuum mechanics of elastic media [2, 11-16].

The second method is now generally accepted as a modelling method of finite irreversible deformations in the material. It is based on the theory of single-crystal plasticity which utilizes the micromechanics of irreversible deformations of a single-crystal to describe large plastic deformations in the material. The corresponding formulations introduce the notion of an intermediate stress-free configuration and use a multiplicative split of a deformation gradient into an elastic part and a plastic part, the theory of nonlinear continuum mechanics of elastic media and classical flow plasticity theories [2, 3, 17-23].

Though it may sound surprising, our ongoing research has shown, that both of the aforementioned methods are just variants of our nonlinear continuum theory of elastoplastic media using an additive decomposition of the displacement field into an elastic part and a plastic part which describes the plastic flow in terms of various instances of a yield surface and stress measures either in the body initial or current configuration. The theory also enables to develop thermodynamically consistent material models in every respect.

The first method using the additive split of the strain rate tensor is moreover constrained when the plastic flow is defined in terms of a Cauchy/Kirchhoff stress tensor based reference yield surface in the body current configuration. As a result, the corresponding plastic part of the spatial strain rate tensor does not have a finite-strain form. That is why contemporary material models based on the additive split of the strain rate tensor look, as if they had a mixed small-strain-finite-strain formulation.

The second method, employing the multiplicative split of the deformation gradient alone also opens a few questions when one considers the definition of the deformation gradient used with it. Since it neglects the elastic/plastic displacement field in the definition of the elastic/plastic part of the deformation gradient representing the factors of the multiplicative split, the formulation of the deformation gradient in this way seems to be rather incomplete and not quite consistent with the theory of nonlinear continuum mechanics. Moreover, it can be shown, that the theory using the multiplicative split would result in the same final formula for the deformation gradient as our theory if the displacement fields were properly considered in the multiplicative split.

We will not present the non-linear continuum theory of elastoplastic media herein, nor will we explore any of its part. Our aim in this paper is to present an alternative $J_2$ material model with isotropic hardening, which as a result of the aforementioned nonlinear continuum theory, allows us to develop thermodynamically consistent alternative material models.

## The Kinematics of Deformation

In order to describe the kinematics of deformation of an elastoplastic media, we assume that the material/Lagrangian displacement field can additively be decomposed into an elastic part and a plastic part $\mathbf{u}=\mathbf{u}^{el}+\mathbf{u}^{pl}$. In this case neither the Green strain tensor $\mathbf{E}=1/2\cdot\left(\mathbf{F}^{T}\cdot\mathbf{F}-\mathbf{I}\right)$ nor the Almansi strain tensor $\mathbf{e}=1/2\cdot\left(\mathbf{I}-\mathbf{F}^{-T}\cdot\mathbf{F}^{-1}\right)$ has a decomposition into an elastic part and a plastic part, but an additive decomposition exists when one evaluates the objective time derivatives of the tensors. The material $\dot{\mathbf{E}}$ and the spatial $\mathbf{d}=\mathcal{L}_{e}(\mathbf{e})$ strain-rate tensors then take the following forms

$$\dot{\mathbf{E}}=\frac{1}{2}\cdot\left(\dot{\mathbf{F}}^{T}\cdot\mathbf{F}+\mathbf{F}^{T}\cdot\dot{\mathbf{F}}\right)=\dot{\mathbf{E}}^{el}+\dot{\mathbf{E}}^{pl}, \tag{1}$$

where

$$\dot{\mathbf{E}}^{el}=\frac{1}{2}\cdot\left[\left(\frac{\partial\dot{\mathbf{u}}^{el}}{\partial\mathbf{X}}\right)^{T}\cdot\mathbf{F}+\mathbf{F}^{T}\cdot\frac{\partial\dot{\mathbf{u}}^{el}}{\partial\mathbf{X}}\right], \tag{2}$$

$$\dot{\mathbf{E}}^{pl} = \frac{\dot{\lambda}}{2} \cdot \left[ \left( \frac{\partial^P \Psi}{\partial \mathbf{P}} \right)^T \cdot \mathbf{F} + \mathbf{F}^T \cdot \frac{\partial^P \Psi}{\partial \mathbf{P}} \right], \tag{3}$$

$$\frac{\partial \dot{\mathbf{u}}^{pl}}{\partial \mathbf{X}} = \dot{\lambda} \cdot \frac{\partial^P \Psi}{\partial \mathbf{P}}, \quad \text{and} \quad \frac{\partial^P \Psi}{\partial \mathbf{P}} \neq \left( \frac{\partial^P \Psi}{\partial \mathbf{P}} \right)^T, \tag{4}$$

$$\mathbf{d} = \mathbf{F}^{-T} \cdot \dot{\mathbf{E}} \cdot \mathbf{F}^{-1} = \mathbf{d}^{el} + \mathbf{d}^{pl}, \quad \mathbf{d}^{el} = \mathbf{F}^{-T} \cdot \dot{\mathbf{E}}^{el} \cdot \mathbf{F}^{-1}, \quad \mathbf{d}^{pl} = \mathbf{F}^{-T} \cdot \dot{\mathbf{E}}^{pl} \cdot \mathbf{F}^{-1}. \tag{5}$$

Here $\mathbf{X}$ denotes the position vector of a material point and $\mathbf{x} = \mathbf{X} + \mathbf{u}$ is the position vector of the corresponding spatial point after deformation. The deformation gradient $\mathbf{F} = \mathbf{I} + \partial \mathbf{u} / \partial \mathbf{X} = \mathbf{I} + \partial \mathbf{u}^{el} / \partial \mathbf{X} + \partial \mathbf{u}^{pl} / \partial \mathbf{X}$ then can either be expressed as a function of the material displacement field $\mathbf{u}$ alone or as a function of its elastic $\mathbf{u}^{el}$ and plastic $\mathbf{u}^{pl}$ parts. The symbols $\dot{\mathbf{E}}^{el}, \dot{\mathbf{E}}^{pl} / \mathbf{d}^{el}, \mathbf{d}^{pl}$ denote the elastic and plastic material/spatial strain rate tensors, in which the plastic flow is defined by Eqn. $(4)_1$ as a product of a plastic multiplier $\dot{\lambda}$ and an appropriate yield surface normal $\partial^P \Psi / \partial \mathbf{P}$ defined in terms of the 1st Piola-Kirchhoff stress tensor $\mathbf{P}$. Here the operator $\mathcal{L}_e(\bullet) = \mathbf{F}^{-T} \cdot \left[ \partial \left( \mathbf{F}^T \cdot (\bullet) \cdot \mathbf{F} \right) / \partial t \right] \cdot \mathbf{F}^{-1}$ is the Lie derivative operator of a spatial strain tensor. Please note, that both the elastic and plastic parts of the strain rate tensor have similar forms as the strain rate tensor. Furthermore, it can be shown, that the plastic flow defined by Eqn. $(4)_1$ is not constrained, resulting in Eqns. (3) and $(5)_3$ respectively, as the only non-degenerated forms of the material and spatial plastic strain rate tensors.

**The Constitutive Equation of the Material**

Proper formulation of a material model for finite-strain elastoplasticity enables to define the constitutive equation of the material in terms of various stress and strain measures or their objective rates in both the body initial and current configurations. As a result, the constitutive equation of a material cannot be unique, but it must have various forms. These forms however have to comply with the principles of material modelling, particularly meet the requirements of material objectivity and moreover be thermodynamically consistent in order that they would defined the same material. Furthermore, because the additive decomposition defined by Eqns. (1), $(5)_1$ exists in rate forms only, the constitutive equation of the material too has to have a rate form. In fact, Eqns. (6)-(9) define a true hypoelastic based elastoplastic material model, which does not have a form in terms of a finite strain measure.

In this research we have modified our former material model capable of imitating ductile-to-brittle failure mode transition of a ductile material at high strain rates [24]. In agreement with the above, the rate form of the constitutive equation of the material can take any of the following forms:

$$\dot{\mathbf{S}} = {}^{mat}\mathbf{C}^{el} : \left( \dot{\mathbf{E}} - x \cdot \dot{\mathbf{E}}^{pl} \right) + {}^{mat}\mathbf{C}^{vis} : \left[ \ddot{\mathbf{E}} - (1-x) \cdot \ddot{\mathbf{E}}^{pl} \right], \tag{6}$$

$$\mathcal{L}_P(\mathbf{P}) = \mathbf{F} \cdot \dot{\mathbf{S}} = \mathbf{F} \cdot \left\{ {}^{mat}\mathbf{C}^{el} : \left( \dot{\mathbf{E}} - x \cdot \dot{\mathbf{E}}^{pl} \right) + {}^{mat}\mathbf{C}^{vis} : \left[ \ddot{\mathbf{E}} - (1-x) \cdot \ddot{\mathbf{E}}^{pl} \right] \right\}, \tag{7}$$

$$\mathcal{L}_O(\boldsymbol{\tau}) = \mathbf{F} \cdot \dot{\mathbf{S}} \cdot \mathbf{F}^T = J \cdot {}^{spat}\mathbf{C}^{el} : \left( \mathbf{d} - x \cdot \mathbf{d}^{pl} \right) + J \cdot {}^{spat}\mathbf{C}^{vis} : \left[ \mathcal{L}_e(\mathbf{d}) - (1-x) \cdot \mathcal{L}_e(\mathbf{d}^{pl}) \right], \tag{8}$$

$$\mathcal{L}_T(\boldsymbol{\sigma}) = J^{-1} \cdot \mathbf{F} \cdot \dot{\mathbf{S}} \cdot \mathbf{F}^T = {}^{spat}\mathbf{C}^{el} : \left( \mathbf{d} - x \cdot \mathbf{d}^{pl} \right) + {}^{spat}\mathbf{C}^{vis} : \left[ \mathcal{L}_e(\mathbf{d}) - (1-x) \cdot \mathcal{L}_e(\mathbf{d}^{pl}) \right], \tag{9}$$

where
$$ {}^{mat}\mathbf{C}^{el} = 2 \cdot G \cdot \mathrm{I} + \lambda^{el} \cdot \mathbf{1} \otimes \mathbf{1}, \quad {}^{mat}\mathbf{C}^{vis} = 2 \cdot G^{vis} \cdot \mathrm{I} + \lambda^{vis} \cdot \mathbf{1} \otimes \mathbf{1}, \tag{10}$$

$$G=\frac{E}{2\cdot(1+\nu)},\ \lambda^{el}=\frac{\nu\cdot E}{(1+\nu)\cdot(1-2\cdot\nu)},\ G^{vis}=\frac{E^{vis}}{2\cdot(1+\nu^{vis})},\ \lambda^{vis}=\frac{\nu^{vis}\cdot E^{vis}}{(1+\nu^{vis})\cdot(1-2\cdot\nu^{vis})},\tag{11}$$

$$^{spat}\mathrm{C}^{el}_{ijkl}=J^{-1}\cdot F_{im}\cdot F_{jn}\cdot F_{ko}\cdot F_{lp}\cdot{}^{mat}\mathrm{C}^{el}_{mnop},\tag{12}$$

$$^{spat}\mathrm{C}^{vis}_{ijkl}=J^{-1}\cdot F_{im}\cdot F_{jn}\cdot F_{ko}\cdot F_{lp}\cdot{}^{mat}\mathrm{C}^{vis}_{mnop}.\tag{13}$$

In Eqns. (6)-(13) the symbols $\mathbf{S},\mathbf{P},\boldsymbol{\tau},\boldsymbol{\sigma},\dot{\mathbf{S}},\mathcal{L}_P(\mathbf{P}),\mathcal{L}_O(\boldsymbol{\tau}),\mathcal{L}_T(\boldsymbol{\sigma})$ denote the 2nd Piola-Kirchhoff stress tensor, the 1st Piola-Kirchhoff stress tensor, the Kirchhoff stress tensor, the Cauchy stress tensor and their objective rates respectively, namely the time derivative of the 2nd Piola-Kirchhoff stress tensor $\dot{\mathbf{S}}$, the Lie derivative of the 1st Piola-Kirchhoff stress tensor $\mathcal{L}_P(\mathbf{P})$, defined in terms of the Lie derivative operator of a mixed spatial-material stress tensor $\mathcal{L}_P(\bullet)=\mathbf{F}\cdot\left[\partial\left(\mathbf{F}^{-1}\cdot(\bullet)\right)/\partial t\right]$, the Oldroyd rate of the Kirchhoff stress $\mathcal{L}_O(\boldsymbol{\tau})$ defined in terms of the Lie derivative operator of a spatial stress tensor $\mathcal{L}_O(\bullet)=\mathbf{F}\cdot\left[\partial\left(\mathbf{F}^{-1}\cdot(\bullet)\cdot\mathbf{F}^{-T}\right)/\partial t\right]\cdot\mathbf{F}^T$ and the Truesdell rate of the Cauchy stress defined in terms of the Truesdell derivative operator of a spatial stress tensor $\mathcal{L}_T(\bullet)=J^{-1}\cdot\mathbf{F}\cdot\left[\partial\left(J\cdot\mathbf{F}^{-1}\cdot(\bullet)\cdot\mathbf{F}^{-T}\right)/\partial t\right]\cdot\mathbf{F}^T$, which actually carries out Lie differentiation, but with rearranged terms. Here the fourth order material elasticity tensor $^{mat}\mathrm{C}^{el}$ and the fourth order material viscosity tensor $^{mat}\mathrm{C}^{vis}$ have similar forms as the fourth order elasticity tensor of the St.-Venant Kirchoff material [25] using two independent material parameters $E,\nu$ and $E^{vis},\nu^{vis}$ respectively. The fourth order spatial elasticity and viscosity tensors $^{spat}\mathrm{C}^{el},{}^{spat}\mathrm{C}^{vis}$ the can be determined in accordance with Eqns. (12) and (13), where $J=\det(\mathbf{F})$ is the Jacobian of the deformation. The variable $x$ denotes the ratio of ductile and total damage increment [24]. Please also note, that the objective rates $\dot{\mathbf{S}},\mathcal{L}_P(\mathbf{P}),\mathcal{L}_O(\boldsymbol{\tau})$, $\mathcal{L}_T(\boldsymbol{\sigma})$ transform in the same way from one of their forms into another as do the stress tensors $\mathbf{S},\mathbf{P},\boldsymbol{\tau},\boldsymbol{\sigma}$, which ensure thermodynamic consistency of the formulation.

### Modelling of the Plastic Flow

Similarly as in the case of the rate forms of the constitutive equation of the material, proper formulation of a finite-strain flow plasticity theory enables to describe the plastic flow in terms of various instances of a yield surface and corresponding stress measures in either the body initial or current configuration. Let the various instances of the yield surface $^S\Psi={}^S\Psi(\mathbf{S},\mathbf{q})$, $^P\Psi={}^P\Psi(\mathbf{P},\mathbf{q})$, $^\tau\Psi={}^\tau\Psi(\boldsymbol{\tau},\mathbf{q})$, $^\sigma\Psi={}^\sigma\Psi(\boldsymbol{\sigma},\mathbf{q})$ be defined in terms of the stress measures $\mathbf{S},\mathbf{P},\boldsymbol{\tau},\boldsymbol{\sigma}$ and a vector of internal variables $\mathbf{q}$. After introducing similar kinematic equations as the ones defined by Eqns. (1)-(5), they serve a basis for the first nonlinear continuum mechanical theory of finite deformations of elastoplastic media. Moreover, since they define the same admissible stress space and the same plastic flow respectively, the instances of the yield surface cannot be independent of each other. In fact $^S\Psi,{}^P\Psi,{}^\tau\Psi,{}^\sigma\Psi$ are related by the following formulas

$$\frac{\partial^P\Psi}{\partial\mathbf{P}}\cdot\mathbf{F}^{-1}=\frac{\partial^\sigma\Psi}{\partial\boldsymbol{\sigma}},\quad\frac{\partial^P\Psi}{\partial\mathbf{P}}\cdot\mathbf{F}^{-1}=\frac{\partial^\tau\Psi}{\partial\boldsymbol{\tau}},\quad\mathbf{F}^T\cdot\frac{\partial^P\Psi}{\partial\mathbf{P}}=\frac{\partial^S\Psi}{\partial\mathbf{S}}.\tag{14}$$

Furthermore, one of the yield surfaces ${}^{S}\Psi, {}^{P}\Psi, {}^{\tau}\Psi, {}^{\sigma}\Psi$ has to be chosen as a reference yield surface to define the material model. It can be shown, that when ${}^{\sigma}\Psi$ or ${}^{\tau}\Psi$ is chosen as a reference yield surface in the current configuration of the body, we recover the contemporary flow plasticity theories.

Crucial part in finite-strain material modelling is thermodynamic consistency of the material model formulation. It ensures, that the material model is independent of the description and the particularities of the model formulation. The thermodynamic consistency of the plastic flow then can be expressed as follows

$$\dot{\mathbf{E}}^{pl} : \dot{\mathbf{S}} \cdot dV_0 = \frac{\partial \dot{\mathbf{u}}^{pl}}{\partial \mathbf{X}} : \mathcal{L}_P(\mathbf{P}) \cdot dV_0 = \mathbf{d}^{pl} : \mathcal{L}_O(\boldsymbol{\tau}) \cdot dV_0 = \mathbf{d}^{pl} : \mathcal{L}_T(\boldsymbol{\sigma}) \cdot dv, \tag{15}$$

where $dV_0$ is an infinitesimal volume element in the body initial configuration and $dv = J \cdot dV_0$ is its spatial counterpart. In terms of the above one can prove, that Eqn. (15) has an equivalent form defined as

$$\frac{\partial {}^{S}\Psi}{\partial \mathbf{S}} : \dot{\mathbf{S}} = \frac{\partial {}^{P}\Psi}{\partial \mathbf{P}} : \mathcal{L}_P(\mathbf{P}) = \frac{\partial {}^{\tau}\Psi}{\partial \boldsymbol{\tau}} : \mathcal{L}_O(\boldsymbol{\tau}) = J \cdot \frac{\partial {}^{\sigma}\Psi}{\partial \boldsymbol{\sigma}} : \mathcal{L}_T(\boldsymbol{\sigma}), \tag{16}$$

which is known as the 'normality rule' and which defines a rate form of a thermodynamically consistent return mapping procedure. The result is of fundamental importance in computational mechanics as it states, how the plastic multiplier ought to be calculated during return mapping when finite-strain elastoplastic analysis is carried out.

## The Reference Yield Surface

As it has been mentioned in the above, the choice of the reference yield surface determines the material model. As a result, alternative material models can be developed. In our research we have generalized the $J_2$ flow plasticity theory with isotropic hardening, where we used the ${}^{P}\Psi = {}^{P}\Psi(\mathbf{P},\mathbf{q})$, Eqn. (17) yield surface as the reference yield surface to define our material model. Please also note, that we have changed the definition of the ${}^{P}J_2(\mathbf{P}) = \mathbf{P}:\mathbf{P}$ invariant, which no longer bases on the deviatoric part of the 1st Piola-Kirchhoff stress tensor $\mathbf{P}$. This is because of the fact, that the 1st Piola-Kirchhoff stress tensor transforms under the change of the observer as $\mathbf{P}^{+} = \mathbf{Q}_R \cdot \mathbf{P}$ and the ${}^{P}J_2(\mathbf{P})$ is the only invariant, which is not affected by the change, i.e. ${}^{P}J_2(\mathbf{P}) = {}^{P}J_2(\mathbf{P}^{+})$, where $\mathbf{Q}_R$ is an arbitrary rotating tensor expressing the relative rotation of the coordinate systems of the observer with respect to the reference coordinate system. As a result, the yield surface is no longer a cylinder, but a sphere.

$${}^{P}\Psi = {}^{P}\sigma_{eq} - {}^{P}\sigma_y \le 0, \tag{17}$$

$${}^{P}\sigma_{eq} = {}^{P}\sigma_{eq}(\mathbf{P}) = \sqrt{{}^{P}J_2(\mathbf{P})} = \sqrt{\mathbf{P}:\mathbf{P}}, \qquad {}^{P}\sigma_y = F_{UT11} \cdot \left[\sigma_y + Q \cdot \left(1 - e^{-b \cdot e^{pl}}\right)\right], \tag{18}$$

$$\dot{e}^{pl} = \dot{e}^{pl}\left(\dot{\mathbf{F}}^{pl}\right) = \sqrt{\dot{\mathbf{F}}^{pl} : \dot{\mathbf{F}}^{pl}} = \dot{\lambda}, \quad e^{pl} = \int_0^t \dot{e}^{pl} \cdot dt, \quad \mathbf{F}^{pl} = \mathbf{I} + \frac{\partial \mathbf{u}^{pl}}{\partial \mathbf{X}}, \quad \dot{\mathbf{F}}^{pl} = \frac{\partial \dot{\mathbf{u}}^{pl}}{\partial \mathbf{X}} = \dot{\lambda} \cdot \frac{\partial {}^{P}\Psi}{\partial \mathbf{P}}. \tag{19}$$

The actual yield stress ${}^{P}\sigma_y$, which is a 1st Piola-Kirchhoff stress measure, determines the radius of the yield surface and is defined by Eqn. $(18)_2$. It is the only nonzero component of a stress tensor $\mathbf{P}_{UT}$ (i.e. ${}^{P}\sigma_y = [\mathbf{P}_{UT}]_{11}$) coming from an uniaxial tensile test of the modelled material, where the operator $\left[(\bullet)\right]_{11}$ extracts the element in the first row and the first column of a 2nd order tensor $(\bullet)$ written as a 3x3 square matrix. The corresponding deformation gradient and the Jacobian of deformation are denoted as $\mathbf{F}_{UT}, J_{UT}$, where $F_{UT11} = \left[\mathbf{F}_{UT}\right]_{11}$ and $J_{UT} = \det\left(\mathbf{F}_{UT}\right) = F_{UT11}$. Please also note, that the only nonzero element of the corresponding 2nd Piola-Kirchhoff stress tensor $\mathbf{S}_{UT}$ determined from the tensile test of the material is $\sigma_y + Q\cdot\left(1-e^{-b\cdot e^{pl}}\right) = \left[\mathbf{S}_{UT}\right]_{11}$. The equation is known as the NOIH rule of isotropic hardening [26], where $\mathbf{P}_{UT} = \mathbf{F}_{UT}\cdot\mathbf{S}_{UT}$. The NOIH rule uses three material parameters, the constant yield limit of the material $\sigma_y$ and $b, Q$, which control the isotropic hardening. The parameters $\sigma_y, Q$ are also 2nd Piola-Kirchhoff stress measures, while $e^{pl}$ is an accumulated plastic strain Eqn. $(19)_2$. One may note here, that we have changed the definition of the accumulated plastic strain rate $\dot{e}^{pl}$ Eqn. $(19)_1$, where $\mathbf{F}^{pl}$ is the deformation gradient of pure plastic deformations in the material whose time derivative is assumed to be in the form of Eqn. $(4)_1$. Other changes in the definitions of the accumulated plastic strain rate $\dot{e}^{pl}$ and the equivalent stress ${}^{P}\sigma_{eq}$ resulted from the need of meeting the requirements of thermodynamic consistency in booth a one-dimensional (1D) stress state and a three-dimensional (3D) stress state in the material.

### Plastic Multiplier Calculation

The calculation of the plastic multiplier is a crucial step in finite-strain elastoplastic stress analysis as it determines the value of the stress rate tensor Eqn. (6)-(9), and the plastic part of the strain rate tensors $\dot{\mathbf{E}}^{pl}, \mathbf{d}^{pl}$ during return mapping. The return mapping procedure moreover has to be thermodynamically consistent, i.e. it has to comply with Eqn. (16). The condition has not yet been met in any formulation so far. The thermodynamically consistent return mapping procedure for plastic multiplier calculation then utilizes the objective differentiation of the yield surface ${}^{P}\Psi$ and it can be expressed as follows

$$\frac{\partial\, {}^{P}\Psi}{\partial \mathbf{P}} : \mathcal{L}_{\mathrm{P}}(\mathbf{P}) - \left[\mathcal{L}_{\mathrm{P}}\left(\mathbf{P}_{UT}\right)\right]_{11} = 0, \tag{20}$$

where $\mathcal{L}_{\mathrm{P}}(\mathbf{P})$ is then replaced by the rate form of the constitutive equation of the material Eqn. (7) and the other term of Eqn. (20) by the form $\left[\mathcal{L}_{\mathrm{P}}\left(\mathbf{P}_{UT}\right)\right]_{11} = F_{UT11}\cdot Q\cdot b\cdot e^{-b\cdot e^{pl}}\cdot \dot{e}^{pl}$. Please also note, that the first term of Eqn. (20) can be replaced with any other term of Eqn. (16) because of the thermodynamic consistency of the formulation.

### The Ratio of Ductile and Total Damage Increment

The idea of the ratio of ductile and total damage increment $x$ was first introduced by Écsi and Élesztős in order to take into account the internal damping of the material properly during plastic deformations, where $x$ allowed us to distribute the plastic flow proportionally between the spring and the damper of a 1D frictional device representing the rheological model of the material [24]. The ratio is determined in an elastic corrector phase during return mapping and its value is then kept constant. Since the return mapping procedure in our material model is carried out in the 1st Piola-Kirchhoff stress space, we had to modify the definition of the ratio as follows

$$x = \frac{\left\langle \mathbf{N} : \mathbf{F} \cdot \left( {}^{mat}\mathbf{C}^{el} : \dot{\mathbf{E}} \right) \right\rangle}{\left\langle \mathbf{N} : \mathbf{F} \cdot \left( {}^{mat}\mathbf{C}^{el} : \dot{\mathbf{E}} \right) \right\rangle + \left\langle \mathbf{N} : \mathbf{F} \cdot \left( {}^{mat}\mathbf{C}^{vis} : \ddot{\mathbf{E}} \right) \right\rangle}, \tag{21}$$

where

$$\frac{\partial^P \Psi}{\partial \mathbf{P}} = \mathbf{N}, \quad \mathbf{N} = \frac{\mathbf{P}}{\sqrt{\mathbf{P} : \mathbf{P}}} = \frac{\mathbf{P}}{\|\mathbf{P}\|}, \tag{22}$$

and

$$\langle y \rangle = \frac{y + |y|}{2} \geq 0 \tag{23}$$

denote the McCauly's brackets, which return zero if $y < 0$ and where we also used the property $\mathcal{L}_P(\mathbf{P}) = \mathbf{F} \cdot \dot{\mathbf{S}}$. Please also note that all terms of Eqn. (21) are objective stress rates, so that the value of $x$ is not affected by the change of the observer.

**Numerical Experiment**

As a numerical experiment, plastic bending of a cantilever of dimensions 50mm x 50mm x 600mm has been studied. One third of the upper surface of the beam was loaded applying $p = 5.0$ MPa constant pressure on it as stepped load using a Heaviside step function using. The beam was initially at rest and the analysis was run as transient-dynamic using $0.5 \cdot 10^{-4}$ s time step size and $0.048$ s calculation end time. Figure 3 depicts the spatially discretized beam, the boundary conditions and the applied pressure with arrows.

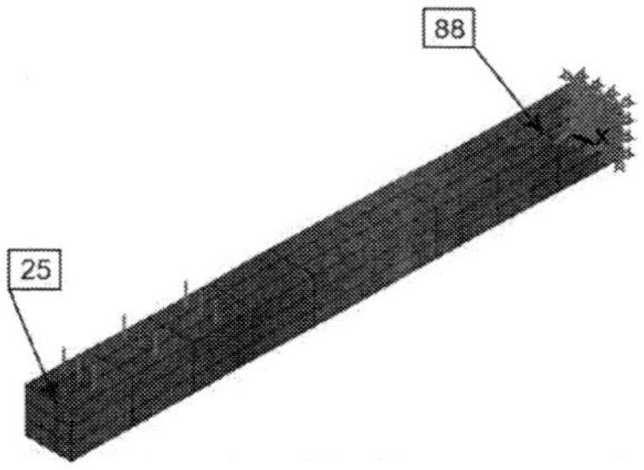


**Fig. 1.** Spatially discretized model of the beam

The cantilever material was a low carbon steel whose material properties are listed in Tab. 1.

**Table 1.** Material properties of the cantilever in the body initial configuration

| | |
|---|---|
| $E$ $[Pa]$ | $2.1 \cdot 10^{11}$ |
| $E^{vis}$ $[Pa \cdot s]$ | $2.1 \cdot 10^{7}$ |
| $\nu = \nu^{vis}[-]$ | 0.3 |
| $\sigma_y$ [Pa] | $350.0 \cdot 10^{6}$ |
| $Q$ [Pa] | $150.0 \cdot 10^{6}$ |
| $b$ [-] | 3.0 |
| $\rho_0$ [kg/m$^3$] | 7800.0 |

In order to simplify the analysis, we assumed, that $\mathbf{F}_{UT} = \mathbf{I}$ and $F_{UT11} = 1$, since we did not have information about the deformation gradient $\mathbf{F}_{UT}$ and its Jacobian $J_{UT}$ coming from the tensile test of the material.

### Numerical Results

Figures 2-5 show a few selected results of the finite element analysis, namely the Von-Mises stress distribution at the end of the analysis, the vertical displacement time history curve at nodes N25 and N88, the vertical velocity and the vertical acceleration time history curves at node N25 (see Fig. 1 for the location of the nodes). As can be seen in the figures, the loading of the body is highly dynamic with maximum deformations around 0.12 m, maximum velocity around 18.0 m.s$^{-1}$ and maximum acceleration around 18500.0 m.s$^{-2}$.

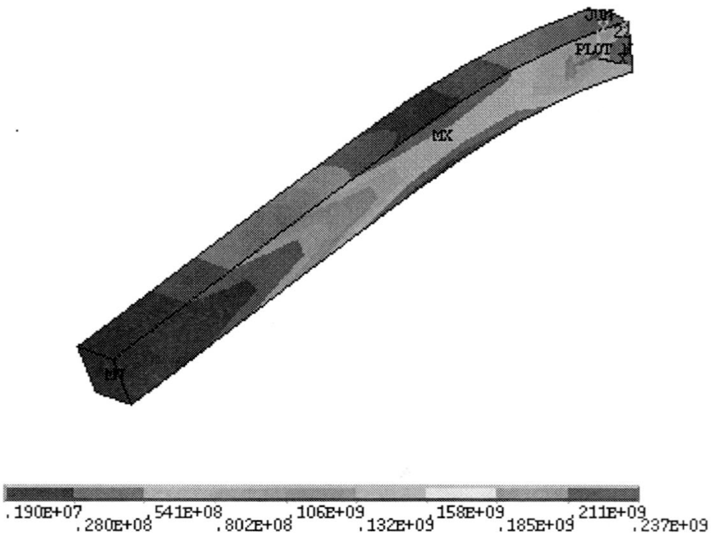


**Fig. 2.** Von Mises stress distribution at the end of the analysis

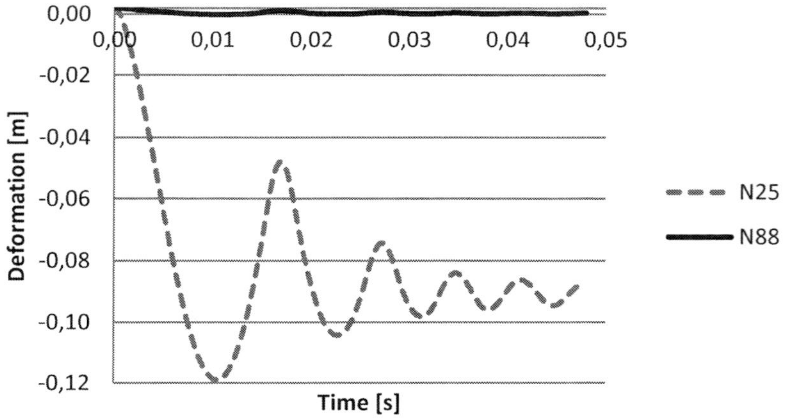


**Fig. 3.** Vertical deformation time history curves at nodes N25 and N88

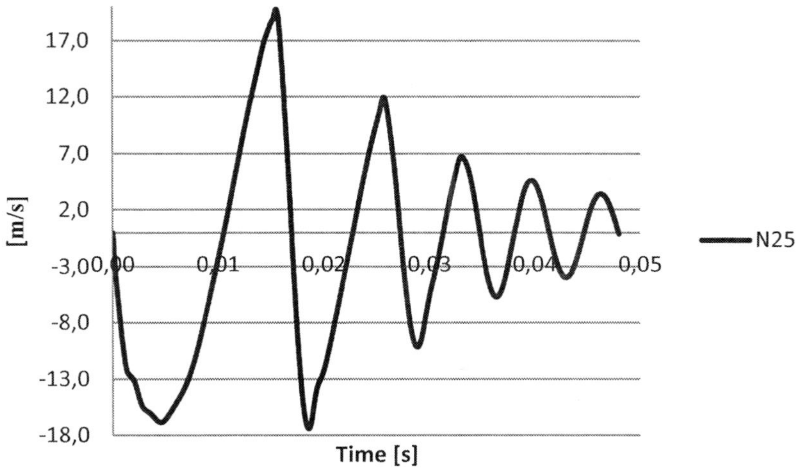


**Fig. 4.** Vertical velocity time history curve at node N25

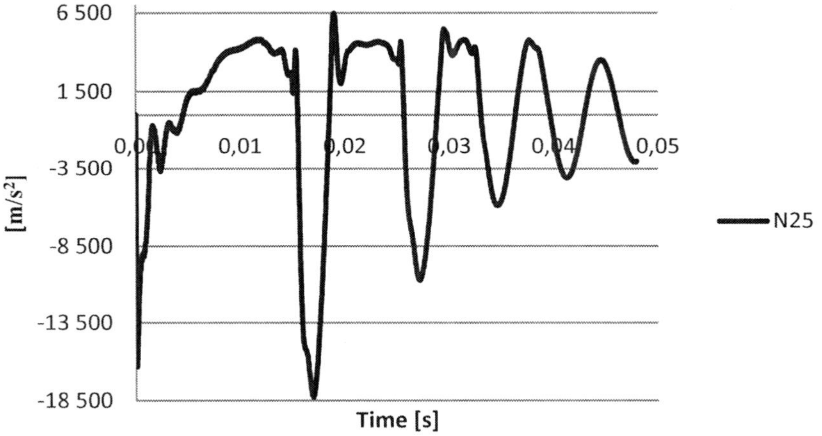


**Fig. 5.** Vertical acceleration time history curve at node N25

## Conclusions

In this paper some recent developments in the solution of structural finite element problems within the framework of finite-strain elastoplasticity was presented. An alternative $J_2$ material model using an improved additive decomposition of the strain rate tensor into an elastic part and a plastic part, capable of modelling ductile-to-brittle failure mode transition of a ductile material at high strain rates, was developed. The model is a result of the first non-linear continuum theory of finite deformations of elastoplastic media which allows to describe the plastic flow in terms of various instances of the yield surface and corresponding stress measures in the body initial and current configurations. The model moreover is thermodynamically consistent. Therefore, it uses constitutive equations, evolution equations and even "normality rules" which can be expressed in terms of power conjugate stress and strain measures or their objective rates either in the body initial or current configuration. Consequently, the analysis results of such a model, are independent of the description used in the model and the particularities of the model formulation.

**Acknowledgement**

This publication is the result of the project implementation: Research of friction stir welding (FSW) application as an alternative to melting welding methods no. 26240220031 supported by the Research & Development Operational Programme funded by the ERDF. Funding from the VEGA grant 1/0740/16 resources is also greatly appreciated.

**References**

[1] G.A. Maugin, The Thermomechanics of Plasticity and Fracture, Cambridge University Press, Cambridge, 1992.

[2] J.C. Simo, T.J.R. Hughes, Computational Inelasticity, Springer, New York, 1998.

[3] E.A. de Souza Neto, D. Perić, D.R.J. Owen, Computational Methods for Plasticity, Theory and Applications, John Wiley & Sons Ltd., Singapore, 2008.

[4] S. Nemat-Nasser, Plasticity, A Treatise on Finite Deformation of Heterogenous Inelastic Materials, Cambridge University Press, Cambridge, 2004.

[5] A. Bertran, Elascity and Plasticity of Large Deformations, An Introduction, 3rd ed., Spinger, Berlin, Heidelberg, 2013.

[6] D.R.J. Owen, E. Hinton, Finite Elements in Plasticity, Theory and Practice, Pineridge Press, Swansea, 1980.

[7] T. Belytschko, W. K. Liu, B. Moran, Nonlinear Finite Elements for Continua and Structures, John Wiley & Sons LTD, Chichester, 2000.

[8] M.A. Crisfield, Non-linear Finite Element Analysis of Solids and Structures, Essentials, Vol. 1, John Wiley & Sons Ltd., Chichester, 2000.

[9] M.A. Crisfield, Non-linear Finite Element Analysis of Solids and Structures, Advanced Topics, Vol. 2, John Wiley & Sons LTD, Chichester, 2001.

[10] K.J. Bathe, Finite Element Procedures, Prentice-Hall INC., Englewood Cliffs, New Jersey, 1995.

[11] C. Truesdell, R.A. Toupin, The Classical Field Theories, Handbook der Physic, III/1, Springer-Verlag, Berlin, 1960.

[12] C. Truesdell, W. Noll, The Nonlinear Field Theories, Handbook der Physic, III/3, Springer-Verlag, Berlin, 1965.

[13] M.E. Gurtin, An Introduction to Continuum Mechanics, Academic Press, Orlando, 1981.

[14] J.E. Marsden, T.J.R. Hughes, Mathematical Foundations of Elasticity, Dover, New York, 1994.

[15] R.W. Ogden, Non-linear Elastic Deformations, Ellis Horwood Ltd., West Sussex, 1984.

[16] P. Cairlet, Mathematical Elasticity. Vol. 1, Three-dimenasional elasticity, North-Holland Publishing Company, Amsterdam, 1988.

[17] R.J. Asaro, Micromechanics of Crystals and Polycrystals, Advances in Applied Mechanics, Vol. 23., Academic Press, New York, 1983..

[18] G.I. Taylor, Analysis of Plastic Strain in a Cubic Crystal, in Stephen Timoshenko 60th aniversary volume, ed., J.M. Lessels, Macmillan, New York, 1938.

[19] R. Hill, Generalized constitutive relations for incremental deformation of metal crystals by multislip, Journal of Mechanics and Physics of solids, 14 (1966) 95-102.

[20] R. Hill, J.R. Rice, Constitutive analysis of elastic-plastic crystals at arbitrary strain, Journal of Mechanics and Physiscs of Solids, 20 (1972) 401-413.

[21] R.J. Asaro, J.R. Rice, Strain localization in ductile single crystals, Journal of Mechanics and Physiscs of Solids, 25 (1977) 309-338.

[22] R.J. Asaro, Geometrical effects in the inhomogeneous deformation of ductile single crystals, Acta Metallurgica, 27 (1979) 445-453.

[23] C. Sansour, J. Bocko, On the Numerical Implications of Multiplicative Inelasticity with an Anisotropic Elastic Constitutive Law, Int. J. Num. Meth. Engrg., 5, 14 (2003) 2131-2160.

[24] L. Écsi, P. Élesztős,: Moving Toward a More Realistic Material Model of a Ductile Material with Failure Mode Transition. "Mat.-wiss. u. Werkstofftech", 5, 43 (2012) 379-387.

[25] J. Bonet, R.D. Wood, Nonlinear Continuum Mechanics for Finite Element Analysis, 2nd ed., Cambridge University Press, Cambridge 2008.

[26] J. Lemaitre, Handbook of Material Behaviour Models, Vol. 1, Deformation of Materials, Academic press, London, 2001.

# Mathematical Setting Eqations of Linear Contact of the Spiral Rope – Phylum VARRINGTON

Eugene Kalentev[1,a], Valerii Tarasov[1,b]
and Aleksandr Korshunov [1,c]

[1] Institute mechanics Ural Branch of the Russian Academy of Sciences,
street T. Baramzina, 34, 426067 Izhevsk, Russian Federation

[a] EugeneDavis@mail.ru, [b] tvv@udman.ru, [c] kai@istu.ru

**Keywords:** set of equations, linear contact, a spiral rope.

**Abstract:** The article deals with the problem of obtaining new solutions of a system of linear contact equations for a spiral rope of the "Varrington" type, which did not require the use of the approximation method. The high complexity of the method under consideration substantially limits its use and requires the use of special software [1, 2]. The question of convergence of the iterative process in the course of computations also usually arises. The main idea is based on the representation of the initial dependencies in the form of series and the subsequent solution of the equations obtained. The proposed procedure can be easily adapted to calculate the geometry of spiral ropes and locks of various types: "Forces", "Varrington-forces" and several others [3,4].

## Introduction

It so happened that steel ropes are traditionally widely used to provide the functioning of a number of samples of modern technology, especially for lifting and transporting machines and mechanisms. Currently, there is a fairly large number of rope sizes, but as the key design factors determining the structure and applicability of a particular type of rope are such as: the cross-sectional profile of an individual wire, the circuit of the contact between the wires in the chain, the order of laying, the contact of the wire in strands [8].
In the usual case, for rope constructions, cables with spiral axes are single- or multi-stage, laid around the central wire of a straight line [5,6]. Ropes of regular laying are also known as twisted ropes. If we continue laying the twisted rope, a so-called double-rope rope will be obtained, where the twisted rope will be called a yarn. It should be noted that for double-laid cables, the central wire is in the form of a spiral line, so you can get a rope of any order. The most common scheme is twisted and double ropes [7].

## Theoretical part

If the wires are stacked in strands in layers with different steps, the layers touch at the points. In the event that the steps of the wire layer are equal, the wires of the upper layer are placed in the grooves formed by the wires of the lower layer, then the threads have a linear contact. For a large number of layers, a combination of point and line contacts between the layers of wire is possible [9]. Consider a new solution to the system of linear contact equations for the spiral rope of the Warrington type. In this case, the use of the approximation method was not required.

On the basis of an approximation method in work [1] the decision of set of equations of linear contact (1) wires a spiral rope of phylum "Varrington" (fig. 1) has been received [11]. High labour input of this method confines its applications and almost always demands any program realisation. Besides there is opened a question concerning convergence of iterative process. Considering above told, we will receive new decisions of system (1).

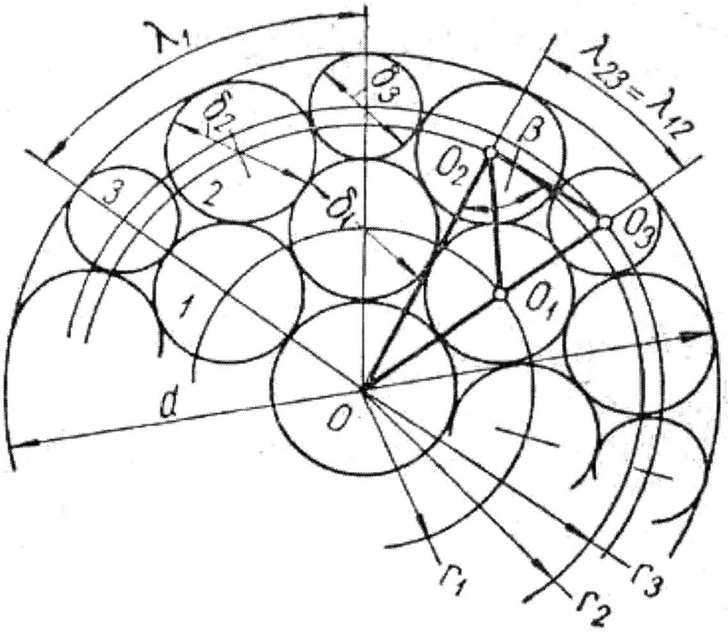


Fig. 1. The Cross-section of a spiral rope of phylum "Varrington"

In drawing it is designated $r_1$, $r_2$, $r_3$ - radiuses of beds wires «1», «2», «3» accordingly, $\delta_1$, $\delta_2$, $\delta_3$ - diameters wires beds «1», «2», «3» accordingly, $\lambda_1$, $\lambda_{12}$, $\lambda_{23}$ - polar angles of contact wires «1», «1-2», «1-3» accordingly.

$$\left.\begin{aligned} &2r_2 + \delta_2 = d, \\ &2r_3 + \delta_3 = d, \\ &2r_1 + 2\delta_1 + 2\delta_3 = d, \\ &\delta_1 = 2r_1 s_1, \\ &\delta_{12}^2 = \left(\frac{\delta_1 + \delta_2}{2}\right)^2 = \Phi_{12}, \\ &\delta_{23}^2 = \left(\frac{\delta_2 + \delta_3}{2}\right)^2 = \Phi_{23}, \end{aligned}\right\} \qquad (1)$$

Where

$$s_1 = \sin\left(\frac{\varepsilon_1}{2}\right)\sqrt{1 + \cos^2\left(\frac{\varepsilon_1}{2}\right)\tan^2(\alpha_1)}$$

Expressions for distance squares, between screw axes linearly contacting wires:

$$\delta_{12}^2 = r_1^2 \tan^2(\alpha_2)\sin^2(\varepsilon_{12}) + r_1^2 + r_2^2 - 2r_1r_2\cos(\varepsilon_{12}),$$
$$\delta_{23}^2 = r_2^2 \tan^2(\alpha_3)\sin^2(\varepsilon_{23}) + r_2^2 + r_3^2 - 2r_2r_3\cos(\varepsilon_{23}) \qquad (2)$$

From drawing 1 it is possible to define:

$$m_1 = m,\ m_2 = 2m,$$
$$\lambda_1 = \frac{2\pi}{m},\ \lambda_{12} = \lambda_{23} = \frac{\pi}{m} \qquad (3)$$

For reception of new decisions we will arrive as follows.

The initial data for calculation: $d$ - diameter of a spiral rope, $h$ - a step-lay, $m_1 = m,\ m_2 = 2m$ - a design (quantity wires on beds). For simplification of calculations we will take over known size $\delta_2$ - diameter wires «2». Thus, it is necessary to define five unknown persons $r_1$, $r_2$, $r_3$, $\delta_1$, $\delta_3$ radiuses of beds wires «1», «2», «3» and diameters wires «1» and «3». We will write down interrelations for angles of lay wires:

$$\alpha_1 = \arctan\left(\frac{2\pi r_1}{h}\right),\ \alpha_2 = \arctan\left(\frac{2\pi r_2}{h}\right),\ \alpha_3 = \arctan\left(\frac{2\pi r_3}{h}\right). \qquad (4)$$

Expressions for the auxiliary equations taking into account (4) will become:

$$\varepsilon_1 = arc\cot\left(\frac{\frac{4\pi^2 r_1^2}{h^2} + \cos\left(\frac{2\pi}{m}\right)}{\sin\left(\frac{2\pi}{m}\right)}\right),\ \varepsilon_{12} = arc\cot\left(\frac{\frac{4\pi^2 r_1 r_2}{h^2} + \cos\left(\frac{\pi}{m}\right)}{\sin\left(\frac{\pi}{m}\right)}\right),\ \varepsilon_{23} = arc\cot\left(\frac{\frac{4\pi^2 r_2 r_3}{h^2} + \cos\left(\frac{\pi}{m}\right)}{\sin\left(\frac{\pi}{m}\right)}\right). \qquad (5)$$

From the first equation of system (1) we will find expression for $r_2$:

$$r_2 = \frac{d - \delta_2}{2}. \qquad (6)$$

Considering (6) we will write down new expressions for $\alpha_2'$ и $\varepsilon_{23}'$ using (4) and (5):

$$\alpha_2' = \arctan\left(\frac{\pi(d-\delta_2)}{h}\right),\ \varepsilon_{23}' = arc\cot\left(\frac{\frac{2\pi^2(d-\delta_2)r_3}{h^2} + \cos\left(\frac{\pi}{m}\right)}{\sin\left(\frac{\pi}{m}\right)}\right). \qquad (7)$$

Auxiliary expression for $S_1$ taking into account (4) and (5) will become:

$$s_1 = \sin\left(\frac{1}{2} arc\cot\left(\frac{\frac{4\pi^2 r_1^2}{h^2}+\cos\left(\frac{2\pi}{m}\right)}{\sin\left(\frac{2\pi}{m}\right)}\right)\right)\sqrt{1+\frac{4\cos^2\left(\frac{1}{2} arc\cot\left(\frac{\frac{4\pi^2 r_1^2}{h^2}+\cos\left(\frac{2\pi}{m}\right)}{\sin\left(\frac{2\pi}{m}\right)}\right)\right)\pi^2 r_1^2}{h^2}}. \qquad (8)$$

Further we will express values of diameters from the first and second expressions of system (1) and we will substitute them in the sixth expression of the specified system, taking into account (2) we will receive:

$$d^2 - 2dr_2 - 2dr_3 + 2r_2r_3 - r_2^2\tan^2(\alpha_3)\sin^2(\varepsilon'_{23}) + 2r_2r_3\cos(\varepsilon'_{23}) = 0. \qquad (9)$$

Let's write down expression (9) in the form of function from $r_3$:

$$f3(r_3) = d^2 - d(d-\delta_2) - 2dr_3 + (d-\delta_2)r_3 - \frac{2(d-\delta_2)^2\pi^2 r_3^2}{h^2\left(1+\frac{\left(\frac{2\pi^2(d-\delta_2)r_3}{h^2}+\cos\left(\frac{\pi}{m}\right)\right)^2}{\sin^2\left(\frac{\pi}{m}\right)}\right)} + \frac{(d-\delta_2)r_3\left(\frac{2\pi^2(d-\delta_2)r_3}{h^2}+\cos\left(\frac{\pi}{m}\right)\right)}{\sin\left(\frac{\pi}{m}\right)\sqrt{1+\frac{\left(\frac{2\pi^2(d-\delta_2)r_3}{h^2}+\cos\left(\frac{\pi}{m}\right)\right)^2}{\sin^2\left(\frac{\pi}{m}\right)}}}. \qquad (10)$$

Let's spread out the yielded function abreast Maklorena with degree of a residual member equal 3:

$$taylor(f3) = d\delta_2 + \left(-d-\delta_2+\cos\left(\frac{\pi}{m}\right)(d-\delta_2)\right)r_3 - \frac{-(d-\delta_2)^2\pi^2\sin\left(\frac{\pi}{m}\right)}{\sqrt{-\frac{1}{-1+\cos\left(\frac{\pi}{m}\right)}}h^2} r_3^2 + O(r_3^3). \qquad (11)$$

Let's reject a residual member and we will resolve expression (11) rather $r_3$:

$$r_3 = -\frac{h}{2\pi^2\sin^2\left(\frac{\pi}{m}\right)\left(d^2 - 2d\delta_2 + \delta_2^2\right)} \times \left(h\cos\left(\frac{\pi}{m}\right)d - h\cos\left(\frac{\pi}{m}\right)\delta_2 - hd - h\delta_2 + vk_1\right), \qquad (12)$$

Where

$$vk_1 = \left( -\left( 4d^3\delta_2\pi^2 - 8d^2\delta_2^2\pi^2 + 4d\delta_2^3\pi^2 - \frac{h^2d^2}{\sin^2\left(\frac{\pi}{m}\right)} + \frac{2h^2d^2\cos\left(\frac{\pi}{m}\right)}{\sin^2\left(\frac{\pi}{m}\right)} - \frac{h^2d^2\cos^2\left(\frac{\pi}{m}\right)}{\sin^2\left(\frac{\pi}{m}\right)} + \frac{2h^2d\delta_2\cos^2\left(\frac{\pi}{m}\right)}{\sin^2\left(\frac{\pi}{m}\right)} - \frac{h^2\delta_2^2\cos^2\left(\frac{\pi}{m}\right)}{\sin^2\left(\frac{\pi}{m}\right)} - \frac{2h^2\cos\left(\frac{\pi}{m}\right)\delta_2^2}{\sin^2\left(\frac{\pi}{m}\right)} - \frac{2d\delta_2h^2}{\sin^2\left(\frac{\pi}{m}\right)} - \frac{h^2\delta_2^2}{\sin^2\left(\frac{\pi}{m}\right)} \right) \sin^2\left(\frac{\pi}{m}\right) \right)^{\frac{1}{2}}.$$

**Discussion**

From the second expression of system (1) it is easily received:

$$\delta_3 = d - 2r_3. \tag{13}$$

Substituting the second and fourth equations of system (1) in the third, we cluster all members in the left part and it is received:

$$2r_1 + 2r_1s_1 - 4r_3 + d = 0. \tag{14}$$

Let's write down expression (14) in the form of function from $r_1$:

$$f1(r_1) = 2r_1 + 2r_1\sin\left(\frac{1}{2}\varepsilon_1\right)\sqrt{1 + \frac{4\cos^2\left(\frac{1}{2}\varepsilon_1\right)\pi^2 r_1^2}{h^2}} + \frac{2}{\pi^2\sin^2\left(\frac{\pi}{m}\right)\left(d^2 - 2d\delta_2 + \delta_2^2\right)} \times$$

$$\times\left(\left(h\cos\left(\frac{\pi}{m}\right)d - h\cos\left(\frac{\pi}{m}\right)\delta_2 - hd - h\delta_2 + vk_1\right)h\right) + d. \tag{15}$$

Let's spread out the received expression abreast Maklorena with degree of a residual member equal 3:

$$taylor(f1) = d + \frac{h}{\pi^2\sin^2\left(\frac{\pi}{m}\right)\left(d^2 - 2d\delta_2 + \delta_2^2\right)} \times \left(h\cos\left(\frac{\pi}{m}\right)d - h\cos\left(\frac{\pi}{m}\right)\delta_2 - hd - h\delta_2 + vk_1\right) +$$

$$+\left(2\cos\left(\frac{\pi}{4} + \frac{1}{2}\arctan\left(\cot\left(\frac{2\pi}{m}\right)\right)\right) + 2\right)r_1 + O(r_1^3). \tag{16}$$

For an illustration of accuracy of used approximating we will result function graphs $f_1(r_1)$ и $taylor(f1)$ (fig. 2).

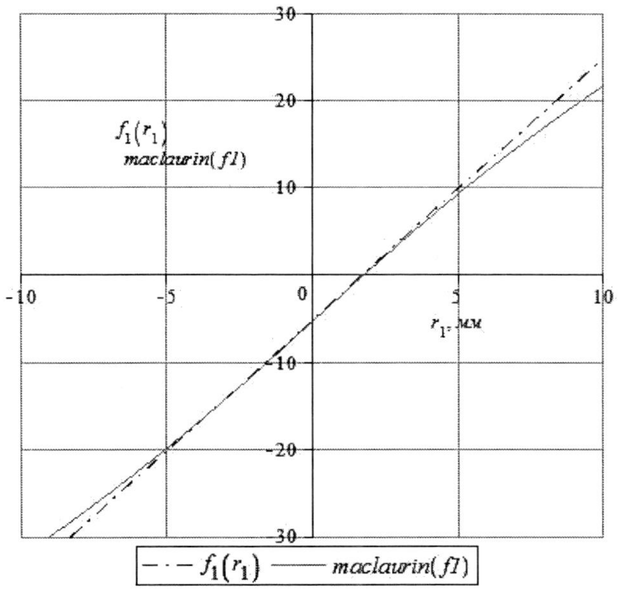


Fig. 2. Function graphs $f_1(r_1)$ и $taylor(f1)$

Having rejected a residual member and having resolved expression (16) rather $r_1$ :

$$r_1 = \frac{-\frac{1}{2}\left(\begin{array}{l} d^3\pi^2\sin^2\left(\frac{\pi}{m}\right) - 2d^2\pi^2\delta_2\sin^2\left(\frac{\pi}{m}\right) + d\pi^2\delta_2^2\sin^2\left(\frac{\pi}{m}\right) + 2h^2 d\cos\left(\frac{\pi}{m}\right) - \\ -2h^2\delta_2\cos\left(\frac{\pi}{m}\right) - 2dh^2 - 2\delta_2 h^2 + 2hvk_1 \end{array}\right)}{\sin^2\left(\frac{\pi}{m}\right)\pi^2\left(vk_2 d^2 - 2vk_2 d\delta_2 + vk_2\delta_2^2 + d^2 - 2d\delta_2 + \delta_2^2\right)}, \quad (17)$$

Where

$$vk_2 = \cos\left(\frac{\pi}{4} + \frac{1}{2}\arctan\left(\cot\left(\frac{2\pi}{m}\right)\right)\right).$$

**Conclusions**

Substituting (17) in (4), (5) we will write down new expressions for $\alpha_1'$, $\varepsilon_1'$ и $s_1'$ , then using the fourth expression of system (1), we find diameter wires an inside layer:

$$\delta_1 = 2r_1 s_1' \quad (18)$$

At last stage it is found the approached value of diameter of the central strand:

$$\delta_0 = 2\left( r_1 - \frac{\delta_1}{2} \right) \quad (19)$$

It is necessary to notice that reception of more exact expression is interfaced to definition of the form of section of a strand in a cross-section of a rope and its orientation.
As a result new decisions of system (1) are received. The yielded procedure can be easily adapted for calculation of geometry of spiral ropes and locks of various phylums: "Forces", "Varrington-forces", etc.

## References


[1] Glushko M. F. Steel elevating ropes. - Kiev, Technics, 1966. – 327p.

[2] J. Wu, The finite element modeling of spiral ropes. International Journal of Coal Science and Technology, Volume 1, Issue 3, 1 September 2014, pp. 346-355

[3] F. Luz, L. Silva, S. Amico and C. Cimini, Tensile behavior of a CFRP wire rope with independent core: Analytical and numerical analyses. 16th European Conference on Composite Materials, ECCM 2014, Seville; Spain; 22 June 2014 through 26 June 2014;

[4] P. Božek, E. Pivarčiová, Flexible manufacturing system with automatic control of product quality. In Strojarstvo. Vol. 55, No. 3 (2013), online, pp. 211-221

[5] G. Fedorko, V. Molnár, Ž. Ferková, P. Peterka, J. Krešák and M. Tomašková, Possibilities of failure analysis for steel cord conveyor belts using knowledge obtained from non-destructive testing of steel ropes. Engineering Failure Analysis, Volume 67, 1 September 2016, pp. 33-45

[6] P. Božek, P. Molda and M. Tóthová. Modelovanie rozbehu kabíny lanovej dráhy. Mathematical analyses of motoring cable synchronization. In Process Control 2004 : 6th International Scientific-Technical Conference `Process Control 2004`, Kouty nad Desnou, Czech Republic, 8.-11.6.2004. Pardubice : Univerzita Pardubice, 2004, pp. R273/1-6.

[7] A. Roduner, A. Vogel and V. Budimir, Spiral rope nets for rockfall protection - Tests, design, application. Regional Symposium of the International Society for Rock Mechanics: Rock Engineering in Difficult Ground Conditions - Soft Rocks and Karst, EUROCK 2009; Dubrovnik, Cavtat; Croatia; 29 October 2009 through 31 October 2009; Code 89361 2010, pp. 735-740

[8] Frankovský, P., Ostertag, O., Ostertagová, E., Trebuňa, F., Kostka, J. and Výrostek, M. (2017). Experimental analysis of stress fields of rotating structural elements by means of reflection photoelasticity. Applied Optics, 56(11), 3064-3070.

[9] Futáš, P., Pribulová, A., Fedorko, G., Molnár, V. Influence of steel scrap in the charge on the properties of gray cast iron. ISIJ International, Volume 57, Issue 2, 2017, pp. 374-379

[10] Ostertagová, E., Frankovský, P. and Ostertag, O. (2016). Application of polynomial regression models for prediction of stress state in structural elements. Global Journal of Pure and Applied Mathematics, 12(4), 3187-3199.

[11] D. Flum, A. Roduner, Findings from large-scale field tests of spiral rope nets for rock protection. 2010, European Rock Mechanics Symposium, EUROCK 2010; Lausanne; Switzerland; 15 June 2010 through 18 June 2010; Code 89434, Pages 625-628

# SECTION 3

## Experiment as a tool for the verification of analytical and numerical methods

EAN 2017
55th conference on experimental stress analysis

# Experimental modal analysis of a free suspended composite plate by using high speed DIC method

Róbert Huňady[1, a] and Peter Pavelka[1, b]

[1]Technical University of Košice, Faculty of Mechanical Engineering, Slovakia

[a]robert.hunady@tuke.sk, [b]peter.pavelka@tuke.sk

**Keywords:** modal analysis, composite plate, digital image correlation, DICMAN 3D

**Abstract.** Experimental modal analysis is a measurement method to determine modal parameters of mechanical structures. It can be used as a tool for the verification of numerical models or to identify the mechanical properties of different material specimens. The aim of this paper is to present the possibilities of using the high speed digital image correlation (DIC) in experimental modal analysis. DIC is non-contact optical method, which for the measurement uses precise digital cameras with high image resolution. The main advantage of this approach is that the results are available as full-field data. On the other hand, the FRF matrix resulting from the measurement is usually relatively large; it may consist of thousands of measurement DOFs. Therefore the authors used a special application DICMAN 3D that was developed to evaluate such type of measurements. The experimental results are correlated with the results of FEM analysis.

## Introduction

Non-contact optical methods allowing full-field measurements belong to the current trends in experimental mechanics. Into this group scanning laser Doppler vibrometry (SLDV), electronic speckle pattern interferometry (ESPI), digital speckle shearography (DSI), and digital image correlation (DIC) can be incorporated [1, 2]. Digital image correlation is a method which has potential for use in vibration analysis [3], for this purpose it is necessary to use of high-speed cameras. Although the DIC method is mainly utilized for testing of components and determination of material properties [4-7], various contributions devoted to vibration analysis as well as modal analysis are published [8-12].

The conception of the method allows observation of various phenomena during deformation and/or movement of an object, and the method is suitable for testing of a wide range of materials. The fundamental of digital image correlation is based on capturing of a stochastic speckle pattern created on an investigated object surface, e.g., by the spraying of black color on a white background. The use of preprinted vinyl foils is also convenient [7, 8]. An observed region is divided into smaller subsets, so-called facets, in such a way that each of facets includes a characteristic part of speckle pattern with sufficient contrast in order to be ensured its uniqueness. The relative displacements are determined by correlation of corresponding facets in a state before and after deformation of the object, or with regard to a reference step. As the system tracks a wide range of points on the object surface via two cameras, the obtained results are in the form of displacement component fields in directions x, y and z [2]. The principle of DIC is explained in detail in the book [13].

As DIC method measures responses in the form of displacements, frequency response functions (FRFs) are in the form of receptance. The sensibility of the measuring system depends on the resolution of CMOS sensor and size of measured surface. The frequency range is limited by the sampling frequency of cameras, i.e. the reciprocal value of a minimal shutter time. Currently there are high-speed cameras with sampling frequencies from several thousand up to several hundred thousand fps at full resolution of a sensor. If the sensor region of interest is decreased, then the maximal frame rate is increased. Measurements at high frame rates require using of an additional source of illumination for ensuring of optimal light conditions. For that purpose, high-power reflectors with specified achromatic light are commonly used. The major advantage of DIC is an ability to capture the responses of all points on the object surface in the same time and under the same conditions as an excitation. When compared to conventional techniques it means a significant time savings in terms of measurement because it is not necessary to relocate the sensors and repeat the measurement several times. However, it is necessary to note that the correlation is relatively time consuming. The density of measuring points (mesh points) and also accuracy of the measurement is dependent on the facet size. Authors [8] investigated the influence of the facet size on modal parameters estimated from DIC measurement. They found that the facet size has no impact on natural frequencies, however, in the case where the facets are too small, an increase in correlation errors which introduce inaccuracies to frequency response functions occurs. The facet size has a significant influence on absolute amplitudes of mode shapes. Depending on the facet size the amplitudes may be changed in tens of percent.

## DICMAN 3D

Since DIC measurement systems usually do not have any software tools for the evaluation of such types of measurement, the use of an external post-processing application is necessary. For that purpose, the authors use a software application called DICMAN 3D [14] that was developed especially for the high-speed digital image correlation system Q-450 Dantec Dynamics.

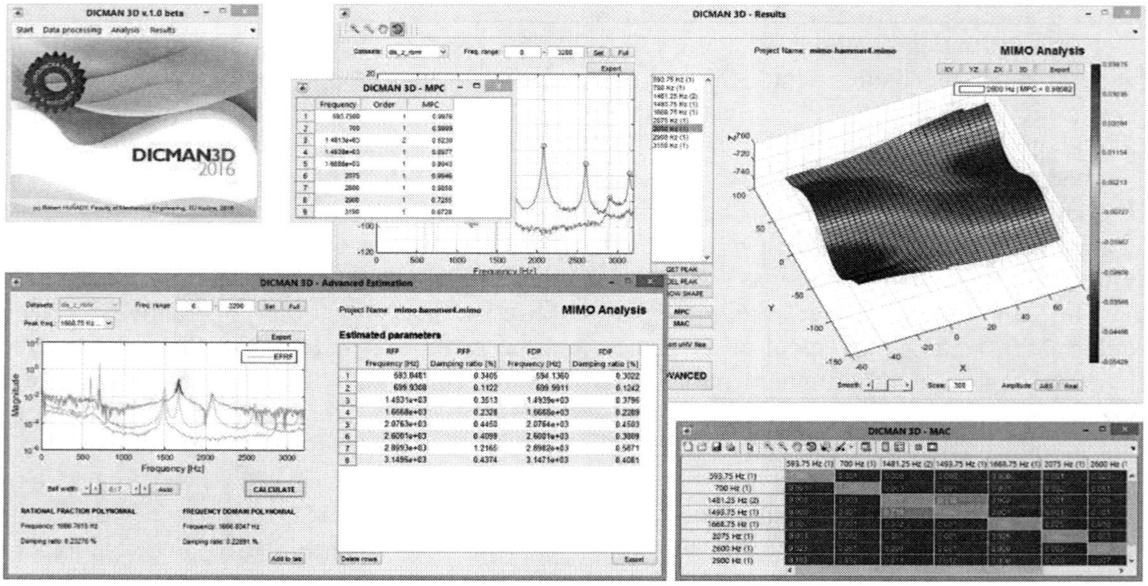


Fig. 1 Graphical user interface of DICMAN 3D.

The system consists of two high-speed cameras, a light source and a powerful notebook with the correlation software Istra4D, which ensures the operation of the measuring device, the cameras calibration and the correlation of images. The measurement is evaluated in Istra4D and the data is saved as HDF5 files. The number of files corresponds to the number of time steps. Each file contains information about the excitation signal and surface point displacements that represent measured responses. The files are subsequently imported to DICMAN 3D. DICMAN 3D has an intuitive graphical user interface (Fig. 1) and is able to evaluate a single-reference as well as a multi-reference measurement. Modal parameters are determined in the frequency domain by means of the following algorithms:

- Complex Mode Indicator Function (CMIF),
- Enhanced Frequency Response Function (EFRF),
- Rational Fraction Polynomial (RFP),
- Frequency Domain Polynomial (FDP).

## Experimental measurement

The measurement object was a free suspended composite plate of a rectangular shape with the dimensions of 130x200 mm and the thickness of 5 mm. The suspension was realized by 4 elastic bands glued on the rear side of the plate. The plate consists of twenty-five plies with an epoxy matrix. Eighteen of the plies were stiffened by the unidirectional carbon fibers (UCF) and the others by the carbon cross-woven (CCW) with the same volume of warp and weft fibers. An overall composition of the composite plate is shown in Table 1.

Table 1 Ply layup of the composite plate.

| Ply | Angle | Reinforc. | Ply | Angle | Reinforc. | Ply | Angle | Reinforc. | Ply | Angle | Reinforc. | Ply | Angle | Reinforc. |
|---|---|---|---|---|---|---|---|---|---|---|---|---|---|---|
| 1 | 45° | UCF | 6 | 90° | CCW | 11 | -45° | UCF | 16 | -45° | UCF | 21 | -45° | UCF |
| 2 | 45° | UCF | 7 | 45° | UCF | 12 | 45° | UCF | 17 | 90° | CCW | 22 | -45° | UCF |
| 3 | 90° | CCW | 8 | 45° | UCF | 13 | 90° | CCW | 18 | 45° | UCF | 23 | 90° | CCW |
| 4 | -45° | UCF | 9 | 90° | CCW | 14 | 45° | UCF | 19 | 45° | UCF | 24 | 45° | UCF |
| 5 | -45° | UCF | 10 | -45° | UCF | 15 | -45° | UCF | 20 | 90° | CCW | 25 | 45° | UCF |

The plate was excited by an impact hammer at two different points. The responses were measured by correlation system Q-450 Dantec Dynamics. The frequency range was set to 6400 Hz. Both measurements were correlated in Istra4D software to determine vibration responses, i.e. displacement fields. Time-dependent data were subsequently exported to DICMAN 3D for the estimation of modal parameters. The experimental configuration for the excitation with an impact hammer is shown in Fig. 2.

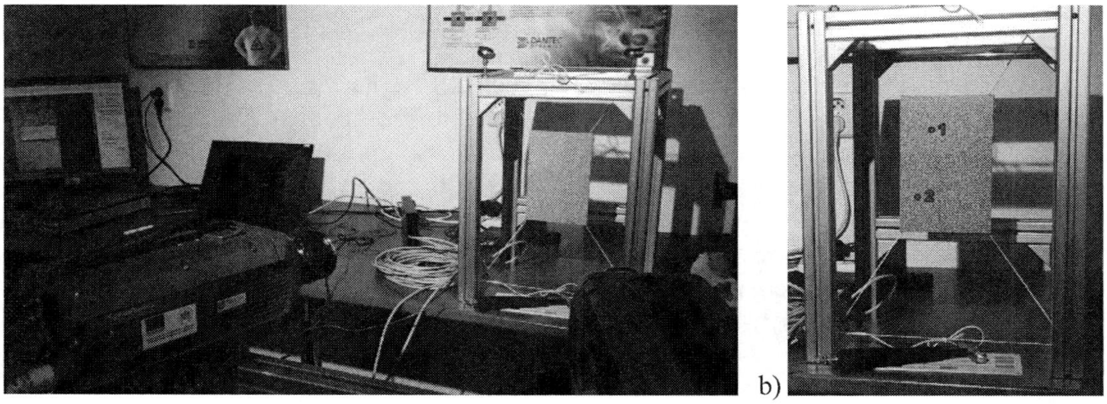
a) b)

Fig. 2 a) Configuration of the experiment, b) Excitation points.

Fig. 3 shows the spectrum of CMIF function that was extracted from the FRF matrix by singular value decomposition. As can be seen, seven modes were identified in the given frequency range. Their natural frequencies and damping ratios estimated by RFP method and FDP method are listed in Table 2. The corresponding mode shapes are shown in Fig. 5.

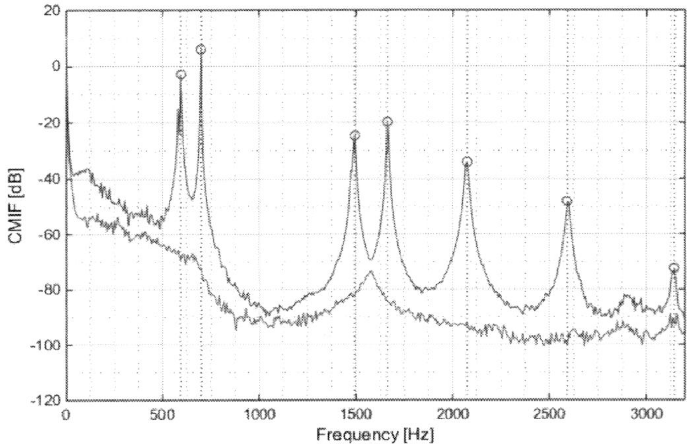


Fig. 3 CMIF plot of the composite plate.

Table 2 Modal parameters of the composite plate determined by DICMAN 3D.

| Mode # | Rational Fraction Polynomial | | Frequency Domain Polynomial | | Modal Phase Collinearity |
|---|---|---|---|---|---|
| | Frequency [Hz] | Damp. ratio [%] | Frequency [Hz] | Damp. ratio [%] | |
| 1. | 593.89 | 0.345 | 594.14 | 0.295 | 0.998 |
| 2. | 699.89 | 0.119 | 699.99 | 0.122 | 0.999 |
| 3. | 1492.9 | 0.362 | 1494.1 | 0.401 | 0.885 |
| 4. | 1666.7 | 0.234 | 1666.8 | 0.228 | 0.994 |
| 5. | 2076.4 | 0.447 | 2076.9 | 0.438 | 0.994 |
| 6. | 2600.2 | 0.419 | 2600.0 | 0.382 | 0.988 |
| 7. | 3149.9 | 0.445 | 3147.6 | 0.401 | 0.841 |

The natural frequencies determined by the RFP method are nearly equal to the frequencies acquired by the FDP method. The differences in the damping ratio values can be considered as acceptable with respect to a different polynomial order of both methods.

## FEM analysis of the composite plate

Finite element (FE) model of the composite plate was created in NX Nastran software. The material properties of the final composite were derived from the material properties of phases forming the individual plies. The continuous phase was epoxy resin. The dispersed phases were created by unidirectional carbon fibers UTS-120 and by carbon woven GG-160T (see Table 1). Linear shell elements QUAD4 of 2 mm length were used to mesh the plate. The additional mass representing the layer of used glue was added to every corner of the model (Fig. 4). It provided the better approximation of dynamic behavior the model. The weight of additional masses was determined by the measurement.

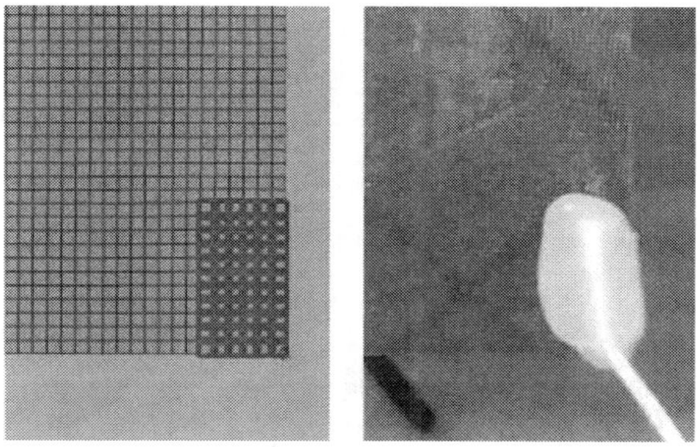

Fig. 4 Additional mass representing the layer of glue.

The eigenvalue analysis was performed in a solver SOL103. There were determined 8 flexible mode shapes of which frequencies are listed in Table 3. The rigid body modes were excluded from the evaluation.

Table 3 Natural frequencies of the composite plate determined by FEM.

| **Mode #** | 1 | 2. | 3. | 4. | 5. | 6. | 7. | 8. |
|---|---|---|---|---|---|---|---|---|
| **Freq. [Hz]** | 567.17 | 677.47 | 1443.9 | 1484.0 | 1595.8 | 2019.4 | 2604.8 | 3119.9 |

### Correlation of the results

It is obvious from Table 3 that one more mode was identified by the FEM analysis. For that reason, the correlation analysis of the results obtained from the experiment and the simulation was performed in NX Nastran. Consistency between the individual mode shapes was determined by Modal Assurance Criterion (MAC). The comparison of the natural frequencies and the mode shapes of the composite plate (Fig. 5) is shown in Table 4.

Table 4 Comparison of modal parameters obtained from the experiment and the simulation.

| **Mode #** | **Natural frequency [Hz]** | | | **MAC** |
|---|---|---|---|---|
| | **Experiment** | **FEM** | **Freq. error [%]** | |
| 1. | 593.89 | 567.17 | -4.499 | 0.784 |
| 2. | 699.89 | 677.47 | -3.218 | 0.885 |
| 3. | *unidentified* | 1443.9 | - | - |
| 4. | 1492.9 | 1484.0 | -0.596 | 0.712 |
| 5. | 1666.7 | 1595.8 | -4.373 | 0.801 |
| 6. | 2076.4 | 2019.4 | -2.679 | 0.879 |
| 7. | 2600.2 | 2604.8 | 0.185 | 0.826 |
| 8. | 3149.9 | 3119.9 | -0.955 | 0.810 |

As can be seen in Table 4, the frequencies of the third and the fourth mode obtained from the simulation are relatively closed. It is likely that in the case of the real plate these two modes overlap each other. If the excitation is applied at the place of a node point of a given mode shape, this mode is not enough excited. If we compare Fig. 2b with Fig. 5-3 we can see that both excitation points lie near the node curve of the third mode shapes. This is the reason why the responses of this mode were not be measured.

Differences between the natural frequencies are within the range of 0.2% to 4.5%. It is acceptable with respect to uncertainty in boundary conditions and material properties. Consistency of the mode shapes given by MAC value shows also a high level.

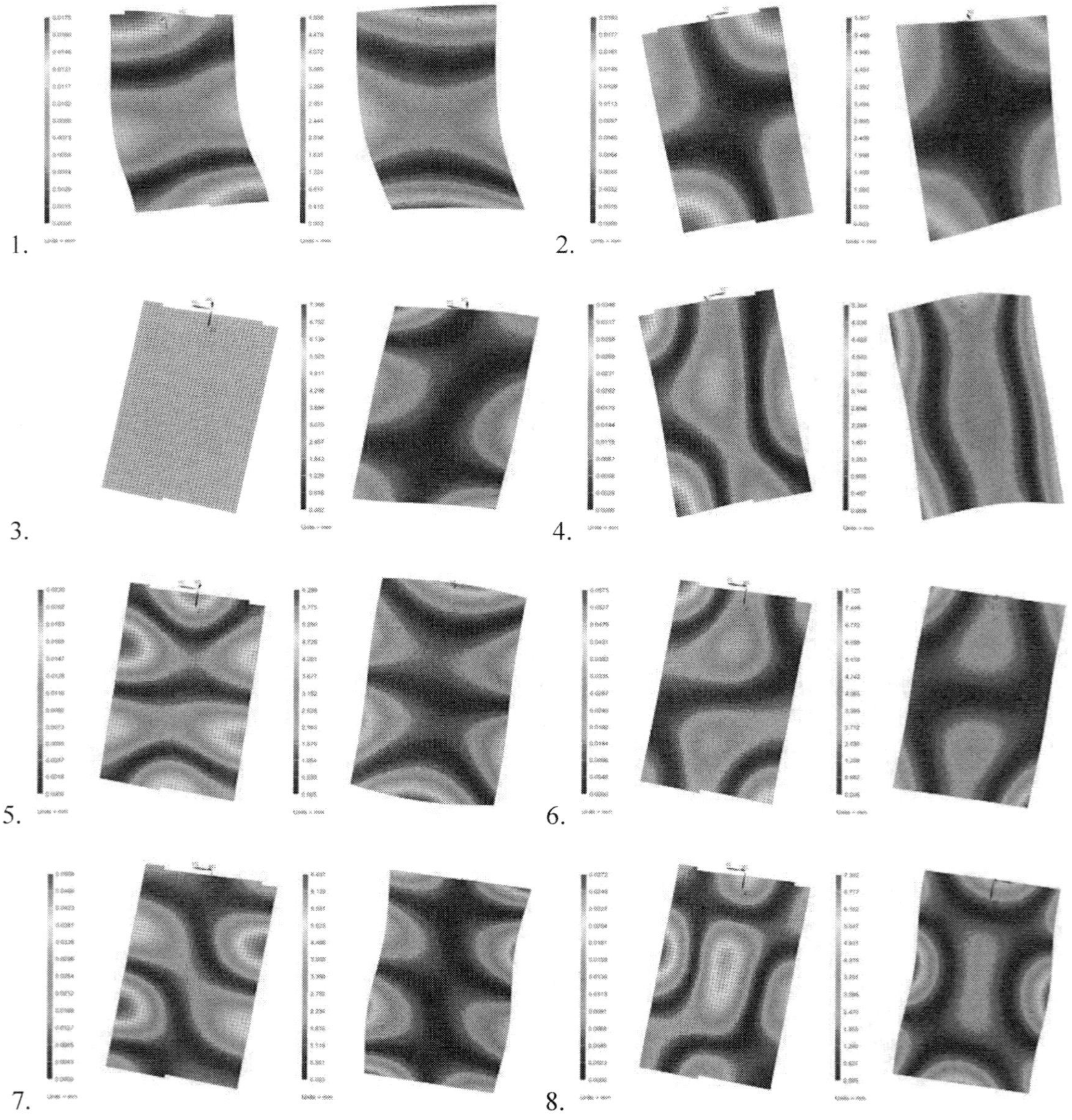

Fig. 5 Mode shapes of the composite plate obtained from the experiment (left) and the FEM simulation (right).

## Conclusions

The paper dealt with the use of high-speed digital image correlation in experimental modal analysis. The paper described the modal analysis of the composite plate that was excited by impact hammer. The multi-reference measurement was performed to obtain modal parameters of the plate. The responses were measured by digital image correlation system Q-450. The measurement was evaluated in the software application DICMAN 3D that was created especially for that purpose. The results of experiment are natural frequencies, damping ratios and shapes of the excited modes. Since DICMAN 3D allows to export the geometric model and mode shapes into UNV file format compatible with I-DEAS applications, the experimental results were compared with the results of FEM simulation.

**Acknowledgement**

The authors would like to express their gratitude to Scientific Grant Agency VEGA MŠ SR for the support of this work under the project no. 1/0751/16 and 1/0393/14.

**References**

[1] F. Trebuňa, F. Šimčák, P. Frankovský, R. Huňady, M. Pástor, *The Application of Optical Methods in Experimental Mechanics 1* (In Slovak), Technical University of Košice, Košice, 2014.

[2] F. Trebuňa, F. Šimčák, R. Huňady, M. Pástor, P. Frankovský, M. Hagara, *The Application of Optical Methods in Experimental Mechanics 2* (In Slovak), Technical University of Košice, Košice, 2015.

[3] F. Trebuňa, F. Šimčák, R. Huňady, *Vibration and Modal Analysis of Mechanical Systems* (In Slovak), Technical University of Košice, Košice, 2012.

[4] K. Genovese, L. Cortese, M. Rossi, D. Amodio, A 360-deg digital image correlation system for materials testing, *Optics and Lasers in Engineering* 82 (2016) 127-134, http://dx.doi.org/10.1016/j.optlaseng.2016.02.015.

[5] B. Pan, L. Yu, Y. Yang, W. Song, L. Guo, Full-field transient 3D deformation measurement of 3D braided composite panels during ballistic impact using single-camera high-speed stereo-digital image correlation, *Composite Structures* 157 (2016) 25-32, http://dx.doi.org/10.1016/j.compstruct.2016.08.017.

[6] E. Gariboldi, K. Naumenko, O. Ozhoga-Maslovskaja, E. Zappa, Analysis of anisotropic damage in forged Al–Cu–Mg–Si alloy based on creep tests, micrographs of fractured specimen and digital image correlations, *Materials Science and Engineering: A* 652 (2016) 175-185, http://dx.doi.org/10.1016/j.msea.2015.11.090.

[7] R. Huňady, M. Hagara, F. Trebuňa, Stress analysis performed in the near surrounding of small hole by a digital image correlation method, *Acta Mechanica Slovaca* 18 (3-4) (2015) 74–81.

[8] R. Huňady, M. Hagara, F. Šimčák, The influence of facet size on the accuracy of modal parameters determined by digital image correlation technique, *Applied Mechanics and Materials* 611 (2014) 496-500.

[9] N.S. Ha, H.M. Vang, N.S. Goo, Modal analysis using digital image correlation technique: an application to artificial wing mimicking beetle's hind wing, *Experimental Mechanics* 55 (5) (2015) 989–998, http://dx.doi.org/10.1007/s11340-015-9987-2.

[10] W. Wang, J.E. Mottershead, A.Ihle, T. Siebert, H.R. Schubach, Finite element model updating from full-field vibration measurement using digital image correlation, *Journal of Sound and Vibration* 330 (8) (2011) 1599–1620.

[11] W. Wang, J. E. Mottershead, T. Siebert, A. Pipino, Frequency response functions of shape features from full-field vibration measurements using digital image correlation,

*Mechanical Systems and Signal Processing* 28 (2012) 333-347, http://dx.doi.org/10.1016/j.ymssp.2011.11.023.

[12] F. Trebuňa, R. Huňady, Z. Bobovský, M. Hagara, Results and experiences from the application of digital image correlation in operational modal analysis, *Acta Polytechnica Hungarica* 10 (5) (2013) 159-174, http://dx.doi.org/10.12700/APH.10.05.2013.5.10.

[13] H. Schreier, J.-J. Orteu, M.A. Sutton, *Image Correlation for Shape, Motion and Deformation Measurements: Basic Concepts,Theory and Applications*, Springer Science & Business Media, 2009.

[14] R. Huňady, M. Hagara, A new procedure of modal parameter estimation for high-speed digital image correlation, *Mechanical Systems and Signal Processing*, Volume 93, 1 September 2017, Pages 66-79.

# Evaluation of glued laminated timber beams reinforced with carbon fiber fabric for use in FEM modelling

Ing. Pavel Klapálek[1, a] and Ing. Lenka Melzerová, Ph.D.[1,b]

[1]Department of mechanics, Faculty of Civil Engineering, CTU in Prague; Thákurova 7, 166 29, Prague, Czech Republic

[a]pavel.klapalek@fsv.cvut.cz, [b]lenka.melezerova@fsv.cvut.cz

**Keywords:** GLT, Glued laminated timber, carbon fiber fabric, destructive, non-destructive.

**Abstract.** Main focus of this article is on glued laminated timber beams reinforced with carbon fiber fabric. In the first part of this article, the tested materials are described, alongside with testing methods used. Several testing methods were used, as there was need for thorough determination of material characteristics. Obtained findings will be used as input data for future FEM models of reinforced GLT beams that should be comparable to real behavior.

## Introduction

In nowadays civil engineering is the timber very popular material, as it is on rise again mainly because of the high demands on energy efficiency of buildings, it is appropriate to devote this material to try to find ways for its optimum utilization. When solid timber ceased to meet the demands of civil engineers, it was necessary to find a way to improve its performance. Accordingly, there was also first thought to remove these drawbacks by means of gluing timber elements to create a composite material with better properties than solid timber. That is glued laminated timber. The prime beginning of glued laminated timber (hereinafter referred to as GLT) dates back to England in 1866, where Josiah George Poole designed and built the still existing roof structure made of glued laminated timber at the University of King Edward in Southampton [1]. A significant shift in technology of GLT came in 1942 when has been introduced the first fully waterproof glue, which allowed the use of GLT in outdoor environment without the risk of degradation of glued joints. GLT is a high value construction material that has excellent properties and can replace solid timber where there no longer sufficient elements can be made of it. It consists of layers of lamellas that are glued together under pressure, using a waterproof adhesive [2]. At present, this material is mainly used in modern timber structures and roof structures of halls. Despite the frequent use of this material, the technology of production is still in development. There are new types adhesives being developed and also methods of reinforcing [5]. For this reason, the main material, which is an essential part of this research is glued laminated timber. The results published in this article are part of a larger project in which we have been focused on beams with rectangular cross section. The beams are made of spruce wood, originally from the Czech Republic and are reinforced with carbon fiber [4].

## Tested material

**Glued laminated timber beams.** Primary tested material is glued laminated timber (GLT) in form of beams, made of spruce wood and classified as GL24h. Length of those beams is 4.6 m, width 0.1 m and height 0.32 m. Each row is made of from so called infinite lamella. Infinite lamella is made of timber parts without knots, those parts are then glued together by finger joints and cut to desired length. Lamellas are glued together with use of melamine-formaldehyde (MF) glue which is very similar to urea-formaldehyde (UF), but has better resistance to water and stability during change of environment. This glue is popular for its light color of hardened joints. Beams were stored under laboratory conditions and the humidity was under 12.0 %. In total, there were six beams (four reinforced with carbon fiber fabric and two beams without reinforcement). Reinforced beams were labelled with numbers 1,4,5 and 6 (Fig.1).

**Figure 1** GLT beams

**Carbon fiber fabric.** Second used material is carbon fiber fabric. This material was used for additional reinforcement of GLT beams on the bottom surface. Type of used fabric is Carbon unidirectional UD CST 200 made of unidirectional carbon fiber with PES grid. It has thickness of 0.4 mm and grammage of 200 g/m$^2$. This fabric was attached to GLT beams with resin based glue epoxide LG 100 with hardener Flex 40. Glues was primary developed for high strength gluing of wooden based materials, mainly because the glue saturates the material and thus becomes part of the composite (Fig 2).

**Figure 2** Carbon fiber fabric

**Testing methods.**

**Destructive testing.** Four-point bending test was chosen for GLT beams as it is most appropriate destructive method for timber, because the most stressed part is without shear force and bending moment is constant. During static loading test were measured vertical deflections and shifts depending on the loading force. Loading of beams was performed in steps, where size of force in each step risen by 4 kN every 240 seconds, until the destruction of specimen. The results of these tests are in form of graph of force and deflection, along with video recordings (Fig. 3).

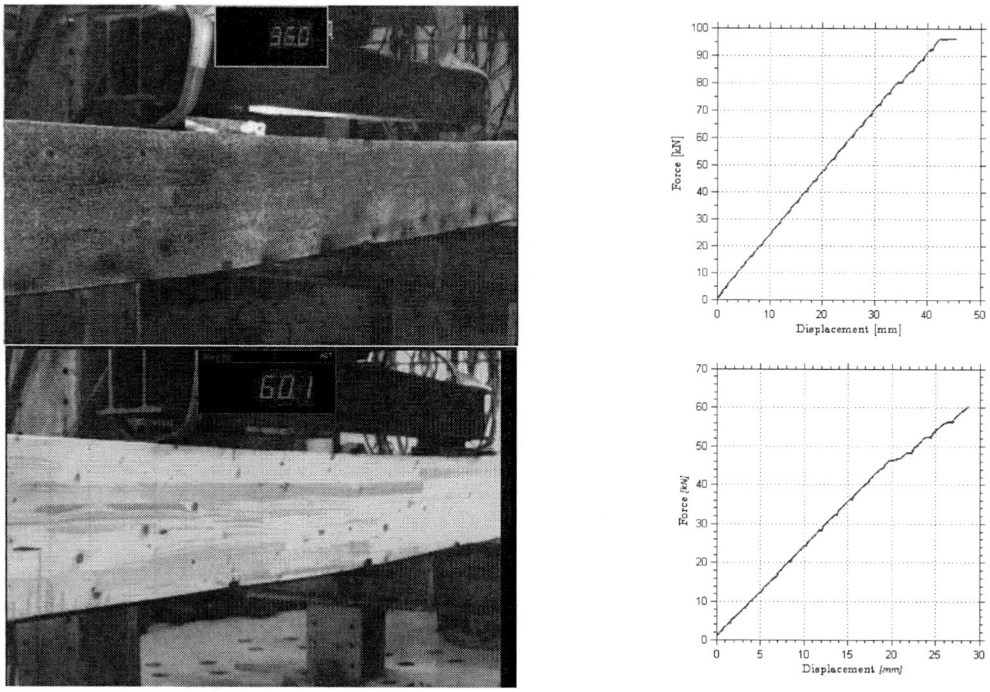


**Figure 3** GLT beams 1 (**top one**) and 6 (**bottom one**) during load with corresponding graph of force and deflection

Real modulus of elasticity was calculated from cleaned values without noise of sensors for all tested beams (Table 1).

| Name | Mean of E [GPa] | Standard error [GPa] |
|---|---|---|
| Beam 1 | 13.5 | 0.8 |
| Beam 4 | 12.9 | 0.7 |
| Beam 5 | 14.4 | 1 |
| Beam 6 | 13.5 | 0.6 |

**Table 1** Values of modulus of elasticity calculated for reinforced beams with standard error

**Non-destructive testing.** To obtain material characteristics in form of modulus of elasticity, two non-destructive methods were used. First one was method of spike. This method is among the available methods of assessing timber elements, without greater mechanical damage to the element, which would alter the mechanical properties of the whole. Method uses device called Pilodyn 6J, that shoots spike with normalized force and the depth of puncture is measured with precision to half of millimeter. Modulus of elasticity is than calculated from those values [5].

Second used method was the pulse method, that is suitable for determining the dynamic elastic properties of flexible materials which are preferably isotropic and homogeneous. The resonant frequency of the elements obtained from such materials, are dependent on the modulus of elasticity. From measured natural frequencies of beams, the modulus of elasticity is then calculated.

## Conclusions

Reinforced glued laminated timber beams were thoroughly explored and tested through various testing methods (destructive four-point bending test and non-destructive method of spike and pulse method). From the four-point bending tests the graphs of progress of loading were obtained, alongside with video recording. Video recordings will be used to observe behavior of reinforced beams during static loading and destruction, mainly a moment before and after maximum load capacity is reached. Calculated values of modulus of elasticity will be used for comparison with FEM models of reinforced beams that will be made. Several variants of FEM models need to be prepared, until the load behavior of model will be compliant with real one. The aim is to create a FEM model, which will behave under load in the same way as the actual beams. Subsequently, it will be possible to change method of reinforcing of the beams (mainly type and thickness of carbon fiber fabric) so that it will be possible to find the ideal reinforcement effect on the beams.

## Acknowledgement

The financial support of this outcome by the Faculty of Civil Engineering, CTU in Prague (SGS project No. SGS17/168/OHK1/3T/11) is gratefully acknowledged. Special thanks also go to company DEKTRADE s.r.o. for providing us with GLT beams and to company GRM systems for providing carbon fiber fabric.

## References

[1] BERGRAM, R., Wood Handbook, Wood as an Engineering Material (AllChapters). Madison, WI: U.S. Department of Agriculture, Forest Service, FPL: 508 p. 2010.

[2] ANGST, V., et al. Handbook 1: Timber structures. 2008. 242 p.

[3] AUGUSTIN, M., et al. Handbook 2: Design of Timber Structures according to EC 5. 2008. 130p.

[4] KUKLÍK, P. and MELZEROVÁ, L. Kompozitní materiály na bázi dřeva. Praha: ČVUT Praha, 2011. 76 s. ISBN 978-80-01-04958-7.

[5] KUKLÍK, P., MELZEROVÁ, L., VÍDEŇSKÝ, J. Závislost modulů pružnosti na penetracích u poškozených a nepoškozených vzorků z lepeného lamelového dřeva. Praha: CIDEAS-Centrum integrovaného navrhování stavebních konstrukcí, 2009, s. 45-46. ISBN 978-80-01-04457-5

# Rotary Table Machine Accuracy Verification

Aleš Richter[1, a], Pavel Klouček[1,b]

[1]VÚTS, a.s., Svárovská 619, Liberec XI-Růžodol I, 460 01, Czech Republic

[a]ales.richter@vuts.cz, [b]pavel.kloucek@vuts.cz

**Keywords:** Rotary Table Machine, Accuracy Verification, Measurement.

**Abstract.** This paper deals with the verification of the operational accuracy of a specially designed rotary table machine.

## Introduction

Growing demands on the production efficiency require very fast and precise machines. In our case, a demand for a rotary table machine (Fig. 1.) with an accuracy of 40 μm and a positioning time 0.5 s was set. Diameter of the table is 1000 mm and total moment of inertia is 100 kg.m$^2$.

The carousel machine serves as a base of a multifunction machine tool. On the carousel plate, 16 beds for clamping workpieces are symmetrically distributed. Within 0.5 s the carousel moves to the exact position (rotation by 22.5°). Then follows a 0.5 s milling period, during which the carousel is stationary. The total cycle of the machine is 1 s. What is important is the positioning accuracy in establishing the plate before machining.

In order to ensure the accuracy of operation and programmable control, a servomotor was chosen [1]. The motor drives the axial cam [2] by a backlash free reduction belt gear. The cam rotates the carousel plate also with a reduction gear (Fig. 1). When the theoretical ideal running, the dependencies of rotation (motor, cam, plate) are as follows: $\varphi_M = 1.25\varphi_C = 40\varphi_P$. Angular velocity is ω, acceleration is α.

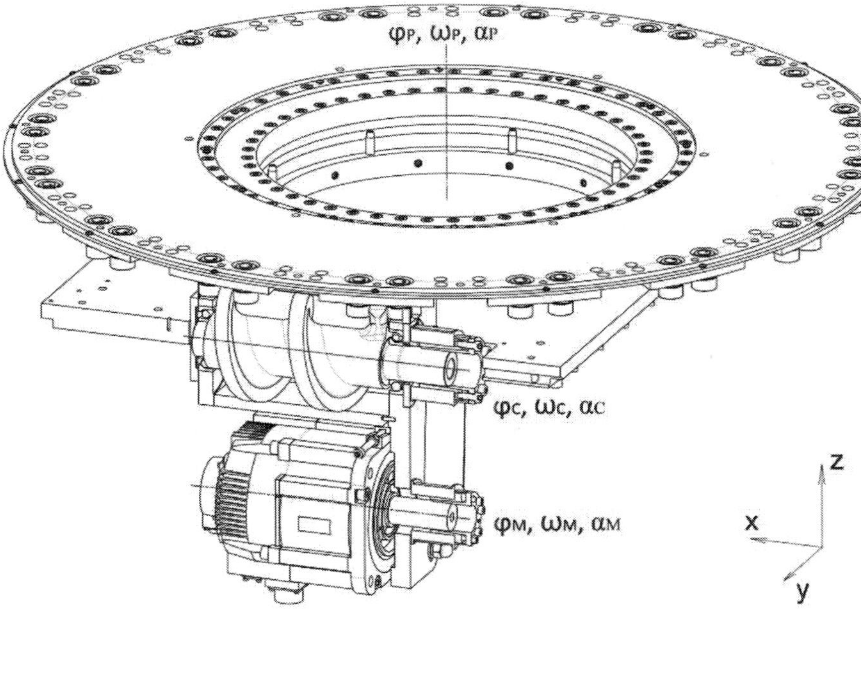

Fig. 1. Model of rotary table machine.

Before carrying out the measurements themselves, it was necessary to develop a mechatronic model of the motor-cam-plate configuration. Furthermore, the selected components were subjected to a stiffness and strength analysis. After many corrections in the calculations and adjustments of the 3D model, a functional sample was made, on which measurements were carried out. [3]

## Measurement

The dynamics of the table, repeating accuracy of the position and static backlash in the transmissions were measured.

**Measurement conditions:** The carousel was positioned on an auxiliary mounting frame loaded with the frame with 16 beds for clamping workpieces. It was driven by a Yaskawa SGMGV-30D3A61 servomotor with an output of 2.9 kW and a nominal torque of 18.6 Nm. Speeds of 60 and 65 steps per minute were tested.
**Measurement points:** Measurement points are shown in Fig. 2 and 3. Angular velocity and angular position of servomotor, worm shaft and working table were measured.
**Encoders:** Incremental optical rotational encoders Heidenhain ROD 1020/3600 imp and Heidenhain ROD 880/36000.
**Used devices:** DMU-PCI plug-in module, A, B, C channels.

Measured quantities were recorded with a Dewetron DEWE-2600 16-channel measurement analyzer equipped with a special DMU-PCI card for precise angular measurements [4]. Evaluation was done using the Weisang FlexPro 9 software.

Fig. 2. Incremental encoders mounted on the output shaft of the motor (bottom) and the worm shaft (top).

Fig. 3. Incremental encoder mounted to the shaft located in the axis of rotation of the working table.

**Method of measurement and evaluation.** As a zero position of the table (rotation angle 0°), the position when bed No. 16 is in the engagement of the driving worm was chosen. In order to verify the dynamics of the table and to determine the position repeatability of particular positions, the table was operated in continuous run at a speed of 60 steps per minute. The cycle comprises the working table motion for about 0.5 second, and a stationary working position for 0.5 second. The angular velocity and position of all three shafts were recorded synchronously with a sampling frequency of 50 kHz. For a comparison of the impact of the speed, three positions of the working table were measured also for a speed of 65 steps per minute (in practice, however, the machine will be operated at speed of 60 steps per minute only).

The static backlash of the working table was determined from the difference of the positions when the driving motor performs a rotation by 10° forwards and then backwards. Several positions of the working table in standard mode with two rollers in contact with the driving worm were measured. Additionally, a measurement with the angularly shifted table, when only one driving roller was in the engagement of the driving worm, was performed.

**Measurement results**

**Dynamics of the working table.** For measuring the dynamics of the working table using an incremental encoder with high resolution (36000 pulses/rev), it was necessary to choose the correct setting of the angular position range. The DMU-PCI plug-in module is equipped with a 16-bit converter, the angle value is then incremented with each edge of a TTL signal of the incremental encoder. If we choose the range of the measured angle of -10° up to +370° so that we can cover the full range of the angle within one revolution with a slight overlap, the quantization step of the converter is 380/65536 = 0.005798°. Thus, we lose the resolution of the encoder because at 144000 levels of the angle per one revolution, we can achieve a resolution 0.0025°. A usage of the full encoder resolution was achieved by reducing the range of the measured angle to 60°, when two or three working positions of the table were recorded in each measurement (the angular step is 22.5° between two working positions). Fig. 4 shows the typical displacement curves for a change of one position of the working table. The angular position curves of the motor, worm and working table are shown. Fig. 5 then displays the curves of the respective angular velocities.

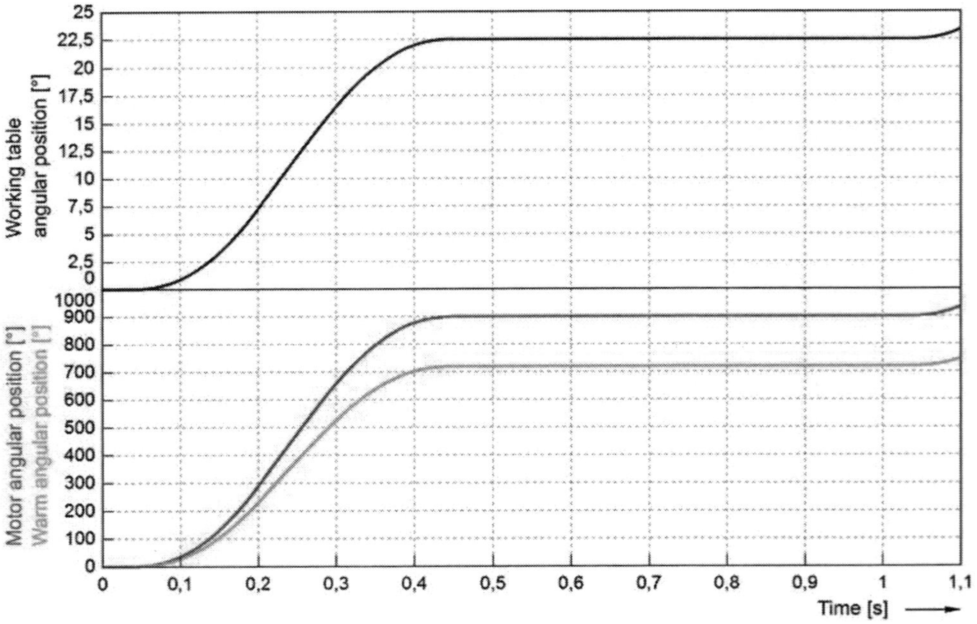


Fig. 4. Displacement/angular position of the working table, driving worm and motor during the transition from position No. 1 to position No. 2, transition and also the dwell part are shown

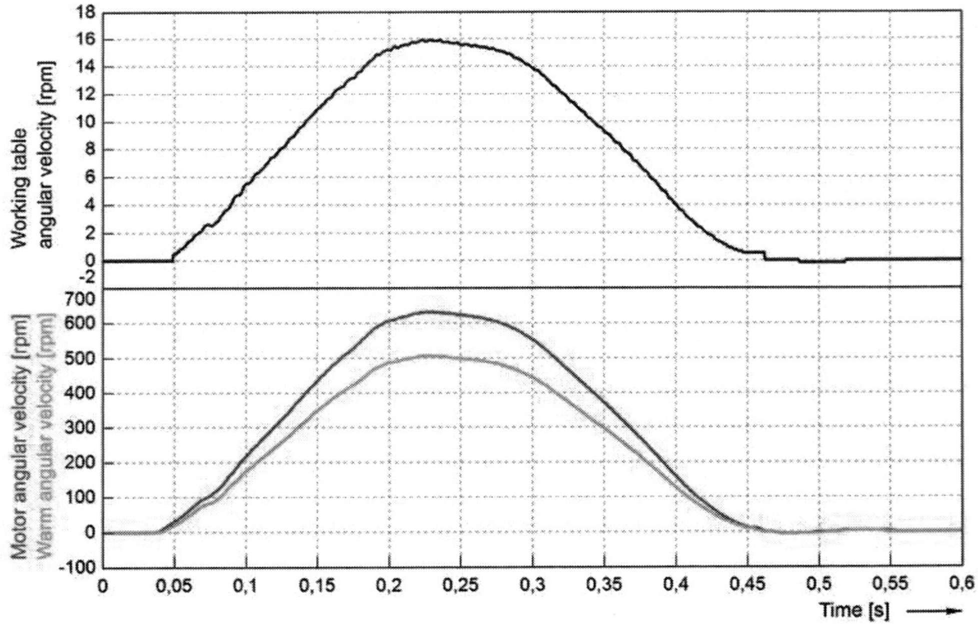


Fig. 5 Angular velocity of the working table, drive worm and motor during the transition from position No. 1 to position No. 2, transition and the beginning of the dwell part are shown

Damping time and overshoot rate of the position are factors affecting the time usable for the actual operation of milling and its accuracy. Therefore, just this segment, when the working table stopped in its new position, was evaluated and compared for all 16 positions of the table. At first, the courses measured by the encoder attached to the working table were assessed. The courses have a stepwise character since the overshoot magnitude is close to couple of quantization steps.. However if we compare the courses with the angular position measured on the driving worm, which can be converted to the motion of the working table with a well-known transmission ratio of 32:1, it can be said that the worm course corresponds to the working table course because backlashes are defined at a unidirectional movement. The incremental encoder on the shaft of the worm has a resolution of 3600 pulses/rev; when converting to the movement of the table and changing the angle size with every edge, it is then 3600 * 32 * 4 = 460800 pulses/rev, i.e. the quantization step of the table movement is 0.00078125°, thus, more than 3 times finer than that from the encoder attached to the table.

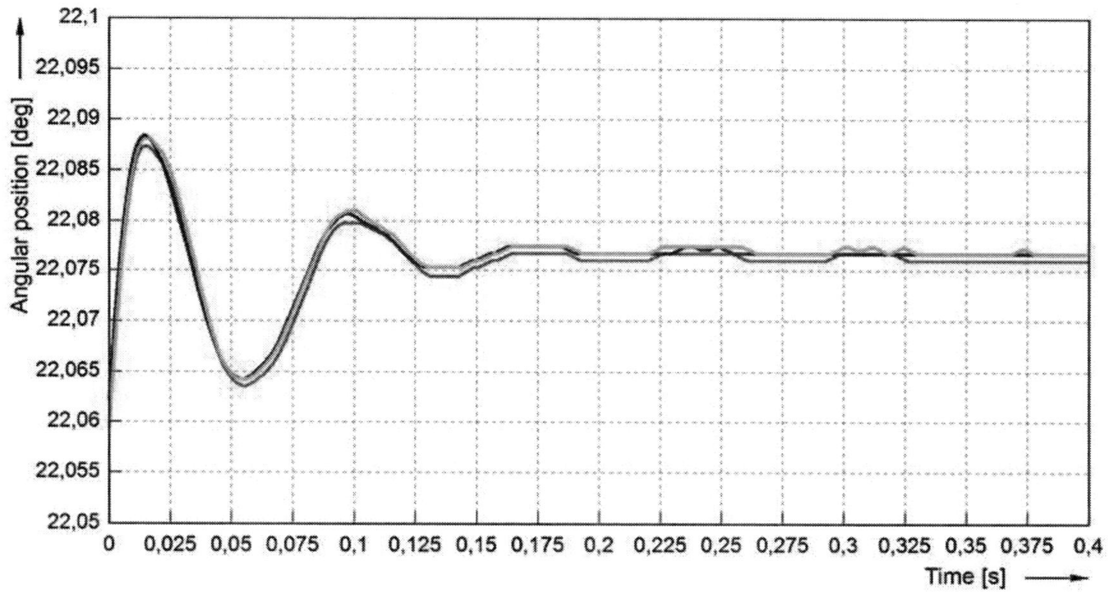


Fig. 6. The course of the angular position of the working table during its stop in the new position when transiting to *position No. 1*. The position measured by the encoder on the shaft of the driving axial cam after conversion by a gear ratio for three consecutive revolutions of the working table (order of the colours: dark blue, red, green).

If we compare the dwell positions in particular positions, , the position repeatability deviation of all positions did never reach a greater value than one quantization step, i.e. ±0.00078125°, during three revolutions of the working table. Should we consider the working table 1 meter in diameter and the maximum position error two quantization steps the maximum peripheral repeatability error is approx. ±13.6 µm. The position error is given by the values conversion, i.e. the output value change will occur only when a specified level is exceeded. It means, the actual value may vary within the full extent of one quantization step and therefore also just under the limit of the following step.. The position overshoot in transition to a new position will reach maximum amplitudes approx. up to 0.015° at a frequency of approx. 14 Hz. Most often, however, a very rapid damping within one or two periods with duration about 0.1 s occurred. After this time segment, the table can be already considered as set in the new position with an accuracy given by the repeatability mentioned above.

For comparison, one measurement of the shifted starting position of the table when only one roller was engaged in the worm at the time of transition to another position was also carried out. Two positions were measured and in both cases, a slight reduction of the overshoot amplitude compared to the operation with two rollers engaged was present. This is surprising as one would expect the contrary.

**Static backlashes.** For particular working positions of the table, backlashes given by the size of the driving rollers were measured. Further, the maximum value of the position overshoot and the table position in the dwell part of the cycle for three consecutive revolutions of the working table were determined for each position.

To determine the static backlash between the driving rollers of the table and the driving worm, an experiment was conducted in which the driving motor rotated at low speed by ten degrees forward and then by the same angle step back. In such a way, the backlash determination in both directions should happen and the run of the table in the return direction should be shorter for the backlash value than in the theoretical run. The measurement was performed for two positions of the table in the standard setting, i.e. with two rollers engaged and consequently with the table rotated in such way so that only one roller should engage in the driving worm in the measurement point.

The backlash values between the working table and the driving worm were determined and the whole backlash of the mechanisms from the driving motor shaft up to the working table. The backlashes were always related to the angular position of the working table. At positions between the worm and the working table, there is a backlash of approx. 0.002° up to 0.004°, which corresponds to a backlash at the circumference of approx. 0.017 up to 0.035 mm, and it also roughly corresponds to the measurements of the manufacturing tolerances of the rollers and the worm. A backlash in the belt transmission between the motor and the driving worm, which is implemented by the so-called toothed belt "without backlash", also contributes to the total static backlash,. The backlash found in this transmission is typically 0.001° with its maximum approx. 0.0025°. It corresponds to the peripheral backlash of approx. 0.022 or 0.008 mm. This backlash is given by both the backlash of the belt teeth in the grooves of the toothed wheels and the inherent flexibility of the belt.

## Conclusions

The function of the stepping mechanism of the milling machine working table was verified by measurement [5]. Relatively high demands are laid on the angular position accuracy because the peripheral fault is quite high for a relatively large table diameter. The position repeatability for all 16 working positions was examined along with the overshoot rate during the transition to a new position, the static backlash of the table in the selected positions.

The angular position and the angular velocity were measured. The results show that the behavior of the worm and the table during the change of the position can be considered almost identical. The positional repeatability error for all positions of the table was at the limit of the resolution of the encoders. The overshoots when transiting to the standing in the new position did not exceed values of 0.015° at a frequency of approximately 14 Hz. They were very quickly damped so that after the first period of the oscillations, the table can be already considered as established in the new position. Additionally, the determined static backlashes were very small.

Some differences between the selected positions, which are related to the dimension of the appropriate rollers, were detected. On the contrary, the influence of the number of rollers (one or two) in the engagement of the worm was almost not reflected in the static backlash.

The maximum peripheral repeatability error is not greater than 30 μm. The static backlash in the transmissions ranged from 17 up to 35 μm.

## References


[1] Souček, P.: Servomechanismy ve výrobních strojích, Vydavatelství ČVUT, ISBN 80-01-02902-6, Praha

[2] Koloc, Z., Václavík, M.: Cam Mechanisms, Elsevier, ISBN 0-444-98664-2, (1993)

[3] Richter A., Ondrášek J.: Rotary Table Machine Input Parameters Optimization. Advances in Mechanism Design II, 2016, pp. 399. ISBN 978-3-319-44086-6

[4] Šidlof P., Svoboda M.: Dynamické vyšetřování mechanismů na základě digitálního měření okamžité úhlové rychlosti, Strojírenství, vol. 34, pp. 267-273, 1984

[5] Klouček P., Braier Z.: Analýza pohybu pracovního stolu frézky, Internal report MER-16/2016, VÚTS, a.s. Liberec

# Experimental Construction of Lamb Wave Dispersion Curves in Plates

ŠOFER Michal[1,a], FERFECKI Petr[1,2,b] and ŠOFER Pavel[3,c]

[1] Department of Applied Mechanics, Faculty of Mechanical Engineering, VŠB-TU Ostrava, 17. listopadu 15/2127, 708 33 Ostrava-Poruba, Czech Republic

[2] IT4Innovations National Supercomputing Center, VŠB-Technical University of Ostrava, Studentská 6231/1B, 708 33 Ostrava-Poruba, Czech Republic

[3] Department of Control Systems and Instrumentation, Faculty of Mechanical Engineering, VŠB-TU Ostrava, 17. listopadu 15/2127, 708 33 Ostrava-Poruba, Czech Republic

[a]michal.sofer@vsb.cz, [b]petr.ferfecki@vsb.cz, [c]pavel.sofer@vsb.cz

**Keywords:** Lamb wave, Dispersion curves, Ultrasound, MATLAB

**Abstract:** The main aim of presented paper is experimental obtaining of Lamb wave dispersion curves in plates using broadband ultrasonic and acoustic emission transducers with additional converters facilitating the measurement of wavelength. The paper contains procedures of experimental investigation of phase as well as group velocity of the particular modes of Lamb wave and presents findings related to the effect of geometry on properties of particular propagation modes of Lamb wave.

## Introduction

Lamb waves have a great deal of interest, especially for flaw detection, material characterization and, last but not least, for inspection of various types of layered structures [1, 2]. Lamb waves are one of the type of guided waves, which can exist only in geometries with finite thickness such as plates, rods or tubes [3]. The excitation and detection of Lamb waves can be realized by a variety of methods such as wedge method or comb-structure method [4] with incorporated piezoelectric transducers. Less frequent, in terms of Lamb wave excitation, is use of optical methods or incorporation of electromagnetic acoustic transducers (EMAT) [5].

Lamb waves exhibit dispersion during their propagation. Dispersion describes a relationship between the phase and group velocity with the frequency. The effect of dispersion is change of overall pulse shape in time domain with related changes in frequency domain. Dispersion can be caused by geometry and/or by viscoelasticity of the material [6]. The dispersion originating from geometry can be described by, so called, dispersion curves, which can be obtained by numerical solving of the Rayleigh-Lamb frequency equation [7].

The aim of this paper is to experimentally determine the phase and as well as group velocity of antisymmetric and symmetric mode of Lamb wave on sheet metals (steel grade 11) of various thicknesses and compare obtained results with numerically computed results. The discussion also contains findings related to propagation characteristics of antisymmetric and symmetric modes of Lamb wave excited in plates of various thicknesses.

## Theory of Lamb waves – solution by method of potentials [7]

Consider plane strain problem of a free traction force surfaces as depicted on Fig. 1. The motion is defined as two-dimensional so that all quantities are independent on $y$ coordinate. The unknown displacement vector $\bar{u}$ will be defined with use of Helmholtz decomposition theorem as follows:

$$\bar{\boldsymbol{u}} = \nabla\boldsymbol{\phi} + \nabla \times \bar{\boldsymbol{\psi}}, \tag{1}$$

where $\phi = \phi(x,z)$ is a potential function and $\bar{\psi} = (0, -\psi(x,z), 0)$ is a vector function, both of them are dependent on $x$ and $z$ coordinates.

*Fig. 1: Geometry of the plate; h is a thickness of the plate*

The equation of motion expressed in terms of displacements can be with use of decomposed displacement vector separated into two wave equation, which represent governing equations for longitudinal (Eq. 2) and transversal waves (Eq. 3):

$$\nabla^2\phi - \frac{1}{c_L^2}\frac{\mathrm{d}^2\phi}{\mathrm{d}t^2} = 0, \tag{2}$$

$$\nabla^2\psi - \frac{1}{c_T^2}\frac{\mathrm{d}^2\psi}{\mathrm{d}t^2} = 0, \tag{3}$$

where $c_L$ is the speed of dilatational waves, $c_T$ is the speed of shear waves. The assumed solutions of Eq. 2, resp. 3 for wave propagating from left to right are:

$$\phi = (K_1\sin(pz) + K_2\cos(pz))e^{\mathrm{i}(kx-\omega t)}, \tag{4}$$

$$\psi = (K_3\sin(qz) + K_4\cos(qz))e^{\mathrm{i}(kx-\omega t)}, \tag{5}$$

where $p^2 = \frac{\omega^2}{c_L^2} - k^2$, $q^2 = \frac{\omega^2}{c_T^2} - k^2$, $K_1$, $K_2$, $K_3$, $K_4$ are constants resulting from boundary conditions, $\omega$ is the angular frequency, $k$ is wavenumber, $t$ is a time variable and $i$ is the imaginary unit. Expressing the displacements $u$ (axis $x$) and $w$ (axis $z$) as well as the stresses $\sigma_{zz}$ and $\sigma_{zx}$ in terms of potentials and applying a boundary condition $\sigma_{zx} = \sigma_{zz} = 0$ at $z = \pm\, 0{,}5\, h$ it is then possible to derive the Rayleigh-Lamb frequency equation for symmetric (Eq. 6) and antisymmetric (Eq. 7) modes:

$$\frac{\tan(\frac{qt}{2})}{\tan(\frac{pt}{2})} = -\frac{4k^2pq}{(q^2-k^2)^2}, \tag{6}$$

$$\frac{\tan(\frac{qt}{2})}{\tan(\frac{pt}{2})} = -\frac{(q^2-k^2)^2}{4k^2pq}. \tag{7}$$

Equations 6 and 7 are further used to obtain the dispersion curves for various modes of symmetric and antisymmetric Lamb wave, namely the dependence between phase velocity and the frequency-thickness product.

The dispersion curves for group velocity $c_g$ as the function of frequency-thickness product can be derived from phase velocity dispersion curves using following definition [7]:

$$c_g = \frac{d(kc_p)}{dk} = c_p + k\frac{dc_p}{dk}. \tag{8}$$

**Experimental measurement of dispersion curves**

The experiments were carried out with the use of AMSY-6 acoustic emission system, manufactured by Vallen Systeme GmbH. The acoustic emission system in itself has been utilized as signal generator and power supply for Vallen AEP5 amplifiers with 34 dB gain (See Fig. 2).

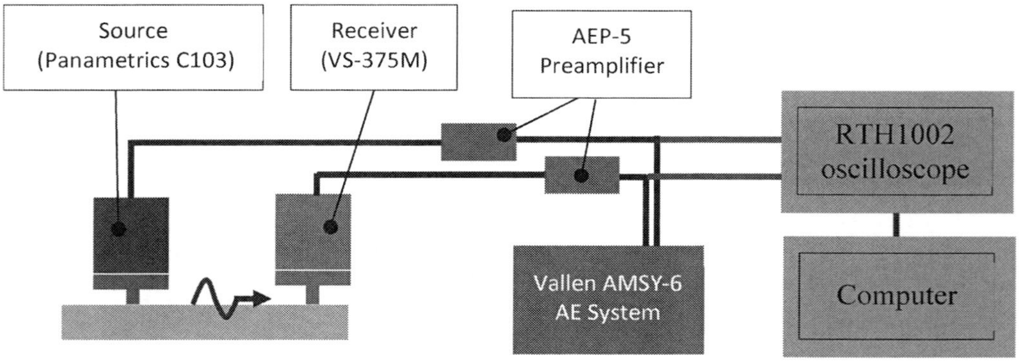


*Fig. 2: Schematic representation of measurement system*

A five-cycle sinusoidal tone burst of 250 $V_{peak\text{-}peak}$ and central frequency of 150 kHz (Fig. 3) has been used as driving signal for the broadband transducer Panametrics C103 1 MHz/0.5", which figured as the source. The signal from propagating waves has been captured by broadband acoustic emission transducer Vallen VS375-M and amplified by Vallen AEP5 amplifier. The signal has been further processed with use of Rohde&Schwartz RTH1002 oscilloscope.

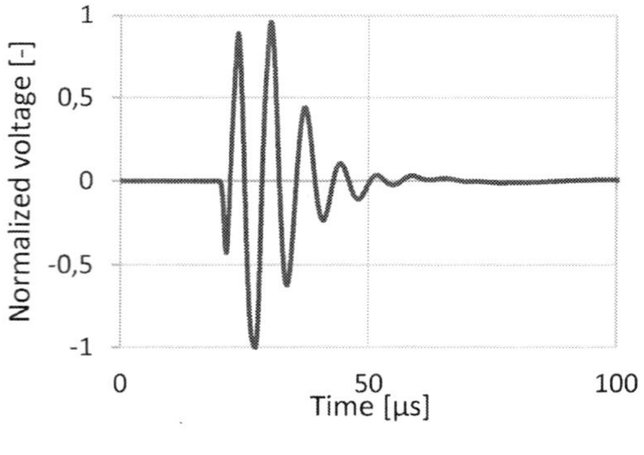


*Fig. 3: Driving signal for wave excitation*

*Fig. 4: Experimental setup for wavelength measurement*

Both transducers were placed on self-manufactured converters (see Fig. 4), which have been mounted onto skids in order to provide sliding on metal sheet including precise measurement of mutual distance between transducer and receiver. The phase velocity has been measured according to principle, depicted in Fig. 5. Namely, the phase difference between points A and B corresponds to $2\pi$, which implicitly results in the offset by the value of one wavelength [4, 7, 8]. The phase velocity can be therefore calculated from following expression:

$$c_p = \lambda f \tag{9}$$

where $\lambda$ denotes wavelength and $f$ denotes frequency of the particular part of the pulse, which wavelength is being measured.

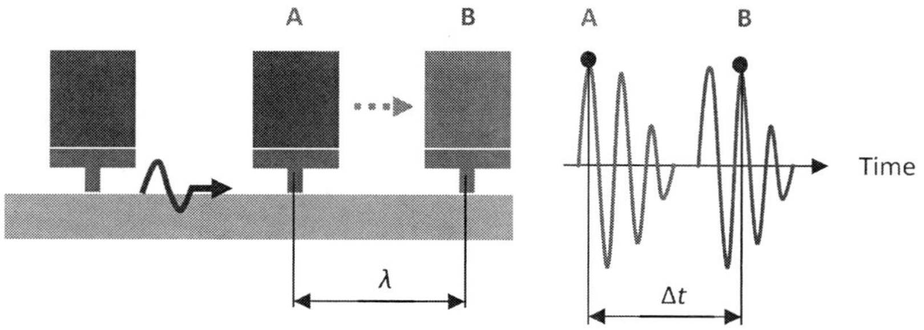


*Fig. 5: Principle of the phase velocity measurement;* $\Delta t$ is the time delay

In case of the group velocity measurement, there has to be used somewhat different approach. The group velocity can be understood as the velocity of energy transportation [7]. In practice, we are measuring the time delay $\Delta t$ corresponding to the distance $\Delta d$ between positions A and B. It has to be noted, that the measurement of the time delay is related to the position of the maximum amplitude of the pulse (see Fig. 6) [7, 8].

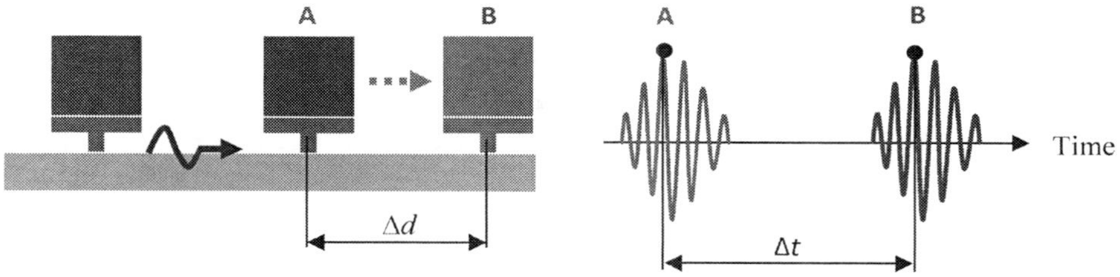


*Fig. 6: Principle of the group velocity measurement*

The group velocity can be therefore calculated as:

$$c_g = \frac{\Delta d}{\Delta t}. \tag{10}$$

## Results and discussion

The phase as well as group velocities were measured on sheet metals (steel grade 11) of thicknesses 1, 2, 3, 4 and 6 millimetres with dimensions ranging from 40x30 cm to 100x70 cm. In case of phase velocity measurement, there has to be earmarked one period of the pulse, which would best possible characterize the entire symmetric (S0) and antisymmetric (A0) Lamb wave mode (see Fig. 7 – orange segment). From chosen period was computed the frequency and according to the procedure illustrated in Fig. 5 was measured the wavelength and finally computed (by Eq. 9) the phase velocity.

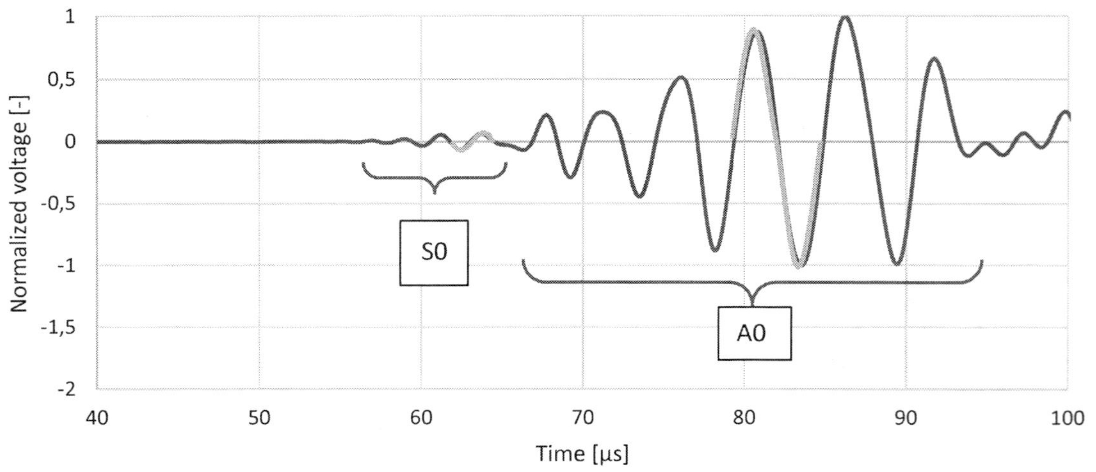


*Fig. 7: A0/S0 mode separation during measurement of group and phase velocity on 1mm thick sheet metal*

The group velocity measurement has been then realized according to the time delay versus propagation distance of the maximum pulse amplitude (see Fig. 6). Measurement of phase and group velocities of A0 flexural mode was easier to implement thanks to its much larger displacements in *z* axis compared to S0 mode within the measured frequency-thickness range. Figs. 8 and 9 display comparison between numerically and experimentally obtained dispersion curves. The numerically obtained dispersion curves were being calculated according to the Eqs. 6 and 7 with use of self-developed code in MATLAB software. The material properties were as follows: $c_L$ = 5900 m/sec, $c_T$ = 3100 m/sec, $\rho$ = 7850 kg/m$^3$. The conformity between numerical and experimental results is, according to the Figs. 8 and 9 very good, the relative error does not exceed 7% in case of all measurements for both modes and both types of velocities.

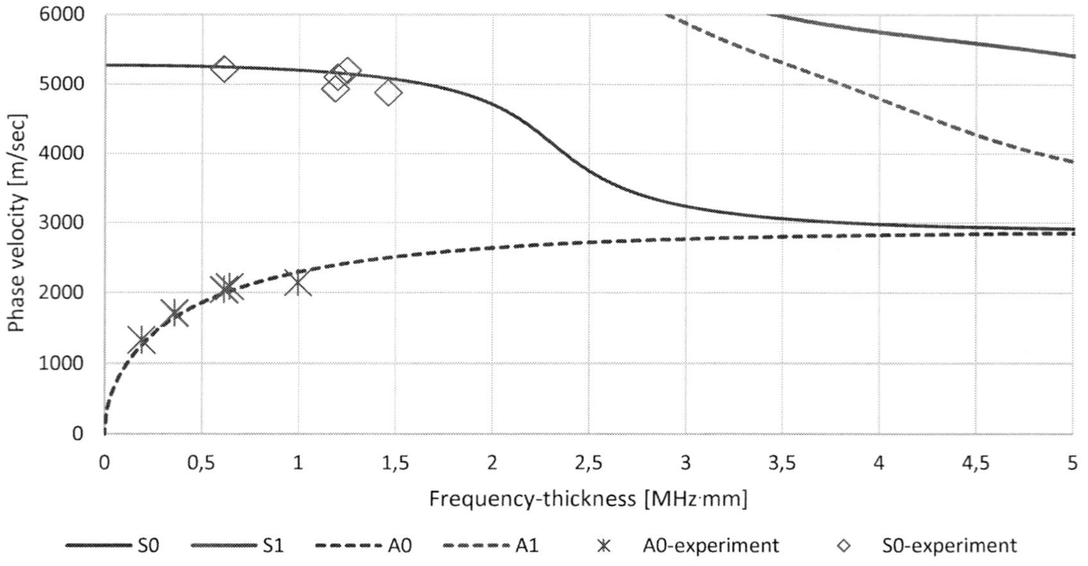


*Fig. 8: Dispersion curves for selected symmetric and antisymmetric Lamb modes*

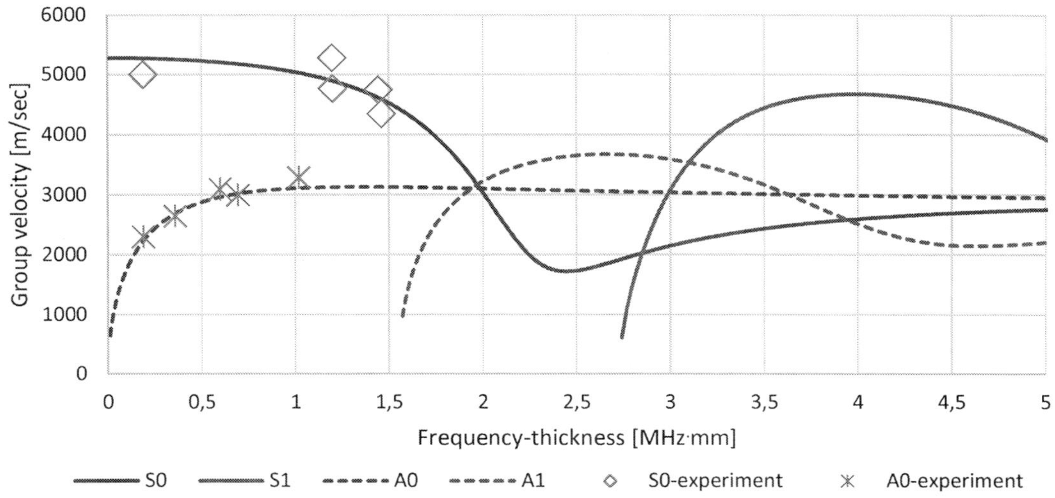


*Fig. 9: Dispersion curves for selected symmetric and antisymmetric Lamb modes*

Presented results show, however, better conformity for A0 Lamb mode. The reason lies in much higher displacement amplitudes in *z* axis compared to S0 mode, which will clearly reflect in the signal obtained from the transducer.

The authors have also focused on dependency of pulse central frequency and wavelength on sheet metal thickness as show Figs. 10-11. It follows, that the A0 mode retains its central frequency (Fig. 10) between 150 and 200 kHz with decreasing character when increasing the plate thickness.

The same trend can be observed for S0 mode (Fig. 10) with the difference of much broader frequency range (614 - 200 kHz versus 200 - 150 kHz in case of A0 mode). This finding is in accordance with the dependence of wavelength on the plate thickness, because of inverse relationship between wavelength and frequency. The origin of these findings is mainly in the wave structure, which fundamentally affects its propagation properties.

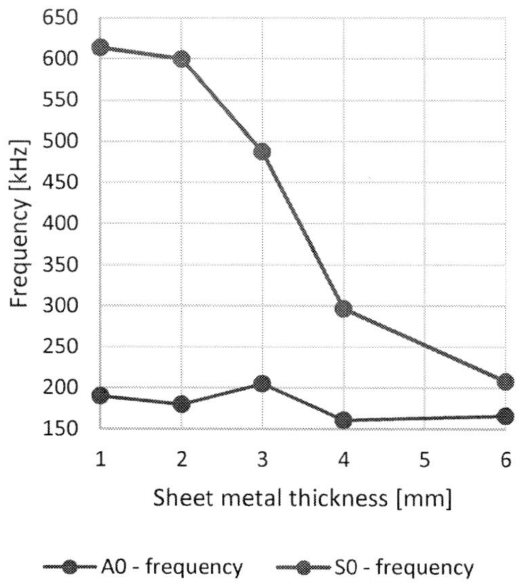


*Fig. 10: Dependency of measured pulse central frequency on sheet metal thickness for A0/S0 Lamb wave modes*

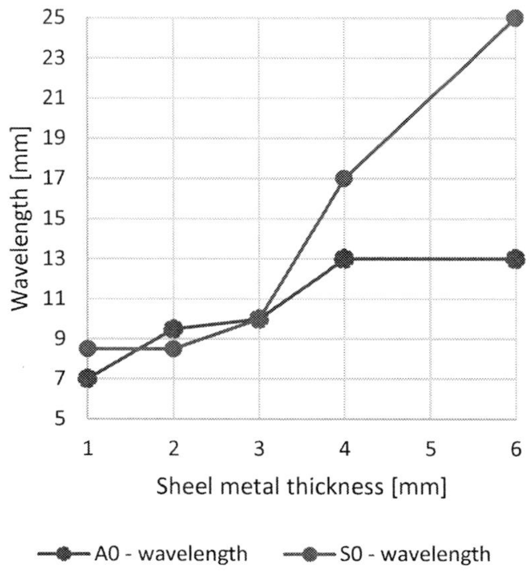


*Fig. 11: Dependency of measured wavelength on sheet metal thickness for A0/S0 Lamb wave modes*

## Conclusions

Despite the relatively challenging evaluation of phase and group velocity of the S0 Lamb mode, the authors have achieved very good conformity between the experimental and numerical results in terms of finding dispersion curves for phase and group velocities of A0 and S0 Lamb mode. Thanks to the presence of modes with higher central frequencies, which were present during the measurement on sheet metals with thickness of 3 and 4 millimetres, it was possible to virtually extent the range of frequency-thickness product up to the value of nearly 1,5.

For future work, authors would like to focus on the obtaining of the dispersion curves for higher values of frequency-thickness product as well as deeper analyse the wave structure and its influence on wave propagation properties.

## Acknowledgement

This work was supported by Specific Research (SP2017/136) and by The Ministry of Education, Youth and Sports from the National Programme of Sustainability II (LQ1602). The support is acknowledged.

## References

[1] X. Jia, Modal analysis of Lamb wave generation in elastic plates by liquid wedge transducers, J. Acoust. Soc. Am. 101 (1997) 834-842.

[2] G. Shkerdin, C. Glorieux, Interaction of Lamb modes with an inclusion, Ultrasonics 53 (2013) 130-140.

[3] T.N. Grigsby, E.J. Tajchman, Properties of Lamb Waves Relevant to the Ultrasonic Inspection of Thin Plates, IRE Trans. Ultrason. Eng. 8 (1961) 26-33.

[4] I.A. Victorov, Rayleigh and Lamb Waves, first ed., Plenum, New York, 1967.

[5] R.B. Thompson, G.A. Alers, M.A. Tennison, Application of direct electromagnetic lamb wave generation to gas pipeline, In: Proceedings of IEEE Ultrasonics Symposium, IEEE New York, 1972, 91-94.

[6] E. Moreno, P. Acevedo, M. Castillo, Pulse propagation in plate elements, Eur. J. Mech. A. Solids 22 (2003) 283-294.

[7] J.L. Rose, Ultrasonic Waves in Solid Media, Cambridge University Press, Cambridge, 2002.

[8] E. Moreno, P. Acevedo, Thickness measurement in composite materials using Lamb waves, Ultrasonics 35 (1998) 581-586.

EAN 2017
55th conference on
experimental stress analysis

# Modal analysis of composite ram

Ing. Milan TANČIN, Ing. Luboš Limberg
University of West Bohemia, Univerzitní 22, 306 14, Plzeň, Czech Republic
University of West Bohemia, Univerzitní 22, 306 14, Plzeň, Czech Republic
milax@rti.zcu.cz, limbi@rti.zcu.cz

**Keywords:** modal analysis, composite ram, eigenfrequency, composite

## Introduction

This topic relate to comparing modal analysis on composite tubes, which are constituent parts the composite ram. These modal analysis are based on modal analysis performed on real part. Three types of modal analysis were performed: 1) modal analysis for freely placed ram, 2) modal analysis for fixed ram on one side, 3) modal analysis for fixed ram on one side with additional mass. The main goal is to create virtual model, which will have almost the same boundary conditions (dimensions, layers, orientations, etc.). Obtained results will be compared with results obtained from real analysis.

## Preparing the model

For preparing the model were available 3D models of whole ram, data from modal analysis [1], including their placement [2] and properties [3] and weight of each part [4].

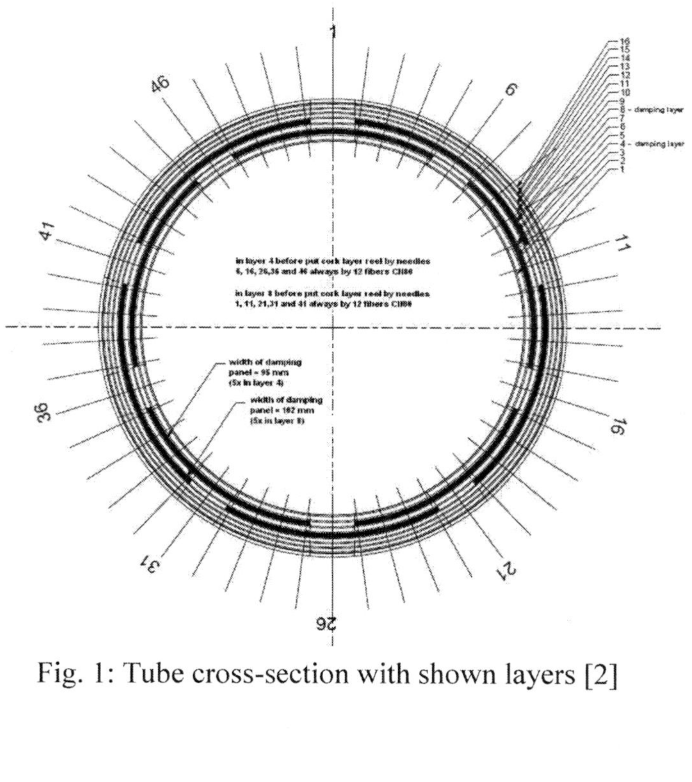


Fig. 1: Tube cross-section with shown layers [2]

Fig. 2: Measuring free oscillation the ram

Fig. 3: 3D model of whole ram

For next operations, only some parts were necessary, concretely model of tube, which is divided into two pieces, further named as *ground tube* and *middle tube* with bigger diameter.

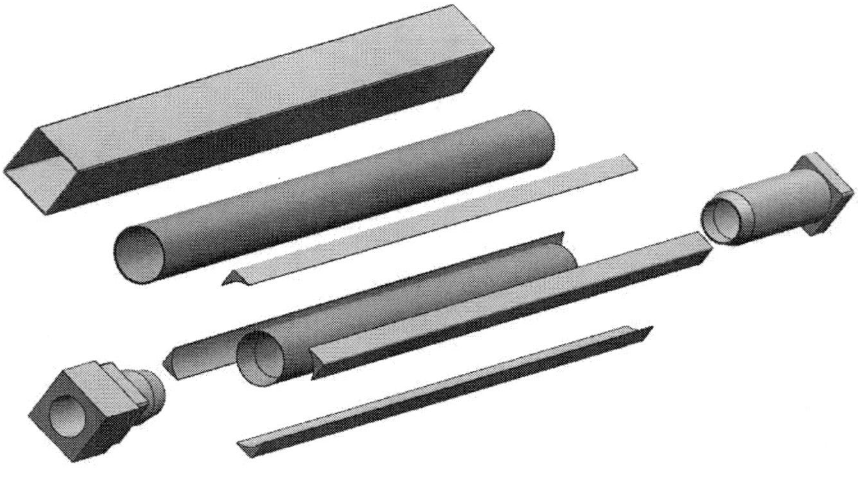

Fig. 4: Exploded view with marked parts (yellow colour), which will be solved

**Selection computational model**

Before preparing computational model is needed to select the best model type for this task type. Exist two main ways, how to model the ram. First way is to model the ram as 3D solid for each composite layer. It means, that individual layers are modelled, meshed and connected to each other. Each layer have defined its properties separately. This way wasn´t prefer, due to demanding preparing the model and connecting layers to each other. In additional, some layers are very thin and this computational model wouldn´t have to be too stable. Second way is to create the model as 2D task and define each layers by laminate modeler. This way was prefer for synoptic layers definition and for better and easier potential future adjusts.

**Model creation**

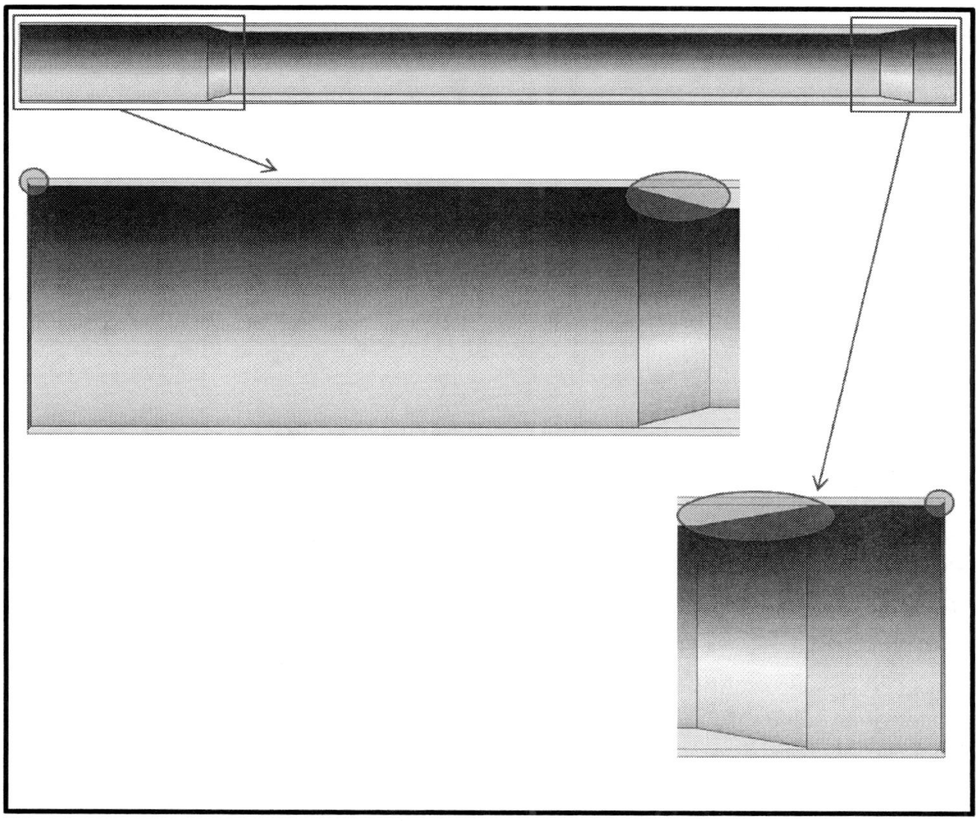

Fig. 5: 3D model of the ram – cross-section with details of chamfers (red marked)

After selection of computational model is possible to model tubes of the ram, which will be computed. This task was a little simplified and certain faces was adjusted to following layer definition. As can be seen on Fig. 1, in certain sectors (black filled), there are cork panels, which have influence on time costs and correct computational approach. Whole 2D model was cut into smaller faces and for each face was defined appropriate layer composition. This approach was selected due to simplify chamfers, which are shown on Fig. 5, where is shown initial model. Consequently, chamfers were represented by cut face, like on Fig. 6.

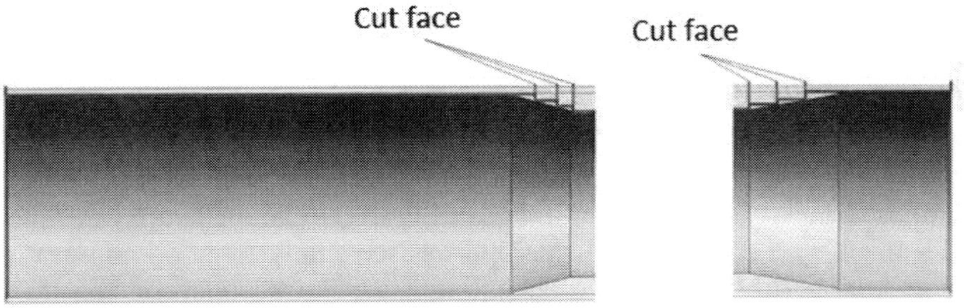


Fig. 6 Simplified chamfer areas – places, where computational 2D model was cut (green marked)

Thanks to these simplifying conditions, 2D model were created, prepared to assign mesh and single layers by laminate modeler. All cut faces will have assign appropriate number of layers.

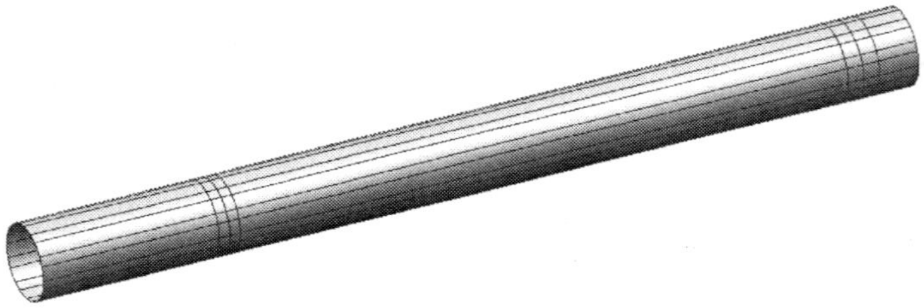

Fig. 7 Final 2D model

**Mesh creation, layer definition**

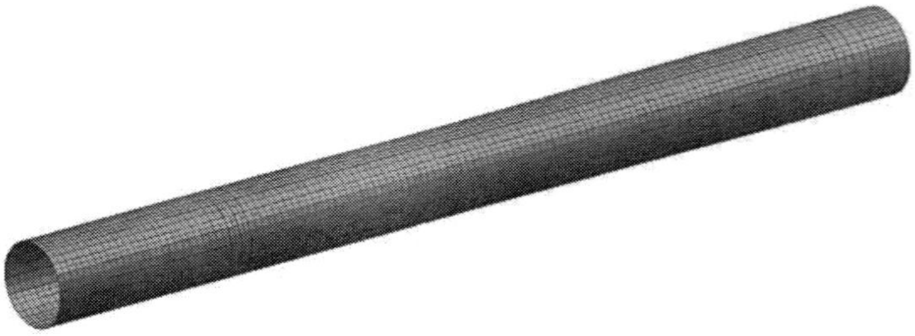

Fig. 8: 2D model of the composite tube

Fig. 9: Layer definition - demonstration example for angle 45°

After creation of computational model, mesh definition follows. Thanks to previous adjustments on 2D model, is it possible to define properties for each face separately, in relation to real layer composition. As was mentioned, this approach was selected mainly due to cork layers. As mesh was selected CQUAD4 with element size 20mm and this mesh was used for all 2D faces. After mesh definition was define layer properties for each face. This stage of definition have the biggest time costs, because by cut caused new 140 faces and each face must be defined separately. It is given due to type of fiber winding, mainly for different angle than 0 and 90 degrees. Fig. 9 describing layer definition for different layers. As it can be seen, each layer with different angle from 0 and 90 degrees must be defined three times. Layers with angle 0 and 90 degrees, are winded completely with this angle. Layers are defined by so-called *layups*. These layups have all information about particular layer. Layups can be united for faces with same properties.

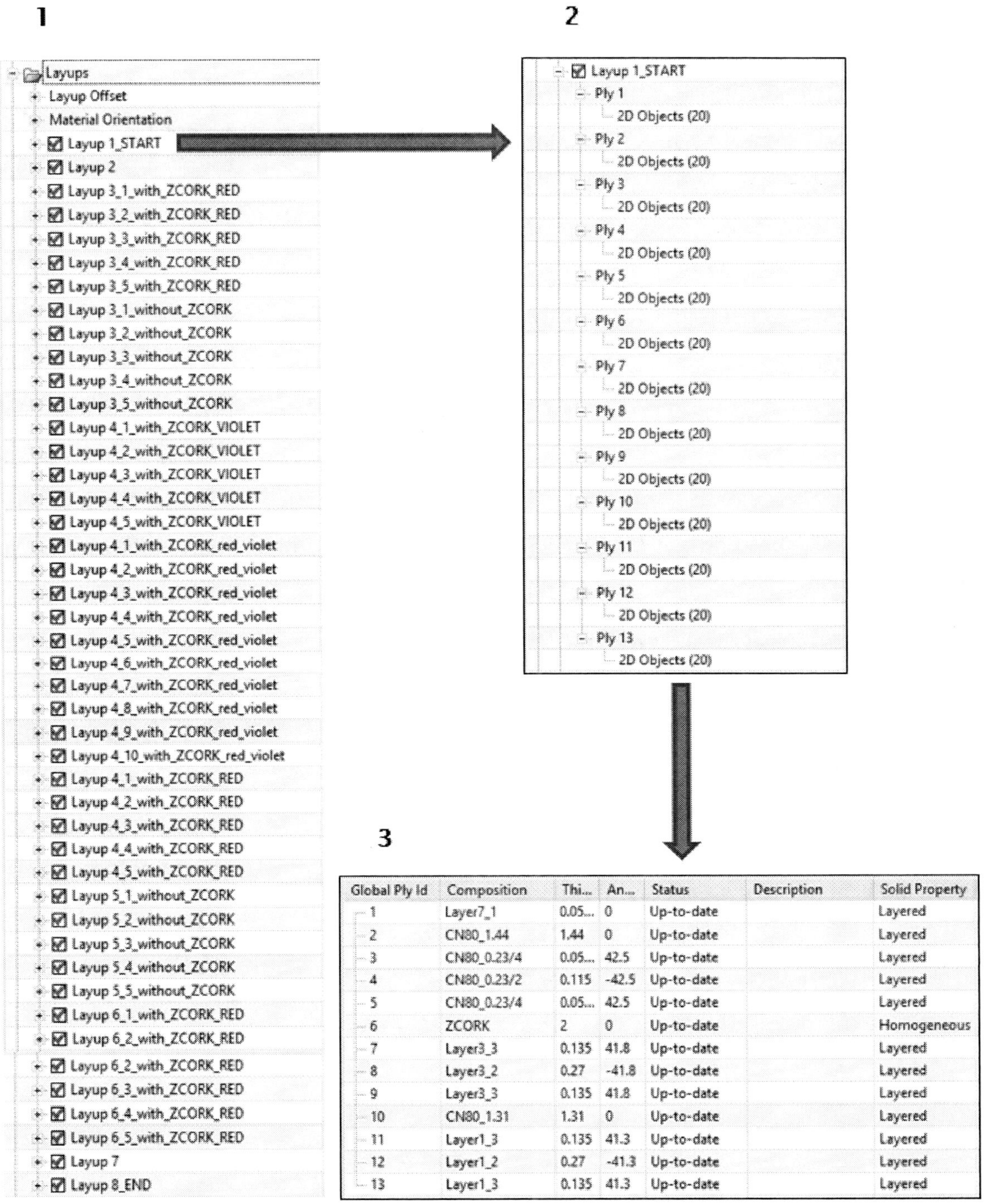


| Global Ply Id | Composition | Thi... | An... | Status | Description | Solid Property |
|---|---|---|---|---|---|---|
| 1 | Layer7_1 | 0.05... | 0 | Up-to-date | | Layered |
| 2 | CN80_1.44 | 1.44 | 0 | Up-to-date | | Layered |
| 3 | CN80_0.23/4 | 0.05... | 42.5 | Up-to-date | | Layered |
| 4 | CN80_0.23/2 | 0.115 | -42.5 | Up-to-date | | Layered |
| 5 | CN80_0.23/4 | 0.05... | 42.5 | Up-to-date | | Layered |
| 6 | ZCORK | 2 | 0 | Up-to-date | | Homogeneous |
| 7 | Layer3_3 | 0.135 | 41.8 | Up-to-date | | Layered |
| 8 | Layer3_2 | 0.27 | -41.8 | Up-to-date | | Layered |
| 9 | Layer3_3 | 0.135 | 41.8 | Up-to-date | | Layered |
| 10 | CN80_1.31 | 1.31 | 0 | Up-to-date | | Layered |
| 11 | Layer1_3 | 0.135 | 41.3 | Up-to-date | | Layered |
| 12 | Layer1_2 | 0.27 | -41.3 | Up-to-date | | Layered |
| 13 | Layer1_3 | 0.135 | 41.3 | Up-to-date | | Layered |

Fig. 10: Created layups – demonstration example for definition one face

### Computing, results

After that material properties was defined, computing can start. Modal analysis was solved by NX NASTRAN solver a type of solver was SOL 103 Real Eigenvalues. Lower border for looking for real eigenvalues was set-up on 1Hz. Totally 3 tasks were solved.

- Freely placed ram

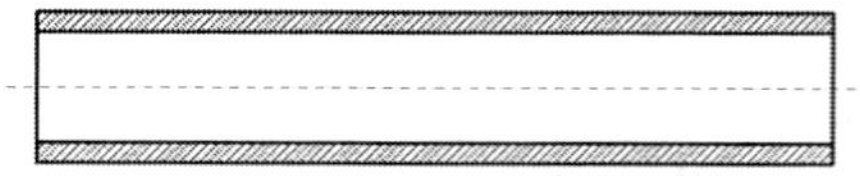

- Fixed ram on one side

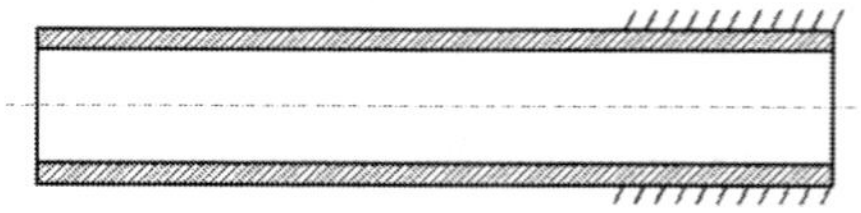

- Fixed ram on one side with additional mass

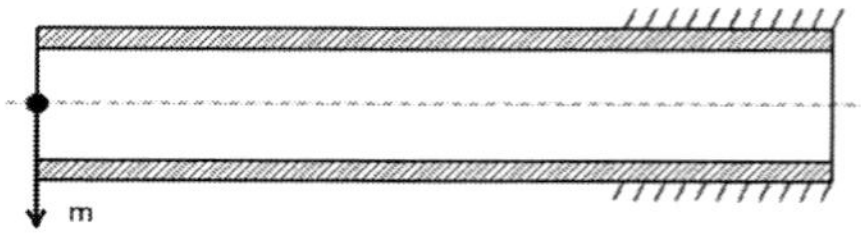


Last two cases have informational character, due to lack of information about length of fixation and size and position the additional mass. For freely placed ram, this set – up is enough and computing can start. For fixed ram on one side is necessary define this fixing, it means, remove degrees of freedom in this place. For fixed ram on one side with additional mass is necessary to define this mass and location.

### Freely placed ram

In case of freely placed ram, results were in compare with experiment about 7 - 38% different. This difference can be caused by simplifying conditions for chamfered faces. Another influences can cause lack of information about real thickness a real ratio between fibers and resin for each layer. This influence can be caused by winding layer on winded previous one. Lower layer can be pressed under nominal value and its properties could be different from reference data.

| Mode | Experiment Frequency (Hz) | Simulation Frequency (Hz) | Difference (%) |
|---|---|---|---|
| 1 | 383 | 356 | **-7.0** |
| 2 | 476 | 359 | **-24.6** |
| 3 | 713 | 575 | **-19.4** |
| 4 | 747 | 578 | **-22.6** |
| 5 | 792 | 590 | **-25.5** |
| 6 | 947 | 592 | **-37.5** |
| 7 | 982 | 663 | **-32.5** |
| 8 | 1060 | 663 | **-37.5** |
| 9 | 1120 | 734 | **-34.5** |
| 10 | 1170 | 735 | **-37.2** |

Table 1: Results for freely placed ram

### Fixed ram on one side

In modal analysis for fixed ram on one side, most modal frequencies were lower than 15%. Fixed area was determined approximately, due to lack of accurate information. Different results can be caused by factors mentioned in chapter above.

| Mode | Experiment Frequency (Hz) | Simulation Frequency (Hz) | Difference (%) |
|---|---|---|---|
| 1 | 273 | 256 | **-6.2** |
| 2 | 392 | 256 | **-34.7** |
| 3 | 400 | 357 | **-10.8** |
| 4 | 572 | 359 | **-37.2** |
| 5 | 606 | 577 | **-4.8** |
| 6 | 653 | 578 | **-11.5** |
| 7 | 684 | 614 | **-10.2** |
| 8 | 775 | 738 | **-4.8** |
| 9 | 809 | 741 | **-8.4** |
| 10 | 970 | 755 | **-22.2** |

Table 2: Results for fixed ram on one side

## Fixed ram on one side with additional mass

Last case, modal analysis for fixed ram on one side with additional mass had similar results in compare with experiment mainly for higher frequencies. For lower frequencies were results more different. As in previous case, there was lack of information about fixing and additional mass. For simulation was used smaller ending. Weight of the ending was known. It was 43kg. On the picture, ending was most probably a little bit moved out from ram. This boundary condition wasn't taken into consideration. Mass is defined as 1D element connected to ram in area, where ending is really placed. On the table 3, it can be seen, that for higher modes, diffence is about 5%, but for lower modes, results are very difficult.

| Mode | Experiment Frequency (Hz) | Simulation Frequency (Hz) | Difference (%) |
|---|---|---|---|
| 1 | 703 | 102 | **-85.5** |
| 2 | 858 | 102 | **-88.1** |
| 3 | 976 | 628 | **-35.7** |
| 4 | 1070 | 737 | **-31.1** |
| 5 | 1160 | 738 | **-36.4** |
| 6 | 1310 | 959 | **-26.8** |
| 7 | 1450 | 959 | **-33.9** |
| 8 | 1540 | 1241 | **-19.4** |
| 9 | 1570 | 1491 | **-5.0** |
| 10 | None | 1491 | |

Table 3: Results for fixed ram on one side with additional mass

## Conclusion

This topic related about realized modal analysis of composite ram. Model was created as divided 2D face and then composite layers were defined. Certain areas of the ram were simplified. After obtain result was shown, that simulation corresponds with experiment only for certain modes. Reasons, why some modes are different, can be more. At first, bottom layers can be stiffer (about 10%), due to way of manufacturing, when each following layer can cause stiffening to previous layer. Some influence can cause not accurate storing layers in little different angle, than reference. Finally, for precise modal analysis is necessary to have more information about precise boundary conditions. This simulation partly confirmed it. To possible improve these results, mainly for modal analysis of freely placed ram, Ansys solver will be used for comparing these results between themselves. For this step should be used shell elements and these new results will be compared with actual ones. Suitable aspect to take into consideration is change material properties for winded layers, placed on lower diameters in the structure. After modal analysis in Ansys, larger model of machining machine and also whole model of the ram (including outer frame, sides and endings) will be created.

## References:


[1] f. a. v. Z. v. P. Katedra mechaniky, „Modální analýza vřeteníku," Plzeň, 2010.

[2] R. Poul, „Smykadlo-TrubkaZaklad-RozvrzeniVrstev-1," 2010.

[3] Unknown, „kompozitni_skladby".

[4] Unknown, „VahyDilu-1".

EAN 2017
55th conference on experimental stress analysis

# Qualitative Analysis of Different Lamb Wave Excitation Methods

ŠOFER Pavel[1,a], ŠOFER Michal[2,b], FERFECKI Petr[2,3,c] and MAREK Martin[3,d]

[1] Department of Control Systems and Instrumentation, Faculty of Mechanical Engineering, VŠB-TU Ostrava, 17. listopadu 15/2172, 708 33 Ostrava-Poruba, Czech Republic

[2] Department of Applied Mechanics, Faculty of Mechanical Engineering, VŠB-TU Ostrava, 17. listopadu 15/2172, 708 33 Ostrava-Poruba, Czech Republic

[3] IT4Innovations National Supercomputing Center, VŠB-Technical University of Ostrava, Studentská 6231/1B, 708 33 Ostrava-Poruba, Czech Republic

[a]pavel.sofer@vsb.cz, [b]michal.sofer@vsb.cz, [c]petr.ferfecki@vsb.cz, [d]martin.marek@vsb.cz

**Keywords:** Lamb wave, Liquid wedge, Comb-structure, Wave excitation

**Abstract:** Guided Lamb waves are widely used for qualitative analysis of damage in various types of structures. An important factor for generally successful analysis is the incorporation of suitable excitation method of the desired mode of Lamb wave. This article is aimed to qualitatively evaluate four different methods of excitation of a particular mode of Lamb wave, namely excitation by longitudinal waves with the use of liquid and plexiglass wedge, excitation by longitudinal waves with the use of comb-structure element and excitation by longitudinal waves with the use of the transducer in itself, placed on the surface.

## Introduction

Guided waves, especially Lamb waves, play important role in the area of qualitative analysis of damage in various types of structures [1,2]. Lamb waves are one of the types of guided waves, which exist only in geometries with a finite thickness such as plates, tubes or layered media [3,4]. The propagation characteristics of Lamb waves are described by dispersion curves, which represent functional dependence between phase/group velocity of a particular mode of Lamb wave on the frequency-thickness product. The dispersion curves for a single as well as multi-layered media can be obtained numerically and also experimentally [5,6]. No less important role during qualitative analysis with use of guided Lamb waves is the method of wave excitation and reception. The excitation of Lamb waves can be realized by two conceptually different methods [4] – in practice more common excitation on the surface of the plate and excitation within the bounding layers of the plate or at the end face/edge of the plate. The generation of a particular mode of Lamb wave in itself can be realized in various ways. The most common way is the use of piezoelectric transducers [4,5]. In recent years, we can notice increasing use of interdigital transducers [6], benefit from a size and multiple capabilities. However, the main drawback of these transducers is their brittleness, which reflects in handling issues when required to place them on curved surfaces [6]. Another possibility is to use electromagnetic acoustic transducers (EMAT) [5], which do not require a direct contact with surface nor any complaint. The main drawback is the limitation to metallic or magnetic structures and the production of lower power than standard piezoelectric transducers [7]. The use of optical methods such as a laser is still not so common with the main drawback of bulky and expensive equipment [8].

The main aim of presented paper is the qualitative assessment of different methods of Lamb wave excitation, which will incorporate classical piezoelectric transducer producing longitudinal bulk waves.

## Derivation of dispersion relation for Lamb waves – solution by method of potentials [7]

Suppose the plane strain problem of a free traction force surfaces as shown in Fig. 1. The motion is defined as two-dimensional so that stresses, strains, displacement and other quantities are independent on $y$ coordinate. The unknown displacement vector $\bar{u}$ can be defined with use of Helmholtz decomposition theorem as follows:

$$\bar{u} = \nabla\phi + \nabla \times \bar{\psi}, \tag{1}$$

where $\phi = \phi(x, z)$ and $\bar{\psi} = (0, -\psi(x, z), 0)$ are potential and vector functions respectively. As mentioned above, both of them are functions of only $x$ and $z$ coordinates.

*Fig. 1: Geometry of the plate; h is the thickness of the plate*

With help of decomposed displacement vector can be the equation of motion, expressed in terms of displacements, separated into two wave equations, one related to longitudinal waves (Eq. 2) and the second one related to transversal waves (Eq. 3):

$$\nabla^2\phi - \frac{1}{c_L^2}\frac{\mathrm{d}^2\phi}{\mathrm{d}t^2} = 0, \tag{2}$$

$$\nabla^2\psi - \frac{1}{c_T^2}\frac{\mathrm{d}^2\psi}{\mathrm{d}t^2} = 0, \tag{3}$$

where $c_L$ is the speed of dilatational waves, $c_T$ is the speed of shear waves and $t$ is time. The assumed solutions of Eq. 2 respectively 3 for the wave propagating from left to right are as follows:

$$\phi = (K_1\sin(pz) + K_2\cos(pz))e^{i(kx-\omega t)}, \tag{4}$$

$$\psi = (K_3\sin(qz) + K_4\cos(qz))e^{i(kx-\omega t)}, \tag{5}$$

where $p^2 = \frac{\omega^2}{c_L^2} - k^2$, $q^2 = \frac{\omega^2}{c_T^2} - k^2$, $\omega$ is the angular frequency, $K_1$, $K_2$, $K_3$, $K_4$ are constants resulting from boundary conditions, $k$ is wavenumber variable and i is the imaginary unit. Expressing the displacements $u$ and $w$ in both axes as well as the stresses $\sigma_{zz}$ and $\sigma_{zx}$ in terms of potentials and applying a boundary condition $\sigma_{zx} = \sigma_{zz} = 0$ at $z = \pm\ 0{,}5h$ it is then possible to finally derive the Rayleigh-Lamb frequency equation for symmetric (Eq. 6) and antisymmetric (Eq. 7) Lamb wave modes:

$$\frac{\tan(\frac{qh}{2})}{\tan(\frac{ph}{2})} = -\frac{4k^2pq}{(q^2-k^2)^2} \tag{6}$$

$$\frac{\tan(\frac{qh}{2})}{\tan(\frac{ph}{2})} = -\frac{(q^2-k^2)^2}{4k^2pq} \tag{7}$$

## Experimental procedure

There have been used four different methods for excitation of symmetric and/or antisymmetric modes of Lamb wave in total. Namely, excitation with the use of comb-structure with transducer producing longitudinal waves, excitation with liquid and plexiglass wedge with transducer producing longitudinal waves and transducer producing longitudinal waves, which has been placed directly on the surface of the plate. In all four cases, Panametrics C103 1MHz/0,5" transducer has been used as a source of acoustic waves. Vallen AMSY-6 acoustic emission system, manufactured by Vallen Systeme GmbH, has been utilized as a signal generator. A five-cycle sinusoidal tone burst of 250 $V_{peak\text{-}peak}$ and the central frequency of 150 kHz (see Fig. 2) was in all four cases used as driving signal for source transducer.

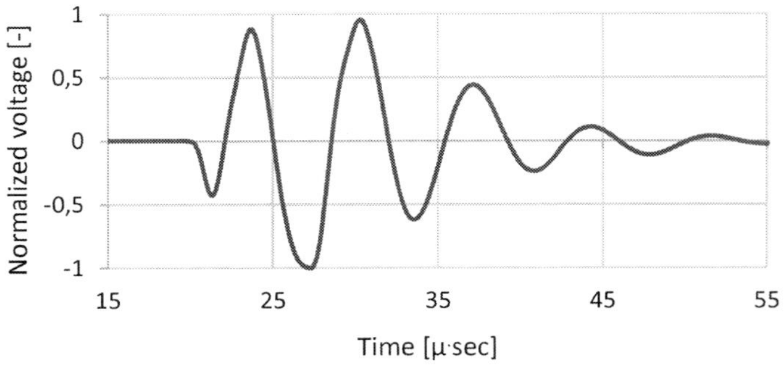


*Fig. 2: Driving signal for wave excitation*

The signal of a wave, propagating from 200 mm distant source, has been detected by Vallen VS 375-M broadband acoustic emission sensor and further amplified and processed with Vallen AEP5 amplifier and Rohde&Schwartz RTH1002 oscilloscope respectively. The experiments were carried out on one-millimetre thick steel sheet of dimensions 100x70 cm (steel grade 11).

The signal, produced by the transducer, which was placed directly on steel sheet surface, was used as a reference for both symmetric and antisymmetric modes of Lamb wave, further excited by different ways. The angles of incidence in the case of wave excitation with the use of liquid and plexiglass wedges we being calculated using Snell´s law and the knowledge of phase velocity of a particular mode of interest and the excitation frequency:

$$\alpha = arcsin\left(\frac{c_{L-wedge}}{c_{phase}}\right), \tag{8}$$

where $\alpha$ is the angle of incidence, $c_{L-wedge}$ is the speed of longitudinal waves in wedge and $c_{phase}$ is phase velocity of a particular mode of interest, which has to be excited. The dispersion curves for phase velocity (Fig. 3) were obtained by solving Rayleigh-Lamb frequency equation [5]. Dispersion curves for group velocity can be obtained with use of phase velocity dispersion curves and the known relation between these two types of velocities [5]:

$$c_g = \frac{\mathrm{d}(kc_p)}{\mathrm{d}k} = c_p + k\frac{\mathrm{d}c_p}{\mathrm{d}k}\,. \tag{9}$$

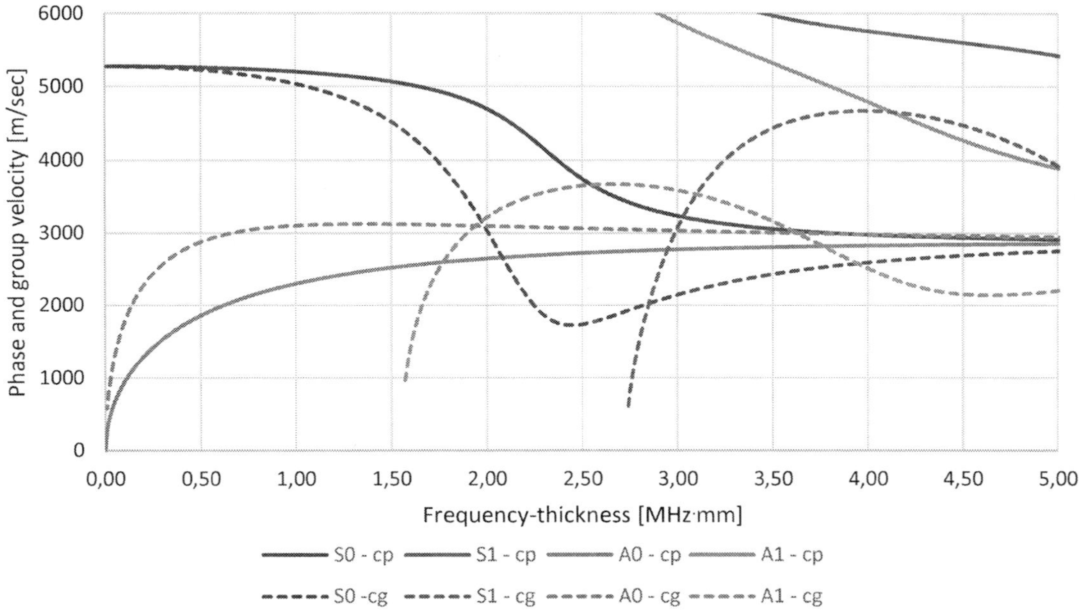


*Fig. 3: Dispersion curves for phase and group velocities in case of 1 mm thick steel sheet ($c_L$ = 5900 m/sec; $c_T$ = 3100 m/sec) produced by self-developed code in MATLAB software; S - symmetric and A - antisymmetric modes of Lamb wave*

**Comb-structure**

The comb-structure method has been originally designed for Rayleigh wave excitation [4]. This method can be however used with success also in the case of Lamb wave generation. According to known excitation frequency and known dispersion curve for phase velocity of mode we want to excite, it is possible to calculate the wavelength, which will then define the spacing of two millimetre wide elements that are in direct contact with the plate (See Fig. 4). In the original concept of comb-structure, there is used a piezoelectric crystal, covering the entire upper part of the comb-structure. In our case, we have used the transducer with a contact surface in the shape of a circle with the diameter of 14 mm. The wavelength of the antisymmetric longitudinal wave with the excitation frequency of 150 kHz is equal to 7,5 mm. The excitation area of the transducer barely covers three segments, however, a certain portion of the acoustic pressure perturbation is delivered to the outer elements thanks to multiple reflections.

*Fig. 4: Comb-structure element*

### Liquid wedge

The incorporation of the liquid wedge for Lamb wave excitation is relatively common technique, capable of delivering reputable results when properly adjusted. The advantage of this method lies in the possibility of excitation of S0 - symmetric and A0 - antisymmetric modes of Lamb wave having relatively small wavenumbers. The cause of this fact lies in the speed of longitudinal waves in water, which is nearly 1500 m/sec. Another considerable advantage is the minimum attenuation in liquid unlike plexiglass and also in existence of only longitudinal waves in the liquid with zero shear modulus. A certain disadvantage is clumsy handling. The self-manufactured liquid wedge uses an inflatable balloon, filled with water, which is together with the transducer mounted in plexiglass holder (See Fig. 5). The liquid wedge is used to generate the symmetric mode of Lamb wave, which results in the inclination angle of 16,3$^{\circ}$ for the S0 phase velocity of 5270 m/sec and frequency of 150 kHz.

*Fig. 5: Liquid wedge*

### Plexiglass wedge

The plexiglass wedge is suitable for generation of S0 - symmetric as well as A0 - antisymmetric modes of Lamb wave. The wedge is 25 mm thick and the radius after which moves the counterpart with the transducer has a radius of 35 mm. In order to set the proper inclination angle, there has to be measured the velocity of longitudinal waves, whose value is 2667 m/sec. For further experiments, there has been additionally measured the speed of transversal waves, which resulted in the value of 1371 m/sec. The measurement of the speed of transversal waves has been realized with use of Panametrics V155 transducer. The plexiglass wedge (Fig. 6) was in our case used to generate the symmetric mode of Lamb wave, which results in the inclination angle of 30,4$^{\circ}$ for the S0 phase velocity of 5270 m/sec and frequency of 150 kHz.

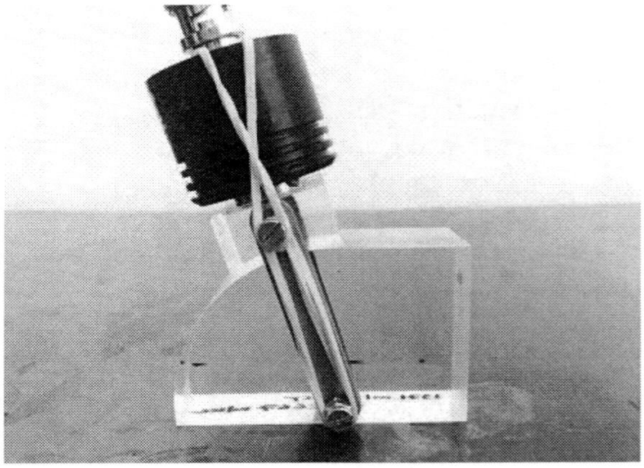

*Fig. 6: Plexiglass wedge*

### Results

The results of detection the symmetric and antisymmetric modes of Lamb wave originating from excitation by the transducer, directly placed on the steel sheet, were used as a reference for the remaining three methods of excitation of a particular mode of Lamb wave. Figure 7 displays us the result of the normalized voltage as the function of time, which belongs to the method when the transducer is directly placed on the steel sheet.

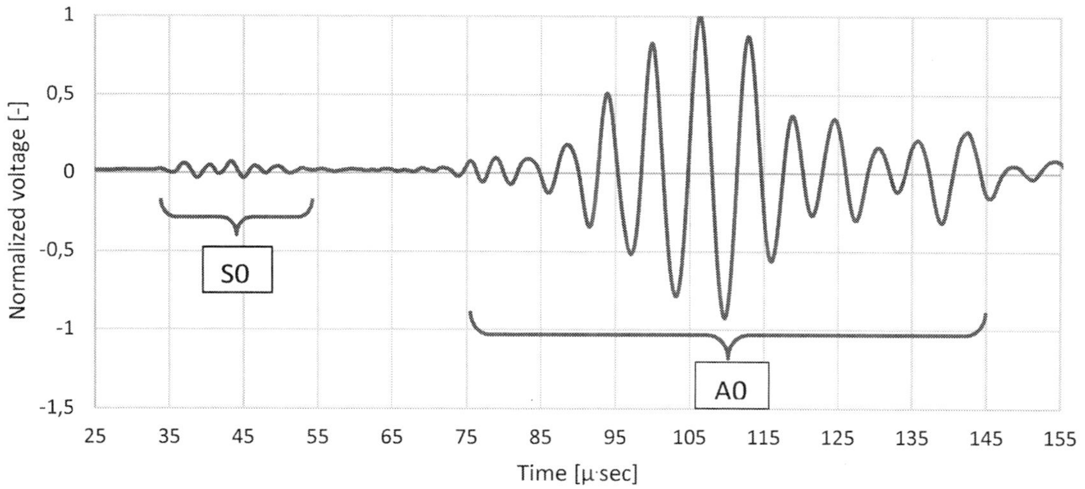


*Fig. 7: Normalized voltage versus time (Transducer directly placed on the steel sheet)*

The methodology of how to evaluate the efficiency can be realized by the way of comparing the amplitude of selected mode, which has to be excited by a particular method to the amplitude of the same mode, which was excited by the transducer directly placed on the steel sheet. The amplitude of A0 and S0 modes of Lamb wave are 96,3 and 72,5 $dB_{AE}$ respectively. $dB_{AE}$ unit expresses the direct signal of the transducer, referenced to the value of 1 µV. The comb-structure was, thanks to its geometry, utilized to generate purely the A0 mode of Lamb wave. The results met the expectations, the increase of the amplitude, compared to the reference value, is 4,6 dB with the present decrease of the amplitude of S0 mode by 10,8 dB. Figure 8 displays the dependency between the normalized voltage and time for the comb-structure method.

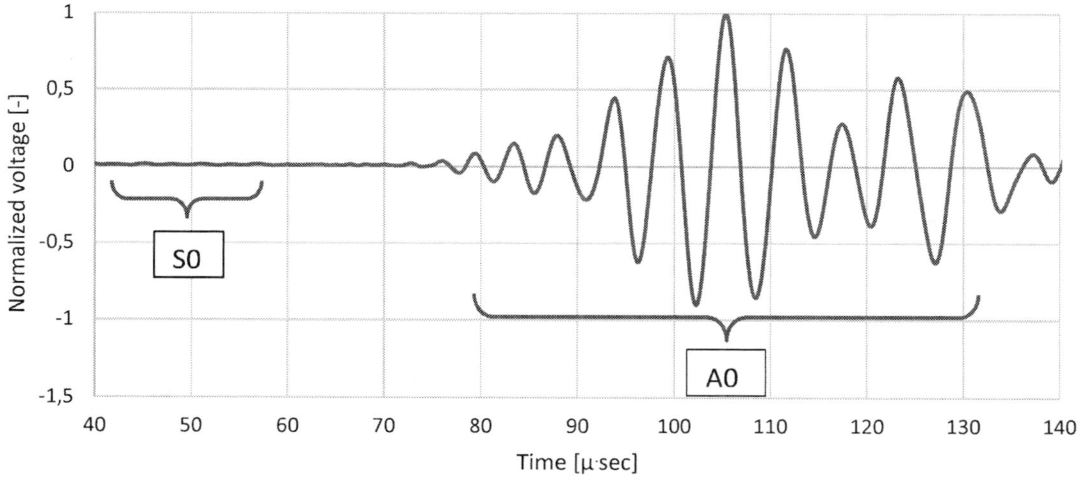


*Fig. 8: Normalized voltage versus time (Comb-structure method)*

The plexiglass wedge, as well as the liquid wedge, had adjusted their inclination angles in order to generate the S0 mode. The increase in amplitude is 1,8 and 2 dB for the case of liquid and plexiglass wedge respectively. Both methods also naturally produced A0 mode. The decrease of the A0 mode amplitude is in the case of plexiglass wedge equal to 16,5 dB and in the case of liquid wedge equal to 5,3 dB. The dependencies between the normalized voltage and time for plexiglass as well as liquid wedge are displayed in Fig. 9.

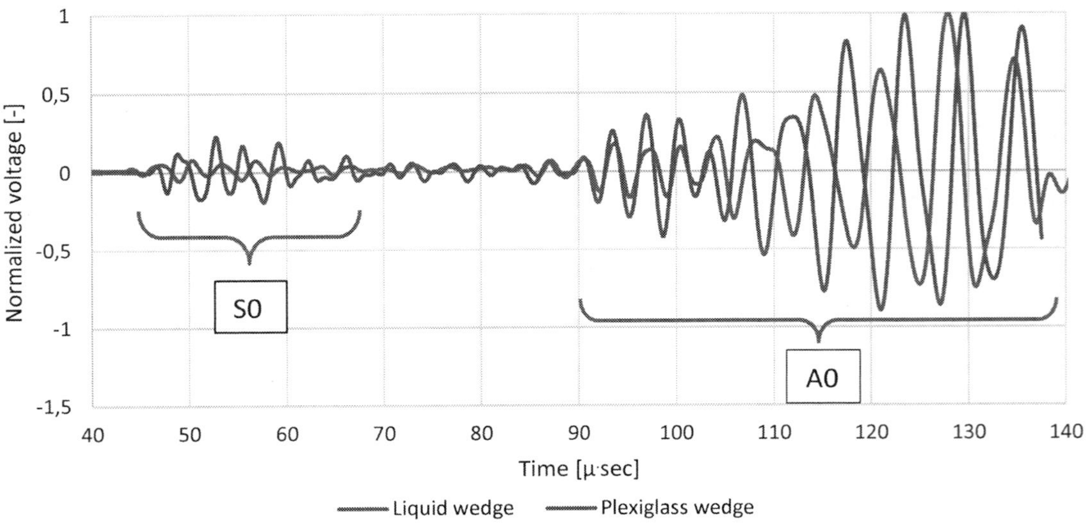


*Fig. 9: Normalized voltage versus time (Liquid wedge and plexiglass wedge method)*

## Conclusion

Four different methods for excitation of symmetric and antisymmetric modes of Lamb wave were used – direct placement of the transducer on the steel sheet, comb-structure method, plexiglass wedge method, and liquid wedge method. In order to compare the methods with each other, the authors have decided to use results originating from the excitation of the S0/A0 modes by direct placement of the transducer on the steel sheet as s reference to which will be the other methods compared. Despite the fact that the comb-structure method, designed for sole excitation of A0 mode, has not used a transducer which will cover the entire area of the comb-structure, has this method reached 4,6 dB higher amplitude of A0 mode compared to the reference value, meanwhile, the amplitude of S0 mode was very well damped. The wedge transducers had adjusted their inclination angles in order to generate the S0 mode. In both cases, the amplitude of S0 mode was approximately 2 dB above the amplitude of reference signal, meanwhile, the amplitude of A0 mode was 5, respectively 16 dB lower than the reference value.

Realized experiments show the specific advantages of used methods. The advantage of the classical approach, when the transducer is directly placed on the plate, is in relatively high amplitudes for both A0 and S0 modes. The comb-structure method belongs to dedicated methods from the perspective of excitation of a particular mode of defined wavelength, which is implicitly bounded to the material properties of the plate including its thickness. The wedge methods proved to be solid performers with the main advantage of the continuous change of the inclination angle, which is directly bounded to various types or modes of waves. On the other hand, the wedge method encounters the presence of leaky waves [4,5], which to a certain extent influence the amplitudes of generated surface wave.

For future work, the authors would like to focus on the design of the liquid as well as plexiglass wedges in order to improve their efficiency.


## Acknowledgement

This research has been elaborated in the framework of the project No: SP2017/106 „Advanced Methods for Machine and Process Control" and No: SP2016/136, both supported by the Ministry of Education, Youth and Sports.

## References

[1] S. Pant, J. Laliberte, M. Martinez, B. Rocha, Derivation and experimental validation of Lamb wave equations for an n-layered anisotropic composite laminate, Compos. Struct. 111 (2014) 566-579.

[2] Ch. Schaal, H. Samajder, H. Baid, A. Mal, Rayleigh to Lamb wave conversion at a delamination-like crack, J. Sound Vib. 353 (2015) 150-163.

[3] T.N. Grigsby, E.J. Tajchman, Properties of Lamb Waves Relevant to the Ultrasonic Inspection of Thin Plates, IRE Trans. Ultrason. Eng. 8 (1961) 26-33.

[4] I.A. Victorov, Rayleigh and Lamb Waves, Plenum, New York, 1967.

[5] J.L. Rose, Ultrasonic Waves in Solid Media, Cambridge, Cambridge University Press, 2002.

[6] T. Stepinski, M. Manka, A. Martinowitz, Interdigital lamb wave transducers for applications in structural health monitoring, NDT & E Int. 86 (2017) 199-210.

[7] P. Fromme, Noncontact Excitation of Guided Waves (A0 Mode) Using an Electromagnetic Acoustic Transducer (EMAT), AIP Conference Proceedings 1706 (2016) 1-8.

[8] Z. Zhang, Y. Pan, Y. Xiao, Z. Zhong, Measurement and analysis of laser generated Rayleigh and Lamb waves considering its pulse duration, Acta Mech. Solida Sin. 28 (2015) 441-452.

EAN 2017
55th conference on experimental stress analysis

# Usage of 3D CAD Software for Verification and Representation of Real Machine Measurements and Results

Ing. Zdeněk Braier[1], Ing. Pavel Šidlof[2], CSc. and Ing. Pavel Klouček[3], Ph.D.

[1,2,3] VÚTS, a.s., Svárovská 619, Liberec XI-Růžodol I, 460 01, Czech Republic

zdenek.braier@vuts.cz, pavel.sidlof@vuts.cz, pavel.kloucek@vuts.cz

**Keywords:** complicated machine, 3D CAD simulation, verification and representation, experiment on real machine, measurement

## Abstract

Many times the connections and transfers between main, driven and end parts are realized by a complicated transmission system with usage of several transfer structures. The contribution describes a complicated sewing machine, which uses several systems, such as tooth belt mechanisms, four-link mechanism, crank mechanism, cam mechanism, tilting mechanism and system shaft inside the hollow shaft etc. for connection of driving and driven mechanism members. A 3D CAD simulation is used for animation and visualization of expected (ideal) status and for verification and representation the measured results.

## Introduction

For reasons of inaccessibility of certain mechanisms (parts) or for their miniature size, it is often not possible to install measuring sensors to the parts. Therefore the movements of these parts are derived from movement measurement of some driving parts of mechanism. The derivation is usually performed by a kinematic or dynamic calculation, but it is advantageous to derivate this form of the movement (stroke) directly from a 3D CAD model of a machine mainly for speed and clarity in some cases. Another beneficial and illustrative way of an unusual use of CAD models for measurement can be a representation of the processed measurement results and their animation. The method requires the use of a sophisticated CAD software for an effective utilization and the data exchange adjustment.

The report shows an example of the method usage for measurement of a relatively complicated real sewing machine. Fig. 1 shows the kinematic schema of machine. One of the main measured results, requested by the customer, was a relative movement of a needle eye 9, a looper 10 and a secondary looper 11 (show at Fig. 2), which affect quality and reliability of the sewing process. The looper is driven by the main shaft 3 through a four-link mechanism 4, from which the movement of the secondary looper is derived by an additional cam mechanism 5. The needle bar 8 is guided by a tilting mechanism 13, which is used for adjustable side bight. The movements of the main shaft 3, the upper (driven) shaft 6, which has irregular motion due to a pulley controlled belt mechanism, the acceleration of the needle bar 8 and the needle bar holder 7, which guides the needle bar, were measured. The measurements were recorded as function of an angular position of the main shaft 3. The data were evaluated in software Wolfram Mathematica. The evaluation of the movement of parts 10, 11, derived from the movement of the shaft 3, was verified by the CAD model. The evaluated relative movements of the needle eye 9, the looper 10 and the secondary looper 11 were compared with their ideal (theoretical) movements within the CAD models.

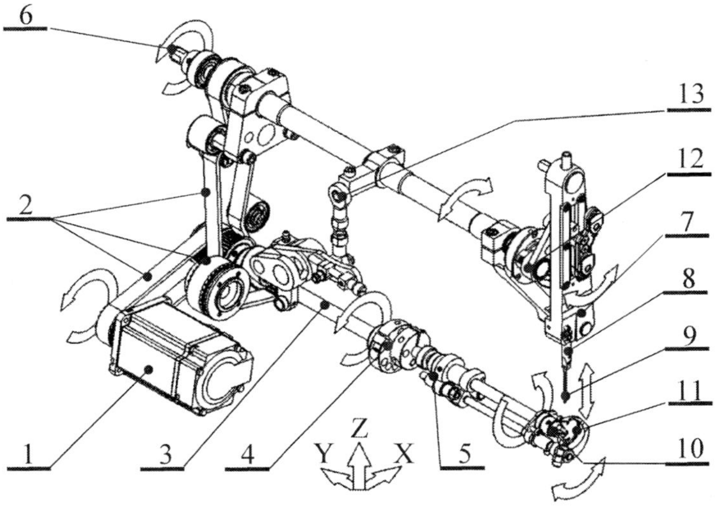

Fig. 1 Kinematic schema of machine

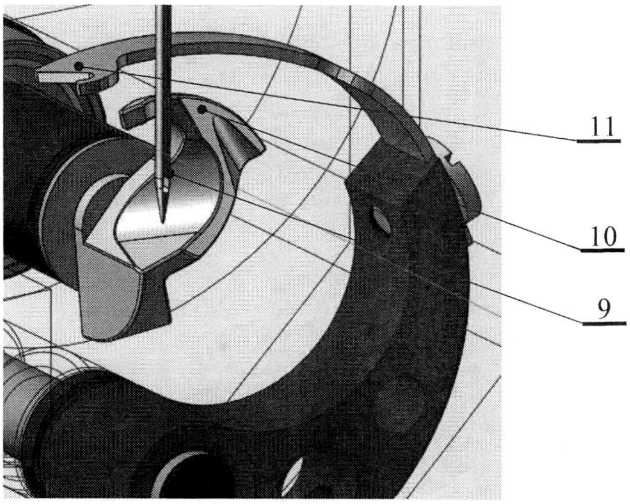

Fig. 2 Detail of needle, looper and secondary looper

Legend of Fig. 1, 2: 1 – servomotor, 2 – belt, 3 – main shaft, 4 – four-link mechanism, 5 – cam mechanism, 6 – upper (driven) shaft, 7 – needle bar holder, 8 – needle bar, 9 – needle (needle eye), 10 – looper, 11 – secondary looper, 12 – crank mechanism of needle bar, 13 – tilting mechanism

## Verification and Representation Methods

CAD designers use ideal single parts models for design of ideal (theoretical) CAD models. The CAD models also have ideal mechanical properties. The kinematic connections and movement are generally realized by stiff linkages without clearances. The software, which allow the dynamic calculation of mechanically complicated machines (e.g. Adams, NX), are hardly accessible for designers due to high cost and the necessity of highly qualified staff. Of course, the calculations by these software are naturally simplified and it is necessary to take measurements. The CAD models may enable the control of measurements, visual presentation of results and animation.

The measured sewing machine has a complex dynamic behavior. There are mechanisms with toothed belts, which have relatively low stiffness. The belt mechanism of the upper shaft is controlled by a tilting mechanism of the needle bar so that during one cycle the rollers vary the lengths of the two branches of the belt and thus modify the angular velocity of the crank shaft and the needle bar movement. The drive of the upper shaft has a considerable influence on the unevenness of the main shaft angular velocity, which influences the looper and secondary looper. In front of the looper, the inserted four-link mechanism significantly changes the looper angular velocity during the cycle. The secondary looper controlled by a cam mechanism performs oscillating movement. The angle and angular velocity of the secondary looper are measured directly by an incremental rotary encoder (IRC). Fig. 3 presents the photos of the machine with the IRC sensor at the main shaft, the upper (driven) shaft (Fig. 3a) and the secondary looper shaft (Fig. 3c). The acceleration sensors positions at the needle bar and the needle bar holder are shown in Fig. 3b. One cannot easily put a sensor on the looper. Because looper drive by the main shaft is relatively stiff, the angle, angular velocity and acceleration of the looper were calculated from the measured angle and angular velocity of the main shaft. The CAD model of the machine was used to check the calculation and correct an initial point of the adjustable looper. Similarly, movement of the needle bar was calculated and checked from the measured movement of the upper (driven) shaft and the needle bar holder.

All variables were simultaneously recorded by a multi-channel Dewetron system that included a special DMU device (a VÚTS system that uses IRC sensors for dynamic measurements of angle, angular velocity and acceleration) enabling angle based records. The calculated speed and acceleration of mentioned parts were calculated from the measured main shaft angle and angular velocity as a time differential.

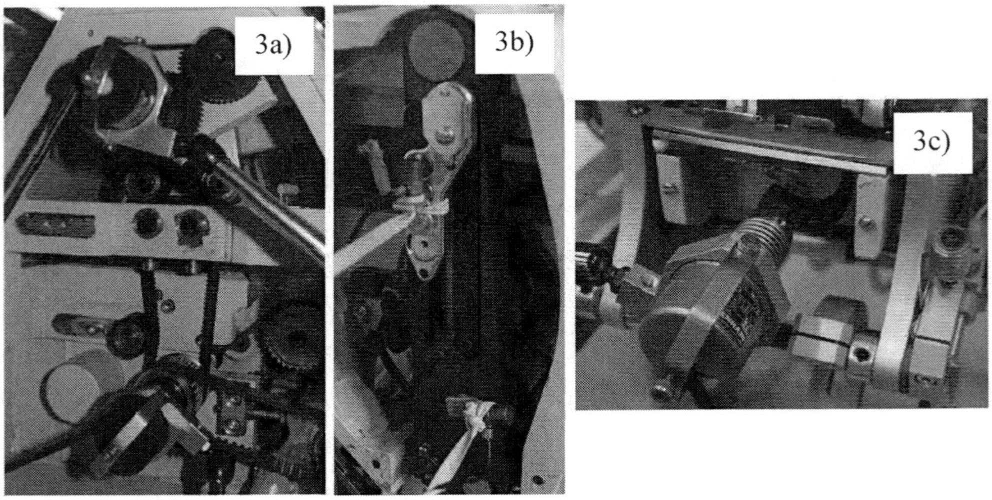


Fig. 3 Photo of real parts with IRC resp. acceleration sensors settings

## Results

Fig. 4 shows the simulation of 3D CAD models in SolidWorks. The simulation shows ideal position of the needle, the looper and the secondary looper in XZ-plane as a function of the main shaft angle from 90° to 280° and from 450° to 640° (second half of the cycle). The figure marked 180° and 540° expresses the lowest dead center of the needle eye and start point of the needle motion upwards (reverse). In range from 90° to 180° the needle moves down.

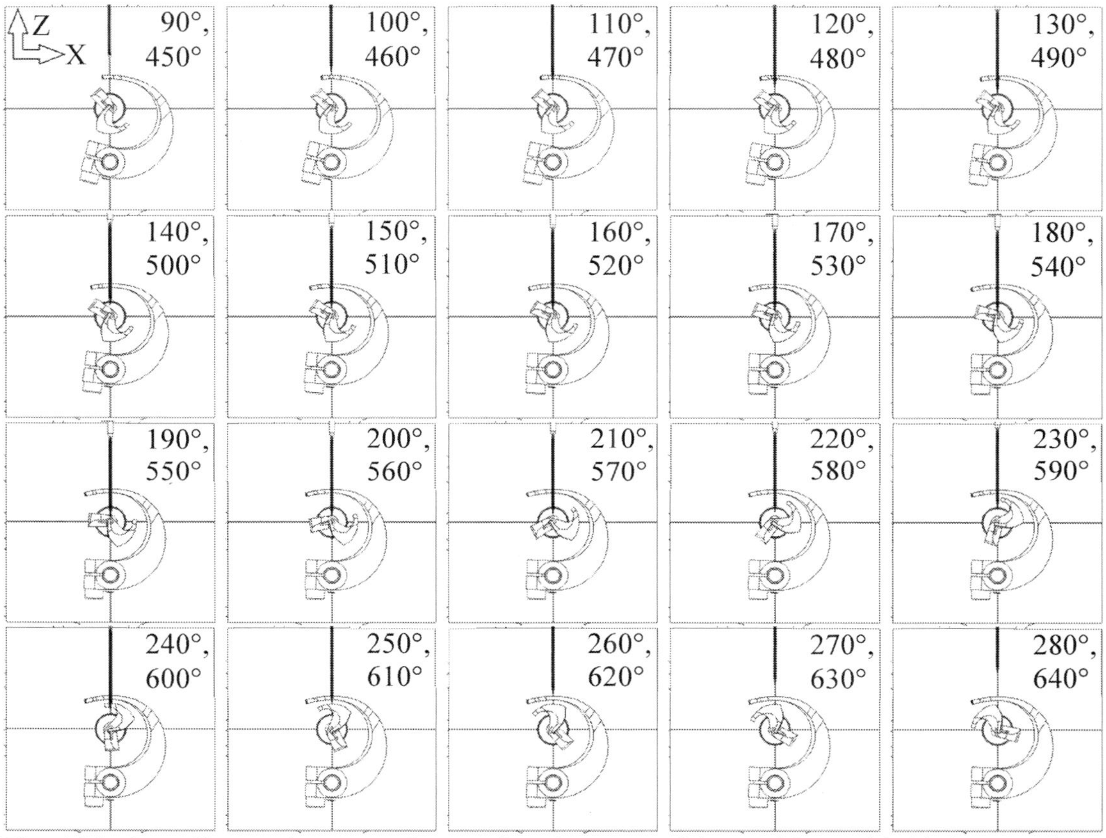


Fig. 4 Ideal needle eye, looper and secondary looper positions without side bight, XZ-plane, 3D CAD models simulation

Fig. 5 shows the mean measured angle of the needle bar tilting without side bight for speeds {1000, 1201, 1401, 1600, 1800, 1899, 1999} [RPM], color marked from green (lowest speed) to red (highest speed). Particular diagrams are mutually shifted by 1 for better clarity, the maximal and minimal values with respect to speed are {{0.05, -0.05}, {0.07, -0.11}, {0.11, -0.16}, {0.13, -0.22}, {0.25, -0.25}, {0.39, -0.33}, {0.45, -0.38}} [deg]. Fig. 6 shows the corresponding mean angular velocity, the diagrams are mutually shifted by 100, the maximal and minimal values are {{3, -4}, {9, -10}, {22, -24}, {26, -26}, {42, -36}, {62, -49}, {66, -56}} [RPM]. Fig. 7 shows the corresponding mean angular acceleration, the diagrams are mutually shifted by 2000, the maximal and minimal values are {{101, -74}, {263, -194}, {496, -567}, {634, -714}, {1179, -1118}, {1560, -1750}, {1872, -1936}} [rad/s$^2$].

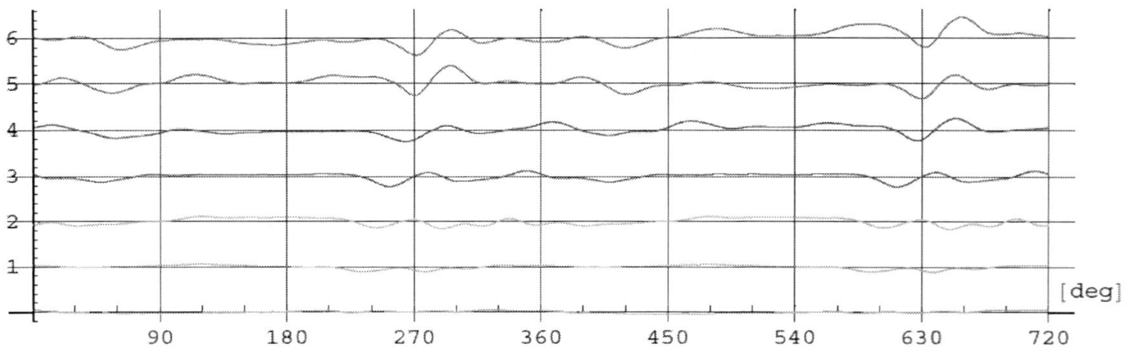

Fig. 5 Mean angle of the needle bar tilting without side bight [deg] vs. main shaft angular position

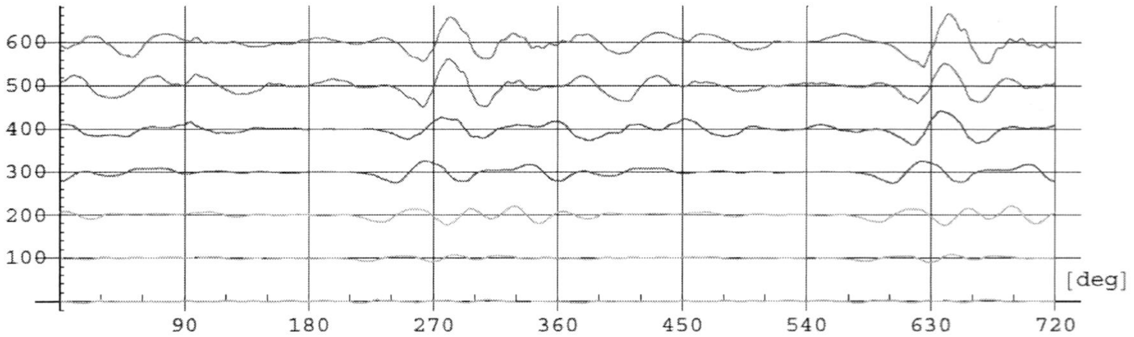

Fig. 6 Mean angular velocity of the needle bar tilting without side bight [RPM] vs. main shaft angular position

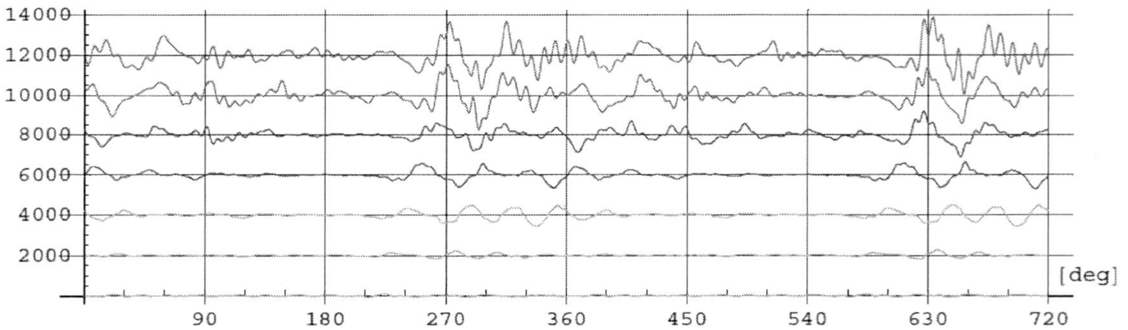

Fig. 7 Mean angular acceleration of the needle bar tilting without side bight [rad/s$^2$] vs. main shaft angular position

Figures 8 and 9 show the corresponding mean velocity and mean acceleration of the looper, which were modulated by transformation of the four-link mechanism. Fig. 8 shows the mean angular velocity of the looper, the diagrams are mutually shifted by 1000, the maximal and minimal values are {{2176, 266}, {2572, 317}, {2973, 373}, {3336, 427}, {3652, 466}, {3938, 490}, {4146, 523}} [RPM]. Fig. 9 shows the mean angular acceleration, the diagrams are mutually shifted by $10^5$, maximal and minimal values are {{23870, -13260}, {34520, -22170}, {48870, -38540}, {61810, -56860}, {89060, -75490}, {100060, -98420}, {116640, -113880}} [rad/s$^2$].

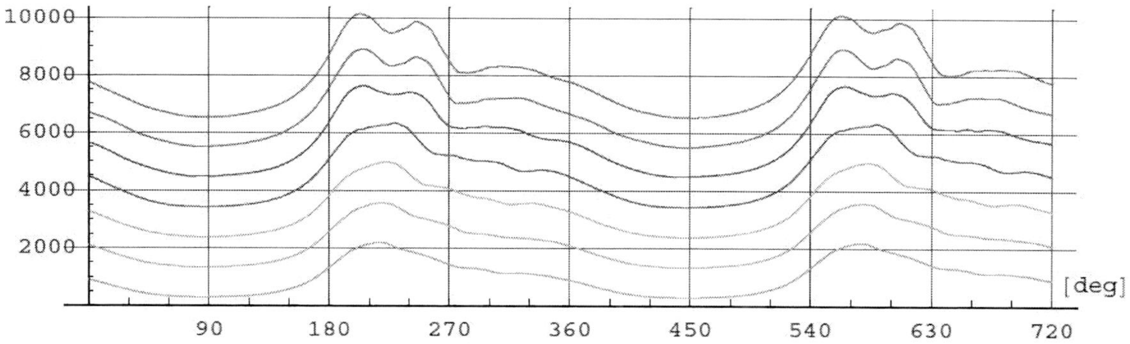


Fig. 8 Mean angle velocity of looper without side bight [RPM] vs. main shaft angular position

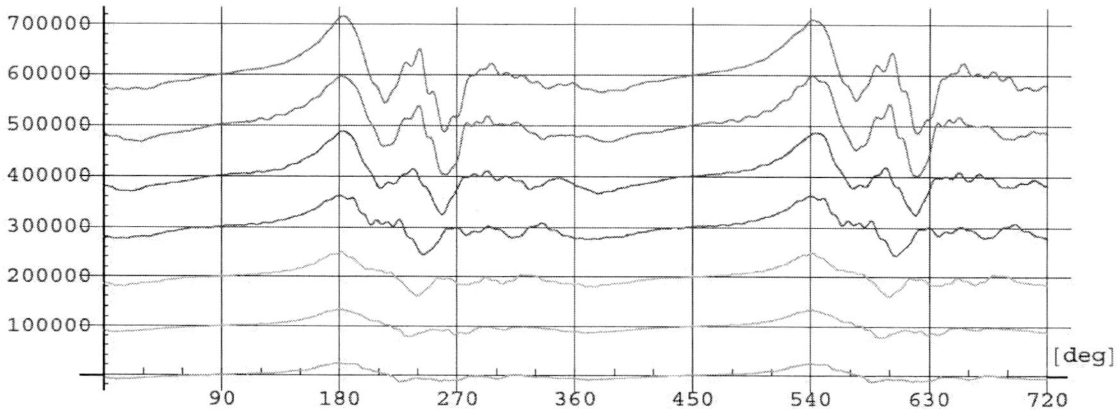


Fig. 9 Mean angular acceleration of looper without side bight [rad/s$^2$] vs. main shaft angular position

Table 1 links the angles and points of Fig. 4, Fig. 10 and Fig. 11. Angle 180° (event. 540°), highlighted by gray background of cells, expresses the lowest dead center of the needle.

Table 1 Links of the angles and points of figures 4, 10 and 11

| | | | | | | | | | | | |
|---|---|---|---|---|---|---|---|---|---|---|---|
| **Fig. 4** | Angle [°] | 110, 470 | --- | 120, 480 | --- | 130, 490 | --- | 140, 500 | --- | 150, 510 | --- |
| **Fig. 10a** | No. [-] | --- | --- | 1 | 2 | 3 | 4 | 5 | 6 | 7 | 8 |
| | Angle [°] | --- | --- | 120 | 125 | 130 | 135 | 140 | 145 | 150 | 155 |
| **Fig. 10b** | No. [-] | 1 | 2 | 3 | 4 | 5 | 6 | 7 | 8 | 9 | 10 |
| | Angle [°] | 470 | 475 | 480 | 485 | 490 | 495 | 500 | 505 | 510 | 515 |
| **Fig. 11a** | No. [-] | --- | --- | --- | --- | --- | --- | --- | --- | --- | --- |
| | Angle [°] | --- | --- | --- | --- | --- | --- | --- | --- | --- | --- |
| **Fig. 11b** | No. [-] | --- | --- | --- | --- | --- | --- | --- | --- | --- | --- |
| | Angle [°] | --- | --- | --- | --- | --- | --- | --- | --- | --- | --- |

| Fig. 4 | Angle [°] | 160, 520 | --- | 170, 530 | --- | 180, 540 | --- | 190, 550 | --- | 200, 560 | --- |
|---|---|---|---|---|---|---|---|---|---|---|---|
| Fig. 10a | No. [-] | 9 | 10 | 11 | 12 | 13 | 14 | 15 | 16 | 17 | 18 |
| | Angle [°] | 160 | 165 | 170 | 175 | 180 | 185 | 190 | 195 | 200 | 205 |
| Fig. 10b | No. [-] | 11 | 12 | 13 | 14 | 15 | 16 | 17 | 18 | 19 | 20 |
| | Angle [°] | 520 | 525 | 530 | 535 | 540 | 545 | 550 | 555 | 560 | 565 |
| Fig. 11a | No. [-] | --- | --- | --- | --- | --- | --- | --- | --- | 1 | 2 |
| | Angle [°] | --- | --- | --- | --- | --- | --- | --- | --- | 200 | 205 |
| Fig. 11b | No. [-] | --- | --- | --- | --- | --- | --- | --- | --- | 1 | 2 |
| | Angle [°] | --- | --- | --- | --- | --- | --- | --- | --- | 560 | 565 |

| Fig. 4 | Angle [°] | 210, 570 | --- | 220, 580 | --- | 230, 590 | --- | 240, 600 | --- | 250, 610 |
|---|---|---|---|---|---|---|---|---|---|---|
| Fig. 10a | No. [-] | 19 | 20 | 21 | 22 | 23 | 24 | 25 | --- | --- |
| | Angle [°] | 210 | 215 | 220 | 225 | 230 | 235 | 240 | --- | --- |
| Fig. 10b | No. [-] | 21 | 22 | 23 | 24 | 25 | 26 | 27 | 28 | 29 |
| | Angle [°] | 570 | 575 | 580 | 585 | 590 | 595 | 600 | 605 | 610 |
| Fig. 11a | No. [-] | 3 | 4 | 5 | 6 | 7 | 8 | 9 | --- | --- |
| | Angle [°] | 210 | 215 | 220 | 225 | 230 | 235 | 240 | --- | --- |
| Fig. 11b | No. [-] | 3 | 4 | 5 | 6 | 7 | 8 | 9 | 10 | 11 |
| | Angle [°] | 570 | 575 | 580 | 585 | 590 | 595 | 600 | 605 | 610 |

Fig. 10 shows the diagrams calculated and evaluated from the measured needle bar acceleration. Fig. 10a) shows the planar detail of the needle eye trajectory without side bight for speed 1600 [RPM], the needle eye vertical coordinate is measured from the looper axes, the main shaft angle range is 120° to 240°, points numbered for 5° steps, X-axis zoomed 10-times. Likewise, Fig. 10b) shows another part of the needle eye trajectory without side bight for the main shaft angle range from 470° to 610° (second part of the cycle). The diagrams show visible influence of vibrations due to the tooth belt connection during the side bight turn-off.

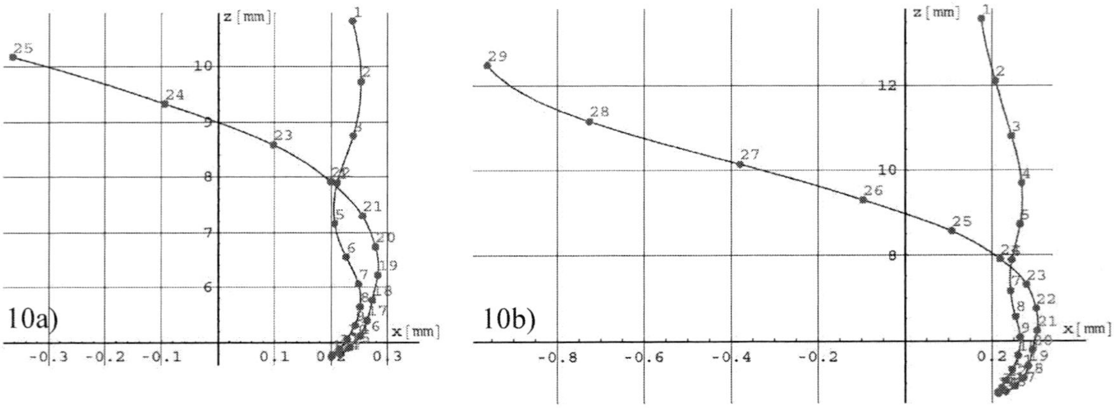


Fig. 10 Needle eye trajectory without side bight, XZ-plane, main shaft angle:
a) 120° to 240°, b) 470° to 610°

Fig. 11 shows the calculated and evaluated results of the looper nose and the needle eye trajectory. Fig. 11a) shows part of the looper nose and the needle eye trajectory without side bight for speed 1600 [RPM] during upwards motion (reverse) of the needle eye, the main shaft angle range is from 200° to 240°, points are numbered for 5° steps. Similarly, Fig. 11b) shows the trajectories for the main shaft angle range from 560° to 610°.

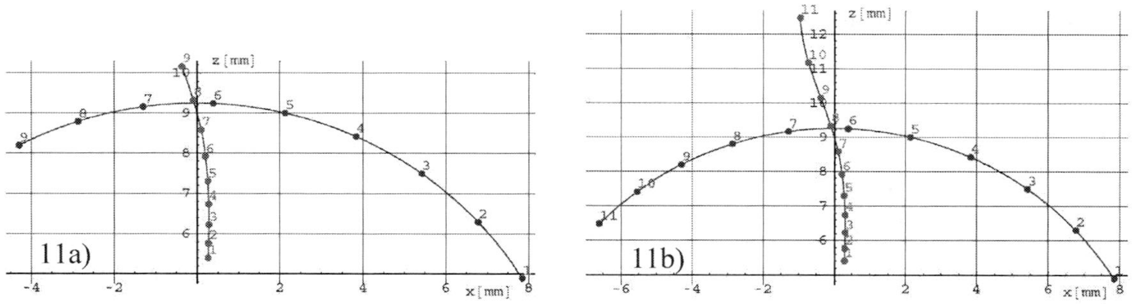


Fig. 11 Looper nose (blue, from right to left) and needle eye (red, upwards) trajectory without side bight, move up (reverse), XZ-plane, main shaft angle: a) 200° to 240°, b) 560° to 610°

Fig. 12 shows the diagrams of the planar detail of the needle eye trajectory with side bight for speed 1600 [RPM]. Fig. 12a) corresponds to the main shaft angle range from 490° to 610° (left stitch), points are numbered for 5° steps, X-axis zoomed 10-times. Similarly, Fig. 12b) shows another part of needle eye trajectory (right stitch), angle range is 120° to 240°, X-axis zoomed equally.

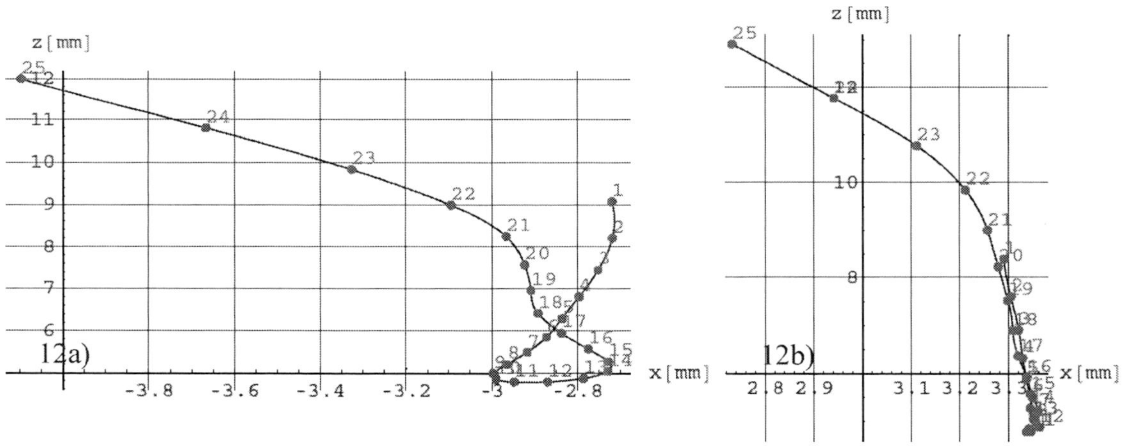


Fig. 12 Needle eye trajectory with side bight, XZ-plane, main shaft angle: a) 490° to 610°, b) 120° to 240°

Fig. 13 shows the calculated and evaluated results of the looper nose and the needle eye trajectory (result of the same measurement). The Fig. 13a) shows trajectories during upwards motion (reverse) of the needle eye, the main shaft angle range is from 560° to 610°, points numbered for 5° steps. Analogously, Fig. 13b) shows trajectories for angle range from 200° to 240°.

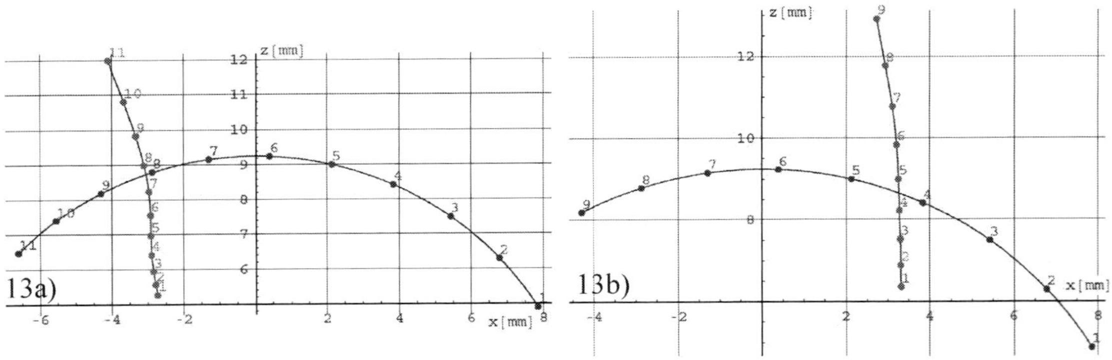


Fig. 13 Looper nose (blue, from right to left) and needle eye (red, upwards) trajectory with side bight, move up (reverse), XZ-plane, main shaft angle: a) 560° to 610°, b) 200° to 240°

The measured results allow us to create animations and visualization of these important movements in the CAD model and assess the reliability of the sewing process and possible errors.

## Conclusions

The contribution shows the example of possible common use of the measurement results of a real machine and simulation results of an ideal (theoretical) 3D CAD model. Measurement, verification and simulation of ideal (theoretical) 3D CAD model vs. measurement of real machine is currently being prepared for an expansion to effective usage for other projects from different branches of industry. The CAD models are used to check the measurement results and illustrative animation and visualization of real processes. It is possible to assess not only the correct operation of the machine, but e.g. vibration impact on motion of machine parts by calculating and imaging the corresponding increased displacements superposed to the parts standard motion.

## Acknowledgments

This work was supported by the Czech Ministry of Education, Youth and Sports, project LO1213.

EAN 2017
55th conference on experimental stress analysis

# Tensile and Shear Tests of a Cork/Rubber Composite

Radek Kottner[1,a], Jiří Kocáb[1,b], Petra Kochová[1,c] and Jan Heczko[1,d]

[1] NTIS - New Technologies for the Information Society, Faculty of Applied Sciences, University of West Bohemia, Technicka 8, 301 00 Pilsen, Czech Republic

[a]kottner@ntis.zcu.cz, [b]kocabj@students.zcu.cz, [c]kochovap@ntc.zcu.cz, [d]jheczko@kme.zcu.cz

**Keywords:** Arcan, biaxial, model parameters, Poisson's ratio, strain rate, uniaxial.

**Abstract.** This paper deals with a composite consisting of cork particles and the rubber matrix. The size of the cork particles is approximately one millimetre. Testing samples were exposed to uniaxial tension, biaxial tension, and shear using special fixture for Arcan samples. Mechanical behaviour of the samples was investigated using different strain rates to provide data for the identification of parameters of a finite-strain viscoelastic constitutive model. The Poisson's ratio was measured under the uniaxial tensile loading.

## Introduction

Knowledge of the mechanical properties of materials is of critical importance for the design of all products. First book devoted to the physical testing of rubbers appeared in 1965 [1]. Since rubbers have a unique combination of the properties, e.g., bulk modulus is thousand times greater than Young's modulus, special testing methods have been developed [2]. Cork is a closed-cell structure. It can be considered, in a first approximation, as a transversally isotropic material. In contrast to rubber, cork has near-zero Poisson's ratio [3].

This paper deals with a composite consisting of cork particles and rubber matrix. The size of the cork particles is approximately one millimetre (Fig. 1). This composite is often used as the core material in constrained damping layer constructions [4, 5]. Testing samples were exposed to uniaxial tension, biaxial tension, and shear using special fixture for Arcan samples [6], (Fig. 2). Mechanical behaviour of the samples was investigated using different strain rates to provide data for the identification of the parameters of a finite-strain viscoelastic constitutive model. The Poisson's ratio was measured under the uniaxial tensile loading. Obtained results supplement mechanical characteristics investigated in [7], where simple tension, simple shear, simple compression, and volumetric compression tests were performed according to recommendations for the Bergstrom-Boyce model calibration [8].

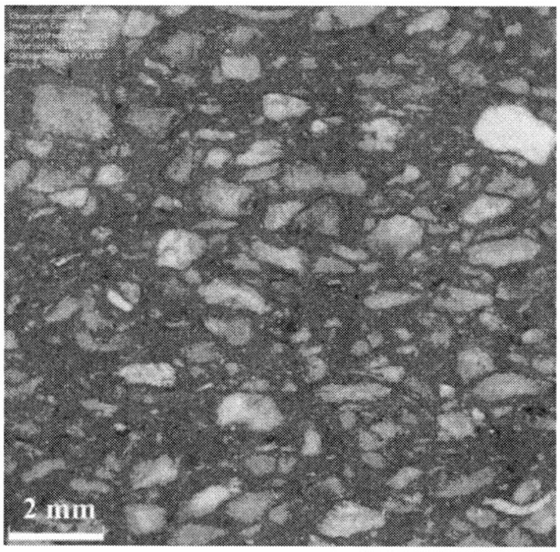


Fig. 1 Detail of cork/rubber particle composite.

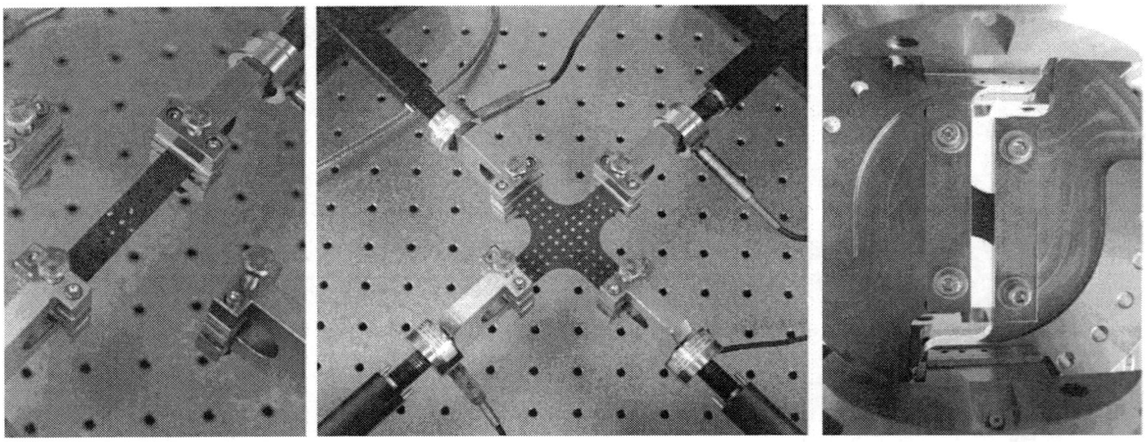

(a) (b) (c)

Fig. 2. Performed tests: uniaxial tension (a), biaxial tension (b), and shear of Arcan sample (c).

**Experiments**

The tests were performed on the electrodynamic planar biaxial test system TestResources 574LE and on the servohydraulic dynamic test system Instron 8850. The temperature was 23±1 °C, the atmospheric moisture was 50±6 %. Experimental samples were cut from 2 mm and 6 mm thick plates using a water jet. The geometry of the samples is shown in Fig. 3 [6, 9]. Displacements of the monitored points, depicted in Fig. 3, were monitored using a real-time digital image correlation system VIC-Gauge 2D. The strain rate range was between 0.001 $s^{-1}$ and 1.00 $s^{-1}$. Three loading/unloading cycles including 60 s of relaxation time were prescribed. A virgin sample was used for each test. At least three samples were tested under identical conditions (the strain rate and maximal nominal strain).

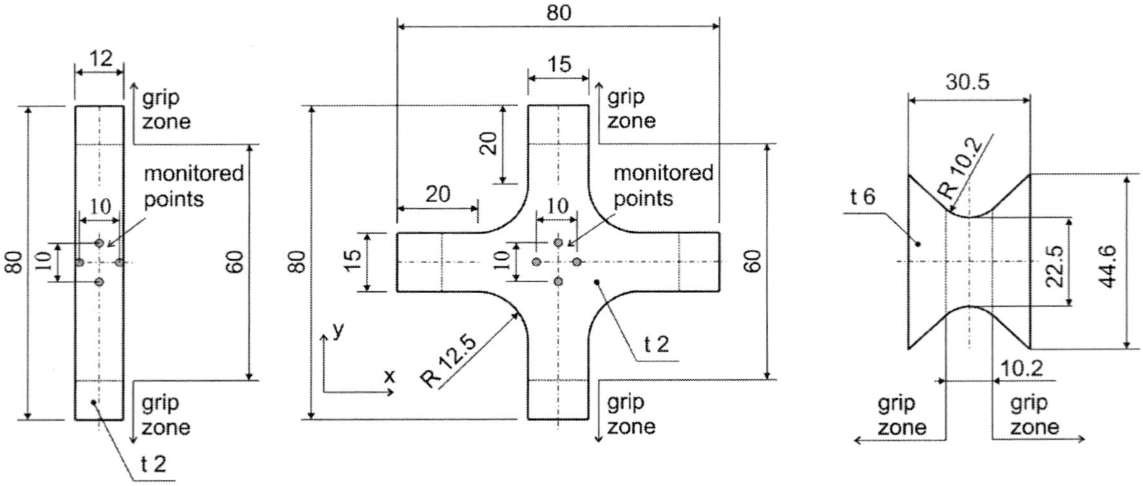


(a) (b) (c)

Fig. 3. Samples for: uniaxial tension (a), biaxial tension (b), and shear (c).

The designation of the samples is obvious from Fig. 4. A symbols denote the type of the test, B symbols denote the strain rate used for loading and unloading, C symbols denote the maximal nominal strain (normal strain related to the initial length between the grips).

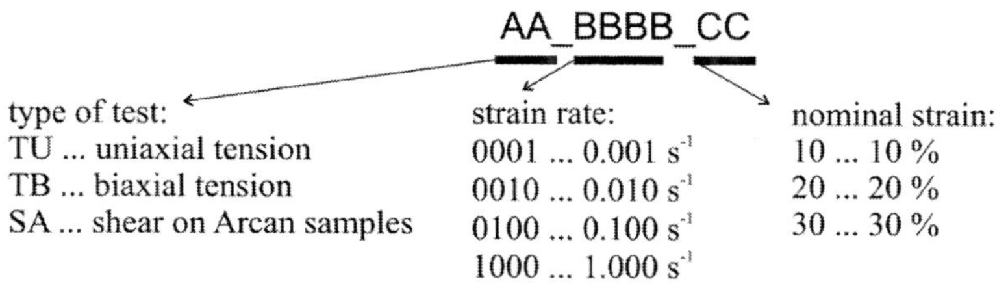


Fig. 4. Designation of samples.

**Uniaxial tension**. The initial sample length between the grips was 60 mm (Fig. 3(a)). The uniaxial tensile (TU) tests were performed with four different crosshead velocities: 0.03 mm·s$^{-1}$, 0.3 mm·s$^{-1}$, 3 mm·s$^{-1}$, and 30 mm·s$^{-1}$. Displacements prescribed to the crosshead are obvious from Fig. 5 (only the first cycle is shown in the case of the strain rate 0.001 s$^{-1}$).

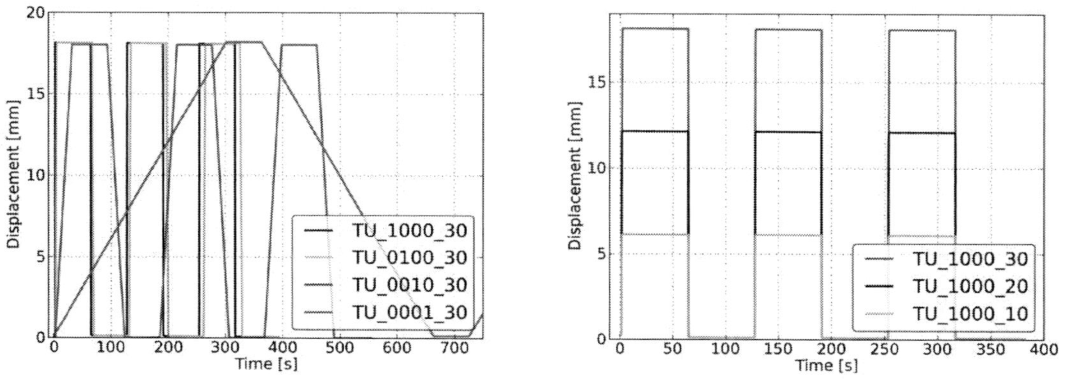


(a) (b)

Fig. 5. Uniaxial tension – prescribed displacements: different strain rate (a), different maximal nominal strain (b).

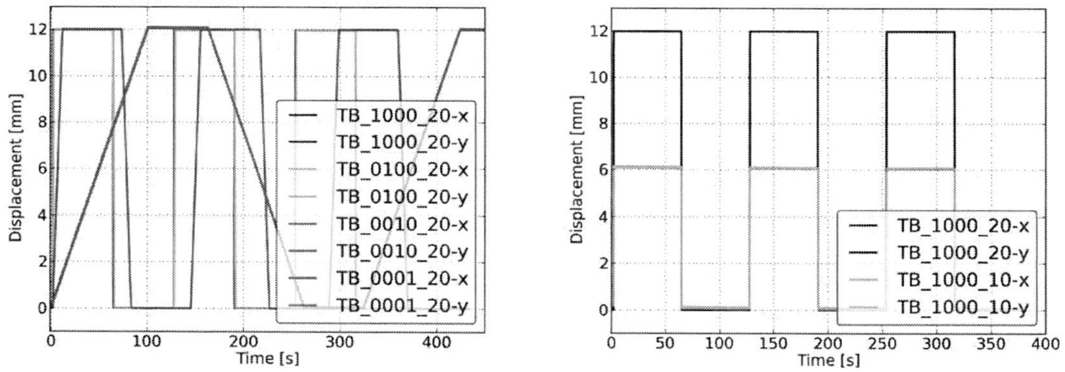


Fig. 6. Biaxial tension – prescribed displacements: different strain rate (a), different maximal nominal strain (b).

**Biaxial tension**. The initial sample length between the grips was 60 mm (Fig. 3(b)). Since the strain rate is not homogeneous in the biaxial samples, the biaxial tensile (TB) tests were performed with higher crosshead velocities than in the case of uniaxial tension. These four velocity values were applied to obtain strain rate values similar to the strain rate values of the TU samples (in the middle of the TB samples): 0.06 mm·s$^{-1}$, 0.6 mm·s$^{-1}$, 6 mm·s$^{-1}$, and 60 mm·s$^{-1}$. Displacements prescribed to the crosshead are obvious from Fig. 6. Since an ultimate failure occurred when the displacements were larger than 12 mm, the TB samples with the maximal nominal strain of 30 % could not be tested.

**Shear**. The initial sample length between the grips was 10.2 mm (Fig. 3(c)). In order to obtain principal strain values similar to the values in TU test, two times higher values of nominal shear strain than the values of nominal normal strain were applied. The shear tests of the Arcan samples (SA tests) were performed with the following four different crosshead velocities: 0.02 mm·s$^{-1}$, 0.2 mm·s$^{-1}$, 2 mm·s$^{-1}$, and 20 mm·s$^{-1}$. Displacements prescribed to the crosshead are obvious from Fig. 7.

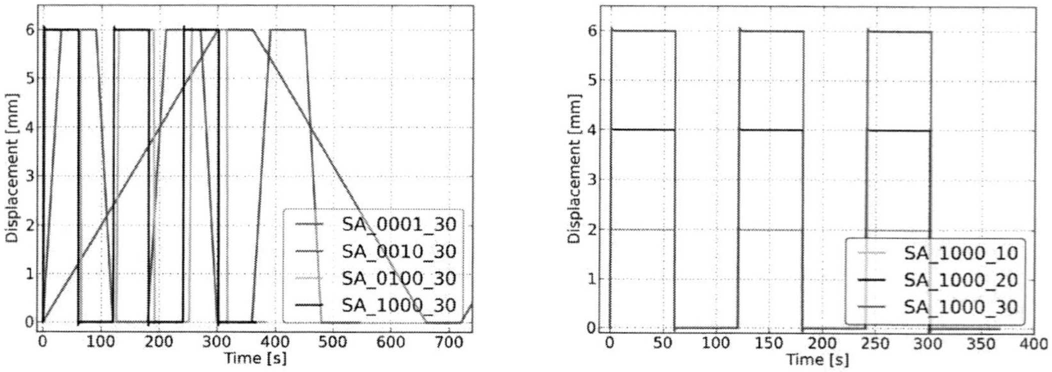


Fig. 7. Shear – prescribed displacements: different strain rate (a), different maximal nominal strain (b).

**Results**

Typical force responses to the prescribed displacements are shown in Fig. 9 – Fig. 11. It is obvious that the forces relaxed significantly when the strain rate was higher than 0.001 s$^{-1}$, especially, when the strain rate was 1 s$^{-1}$. Since the thickness of the tensile samples was small, the samples lost stability during unloading (this phenomenon in the case of the TB sample is shown in Fig. 8). Therefore, significant negative force values were not registered in cases of the TU and TB tests. Moreover, obtained displacements of the monitored points, when the crosshead displacements were close to zero, were not correct.

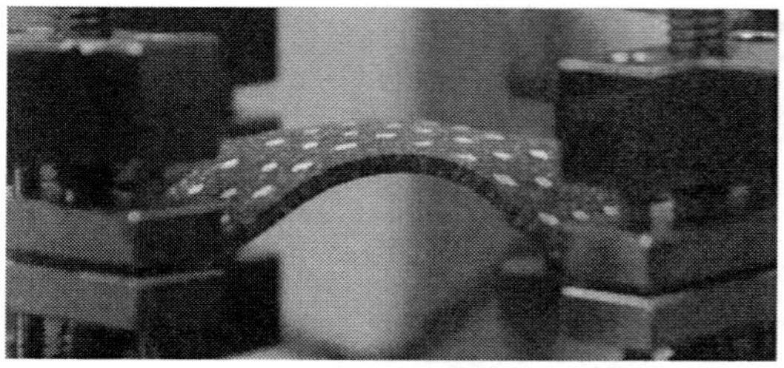

Fig. 8 Buckling of TB samples

Fig. 9(b) –- Fig. 11(b) demonstrate the force-time curves of the tests when the force values of each test are divided by the maximal force value of the test. If the curves for tests with different maximal nominal strain are identical, the viscoelasticity is linear. It is obvious that the curves were very similar in the case of the tensile tests, however, the differences in the case of the shear test are significant. The reason of that can be different behaviour of the cork in tension and compression (cork is a closed-cell foam).

Anisotropy of some samples was proved in the tensile tests (see e.g. Fig. 10, sample TB_0001). Besides the fact that cork is anisotropic, cork particles exhibited a preferred orientation in some samples (see longitudinally and transversally oriented particles of TU samples in Fig. 12). Significant difference between tensile curves of samples loaded in the longitudinal direction and the transverse direction is demonstrated in Fig. 13(a).

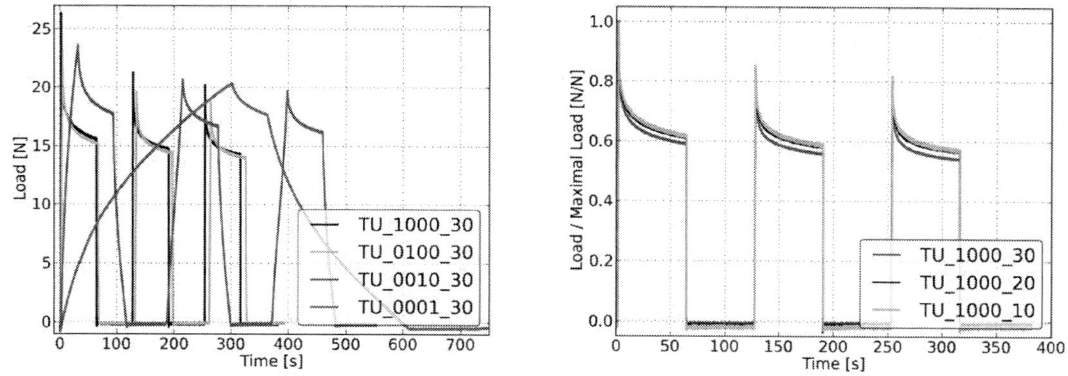


(a) (b)

Fig. 9. Uniaxial tension – force responses: different strain rate (a), different maximal nominal strain (b).

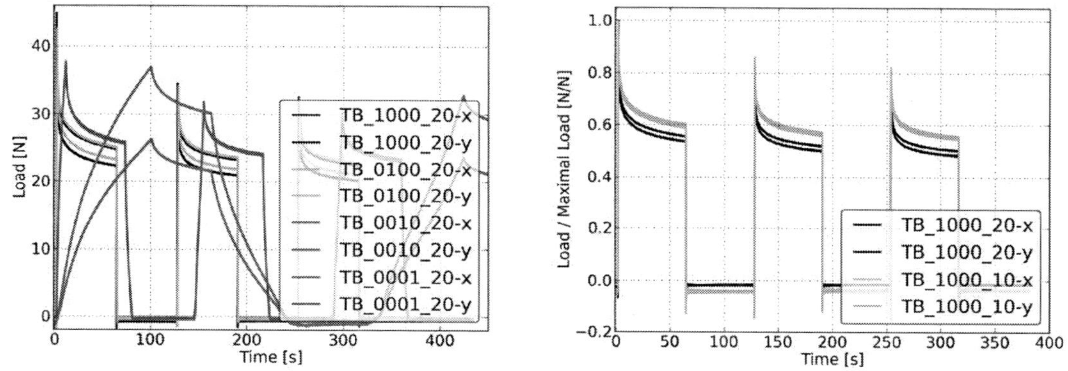


(a) (b)

Fig. 10. Biaxial tension – force responses: different strain rate (a), different maximal nominal strain (b).

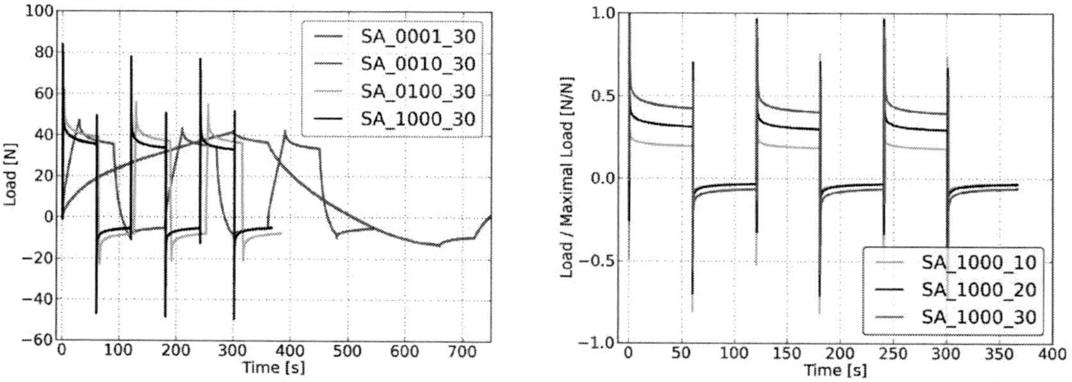


(a) (b)

Fig. 11. Shear – force responses: different strain rate (a), different maximal nominal strain (b).

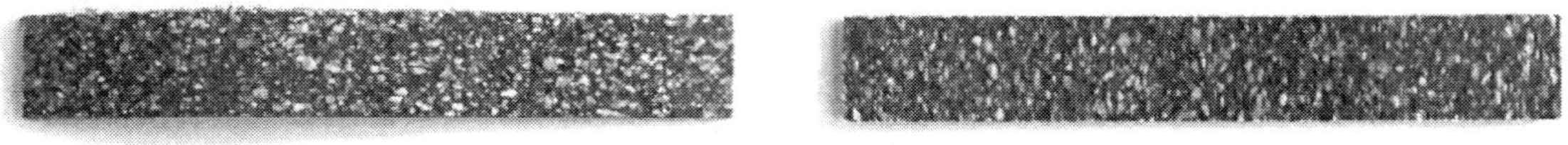

longitudinal orientation transverse orientation

Fig. 12. Longitudinally and transversally oriented particles of TU samples.

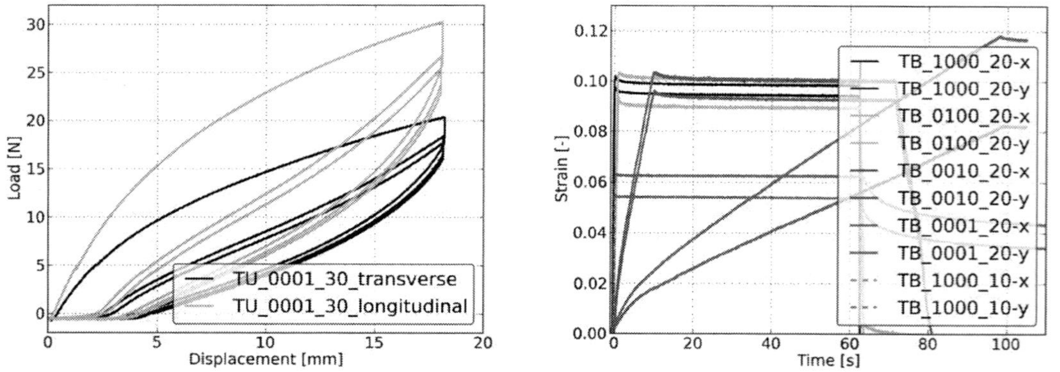


(a) (b)

Fig. 13. Uniaxial tension of samples with oriented particles (a). Biaxial tension – strains in the middle of samples (b).

The anisotropy and the shape (orientation) of the cork particles probably also influenced the results of the Poisson's ratio measurement (calculated from the displacements of the monitored points). Since area corresponding to the monitored points is relatively small (comparing to the size of the cork particles), the Poisson's ratio was in the range shown in Fig. 14. Nevertheless, the results show a trend that the Poisson's ratio decreased with the strain rate.

Fig. 13(b) shows normal strains $\varepsilon_x$ and $\varepsilon_y$ in the middle of TB samples (calculated from the displacements of the monitored points). It is obvious that the normal strain in this area reached approximately half of the values of the nominal normal strain.

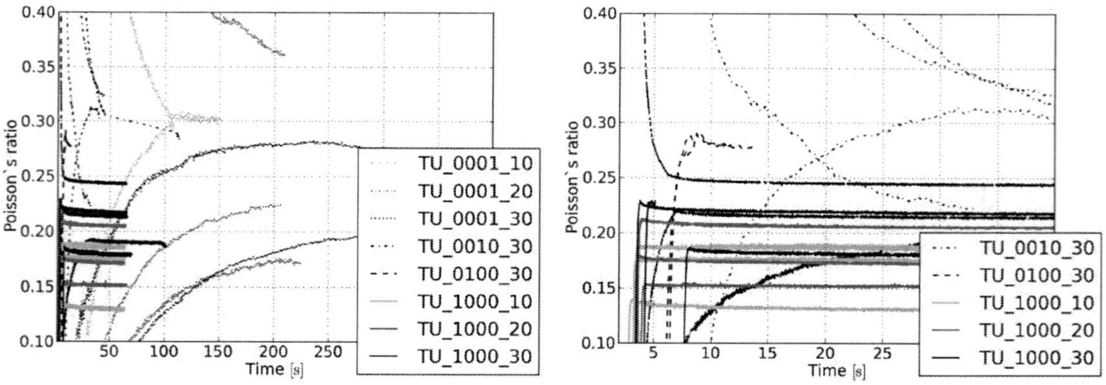


Fig. 14. Uniaxial tension – Poisson's ratio.

## Conclusions

The results of the Poisson's ratio measurement showed a trend that the Poisson's ratio decreased with the strain rate. The Poisson's ratio was in the range of 0.13 to 0.36. As expected, the maximal forces increased with the strain rate. Although the tensile tests (uniaxial and biaxial) suggested linear viscoelastic behaviour, the shear response was nonlinear with respect to the prescribed strain history.

Mechanical properties of the investigated cork/rubber composite depended on the location of a sample in a plate from which the sample was cut (based on shape and anisotropy of the cork particles). This fact complicates obtaining of reliable experimental data for the calibration of a finite-strain viscoelastic constitutive models. Testing of large-scale samples, which would correspond better to the constrained damping layers of structures, can improve the experimental data reliability (the unification of experimental results).

## Acknowledgement

This publication was supported by the project LO1506 of the Czech Ministry of Education, Youth and Sports.

## References

[1] J.R. Scott, Physical Testing of Rubbers, Maclaren & Sons Ltd., London, 1965.

[2] R. Brown, Physical Testing of Rubber, fourth ed., Springer, USA, 2006.

[3] S.P. Silva, M.A. Sabino, E.M. Fernandes, V.M. Correlo, L.F. Boesel, R.L. Reis, Cork: properties, capabilities and applications, International Materials Reviews 50 (2005) 345-365.

[4] Information on http://www.corkhouse.co.il/WEB/8888/NSF/Web/8030/ACM-87.pdf

[5] R. Kottner, J. Vacík, R. Zemčík, Improvement of the damping properties of carbon-fibre-reinforced laminated plastics using damping layers, Materials and Technology 47 (2013) 189-193.

[6] W.G. Knauss, I. Emri, H. Lu, Mechanics of polymers: Viscoelasticity, in: W.N. Sharpe, Jr. (Ed.), Springer Handbook of Experimental Solid Mechanics, Springer, New York, 2008, pp. 49-95.

[7] R. Kottner, J. Kocáb, J. Heczko, J. Krystek, Investigation of the mechanical properties of a cork/rubber composite, Materials and Technology 50 (2016) 579-583.

[8] Information on https://polymerfem.com/

[9] J.O.V. Delgadillo, S. Delorme, R. El-Ayoubi, R. DiRaddo, S.G. Hatzikiriakos, Effect of freezing on the passive mechanical properties of arterial samples, Biomedical Science and Engineering 3 (2010) 645-652.

EAN 2017
55th conference on
experimental stress analysis

# Influence of Distribution of Finger Joints and Timber Flaws on the Damage Evolution of Laminated Glued Timber Beams

MELZEROVÁ Lenka[1, a], ŠEJNOHA Michal[1,b]

[1] CTU in Prague, Faculty of Civil Engineering, Department of Mechanics, Thákurova 7, 16629, Prague 6, Czech Republic

[a]melzerov@fsv.cvut.cz, [b]sejnom@fsv.cvut.cz

**Keywords:** Four-point bending, laminated glued timber beams, knots, timber flaw, damage.

**Abstract.** A group of six glued laminated timber beams was tested in four-point bending until failure. Both standard measuring devices mounted to the beams and digital cameras were employed to provide for a continuous measuring of displacements and strains as well as visualization of damage evolution and subsequently for quantification of damage mechanisms leading to failure of individual beams. This was accompanied by identification of positions of all visible knots and finger joints. It is shown that their distribution plays an important role in the onset of damage evolution and final failure pattern.

## Introduction

All six tested beams were commercially manufactured adopting standard manufacturing process. They were made of spruce having the same geometry and number of layers. However, the number of segnents, their length and arrangement and thus the location of finger joints varied from beam to beam. No particular attention has been paid to the selection of individual laminations. This resulted in a considerable spread of the ultimate strength in between beams. In particular, the ultimate loading ranged from 68kN do 148kN. The objective of this paper is to reconcile such differences by drawing parallel with the location and density of individual finger joints, knots and other visible timber flaws in the region of the maximum stress, and corresponding strength of local segments of timber. This might help in formulation of various structural recommendations leading to an increase of bearing capacity of laminated timber beams loaded primarily in bending. Based on the available graphical documentation of failure, it is possible to tie the onset of damage evolution to a particular flaw, i.e. the finger joint or a secondary flaw such as knot.

## Monitoring Crack Evolution

The loading was introduced with the help of hydraulic cylinders. The location of points of load application, beam supports and the specimen geometry is evident from Fig. 1, see also [1]. A step-wise increase of loading forces with a constant increment of 4kN and constant hold period prior to a new load increment as plotted in Fig. 1 was maintained. A gradual monitoring of the failure process was accomplished by employing a high-speed digital camera. However, it was impossible to track every detail of the failure process owing to the load control regime and a relatively short time between the first crack appearance and final failure. In some cases, this took less than 1s. Nevertheless, it was still possible to get a clear insight into various failure mechanisms. The final collapse was always initiated by the crack

developed at the most stressed central part of the beam due to the presence of a finger joint or a large knot.
To this end, two groups of beams were examined in this study. Four beams in particular were manufactured from one set of wood while other two beams from a different one. Both beams in the second set were also subjected to DIC (Digital Image Correlation) measurements to assess the evolution of displacements and potentially also strains and damage. However, recall the load control regime which precluded a direct exploitation of these measurements in this study. Nevertheless, they provided useful information utilized in statistical analysis of effective elastic properties within the framework of Bayesian inference [4]. In this study we only present averages of longitudinal Young's moduli in GPa pertinent to individual laminations. These were determined from a large set of measurements using the Pilodyn 6J measuring device [2,3].

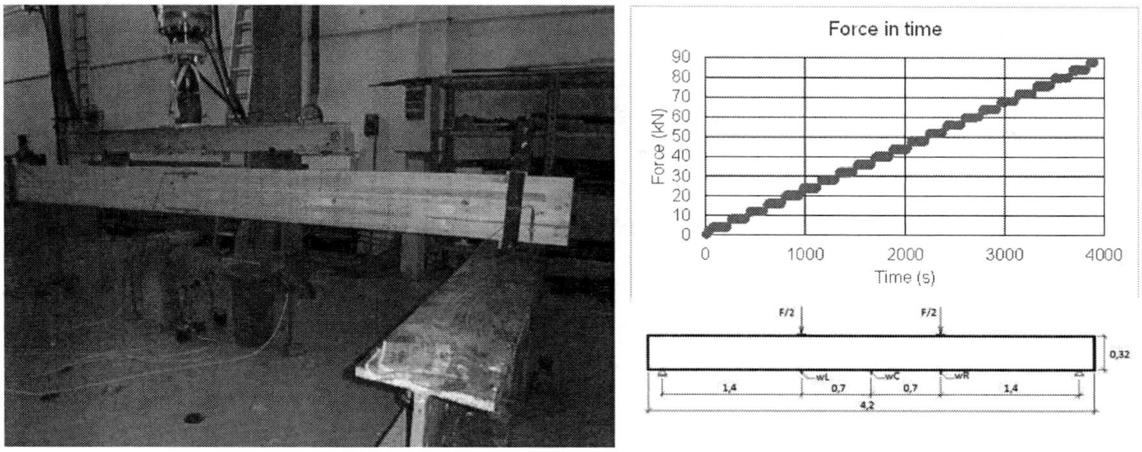

**Figure 1**: Example of tested beam including the test setup and time variation of the applied load

**Beam1.** The beam No. 1 experienced loading out of the original loading plane just prior to the final collapse. This might explain its relatively low bearing capacity equal to 68kN only. The onset of failure can be attributed to a pair of large knots in the middle part of the beam having the distance of 35mm only and spreading the entire thickness of the second and third bottom plies, respectively, see Fig. 2.

These partial results suggest the following rather intuitive recommendations. First, large knots spreading the whole lamination thickness should be removed. If that is not possible, we should avoid grouping of knots. One should also pay attention to the location of finger joints and if possible to avoid their grouping at least in the most stressed section of the beam as well.

**Beam 2.** The beam No. 2 experienced the first failure in the finger joint which happened to be located exactly in the center of the bottom ply. The failure then propagated to the right up to the second bottom ply. Next crack opened in the third bottom ply, which was found to have the smallest average stiffness over the entire length, see Fig. 2. The next failure appeared in a poor-quality joint in the fifth ply from the bottom located 25cm to the right from the first failed joint. This was followed by the total collapse connecting all preceding cracks. This beam reached the collapse load equal to 88kN. As seen from Fig.2 it contained five finger joints located in the central part of the beam with one not suitably placed in the beam center. The partial design recommendation thus suggest to avoid application of poor-quality joints, to divide the most loaded bottom ply into the least number of laminations and to distribute all joints evenly throughout the beam.

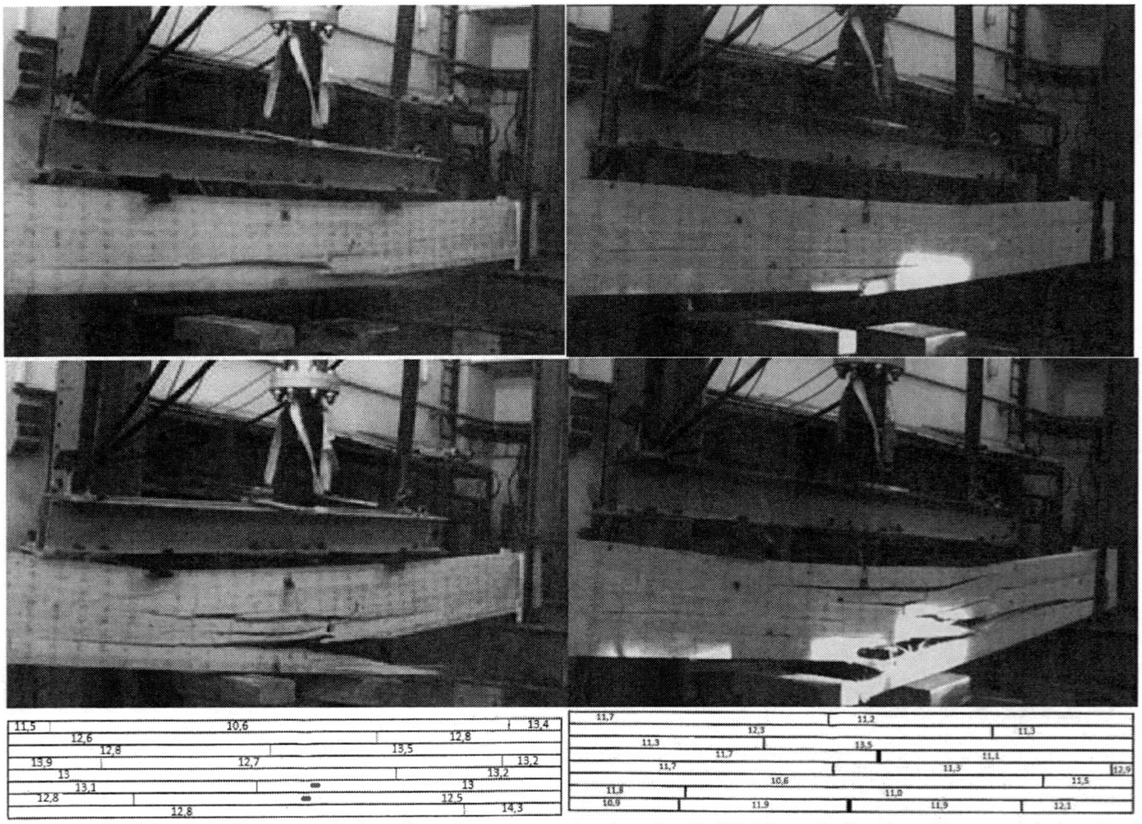

**Figure 2**: Final collapse of the first beam at the load of 68kN and final collapse of the 2nd beam at the load of 88kN

**Beam 3.** This beam achieved the same bearing capacity as the previous one. Even here the distribution of finger joints is completely unsuitable. Each ply contains one finger joints found within a 60cm wide band around the beam central part. Five of them, all located in first five bottom plies, were damaged upon the final failure of the beam. The failure mechanism is therefore rather predictable.

As shown in Fig. 3, the beam broke from bottom up due to crack propagation from one joint to the other. In this case, the distribution of large knots and an average stiffness of individual laminations had no effect on the beam failure, even though the two laminations with the smallest stiffness were placed in the bottom ply.

**Beam 4.** This beam reached a bit higher load bearing capacity equal to 96kN. The failure mechanism was, however, similar to the previous one. Herein, the first crack initiated from the damaged finger joint in the bottom ply and propagated towards the finger joint in the third bottom ply. Most probably, this region contained a hidden defect leading to fracture of the fourth ply in the same location. This indicates the need for identification and removal of such defects prior to gluing. The crack then propagated within this ply in the longitudinal direction. Yet before the final collapse another crack initiated in the fifth ply in the section having a significantly smaller stiffness of about 11.4GPa in comparison to the surrounding, see Fig. 3. Similarly to the previous cases, a finger joint located in the center of the beam caused the final failure. This further supports the need for not having joints in the most stressed regions. Apart from knots and finger joints, we might also want to concentrate on at least approximate knowledge of elastic moduli in individual laminations. These should be examined prior to manufacturing to avoid placing laminations of a large stiffness difference next to each other. This is certainly achievable for machine-wise classification of wood.

**Figure 3**: Gradual evolution of cracks. Location of finger joints measured in mm from the right and final collapse of the first beam at the load of 96kN.

**Beam 5.** A totally different failure mechanism has been observed for the last two tested beams having a considerably higher load bearing capacity. This beam, the last before one in terms of bearing capacity, collapsed at the load of 128kN, which is more than 30% higher load than encountered for the previously discussed beams. On the other hand, the actual progress of damage resembled almost a brittle failure. The first two photographs in Fig. 4 were taken 1s apart. Only one finger joint, highlighted also in Fig.4, failed in this case. Again it was damaged not until propagation of longitudinal cracks. The sequence of photographs in Fig. 4 further indicates that no finger joint contributed to the beam failure as three longitudinal cracks developed in the middle most loaded section of the beam. The first crack initiated in the third bottom ply having the lowest stiffness equal to 10.8GPa. This was also the lowest value of the Young modulus among all tested laminations. The next crack appeared one ply above with the average value of the Young modulus equal to 11.6GPa. Occasionally, the crack propagation was terminated at the finger joint. These joints, however, were always found out of the most loaded central part of the beam. Fracture of the bottom ply approximately in the middle of the beam took place at the final stage of the total collapse. Reading the damage pattern was simplified through a regular grid of points, where indentation measurements of Young's modulus took place.

**Beam 6.** The last beam with the highest load bearing capacity failed at the load level of 148kN. The failure mechanism was similar to the beam No. 5 so that the beam failed again due to the massive evolution of longitudinal cracks as seen in Fig. 4. In this case the first crack initiated in the bottom ply below the left force and propagated through the corresponding lamination to the right up to the point of force application. The first failure was followed by the evolution and propagation of cracks in the second and third bottom plies starting in the vicinity of two large knots below the right force. The whole fracturing process is documented via a series of photographs in Fig. 4. Checking the average values of Young's modulus we notice a smaller value of longitudinal modulus in the first two bottom plies in comparison to the third ply. The lower value of the elastic modulus suggests also a lower wood density and consequently a lower strength. Nevertheless, these moduli are still of about 1 to 2 GPa higher when compared to the beam No. 5. This explains, together with a better redistribution of finger joints in the most stressed section, its larger load bearing capacity. The two finger joints in the bottom ply played no role as they were located out of the most stresses section. By coincidence the second bottom ply was compact made of one lamination and the third bottom ply had a finger joint close to the support. On the contrary, there was a finger joint in the center of the top ply. It thus can be expected that flipping the beam upside down would result in a much lower load bearing capacity. So apart from grouped large knots in the most stressed section the position of the beam should be correctly selected.

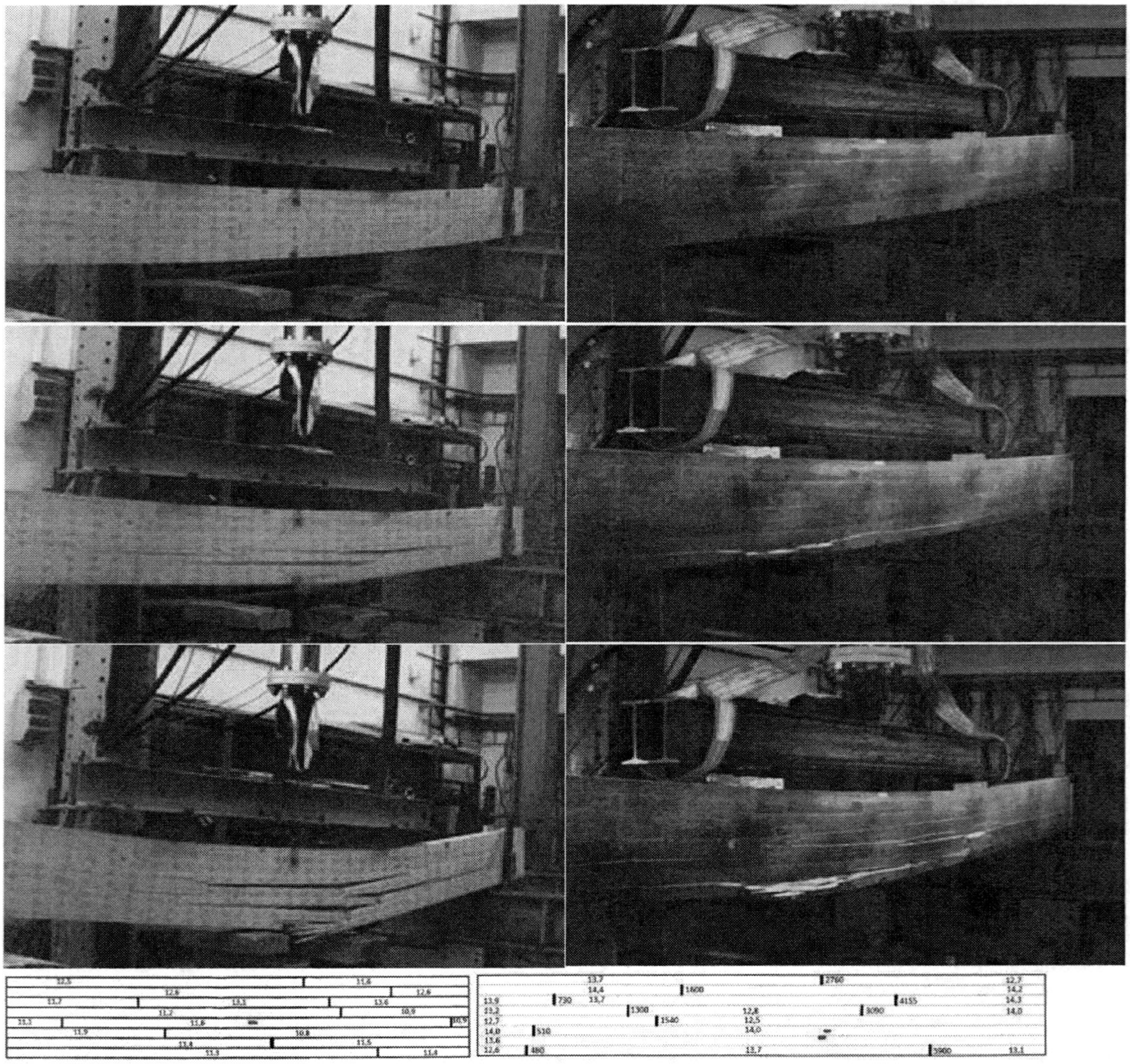

**Figure 4**: Beam with load bearing capacity equal to 128kN with identified location of a large knot and damaged finger joint and beam with load bearing capacity equal to148kN and the location of pair of knots.

## Discussion

The presented study suggests several import design recommendations for glued laminated timber beams. Even though the finger joints are processed according to the current standards, they still present the weakest point in the beam when loaded in four-point bending. All three beams with such an arrangement, recall beams No. 2-4, experienced the first failure in the finger joint. On the contrary, the longitudinal interfacial joints played no role. Cracks have always propagated through individual laminations and their transition from one ply to the other was smooth regardless of the horizontal glued surfaces. Thus suitably arranging the finger joints may considerably increase the beam load bearing capacity, which amounted to more than 30% in our particular case.

From the presented results we may identify two particular failure mechanisms. In all beams with an unsuitably located finger joint in the most stressed section the final fracture took place across this joint leading to a relatively low ultimate load bearing capacity. In the last two beams with a higher capacity the final failure was manifested by a longitudinal fracture of one of the bottom lamination having the lowest stiffness compare to its surrounding or containing large knots.

## Conclusions

To increase the load bearing capacity of glued laminated timber beams it is crucial to ensure a suitable redistribution of finger joints. Avoiding their placement in the most stressed section increased the bearing capacity in our study from about 30 up to 50%.

Similar to finger joints the large knots should also be excluded from the most stressed sections. Regions containing a group of such knots are also source of bearing capacity reduction.

Next step should ensure a correct position of individual laminations within individual plies of the beam from their stiffness point of view. It has been suggested that neighboring laminations should not have a significant difference in their stiffness.

When fulfilling all these recommendations we most probably encounter a collapse mechanism initiated by the evolution of longitudinal crack in one the bottom most stressed laminations.

## Acknowledgement

The support by the GAČR grant No. 15-10354S is gratefully acknowledged.

## References

[1] L. Melzerová, M. Šejnoha, P. Klapálek, Bending test and simulations of GLT beams. Applied Mechanics and Materials. 611 (2014) 54-59.

[2] L. Melzerová, M. Šejnoha, Results of penetration tests performed on timber GLT beams. WCTE 2014 World Conference on Timber Engineering, Proceedings. (2014) 49-50.

[3] L. Melzerová, M. Šejnoha, M, Influence of distribution of finger joints and timber flaws on the damage evolution of laminated glued timber beams during four point bending. WCTE 2016 CD-ROM Proceedings. (2016).

[4] M. Šejnoha, T. Janda, L. Melzerová, V. Nežerka, J. Šejnoha, Modeling glulams in linear range with parameters updated using Bayesian interface, Engineering Structures. 138 (2017) 293-307.

# Automated tool for simulation of standardized tests of cohesive interfaces

Petr Hanzlik[1,a], Tomas Kroupa[2,b]

[1]University of West Bohemia, Univerzitní 8, 301 00 Pilsen, Czech Republic

[2]University of West Bohemia, NTIS – New Technologies for the Information Society, Univerzitní 22, 301 00, Plzeň, Czech Republic

[a]petrhan@kme.zcu.cz, [b]kroupa@kme.zcu.cz

**Keywords:** interface failure, Abaqus 6.14, Hunstman Araldite 2021, Gurit Spabond 345, epoxy inner layer of laminated composite, DCB, ENF, MMF, strain energy release rate $G^c$, finite element method, traction separation law, quadratic stress criterion, power law criterion, laminated composite, linear elastic orthotropic material model

## Introduction

This work deals with determination of material parameters of cohesive interface using finite element models. Cohesive interface means conjunction of two deformable materials with relatively high stiffness compared to the interface. This conjunction or adhesive bonding determines overall strength of this system for certain loading types. In general, failure of such interface is caused by normal and shear stresses. Resulting failure is in fact a case of the crack growth and can be determined by parameters used in fracture mechanics. Strain energy release rate (SERR) $G$ was selected as a basic parameter for evaluation of cohesive interface failure for this reason. Numerical analyses are performed in FE software Abaqus 6.14. Automated tool for identification of material parameters of two of the mentioned basic modes was created using Python 2.7. programing language. Material parameters of two adhesives, Hunstman Araldite 2021 and Gurit Spabond 345, and an inner layer of laminated composite are identified through secondary material parameters: stiffness $k$, displacement $\delta^0$ (damage initiation) and $\delta^c$ (failure determination).

## Analytical models

General behaviour of loaded cohesive interface can be described using decomposition to three basic modes described on figure 1: Mode I – interface is loaded in the normal direction to a crack plane, Mode II – interface is loaded in the crack plane perpendicularly to a crack front, Mode III – load is in the direction of the crack front in the crack plane.

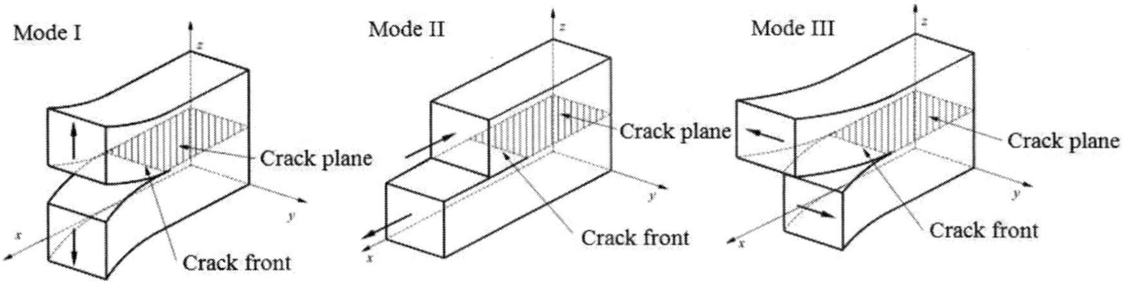

Figure 1: Basic modes of interface load [4]

As mentioned before, strain energy release rate (SERR) $G$ was selected as a basic parameter for evaluation of cohesive interface failure. This material parameter can be mathematically defined through strain energy $\pi$ as

$$G = -\frac{1}{b}\frac{d\pi}{da}, \tag{1}$$

where $b$ is width of thin-walled specimen of standardized test and $a$ is length of interface crack. In case of cantilever beams equation 1 can be transformed into

$$G = \frac{F^2}{2b}\frac{dC}{da}, \tag{2}$$

where $C$ is compliance of cantilever beam mathematically defined by displacement $\delta$ measured in place where load $F$ is applied as

$$C = \frac{\delta}{F}. \tag{3}$$

Crack growth is evaluated through Griffith criterion that can be summed up into four points [2]:

1. Crack grows when SERR is equal or exceeds its critical value $G^c$
2. At the moment of the crack growth are stresses in the area of growth constant.
3. Critical value of SERR is constant.
4. Crack grows in the direction of maximal component SERR.

Critical value of SERR is determined via analytical equations obtained by standardized tests. For load mode I the double cantilever beam test was used. For mode II end notched flexure test was used. For combination of mode I and II MMF test was used.

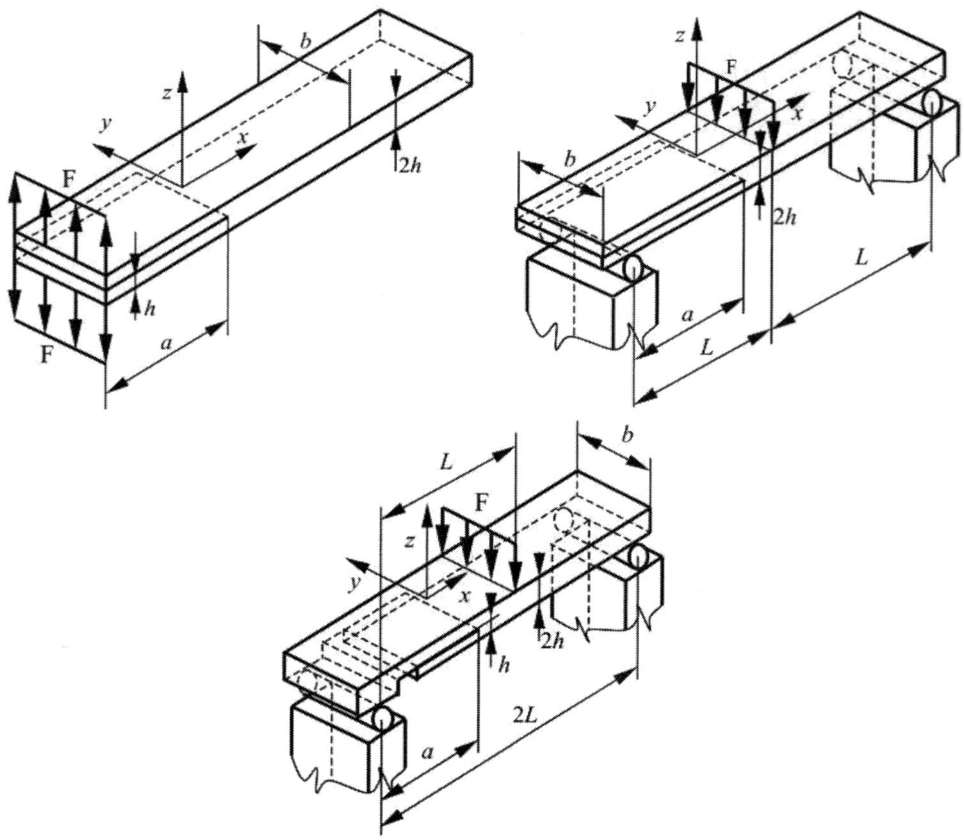


Figure 2: Standardised tests DCB (top left), ENF (top right) and MMF (bottom) [4]

Critical value of SERR for DCB test can be obtained using Euler-Bernoulli beam theory as

$$G_{\mathrm{I}}^{c} = \frac{12F^2a^2}{bh^3E_x} = \frac{3F\delta}{2ba}, \tag{4}$$

where $E_x$ is an effective Young's modulus in the direction of X-axis. For ENF test is the critical value mathematically described as

$$G_{\mathrm{II}}^{c} = \frac{9F^2a^2}{16b^2h^3E_x} = \frac{3Fa^2\delta}{2b(3a^3+2L^3)}. \tag{5}$$

Analogically for test MMF

$$G_{\mathrm{I/II}}^{c} = \frac{21F^2a^2}{16b^2h^3E_x} = \frac{672Fa^2\delta}{b(448a^3+121L^3)}. \tag{5}$$

## Finite element models

Finite element model of standardized tests created in Abaqus CAE consist of two distinct parts: joined components and cohesive interface. Joined components are considered as linear elastic orthotropic material with following constitutive equation written for Cartesian material coordinate system as

$$\begin{bmatrix}\sigma_1\\\sigma_2\\\sigma_3\\\sigma_4\\\sigma_5\\\sigma_6\end{bmatrix}=\begin{bmatrix}\frac{E_1(1-\nu_{23}\nu_{32})}{\det(\boldsymbol{\nu})} & \frac{E_1(\nu_{21}+\nu_{23}\nu_{31})}{\det(\boldsymbol{\nu})} & \frac{E_1(\nu_{31}+\nu_{32}\nu_{21})}{\det(\boldsymbol{\nu})} & 0 & 0 & 0\\ \frac{E_1(\nu_{21}+\nu_{23}\nu_{31})}{\det(\boldsymbol{\nu})} & \frac{E_2(1-\nu_{13}\nu_{31})}{det(\boldsymbol{\nu})} & \frac{E_2(\nu_{32}+\nu_{31}\nu_{12})}{\det(\boldsymbol{\nu})} & 0 & 0 & 0\\ \frac{E_1(\nu_{31}+\nu_{32}\nu_{21})}{\det(\boldsymbol{\nu})} & \frac{E_2(\nu_{32}+\nu_{31}\nu_{12})}{\det(\boldsymbol{\nu})} & \frac{E_3(1-\nu_{12}\nu_{21})}{\det(\boldsymbol{\nu})} & 0 & 0 & 0\\ 0 & 0 & 0 & G_{23} & 0 & 0\\ 0 & 0 & 0 & 0 & G_{13} & 0\\ 0 & 0 & 0 & 0 & 0 & G_{12}\end{bmatrix}\begin{bmatrix}\varepsilon_1\\\varepsilon_2\\\varepsilon_3\\\varepsilon_4\\\varepsilon_5\\\varepsilon_6\end{bmatrix}, \tag{6}$$

where $\sigma_i$ is normal or shear stress, $\varepsilon_i$ is normal or shear strain, $E_i$ is Young's modulus, $G_{ij}$ is shear modulus and $\nu_{ij}$ is Poisson ratio [1]. Matrix $\boldsymbol{\nu}$ can be written as

$$\boldsymbol{\nu}=\begin{bmatrix}1 & -\nu_{12} & -\nu_{13}\\ -\nu_{12} & 1 & -\nu_{23}\\ -\nu_{13} & -\nu_{23} & 1\end{bmatrix}. \tag{7}$$

Cohesive interface is created by cohesive contact that is defined by a *bilinear traction-separation law* (visualized on figure 3), *quadratic stress criterion* (Eq. 8) and a *power law criterion* (Eq. 9) in each node of a contact pair.

$$\left(\frac{t_{nn}}{t_{nn}^0}\right)^2+\left(\frac{t_{ss}}{t_{ss}^0}\right)^2+\left(\frac{t_{tt}}{t_{tt}^0}\right)^2=1, \tag{8}$$

$$\left(\frac{G_{nn}}{G_{nn}^c}\right)^\alpha+\left(\frac{G_{ss}}{G_{ss}^c}\right)^\alpha+\left(\frac{G_{tt}}{G_{tt}^c}\right)^\alpha=1. \tag{9}$$

*Traction-separation law* prescribes relation between stress component (normal/shear) and corresponding component of displacement. Surface under traction-separation curve is equal to critical value of SERR. *Quadratic stress criterion* defines failure of cohesive interface and *power law criterion* determines its damage initiation [3]. When power law criterion is fulfilled stiffness of interface is decreased (plastic strain is not present). For better understanding is this phenomenon described on figure 4.

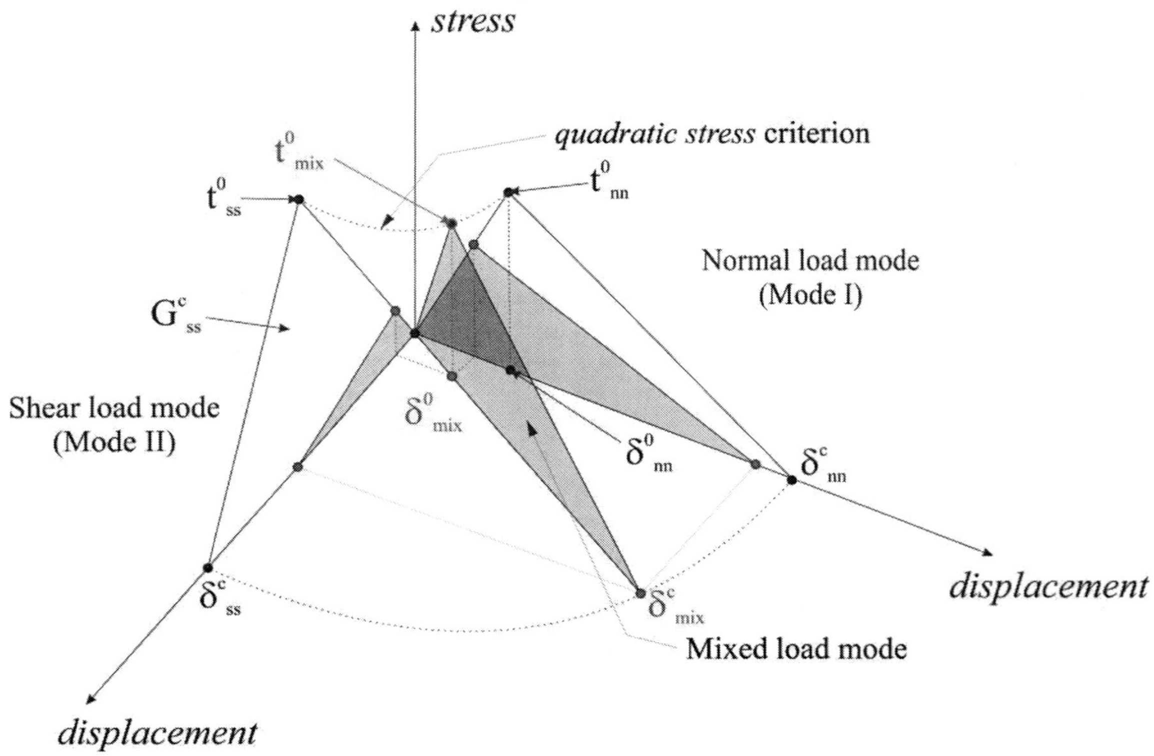


Figure 3: Bilinear traction-separation law for combination of mode I and II

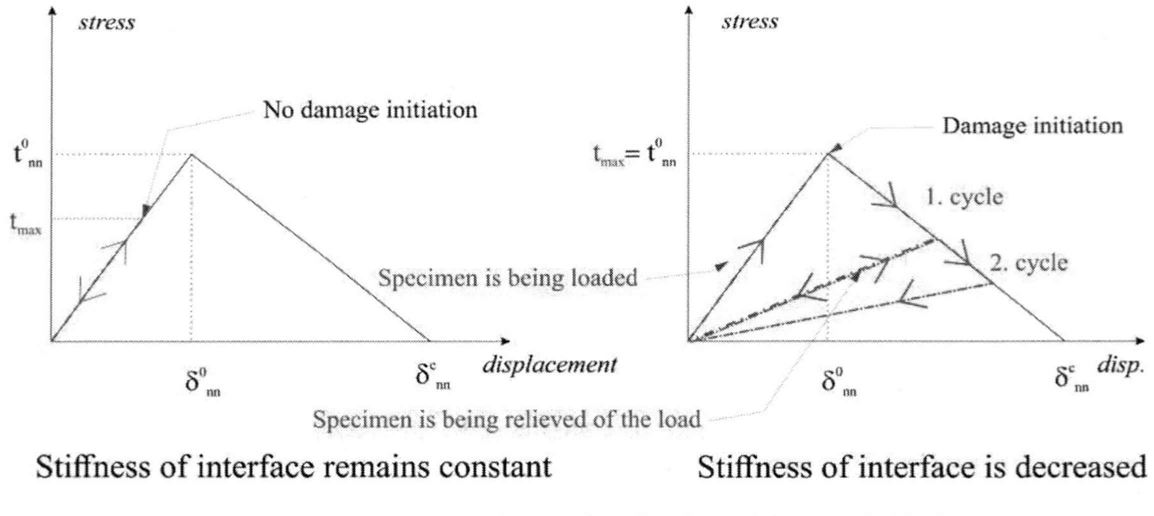


Figure 4: Power law criterion – visualization of damage initiation

## Realization

Material parameters of cohesive interface were determined by standardized tests of an unidirectional carbon fibre-reinforced epoxy beams: DCB, ENF and MMF (mixed mode flexure). These three types of standardized test were performed to obtain dependence of force $F$ and displacement $\delta$ from loading point. As mentioned before failure analysis is based on identification of critical value of SERR $G^c$. Critical value of SERR extracted from experimental data is determined via analytical equations set for standardised tests stated above in section analytical model. Determination of load force $F$ is depicted on figure 5.

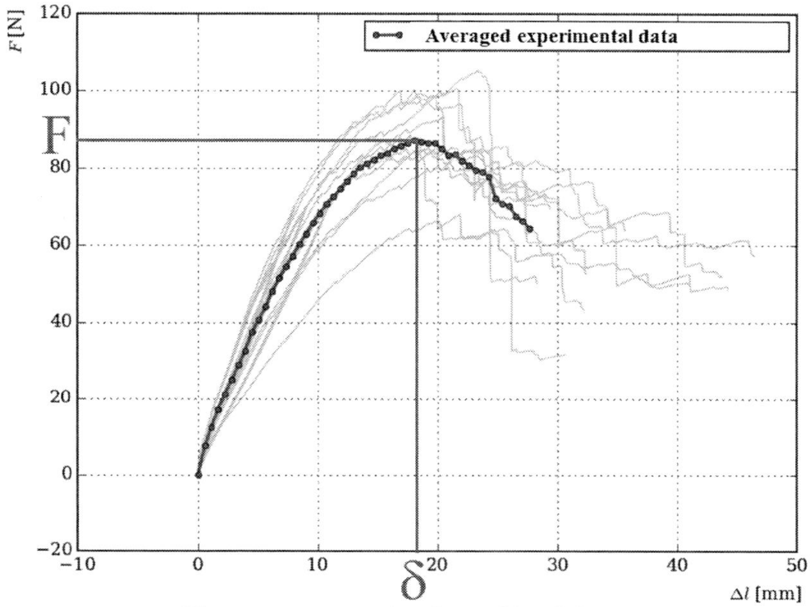


Figure 5: Determination of load force

Critical value of SERR obtained from FEA is determined through secondary material parameters of cohesive interface: stiffness $k$, displacement $\delta^0$ (damage initiation) and $\delta^c$ (failure determination). Secondary material parameters of cohesive interface were identified by comparison of force-displacement curves obtained from standardized tests and FE analysis. Comparison of force-displacement experimental and fitted FEA curves and visualisation of contact status change during load for corresponding model for DCB and ENF specimen with cohesive interface made of adhesive Araldite Hunstman 2021 is displayed on figure 6.

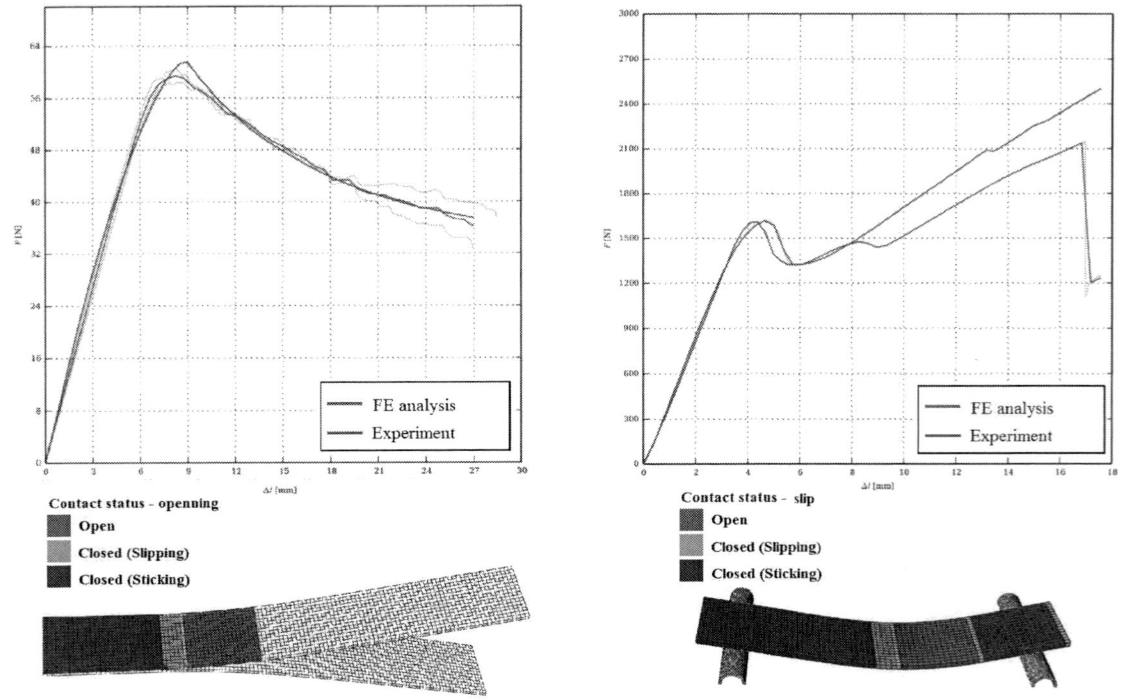


Figure 6: Comparison of force-displacement curves for DCB (left) and ENF specimen

This system of determination was chosen for its relative simplicity, because every parameter influences force-displacement curve in its own way. For e. g. effect of rising stiffness $k_{normal}$ on critical value of SERR and its effect on force-displacement curve (while remaining material parameters are constant) is depicted on figure 7 and effect of the $\delta^0_{shear}$ growth is depicted on figure 8.

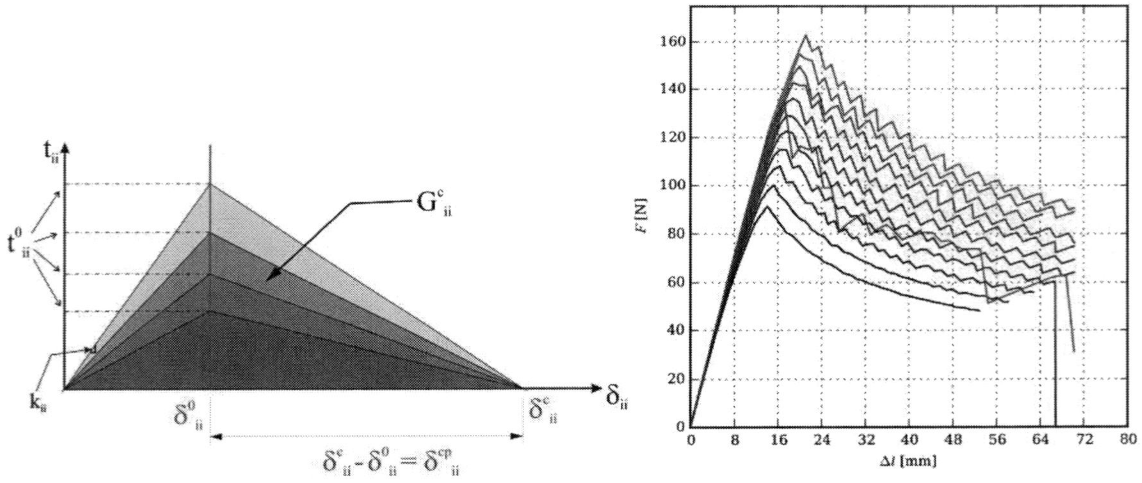


Figure 7: Influence of rising stiffness $k_{normal}$ on determination of $G^c$ for DCB test (curves: red – FEA, green – experimental data)

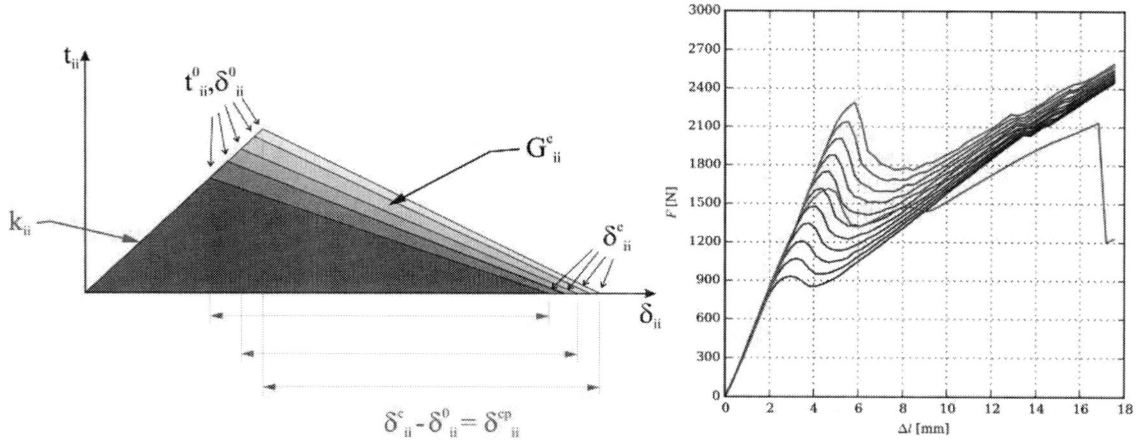


Figure 8: Influence of $\delta^0_{shear}$ on determination of $G^c$ for ENF test
(curves: red – FEA, green – experimental data)

As mentioned before, material parameters of two adhesives (Hunstman Araldite 2021, Gurit Spabond 345) and inner layer of laminated composite were identified. Resultants for secondary material parameters can be found in table 1. Comparison of experimental and FEA critical value of SERR $G^c$ resultants is recorded in table 2.

Table 1: Resultants for secondary material parameters

| Mat. parameter | | Araldite | Spabond | Laminate |
|---|---|---|---|---|
| $k_{normal}$ | [GPa/m] | 32.0 | 42.0 | 24.0 |
| $k_{shear}$ | [GPa/m] | 72.0 | 100.0 | 100.0 |
| $\delta^0_{normal}$ | [μm] | 85.0 | 48.0 | 59.0 |
| $\delta^0_{shear}$ | [μm] | 155.0 | 220.0 | 30.0 |
| $\delta^c_{normal}$ | [μm] | 835.0 | 295.0 | 769.0 |
| $\delta^c_{shear}$ | [μm] | 395.0 | 350.0 | 450.0 |

Table 2: Comparison of experimental and FEA critical SERR resultants

| Experimental / FEA resultants of $G^c$ [J/m$^2$] | DCB ($G_I^c$) | ENF ($G_{II}^c$) | MMF ($G_{I/II}^c$) |
|---|---|---|---|
| **Adhesive Gurit Spabond 345** | 385/301 | 3750/3850 | --- |
| **Adhesive Hunstman Araldite 2021** | 1353/1135 | 2045/2204 | --- |
| **Inner layer of composite** | 535/544 | --- | 730/--- |

**Conclusion**

Standardised tests DCB and ENF seems to be reliable tools for identification of failure defining material parameters for previously mentioned cohesive interfaces. MMF test alone is insufficient for determination of material parameters describing load mode II. Critical values of SERR, obtained from analytical equations and FEA, show agreement.

**Acknowledgement**

This publication was supported by the project LO1506 of the Czech Ministry of Education, Youth and Sports.

## References

[1] Laš, V., 2008. *Mechanika kompozitních materiálů.* University of West Bohemia, Faculty of Applied Sciences, Pilsen, Czech Republic.

[2] Li C.F., Hu N., Yin Y.J., Sekine H.,Fukunaga H., 2002. *Low-velocity impact-induced damage of continuous fiber-reinforced composite laminates. Part 1. An FEM numerical model.*, Composites Part A str. 1055 - 1062

[3] Reeder, J. R., 2006. *3D Mixed-Mode Delamination Fracture Criteria – An Experimentalist's Perspective.* NASA Langley Research Center, Hampton.

[4] Szekrényes A., 2005. *Delamination of composite speciments*, Ph. D. dissertation, Budapest University of Technology and Economics, Budapest, Hungary.

EAN 2017
55th conference on experimental stress analysis

# Effective Service Life Testing of the Cam Mechanisms

Hejnova Monika[1, a]

[1]VUTS, a.s., Svarovska 619, Liberec, Czech Republic

[a]monika.hejnova@vuts.cz

**Keywords:** service life, experimental testing, rolling contact fatigue.

**Abstract.** This paper is dealing with the description of the effective service life testing of material suitable for production of cams and other parts loaded by contact stress. An aim is to determine the fastest procedure for obtaining of necessary information about phenomena at RCF (rolling contact fatigue). This procedure allows us to determine the suitable surface treatment or we can suggest the changes in the loading force and we can determine the real service life of the contact surfaces. In the next step we can suggest the correction of the theoretical model. By this model we usually estimate the service life of the contact surfaces.

## Introduction

A part of the cam mechanism design is the service life estimation, respectively the determination of the suitable material for production. Very often we use only the theoretical estimation. Previous research shows that theoretical and real service life are considerably different [1]. The aim of this paper is to give a concrete procedure, which enables the effective testing of the service life and get the maximum of information about examined material. Important factor at the service life testing is time. Required service life is mostly $10^7$ to $10^8$ cycles, so the testing is very time consuming. Methodic given in this paper can be used by the service life determination of all other machine parts which are loaded by contact stress.

## Testing procedure

The complete testing procedure we can divide into four phases. These are described below.

### Phase 1: Theoretical service life

A value of the theoretical service life we determine according to appropriate theory. They are available in the literature e.g. [3], [6]. Usually, we are using two theories. The first theory [3] is based on the service life estimation by means of the surface hardness. The calculation is simple but the surface hardness does not describe all the mechanical properties of the examined material. The second theory [6] is based on two material parameters. These parameters better describe the mechanical properties of the material but the database of these parameters is limited. It is very expensive and time consuming to obtain the parameters for materials outside the database.

The comparison of these theories results shows that the service life determined by each theory can be very different [1]. Therefore we need to find a suitable experimental procedure which can help verify the accuracy of the theories respectively suggest their correction.

### Phase 2: Testing preparation

The first step is the preparation of the experiment in the field of the suitable material selecting. This choice is necessary to make in according to the results from the calculations (FEM, analytic methods). It is also necessary to control the supplied material, its chemical composition and micropurity. In the next step we should determine exactly the parameters of heat, respectively chemical heat treatment. The parameters of these technological processes are very important for ensuring of required properties of the surface layers of the examined component. These are parameters of carburizing, heating etc.

It is necessary to take into account not only the material properties, but the influence of the working conditions on the RCF forming in the experiment too. The most significant is lubricant and its interaction with the contact surfaces and the shape of the contact surfaces too. Information about the interaction between oil and the contact surfaces is important because the behavior can change depending on change of the contact surface properties (change of surface tension, chemical interaction etc.). The properties of the contact surfaces can affect the RCF in two ways. The first one says that if there are cracks on the surface, it can lead to pitting [7]. The second problem can be, when the contact surfaces are too smooth, then can be insufficient space there for oil film [8].

After verification of the properties of the relevant steel, we can produce the test specimen. We produce two types of specimens, the first one is for testing the service life and the second type is for testing the material properties. By these specimens we can define quality and depth of the heat treatment (respectively carburizing, coating, weld deposit etc.). The depth of the carburizing layer can be determined by removing thin layers and then by using basic chemical analysis to determine the carbon content. Other parameters which are useful we can obtain by nanoindentation measuring. We can get information about mechanical properties of the examined specimens, hardness, modulus, visco-elastic properties and other in the dependence on depth, very detailed in the area of interest.

**Phase 3: Testing**

After choosing the suitable material, production of specimens and determining the right test conditions the specimens can undergo the service life tests. More about these tests in [2].

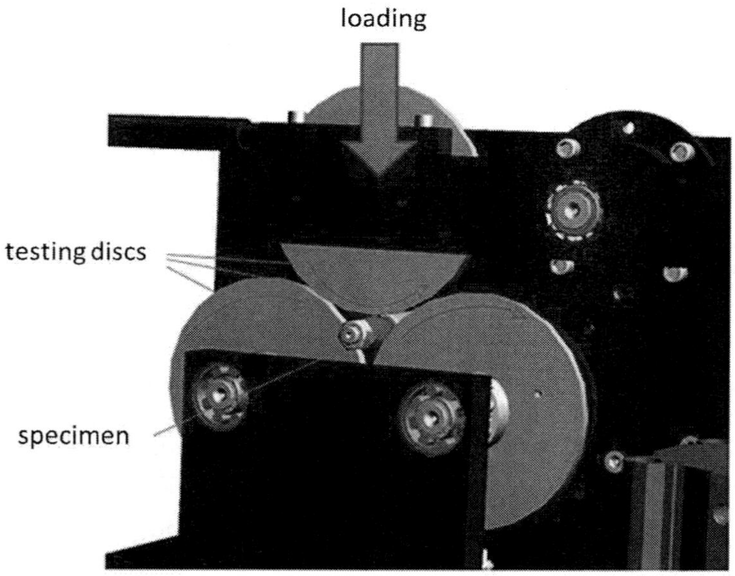


**Fig. 1** Test rig

There is shown the test rig in Fig. 1. The basis of the test is that the specimen is inserted between the three discs and it is loaded by the force determined by means of the theoretical estimation or from previous tests. This way of testing allows the shortening of the time testing to 1/3. This is significant time saving because we need to test every specimen for $10^7$ or $10^8$ cycles. Tests are running to detection of defect or to achieving the desired number of cycles.

**Phase 4: Results evaluating**

After service life testing the tested surface can be observed by stereomicroscope (or other suitable microscope) by low magnification in the first step. The task is to obtain information about surface damage from a macroscopic view.

When we look through the specimens on the stereomicroscope, we can analyze the specimens on the electron microscope. There is picture obtained on a Tescan electron microscope in Fig. 2 (shown pitting on specimen).

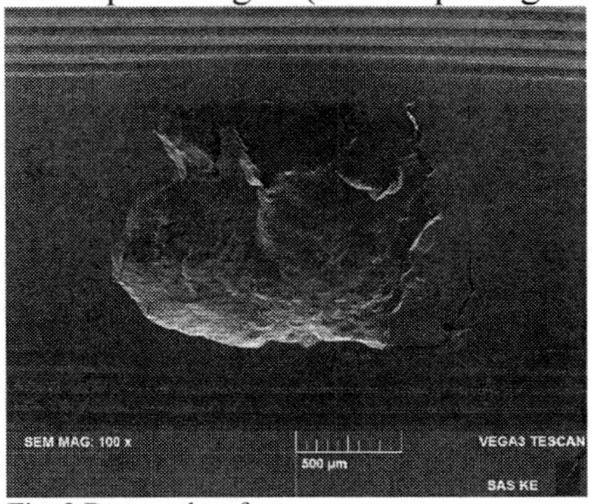

**Fig. 2** Damaged surface

There are other possibilities of the specimen analysis on electron microscopes. We used Jeol JSM 7000F. This microscope works with high magnification and this allows detailed monitoring of the cracks. Except standard display allows display in regimes COMPO and TOPO (see Fig. 3 and Fig. 4). Thanks to these regimes we are able to find out more information about the examined specimen. The regime COMPO allows us to identify inhomogeneity on the basis of the differences in chemical composition. It shows grayscale in the dependence of atomic numbers. The regime TOPO shows topographic view at the examined surface.

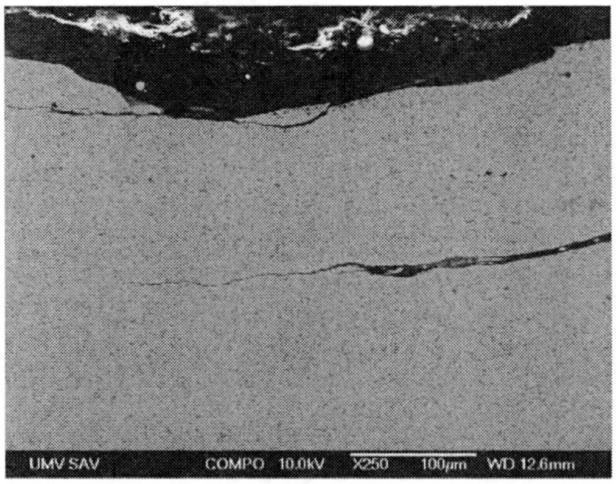

**Fig. 3** Detail of the damaged part shown in regime COMPO

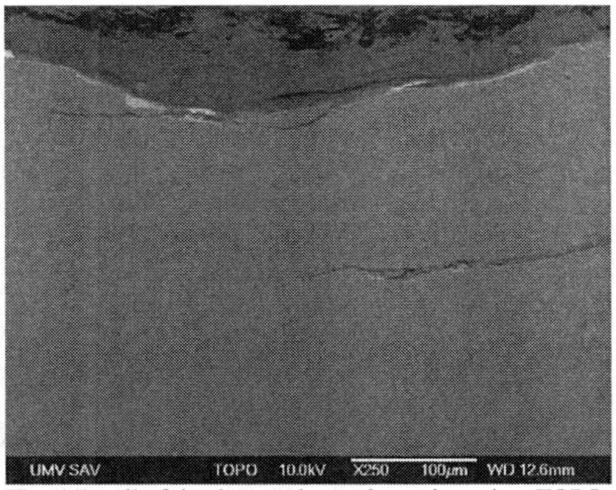

**Fig. 4** Detail of the damaged part shown in regime TOPO

There is marked a maximal depth of cracks presence in Fig. 5. From this image we obtained a value of 0,33mm. The depth of maximal shear stress is 0.77 mm. This value was determined by analytical calculation [5].

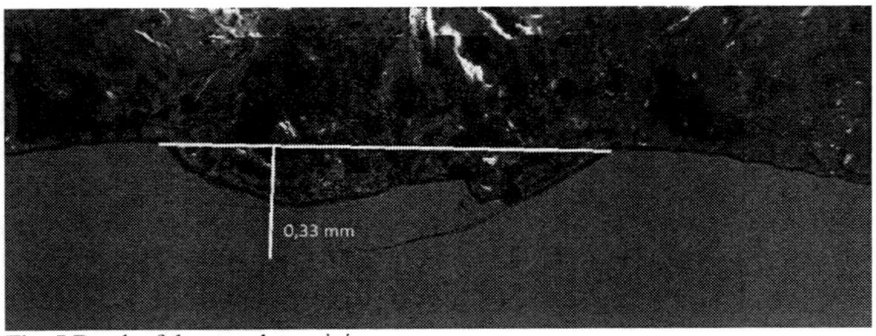


**Fig. 5** Depth of damage determining

From microscopic analyses we can obtain information about observing the technological process, determining of depth and uniformity of the heat treatment [4]. In Fig. 6 is shown image with a marked depth of the carburised layer (obtained on optical microscope).

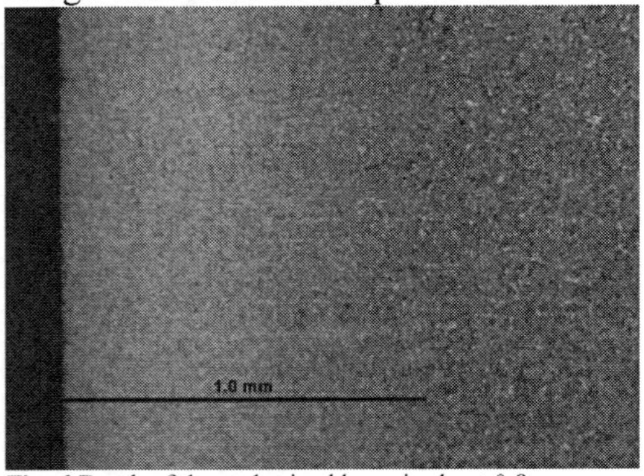


**Fig. 6** Depth of the carburized layer is about 0,8 mm

In Fig. 7 and Fig. 8 shows a detail of the crack root. From these images it is clear that the crack does not form on the grain boundaries. There is not the increase presence of impurities, inclusions, cavities in the examined steel. In Fig. 7 globular carbides are clearly visible, their chemical analysis was performed. We can say that the crack did not form from any inclusion or impurity or cavity. It was formed only by the exhaustion of mechanical properties of the examined steel.

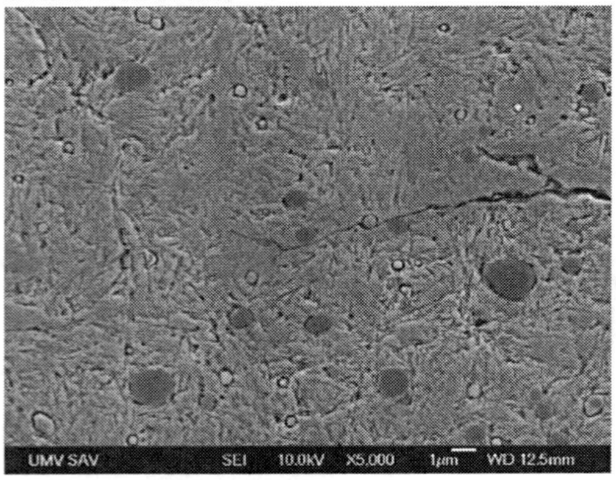


**Fig. 7** Crack root shown with carbides

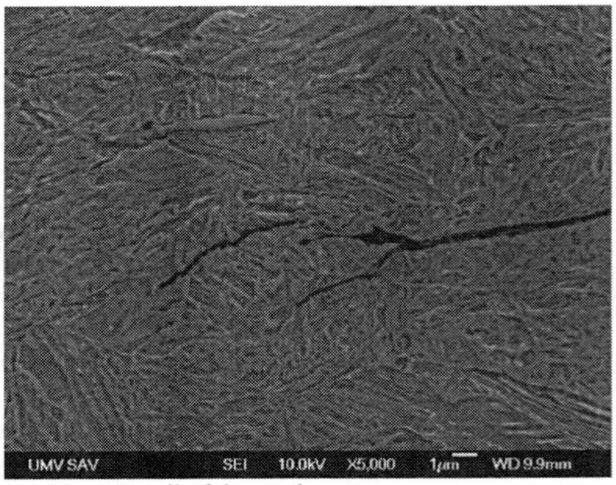


**Fig. 8** Detail of the crack root

From analysis can we say whether the structure is uniform, micropurity of the examined material is convenient and to determine the content of the inclusions and cavities. We can determine if the damage occurred in the material and the cracks are formed on the basis of exhaustion of the mechanical properties of the examined steels or on the basis of non-uniform structure. To have complete information about the examined material, it is useful to measure microhardness and specimens continually (at service life testing) undergo the defectoscopic tests (detection of surface and subsurface defects).

## Conclusions

From results of the tests we can say that the examined material will meet conditions for the service life respectively for loading force. After the evaluation of experimental testing we will suggest an adjustment of the mathematical model and possibly technologic parameters of the heat treatment (mainly depth of the carburized layer and carbon concentration).

The most suitable solution of the material problematics of the rolling contact fatigue is to choose the material which will be loaded in its elastic area. This is possible to achieve either material with high yield strength, with sufficient heat or chemical heat treatment surface layer or with weld deposit.

## Acknowledgments

The research work reported here was made possible by LO1213–project supported by the Ministry of Education, Youth and Sports of the Czech Republic.

## References

[1] Hejnová, M. and Ondrášek, J.: Life Estimation of the Contact Surfaces, In Advances in Mechanism Design II. Proceedings of the XII International Conference on the Theory of Machines and Mechanisms. Springer, 2016. ISBN 978—3—319—44086—6. p. 43-50.

[2] Hejnová. M.: Problematika životnosti vačkových mechanismů. In Mechatronika pohonů pracovních členů mechanismů. Liberec: VÚTS, 2015. Chapter 21, p. 205-208.

[3] Koloc, Z. — Václavík, M.: Vačkové mechanismy. Praha: SNTL—Nakladatelství technické literatury, 1988. 379 p.

[4] Koutský, J. — Šmíd, J.: Makroskopické a mikroskopické zkoušení oceli. Praha: SNTL—Nakladatelství technické literatury, 1964, 126 p.

[5] Mesys: Calculation of contact stress [online]. Available at: https://www.mesys.ag/?page_id=1220

[6] Norton, R. L.: Cam Design and Manufacturing Handbook. Second edition. New York: Industrial Press, 2009. ISBN: 978—0—8311—3367—2. 592 p.

[7] Požár, R.: Rozvoj materiálového poškození při záběru ozubených kol a jeho vizualizace. Brno, 2009. 76 p. Diplomová práce, 2009. Mendelova zemědělská univerzita v Brně, Agronomická fakulta. Available at: http://is.mendelu.cz/zp/portal_zp.pl?prehled=vyhledavani;podrobnosti=30442;download_prace=1

[8] Vrbka, M. et al.: Effect of surface texturing on lubrication film formation and rolling contact fatigue within mixed lubricated non-conformal contacts [online]. In Meccanica (2011) 46:491-498. DOI 10.10007/s11012-010-9288-x. Available at: http://link.springer.com/article/10.1007/s11012-010-9288-x

EAN 2017
55th conference on experimental stress analysis

# Measurement of Cyclic Creep by Digital Image Correlation Method

R. Halama [1, 2, a], L. Lipinová[1, b], R. Páleník[1, c]

[1] Faculty of Mechanical Engineering, VŠB – Technical University of Ostrava, 17. listopadu 2172/15, 708 33 Ostrava-Poruba, Czech Republic

[2] IT4Innovations, VŠB – Technical University of Ostrava, 17. listopadu 2172/15, 708 33 Ostrava-Poruba, Czech Republic

[a]radim.halama@vsb.cz, [b]ladislava.lipinova.st@vsb.cz, [c]radek.palenik.st@vsb.cz

**Keywords:** low cycle fatigue, ratcheting, digital image correlation, full-field analysis

**Abstract.** This contribution presents a new methodology of cyclic creep investigation based on digital image correlation application. The full-field strain analysis brings the possibility of strain accumulation evaluation on a curved part of standardly used specimen in multiaxial fatigue research. Resulting cyclic creep curves can subsequently serve for evaluation of strain accumulation prediction, what can positively influence future research in cyclic plasticity domain. The new methodology saves amount of material and leads to a shorter experimental time.

## Introduction

Digital Image Correlation Method (DICM) is a suitable tool for 3D full-field strain analysis in practice as well as material research [1]. DICM makes possible to measure complete history of strain tensor in examined point of the specimen in a large scale of strains. Thus, it is an appropriate optical method for effective evaluation of fatigue tests. In the previous study [2], DICM was compared with standard uniaxial extensometer to verify its applicability for cyclic creep (ratcheting) strain measurement with acceptable accuracy. Indeed, the power of DICM is in 3D full-field analysis. There is a question, if strain analysis in the part of specimen with increasing cross-section area might be useful.

This paper presents some new results of deformation response on cyclic loading in tubular specimen measured by DICM. The research was conducted within the framework of meetings with students involved in the program for talented students of the Faculty of Mechanical Engineering at VŠB - Technical University of Ostrava.

## Experimental setup

Specimens were made of R7T steel, which is commonly used material of railway wheels. All tests were realized on biaxial testing machine LabControl 100kN/1000Nm at the VŠB – TU Ostrava, see Fig. 1. For DIC measurements the Mercury RT® system (2x2.3Mpx@40Hz), provided by Sobriety s.r.o. company, was used. Averaged strains were measured by EPSILON3550 extensometer with 25mm gauge length in order to verify the measured strain response by DICM in the testing part of specimen. An optical contrast coating known as pattern was prepared on each specimen for application of DICM.

Fig. 1. A photo from realized measurements

The specimen geometry considered for all measurements is shown in the Fig. 2. The tubular specimen with outer diameter 12.5mm and inner diameter 10mm was used. The proportional loading was investigated exclusively in this experimental study. More precisely two stress states were investigated: tension-compression and pure torsion. However, it should be mentioned, that the new methodology is applicable also to other types of loading, including non-proportional cases.

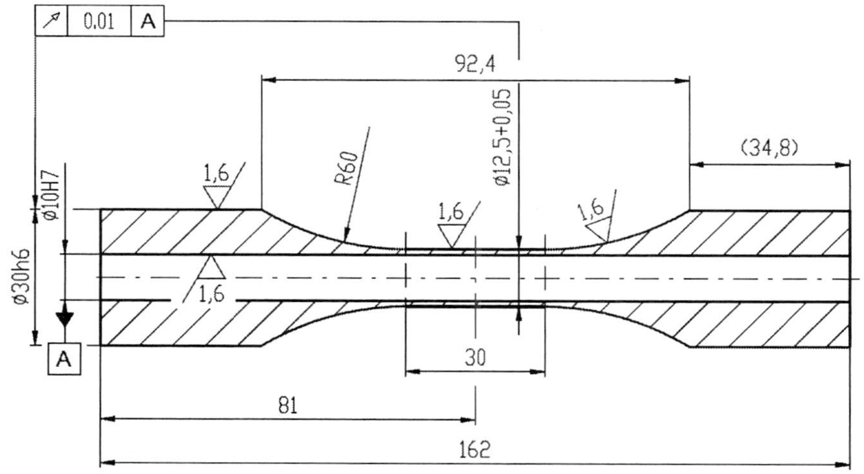


Fig. 2. A scheme of specimen's geometry

**Results of cyclic uniaxial tests**

First two fatigue tests were realized under uniaxial loading. In order to investigate the stress-strain curve of the material considered, a sequential test was performed under strain control. The comparison of static stress-strain curve with cyclic stress-strain curve is done graphically in the Fig. 3. The reader can observe a cyclic softening behaviour for small strain amplitudes and a cyclic hardening behaviour for higher strain amplitudes.

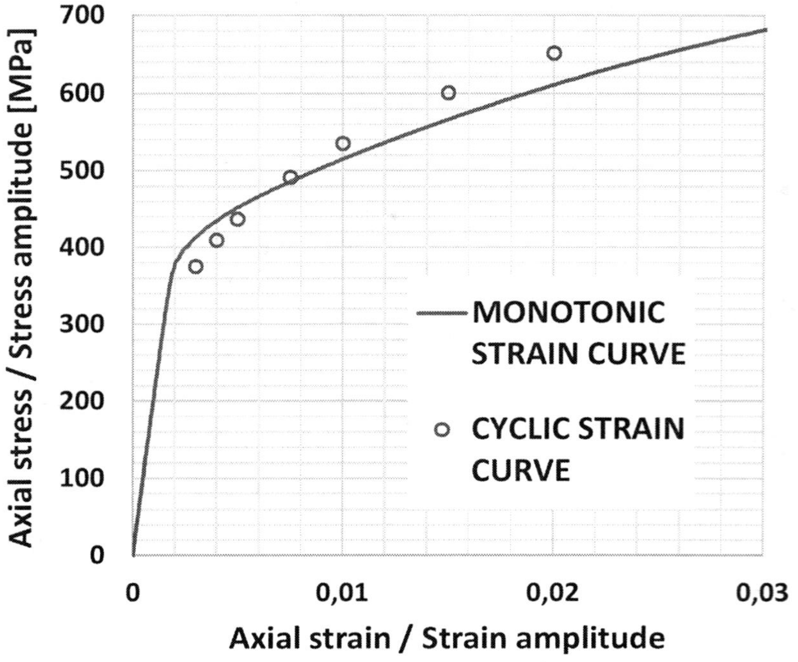


Fig. 3. Stress-strain curves of investigated material obtained from a sequential uniaxial test

The next conducted experiment was done with emphasis on cyclic creep investigation. The harmonic loading with the stress amplitude of 450MPa considering mean stress of 50MPa was applied (frequency 0.1Hz up to 200 cycles, then increased to 0.5Hz). The test was conducted until fracture. Accumulation of total strain is apparent from Fig. 4. Number of cycles to fracture was 1006 cycles. The material shows a ratcheting behaviour with steady state after stabilization of strain response, see Fig. 5.

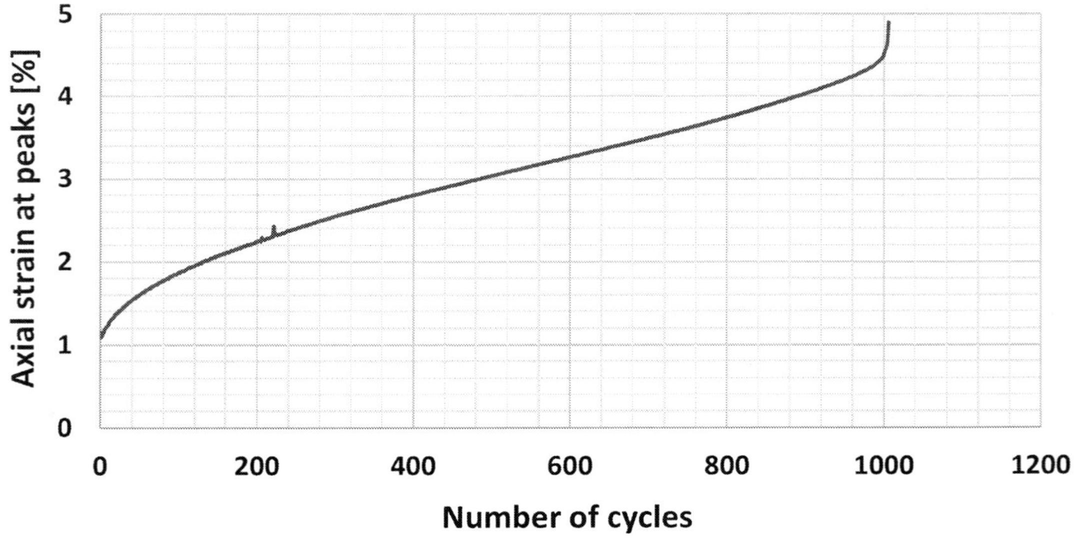


Fig. 4. Uniaxial ratcheting strain measured at peaks as a function of number of cycles

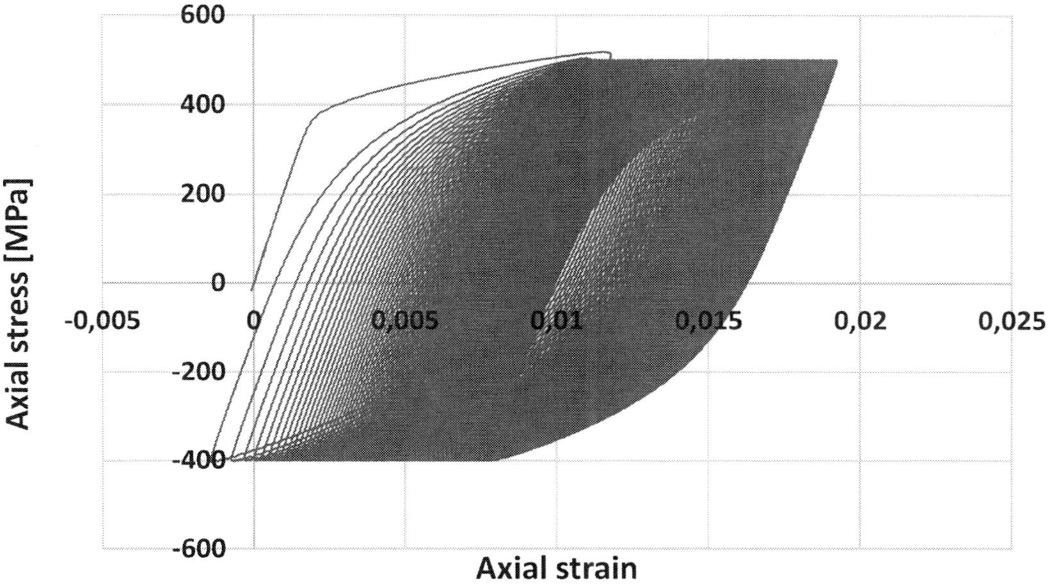


Fig. 5. Stress-strain behaviour of the material in uniaxial ratcheting test

Initial cycles were measured also by DICM. Points selected for evaluation of strains are shown in the Fig. 6 (left). Corresponding uniaxial strain accumulation evaluated at peaks are compared in the graphical form in the Fig. 6 (right). The larger cross-section the less axial strain accumulation was observed as expected.

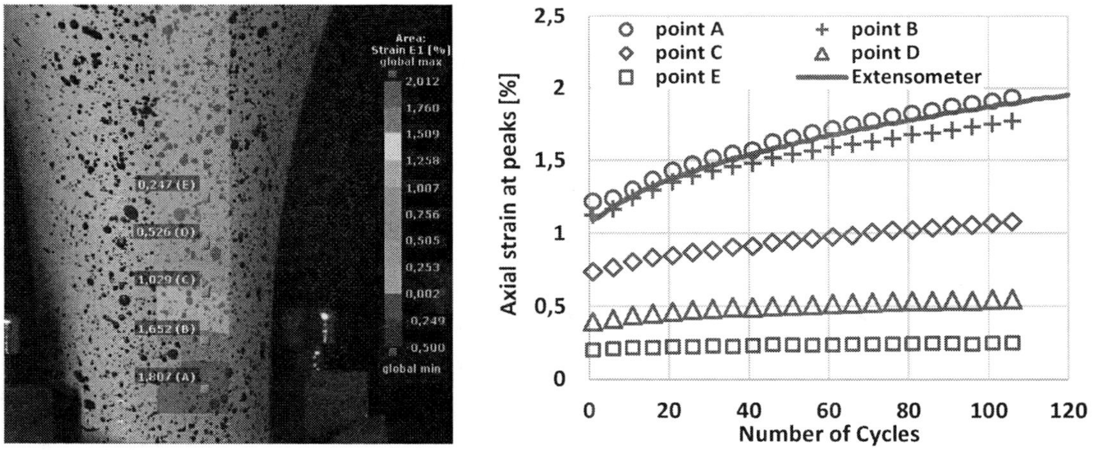


Fig. 6. Contours of first principle strain (left) and axial strain accumulation (right) evaluated from DIC measurement in uniaxial ratcheting test.

## Results of cyclic torsion test

The last experiment realized in this study was done in cyclic torsion. The harmonic loading was applied considering shear stress amplitude of 233MPa with mean shear stress of about 30MPa (frequency 0.1Hz up to 322 cycles, then increased to 0.5Hz). The test was realized until fracture. Shear strain accumulation measured by extensometer during the whole test is presented in Fig. 7. Small decrease of shear strain was caused by interruption of the test in order to change the loading frequency. Number of cycles to fracture was 18780 cycles.

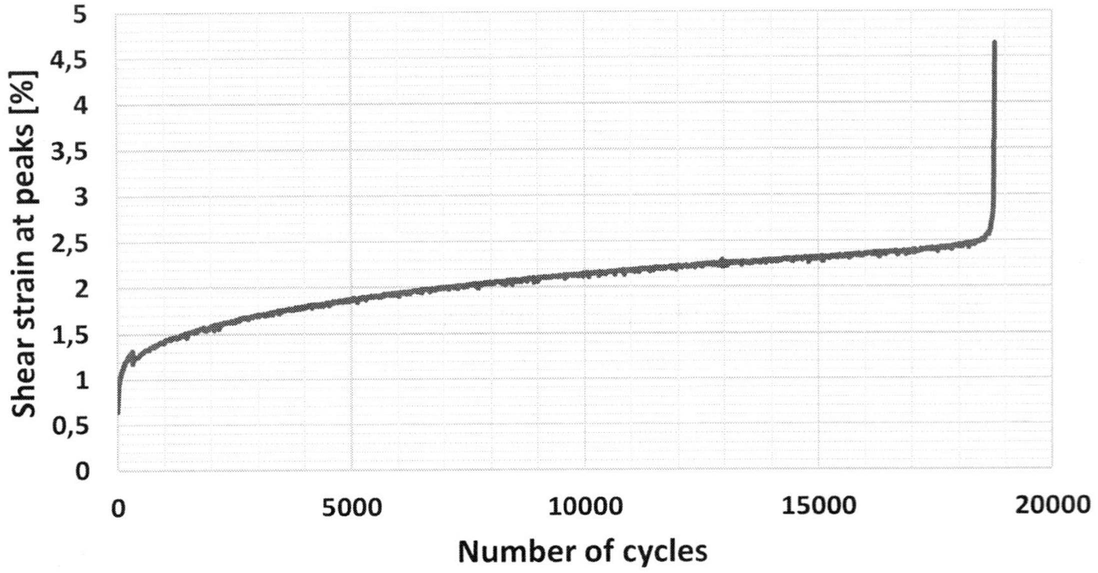


Fig. 7. Shear strain accumulation measured by extensometer in torsional ratcheting test.

Examined material again shows a ratcheting behaviour with steady state after saturation of opened hysteresis loop as can be seen in the Fig. 8, where stress-strain history for selected cycles is presented. Equivalent stress and equivalent strain were evaluated as

$$\sigma_{eqv} = \sqrt{3}\tau, \quad (1)$$

$$\varepsilon_{eqv} = \frac{\gamma}{\sqrt{3}}, \quad (2)$$

where $\tau$ and $\gamma$ are the shear stress and the shear strain respectively. The stabilized behaviour occurs after about 10000 cycles in this case as can be seen in the Fig. 8.

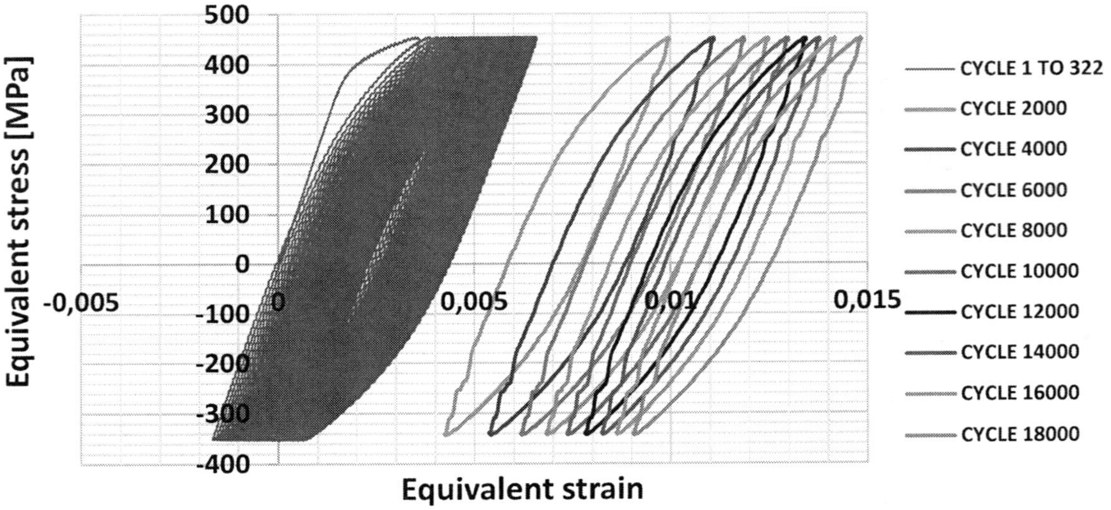


Fig. 8. Stress-strain behaviour of the material in torsional ratcheting test

Shear strains were simultaneously measured in initial cycles by DICM. Evaluation points are shown in the Fig. 9 (left). Shear strain accumulation evaluated at peaks in mentioned points are compared in the Fig. 9 (right). The larger cross-section the less amount of plastic strain can be seen again.

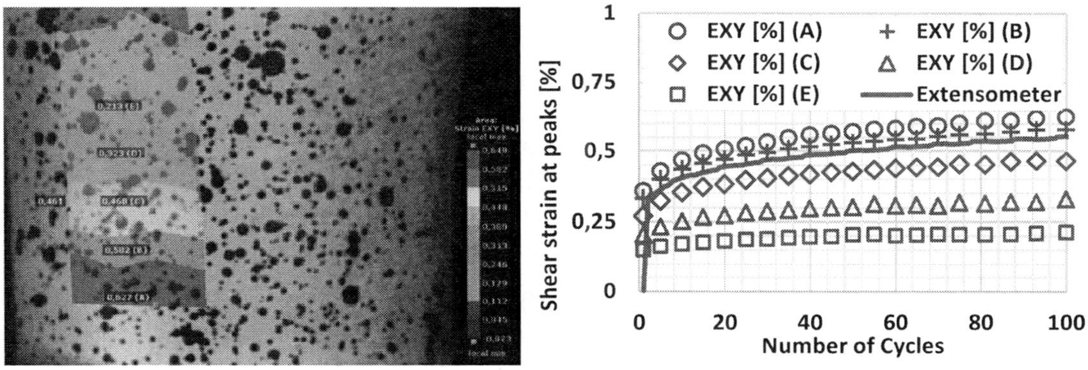


Fig. 9. Contours of shear strain (left) and shear strain accumulation (right) evaluated from DIC measurement in torsional ratcheting test.

## Conclusion

As it is clear from presented results (Fig. 6 and Fig. 9), DICM gives comparable results as extensometer for the point A placed in the testing part. The difference is caused by stress concentration in the area close to the border between testing part and the part with curved surface. The stress concentration was obtained in a FE simulation too. The accumulation of plastic deformation is decreasing with increasing cross-sectional area as could be expected. The strain responses at the point E in both cases (uniaxial as well as torsional) show almost plastic shakedown. Obtained experimental data will be used for evaluation of cyclic plasticity models [3,4] developed for accurate ratcheting prediction of metallic materials.

## Acknowledgement

This work was supported by the Czech Science Foundation (GA15-18274S), by the Specific Research Project (SP2017/136) and by the Institutional Development Project (IRP/2016/60). The support is gratefully acknowledged.

## References

[1] F. Trebuňa, R. Huňady, M. Hagara. *Experimental methods of mechanics - Digital image correlation.* Košice: Technická univerzita v Košiciach, 2016. (in Slovak)

[2] R. Halama, M. Fusek, F. Fojtík, J. Brumek, Strain Measurement in Ratcheting Tests and Prediction by FEM, In Proceedings of the 47th International Conference Experimental Stress Analysis, Czech Republic, Sychrov, June 8-11, 2009, Ed. Marvalová, B., Petríková, I., Čapek, L., Liberec: Technical University in Liberec, 2009, p.115-119.

[3] R. Halama, M. Fusek, Z. Poruba, Influence of Mean Stress and Stress Amplitude on Uniaxial and Biaxial Ratcheting of ST52 Steel and Its Prediction by the AbdelKarim–Ohno Model, Int J Fatigue 91 (2), 2016, 313-321.

[4] R.Halama, M. Fusek, M. Šofer, Z. Poruba, P. Matušek, R. Fajkoš. Ratcheting Behavior of Class C Wheel Steel and Its Prediction by Modified AbdelKarim-Ohno Model, In Proceedings of the 10th International Conference on Contact Mechanics CM2015, Colorado Springs, Colorado, USA, August 30 – September 3, 2015.

EAN 2017
55th conference on experimental stress analysis

# Construction machinery life estimation - Machine load parameters monitoring

Karel Weigel, Ivan Jeřábek, Lukaš Horký

**Doosan Bobcat Engineering s.r.o., Czech Republic**

e-mail: karel.weigel@doosan.com

e-mail: ivan.jerabek@doosan.com

e-mail: lukas.horky@doosan.com

**Keywords:** brittle coat, strain gauge, life estimation.

Essential component of successful product on a market is accurate knowledge of machine life cycle. This knowledge is based on several components – machine properties, customer's requirements and usage. At Doosan Bobcat Engineering, Doosan Company R & D facility located at Dobříš, Czech Republic, lifetime monitoring of components, assemblies or whole machines is done in two phases:

- Laboratory tests: static, short term and/or long term cyclic tests, noise test, vibrations, etc.
- Field tests – real life tests of machines during actual landscaping on special test polygon

The goal for all tests is accurate monitoring and analysis of specimen properties and prediction of behaviour over time – life cycle definition. Stress analysis and lifetime prediction calculation is one of very essential task.

Finite element analysis (FEA) provides basic list of critical points on the structure. Next step is static test using brittle coat (special paint) which enables to verify critical points location provided by FEA and also serves as input to improve mathematical models. Major disadvantage of brittle coat is high temperature sensitivity and its application is very expensive and time consuming. Two special chambers with temperature and humidity control are needed to successfully apply brittle coat. Both chambers are available in Doosan Bobcat Engineering Innovation Center in Dobříš.

Fig. 1 Brittle coat results - critical point mark, each colour represents different load case

Final positions of strain gauges are determined after careful result analysis from both methods – brittle coat and FEA. Strain gauges are supposed to monitor all critical point on the structure. Additional reference points are added for mathematical model validation purposes.

Fig. 2 Strain gauges installation ready for tests

Additional machine parameters are recorded aside of mechanical stress in the structure. For example hydraulic pressure, engine speed, positions, vibrations, etc. All parameters are recorded synchronously at rate of several hundred samples per second. This way of data acquisition is essential for stress analysis, machine behaviour and life estimation calculations. Tests are performed either in simulated laboratory environment or in real life landscaping on test polygon. Measurement units are configured as "measurement nest". Measurement units are located at various locations over the machine and are using common synchronization channel and wireless configuration and control. Synchronization is at the moment via cable, but we are working on wireless variant, that would allow us more flexibility with "measurement nest" locations over the machine. Synchronization is based on value change detection signal.

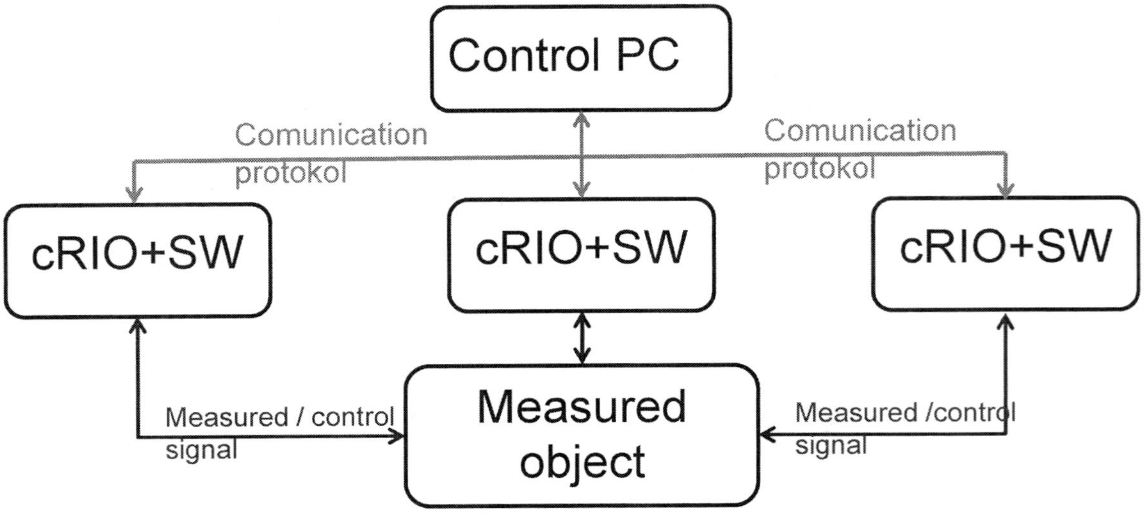

Fig. 3 Measurement system architecture

This approach is very flexible and can be used to synchronize several dozens of channels on different measurement units, different inputs and different sampling frequencies. Major advantage of this is detailed knowledge of stress and other monitored parameters. Drawback is large amount of data (several Gigabytes) that has to be processed and analysed.

Result of life estimation analysis is simple table (Tab. 1). Free capacity of department is available for external clients.

| Gauge name | Target life [Hr] | Life Estimation [Hr] |
|---|---|---|
| A01 | 5 000 | 11 355 |
| A13 | 5 000 | 29 358 |
| A04 | 5 000 | 41 547 |
| A02 | 5 000 | 96 096 |
| B03 | 5 000 | 139 052 |
| H11 | 5 000 | 206 769 |

Tab. 1 Life estimation analysis results

**References**

[1] K. Hoffmann, An Introduction to Stress Analysis and Transducer Design using Strain Gauges, e-book, www.hbm.com.

[2] http://test.bobcatdobris.cz/en

[3] Information on http://www. www.stresskote.com

[4] Information on http://www.ni.com

[5] Information on http://www. www.dewetron.com

[6] Information on http://www. www.hbm.com

# Absorbing materials properties in experiment and simulation

Ing. Martina Drdlová[1, a], Ing. Miloslav Popovič[2,b] Ing. Ondřej Koutný[3,c]

[1]Výzkumný ústav stavebních hmot, a. s., Hněvkovského 65, 617 00 Brno; CZ

[2]SVSFEM s.r.o., Škrochova 42; 615 00, Brno; CZ

[3]BOGGES spol. s.r.o., Rudice 1, 679 06 Rudice; CZ

[a]drdlova@vustah.cz, [b]mpopovic@svsfem.cz, [c]ondrej.koutny@bogges.cz

**Keywords:** Split Hopkinson Pressure Bar, blast load, Newton Cradle test, absorbing materials.

## Introduction

A blast loading applied to different kinds of materials and structures is becoming more important task every year because of actual terrorist threats. An absorption capacity of soft, viscoelastic materials at high strain rates is included in wide range of practical applications. One of the critical questions in any similar analysis is setup of material properties including all physical constants. There are many kinds of tests analysing dynamic properties of absorbing elements. In many cases testing and also numerical simulations are influenced by many types of measurement and simulation conditions. This investigation was focused on relations between particular testing results in their complexity and related numerical simulations as well.

The goal of this investigation is to find a numerical material model of absorbing material suitable for explicit numerical simulation especially focused on blast load. Number of variations in real testing is usually strictly limited because of experiment costs. On the contrary numerical simulations have no limits in number of analysed variants but they have handicap in correlation with reality. This should be solved by combined development of requested products. This analysis used four kinds of real tests with different strain rates as consistent and verified points that numerical models should confirm and fill in unknown gaps between them.

Those testing techniques were applied to materials composed of porous glass (Liaver) or ceramic (Liapor) filler and polymeric binder, with density of 125 - 300 kg/m3 and particle size in range of 0,25 – 2 mm.

| Material filler | Particle size | Material Density [kg/m3] |
|---|---|---|
| Liaver | 0,25 – 0,5 | 381 |
| | 0,5 – 1 | 343 |
| | 1 – 2 | 334 |
| Liapor | 0,25 – 0,5 | 785 |
| | 0,5 – 1 | 752 |
| | 1 – 2 | 644 |

Table 1: Analyzed materials

## Set of Experiments

**Quasistatic compression** in closed space is the simplest test of analyzed materials. In this test absorbing materials exhibit typical behavior consisting of three phases. In the first phase the material exhibits quite high values of stiffness at the beginning of deformation. This phase is followed by a process of permanent destruction of bonds between individual elements. In the last phase the material is maximally compressed and the value of force necessary even for small increment of compression rises significantly.

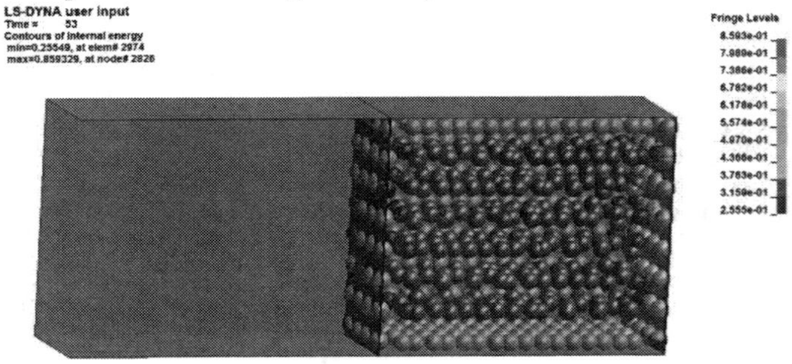


Figure 1: LS-DYNA simulation of quasistatic compression

This test was performed with 40x40x80 mm specimens. The measured data were used to create numerical material models of both types of absorbing materials. Material model from LS-Dyna database which is dedicated to modeling crushable foam with optional damping and tension cutoff was used. Unloading is fully elastic. Tension is treated as elastic-perfectly-plastic at the tension cut-off value.

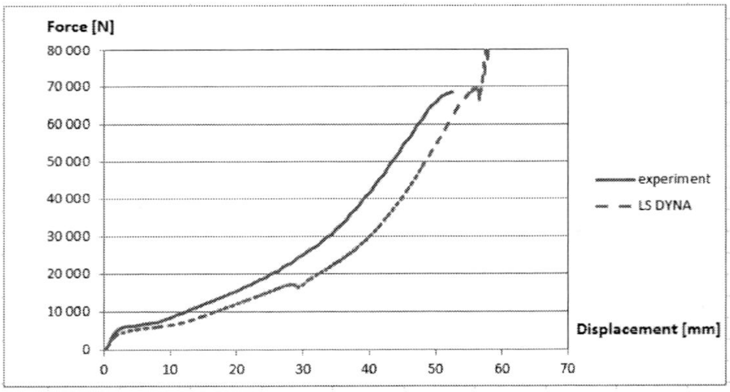


Figure 2: Comparison between numerically and experimentally found material behaviors (quasistatic compression)

**The Newton cradle principle** was used for dynamic test. The impact test of absorbing material with sample size of 50 mm was performed. The impact velocity 3.5 m/s is relatively low but obtained testing results present significantly different process of deformation regarding to static pressure test. The energy absorbed in sample can be simply evaluated by residual potential energy of a punch body.

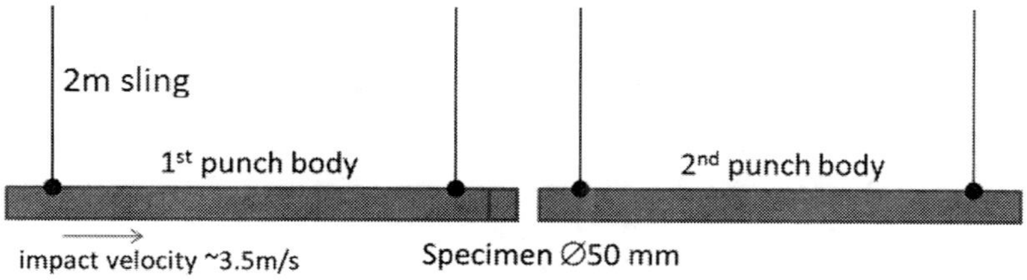


Figure 3: Scheme of "Newton Cradle" experimental setup

Figure 4: "Newton Cradle" numerical model

Figure 5: "Newton Cradle" experimental setup

The residual velocity of both hitting rods from an experiment and simulation were used for a verification of the numerical model. A maximum difference under 5% in velocity values was observed. This satisfies model from an energy point of view. However a process of sample destruction should be analysed in next step with regards to simulation boundary conditions.

**The Split Hopkinson Pressure Bar (SHPB)** test was performed with goal to evaluate stress-strain in higher strain rate values (about 2000 $s^{-1}$). Of course higher strain rate in SHPB experiment caused higher deviation on stress-strain results than impact test presents. However an evaluated absorption of analysed materials was found in sufficient precision.

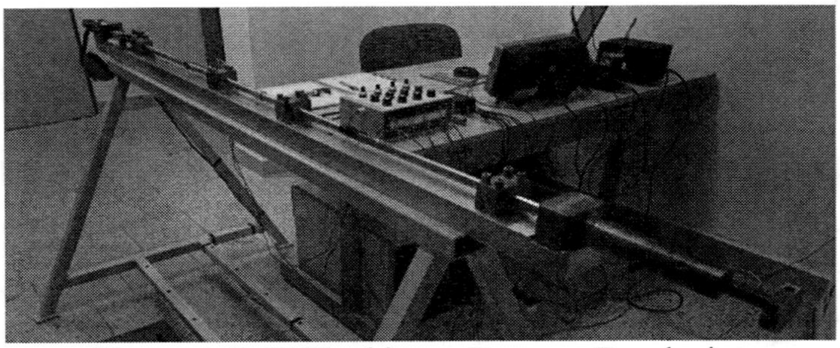

Figure 6: Split Hopkinson Pressure Bar device

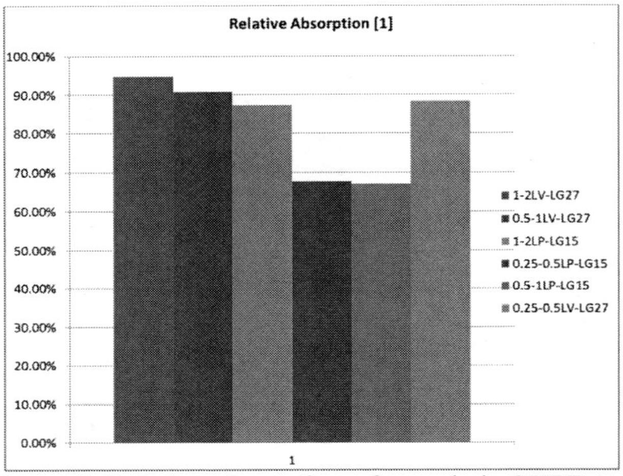


Figure 7: Relative absorption of sample in SHPB test

**Blast load.** All previously described tests enables testing of relatively big number of samples (more than 200). A real blast load limits strongly number of successful tests. A blast load was analysed in two kinds of experiments. The first one investigated relative absorption of analysed materials in simplified one dimension experimental setup. The absorption was evaluated from an elastic wave measured in steel rod, which had a sample attached to its front face, during an explosion (see figure below).

For the explosive load test there was a steel rod with diameter of 50 mm and length of 2000 mm freely hanged on the construction (see Fig. 9). The hanging enabled the rod to move freely in axial direction. There was a specimen fastened to the front face of the rod. The specimen had the same diameter as the rod (50 mm) and it was 40 mm long. Right on the other face of the specimen there was a cylinder shaped, 40 mm long explosive attached to it. The axial deformation caused by the shock wave from the explosion was measured in the middle of the rod's length with resistance strain gauges.
The comparison on fig. 10 shows good correlation between experimentally measured data and numerical simulations based on material curve acquired from SHPB test.

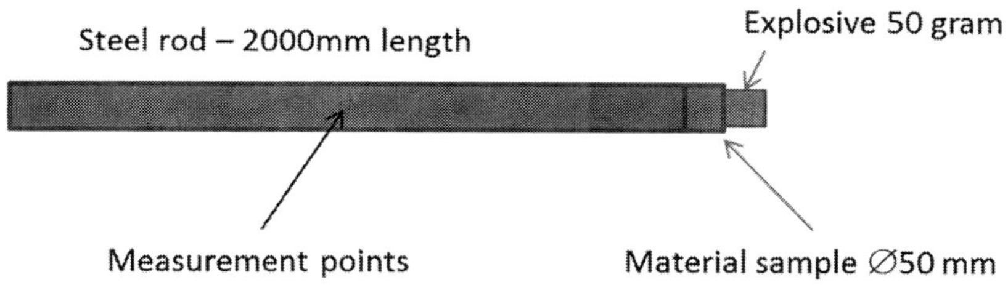


Figure 8: Scheme of explosive load experimental setup

Figure 9: Explosive load experimental setup

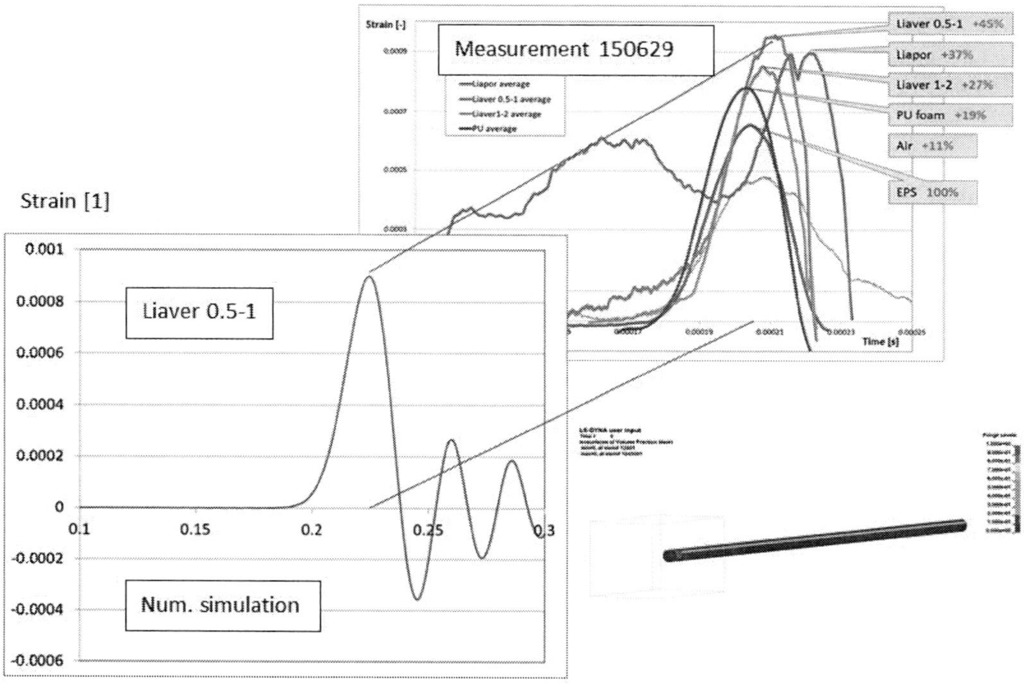


Figure 10: Comparison of numerically and experimentally found results of explosive load testing

The second experiment with blast load was prepared as explosion of 100 g of TNT placed in limited distance above a square sample with size of 500x500 $mm^2$ supported around in fixed frame.

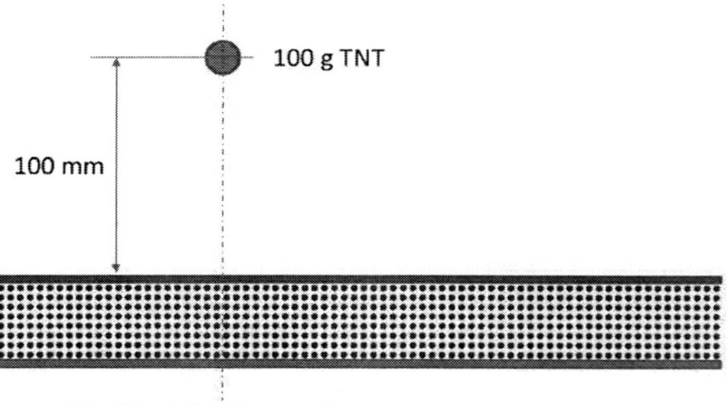


Figure 11: Blast load experiment setup

The response of the absorbing material was evaluated by a residual deformation of the sample, especially deformation of a back sheet.

## Optimization

In order to create universal material model, that would be able to describe behaviour of analysed materials when variously loaded, the process of optimization was utilized. For this purpose optimization software optiSLang was used. The project, that was created to optimize material model to behave like the real material, consisted of one input file and several solvers. Each solver represented one type of testing that was performed on the material before. The input file was a file which contained material model. Therefore the optimization was able to design a material model and then use this model in several numerical simulations representing various material tests. Then in these simulations the same kind of data as in the real tests were collected and compared with experimentally measured data. The differences between numerically obtained data and experimentally measured data were responses of the process. The aim of the optimization was to alter the input in such way that it would minimize these differences.

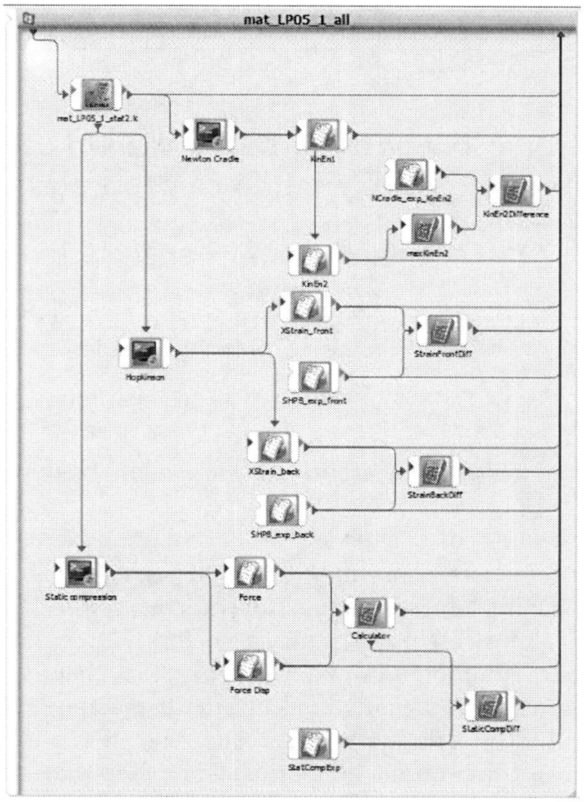


Figure 12: Optimization project

**Conclusion**

The aim of this project is to find universal material model of absorbing materials composed of porous glass/ceramic filler and polymeric binder.

The wide range of experiments on selected absorbing materials brings a general numerical models matching expected behavior of this material in real applications. The numerical model was based on stress-strain dependency analyzed in quasistatic commpression test and then altered according to stress-strain dependency found in SHPB testing. This relation was implemented in crushable foam material model. Together the absorption capacity of this numerical material model was compared with results from the blast test and "Newton Cradle" experiment. Performed real experiments represent very wide spectrum of material behavior relating to strain rate. At the point of designing a real application this model is sufficient for the first response assessment of analyzed absorbing system. It is obviously necessary to experimentally verify the final design of the application.

The final numerical material model is actually used in design of blast resistant litter bin and other application as anti-blast wall and so on.

**Funding Statement**

The authors wish to express their gratitude and sincere appreciation to the authority of The Technology Agency of the Czech Republic, project No. TA02000162 for financial support.

EAN 2017
55th conference on experimental stress analysis

# Study of bending properties of thin walled carbon and flax composite rods winded of prepreg fibers

P. Kulhavý[1, a], P. Lepšík[1,b]

[1] Technical University of Liberec, Faculty of Mechanical Engineering, Studentská 2, 461 17 Liberec 1, Czech Republic

[a] Petr.Kulhavy@tul.cz, [b] Petr.Lepsik@tul.cz

**Keywords:** Prepreg, composite, flax, carbon, acp simulation, bending

**Abstract.** Mechanical testing of composite parts and materials in order to obtain basic mechanical parametres is a time consuming often difficult process. However, due to rising prices of materials and the high energy demand plays reduction of weight of final products an important role through almost all manufacturing sectors [1]. When replacing conventional materials by composites it is important to know the exact mechanical properties, because just with an appropriate combination of individual layers, dispersion and matrix we can achieve the solution tailored to the specific application. Therefore the aim of the carried work was experimentally and by using appropriate numerical and material models describe the behavior of the two kinds of composite rods during classical three-point bending test. A difference in the behavior of both structures was clearly noticeable during the experiment. The flax material endured lesser strain and also the arising deformations had distinctly different character. At the same time in the ACP module of ANSYS was created model as a tool for verification of the obtained data. The initial data obtained from the model and experiment were in quite good agreement until the first peak of maximal strength, when in a real case occurs warping of the rod shape, fiber rupture and delamination. This in the experiment caused significant decrease of the strength, but in the model, however, the force still continues to grow. For this reason is for advanced modeling necessary to supplement the model of cohesive interlayers and failure criterion.

## Introduction

Composite materials are used especially because of their low weight, very high strength and offer also solution for applications subjected to fatigue loadings. Whereas metal materials show one failure mode i.e. cracking, composites can exhibit one or a combination of failure modes, including fiber rupture, matrix cracking, delamination, interface debonding and void growth [2].

Mechanical testing is, however essential process, and according to Camanho [2] can be simplified by testing of simple structures, such as flat coupons. With this absolutely did not agree Wang [3] who mentioned that the laboratory samples and real parts show considerable differences in the obtained mechanical properties. The use of flexural tests to determine the mechanical properties is widespread to the relative simplicity of the test method and required equipment. Using a short beam it is also possible to make flexure tests to determine the interlaminar shear strength of a laminate. The stiffness of anisotropic composite plates depends on several factors, i.e. fiber orientation, laminate stacking, surface waviness and molding temperature. The modes of failure in a bending with large deflection and destruction of the samples describe [4, 5]. In their works is possible to find description how in fiber composites begins damage at micro-scale with matrix cracking, fibre matrix debonding and fibre failure. This is followed by meso-scale damage such as intra-yarn cracking and inter-ply delamination.

## Materials

Plastics materials reinforced by long fibers are widely used because of their high strength and modulus to density ratio [6]. The high strength composite materials are capable to transform stress only in the fiber direction (Fig. 1). Tensile strength in the direction perpendicular to the direction of longitudinal fibers is even lower than the ultimate strength of the matrix itself, which is caused by the concentration of local stress on their interface [1]. The main constituent of tested samples are carbon and flax fibers (Tab.1) pre-impregnated with resins. The used material is an epoxy UD prepreg with thickness of 0.2 mm. According to conducted measurements and to [7] is the final thickness after polymerization approximately 85 % of the original one. Three plies (45/0/-45) was layered on a nonbearing mandrel with the outer diameter 24 mm.

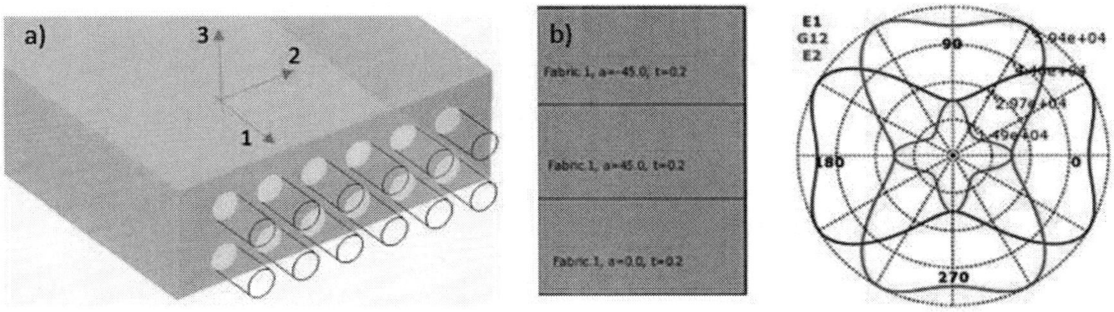


**Fig. 1** UD composite: a) Convention of the global coordinate axis b) Polar properties of the created carbon stackup (45, 0,-45)

Tab. 1 material properties of used UD prepreg materials

| | E1 [MPa] | E2 [MPa] | ρ [g/cm$^3$] | μ [-] | Ductility [%] |
|---|---|---|---|---|---|
| Flax | 10 100 | 6 000 | 1,5 | 0,21 | 2 |
| Carbon | 101 000 | 9 000 | 1,9 | 0,25 | lim -> 0 |

In the view of homogenization a transversally isotropic behaviour is commonly assumed for the composite [8]. Consequently, the stiffness could be described by a set of five parameters: young-moduli ($E_{11}$and $E_{22}$), poisson's ratios ($\nu_{12}$ and $\nu_{23}$) and a shear-modulus ($G_{12}$). With the help of homogenization techniques these five parameters describing the stiffness behaviour of the composite can be derived. It is necessary to mention that this idea is valid only with an assumption of a really perfect alignment of fibres. Then the compliance matrix can be rewritten in terms of the engineering constants (1).

$$
\begin{bmatrix} \sigma_1 \\ \sigma_2 \\ \sigma_3 \\ \tau_{23} \\ \tau_{31} \\ \tau_{12} \end{bmatrix} = \begin{bmatrix} \frac{1}{E_1} & -\frac{\mu_{12}}{E_1} & -\frac{\mu_{13}}{E_1} & 0 & 0 & 0 \\ -\frac{\mu_{21}}{E_2} & \frac{1}{E_2} & -\frac{\mu_{23}}{E_2} & 0 & 0 & 0 \\ -\frac{\mu_{31}}{E_3} & -\frac{\mu_{32}}{E_3} & \frac{1}{E_3} & 0 & 0 & 0 \\ 0 & 0 & 0 & \frac{1}{G_{23}} & 0 & 0 \\ 0 & 0 & 0 & 0 & \frac{1}{G_{31}} & 0 \\ 0 & 0 & 0 & 0 & 0 & \frac{1}{G_{12}} \end{bmatrix} \begin{bmatrix} \varepsilon_1 \\ \varepsilon_2 \\ \varepsilon_3 \\ \gamma_{23} \\ \gamma_{31} \\ \gamma_{12} \end{bmatrix} \quad (1)
$$

**Experiment**

The flexural strength of a material is the maximum stress that material subjected to bending load could resist before failure [9] Chung. The aim of the carried experiment was to study behaviour of two kind of tubes in flexural stress. The applied quasi static loading had increased with step 0,5 mm /s [10] until caused the final 80 mm displacement of used indenter - measured from the point of a first touch with the tube (Fig. 2). The distance *l* between the supports is in our case equal to 380 mm.

The basic force equilibrium equation at a point (x, z) in the beam is given by (2).

$$
\frac{d\sigma}{dx} - \frac{d\tau}{dz} = 0 \quad (2)
$$

Where σ and τ are the respectively normal and shear stresses.
Because for composite the mid plane is not exactly in the center of a tested sample, we need to distinguish h measured from the neutral plane to the bottom of the beam and ht to the top in equation (3).

$$
\int_{-h_b}^{0} \sigma_T \, dz + \int_0^{h_t} \sigma_c \, dz' = 0 \quad (3)
$$

Where indexes T and C denote tension and compression, and $h_b + h_t$ are equal to the outer diameter of tested rod.

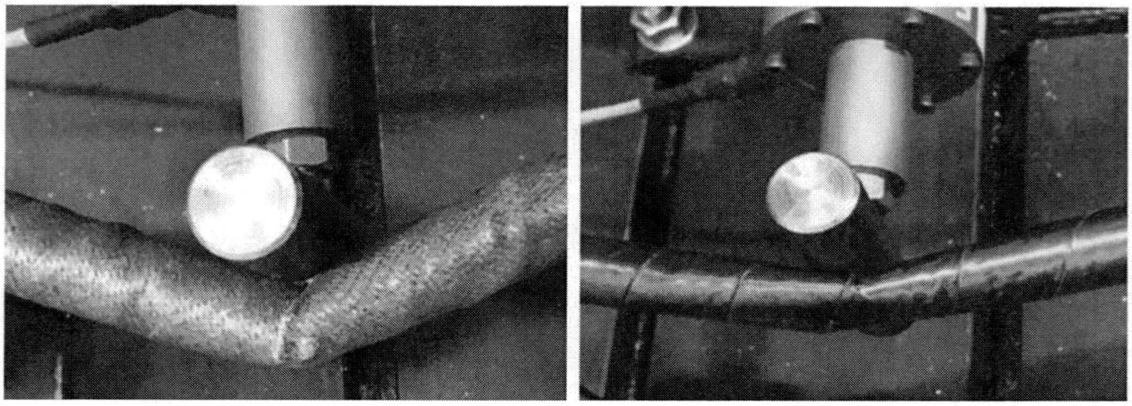

**Fig. 2** Shape of the deformed samples: a) flax fiber tube, b) carbon fiber tube

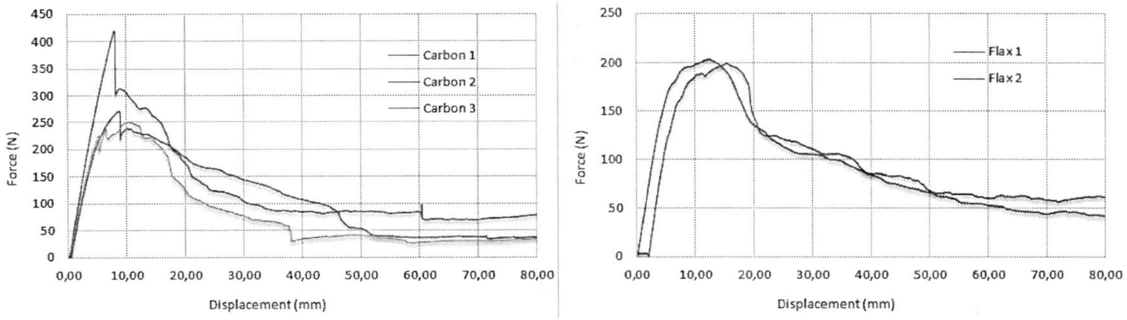


**Fig. 3** Experiment, Force – Displacement relation for the tested rods a) Carbon b) Flax

Based on the measured deformation, it is possible to determine the flexion stress (4).

$$\sigma_O = \frac{M_{Omax}}{Wo} = \frac{F_{max} \cdot l}{4 \cdot Wo} \tag{4}$$

**Numerical model of the solved problem.**

Finite-element method (FEM) is a powerful tool without is today not possible to efficiently design composite parts. The prediction of their mechanical behaviour is very complex problem, because the process induces fiber orientation, interface of plies etc. Numerical analysis allows us to derive the different strain energies stored in the material directions of the constituents of composite materials [11]. Advanced methods could even describe the entire damage process from its initiation through evolution to a complete failure of a composite structure [12]. An unanswered question is, how accurate the simulation should be to be suitable [8].

Because the modeling of contact like in our case the between layered shell and solid element is very problematic. It is possible to find various simplifications, e.g. Gruber [8] created and simplified static model without use of any contact. The boundary conditions were applied just to the nodes as could be seen in Fig. 4a. During the test are the boundary of the plate considered to be free as the plate is supported at interior points, and no special treatment is required [13].

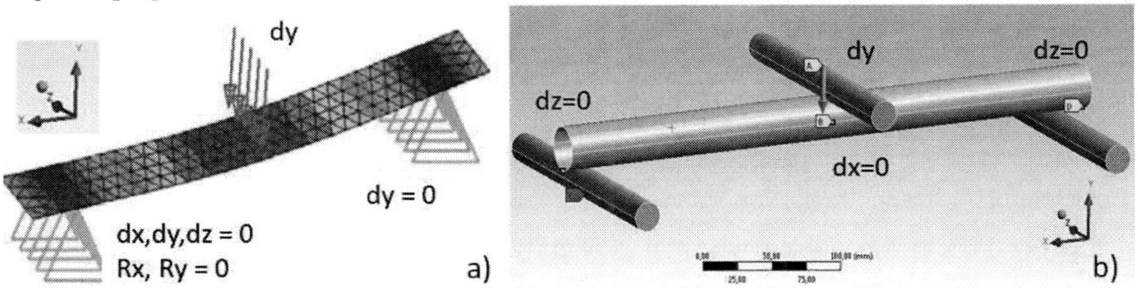


**Fig. 4** Model of bending test: a) Simplified according to [8], b) Fully defined contact model

In our case, the model was solved as a fully contact task, Fix. 4b. The tested part was laid on two fixed supports. According to Sonner [14] it is better to use fixed constraint on the domain level than on the boundary. Then it is not necessary to solve for a lot of zeros in the internal of these domains. The frictional support with asymmetric behavior has been set.

For combination of solid and shell elements the pure penalty formulation with nodal-normal detection of integration points was used. The chosen basic normal stiffness 1e-002 could be additionally adjusted by program. The simplest way of handling an initially unconstrained model was to add weak springs. The spring constant can be made dependent on the load parameter, so that the spring has effect only in the beginning of the simulation [15].

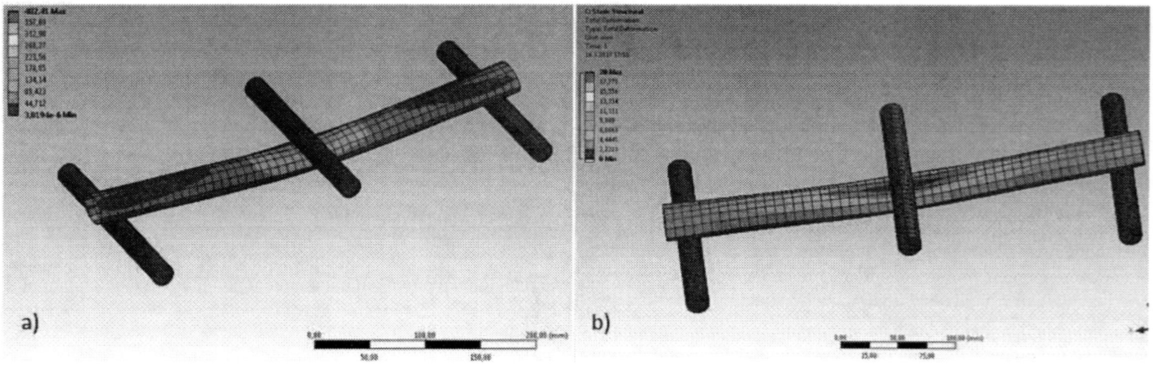


**Fig. 5** Results of the carried simulation for the stacked layer a) The Equivalent stress, b) Directional displacement

**Results and discussions.**

The carried model (Fig. 5) was in a good agreement for approx. first 10 - 20 mm, it means to the point where the maximal strength and subsequent deformation of the tested parts are. Then, in the real case start the force decreasing because of arising nonlinear deformations, delamination, rupture etc. meanwhile the force in model is still exponencially increasing. One example of the experimental results and deformed model are shown in Fig. 2, 3. The deviation between the experiment and model was higher for the flax tube (Tab. 2). This could be mainly caused that the material model of flax prepreg is not sufficiently precise. With this idea also agrees the fact that for structure designing and numerical simulation is flax material not too common as e.g. the carbon one.

Tab. 2 Resultant values from experiment compared with the numerical model

| | **Displacement** | $F_{max}$ | | $\sigma_{omax}$ | |
|---|---|---|---|---|---|
| | at Fmax | Experiment | Model | Experiment | Model |
| **Carbon** | 8,2 mm | 305 N | 338 N | 98 MPa | 134 MPa |
| **Flax** | 13 mm | 220 N | 308 N | 68 MPa | 104 MPa |

**Conclusions**

During the experiment we have obtained significantly different behaviour of the two tested kind of prepregs. There was not only the expected earlier rupture of flax fibers, but also fundamentally different fracture mechanisms, delamination and the resulting deformation of the entire test rods. The significant influence of these elements was evaluated using numerical model, created in ACP preprocessor. Until a moment of the experimentally found maximal strength, were results of the carried model and experiment in a very good agreement. But without an additional function of sophisticated post-procesor, we are not able to recognize the maximal values and the force would still just increase. It means, if we would like to know the entire process of composite deformation is the classical nonlinear static model, used e.g. for conventional steel parts insufficient. It is necessary to change the actually used shell composite model to a solid model suplemented by an interphase, cohesive elements, failure criteria etc. which would be the next step in our future work.


## Acknowledgement

This publication was written at the Technical University of Liberec as part of the project "Innovation of technical systems structures with the use of composite materials" with the support of the Specific University Research Grant, as provided by the Ministry of Education, Youth and Sports of the Czech Republic in the year 2017.


## References


[1] P. Kulhavý, J. Petřík, P. Srb, P. Lepšík: Vibration response of composite structures, 54th Int. conference on Experimental Stress Analysis, University of West bohemia, (2016)

[2] P. Camanho, C. G. Davila, M. F. de Moura, Numerical Simulation of Mixed-Mode Progressive Delamination in Composite Materials, J. of Comp. Materials, Vol. 37, (2003)

[3] W. T. Wang, T. Y. Kam, Material characterization of laminated composite plates via static testing, Composite Structures 50/4 (2000) doi: 10.1016/S0263-8223(00)00112-4

[4] A. Azzam, W. Li, An experimental investigation on the three-point bending behavior of composite laminate, Global Conference on Polymer and Composite Materials, Materials Science and Engineering, Vol. 62 (2014)

[5] H. Ullah, A. R. Harland, V. Silberschmidt, Experimental and Numerical Analysis of Damage in Woven GFRP Composites Under Large-deflection Bending, Appl Compos Mater, Vol. 19, (2012)

[6] Y. Suganuma, H. Fukuda, Applicability of compression bending test to measure compressive failure strain, 16th conference on composite materials, Kyoto, Japan, (2007)

[7] J. M. Hodgkinson, Mechanical testing of advanced fibre composites, Woodhead Publishing, (2000)

[8] G. Gruber, S. Wartzack, Three-point bending analyses of short fiber reinforced thermoplastic: Comparison between simulation and test results. SAS Tech Journal, (2013)

[9] S. M. Chung, A. U. J. Yap, S. P. Chandra, and C. T. Lim, Flexural strength of dental composite restoratives: Comparison of biaxial and three-point bending test. J. Biomed. Mater. Res., (2004)

[10] Nasa technical note, D-7572 Analysis of the three-point-bend test for materials with unequal tension and compresion properties, (1974)

[11] Y. Zou, L. Tong, G. P. Steven, Vibration based model dependent damage identification and health monitoring for composite structures. Journal of Sound and vibration 230, (2000), 357-378, doi:10.1006/jsvi.1999.2624

[12] H. Ullah, A. R. Harland, V. Silberschmidt, Damage modelling in woven-fabric CFRP laminates under large-deflection bending, Computational Materials Science, Vol. 64, (2012)

[13] Kirstein AF, Woolley RM. Symmetrical bending of thin circular elastic plates on equally spaced point supports. Washington,DC: National Bureau of Standards; (1966), p 1–9.

[14] H. Sönnerlind, Comsol user manual, Comsol engineering, Sweden, 2013

[15] Mechanical Structural Nonlinearities - Advanced Contact, Training Material, ANSYS, Inc. Proprietary, (2010)

EAN 2017
55th conference on experimental stress analysis

# Experimental Analysis of Efficiency of Profiled Dimpled Membranes in Terms of Wall Moisture Reduction

HÁJKOVÁ Eva[1, a]

[1]Department of Building Structures, Faculty of Civil Engineering, Czech Technical University in Prague, Czech Republic

[a]eva.hajkova@fsv.cvut.cz

**Keywords:** profiled dimpled membrane, moisture, masonry, efficiency, rehabilitation.

**Abstract.** One of the major problems in the reconstruction of existing buildings is an additional protection of substructure against the negative effects associated with increased moisture levels in building structures. Currently, one of the most frequently rehabilitation method is the use of profiled dimpled membranes. The efficiency of this rehabilitation method has not yet been precisely determined. This rehabilitation system is mostly designed based on practical experience and the effect of the proposed measures on the rehabilitated structure may be insufficient. This paper analyses the efficiency of the dimpled membrane in terms of its ability to reduce moisture in the structure compared to a reference model and theoretical calculation models.

## Introduction

One of the most important measures ensuring structural lifetime is adequate protection of substructure from negative effects associated with increased moisture in building structures. Increased moisture levels result in changes in mechanical and physical properties [1] (e.g., modulus of elasticity and strength), which may lead to reduced load carrying capacity of masonry structures [2]. There are many types of rehabilitation methods for adequate protection of buildings from effects of groundwater and water rising into walls from below the foundations [3]. One of the most common and most affordable methods is the use of profiled dimpled membranes.

A correct design of an appropriate rehabilitation method requires the efficiency of the proposed measure (including all of its weaknesses), reliability and impacts of marginal conditions on the resulting waterproofing effect. These rehabilitation systems are currently mostly designed based on practical experience. Their efficiency is usually only verified using results originating from field surveys. The disadvantage of these field surveys is the fact that moisture measurement can be done destructively (exactly) before the rehabilitation intervention, which is no longer possible once the rehabilitation measure is taken. Therefore, objective comparison of the resulting measured moisture levels before and after the rehabilitation intervention is impossible. For these reasons, experimental analysis using a model under laboratory conditions is beneficial. This paper analyses a structure that uses a profiled dimpled membrane as a rehabilitation measure against water and moisture. The objective of the experiment is to objectively determine the efficiency of the dimpled membrane in terms of its ability to reduce moisture in the structure compared to a reference model and theoretical calculation models (which can be fed with exact parameters). The advantage of these laboratory tests is the direct comparability under identical conditions (reference model vs. model with rehabilitation method applied) and the ability to carry out destructive tests (more exact measurement).

### Profiled Dimpled Membrane

The principle of this rehabilitation method consists in prevention of direct contact of moist earth with the building structure. The dimples in the membrane produce an air gap between the structure and the adjacent earth. The system of air cavities enables ventilation (and evaporation) of water vapour from adjacent building structures [4]. The dimpled membrane has a number of uses in both new construction, where it is used primarily as drainage and mechanical protection for the foundation structure, and renovation projects, where it is used primarily for ventilation and moisture reduction of existing building structures. The use of dimpled membrane for solving the problem of rehabilitation of moist walls has the advantages of very quick installation and monetary savings. However, the actual effect of dimpled membrane on the rehabilitated structure has not been determined exactly.

### Experimental Analysis

For the purposes of objective determination of the efficiency of dimpled membrane, we made two laboratory models on a scale of 1 : 2 (Fig. 1). These models simulated a segment of perimeter brick wall of a building with a cellar. Profiled dimpled membrane was applied to one of the models as a protection against water rising into the wall from below the foundation; the other model (reference) was left without any rehabilitation measures.

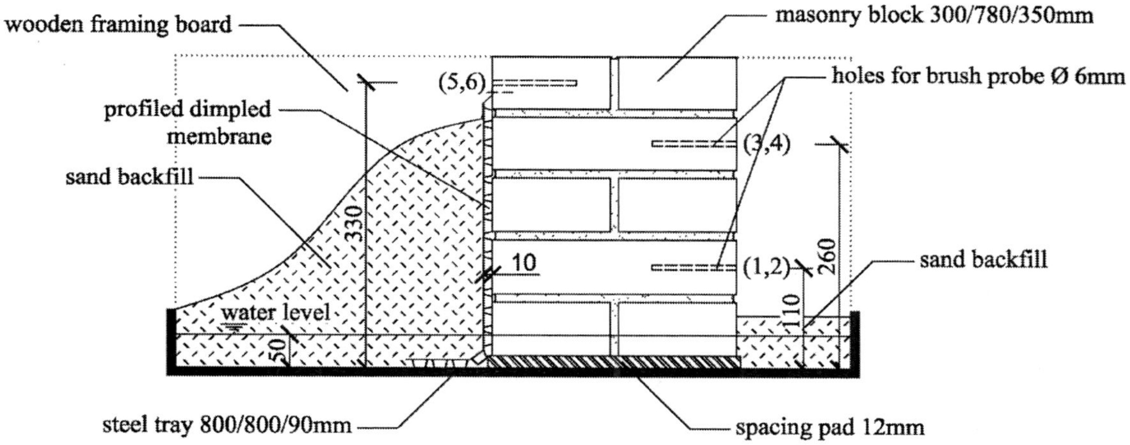


Fig. 1 Scheme of a laboratory model with profiled dimpled membrane in a scale of 1:2

The test models were bricked in steel trays, in which a constant water level was maintained throughout the experiment duration. The laboratory models were built of solid fired bricks with dimensions of 290 x 140 x 65 mm. Lime mortar was used as the binder. To achieve efficient saturation of the brickwork with water, the masonry blocks were built on spacing pads 12 mm high. The moisture measurements were made using a resistance moisture meter. That was why holes for the device brush electrodes 6 mm in diameter had to be drilled in the brickwork. Each masonry model contained 6 measuring points. Each measuring point consisted of two holes for electrodes. The location of measuring points can be seen in Figure 2. The brush probes up to 100 mm in depth were used for the measurement. The length of the probes was sufficient in this case. It is assumed that the highest moisture values due to capillary attraction are achieved at the centre of the brickwork [5]. To be able to compare the results obtained from the laboratory measurements and the results of the numerical simulation, the brickwork segment model had to comply with the parameters of the 2D problem. For this reason, the masonry blocks were wrapped in impermeable film on the lateral sides and additionally were covered with wooden boards. The boards were used as framing for the sand backfill, simulating the surrounding earth adjacent to the brickwork.

Fig. 2 Designation of measuring points on the laboratory model

**Measuring Devices.** The moisture measurements of the brickwork used an advanced Greisinger GMH 3810 electrical resistivity moisture meter. After the experiment was completed, the gravimetric method was employed to verify the results measured with the resistance device. Control samples were collected from the points where the measuring end of the brush probe had been located during the measurement.

**Laboratory Testing.** In the first phase of the experiment, the models were left in the laboratories for a period of 3 months, during which the moisture of the brickwork was not measured. In this period, the masonry structure was saturated with rising water. Moisture measurements of brickwork were made in each measuring point during the next ten months. The measuring interval was 1 month. The bulk moisture of masonry materials is also influenced by the relative humidity of the air in the area where they are located. Therefore, the relative air humidity and the air temperature were measured in the laboratory. Both models were in the same environment, exposed to identical conditions. The influence of the salts was not considered in this laboratory experiment. Salts may negatively affect the masonry surface structure in the course of evaporation of salt-containing water.

**Results and Discussion.** The efficiency of the dimpled membrane type used was determined based on the moisture readings. The efficiency was calculated based on differences between the moisture readings at individual points of both models and is expressed as a percentage reduction in the mass moisture in the brick block with the dimpled membrane compared to the reference model. The following charts show the values of the moisture readings w [% weight] at individual measuring points during a period of 10 months. Thus, the charts show the rate of moisture decrease in the individual measuring points (Fig. 3).

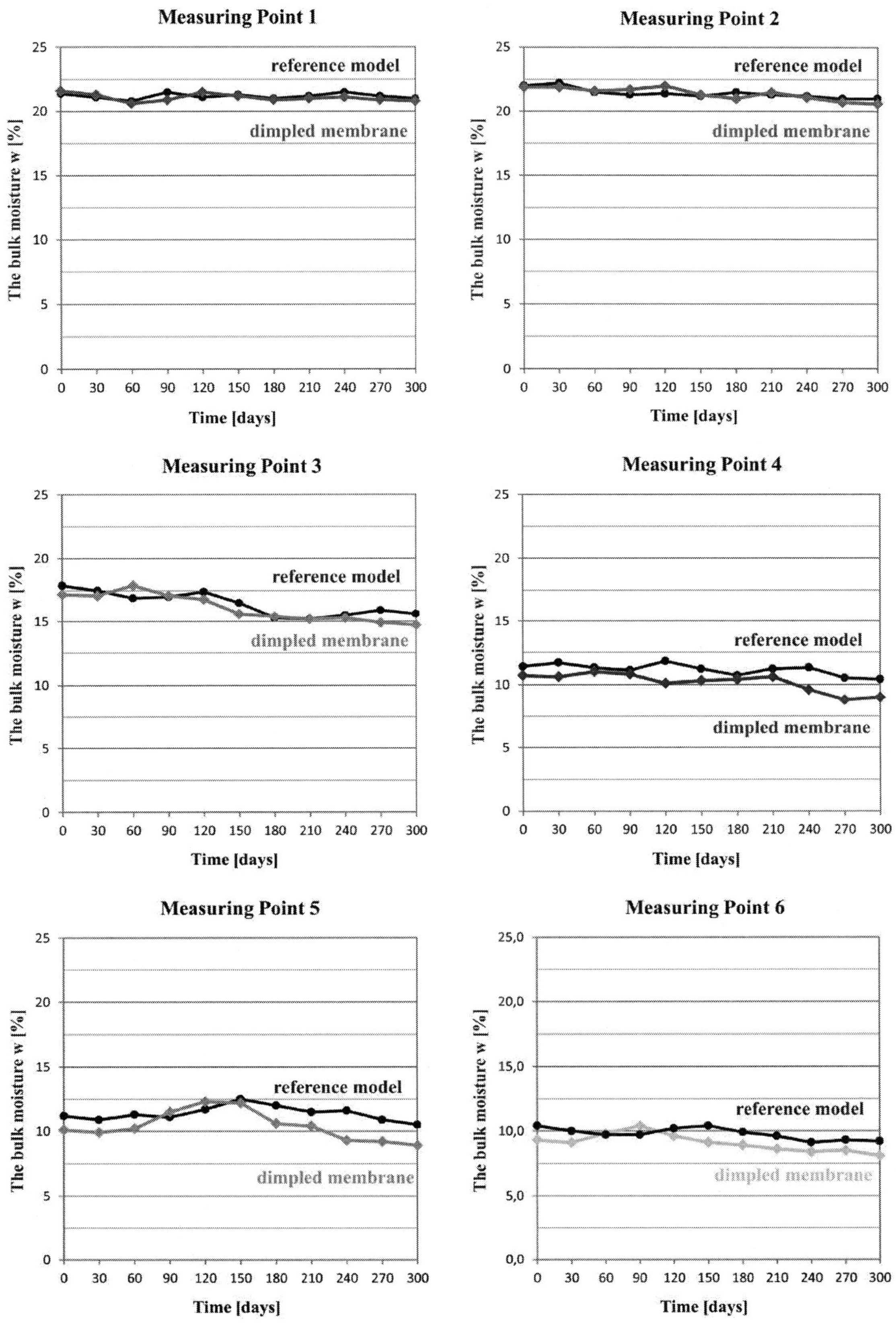


Fig. 3 The measured values of moisture in the individual measurement points

Based on the measured moisture values, it can be concluded that the efficiency of the dimpled membrane is very limited. Values of the moisture in the model with dimpled membrane are almost identical compared to the reference model. Depending on differences season, slight fluctuations of moisture values are evident (increase moisture values in the summer months). However, the bulk moisture of the brickwork does not decrease significantly.

The final bar charts summarize the average values of moisture measured w [% weight] at individual points for the reference model and the model with dimpled membrane. The first chart shows the initial moisture values in the masonry blocks after saturation with rising water for a period of 3 months (Fig. 4). The second chart shows the final values of moisture after 10 months from the beginning of measurement (Fig. 5). The results of laboratory testing showed that the moisture in the masonry had decreased considerably for both models due to the evaporation of moisture from the structure. The moisture in the wall with dimpled membrane was reduced by 8% on average compared to the reference model.

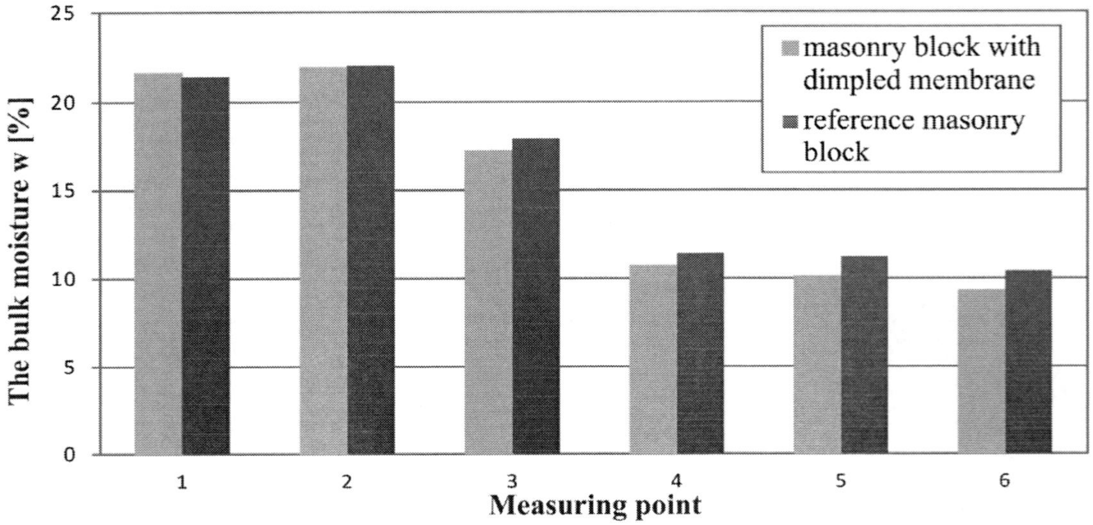


Fig. 4 The measured values of moisture at the beginning of measurement

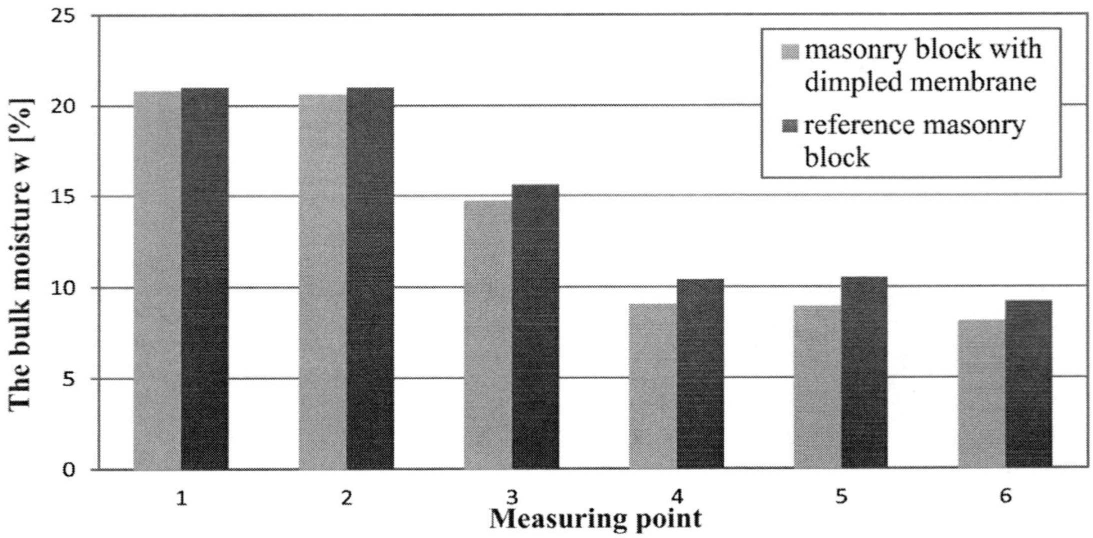


Fig. 5 The final values of moisture after 10 months from the beginning of measurement

## Numerical Simulation

**Simulation Program used and Validation.** The experimental study was afterwards verified using a numerical simulation made in the WUFI 2D software. This software was developed at the Fraunhofer Institute for Building Physics in Germany based on the thesis of H. M. Künzel [6]. WUFI 2D enables to create a model of dynamic two-dimensional heat and moisture transfer in structures and materials depending on the boundary conditions [7]. It is also necessary to know the thermal and moisture-related properties of the materials used. In this case, we used the thermal and moisture-related material properties determined experimentally in the laboratory study. Boundary conditions entering into the computation as e.g. the test duration, the relative air humidity and air temperature in the laboratory were the same as in the laboratory experiment. The numerical simulation was made for a reference model and a model with dimpled membrane.

**Simulation Results and Discussion.** The results of the numerical simulation confirmed the results obtained from the laboratory measurements. Figure 6 shows the total content of water in the masonry block with dimpled membrane and in the reference masonry block at the end of the simulation. Compared to the reference model, the total content of water in the model with dimpled membrane was fractionally lower. We can therefore conclude that the numerical simulation confirmed the behavior of test blocks determined by experimental measurement.

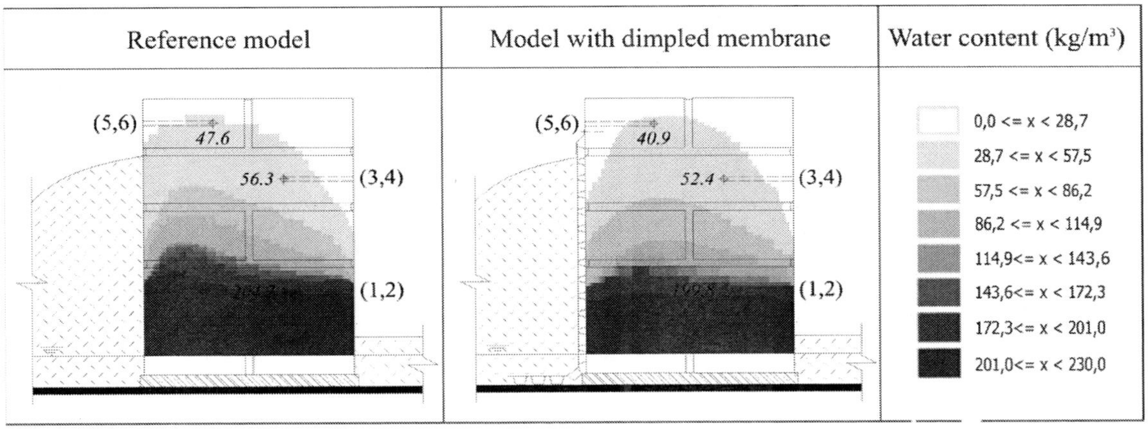


Fig. 6 The results of the numerical simulation – water content in reference model and model with dimpled membrane (created in WUFI 2D)

## Conclusion

Based on the results of the laboratory tests, it can be concluded that the efficiency of this measure is very limited. According to data from the manufacturers' data sheets, dimpled membrane is suitable for removing moisture from building walls and permits air circulation as well. The laboratory experiment demonstrated that the moisture in the wall with dimpled membrane was reduced by 8% on average compared to the reference model. Obviously, no air circulation occurs without provision of air inflow and exhaust. Dimpled membrane is only suitable as protection against water entering the walls from the sides and water that flows down freely. As can be seen from the results, it is not effective against water rising into the wall from below the foundations by itself; it has to be accompanied by an additional rehabilitation measure. The results of the numerical simulation showed agreement with the data obtained from the laboratory measurements: like in the laboratory experiment, the total content of water in the brick block with the dimpled membrane was lower. The partial results of the experiment verification will be presented in this article and at a conference.

**Acknowledgement**

The article was created with support from the research project no. SGS17/117/OHK1/2T/11 “Experimental verification of the reliability of selected systems for the protection of substructure against water”.

**References**

[1] D. D'Agostino, Moisture Dynamics in a Historical Masonry Structure: The Cathedral of Lecce (South Italy), Build Environ. 63 (2013) 122-133.

[2] J. Sykora, T. Krejci, J. Kruis and M. Sejnoha, Computational Homogenization of Non-Stationary Transport Processes in Masonry Structures, J Comput Appl Math 236(18) (2012) 45-55.

[3] E. Grytli, L. Kværness, L. Rokseth and K. Ygre, The Impact of Energy Improvement Measures on Heritage Buildings, Journal of Architectural Conservation 18(3) (2012) 89-106.

[4] M. Balik et al, Odvlhčování staveb, second ed., Grada, Prague, 2008.

[5] V. P. de Freitas, V. Abrantes and P. Crausse, Moisture Migration in Building Walls - Analysis of the Interface Phenomena, Build Environ. 31 (2) (1996) 99-108.

[6] H. M. Künzel, Simultaneous Heat and Moisture Transport in Building Components. One- and Two-dimensional Calculation using Simple Parameters, PhD Thesis, Fraunhofer Institute of Building Physics, Stuttgart, 1995.

[7] D. de Vries, The Theory of Heat and Moisture Transfer in Porous Media Revisited, Journal of Heat and Mass Transfer 30 (7) (1987) 1343-50.

EAN 2017
55th conference on experimental stress analysis

# Experimental Verification of the Demountable Structure Prototype

POLÁK Aleš[1, a]

[1]Department of Building Structures, Faculty of Civil Engineering, Czech Technical University in Prague, Czech Republic

[a]ales.polak@fsv.cvut.cz

**Keywords:** precast system, demountable joints, experimental verification.

**Abstract.** Demountable steel joints allow easy assembly and disassembly of precast reinforced concrete elements. Traditional prefabricated joints use grout mortar and steel reinforcement to achieve the required rigidity of the structure. Demountable structures are characterized by discrete joining of individual elements through contact, usually articulated, joints. This type of joints brings different demands for ensuring of the structures rigidity. This paper is focused on experimental verification of the demountable structure's prototype and its demountable steel joints. Both static and dynamic load tests were carried out on this prototype.

## Introduction

The negative impact of reinforced concrete structures on the environment, a high amount of energy associated with their production and carbon dioxide released into the atmosphere can significantly eliminated by frequent use of demountable structures. Demountable structures consume only a fraction of energy during disassembly compared to the energy required for demolition. In addition, the individual elements of demountable structures can be reused. This method of construction is compared with conventional concrete structures very environmentally friendly and saves natural resources [1].

The goal of the project TA02010837 was research and development of a new multipurpose demountable prefabricated reinforced concrete building system with controlled joints' properties and possibility of repeated use. The demountable structure prototype was developed under this project.

## Prototype of the Demountable Structure

The prototype was designed for experimental testing the feasibility of the system and verifying the functional properties of individual demountable joints.

The proposed arrangement of the load-bearing structure included characteristic parts of the column system, column system with embedded flexible diaphragms, pillar system and combination systems within the so-called integrated systems, the system of spatial units, basic types of precast reinforced concrete elements and characteristics bearing joints.

The prototype was designed as a two-storey building with a usable area 160 m$^2$. The realization itself was divided into two phases. The structure of a rectangular shape with plan dimension 12.75 x 6.25 m with the same structural height of both floors 3.3 m was conducted in the first phase. The experimental verification was carried out on this structure, after that the second realization phase followed.

The ceiling structure was formed hollow core prestressed ceiling panels Partek thickness of 200 mm. Ceiling panels were placed on the girders and walls via steel pins. These pins were screwed into the embedded steel anchor elements in the hollows of the panels during assembly. Anchor elements were inserted into the freshly cut ceiling panels and concreted by self-compacting concrete through the preformed hole in the upper face of the panel [2].

## Experimental Verification

The aim of the experiment was to verify properties of the ceiling plate composed of prestressed ceiling panels. The structure was loaded with a horizontal force to verify critical issues related to spatial rigidity of the system with demountable joints.

Fig. 1 Horizontal static loading in transverse (left) and longitudinal direction (right)

The static test examined the response of the experimental prototype (superstructure) on the horizontal load acting in the longitudinal and transverse directions. The loading was applied via hydraulic press at the level of second storey ceiling. Supporting wooden structure allowed imposing the hydraulic press at the required height. The static loading was applied by means hydraulic press, a hand pump with manometer and a valve. The loading with a horizontal force was applied into the structure by the hydraulic press in 9 kN steps. The maximum force reached the value of 54 kN.

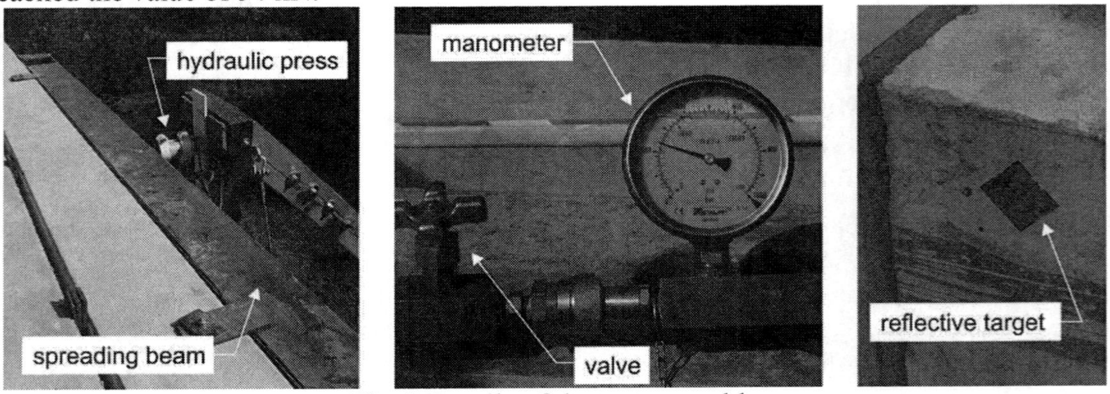

Fig. 2 Details of the test assembly

Geodetic measurements method was chosen for measuring deformations. This method (in contrast to the other methods of measuring deformation and strain) did not require complicated auxiliary structure for the installation and monitoring of measuring sensors. Deformation measurement were carried out by geodetic total station Nikon DTM 430. The mean error of changing position of the monitored point was less than 0.25 mm. Four points at the top edge of the object were chosen and fitted with reflective targets for monitoring horizontal displacement of the load-bearing structure.

Verification of the prototype structure to the effects of dynamic loading was carried out between individual loading cycles. This loading was induced by dynamic vibration generator positioned at the top center of the ceiling slab. Six accelerometers with high sensitivity was used to measure the response of the structure. Accelerometers Wilcoxon Research Model 731A with high sensitivity of 10 V / g were placed according to the scheme in Fig. 3.

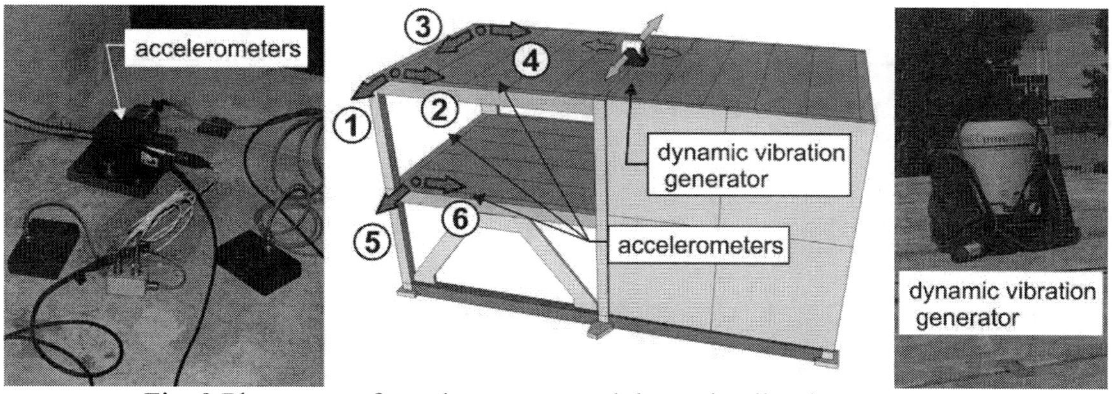

Fig. 3 Placement of accelerometers and dynamic vibration generator

Accelerometers measure the vibrations reliably from -10 ° C to + 65 ° C, and have an integrated low-pass filter, which can eliminate the high frequencies resulting from changes in acoustic pressure in the area. Accelerometers are connected to the sixteen-channel analog computer Dewetron DEWE 43 with a 24-bit AD converter and simultaneous sampling of 200 kS/s. Input voltage can be set for four ranges of ± 10V, ± 1V, ± 100 mV and ± 10mV. One accelerometer was placed on the movable part of the dynamic vibration generator and measured movement of the mass. The obtained data were evaluated by own procedures in MATLAB.

## Measured Values

Resonant frequencies can be read from the spectral densities (Fig. 4) taken from the Fourier transform of the autocorrelation functions obtained from the structure excited by pulsed load –rubber hammer blow on the structure in the direction of the sensors. The hammer blow was conducted for the transverse and longitudinal direction.

The identification test was followed by the horizontal cyclic load (in longitudinal and transverse direction), that was applied between individual static load cases. The resonant frequencies of the concrete structure and the loading parameters are shown in Table 1, together with the number of counted oscillations during 1 oscillation hour.

| Loading cycle | Direction of loading | Mobile mass [kg] | Oscillation frequency [Hz] | Number of oscillation | Dynamic force [kN] |
|---|---|---|---|---|---|
| 1 | Longitudinal | 13.2 | 29.14 | 105 000 | 4.926 |
| 2 | Longitudinal | 13.2 | 29.14 | 110 000 | 4.873 |
| 3 | Transverse | 13.2 | 30.21 | 108 800 | 5.079 |
| 4 | Transverse | 13.2 | 30.21 | 109 100 | 5.143 |

Table 1 Resonant frequencies with the appropriate parameters for the structure excitation

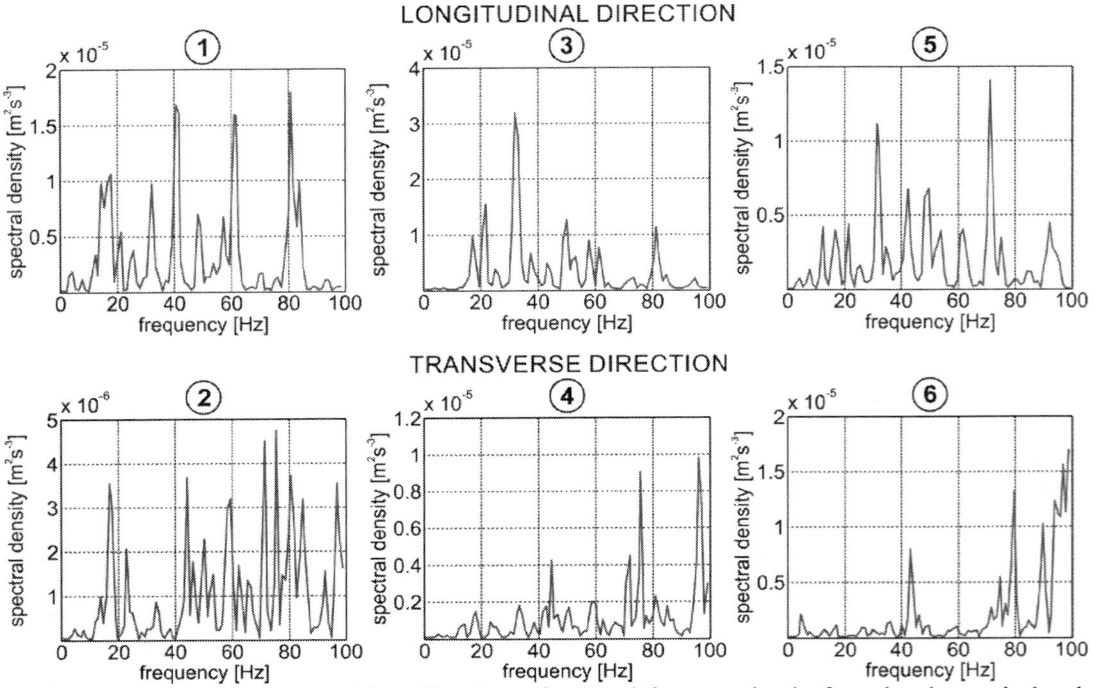


Fig. 4 Response spectrum identification of natural frequencies before the dynamic loads

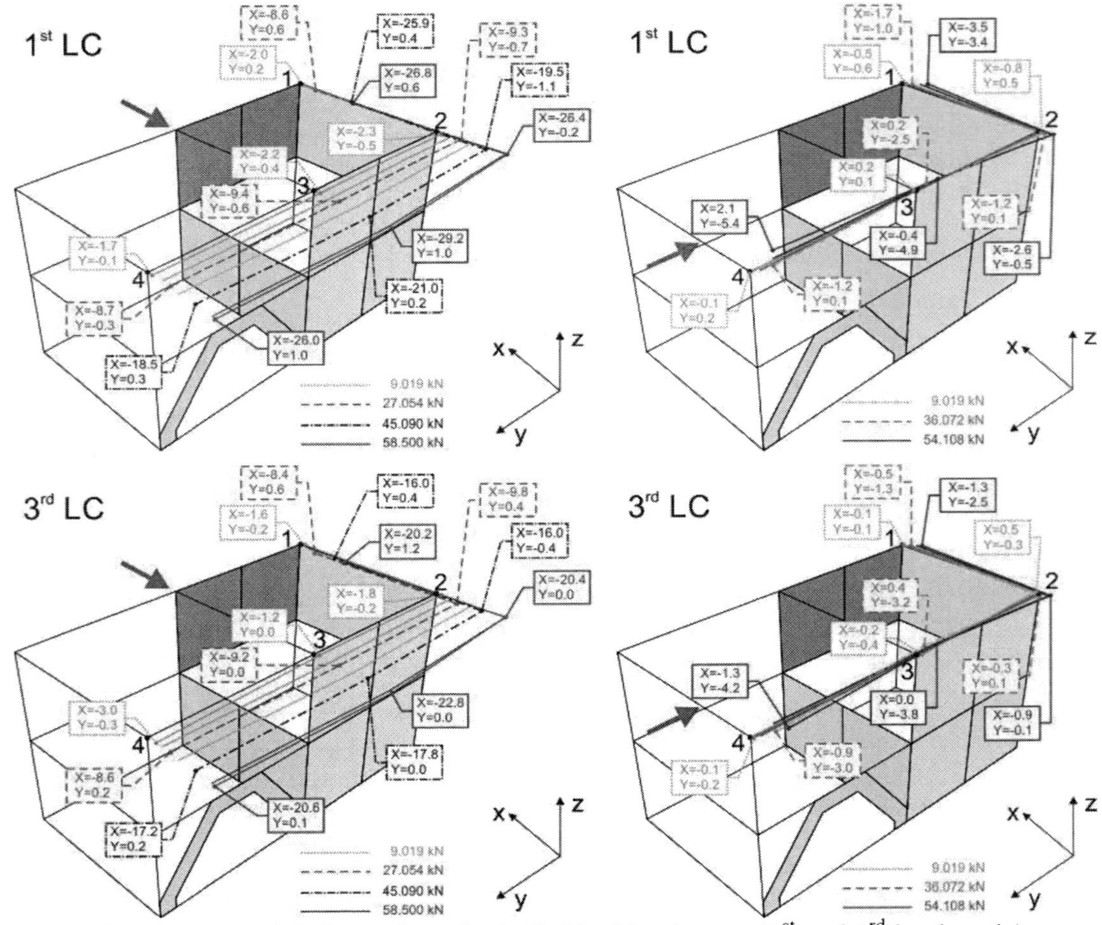


Fig. 5 Measured deformations for individual load steps (1st and 3rd load cycle)

Different values of deformation between the static load cases were caused by abutment of individual steel joints, dynamic load and tolerances of individual joints and prefabricated elements. Table 2 shows the resulting experimentally determined deformations' values of the ceiling slab. Percentage values are relative to the dimensions of the ceiling board in the second direction.

| Direction of loading | Loading cycle | Deformation of flatness | |
|---|---|---|---|
| | | [mm] | [%] |
| Transverse | 1 | 2.7 | 0.020 |
| Transverse | 2 | 1.0 | 0.012 |
| Transverse | 3 | 1.3 | 0.011 |
| Longitudinal | 1 | 2.0 | 0.030 |
| Longitudinal | 2 | 1.5 | 0.025 |
| Longitudinal | 3 | 1.65 | 0.027 |

Table 2 Experimentally determined deformations of the ceiling slab

## Conclusion

The analysis shows that experimentally determined deformations' values of the ceiling slab when the maximum horizontal force (acting in the longitudinal or transverse direction) reached 54 kN are in the range from 0.011 to 0.030%. Based on these percentage expressed deformations' values of ceiling slab can be stated that the prefabricated concrete slab with a discrete steel joints meets the requirements of a rigid horizontal beam and ensures spreading of the effects of the horizontal load on the individual vertical support elements in proportion to their stiffness.

## Acknowledgement

The paper was written with support from TA02010837 "Multipurpose dismantleable prefabricated reinforced concrete building system with controlled joint properties and possibility of repeated use". Lead researcher of the project is prof. Ing. Jiří Witzany, DrSc. Ing. Radek Zigler, Ph.D., doc. Ing. Tomáš Čejka, Ph.D., Ing. Jiří Karas, CSc., Ing. Pavel Kokeš and Ing. Aneta Maroušková contribute significantly to this project.

## References

[1] J. Witzany, T. Čejka, R. Zigler, A Precast Reinforced Concrete System with Controlled Dynamic Properties, in: ISEC Press, Sustainable Solutions in Structural Engineering and Construction, Kasetsart University, Bangkok, 2014, pp. 227–232.

[2] J. Witzany et al., A Demountable Precast Reinforced Concrete Building System of Multi-Story Buildings, in proc.: Implementing Innovative Ideas in Structural Engineering and Project Management, Fargo: ISEC Press, Sydney, 2015, pp. 219–224.

# Validation of high-speed impact numerical simulations on flat test panels for real structural analysis

Radek Doubrava[a], Martin Oberthor[b] and Petr Bělský[c]

[1]Výzkumný a zkušební letecký ústav, a.s., Beranových 130, Prague - Letnany, Czech Republic

[a]doubrava@vzlu.cz, [b]oberthor@vzlu.cz, [c]belsky@vzlu.cz

**Keywords:** hail strike, FE, SPH

**Abstract.** This paper aims to integrate numerical analyses with high-speed impact tests on flat panels for application in real structural design.
The work described in this paper and the research leading to these results was funded through the European Community's Seventh Framework Programme FP7/2007-2013 under grant agreement n° 604013, AFLONEXT project.

## Introduction

Innovative designs of aircraft structures must be validated to meet airworthiness requirements (e.g., [1]). The most commonly used design approach for such structures involves making simulations based on numerical simulations, which require the use of reliable validation techniques, especially when using anisotropic materials, such as fibre-reinforced composite.

## Proposed technique

The technique involves validation of FE models based on tests of a simplified structure. The simplification involves the use of flat test panels that ensure:

- Fast automatic meshing with guaranteed element quality
- Application of real boundary conditions according test conditions
- Rapid modification of FE model parameters
- Ease of test specimen production (flat panels)
- Comparability between test results

The following measurement and investigated parameters are used as a standard for high-speed impact:

- High-speed cameras
- NDT before and after impact
- Reaction force measurements

Figure 1 shows a schema of the proposed technique.

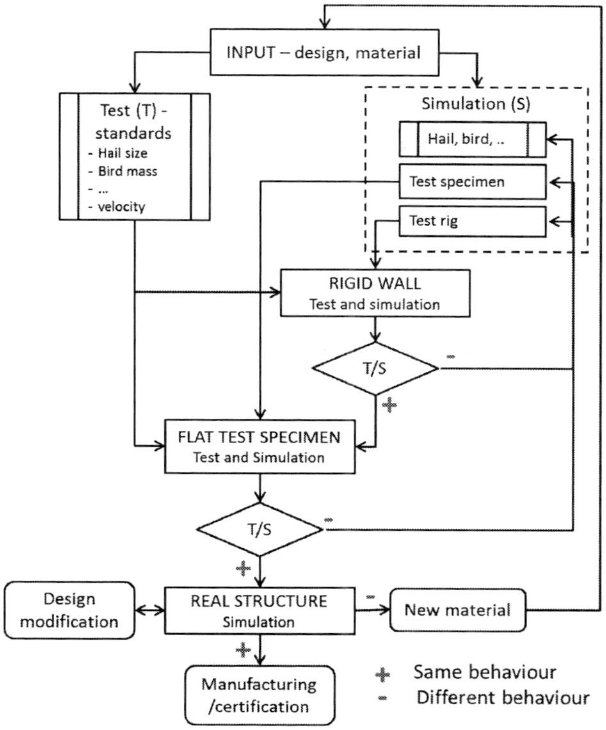


Fig. 1 – Proposed technique for the validation of numerical simulations of high-speed impact

## Hail strike tests

The tests were performed in VZLÚ (Aerospace Research and Test Institute) according ASTM standards [2].
Tests were carried out on a flat composite plate that was representative of Krueger flap skin (see figure 2 - left). The dimensions of the test specimen are shown in figure 2 (right).

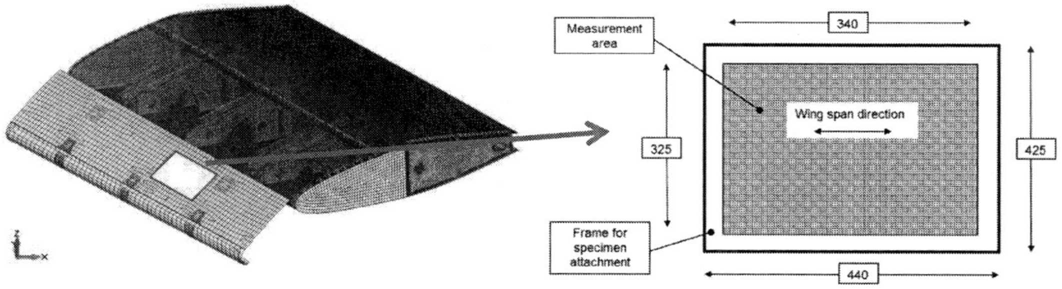


Fig. 2 – Global FE model of the demonstrator with a composite Krueger flap (left) and the proposed flat test specimen for hail strike tests (right)

The flat test specimens were manufactured based on a flap design to transfer the results from the flat test specimen on the real structure through the analysis. The test specimens were manufactured in Polytechnic Warsaw (see figure 3). Two types of composite test specimens were used:

- Skin with 10-layered CFRP composites (test specimen No. 1, 4, and 5)
- Sandwich with two 10-layered CFRP composites and 30 mm thick foam (test specimen No. 2 and 3)

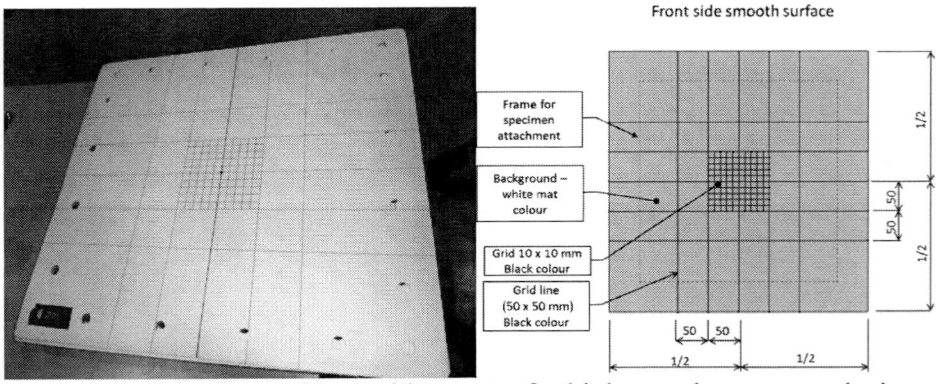


Fig. 3 – Flat test specimen with a raster for high-speed camera analysis

Figure 4 shows the hail strike test arrangement at the VZLU test hall.

Fig. 4 – Hail strike test arrangement at VZLU

Two high-speed video cameras were used to record the hail impact. One camera was positioned in front of the test specimen, and one camera was positioned from behind the test specimen.
An impactor with a 25.4-mm ice ball diameter was used for the hail strike tests (see figure 5).

Fig. 5 – Rubber mould and ice ball used for the hail strike tests

The ice ball was accelerated through an air gun with a smooth borehole barrel up to the specified velocity using expanding compressed air (see Figure 6).

Fig. 6 – Air gun with pressure vessel and loading chamber

A high-speed camera was used to make impact speed measurements.

**Test results**

The test results were analysed based on:

- High-speed camera pictures
- Plastic deformation and damage patterns
- Reaction forces drawn from load cell measurements

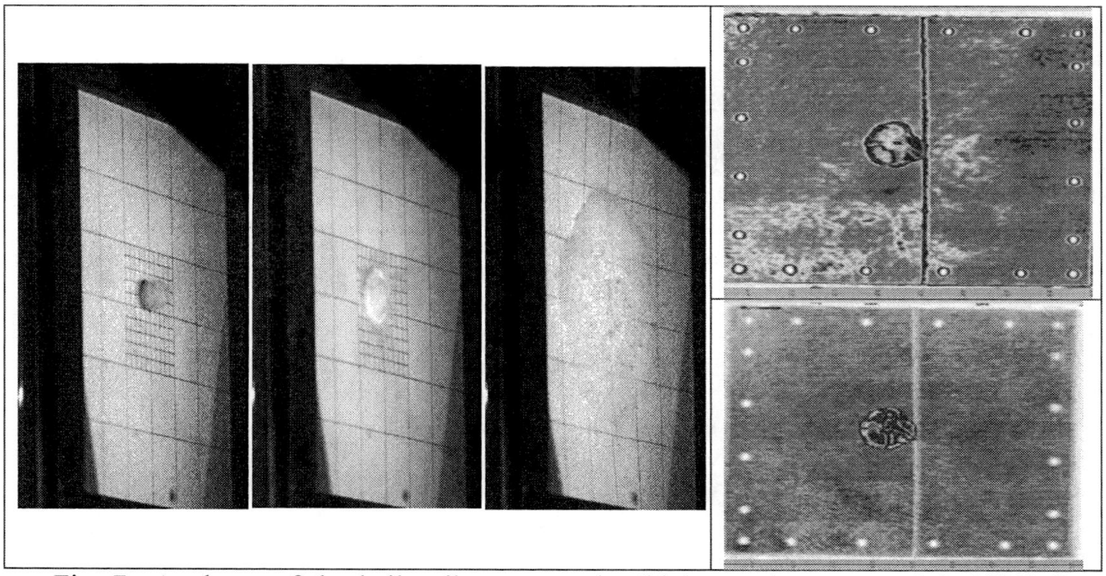

Fig. 7– Analyses of the hail-strike test results (high-speed camera – left, NDT C-scan – right)

Figure 8 shows damage measurements drawn from the C-scan results. The perpendicular hail strike of 640 km/h created an approximately 70-mm dent.

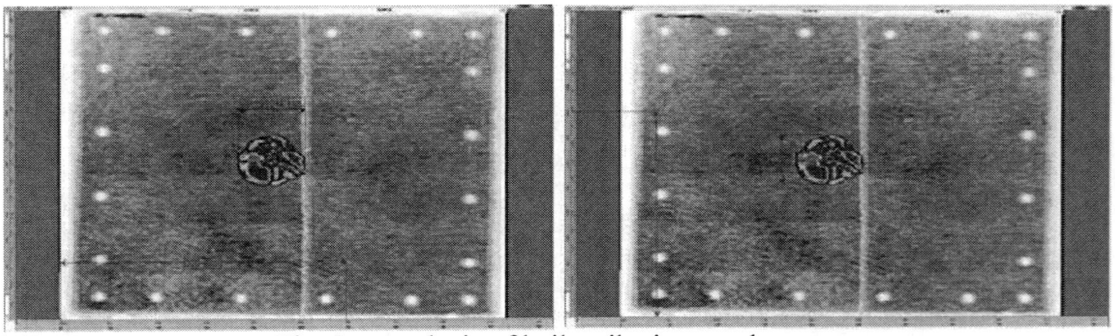

Fig. 8 – Analysis of hail-strike impact damage

Table 1 presents a summary of the test measurement results.

| test No. | test specimen | Hail diameter | impact speed | cameras measurement | | | NDT C-scan | | |
|---|---|---|---|---|---|---|---|---|---|
| | | | | front | speed measurement | back | horizintal size | vertical size | average |
| | | mm | km/h | fps | | | mm | | |
| 1 | BSP01 | 25.4 | 545 | 8000 | 3000 | 500 | N | N | N |
| 2 | BSPS01 | | 540 | 10000 | 3000 | 500 | N | N | N |
| 3 | BSPS02 | | 600 | 10000 | 3000 | 500 | 55.3 | 52.1 | 53.7 |
| 4 | BSP02 | | 525 | 10000 | 3000 | 500 | N | N | N |
| 5 | BSP03 | | 640 | 10000 | 3000 | 500 | 70.4 | 72.3 | 71.4 |

Tab.1 – Summary of the hail strike test results (N - without damage)

The reaction force was analysed from load cells measurements during the impact test. Measurements were made on 3 load cells between the moving part of the test rig and frame connected to the ground.

**Simulation**

The FE simulation was performed using the ABAQUS FE software program. An explicit solver with double precision was used for the analysis. The aim of this analysis was to tune unknown inputs and computational parameters to obtain the same behaviours between the test and simulation, i.e., energy absorption and damage. The same set up was then used for analysis of the entire structure.

In terms of the impactor's (hail) material properties and features, ice is a complex material presenting high degrees of variability. The material density of ice and hail varies and is subject to a range of factors, such as the weather systems through which it was created. Standard ice has a density of 917 kg/m3; while this value slightly increases as temperature declines, it never reaches the density of liquid water.

Ice may exhibit two forms of non-elastic behaviour under stress. Under low-velocity deformation, ice exhibits ductile behaviour and yet as velocity increases, the material become brittle.

The method used to adjust hail strike simulations is certified by the Czech Republic Civil Aviation Authority [3]. Figure 10 shows the geometry and FE model of the impactor.

Fig. 10 – Geometry, FE mesh and SPH elements of a ball representing the hail impactor

For the simulation of the composite structure, we used 4-node shell elements (S4R) of different mesh sizes. For the damage analysis of the composite material, Hashin's damage material model was used.

Figure 11 compares the test results (high-speed camera) with the numerical simulation of hail striking a flat composite panel representing a lay-up of a composite flap.

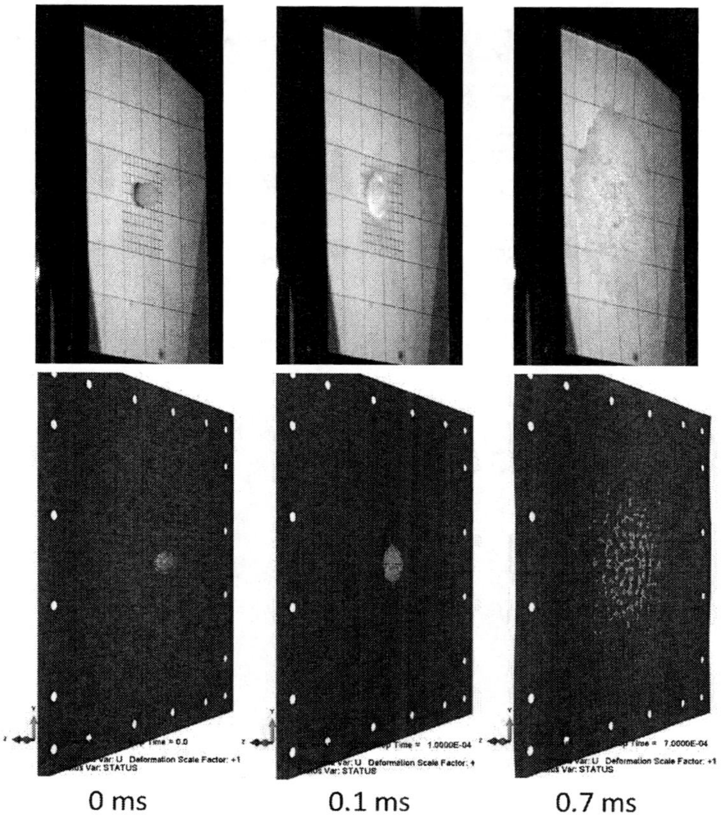


Fig. 11 – Comparison of impactor hail damage between the hail strike test and simulation

Four different impact area FE mesh sizes were used for the influence analysis. Figure 12 shows the results of this analysis.

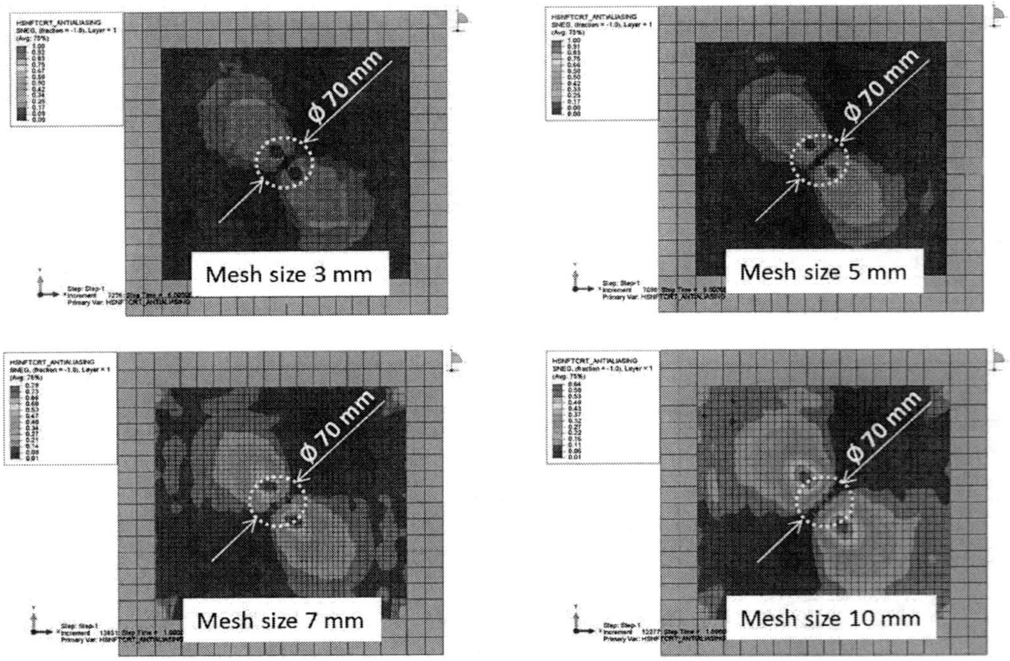

Fig. 12 – Analysis of the FE mesh size effects on damage predictions

Figure 13 compares the test and simulation results in terms of reaction forces during impact.

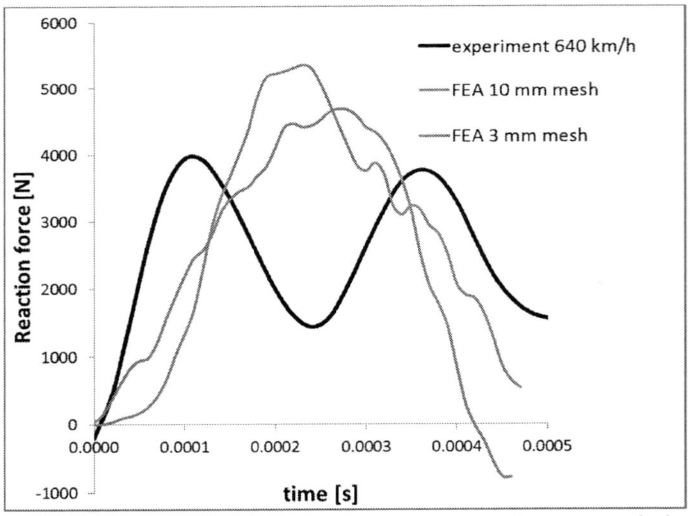

Fig. 13 – Comparison of reaction forces between the test and simulation

The peaks of the test reaction forces (figure 13) are influenced by the load cell scanning frequencies.

The verified simulation of 1" hail strikes (25.4 mm) was used to determine the airworthiness required (e.g., EASA airworthiness CS E 790 [1]) for 2" hail (50.8 mm). Figure 14 compares the above mentioned hail size with the estimation of speed limits for the defined composite material and lay-up for perpendicular impact.

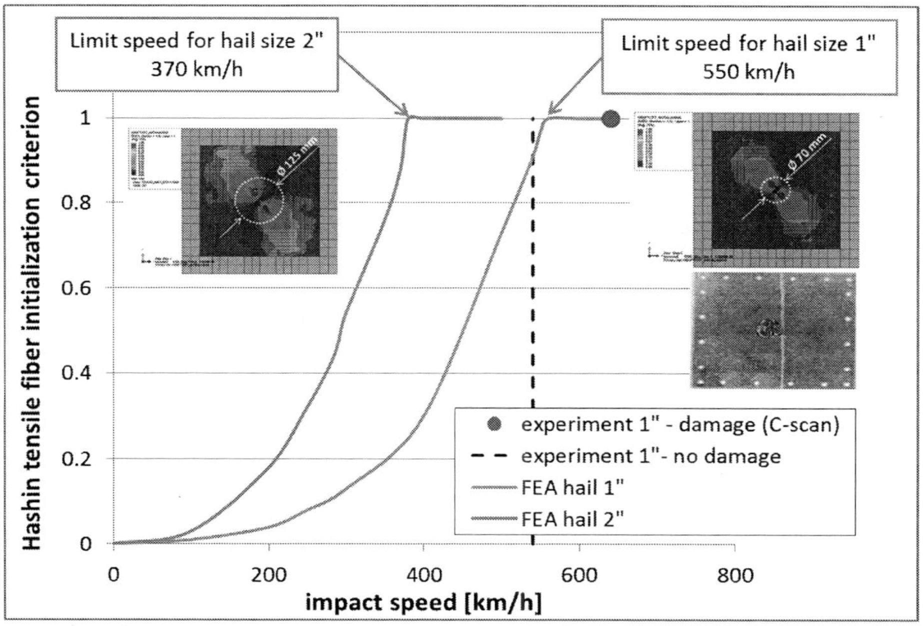


Fig. 14 - Estimation of the impact resistance of 10 layered CFRP composites for the perpendicular impact of 1" and 2" diameter hail.

## Conclusions

A comparison between the test and simulation results reveals consistency in predictions of damage initialization and propagation.

These tests provide important information for design and computational analysis. Test and numerical model verification based on different impact speeds also provide important generalizations for the application and optimization of composite structures.

## References

[1] European Aviation Safety Agency, CS-E Certification Specifications for Engines, Amendment 3, 2010

[2] ASTM international, ASTM F320-05 Standard Test Method for Hail Impact Resistance of Aerospace Transparent Enclosures, 2005

[3] Doubrava, R.: SPH Technique Application for Fast Impact Simulation, report R-5841, VZLU, Prague, 2013

EAN 2017
55th conference on experimental stress analysis

# The Vibrating Bowl Feeder Dynamic Model

M. Sturm [1a], L. Pešík[2b]

[1]Faculty of Mechanical Engineering, University of Applied Sciences Zittau/Görlitz, Germany

[2]Faculty of Mechanical Engineering, Technical University of Liberec, Czech Republic

[a]m.sturm@hszg.de

[b]lubomir.pesik@tul.cz

**Keywords:** Vibrating conveyor, bowl feeder, resonance, dynamic model,echanical parameters.

**Abstract:** Vibrating conveyors also named bowl feeders are a common equipment for conveying goods into production systems. These systems are used for the supply of a certain number of goods to an individual designed inter-face and simultaneously arranging a correct orientation of the goods for the following manufacturing step conveyed by the same time. This type of conveyors are used in various industries, such as automotive industry, electronic industry and medical industry. The target of this article is to determine a dynamic model and mechanical parameters by means of testing, and a numerical simulation of a ready-to-operate conveyor under standard working conditions.

## Introduction

The bowl feeder, which is used for determination of the dynamic model and mechanical parameters for future testing and numerical simulations, originally has been built to forward parts used in the automotive industry.

**Fig. 1** Bowl feeder for testing

The technical specification of the bowl feeder was as follows in Tab. 1 and on Fig. 2:

**Tab. 1** Parameters of the bowl feeder

| Top Diameter | Bowl hight | Distance bottom to outlet | Foot diameter |
|---|---|---|---|
| 700 mm | 450 mm | 400 mm | 535 mm |
| **Spring length** | **Total mass** | **Capacity** | **Moving mass** |
| 130 mm | 123 kg | 12 kg | 33 kg |

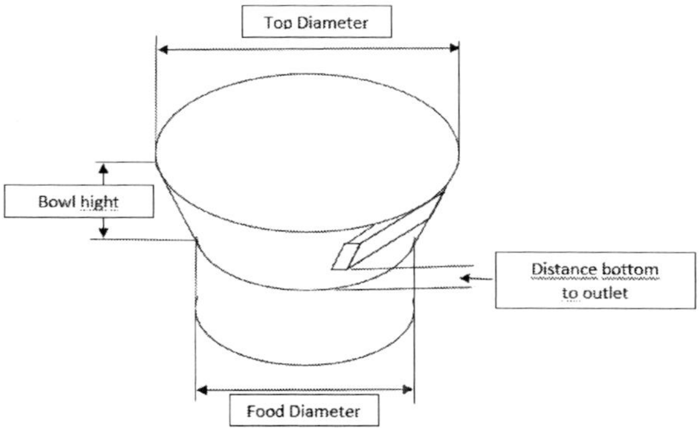


**Fig. 2** Parameters of the bowl feeder

The spring material of the bowl feeder is a fiber-reinforced synthetic material. Each of the three springs (flat spring packages) is installed under 15° to a vertical boundary line. Each package consists of 8 flat spring elements, screwed together using 1mm thick distance discs. The excitation device consists of three electromagnetic vibration exciters. Each exciter is situated close to a spring package (Fig. 3).

**Fig. 3** Springs package with excitation device

## 1. Determination of the bowl feeder dynamic model

The assembly of a bowl feeder includes two principal parts, the conveying element with a conical form and a supporting cylindrical element. The conveying element is indexed by character $N$ and the supporting element by character $S$. Both parts are connected by springs, which are uniformly distributed around their rotational axes $y_N$ and $y_S$. In this case, three flat springs packages are assembled. Each package is declined by angle $\beta_{NS}$ to the axis plane, it has its stiffness and the total values of all springs packages are defined as $k_{NS}$. The total damping coefficient is described by value $b_{NS}$.

The conveying element is defined by its mass $m_N$ and by its moment of inertia $J_{Ny}$ around rotational axe $y$. The supporting element has the mass $m_S$, the moment of inertia $J_{Sy}$, and it is placed on rubber springs with total stiffness $k_{SRx}$ and $k_{SRy}$, and the total damping coefficients $b_{SRx}$ and $b_{SRy}$. The whole system is fixed by a rigid connection to the frame.

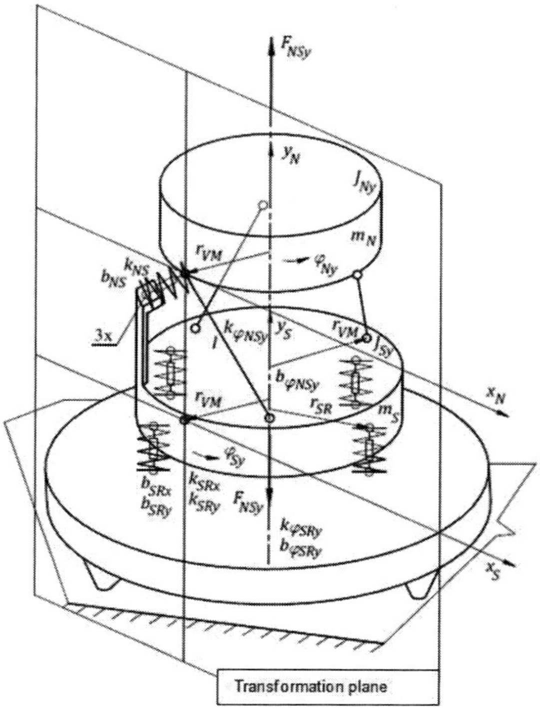


**Fig. 5** Mechanical model of the bowl feeder with transformation plane

For the dynamic model of the system, each flat springs package is replaced by a rod and a spring with a perpendicular effect. All three rods creating a joint mechanism between two parts of the bowl feeder.

The motion of the bowl feeder system can be described by the help of four coordinates, two translating motions $y_N$, $y_S$ and two rotating motions $\varphi_{Ny}$, $\varphi_{Sy}$. The relative motion of both parts is defined:

$$y_{NS} = y_N - y_S \tag{1}$$

and

$$\varphi_{NSy} = \varphi_{Ny} - \varphi_{Sy} \, . \tag{2}$$

Based on these intentions, four differential equations describing the movement of the bowl feeder system can be formed:

$$m_N \ddot{y}_N + F_{bNSy} + F_{kNSy} + F_{VM} \cos \beta_{NS} = F_{NSy} \, , \tag{3}$$

$$J_{Ny} \ddot{\varphi}_{Ny} + M_{bNSy} + M_{kNSy} - r_{VM} F_{VM} \sin \beta_{NS} = 0 \, , \tag{4}$$

$$m_S \ddot{y}_S + b_{SRy} \dot{y}_S - F_{bNSy} + k_{SRy} y_S - F_{kNSy} - -F_{VM} \cos \beta_{NS} = -F_{NSy} \tag{5}$$

and

$$J_{Sy} \ddot{\varphi}_{Sy} + b_{\varphi SRy} \dot{\varphi}_{Sy} - M_{bNSy} + k_{\varphi SRy} \varphi_{Sy} - M_{kNSy} + +r_{VM} F_{VM} \sin \beta_{NS} = 0 \, , \tag{6}$$

The force $F_{VM}$ is a connection force between the conveying part and the supporting part which is transferred by the joint mechanism of both parts. The force $F_{NSy}$ is the exciting force between the conveying part and supporting part.

The other dynamic forces can be calculated as follows:

$$F_{bNSy} = b_{NS}\dot{y}_{NS} , \tag{7}$$

$$F_{kNSy} = k_{NS}y_{NS} , \tag{8}$$

$$M_{bNSy} = F_{bNSx}r_{VM} = b_{NS}r_{VM}^2\dot{\varphi}_{NSy} , \tag{9}$$

and

$$M_{kNSy} = F_{kNSx}r_{VM} = k_{NS}r_{VM}^2\varphi_{NSy} , \tag{10}$$

where

$$F_{bNSx} = b_{NS}\dot{x}_{NS} , \tag{11}$$

and

$$F_{kNSx} = k_{NS}x_{NS} . \tag{12}$$

Torsional damping coefficient $b_{\varphi SRy}$ and torsional stiffness $k_{\varphi SRy}$ are:

$$b_{\varphi SRy} = b_{SRx}r_{SR}^2 \tag{13}$$

and

$$k_{\varphi SRy} = k_{SRx}r_{SR}^2 . \tag{14}$$

The relation between relative coordinates $y_{NS}$ and $\varphi_{NSy}$ is done by the joint mechanism:

$$y_{NS} = \varphi_{NSy}r_{VM}\tan\beta_{NS} . \tag{15}$$

In order to make a simplification of the calculation of equations (3) to (6), it is possible to write the balance of forces in the transformation plane (Fig. 6) as follows:

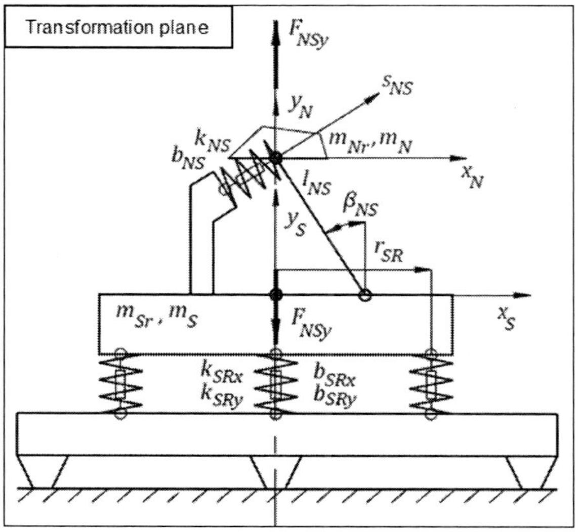


**Fig. 6** Dynamic model in the transformation plane

$$x_{Nr} = r_{VM}\varphi_{Ny}\,, \tag{16}$$

$$x_{Sr} = r_{VM}\varphi_{Sy}, \tag{17}$$

$$x_{NSr} = r_{VM}\varphi_{NSy} \tag{18}$$

$$y_{Nr} = y_{N}, \tag{19}$$

$$y_{Sr} = y_{S}\,, \tag{20}$$

$$y_{NSr} = y_{NS} \tag{21}$$

$$J_{Ny} = m_{Nr}r_{VM}^2\,. \tag{22}$$

$$J_{Sy} = m_{Sr}r_{VM}^2\,. \tag{23}$$

By introducing the coordinate $s_{NSr}$ , the number of unknown functions decreases to three:

$$x_{NSr} = s_{NSr}\cos\beta_{NS} \tag{24}$$

and

$$y_{NSr} = s_{NSr}\sin\beta_{NS} \tag{25}$$

After the adjustment of the equations (3) to (6), it is possible to write in the matrix form:

$$\bar{M}\ddot{\bar{q}} + \bar{B}\dot{\bar{q}} + \bar{K}\bar{q} = \bar{Q}\,. \tag{26}$$

where are the mass matrix is:

$$\bar{M} = \begin{vmatrix} m_N\sin^2\beta_{NS} + m_{Nr}\cos^2\beta_{NS} & m_{Nr}\cos\beta_{NS} & m_N\sin\beta_{NS} \\ m_{Nr}\cos\beta_{NS} & m_{Sr} + m_{Nr} & 0 \\ m_N\sin\beta_{NS} & 0 & m_S + m_N \end{vmatrix}, \tag{27}$$

the damping matrix is:

$$\bar{B} = \begin{vmatrix} b_{NS} & 0 & 0 \\ 0 & b_{SRr}\frac{r_{SR}^2}{r_{VM}^2} & 0 \\ 0 & 0 & b_{SRy} \end{vmatrix}, \tag{28}$$

and the stiffness matrix is:

$$\bar{K} = \begin{vmatrix} k_{NS} & 0 & 0 \\ 0 & k_{SRr}\frac{r_{SR}^2}{r_{VM}^2} & 0 \\ 0 & 0 & k_{SRy} \end{vmatrix}, \tag{29}$$

the coordinates vector is:

$$\bar{q} = \begin{vmatrix} s_{NSr} \\ x_{SR} \\ y_{SR} \end{vmatrix} \tag{30}$$

and the exciting vector is:

$$\bar{Q} = \begin{vmatrix} F_{NSy}sin\beta_{NS} \\ 0 \\ 0 \end{vmatrix}. \tag{31}$$

From the matrix equation (26) it is possible to calculate the time vector function $\bar{q}$. The parameters of the bowl feeder system are either known or can be measured. The mass parameters of the conveying and supporting parts are defined in the production documentation. The stiffness and damping parameters are difficult to measure and some differences can be expected between the parameters measured in a static or a dynamic way.

## 2. Experimental determination of dynamic parameters

Various dynamic parameters can be taken from the technical documentation of the bowl feeder (Tab. 1), others have to be calculated or measured during the correct excited oscillation.

The primary task of the bowl feeder is the generation of micro throws, and by doing this, the conveyance of goods. Therefore the excitation of bowl feeder free oscillating is generated by a rubber coated hammer in the tangential direction and the oscillation iss observed in the vertical direction. For this test, a sensor is mounted vertically under the spiral path of the bowl.

The base for the experimental determination of the dynamic parameters is the measurement of free oscillation kinematic values. The results of this step are time and frequency functions of acceleration (Fig. 7 and Fig. 8).

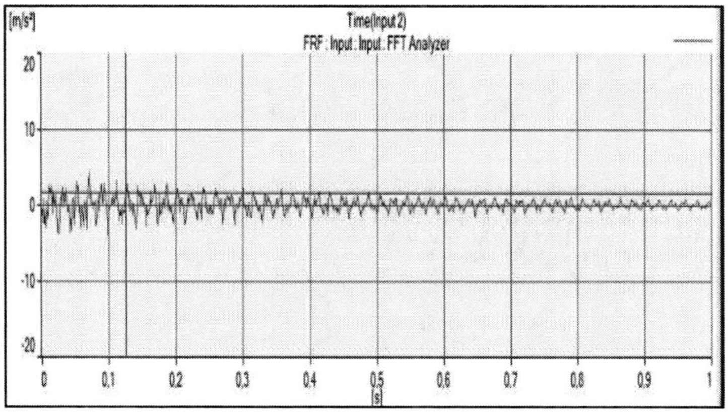


**Fig. 7** Time function of acceleration in vertical direction by free oscillation

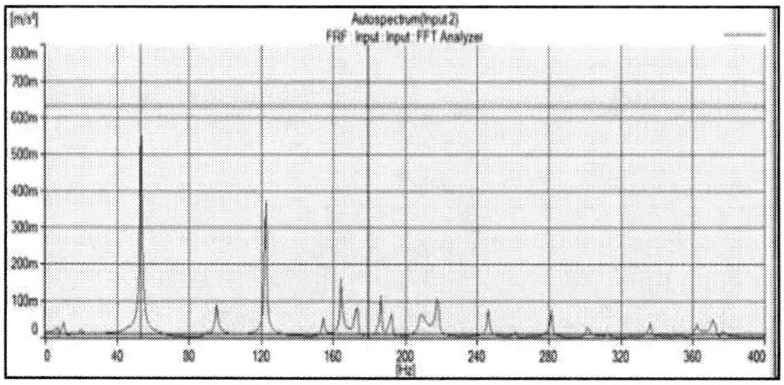

**Fig. 8** Frequency function of acceleration in vertical direction by free oscillation

The data supplying the necessary information on amplitudes decrease and frequencies. Based on the time function of acceleration, it is possible to find the damping parameters of the bowl feeder system. The frequency function supplies the information on the natural frequencies of the system.

In the analysed case of the particular bowl feeder, relative weak damping and three principal natural frequencies have been determined, which corresponding to three coordinates of vector function $\bar{q}$.

The identification of the dynamic parameters is simplified by a reduction of the system. An example for this is a system with only one degree of freedom. It is considered that the stiffness of the rubber springs having an infinitive value. In reality, the supporting part is fixed to the frame and then the equation (26) in case of the free oscillation is reduced to the form:

$$\left(m_N \sin^2 \beta_{NS} + \frac{J_{Ny}}{r_{VM}^2} \cdot \cos^2 \beta_{NS}\right) \ddot{s}_{NS} + b_{NS}\dot{s}_{NS} + +k_{NS}s_{NS} = 0\,. \qquad (32)$$

Based on this equation (32), the values of damping coefficient $b_{NS}$ and stiffness $k_{NS}$ are determined by the help of comparing measured and calculated time function $\ddot{s}_{NS}$.

The simplified system is excited and brought in free oscillating. Measured acceleration $\ddot{s}_{NS}$ of coordinate $s_{NS}$ is shown on Fig. 9 as the time function.

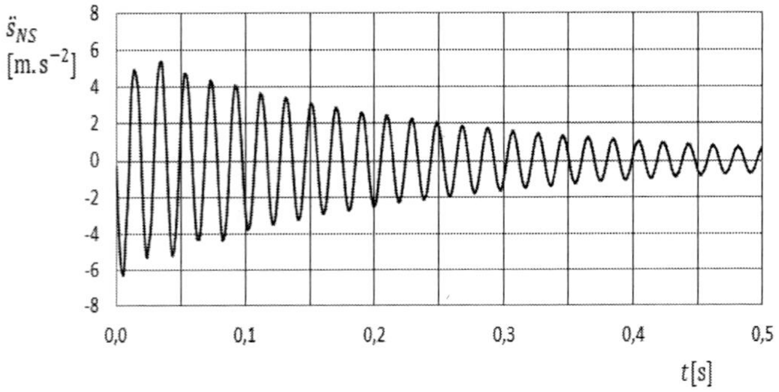


**Fig. 9** Time function of acceleration measured on the simplified system

The values of parameters $b_{NS}$ and $k_{NS}$ are chosen in the way that the measured and calculated time function will show the same or similar curve (Fig. 10).

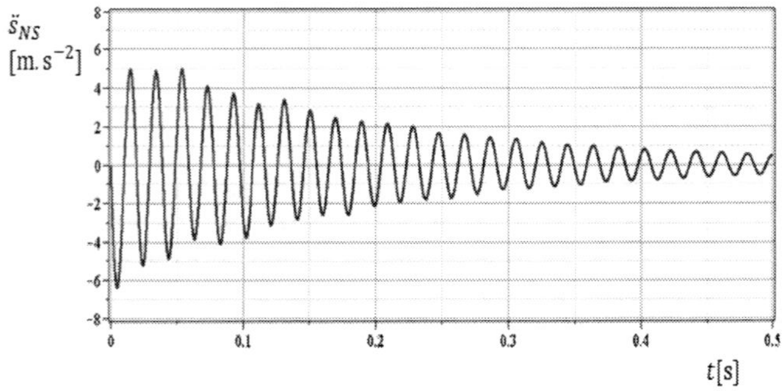


**Fig. 10** Time function of acceleration calculated for the simplified system

After obtaining of the parameters $b_{NS}$ and $k_{NS}$, the bowl feeder system can be returned into a more general configuration as illustrated in Fig. 6. The method then will be repeated for the parameters $b_{SRx}$, $b_{SRy}$, $k_{SRx}$ and $k_{SRy}$.

## Conclusions

This paper deals with the determination of dynamic parameters of a bowl feeder conducted with a comparison method of simulation and testing results.

The behavior of the dynamic bowl feeder system is simulated by free oscillation. The results of this simulation are kinematic functions of the bowl feeder oscillation, above all the acceleration time function with data about natural frequencies and the damping of the system. It is advantageous to simplify the dynamic system and thus the kinematic functions. In addition, its parameters have to be determined and advanced systematically as far as the full system of the bowl feeder. The same time functions are measured by tests done on the bowl feeder in a corresponding configuration of the system. By identical behavior of simulated and measured functions, the dynamic parameters are defined in equations of motion.

## References


1. **Dresig H., Holzweißig F.** (2008), *Maschinendynamik 8, Auflage, Springer Verlag,* Berlin.
2. **Harris C.M.** (2005)**,** *Shock and Vibration Handbook, Fifth edition, McGraw-Hill***,** New York.
3. Rhein-Nadel Automation GmbH**:** *Zufuhrtechnik. [online].[Cit. 5. 2. 2013]. Available on WWW: http://rna.de/Zufuehrsysteme_Basis-geraete.mfpx?ActiveID=*
4. **Nendel, K.** (2008), *Zweidimensionale Bewegungsformen bei Vibrationsförderern [online].. Available on WWW: http://www.vibrationsfoerdertechnik.de/ Download/ 2D-Bewegungsformen_WGTL_Tagungsband08.pdf.*
5. **Risch, T.** (2008), *2D-Vibrationsformen bei Vibrationsförderern. [online].. Available on WWW: http://www.vibrationsfoerder-technik.de/Download/2D-Bewegungsformen_WGTL_Poster08.pdf*
6. **Buja, H.O.** (2007): *Praxishandbuch Ramm- und Vibrationstechnik. Bauwerk Verlag Berlin.*
7. **Pešík, M.** (2012), *Kinematics of Objects in Vibratory Conveors. In ACC JOURNAL. 1st edition. Liberec: Technical University of Liberec. Available on WWW: http://acc-ern.tul.cz/cs/journal/item/root/acc-journal-xviii-12012.*

EAN 2017
55th conference on experimental stress analysis

# Testing and numerical analysis of mechanical stress mobile flood barriers

Michal Petrů[1, a] Pavel Srb[1, b] Ladislav Ševčík[2, c] and Aleš Lufinka[2,d]

[1]Technical University of Liberec, Institute for Nanomaterials, Advanced Technology and Innovation, Studentská 2, 461 17 Liberec 1, Czech Republic

[2]Technical University of Liberec, Faculty of Mechanical Engineering, Studentská 2, 461 17, Liberec 1, Czech Republic

[a]michal.petru@tul.cz, [b]pavel.srb@tul.cz, [c]ladislav.sevcik@tul.cz, [d]ales.lufinka@tul.cz

**Keywords:** flood, barrier, experimental measurement, numerical simulation

**Abstract.** This paper deals with experimental testing and numerical simulating of flood barriers. Floods are an integral part of the natural water cycle and can cause great loss of life and great property damage. The Czech Republic has been hit by several very devastating floods in past years. The development of new types of flood barriers is very important. The currently developed mobile barrier is characterized by its easy assembly on different surfaces. During the development of the mobile barrier several experiments have to be carried out. Based on the results from these experiments numerical simulations were created using Abaqus software. The results of simulations show good agreement with the experiment. These numerical simulations will be used as an effective replacement for experimental tests, which will lead to shorter development times and lower costs.

## Introduction

Nowadays flood barriers with concrete substructure reliably protect against flood. These barriers are used mainly in the large port cities or cities with large rivers. The issue, however, is the construction of flood barriers in the villages and smaller towns. Inhabitants of small towns and villages do not want barriers with substructure because of landscape distorting, damaging vegetation and complicate agricultural work in the fields. Therefore, sandbags are still used for flood protection and removed again after finishing danger. Development of flood systems leads to the mobile flood barriers which can be quickly installed and uninstalled after the flood. Constructional prototypes of these barriers must be tested in test channel at the different heights of water level, against log impact [1, 2] also for different types of terrain. These new barriers must be low maintenance and storage demands.

## Experimental measurement

The mobile barriers are tested in real flood conditions with floating log. Several measurements were carried out where the aim was to achieve statistically significant values. Therefore the type 1 measurements (artificial test channel) Fig.1 (left) were carried out first, for required water velocities from 1 to 4 $ms^{-1}$ with a step of 0.5 $ms^{-1}$ at the moment of impact, as published in [3]. Subsequently the type 2 measurements (river bed) Fig. 1 (right) were carried out. The water was released from reservoir and then went through the test track with a tidal wave. In the type 1 measurement the direction of log movement was defined by its anchorage by pulleys on the guide wire. In the type 2 measurement the direction of movement of the log was not limited; the log movement was given by the water flow.

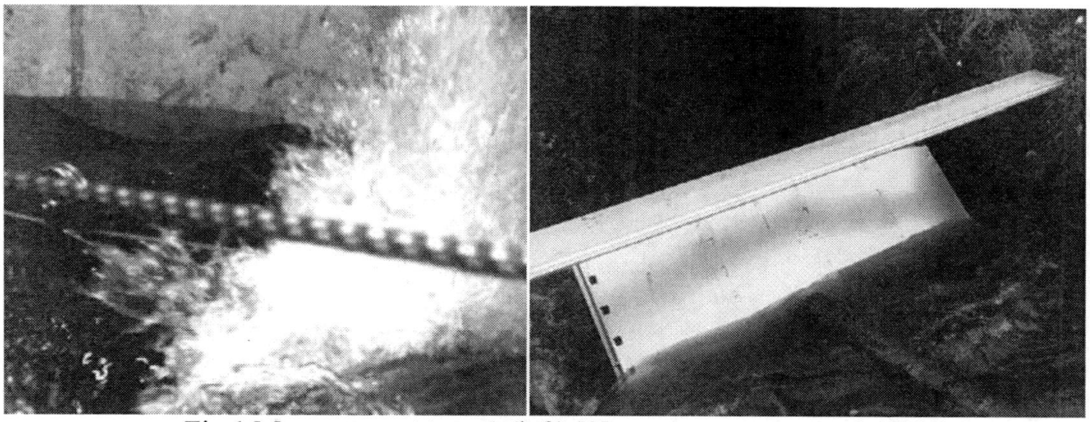

Fig.1 Measurement type 1 (left) [3], measurement type 2 (right)

**Used instrumentation:**

- 2x contactless laser sensor for measuring barrier movement
- incremental motion sensor for measuring log movement (position + speed)
- uniaxial accelerometer for measuring log acceleration
- 1x high speed video camera at the point of impact of the log into the barrier

The speed of the log movement was calculated in real time as a derivation of the incremental position sensor signal. The barrier and the log was recorded by high speed video camera. Laser sensors measured deformation of the barrier. The layout of experiment of flood system without substructure in real geological conditions is shown in Fig.2

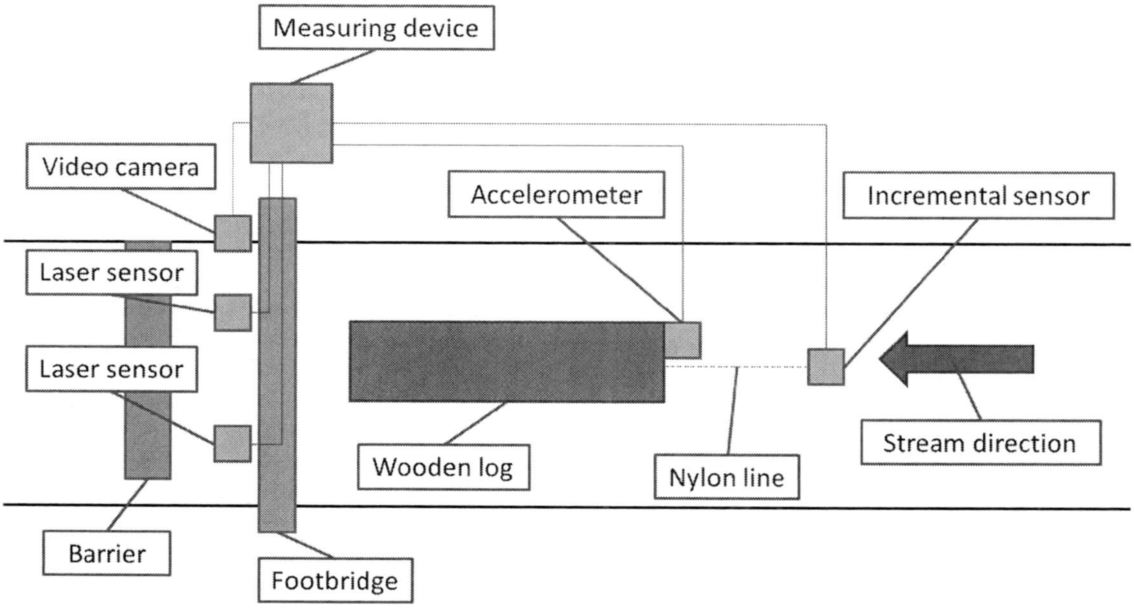


Fig. 2 The layout of experiment.

## Numerical modeling

A simplified CAD Fig.3 (left) model of flood barrier, wooden log and water space, was used to define the numerical simulation. For the simulation the barrier was replaced by simple plate of 1.5x1 m. The volume where water can move was created as a block of 7.5x1.5x1.5 m. The initial volume of water was created as block of 5x1.5x0.5 m. A log was created as cylinder with a diameter of 0.2 m and 3 m length with density of 900 $kgm^{-3}$ (weight 85 kg), which corresponds approximately to the log used in real experiment.

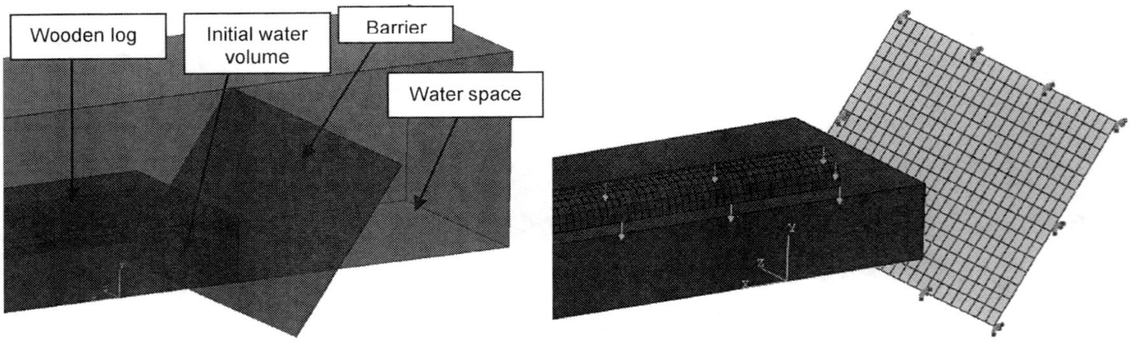


Fig. 3 CAD model of barrier (left), numerical simulation assembly (right).

The mesh used in the simulation was quite coarse. The water volume was meshed by 16,000 brick elements with edge size 0.1 m, the wooden log contains 4,800 brick elements with the shortest edge 0.028 m and the barrier contains 375 shell elements with edge length 0.1 m.

The gravitational acceleration of 9.81 $ms^{-2}$ was applied on the whole assembly. On the barrier were applied fixed constraints at the points corresponding to footholds of real barrier Fig. 3 (right). The initial water velocity was set to match the tidal wave velocity measured in the test channel.

## Results

The first simulation was carried out mainly to prove that the FEM method can be used for water modeling in such time and spatial range. From a visual comparison of simulation and the record of high speed camera it can be concluded that both waveforms are very similar Fig.4. Subsequently refined simulations were carried out. Results of these simulations are relatively accurate and are an auxiliary tool in developing new barriers.

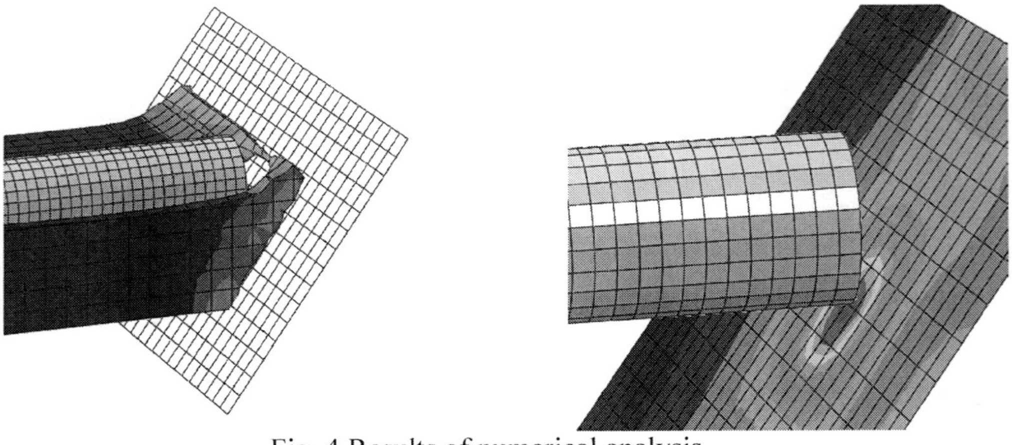

Fig. 4 Results of numerical analysis

Tidal wave speed of 1 $ms^{-1}$ was achieved in a real experiment. At simulation was achieved higher speed of tidal wave, which can be about 3 $ms^{-1}$ during the flood. Fig. 4 shows the course of deformation of the barrier at different tidal wave speed.

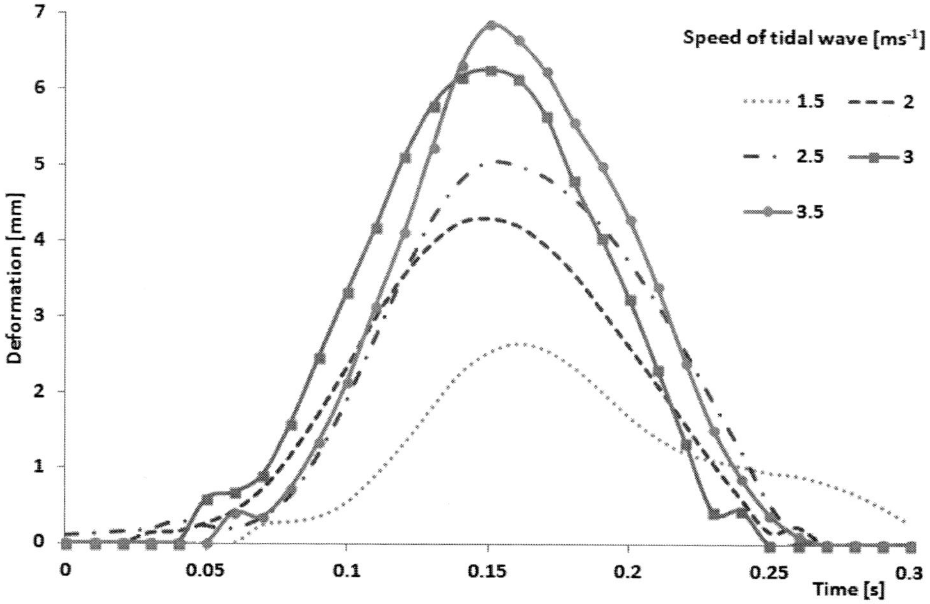


Fig. 5 Deformation of the barrier

## Conclusion

In case of a sudden flood, it is necessary to divert water out of the inhabited area as quickly as possible. The solution may be the barriers with substructure, but these are quite expensive. Therefore, use of mobile barriers may be a better solution in some cases. Mobile flood barriers have to meet many requirements, like easy installation, low weight, low price, low storage demands and so on. All these requirements must be considered for development of new design of mobile flood barriers. During the development of barrier many experiments have to be done. FEM and computing capabilities are now at a high enough level to allow numerical simulation of relatively complex phenomena such as fluid and structure interactions. These numerical simulations can successfully replace and complements some of the experimental tests with sufficient accuracy.


## Acknowledgements

The results of this project VI20152018005 were obtained with co-funding from the Ministry of the Interior of the Czech republic as part of targeted support from the project "Vývoj protipovodňových systémů pro zvýšení ochrany obyvatelstva a infrastruktury".


## References


[1] (2013) Approval Standard for Flood Abatement Equipment Class Number 2510, FM Approvals LLC https://www.fmglobal.com/assets/pdf/fmapprovals/2510.pdf

[2] Muttray, M.O., Oumeraci, H. Theoretical and experimental study on wave damping inside a rubble mound breakwater (2005) Coastal Engineering, 52 (8), pp. 709-725.

[3] Petrů, M. Ševčík, L., Lufinka, A., Martinec, T. Analysis of stress and deformation of the flood barrier , 54th International Conference on Experimental Stress Analysis, EAN 2016; Hotel SrniSrni; Czech Republic

EAN 2017
55th conference on experimental stress analysis

# Failure Mechanism of Compressed Masonry Column

MAROUŠKOVÁ Aneta[1, a]

[1]Czech Technical University in Prague, Faculty of Civil Engineering, Thákurova 7, Prague 6, 166 29, Czech Republic

[a]aneta.marouskova@fsv.cvut.cz

**Keywords:** Masonry, Column, Compression, Numerical Analysis, FEM.

**Abstract.** This paper deals with numerical analysis of masonry column under concentric compressive load. The heterogeneous model with the possibility of describing the contact between distinct materials was chosen for the numerical analysis. Mutual interaction between masonry unit (in this case solid burnt bricks) and mortar joints results in additional tensile stresses of the material with higher value of Young's modulus. Tensile vertical cracks following by loss of the structure's stability are caused not only from the contraction but also from the mutual interaction in contrast with compressed concrete pillars. For all simulations the commercial software package ABAQUS was used. Theoretically and experimentally obtained results are mutually discussed.

## Introduction

The determination of required material characteristics entering the computation of masonry structure is very complex. Masonry consists at least of two components, in this case solid burnt bricks and one type of mortar. These components are considered as heterogeneous materials with huge spread of mechanical characteristic. Experimental part has proved quasi-brittle behaviour of mortar and brittle behaviour of solid burnt bricks in three-point bending test. For bricks, after the value of failure stress was reached the softening curve has directed straight down, on the contrary of mortar when the softening curve had gradient expressing the brittleness. Experimentally detected working diagram in compression of brick and mortar were characterized after first crack's appearance by mild hardening following by softening [1]. However even if all mechanical characteristic of the individual components are known, behaviour of masonry column as compact structure is influenced by several factors, such as manufacturing, the size and the shape of masonry unit, damage or degradation. To eliminate unknowns, the linear – elastic behaviour for all components simulating the state before damage were chosen for verification of the character of behaviour masonry column under concentric compressive load.

The modelled column with plan dimensions of 0.3 m x 0.3 m and a height of 1 m matches with the real column tested in laboratory. The column was constructed using P20 solid burnt bricks with dimensions of 0.29x0.14x0.065 m. Mortar joints with a thickness of 0.02 m were made with M2 mortar. The experimental program is part of a research project NAKI, which is being carried out at the Faculty of Civil Engineering, CTU in Prague.

## Failure Mechanism of Masonry Column

Failure mechanism of compressed brick masonry column is characterized by the appearance and development of vertical tensile cracks passing in the direction of principal stresses and is accompanied by progressive growth of horizontal deformations. These cracks are caused by contraction and interaction between two materials with different mechanical characteristics [2, 3, 4]. As a result of this mutual interaction, mortar that has a lower Young´s modulus and has tendency to greater lateral strain, is transversely "pressed" and masonry units, on the contrary are transversely "stretched".

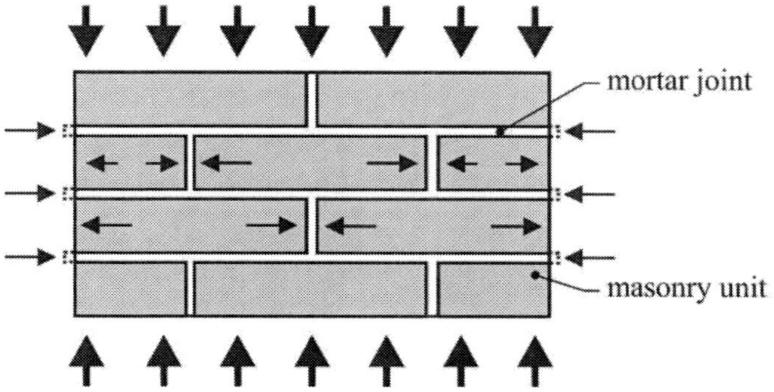


Fig. 1: Mortar-brick interaction

Intensive masonry unit-mortar interaction does not occur for example regular ashlar masonry with small differences in mechanical properties between masonry unit and binder. In that case, the failure mechanism is characterized by gradual local damage of masonry in compression.

**Numerical Model**

The heterogeneous model (all components are geometrically and materially accurate described) was chosen for the analysis under the assumption of linear-elastic behaviour for all components simulating the state before damage. Linear-elastic parameters are shown in table 1.

Tab. 1: Linear – elastic parameters of materials.

| Material | Young Modulus E [GPa] | Poisson's coefficient ν [-] |
|---|---|---|
| Masonry unit | 2.5 | 0.20 |
| Mortar | 0.5 | 0.15 |
| Steel | 210 | 0.3 |

The maximal size of finite element's mesh was a third of masonry unit's height with refining for mortar joints matches with two finite element per mortar joint thickness. The chosen type of finite elements were three dimensional solid elements C3D8 – 8-node linear brick. For brick-mortar interface, "hard" contact was assumed in normal direction and frictional behaviour (with a friction coefficient of 0.6) in tangential direction.

The boundary conditions were applied on two steel plates positioned at the head and the bottom of the masonry column. Vertical load were applied in a form of displacement at the top surface of steel plate at the column's head and lower surface of bottom plate were rigidly fixed.

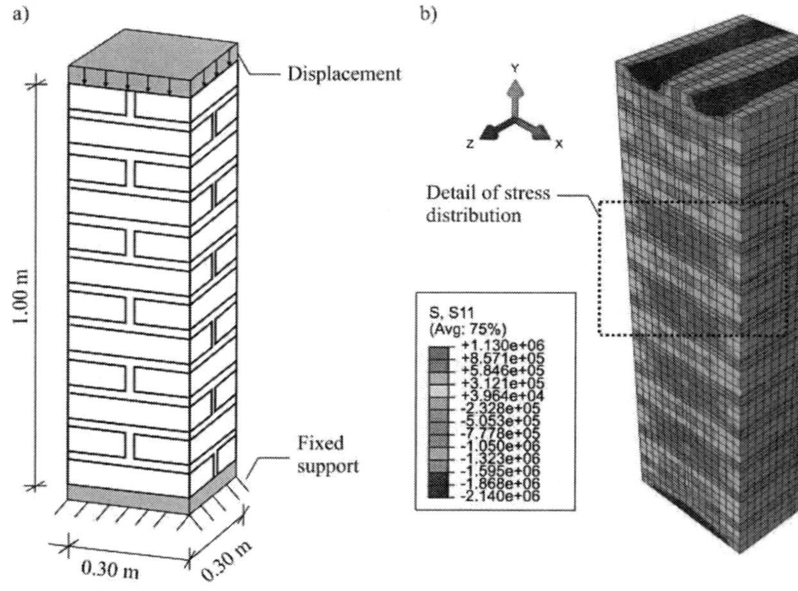


Fig. 2: Numerical model's scheme (a), horizontal stress distribution (b)

**Discussion of Results**

The horizontal stress (s11) distribution is evident from the results of numerical analysis. Figure 3 shows in detail one of the horizontal stresses, the second (s33) one is symmetric. The maximal principal stress's trajectories (combining primarily both horizontal stresses s11 and s33) run along masonry units and cause transversal cracking of solid burnt bricks. Direction of the crack thus matches with the middle principal stress's trajectories. Along the height of the column, cracks run in the direction of minimum principal stresses (almost corresponding to the vertical stress s22).

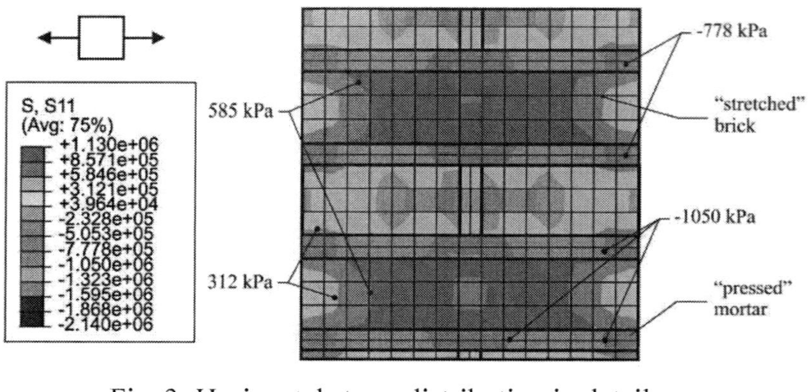


Fig. 3: Horizontal stress distribution in detail

Results from experimental testing are consistent with theoretically findings. Hairline cracks had appeared at the brick-butt joint interface. The vertical cracks approx. in a half of brick had firstly appeared in the middle part of the column. The next loading were accompanied by the appearance of other crack in the upper part and growth of cracks in the middle third of the column. Development of the cracks has resulted in large horizontal deformation and formation of several main continuous cracks causing the splitting of the column to the sub-parts.

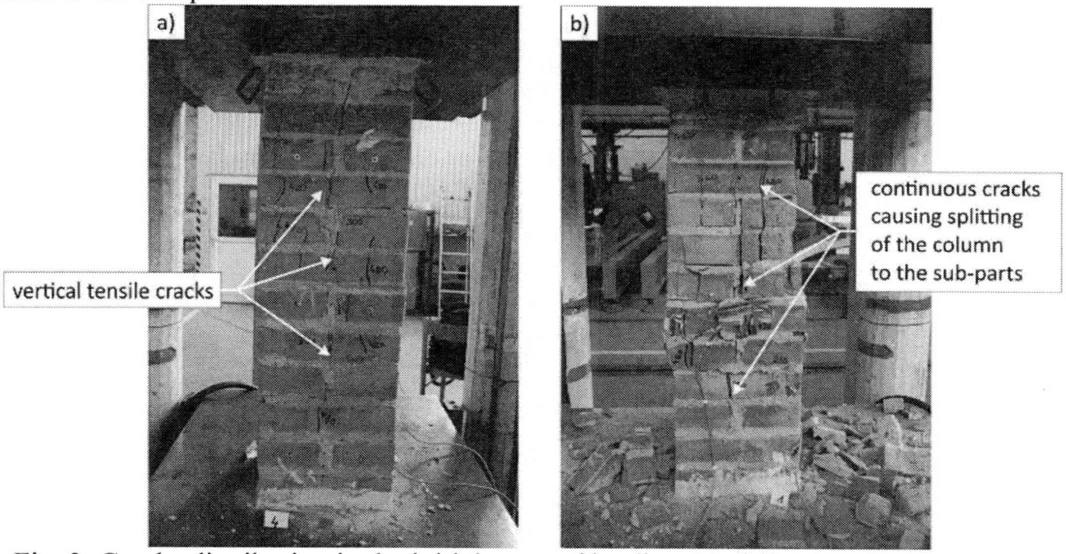


Fig. 3: Cracks distribution in the initial stage of loading (a), failure of the column after reaching the maximal stress (b)

## Conclusions

The formation and growth of vertical tensile cracks in masonry units (bricks) due to the additional tensile stresses caused by lateral strain of more compliant mortar in bed joints and contractions precedes redistribution of stresses and failure of the structure. This type of failure mechanism corresponding with presented brick work masonry column under concentric compressive load was verified by stress distribution from numerical simulation and experimental testing.

## Acknowledgement

The results were obtained with support from research project DF12P01OVV037 - "Progressive Non-Invasive Methods of the Stabilization, Conservation and Strengthening of Historic Structures and their Parts with fibre- and nanofibre-based Composite Materials".
The grant researcher is prof. Ing. Jiří Witzany, DrSc. dr.h.c.

## References

[1] A. Maroušková, Inelastic material models for numerical analysis of unreinforced compressed masonry columns, Key Engineering Materials 677 (2016) 197-202, doi: 10.4028/www.scientific.net/KEM.677.197.

[2] J. Witzany, R. Zigler, Stress State Analysis and Failure Mechanisms of Masonry Columns Reinforced with FRP under Concentric Compressive Load, Polymers 8 (2016), 176, doi: 10.3390/polym8050176.

[3] J. Witzany, T. Čejka, R. Zigler, Failure Mechanism of compressed short brick masonry columns confined with FRP strips, Construction and Buildings Materials 63 (2014) 180-188, doi: 10.1016/j.conbuildmat.2014.04.041.

[4] A. Maroušková, Masonry Column Reinforced by FRP Wrapping: Behavior and Numerical Analysis, Applied Mechanics and Materials 825 (2016) 27-30, doi:10.4028/www.scientific.net/AMM.825.27.

EAN 2017
55th conference on experimental stress analysis

# Comparison of ESPI and DIC Strain Contours Measurements under Three Loading Cases

Radim Halama[1,2, a], Lukáš Horňáček[1, b], Jakub Šmiraus[1, c]

[1] Faculty of Mechanical Engineering, VŠB – Technical University of Ostrava, 17. listopadu 2172/15, 708 33 Ostrava-Poruba, Czech Republic

[2] IT4Innovations, VŠB-Technical University of Ostrava, 17. listopadu 2172/15, 708 33 Ostrava-Poruba, Czech Republic

[a]radim.halama@vsb.cz, [b]lukas.hornacek@vsb.cz, [c]jakub.smiraus@vsb.cz

**Keywords:** Electronic Speckle Pattern Interferometry, Digital Image Correlation, Three loading cases, strain contours

**Abstract.** This article describes strain and displacement contours comparison of Electronic Speckle Pattern Interferometry and Digital Image Correlation. Both methods were used in tension mode, shear mode and combination of these modes.

## Introduction

Electronic Speckle Pattern Interferometry (ESPI) is widely used contactless method based on the principle of optical interferometry. For instance, Dantec Dynamics Q100 ESPI system allows to measure the displacement field, which can hold sensor, approximately 50x50 mm. During measurement it is necessary to proceed step by step and always keep the load value for as long as necessary. The principle of measurement is based on the lighting pattern from four different directions. The reflection of light is sensed by one central CCD camera [1]. In contrast Digital Image Correlation Method (DICM) is more progressive than ESPI, because it allows continuous scanning of surface without making load steps and less sensitivity to vibrations. The principle of DCIM is based on scanning pattern on the specimen by two cameras with external lighting. Differences in displacements obtained by an appropriate correlation algorithm for each evaluation points captured by both cameras are evaluated as the displacement field [2]. This field can be many times larger than the ESPI's field (considering Dantec Dynamics Q100 equipment) [3]. In case of composites more attention must be paid to find out suitable methods for deformation measurements and precise execution of the experiment [4].

This article presents comparison of strain fields and displacement fields measured on the sheet specimen surface under shear, tension and combined loading by ESPI and DICM methods in the area of small strains.

## Experimental setup

Specimen was made of PP50GF composite material, which is a commonly used material in automobile industry. Glass fibers in specimen are oriented in angle 45° from longitudinal axis. All tests were realized on the universal testing machine Testometric 500-50CT at the VŠB –TU Ostrava. For DIC measurements the Mercury®RT equipment (2x2.3Mpx@40Hz) provided by Sobriety s.r.o. company was used [5]. For ESPI measurements Dantec Dynamics

Q100 equipment was used. Creation of optical contrast coating known as pattern was prepared on one specimen side for full field DIC measurements. The opposite side was sprayed by a white colour to prepare optically sensitive layer for ESPI method application. Both measurements were done on the same specimen in the same time, during test, so all measurements were performed simultaneously with DIC system and ESPI equipment.

In all measurements ESPI Q100 sensor was fixed to the testing jig. Therefore the coordinate system remained unchanged for all measurements. DIC equipment Mercury®RT was mounted on a tripod. For purposes of possible comparison of ESPI and DIC, the coordinate system for evaluation was always changed in postprocessing to be analogous to the ESPI coordinate system. Additionally, when the ESPI measurement was evaluated, the displacement offset was set to one selected point based on the DIC measurement. Otherwise, it would be impossible to make a quantitative comparison of the two methods, because ESPI sensor would be shifted with the test jig. All measurements were performed after preload of 100N.

a) b)

Fig. 1. Photographs of experimental setup: a) combined mode, b) shear mode

**Results for tension mode**

When the specimen was subjected to pure tension, load steps have been set to value 50N and subsequently decreased to the value of 30N. The measurement was performed up to a 300N load from zero state with established preload. Evaluation of the measurements of both methods is shown at Fig. 2. Displacement fields obtained by DIC (Fig.2a) and ESPI (Fig.2b) are comparable. Both methods also show strain concentration in the area close to notches (Fig.2c-d).

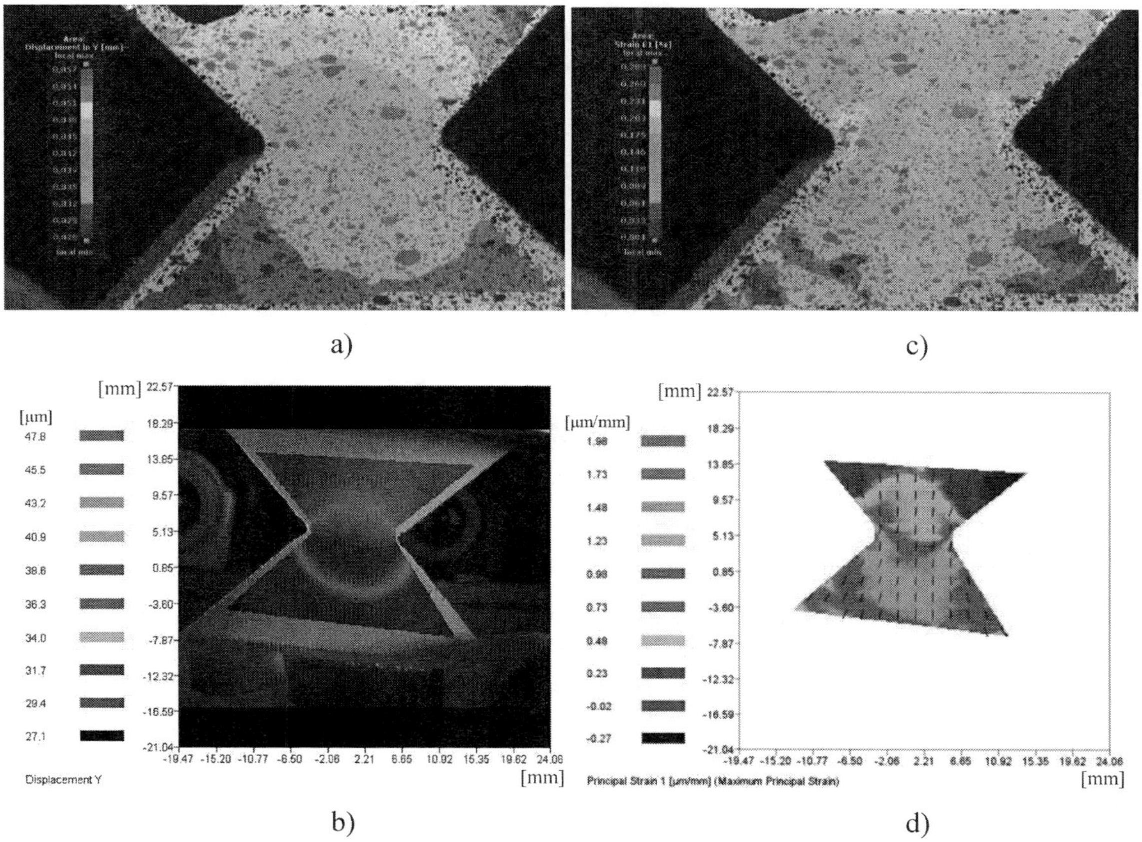


Fig. 2. a) Displacement UY contours - DIC, b) Displacement UY contours - ESPI, c) first principal strain contours - DIC, d) first principal strain contours - ESPI

## Results for shear mode

For the case of shear mode, load steps have been set to value 10 N. The measurement was performed up to a 155N load from zero state with established preload. An example of evaluation of the measurements by both methods is shown at Fig. 3 and Fig. 4.

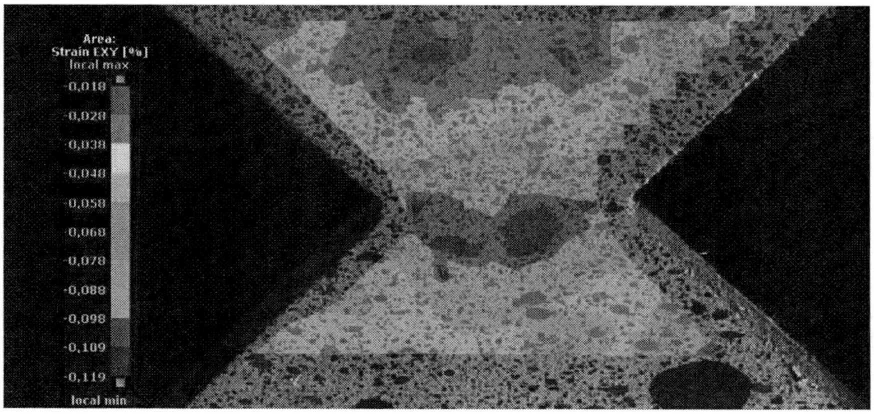


Fig. 3. Shear strain contours - DIC

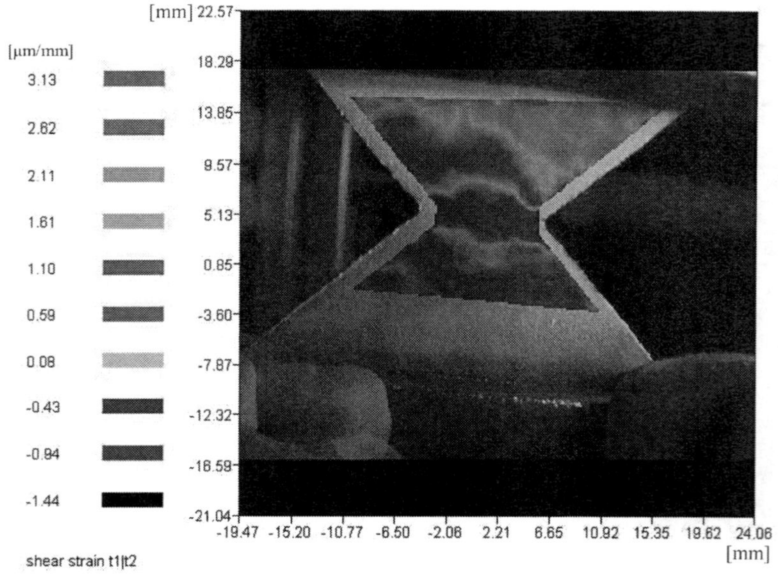


Fig. 4. Engineering shear strain contours - ESPI

It should be mentioned, that the ESPI equipment is configured to display the value of engineering shear strain, whereas the DIC equipment shows values half as ESPI, because the shear strain value is computed as a component of the strain tensor not as the engineering shear strain.

### Results for combined mode

When the testing jig is used in such a way, that specimen's axis is situated in 45° angle to loading force direction, the stress state corresponds to the combination of tension and shear mode. In this case, load steps have been set to value 40 N and subsequently decreased to the value of 30N. The measurement was performed up to a 370 N load from zero state with established preload. Comparison of results gained by both measurement methods is possible in Fig. 4 and Fig. 5.

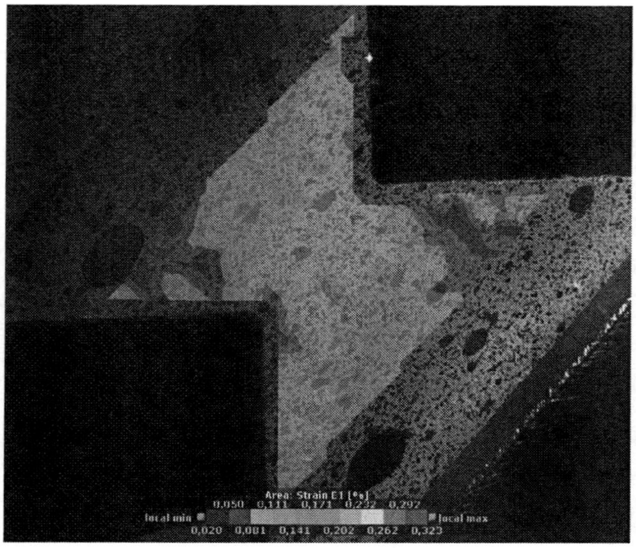

Fig. 4. First principal strain contours - DIC

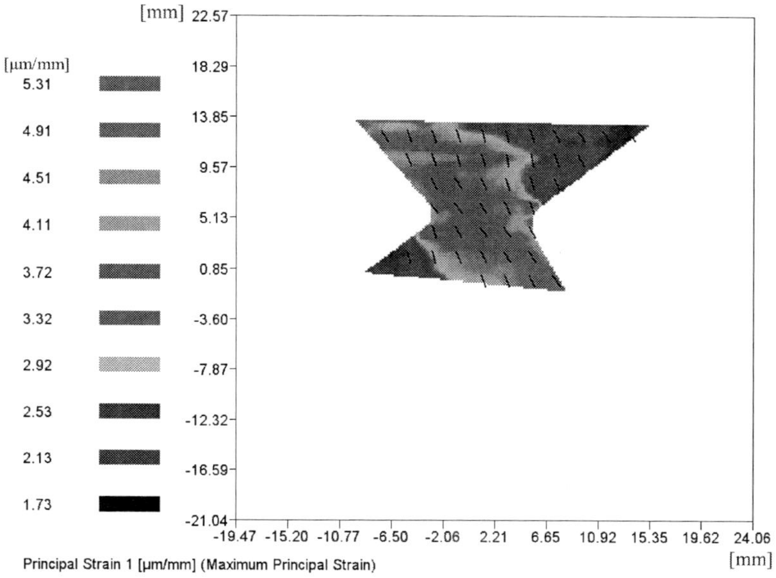


Fig. 5. First principal strain contours - ESPI

## Conclusion

Two progressive methods of experimental stress analysis (Electronic Speckle Pattern Interferometry and Digital Image Correlation Method) were evaluated in this study. The measured results are qualitatively comparable for all three loading modes.

There are presented comparison of measurements results (displacement contours, first principal strain contours, shear strain contours) for tension, shear and combined measurements. From the comparison follows, that both methods are suitable for measurement of very small displacements and strain values. The ESPI method is more accurate for the case of small strain analysis [3]. The disadvantage of Q100 equipment in necessity of load increase interruption is not so important in investigated case.

The measured data will be used to calibrate a material model in a finite element software to describe properly the stress-strain behavior of investigated composite material in a 3D finite element simulations.

## Acknowledgement

This work was supported by The Ministry of Education, Youth and Sports from the National Programme of Sustainability II (LQ1602) and from the Specific Research Project (SP2017/136) and by the Grant Agency of the Czech Republic (GACR) project No. 15-18274S.

## References

[1] Pečenka L., Kalina M., Halama R., Fojtík F. Verification of ESPI Stress Analysis by Means of FEM, Transactions of the VŠB - TU of Ostrava, Mechanical Series, 2013, Vol.59, No.1, p.65-68.

[2] Information on https://www.dantecdynamics.com

[3] Hagara, M., Šimčák, F., Kalina, M. The knowledge acquired by using of optical methods by strain fields investigation. Applied Mechanics and Materials, Vol.486, 2014, p.141-146.

[4] Information on http://www.sobriety.cz/solutions-mercury.htm

[5] Hodgkinson J. M. (editor). Mechanical Testing of Advanced Fibre Composites. CRC Press 2000, 362p.

EAN 2017
55th conference on experimental stress analysis

# Modeling of friction stir welding of ferritic steels by considering phase transformations

Daniel Kaločany[1,a], Pavel Élesztős[2,b], Ladislav Écsi[3,c], Roland Jančo[4,d]

[1] EDAG Engineering CZ, spol. s. r. o., Praha 5, CZ

[2,3,4]SjF STU in Bratislava, Bratislava, SK

[a]daniel.kalocany@gmail.com, [b]pavel.elesztos@stuba.sk, [c]ladislav.ecsi@stuba.sk, [d]roland.janco@stuba.sk

**Keywords:** FEM, friction stir welding, SYSWELD.

**Abstract.** The paper content deals with readers familiarisation with the numerical simulation of friction stir welding issues and computation verification according to the experimental measured data. In order to achieve results as close to the reality as possible, microstructure transformation of the weld joint and heat affected zone, and also temperature dependent material nonlinearities must be taken into account. Temperature field distribution in time depends on the type of heat source and parameters of the welding process. Also, microstructure and state transformations have influence on temperature field distribution. Temperature field distribution has significant influence on stress and strain development during and after welding.

## Introduction

Welding joints belongs to the most commonly used technologies of permanent joint manufacturing processes. The weld is created by local temperature increase of joined materials to the temperature that is usually higher than its melting temperature. Friction welding methods, including friction welding with mixing, do not reach a temperature higher than the melting point. Rapid local temperature changes cause high temperature gradient slopes, which causes temperature loads, phase and state transformations. Clamping condition loads during welding and the nonlinear material behaviour make welding simulations a very complex problem. Phase distribution in the weld and a heat affected zone develops during cooling after welding and has an influence on residual stresses distribution in the welded specimen. The combination of residual stresses and external loads have an influence on fatigue life of the welded constructions or on the development of discontinuities in the welds. In order to increase fatigue life of the welded structures it is necessary to keep the residual stresses on as low values as possible. A numerical simulation based on the finite element method can be a very useful tool for predicting residual stresses development during and after welding. The current state of the finite element method development leads to a tendency to simulate complex problems as technological processes. . With the use of the finite element method it is possible to transform a system of partial differential equations for finite amount of discrete points into a system of algebraic equations. These algebraic equation systems can be solved and evaluated effectively.

## Problem description

Friction stir welding is a method of joining materials permanently in a solid state. That means that temperature in the welding location will not be increased over the melting temperature of the welded materials. The material is heated by friction of the rotation tool with the shape of a pin, which is plunged by force into the specimen that will be welded. By sufficient heating the material becomes more easily plastic-formable. Material retreat and stir are then caused by the rotational movement of the tool. After the rotating tool is fully plunged into the material, the tool will start to move along the weld trajectory by translational movement. Blinding of the hole after welding is provided by the tool's shoulder that also has the function of protecting the weld. This welding process was primary developed for aluminium welding in the airospace industry with this welding method high-quality, high-strength, and low-distortion welds can be achieved. Another advantage of this method is the absence of additional materials and harmful fumes. These undisputed advantages lead to the acceptance and willingness of applying friction stir welding in the commercial sphere. Nowadays numerical simulations have become a common part of industry. In order to fully integrate friction stir welding into industry, numerical simulation of the whole process with cooling, taking in account phase transformation, need to be created. The current level of commercially available tools for friction stir welding simulations, FSW Toolbox for SYSWELD, have limited options. The use of this toolbox is primary intended for friction stir welding of aluminium alloys and only for quasi-unstable time step. The content of the presented paper deals with the simulation results after implementation of ferritic steel material models. These material models are taking into account phase transformations. That makes friction stir welding a very complex problem, which is coupled with the task of fluid flow, heat transfer, metallurgy and mechanics. In this case, metal is considered as an incompressible liquid that behaves according to the Norton-Hoff law.

## Computation of quasi-unstable time step for friction stir welding of ferritic steels

In the computation of quasi-unstable time step temperature depending material properties of steel S355J2G3 (STN 11 503) were used. The sequence scheme of algorithms for quasi-stable time step computation is shown in Fig. 1.

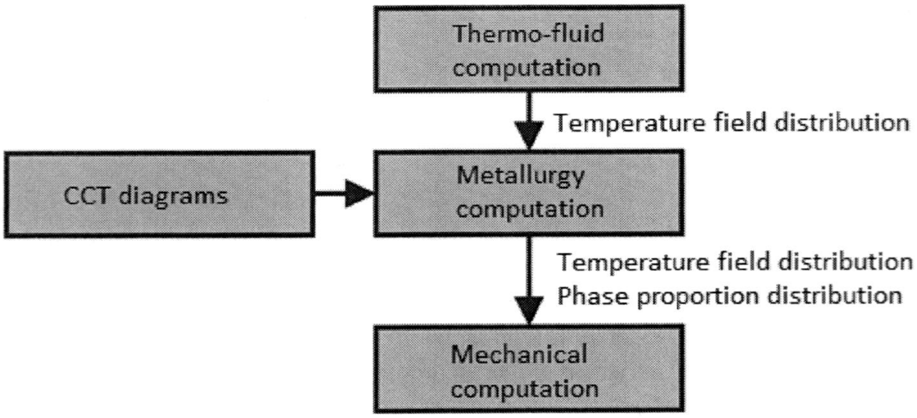


Fig.1. - Sequence scheme of friction stir welding computation

The first step is thermo-fluid computation for simulation of temperature field distribution, which is generated by the friction of the rotating tool. The weld metal is considered as an incompressible liquid that behaves according to the Norton-Hoff law and it depend only on the deformation speed rate. Heat transfer is considered according to the Fourier's law of heat conduction. As boundary conditions are considered convective temperature losses to the surroundings and technological parameters of the welding process, which have influence on the shape and efficiency of the heat source. The results of thermos-fluid computation are distributions of temperature field and velocity vectors for nodes. The temperature field distribution does not take into account temperature losses due to a phase or state transformation.

Metallurgy computation is used to determine the phase proportion distribution and takes into account temperature losses caused by phase and state transformation. For metallurgy computation the Johnson-Mehl-Avrami transformational kinetic law modified by Leblond is used, which takes into account CCT diagrams. Martensitic bi-difusional transformation is computed according to the Marburger's equation. These are semi-empirical equations. The results of metallurgy computation are phase proportion distributions and temperature field distributions with additional temperature losses caused by phase and state transformation.
Temperature field and phase proportion distributions are input loads for mechanical computation. Boundary conditions for this step of computation are clamping conditions and their stiffness. In general, for total strain rate computation Leblond's model is used in welding and heat treatment simulations.

$$\dot{\varepsilon} = \dot{\varepsilon}_e + \dot{\varepsilon}_p + \dot{\varepsilon}_{th} + \dot{\varepsilon}_{tp} \qquad (1)$$

| | | |
|---|---|---|
| $\dot{\varepsilon}$ | - | total strain rate [-] |
| $\dot{\varepsilon}_e$ | - | elastic strain rate [-] |
| $\dot{\varepsilon}_p$ | - | plastic strain rate [-] |
| $\dot{\varepsilon}_{th}$ | - | temperature and metallurgical strain rate [-] |
| $\dot{\varepsilon}_{tp}$ | - | transformation plasticity [-] |

**Tuning of temperature field distribution**

For the experimental measurement of temperature field distribution a non-contact thermal-vision measurement method with FLIR® sc660 thermal imaging camera was chosen. In order to avoid rough mistakes during measurement, it is necessary to keep the emissivity of measured specimens on the same value during the measurement. For this reason, special paint, ThermaSpray 800, with known emissivity $\varepsilon = 0{,}96$ guaranteed to the value of 600 °C was spread on the specimens. Temperature profiles on a line from a simulation were compared to those of an actual measurement. Both lines were located 20 mm parallel to the weld trajectory to avoid rough mistakes caused by emissivity changes during the measurement.

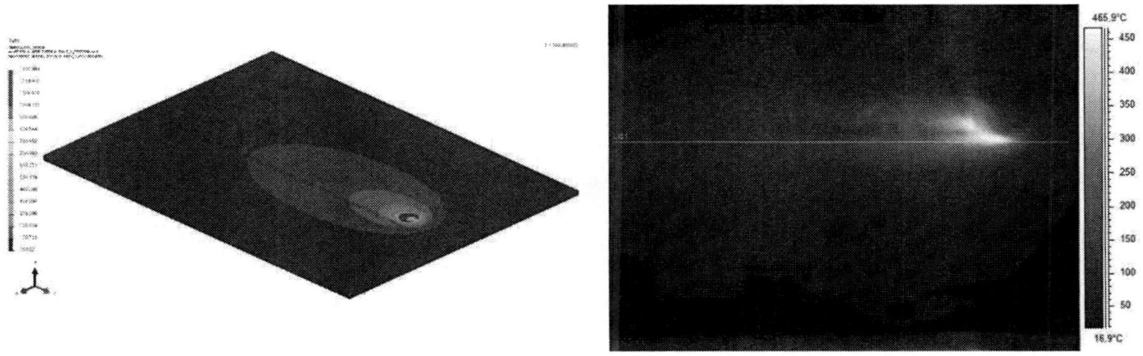


Obr. 2 – Comparison of temperature distribution of a simulation and measurement in time t = 374 s

Tuned parameters for determining the temperature field distribution were heat source efficiency $H$, friction coefficient $\mu$ and exponent $R$. Values of these coefficients are shown in table Tab.1. Heat exchange coefficients to the environment were tuned for each specimen's areas, because of different cooling conditions. Fig.3. is shows the comparison of the temperature profiles obtained by simulation and measurement.

Tab. 1

| | $H$ | $\mu$ | $R$ |
|---|---|---|---|
| Default value | 0,5 | 1 | 1 |
| Tuned value | 0,210 | 0,315 | 1 |

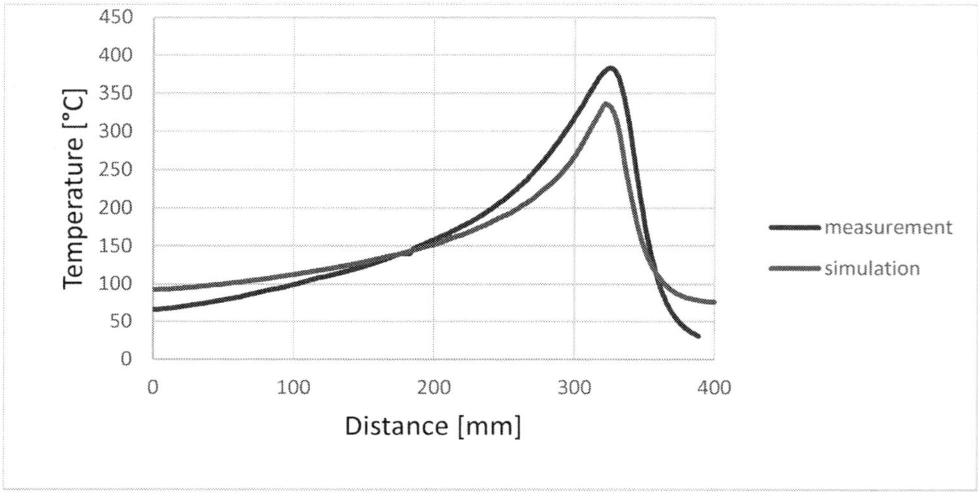


Fig. 3 Comparison of the temperature profiles obtained from simulation and measurement

From the temperature field distribution phase proportions can be determined. In this case for the quasi-unstable time step in time t = 374 s. This time value is derived from the translational speed of the tool and from the weld line length. Phase proportion distributions are shown in Fig. 4.

Austenit max. 100 %

Bainit max. 75 %

Ferit max. 100 %

Martenzit max. 32 %

Fig.4. - Phase proportion distribution in time t = 374 s

### Validation of Mechanical computation results

As boundary conditions for the mechanical computation clamped areas for nodes at certain localities were considered, where the specimen were clamped during the measurement. At a certain location, on the top surface of the measured specimen, a strain rosette was placed. Obtained data were processes in order to obtain results of the stress state according to Von Mises theorem. This Results were compared with the simulation results of stress state at a certain locality in time t = 374 s. Measurement locality is shown on the left in the red rectangle in Fig. 5 together with the stress distribution obtained by simulation at the same locality.

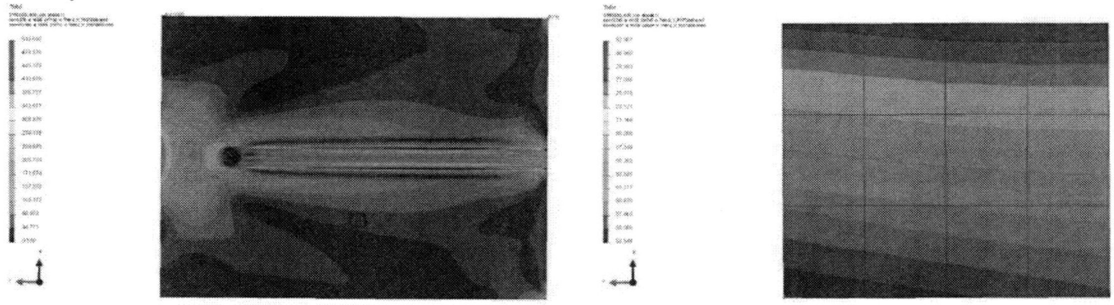

Fig.5. - Von Mises stress distribution in time t = 374 s, right – detail of measured locality $\sigma_{min}$ = 53,55 MPa, $\sigma_{max}$ = 82,91 MPa

Stress state results in time according to the Von Mises theorem obtained from the measurement in locality highlighted in Fig.5. are shown in Fig.6. Values of Von Mises stress in time t = 374 s are approximately around 60 MPa.

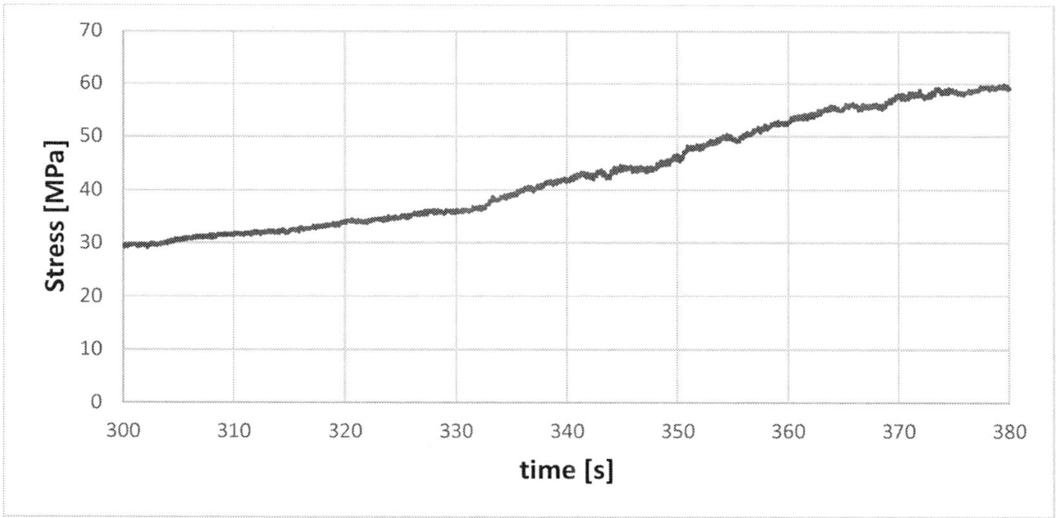


Fig. 6. – Stress state according to the Von Mises theorem obtained from measurement

### Summary

This paper deals with friction stir welding simulations of ferritic steels undergoing phase transformations. The use of temperature dependent material models for austenite, bainite, ferrite, perlite, and martenzite were verified by experiment. The tuning of boundary conditions, the heat source shape and parameters have shown that this simulation can be successfully used for ferritic steels as well.

## Conclusions

Further research has to be carried out to determine the final phases and the residual stress distribution over the body when the cooling process takes place.

## Acknowledgements

This publication is the result of the project implementation: Research of friction stir welding (FSW) application as an alternative to melting welding methods no. 26240220031 supported by the Research & Development Operational Programme funded by the ERDF. Funding from the VEGA grant 1/0740/16 is also greatly appreciated.

## References

[1] Feulvarch, E., Robin, V., Boitout, F., Bergheau, J. M.: 3D Modelling of thermofluid flow in friction stir welding including metallurgical and mechanical censequences. ESI Group, (2008)

[2] SYSWELD Reference Manual. ESI Group, (2013)

[3] Friction Stir Welding Modeling using SYSWELD V2008. ESI Group, (2008)

[4] Mishra, R. S., Mahoney, M. W.: Friction Stir Welding and Processing. 1. Vyd. AMS International, (2007)

[5] Slováček, M.: Numerické simulace zvařování. Výpočet a hodnocení dostorzí a zbytkových napětí. Dissertation thesis. Brno: Fakulta vojenskych technologii univerzity obrany v Brne, (2005) in Slovak

[6] Shercliff, H. R., Colegrove, P.A.: Modeling of Friction Stir Welding. Mathematical Modelling of Weld Phenomena 6, 927-974, (2002)

[7] Goldak, J., Akhlaghi, M.: Computational welding mechanics. Springer, (2005)

[8] Jančo, R., Écsi, L., Élesztös, P.: FSW Simulation of Aluminium plates by SYSWeld – Part I. The International Conference on Numerical and Experimental Solution of Welding, 15 – 19, (2014)

[9] Trebuňa, F., Šimčák, F., Bocko, J.: Using of numerical and experimental methods of mechanics to safety assessment of sluice equipments. In: Technológ. Roč. 5, č. 4 (2013), s. 19-22. in Slovak

[10] Jančo, R., Écsi, L., Élesztös, P.: FSW Simulation of Aluminium plates by SYSWeld – Part II. The International Conference on Numerical and Experimental Solution of Welding, 21 – 28, (2014)

[11] Kaločány, D.: Numerická a experimentálna analýza deformačného a napäťového stavu pri trecom zváraní s premiešavaním. PhD thesis. Bratislava: SjF STU, (2016) in Slovak

EAN 2017
55th conference on experimental stress analysis

# Tuning the material models for the nonlinear dynamic simulations by considering the exact material properties

Pavel Élesztős, jr. [1,a], Juraj Máčaj [1,b], Pavel Élesztős[2,c]

[1] TEN SLOVAKIA, s.r.o., Gazdovský rad 49/A, 931 01 Šamorín, SK

[2] SjF STU v Bratislave, námestie Slobody 17, 812 31 Bratislava, SK

[a] p.elesztos@ten.sk , [b] j.macaj@ten.sk

[c] pavel.elesztos@stuba.sk

**Keywords:** *material models, strain-rates, Krupkowski law, dynamic bending test, simulations*

## Introduction

In the automotive and ground transportation industry simulations have been replacing prototype tests for quite some time. The "digital twin" is required to acredibly copy the physical properties of the actual prototype. For the non-linear dynamics analyses it is necessary to choose the realistic physical properties of elastic-plastic materials. The goal is to have the simulation achieve the best correlation compared with the results of the experimental tests.

In this article we will show how optimizing the material parameters can significantly reduce the computational time while offering the same quality results. The aim of this article is to compare the results obtained when using two highly advanced material models, the hardening curves and the Krupkowski law [3]. The results of the simulations in both cases were compared to the results of the experimental dynamic bending tests carried out on test samples with a cross section size of 50x50 [mm], wall thickness of 5 [mm], where the material is steel class S355 and it is manufactured by hot rolling.

## 1. Creation of material cards

The first step of the „digitization" of the material properties for the numerical simulation purposes is the acquisition and adjusting of the basic mechanical properties of the given material. Those we obtain by means of static tensile tests according to the STN EN ISO 6892-1 standards. At this stage it is necessary to create a model of the tensile specimen with a fine meshing into a finite elements and to carry out the tensile test with the given simulation software. In this case, it is best to use the 4 nodes shell or 10 nodes tetrahedron elements, depending on whether the material is tuned for the volume or shell elements. In case of volume mesh it is necessary to check the numerical material model thanks to which we intend to count elastoplastic behaviour of steel. Some elements have the tendency to be too soft (supple – for example hexahedrons), others can exhibit very high stiffness resulting in locking. At any case, the real and computed tensile curve must show compliance in the whole elastic-plastic domain. If this is not the case, it is necessary to adjust the characteristic variables (parameters) of the material models so that the difference is in the range of acceptable values.

Next, it is necessary to proceed with the tuning of the material properties to the size of the computation mesh. Although the speed at which computers perform is constantly improving , currently we are not able to calculate large models and large-scale tasks with a fine mesh size. To hammer out the results within a reasonable time, it is necessary to use a coarser mesh grid that is determined by the optimal characteristics of the meshing. In practice, for steel it means that instead of 5 mm mesh size for the dynamics tasks (with large deformations), a mesh with a size of 10 to 20 mm is used. Of course it is necessary to adjust and tune the material cards for that computational mesh size. The basic parameters of the material cards for the computational mesh size are also the tensile test results. In this step we prepare another set of the tensile test samples but this time with the element size that corresponds to the computational mesh size. At this stage it is necessary to adjust the parameters of the material model in such a way so that the tensile curves of both simulations (thus fine and real – computational mesh size) are in compliance. It should be emphasized that in all cases we have to compare the real, and not the engineering tensile curves.

## 2. Elastic-plastic material models

In the case of dynamic tasks most often two material models are used – the Krupkowski law and the strain-rate curves for the elastic-plastic description of the material behaviours.

The Krupkowski material model characterizes the behaviour of the material in the elastic-plastic domain which also implies isotropic and kinematic behaviours of the flexible bodies. The material model is described by the following equation:

$$\sigma = K(\varepsilon_0 + \varepsilon)^n \tag{1}$$

where:

$K$ – material constant (hardening coefficient)
$n$ – strain hardening exponent, which has a value of <0,1>
(0 – isotropic hardening, 1 – kinematic hardening)
$\varepsilon_0$ – initial strain on the limit of elasticity

Constants $K$ and $n$ of the material model are obtained through the approximation of the tensile diagram, while engineering stress and strain values are recalculated into real values and thus construct a curve. We do this because the Krupkowski equitation better approximates the actual curve of stress and strains than the measured engineering curve value. The main benefit of the Krupkowski material model is the correct description of the material behaviour between the limits of proportionality and strength. Therefore for the search of constants of this material model we use that part of the tensile curves which describes this plastic area. The course of the real and engineering tensile curve is shown in figure Nr. 2.

Within the parameter settings the goal is to find the constants thanks to which the mathematical approximation is as close as possible to reality. Based on the calculated constants a curve is generated together with the plastic part of the real tensile curve are described in figure Nr. 3. The Krupkowski material law is tuned for a **specific loading speed**.

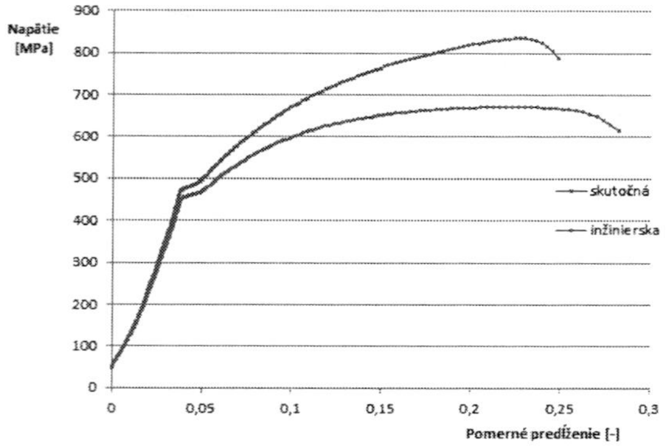


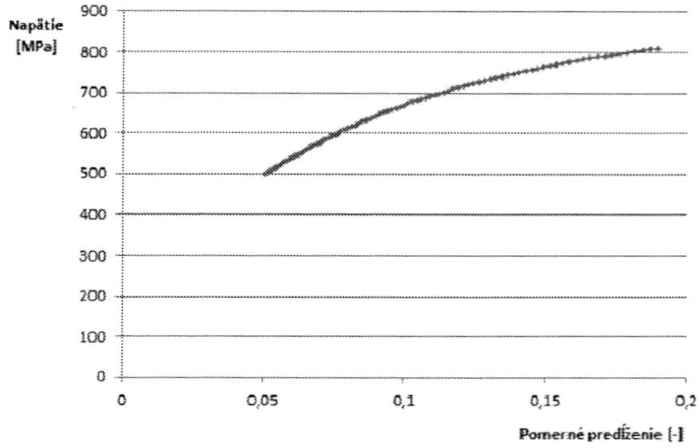

*Fig. 1 The real and engineering tensile curve (left) and the plastic part of the real tensile curve (right)*

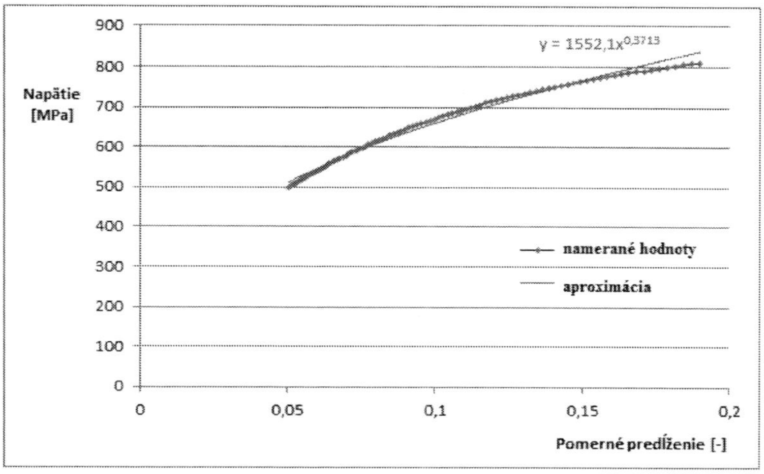


*Fig. 2 the plastic part of the real tensile curve (blue) and its comparison with the computation by using the tuned Krupkowski law and given mesh size (red)*

Another option to describe the material behaviour in the whole elastic-plastic domain is to use the strain-rate curves for a given range of speeds. This means that in this case it is necessary to experimentally measure the elastic-plastic behaviour of the material for different load speeds. These curves describe the strain rate influence on the plastic strain response for the given load speed. The number of curves that can be used as an input for a given material is usually limited by the specific software only – in the case of Virtual Performance Solution of ESI it is possible to define 8 different curves. These curves describe the strains $\varepsilon$ depending on the strain rates $\dot{\varepsilon}$ the software interpolates the values $\sigma\ (\varepsilon, \dot{\varepsilon})$ from the specified curves. The first curve representing the basic stress-strain law $\dot{\varepsilon} = 0$ must always be defined.

## 3. Three-point dynamic bending

Static tensile tests and their execution by means of simulations help us create the basic material cards. For the dynamic calculations it is necessary to check the behaviour of a material under a dynamic loading of which impact velocity is typical for this particular problem. In case of elastic-plastic materials such verification can be done by using the dynamic bending test.

The dynamic bending test is characterized by a load of known weight *m*, which is impacting the test sample from a predetermined height. Since the value of the track of the free fall is known, the impact speed can be determined with a very high precision. The system must be set so that the impact velocity replicated that speed we require to perform the virtual test.

The weight of the impactor can be selected so that permanent deformation of the test specimen is approaching the expected deformation of the realtest. The scheme of the dynamic bending test si shown in figure nr. 4.

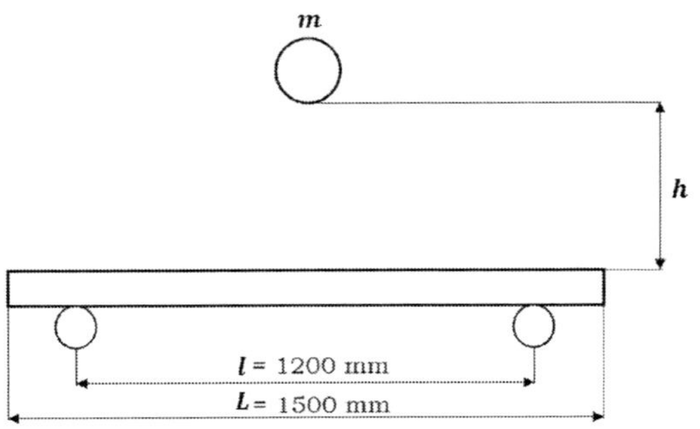


*Fig. 3 The scheme of the dynamic bending test*

After the impact, the impactor of a given mass $m$ falls onto the test piece which bents and then after the impact and springback it results in a permanent deformation $y_d$. The magnitude of this distortion can be expressed by the following formula:

$$y_D = \frac{G_D \cdot l^3}{48 \cdot E \cdot Jz} \qquad y_d = \frac{G_D . l^3}{48 E J_z} \qquad (2)$$

An established method of obtaining the material data necessary for the simulation is performing dynamic bending tests on real samples and also on a computer model. Then these tests are compared to see whether they match up. The acceptable tolerance band is is up to 10%.

## 4. Experimental tests

Our experimental tests were carried out on test specimen of the square cross-section with dimensions of 50x50 [mm], with a wall thickness of 3 [mm] and a length of 1500 [mm]. The test specimen were made of steel of class S355. The impactor mass is 60 [kg] and its impact speed pn the test specimen were $v_d = 8$ [m.s$^{-1}$].

Substituting the appropriate values into the formula (2) we can estimate that the theoretic value of the maximal bending deformation after springback, without taken into account the hardening of the material, is

$$y_D = 56.14\ [mm] \qquad (3)$$

In our case it is very difficult to measure the maximum deformation of the test specimen. Although the whole course of the test is recorded by a high-speed camera with a speed of 1 picture for a millisecond, the maximum deformation under our conditions can on;y be read and changed only with an accuracy of two-tenths of a millimetre. Hence we decided to evaluate the value of the permanent deformation of the test specimen. The actual experiment we conducted on 10 identical test specimen, while residual deflection was measured in 3 places of each one sample, The first measurement point was right in the middle of the sample, the second and the third were at a distance of ¼ L from the underpinning points towards the centre of the test piece. For the footage of f the carried out experiment, see Figure no. 5.

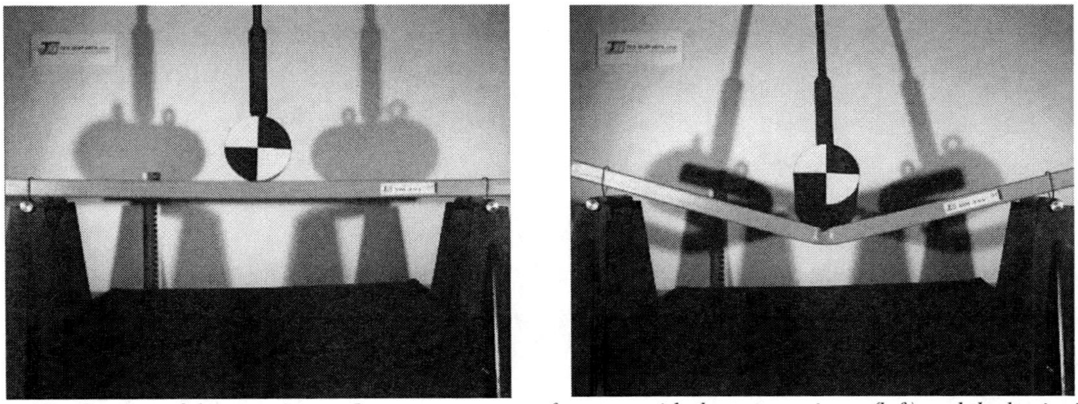

*Fig. 4 Realization of the experiment: Impactor moment of contact with the test specimen (left) and the beginning of the springback*

We used an identical approach in the case of virtual tests. In the modul VisualMESH™ of the software Virtual Performance Solution™ we prepared a finite element model with the identical bending test conditions. The model was prepared using mid surfaces, then we have given geometry divided into a finite network. For the discretization we used 4-nodes shell elements. Subsequently onto the prepared model we applied the initial and boundary conditions coincide with the real conditions of the experiment.

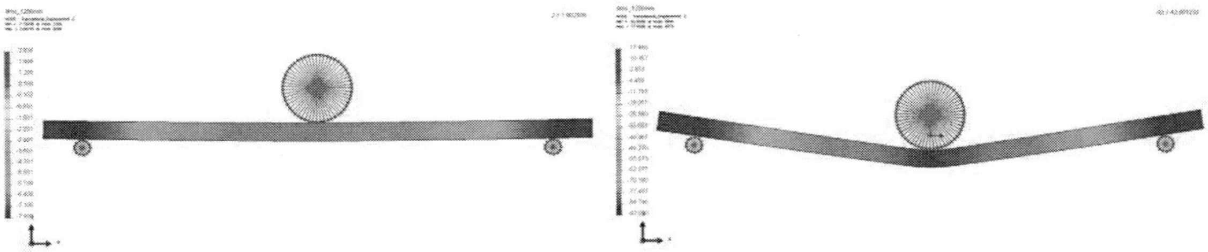

*Fig. 5 Outputs of the simulation: Impactor moment of contact with the test specimen (left) and the beginning of the springback*

## 5. Obtained results

The aim of this work was to analyze the extent to which it is possible to use simplified material models for complex calculations of dynamic events. For a more complete picture, we compared the two most widely used material models with experimental results in 3 characteristic points of the test specimen. The test results have been included in the following table:

| The used test method | Deformations in the characteristic point of the test sample | | |
|---|---|---|---|
| | 1 | 2 | 3 |
| Measured experimental data | -8,1 | -45,6 | -8,2 |
| Computation through strain-rate curves | -8,4 | -47,9 | -8,4 |
| Computation using the Krupkowski law | -8,8 | -48,7 | -8,7 |

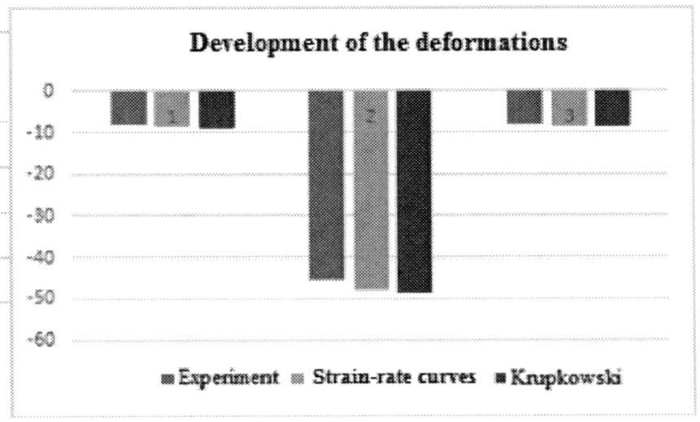


*Fig. 6 Comparison of the results of the experiment and computations*

Based on the obtained results we can conclude, that the maximum value of the percent deviation of the experiment and calculation when using the Krupkowski law is 8.6%. If we compare simulation results with the use of strain-rate curves, the maximum value of the percent deviation of the experiment and calculation is 5.0% .

If we compare the results obtained by calculations with two different material models, we can conclude that the maximal percent deviation is up to 2%.

## Conclusions

The present article discusses the use of various flexible-plastic material models for simulation of nonlinear dynamic events. The subject is to optimize the material parameters to minimize the computational time in compliance with the quality requirements of the numerical calculation results. Based on the obtained results we can conclude, the two methods are fully applicable, and provide the result in the frame of the permitted tolerance margin percentage deviation. In areas with a strong plastic deformation the difference of the results of the computations using the Krupkowski law and the strain-rate curves is only 1.6%, while the latter method is financially much more expensive. Experiments to measure the strain-rate curves have of an order of higher costill there?
s. The computation time for the solution with the Krupkowski law is only 6% longer in comparison of the computation using the strain-rate curves.

Using the Krupkowski material model, it is possible to finalize the results quickly and efficiently, and the results thus obtained are consistent with the results obtained by experimental tests.

## References


[1] Norma STN EN ISO 6892-1:2010.

[2] Virtual Performance Solution 2015 Solver Reference Manual, ESI Group, 2016.

[3] Andrej Hrivňák, Emil Evin: Lisovateľnosť plechov, ELFA, Košice, 2004.

[4] R. Hasa, J. Danko, T. Milesich, Ľ. Magdolen: Modeling of semi-active vehicle suspension with magnetorhological damper. In Scientific Proceedings Faculty of Mechanical Engineering STU Bratislava, Vol. 22, 2014, s. 35-40.

# SECTION 4

## Experimental research and prediction of strength, life and operational reliability of structures and devices

EAN 2017
55th conference on experimental stress analysis

# Resistance Welding Simulation

Dub Martin[1,a], Lopot František[1,b], Berka Ondřej[1,c] and Marek Sýkora[2,d]

[1]CTU in Prague, Faculty of Mechanical Engineering, Department of Designing and Machine Components, Technická 4, 166 07, Prague 6, Czech Republic

[2]KOPR spol. s.r.o., Michelská 367/4, 140 00, Prague 4, Czech Republic

[a]martin.dub@fs.cvut.cz, [b]flopot@seznam.cz, [c]ondrej.berka@fs.cvut.cz, [d]marek.sykora91@gmail.com

**Keywords:** resistance welding; simulation, fem, fea, temperature dependency

**Abstract.** Simulation of technological processes are very useful for the assessment of its parameters. Conveniently chosen parameters can do process more effective, economic and results more quality. This contribution deals with the simulation of resistance welding process. The simulation is carried out for the welding of the rod and connecting-rod for car shock-absorber. The aim of this simulation is to show the methodics of coupled thermal-electric analysis and the temperature dependency of material constants is taken into account.

## Introduction

Simulation of technological processes are very useful for the assessment of its parameters. Conveniently chosen parameters can do process more effective, economic and results more quality. This contribution deals with the simulation of resistance welding process. The simulation is carried out for the welding of the rod and connecting-rod for car shock-absorber, see Fig. 1. The aim of this simulation is to show the methodics of coupled thermal-electric analysis and the temperature dependency of material constants is taken into account. The welded parts are manufactured of the carbon steel C40.

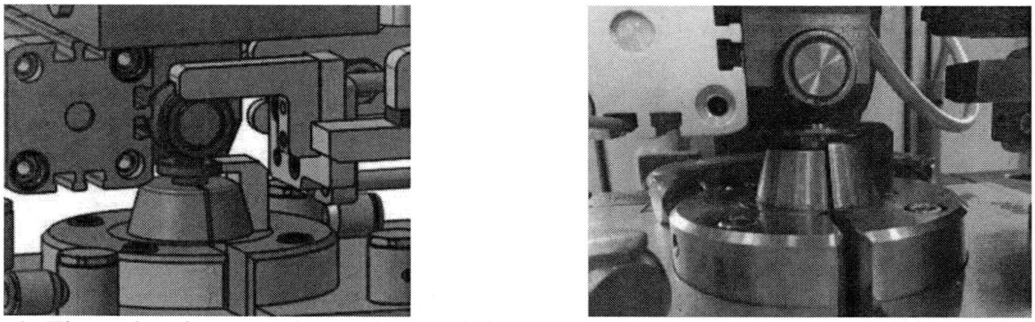

Fig.1: The rod and connecting-rod on welding press – CAD model and real situation, [1]

## Temperature dependency of material constants

In case of welding the material is exposed to a large range of temperatures (from 20°C to melting point approx. 1500°C). Thus, the temperature dependency of material constants such as modulus of elasticity, thermal conductivity or electrical resistivity should be taken into account. The temperature behaviour for modulus of elasticity can be seen in Fig. 2. There is evident decrease of the modulus of elasticity round the temperature 700°C, which corresponds to the crystal structure change from ferrite to austenite.

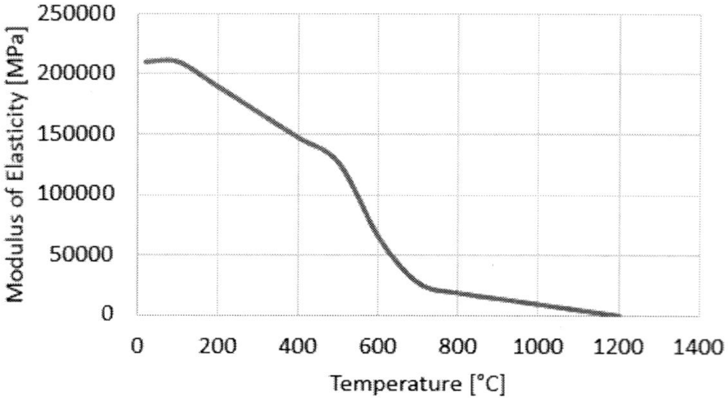

Fig. 2: Temperature behaviour of modulus of elasticity (steel C40), taken from [2]

**Analytical verifications**

Some subtasks of complete thermal-electric analysis with mechanical loading are carried out independently and they are verified by analytical calculations, for example stationary heat transfer with electrical heat source. The heat power $P$ and the volumetric heat source $\dot{Q}$ is defined by the Eq. 1 and Eq. 2 respectively.

$$P = RI \tag{1}$$

$$\dot{Q} = \frac{P}{V} \tag{2}$$

$R$ is the electrical resistance, $I$ is the current and $V$ is the volume of the part. One-dimensional analysis of the electrical heating of the block is performed and it is compared with FEA results. The schema of this verification including the boundary conditions can be seen in Fig. 3

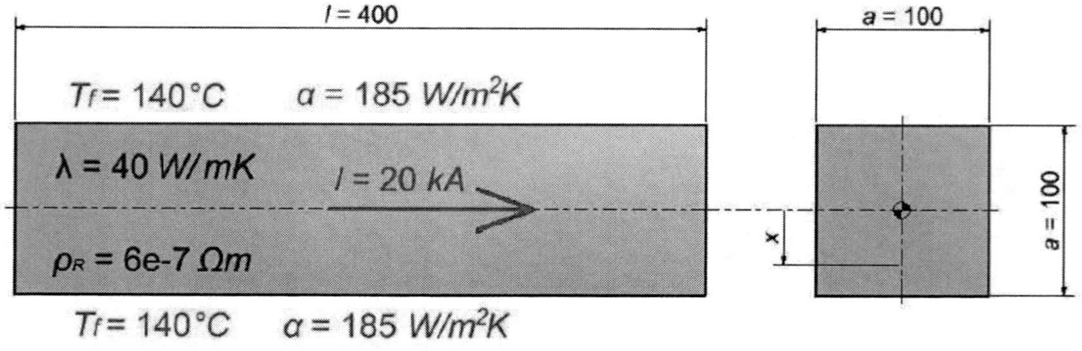

Fig. 3: The schema of the verification example, [1]

The heat flow is described by the Fourier heat equation in Eq. 3

$$\lambda \frac{d^2T}{dx^2} + \dot{Q} = 0 \tag{3}$$

Where $\lambda$ is the thermal conductivity of the material. The result equation for the temperature follows.

$$T(x) = T_f + \frac{\dot{Q}}{2\lambda}\left(\frac{a}{2}\right)^2\left[1 + \frac{4}{Bi} - \left(\frac{2x}{a}\right)^2\right] \tag{4}$$

$T_f$ is the outer temperature and $a$ is the transversal dimension of the block. The Biot number $Bi$ is defined as follows, $\alpha$ is the heat transfer coefficient.

$$Bi = \frac{\alpha a}{\lambda} \tag{5}$$

The FEA was performed and obtained results (maximum and minimum temperature) were compared, see Table 1.

Table 1: Comparison of analytical and FE solution

| Quantity | Analytical solution | FEA | Ratio |
|---|---|---|---|
| $T_{max}$ [°C] | 863.6 | 863.6 | 1.0 |
| $T_{min}$ [°C] | 788.6 | 788.6 | 1.0 |

Further subtasks, such as non-stationary heating by the passage of electric current and non-stationary analysis of free cooling, were performed and verified by analytical solution.

**Simulation model**

Simulation model is created in *FEM* software *Abaqus CAE*. The assembly of the rod and connection-rod and derived FE model is shown in Fig. 3. The hexagonal finite elements for space deformations, electric potential and temperature Q3D8 are used.

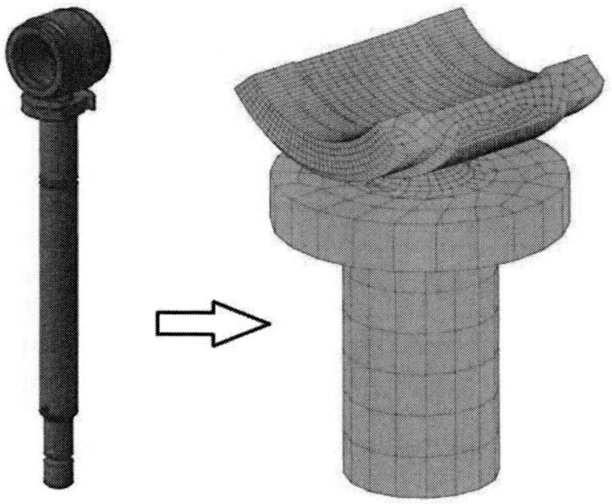

Fig. 3: The assembly of the rod and connection-rod and derived FE model, [1]

The rod and connection-rod are coupled by the *Surface to Surface Contact* and they are pressed to each other in *Static, General Step*. For the combination of thermal-electric analysis and mechanical loading (pressing of both parts to each other) the *Coupled Thermal-Electrical-Structural Step* is used, [3]. The temperature dependent data for steel C40 are used. The current is modeled as alternating with frequency *50 Hz*. Thus the direction of the current density is oscillating as can be seen in Fig. 4.

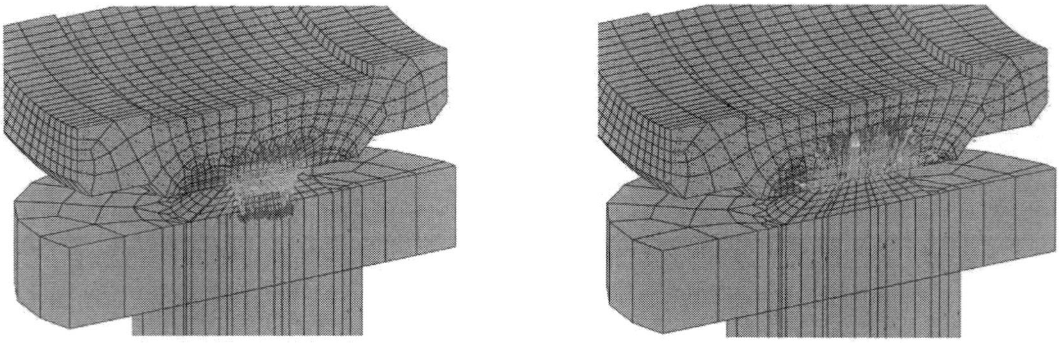

Fig. 4: Oscillating current density, time *t=0.005 s* (left) and time *t=0.015 s* (right), [1]

The maximum contact pressure at time *t=0 s* is $\sigma_{t=0}^{max} = 3140\ \mathrm{N/mm^2}$ and at time t=0.54 s $\sigma_{t=0.54}^{max} = 1820\ \mathrm{N/mm^2}$. That is the result of decreasing material stiffness due to the temperature. The result thermal field is shown in Fig. 5.

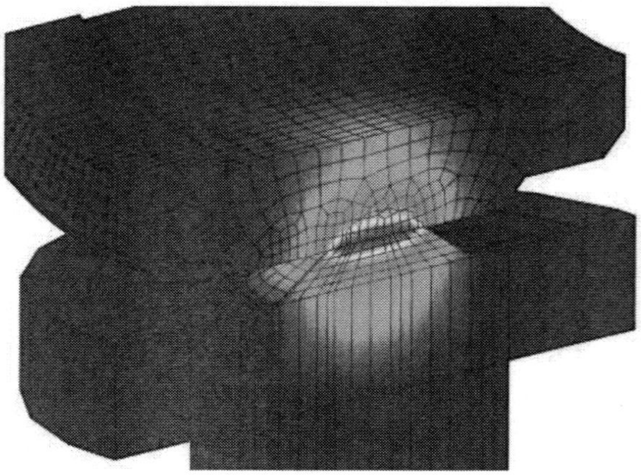

Fig. 5: Result thermal field in welded parts, [1]

**Results and conclusions**

Specialized software is usually used for this kind of simulations. This contribution shows the methodics for simulation in non-specialized engineering software *Abaqus CAE*. The simulation also gives the general information about required welding parameters for quality welds such as pressure, current and time. Because of the large temperature range, the necessity of temperature dependent material model is shown.

**Acknowledgements**

The work was supported by the company KOPR spol. s.r.o.

**References**

[1] M. Sýkora, Redesigning of Resistance Welding Press, CTU Prague (2015).

[2] Information on http://www.matweb.com/

[3] O. Berka, M. Dub, F. Lopot, M. Janda, P. Burda and V. Dynybyl, Determination of Transmission Mechanical Efficiency by Strain Gauges Measurement, EAN 2016 - 54th International Conference on Experimental Stress Analysis (2016).

# Influence of Temperature on Selected Mechanical Properties of Geopolymer Composites

Jan Krystek[1,a], Vladislav Laš[2,b], Vilém Pompe[3,c] and Pavlína Hájková[4,d]

[1,2,3] European Centre of Excellence, NTIS – New Technologies for Information Society, University of West Bohemia, Univerzitní 8, 301 00 Plzeň, Czech Republic

[4]Unipetrol Centre for Research and Education a.s., Revoluční 1521/84, 400 01 Ústí nad Labem, Czech Republic

[a]krystek@kme.zcu.cz, [b]las@kme.zcu.cz, [c]pompe@ntis.zcu.cz, [d]pavlina.hajkova@unicre.cz

**Keywords:** Bending; Composite; Geopolymer; Temperature conditioning; Tension.

**Abstract.** The composites with polymer matrix cannot be used in high temperature environment, while geopolymers excel in high temperature resistance. This paper presents analysis of influence of temperature conditioning on selected mechanical properties of geopolymer composites. Prismatic specimens with two types of geopolymer matrix were tested in tensile or bending tests. The dependencies of the selected mechanical properties on the temperature conditioning were identified for both types of geopolymer matrix.

## Introduction

Currently, composite materials are applied in many industrial areas. It is due to their very good mechanical properties, for example low weight, high strength and high stiffness. Usually, these composites are made with carbon or glass fibers and polymer matrix, for example epoxy matrix. A disadvantage of these composites is that they cannot be used in high temperatures, because their mechanical properties significantly degrade with increasing temperature. This shortcoming can be removed using a geopolymer matrix. Geopolymer matrix is an inorganic polymer material. Preparation of this material is based on aluminosilicate alkali activation. The bond Si-O-Al-O is formed by the polymerization usually under normal temperature and pressure. Geopolymers excel in many properties, primarily high temperature resistance [1, 2], frost resistance or resistance against acids and organic solvent agents. The temperature resistant matrix does not have good mechanical properties such as strength, stiffness or impact resistance, therefore fiber reinforcement is used in order to improve these properties [3-7].

The paper presents analysis of variation of mechanical properties of geopolymer composite with carbon fibers subjected to high temperature. Two types of geopolymer matrix were used. Tensile and bending tests were performed for identification of mechanical properties.

## Materials and specimens

Carbon fibers (Table 1) and two types of geopolymer matrix were used in this work. Geopolymer matrix FC4 consists of potassium water glass, potassium hydroxide (KOH), silica fume, material with high content of metakaolinite, and boric acid. Molar ratios of the components are presented in Table 2. Geopolymer matrix B3P1 consists of potassium water glass, material with high content of metakaolinite, and ingredients with calcium. Material with metakaolonite Mefisto $L_{05}$ (produced by České lupkové zavody, a.s.) was used for both types of matrix.

Table 1: Properties of carbon fabric.

| Material of fibers | Binding | Area weight | Thickness | Density |
|---|---|---|---|---|
| Toray 3K 200 tex | plain | 200 [g/m$^2$] | 0.32 [mm] | 1 760 [kg/m$^3$] |

Table 2: Molar ratios.

| | alkali activator modulus ($SiO_2/M_2O$) | Si:Al | M:Al | Ca:Al | $H_2O$:Al |
|---|---|---|---|---|---|
| **FC4** | 1.08 | 17.07 | 4.35 | - | 24.62 |
| **B3P1** | 1.56 | 1.80 | 1.00 | 0.12 | 6.01 |

The composite plates were made from 10 layers of plain weave carbon fabric (200 g/m$^2$). The prismatic specimens were cut using diamond blade. The specimens were subjected to conditioning temperature at 23 °C, 200 °C, 400 °C or 600 °C. Afterwards, the tensile or bending tests were performed using universal testing machine *Zwick/Roell Z050* at room temperature. Designation of specimens is presented in Fig. 1.

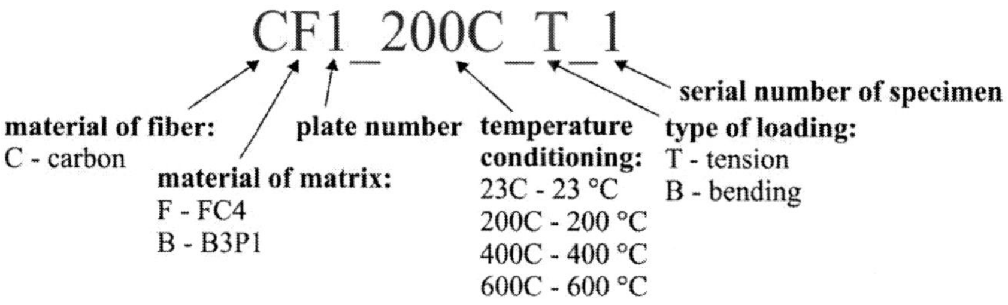


Fig 1: Designation of specimens.

**Tensile test**

The force–displacement dependencies were obtained from tensile test complying with *ASTM D 3039*. The specimen size was modified according to possibilities resulting from the plate size. Total length of the tensile specimens was $l_c = 150$ mm. An initial grip distance was $l_j = 100$ mm. An extensometer was used for measuring the displacement (gage length was $l_e = 60$ mm). The geometric parameters of specimens (width $W$, thickness $H$, weight $M$ and fiber volume ratio $V_f$) are presented in Table 3. The load velocity (crosshead displacement) was $v_T = 2$ mm/min.

The specific load capacity–strain dependencies were calculated. The specific load capacity $f$ is the load, which is carried by one layer of fibers having a width of one meter

$$f = \frac{F_{max}}{W \cdot n_L}, \quad (1)$$

where $F_{max}$ is maximum force, $W$ is width of specimen and $n_L$ is number of fiber layers in specimen. Values of maximum force $F_{max}$, maximum tensile stress ($\sigma_{max} = F/(W \cdot H)$), strain for maximum tensile stress $\varepsilon_{max}$, specific load capacity $f$ and effective moduli $E$ are presented in Table 4. The effective moduli was identified on interval of strain $\varepsilon \in \langle 0.0005, 0.001 \rangle$. The dependencies of the specific load capacity on temperature conditioning are shown in Fig. 2 for both types of the geopolymer matrix. The dependencies of the effective moduli on temperature conditioning are shown in Fig. 3.

Table 3: Weight, geometric parameters and fiber volume ratio of tensile specimens.

| specimen | $W$ [mm] | $H$ [mm] | $M$ [g] | $V_f$ [%] | specimen | $W$ [mm] | $H$ [mm] | $M$ [g] | $V_f$ [%] |
|---|---|---|---|---|---|---|---|---|---|
| CF1_23C_T_1 | 25.7 | 2.9 | 18.45 | 39.2 | CB4_23C_T_1 | 25.9 | 3.5 | 18.76 | 32.5 |
| CF1_23C_T_2 | 25.7 | 3.0 | 18.74 | 37.9 | CB4_23C_T_2 | 25.7 | 3.6 | 18.57 | 31.6 |
| CF1_23C_T_3 | 25.9 | 3.0 | 19.92 | 37.9 | CB4_23C_T_3 | 26.2 | 3.7 | 19.28 | 30.7 |
| CF1_200C_T_1 | 25.7 | 2.9 | 18.19 | 39.2 | CB4_200C_T_1 | 25.9 | 3.6 | 18.66 | 31.6 |
| CF1_200C_T_2 | 25.8 | 2.9 | 18.65 | 37.9 | CB4_200C_T_2 | 26.0 | 3.6 | 19.03 | 31.6 |
| CF1_200C_T_3 | 25.7 | 3.0 | 18.95 | 37.9 | CB4_200C_T_3 | 26.1 | 3.6 | 19.02 | 31.6 |
| CF1_400C_T_1 | 26.2 | 3.0 | 19.53 | 36.7 | CB4_400C_T_1 | 26.2 | 3.6 | 19.19 | 31.6 |
| CF1_400C_T_2 | 25.9 | 3.1 | 20.08 | 37.9 | CB4_400C_T_2 | 26.2 | 3.7 | 19.63 | 30.7 |
| CF1_400C_T_3 | 25.9 | 3.0 | 19.79 | 37.9 | CB4_400C_T_3 | 25.8 | 3.6 | 19.12 | 31.6 |
| CF1_600C_T_1 | 25.7 | 3.0 | 19.17 | 37.9 | CB4_600C_T_1 | 25.9 | 3.6 | 19.11 | 31.6 |
| CF1_600C_T_2 | 26.0 | 3.0 | 19.53 | 37.9 | CB4_600C_T_2 | 25.7 | 3.5 | 18.58 | 32.5 |
| CF1_600C_T_3 | 26.0 | 3.0 | 19.30 | 37.9 | CB4_600C_T_3 | 25.7 | 3.5 | 18.32 | 32.5 |

Table 4: Maximum force, stress and strain, specific load capacity and effective tensile moduli.

| specimen | $F_{max}$ [N] | $\sigma_{max}$ [MPa] | $\varepsilon_{max}$ [%] | $f$ [kN/m] | $\bar{f}$ [kN/m] | E [GPa] | $\bar{E}$ [GPa] |
|---|---|---|---|---|---|---|---|
| CF1_23C_T_1 | 19388 | 260.1 | 1.00 | 94.3 | | 44.9 | |
| CF1_23C_T_2 | 17354 | 225.1 | 0.94 | 84.4 | 92.77 | 41.0 | 42.67 |
| CF1_23C_T_3 | 20631 | 265.5 | 1.10 | 99.6 | | 42.1 | |
| CF1_200C_T_1 | 17553 | 235.5 | 1.01 | 85.4 | | 38.6 | |
| CF1_200C_T_2 | 17175 | 229.5 | 1.05 | 83.2 | 83.30 | 36.4 | 37.47 |
| CF1_200C_T_3 | 16707 | 216.7 | 0.91 | 81.3 | | 37.4 | |
| CF1_400C_T_1 | 15001 | 190.9 | 0.99 | 71.6 | | 25.0 | |
| CF1_400C_T_2 | 11773 | 146.6 | 0.92 | 56.8 | 63.50 | 20.0 | 21.83 |
| CF1_400C_T_3 | 12861 | 165.5 | 0.94 | 62.1 | | 20.5 | |
| CF1_600C_T_1 | 10227 | 132.6 | 0.70 | 49.7 | | 25.6 | |
| CF1_600C_T_2 | 9918 | 127.2 | 0.58 | 47.7 | 48.73 | 28.6 | 25.97 |
| CF1_600C_T_3 | 10159 | 130.2 | 0.64 | 48.8 | | 23.7 | |
| CB4_23C_T_1 | 19147 | 211.2 | 1.23 | 92.4 | | 19.2 | |
| CB4_23C_T_2 | 18520 | 198.6 | 1.08 | 89.4 | 91.03 | 21.3 | 19.47 |
| CB4_23C_T_3 | 18769 | 197.4 | 1.12 | 91.3 | | 17.9 | |
| CB4_200C_T_1 | 17769 | 190.6 | 1.13 | 85.8 | | 19.2 | |
| CB4_200C_T_2 | 17461 | 186.6 | 1.09 | 83.9 | 84.97 | 16.8 | 18.57 |
| CB4_200C_T_3 | 17799 | 189.4 | 1.08 | 85.2 | | 19.7 | |
| CB4_400C_T_1 | 14092 | 149.4 | 0.86 | 67.2 | | 10.3 | |
| CB4_400C_T_2 | 14707 | 151.7 | 0.90 | 70.2 | 68.43 | 10.3 | 9.97 |
| CB4_400C_T_3 | 14015 | 150.9 | 1.13 | 67.9 | | 9.3 | |
| CB4_600C_T_1 | 14120 | 151.4 | 1.07 | 68.1 | | 9.0 | |
| CB4_600C_T_2 | 13770 | 154.3 | 1.05 | 67.5 | 67.33 | 8.0 | 8.30 |
| CB4_600C_T_3 | 13590 | 151.1 | 1.01 | 66.1 | | 7.9 | |

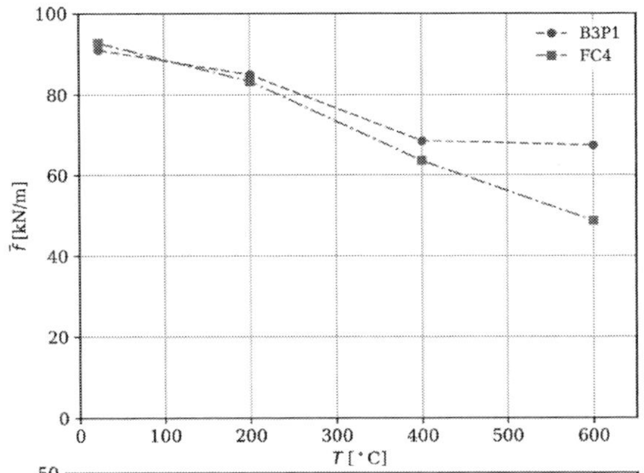


Fig. 2: Dependencies of the specific load capacity on temperature conditioning.

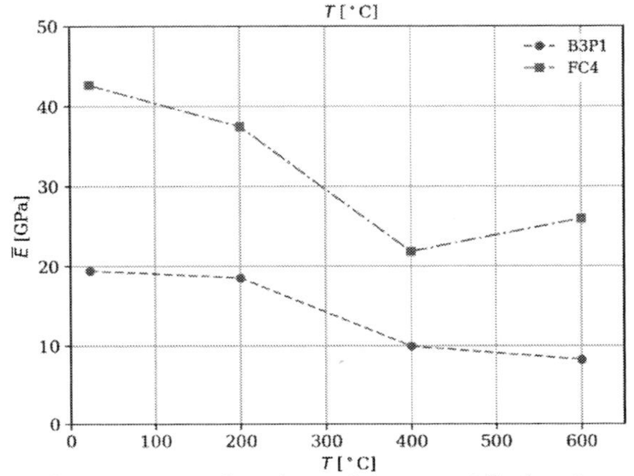


Fig. 3: Dependencies of the effective moduli on temperature conditioning.

**Bending test**

The force–displacement dependencies were obtained from bending test complying with *ČSN EN 2562*. Total length of the tensile specimens was $l_c = 150$ mm. The support span was $l_s = 80$ mm. An extensometer was used for measuring the displacement of loading nose. The geometric parameters of specimens (width $W$, thickness $H$), weight $M$ and fiber volume ratio $V_f$ are presented in Table 5. The load velocity (crosshead displacement) was $v_B = 5$ mm/min.

Table 5: Weight, geometric parameters and fiber volume of bending specimens.

| specimen | $W$ [mm] | $H$ [mm] | $M$ [g] | $V_f$ [%] | specimen | $W$ [mm] | $H$ [mm] | $M$ [g] | $V_f$ [%] |
|---|---|---|---|---|---|---|---|---|---|
| CF1_23C_B_1 | 26.1 | 3.0 | 19.44 | 37.9 | CB4_23C_B_1 | 25.6 | 3.5 | 18.58 | 32.5 |
| CF1_23C_B_2 | 25.7 | 2.8 | 17.74 | 40.6 | CB4_23C_B_2 | 25.6 | 3.5 | 18.75 | 32.5 |
| CF1_23C_B_3 | 25.7 | 2.8 | 17.20 | 40.6 | CB4_23C_B_3 | 25.8 | 3.5 | 18.72 | 32.5 |
| CF1_200C_B_1 | 25.8 | 2.9 | 19.02 | 39.2 | CB4_200C_B_1 | 25.7 | 3.5 | 18.80 | 32.5 |
| CF1_200C_B_2 | 25.9 | 2.9 | 18.51 | 39.2 | CB4_200C_B_2 | 26.2 | 3.6 | 19.05 | 31.6 |
| CF1_200C_B_3 | 25.8 | 2.8 | 17.96 | 40.6 | CB4_200C_B_3 | 25.9 | 3.6 | 18.84 | 31.6 |
| CF1_400C_B_1 | 26.0 | 2.8 | 18.19 | 40.6 | CB4_400C_B_1 | 25.8 | 3.6 | 18.85 | 31.6 |
| CF1_400C_B_2 | 26.1 | 2.8 | 18.74 | 40.6 | CB4_400C_B_2 | 26.0 | 3.6 | 19.33 | 31.6 |
| CF1_400C_B_3 | 26.0 | 2.9 | 17.53 | 39.2 | CB4_400C_B_3 | 26.2 | 3.6 | 19.43 | 31.6 |
| CF1_600C_B_1 | 26.1 | 2.9 | 19.39 | 39.2 | CB4_600C_B_1 | 25.9 | 3.6 | 19.13 | 31.6 |

| | | | | | | | | | |
|---|---|---|---|---|---|---|---|---|---|
| CF1_600C_B_2 | 26.1 | 2.9 | 18.94 | 39.2 | CB4_600C_B_2 | 25.8 | 3.5 | 18.31 | 32.5 |
| CF1_600C_B_3 | 26.1 | 2.9 | 19.15 | 39.2 | CB4_600C_B_3 | 26.2 | 3.6 | 19.35 | 31.6 |

Values of maximum force $F_{\max}$, maximum tensile stress

$$\sigma_{\max} = \frac{3 \cdot F_{\max} \cdot l_s}{2 \cdot W \cdot H^2}, \tag{2}$$

and displacement for maximum force are presented in Table 6. The dependencies of the maximum stress on temperature conditioning is shown in Fig. 4 for both types of the geopolymer matrix. Experimental setup and typical failure after the bending test is shown in Fig. 5.

Table 6: Maximum force, maximum stress and displacement for maximumforce - bending.

| specimen | $F_{\max}$ [N] | $\sigma_{\max}$ [MPa] | $\overline{\sigma_{max}}$ [MPa] | $u_{\max}$ [mm] | specimen | $F_{\max}$ [N] | $\sigma_{\max}$ [MPa] | $\overline{\sigma_{max}}$ [MPa] | $u_{\max}$ [mm] |
|---|---|---|---|---|---|---|---|---|---|
| CF1_23C_B_1 | 210 | 107.4 | | 2.05 | CB4_23C_B_1 | 136 | 52.2 | | 1.32 |
| CF1_23C_B_2 | 194 | 115.7 | 116.47 | 2.37 | CB4_23C_B_2 | 146 | 55.8 | 53.57 | 2.02 |
| CF1_23C_B_3 | 212 | 126.3 | | 2.40 | CB4_23C_B_3 | 139 | 52.7 | | 1.57 |
| CF1_200C_B_1 | 170 | 93.8 | | 2.61 | CB4_200C_B_1 | 77 | 29.2 | | 1.69 |
| CF1_200C_B_2 | 164 | 90.2 | 92.73 | 1.90 | CB4_200C_B_2 | 79 | 27.8 | 28.07 | 1.56 |
| CF1_200C_B_3 | 159 | 94.2 | | 1.90 | CB4_200C_B_3 | 76 | 27.2 | | 1.76 |
| CF1_400C_B_1 | 123 | 72.4 | | 1.83 | CB4_400C_B_1 | 35 | 12.7 | | 2.76 |
| CF1_400C_B_2 | 136 | 79.9 | 72.8 | 2.37 | CB4_400C_B_2 | 40 | 14.1 | 13.43 | 2.15 |
| CF1_400C_B_3 | 120 | 66.1 | | 1.68 | CB4_400C_B_3 | 38 | 13.5 | | 1.67 |
| CF1_600C_B_1 | 200 | 109.2 | | 1.66 | CB4_600C_B_1 | 31 | 11.2 | | 1.58 |
| CF1_600C_B_2 | 192 | 104.9 | 108.37 | 1.74 | CB4_600C_B_2 | 29 | 11.1 | 11.23 | 1.44 |
| CF1_600C_B_3 | 203 | 111.0 | | 1.76 | CB4_600C_B_3 | 32 | 11.4 | | 1.57 |

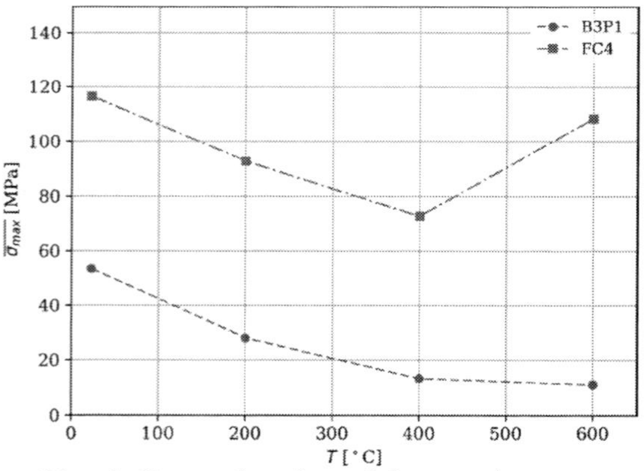


Fig. 4: Dependencies of the maximum stress on temperature conditioning – bending test.

Fig. 5: Typical failure after bending test.

## Conclusion

The composite specimens were composed of carbon fibers and geopolymer matrix. Two types of geopolymer matrix were used. The specimens were subjected to conditioning temperature at 23 °C, 200 °C, 400 °C or 600 °C. Afterwards the tensile or bending tests were performed at room temperature.

The dependence of the specific load capacity and the effective moduli on the conditioning temperature were determined in case of tensile test. The specific load capacity decreases with increasing temperature conditioning for both types of geopolymer matrix. The effective moduli in tension for the specimens with matrix B3P1 is approximately half that for the specimen with FC4 matrix.

In case of bending test, maximum stress for the specimens with matrix B3P1 is lower than for the specimen with FC4 matrix. The maximum stress decreases with increasing conditioning temperature for matrix B3P1. In case of matrix FC4, the maximum stress decreases with increasing conditioning temperature only up to 400 °C.


## Acknowledgment

This publication was supported by the project LO1506 of the Czech Ministry of Education, Youth and Sports. The publication is a result of the project Development of the UniCRE Centre (project code LO1606) which was financially supported by the Ministry of Education, Youth and Sports of the Czech Republic under the National Programme for Sustainability I.


## References


[1] C. Papakonstantinou, P. Balaguru, R.Lyon, Comparative study of high temperature composites, Composites Part B: Engineering 32 (8) (2001) 637-649.

[2] J. Mills-Brown, K. Potter, S. Foster, T. Batho, Thermal and tensile properties of polysialate composites, Ceramics International 39 (2013) 8917-8924.

[3] D. Pernica, P. Reis, J. Ferreira, P. Lauda, Effect of test condition on bending strength of geopolymer-reinforced composite, J Mater Sci (2010) 45:744-749.

[4] M. Alzeer, K. MacKenzie, Synthesis and mechanical properties of novel composites of inorganic polymers (geopolymers) with unidirectional natural flax fibers (phormium tenax), Applied Clay Science 75-76 (2013) 148-152.

[5] A. Natali, S. Manzi, M. Bignozzi, Novel fiber-reinforced composite materials based on sustainable geopolymer matrix, Procedia Engineering 21 (2011) 1124-1131.

[6] S. Samal, B. Marvalová, I. Petriková, K. Vallons, S. Lomov, H. Rahier, Impact and post impact behavior of fabric reinforced geopolymer composite, Construction and Building Materials 127 (2016) 111-124.

[7] S. Samal, et al. Impact and post impact behavior of fabric reinforced geopolymer composite. Construction and Building Materials, 2016, 127: 111-124

# Usage of Illite-kaolinitic Clay for Rammed Earth Manufacturing and Its Drying Rate

Tereza Otcovská[1, a],Barbora Mužíková[1,b] and Pavel Padevět[1,c]

[1]CTU in Prague, Faculty of Civil Engineering, Department of Mechanics, Thakurova 7, Prague 6, Czech Republic

[a]tereza.otcovska@fsv.cvut.cz, [b]barbora.muzikova@fsv.cvut.cz, [c]pavel.padevet@fsv.cvut.cz

**Keywords:** Clay, rammed earth, water coefficient, drying rate, coefficient ratio.

**Abstract.** This submission concerns by unburned clay as a construction material. It briefly summarize basic reasons why to deal with an earth construction and its main goal is determine rammed earth drying speed. The speed of drying is crucial because it is only process with which unburned clay obtains its strength. A main part of the submission is dedicated to description of carried out experiments and their evaluation. There was one earth mixture tested from which two testing body sizes were manufactured by technology of ramming. The speed of drying was determined by measuring of moisture drop.

## Introduction

An unburned clay was commonly used construction material of the past. However, with the boom of new modern construction materials, unburned clay was pushed out from the business, and therefore the necessary scientific research of unburned clay mechanical properties was never performed [1–3].

These days unburned clay returns to focus again generally for three reasons. First reason is reconstruction of already built clay buildings[4, 5]. Second, the unburned clay constructions are compatible with the principles of sustainable development [6, 7]. Third, unburned clay has a blessed impact on the internal micro-climate of the building and consequent improvement in the health of inhabitants [1, 8].

For proper unburned clay construction design it is necessary to employ construction methods that are properly based on scientific research. Such methods are not available nowadays. This article concerns especially with a method of unburned rammed earth test bodies manufacture and theirs drying speed. The manufacturing method is critical because it considerably influence properties of final product such as compressive strength, bending tensile strength,etc. [1, 9].

The drying speed of unburned clay is important because drying is the only process with witch unburned clay receive its strength. It is so necessary to know drying speed to be able to estimate the time that must pass from a manufacture to full construction load. [1, 7].

## Manufacturing Method

An earth appropriate for building constructions consists from three elementary components: sand, clay and water. The clay fulfils a function of binder, the sand a function of filling agent and the water serves for activation of bonding properties of the clay and for good earth mixture processing.

An earth mixture for testing bodies manufacture was proposed. A sand to clay ratio was 80/20 and water to clay coefficient (water coefficient) was 0.37. The components were mixed and mechanically and manually kneaded until homogenous clay mixture was not obtained. The used clay was of illite-kaolinitic type.

The clay mixture was rammed to steel moulds and the testing bodies was created. It was totally manufactured 5 test bodies of size 4 x 4 x 16 cm and 12 testing bodies of size 2 x 2 x 10 cm (see Fig. 1). Ready-made testing bodies were right after unmoulding inserted into air-conditioned chamber (see Fig. 2). The environment properties in air-conditioned chamber were set to 20 °C and 50 % relative air humidity. Test bodies composition is stated in Table 1.

Fig. 1 Manufacture of test bodies

Fig.2 Test bodies in an air-conditioned chamber

Table 1 Composition and quantity of test bodies

| Size of test bodies [cm] | Number of test bodies [ks] | Kind of clay | Sand/clay ratio [-] | Water clay ratio [-] |
|---|---|---|---|---|
| 4x4x16 | 5 | illite-kaolinitic | 80/20 | 0.37 |
| 2x2x10 | 12 | illite-kaolinitic | 80/20 | 0.37 |

**Measured Results**

Test bodies put into air conditioned chamber were periodically weighted and a drop of their moisture was observed. Average weight drop tested bodies in time is stated in Table 2.

The measured mass decrease was used for mass moisture content (see Table 3, Fig. 3) and mass moisture content to test body form coefficient (volume to surface ratio) (see Table 4 Fig. 4).

Table 2 Weight drop of tested bodies in time

| Size of test bodies [cm] | Date of measuring | | | | | | |
|---|---|---|---|---|---|---|---|
| | 27.2.2017 | 28.2.2017 | 1.3.2017 | 3.3.2017 | 7.3.2017 | 10.3.2017 | 14.3.2017 |
| 4x4x16 | 603.4 | 567.9 | 566.6 | 565.7 | 565.5 | 565.5 | 565.4 |
| 2x2x10 | 90.2 | 84.3 | 84.1 | 83.5 | 83.5 | 83.4 | 83.3 |

Table 3 Moisture content

| Size of test bodies [cm] | Date of measuring | | | | | | |
|---|---|---|---|---|---|---|---|
| | 27.2.2017 | 28.2.2017 | 1.3.2017 | 3.3.2017 | 7.3.2017 | 10.3.2017 | 14.3.2017 |
| 4x4x16 | 7.4 | 3.1 | 0.9 | 0.7 | 0.7 | 0.7 | 7.4 |
| 2x2x10 | 7.4 | 1.0 | 0.9 | 0.8 | 0.7 | 0.6 | 7.4 |

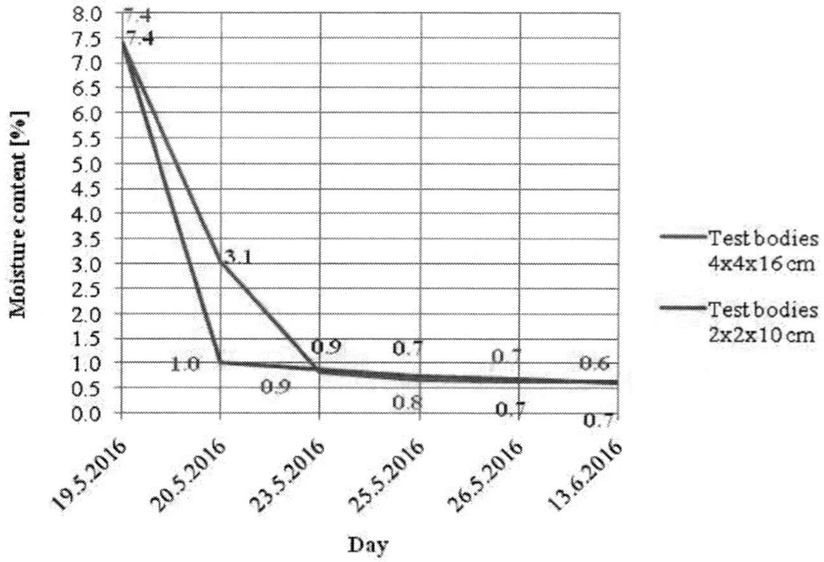


Fig. 3 Moisture content

Table 4 **Moisture content and form coefficient ratio**

| Size of test bodies [cm] | Date of moisture content and form coefficient ratio measure | | | | | | |
|---|---|---|---|---|---|---|---|
| | 27.2.2017 | 28.2.2017 | 1.3.2017 | 3.3.2017 | 7.3.2017 | 10.3.2017 | 14.3.2017 |
| 4x4x16 | 8.3 | 3.4 | 1.0 | 0.8 | 0.7 | 0.7 | 8.3 |
| 2x2x10 | 16.3 | 2.2 | 1.9 | 1.7 | 1.5 | 1.3 | 16.3 |

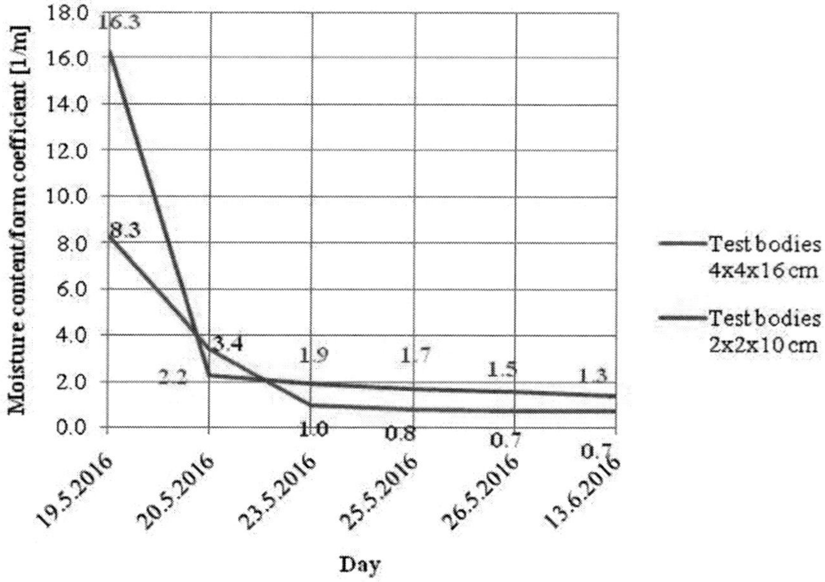


Fig. 4 Moisture content and form coefficient ratio

## Conclusion

We can establish that the drying speed is dependent on ratio of total body volume and body surface. The tested bodies'mass moisture content unified after two day from a manufacture.

We consider as appropriate to measure speed of drying with a use of the form coefficient. However, for proper assessing of speed of drying it is necessary carry out described measurement on another test bodies of other sizes (other value of form coefficient) and measure drop of test bodies' weight in shorter time intervals particularly in first days after manufacture. Proper interval for the measurement just after manufacture is about 1 hour. That experiment we plan to carry out in following months.


## Acknowledgement

The financial support of this work by the Faculty of Civil Engineering, Czech Technical University in Prague (SGS project No. SGS16/201/OHK1/3T/11) is gratefully acknowledged.

We would like to express our thanks to LB MINERALS, s.r.o. company for free supply of material necessary for experimental measurement and we would also like to express thanks to HELUZcihláhřskýprůmyslv.o.s. for giving statistical data and information.


## References


[1] MINKE, Gernot. *Building With Earth* [online]. B.m.: Ökobuch Verlag, Staufen, 2006 [vid. 2015-08-18]. ISBN 978-3-0346-0822-0. Dostupné z: http://archive.org/details/Gernot_Minke-Building_With_Earth

[2] NORMA. *ČSN 1168-1939 - ODMÍNKY PRO ZEDNICKÉ A PŘIDRUŽENÉ PRÁCE POZEMNÍCH STAVEB*. 1951. bibtex: csn_csn_1951

[3] JAQUIN, Paul a Charles AUGARDE. *Earth building : history, science and conservation* [online]. Report. B.m.: Bracknell : IHS BRE Press. 2012 [vid. 2017-02-16]. Dostupné z: http://www.envia.bl.uk//handle/123456789/4131

[4] HÁJEK, Václav. *Lidová stavení: opravy a úpravy*. 1. vyd. Praha: Grada, 2001. Stavitel. ISBN 80-247-9054-8.

[5] KOVÁŘŮ, Věra. Hliněný dům, problematika jeho památkové obnovy a využití. In: *Zděná architektura v Čechách, na Moravě a ve Slezsku*. 2004, s. 31 – 33.

[6] *Agenda 21 - United Nations Environment Programme (UNEP)* [online]. [vid. 2017-02-14]. Dostupné z: http://www.unep.org/Documents.Multilingual/Default.asp?documentid=52

[7] RŮŽIČKA, Jan. Building structures in environmental context, mechanical physical properties of unburned bricks by impact of humidity. *Mezinárodní vědecký seminář Zdravé domy*. 2005.

[8] ŽABIČKOVÁ, Ivana. *Přestavby hliněných budov*. Brno: Sdružení hliněného stavitelství, 2009. ISBN 978-80-254-3907-4.

[9] ŽABIČKOVÁ, Ivana. *Hliněné stavby* [online]. Brno: Era 21, 2002 [vid. 2015-08-18]. ISBN 80-86517-21-7. Dostupné z: http://www.kosmas.cz/knihy/107760/hlinene-stavby/

# Impact of the Heat Load of Concrete on the Propagation of Ultrasound Waves

PAZDERA Luboš[1, a], TOPOLÁŘ Libor[1,b], HODULÁKOVÁ Michaela[1,c], MIKULÁŠEK Karel[1,d] and SMUTNÝ Jaroslav[1,e]

[1] Brno, University of Technology, Faculty of Civil Engineering, Czech Republic

[a]pazdera.l@fce.vutbr.cz, [b]libor.topolar@vutbr.cz, [c]hodulakovam@study.fce.vutbr.cz, [d]mikulasek.k@fce.vutbr.cz, [e]smutny.j@fce.vutbr.cz

**Keywords:** ultrasound waves, heat load, impact, concrete, acoustic emission

**Abstract.** Due to fires of building structures, there has been a growing interest in investigating the properties of concrete at high temperatures. In the fire, a temperature of up to 1400 °C may be measured. Such high temperatures will bring about changes in the structures of the entire construction. Ultrasonic methods such as acoustic emission method can be used to monitor changes in the structure caused by a fire. Two similar concrete mixtures after thermal degradation are evaluated using an impulse ultrasonic method.

## Introduction

Interest in the behaviour of concrete under high temperatures has recently been fuelled by fires of industrial, government- or private-owned buildings, tunnels, and other building structures. In a fire, the temperature may reach up to 1400 °C [1]. Due to such heat, changes occur in structures that may alter the mechanical properties and structures of materials through macro- and micro-cracks.

The resistance to fire of reinforced concrete may be defined as its capacity, in the event of a fire, to keep the concrete's native function, its structural properties, to protect the steel reinforcements, and protect the immediate environment against toxic gases [2]. Damage to concrete due to a fire is not regarded as a degradation mechanism, but rather as an accident and direct external influence. However, the driving forces that cause explosive crumbling of concrete exposed to a fire may be related to the microstructure and transport mechanisms of the cement material, as is the case with other degradation mechanisms [3]. A major difference is that the damage results in a much shorter time and that concrete may be destroyed within an hour or even a shorter interval [4].

The mechanisms causing explosive crumbling or disintegration in the event of fire include concrete humidity and material microstructure. At 100° C, the water contained in concrete will turn to steam. As the temperature grows, the pressure of the steam in concrete will mount, too. With the concrete microstructure being rather open, i.e., with an interconnected system of pores (high water coefficient), the steam pressure will drop as the steam is leaking rather fast. If, however, the concrete structure is rather dense, the steam pressure may reach values exceeding 3 N/mm$^2$. As a result of high internal pressure, a thin surface layer of concrete may abruptly be torn off, which is called explosive crumbling [5].

The paper aims to evaluate the concrete properties monitored at different temperatures of up to 1200 °C [6]. The observed changes in the propagation of ultrasound waves through degraded concrete material may be used to quickly assess the quality of a concrete structure after being exposed to heat from a fire [7]. Acoustic or ultrasound methods are among the non-destructive methods that may often enable an early detection of material failures in a structure [8,9].

## Experimental set up

This paper presents evaluation of ultrasound signals when testing their passage through the material of concrete specimens after a heat load. The used specimens were a prism of 100 mm by 100 mm and length of 400 mm, Each specimen was placed in an oven to be exposed for one hour to one of the temperatures of 200 °C, 400 °C, 600 °C, 800 °C, 1000 °C, and 1200 °C. Two similar mixture was made - cement CEM I 42.5R of 350 kg/m$^3$, sand 0/4 of 820 kg/m$^3$, water of 160 kg/m$^3$, super plasticizer of 3 kg/m$^3$. The mixture A contains coarse aggregate 8/16 which mixture B makes compensate of coarse aggregate 11/22 of 1000 kg/m$^3$.

The PUNDIT ultrasound device with a classic ultrasound probe placed on a 100 mm by 100 mm square wall was generated an impulse signal. Four sensors were pick up ultrasound signals as shown in Fig. 1. The two sensors are placed of 50 mm from the opposite wall of the exciter.

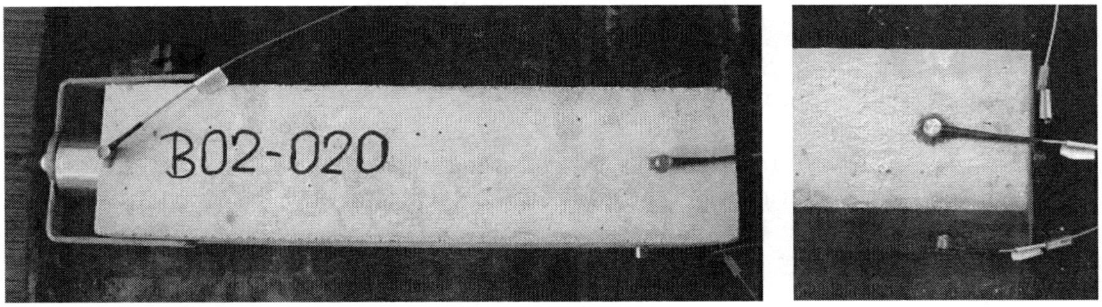


Figure 1 Location of the exciter and the sensors on specimens.

By impact acoustic emission tests, for each temperature degradation, twenty specimens were put to an impact acoustic emission test. Each specimen was loaded with three different pulse amplitudes of 250 V, 500 V, and 1200 V, respectively. The errors of the values measured are below five percentage points.

## Monitored properties

The values measured of selected material properties obtained from destructive and non-destructive tests are illustrated in Fig. 2, 3 and 4 and in Table 1. The bar diagrams show the temperature-dependent trends between the compressive strengths, the flexural bending strengths, maximal amplitudes of the impact acoustic emission and the linearity of increasing loading impulse. Similar trends were observed between the monitored properties and the temperature, that is, with an increasing temperature, the value measured will drop down to a degrading temperature of 1000 °C at which the sample structure begins to change significantly.

The compressive strengths of both mixtures are similar except at degraded usually temperature as shown in Fig. 2 left graph. However, the flexural bending strengths shown in Fig. 2 right graph are a little different; here mixture B contains higher values except for degradation temperature of 1200 °C.

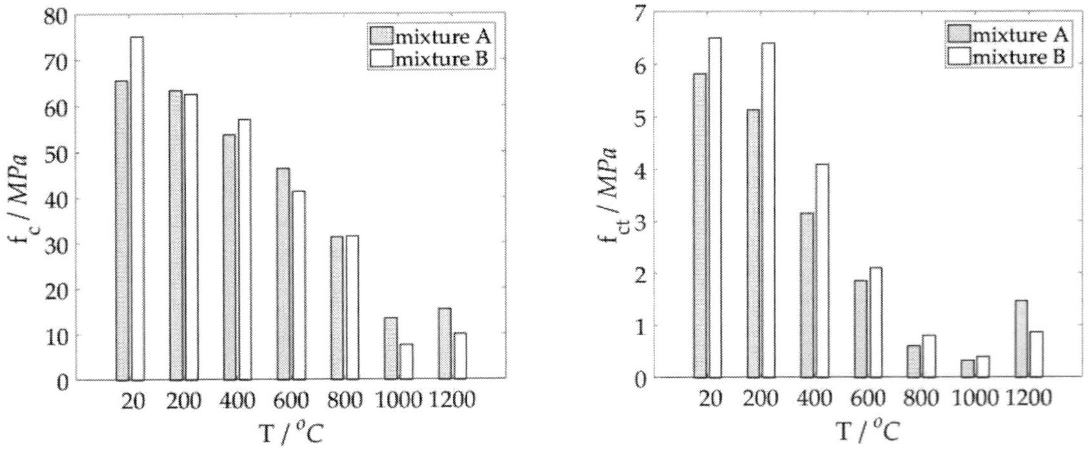


Figure 2 The compressive strength $f_c$ and the flexural bending strength $f_{ct}$ at the different degradation temperatures T

To describe the response emission signals, which are formed during the impact test in the specimens, we focused on the selected parameters of these signals such as maximal amplitude. In Fig. 3 frequency spectral responses of sensor No 4 of mixture A (left graph) and of mixture B (right graph) dependent on the degraded temperature at the loaded impulse of 1200 V will be evaluated in a next article. A response sample were recorded by 4 MHz, therefore the frequency are up to 200 kHz contains higher values of spectra. This sampled frequency was applied for point of view of higher time resolution.

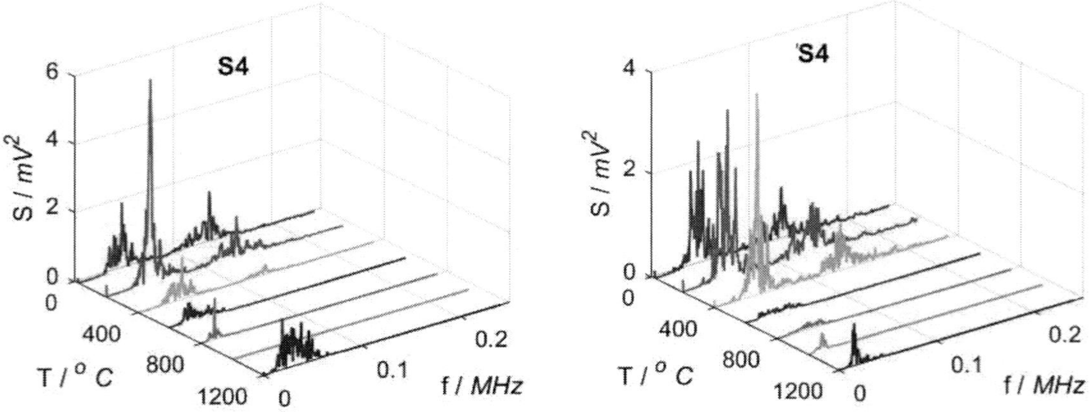


Figure 3 The spectral response S at the different degradation temperature T and the frequency f for specimens of mixture A in the left and of mixture B in the right

The bar graphs in the Figure 4 show seven groups of columns. Each group of six columns shows pairs, that is, for each mixture, different excitation voltages, the first pair 250 V, the second 500 V and the third 1200 V. The degrading-temperature-dependent character of the resizing of the maximum amplitude (see Fig. 4) is similar to that in the case of changes in the compressive strength or the flexural bending strength (see Fig. 2).

There are two areas of interest in the range of temperatures of up to about 200°C and around 1000°C. In the first temperature region of up to 200°C, the maximum amplitude increases, thus the compressive strength does not change significantly. The minimum value is reached both by the maximum pulse amplitude of the impact acoustic emission and the compressive strengths and flexural bending strengths at temperatures around 1000°C, at which the structure of cement mixtures changes.

The changes up to temperature of 200°C are the effect of loss of free water occurs at about 100 °C. Here is formatted Calcium-Silicate-Hydrate (C-S-H) and Calcium Hydroxide $Ca(OH)_2$ - Portlandite. During further raising temperatures of up to 200 °C begins dehydration cementing compound. The main phenomena for decreasing properties until about temperature of 600 °C are decomposition of calcium hydroxide ($Ca(OH)_2 \rightarrow CaO + H_2O$) takes place at about 450 °C and the crystal structure of quartz transforms at temperature of 573 °C from α- to the β- form and at temperature of 900°C calcite starts shrinking due to decomposition. The next phenomena, that affects the mechanical properties, is that the volume and surface of pores increase up to a temperature of 500°C and decrease with further increase of temperature. Whenever the temperature was increased to of 800 °C were the second phase decomposition C-S-H took place and also decomposition of calcium carbonate ($CaCO_3 \rightarrow CaO + CO_2$). When the temperature is above 1000 °C, it changes the overall structure of the concrete mixture to rocks melt, therefore a structural change, accompanied by the creation of new crystal phases (Wollastonit β - $CaO.SiO_2$).

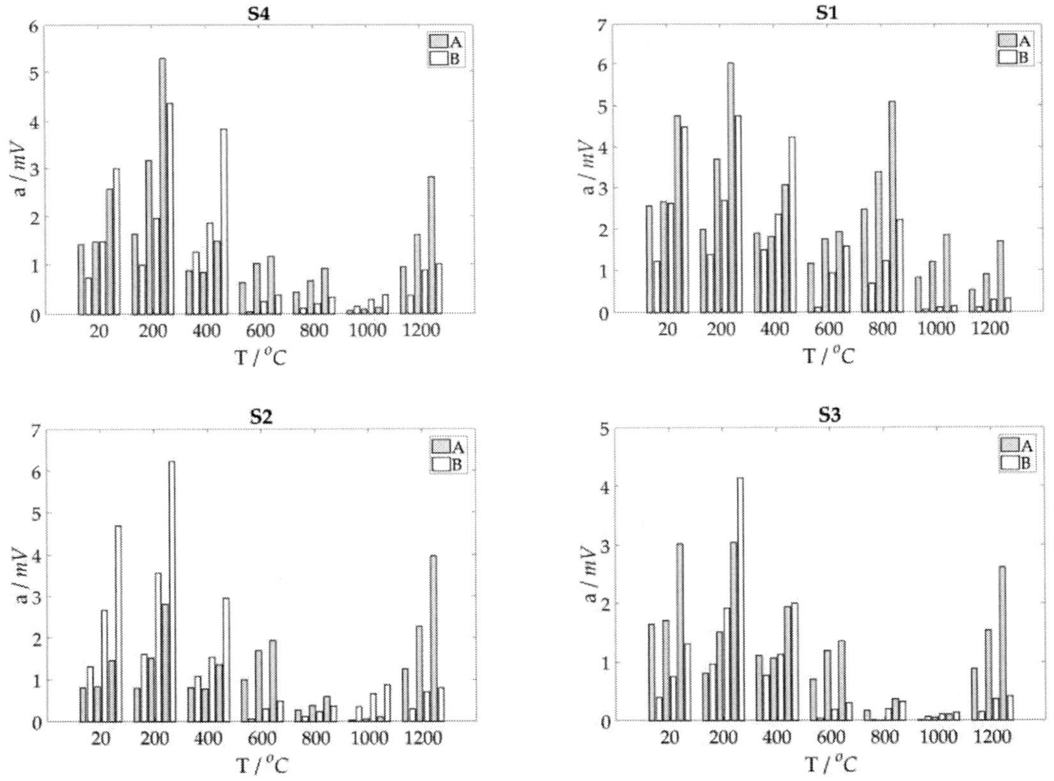


Figure 4 Maximal amplitudes a depending on degraded temperature T and excited voltage (each group contained 250 V, 500 V and 1200 V)

The correlation coefficients of the linear increase in the maximum amplitude due to the rising pulse excitation voltage according to Table 1 are very close to one, which means that the behaviour of the sample is linear. Lower values of the correlation coefficients occur at degradation temperatures of above 600 °C.

Table 1 The correlation coefficients

| | Degraded temperature [°C] | 20 | 200 | 400 | 600 | 800 | 1000 | 1200 |
|---|---|---|---|---|---|---|---|---|
| Sensor 1 | Mixture A | 0.977 | 0.984 | 0.951 | 0.901 | 0.996 | 0.994 | 0.998 |
| | Mixture B | 0.983 | 0.991 | 0.998 | 0.945 | 0.995 | 0.915 | **0.786** |
| Sensor 2 | Mixture A | 0.976 | 0.995 | 0.953 | 0.906 | 0.999 | 0.998 | 0.993 |
| | Mixture B | 0.988 | 0.984 | 1.000 | 0.940 | 0.979 | 0.939 | 0.811 |
| Sensor 3 | Mixture A | 0.977 | 0.998 | 0.954 | 0.907 | **0.752** | 0.982 | 0.992 |
| | Mixture B | 0.990 | 0.999 | 1.000 | 0.944 | 0.921 | 0.923 | 0.805 |
| Sensor 4 | Mixture A | 0.997 | 0.985 | 0.952 | 0.905 | 0.970 | 0.981 | 0.995 |
| | Mixture B | 0.998 | 1.000 | 1.000 | 0.931 | 0.981 | 0.929 | 0.824 |

## Conclusions

The paper describes an analysis of the impact acoustic emission tests carried out on temperature degraded concrete specimens. The evaluations of the experiments imply the possibility of using the impulse acoustic emission method to describe the changes of concrete structure after its fire degradation. The outputs of the impulse acoustic emission method can help for more accurate input parameters for numerical models. As can be seen in Table 1 for mathematical modelling, it can be assumed that increasing the excitation amplitude is a linear process.

The replacement of the coarse aggregate 8/16 by the coarse aggregate 11/22 can improve the fire resistant properties of concrete.

Based on the above results it can be concluded that the monitoring history of concrete using the impulse acoustic emission method can approximately determine the degradation of a structure due to thermal loading. However, in practice, there are many more factors influencing the transport of waves through an observed structure, especially a heterogeneous one.

The above method of acoustic emission combines the advantages of the ultrasound impulse method and the signal recording by the acoustic emission method with valuable information about the longitudinal, transversal, and surface wave obtained from a single impulse. The change in the input pulse shape and the the different levels of the excitation voltage yield comprehensive information on the observed structure.

## Acknowledgement

This paper has been written under the project GACR No.16-02261S supported by Czech Science Foundation and the project No. LO1408 "AdMaS UP - Advanced Materials, Structures and Technologies", supported by Ministry of Education, Youth and Sports under the „National Sustainability Programme I" and under the projects No. FAST-S-16-2967 and No. FAST-J-17-4554 both supported by the Brno University of Technology.

**References**

[1] J. Piasta, Heat deformations of cement paste phases and the microstructure of cement past. Materials and Structures, 17(6) (1989), 415–420

[2] PJ Burke, LA Bisby, MF Green, Effects of elevated temperature on near surface mounted and externally bonded FRP strengthening systems for concrete, Cement & Concrete Composites, 35(1) (2013), 190-199

[3] SJ. Park, HJ. Yim, Evaluation of residual mechanical properties of concrete after exposure to high temperatures using impact resonance method, Construction and Building Materials, 129 (2016) 89-97

[4] C Maraveas, AA Vrakas, Design of Concrete Tunnel Linings for Fire Safety, Structural Engineering International, 24(3) (2014), 319-329

[5] L. Bodnarova, D. Horak, J. Valek, R. Hela, L. Sitek, Possibilities of observation of behaviour of concrete- and cement-based composite materials exposed to high temperatures, Materials Research Innovations, 19 (2015), 936-940

[6] L. Topolar, L. Pazdera, P. Cikrle, Acoustic Emission Monitoring during Static Modulus Elasticity Test of Concrete Specimen ,Applied Mechanics and Materials, 486 (2014), 267-272

[7] A. Benedetti, On the ultrasonic pulse propagation into fire damaged concrete, ACI Structural Journal, 95(3) (1998) 259-271

[8] M. Korenska, M. Manychova, L. Pazdera, Experimental Study of the Nonlinear Effects Generated in a Concrete Structure with Damaged Integrity, Russian Journal of Nondestructive Testing, 49(9)(2013), 530-537

[9] SJ Park, HJ Yim, HG. Kwak, Nonlinear resonance vibration method to estimate the damage level on heat-exposed concrete, Fire Safety Journal, 69 (2014) 36-42

EAN 2017
55th conference on experimental stress analysis

# Water Absorption Properties of Rammed Earth Material with Montmorillonite Clay

BARBORA MUŽÍKOVÁ[1, a], TEREZA OTCOVSKÁ[2,b] and PAVEL PADEVĚT[3,c]

[1,2,3] Thákurova 7, 166 29 Prague, Czech Republic

[a]barbora.muzikova@fsv.cvut.cz, [b]tereza.otcovska@fsv.cvut.cz, [c]pavel.padevet@fsv.cvut.cz

**Keywords:** Rammed earth, water absorption, clay, building.

**Abstract.** This temple presents water absorption properties of rammed earth material. Earth material is very sensitive to moisture and water. Specimens of two recapture were produced in the laboratory using montmorillonite clay. Specimens were rammed in the moulds. Then they were putted in the covering of nylon. They were settled in the box with soft foam that was moistened. Fifteen pieces of them of each recipe were tested. The level of moistening was constant. The specimens were regularly measured and weighted. The weight growth is 11.21 % for GEM I and 13.05 %. The amount of the clay had an influence to the water absorption characteristics. The more clay is used, the more water can the final product absorb.

## Introduction

This temple presents earth material as a modern building material based on the long history of using, but in last centuries it was replaced by modern materials. Due to the replacement the material characteristics were never fully researched. The demand for the earth material is increasing nowadays following the principles of sustainable buildings development [4, 5].

The main advantages of using earth as a building material are good air humidity balance, lower costs for material and construction, and also the fact that the earth based material has a positive effect on human health. Rammed earth is accessible, environmentally friendly and suitable for low-cost building in person. According to the sustainable buildings development not only the mechanical properties are important in the structure design. The quality of building material is also given by economic and socio-cultural criteria. For example the rammed earth primary energy consumption connected with production of building material is about 44 kWh/m$^3$. This is 18 times less than for prefabricated concrete material [1-3, 6, 7].

Table 1 Recipes of used specimens.

| | Sand [%] | Clay [%] | Water –clay ratio [-] |
|---|---|---|---|
| GEM I | 80 | 20 | 0.37 |
| GEM II | 75 | 25 | 0.37 |

The mechanical properties of the unfired earth depend mainly on the composition and the method of processing. The basic methods of using earth are mudwalls, rammed earth, earth bricks, earth infill in a timber frame construction etc. This research focuses on the rammed earth material. Nowadays, there are no standards for using earth as a building material in the civil engineering. It is necessary to search for more unfired earth characteristics in order the engineers could design the earth constructions more than today [2, 3, 6].

## Testing of Unfired Earth Material

**Rammed Earth and Water.** The earth material is very sensitive to moisture and water that is why it is an important characteristic that needs to be examined. The structure of earth is widely un-homogeneous, porous and opened. It is capable to absorb water and transport it in the material itself.

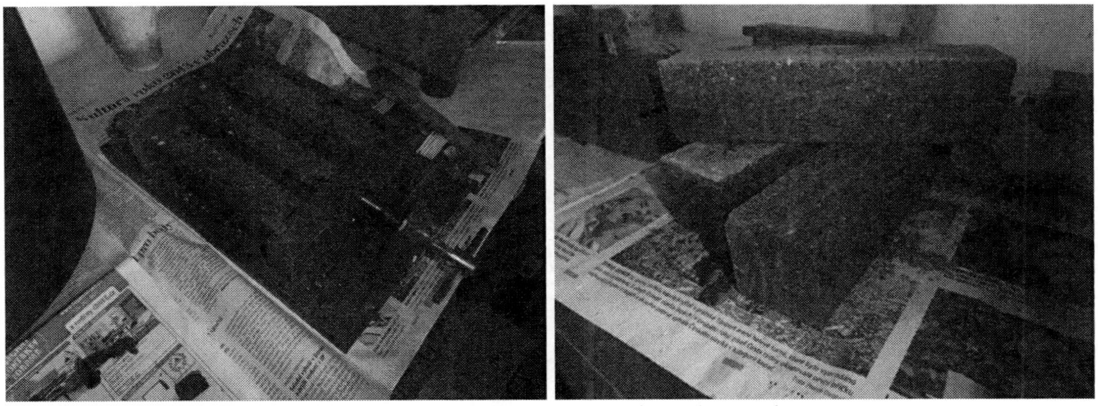

Fig.1 The production of rammed earth specimens.

**Preparing Material.** The recipe for the specimen was designed of montmorillonite clay, sand and water. There are two recipes GEM I and GEM II, both of montmorillonite, but they differ in the sand/clay ratio as it is shown in Table 1. The amount of the components is expressed by weight per cent. The amount of water was settled as a water/clay ratio, same at both recipes. Fifteen specimens of each recipe were tested.

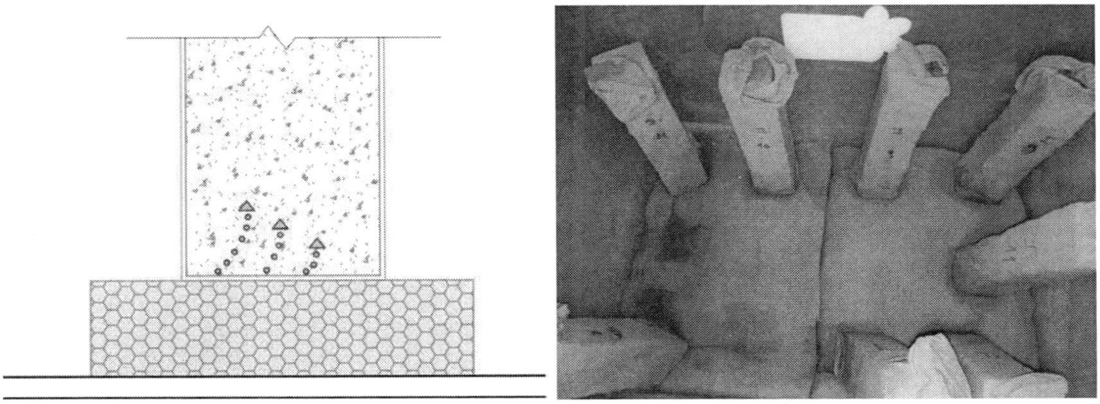

Fig.2 Theoretical process of moisture absorbing and the scheme of testing.

First of all, the sand was well mixed with two thirds of water. Than the clay and the rest of water was added. It was mingled by drilling machine with a special ending. Specimens of size 40 x 40 x 160 mm were prefabricated for the testing in the laboratory. The production was made by ramming into moulds (Fig. 1) by hand and by drilling machine. The moulds were wiped by oil and the earth was rammed in four to five layers. The last one was always made by hand.

**Practical Tests.** Specimens were tested after four weeks from production. By this time the humidity from the prefabrication is stable and is not changing any more. The hardening process is based only on the drying out the mixed water and it is almost over.

Specimens were putted in the covering of nylon this was made not to lose the material during manipulation when the specimens were weighted. Specimens were settled in the box with soft foam that was moistened. The level of moistening was constant, the water was added during the test, but it was not for the purpose to have too much water, the specimens cannot stay in the height level of water, they just absorbed the moisture from the soft foam (Fig.2). The specimen were regularly measured and weighted.

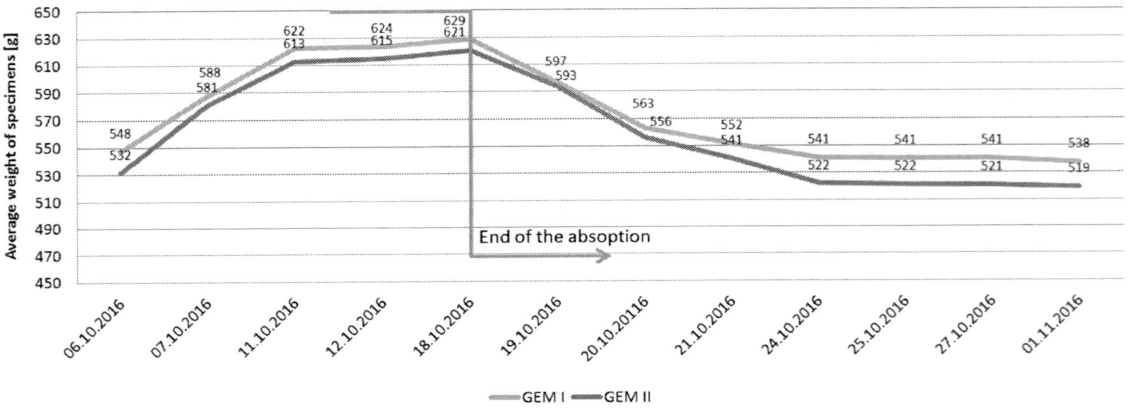


Fig. 3 Development of the weight in time (no broken specimen).

Specimens were measured and weighted then they were settled in the box. Regularly there was made a weighting and a measuring of dimensions. The test took place for twelve days, data was recorded. Some of the specimens were broken during the test, they were removed from the testing but they were still weighted because the leaving moisture was recorded too.

### Evaluation of the Measurement

**The Course of the Test**. The weight and the sizes of the specimens were recorded. In the Fig. 3 the development of the weight in time is shown. These data represent the specimens that were not broken during the whole testing. In the same time they had the same conditions. At 6.10. the specimens were settled in the box with water and 18.10. were taken out.

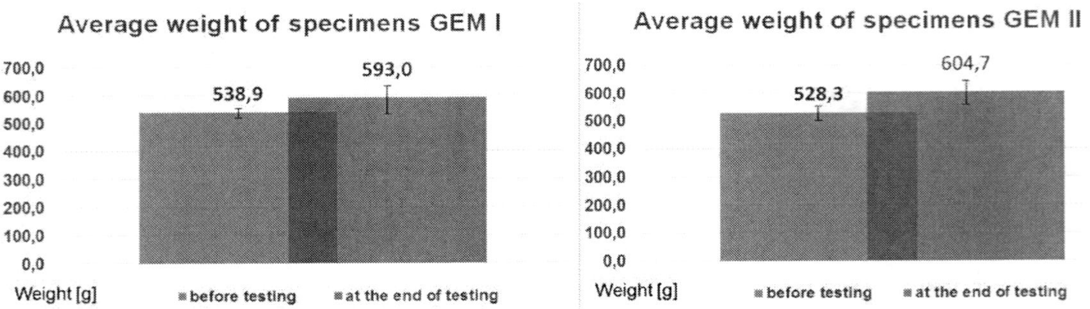


Fig. 4 Average weight of specimens GEM I and GEM II.

**Results from Testing**. The average weight for each recipe was calculated from the maximal weight of the specimen. If the specimen was broken and taken earlier from the testing box, the maximal weight in the moment was considered. It is shown in the Fig. 4. The blue column represents the average starting weight of GEM I and GEM II and the red one is the average weight at the taking out. The black line segments represent the maximal and minimal weight in the relevant category, the range of scatter of the measured data can be seen from this. At Fig. 5 is shown the bulk density in the same way. The starting weight and also the bulk density are higher for GEM II than GEM I.

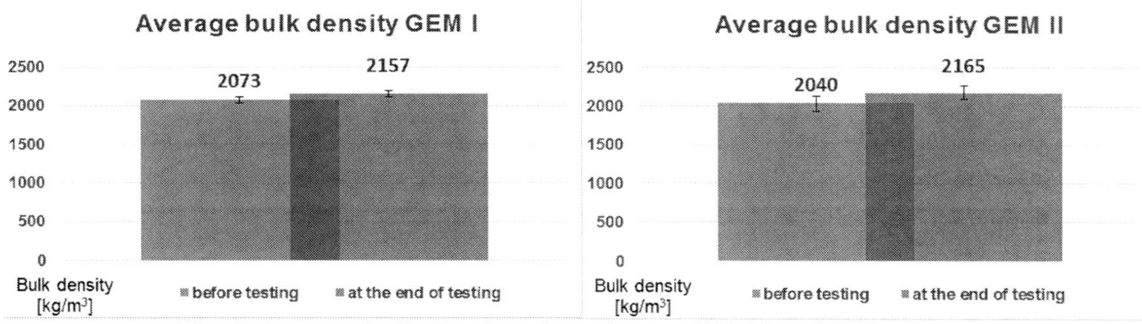


Fig. 5 Average bulk density of specimens GEM I and GEM II.

**Evaluation of Measured Data**. The average weight of GEM I specimens at the start was 539 g, at the end of water absorption the weight is 593 g, for GEM II the weights are 528 g and 605 g. The weight growth is 11.21 % for GEM I and 13.05 % for GEM II. The percentage growth of the bulk density is 4.06 % for GEM I and 6.12 % for GEM II. The capability for water absorbing is higher at the recipe GEM II with higher volume of clay. In this way the clay negatively influences the characteristic of the final product despite the fact that for the bulk density it is reversely.

## Conclusion

Earth material is very sensitive to moisture and water, that why it is an important characteristic. The recipe with the higher volume of clay have higher lever of bulk density. The weight growth is 11.21 % for GEM I and 13.05 % for GEM II. This number is not negligible. The amount of the clay has an influence to the water absorption characteristics. The more clay is use, the more water can the final product absorbs.


## Acknowledgement

The financial support of this experiment by the Faculty of Civil Engineering at CTU in Prague (SGS project No. 16/201/OHK1/3T/11) is gratefully acknowledged.


## References


[1] Little, B., Morton M., Building with Earth in Scotland: Innovative Design and Sustainability. (2001) 1-27.

[2] Norton, J., Building with Earth, A Handbook. Berlin, 2006.

[3] Minke, G., Building with Earth - Design and Technology of Sustainable Architecture. Berlin, 2006 pp. 11-18, 158-160.

[4] Suske, P., Hlinené domy novej generacie. Bratislava, 1991. pp. 7-10, 90-98, 121-125.

[5] Suske, P., Nepálená hlína v moderní architektuře. Dokument Era 21: ekologie, realizace, architektura, 2004. pp. 69-70.

[6] Žabičková, I., Hliněné stavby. Brno, 2002. pp. 5-14.

[7] AGENDA 21 pro udržitelnou výstavbu. Praha, 2001.

EAN 2017
55th conference on experimental stress analysis

# The Analysis of Acoustic Emission Signals Generated in Cement-based Composites During the Cyclic Loading Test

TOPOLÁŘ Libor[1, a], KOCÁB Dalibor[1, b], MISÁK Petr[1, c], PAZDERA Luboš[1, d], HODULÁKOVÁ Michaela[1, e] and KUCHARCZYKOVÁ Barbara[1, f]

[1]Brno University of Technology, Faculty of Civil Engineering, Veveří 331/95, Brno, Czech Republic

[a]Libor.Topolar@vutbr.cz, [b]Dalibor.Kocab@vutbr.cz, [c]Petr.Misak@vutbr.cz, [d]Pazdera.L@fce.vutbr.cz, [e]M.Hodulakova@seznam.cz, [f]Barbara.Kucharczykova@vutbr.cz

**Keywords:** Acoustic Emission Method, Cyclic Loading Test, Concrete, Cumulative Counts

**Abstract.** Acoustic Emission Method (AE) is a non-destructive testing with potential applications for locating and monitoring cracks during structural health management. In this paper, the AE signal properties for identifying the presence of an initial crack are used. AE signals generated under different loading patterns can provide valuable information concerning the structural integrity of a material. The experimental investigation from uniform cyclic loading tests performed on concrete specimens indicated that onset of crack can be identified of AE measurement. Each test specimen was subjected to cyclic loading with the loading force equivalent to one-third of the expected compressive strength value. The total number of loading cycles was one thousand. The AE method has significant potential to be used for in-situ monitoring and evaluation of the current state of structures.

## Introduction

Non-destructive testing techniques (NDT) for concrete structures are recently required for maintenance purposes. This results from the fact that concrete constructions are no longer maintenance-free. NDT used to estimation the current state of the concrete structures is extensively investigated and are practically under development [1].

In concrete, the damage of defects and cracks is primarily nucleated due to chemical reaction, mechanical stress, and fatigue. To estimate the damage of a concrete construction in service, a detailed inspection is generally conducted by taking the diamant-drilled cores from the structure. Physical properties are generally measured. As for mechanical properties, compressive strength and Young's modulus are usually determined by conducting an uniaxial compression test [2].

These obtained mechanical properties are then compared, if possible, with the standard specifications. Measurement of AE acoustic emission activity during the cyclic loading in the uniaxial compression test of the cylindrical specimens have been proposed. A cyclic loading test generally consists of several load cycles on the material or structure of interest. Each cycle includes a loading phase and an unloading phase. The AE activity recorded during the unloading phase of the cycling loading procedure increase when the damage of the specimen is in progress. The increased AE activity observed during the unloading process is generally an indication of structural instability [3-5]. This is consistent with the Kaiser effect for dilatant microcracks and implies that shear cracks do not form until near the macroscopic structural failure. Damage evaluation in concrete by AE activity of a drilled core in a compression test was under investigation [1].

AE parameters obtained by a conventional system are the number of counts, the number of events, an amplitude, the energy of AE signals, rise time, duration etc. The RA value and the average frequency can be defined from AE parameters as [6,7]:

RA value = the rise time / the maximum amplitude (1)

Average frequency = the number of counts / the duration time (2)

By means of these two AE indices, cracks could be classified into tensile and shear cracks as referring to Fig. 1.

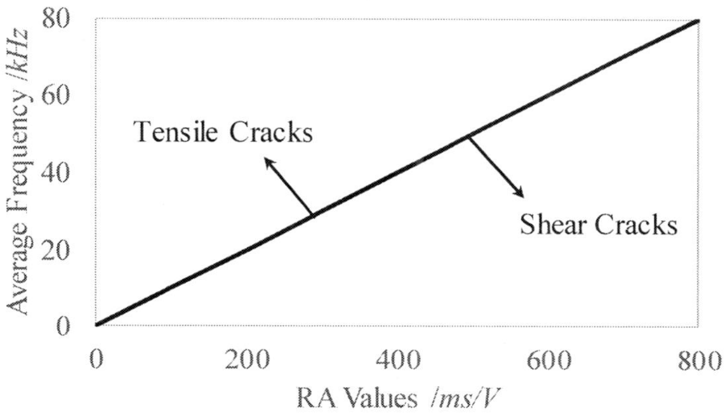


Fig. 1. Relationship between average frequency and RA values for crack classification [7].

## Materials and Testing Methods

Fresh concrete was made so that the quality of the specimens could affect the experiment results to as little a degree as possible. The concrete was mixed in a mixer with the volume of approx. 1 $m^3$ in a concrete plant and stored in polyurethane moulds of a single type. This method was chosen with the purpose of producing concrete as uniform as possible. In order to eliminate the factor of concrete aging during the experiment, the concrete was tested at an age of more than one year.

Fresh concrete composition and properties of the concrete in the fresh and hardened state is in Tab. 1. The following specimens were made: 6 cylinders with the diameter of 150 mm and height of 300 mm, 3 cubes with the size of 150 mm and 3 beams with the dimensions of 100 x 100 x 400 mm. The cubes and beams were used for control tests of bulk density and 28-day flexural and compressive strength (cf. Tab. 1).

The loading was performed in accordance with method B described in EN 12390-13 [8]. The testing procedure was configured in a way that allowed performing as many loading cycles as possible within several dozens of hours, while still performing a static loading test. The goal of the experiment was to observe the behaviour of concrete during loading in the elastic region of the stress-strain curve; for this reason, the upper loading stress was set to $\sigma_a = f_c/3$, where $f_c$ is compressive strength measured on reference specimens of the same shape and dimensions as the test specimens. According to EN 12390-13 [8], the basic loading force should be within $0.1 \cdot f_c \leq \sigma_b \leq 0.15 \cdot f_c$. This experiment used a lower loading stress of the highest possible value, i.e. $\sigma_b = 0.15 \cdot f_c$. The reason for this was to save time when transferring between the loading stresses. For the same reason, the highest allowed loading rate was chosen to be 0.80 MPa/s with a holding time of 3 s. The cyclic loading was performed using a mechanical testing press LaborTech with a loading force range of 0–1000 kN; the loading rate was set by a compressive force increment.

Tab. 1. Fresh concrete composition

| Material | Amount per 1 $m^3$ of fresh concrete /*kg* |
|---|---|
| CEM I 42.5 R | 338 |
| Bratčice 0-4 mm | 905 |
| Olbramovice 4-8 mm | 183 |
| Olbramovice 8-16 mm | 667 |
| Sika ViscoCrete 4035 | 1.77 |
| w/c /– | 0.48 |
| Flow table test (fresh concrete) /*mm* | 340 |
| Slump-test (fresh concrete) /*mm* | 70 |
| Air content (fresh concrete) /*%* | 4.5 |
| Density (fresh concrete) /$kg/m^3$ | 2 270 |
| Compressive strength (hardened concrete) /$N/mm^2$ | 52.2 |
| Tensile strength (hardened concrete) /$N/mm^2$ | 5.6 |

The three acoustic emission sensors (IDK-09) with preamplifiers (AS3K with 35 dB gain) were attached to each specimen using beeswax. Acoustic emission signals were taken by measuring equipment DAKEL XEDO. Universal measurement and diagnostic system DAKEL-XEDO was developed by ZD Rpety-Dakel company. XEDO is a modular system. One communication card and up to 15 various input cards can be located in a metal box. Communication between cards within a box is realized by hi-speed bus. An allows sampling of the signal from one sensor (speed up to 8 MSamples/sec), enumerates standard acoustic emission parameters, process emission events parameters for possible emission source localization. Elimination of the noises is achieved by simply setting the threshold level (200 mV) over the noise level, or by filtering during a post-analysis of the data.

**Results and Discussions**

The measurements results presented in Fig. 2 shown, that during the loading the Kaiser effect was detected. The Kaiser effect is a special behavior of the material under uniaxial compression [9]. It results from the fact that cracks which were created due to the previous loading do not propagate until the load exceeds the former stress level in the next loading. In usual cases, AE activity under uniaxial compression is observed from low-stress level. Some critical microcracks in a loading path usually propagate due to stress applied. When concrete contains a number of critical microcracks, active AE generation is expected from low-stress level due to new crack propagation. In contrast, AE activity in the concrete of few microcracks is considered to be stable and low at low-stress level. Thus, a number of critical microcracks in concrete could be evaluated by monitoring AE activity under uniaxial compression stress.

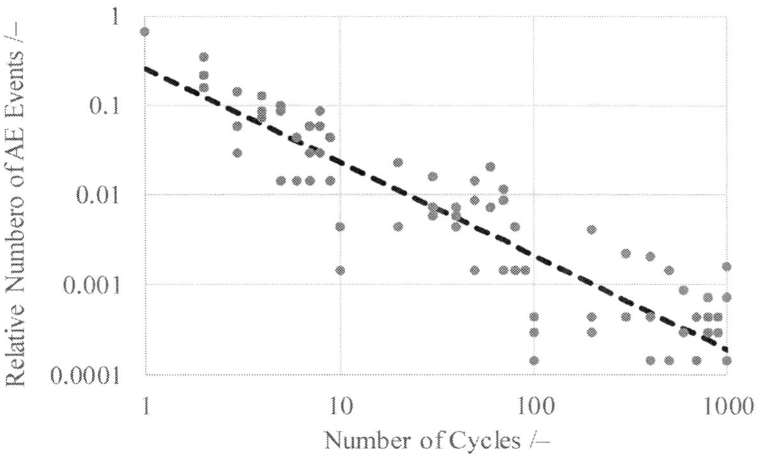


Fig. 2. Kaiser effect in a cyclically loaded concrete specimen

The fundamental approach to the crack classification is the possibility to determine the tensile crack and shear crack in the structure. In crack classification of composite material, a conventional method based on only one parameter is inadequate [10]. Utilization of several AE parameters such as rise time, count, amplitude and duration are vital to access the crack location and classification. The classification of cracks namely tensile crack and shear crack based on relationship between average frequency and RA value has been established by Ohtsu et al. [7,11] as shown in Fig. 1. Shear crack occurred when AE signal has high RA value and low average frequency. Meanwhile, tensile crack occurred when AE signal reflects low RA value and high average frequency. Shear events are characterized by longer rise time and higher amplitude than those of tensile events [12].

The relationship between average frequency and RA value (Fig. 3) exhibit strong sensitivity to the damage modes and the classification enables a warning against final failure. In this paper, the application of average frequency versus RA value is limited to the concrete cylinder subjected to static loading. If the real phenomenon is taken into account, most of the structures are subjected to dynamic loading as well as fatigue loading. It is a good opportunity for information if the fatigue loading can be applied to the concrete cylinder for damage modes classification. Both the AE waveform at low signal and high signal were considered. In the case of fatigue test, the high AE waveform occurred at a few cycle of loading and the AE waveform reduces as constant load is continuously applied. Therefore, the low AE signal is more dominant. If high and low AE signal are considered, the point in the relationship between average frequency and RA value tends to position itself in the low RA value region (Tab. 2.).

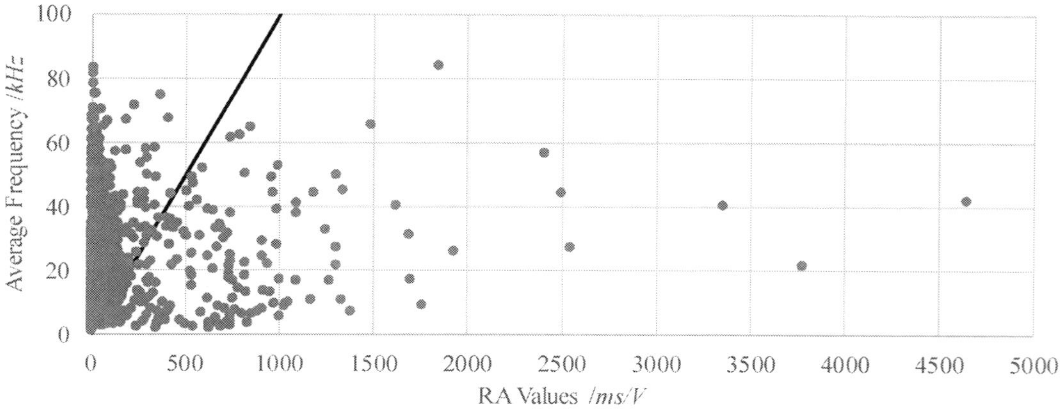


Fig. 3. Types of cracks are shown from the relationship between RA value and the average frequency

Tab. 2. The types of cracks during the first load cycle

| Phase of the cycle | Time interval /s | Total Number of AE events /– | Tensile Cracks /% | Shear Cracks /% |
|---|---|---|---|---|
| Loading | 30 | 732 | 80.2 | 19.8 |
| Stress level $\sigma_a$ | 3 | 191 | 89.0 | 11.0 |
| Unloading | 30 | 562 | 88.1 | 11.9 |
| Stress level $\sigma_b$ | 3 | 121 | 91.7 | 8.3 |

## Conclusions

For quantitative estimation of the damage in structural concrete, AE measurement is applied to a uniaxial compression test of a concrete specimen. AE reflected behavior, which is closely associated with the damage inside concrete. This paper describes the application of a relatively new type of analysis of AE signals (a relationship between average frequency and RA value) which propagate in concrete‘s cylindrical specimens during the cyclic loading compressive uniaxial test. Although, it deals with a pilot measurements and the conclusions can be not generalized. The method of analysis appears to be very promising and suitable to practical applications.

## Acknowledgement

This outcome has been achieved with the financial support of the Czech Science Foundation under project No. 17-14302S "Experimental analysis of the early-age volume changes in cement-based composites" and under the project No. J-17-4554 supported by Faculty of Civil Engineering of Brno University of Technology.

## References

[1] M. Ohtsu, H. Watanabe, Quantitative Damage Estimation of Concrete by Acoustic Emission, Construction and Building Materials 15(5-6) (2001) 217-224.

[2] M. Ohtsu, T. Suzuki, Quantitative damage estimation of concrete core based on AE rate process analysis, Journal of acoustic emission, 22 (2005) 30-38.

[3] S. Yuyama, Z.W. Li, M. Yoshizawa, T. Tomokiyo, T. Uomoto, Evaluation of fatigue damage in reinforced concrete slab by acoustic emission, NDT & e International 34(6) (2001) 381-387.

[4] S. Colombo, M.C. Forde, I.G. Main, M. Shigeishi, Predicting the ultimate bending capacity of concrete beams from the "relaxation ratio'' analysis of AE signals, Construction and Building Materials 19 (2005) 746-754.

[5] F.F. Barsoum, J. Suleman, A. Korcak, E.V. Hill, Acoustic emission monitoring and fatigue life prediction in axially loaded notched steel specimens, Journal of acoustic emission 27 (2009) 40-63.

[6] JCMS-III B5706, Monitoring Method for Active Cracks in Concrete by Acoustic Emission, Federation of Construction Materials Industries, Japan, 2003.

[7] M. Ohtsu, I. Toshiro, T. Yuichi, Acoustic emission techniques standardized for concrete structures, Journal of Acoustic Emission 25 (2007) 21-32.

[8] ČSN EN 12390-13 Testing hardened concrete, (the Czech version of the EN 12390), ÚNMZ, (2014)

[9] K. Ono, Structural Integrity Evaluation by Means of Acoustic Emission, in: A. Carpinteri, G. Lacidogna (Eds.), Acoustic Emission and Critical Phenomena, Taylor & Francis, 2008, pp. 13-27.

[10] S. Momon, N. Godin, P. Reynaud, M. R'mili, G. Fantozzi, Unsupervised and supervised classification of AE data collected during fatigue test on CMC at high temperature, Composites A 43 (2012) 254–260.

[11] M.N. Noorsuhada, An overview on fatigue damage assessment of reinforced concrete structures with the aid of acoustic emission technique, Construction and Building Materials 112 (2016) 424-439.

[12] T. Shiotani, Evaluation of repair effect of deteriorated concrete piers of intake dam using AE activity, Advanced Material Research 13–14 (2006) 175–180.

EAN 2017
55th conference on experimental stress analysis

# Thermal FEM analysis of hybrid structure

Luboš LIMBERG[1, a], Milan TANČIN[1, b]

[1] The Faculty of Mechanical Engineering, The University of West Bohemia, Univerzitni 8, 306 14 Pilsen, CZ

[a] limbi@rti.zcu.cz, [b] milax@rti.zcu.cz

Keywords: Polymer concrete; thermal FEM analysis; stress; deformation; comparative of material, hybrid structure.

**Abstract.** This article is focuses on mechanical properties of hybrid structure, which is loaded, by a heat. The type of hybrid structure is the hybrid structure of polymer concrete in combination with steel and cast iron. There will be presented a basic information about polymer concrete, steel and iron cast and their properties. In the article is also described the thermal FEM analysis of hybrid structure.

## Introduction

Polymer concrete is increasingly used material that is an alternative to cast-iron and steel in the design of beds of machine tools. It is a composite material composed of minerals, which serve as a filler. The matrix consists of resin and a hardener. This material provides interesting and design-advantageous properties, e.g. damping, dimensional and thermal stability, high stiffness, corrosion resistance, etc.

Thanks to high thermal capacity in combination with low thermal conductivity, the polymer concrete responds slower to temperature changes in its surroundings in comparison with cast-iron.

## Meaning of analyzes

Polymer concrete has different properties than steel and other metals. Two of them are thermal expansion coefficient and thermal conductivity. Differences of these material properties can cause a tension in the transition layer of hybrid structure. If used high temperatures it can produce a critical stress and then permanent deformations. Polymer concrete is very susceptible to a tensile stress. Therefore, relatively small thermal loads can cause a damage of hybrid structures.

The purpose of the analyses is finding the critical temperature loads. These analyses show the stress in the transition layer of the materials if used different thermal loads.

## Mesh

There was used 3D swept mesh (CHEXA (8)) with element size 5mm. 2D mapped mash (CQUAD4) was used for creating the 3D mesh (Fig. 1). Material properties were used according to the table (Table 1).

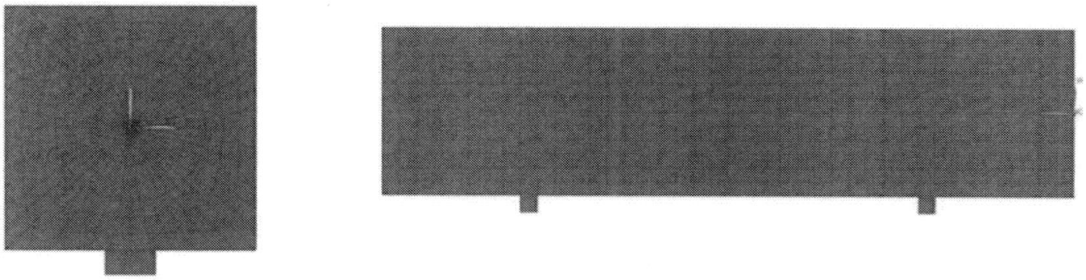

Fig. 1: Meshing model

Table 1: Material properties

| | POLYMERCONCRETE | STEEL | IRON CAST G25 |
|---|---|---|---|
| Mass density [kg/mm$^3$] | $2.3*10^{-6}$ | $7.85*10^{-6}$ | $7.15*10^{-6}$ |
| Young´s modulus E [N/mm$^2$] | 35 000 | 210 000 | 90 000 |
| Thermal Expansion Coefficient $\alpha$ [1/K] | $16*10^{-6}$ | $12*10^{-6}$ | $10.5*10^{-6}$ |
| Thermal Cond. [W/mK] | 2 | 50 | 58 |
| Specific Heat [μJ/kgK] | 943 000 000 | 500 000 000 | 434 000 000 |

**Boundary conditions**

The connection between materials (steel, iron cast + polymer concrete) is formed using gluing function. The function makes perfectly rigid connection. Perfectly rigid connection can skew the results, but that is the most affordable way to usable results. There are two different thermal loads (with cooling and without cooling).

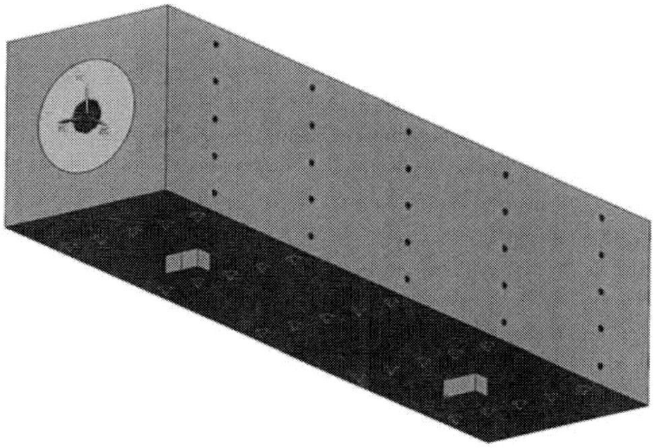

**Analysis results**

Fig. 2: Boundary conditions

**Conditions I.**

There is thermal load without cooling. Size of thermal loads are 20°C, 50°C, 100°C for the entire model.

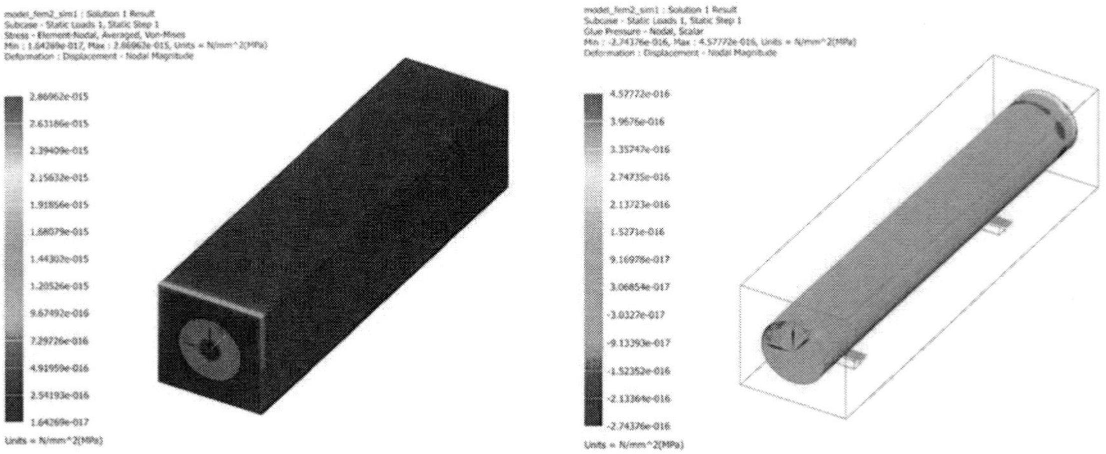


Fig. 3: Overall stress and stress at the transition layer for 20°C / no cooling

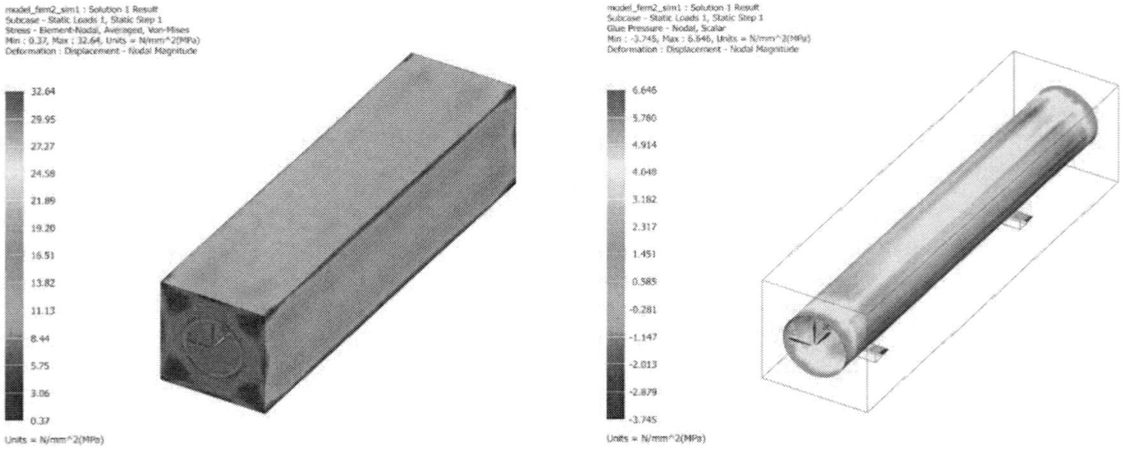


Fig. 4: Overall stress and stress at the transition layer for 50°C / no cooling

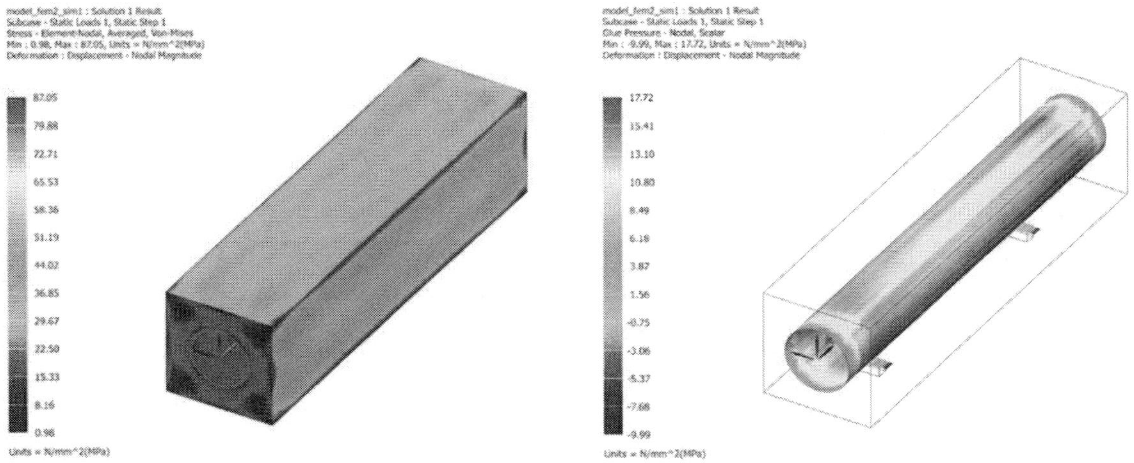


Fig. 5: Overall stress and stress at the transition layer for 100°C / no cooling

Table 2: Conditions I. results

| Temperature for the entire model (no cooling) | 20 °C Steel \| Cast iron G25 | | 50 °C Steel \| Cast iron G25 | | 100 °C Steel \| Cast iron G25 | |
|---|---|---|---|---|---|---|
| Overall stress [MPa] | 2,869e-15 | 2,297e-15 | 32,64 | 18,47 | 87,05 | 49,26 |
| Stress at the transition layer [MPa] | 4,677e-15<br>-2,743e-15 | 3,402e-16<br>-2,080e-16 | 6,64<br>-3,74 | 5,36<br>-3,31 | 17,72<br>-9,99 | 14,29<br>-8,83 |
| Deformation [mm] | 1,400e-18 | 1,189e-18 | 0,044 | 0,035 | 0,117 | 0,095 |

## Conditions II.

There is thermal load in steel / cast iron with cooling (20 °C) on sides of polymer concrete. Size of thermal loads are 50°C, 100°C.

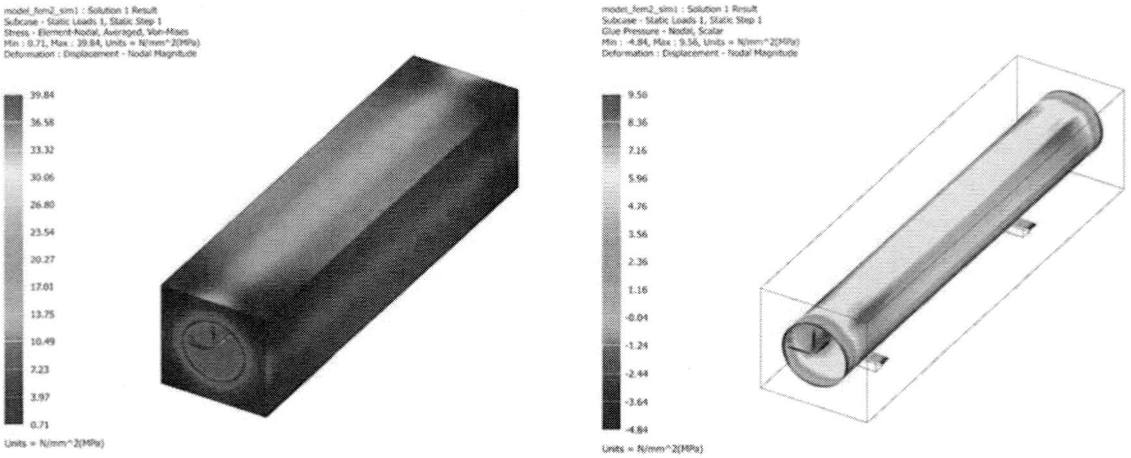


Fig. 7: Overall stress and stress at the transition layer for 50°C / with cooling

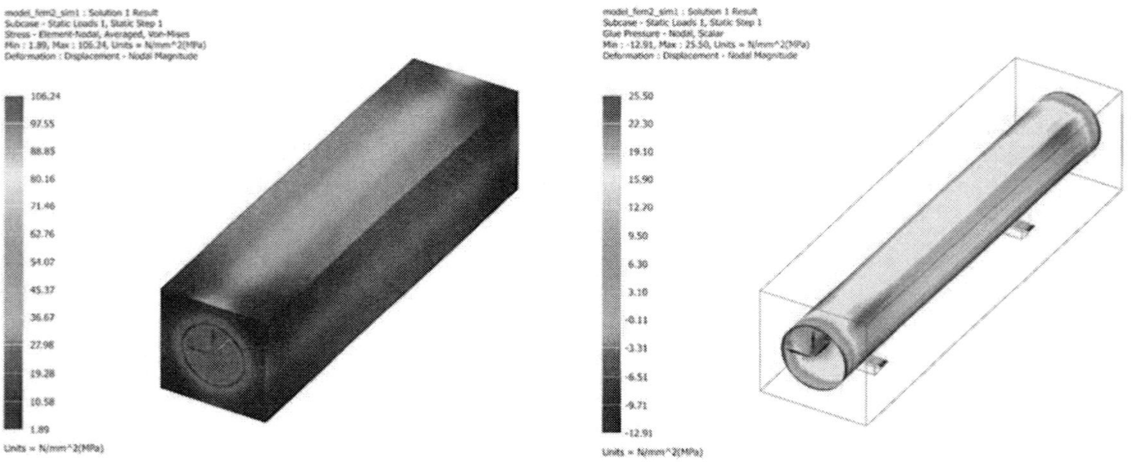


Fig. 6: Overall stress and stress at the transition layer for 100°C / with cooling

Table 3: Conditions II. Results

| Temperature in steel/cast iron (cooling) | 20 °C | | 50 °C | | 100 °C | |
|---|---|---|---|---|---|---|
| | Steel | Cast iron G25 | Steel | Cast iron G25 | Steel | Cast iron G25 |
| Overall stress [MPa] | 2,869e-15 | 2,297e-15 | 39,84 | 22,59 | 106,24 | 60,25 |
| Stress at the transition layer [MPa] | 4,677e-15<br>-2,743e-15 | 3,402e-16<br>-2,080e-16 | 9,56<br>-4,84 | 8<br>-4,55 | 25,5<br>-12,91 | 21,33<br>-12,15 |
| Deformation [mm] | 1,400e-18 | 1,189e-18 | 0,040 | 0,029 | 0,107 | 0,079 |

## Conclusions

There are two types of thermal loads. The first is constant thermal load where both of the parts of the hybrid structure have same temperatures. The second is analysis where one of the parts is cooling to 20 °C. The results of these analyses show the size of deformations and stresses in hybrid structures.

Overall stress of conditions I (no cooling) is lower than overall stress of conditions II (cooling). Maximal tensile stress at the transition layer is around 13 MPa. The critical tensile stress for polymer concrete is around 10-15 MPa so 100 °C load of condition II can be critical for maintaining the integrity of the polymer concrete.

This paper is based upon work sponsored by TAČR TA project TE01020075

## Reference

[1] Jančář, J.: Úvod do materiálového inženýrství polymerních kompozitů. Vyd.1. Brno : Vysoké učení technické v Brně, Fakulta chemická, 2003. 193 s. ISBN 8021424435

[2] Schneeberger.com [online]. 2011 [cit. 2011-05-24]. Mineral casting - Schneeberger. <http://www.schneeberger.com/products/mineralcasting>.

[3] Daďourek, K.: Kompozitní materiály – druhy a jejich využití, Technická univerzita v Liberci, 1. vydání, 113 s., 2007, ISBN 978-80-7372-279-1

[4] Kosnar,M.,Lašová,V. a kol : Závěrečná zpráva projektu 1.2.2, Praha 2005

EAN 2017
55th conference on experimental stress analysis

# Cyclic Tensile Tests of Epoxy Resin and the Experimentally Obtained Data Processing

Hana Srbová[1, a], Tomáš Kroupa[2, b]

[1, 2] University of West Bohemia, NTIS – New Technologies for the Information Society, Univerzitní 22, 306 14, Plzeň, Czech Republic, tel.: +420 377 63 2308

[a] hsrbova@kme.zcu.cz, [b] kroupa@kme.zcu.cz

**Keywords:** tensile test, data processing, loading cycle, envelope curve, damage factor, hardening curve

Material parameters of epoxy resin subjected to cyclic tensile tests are derived from the experimentally obtained data. The cyclic tensile tests are processed to obtain material parameters necessary to define a material model in a finite element software including damage and plastic behavior. A library of functions was written in programming language *Python* for purposes of the data processing.

## Introduction

In order to examine mechanical response of one type of composite matrix, specimens made of epoxy resin were subjected to cyclic tensile loading with increasing amplitude. Low molecular weight epoxy resin CHS-EPOXY 520 was hardened with CHS-P 11 and cured in a shape of a plate at room temperature before cutting the specimens. The manufacturing process was performed according to standard test method ASTM D638-10 [1]. Three groups of specimens varied in the curing time, from one week to 9 months.

The paper includes description of obtaining the key information from the experimental data, their averaging and further data processing for the use in FE or semi-analytical isotropic elasto-plastic-damage material model. The data are processed with the aim to obtain quantities or information such as Young's modulus, hardening curve or dependency between damage factor and energy necessary to damage level increase.

## Experiments

Specimens were manufactured according to standard test method ASTM D638-10 Tensile Properties of Plastics and divided into 3 groups according to their curing time:

1. 7 days
2. 152 days
3. 255 days

Both, the manufacturing and the curing process were performed at room temperature.

Tensile tests were performed on test machine ZWICK ROELL/Z050 with clip-on biaxial extensometer BTC-EXACLBI.001. Specimens were subjected to cyclic tensile loading with increasing amplitude until failure (Fig. 1). Unloading in each cycle started when gage area elongation exceeded a multiple of $\Delta l$ = 0.02 mm. Loading in the corresponding cycle started when tensile force decreased under 30 % of force value at the start of the unloading.

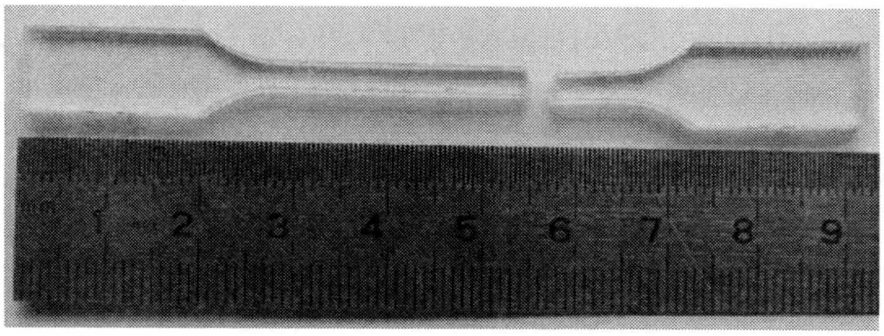


**Figure 1:** Cracked specimen.

## Data Processing

Data from the experiments were further processed using *ploTRA* - library of functions developed for purposes of processing experimental data written in *Python* programming language. Loading force $F$, elongation $\Delta l$ of gage area $l_g = 25$ mm and transverse strain of the specimen were recorded during the tests.

Engineering stress $\sigma$ and strain $\varepsilon$ were defined as

$$\sigma = \frac{F}{A_0} \qquad \text{and} \qquad \varepsilon = \frac{\Delta l}{l_g}. \tag{1}$$

The cross-section area $A_0$ change is neglected. In case of finite strain, true stress $\sigma_{true}$ and strain $\varepsilon_{true}$ would be considered.

Specific points of all loading cycles in shape of force-displacement dependency $F(\Delta l)$ and stress-strain dependency $\sigma(\varepsilon)$ (Fig. 2) were identified. The exact position of the intersection point $P_{int}$ was found and maxima — $P_{max}$ and minima — $P_{min}$ of the loading cycles were defined. In case of stress-strain dependencies, all the points were identified in terms of strain $\varepsilon$ and in case of force-displacement dependencies, the $P_{min}$ points were identified in terms of displacement $\Delta l$ and the $P_{max}$ points in terms of force $F$. To obtain a representative behavior of all the specimens within the set of specimens with the same curing time, the specific points — $P_{min}$, $P_{int}$ and $P_{max}$ were averaged for each group of specimens in terms of both coordinates – stress–strain and force–displacement.

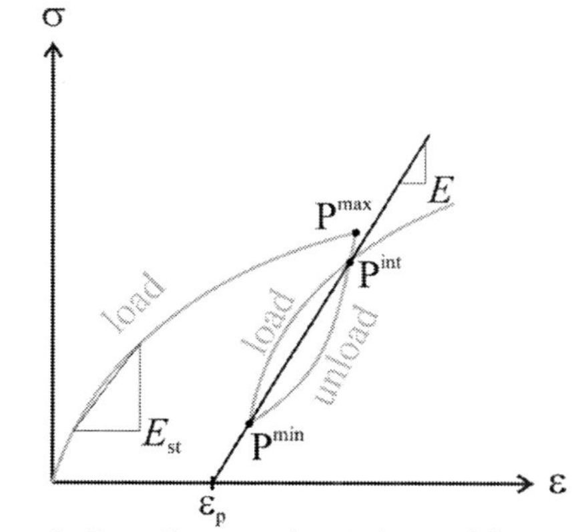


**Figure 2:** Loading cycle and specific points.

The averaged specific points were further used for definition of averaged envelope curve for each set of experiments and the tangents of the loading cycles. An envelope curve was set as the curve connecting points $P_{max}$. Tangents of the cycles, which can be considered as moduli $E$ are identified as the slope of a line connecting given averaged cycle intersection point and minimum point ($P_{int}$ and $P_{min}$). In general, Young's modulus is defined in standard

ČSN EN ISO 527-1 [2] from the difference between two strain points $\Delta\varepsilon$ (0.05% and 0.25%) and the difference in applied stress between the two strain points as $E_{st} = \Delta\sigma/\Delta\varepsilon$ (Fig. 2). It can be more convenient, however, to determine it as the slope of a linear regression line from the data points on the given strain interval. Values of the identified Young's moduli $E_{st}$ and other parameters obtained from the averaged data are in Tab. 1. Plastic strain $\varepsilon_p$ for reached strain $\varepsilon$ given by the point $P_{max}$ is obtained as the value of intersection of the line connecting $P_{int}$ and $P_{min}$ and the axis $\varepsilon$.

**Table 1**: Averaged data for sets of specimens obtained from experimental tests.

| Curing time $t$ [days] | Young's modulus $E_{st}$ [MPa] | Max. stress $\sigma_{max}$ [MPa] | Max. strain $\varepsilon_{max}$ [%] | Achieved cycles |
|---|---|---|---|---|
| 7 | 3.73 | 40.50 | 1.20 | 14 |
| 152 | 3.40 | 59.37 | 2.77 | 32 |
| 255 | 3.41 | 63.28 | 3.26 | 40 |

The envelope curves

$$\sigma = \sigma(\varepsilon) \tag{2}$$

for the sets of experiments and the slopes of averaged loading cycles are shown in Fig. 3. The curves were resampled and point $[\varepsilon, \sigma] = [0, 0]$ for the unloaded state was added. Values of strain $\varepsilon$, where the maximum stress $\sigma_{max}$ was reached is different for engineering stress $\sigma$ and for effective stress $\bar{\sigma}$, which is the stress in the undamaged part of the cross-section.

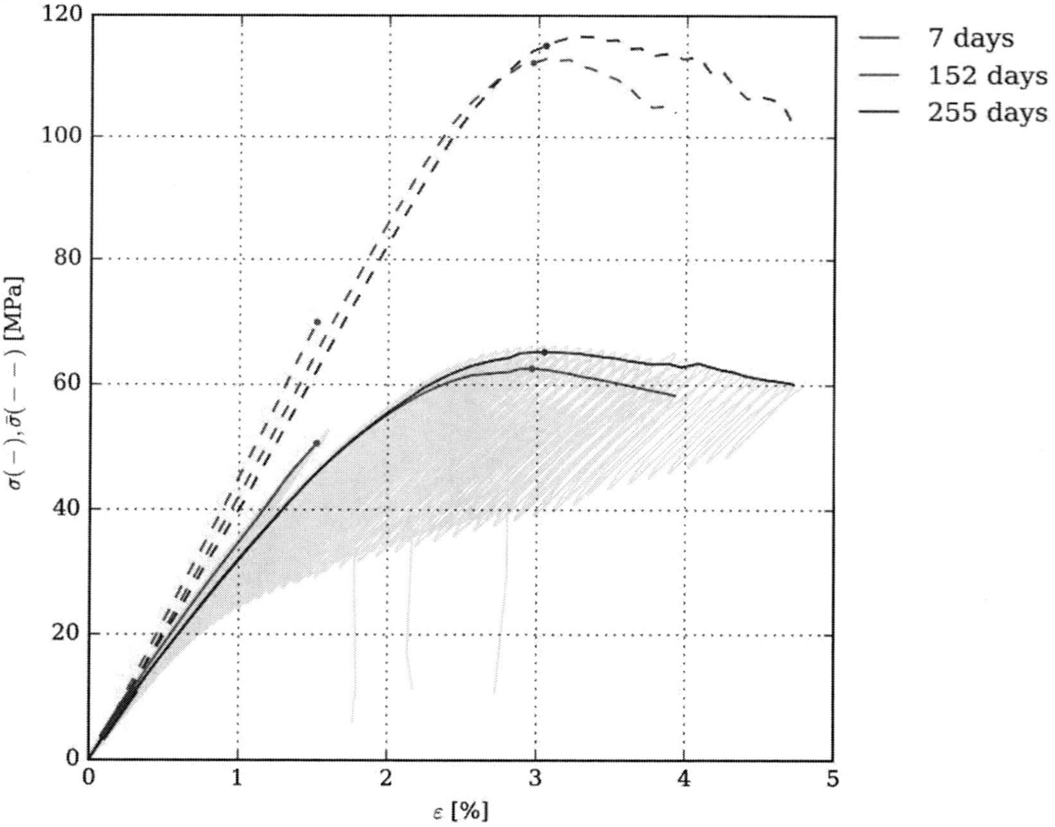


**Figure 3:** Envelope curves for sets of specimens, points of the highest reached stress $\sigma_{max}$, effective stress-strain dependencies.

Effective stress depicted in Fig. 3 was obtained as

$$\bar{\sigma}(\varepsilon) = \frac{\sigma(\varepsilon)}{1-D(\varepsilon)}, \tag{3}$$

where $D$ is damage factor defined as

$$D(\varepsilon) = 1 - \frac{E(\varepsilon)}{E(0)}, \tag{4}$$

and $E(\varepsilon)$ is modulus of the loading cycle for strain $\varepsilon$ defined at the $P_{max}$ point and $E(0)$ is initial modulus. It is the coordinate of point $[\varepsilon, E] = [0, E(0)]$ of a line connecting the first two moduli (moduli of the first two cycles) of modulus-strain dependency (Fig. 4).

$$E = E(\varepsilon). \tag{5}$$

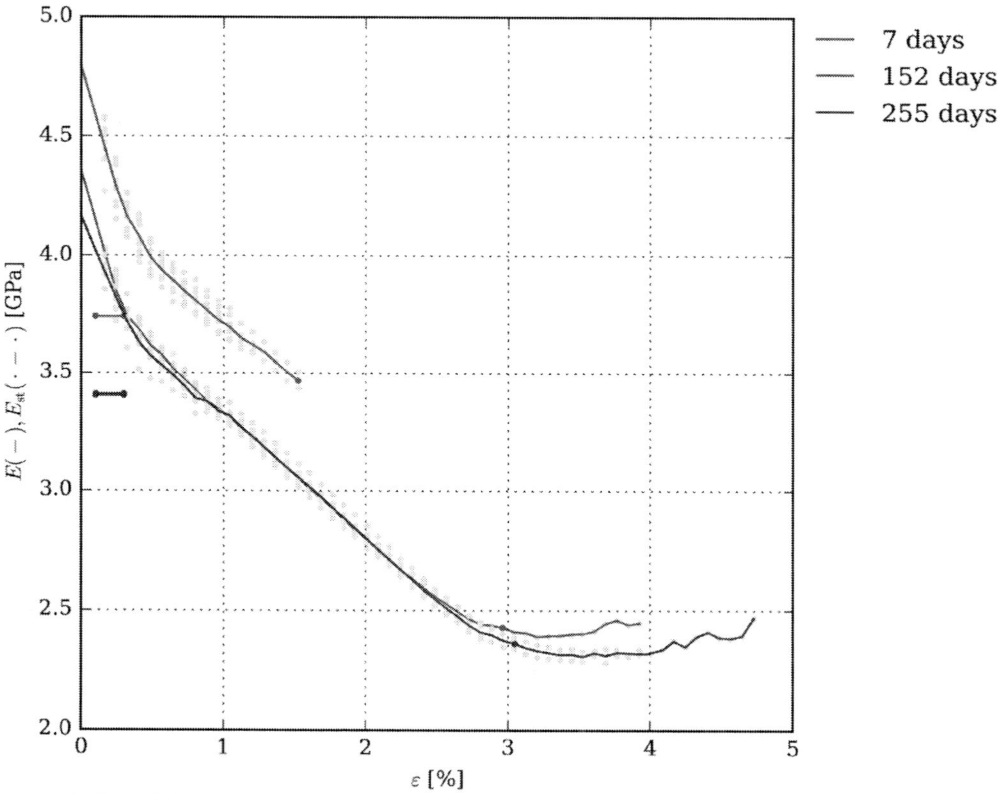


**Figure 4:** Modulus-strain dependencies, Young's moduli $E_{st}$ depicted on the strain interval defined in standard, and points of the highest reached stress.

The epoxy resin showed damage behavior, since the values of the cycle moduli $E$ are a decreasing function of the maximum reached strain $\varepsilon$. In case of the specimens cured for 152 and 255 days, the values of the cycle tangents slightly increased just before failure. It is caused by the definition of loading cycles leading to incorrect values for the advanced cycles. The material cured for 7 days has the highest stiffness, since the moduli of cycles $E$ as well as the Young's modulus $E_{st} = 3.7$ GPa identified according to standard are the highest.

The material also showed plastic behavior. Dependency of plastic strain on the strain reached at point $P_{max}$

$$\varepsilon^P = \varepsilon^P(\varepsilon) \tag{6}$$

is shown in Fig. 5. The plastic strain was evaluated for each cycle as the intersection of the cycle tangent (line through points $P_{int}$ and $P_{min}$) and the strain axis (Fig. 2). The strain values $\varepsilon$ are the strains at the $P_{max}$ points, where unloading for each cycle begins. The point for the unloaded state $[\varepsilon, \varepsilon^P] = [0, 0]$ was added and the function was resampled with the same frequency as the envelope curves (Fig. 3). The damage behavior is also noticeable from the dependency of plastic strain on maximum reached strain. The dependency $\varepsilon^P(\varepsilon)$ is shown in Fig. 5. The ability to undergo large plastic deformation (plastic strain $\varepsilon^P$) increases with increasing curing time.

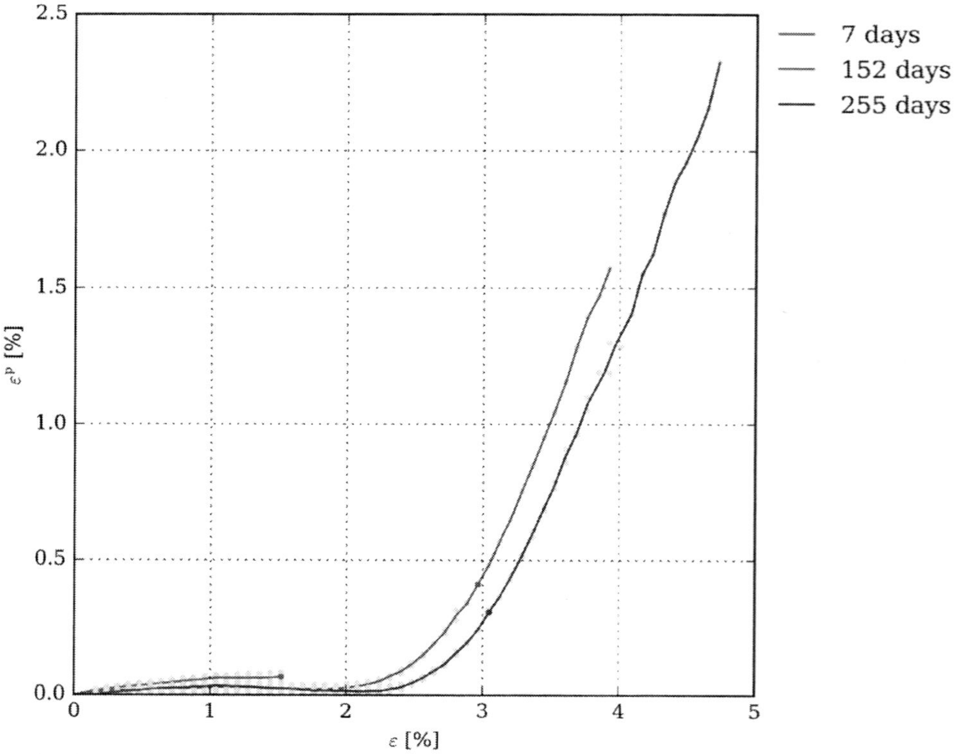


**Figure 5:** Plastic strain-strain dependencies, points of the highest reached stress $\sigma_{max}$.

The specimens cured for 7 days cracked by brittle fracture. On the other hand, the specimens cured for 152 and 255 days withstood loading cycles even after reaching the highest stress $\sigma_{max}$.

Damage functions in form of dependency of the damage factor on strain $D(\varepsilon)$ or elastic strain $D(\varepsilon^e)$ are shown in (Fig. 6). The graphs show that the damage function in case of material which is not failed at the maximum reached stress is desirable to be described in the form $D(\varepsilon)$. Damage function in form of $D(\varepsilon^e)$ for the first set of experiments with a different definition of the damage factor $D$ is presented in [3].

The damage associated energy is a function describing the damage evolution in the material. It is defined by integral of the effective stress (3) as

$$B(\varepsilon) = \int_0^{\varepsilon} \bar{\sigma}(\epsilon) d\epsilon. \quad (7)$$

By expressing the damage evolution dependency $B$ on the damage factor $D$ as

$$B = B(D), \quad (8)$$

a dependency used for material model in *Abaqus CAE* is obtained (Fig. 7).

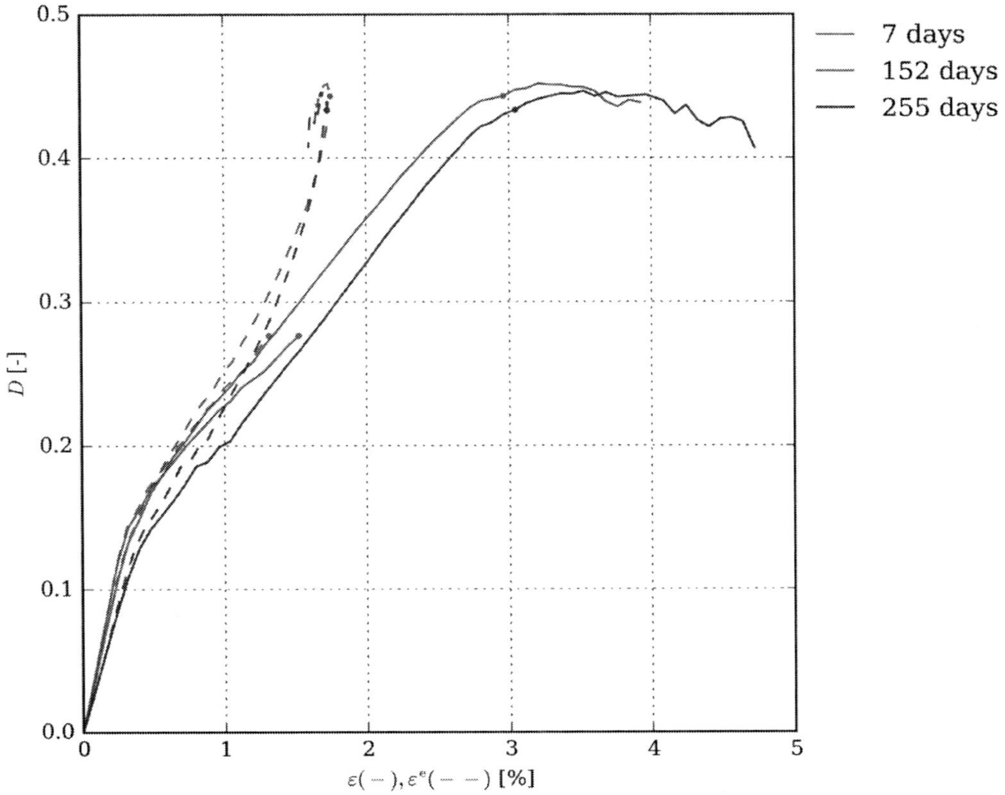


**Figure 6:** Damage factor-strain and damage factor-elastic strain dependencies, points of the highest reached stress $\sigma_{max}$.

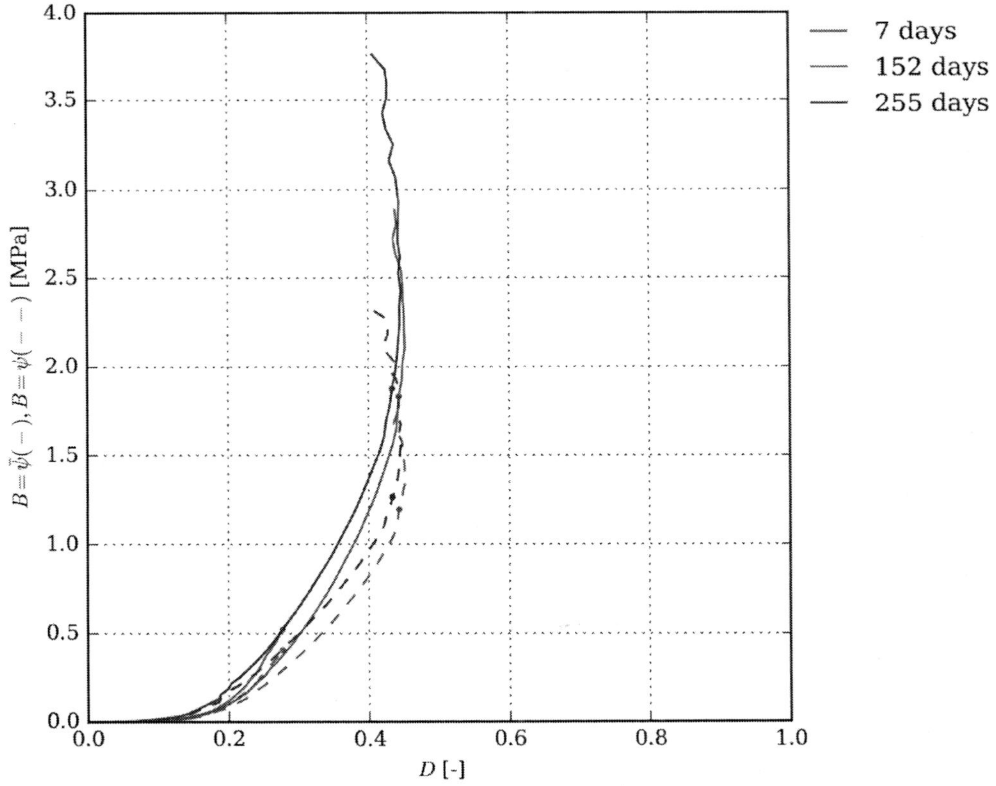


**Figure 7:** Damage evolution energy-damage factor dependency.

The dependency of damage evolution energy on damage factor shows that the epoxy resin with the shortest time of curing can absorb the lowest amount of damage energy (Fig. 7). The epoxy cured for 7 days is the most brittle material. The parts of the dependencies, where the value of the damage factor has two or more values of the energy are removed before assigning to the material model.

Next material characteristic for material model with plastic behavior is hardening curve displayed in Fig. 9. Hardening curve is a yield stress function of equivalent plastic strain $R = R(\bar{\varepsilon}^{\mathrm{p}})$. For material model considering damage behavior it is also advantageous to define dependency (Fig. 9)

$$\bar{R} = \bar{R}(\bar{\varepsilon}^{\mathrm{p}}), \tag{9}$$

where for a uniaxial stress $\bar{R}$ is equivalent stress $\bar{\sigma}$ and the equivalent plastic strain is equal to the plastic strain $\varepsilon^{\mathrm{p}} \equiv \bar{\varepsilon}^{\mathrm{p}}$.

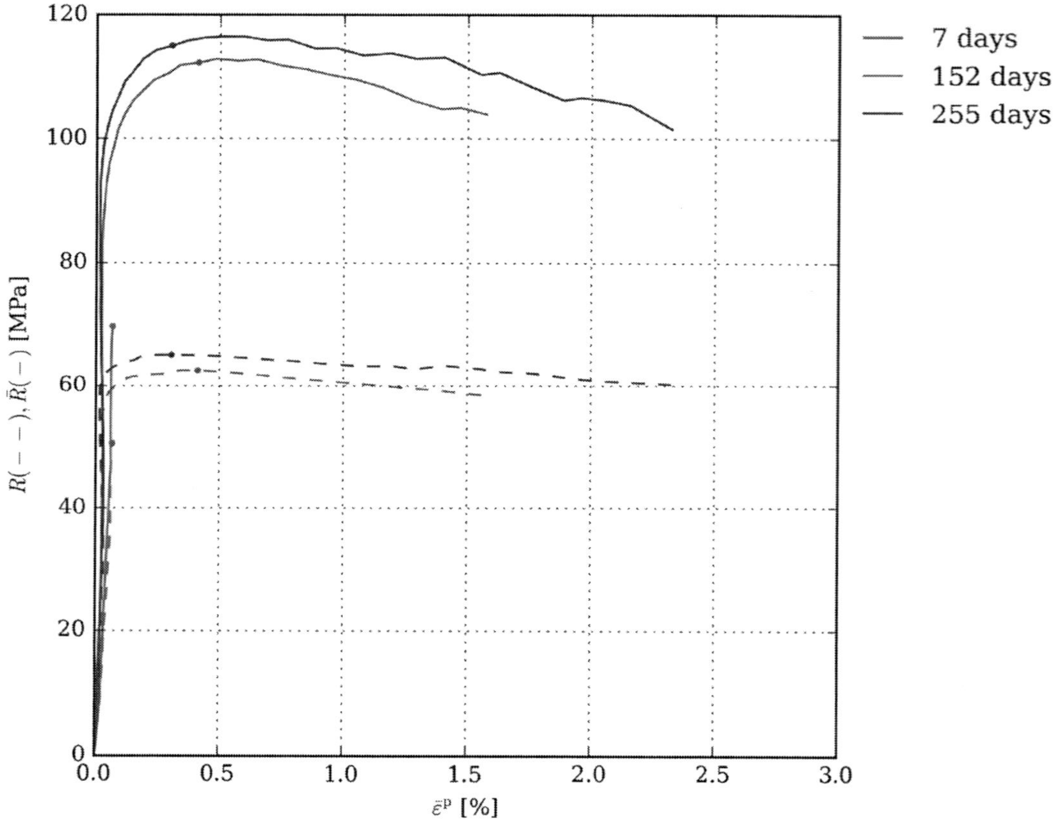


**Figure 9:** Equivalent stress-equivalent plastic strain dependency and yield stress-equivalent plastic strain dependency (hardening curve).

The parasitic effect seen in Fig. 5 where the plastic strain is not monotonically increasing function of maximum reached strain has the consequence that the equivalent plastic strain is not an increasing function (Fig. 9). This experimentally obtained error should be avoided in correct material model.

## Conclusions

Library of functions for processing experimentally obtained data was created. It was used for obtaining crucial parameters from three sets of experiments on epoxy resin. The effect of curing time of epoxy resin was expressed by evaluating material parameters averaged within the sets as well as by depicting the averaged stress strain dependencies.

The tool built in the work enables to process a large number of experiments and visualize them in terms of various dependencies. Sets of data may be averaged in different sets according to the user defined aspect. The data may be either expressed graphically or numerically. The proposed library also enables the user to save the experimental and processed data in a structured data file with all desired information. A structure system exploiting data type with keys designating all the saved data was proposed for the structure of the data organization. This system enables synoptical data sorting and clear storage system.

The processed data may be further used in finite element software. The dependencies may be used in finite element software either to define material properties or in identification processes (e.g. material model calibration). Data from the cyclic tensile tests of the epoxy resin will be further used for defining material models of a substituent (matrix) of carbon/epoxy long-fiber composite materials.

In further authors' work, the cyclic tensile tests will be performed on specimens made of thermoplastic material. The value of the reloading force (stress) at point $P_{min}$ will be set a constant in next author's work. The recently defined increasing value of the start of the loading process causes inaccuracies in further data processing.

Elasto-plastic material model with damage proposed in [3] will be validated with the experimental data processed by the proposed approach with use of the library of functions *ploTRA*.

## Acknowledgement

This publication was supported by the project LO1506 of the Czech Ministry of Education, Youth and Sports.

## References

[1] ASTM D638-10, Standard Test Method for Tensile Properties of Plastics, ASTM International, 2010.

[2] ISO 527-1, Plastics – determination of tensile properties part 1: General principles. International Organization for Standardization, Geneva, Switzerland, 2012.

[3] Kroupa, T., Srbová, H., Klesa, J. One-Dimensional Elasto-plastic Material Model with Damage for Quick Identification of Material Properties. Materials and Technology, 51(2), 2017. ISSN 1580-2949.

EAN 2017
55th conference on experimental stress analysis

# Runtime Testing of Casehardened Gears

K. Petr[1, a]

[1]Czech Technical University in Prague, Faculty of Mechanical Engineering, Technicka 4, 166 07 Prague 6, Czech Republic

[a]karel.petr@fs.cvut.cz

**Keywords:** runtime testing; casehardened gears; annealing; forged bar; gears-mesh; thickness; FEM.

## Introduction

Experimental testing of gearing (gears) [1] can be done in several ways, e.g. by pulsating [2] or runtime. This article describes the runtime tests of carburized gears with different thickness of casehardened depth (CHD) and casehardening technologies, i.e. addition or removal of certain steps during the casehardening process of the tooth flank and tooth root. Results of tests were compared with respect of results of metallographic structure.

## Description of the Tests and Parameters

Runtime test was based on the following principle: gears-mesh of two gears with described parameters (torque preload, rpm…). For these tests were used test equipment with mechanical closed loop (Fig. 1) – $M_k$=4500 Nm, n=1500min$^{-1}$. Testing parameters are shown in Table 1.

Fig. 1. Testing device for runtime test with variable load spectrum.

In this type of test may be tested by contact failures of gears, such as pitting, micropitting, wear and scuffing, fractures of teeth, etc. The test cycle consists of the process running, which is necessary for elimination of damage of teeth (scuffing from test cycle). Running-in cycle has been chosen so that have minimal impact on fatigue gearing during operational tests.

Table 1. Parameters of runtime tests.

| *Testing phase* | *Rpm [min$^{-1}$]* | *Torque [N m]* | *Time [min]* |
|---|---|---|---|
| ***Running-in*** | *500* | *500* | *60* |
| | *1 000* | *500* | *60* |
| | *1 500* | *1 000* | *60* |
| | *1 000* | *2 000* | *90* |
| ***Test*** | *750* | *3 475* | *Acc. to lifetime* |

During runtime test were tested four types (see Table 2) of gear set (three gearing in the set). These experiments were aimed at the comparison between sets of gears in terms of the growth of pitting (see Table 3) and flank breakage (see Table 4). Semi-product was the same for all gear set (forged bar). Within four sets of gear set were compared different CHD and within the same CHD were realized different parameters of casehardening process.

Table 2. Pressure angel ($\alpha = 20°$ and $\alpha = 25°$).

| *Indication* | *Semi-product* | *Casehardening technology* | *CHD [mm]* |
|---|---|---|---|
| ***CHD04-St*** | *Forged bar* | *Standard* | *0.4 ÷ 0.6* |
| ***CHD07-St*** | | *Standard* | *0.7 ÷ 0.9* |
| ***CHD07-WA*** | | *Without annealing* | *0.7 ÷ 0.9* |
| ***CHD07-C*** | | *More % of carbon (C)* | *0.7 ÷ 0.9* |

**Comparison of Gears with Different CHD and Casehardening Technologies**

From a comparison of results of runtime tests ($\alpha$ = 20°) can be concluded, that worst was gearing with lower hardened layer. There was pitting almost immediately, and there was flank breakage very soon too.

Table 3. Example of growth pitting on gears ($\alpha = 20°$ and $\alpha = 25°$).

| *Indication* | *Origin of pitting* | *Growth of pitting* | *Final pitting* |
|---|---|---|---|
| ***CHD07-C-20°*** |  |  |  |
| ***CHD07-C-25°*** | 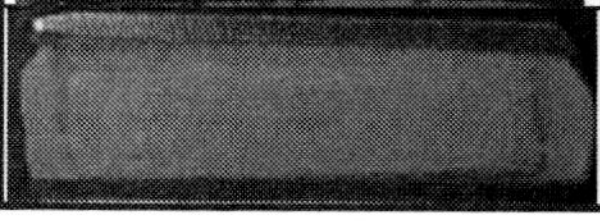 |  |  |

For gearing with the same CHD was achieved best results for gearing with higher percent of carbon (C). For the geometry of gearing with $\alpha$ = 25° were created two sets of gearing with different CHD = (0.4 – 0.6) mm and CHD = (0.7 – 0.9) mm. Among the results was a big difference.

Table 4. Sample fractures surface on the tooth and gear from runtime testing ($\alpha = 20°$).

| *Indication* | *Photo of fracture surface on the gear* | *Photo of fracture surface on the tooth* |
|---|---|---|
| *CHD07-St-20°* |  |  |
| *CHD07-C-25°* |  |  |

The important criterion for evaluation of the properties of hardened layers were their metallographic structure, progress of hardness and progress of residual stresses on the surface and under the surface of case hardened layer. Fig. 2 shows results of residual stresses for each set of gearing (measured by the method of Bauhausen noise). By means of this methodology, was compared magnetoelastic parameter where did not desirable higher values. The best was set of gearing with higher percent of carbon in the carburizing layer.

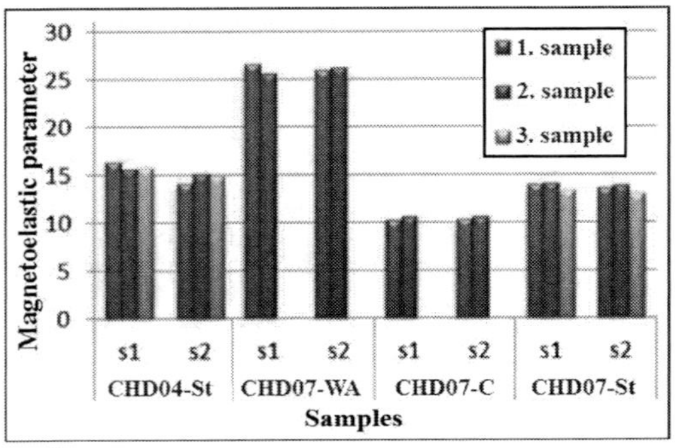


Fig. 2.: Magnetoelastic parameter.

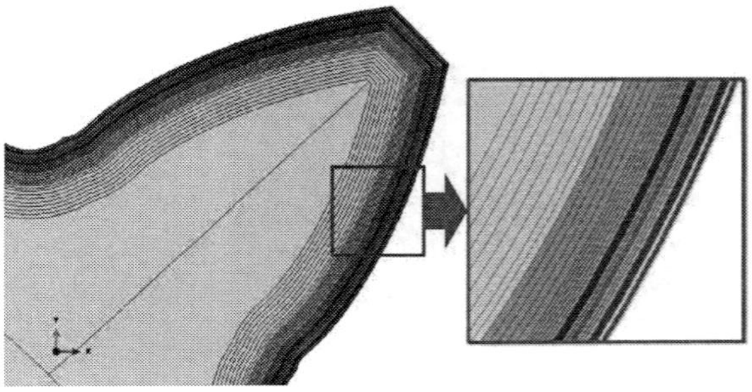

Fig. 3.: Detail of the "part of the layers - shells" casehardened layer.

## Comparison of FEM Simulations and Experimental Tests of Gears with Respect of Different CHD and Pressure Angle

FEM model and calculation of gear-mesh is the difficult contact problem. The solution of this problem consists in creating contact between teeth in the gearing (2D FEM model, see Fig. 3) [3]. Simulation of gear mesh is performed quasi-statically during three steps. On pinion and gear there have been created only three teeth. Three teeth are sufficient for the simulation of gear-mesh across the tooth (middle tooth) [4].

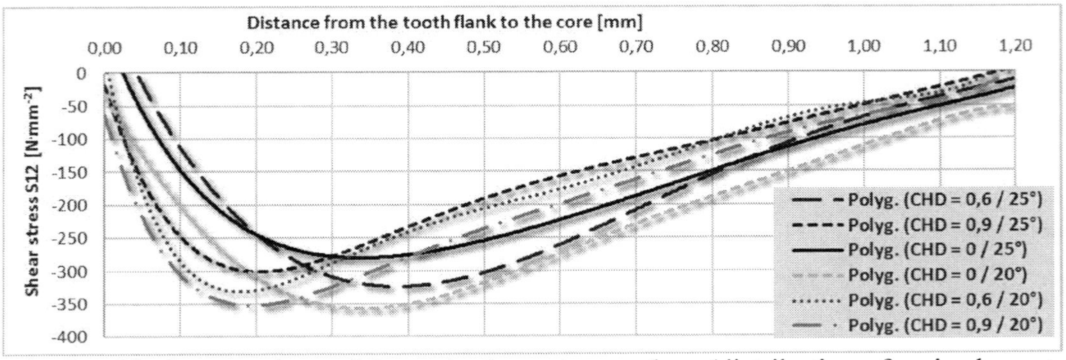


Fig. 4.: Graphical comparison of the main shear stress values (distribution of main shear stress

over the depth – CHD 0; 0.6; 0.9 mm – 20° and 25°).

From distribution of main shear stress (S12) according to Fig. 4 is obvious that with greater thickness of CHD the maximal value S12 is moved nearer to the tooth flank. For pressure angle 20° the area with maximal shear stress is greater than for pressure angle 25° and is placed deeper under casehardened layer. Fig. 4 shows distribution of main shear stress over the depth (path of depth is in the direction of normal line) for $CHD_{20° and\ 25°} = 0$; $CHD_{20° and\ 25°} = 0.6$; $CHD_{20° and\ 25°} = 0.9$ mm.

## Conclusions

From a comparison of results of tested gears with respect to different CHD we can say follows. With respect to resistance of pitting and flank breakage it was undesirable lower CHD. Furthermore, it is seen that the excessive increase of carburizing depth was also not efficient.

A positive result has been used higher percent of carbon (C) in the hardened layer with respect to loading capacity (damage to pitting). This positive step was confirmed by analysis of residual stresses.

FEM simulation is very useful tool for verification of experimental and calculated data. This analysis suggests that when using greater pressure angle than 20°, it is necessary to calculate with the fact that position of the maximum main shear stress is moved towards from tooth flank to core. On this problem is necessary to respond by greater CHD. With sufficient increase of CHD the area of critical values of shear stress is reduced. A positive aspect of greater pressure angle is greater resistance to bending stress.


## Acknowledgments

The research work reported here was made possible by Ministry of Industry and Trade of the Czech Republic grant No. FR-TI4/054.


## References


[1] K. Petr, V. Dynybyl, Methods of testing gearboxes of rail vehicles, In Applied Mechanics and Materials. Uetikon-Zurich: Trans Tech Publications, 2014, p. 277-282. ISSN 1660-9336. ISBN 978-3-03785-977-3.

[2] K. Petr, Experimental mechanical pulsation tests of gears with different case-hardened depth. Uetikon-Zurich: Trans Tech Publications, 2015, p. 283-286. ISBN 978-800105734-6.

[3] K. Petr, V. Dynybyl, M. Dub, FEM Simulation of Flank Breakage on Tooth of Gears and Experimental Photos. In Modern Methods of Construction Design. Cham: Springer International Publishing AG, 2014, 291-297. ISSN 2195-4356. ISBN 978-3-319-05203-8. http://dx.doi.org/10.1007/978-3-319-05203-8_41.

[4] Petr, K., Kanaval, J. and Dynybyl, V. Modification and Non-standard Methods of Increasing Tooth Flank Resistance of Gears, in: R.B. Höhn, ed. Proceedings of International Conference on Gears, VDI-Society for Product and Process Design and TUM, Munich, VDI Verlag GmbH, Düsseldorf, 2010, pp.1439-1142.

EAN 2017
55th conference on experimental stress analysis

# Determination of condition of concrete samples of different mixture degraded by high temperatures by acoustic non-destructive method Impact-echo

DVOŘÁK Richard[1, a], HODULÁKOVÁ Michaela[1,b] and TOPOLÁŘ Libor[1,c]

[1]Brno University of Technology, Veveří 331/95, 602 00 Brno, Czech Republic

[a]dvorak.r1@fce.vutbr.cz, [b]135542@vutbr.cz, [c]Libor.Topolar@vutbr.cz

**Keywords:** fire resistance, concrete, high temperature degradation, impact-echo method, non-destructive testing, transverse waves, longitudinal waves.

**Abstract**

The advantage of concrete over other building materials is its fire-resistive properties. The civil structures must be able to withstand dead and live loads without collapse even though the rise in temperature causes a decrease in the strength and modulus of elasticity for concrete. At current state determination of damage in concrete is performed mainly by microscopic analysis which is a destructive method. It requires a significant time interval to produce results and by simple compressive strength tests which also requires sample removal. On the other hand, the inspection by Acoustic Non-Destructive Testing methods [1], [2], [3]. (Impact-echo, ultrasound measurement etc.) can be applied, when macro-cracking or another form of damage appears.

Impact echo method is one of the oldest acoustic testing method in civil engineering [4]. We excite the test specimen by mechanical impulse and then listen to response [5]. Skillful observer then can assess the condition of construction. Right interpretation of measured data is crucial part for this acoustic non-destructive method. In this paper, the possible usage of impact echo method used for evaluation of condition of high temperature degraded concrete structures [6] is examined. It is focused on two mixtures different by coarse aggregate 8/16 and 11/22 mm. The temperature load is divided to 200, 400, 600, 800, 1000 and 1200 °C sets. The signals were excited by mechanical hammer, and response in form of mechanical wave was recorded by piezoelectric sensor. The signal was then converted to frequency spectrum by Fast Fourier Transform [7].

We assessed two testing positions of exciter and receiver, thus two frequency spectrums for each temperature set. A comparison of longitudinal and transverse wave is presented and as a guideline for material properties destructive tests have been made. A correlation between condition of temperature degraded concrete samples obtained by destructive tests with dominant frequency peaks obtained by non-destructive method is discussed and evaluated.

## Introduction

Impact echo method was invented at the U.S: National Bureau of Standards in 1980 and developed at Cornell University, in Ithaca, New York. Professor Mary J. Sansalone was principal inventor of the impact-echo method and a leading authority on the use of transient stress waves for non-destructive evaluation of heterogenous materials [8].

Impact-echo method belongs to stress wave propagation non-destructive methods based on the use of transient stress waves. A transient stress pulse is introduced into a structure by mechanical impact at a point on the surface. This pulse travels into the structure as dilatational (P) and distortional (S) waves and along the surface as a Rayleigh (R) wave (Figure 2.). The P- and S-waves propagate into the structure along spherical wavefronts and are reflected by internal cracks, voids, interface between different mediums and by external boundaries of the structure [9]. A P-wave is associated with normal stress, while an S-wave is associated with shear stiffness. A Rayleigh wave, also called a surface wave or R-wave, propagates along the surface of a solid, and particle motion is retrograde elliptical [5].

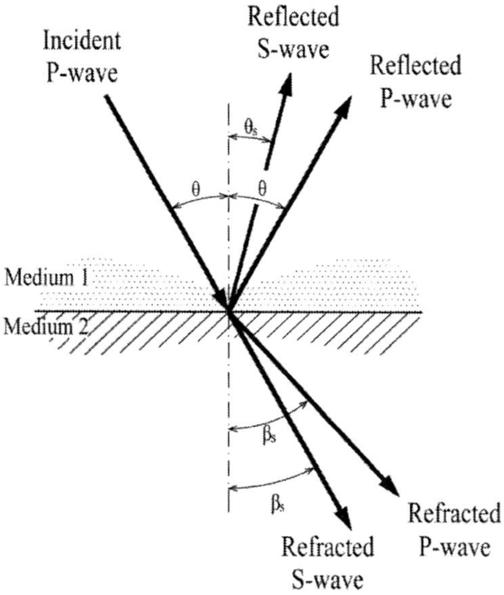


Figure 1. Reflection and refraction of P-wave [5].

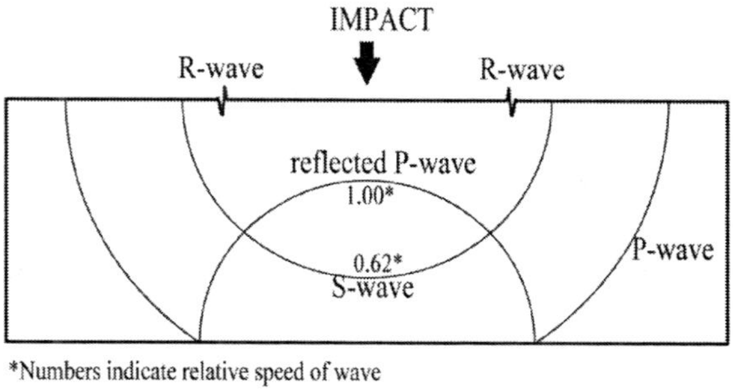


Figure 2. Different types of waves [4].

When a P- or S-wave front is incident on a boundary between dissimilar media, "specular" reflection occurs (Figure 1.). At a boundary between two different media only portion of a stress wave is reflected. The remainder penetrates into the underlying medium (wave refraction). The angle of refraction, β, is a function of the angle of incidence, θ, and the ratio of wave speed, $C_2/C_1$, in the different media, and is given by Snell's Law. Depending on the angle of incidence, a P-wave can be partially reflected as both P- and S-waves and can be refracted as both P- and S- waves.

The portion of an incident ray of a P-wave that is reflected at an interface between two media depends on the specific acoustic impedances of each medium (equation 1.1). Typical values of acoustic impedances are shown at Table 1. [5].

$$Z = \rho \cdot C_p \tag{1}$$

Table 1. Specific acoustic Impedances [5].

| Material | Density [kg/m$^3$] | P-wave Speed [m/s] | Specific Acoustic Impedance [kg/m$^2$·s] |
|---|---|---|---|
| Air | 1.205 | 343 | 0.413 |
| Concrete* | 2300 | 3000-4500 | 6.9-104 x $10^6$ |
| Granite | 2750 | 5500-6100 | 15.1-16.8 x $10^6$ |
| Steel | 7850 | 5940 | 46.6 x $10^6$ |

**Concrete density depends on its mixture.*

## Material properties

Concrete is heterogenous composite made by cement as a binder, fine aggregate and coarse aggregate as filler and additives. In most static bearable structures concrete is also densely reinforced with steel bars. When mechanical P-wave travel through the concrete specimen, on every disparity such as cracks, air voids, defects, external boundaries or interface between concrete and reinforcement, it reflects and refracts and behave as a new source of P- and S-waves.

Table 2. Review of changes in microstructure that occurs during temperature load [10].

| Temperature | Changes |
|---|---|
| 20 – 200 °C | Loss of physically bounded water and steam expansion, thus loss in binding force in concrete. Beginning of dehydration of CSH gel.<br>- 80 °C – 150 °C dehydration of ettringite.<br>- 150 °C – 170 °C decomposition of gypsum $CaSO_4 \cdot 2H_2O$. |
| 300 – 400 °C | Cracking of silicate aggregate (350 °C).<br>Critical temperature for water (374 °C) – after this temperature, there can be no physically bounded free water. |
| 400 – 500 °C | Decomposition of portlandite $Ca(OH)_2 \rightarrow CaO + H_2O$. |
| 500 – 600 °C | Quartz change from β phase to α phase at 573 °C. |
| 600 – 800 °C | Second decomposition phase of CSH gel's and production of β-$C_2S$. |
| 800 – 1000 °C | Beginning of ceramic bond, which substitutes hydraulic bonds accompanied by production of $CO_2$.<br>- Decomposition of dolomitic limestone at 840 °C.<br>- Decomposition of calcite $CaCO_3 \rightarrow CaO + CO_2$ at 930 °C – 960 °C. |
| 1000 – 1200 °C | Production of Wollastonite β ($CaO \cdot SiO_2$).<br>- Melting of basalt starts at 1050 °C. |
| ≥ 1300 °C | Total degradation of concrete, loss of mechanical properties. Some compounds start to melt and partial sintering occurs. |

Intact healthy well compacted concrete elements have very few defects, and the measured signal is clear from noise and easily readable. Vice versa damaged concrete elements have cracks, bigger air voids, and larger amount of disparities. In high temperature degradation, material changes occur, and material properties such as density, compressive strength and flexural bending strength are changing.

During temperature load many changes occur in concrete, as seen in Table 2. First critical section of temperature load is between 400 °C and 600 °C, where physical free and bounded water and chemical bounded water vaporize and gives rise to pressure in pores and cracks, incapable of leaving the concrete. This leads to spalling for low performance concrete and explosive spalling for high performance concrete [11], [2].

The second critical section is around 1000 °C, where concrete reaches the bottom of its mechanical properties and ceramic bounds do not appear yet. Concrete degraded by this temperature is fragile, aggregate and cement matrix almost lost its binding force.

After 1100 °C partial sintering and ceramic bonds occurs, which leads to sintered structure and thus slight improvement in mechanical properties.

## Mixture and degradation

For test specimens, two different mixtures were designed with the usage of different coarse aggregate (Table 3). Coarse aggregate was chosen in order to test concrete of different quality, and observe changes in measured dominant frequencies and destructive tests.

Table 3. Mixture designs.

| Compounds | Weight for 1 m$^3$ [kg] | |
|---|---|---|
| | Mixture A | Mixture B |
| Portland cement CEM I 42.5 R Mokrá | 345 | |
| Fine aggregate 0/4 mm (Žabčice) | 813 | |
| Coarse aggregate 8/16 mm (Olbramovice) | 980 | – |
| Coarse aggregate 11/22 mm (Olbramovice) | – | 980 |
| Super-plastizer additive SicaViscocrete 2030 (0,8 % mc) | 3.1 | |
| Water (w/c = 0,5) | 176 | |

In this experiment beams of dimension 0.1 x 0.1 x 0.4 m were used as test specimens. Test specimens were created in steel form and after demolding, they were stored in water for 28 days. After that, they were stored at laboratory conditions of 21 °C and relative humidity 60 – 80 % for 7 days. Test specimens were dried in laboratory dryer for 72 hours at 110 °C. By this we removed most of the physically bounded water, which could otherwise lead to explosive spalling and potentially damaging the furnace.

Each mixture was then divided into one intact reference set (kept at 20 °C) and temperature sets which were degraded at 200 °C, 400 °C, 600 °C, 800 °C, 1000 °C and 1200 °C. After drying, the test specimens were burn in ceramic furnace Rhode KE 130B at the heating rate of 5 °C/min. Selected temperatures were maintained for 60 minutes. Test beams were then spontaneously cooled in furnace to laboratory temperature.

## Measuring process

Studies have shown, that impact-echo method can be very reliable and effective tool for evaluating fire damaged concrete structures [12], [6], [2].

Effectiveness and reliability of this method depend on many factors but in particular on the skillful observer and right choice of measuring system.

In the experiment, each test specimen was tested by impact-echo method in two positions exciter – sensor. For each position as an exciter a steel spherical hammer of 70.5 g was used. Hammer is attached to test equipment, so each fall and hit on the body of the test specimen is identical.

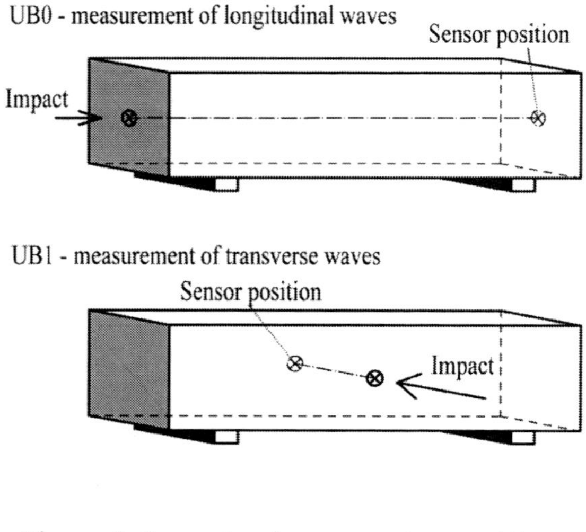


Figure 3. Impact-echo measuring positions.

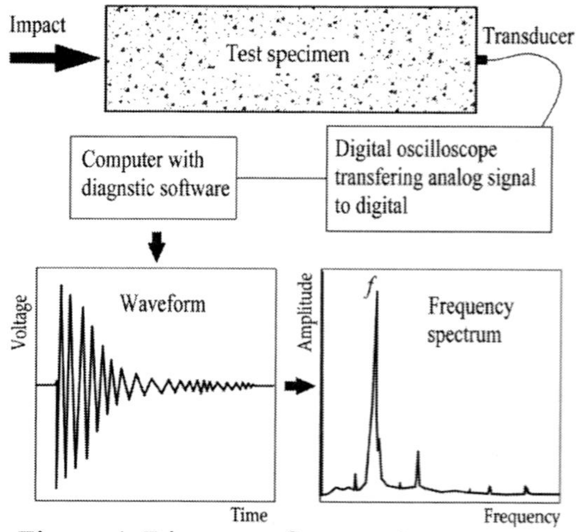


Figure 4. Diagram of measuring process.

The measuring of longitudinal waves is labeled UB0 (see Figure 3. - upper part) and the measurement of transverse waves is labeled UB1 (see Figure 3. - lower part). The diagram of the measuring and Fast Fourier Transform process is shown in Figure 4. As transducer, a piezoelectric sensor is used attached to surface of concrete by beeswax. Signal is recorded by digital oscilloscope Handyscope HS3 with the sampling speed of 10 MHz. For analyzing the signal a TiePie software is used. The waveform is transferred to frequency spectrum by fast furrier transformation [4].

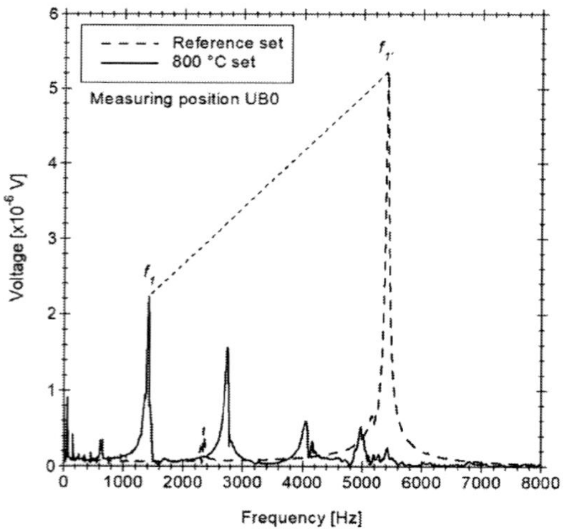


Figure 5. Shift of dominant frequency in longitudinal way of measurement.

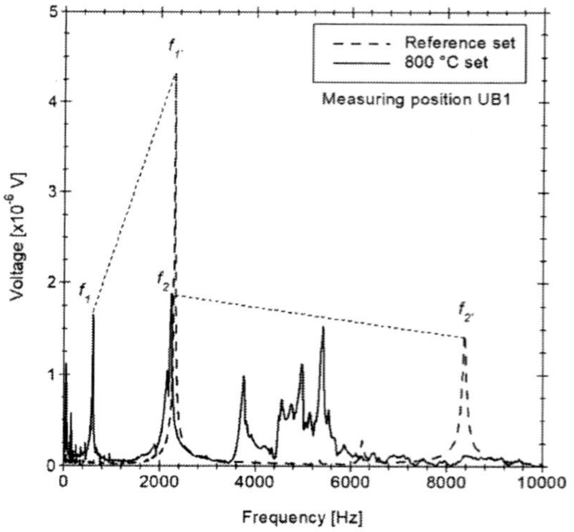


Figure 6. Shift of dominant frequency in transverse way of measurement.

The shift of measured dominant frequency is shown in Figure 5 for longitudinal waves and in figure 6 for transverse waves. In the measuring position UB0, we focus on one dominant frequency $f_1$. In measuring position UB1, we focus on two dominant frequencies $f_1$ and $f_2$.

After degradation of concrete samples by the high-temperature process, the dominant frequencies are shifted from higher to lower frequencies. We compare this change with compressive strength [13] and flexural bending strength [14] which results from destructive tests.

### Measured data

On Figure 7. values of dominant frequency in a longitudinal way for mixture A and B are presented. Both mixtures A and B are shifting to lower frequencies as temperature of degradation rises. The significant change appears from 200 to 800 °C for mixture A. Mixture B shows to be more resistance to high-temperature when its downfall appears at 600 °C to 1000 °C. Both mixtures reach lowest frequencies at 1000 °C. Mixture B have a bigger dispersion of measured values than mixture A at 400 °C, 600 °C and 1000 °C. Dominant frequency of both mixtures shows increase of dominant frequency at 1200 °C. This is mainly due to partial sintering, when ceramic bonds start to substitute the damaged hydraulic bonds.

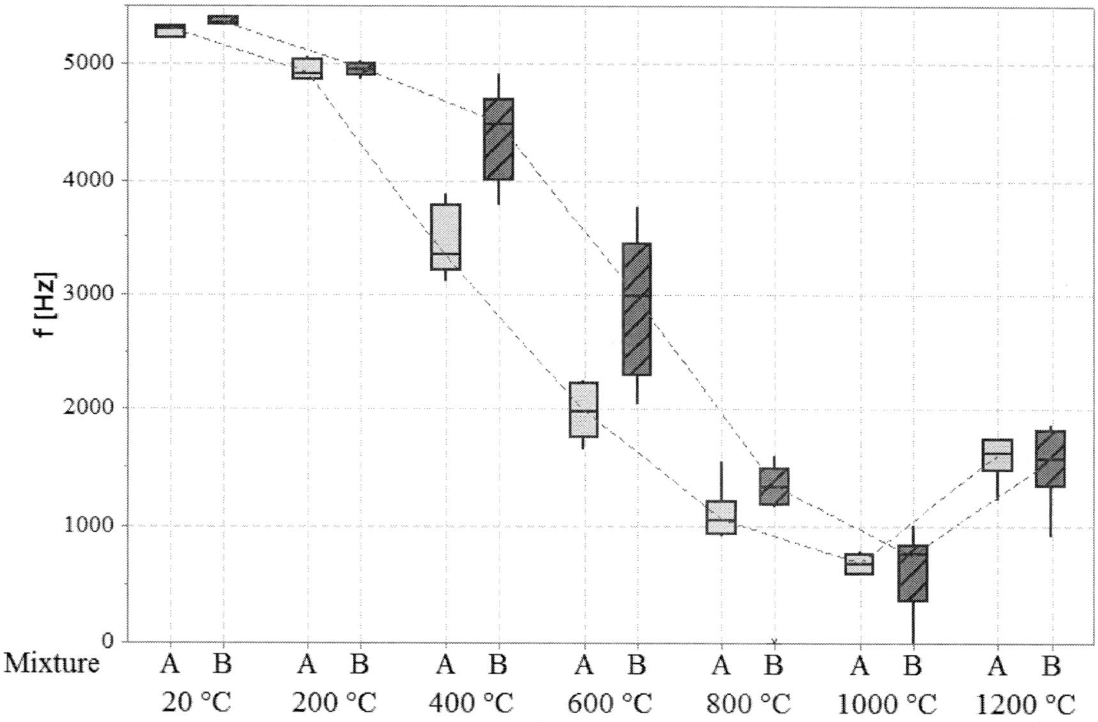


Figure 7. Measured dominant frequencies for mixture A and B in measurement position UB0.

Most rapid changes around 400 and 600 °C are mainly caused by decomposition of portlandite $Ca(OH)_2$ and loosing rest of chemically bounded water. These temperatures were also critical in term of steam pressure rising in pores. In one case, rising of steam pressure led to explosion of test specimen and damaging the furnace, even all of the test specimens were pre-dried. When we compare this results with destructive test, we can see correlation between measured dominant frequency and compressive strength and flexural bending strength. Figure 8. shows, that decrease in strength start already at 200 °C and continues for both mixtures to 1000 °C. At 1200 °C mixture A shows higher values of dominant frequencies and also values of compressive strength and flexural bending strength than mixture B. This is due to smaller fraction of aggregate 8/16 mm. The smaller fraction can absorb more heat energy during burning, and change its structure more easily that bulky 11/22 aggregate, which would take much more energy to change its structure. The granodiorite and quartz sand was used. The granodiorite contained as major components feldspar, quartz and biotite. Feldspar has a melting point about 1170 °C. The quartz has melting point at 1720 °C, but in the presence of alkaline substances (e.g. feldspar, CaO, carbonates, FeO, MgO, etc.), the melting point decreases [15], [16].

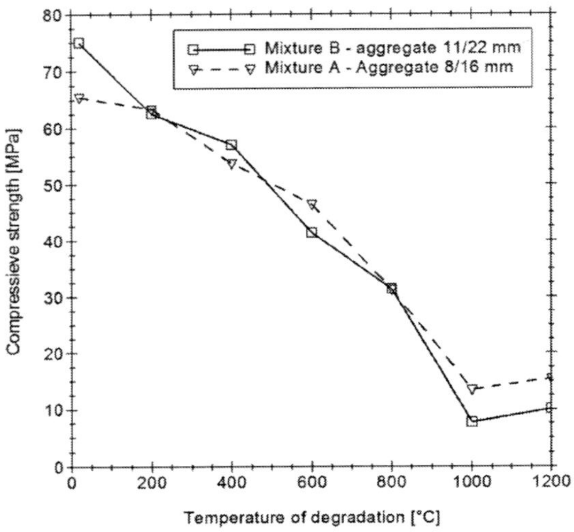

Figure 8. Compressive strength test.

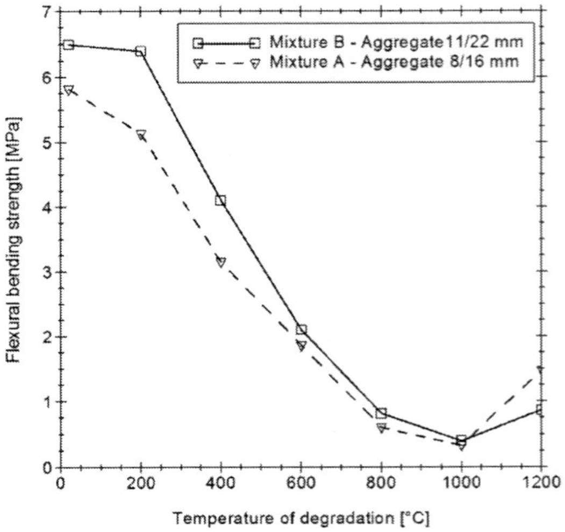

Figure 9. Flexural bending strength test.

The shift of dominant frequencies in a longitudinal way is shown in Figure 10 and shift of two observed frequencies in transverse way is shown in Figure 11. We can see similar trend in both figures. Transverse frequencies $f_1$ and $f_2$ envelope longitudinal frequency. Signals of some test specimens were hard to read, and results from longitudinal or transverse waves helped determination of specific frequency. This approach proved to be successful in distinguishing dominant frequency from frequency noise.

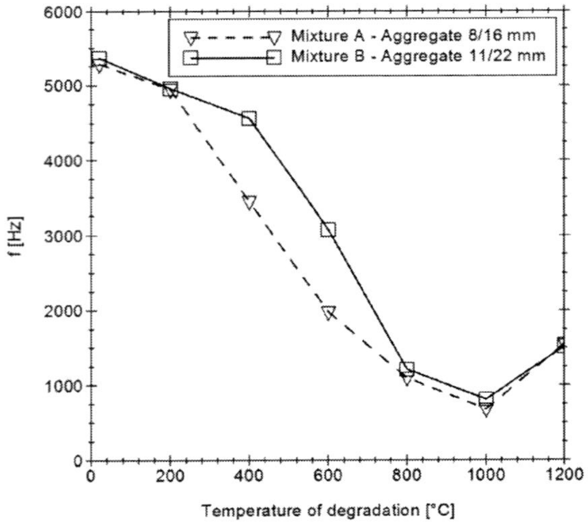

Figure 10. Dominant frequency for each mixture in longitudinal way.

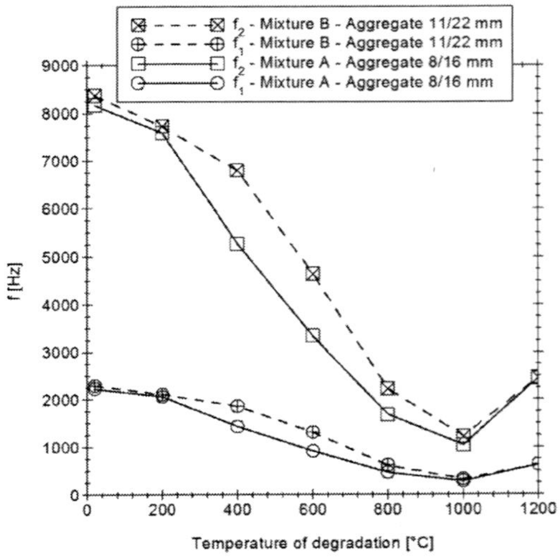

Figure 11. Dominant frequency for each mixture in transverse way.

The significant change in the percentage ratio of flexural bending strength to compressive strength due to the effect of temperature was observed (see Figure 8 and 9). For normal concrete (without steel fibers) the percentage ratio of flexural bending strength to compressive strength is about 8 or 10 % at temperature 20 °C. The higher the temperature had a more destructive influence on flexural bending strength than on compressive strength. Concrete became brittle at high temperature, but at a temperature of 1200 °C the flexural bending strength increased.

## Conclusions

This article shows other possibilities of Impact echo method as rapid non-destructive method for evaluating temperature degraded concrete. We've tested test specimens with dimensions of 0.1 x 0.1 x 0.4 m of two mixtures – A with usage of coarse aggregate 8/16 mm and B with usage of coarse aggregate 11/22 mm. Specimens were tested up to 1200 °C by 200°C steps. We observed a shift of dominant frequencies toward the lower frequencies for longitudinal way of measurement and as well for transverse way of measurement. The value of dominant frequencies decreases up to 1000 °C, from this point a slight increase appeared up to 1200 °C. These results are in strong correlation with the values of compressive strength of these samples. Mixture A had flexural bending strength $f_{ct}$ for reference set 5.82 MPa and mixture B had 6.50 MPa. For 1000 °C mixture A had $f_{ct}$ 2.44 MPa with measured dominant frequency 680 Hz in longitudinal way and mixture B had compressive strength of 5.19 MPa with measured dominant frequency 810 Hz. By impact echo results, we can then distinguish different mixture up to 1000 °C. At temperature 1200 °C measured dominant frequencies are 1560 Hz for mixture A and 1520 Hz for mixture B. These measurements show that, for high temperature degraded concrete, Impact-echo results can distinguish level of degradation, on which temperature were specimens tested and thus condition of concrete up to 1200 °C. At 1200 °C melting of basalt and $SiO_2$ starts and ceramic bonds appears. From these results, we can conclude that Impact echo method is suitable nondestructive method for rapid non-destructive testing of concrete structures and elements affected by fire. However, the testing environment is considered to be a laboratory regulated environment. For specific application in civil engineering for diagnostic purposes, where rise in temperature is much steeper, and temperature last only for brief moment, more experiments had to be done.

## Acknowledgement

This paper has been worked out under the project No. LO1408 "AdMaS UP - Advanced Materials, Structures and Technologies", supported by Ministry of Education, Youth and Sports under the „National Sustainability Programme I" and under the project GAČR No.16-02261S supported by Czech Science Foundation and under the project FAST-J-17-4585 supported by Faculty of Civil Engineering of VUT.

## Reference

[1] M. Lunak and I. Kusak, Electric and Temperature Properties of Concrete Setting Measurement Set, Proc. of the Defektoskopie 2009 - NDE for Safety. Praha, BETIS, 2009, 269–274.

[2] RICHTER, R., F. DEHN, J. SCHMIDT, M. JUKNAT, C.U. GROSSE, P. PIMIENTA a F. MEFTAH. Measurement systems to detect the time-dependant development of concrete spalling under fire exposure. MATEC Web of Conferences [online]. 2013, 6, 03006- [cit. 2017-04-04]. DOI: 10.1051/matecconf/20130603006. ISSN 2261-236x.

Available also from: http://www.matec-conferences.org/10.1051/matecconf/20130603006

[3] WHELAN, Matthew J., Brett Q. TEMPEST a David B. SCOTT. *Post-Fire Nondestructive Evaluation of a Prestressed Concrete Double-Tee Joist Roof.* Journal of Performance of Constructed Facilities [online]. 2015, 29(2), 04014055- [cit. 2017-04-04]. DOI: 10.1061/(ASCE)CF.1943-5509.0000543. ISSN 0887-3828. Available also from: http://ascelibrary.org/doi/10.1061/(ASCE)CF.1943-5509.0000543

[4] CARINO, N. THE IMPACT-ECHO METHOD: AN OVERVIEW. *Proceedings of the 2001 Structures Congress & Exposition.* Gaithersburg, USA: National Institute of Standards and Technology, 2001, **18**, 1-19.

[5] MALHOTRA, V. a Nicholas CARINO. Handbook on nondestructive testing of concrete. 2nd ed. Boca Raton, Fla.: CRC Press, 2004. ISBN 08-031-2099-0.

[6] HAGER, I. Behaviour of cement concrete at high temperature. Bulletin of the Polish Academy of Sciences: Technical Sciences [online]. 2013, 61(1), - [cit. 2016-11-09]. DOI: 10.2478/bpasts-2013-0013. ISSN 0239-7528. Available from: http://www.degruyter.com/view/j/bpasts.2013.61.issue-1/bpasts-2013-0013/bpasts-2013-0013.xml

[7] BRACEWELL, R. The Fourier Transform & Its Applications. 3. USA: McGraw-Hill Science/Engineering/Math, 1999. ISBN 0073039381.

[8] SANSALONE, Mary. a William B. STREETT. *Impact-echo: non-destructive evaluation of concrete and masonry.* Ithaca, N.Y.: Bullbrier Press, 1997. ISBN 978-0961261061.

[9] SANSALONE, M., J.M. LIN a W.B. STREET. A procedure for Determining P-Wave Speed in Concrete for Use in Impact-Echo Testing Using a P-wave Speed Measurement Technique. In: ACI Materials Journal. 1997, s. 9.

[10]HAGER, I. *Behavior of high performance concretes at high temperature – evolution of mechanical properties.* French, 2004. PhD thesis. Ecole Nationale des Ponts et Chaussées.

[11]OZAWA, Mitsuo, Shinya UCHIDA, Toshiro KAMADA a Hiroaki MORIMOTO. Study of mechanisms of explosive spalling in high-strength concrete at high temperatures using acoustic emission. *Construction and Building Materials.* 2012, **37**, 621-628. DOI: 10.1016/j.conbuildmat.2012.06.070. ISSN 09500618. Available also from: http://linkinghub.elsevier.com/retrieve/pii/S095006181200462X

[12]EPASTO, Gabriella, E. PROVERBIO a V. VENTURI. Evaluation of fire-damaged concrete using impact-echo method. *Materials and Structures.* 2010, **43**(1-2), 235-245. DOI: 10.1617/s11527-009-9484-0. ISSN 1359-5997. Available also from: http://www.springerlink.com/index/10.1617/s11527-009-9484-0

[13]ČSN EN 12 390-3. Testing hardened concrete – Part 3: Compressive strength of test specimens. Praha: Český normalizační intitut, 2009.

[14]ČSN EN 12 390-5. Testing hardened concrete – Part 5: Flexural strength of test specimens. Praha: Český normalizační institut, 2009.

[15]L. Sitek, L. Bodnárová, , J. Válek, M. Zeleňák, J. Lich, , J- Foldyna, M. Novotný, M. Effects of water jet on heat-affected concretes, 11th International Scientific Conference on Modern Building Materials, Structures and Techniques, MBMST 2013; Vilnius; Lithuania; 16 May 2013 through 17 May 2013; Procedia Engineering , Vol. 57, 2013, pp. 1036-1044.

[16]L. Sitek, L. Bodnarova, K. Souček, L. Staš, L. Gurkova, Analysis of inner structure changes of concretes exposed to high temperatures using micro X-Ray Computed Tomography. Acta Geodynamica Et Geomaterialia. 2015. 12(1). p. 79-89. ISSN 1214-9705

# Motorcycle Suits: Uniaxial and Biaxial Tensile Tests of Leather

BOŃKOWSKI Tomasz[1, a], ŠOLTÉS Lukáš[1, b], KOCHOVÁ Petra[1, c], KOTTNER Radek[1, d] and HYNČÍK Luděk[1, e]

[1]NTIS - New Technologies for the Information Society, Faculty of Applied Sciences, University of West Bohemia, Technicka 8, 301 00 Pilsen, Czech Republic

[a]tomasz@ntis.zcu.cz, [b]soltesl@ntis.zcu.cz, [c]kochovap@ntc.zcu.cz, [d]kottner@kme.zcu.cz, [e]hyncik@ntc.zcu.cz

**Keywords:** motorcycle suit, leather, testing, uniaxial, biaxial, tension

**Abstract.** This paper describes uniaxial and biaxial tests of leather samples. The tests were conducted on non-standardized samples. The uniaxial test was performed on 100 samples. The biaxial test was done on 20 samples. The aim of the uniaxial test was to obtain a reliable value of the leather ultimate strength, on a large population of samples. Additionally, this test revealed the differences in the ultimate strength between two relative perpendicular directions of cut. The biaxial test was used to obtain the stress-strain relationship and the ultimate strength. The stress-strain relationship will be used in the future to improve the constitutive material model of the leather. The biaxial test was conducted under four logarithmically graded clamps separation speeds (1, 10, 100, 1000 mm/min).

## Introduction

A natural leather is the most commonly used material in motorcycle suit industry [1]. Potential reduction of the rider's injuries by personal protective equipment makes the assessment of the drivers clothing a very important topic in powered two wheelers safety research [2]. The manufacturers and developers of motorcyclist outfits often follow the standardized test procedures [3] during the examination of materials provided by their suppliers. However, these standards do not take into account non-isotropic nature of the biological material. In previous research, the authors identified Ogden's material parameters of the leather [4]. This paper presents the progress in the author's leather tensile testing procedure.

## Tensile test methodology

The tensile tests were conducted on the 574LE2 Biomechanics and Tissue Tester from TESTRESOURCES Company. The tests were held in temperature equal to 23±1 °C and relative humidity 50±6 %.

The uniaxial test was designed to measure differences in the ultimate strength between relative perpendicular material directions. The five standardized samples [3] (S samples) were cut randomly from the motorcycle suit. Then, 20 samples were cut from each S sample. The first half of them were cut along the S sample axis (0 degrees angle), the rest were cut in the perpendicular direction (90 degrees angle). The uniaxial samples (U samples) had dimensions 40 x 4 mm with average thickness 1.24 mm. The U samples (Fig. 1) were exposed to a uniaxial strain described by Fig. 2 (the loading continued up to the ultimate strength).

Fig. 1 Clamped uniaxial sample.

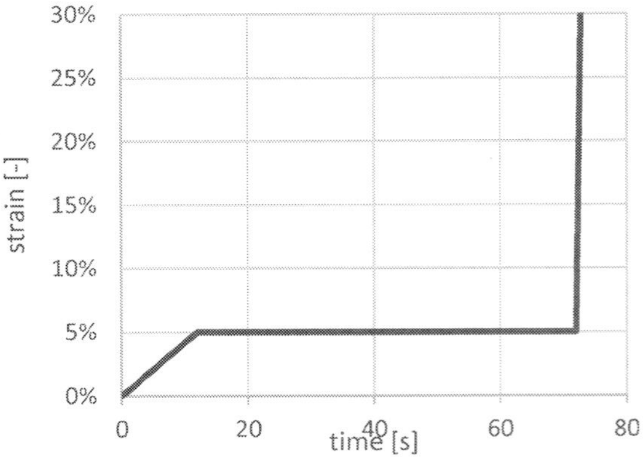


Fig. 2 The testing machine control signal.

To obtain data for the more reliable calibration of the hyperelastic material model, the biaxial test was performed (Fig. 3). The biaxial samples (B samples) were cross-shaped with 12.5mm radius in the corners (Fig. 4). This shape allowed to avoid the stress concentrations in the cross corners. The B samples were cut randomly from a motorcycle garment leather. The B samples thickness wavered between 1.017 mm and 1.468 mm, with an average of 1.283 mm.

The group of 20 B samples was divided into four sub-groups. Each sub-group was exposed to biaxial strain with various strain rate. The elongation control signal had the ramp shape. The slope of the ramp was logarithmically graded (1, 10, 100, 1000 mm/min). During this test, four independent force-cells were used (one force cell for each clamp).

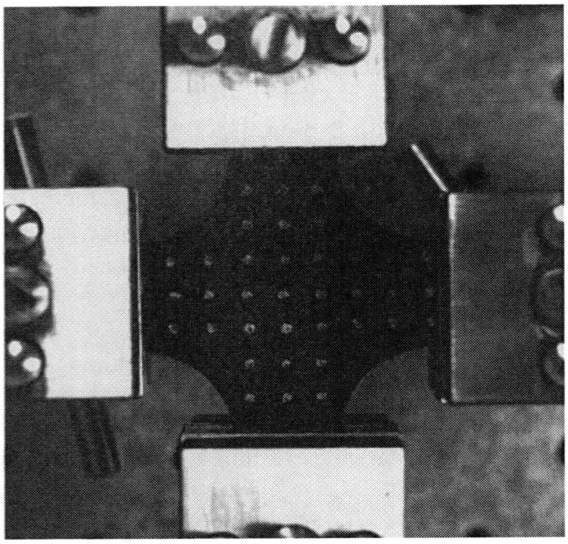

Fig. 3 Clamped B sample before the test.

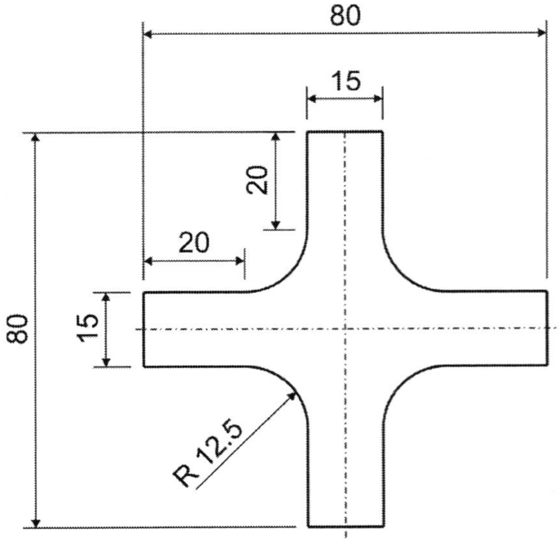


Fig. 4 The B sample dimensions.

**Uniaxial tensile test results**

The conducted test brought the ultimate damage force with last 20 data points before the samples damage, for the each U sample. These data were presented in the charts below (Fig. 5- 9). Each S sample had a specific color (black, green, blue, red and white). Following this nomenclature, figures 5-9 had the corresponding color. We assumed that the color corresponds to a different cowhide. On the Fig. 5- 9 solid lines represents U samples cut along the S sample axis, while dashed lines represent the U samples cut in the perpendicular direction. The samples cut from white S sample (Fig. 9) shows the smallest dispersion in ultimate damage force.

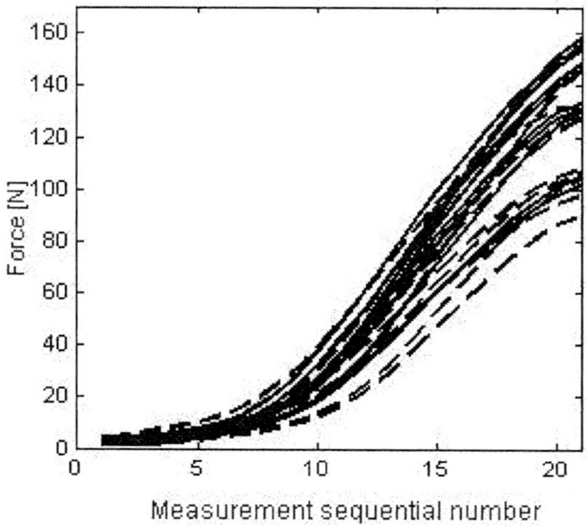


Fig. 5 Last 20 measurements for black samples.

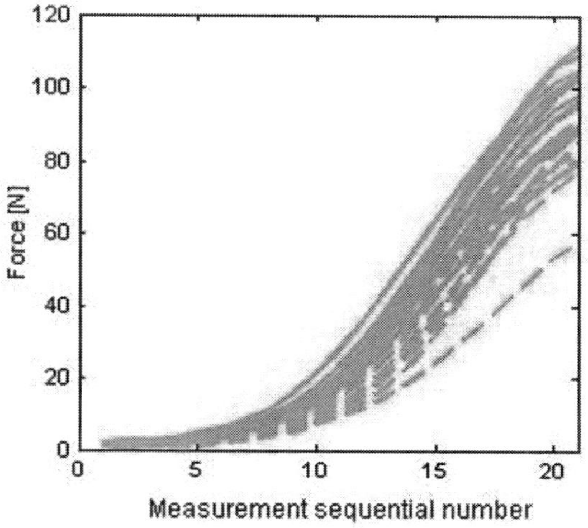


Fig. 6 Last 20 measurements for green samples.

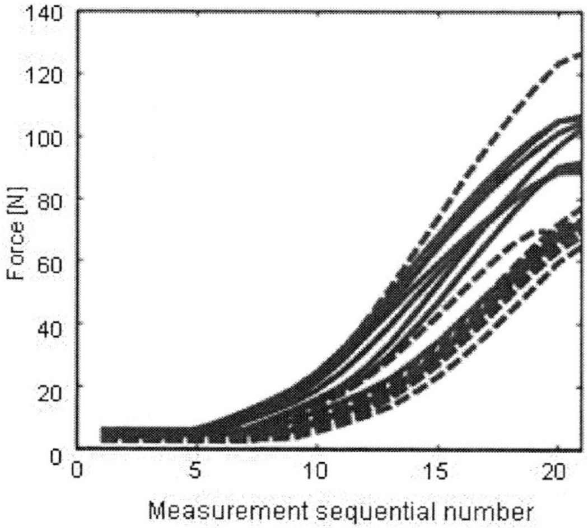


Fig. 7 Last 20 measurements for blue samples.

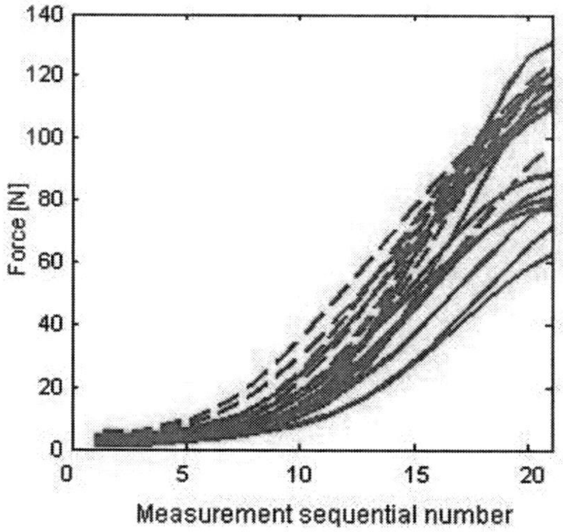


Fig. 8 Last 20 measurements for red samples

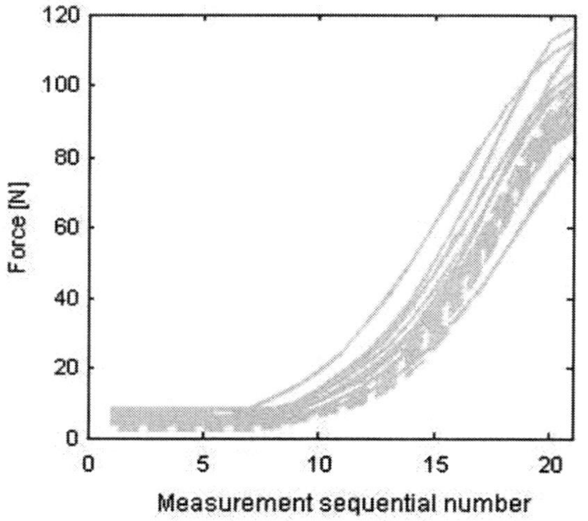

Fig. 9 Last 20 measurements for white samples.

Based on obtained data (Fig. 5- 9) and real dimensions of the U sample, the ultimate strength was calculated (fig 10). The ultimate strength was in the range 11.58- 31.8 MPa. The average ultimate strength difference between the U samples in the two directions was in the range 7– 28 % (Table 1).

Table 1 Average ultimate strength of uniaxial samples [MPa].

| Samples belonging to S sample | Black | Blue | Red | White | Green |
|---|---|---|---|---|---|
| 0 deg | 27.45 | 18.35 | 18.20 | 20.14 | 19.79 |
| 90 deg | 24.16 | 15.35 | 23.36 | 18.71 | 16.80 |
| difference | 12.01% | 16.33% | 28.30% | 7.07% | 15.09% |

The Fig. 10 shows the dispersion in the ultimate strength through all U samples. This graph illustrates that the blue and white samples, cut in the perpendicular direction, are the most homogenized group. Only two measurements (one BL90 and one G90) were out of 99.3 % of result distribution.

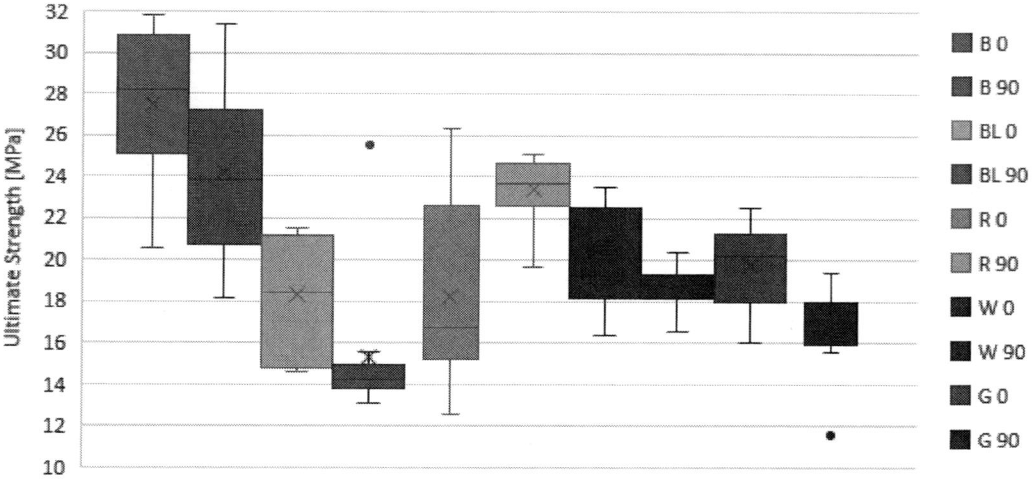

Fig. 10 Box plot of U samples ultimate strength [MPa].

**Biaxial tensile test results**

The biaxial test was designed to obtain leather stress-strain relationship (Fig. 11). The biaxial load characteristics are necessary for reliable hyperelastic model fitting in terms of Ogden parameters [5]. In paper [4], the authors used the load characteristics from the uniaxial standardized test with the clamp separation speed of 100 mm/min, therefore the biaxial test with the same speed was desirable (Fig. 11).

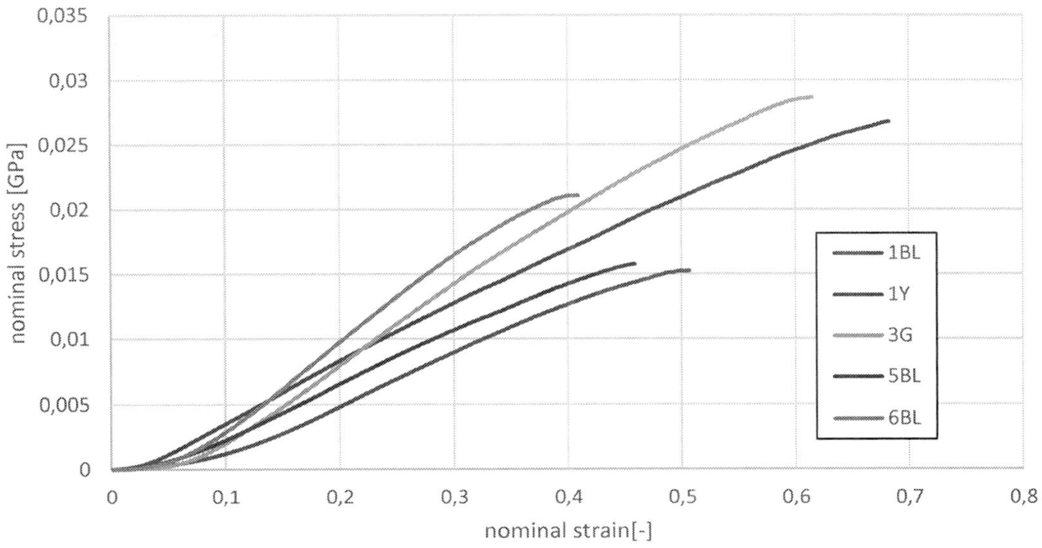


Fig. 11 Nominal stress vs. nominal strain under speed of 100 mm/min

Additionally, the ultimate damage force in the different jaw separation speeds was examined (Fig. 12). The following graph shows that under the speed of 1, 100, 1000 mm/min the average ultimate force is around 500 N. Only for speed 10 mm/min higher average ultimate with the value of 625 N was found.

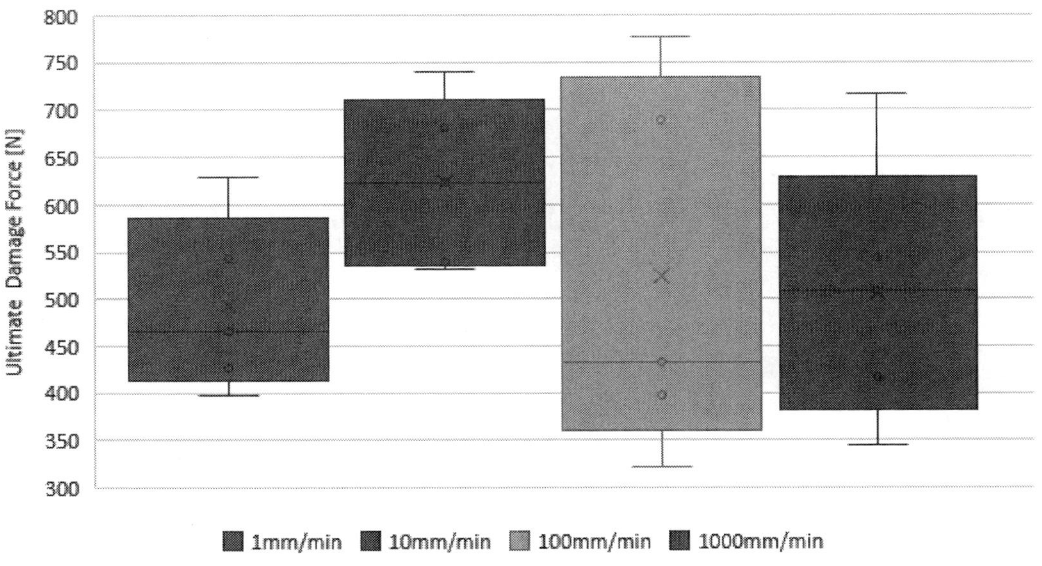


Fig. 12 Box plot of the B samples ultimate force [N].

## Conclusions

The ultimate tensile strength under uniaxial loading was in the range 11.58- 31.8 MPa. The average ultimate strength difference between the U samples in the two perpendicular directions was in the range 7– 28 %. The nominal stress–nominal strain relationship and the ultimate damage forces under biaxial load were obtained. The nominal stress-strain relationship will be used in the future to improve the leather constitutive model in terms of Ogden parameters.


## Acknowledgment

The research leading to these results has received funding from the People Programme (Marie Curie Actions) of the European Union's Seventh Framework Programme FP7/2007-2013/ under REA grant agreement no 608092. This work was supported by the internal grant project SGS-2016-059 and by the project LO1506 of the Czech Ministry of Education, Youth and Sports under the program NPU I. The authors would like to thank PSí Hubík for providing leather samples.


## References


[1] MAIDS – In-Depth Investigation of Motorcycle Accidents, The Association of European Motorcycle Manufacturers (ACEM) (2009).

[2] L. Meredith, C. Hurren, E. Clarke, M. Fitzharris, M. Baldock, L. de Rome, J. Olivier, J. Brown, Validation of the abrasion resistance test protocols and performance criteria of EN13595: The probability of soft tissue injury to motorcycle riders by abrasion resistance of their clothing, Journal of Safety Research 61 (2017) 1-7, ISSN 0022-4375.

[3] Leather - Physical and mechanical tests - Determination of tensile strength and percentage extension, ISO 3376:2002(E).

[4] T. Bońkowski, R. Kottner, J. Krystek, L. Hubík, T. Görner, L. Hynčík, Tensile test of motorcycle garment leather. In 54th International Conference on Experimental Stress Analysis (EAN 2016) (2016) 23-29, ISBN: 978-80-261-0624-1.

[5] Virtual Performance Solution 2016 - Solver Reference Manual Volume III, (2016).

EAN 2017
55th conference on experimental stress analysis

# Assessment of serviceability criteria in Eurocodes

Markova Jana[1,a] and Holicky Milan [1,b]

[1]Klokner Institute, Czech Technical University in Prague, Czech Republic

[a]jana.markova@cvut.cz, [b]milan.holicky@cvut.cz

**Keywords:** serviceability constraints; reliability assessment; target reliability.

**Abstract.** The load-bearing capacity, serviceability or durability of a structure designed in accordance with different standards or nationally implemented Eurocodes might be expected within a broad range. The actual structural resistance depends not only on used theoretical models and selected reliability elements for calculation of structural resistance but also on various prescriptive rules recommended in applied current standards including structural detailing. Moreover, in some cases the theoretical models given in various standards for determining structural resistance or serviceability provide considerably different probability of over-crossing the specified design value

## Introduction

Construction works are designed using methods provided in national, European or international standards. The Eurocodes for structural design allow the national selection of more than 1500 Nationally Determined Parameters (NDPs) including alternative design approaches, combinations of actions, values of partial factors and other reliability elements and also serviceability constraints. The actual reliability of a designed structure depends on applied national standards or selected NDPs in the National annexes to Eurocodes of the CEN Member countries.

The reliability of structural members designed for the serviceability limit states according to current National Annexes to Eurocodes may have a considerable scatter and during the current evolution of Eurocodes should be further harmonised.

## Serviceability limit states

The serviceability limit states concern the functioning of a structure or structural members, comfort of people and appearance of the construction works. Taking into account the time dependency of load effects, two types of serviceability limit states should be distinguished, as illustrated in Figure 1.

- Irreversible serviceability limit states (see Figure 1(a)), which are those limit states that remain permanently exceeded even when the actions that caused the infringement are removed; the failure domain is the total time following the first passage of the limiting value.
- Reversible serviceability limit states (see Figure 1(b)), which will not be exceeded when the actions which caused the infringement are removed; the failure domain consists of all parts where the response is above the limiting value.

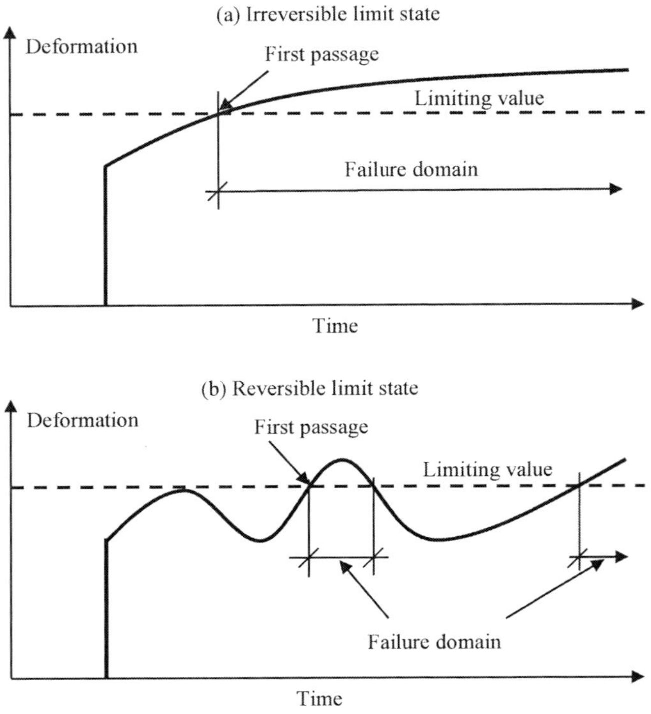

Figure 1: Irreversible and reversible limit states

For irreversible limit states the design criteria are similar to those of ultimate limit states, but with reduced reliability. The first passage of the limit state is decisive. For reversible limit states the first infringement (first passage) does not necessarily lead to the loss of serviceability.

Various serviceability requirements can be formulated taking into account the acceptance of infringements, their frequency and their duration. Generally, three types of serviceability limit states are applicable as follows:

- no infringement is accepted,
- specified duration and frequency of infringements are accepted,
- specified long-term infringement is accepted.

The serviceability criteria are then associated as appropriate with the characteristic, frequent and quasi-permanent values of variable actions.

What must be verified is that the design values of the actions' effects, $E_d$, specified via the criterion in question and calculated on the basis of suitable combinations, remain lower than the corresponding design limit values $C_d$ of the relevant serviceability criterion

$$E_d \leq C_d \quad (1)$$

Conceptually, all such serviceability limit state checks conform to this general relation, though, they may concern aspects that are quite distinct one from the other, such as limitations to deflections, or crack widths in reinforced concrete structures.

Presently the recommended serviceability constrains are not provided in EN 1990 [5] for basis of structural design. It is expected that selected recommendations will be given in new, revised EN 1990. Limiting design values of vertical deflections based on several National Annexes to Eurocodes of the CEN Member countries, and also currently recommended in the latest draft of new EN 1990 by the Subcommittee CEN/TC 250/SC10 are given in Table 1, for natural frequencies in Table 2 and for crack widths limits in Table 3.

Table 1. Recommended limiting values of deflections for selected structures

| Structures | National Annexes | EN 1990 |
|---|---|---|
| Roofs non-accessible | $w_{max} \leq L/200$ up to $\leq L/300$ | rigid roofing:<br>$w_2+w_3 \leq L/250$<br>resilient roofing:<br>$w_2+w_3 \leq L/125$ |
| Roofs accessible | $w_{max} \leq L/250$ up to $\leq L/300$ | partition walls:<br>- with openings:<br>$w_2+w_3 \leq L/1000$<br>- no opening:<br>$w_2+w_3 \leq L/500$<br>reinforced walls:<br>$w_2+w_3 \leq L/350$<br>removable walls:<br>$w_2+w_3 \leq L/250$ |
| Floors | $w_{max} \leq 15$ up to 28 mm; $L/400$; $L/250$<br>$L/400$ main girders; supporting plasters $L/250$; brittle materials 15 mm | |
| Structures supporting crane runways | $w_{max} \leq L/600$ up to 700; $w_{max} \leq 25$ mm | $w_{max} \leq L/600$;<br>$w_{max} \leq 25$ mm |
| Appearance | $w_{max} \leq L/250$ | $w_{max} \leq L/250$ |

For the serviceability limit state of a structure not to be exceeded when subjected to vibrations, the natural frequency of vibrations of the structure should be kept above appropriate values which depend upon the function of the building and the source of the vibration, and agreed with the client or the relevant authority.

Table 2. Appropriate values of natural frequencies for selected structures

| Structures | National Annexes | MC 2010 | EN 1990 |
|---|---|---|---|
| Gymnasia and sports halls | > 3 up to > 6,0 Hz | 8 | > 8,0 Hz |
| Concert halls without permanent seats | > 8,0 Hz | 7 | > 7,0 Hz |
| Concert halls with permanent seats | > 8,0 Hz | 3,4 | > 5 Hz |
| Floors and staircases of public buildings | > 5 up to > 6 Hz | | > 5 Hz |
| Floors of residential or office buildings | > 3 up to > 5 Hz | 1,4-4; 4 | > 3 Hz |

Recommended crack width limits provided in EN 1992-1-1 are given in Table 3.

Table 3. Crack width limits in EN 1992-1-1 [5]

| Exposure class | Quasi-permanent combination | Frequent load combination |
|---|---|---|
| X0, XC1 | 0,4 mm | 0,2 mm |
| XC2, XC3, XC4 | 0,3 mm | 0,2 mm |
| XD1, XD2, XS1, XS2, XS3 | | Decompression |

MC 2010 [9] recommends for reinforced concrete members in exposure classes XC, XD, XF, XS the crack width limit $w_{lim} = 0,3$ mm under the quasi-permanent combination of actions.

## Probabilistic verification of structural reliability

The knowledge of the reliability level of the structure designed according to the national standards or nationally implemented Eurocodes and also the reliability (credibility) of prescriptive analytical models or serviceability constrains in standards can be used for optimisation of design procedures or for further harmonisation of standards. The structural member or theoretical model may be considered as reliable, if the condition $p_F < p_t$ (or $\beta > \beta_t$) is satisfied where the probability of failure $p_F$ is given as

$$P_f = P\{g(X) < 0\} = \int_{g(X)<0} \varphi_X(x)\,dx\,. \tag{2}$$

The failure probability $p_F$ may be expressed by reliability index $\beta = -\Phi^{-1}(p_F)$, where $\Phi$ is the distribution function of standardised normal variable. The probability of failure $p_t$ and reliability index $\beta_t$ are the specified (target) values that should not be exceeded during the intended reference period.

The reliability differentiation of structures in EN 1990 [4] is based on three different levels of failure consequences with respect to the ultimate limit states (consequence classes CC1 to CC3). However, for the serviceability limit states similar differentiation is not provided yet. In some cases this differentiation of structures in serviceability limit states might also be useful for distinguishing the severity of potential consequences. Presently EN 1990 [4] recommends for the reversible limit states the target reliability index $\beta_t = 0$ and for the irreversible serviceability limit states $\beta_t = 1{,}5$ (for the fifty years reference period).

Some further recommendations for the target values of reliability indices in the serviceability limit states are given in the JCSS Probabilistic Model Code [6] where the target indices $\beta_t$ are recommended in a range from 1,3 to 1,7 for CC1 to CC3.

The reliability analysis of structural members for the ultimate or serviceability limit states can be determined through the probability $p_{F1}$ of the action effects $E(\boldsymbol{X})$ randomly exceeding the structural resistance $R(\boldsymbol{X})$ according to the following relationship

$$p_{F1} = P\{(\xi_R R(\boldsymbol{X}) - \xi_E E(\boldsymbol{X})) < 0\}\,. \tag{3}$$

where $\boldsymbol{X}$ is a vector of basic variables and $\xi_R$ and $\xi_E$ are the model uncertainties of a resistance and action effects.

The reliability (credibility) of theoretical models given in current standards can be examined by means of the credibility of specified design value $v_d(\boldsymbol{x}_d)$ (e.g. design stress, design deflection, design crack width) based on the vector of design variables. The probability $p_{F2}$ of exceeding the design value $v_d(\boldsymbol{x}_d)$, which was determined according to relevant theoretical formulae and recommendations of prescriptive design procedure, might be analysed as

$$p_{F2} = P\{(v_d(\boldsymbol{x}_d) - \xi_E\, v(\boldsymbol{X})) < 0\}\,. \tag{4}$$

where $\boldsymbol{X}$ is a vector of random basic variables and $\xi_E$ represents the uncertainties of the effects of basic variables and model uncertainties in the considered limit state.

The serviceability requirements in the limit states of crack width are analysed for an example of a reinforced concrete member as follows.

## Verification of structural reliability with respect to crack width

The time-independent reliability analysis of a reinforced concrete slab for the limit state of crack width is dealing with the probability $p_{F1}$ of the random crack width $w(X)$ over-crossing the required constraint $w_{lim}$ expressed by

$$p_{F1} = P\{(v_{lim}(\boldsymbol{x}_d) - \xi_E\, v(\boldsymbol{X})) < 0\}\,. \tag{5}$$

where $\boldsymbol{X}$ is a vector of basic variables and $\xi_E$ a model uncertainty for action effects.

The probabilistic models of basic variables entering the equation (5) are based on recommendations of JCSS and reliability analyses developed in the Klokner Institute CTU, and they are listed in contributions [7,8].

The theoretical models assume different probabilities of exceeding the characteristic value of crack width, or maximum crack spacing. The probability of over-crossing the characteristic crack width $w_k$ is 5 % according to Eurocodes, MC 2010 [9] and CSN 73 1201 [3], 10 % in CEP FIP Model Code 1990, 20 % in BS 8110 [2].

The results of reliability analysis of the reinforced concrete slab (height from 0,19 to 0,29 m) determined using the software Comrel are illustrated in Figure 2.

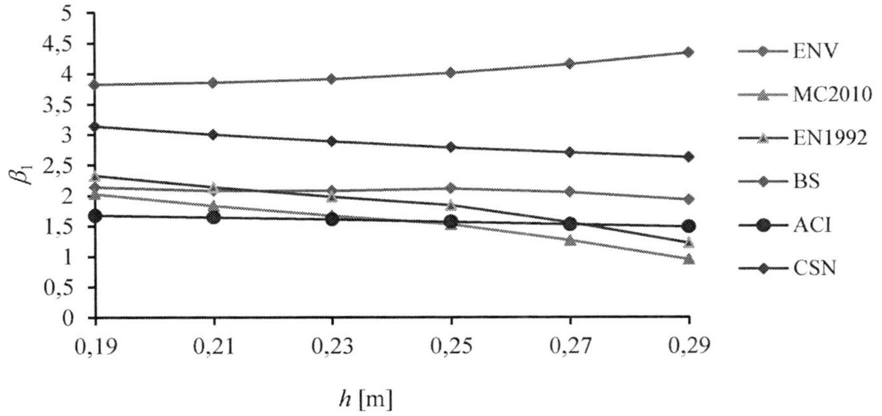


Figure 2: Reliability of a reinforced concrete slab for the limit states of crack width

The results of reliability analyses show that the reliability of structural member depends on used theoretical models for crack width. The reliability index $\beta_1$ is in a broad range from 1 to 4,5. The reliability of the slab seems to be rather high according to British (BS) and also Czech (CSN) national standards.

The credibility (reliability) of the specified characteristic value of crack width $w_k$ is also verified. The probability $p_{F2}$ of over-crossing the crack width $w_k$ according to relationship (4) for the slab expressed here by reliability index $\beta_2$ is illustrated in Figure 3.

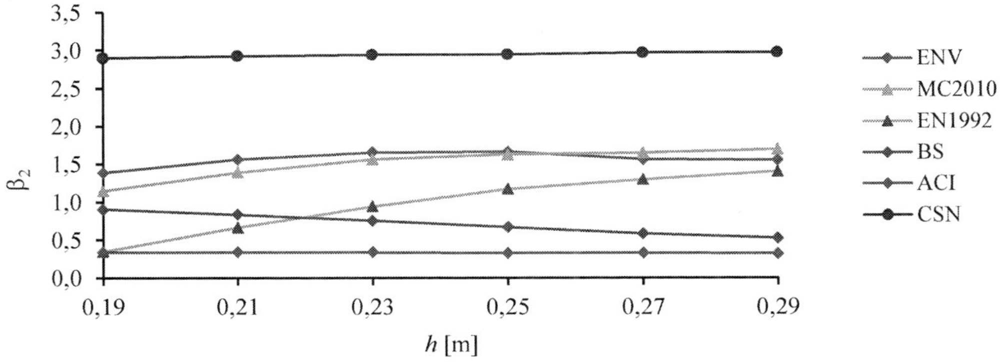


Figure 3: Credibility of specified characteristic value of crack width $w_k$

Analysis of the credibility of specified crack width $w_k$ indicates that the reliability index $\beta_2$ appears to be low for the theoretical models introduced in national American and British standards and rather high in Czech standard (about 2,9). The credibility of theoretical models seems to be sufficient in Eurocodes and Model Code 2010 [9], satisfying the target reliability index $\beta_t$ = 1,5.

## Concluding remarks

Presently the basic Eurocode EN 1990 introduces general recommendations for the verification of the serviceability limit states only. The development of supplementary provisions is needed including classification of structures based on the consequences of failure. An example of verification of a reinforced concrete slab with respect to the limit state of crack width indicates that the same limiting serviceability constraint (crack width limit) is compared with characteristic values of crack width obtained on the basis of a broad range of normative recommendations and also having different statistical meaning.

Reliability indices $\beta$ assessed in the analysis of the credibility of the analytical crack width formulae and the reliability of reinforced concrete slab with respect to limit crack width have a significant scatter and in some cases seem to be rather low.

It appears that the probabilistic methods can be effectively used for the calibration of serviceability constrains used in design or verification of structures. They might be applied for further harmonisation of national standards or parameters NDPs in the second generation of Eurocodes.

## Acknowledgments

This is an outcome of the research projects 16-04132S supported by the Czech Science Foundation and LD15037 supported by the Ministry of Education.

## References

[1] ACI 318-89. 1989. Manual of Concrete Practice, reported by ACI Committee 301. American Concrete Institute. Detroit

[2] BS 8110. 1989. Structural Use of Concrete. Part 2: Code of Practice for Design and Construction, British Standards Institution.

[3] ČSN 73 1201. 1986. Design of Concrete Structures and Amendments.

[4] EN 1990. 2002. Basis of Structural Design, CEN

[5] EN 1992-1-1. 2004. Design of Concrete Structures - Part 1: General Rules and Rules for Buildings, CEN

[6] JCSS Probabilistic Model Code. 2002.

[7] Marková J., Holický M. 2000. *Probabilistic Analysis of Crack Width.* Acta Polytechnica. Vol. 40. No. 2/2000, Prague, Czech Republic. pp. 56-60

[8] Marková J., Holický M. 2009. *Probabilistic Assessment of Crack Width for Existing Bridges,* in Esrel 2009, Prague, Czech Republic

[9] Model Code 2010, Design Code, fib, 2010

EAN 2017
55th conference on experimental stress analysis

# Inspection of Generator Shield Heat Treatment by Hole Drilling Method

Otakar Weinberg[1, a], Petra Hájková[2, b], Stanislav Holý[3, c], Jiří Janovec[4,d] and Jaroslav Václavík[1, e]

[1]Výzkumný a zkušební ústav Plzeň s.r.o., Tylova 1581/46, 301 00 Plzeň, Česká republika

[2]ČEZ, a. s., Jaderná elektrárna Temelín, 373 05 Temelín, Česká republika

[3]Ústav mechaniky, biomechaniky a mechatroniky, Strojní fakulta, ČVUT Praha, Karlovo náměstí 13, 121 35 Praha, Česká republika

[4]Ústav materiálového inženýrství, Strojní fakulta, ČVUT Praha, Karlovo náměstí 13, 121 35 Praha, Česká republika

[a]weinberg@vzuplzen.cz, [b]petra.hajkova@cez.cz, [c]stanislav.holy@fs.cvut.cz, [d]jiri.janovec@fs.cvut.cz, [a]vaclavik@vzuplzen.cz

**Keywords:** Residual stress, hole-drilling, casting, generator shield.

**Abstract.** In this article, the level of residual stresses on the set of several large-size castings of turbo-generator shield block is investigated to optimize the heat treatment and machining using hole-drilling method. Heat treatment of castings and methods and arrangements of the investigations including results are described as well.

## Introduction

The residual stresses, put to the structure during the incorrect casting process, can lead to change of dimensions, problems with the assembly, induce higher operational stresses and subsequently may cause the fatigue damage of the structure.

The level of residual stresses on the set of several large-size castings of turbo-generator shield block was investigated to optimize the heat treatment and machining using hole-drilling method. Heat treatment of castings and methods and arrangements of the investigations including results are described as well.

## Description of the Problem

**The Goal of Performed Investigations.** The measurement of residual stresses of six sets of the generator shield from the turbine side which should determinate the necessary heat treatment for minimum costs and minimum residual stresses was the goal of the investigation. The producer effort also was to avoid the stress relief annealing (SRA).

The shield is built from bottom and upper part; both halves are designed as the ø3780 mm semicircular plates with stiffening ribs and flanges, mutually connected with pins and bolts (Fig. 1a, 1b). The masses of the bottom an upper parts are 9500 kg and 6500 kg, used material GS-30Mn5 with $R_{p0.2}$ = 450 MPa and $R_m$ = 600 MPa.

The problems were started at the moment, when both finish-machined parts were replaced for new ones. The assembly of both parts with dowel pins was impossible due to the different dimensions overlap at the assembly plane. It was estimated, that the cause of this feature can be high level of residual stresses, lead to the both parts during casting, heat treatment or raw machining. Therefore a comprehensive program of production conditions to the shields stability was performed.

Fig. 1a Generator shield, face side

Fig. 1b Generator shield, internal side

**Heat treatment.** The positions for the measurement were selected according the calculation model. Both the distribution of temperature stresses coming from the controlled heat treatment and acting operational stresses were taken in considerations. The final decision of the places was performed according measurement limits of hole-drilling method.

Meeting the criteria of stated heat treatment conditions is critical for keeping the calculated mechanical parameters of castings. The used procedure comes therefore from the analysis of the chemical compositions of the used castings. Subsequently the individual phases of heat treatment are performed. It concerns the normalization, hardening, tempering and the stress relief annealing (SRA). Setting and controlling the particular operations of heat treatment is performed according theoretical and empirical knowledge of the producer. According this scheme the heat treatment of investigated parts was performed. The basic parameters for heat treatment were: normalization at 920°C/8÷12 hours in the air, hardening at 920°C/8÷12 hours to water, tempering 570÷620°C/6÷8 hours in the air and the annealing for residual stress removing 550÷600°C/8 hours in the air.

## Used Method and Measurement Positions

**Measurement Method.** The hole drilling method was used for residual stress investigation. Using the standard equipment for hand-drilling, the hole ø4 mm was drilled up to 4 mm depth with 2-edge eccentric end mill ø 4 mm and the residual stresses were evaluated according the own procedure, which is in conformity with the methodology, used in [1]. The advantage of this procedure is the possibility of measurement of residual stresses in large depths. The used strain gauge rosette HBM 3/120 RY21 was pre-drilled with the end mill ø2 mm at each step. It was verified that using this procedure no significant stresses are induced during the low speed drilling.

**Measurement Positions.** Two measured positions were chosen at each casting, one measuring point on the face side marked "1" and one on the internal back side marked "2" (see Fig. 1 and 2). On the bottom part face side the position was chosen on the lug side of the bearing housings. On the upper part face side the position was chosen in the cavity of the bush for shaft leading. The measurement positions on the internal shield sides were at the similar positions at the annual area for the shaft leading.

Fig. 2 On-site measurements of the shield

## Performed tests and results

**Performed Tests.** Six sets of shields were chosen for the test from which results from three sets are presented here. Each set differs with heat treatment and machining parameters. The pulsed pressuring with 0.9MPa (PP) was also investigated to be able to lower the residual stresses.

The first shield set was as delivered from the producer on final mechanical conditions – at these parts the problem has occurred. The expectations of high residual stresses were confirmed. The measured residual stresses reached the levels higher than 500 MPa in compressive sense. No other heat treatment investigations were performed with this set. Probably no stress relief annealing was performed by the producer (Fig. 3a, 3b).

The second shield set was new product with all heat treatment steps including the annealing for residual stress removing. The manufacture was followed by the customer. The shield has left only the finish allowance. The measured residual stresses were very low < 25 MPa. The second step was the investigation residual stress change after the final machining and (PP). Both procedures have insignificant influence to residual stress level, the measured residual stresses were very low < 30 MPa (Fig. 4a, 4b).

The third shield set was delivered from metallurgical works with only standard heat treatment. The measures residual stresses were again high compressive about 300 MPa, but lower in comparison with the original set. After this step the rough machining (RM) was performed. The raw machining also has no significant influence to the level of residual stresses; measured values were again compressive about 300 MPa. The final step was the stress relief annealing. The residual stresses after this step fall down significantly down to 30 MPa (Fig. 5a, 5b). All measured residual stresses for all sets were compressive.

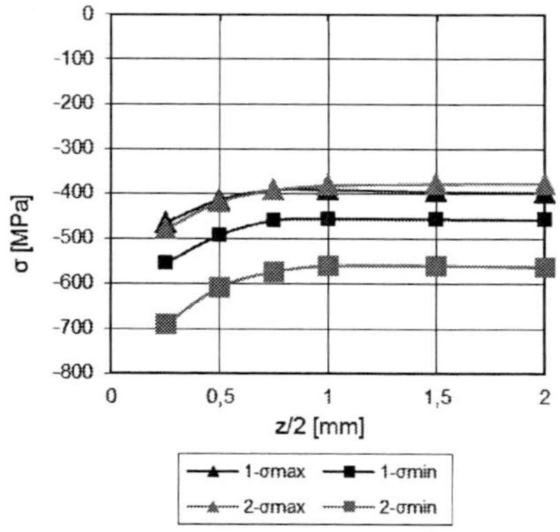


Fig. 3a 1st set, upper part, without SRA

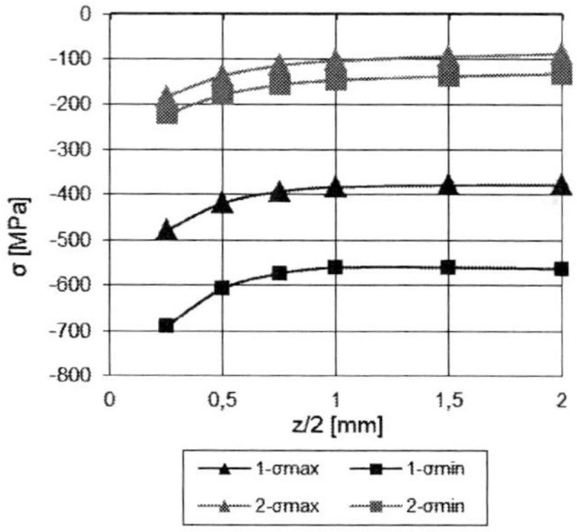


Fig. 3b 1st set, bottom part, without SRA

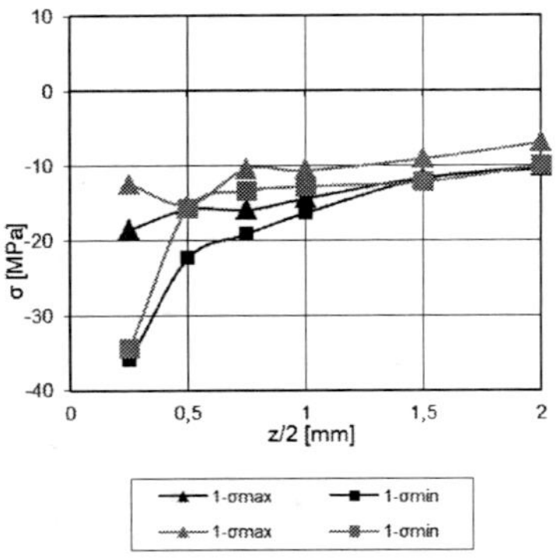


Fig. 4a 2nd set, bottom part, SRA+PP

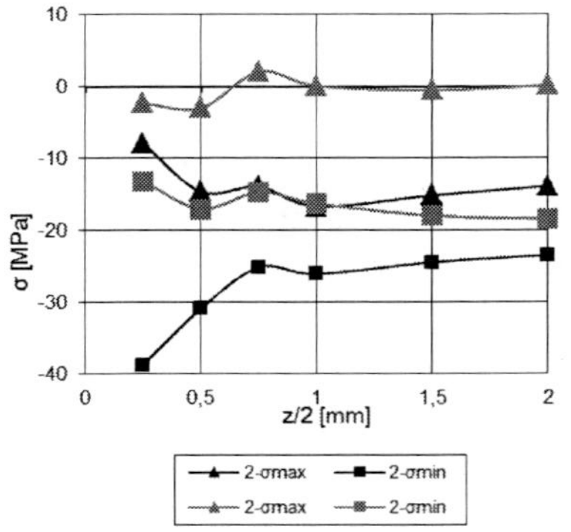


Fig. 4b 2ndset, bottom part, SRA+PP

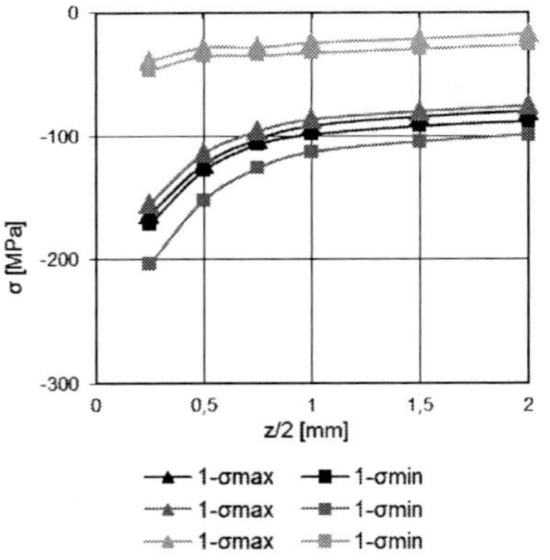


Fig. 5a 3nd set, bottom part, RM+SRA

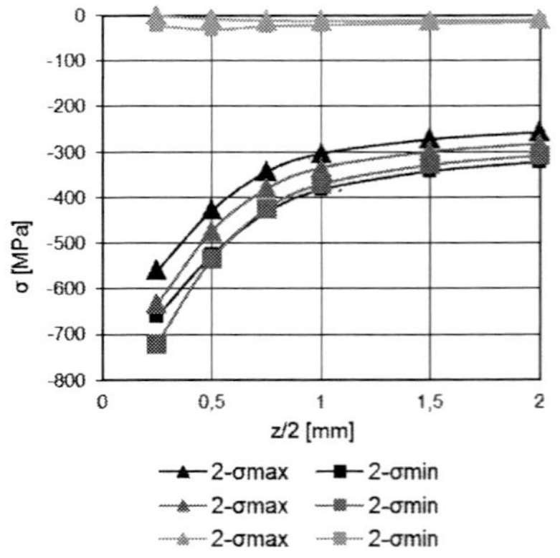


Fig. 5b 3nd set bottom part, RM+SRA

## Conclusions

The resulting residual stresses for all sets of investigated pairs of turbo generator shield did not fulfill the condition for low values for the technological operations, where as the final technological operation the annealing for residual stresses was not presented. The exact complex tuning of this heat treatment (heat velocity, temperature holding time, the velocity of cooling, kinds of cooling) influence significantly the required controlled features. The technological operation of pulsed pressuring has only marginal effect and does not considerably influence the level of residual stresses. Its account is only in slight redistribution of residual stresses.

The article has originated in the framework of the institutional support for the long-term conceptual development of the research organization.

## References

[1] ASTM E837-13a, Standard test method for determining residual stresses by hole-drilling strain-gage method. September 2013.

[2] J. Janovec, J., Analysis of heat treatment o generator shields, sets No. 2, Final report, Technical report No. 25-15, ČVUT Prague, 2015 (in Czech).

EAN 2017
55th conference on experimental stress analysis

# Destructive Testing of Cement Paste with Micronized Waste Marble Powder as Partial Replacement for Cement – Long Time Development of Mechanical Properties

PROŠEK Zdeněk [1,2,a], TOPIČ Jaroslav [1,b] and TESÁREK Pavel [1,c]

[1]CTU in Prague, Faculty of Civil Engineering, Thákurova 7, 166 29 Praha 6, Czech Republic

[2]CTU in Prague, UCEEB, Trinecka 1024, 273 43 Bustehrad, Czech Republic

[a]zdenek.prosek@fsv.cvut.cz, [b]jaroslav.topic@fsv.cvut.cz, [c]tesarek@fsv.cvut.cz

**Keywords:** Cement Pastes, Compressive Strength, Flexural Strength, Micronized Marble Powder

**Abstract.** This article deals with using waste marble powder as partial replacement for cement. In particular, we focus on mechanical properties of the cement pastes. We used five mixtures of cement pastes containing 5, 10, 15 and 50 wt. % of micronized marble powder as partial replacement for cement. The investigated mechanical properties were flexural strength and compressive strength for the 28 and 377 days old samples. The results obtained from these materials were compared with reference material.

## Introduction

Leaving the waste materials from marble industry to the environment directly can cause environmental problem [1]. Hence, many countries have been working on how to reuse the waste material so that they reduced hazards to the environment [2]. One possible utilization of this waste is used in the production of blended cement [3]. There are several works that deal with use the marble sludge as replacement for cement in creating the blended cement. The first study found that the addition of the marble sludge to the cement composite is changed CH crystals ($Ca\ (OH)_2$) [4]. The second study showed the morphology of the internal structures of cement pastes with and without marble sludge by using a scanning electron microscopy. The morphology was identical [5]. In all case, marble sludge in a small amount had a good influence on compressive strength [6,7]. Moreover, according to the study, which was written by Menendez et al. [8] marble and limestone dust resulted in slight decrease compressive strength after 90 days compared to the value of compressive strength 28 days old samples .Reduction of compressive strength was explained by measurement uncertainty.

New options to possibly confirm or deny the allegations about the influence of the marble sludge on long-term mechanical properties of cement pastes is using the method of micro milling by high-speed mill, because the resulting micronized marble powder has a higher specific surface area and thus possible higher reactivity [9].

## Materials and Samples

The mixtures composed of Portland cement CEM I 42.5R (Radotín) and micronized marble powder with fractions 0-40 μm has been prepared for investigating the impact of micronized marble powder on long time mechanical properties. Micronized marble powder was obtained from the processing of limestone marble and it is mixture of marble dust and sludge from Spain, Italy, Turkey, West Bank of Jordan and Egypt. Micronized marble powder was micronized by a high-speed mill from Lavaris Ltd. (Libčice – Czech Republic). The figure 1 show particle size distribution of Micronized marble powder. The mixtures were designed to maintain the same workability of the fresh mixture. Consistency of the mixtures was determined using a flow expansion test after 10 ($d_{10}$) and 20 ($d_{20}$) impulses (Tab. 1). Individual material samples differed in the quantity of micronized marble powder. The cement was replaced by the waste micronized marble powder in an amount of 5, 10, 15 and 50 weight percent (Tab. 1). The results obtained from these materials were compared with reference.

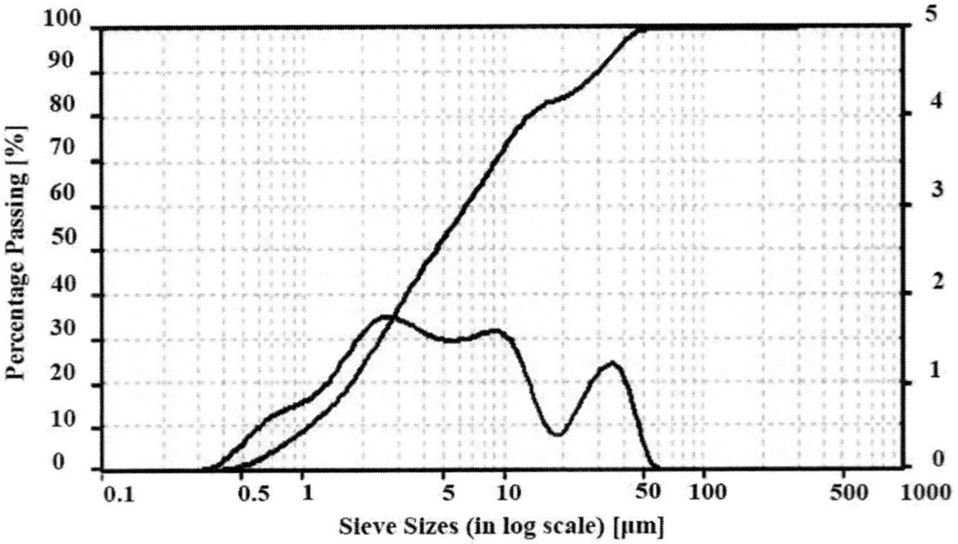


Fig. 1: Particle size distribution curve of Micronized marble powder

Mixtures prepared according to Tab. 1 were cast in the rectangular mold with dimensions equal to 40 × 40 × 160 mm. All the rectangular molds were measured their dimensions before the mixture was applied to the mold. These values were used to calculate approximate shrinkage (Tab. 1). After casting, these specimens were kept in the molds for 24 h at room temperature of 22 ° C. The six samples from each mixture. After demolding, these specimens were stored in a laboratory environment at 22 ± 1 ° C and relative humidity 50 ± 2 %.

Tab. 1: Composition of the mixtures.

| Set/ Material | Cement [wt. %] | Marble powder [wt.%] | Water ratio [-] | Flow expansion test $d_{10}$ [mm] | Flow expansion test $d_{20}$ [mm] | Bulk density [$kgm^{-3}$] | Shrinkage [%] |
|---|---|---|---|---|---|---|---|
| Ref. MP0 | 100 | 0 | 0.35 | 160 | 190 | 1919 ± 5 | 0.19 ± 0.06 |
| MP5 | 95 | 5 | 0.317 | 160 | 190 | 1915 ± 6 | 0.27 ± 0.11 |
| MP10 | 90 | 10 | 0.315 | 160 | 190 | 1920 ± 3 | 0.14 ± 0.05 |
| MP15 | 85 | 15 | 0.31 | 160 | 190 | 1879 ± 6 | 0.21 ± 0.07 |
| MP50 | 50 | 50 | 0.305 | 158 | 190 | 1760 ± 8 | 0.45 ± 0.16 |

## Experimental Methods and Results

The values of the compressive and the flexural strength of the samples were determined at the 28th day and 377th day. Three samples for each series and each period were used for the determination of the flexural strength on beams of dimensions equal to 40 × 40 × 160 mm and the uniaxial compressive test was performed on the broken halves of the specimens with effective dimensions of 40 × 40 × 40 mm. The load was applied at a constant rate of 0.1 mm/s in the case of the three-point bending test and 0.3 mm/s in the case of the compressive test. The distance between supports for the three-point bending test was equal to 100 mm.

Samples MP5, MP10 and MP15 had approximately same shrinkage as reference samples and value of shrinkage was about 0.20 % (Tab. 1). Samples MP50 had values of shrinkage two times higher than others, namely 0.45 ± 0.16 %. Increased shrinkage of samples was due to the addition of fine particles of micronized marble powder with high specific surface area. Also the sample MP5 and MP10 had the same bulk density as the reference sample. It is the same bulk density and marble powder is lighter than cement paste so, part of micronized marble powder must form a microfiller.

The figure 2 show results from mechanical tests. The results show the effect of long time on the mechanical properties of the cement pastes. The flexural strength was increased after 377 days for all samples. The largest flexural strength had set composed with 50 wt. % micronized marble powder, directly 5.95 ± 0.42 MPa. Compressive strength of cement pastes with micronized marble powder decreased after 377 days. The largest compressive strength after 28 days had mixture with 5 wt. % of micronized marble powder, directly 87.7 ± 6 MPa, but the value decreases after 377 days to 55.6 ± 4 MPa.

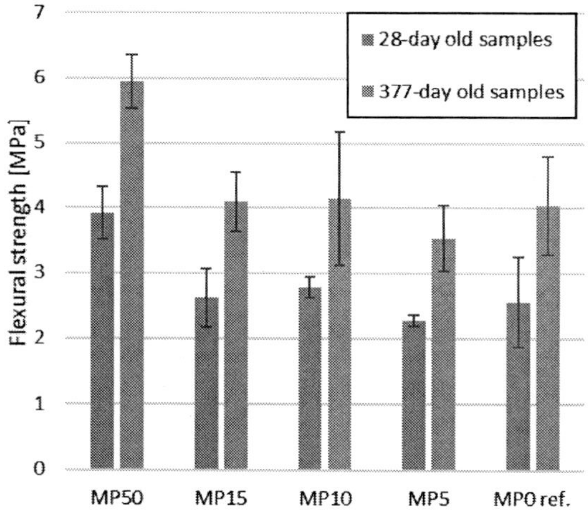


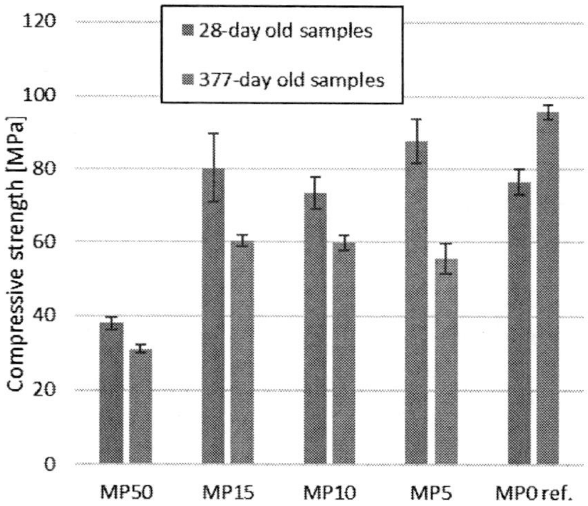


Fig. 2: Comparison of mechanical properties of cement pastes, flexural strength – left, compressive strength – right

## Conclusions

Using micronized marble powder was in initial stage enhances compressive strength of cement pastes, but after 377 days impairment compressive strength by about 20%. The results show that micronized marble powder may effect on hydration process of cement. In the future, we will focus on determining of effect of micronized marble powder in the hydration process by using chemical analysis and for final confirmation or refutation of the possibility of the involvement of the micronized marble powder in hydration of Portland cement.

**Acknowledgement**

This outcome was supported by the Czech Technical University in Prague under No. SGS16/201/OHK1/3T/11 and by the Ministry of Education, Youth and Sports within National Sustainability Programme I, project No. LO1605. The authors also thank the Center for Nanotechnology in Civil Engineering at the Faculty of Civil Engineering, Czech Technical University in Prague.

**References**

[1] H. Yilmaz, M. Guru, M. Dayi, I. Tekin, Utilization of waste marble dust as an additive in cement production, Materials and Design 31 (2010) 4039-4042.

[2] M.S. Ibrahim, et al., Health Risk Assessment of Marble Dust at Marble Workshops, Nature and Science, 9 (2011) 144-154.

[3] H. Binici, H. Kaplan, S. Yilmaz, Influence of marble and limestone dusts as additives on some mechanical properties of concrete, Scientific Research and Essay, 2 (2007) 372-379.

[4] B. Demirel, The effect of the using waste marble dust as fine sand on the mechanical properties of the concrete, International journal of physical sciences 5 (2010) 1372-1380.

[5] A. A. Aliabdo, A. E. M. Abd Elmoaty, E. M. Auda, Re-use of waste marble dust in the production of cement and concrete, Construction and Building Materials 50 (2014) 28-41.

[6] A. Ergün, Effects of the usage of diatomite and waste marble powder as partial replacement of cement on the mechanical properties of concrete, Construction and Building Materials 25 (2011) 806-812.

[7] V. M. Shelke, P. Pawde, R. Shrivastava, Effect of marble powder with and without silica fume on mechanical properties of concrete, J Mech Civ Eng 1 (2012) 40-45.

[8] G. Menéndez, V. Bonavetti, E.F. Irassar, Strength development of ternary blended cement with limestone filler and blast-furnace slag, Cement and Concrete Composites 25 (2003) 61-67.

[9] Z. Prošek, J. Topič, K. Šeps, P. Tesárek, Influence of Waste Crushed Limestone and Waste Micronized Marble Powder on Mechanical Properties of Cement Composite, Key Engineering Materials 722 (2017) 222-227.

EAN 2017
55th conference on experimental stress analysis

# Influence of Different Admixtures on Hydration Process and Mechanical Properties of Cement Paste

TOPIČ Jaroslav[1,a], PROŠEK Zdeněk[1,b], PLACHÝ Tomáš[1,c] and TESÁREK Pavel[1,d]

[1]CTU in Prague, Faculty of Civil Engineering, Thákurova 7, 166 29 Praha 6, Czech Republic

[a]jaroslav.topic@fsv.cvut.cz, [b]zdenek.prosek@fsv.cvut.cz, [c]plachy@fsv.cvut.cz, [b]pavel.tesarek@fsv.cvut.cz

**Keywords:** Cement paste; recycled concrete powder; micro-milled silica sand; hydration process; mechanical properties.

**Abstract.** Sustainable developments requires massive use of recycled materials but recycling of fines part of cementitious materials, which mostly composed of cement paste, proves to be very complicated. Therefore reactive properties of recycled concrete powder due to presence of non-hydrated cement grains are investigated. To eliminate influence of larger nucleation area (caused by presence of fine particles) on hydration process the inert micro-milled silica sand was used to comparison. Calorimetric measurement proves the presence of exposed non-hydrated particles in recycled concrete powder that can react again. But lower density of old cement paste in RCP overweight the mentioned potential of recycled concrete powder and mechanical properties are decreased compared with reference cement paste and cement paste with micro milled silica sand.

## Introduction

Sustainable developments requires massive use of recycled materials. Due to high consumption of cementitious materials in construction is necessary to focus on complete recycling of it. Recycling process of fines part of cementitious materials, which mostly composed of cement paste, proves to be very complicated. In previous papers where the recycled concrete powder (RCP) is used as partial cement replacement and microfiller the reactivity and hydration properties are examine [1,2]. The aim of this paper is to compare hydration and mechanical properties of RCP with inert filler with similar granulometric curve. Micro-milled silica sand was chosen to comparison. Due to his crystalline character, the pozzolanic activity of $SiO_2$ should be minimal compared with amorphous $SiO_2$, which is used as reactive additive [3,4]. Thanks to this comparison, the possible binder properties of RCP could be highlighted.

Measurement of heat flow during hydration were supplement by testing of mechanical properties. This results should be helpful to understand influences of changes in hydration process on final mechanical properties of cement paste with different admixtures.

## Materials and Samples

The Portland cement CEM I 42.5 used in this experiment was produced in Radotín (Czech Republic). Old railway sleepers of type PB2 and SB8 were used to produce recycled concrete powder (RCP) by high speed mill from LAVARIS Ltd. As micro-milled silica sand (MSS) was used product ST2 from company Sklopísek Střeleč, Inc.

For calorimetric measurement, three different mixtures were made. First mixture (denote as CEM) was reference cement past. Second mixture (denote as CEM+RCP 50) contained 50 % of cement and 50 % of RCP and third mixture (denote as CEM+MSS 50) contained 50 % of cement and 50 % of MSS. For this measurement had those three samples same water/mixture ratio equal to 0.45.

One reference cement paste without any admixture and four mixtures with different admixtures were made for measurement of mechanical properties. Two mixtures contain cement with 10 and 50 % of recycled concrete powder. Last two mixtures contain cement with 10 and 50 % of micro-milled silica sand. Mixtures had different water/mixture ratio from 0.35 to 0.40 to compensate the deteriorated workability caused by increasing amount of RCP and MSS. Composition of the samples is shown in Table 1. Each set contained 6 prismatic samples having dimensions of 40 × 40 × 160 mm. The samples were removed from their casts after 2 days. After that the samples were cured for 28 days in water at the temperature of 21 ± 2 °C.

Tab. 1: Composition of the tested samples.

| Mixture | Cement [g] (CEM I 42.5 R) | Recycled concrete powder (RCP) [g] | Micro-milled silica sand (MSS) [g] | Water/mixture ratio |
|---|---|---|---|---|
| **CEM** (ref) | 1500 | - | - | 0.35 |
| **CEM+RCP 10** | 1350 | 150 | - | 0.36 |
| **CEM+RCP 50** | 750 | 750 | - | 0.40 |
| **CEM+MSS 10** | 1350 | - | 150 | 0.36 |
| **CEM+MSS 50** | 750 | - | 750 | 0.40 |

## Experimental methods

Calorimetric measurement was conducted on an isothermal TAM Air calorimeter for measurement of heat flow. Mixtures were tested for 7 days at a constant temperature of 20 °C and were stored in sealable plastic containers, each containing from 32 g to 36 g of mixture. Based on the weight of each sample (measured before testing) the results of the heat flow were related to 1 g of cement for a better detection of changes caused by admixtures.

Dynamic Young's modulus were monitored by using the non-destructive resonance method. This method is based on measuring the natural frequency of the prismatic 40 × 40 × 160 mm samples. For evaluation of dynamic Young's modulus the basic longitudinal and flexural natural frequency, dimension and weight of the samples were used.

Flexural and compressive strength were determined on 28 day old samples using a model FP100 Heckert device. The testing was displacement controlled at a constant rate of 0.1 mm/s in the case of three-point bending and 0.3 mm/s during the compression test.

## Experimental results

Results of calorimetric measurement are shown in Fig. 1. By relating heat flow to 1 g of cement and not an entire RCP or MSS mixture it was possible to highlight differences in the hydration process caused by RCP. In case of mixture CEM+MSS 50 the increase of heat flow compared with reference cement is caused by larger nucleation area due to MSS addition. Increase of heat flow of mixture CEM+RCP 50 is partly (as CEM +MSS 50) caused by larger nucleation area, but mostly by presents of non-hydrated cement grains in RCP. Increase of heat flow of mixture with RCP compared with mixture CEM proves that exposed cement grains in RCP can react again and comparison with CEM+MSS 50 proves that increase of heat flow wasn't caused only by larger nucleation area.

Bulk density of the samples was measured before dynamic Young's modulus testing. Reference cement samples had bulk density 1919 ± 9 kg/m$^3$. Bulk density of the samples with 10 % RCP was 1851 ± 1 kg/m$^3$ and with 50 % of RCP was 1566 ± 3 kg/m$^3$. Bulk density of the samples with 10 % MSS was 2026 ± 11 kg/m$^3$ and with 50 % of MSS was 1964 ± 3 kg/m$^3$.

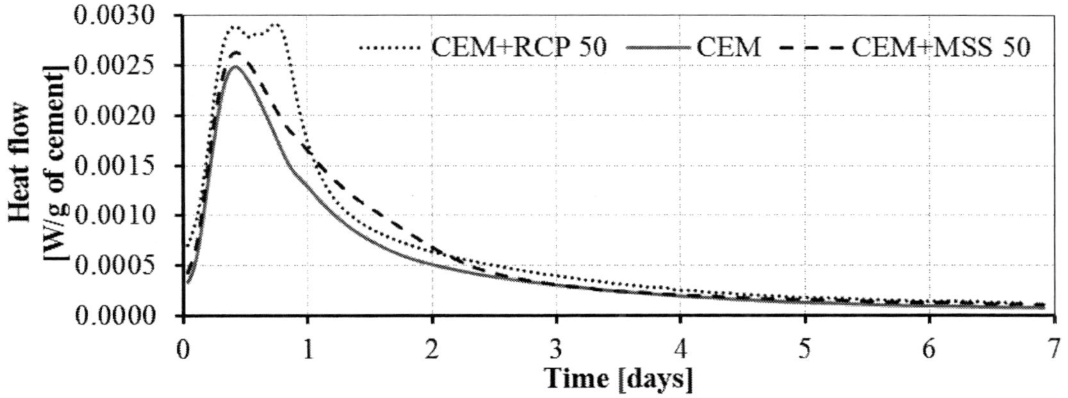


Fig. 1: Heat flow of cement paste and cement with 50 wt. % of recycled concrete powder (CEM+RCP) and micro-milled silica sand (CEM+MSS).

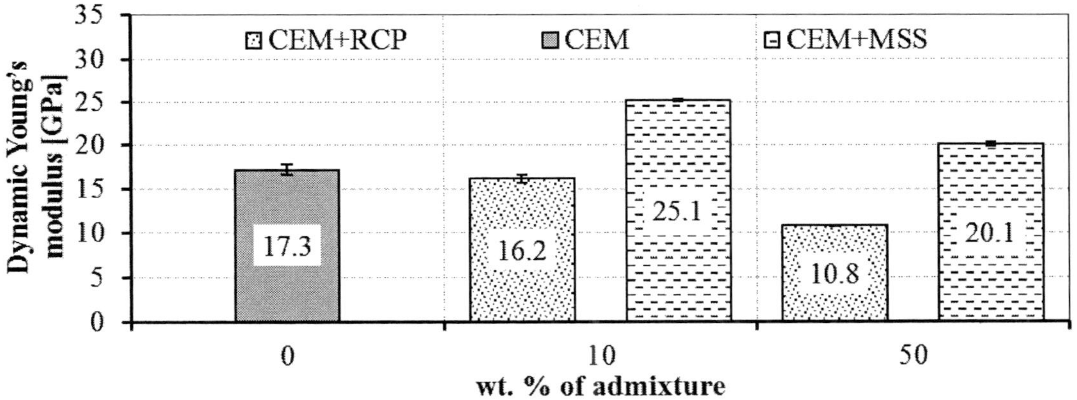


Fig. 2: Dynamic Young's modulus of cement paste and cement with 10 and 50 wt. % of recycled concrete powder (CEM+RCP) and micro-milled silica sand (CEM+MSS).

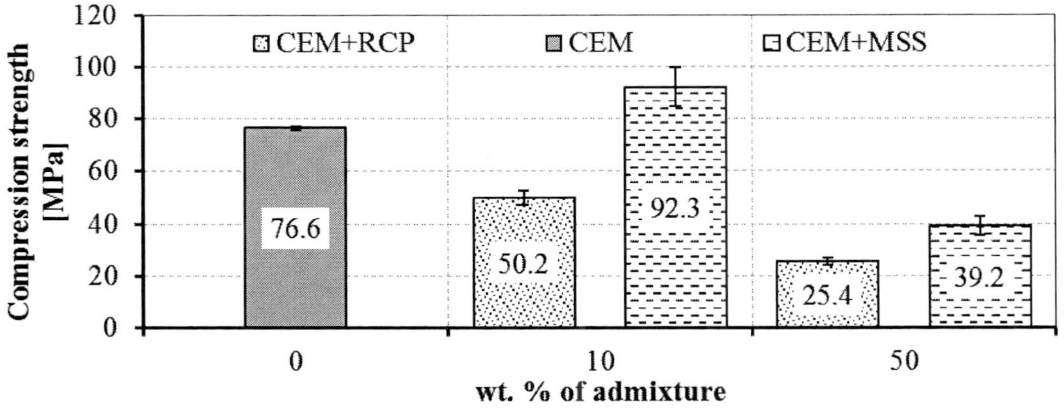


Fig. 3: Compression strength of cement paste and cement paste with 10 and 50 wt. % of recycled concrete powder (CEM+RCP) and micro-milled silica sand (CEM+MSS).

Dynamic Young's modulus of the samples illustrated in Fig. 2 and compression strength illustrated in Fig. 3 are proportional to bulk density. As can be seen, the mechanical properties of samples with RCP are despite the hydration of cement grains in RCP decreased compared with other samples. This decrease is probably caused by lower density of old cement paste in RCP compared with MSS and new hydration products.

Based on this results, the potential improvements of mechanical properties could be achieved by using small amount of RCP in cement mortars or concrete where aggregate is present. In this combination, the RCP would be fulfills the role of micro filler and partially the role of binder and lower density of RCP would not be obstacle as in case of cement paste. Those experiments will follows in further research.

## Conclusions

This paper is focused on hydration and mechanical properties of cement paste with addition of RCP (filler and possible binder replacement) and MSS (inert filler). Based on results the following findings are summarized:

- Based on heat flow during hydration process, the presence of exposed non-hydrated cement grains in RCP has been proved and influence of larger nucleation area was define as minimal thanks to comparison with influence of MSS.
- Despite the additional hydration of cement grains in RCP, the mechanical properties of cement paste with RCP were negatively influenced by lower density of cement paste in RCP compared with MSS and new hydration products.
- Probably only solution how micro filler and binder properties of RCP could be utilized to improve the mechanical properties is combination of small amount of RCP with cement composites where aggregate is present. Then the RCP would be fulfills the role of micro filler and partially the role of binder and lower density of RCP would not be obstacle.

## Acknowledgement

The authors wish to express their gratitude for financial support to the Faculty of Civil Engineering, Czech Technical University (CTU) in Prague under SGS project No. SGS16/201/OHK1/3T/11 and Czech science foundation GAČR under project No. 17-06771S. The authors also thanks to the Center for Nanotechnology in Civil Engineering at the Faculty of Civil Engineering, CTU in Prague.

## References

[1] J. Topič, Z. Prošek, Properties and microstructure of cement paste including recycled concrete powder, Acta Polytechnica 57 (2017) 49-57.

[2] J. Topič et al, Properties and microstructure of cement paste including recycled concrete powder, In: Central Europe towards Sustainable Building 2016 - Innovations for Sustainable Future (2016) 1320-1324.

[3] J. Bensted, P. Barnes, Structure and Performance of Cements, Second ed., Spon Press, London, 2002.

[4] S.S. Shebl, H.S. Seddeq, H. A. Aglan, Effect of micro-silica loading on the mechanical and acoustic properties of cement pastes, Construction and Building Materials 25 (2011) 3903-3908.

EAN 2017
55th conference on experimental stress analysis

# Analysis of tensile strength in bending of plasterboard with use of optical correlation

Ondřej Zobal [1,2,a], Jiří Litoš [1,2,b], Jakub Antoš [1,c]

[1]CTU in Prague, Faculty of Civil Engineering, Department of Mechanics, Thákurova 7, 166 29 Prague 6, Czech Republic

[2]University Centre for Energy Efficient buildings, Czech Technical University in Prague, Třinecká 1024, 273 43 Buštěhrad, Czech Republic

[a]ondrej.zobal@fsv.cvut.cz, [b]litos@fsv.cvut.cz, [c]jakub.antos@fsv.cvut.cz

**Keywords:** plasterboard, tensile strength in bending, optical correlation.

**Abstract.** One of the basic observed parameters for plasterboards is the tensile strength in bending and sag. Therefore, tensile strength and sag is regularly verified in the development of new boards and subsequent certification. This paper shows a classic destructive standardized test extended by optical correlation of stress.

## Introduction

Plasterboards are formed by combining pressed gypsum and two paper cartons and are used for dry construction. They have application in renovations and for new buildings for the production of walls, floors, ceilings and roofs and their advantage is easy and fast usage and protection against noise, moisture and fire [1]. The standard ČSN EN 520 + A1 (2010) prescribes marking, defining the properties and test methods of plasterboard and additionally unifies the requirements and designations with others countries in Europe. Declared tensile strength in bending and sag of are among the fundamental requirements [2-5]. Another definition testing of tensile strength in bending is described in standard ASTM C 473 from USA [6]. Measurement of tensile strength in bending can be carried out as a three-point bending [7] or four-point bending as described in this paper. Mechanical tests were supplemented optical correlation of digital image (DIC).

## Methodology and evaluation of measurement

**Specimens.** Test of tensile strength in bending was carried out. Two specimens with dimensions 600×300×12,5 mm were cut of each plate faceup, one in the transversal and second in the longitudinally direction according to standard. Altogether for test were available 4 samples of one type of board. Designation of specimens and their geometry are shown in Table 1.

Table 1 Designation of specimens and their geometry.

| Designation | Cut | Length [mm] | Width [mm] | Thickness [mm] | Basis weight [kg/m$^2$] |
|---|---|---|---|---|---|
| D1-h | transverse | 600 | 301 | 12.5 | 13.5 |
| D1-v | longitudinal | 600 | 302 | 12.5 | 13.5 |
| D2-h | transverse | 602 | 302 | 12.5 | 13.5 |
| D2-v | longitudinal | 603 | 300 | 12.5 | 13.5 |

**Four-point bending.** All specimens were loaded with increment deformation constant speed of 2 mm/min until the collapse. The static scheme of the test shows Fig. 1 and photo of measuring assembly is on Fig. 2.

Tensile strength in bending (four-point) is given by the following relationship:

$$f_{cf} = \frac{F.l}{d_1 .d_2^2}, \tag{1}$$

where $f_{cf}$ is tensile strength in bending [MPa], $F$ is maximal load [N], $l$ is distance between supports [mm], $d_1$ a $d_2$ are dimensions transverse cut of specimen [mm]. Sensors for monitoring the deflection were placed at specified points.

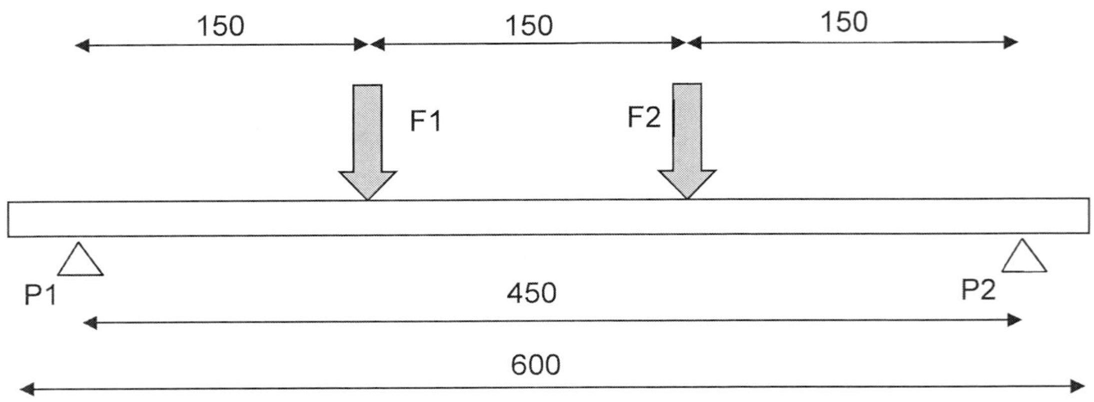


Fig. 1: Static scheme of testing.

Fig. 2: The ready measuring assembly for a four-point flexural test.

**Evaluation of tensile strength in bending and sag.** The values of tensile strength of individual specimens and the average values are given in Table 2. A symbol of sample orientation relative to the plate is always stated in designation of specimen (h - horizontal, longitudinal, v - vertical, transverse). Typical working diagrams of loading tests show Fig. 3 and 4.

Table 2: Measured and calculated values of strength for each sample and their average.

| Designation | m [kg] | F [kN] | $f_{ct}$ [MPa] | Sag [mm] |
|---|---|---|---|---|
| D1-h | 2.4 | 0.3 | 2.9 | 2.9 |
| D2-h | 2.5 | 0.3 | 2.9 | 6.9 |
| average | **2.5** | **0.3** | **2.9** | **4.9** |
| D1-v | 2.4 | 0.8 | 7.6 | 14.6 |
| D2-v | 2.4 | 0.8 | 7.7 | 15.0 |
| average | **2.4** | **0.8** | **7.7** | **14.8** |

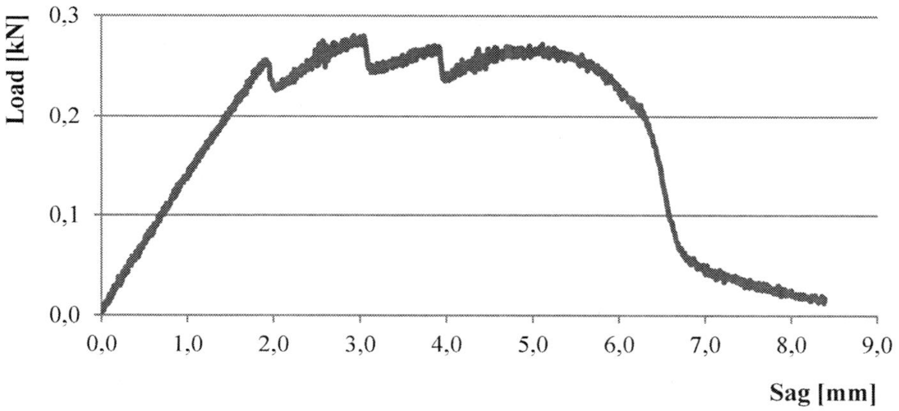


Fig. 3: Working diagram of specimen D1-h from the tensile test in bending at speed of load of 2 mm/min, $F_{max} = 0.3$ kN $u_{Fmax} = 2.9$ mm.

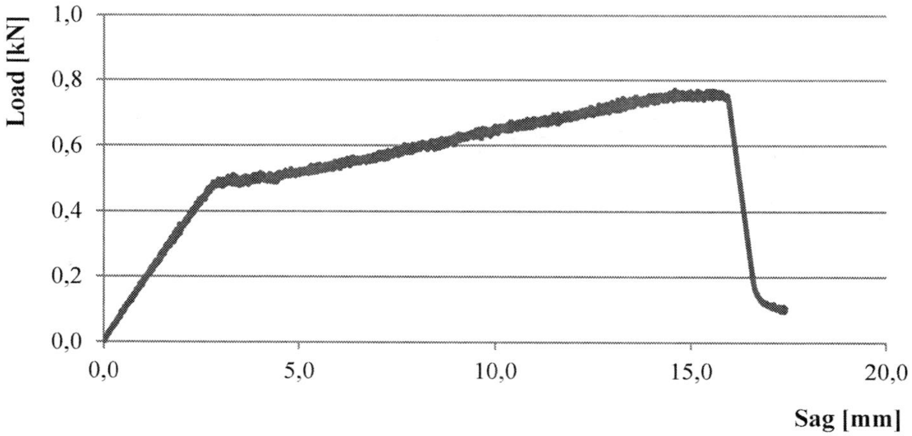


Fig. 4: Working diagram of specimen D1-v from the tensile test in bending at speed of load of 2 mm/min, $F_{max} = 0.8$ kN $u_{Fmax} = 14.6$ mm.

**Optical correlation**. The method of optical digital image correlation (DIC) can be successfully used for measurement of displacements, relative deformation, the changes in shape, material properties, stress and modal harmonic analysis and motion analysis. The universality of this method also lies in its independence from the scale, ie. it can be used from micromechanics, through mechanics and testing samples of standard sizes according to the CSN and the EU to analyze the whole structure (eg. lattice girders). The principle of this method is to acquire a reference image the examined surface on which is applied a contrasting pattern and in a sequence of images during the destructive testing. These digital images were then evaluated using a special software which is able to compare how between images transformed contrasting pattern and consequently to calculate therefrom the relative deformation at those locations [8]. Canon 90D DSLR camera with a remote trigger and evaluation software Ncorr was used for testing. The main deformation is shown in Fig. 5. Photos on Fig. 6 and 7 show the state of measured boards at the beginning and end of the test.

Fig. 5: Optical correlation - the main strain.

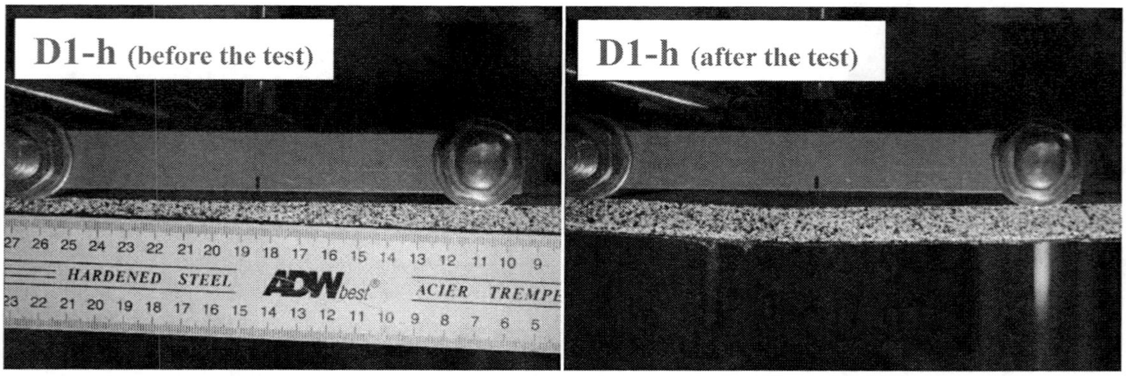


Fig. 6: State of measured boards at the beginning and end of the test.

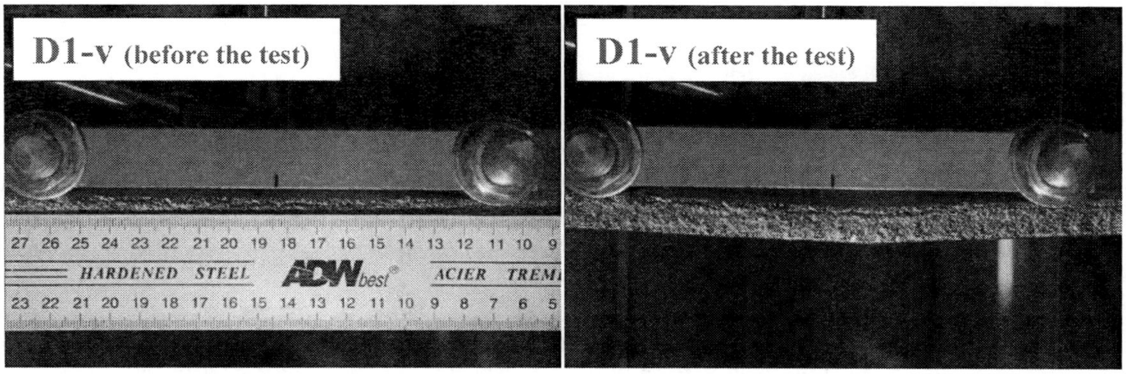


Fig. 7: State of measured boards at the beginning and end of the test.

## Conclusions

Tensile strength in bending was obtained by four-point bending test. This is a classic method of measurement. A supplementary optical digital image correlation (DIC) gives except comparison to specific values detailed information about the behavior of the material during loading, which can be used to better analyze the properties and measurement of deflection.


## Acknowledgement

This paper has been supported by the Faculty of Civil Engineering, Czech Technical University in Prague (SGS project No. SGS16/201/OHK1/3T/11) and by the Ministry of Education, Youth and Sports within National Sustainability Programme I, project No. LO1605.


## References


[1] Gypsum Unique Properties in Buildings, Information on - http://www.eurogypsum.org/about-gypsum/properties/.

[2] Sádrokartonové desky podle evropských norem - Information on http://imaterialy.dumabyt.cz/rubriky/drevene-a-montovane-konstrukce/sadrokartonove-desky-podle-evropskych-norem_101335.html (in Czech).

[3] ČSN EN 520+A1 Gypsum plasterboards – Definitions, requirements and test methods, Prague, in Czech, (2010).

[4] D. Supik, Analýza vybraných mechanických vlastností konstrukčních materiálů, diplomová práce, Mendelova Univerzita v Brně, in Czech, (2009).

[5] P. Crescenzo, G- Magliulo and G. Manfredi, Mechanical properties of plasterboards: experimental tests and statistical analysis, Journal of Materials in Civil Engineering 28.11 (2016).

[6] Gypsum Association, Gypsum board typical mechanical and physical properties, GA 235 10 (2010).

[7] I. Rahmanian,Thermal and mechanical properties of gypsum boards and their influences on fire resistance of gypsum board based systems, Ph.D. thesis, (2011).

[8] V. Nežerka, V., J. Antoš, P. Tesárek, Open Source 2D DIC Tools for Applications in Solid Mechanics, In: 53rd conference on experimental stress analysis. Praha: Czech Technical University in Prague, (2015), pp. 119-120

EAN 2017
55th conference on experimental stress analysis

# Effects of Advanced Laser Processing on the Microstructure and Residual Stresses of H13 Tool Steel

Karel Trojan[1,a], Václav Ocelík[2,b], Nikolaj Ganev[1,c], Stanislav Němeček[3,d] and Jiří Čapek[1,e]

[1]CTU in Prague, Department of Solid State Engineering, Faculty of Nuclear Sciences and Physical Engineering, Trojanova 13, 120 00 Prague 2, Czech Republic

[2]Department of Applied Physics, Zernike Institute for Advanced Materials, Faculty of Science and Engineering, University of Groningen, Nijenborgh 4, 9747 AG, Groningen, The Netherlands

[3]RAPTECH s. r. o. ,U Vodárny 473, 330 08 Zruč-Senec, Czech Republic

[a]karel.trojan@fjfi.cvut.cz, [b]v.ocelik@rug.nl, [c]nikolaj.ganev@fjfi.cvut.cz, [d]nemecek@raptech.cz [e]jiri.capek@fjfi.cvut.cz

**Keywords:** laser processing; cladding; residual stresses; microstructure; H13 tool steel.

**Abstract.** The aim of this paper is to describe the effects of laser processing on the microstructure and residual stresses of laser cladded H13 tool steel on the classical construct steel S355 substrate. This research paper concludes that in this case of laser cladding, phase transformation and not shrinkage is likely to be a dominant effect on the formation of compressive residual stresses along the clad. Furthermore, martensitic structure and unequal concentration of alloying elements was observed on the cross-section of the clad using electron backscattering diffraction and energy-dispersive X-ray spectroscopy.

## Introduction

Processing by high power lasers is nowadays the standard in many different industrial fields which require precision, accuracy and high production efficiency. Therefore, this paper outlines the capability of the advanced laser processing for additive manufacturing in relation to Telesang et al. (2015) [1]. Hot working tool steel H13 is one of the most common die material used in metal and casting industries. Dies suffer great damage due to wear and thermo-dynamic stresses during their lifetime [1]. Therefore, various methods have been developed for its repair. A great benefit of laser cladding in this field is a high productivity along with minimal influence on surrounding material by thermal stress [2]. This study reports an investigation of residual stresses (RS) by X-ray diffraction (XRD) and microstructure (chemical and phase composition, crystallite orientation) by orientation imaging microscopy (OIM) using electron backscatter diffraction (EBSD) and energy-dispersive X-ray spectroscopy (EDS) with scanning electron microscopy (SEM) of laser cladded tool steel AISI H13 on the classical construct steel S355 substrate. The research aims to understand the residual stresses and microstructural characteristics that are one of the most important factors that influence the products behaviour, especially the fatigue life [3].

## Experimental

Laser cladding was carried using a diode laser *Laserline* 5,5kW with optical head *Precitec Y52*. Laser power density 108 J/mm$^2$ was applied to form a single clad with the length of 137 mm, width of 6 mm and height of approx. 1 mm above substrate. The powder of the AISI H13 tool steel was used with an average particles diameter of 53.1 ± 15.9 μm.

X-ray diffraction analysis was carried out using an *X'Pert PRO MPD* diffractometer with chromium tube anode and pinholes determining irradiated area of primary beam, with size of 1×0.5 mm$^2$ in longitudinal clad direction L and 2×0.25 mm$^2$ in transversal clad direction T, respectively. Diffraction angle $2\theta^{211}$ was taken as a centre of gravity of the {*211*} diffraction doublet *CrKα*. For residual stress evaluation, $\sin^2\psi$ method and X-ray elastic constants $½s_2$ = 5.76 TPa$^{-1}$ and $s_1$ = −1.25 TPa$^{-1}$ were used.

In order to determine the phase composition of the surface layer, diffraction patterns were obtained by measuring the clad on the instrument *X'Pert PRO MPD* in classical Bragg – Brentano (BB) focusing configuration with cobalt tube anode and 1×0.25 mm$^2$ pinholes. Measured diffraction diagrams were processed with the program *X'Pert HighScorePlus* and crystalline phases were identified using a *PDF-2* database. Quantitative analysis was evaluated using the Rietveld analysis in the *TOPAS 4.2* software. The effective depth of penetration ($T^{eff}$) is corresponding to the thickness of a surface layer that provides about 63 % of the diffracted intensity. In the case of the used wavelength, the $T^{eff}$ is about 5 μm.

The OIM data was collected using a *Philips XL 30 FEG* scanning electron microscope equipped with a *TSL OIM system* based on *DigiView 3* camera and *EDS EDAX SUTW+* detector. The accelerating voltage of 25 kV and 0.4 μm step size of scanning were used. A grain boundary is defined in the microstructure as a boundary between two neighbouring scanning points having crystallographic misorientation larger than 5°. All EBSD data were analysed with *TSL OIM Analysis 7.3.0* software and only data points with *Confident index* higher than 0.05 were used. *Confident index* is based on voting scheme during automated indexing of the diffraction pattern where it is counted as a ratio of votes for the best solution minus votes for the second best solution divided by total possible number of votes from the detected Kikuchi bands. Certain crystallographic orientation receives a vote when observed angles between the three bands are the same as table values of given orientation. *Confident index* 0.05 corresponds on a face-centred cubic material to approx. 70% probability of correct indexing.

## Results and discussion

Table 1 gives the range of the chemical composition of H13 steel determined quoted by ISO 4957 [4], the next row shows average chemical composition and standard deviation obtained by EDS on a cross-section of four randomly selected particles of the used powder obtained by EDS on a cross-section; the bottom row shows the average chemical composition and standard deviation in four different clad cross-section areas of dimensions approx. 500×250 μm$^2$. Table 1 shows that only chromium in the clad is on average about one weight percentage less than prescribe standard. Manganese has been observed on the contrary of 0.1 % or more. SEM image of the cross-section of the single clad with used directions and an area 200×200 μm$^2$ which was observed using OIM can be seen in Fig. 1. Lack of the fusion on the sides of the clad could be observed, it is probably the result of improper setting of laser power density.

Table 1: Chemical composition of the steel AISI H13 according ISO 4957 [4], the powder AISI H13 and the clad.

| | Cr [Wt.%] | Mo [Wt.%] | Si [Wt.%] | V [Wt.%] | Mn [Wt.%] | C [Wt.%] |
|---|---|---|---|---|---|---|
| EN ISO 4957 | 4.80 – 5.50 | 1.20 – 1.50 | 0.80 – 1.20 | 0.85 – 1.15 | 0.25 – 0.50 | 0.35 – 0.42 |
| Powder H13 | 5.10 ± 0.07 | 1.79 ± 0.08 | 0.92 ± 0.11 | 1.08 ± 0.02 | 0.45 ± 0.07 | |
| Clad | 3.77 ± 0.05 | 1.21 ± 0.05 | 0.72 ± 0.01 | 0.84 ± 0.03 | 0.62 ± 0.03 | |

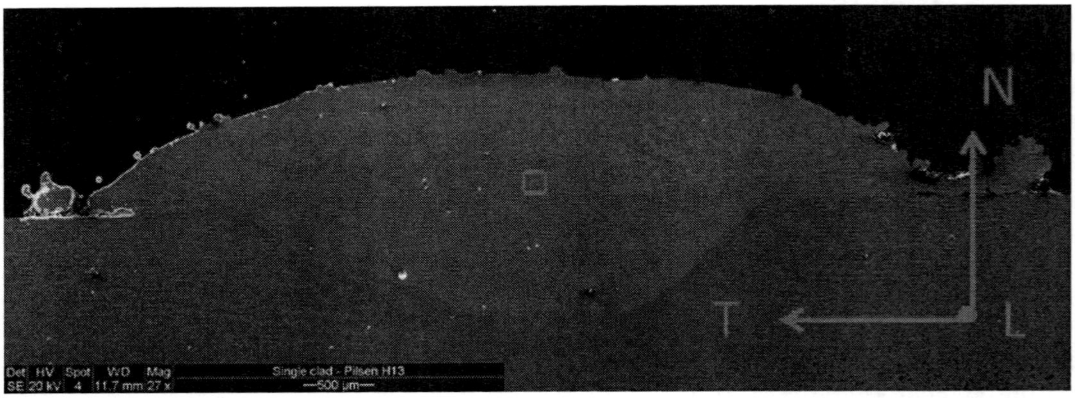


Fig. 1: SEM image of the cross-section of the single clad AISI H13 tool steel with marked directions and area which was observed using OIM.

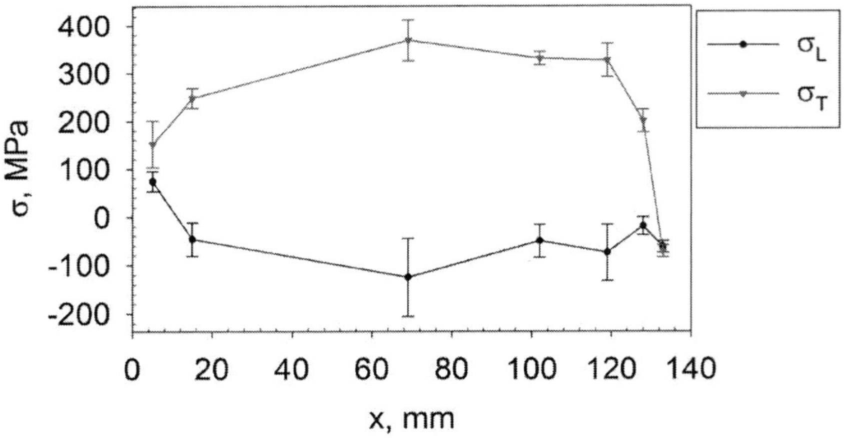


Fig. 2: Surface RS on the top of the clad in direction L and T along the length of laser track, where $x$ is the distance from the beginning of the clad.

The comparison of surface RS obtained by XRD in direction L and T is plotted in Fig. 2. RS at the beginning and the end of the clad are comparable with the experimental error. In the direction L, RS exhibit slight compressive stresses along the clad. However, in the direction T, tensile RS were obtained. This dependence is opposite to what was generally observed in the literature [1,5]. However, there were published studies confirming tensile residual stresses in the transversal direction as well [6].

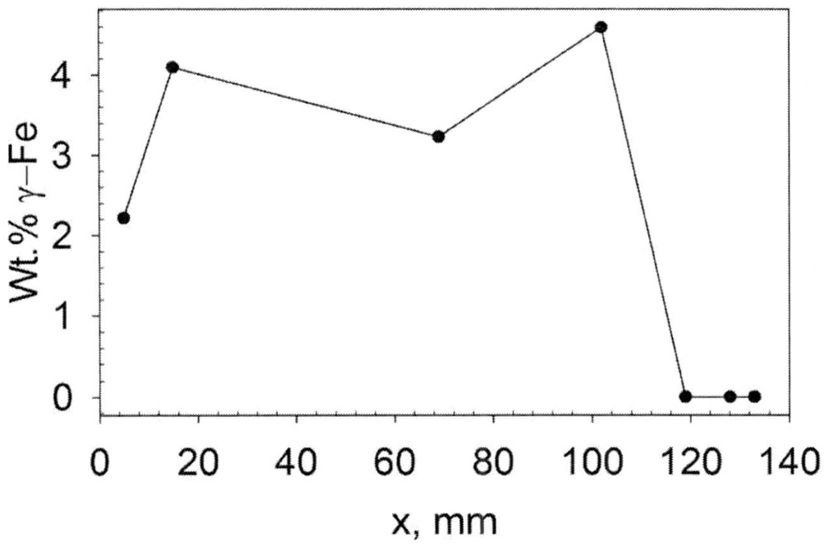


Fig. 3: Mass weight content of γ-Fe in the surface layer obtained by XRD.

Mass weight content of γ-Fe in the surface layer depending on the distance from the beginning of the clad is plotted in Fig. 3. It can be seen that at the beginning of the clad a γ-Fe content is increasing to the value of around 4 %, and in the end no retained austenite was observed by XRD. This could be due to slower cooling rate at the end as the heating of the substrate occurred. After finding these results, the real structure and the microstructure of the single clad have been taken to deeper evaluation by OIM on the cross-section 5 mm from the beginning of the clad.

The main phase detected in the clad was body centred-cubic iron. Fig. 4 shows inverse pole figure maps of α-Fe in the selected area, wherein individual colours correspond to the normal vectors of crystallographic planes that are parallel to the given sample direction. The original austenite grains with a characteristic size of 20-50 μm, which were established during the transition of the melt into a solid phase and whose were subsequently transformed into martensitic laths are clearly seen in the figure. It has to be noted that EBSD technique is not able directly distinguish ferrite and martensite. Carbides of alloying elements were not observed using EBSD. However, in the analysed area, 1.1 % of retained austenite was confirmed. The difference compared to the XRD value could be caused by an error of each method and by larger volumes analysed by XRD, especially along the clad. At the beginning of the clad, proportion of retained austenite by XRD is growing.

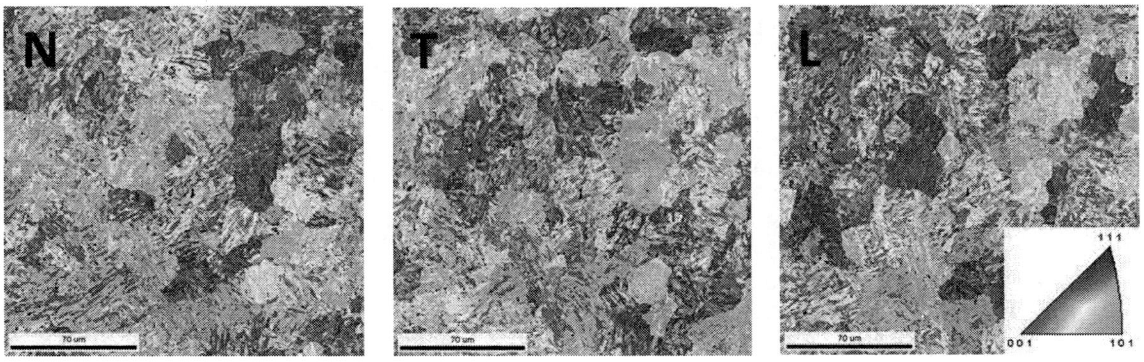


Fig. 4: Orientation maps of α-Fe for different directions in the selected area.

Fig. 5 gives an idea of the relative distribution of individual alloying elements in the selected area obtained by energy-dispersive X-ray spectroscopy mapping. Darker colours indicate higher concentration of the element.

Subsequently, the distribution of the misorientation angle in the selected area is plotted in Fig. 6. In the article [7], there was shown that misorientation angles between the variants that satisfy the Kurdjumov-Sachs orientation relationship of phase transformation from face-centred cubic phase to body-centred cubic phase are up to 21.06° and then higher than 47.11°.

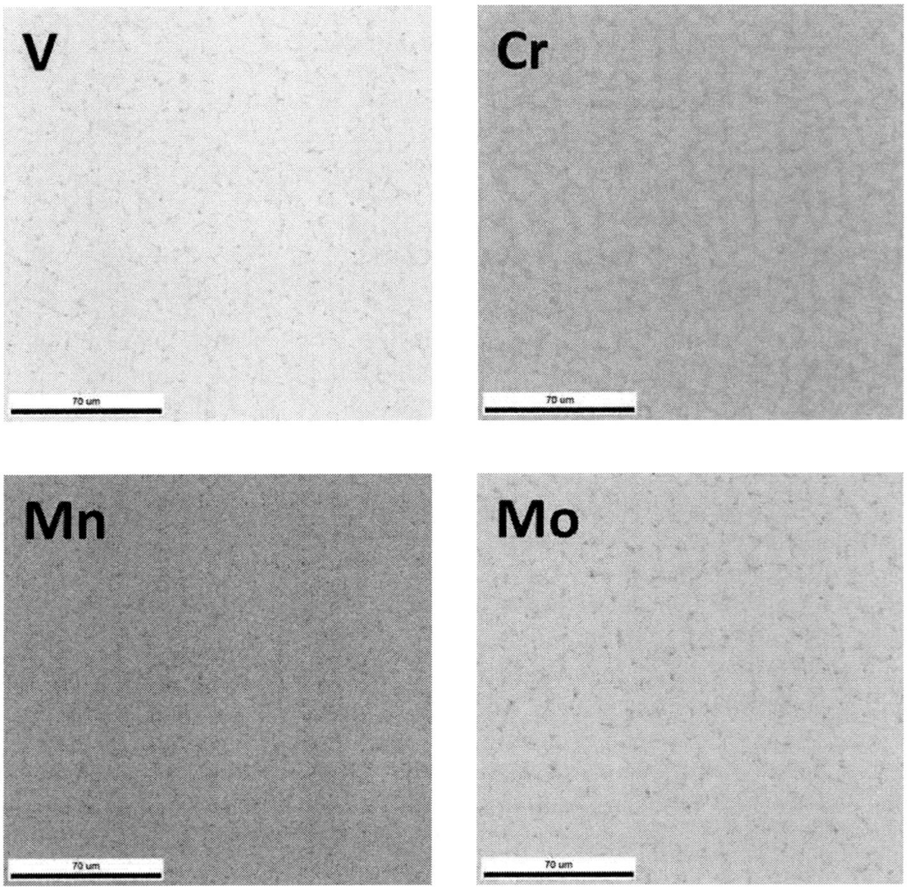


Fig. 5: Relative representation of content of alloying elements in the selected area acquired by EDS.

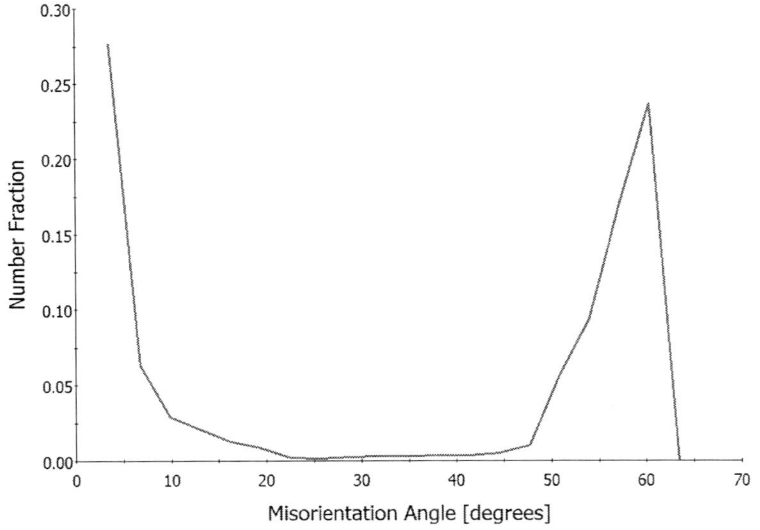


Fig. 6: Misorientation angle distribution in the selected area.

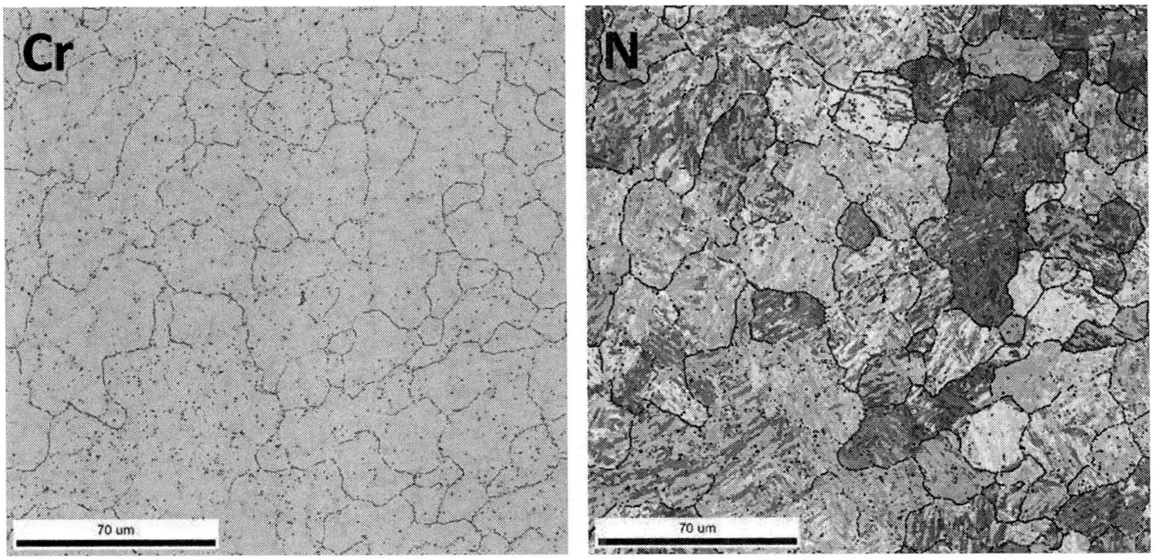


Fig. 7: Comparison of the map of relative representation of content of chromium and orientation map with highlighted grain boundaries between 21.2° and 47°.

Thus, Fig. 7 compares the map of relative representation of content of chromium and grain orientation map with highlighted grain boundaries between 21.2° and 47°. It can therefore be clearly seen that higher concentration of chromium corresponds with the random grains boundaries between 21.2° and 47°. These boundaries relate to the grain boundaries of the original austenitic grains which were formed during cooling of the melt. The unequal concentration of alloying elements is caused by the lever rule during solidification. This could induce inferior mechanical properties of the steel as carbides of alloying elements were not observed.

According to the linear equation developed by Andrews [8], $M_s$ temperature is 316 °C for the lower limit of alloying elements and 268 °C for upper limit of alloying elements [9]. So if we assume that the coefficient of thermal expansion for steel H13 is $12.5\times10^{-6}$ $K^{-1}$, relative shrinkage 0.0034 results from cooling from the average $M_s$ to the room temperature [10]. On the other hand, the volume changes during phase transformation in steels are a function of carbon content. Dimensional change during transformation from spheroidite to martensitic microstructure is 0.0026×(Wt.% of C) [9]. For steel H13, the dimensional change is therefore between 0.0020 and 0.0024. Shrinkage and expansion are therefore the same order of magnitude and depends on the specific situation that prevails. Consequently, it depends on the particular situation which effect would be dominant.

The results above suggest that there has been a rapid cooling of the melt to form a martensitic structure. Moreover, no carbides and a half of the value of retained austenite in comparison with XRD were observed using EBSD. Therefore, it may be concluded that in this case of laser cladding, phase transformation and not shrinkage is likely to be a dominant effect on the formation of residual stress in the direction L along the clad. This caused the tensile RS in the perpendicular direction T.

## Conclusions

Laser deposition of the H13 tool steel showed a great application potential. Martensitic structure dominates in the single clad. This suggests high cooling rates and quenching of the clad. Laser deposition of H13 induced significant residual stresses. Residual stresses at the beginning and at the end of the track are almost equal within the error. In the longitudinal direction RS exhibit slight compressive character along the clad. However, in the transversal direction tensile RS were detected. The research topic and primarily the main origin of residual stresses has not yet been accurately described and understood, so further measurements and adjustments of the deposition parameters will be needed.

## Acknowledgement

Measurements were supported by the project TH02010664 of the Technology Agency of the Czech Republic and by University of Groningen. This work was supported by the Student Grant Competition CTU in Prague grant No. SGS16/245/OHK4/3T/14.

## References

[1] G. Telasang, J. Majumdar, N. Wasekar, G. Padmanabham, I. Manna, Microstructure and Mechanical Properties of Laser Clad and Post-cladding Tempered AISI H13 Tool Steel. Metall. Mater. Trans. A. 46A (2015) 2309-2321.

[2] M. Vedani, B. Previtali, G.M. Vimercati, A. Sanvito, G. Somaschini, Problems in laser repair-welding a surface-treated tool steel, Surf. Coat. Tech. 201 (2007) 4518-4525.

[3] G.E. Totten, M. Howes, T. Inoue, Handbook of residual stress and deformation of steel, ASM International, Materials Park, 2002.

[4] ISO 4957:1999. Tool steels.

[5] J. Chen, S. Wang, L. Xue, On the development of microstructures and residual stresses during laser cladding and post-heat treatments, J. Mater. Sci. 47.2 (2012) 779-792.

[6] A. Suárez, J.M. Amado, M.J. Tobar, A. Yáñez, E. Fraga, M.J. Peel, Study of residual stresses generated inside laser cladded plates using FEM and diffraction of synchrotron radiation, Surf. Coat. Tech. 204 (2010) 1983-1988.

[7] H. Kitahara, R. Ueji, N. Tsuji, Y. Minamino, Crystallographic features of lath martensite in low-carbon steel, Acta. Mater. 54 (2006) 1279-1288.

[8] K.W. Andrews, Empirical Formulae for the Calculation of Some Transformation Temperatures, Journal of the Iron and Steel Institute 203 (1965) 721–727.

[9] G. Krauss, Steels: Processing, Structure, and Performance, ASM International, Materials Park, 2005.

[10] D. Klobčar, J. Tušek, B. Taljat, Thermal fatigue of materials for die-casting tooling, Mater. Sci. Eng. A 472.1 (2008) 198-207.

# Creep of Unfired Clay Mortar Based on Illitic Clay

PADEVĚT Pavel[1, a] and BITTNAR Petr[1,b]

[1]Czech Technical University in Prague, Faculty of Civil Engineering, Thákurova 7, Prague 6, 166 29, Czech republic

[a]pavel.padevet@fsv.cvut.cz, [b]petr.bittnar@fsv.cvut.cz

**Keywords:** Unfired clay, rammed clay, creep, shrinkage.

**Abstract.** Paper focuses attention on production and preparation of test specimens of unfired clay. The material of unfired clay consists of illitic clay, sand and a specified amount of water. The shrinkage and creep measurements are started the second day after production. The result of measurement is to determine the size of the length changes of unfired clay and monitoring long-term behavior of the material. The resulting creep and shrinkage curves are related to temperature changes in the laboratory.

## Introduction

Problems of the properties of unfired clay have not preference like a other modern material, today. Buildings made of unfired clay accompany mankind since time immemorial [1]. Yet there are no reliable data for their safe design. The absence of information can be one of the reasons why earthen structures are not space for their implementation. Buildings made of unfired clay have their benefits, such as high-quality indoor environment, environmentally friendly materials and good availability [2].

One of the material information is a behavior of the structure, or the material from the perspective of rheology [3]. The size of creep and shrinkage are very important in terms of the design of structures made of unfired clay. This issue is dedicated the article. Specifically, it is focused on rheology rammed clay which was prepared from the illitic clay [4].

Fig.1 The specimens used for measurement of creep and shrinkage, prepared from illitic clay.

## Material testing

Tested material was prepared from illitic clay, sand and water. The sand was composed of two fractions, 0-1 and 1-2 mm. Smooth line granularity was created using of both fractions in the ratio of 60 : 40 (fine and coarse part). Both fractions of sand were mixed before production of the dry clay mixture. The clay was in a mixture represented 25 % of weight, and sand accounted for 75 % of weight of the dry mixture. Water ratio is defined as the relation of the weight of mixing water and clay. Its size was 0.37 for the mixture. The mixture was compacted - rammed into the molds and then tested and non-fired [5].

The start of the measurement was the second day after the made of specimens. Measurement time was 34 days. During the measurement, a steady temperature with a deviation of ±1° C was maintained in the laboratory as shown in Figure 2.

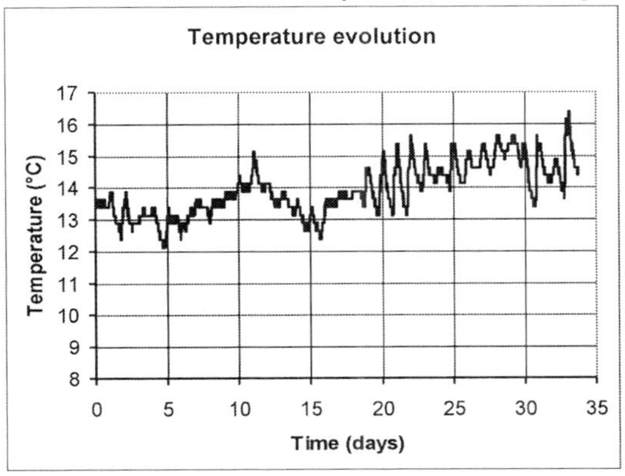


Fig.2 Evolution of temperature during the test.

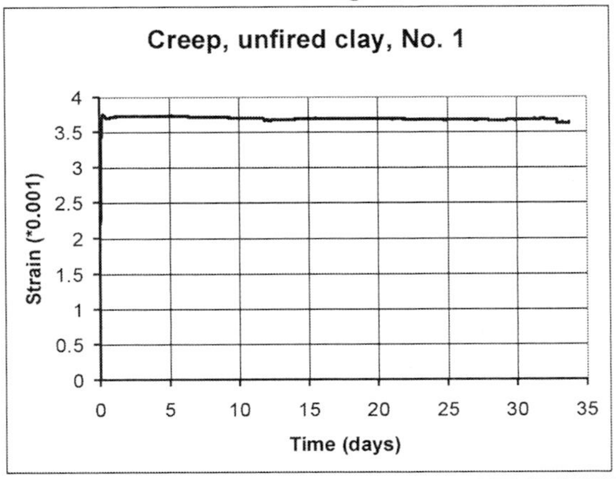


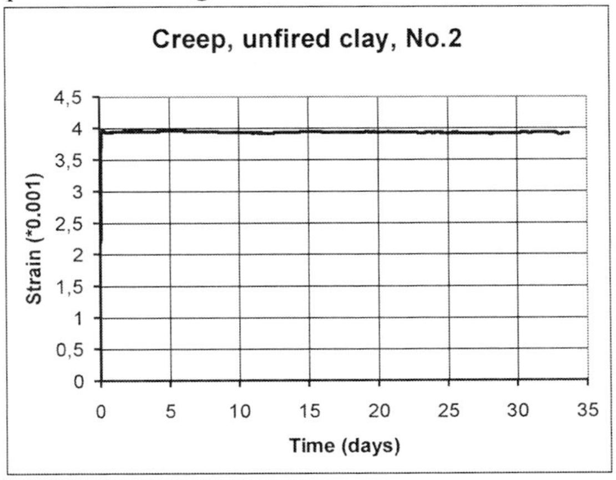


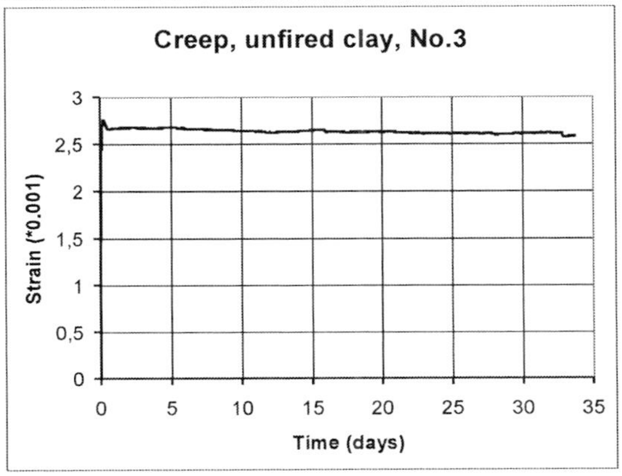


Fig.3 The curves of the creep measured on the specimens 1-3 prepared from described clay.

A temperature increase of 1° C occurred after 20 days of measurement. This increased ambient temperature in the laboratory did not affect creep curves, as shown in Figure 3.

The test was carried out with three specimens for measurement of creep and 3 of the body for measuring the shrinkage. Used specimens are displayed on Figure 1. The base of bodies was octagonal. The body has a length of 70 mm, which is suitable for measuring creep and shrinkage in lever mechanisms [6]. Load lever mechanism having a size 59N. Bodies intended for measuring creep was so loaded. The load represents approximately 10% load capacity of perpendicularly to the ramming of rammed clay. The moisture from the body during measurement can freely escape [7]. The transport of moisture was not blocked.

Shrinkage measurements were also carried out in lever mechanisms, but without load. The shrinkage measurement results are shown in the graphs in Figure 4.

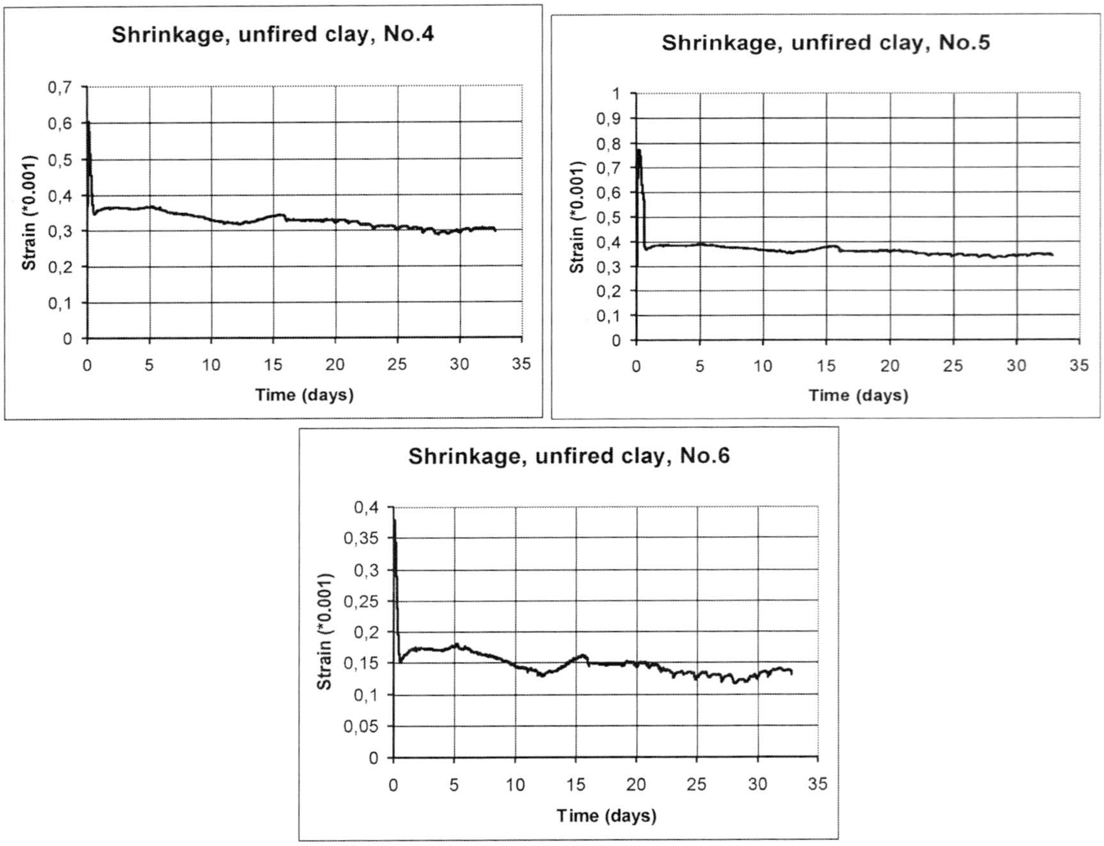


Fig.4 The curves of the shrinkage on the specimens 4-6 prepared from described clay.

During the measurement, a possible change of weight with changes in the humidity in the laboratory was compared using control bodies in the temperature chamber. The same bodies were used to compare. Three of them were placed near the measurement and three bodies in the temperature chamber. Weight loss is caused by moisture removal from clay. The course of weight loss is shown in Figure 6. Bodies 1-3 were located at the creep measurement point, and body 4-6 was located in the temperature chamber. The temperature chamber maintained 50% humidity and 20° C. By comparing the shape of the curves, it appears that a stable temperature and humidity environment was maintained during the measurement [8]. Measurement of shrinkage and creep as was done in a steady environment, and the results are affected only by the behavior of the material.

The greatest loss of moisture occurs during the first two days. At the time of the loss of free water from the bodies, the strength increases. The rammed clay hardened at that time.

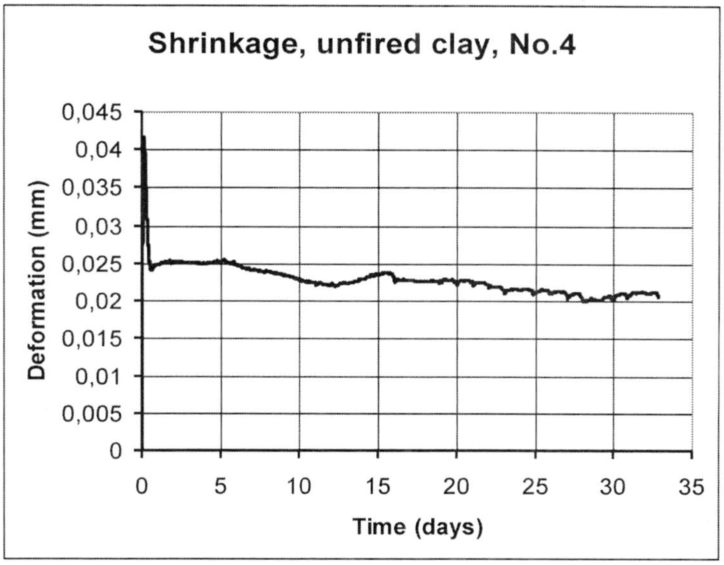


Fig.5 The imported curve of evolution of temperature during the test.

## Results

Only body No.3 showed a decreasing value over time of creep. For explanation, it means increasing of strain the shrinkage and diminishing the values of strain the expansion. Bodies 1 and 2 practically did not change the initial creep value throughout the test period. The increase in shrinkage occurs almost only in the beginning, after loading the specimens. Size creep is around 3 milistrain. Creep results include the effect of shrinkage.

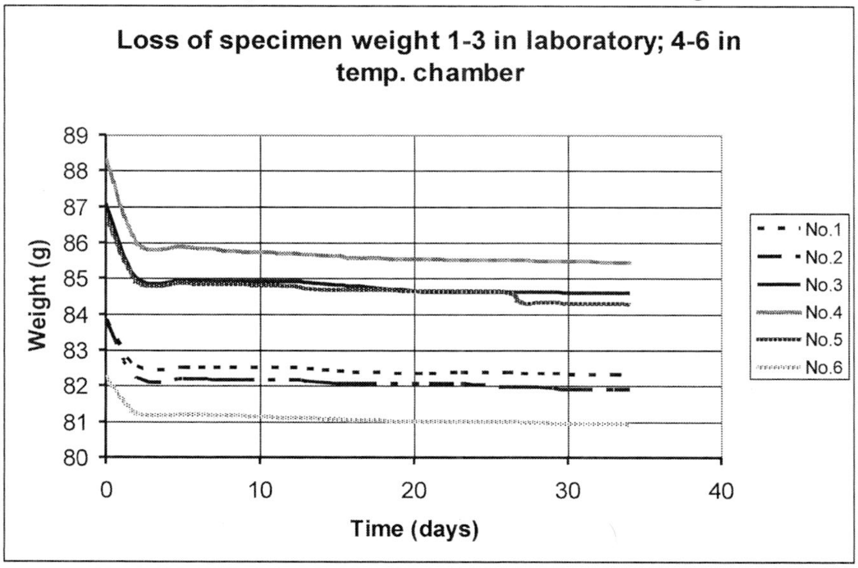


Fig.6 The change of the weight of specimens during the test.

In contrast, the shrinkage changes throughout, as shown in Figure 4. The strain diminished by 0.05 milistrain in all three bodies. This means that during the entire test, the specimen length was increased under steady temperature and humidity conditions. The reason is probably the ability of the illitic clay to expand even at steady air humidity. Very well, this is shown in Figure 5, where the deformation of the 70 mm bead body is plotted on the vertical axis of the graph. It is seen that the elongation value is 5 microns. Initial deflection of the strain is caused by clamping the specimens into the measuring devices.

## Conclusions

Rammed clay has a very interesting perspective creep properties. The main change of creep occurs in the early stages of measuring after the loading. Totally interesting result is the size of shrinkage after 30 days. This reached a value of $0.05 \times 10^{-3}$ milistrain in the direction of expansion. It can be confirmed that this behavior is not random, but corresponds to the behavior of unfired and rammed clay [1]. The results of testing 3 specimens for creep and 3 elements for shrinkage are very consistent even though they are the rammed clay. It can be deduced from the results that illitic clay tends to expand even at steady air humidity.


## Acknowledgement

This research has been supported by project SGS project No.16/201/OHK1/3T/11 Advanced from design of materials to the constructions.


## References


[1] G. Minke, Building with Earth [online], available from:

http://archive.org/details/Gernot_Minke-Building_With_Earth

[2] F. Pacheco-Torgal, S. Jalali, Earth constructions: Lessons from the past for future eco-efficient constructions, Constructions and Building Materials, Vol. 29, April 2012, pp. 512-519.

[3] P. Padevět, P. Bittnar, B. Mužíková, Smrštění nepálené hlíny připravené z montmorillonitu, Proceedings of the 14th International Conference on New Trends in Statics and Dynamics of Buildings October 13-14, 2016 Bratislava, Slovakia (in Czech).

[4] S. Ferrari, A.F. Gualtieri, The use of illitic clays in the production of stoneware tile ceramics, Applied Clay Science, Vol. 32, Issue1-2, April 2006, pp. 73-81.

[5] J.M. Kinuthia, J.E. Oti, Designed nom-fired clay mixes for sustainable and low carbon use, Applied Clay Science, Vol. 59-60, May 2012, pp 131-139.

[6] P. Padevět, P. Bittnar, Measuring the creep and material properties of cement paste specimens, WSEAS Transaction on Applied and Theoretical Mechanics pp. 81-90, 2010.

[7] M. Maddison, T. Mauring, K.Kirsimäe, U. Mander, The humidity buffer capacity of clay-sand plaster filler with phytomass from treatment wetlands, Building and Environment, Vol. 44, Issue 9, September 2009, pp. 1864-1868.

[8] M Jankula, T. Hulan, I. Štubňa, J. Ondruška, R. Podoba, P. Šín, P. Bačík, A. Trnik, The influence of heat on elastic properties of illitic clay Radobica, Nippon Seramikkusu Kyokai Gakujutsu Ronbunshi/Journal of the Ceramic Society of Japan, Vol. 123, Issue 1441, 2015, pp. 874-879.

EAN 2017
55th conference on experimental stress analysis

# Calculation of B-Basis Values from Composite Material Strength Parameters Obtained from Measurements of Non-Identical Batches

Lukáš Bek [1, a], Radek Kottner [1, b], Jan Krystek [1, c], Vladislav Laš [1, d]

[1] NTIS - New Technologies for the Information Society, Faculty of Applied Sciences, University of West Bohemia, Technicka 8, 301 00 Pilsen, Czech Republic
[a]lukasbek@kme.zcu.cz, [b]kottner@kme.zcu.cz, [c]krystek@kme.zcu.cz, [d]las@kme.zcu.cz

**Keywords:** Carbon fibre composite, Anderson-Darling, Analysis of variance, B-basis value.

**Abstract.** Experimental investigation of stiffness and strength parameters of unidirectional carbon fibre plates produced in an autoclave using the Vacuum Assisted Process was performed. Parameters belonging to five production batches were subjected to the statistical evaluation. The Maximum Normed Residual test was used for the identification of outliers and the k-sample Anderson-Darling test was used in order to determine whether the parameters can be processed as belonging to one big group of data. Due to the previous work and negative result of the test for strength parameters, the scope of the work was focused to the evaluation B-basis values using the Analysis of variance method.

## Introduction

High stiffness and strength to weight ratios of composite materials cause increasing use of these materials in various applications in sport, automotive, marine, and aerospace industries. However, manufacturing process is a very complex process consisting of a fiber placement done by a worker or a robot, a potential resin infusion and from a curing process. These process critical points may lead to batch-to-batch variability caused by process variability. Except of listed process critical points, also raw materials batch-to-batch variability and testing variability may lead to a final material variability [1]. This material variability results in stiffness and strength variability of a final composite product. Therefore, it is essential to use a statistical evaluation of experimentally investigated material parameters. These parameters should be obtained from measurements of samples belonging to different batches and later the hypothesis, whether it can be processed as belonging to one group, should be tested.

In case of the positive test, a statistical distribution can be usually used for the approximation. In [1], the Normal distribution, the Lognormal distribution and the 2-parameter Weibull distribution are recommended as the most suitable for the approximation of material parameters of composites. The Normal distribution was used in [2] for the approximation of the stiffness and strength parameters of carbon fibre composite. The Weibull distribution was used in [3], [4], and [5]. These three papers were focused on the approximation of the strength parameters of unidirectional carbon fibre samples, of carbon/epoxy laminate having the layup $[0/90]_{2S}$, and of bamboo fibres, respectively. However, none of the papers mention whether the material parameters were measured using samples from one or more batches. Therefore, the test for the hypothesis, whether the parameters can be processed as belonging to one group, was not mentioned. Due to the previous work of the author [6], it was observed that the material strength parameters tend not to confirm the hypothesis of behaving as belonging to one group (non-identical batches).

Therefore, this work was focused on the extension of the previous work in form of the batch count increase and on the evaluation of B-basis values from material strength parameters measured from non-identical batches.

## Theoretical background

### The Maximum normed residual test

The Maximum normed residual test (MNR) is a test for the identification of outliers in a set of unstructured data $x_1, x_2, \ldots, x_n$. The MNR can be calculated as:

$$\mathrm{MNR} = \max_i \frac{|x_i - \bar{x}|}{s}, \qquad i=1, 2, \ldots, n \tag{1}$$

where $\bar{x}$ denotes the sample mean and $s$ the standard deviation. This number is compared to the critical value (CV) calculated for the sample size $n$ as:

$$\mathrm{CV} = \frac{n-1}{\sqrt{n}} \sqrt{\frac{t^2}{n-2+t^2}}, \tag{2}$$

where

$$t = [1 - \alpha/2n]. \tag{3}$$

$\alpha$ is the significance level and recommended $\alpha = 0.05$ for this test. If the MNR is lower than CV, then it can be concluded that no outlier was detected. Otherwise, the sample associated with the highest $|x_i - \bar{x}|$ is detected as an outlier. Once the outlier is detected, it has to be removed from calculations and the MNR test must be repeated [1].

### The k-sample Anderson-Darling test

The k-sample Anderson-Darling test is used for the verification of the hypothesis whether the samples from different batches can be combined into one group of observations $z_1, z_2, \ldots, z_L$. This hypothesis is confirmed if the ADK number is lower than critical value ADC. ADK can be calculated as [1]:

$$\mathrm{ADK} = \frac{n-1}{n^2(k-1)} \sum_{i=1}^{k} \left\{ \frac{1}{n_i} \sum_{j=1}^{L} h_j \frac{\left(nH_{ij} - n_i H_j\right)^2}{H_j(n-H_j) - nh_j/4} \right\}, \tag{4}$$

where

| | |
|---|---|
| $n$ | is the number of combined samples, |
| $k$ | is the number of batches, |
| $L$ | is the number of observations, |
| $h_j$ | is the number of values in the combined samples equal to $z_j$, |
| $H_j$ | is the number of values in the combined samples less than $z_j$ plus one half the number of values in the combined samples equal to $z_j$, |
| $F_{ij}$ | is the number of values in the i-th group which are less than $z_j$ plus one half the number of values in this group which are equal to $z_j$. |

The critical value ADC can be evaluated as:

$$\mathrm{ADC} = 1 + \sigma_n \left[1.645 + \frac{0.678}{\sqrt{k-1}} - \frac{0.362}{k-1}\right], \tag{5}$$

where

$$\sigma_n^2 = \mathrm{Var(ADK)} = \frac{an^3+bn^2+cn+d}{(n-1)(n-2)(n-3)(k-1)^2}, \tag{6}$$

$$\begin{aligned} a &= (4g-6)(k-1)+(10-6g)S, \\ b &= (2g-4)k^2+8Tk+(2g-14T-4)S-8T+4g-6, \\ c &= (6T+2g-2)k^2+(4T-4g+6)k+(2T-6)S+4T, \\ d &= (2T+6)k^2-4Tk, \end{aligned} \tag{7}$$

$$S = \sum_{i=1}^{k} \frac{1}{n_i}, \tag{8}$$

$$T = \sum_{i=1}^{n-1} \frac{1}{i}. \tag{9}$$

**The One-way Analysis of variance**

The one-way Analysis of variance (ANOVA) can be used for the calculation of the B-basis value in case of negative result of the k-sample Anderson-Darling test. The analysis can be used only if following assumptions are fulfilled [1]:

1. The data from each batch are normally distributed,
2. The within-batch variance is the same from batch to batch,
3. The batch means are normally distributed.

Currently, there is no available test for the first assumption. For the second assumption, there is advised to use the Levene's test for equality of variances. Also for the third assumption, there is no useful test if the batch number is not sufficient high (twenty or higher) [1].

For the calculation of the B-basis value, following calculations must be performed:

$$n^* = \sum_{i=1}^{k} n_i^2/n, \tag{10}$$

$$n' = (n-n^*)/(k-1), \tag{11}$$

$$\bar{x} = \sum_{i=1}^{k} n_i \bar{x}_i/n, \tag{12}$$

$$MSB = \sum_{i=1}^{k} \frac{n_i(\bar{x}_i-\bar{x})^2}{k-1}, \tag{13}$$

$$MSE = \frac{1}{n-k} \sum_{i=1}^{k} (n-k)s_i^2, \tag{14}$$

where $n_i$ is the number of samples in $i$-th batch, $n$ denotes the total number of samples across $k$ batches. $n'$ is the effective batch size, $\bar{x}_i$ and $s_i$ are the batch mean and batch standard deviation, respectively. $\bar{x}$ denotes the overall mean. $MSB$ is called between-batch mean square and the $MSE$ is called the within-batch mean square.

Furthermore, it is necessary to calculate the population standard deviation:

$$S = \sqrt{\frac{MSB}{n'} + \left(\frac{n'-1}{n'}\right) MSE}, \tag{15}$$

and also the ratio of mean squares defined as:

$$u = \frac{MSB}{MSE}. \tag{16}$$

The tolerance limit factor can be calculated as follows:

$$T = \frac{k_0 - \frac{k_1}{\sqrt{n'}} + (k_1 - k_0)w}{1 - \frac{1}{\sqrt{n'}}},$$

where

$$w = \sqrt{\frac{u}{u+n'-1}} \tag{17}$$

and parameters $k_0$ and $k_1$ are tolerance limit factors for a simple random sample from a normal distribution with sample size $n$ and size $k$, respectively. Mentioned parameters can be found in tables [1].

Thus, B-basis value can be calculated using formula:

$$B = \bar{x} - TS. \tag{18}$$

**The Levene's test for equality of variances**

One of the assumptions for the applicability of the ANOVA method is the equality of batch variances. For this purpose, the Levene's test is widely used. For this test, data must be transformed as follows:

$$w_{ij} = \left| x_{ij} - \tilde{x}_i \right|, \tag{19}$$

where $\tilde{x}_i$ is the median of the $n_i$ values of the *i*-th batch. Using transformed data, the F-test must be performed. If the test statistics is greater than or equal to the tabulated F-distribution quantile, the variances are significantly different [1].

**The F-test**

The F-test is used for testing of batch mean (or variance) equality. In order to test the hypothesis if the sample batches have the same mean, the statistics must be calculated [1]:

$$F = \frac{\sum_{i=1}^{k} n_i(\bar{x}_i - \bar{x})^2/(k-1)}{\sum_{i=1}^{k}\sum_{j=1}^{n_i}\left(x_{ij} - \bar{x}_i\right)^2/(n-k)} \tag{20}$$

where $\bar{x}_i$ is the mean of $n_i$ values in the *i*-th group and $\bar{x}$ is the mean of all $n$ observations. If the $F$ statistics is greater than $1-\alpha$ quantile of the F-distribution, it can be concluded (with $\alpha$ percent risk of error), that the means of $k$ groups are not equal.

**Test Samples**

For purposes of the test sample creation, five carbon fibre plates with unidirectional (UD) fibre orientation were produced by the author in the ASC Econoclave autoclave (Fig. 1). Plates were produced from non-crimped fabric (NCF) Saertex with fibres TENAX-J IMS60 E13 24K and from epoxy resin MSG L285 using the Vacuum Assisted Process (VAP®). VAP® is the process patented the Airbus Group and it is using the VAP® membrane made out of gore-tex®. This membrane is used for separation of the air (vacuum) from the resin. Thus, the amount of air bubbles contained in resin is reduced and nearly optimal fibre to resin ratio is achieved.

Test samples were cut from produced plates using water jet cutter. Samples for testing of tensile and shear material parameters were cut from each plate. Therefore, obtained samples had fibre orientation 0°, 45° and 90° with respect to the longer edge of a sample. The dimensions followed recommendations of ASTM D3039 and were determined as 200 mm × 25 mm × 2.4 mm (Fig. 2). Six samples per each batch and fibre orientation were cut and numbered to guaranty the full traceability.

All samples were measured and weighted for purposes of the fibre volume ratio calculation. Resulting means and standard deviations of the fibre volume ratios for each batch are listed in Table 1.

Table 1: Mean fibre volume ratios and standard deviations of tested batches.

| Batches | Mean fibre volume ratio [%] | Standard deviation of fibre volume ratio [%] |
|---|---|---|
| 1 | 60.7 | 0.9 |
| 2 | 64.0 | 0.9 |
| 3 | 61.6 | 0.7 |
| 4 | 61.6 | 0.5 |
| 5 | 62.6 | 0.4 |

Fig. 1: VAP® manufacturing process of UD carbon fibre plates.

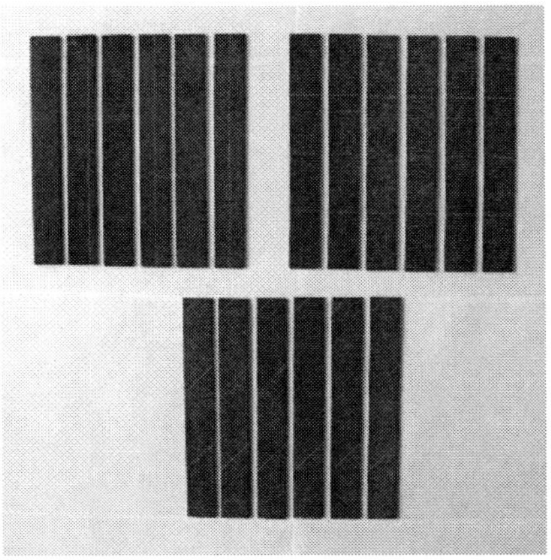

Fig. 2: Example of sample batch with fibre orientation 0°, 90° and 45°.

**Experiments and statistical evaluation**

Experiments were carried out using the universal testing machine Zwick/Roell Z050. The loading rate of 2 mm/min was selected based on recommendation of the ASTM D3039. The load was measured using a load cell and the displacement of the sample was measured using an extensometer. The gage length was set to 60 mm. All samples were meant to load up to the failure.

However, it was observed that the load and also the clamp capability of the testing machine were not sufficient for the test samples having the fibre orientation of 0°. Therefore, it was not possible to measure properly the stiffness and strength parameters of these samples and they were removed from the evaluation.

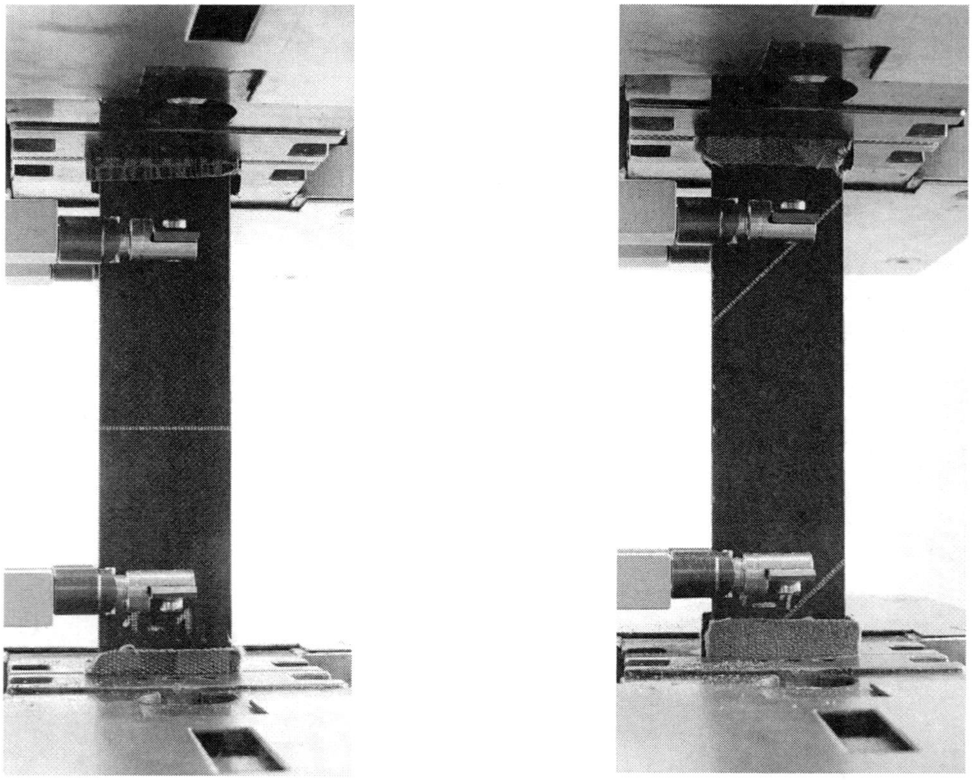

Fig. 3: Experiments with samples having 90° (left) and 45° (right) fibre orientations.

Obtained load-displacement curves were recalculated into the engineering stress-strain curves and stiffness and strength parameters were evaluated. These parameters were subjected to the MNR test. Based on the results, sample no. 2 having the 45° fibre orientation from the batch 2 and sample no. 3 having the 45° fibre orientation from the batch 5 were detected as outliers from the strength point of view. Therefore, they were removed from other evaluations. Without mentioned outliers, all batches passed the MNR test. Resulting critical values are listed in Table 2 and Table 3. Based on the parameters for each sample and batch, mean values and standard deviations were calculated for stiffness and strength parameters. Parameters related to the samples having fibre orientation of 90° and 45° are listed in Table 4 and Table 5, respectively.

Table 2: MNR and CV values for stiffness and strength parameters for samples having 90° fibre orientation

| Batch | MNR for stiffness parameters [-] | CV for stiffness parameters [-] | MNR for strength parameters [-] | CV for strength parameters [-] |
|---|---|---|---|---|
| 1 | 1.594 | 1.887 | 1.484 | 1.887 |
| 2 | 1.239 | 1.887 | 1.310 | 1.887 |
| 3 | 1.271 | 1.887 | 1.314 | 1.887 |
| 4 | 1.620 | 1.887 | 1.491 | 1.887 |
| 5 | 1.415 | 1.887 | 1.534 | 1.887 |

Table 3: MNR and CV values for stiffness and strength parameters for samples having 45° fibre orientation

| Batch | MNR for stiffness parameters [-] | CV for stiffness parameters [-] | MNR for strength parameters [-] | CV for strength parameters [-] |
|---|---|---|---|---|
| 1 | 1.586 | 1.887 | 1.514 | 1.715 |
| 2 | 1.517 | 1.887 | 1.720 | 1.887 |
| 3 | 1.881 | 1.887 | 1.591 | 1.887 |
| 4 | 1.305 | 1.887 | 1.676 | 1.887 |
| 5 | 1.717 | 1.887 | 1.200 | 1.715 |

Table 4: Mean values and standard deviations of stiffness and strength parameters for samples having 90° fibre orientation

| Batch | Stiffness mean [MPa] | Stiffness standard deviation [MPa] | Strength mean [MPa] | Strength standard deviation [MPa] |
|---|---|---|---|---|
| 1 | 7685.2 | 114.2 | 43.1 | 1.0 |
| 2 | 7602.8 | 148.4 | 39.7 | 0.9 |
| 3 | 7822.3 | 244.7 | 40.4 | 1.2 |
| 4 | 7760.8 | 238.0 | 40.7 | 1.4 |
| 5 | 7946.1 | 126.6 | 41.9 | 1.5 |

Table 5: Mean values and standard deviations of stiffness and strength parameters for samples having 45° fibre orientation.

| Batch | Stiffness mean [MPa] | Stiffness standard deviation [MPa] | Strength mean [MPa] | Strength standard deviation [MPa] |
|---|---|---|---|---|
| 1 | 11247.3 | 401.6 | 67.0 | 2.5 |
| 2 | 11474.8 | 537.7 | 62.5 | 1.9 |
| 3 | 10875.4 | 561.0 | 63.1 | 2.5 |
| 4 | 11814.8 | 887.0 | 63.3 | 1.9 |
| 5 | 11568.1 | 357.6 | 64.7 | 2.0 |

Furthermore, parameters from all batches were subjected to the k-sample Anderson-Darling test in order to test the hypothesis whether parameters can be processed as belonging to one big group of samples (further referred to as pooled parameters). This hypothesis was confirmed for the stiffness parameters for samples having the fibre orientation of 90° and also of 45° (Table 6). The further evaluation of pooled parameters was the scope of the previous work and it is described in [6]. However, for the strength parameters, this hypothesis was rejected. Therefore, it was necessary to perform the Levene's test for equality of variances, which is involving the F-test, in order to test whether one of the requirements for the application of ANOVA method for calculation of B-basis values is fulfilled (Table 7). Listed critical value is the 0.95 quantile of the F-distribution.

Table 6: Results for k-sample Anderson Darling test.

| Samples | ADK for stiffness param. [-] | ADC for stiffness param. [-] | ADK for strength param. [-] | ADC for strength param. [-] |
|---|---|---|---|---|
| 90° fibre orientation | 1.6123 | 1.6412 | 2.4490 | 1.6412 |
| 45° fibre orientation | 1.5896 | 1.6412 | 2.5584 | 1.6356 |

Table 7: Results for the Levene's test for equality of variances

| Sample strength parameters | $F$ [-] | Critical value [-] |
|---|---|---|
| 90° fibre orientation | 0.17 | 2.76 |
| 45° fibre orientation | 0.30 | 2.80 |

Based on the results from the Table 7, it was concluded that the ANOVA method can be used for the calculation of B-basis values for the strength parameters. Using the eq. 10 through eq. 18, B-basis values and population standard deviations were calculated and listed in the Table 8.

Table 8: B-basis values and population standard deviations for the strength parameters

| Sample strength parameters | B-basis value [MPa] | Population standard deviation [MPa] |
|---|---|---|
| 90° fibre orientation | 36.3 | 1.7 |
| 45° fibre orientation | 57.4 | 2.6 |

## Conclusion

The statistical evaluation of the strength material parameters of the produced carbon fibre plates was performed. Two outliers from the strength point of view were removed from batches and the B-basis values were calculated in case of the negative k-sample Anderson-Darling test. The investigation of measurement possibilities for the high strength samples having the 0° fibre orientation, the evaluation of the B-basis values for these samples and the use of calculated values for the research of bolted joint strength will be the scope of the future work.

## Acknowledgement

This publication was supported by the project LO1506 of the Czech Ministry of Education, Youth and Sports.

## References

[1] U.S. Department of Defense. Military Handbook - MIL-HDBK-17-1F: Composite Materials Handbook; Volume 1. Polymer matrix composites guidelines for characterization of structural materials. U.S. Department of Defense, 2002.

[2] W. Lian and W. Yao. Fatigue life prediction of composite laminates by FEA simulation method. Internal Journal of Fatigue, 2010.

[3] M.H. Dirikolu, A. Aktas, and B. Birgören. Statistical Analysis of Fracture Strength of Composite Materials Using Weibull Distribution. Turk. J. Eng. Environ. Sci, Vol. 26, 2002.

[4] E. Barbero, J. Fernádez-Sáez, and C. Navarro. Statistical analysis of the mechanical properties of composite materials. Composites Part B: Engineering, Vol. 31, 2000.

[5] F. Wang and J. Shao. Modified Weibull Distribution for Analysing the Tensile Strength of Bamboo Fibers. Polymers, Vol. 6, 2014.

[6] L. Bek, R. Kottner, J. Krystek, V. Laš, Statistical based approach to material identification of composite materials, 54th International Conference on Experimental Stress Analysis , Srni, 2016, pp 1-8, ISBN: 978-80-261-0624-1.

EAN 2017
55th conference on experimental stress analysis

# Geometry influence on strength of thermal drilled threaded holes

M. Kováčik[1], R. Bělík[1]

[1]Department of Applied Mechanics, Faculty of Mechanical Engineering, VŠB-Technical University of Ostrava, Czech Republic

**Keywords:** flowdrill; thermal drilling; thread stripping; DOCOL; bolted joint.

**Abstract.** This paper focuses on thermal drilling technology for fabricating threaded holes. Furthermore, experiments presented here focus on the influence of geometrical constraints in such bolted joints and their effect on the joint strength. The experimental results are analysed and compared to different types of geometry. The experiments are done on both mild and high strength steel types ensuring a further set of variables to the problem.

## Introduction

There are not many standards regarding thermally drilled threaded holes. It is therefore difficult to predict strength and reliability in general of the joints with bolts screwed in thermally drilled threaded holes. Some of the assumptions from ISO 898 [1] can be extended to the case as it was done in a previous study [2] to this matter by these very authors, even though the ISO standard presents predictions on standard bolted joints (bolt and a nut).

According to ISO 898 [1], bolted joints failures are categorized to three types: Failure in the bolt shank, thread stripping on the bolt and thread stripping in the nut (threaded hole in this case). Any joint failure is undesirable; however, if a bolt breaks in the shank, it can be easily replaced. If thread stripping on the bolt occurs, it may not be as easily detected and it may damage the thread in the hole. However, the least desirable failure is the thread stripping in the hole, thus damaging the part as a whole.

According to studies by Alexander [3] and Hagiwara [4], a sufficient length of engagement of the thread has to be achieved in order to raise the bolt stripping strength above the bolt shank strength. This way, the bolt shank breaks always before thread stripping occurs.

Achieving sufficient length of engagement is particularly difficult for thin plates, where the ratio of the thread size and depth of the hole is poor. The thermal friction drilling technology [5] can partly overcome this difficulty by extruding a sleeve from the removed material while creating hole, thus extending the depth of the hole and the effective length of engagement with it. It was discovered in the previous work [2] that there is another often occurring type of failure typical for this type of technology characterized as a breakage of the extruded sleeve, which splits from the base material.

Previous study [2] showed also a relatively large deflection of the plates which apparently influenced the effective length of engagement and thus lowered the joint bearing capability. The purpose of this work is to study if tighter constraints could result in strength of such joints and if so, how is the strength by the geometry influenced.

## Method of Evaluation

According to the standards for bolted joints [1], the measure of such joint strength is the amount of axial tensile force at failure for each size and property class. Since this paper focuses mainly on thread stripping, the property class was not an object of variations and the highest property class of 12.9 was used to maximize the bolt shank strength. There are three flange material types considered in this study. Two mild steel grades S235JRG and S355J0 according to EN 10025 [6] and a DOCOL 1200M high strength steel acc. to its datasheet [7]. The reference values of their tensile strengths are used from previous paper [2], where they were measured additionally.

Thickness of the base material before hole fabrication plays a major role as well on the resulting depth of the manufactured hole. 2mm thick specimen were used for all steel types, and 3mm thick specimen for mild steels (high strength steel of 3mm thickness was not standardly available).

To extend the results in the previous paper [2] for comparison, the same thread sizes were assumed, e.g.M5x0.8, M6x1, M8x1.25. All these parameters listed above were included into the specimen fabrication.

## Experimental setup

A tensile testing device is a convenient device for axial tensile force measurements. It was already mentioned above, that this paper focuses on the geometrical constraints. In order to isolate the dependency, the same clamping adapter as the one used earlier [2] was used with different hole dimensions, see fig 1. It is assumed, that a tighter flange would suppress the deformation of the hole more and therefore would result in higher strength.

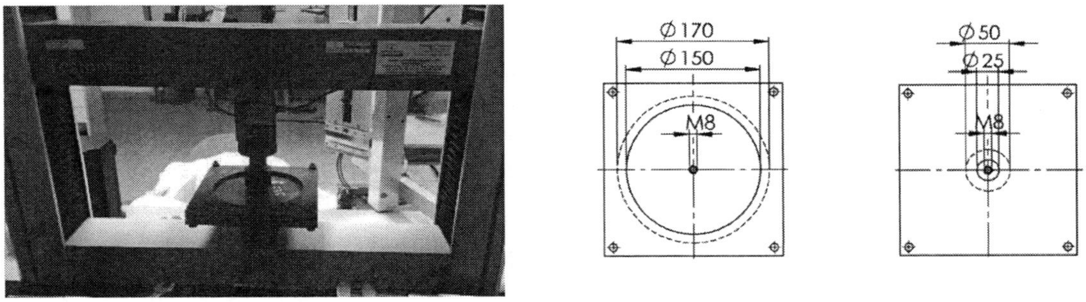


a) test rig b) original geometry c) altered geometry

Fig. 1: The tensile testing device with the interface for a clamped specimen.

## Results

The tabled values of a wider geometry are used from the previous paper [2]. The tighter flange clamp results are used from the present study. Each failure is marked with its type marked below.

Tab. 1: Average of measured values of joint strength for base material thickness 2mm

| Geometry | Original | | | Altered | | |
|---|---|---|---|---|---|---|
| Material | S235JRG | S355J0 | DOCOL 1200M | S235JRG | S355J0 | DOCOL 1200M |
| M5 | 6 700* N | 8 110 N | 12 630 N | 7 351* N | 9 949 N | 13 451 N |
| M6 | 8 420* N | 9 140 N | 15 000 N | 8 402* N | 10 828** N | 15 877 N |
| M8 | 9 580* N | 11 600** N | 16 720* N | 10 795* N | 13 657** N | 18 583 N |

** failures with dominant internal thread stripping*
***failures with dominant sleeve breakage*

Tab. 2: Average of measured values of joint strength for base material thickness 3mm

| Geometry | Original | | Altered | |
|---|---|---|---|---|
| Material | S235JRG | S355J0 | S235JRG | S355J0 |
| M5 | 11 270* N | 12 720 N | 13 132* N | 15 819** N |
| M6 | 13 420* N | 15 250 N | 15 247* N | 17 574** N |
| M8 | 14 200* N | 18 510** N | 17 737* N | 21 213** N |

** failures with dominant internal thread stripping*
***failures with dominant sleeve breakage*

The improvement using the tighter geometry is evaluated below as the ratio of results from present study (altered geometry) to the results from the previous study (Original geometry).

Tab. 3: Percent improvement due to tighter geometry

| Thickness | 2 mm | | | 3 mm | |
|---|---|---|---|---|---|
| Material | S235JRG | S355J0 | DOCOL 1200M | S235JRG1 | S355J0 |
| M5 | 10% | 23% | 7% | 16% | 24% |
| M6 | 0% | 18% | 6% | 14% | 15% |
| M8 | 13% | 18% | 11% | 25% | 15% |

The values resulting from the altered geometry are compared to standardized nut values from ISO 898-2 [1] for different strength classes. The table 4 summarizes values equivalent to the strength class for each configuration.

Tab. 4: Equivalent nut strength class (acc. to Table 4 in [1])

| Thickness | 2 mm | | | 3 mm | |
|---|---|---|---|---|---|
| Material | S235JRG | S355J0 | DOCOL 1200M | S235JRG1 | S355J0 |
| M5 | 05 | 6 | 9 | 9 | 10 |
| M6 | 04 | 04 | 6 | 6 | 8 |
| M8 | <04 | < 04 | 04 | 04 | 05 |

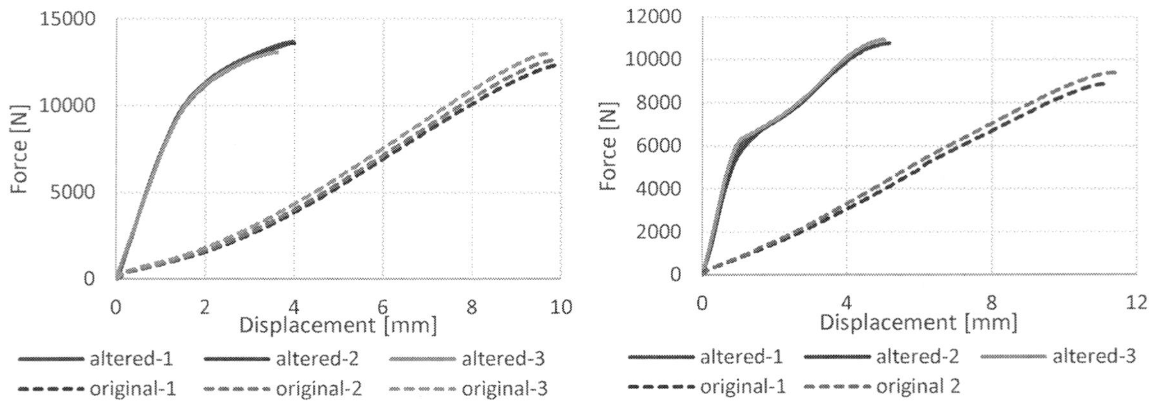


a) DOCOL 1200M, M5 size, Thickness 2mm   b) S355J0, M6 size , Thickness 2mm

Fig. 2: The tensile testing device with the interface for a clamped specimen.

## Conclusions

A tighter flange constraint results clearly in an increase of joint strength. However the effect is non-linear in terms of different configurations. It seems quite difficult at this point to find any dependence among the improvements from table 3. It can be observed from table 4 that the equivalent strength class for which can be nuts replaced with the threaded holes, gets better when moving to the right top in the table, while it is getting poor when moving to the opposite direction. This can be concluded as the ratio of the flange thickness to thread size still plays the major role on the strength. The material does as well but not as much as it is evident from values for M8 size.

Again DOCOL high strength steel specimens result now always in bolt thread stripping, which indicates that the flange material Yield strength is near the yield strength of the bolt. It is evident that opposed to the case of last study, the failure modes are now more consistent, because nearly all S355J0 specimens have a sleeve failure, all S235JRG have hole thread stripping and all DOCOL specimens cause a bolt thread stripping indicating that the Alexander's theory [3] might be extended to this case as well, since there is only one discrepancy at S355J0-M5-2mm. The deformation in the hole is in this case is apparently still large enough to damage the bolt thread before damaging the hole itself.

The Force-displacement relationship shows now a very similar shape to an uniaxial material testing curve with spots possibly indicating yield points as well as plastic and elastic regions which were not between examined phenomena. Future studies to this matter may focus to pretensioned specimens or circularly non-symmetrical specimens.

## Acknowledgements

The authors would like to kindly show their gratitude to company Borcad CZ s.r.o. for fabrication of the specimen and fabrication of the interface needed for actual measurement including the material.

## References

[1] ISO 898-2: 1992, Mechanical properties of fasteners-Part 2: Nuts with specified proof load values - coarse thread.

[2] R. Bělík, M. Kováčik: Strength of thermal drilled threaded holes, EAN 2016 - 54th International Conference on Experimental Stress Analysis, University of West Bohemia, 2016. ISBN 978-802610624-1.

[3] Alexander, E. M., Analysis and Design of Threaded Assemblies, SAE Papers, No. 770420 (1977)

[4] Hagiwara, M., Sakai, H., Verification of the Design Concept on Nuts in Bolt/Nut Assembly for the Revision of ISO 898-2 and ISO 898-6, Journal of Advanced Mechanical Systems, and Manufacturing, Vol. 1, No. 5 (2007), pp. 755-762

[5] Flowdrill technology from http://www.flowdrill.com/eu_en/, 14-04-2016

[6] EN 10025-2:2004 - Hot rolled products of structural steels

[7] Data sheet 8251 Docol 1200M http://www.ssab.com/api/sitecore/ProductPdfRender/PrintProductSheet?pid=F804FE49BF2F496186D8F3B8F90220E6, 24-06-2014

# Experimental verification of the mechanical properties of the material of the rotor shaft of the turbine damage

M. Pástor [1,a], F. Trebuňa [1], F. Šimčák [1], M. Buršák [2]

[1] Technical University of Košice, Faculty of Mechanical Engineering, Department of Applied Mechanics and Mechanical Engineering, Letná 9, Košice, 04200, Slovakia

[2] Technical University of Košice, Faculty of Mettalurgy, Institute of Materials, Letná 9, Košice, 04200, Slovakia

[a] miroslav.pastor@tuke.sk, [b]frantisek.trebuna@tuke.sk, [c]frantisek.simcak@tuke.sk, [d]marian.bursak@tuke.sk

**Keywords:** mechanical properties, fatigue test, lifespan.

**Abstract.** Analysis of lifespan of machines and equipment is connected with good knowledge of mechanical properties of used materials. In case of failure is necessary, beside others, to verify, if the material used in structure fulfill all demanded mechanical as well as fatigue properties. In the paper is described a procedure for experimental verification of mechanical properties of material of steam turbine rotor shaft after its damage. The results of measurements document that shaft material correspond to declared quality and mechanical as well as fatigue properties correspond to state of heating processing of given steel.

## Introduction

The heating turbines work in cycle, where working medium changes its phase. The most spread is R-C cycle, where the working medium is water. In that case, in turbine expands water steam. The steam turbines are always parts of bigger technological unit which has a specific purpose (production of electricity, heat, work on the shaft, steam for industrial purposes and so on). From the purpose and properties of technological unit results basic parameters for design of turbine, e.g. power, rotations per minute, or special demands like power characteristics of turbine and so on [1,2].

The structural design of steam turbine includes problems of strength, stiffness and shape, oil systems, foundation of turbine, clutches, gear boxes. Very important are questions connected with purpose, using and service of the turbine. The steam turbines are equipped by hydraulic and electric equipment that serve for regulation, measurement and diagnostics of turbine. Design of turbine consists of many classical machine parts (bearings, oil system, clutches, valves, bolted joints, etc.) that can be computed by standard methods, they exhibit high level of unification and they are bought by turbine producers on the market.

One of critical part of steam turbine is a rotor shaft. From the design point of view, the rotor can be made with assembled rotors, forged as one piece, made as welded disc, welded drum, or combined [3]. The rotor with assembled wheels has simple structure from the point of design as well as technology. Its dimensions are obviously limited by shaft length and sometimes such a rotor cannot be used. The strength of turbine shaft is checked for torque which is the biggest one in location of connection between turbine and generator or powered machine. In the location of joint between turbine and electric generator, the stress in shaft is checked under conditions of short circuit of generator, when the torque increases in comparison to common operation several times [4,5]. Failure analysis of steam turbines is performed in terms of microstructure and mechanical properties of the material, fatigue tests, fractography, and so on [6,7,8].

## Description of turbogenerator damage

The condensation steam turbogenerator before damage is shown in Fig. 1. Analyzed turbine rotor is made of alloy steel, forged as one piece (Fig. 2). The blades processed from anti-corrosive steel are assembled to the discs by special joints.

Fig. 1 Condensation steam turbogenerator [3].

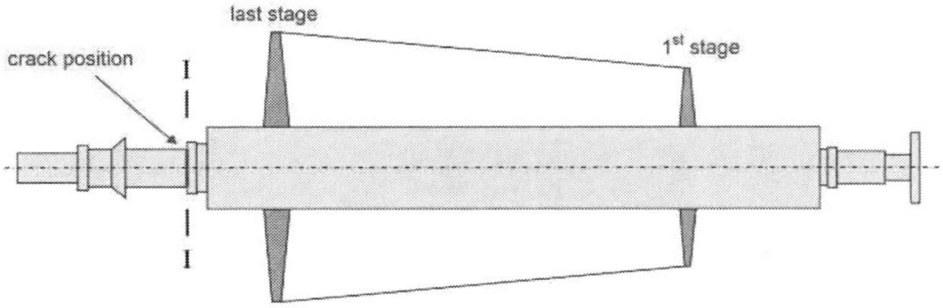


Fig. 2 Shaft of turbine rotor.

On the side of steam inlet on the end of rotor, the shaft consists of mechanical equipment for regulation of rotations. On the outer side of rotor, near mechanical equipment for rotation regulation are positioned cams of starting equipment for axial shifting of rotor. In case of excessive rotor shifting due to stochastic loading, or due to excessive wearing of axial bearing, these cams act as starting equipment and they activate stopping of the equipment. The rotor is positioned and led in two radial bearings and one axial bearing with pressure lubrication in oil bath. The turbine rotor is connected with the generator by reduction transmission box.

Cross-section of rotor in location of damage is loaded during operation of steam turbine by bending moment due to weight of rotor as well as by torque. The torque in the location of damage was caused by torsion moment of steam turbine. For the nominal rotations 8200 rpm, guaranteed power on generator (on terminals) 6 MW and efficiency 0.8, is the power of steam turbine after considering power lost on the rotation reductor equal cca to 7,5 MW. For nominal rotations 8200 rpm is the torque 8,74 kNm. Maximum stress due to torsion under nominal power in location of notch is 37,4 MPa. It results that such values of stress cannot lead to the failure state. State of equipment – steam turbogenerator after damage is given in Fig. 3.

Fig. 3 View to steam turbogenerator after failure.

Mechanical energy was transmitted from turbine to the reduction gear by shaft connected with flange of turbine rotor shaft by bolts (Fig. 4).

a)  b) 

Fig. 4 a) Shaft connecting turbine rotor and reduction gear, b) view to the side of the shaft consists of a pipe.

The right side in Fig. 4a represents broken part of turbine rotor shaft from the location damaged by fatigue fracture, while the outer diameter of full shaft is 120 mm in the location of fracture. The left side shows a shaft that consists of a pipe (Fig. 4b) on the end of which is a flange of clutch. View to the second part of clutch (from the side of gear) under aluminium cover is shown in Fig. 5.

Fig. 5 Part of clutch from the side of reduction gear.

On the basis of photographs taken during inspection (Fig. 6) as well as views to damaged part of shaft with fractured surface (Fig. 7), plastic deformation near fracture location (Fig. 8), detail view to fracture surface (Fig. 9), but also shape of fracture surface can be stated that the failure of shaft is connected with fatigue process (concluded also by bright conchoidal fracture).

Fig. 6 View to broken part of turbine rotor shaft.

Fig. 7 View to fracture surface on the shaft on the side of turbine.

Fig. 8 Shaft with fracture surface.

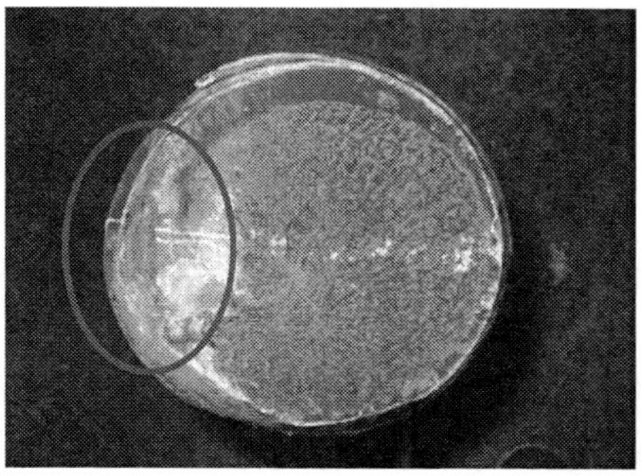

Fig. 9 Detail view to fracture surface – with marked area of macrocrack.

Creation of circular part of fractured surface concludes that it is resulting from combination of bending and torsion where the initial crack was caused not only by high level of equivalent stresses resulting from combination of bending and torsion, but also by stress concentrator due to defective structure of material resulting from smithing, Fig. 9. Such defective part together with sharp angle of relatively non-traditional shape (Fig. 10) localizes the area of first failure of turbine rotor shaft. This area was selected for further analysis of shaft material.

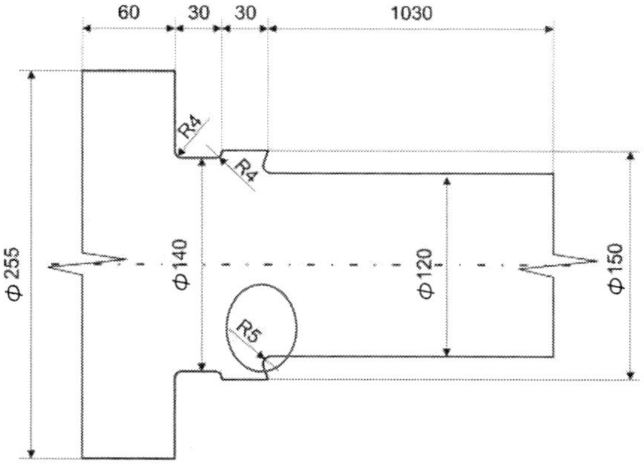


Fig. 10 Dimensions of turbine rotor shaft.

Fig. 8 documents that during crack spreading the pressing of material occurred which were invoked by cyclic loading resulting from rotation of shaft and axial deformation of flange clutch. In Fig. 11a is shown the separated part of turbine rotor from which a material for tests was taken. The material for mechanical and metallographic tests, see Fig. 11a, was taken according to plan given in Fig. 11b.

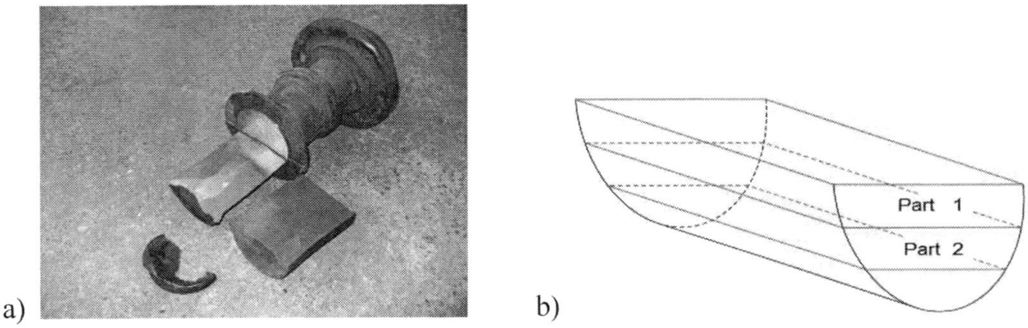


Fig. 11 a) Location on turbine rotor shaft from which was taken the material for tests, b) plan for preparation of specimens.

The segment of material separated by perpendicular section is shown in Fig. 12.

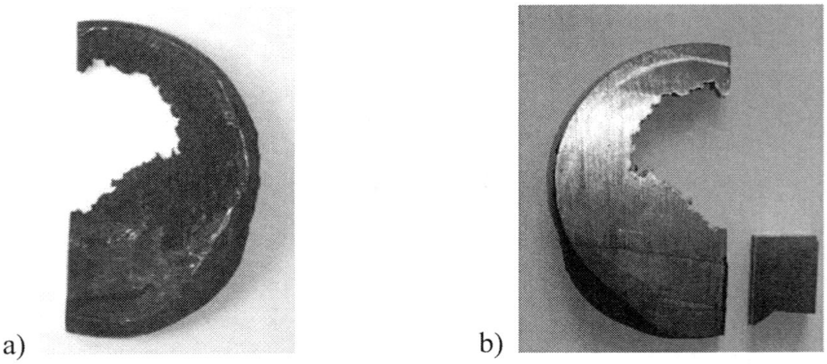


Fig. 12 Segment separated from damaged part of rotor.

On the left side of Fig. 12a is apparent area of fatigue failure and final crack caused by force. On the right side is a crack (Fig. 12b, failure in material) resulting from imperfect technology of shaft production. Strong plastic deformation along circumference of failure cross-section shows that in transversal macrocrack at the same time acted high normal stresses leading to plastic deformations during cyclic loading caused by rotation of turbine rotor.

**The analysis of the mechanical properties of the material of the turbine shaft**

The material for mechanical and metallographic tests was taken according to plan given in Fig. 13.

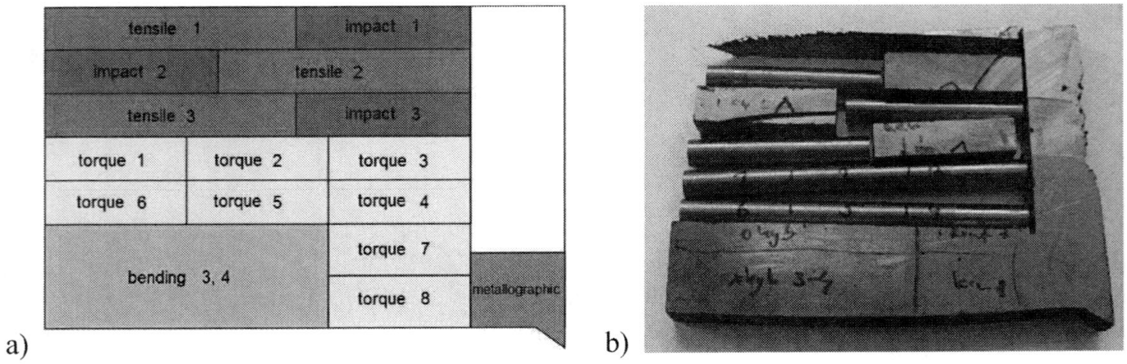


Fig. 13 Plan for preparation of specimens.

**Tensile test**

Basic mechanical properties were determined by static tensile test with specimens 6 x 30 STN EN 1000 2 -1 on the test machine ZWICK 1387 in accordance with treatment given in Standard STN EN 1000 2 -1. The measured results for given analysed material taken from broken shaft as well as properties of steel STN 41 6431.6 are given in Tab. No. 1. The testing velocity until reaching yield point of material $R_{p0,2}$ was 0,5 mm/min.

Table No. 1. Measured and prescribed mechanical properties of shaft material

| Material 16431 | Specimen | $R_{p0,2}$ [MPa] | $R_m$ [MPa] | $A_5$ [%] | Z [%] |
|---|---|---|---|---|---|
| Measured value | 1 | 645 | 777 | 22,3 | 71,4 |
| | 2 | 662 | 769 | 20,7 | 72,3 |
| | 3 | 613 | 789 | 22,3 | 73,4 |
| Prescribed value | - | min. 569 | 686-834 | 20 | - |

Average value of yield point $R_{p0,2}$ was 640 MPa and strength $R_m$ was 778 MPa. Shapes of tensile diagrams for specimens 2 and 3 are given in Fig. 14a, the test specimens after tests are shown in Fig. 14b.

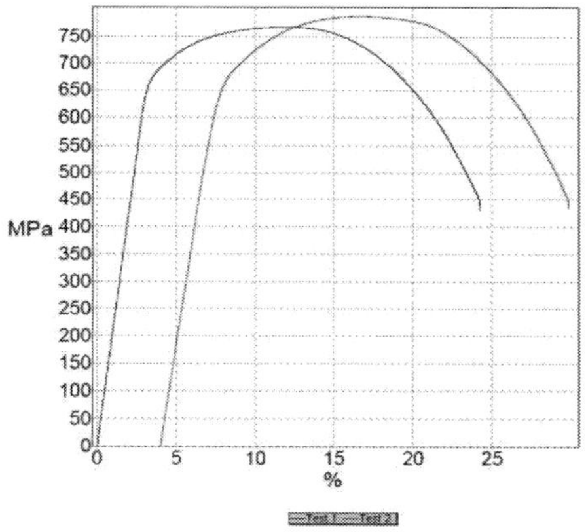


a)

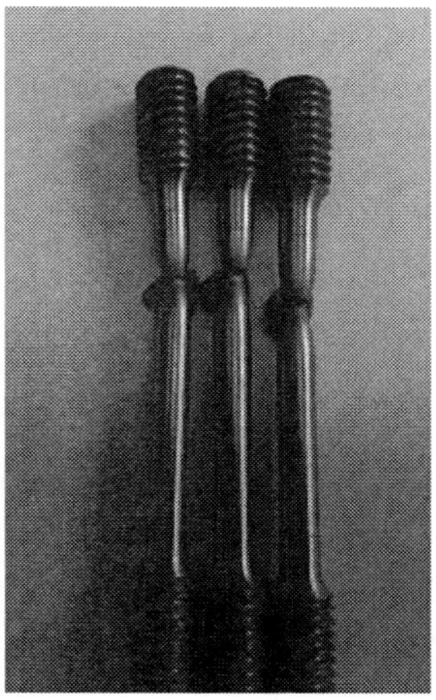

b)

Fig. 14 a) Tensile test diagram of specimens 2, 3, b) damaged test specimens 1, 2, 3.

***Bending impact test***

Toughness of material of damaged shaft was measured by impact test by bending for small bodies on the pendulum hammer PSW300 (Fig. 15) according to procedure given in STN EN 10045-1. The notches of shape Charpy „V“ and Mesnager „U2“ were used for crack initiation. The results of measurements are given in Tab. No. 2, which is completed with prescribed values for material 16 431.

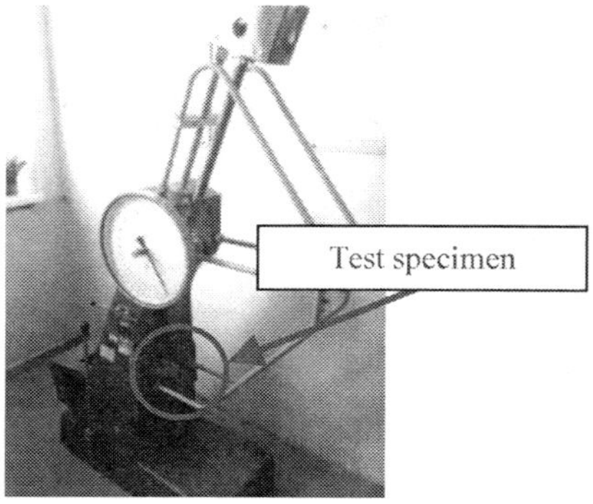


Fig. 15 Impact hammer PSW300 for toughness' test in bending for small specimens.

Table No. 2 Impact toughness of shaft material

| Specimen | Impact work KV [J] | Toughness KCV [$Jcm^{-2}$] | Specimen | Impact work KU [J] | Toughness KCU [$Jcm^{-2}$] |
|---|---|---|---|---|---|
| 1 | 169 | 211 | 4 | 201 | 251 |
| 2 | 186 | 233 | 5 | 196 | 245 |
| 3 | 167 | 209 | 6 | 194 | 243 |
| Prescribed value | - | - | - | - | 78 |

Average value of toughness with notch „V" is KCV=218 $Jcm^{-2}$ and with notch "U2" is KCU = 248 $Jcm^{-2}$. Damaged test specimens after impact bending test are shown in Fig. 16.

Fig. 16 Damaged test specimens after impact test in bending.

The measured results show that the toughness of material fulfills demands given in material standard for the steel STN 41 6314.

## Hardness

Hardness of shaft material was assessed by measurement of hardness according to Brinell by procedure given in STN EN 10003 – 1 (STN 42 0371). Prescribed value of the hardness is according to catalogue of steel STN 41 6314 HBS 208-253. Measured values were in range from 247 to 257 HBS.

## Metallographic documentation

The samples were taken from damaged shaft for the metallographic scratch pattern in order to have possibility to investigate structure along circumference of shaft, or structure perpendicular to the shaft axis (see Fig. 12b, Fig. 13a). The investigation was accomplished on light microscope LEITZ. Microstructure of shaft material is given in Fig. 17a and Fig. 17b.

a) 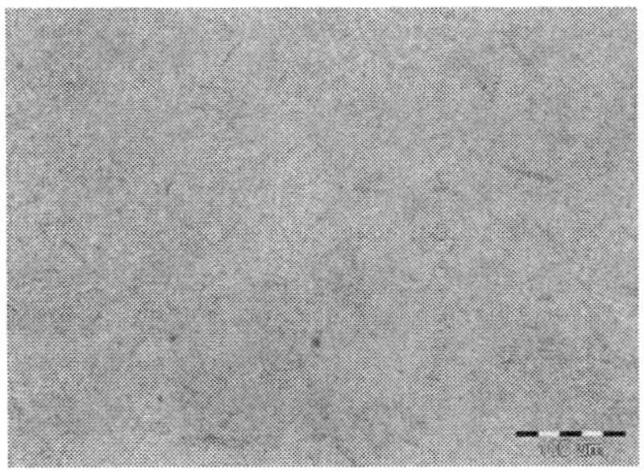 b) 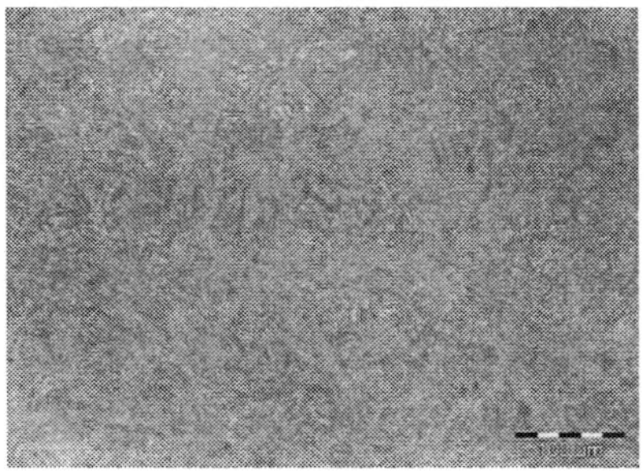

Fig. 17 a) View to microstructure in direction of shaft, b) view to microstructure in direction perpendicular to the axis of shaft.

Microstructure consists of tempered martensite-sorbit that arose out during tempering of martensite over temperature 600°C. In general, the highest fineness of sorbit, the highest strength and hardness [9]. The documented state shows that the technology of temperature processing was appropriate.

**Fatigue**

The characteristics of fatigue properties were assessed by fatigue equipment PWO 0192N, product of company Carl Schenk, Fig. 18. The fatigue characteristics such as fatigue limit were assessed in conditions of torsion and plane bending. The shape of used test specimens is given in Fig. 19 for torsion and in Fig. 20 for plane bending. The diameter of test specimen for torsion test was 6 mm. The thickness of test specimen for bending test was 4 mm and its width 10 mm. The coefficient of cycle asymmetry was R = -1. The test was accomplished by Standard STN 42 0363.

Fig. 18 View to microstructure in direction of shaft.

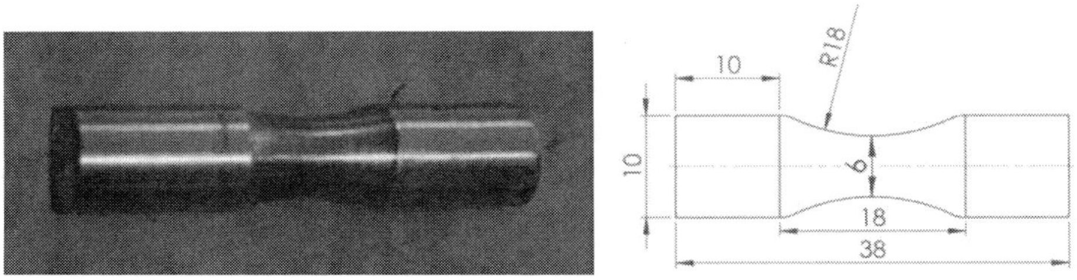


Fig. 19 Test specimens for torsion.

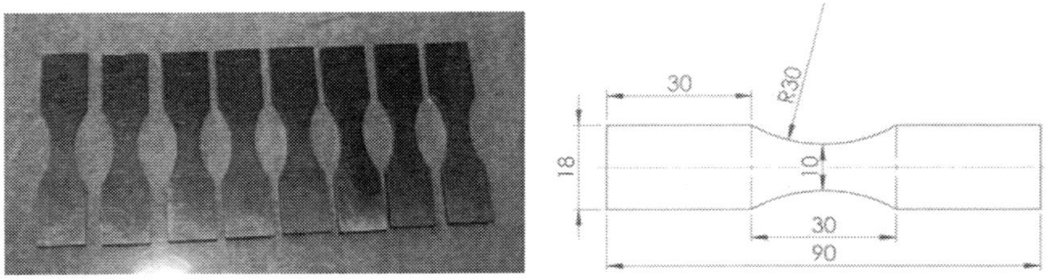


Fig. 20 Test specimens for plane bending.

Measured fatigue characteristics for torsion test are given in Tab. No. 3, and for the plane bending test in Tab. No. 4.

Table No. 3 Number of cycles to failure gained from measurements of torsional loading.

| No. specimen | **2** | **6** | **7** | **4** | **1** | **9** | **5** | **3** | **8** | **10** |
|---|---|---|---|---|---|---|---|---|---|---|
| Stress [MPa] | 150 | 174 | 181 | 186 | 205 | 214 | 237 | 258 | 274 | 291 |
| Number of cycles [N] | $10^7$ | $10^7$ | $10^7$ | 981800 | 845600 | 335600 | 2081300 | 648600 | 742300 | 342600 |

Table No. 4 Number of cycles to failure gained from measurements of plane bending.

| No. specimen | **2** | **5** | **4** | **1** | **3** | **7** | **8** | **6** |
|---|---|---|---|---|---|---|---|---|
| Stress [MPa] | 318 | 330 | 348 | 368 | 393 | 405 | 411 | 427 |
| Number of cycles [N] | $10^7$ | $10^7$ | 473000 | 731200 | 522200 | 412000 | 254000 | 174800 |

Data scattering can be joined with specific location of specimen take off from the turbine rotor shaft, which was loaded over yield point of material (see Fig. 8 and Fig. 13).

**Summary findings from the experimental determination of the mechanical properties of the material of the turbine rotor**

Mechanical properties were determined by tensile tests. The tensile tests were realized on three specimens and there were determined magnitudes of yield point $R_{p0,2}$, strength $R_m$ as well as plastic properties A5 and Z. Average values from three specimens are $R_{p0,2}$=640 MPa and $R_m$=778 MPa. These values created a base for the assessment of shaft strength under normal and emergency conditions.

Further test was realized for impact in bending. The tests were accomplished with notch of shape V and U. The tests were provided on three specimens for every of notch type. Material toughness of specimens with notches V and U did not exceed 209 J/cm$^2$, so the values prescribed by standards are fulfilled.

The hardness was in interval of prescribed values, too. Microstructure of material was determined in axial and transversal direction of shaft. Microstructure is created by tempered martnesite-sorbit that arise from steel tempering over 600 °C. In general, the highest fineness of sorbit, the highest strength and hardness. The documented state shows that the technology of temperature processing was appropriate.

The fatigue limit was measured for variable symmetric bending and torsion loadings. For the measurement were used 10 specimens for each type of loading and from the maximum measured amplitudes was determined number of cycles to damage. The measured values were used for creation of Wöhler curves from which is apparent fatigue limit for bending $\sigma^*_{co} = 330$ MPa and fatigue limit for torsion $\tau_c$ = 181 MPa. The ratios $\sigma^*_{co}/R_m$ and $\tau_c/R_m$ corresponds to ratios determined by Research Institute of Materials in Prague [3].

On the basis of realized material analysis can be stated that:

a) Structure of damaged shaft corresponds to steel 16 431.6, while the prescribed technological procedures for tempering of given steel were obeyed [10].
b) Mechanical and fatigue properties correspond to the demands of thermal processing of given steel.
c) One of the reasons of failure can be a defect – inclusion in material that is a result of faults in production technology of shaft. This results to the stress concentration and initiation of fatigue process of failure of turbine rotor shaft in location where cumulated influences of notch as well as inclusion in material were.

## Conclusion

The conclusions concerning reasons of failures and the answers to individual questions were elaborated on the basis of several visual inspections, analysis of data given by operator, but mainly on the basis of extensive experimental tests, chemical composition of material, determination of mechanical properties, assessment of microstructure, determination of mechanical properties under fatigue tests, extensive numerical and analytical modelling of tha shaft loading in location of damage. Definitely, the authors came out from general rules used for the design of steam turbines. The knowledge of educational institutions as well as production firms were used in formulation of conclusions.

From the analysis of fracture surface from the point of view of failure mechanism can be stated that despite of variable loading by torsion there was no clam-shell structure in the spreading crack in critical cross-section of shaft at circumferential direction, but the crack was non-concentric. From detailed analysis of fractured surface was found out that the location with the highest deepness of fatigue failure contains a crack in circumferential direction that con not be connected with loading, but it is joined with inappropriate production technology of turbine rotor shaft. With the highest probability, there was a failure as residuum from smiting. This error in the structure led to the crack towards surface and accordingly led to sharp notch which was reason of turbine shaft damage.

## Acknowledgement

This paper was supported by Projects VEGA No. 1/0751/16 and VEGA No. 1/0393/14.

## References

[1] P. Heinz, B. Murari, P. Singh, Steam Turbines: Design, Applications, and Rerating, Second Edition. New York, 2009 The McGraw-Hill Companies, Inc., ISBN 9780071508216.

[2] D. Japikse, Introduction to turbomachinery, 1997. 2. vydání. Oxford: Oxford University Press, ISBN 0–933283-10-5.

[3] F. Trebuňa et al., Posúdenie technických a prevádzkových parametrov turbogenerátora, vyhodnotenie príčin kolapsu, materiálová analýza súčiastok a konzultácia budúcich technických riešení, Final Report, TUKE SjF, Košice, 2015, 137 p.

[4] F. Trebuňa, P. Bigoš, Intensification of technical competence heavy supporting structures. Vienala, Košice, 1998.

[5] F. Trebuňa, F. Šimčák, Odolnosť prvkov mechanických sústav. EMILENA, Košice, 2004.

[6] S. Barella, M. Bellogini, M. Boniardi, S. Cincera, Failure analysis of a steam turbine rotor. Engineering Failure Analysis 18 (2011) 6, 511-519.

[7] Z. Mazur, R. G. Illescas, J. A. Romano, N. P. Rodrigue, Steam turbine blade failure analysis. Engineering Failure Analysis 15 (2008) 1 - 2, 129-141.

[8] E. Poursaeidi, M. R. Mohammadi Arhani, Failure investigation of an auxiliary steam turbine. Engineering Failure Analysis 17 (2010) 6, 1328 - 1336.

[9] F. Trebuňa, M. Buršák, Medzné stavy – lomy. Grafotlač Prešov, Prešov, 2002.

[10] I. Fűrbacher, K. Macek, J. Steidl, Lexikon technických materiálů. Svazek 1, Verlab Dashőfer, 2005.

# The Shrinkage of Fine Recycled Aggregate Concrete

ŠEFFLOVÁ Magdaléna[1, 2 a]

[1]Czech Technical University in Prague, Faculty of Civil Engineering, Department of Building Structures, Thákurova 7, 166 29, Prague, Czech Republic

[2]Czech Technical University in Prague, University Centre of Energy Efficient Buildings, Třinecká 1024, 273 43, Buštěhrad, Czech Republic

[a]magdalena.sefflova@cvut.cz

**Keywords:** recycled aggregate, concrete, durability, shrinkage.

This paper is focused on the shrinkage of fine recycled aggregate (FRA) concrete. The durability of FRA concrete is connected with many uncertainties and doubts. This paper presents results of shrinkage measuring of FRA concrete. The FRA was originated from crushed old concrete structures. There were prepared a total four concrete mixtures. The first mixture (REF) was prepared with natural sand. In other concrete mixtures, natural sand was replaced by the FRA in various replacement ratios. There were tested the shrinkage of FRA concrete. The shrinkage was tested according to ČSN 73 1320. It is possible to say that the use of FRA influences the concrete shrinkage. The shrinkage of FRA concrete depends on the amount of FRA in concrete.

## Introduction

Concrete is one of the most used building materials in the world. Concrete is popular for good mechanical properties, durability and simultaneously cost effectiveness. Concrete´s world production is estimated around 10 billion cubic meters per year [1]. With increasing production of new concrete elements, the amount of construction and demolition (C&D) waste on the landfills increases too. The C&D waste constitutes 1/3 of total waste production in the European Union. The targets for the reduction reuse and recycling of C&D waste were determined in connection with this problem. Minimally 70 % of the produced C&D waste must be reused or recycled by 2020 [2].

The use of recycled aggregate concrete, made from crushed C&D waste, is one of possible ways how save natural resources and simultaneously reduces the amount of C&D waste on landfills. Unfortunately, the use of recycled aggregate is connected with uncertainties and doubts [3]. The shrinkage is one of the most unknown aspects of recycled aggregate concrete.

## Experimental research program

**Materials.** Portland cement CEM I 42.5 R, natural sand, water and the FRA were used in the experimental part. The FRA was obtained from recycling plant in the Czech Republic. The FRA originated from demolished concrete structures. The FRA were crushed from fraction 32/64 mm to fraction 0/16 mm and sieved to fraction 0/4 mm. There were tested geometrical and physical properties of the FRA. The physical properties were tested by pycnometric method according to ČSN EN 1097 – 6 [4]. The test results were compared with natural sand. The test results of physical properties are given in Table 1.

Table 1: Physical properties of aggregate

| Aggregate | Fraction | Density [kg/m$^3$] | Water absorption capacity [%] |
|---|---|---|---|
| Natural aggregate | 0 – 4 mm | 2600 | 2.00 |
| Fine recycled aggregate | 0 – 4 mm | 2090 | 8.29 |

**Concrete mix proportion.** A total four concrete mixtures were designed and prepared. All concrete mixtures were designed with the same w/c ratio and the same amount of cement. The amount of cement and water- cement ratio was designed according to the Czech standard CSN EN 206 [5]. The first concrete mixture was reference (FRA) which did not include the FRA. Natural sand was replaced by the FRA in concrete mixtures REC10, REC20 and REC30 in varying replacement ratio (10 %, 20 %, 30 %). Details of concrete mix proportion are given in Table 2.

Table 2: Concrete mix proportion, per cubic meter

| Designation | REF | REC10 | REC20 | REC30 |
|---|---|---|---|---|
| Cement [kg] | 486 | 486 | 486 | 486 |
| Water [kg] | 243 | 243 | 243 | 243 |
| Sand [kg] | 1458 | 1312 | 1166 | 1021 |
| FRA [kg] | 0 | 146 | 292 | 437 |

**Methodology.** There were tested shrinkage and the development of compressive strength. The shrinkage was tested according to ČSN 73 1320 [6]. The shrinkage drains of dimensions 100x60x1000 mm u-shaped stainless steel profile were used for testing (see Fig. 1). The drains are covered with a removable neoprene sheet to avoid wall friction. On one side a removable anchor is fixed. On the other side this anchor is movable and sliding on three wheels. The motion of this anchor is registered by a high sensitive digital probe. Digital probes, connected over a digital bus system, were used for measurement. Concrete samples were kept in constant environment. The shrinkage was measured during 72 hours. The air temperature and humidity were monitored during the measuring period.

The compressive strength was tested according to ČSN EN 12390 – 3 [7]. Beams of dimensions 160x40x40 mm and 100 mm cubes were used for the testing. The compressive strength was tested at the age 7, 14, 28 and 90 days.

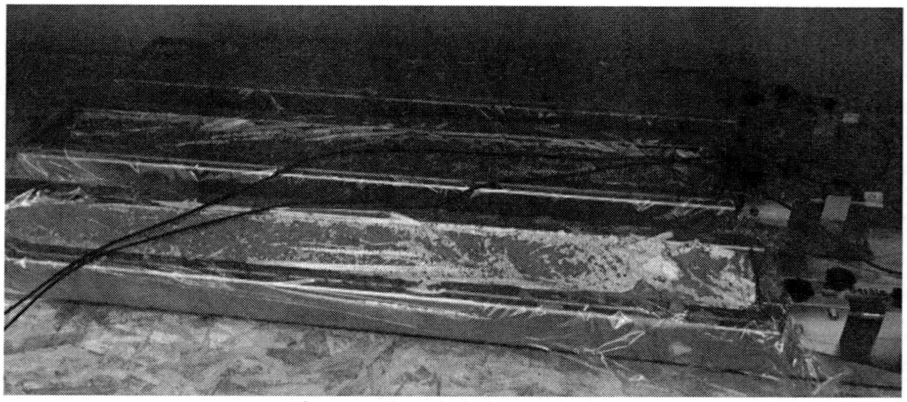

Fig. 1: The shrinkage drains

**Test results.**

The graph in Fig. 2 shows the development of shrinkage of FRA concrete. The use of FRA in concrete has negative influences on shrinkage of concrete. With increasing amount of FRA, shrinkage increases too.

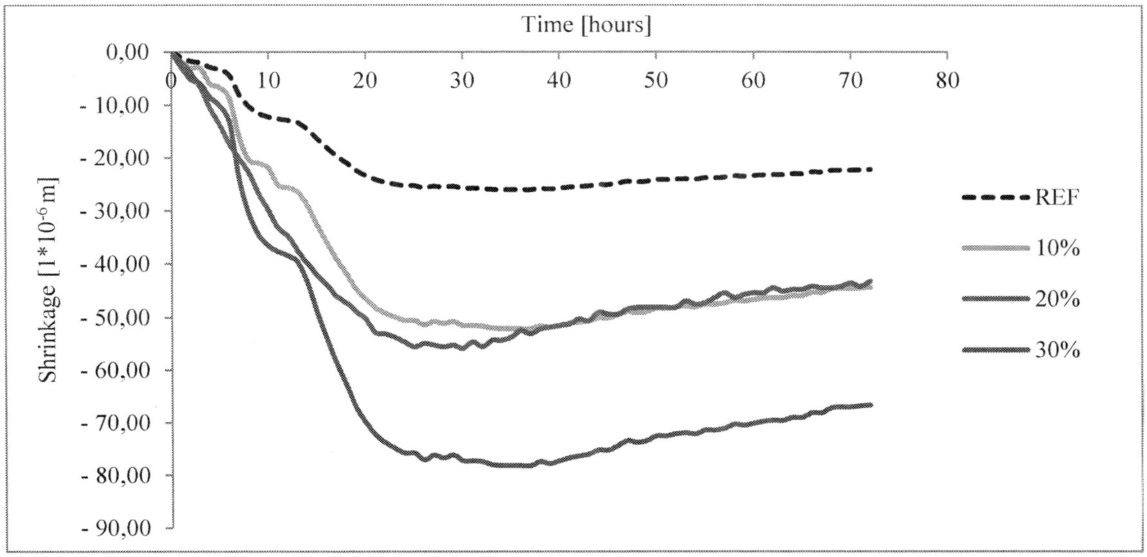

Fig. 2: The shrinkage development of concrete

The graph in Fig. 3 shows the development of the compressive strength at the age 7, 14, 28 and 90 days. Concrete samples with FRA show a mild improvement of the compressive strength of FRA concrete with comparison of reference concrete. However differences of the compressive strength between each FRA concrete mixtures are minimal.

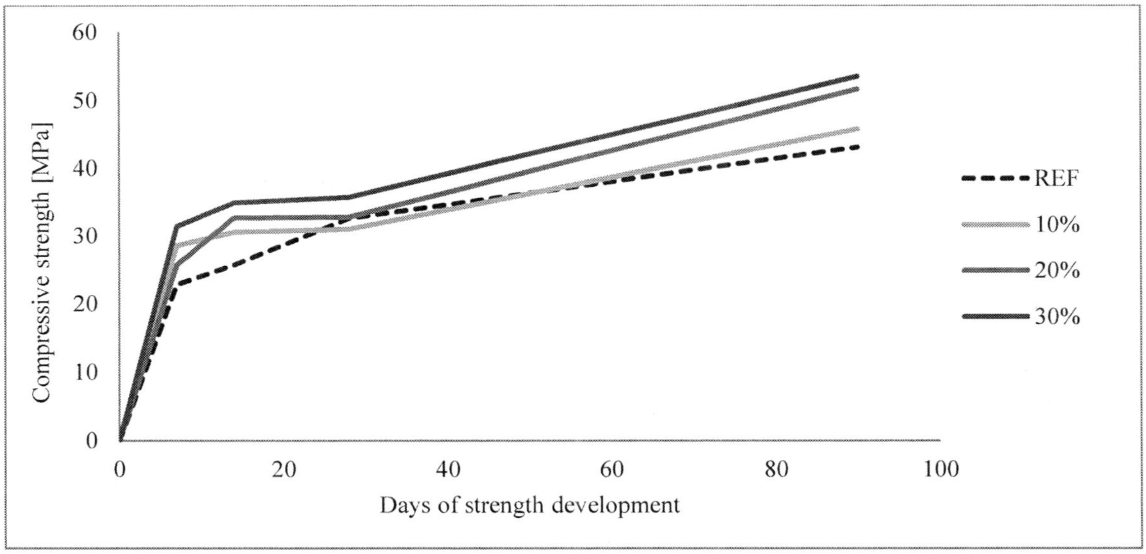

Fig. 3: The development of compressive strength

## Conclusions

In this paper, experimental results of the shrinkage testing of FRA concrete are presented and discussed. From the test results, the following conclusions are drawn:

- The use of FRA as partial replacement of natural fine aggregate influences the resultant properties of concrete.
- Concrete samples with FRA show a mild improvement of compressive strength with comparison of reference concrete.
- By the way of the contrast, the use of FRA in concrete negative influences the shrinkage of concrete.
- This influence depends on the amount of FRA in concrete.
- With the increase of the amount of FRA, the shrinkage of recycled aggregate concrete increases too.

However, FRA concrete has a big potential for the use in the manufacturing of building structures, but it is necessary to test durability and lifespan of FRA concrete.

## Acknowledgement

This work has been supported by SGS15/182/OHK1/3T/11 Verification of durability and lifespan of cement composites and recycled concrete.

## References

[1] C. Meyer, The greening of the concrete industry, in: Cement and Concrete Composites 31, 2009, pp. 601 - 605.

[2] BIOIS. Service contract on management of construction and demolition waste - SR1. Draft final report task 2, Management. Paris; 2010.

[3] A. Katz, Properties of concrete made with recycled aggregate from partially hydrated old concrete,Cement and Concrete Research 35 (2003) 703–711.

[4] ČSN EN 1097 – 6. *Tests for mechanical and physical properties of aggregates – Part 6: Determination of particle density and water absorption.* Prague: Czech Office for Standards, Metrology and Testing, 2013. (in Czech)

[5] ČSN EN 206. *Specification, performance, production and conformity.* Prague: Czech Standard Institut, 2014. (in Czech)

[6] ČSN 73 1320. *Determination of volume changes of concrete.* Prague: Czech Standard Institut 1987. (in Czech)

[7] ČSN EN 12390 - 3 (731302). *Testing hardened concrete – Part 3: Compressive strength of test specimens.* Prague: Czech standard institute, 2009. (in Czech)

# Fatigue Behaviour of Bolted Joints without Load Transfer

Martin Kadlec[1, a], Ivan Mlch[1,b] and Roman Růžek[1,c]

[1]VZLU – Aerospace Research and Test Establishment, Beranových 130, Prague, Czech Republic

[a]kadlec@vzlu.cz, [b]mlch@vzlu.cz, [c]ruzek@vzlu.cz

**Keywords:** Aluminum alloy, fatigue testing, joint, wing, aeronautics

**Abstract.** The fasteners affect the fatigue performance which significantly influences lifetime of aeronautical structures. The aim of the experiments was to determine and compare the fatigue characteristics of typical joints with various fastener types applied in L 410 NG airframe structure. The bone shape specimens with hole filled by fastener were used. Fatigue behaviour of two materials 7475-T7351 and of 2124-T851 and two types of Hi-Lite bolts and LeN 3366.5 rivets were investigated. Constant force amplitude loading with stress ratio of 0.05 and frequency of 5 Hz were applied. Four types of failure modes were described for each specimen according to the crack initiation site. Due to various invalid failure modes, no difference among the configurations was determined as significant.

## Introduction

Bolts and rivets are typically used for joining of metallic parts used in aircraft wing structures. The fasteners as well as joint design, production quality and stress level affect the fatigue performance which is fundamental for aeronautical structures [1, 2]. The fatigue properties of joints can be critical for several reasons: stress concertation, fretting corrosion, eccentricity producing secondary bending [3]. Fatigue behaviour of an aircraft is typically determined on specimens with stress concentration that is representative for the real structure. The typical stress concentration for a riveted small commuter aircraft is defined by value of 2.6. Moreover additional technological factors such as riveting, hole expansion, or tightening torque which can significantly affect the fatigue life have to be evaluated. Therefore, specimens with a hole filled by the actual fastener are used. The aim of this paper was to evaluate fatigue behavior of the joints without load transfer. The results are to be utilize in the framework of life enhancement of principal critical areas in a wing structure.

## Materials and Methods

The experiments were performed on joints consisting of a bone shape main part and two blocks presenting ends of a joined part (Fig. 1a). Test matrix in Table 1 shows two types of materials that were used: 76.2 mm thick plate of 7475-T7351 and 100 mm thick plate of 2124-T851. Two types of fasteners were used: Hi-Lite bolts and rivets LeN 3366.5. For the Hi-Lite bolts, hermetic was used [4, 5]; for the LeN rivets, no hermetic was used. The joints were loaded by constant amplitude with $R = 0.05$ and frequency of 5 Hz at room temperature. Various plate thicknesses from 2.2 mm up to 11.3 mm and different rivet diameter combinations were analysed based on the structure of L 410 NG airplane. The fatigue tests were performed on INOVA ZUZ 100 kN and Hydropuls Schenck 250 kN.

The photo documentation of the fracture surface was obtained by Canon EOS 500D using side halogen lighting.

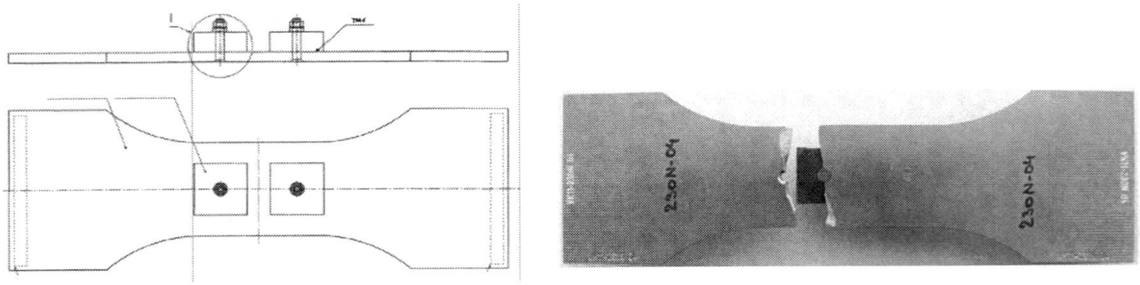


a) Schematic drawing of unloaded joints. b) Valid failure from rivet hole

Fig. 1: Unloaded joint schematic and a failure mode

## Results

**Regression model evaluation.** First, a linear regression model and then a nonlinear model in compliance with MMPDS-11 [6] were applied for series with at least four specimens with valid failure mode.

A nonlinear regression model is represented by the expression:

$$\log(N) = A_1 + A_2 \cdot \log((\sigma_{br})_{MAX} - A_4), \quad (1)$$

where $A_1$, $A_2$ and $A_4$ are the regression constants and $(\sigma_{br})_{MAX}$ is the maximal stress level of the load cycle applied to the specimen gross cross section. The linear model was used as Eq. 1 with $A_4 = 0$. Fig. 1b shows the obtained two valid Wöhler curves. The other curves were not evaluated due to invalid failure modes.

The fatigue data are also marked according to failure type that is described in the following paragraph. No point crossing means valid crack initiation from the hole. Fig. 2 shows summary of the fatigue lives including linear and non-linear regressions. The linear regression was split into two parts because of different slope for load below 120 MPa.

Table 1: Specimen parameters.

| Joint series ID | Qty. | Main part | | Opposite part | | Fastener spec. | |
|---|---|---|---|---|---|---|---|
| | | $t_1$ (mm) | Material | $t_2$ (mm) | Material | ID | dia. (mm) |
| **224N** | 12 | 7.5 | 7475-T7351 | 10 | 2124-T851 | HST10AG8 | 6.3 |
| **227N** | 12 | 10 | 2124-T851 | 7.5 | 7475-T7351 | HST10AG8 | 6.3 |
| **230N** | 12 | 5.7 | 7475-T7351 | 11.3 | 2124-T851 | HST11AG6 | 4.8 |
| **233N** | 12 | 11.3 | 2124-T851 | 5.7 | 7475-T7351 | HST11AG6 | 4.8 |
| **236N** | 12 | 7.2 | 2124-T852 | 3.5 | 7475-T7352 | 5Du Zk 5x | 5 |
| **239N** | 12 | 3.5 | 7475-T7354 | 2.2 | 2124-T854 | 5Du Zk 4x | 4 |
| **242N** | 12 | 2.2 | 2124-T854 | 3.5 | 7475-T7354 | 5Du Zk 4x | 4 |

**Failure modes.** Failure modes were described for according to the crack initiation site as follows:

1) Fracture initiated from the fastener hole which was considered as the only valid failure. The top surface view is included in Fig. 1b. The fracture surface in Fig. 3a shows radial crack growth from the fastener countersunk;

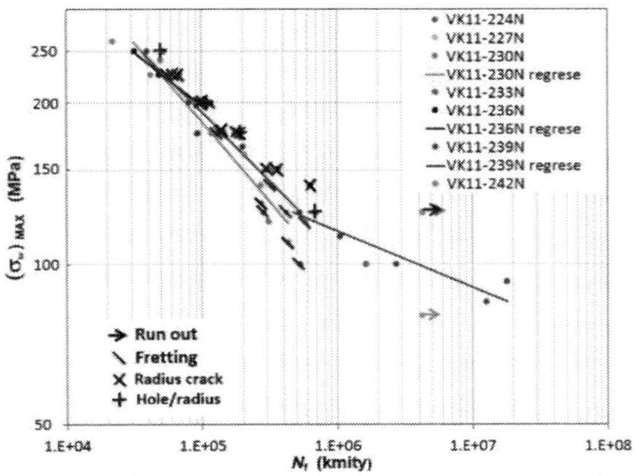


a) Linear regression model.

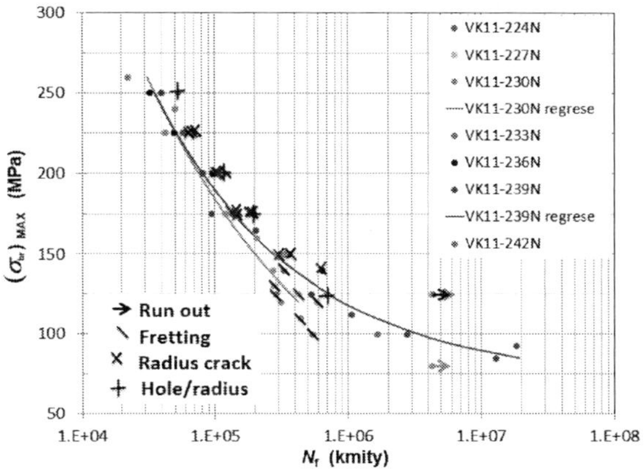


b) Non-linear regression model.

Fig. 2: Results of fatigue crack growth of unloaded joints.

2) Fretting corrosion between main and opposite part was observed only for the Hi-Lite fasteners (Fig. 3b). The fretting initiated not only from the opposite block corner but also from its edges.
3) Specimen radius crack initiation, more preferentially for higher load levels over 150 MPa (Fig. 3c).
4) Longitudinal cracking. It was observed for specimens with the radius initiation and it appeared only for main parts extracted in LS direction.

Fig. 3: Typical failure observed during testing: a) hole initiation, b) fretting corrosion, c) radius initiation.

## Discussion

There is a possibility that some of radius failed specimens had already hole initiated crack. Detailed non-destructive inspection of holes (after joint disassembly) was not performed. This idea is supported by the fact that the radius initiated fatigue lives were not significantly different from the valid failure modes. However, a combination of radius initiation and fretting corrosion was confirmed on several specimens. Testing of series with high ratio of the invalid failure modes was interrupted. The specimens with other than hole crack initiation were not included to the evaluation.

At least 10 specimens are needed for a valid Wöhler curve according to standard [2] that was met only for series 239N. Therefore, only the 239N curve parameters can be used for modelling and no comparison of curves is possible. It is recommended to complete number of the regular results by means of additional experiment. For additional test design and realization following recommendation are suggested:

1) Design modified specimens with a large or double radius (larger radius near the gage section) to prevent radius crack initiation.

2) Use lower cycling frequency for Hi-Lite bolt specimens to prevent fretting which can be done on the remaining specimens. Some additional research activities are needed.

## Conclusion

Several Wohler's curves for various configurations was determined for various combination of thickness and fastener type. Due to various invalid failure modes, no difference among the configurations was determined as significant. To clarify the fatigue data trends, two regression models were used (linear and non-linear equation). There were several types of invalid failure modes. Fretting corrosion between main and opposite part occurred for Hi-Lite fasteners. The fretting corrosion initiated not only from the opposite block corner but also from its edges. Especially for load levels, fatigue crack initiated from the radius surface. Additionally, longitudinal cracking was observed for specimens with the radius initiation and it appeared only for main parts extracted in LS direction.

The two valid Wöhler curves will be used for modelling. For additional test design and realization, modified specimens with a large or double radius to prevent radius crack initiation are recommended. Using lower cycling frequency for Hi-Lite bolt specimens should prevent fretting corrosion.

## Acknowledgement

The project TE02000032 / Advanced Aerostructures Research Centre is solved with the financial support of the Technology Agency of the Czech Republic (TA ČR).

## References

[1] A. Skorupa, M. Skorupa, T. Machniewicz, A. Korbel, Fatigue crack location and fatigue life for riveted lap joints in aircraft fuselage, Int. J. Fatigue. 58 (2014) 209–217.

[2] M. Skorupa, T. Machniewicz, A. Skorupa, A. Korbel, Fatigue life predictions for riveted lap joints, Int. J. Fatigue. 94 (2017) 41–57. doi:10.1016/j.ijfa

[3] J. Schijve, Fatigue of structures and materials in the 20th century and the state of the art, Int. J. Fatigue. 25 (2003) 679–702. doi:10.1016/S0142-1123(03)00051-3.

[4] L. Boni, A. Lanciotti, Fatigue behaviour of double lap riveted joints assembled with and without interfay sealant, Fatigue Fract. Eng. Mater. Struct. 34 (2011) 60-71.

[5] M. Kadlec, P. Kucharský, Fastener stiffness measurement methodology, in: F. Plánička, J. Krystek (Eds.), EAN 2016 - 54th International Conference on Experimental Stress Analysis, Srní (CZ), 30 May – 2 June 2016.

[6] MMPDS-11, Metallic material properties development and standardization (MMPDS), Federal Aviation Administration, United States, 2016.

EAN 2017
55th conference on experimental stress analysis

# Effect of Poly(vinyl-alcohol) Amount on Flexural Strength of Cement Mortars with Micronized Recycled Concrete

TESÁREK Pavel[1,a], PROŠEK Zdeněk[1,2,b] and FLÁDR Josef[1,c]

[1] Faculty of Civil Engineering, Czech Technical University in Prague, Thákurova 7, 166 29 Prague, Czech Republic

[2]CTU in Prague, UCEEB, Trinecka 1024, 273 43 Bustehrad, Czech Republic

[a]tesarek@fsv.cvut.cz, [b]zdenek.prosek@fsv.cvut.cz, [c]josef.fladr@fsv.cvut.cz

**Keywords:** Cement Composites, Poly(vinyl-alcohol), Micronized Recycled Concrete, Flexural Strength, Curing Conditions

**Abstract.** Main goal of the presented study was to investigate the influence of different curing conditions and different amount of PVA (Poly(vinyl-alcohol)) on flexural strength of cement-PVA composites. Resulting cement-PVA composites was composed of Portland cement CEM I 42.5 R, micronized recycled concrete and PVA with different percentage amounts (0 wt. %, 1.4 wt. %, 2.8 wt. %, 4.0 wt. % and 5.6 wt. %). Two sets of samples were made from each mixture. The first set of samples was stored in water after their removal from molds, while the second set was stored exposed to air at a constant temperature and relative humidity. The flexural strength was measured on samples of size 40 × 40 × 160 mm using a three-point bending test. The results obtained from these materials were compared with reference material.

## Introduction

Polymers in cement composites have been used since the 20th century. Since that time many types and forms of the polymers been used for cement modification. Best results in cement composites been achieved by adding latex and polymer resins. Because of the massive consumption of cement composites in the construction industry, growing demands on the mechanical properties and the high cost of latex and resins the other alternative polymers as water soluble polymers are interested [1].

Poly(vinyl-alcohol) (PVA) is a water soluble polymer, which is used for modification of cement composites for construction concretes [2, 3]. Material properties of composites based on cement and PVA achieved better characteristics, e.g. hygric properties, durability, shrinkage or tensile, respectively flexural, strength [4]. On the other hand values of mechanical properties as compressive strength and Young's Modulus are reduced, this phenomena related with higher values of bulk density because supplement of PVA create higher total porosity [5]. Knapen et al. [6, 7] conducted a study in which he tested the influence of water on cement composite with PVAA (polyvinyl alcohol-acetate), some mechanical properties and changes in hydration process was. He observe changes towards to retardation of hydration process and decrease content of $Ca(OH)_2$. Morlat et al. [8] conducted a study in which he used mixture of cement and PVA and examined its compression strength, flexural strength and Young's modulus of elasticity of cement paste. The flexural strength was increased and total fracture energy was increased by a factor of three by adding PVA less than 4 wt. % of cement. The purpose of this paper is to investigate the influence of the recycled concrete powder and conditions of curing after 28 days of hardening on the mechanical properties of PVA-cement composites.

## Materials and Samples

Portland cement CEM I 42.5 R Mokrá (65 wt. %), micronized recycled concrete (from concrete gutters with a fraction of 0-0.250 mm (35 wt. %), water and PVA (16 wt. % water solution) were used for preparing of the specimens with different values of PVA. PVA was made by alkaline hydrolysis of polyvinyl acetate in methanol in form of water solution SLOVIOL® R produced by Fortischem, a.s. The recycled concrete was created from the gutters. A high speed mill from LAVARIS Ltd. was used for micronizing of the recycled concrete. Individual material specimen differed in the values of solid PVA (Tab. 1). The six samples with dimensions of 40 × 40 × 160 mm were prepared for each material. One-half of the specimen were placed into a water bath immediately after their removal from molds (W) (24 hours after they were cast) while the other was kept at laboratory conditions (D). Standard laboratory conditions were in a laboratory environment at 22 ± 1 ° C and relative humidity 50 ± 2 %. The composition of individual mixtures can be seen in Table 1. Water ratio was constant, equal to 0.35.

Tab. 1: Composition of the tested specimens.

| Samples | Cement [g] | Recycled concrete powder [g] | PVA [g] | Water ratio = 0.35 | |
|---|---|---|---|---|---|
| | | | | Water from solution PVA [g] | Water [g] |
| D/W 0.0 % | 2275 | 1225 | 0 | 0 | 1225 |
| D/W 1.4 % | 2275 | 1225 | 49 | 257 | 968 |
| D/W 2.8 % | 2275 | 1225 | 98 | 514 | 711 |
| D/W 4.0 % | 2275 | 1225 | 140 | 753 | 472 |
| D/W 5.6 % | 2275 | 1225 | 196 | 1026 | 199 |

## Methods and Experimental Results

A comparison of basic material properties as workability of fresh mixtures, bulk density, total porosity and shrinkage after 24 hours shows table 2, the tested properties depend on quantity of used PVA. The workability of fresh mixtures defined by the spillage after 20 impulses. The destructive testing was performed using standard procedures. The flexural strength were determined on the 28 days old samples using the Heckert device model FP100. The testing was controlled by a constant rate of 0.1 mm/s, a configuration of a three-point bending was in our case used. The distance between supports for three-point bending test was equal to 100 mm. The results of basic material properties show influence of PVA. Adding PVA up to 4.0 wt. % of mixture worsens the workability of the fresh mixtures by reducing the spillage. As a result, there is an increase number of technological pores and thus the total porosity, which results in a reduction of bulk density (Tab. 2).

Tab. 2: Comparison of the material properties (with standard deviations).

| Samples | Spillage [mm] | Bulk density [kg/m$^3$] | Total porosity [-] | Shrinkage (after 24 hours) |
|---|---|---|---|---|
| D/W 0.0 % | 240 ± 5 | 1820 ± 40 | 0.24 ± 0.02 | 0.18 ± 0.01 |
| D/W 1.4 % | 230 ± 4 | 1620 ± 35 | 0.29 ± 0.01 | 0.13 ± 0.02 |
| D/W 2.8 % | 190 ± 5 | 1590 ± 38 | 0.31± 0.02 | 0.12 ± 0.01 |

| | | | | |
|---|---|---|---|---|
| D/W 4.0 % | 179 ± 4 | 1540 ± 31 | 0.33 ± 0.01 | 0.09 ± 0.01 |
| D/W 5.6 % | 136 ± 3 | 1663 ± 50 | 0.27± 0.01 | 0.24 ± 0.01 |

Fig. 1 presented results of flexural strengths. The effect of PVA quantity is clearly visible for the dry samples (D), a value of the dry samples with 5.6 wt. % of PVA is two time higher that ones for the reference samples without PVA. Decrease of flexural strength corresponded with lower quantity of PVA for the other values. The different results were obtained for the wet samples, where the values of the reference samples and the samples with 5.6 wt. % are similar.

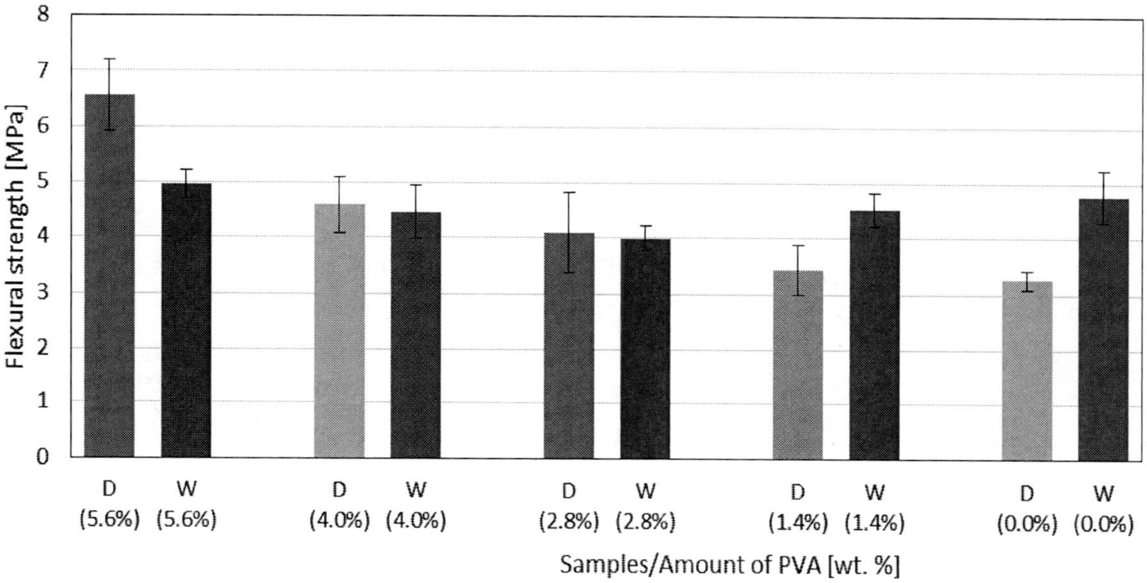


Fig. 1: Comparison of flexural strength (with standard deviations).

## Conclusions

This work is focused on influence of PVA on flexural strength of cement composites. The resulting cement composites was composed of Portland cement, recycled concrete powder and PVA with different amount (namely 0, 1.4, 2.8, 4.0 and 5.6 wt. % of cement and recycled concrete). Based on the results it can be concluded that:

- PVA solution resulted in a decrease of the workability of fresh mixture because of its high viscosity,
- increasing the amount of PVA up to 4.0 wt. % resulted in increasing porosity. Due to the increase of porosity was reduced the bulk density,
- the value of flexural strength of the cement composites increases with increasing the volume of the PVA despite the higher porosity in the case of dry samples (D),
- the condition of curing under water level had bad influence on the value of flexural strength of cement composites with PVA,
- the best average value of flexural strength had samples with 5.6 wt. % PVA at standard laboratory conditions, namely 6.6 ± 1.1 MPa.

In the future, we will focus the possibility of involvement of PVA and recycled concrete powder in the hydration process, to follow the studies of e.g. [5]; and PVA will be tested for modification of historic mortars [9] or gypsum composites [10] in solid form or as an alternative as nanotextile [11, 12].

**Acknowledgement**

This was financially supported by Czech Technical University in Prague – project SGS16/201/OHK1/3T/11, by the Grant Agency of the Czech Republic research project 17-06771S and by the Ministry of Education, Youth and Sports within National Sustainability Programme I, project No. LO1605.

The authors also thank the Center for Nanotechnology in Civil Engineering at the Faculty of Civil Engineering CTU in Prague.

**References**

[1] Y. Ohama, Handbook of polymer-modified Concrete and Mortars, Noyes Publications, New Jersey, 1995, pp. 45-182.

[2] M. Lidmila et al., Mechanical Properties of Recycled Binder/Micro-Filler Cement-Based Material, Advanced Materials Research 2014, 1054 (2014) 234-237.

[3] M. Lidmila et al., Utilization of Recycled Fine-Ground Concrete from Railway Sleepers for Production of Cement-Based Binder, Appl. Mechanics and Materials 486 (2014) 323-326.

[4] M. Lidmila et al., Composite Material Based on Cement and PVA: Evolution of Mechanical Properties during First 28 Days, Advan. Mat. Res. 1054 (2014) 215-220.

[5] J. Topič et al., Effect of PVA modification on the properties of cement composites, Acta Polytechnica 55 (2015) 64-75.

[6] E. Knapen and D. V. Gemert, Effect of under water storage on bridge formation by water-soluble polymers in cement mortars, Construction and Buildings Materials 23 (2009) 3420-3425.

[7] E. Knapen and D. V. Gemert, Cement hydration and microstructure formation in the presence of water-soluble polymers, Cement and Concrete Research 39 (2009) 6-13.

[8] R. Morlat, G. Orange, Y. Bomal, P. Godard, Reinforcement of hydrated Portland cement with high molecular mass water-soluble polymers. Journal of Materials Science 42 (2007) 4858-4869.

[9] V. Nezerka et al., V. Investigation of crushed brick-matrix interface in lime-based ancient mortar by microscopy and nanoindentation, Cement & Concrete Composites 55 (2015) 122-128.

[10] P. Tesárek et al., Microstructural and micromechanical study of gypsum, Chemicke listy 105(S) (2011) S852-S853.

[11] P. Tesárek, et al., Macro Mechanical Testing of Nanofibers: Tensile Strength, in: M. Ruzicka et al. (Eds.), Proceedings of the 50th annual conference on Experimental stress analysis, 2012, CTU Prague, pp. 465-468.

[12] P. Tesárek, et al., Mechanical Properties of Single and Double-Layered PVA Nanofibers, Key Engineering Materials 586 (2014) 261-264.

EAN 2017
55th conference on experimental stress analysis

# Experimental verification of thickness effects on fatigue crack growth in an AA 7475 plate

RŮŽEK Roman[1, a], HOMOLA Petr[1,b] and KAKOS Martin[1,c]

[1]Aerospace Research and test Establishment (VZLÚ),
Strength of Structure Department,
Beranových 130, 199 05 Praha – Letňany, Czech Republic

[a]ruzek@vzlu.cz, [b]homola@vzlu.cz, [c]kakos@vzlu.cz

**Keywords:** Crack growth, thickness, constant amplitude, variable amplitude, fatigue.

**Abstract.** Fatigue crack propagation behaviours under constant and variable amplitude loading while taking into account thickness effects is investigated. A plate aluminium alloy 7475 T351 with a thickness of 76 mm is analysed. Thicknesses of 2 to 8 mm are examined. The significant differences between fatigue crack behaviours under constant and variable amplitude loading are defined. While different thicknesses have no significant effects under constant amplitude loading, significant differences are identifiable in cases of variable amplitude loading. A thickness of 2 mm results in a 2.6-fold longer lifetime compared to the lifetime corresponding to a thickness of 8 mm.

## Introduction

Fatigue crack propagation in metals has generally been quantified in terms of stress intensity factors. The magnitude of higher order terms in the near-tip asymptotic elastic stress field is dependent on geometric features. It has been found experimentally that a fatigue crack can remain closed for part of the fatigue cycle and in cases of tensile-tensile loading. The portion of the fatigue cycle for which the crack is open provides the driving force for crack growth. Closure ratios have been defined for various alloys in consideration of constant amplitude loading. Based on this knowledge, different analytic and numerical models have been derived. Prediction models are based on crack closure phenomena and stress-strain condition development in the area surrounding a crack tip. The most well-known models used in aerospace include the following: Willenborg, Wheller, ONERA, CORPUS, AFGROW and NASGRO ([1]-[5]). All of these models consider material characteristics evaluated under constant amplitude loading. The parameters of prediction models are usually defined through experiments conducted under variable amplitude loading.

The new concept of an airframe structure has been adopted throughout the development of new versions of the L-410 commuter aircraft. Integral panels are used in the airframe structure. These thin curvature panels are composed of thick plates that are milled to thin parts with integrated stiffeners referred to as stringers [6]. Unfortunately, some significant discrepancies between actual and predicted fatigue crack propagation patterns under variable amplitude loading conditions have been identified [7]. Several works have focused on the enhancement and verification of prediction tools [6 - 10]. All of these studies confirm problematic propensities for current models to describe thickness changes via standard procedures.

This paper describes an experimental investigation of fatigue crack propagation behaviours under constant and variable amplitude loading conditions while taking into account thickness effects on 7475 T351 aluminium alloys. The aim of this work was to address the following questions:

- Do thick specimens composed of thick plates affect the material characteristics defined under constant amplitude loading conditions?
- Do specimens composed of thick plates affect fatigue crack propagation behaviours under variable amplitude loading conditions?

## Material and specimen configuration

An ALCOA 7475 T7351 plate material was experimentally studied as a typical structure material used in L410NG airframes. Plate semi products with a nominal thickness of 76 mm were used. During airframe manufacturing, 76 mm-thick plate materials were machined into final products of different thicknesses (of typically 2 to 8 mm). This paper thus only considers data for this typical thickness range. Two extreme thicknesses of 8 and 2 mm were examined under constant amplitude loading to answer the first question listed above. Three nominal thicknesses of 2, 4 and 8 mm were examined under variable amplitude loading conditions.

The M(T) specimens shown in Fig. 1 were used for the crack propagation investigation. A specimen width of 100 mm was applied.

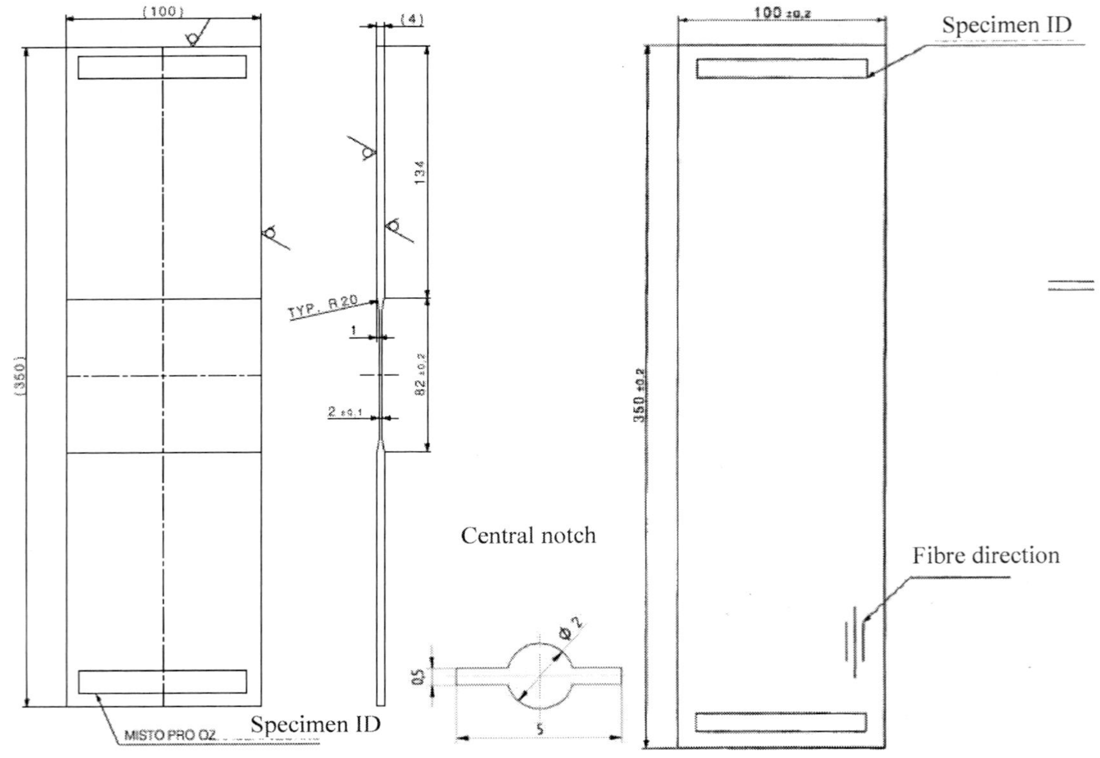


Fig. 1 M(T) specimen geometry (left - thickness of 2 mm, right - thicknesses of 4 and 8 mm)

## Test and loading conditions description

Fatigue crack growth tests were conducted using a hydraulic SCHENCK load frame with a capacity of 250 kN. The SCHENCK load frame was controlled by the INSTRON FastTrack 8800 test control system. It provides real-time closed loop control, including transducer conditioning and function generation. Specimens were clamped into the hydraulic test frame with mechanical grips. The test set up, i.e., the M(T) specimen, load frame, mechanical grips and microscopes, is shown in Fig. 2.

Fig. 2M(T) fatigue crack growth test configuration

The test procedure was conducted in agreement with ASTM E-647 standard requirements [20]. Fatigue crack initiation and propagation were monitored via a visual method (VT) using an Olympus SZ40 light stereomicroscope with a maximum magnification of 40x. Crack length measurements were carried out on both surfaces and sides relative to the longitudinal axes of the specimens.

A harmonic loading (sinusoidal load cycle) with a constant force amplitude was used in the crack initiation phase. The effects of thickness were studied for the constant and variable (randomized flight-by-flight sequence) amplitude loading. A harmonic loading (sinusoidal load cycle) with a constant force amplitude and stress ratio of 0.05 was used for basic material characteristic definition. A typical randomized flight-by-flight sequence with a variable amplitude representative of the wing of a small SQ_0060 commuter aircraft was used. The sequence consists of 3,000 flights and represents stress spectra of an L 410 NG airplane bottom wing surface panel skin in section 21. The SQ_0060 loading sequence contains 174.975 cycles. One flight corresponds to 1 flight hour.

## Experimental results and discussion

Basic fatigue crack growth material characteristics of the plate material were defined using 8 mm thick specimens. These material characteristics have been used as a basis for crack growth predictions generally. Figs. 3 and 4 compare data obtained under the constant amplitude loading (specimen thickness of 8 vs. 2 mm). Fig. 3 compares the crack rate vs. effective stress intensity factor data for the 8 mm thick specimens (data includes three different stress ratios (R = 0.02, 0.2 and 0.6)) with those of the 2 mm thick specimens (stress ratio of 0.02). The effective stress intensity factor was calculated using the Schijve equation [5]. The crack growth data measured for the 2 mm specimens overlap with data obtained on the 8 mm specimens. The 2 mm thickness data shown in the upper scatter data range of the 8 mm thickness data in terms of crack growth rates range between $10^{-8}$ and $10^{-7}$ mm.cycle$^{-1}$. This implies that the crack propagation data corresponding to a thickness of 2 mm are faster than those of the 8 mm data under the same loading conditions. This finding is illustrated in Fig. 4. Fig. 4 also shows scatter data for two 8 mm specimens loaded under the same loading conditions. Relatively high scatter levels in crack growth data under constant amplitude loading conditions are shown (±15%). It is recommended that data scatter results are confirmed and analysed using a larger number of specimens.

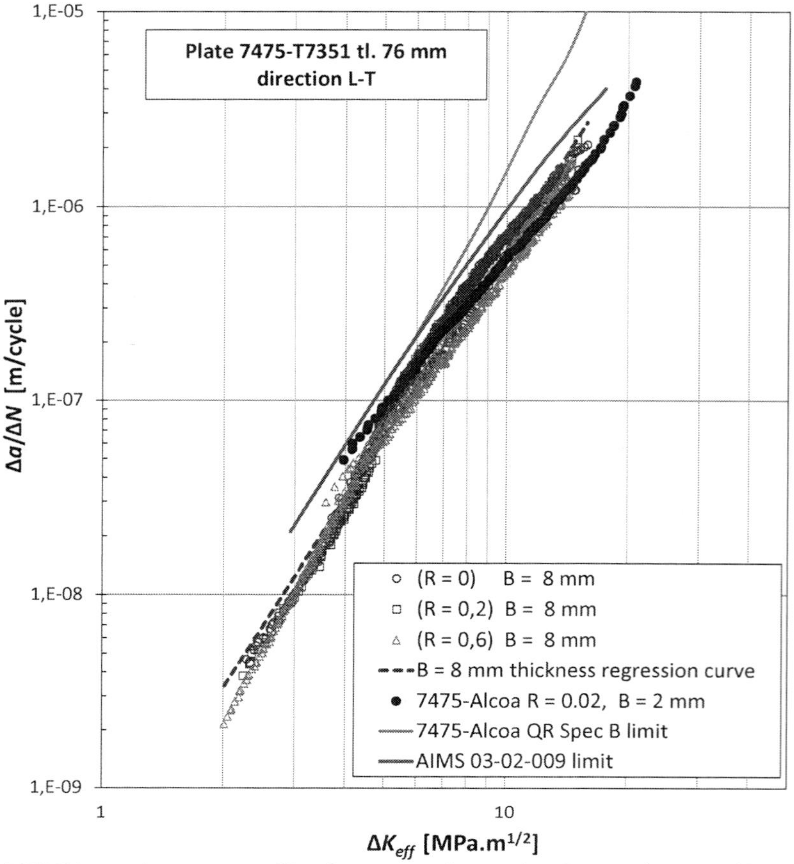


Fig. 3 7475 T7351 crack rate vs. effective stress intensity factor data – effect of thickness

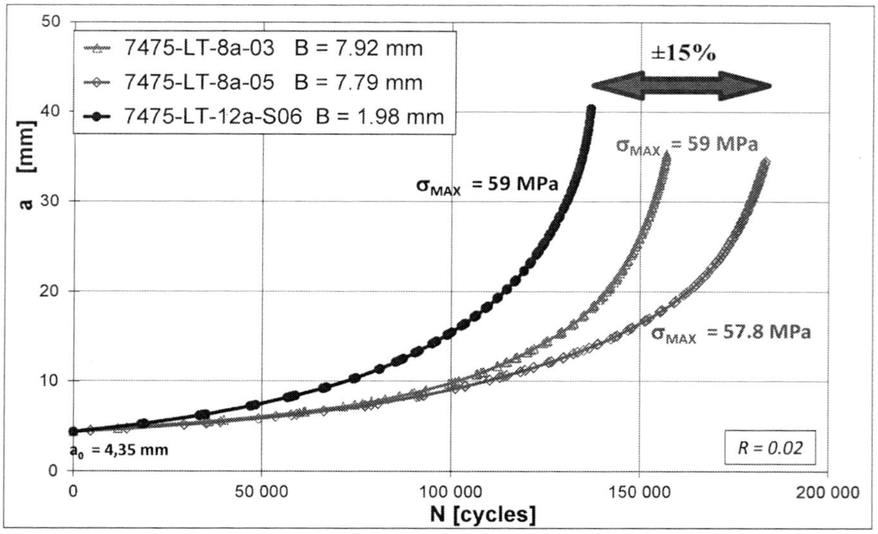


Fig. 4 7475 T7351 crack propagation data under constant amplitude loading – effect of thickness

Figs. 5 presents fatigue crack propagation data measured for the three different thicknesses (2, 4 and 8 mm) under variable amplitude loading. All of the specimens were loaded using the same stress loading parameters – the maximum stress value in the sequence of 89 MPa. Thickness effects on crack propagation behaviours are evident. A thickness of 8 mm gives the shortest crack propagation lifetime while a thickness of 2 mm gives the longest crack propagation lifetime. This difference is significant. A thickness of 2 mm results in a 2.6-fold longer lifetime than that corresponding to a thickness of 8 mm. This difference cannot be explained by the presence of potential differences in basic material characterisation data alone. Moreover, the material data under constant amplitude loading show the opposite relationship. Significant changes in crack propagation mechanisms associated with plane stress states, plane strain states and transition modes between both states must be taken into account in prediction models. This may explain why no present prediction model is able to predict this feature. Further data on shear lip development patterns are therefore necessary for this feature's description. Additional experimental and numerical research must therefore be conducted to fully understand the effects of thickness levels on fatigue crack propagation behaviour.

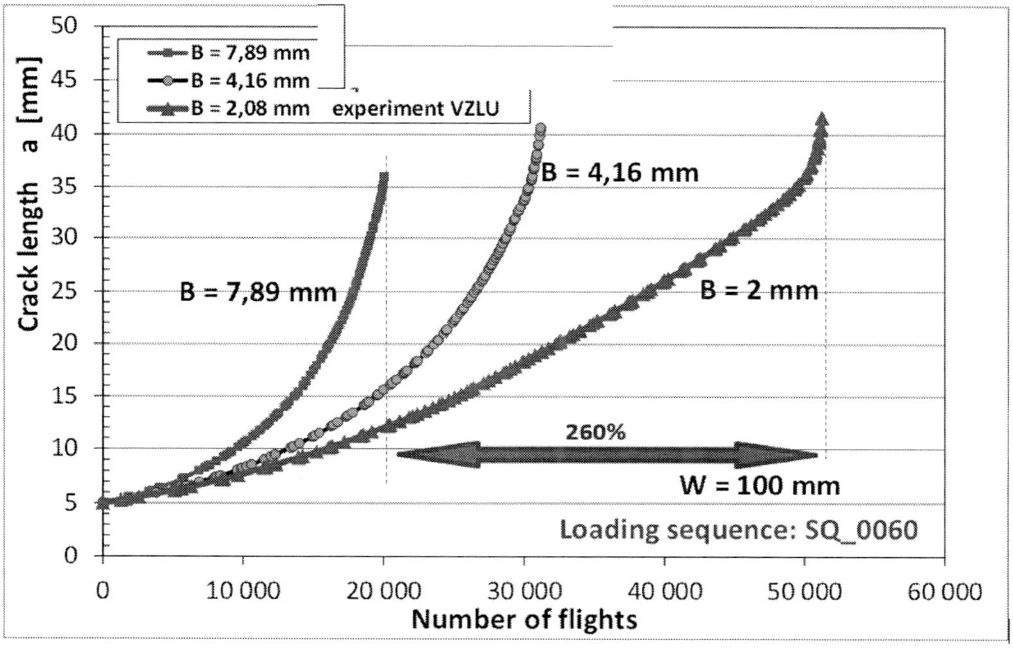


Fig. 5 7475 T7351crack propagation data comparisons under variable amplitude loading conditions $\sigma_{max}$ = 89 MPa – effect of thickness

## Conclusions

This paper focuses on an experimental investigation of fatigue crack propagation behaviours under constant and variable amplitude loading conditions that considers the effects of thickness on ALCOA 7475 T351 aluminium alloy.

Thickness effects were studied under constant and variable amplitude loading conditions. We studied 7475 T7351 plate materials with a nominal thickness of 76 mm. Three final machined thicknesses of 2 to 8 mm were examined. A significant difference between fatigue crack behaviours under constant and variable amplitude loading conditions was found. While different thicknesses had no significant effect under conditions of constant amplitude loading, significant differences were found in cases of variable amplitude loading. This implies that the existing prediction models are not able to predict crack propagation patterns solely based on material characteristics defined under constant amplitude loading conditions without defining additional model parameters.


**Acknowledgements**

The research leading to these results was performed with the support of funding from the Technology Agency of the Czech Republic under "The advanced aerostructure research centrum" project no. TA0200003.


**References**


[1] J.G. Willenborg, R.M. Engle, H.A. Wood: A crack growth retardation model using an effective stress concept. AFFDL-TM-71-1 FBR, Air Force Flight Dynamics Lab. (1971).

[2] D. Aliaga, A. Davy, H. Schaff: A simple crack closure model for predicting fatigue crack growth under flight-simulation loading. ASTM STP-982, American Society for testing and material, (1988) pp.491-504.

[3] A.U. DeKoning: A simple crack closure model for prediction of fatigue growth rates under variable-amplitude loading. ASTM STP-743 (1981) pp.63-85.

[4] U.H. Padmadinata, J. Schijve, Prediction of fatigue crack growth under flight-simulation loading with the modified CORPUS model, Report LR-760, Delft University of Technology, the Netherlands (1994).

[5] R. Růžek: Fatigue crack growth in aerospace structures under complex loading conditions. Doctoral dissertation, Czech Technical University, Faculty of Nuclear Sciences and Physical Engineering, Prague (1993).

[6] J. Šedek, R. Růžek, J. Raška, J. Běhal: Comparative study of prediction methods for fatigue life evaluation of an integral skin-stringer panel under variable amplitude loading. Procedia Engineering, 114 (2015) pp.124 – 131, doi: 10.1016/j.proeng.2015.08.050

[7] J. Běhal: Comparison of crack growth prediction models implemented into the AFGROW software. Technical report R-6592, VZLÚ, Praha (2016).

[8] J. Běhal, P. Homola, R. Růžek.: Effect fracture resistance on retardation of crack growth. International Journal of Structural Integrity, Vol. 6 No. 4 (2015) pp.1-13, doi: 10.1108/IJSI-10-2014-0053

[9] J. Běhal, P. Homola,: Simulation of Operational Loading Spectrum and its Effect on Fatigue Crack Growth, Procedia Engineering 101 ( 2015 ) pp.26-33, doi: 10.1016/j.proeng.2015.02.005

[10]J. Běhal, L. Nováková: Stress state factor evaluation based on fractographic analysis for use in crack growth FASTRAN retardation model of AFGROW computing code, Engineering Failure Analysis, Vol. 35 (2013), pp.645-651, doi:10.1016/j.engfailanal.2013.06.014

EAN 2017
55th conference on experimental stress analysis

# Residual stress analysis after laser welding

Mária Blatnická[1,a], Michal Šajgalík[2,b], Milan Sága[3,c]

[1]Department of Applied Mechanics, Faculty of Mechanical Engineering, University of Žilina, Univerzitná 1, 010 26 Žilina, Slovak Republic

[2]Department of Machining and Manufacturing Technologies, Faculty of Mechanical Engineering, University of Žilina, Univerzitná 1, 010 26 Žilina, Slovak Republic

[3]Department of Applied Mechanics, Faculty of Mechanical Engineering, University of Žilina, Univerzitná 1, 010 26 Žilina, Slovak Republic

[a]maria.blatnicka@fstroj.uniza.sk, [b]michal.sajgalik@fstroj.uniza.sk, [c]milan.saga@fstroj.uniza.sk

**Keywords**: residual stress, laser welding, X-ray method, high strength steel Domex

**Abstract.** The paper deals with experimental measurement of residual stress in high strength steel sample by x-ray method before and after laser welding. All the data was collected and then processed in program Matlab to create surfaces of stress distribution in the sample. Polynomial functions were fit to the data to smooth them and achieve better understanding of results. Finally, results of residual stress in welded sample were compared to those in non-welded sample for determining the effect of welding on residual stress.

## Introduction

Nowadays, the use of high strength steels is being developed, yet their usage is sometimes limited by their fatigue critical welds [1]. One common problem associated with welding is formation of residual stress, which is defined as any stress that would exist in a continuum body if all external loads were removed [2]. Heating, melting and cooling of the weld and nonuniform temperature distribution are causing plastic thermal strain formation, which results in permanent deformation and residual stress near the weld and its heat-affected zone and can be high enough to cause cracking without any applied loads. The welding-induced tensile residual stresses have harmful effect on fatigue lifetime of structures, corrosion resistance and other mechanical properties. On the other hand, compressive residual stress can improve fatigue resistance of material [7]. This issue has been studying for many years by researchers and is still underway. According to fatigue resistance it is of utmost importance to investigate detailed distribution of welding-induced residual stresses [8,9].

## Residual stress determination

Calculation of residual stress requires the introduction of many simplifications due to vast number of variables included. These assumptions enable a mathematical description of processes, but also simplify their physical nature; calculation of the stress is complicated nevertheless. Therefore, the residual stresses, despite huge progress in computer science and good results in the simulation of the process, are still largely determined by application of experimental measurements.

Traditionally, residual stresses can be measured by destructive or non-destructive methods [10]. The former contain for instance cutting and hole-drilling techniques, the latter neutron and x-ray diffraction techniques. X-ray diffraction method obtains strains through change of micro-structural crystal lattice spacing in material; and from this quantity, the total stress can be acquired [3]. When x-ray beam is cast over a solid material, it is sprinkled by the atoms forming the material. The diffracted (sprinkled) waves lead to interferences and the beam can be collected to reach the diffraction intensity peak according to the Bragg's Law (Eq. 1).

$$2D_0 sin\theta_0 = n\boldsymbol{\lambda}\ (n = 1, 2, 3 \ldots) \quad (1)$$

Where, $D_0$ is original distance between lattice planes of adjacent crystal surfaces at the non-stress state,

$\theta_0$ is original diffraction angle between the x-ray beam and crystal surface at the non-stress state,

$\boldsymbol{\lambda}$ is the wavelength of the x-ray beam,

$n$ is order of diffraction.

However, the x-ray beam can penetrate into the material usually not deeper than 40μm, hence this method assumes plane-stress condition i.e. stress perpendicular to the surface is assumed to be zero. For many applications, particularly where fatigue failure is involved, this is not a serious disadvantage [4].

## Welding induced residual stress

A number of factors influence residual stress in a welded structure. Shrinkage of the weld metal, thermal expansion and contraction of the base metal, quenching effects, phase transformation and fixtures of structure being welded are the most important.

The diagram in Fig. 1 shows the influence of increasing heat input on the distribution of transverse residual stresses along a line perpendicular to the weld seam. In the case of low heat input (Curve 3), one can see a reduction of tensile residual stress or even compressive residual stresses. In the case of high heat input (Curve 1), one can expect a buildup of high tensile stresses across the weld seam [6]. On account of this fact, welding the sample by means of laser technology was selected.

Laser welding has become popular recently owing to high power density, which has the possibility of focusing the beam power to a very small spot diameter resulting in small heat affected zone. This is a huge perquisite in contrast to the conventional welding processes, e.g. arc welding as it means also lower distortions, residual stresses and strains [5, 2].

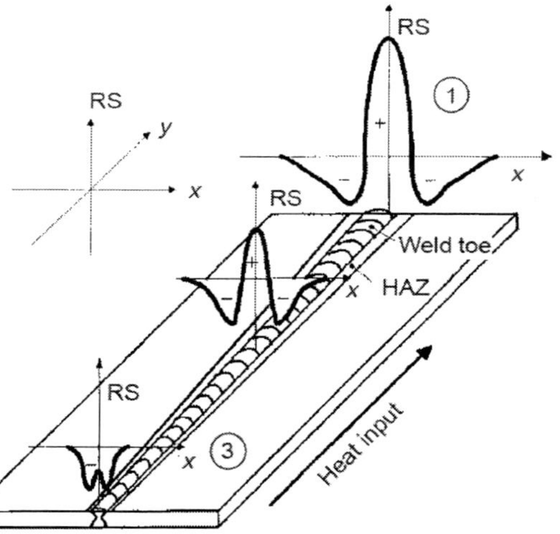


Fig. 1: Influence of heat input on transverse residual stress distribution (RS = residual stress) [6]

## Measurement results

The measurement of residual stress in a sample (Fig.2) from high strength steel Domex 700 MC was carried out. Domex 700 MC is a hot-rolled, low-alloy high strength sheet steel with 700 $N/mm^2$ yield strength. The sheet was welded by laser and then cut into individual samples. The measurement of transverse residual stress was conducted first in base metal sample and subsequently in welded sample on the cut surface (highlighted according to Fig.2) with dimensions (6×6) mm.

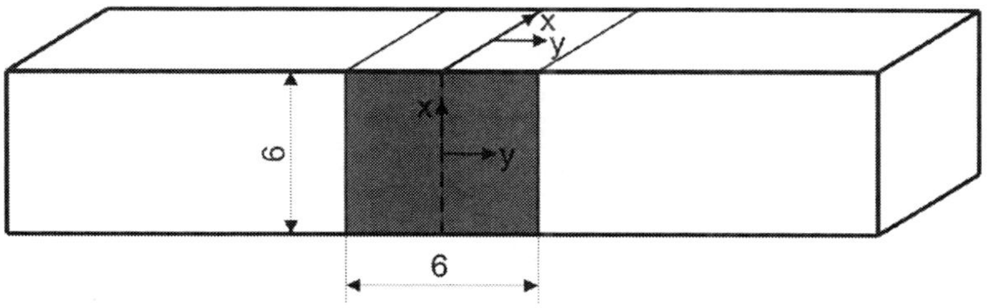


Fig. 2: Schematic representation of a measured sample – highlighted cut surface (transverse)

In Fig. 3, there can be seen high compressive residual stresses (in $N/mm^2$) in "x" direction (see also Fig. 2) on non-welded cut surface caused by rolling of the material. In Fig. 3b, there is distribution of residual stress along a line y = 0. Fig. 4 shows distribution of normal residual stress in the same direction on welded cut surface and a comparison of these two distributions of residual stress on non-welded and welded cut surface can be seen in Fig. 8a. It is apparent that high compressive residual stresses caused by rolling were released by welding; specifically the highest value of residual stress in "x" direction in non-welded sample cut surface was (according to Fig. 3b) ${}_{trans}\sigma_x^{nw} = -612$ MPa and the same in welded sample was (according to Fig. 4b) ${}_{trans}\sigma_x^{w} = -495$ MPa, which means that stress was reduced nearly by 20%. Compressive nature of resulting residual stress after laser welding can be explained by low heat input (see Fig. 1) typical for laser welding.

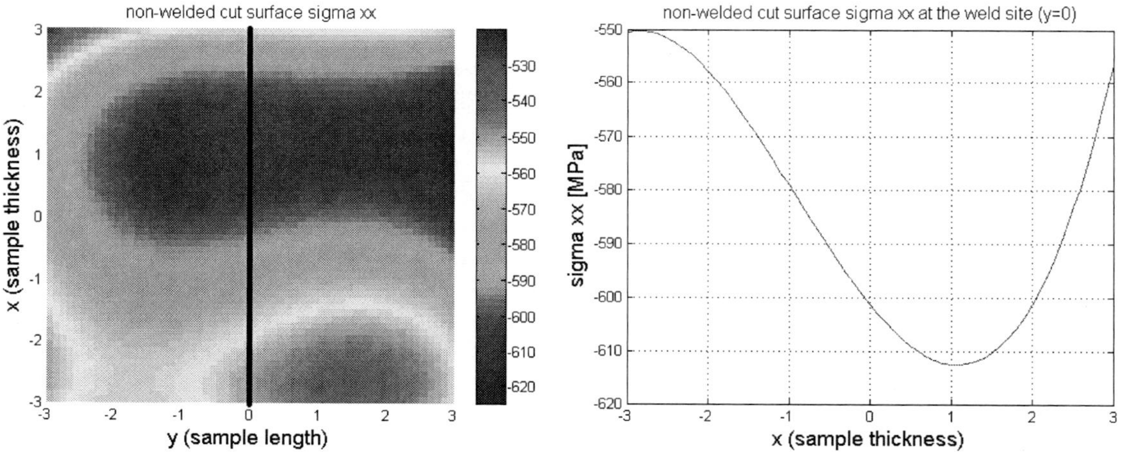


Fig. 3: a) Distribution of normal residual stress in "x" direction on non-welded cut surface and b) distribution of normal residual stress in "x" direction on line y = 0 on non-welded cut surface

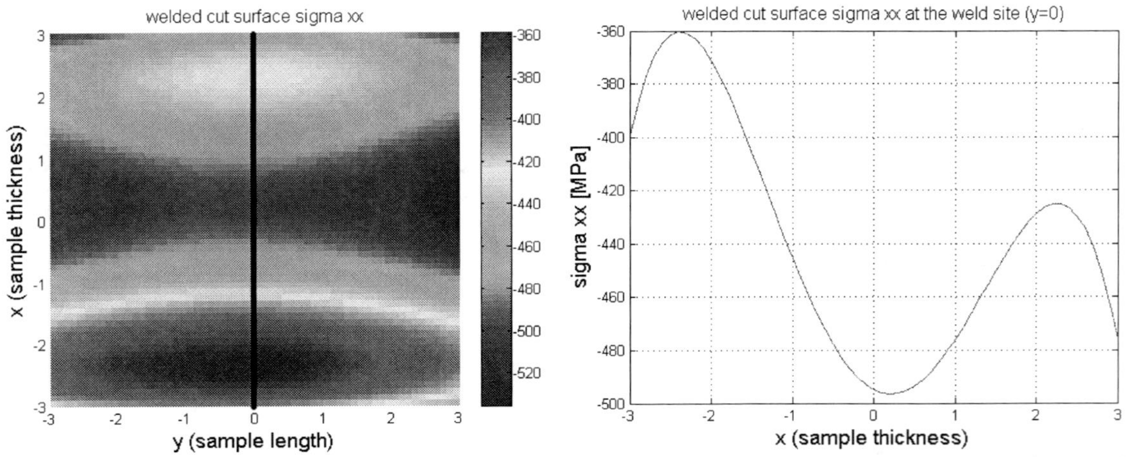


Fig. 4: a) Distribution of normal residual stress in "x" direction on welded cut surface and b) distribution of normal residual stress in "x" direction on line y = 0 on welded cut surface

The next step was measurement of longitudinal residual stress, again first in base metal sample and thereafter in welded sample on the unworked surface (highlighted according to Fig.5) with dimensions (6×6) mm.

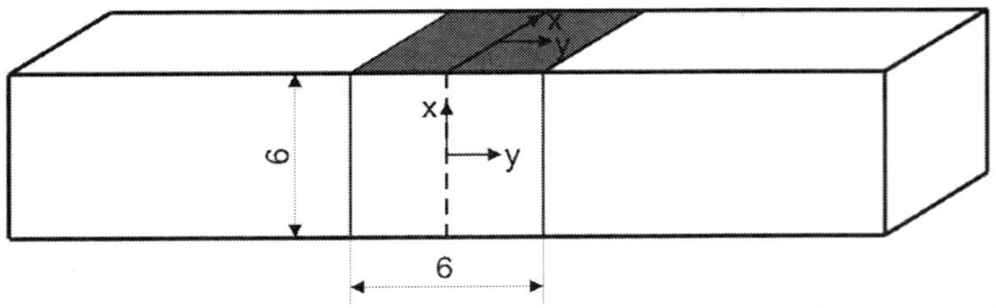


Fig. 5: Schematic representation of a measured sample – highlighted unworked surface (longitudinal)

In Fig. 6, one can see approximately zero normal residual stress in "x" direction on non-welded unworked surface near the middle of sample thickness and low values of compressive residual stress near the edges, which can be caused by cutting the sheet into samples.

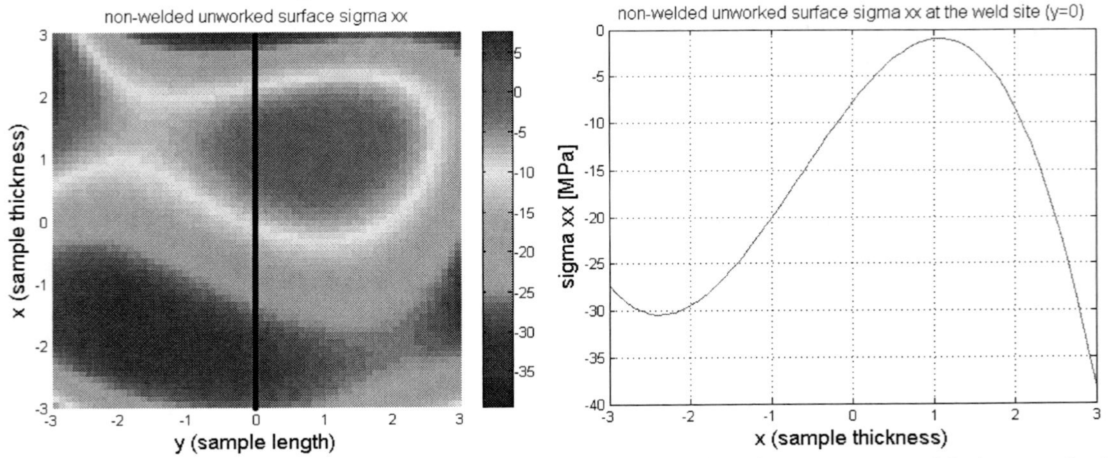


Fig. 6: a) Distribution of normal residual stress in "x" direction on non-welded unworked surface and b) distribution of normal residual stress in "x" direction on line y = 0 on non-welded unworked surface

In Fig. 7, there can be seen relatively high values of compressive residual stress after welding, mainly near the middle of sample length and also thickness – in the weld spot. Comparison of residual stress before and after welding is illustrated in Fig. 8b. In contrast to the above results of transverse residual stress in cut surface, there can be seen increase of compressive residual stress by welding from nearly zero to maximum value of -192 MPa.

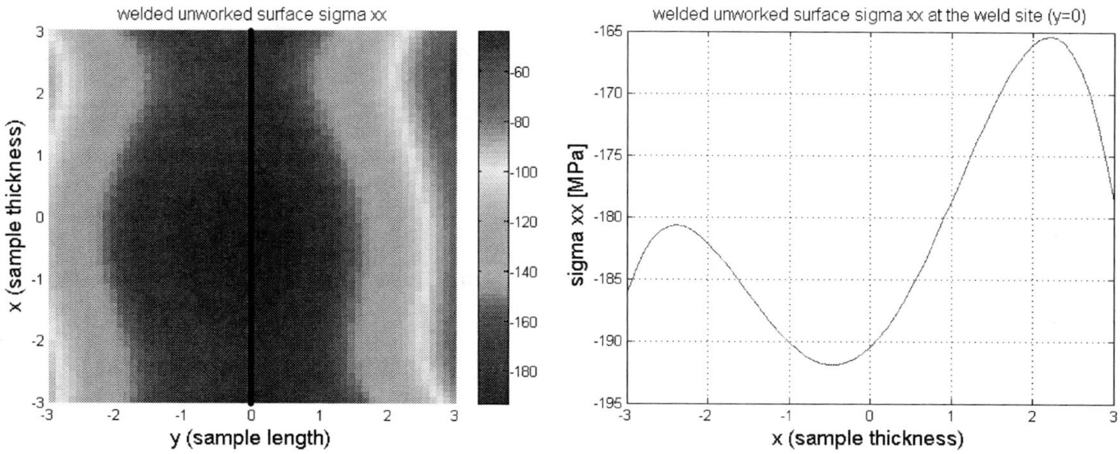


Fig. 7: a) Distribution of normal residual stress in "x" direction on welded unworked surface and b) distribution of normal residual stress in "x" direction on line y = 0 on welded unworked surface

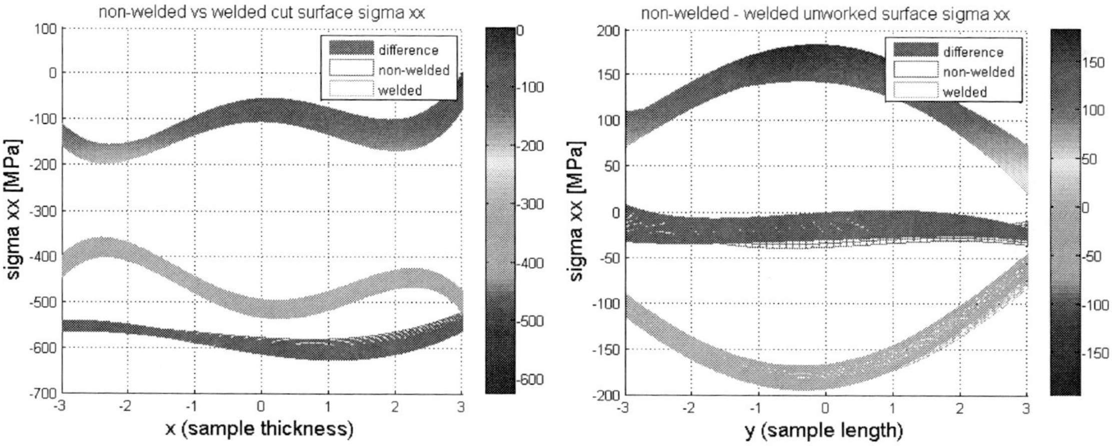


Fig. 8: a) Comparison of normal residual stress in "x" direction on non-welded and welded cut surface and b) comparison of normal residual stress in "x" direction on non-welded and welded unworked surface

In Fig. 8, there are two comparisons – of transverse and longitudinal residual stress before and after welding. It should be noted that in Fig. 8a, stress distribution varies in a direction of the sample thickness whilst in Fig. 8b, stress distribution changes with sample length.

## Conclusion

The aim of this paper was to conduct experimental measurement of residual stress in high strength steel sample before and after laser welding by X-ray method, process the data and compare and evaluate the results. Residual stress was determined in transverse and longitudinal direction across the sample thickness before and after laser welding; data was processed in program Matlab, where surfaces of stress distributions were created and smoothened by polynomial functions. Before welding, transverse residual stress was compressive with maximum value of -612 MPa resulting from rolling of the material and was reduced after welding to maximum value -495 MPa. This means that welding caused nearly 20% decrease of compressive transverse residual stress. Longitudinal residual stress before welding was negligible and after welding was increased to maximum value of -192 MPa. Results showed compressive character of normal residual stress both in transverse and longitudinal direction after laser welding due to the low heat input typical for this technology. This fact is favourable as compressive residual stress can enhance fatigue resistance of material, which will be verified in later research.

**Acknowledgement**

This work has been supported by the Slovak Research and Development Agency under the contract No. APVV-14-0096.

**References**

[1] X. Cheng, J. W. Fisher, H. J. Prask, T. Gnäupel-Herold, B. T. Yen, S. Roy, Residual stress modification by post-weld treatment and its beneficial effect on fatigue strength of welded structures, International Journal of Fatigue 25 (2003) 1259-1269.

[2] Z. Feng, Processes and mechanisms of welding residual stress and distortion, Woodhead Publishing Limited and CRC Press LLC, 2005.

[3] J. Lin, N. Ma, Y. Lei, H. Murakawa, Measurement of residual stress in arc welded lap joints by cosα X-ray diffraction method, Journal of Materials Processing Technology 243 (2017) 387-394.

[4] G. E. Dieter Jr., Mechanical Metallurgy, McGraw-Hill Book Company, New York, 1961

[5] G. A. Moraitis, G.N. Labeas, Residual stress and distortion calculation of laser beam welding for aluminum lap joints, Journal of Materials Processing Technology, 198 (2008) 260-269.

[6] G. Totten, M. Howes, T.Inoue, Handbook of Residual Stress and Deformation of Steel, Materials Park, Ohio, 2002.

[7] M. Vaško, M. Blatnický, P. Kopas, M. Sága, Research of weld joint fatigue life of the AlMgSi07.F25 aluminium alloy under bending-torsion cyclic loading, Metalurgija 56(1-2) (2017) 94-98.

[8] P. Kopas, M. Blatnický, M. Sága, M. Vaško, Identification of mechanical properties of weld joints of AlMgSi07.F25 aluminium alloy, Metalurgija 56(1-2) (2017) 99-102.

[9] P. Kopas, L. Jakubovičová, M. Vaško, Fatigue resistance of reinforcing steel bars, Procedia Engineering 136 (2016) 193-197.

[10] M. Neslusan, A. Czan, U. Zuperl, Analysis of the heat distribution when grinding of a VT 9 titanium alloy and its relation to residual stresses, Strojniski vestnik-journal of mechanical engineering 48 (10) (2002) 557-564

EAN 2017
55th conference on experimental stress analysis

# Contribution to multiaxial fatigue analysis of aluminium alloy EN AW 6063.T66

Milan Sága[1,a], Milan Uhríčik[2,b] and Milan Vaško[1,c]

[1]University of Žilina, Faculty of Mechanical Engineering, Department of Applied Mechanics, Univerzitná 1, 010 26 Žilina, Slovak Republic

[2]University of Žilina, Faculty of Mechanical Engineering, Department of Materials Engineering, Univerzitná 1, 010 26 Žilina, Slovak Republic

[a]milan.saga@fstroj.uniza.sk, [b]milan.uhricik@fstroj.uniza.sk, [c]milan.vasko@fstroj.uniza.sk

**Keywords:** multiaxial fatigue criteria, bending-torsion loading, aluminium alloy

## Introduction

The experimental knowledge of multiaxial fatigue life plays an important role in the design of machine components loaded cyclically. The stress states of components are typically multiaxial [1, 2]. These components are often loaded by combination of bending-torsion [3, 4, 5]. The aim of this article will be a presentation of experimental and numerical research of the multiaxial fatigue criteria for aluminium alloy EN AW 6063.T66. In the centre of attention will be the analysis and comparison of selected multiaxial fatigue criteria, which are confronted with experimental results [1, 6]. Fatigue life surfaces of aluminium alloy (when bending-torsion loading and phase shift $\varphi = 0°$ and $90°$ were combined) have been prepared on this basis [1].

Aluminium is the world's most abundant metal and is the third most common element, comprising 8% of the earth's crust. The versatility of aluminium makes it the most widely used metal after steel [7, 8, 9]. Pure aluminium is soft, ductile, corrosion resistant and has a high electrical conductivity. It is widely used for foil and conductor cables, but alloying with other elements is necessary to provide the higher strengths needed for other applications [8, 9]. Aluminium is one of the lightest engineering metals, having strength to weight ratio superior to steel [7]. By utilising various combinations of its advantageous properties such as strength, lightness, corrosion resistance, recyclability and formability, aluminium is being employed in an ever-increasing number of applications. This array of products ranges from structural materials through to thin packaging foils.

## Experimental equipment for modelling of bending-torsion combination

Department of Applied Mechanics at the University of Žilina has developed the unique testing device that allows performing fatigue tests under cyclic loading on specimens [1]. The working part of the device is shown in Fig. 1 (left). This experimental equipment is based on mechanical principle. Power of device is secured by two synchronic electromotors with frequency converters from 0.5 Hz to 100 Hz. Loading frequencies are identical with frequency of rotation drive. Synchronization of the electromotors is secured using by electronics and allows synchronization of loading amplitude. Synchronization of electomotors also allows setting phase shift for individual loading levels. The constant rotation is generated by excenter and linkage mechanism. By changing of eccentric magnitude it is possible to change a loading magnitude. Also if we change a length of connecting crank on the experimental equipment, there will be change in a loading cycle character. A sinusoidal waveform was used as command signal. The fatigue tests were conducted with constant strain amplitudes, at room temperature, in air. For evaluation of fatigue curves it needs to know stress and strain conditions on individual loading levels.

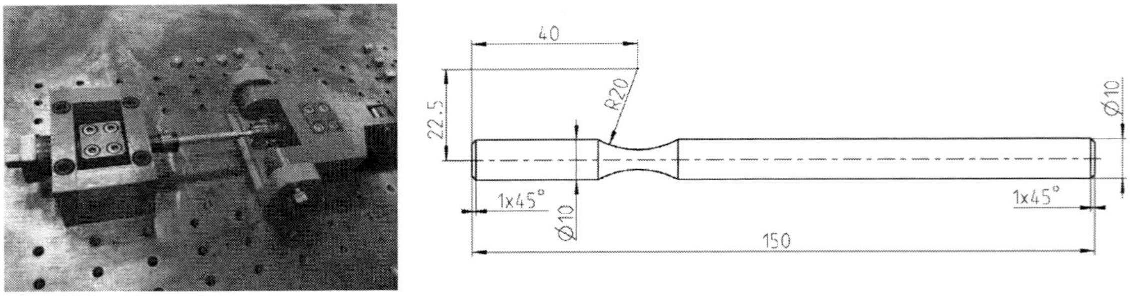


Fig. 1 Unique testing device (left) and geometry of the test specimen (right)

The geometry of the test specimen is shown in Fig. 1 (right). Stress concentrator parameters were determined by calculation according to the manufacturing process and the initial dimensions of the material [6, 7].

## Testing material characteristics

As an experimental material to perform multiaxial fatigue tests were chosen the EN AW 6063.T66 aluminium alloy with a normalised chemical composition (Table 1). The material used in this research was delivered in the form of cylindrical shape with a diameter 10 mm. The length of cylindrical bars was 150 mm. The material was in rolled state. The EN AW 6063.T66 is a medium strength alloy, suitable for applications where no special strength properties are required. Simple to complex shapes can be produced with very good surface quality characteristics, and suitable for many coating operations such as anodizing and powder coating. The T66 treatment corresponds to solution heat-treated and then artificially aged (precipitation hardened) to a higher level of mechanical properties through special control of manufacturing process [8].

Tab. 1 Chemical composition of the EN AW 6063.T66 aluminium alloy (weight in [%]) according to EN 573-3

| **Si** | **Fe** | **Cu** | **Mn** | **Mg** | **Cr** | **Zn** | **Ti** | **Other** | | **Al** |
|---|---|---|---|---|---|---|---|---|---|---|
| | | | | | | | | **Each** | **Total** | |
| 0.20-0.60 | 0.35 | 0.10 | 0.10 | 0.45-0.90 | 0.10 | 0.10 | 0.10 | 0.05 | 0.15 | rest |

Static tensile test with standard specimens had been carried out. Results are summarized in Table 2. The stress-strain diagram contains stresses and true stresses depending up strain (Fig. 2). For FEM analyses by ANSYS were used true stresses [6].

Tab. 2 Mechanical properties of the EN AW 6063.T66 aluminium alloy

| | |
|---|---|
| Young modulus | 62 500 MPa |
| Ultimate tensile strength | 247 MPa |
| Tensile yield strength | 212 MPa |

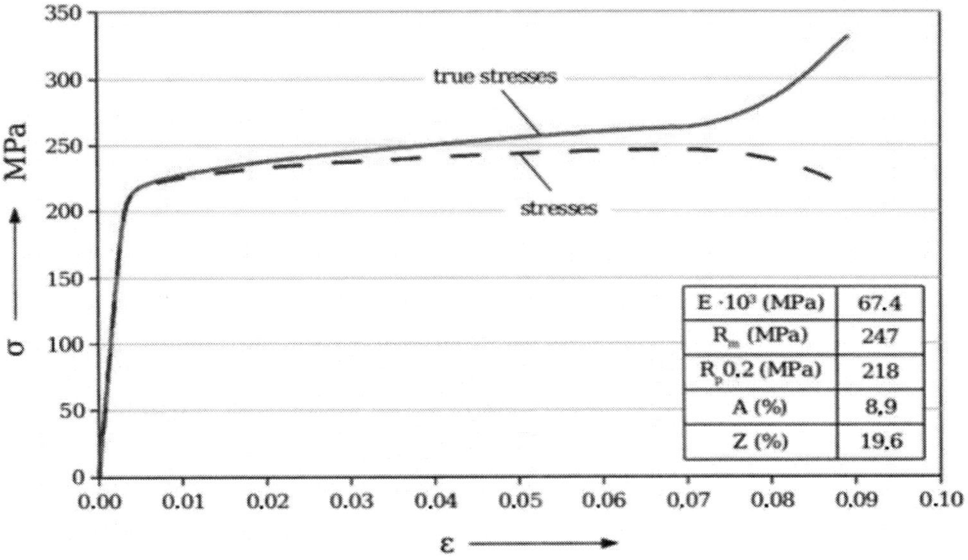


Fig. 2 Stress-strain diagram for EN AW 6063

## Results

Cyclic loading of the test specimen was done under controlled amplitude of deformation with a zero mean deformation component. The deformation amplitude was characterised by the value $\varepsilon_{ac}$ ($\varepsilon_{xx}$) in the case of alternating bending, and $\gamma_{ac}$ ($\gamma_{xy}$) in the case of alternating torsion. The loading systems are synchronized to each other with a phase shift of $\varphi = 0°$ and 90° at the same frequency. Multiaxial fatigue tests under bending-torsion loading and with phase shift $\varphi = 0°$ were made in the first phase of the experiment to determine the fatigue life of the specimens. The measured values of the number of cycles to fracture are shown in the fatigue surface $\varepsilon_{xx}$-$\gamma_{xy}$-$log\ N_f$ (Fig. 3) [1].

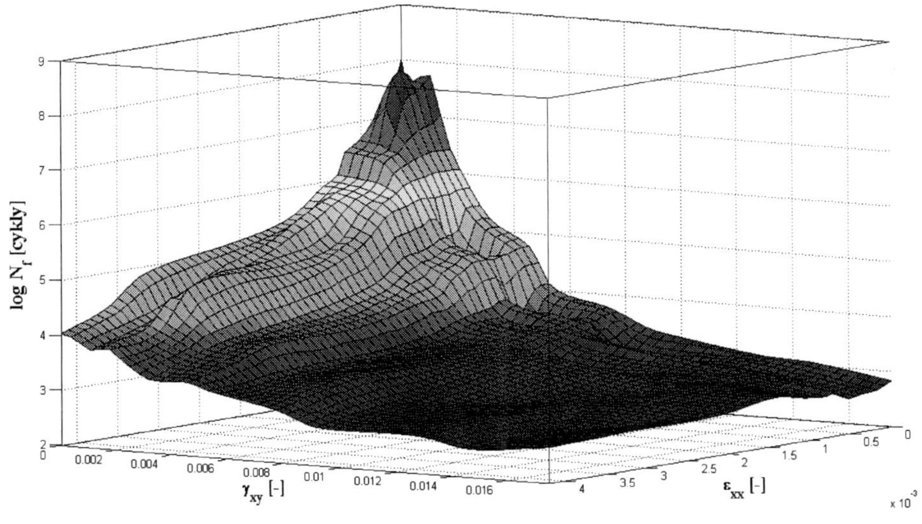


Fig. 3 The original fatigue surface $\varepsilon_{xx}$-$\gamma_{xy}$-$log\ N_f$

Measured fatigue surface is necessary to approximate by a polynomial function for the mathematical description of the fatigue properties (Fig. 4) [1].

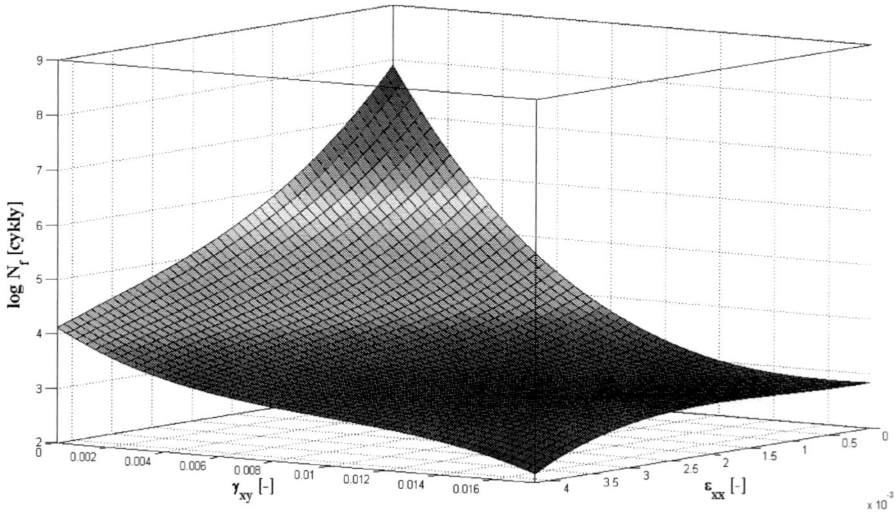


Fig. 4 The fatigue surface $\varepsilon_{xx}$-$\gamma_{xy}$-$log\ N_f$ after the approximation of the third degree polynomial

The function of the third degree with a correlation coefficient of 0.9656 was used for the approximation (Eq. 1). Coefficients of the polynomial function are shown in Table 3 [1].

$$log\ N_f = k_1 + k_2.\varepsilon_{xx} + k_3.\gamma_{xy} + k_4.\varepsilon_{xx}^2 + k_5.\varepsilon_{xx}.\gamma_{xy} + k_6.\gamma_{xy}^2 + k_7.\varepsilon_{xx}^3 + +k_8.\varepsilon_{xx}^2.\gamma_{xy} + k_9.\varepsilon_{xx}.\gamma_{xy}^2 + k_{10}.\gamma_{xy}^3 \ . \quad (1)$$

Tab. 3 Coefficients of the polynomial function for equation (1)

| | | | | | | | | | |
|---|---|---|---|---|---|---|---|---|---|
| $k_1$ | 7.902 | $k_3$ | -819 | $k_5$ | $1.894\cdot10^5$ | $k_7$ | $-2.967\cdot10^7$ | $k_9$ | $-5.084\cdot10^6$ |
| $k_2$ | -1781 | $k_4$ | $3.333\cdot10^5$ | $k_6$ | $4.307\cdot10^4$ | $k_8$ | $-1.388\cdot10^7$ | $k_{10}$ | $-7.298\cdot10^5$ |

The fatigue surface arose from experimental values by polynomial function approximation for recalculated stresses obtained by using FEM (Fig. 5). The function of the third degree with a correlation coefficient of 0.9651 was used for the approximation (Eq. 2). Coefficients of the polynomial function are shown in Table 4 [1].

$$log\ N_f = k_1 + k_2.\sigma_{xx} + k_3.\tau_{xy} + k_4.\sigma_{xx}^2 + k_5.\sigma_{xx}.\tau_{xy} + k_6.\tau_{xy}^2 + k_7.\sigma_{xx}^3 + +k_8.\sigma_{xx}^2.\tau_{xy} + k_9.\sigma_{xx}.\tau_{xy}^2 + k_{10}.\tau_{xy}^3 \ . \quad (2)$$

Tab. 4 Coefficients of the polynomial function for equation (2)

| | | | | | | | | | |
|---|---|---|---|---|---|---|---|---|---|
| $k_1$ | 4.75 | $k_3$ | -0.9301 | $k_5$ | 0.5524 | $k_7$ | -0.08524 | $k_9$ | 0.07708 |
| $k_2$ | -0.6001 | $k_4$ | 0.1809 | $k_6$ | -0.4967 | $k_8$ | -0.1529 | $k_{10}$ | -0.1897 |

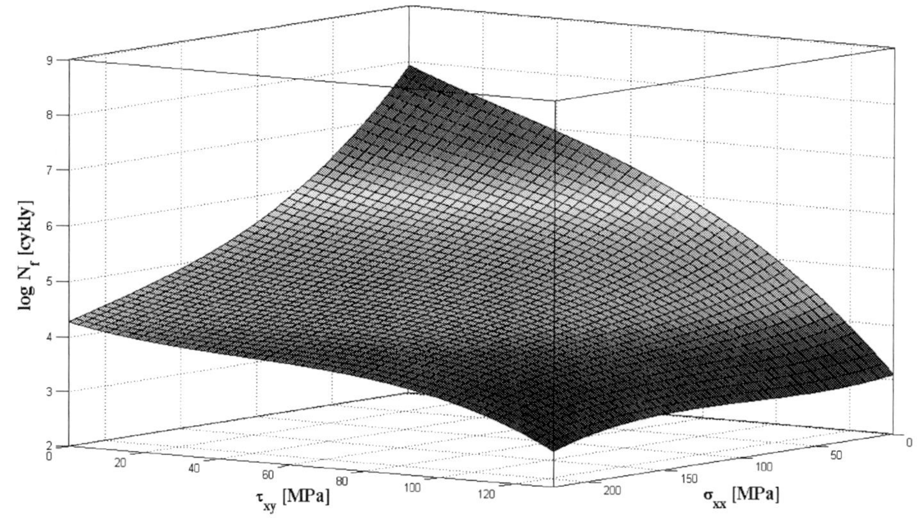


Fig. 5 The fatigue surface $\sigma_{xx}$-$\tau_{xy}$-$log\ N_f$ of the third degree polynomial function for $\varphi = 0°$

Multiaxial fatigue tests under bending-torsion loading and with phase shift $\varphi = 90°$ were made in the second phase of the experiment to determine the fatigue life of the specimens. Like in previous case, the measured values of the number of cycles to fracture are shown in the fatigue surface $\varepsilon_{xx}$-$\gamma_{xy}$-$log\ N_f$.

The experimentally obtained fatigue surface is shown in Fig. 6 [1].

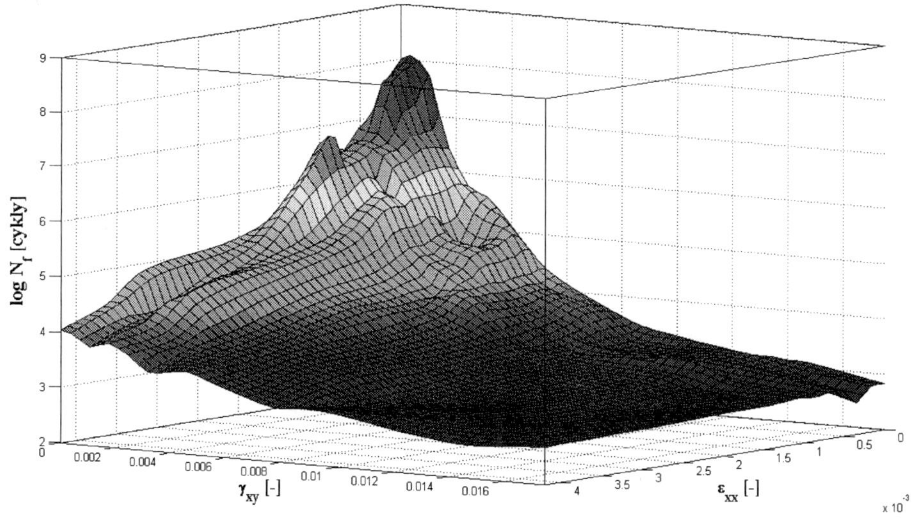


Fig. 6 The fatigue surface $\varepsilon_{xx}$-$\gamma_{xy}$-$log\ N_f$ for the bending-torsion cyclic loading with phase shift $\varphi = 90°$

Obtained fatigue surface is necessary to approximate by a polynomial function of the third degree with a correlation coefficient of 0.9770. The equation of the polynomial function has the following form and its coefficients are listed in the Table. 5 [1].

$$\begin{aligned} log\ N_f &= k_1 + k_2.\varepsilon_{xx} + k_3.\gamma_{xy} + k_4.\varepsilon_{xx}^2 + k_5.\varepsilon_{xx}.\gamma_{xy} + k_6.\gamma_{xy}^2 + k_7.\varepsilon_{xx}^3 + \\ &+ k_8.\varepsilon_{xx}^2.\gamma_{xy} + k_9.\varepsilon_{xx}.\gamma_{xy}^2 + k_{10}.\gamma_{xy}^3\ . \end{aligned} \tag{3}$$

Tab. 5 Coefficients of the polynomial function for equation (3)

| | | | | | | | | | |
|---|---|---|---|---|---|---|---|---|---|
| $k_1$ | 8.052 | $k_3$ | -843.9 | $k_5$ | $1.806 \cdot 10^5$ | $k_7$ | $-1.169 \cdot 10^7$ | $k_9$ | $-5.559 \cdot 10^6$ |
| $k_2$ | -1526 | $k_4$ | $1.884 \cdot 10^5$ | $k_6$ | $4.488 \cdot 10^4$ | $k_8$ | $-9.121 \cdot 10^6$ | $k_{10}$ | $-7.64 \cdot 10^5$ |

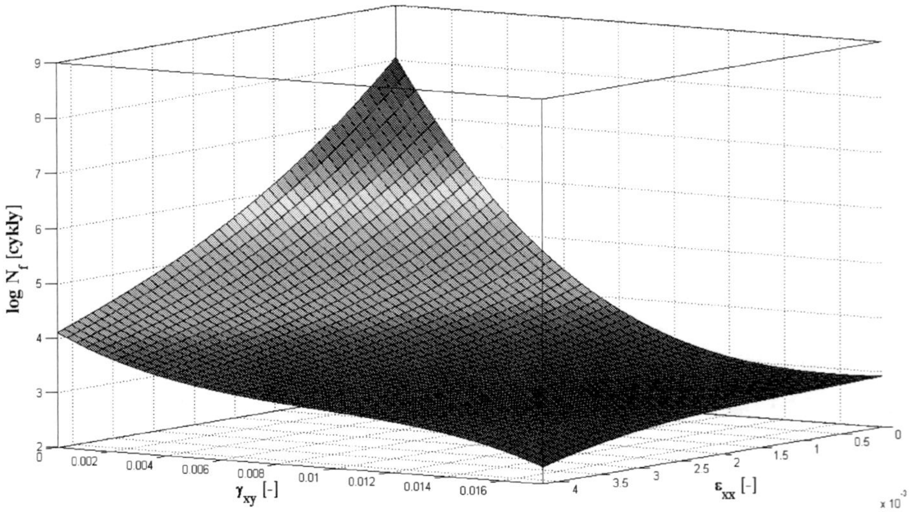


Fig. 7 The fatigue surface $\varepsilon_{xx}$-$\gamma_{xy}$-$log\ N_f$ of the third degree polynomial function for $\varphi = 90°$

The fatigue surface arose from experimental values by polynomial function approximation for recalculated stresses obtained by using FEM (Fig. 8). The function of the third degree with a correlation coefficient of 0.9604 was used for the approximation (Eq. 4). Coefficients of the polynomial function are shown in Table 6 [1].

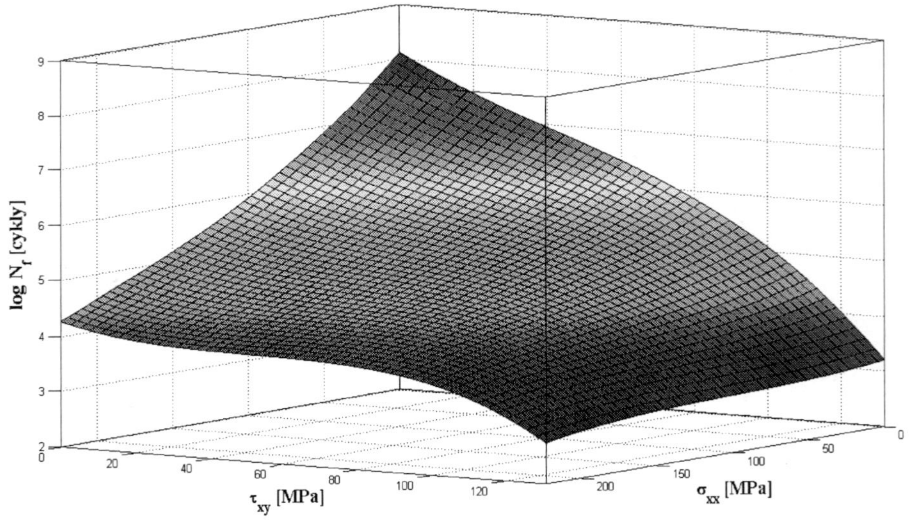


Fig. 8 The fatigue surface $\sigma_{xx}$-$\tau_{xy}$-$log\ N_f$ of the third degree polynomial function for $\varphi = 90°$

$$log\ N_f = k_1 + k_2.\sigma_{xx} + k_3.\tau_{xy} + k_4.\sigma_{xx}^2 + k_5.\sigma_{xx}.\tau_{xy} + k_6.\tau_{xy}^2 + k_7.\sigma_{xx}^3 + \\ + k_8.\sigma_{xx}^2.\tau_{xy} + k_9.\sigma_{xx}.\tau_{xy}^2 + k_{10}.\tau_{xy}^3 . \quad (4)$$

Tab. 6 Coefficients of the polynomial function for equation (4)

| | | | | | | | | | |
|---|---|---|---|---|---|---|---|---|---|
| $k_1$ | 4.906 | $k_3$ | -0.9345 | $k_5$ | 0.5095 | $k_7$ | -0.04345 | $k_9$ | 0.03901 |
| $k_2$ | -0.6343 | $k_4$ | 0.1164 | $k_6$ | -0.5229 | $k_8$ | -0.08765 | $k_{10}$ | -0.2705 |

## Analysis of selected fatigue criteria

The number of cycles was determined using the program Fatigue Calculator after realizing all experimental measurements [10]. The Fatemi-Socie, Smith-Watson-Topper (SWT), Brown-Miller and Liu criteria were used. The Fatigue Calculator software was created at the University of Illinois. The number of cycles at which fatigue fracture will occur can be detected using this program [1]. This program works with low cycle and high cycle fatigue.

Multiaxial fatigue tests under bending-torsion loading were made to determine the fatigue life of the specimens. The same conditions of combined loading were chosen (sinusoidal cyclic loading) for the purpose of comparison of experimental obtained results with calculated results. The strain values $\varepsilon_{xx}$ and $\gamma_{xy}$ input into the calculation and it is working with the symmetric frequency of loading cycles (30 Hz) and a phase shift $\varphi = 0°$ and $\varphi = 90°$.

The Fatigue Calculator after the calculating the fatigue life shows the number of cycles to fracture for the various models of damage ($N_{f(Fatemi\text{-}Socie)}$, $N_{f(SWT)}$, $N_{f(Brown\text{-}Miller)}$ a $N_{f(Liu\ I\ a\ Liu\ II)}$). The calculated values for the multi-axial cyclic loading (bending-torsion) with the phase shift of $\varphi = 0°$ are processed in the spatial fatigue surfaces $\varepsilon_{xx}$-$\gamma_{xy}$-$log\ N_f$. Fig. 9 shows the fatigue surface using the fatigue criteria Smith-Watson-Topper (SWT) with phase shift $\varphi = 0°$.

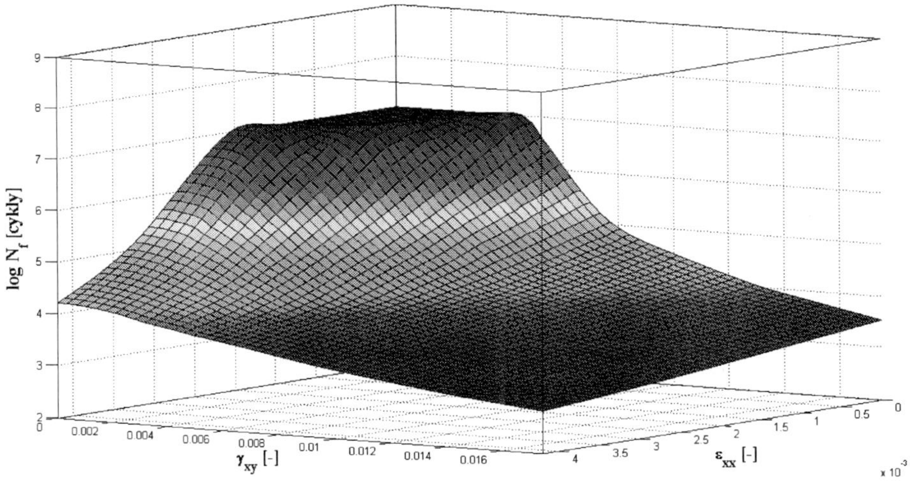


Fig. 9 The fatigue surface $\varepsilon_{xx}$-$\gamma_{xy}$-$log\ N_f$ using the fatigue criteria SWT with phase shift $\varphi = 0°$

The calculation of the multi-axial cyclic loading (bending-torsion) with the phase shift $\varphi = 90°$ and with the same input values was performed subsequently. Like in previous case, the number of cycles to fracture are shown in the fatigue surface $\varepsilon_{xx}$-$\gamma_{xy}$-$log\ N_f$ (Fig. 10).

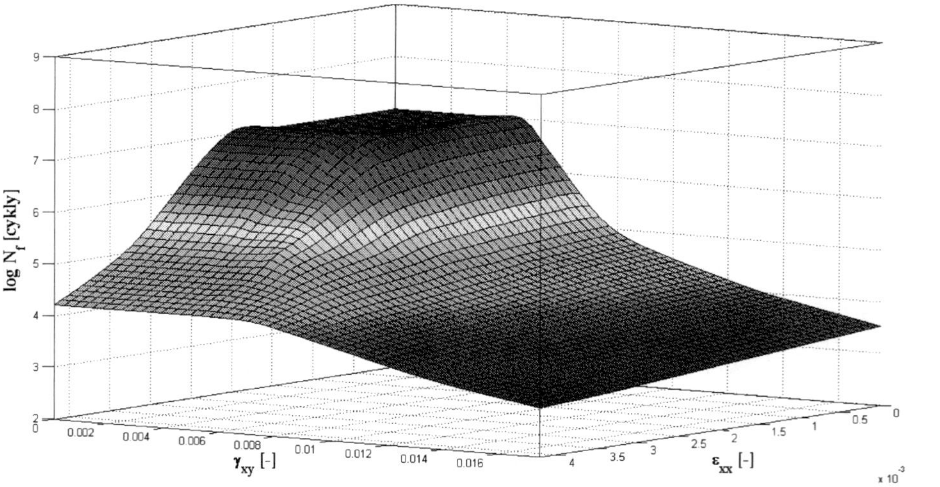

Fig. 10 The fatigue surface $\varepsilon_{xx}$-$\gamma_{xy}$-$log\ N_f$ using the fatigue criteria SWT with phase shift $\varphi = 90°$

The calculation of the number of cycles to fracture using the Fatigue Calculator was made in addition to the experiment [1]. The numbers of cycles were evaluated by fatigue surfaces $\varepsilon_{xx}$-$\gamma_{xy}$-$log\ N_f$ for the various fatigue criteria (Fatemi-Socie, SWT, Brown-Miller and Liu I, II). Fatemi-Socie model shows a lower number of cycles than experiment and Liu model I and II indicates conformity with the experiment results in the lower number of cycles and at higher values of the deformation under torsion (from $\gamma_{xy} \approx 0.008$). Brown-Miller method gives a lower lifetime under all loading amplitudes of both deformations and at the low number of cycles (to the $N_f \approx 10^5$ cycles). SWT method determines a longer lifetime for the whole fatigue surface [1]. Similar results were obtained even with a phase shift $\varphi = 90°$.

## Conclusions

The article presents multiaxial fatigue resistance of samples from aluminium alloy EN AW 6063.T66 in different variants, using a combined computationally – experimental method. From experimentally measured values of number of cycles to failure was created three-dimensional fatigue surface $\varepsilon_{xx}$-$\gamma_{xy}$-$log\ N_f$ for phase shift 0° and 90°.

For further use of these fatigue surfaces was necessary to approximate by appropriate polynomial function. This calculation was carried out using a program MATLAB [1]. Equations of polynomial functions incurred by an approximation, may be used in the compilation of own fatigue criteria for calculating fatigue lifetime, which are based on the experimental obtained results of number of cycles to fracture [1].

Fatigue Calculator software was used for another – computational analyses. This program can quickly calculate fatigue lifetime of selected material. In our calculation we considered with multiaxial criterion SWT described above. From these calculated values of number of cycles to failure were created three-dimensional fatigue curves for phase shift 0° and 90°. However, this criterion do not specify the lifetime more than $10^7$ cycles.

Multiaxial criterion SWT applied to fatigue lifetime calculation and also values of number of cycles to failure from experiments for specimens of aluminium alloy EN AW 6063.T66 increases with decreasing strain amplitude continuously in the cycles of number region [1]. Comparing three-dimensional curves is evident that criterion SWT give higher lifetime than experiment in the whole area of the number of cycles at the same load amplitudes. This may be caused by different material parameters, which were used for each models of damage. They probably do not include all real parameters and properties of the comparison of the experimental material that probably affected the sensitivity of the numerical calculation, especially at the lower values of amplitude deformation, which was applied in experiment. Our calculations can give a higher value of fatigue lifetime in comparison with experiments in the area of loads.

**Acknowledgements**

This work has been supported by the Slovak Research and Development Agency under the contract No. APVV-14-0096 and KEGA project No. 015ŽU-4/2017.

**References**

[1] M. Uhríčik, Analysis, algorithmisation and experimental verification of multiaxial fatigue criteria of construction elements from aluminium alloys, PhD thesis, Žilina, 2013.

[2] F. Trebuňa, F. Šimčák, Odolnosť prvkov mechanických sústav, 1st ed, Košice, Vienala, 2004. 980 p.

[3] L.M.P. Abreu, J.D. Costa, J.A.M. Ferreira, Fatigue behaviour of AlMgSi tubular specimens subjected to bending-torsion loading, International Journal of Fatigue 31 (2009).

[4] A. Fatemi, D.F. Socie, A critical plane approach to multiaxial fatigue damage including out-of-phase loading, Fatigue Fract. Eng. Mater. Struct. 11 (1988) 149-165.

[5] D.F. Socie, G.B. Marquis, Multiaxial fatigue, Warrendale, SAE International, 1999. 502 p.

[6] P. Kopas, M. Sága, M. Uhríčik, Three-dimensional S-N curve method to estimate fatigue life of aluminium alloy under combined loading and comparing with fatigue criteria, Technológ 5 (2013) 231-234.

[7] M. Sapieta, V. Dekýš, A. Sapietová, Thermal-stress analysis of beam loaded by 3 point bending, Procedia Engineering 136 (2016) 216-219.

[8] A. Czan, M. Sajgalik, A. Martikan, V. Kuzdak, Identification of residual stress and structure properties in surface and subsurface layers of austenitic steel designed for nuclear energetics, Applied Methods of the Analysis of Static and Dynamic Loads of Structures and Machines II, 827 (2016) 125-128.

[9] V. Dekýš, M. Sapieta, M. Uhríčik, Experimental measurement of stress distribution, Applied Mechanics and Materials 816 (2015) 416-420.

[10] Fatigue Calculator, online at: www.efatigue.com

# Determining the Strength Properties of Castellated Girders by Experimental and Numerical Modeling

Peter Sivák[1,a], Ingrid Delyová[1,b] and Tomáš Kaluja[1,c]

[1]Department of Applied Mechanics and Mechanical Engineering, Faculty of Mechanical Engineering, Technical university of Košice, Košice, Slovakia

[a]peter.sivak@tuke.sk, [b]ingrid.delyova@tuke.sk, [c]tomas.kaluja@gmail.com

**Keywords:** Castellated girder (beam), lateral torsional buckling, experimental and numerical modeling.

**Abstract.** Castellated girders are relatively efficient alternative to the classical beams made from rolled sections. The essence of the production process of castellated girders consists in cutting the beam for example zigzag incision and then welding the cut parts of the original beam. This will provide greater heights beam with hexagonal holes. Vertical height of the beam can be increased further by inserting short webs to form octagonal openings. Castellated girder from the original beam has greater strength and rigidity. The features can be related mainly to the weight. However, it is more prone to lateral torsional buckling, local buckling of web and rupture of weld in the web post. Other advantages of castellated girders include the possibility of plumbing, wiring and other installations through openings. The paper presents stress and strain analysis of critical points of castellated girders with circular and hexagonal holes. There were applied three modifications of standard IPE profile. For the purpose of stress and strain analysis were applied the methods and means of experimental and numerical modeling.

## Introduction

Castellated beams or castellated girders are known as an effective alternative to standard profiles, especially to IPE profiles. They are made from rolled and welded sections, when the web is cutting on two symmetric sides and subsequent is welded (Fig. 1). Following such an adjustment of beam it is increasing its height with created holes in the middle of the web, which contributes to the bending stiffness while maintaining low weight. Course there is fact, that such structural update is positively reflected in the final price of the beam. Castellated beams have application in various technical works for example roof constructions, the structures for the portal cranes, bridges etc. (Fig. 2). They are application wherever, where are the emphasis on the maximum capacity with minimum weight [5,6].

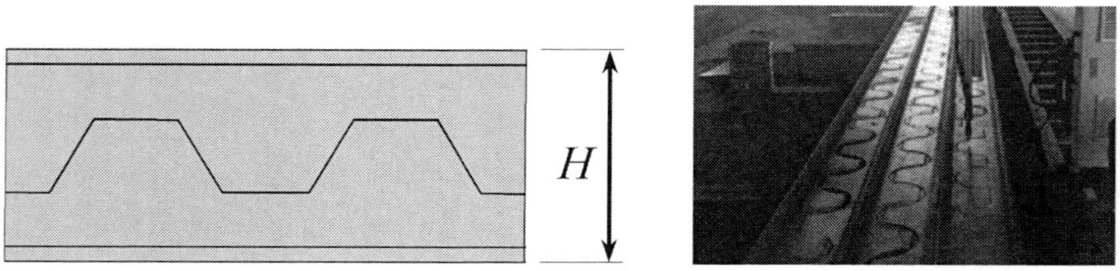


Fig. 1 Creating castellated girders using cutting and welding [1]

Fig. 2 Various applications of castellated girders [2,3,4]

Castellated girders also more effective in terms of passing service wires, pipes and ventilating ducts through opening. Castellated girders deliver increasing value of the bending section modulus $W_{ox}$ according Fig. 3 [7] with value $z$ according Fig. 4c.

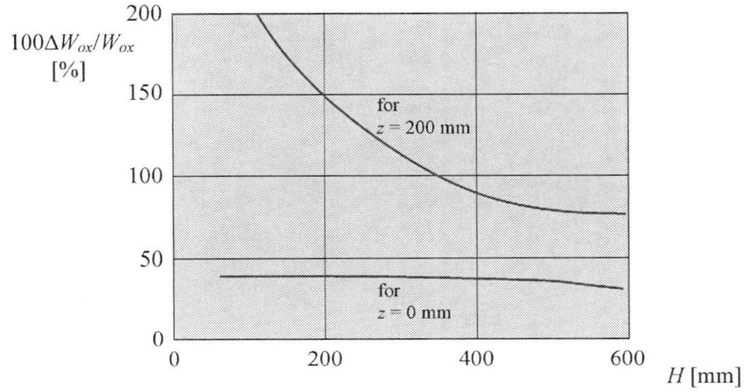


Fig. 3 Increasing value of the bending section modulus

**Typology of castellated girders**

Standardly produced types of castellated girders represent the following groups of shapes: Castellated symmetrical girder with a hexagonal holes (CGH) (Fig. 4a) making from IPE profile using zigzag cutting. This base type beam predominantly used as a bending stress component of steel structures. Castellated asymmetric girder with a hexagonal holes (Fig. 4b). This type of beam has application in cases, where the stress of the two flanges is different and not symmetrical. This is the case for example of composite steel structures that are not protected against lateral buckling. Castellated symmetrical girder with a hexagonal holes and with inserting element (Fig. 4c). Castellated symmetrical girder with a circular holes (CGC) (Fig. 4d). This type of beam is particularly suitable for uniform load because it has less resistance for shear loading [5]. Very special shape of castellated girder is tapered shape (Fig. 4e) and curved shape (Fig. 4f).

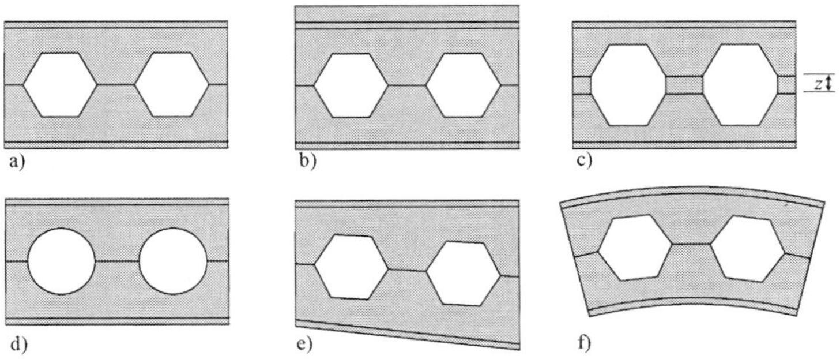


Fig. 4 Basic and special shapes of the castellated girders

## The possibility of failure of castellated beams

Failure to properly design of the shape or geometry of castellated girders may be permanent damage to such proposed beams. Failure of web opening depends on opening shape, opening dimensions, the distance between opening, web post thickness. Several modes of Failure are [8]:

1. flexural failure of the section;
2. lateral torsional buckling of beam;
3. local buckling of web or flange (Fig. 5);
4. rupture of weld in the web post;
5. Vierendeel failure of perforation of section;
6. web post buckling.

Fig. 5 Collapsed castellated girder - local buckling [5,9]

## Calculation and stress analysis

Calculation, strain and stress analysis of the castellated girder can be performed by analytical, experimental and numerical modeling in the comparative, complementary and verification connections.

## Analytical solution

Static analytical calculation of the castellated beams can be performed in the same manner as is used in calculating Vierendeel beam. The basis for calculation of the beam (Fig. 6) is element obtained by imaginary cutting. In place of the envisaged joints are internal variables replaced with outside variables $F/2$, $N_L$, $N_R$ and $Q$ as shown on Fig. 7.

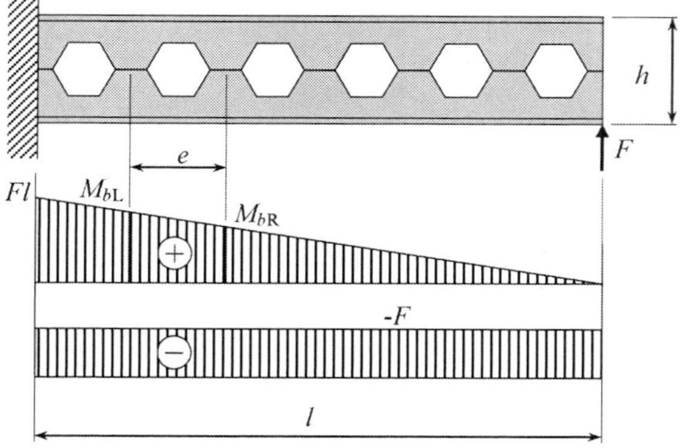


Fig. 6 Loading force $F$, bending moment $M_b$ and shear force along the castellated beam

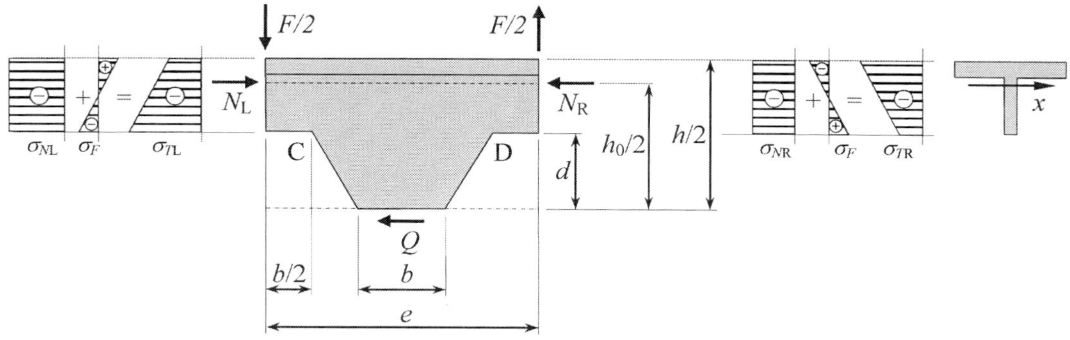


Fig. 7 Preparing of the castellated beam CGH to analytical solution

The forces $N_L$, $N_R$ and $Q$ are expressed:

$$N_L = \frac{M_{bL}}{h_0},\ N_R = \frac{M_{bR}}{h_0},\ Q = \frac{F \cdot e}{h_0}. \tag{1}$$

Total sum of normal stress in the point C is:

$$\sigma_{TC\max} = \sigma_N + \sigma_{F\max} = \frac{M_{bL}}{h_0 A} + \frac{Fb}{4W_{ox}}, \tag{2}$$

where $A$ is the area and $W_{ox}$ is the bending section modulus of the cross section of the upper belt [7].

**Experimental solution**

During the experiment was used measuring system as shown in the Fig. 8. The main parts of the measuring chain were: the active and compensation resistance 120-ohm strain gauges (single and 90-degree rosette) (Ekolab, Micro Measurements Group Vishay and Baldwin Lima Hamilton) in quarter and half-bridge circuits, the digital strain indicator Model P-3500 (Micro Measurements Group Vishay), the switch & balance unit SB-10 (Micro Measurements Group Vishay) and respective loading units and elements. The models P-3500 and SB-10 are portable, battery-powered high precision instruments for use with resistive strain gages and transducers, primarily intended for the experimental stress analysis (ESA).

Fig. 8 Experimental model of the IPE 100, CGC, CGH and respective experimental chain and loading unit

The position of the electric resistance strain gauges on the IPE 100 shows Fig. 9, on the CGC shows Fig. 10 and on the CGH shows Fig. 11.

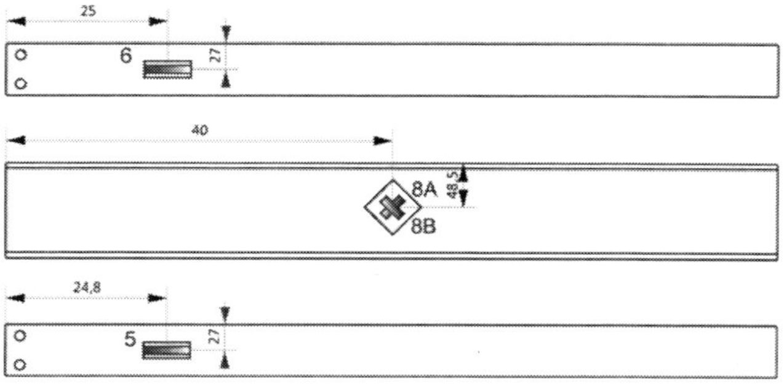

Fig. 9 Position of the resistance strain gauges on the IPE 100

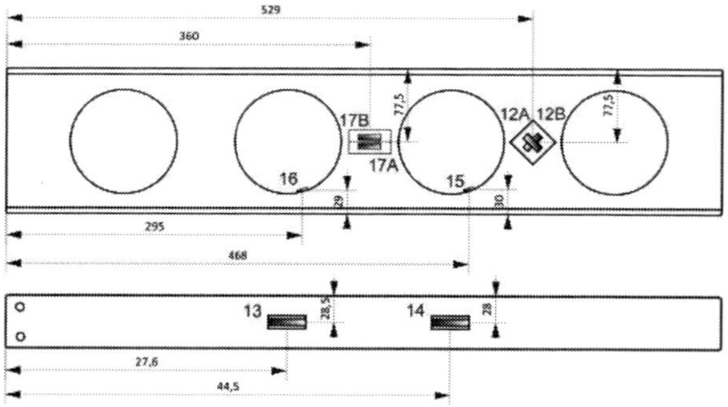

Fig. 10 Position of the resistance strain gauges on the CGC

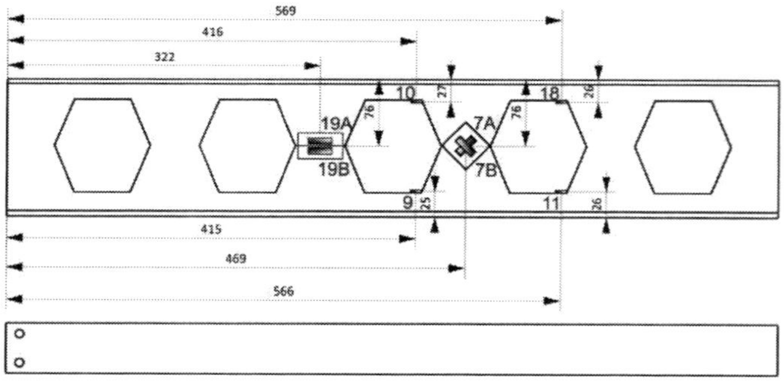

Fig. 11 Position of the resistance strain gauges on the CGH

The experimental solution has included also ultrasonic check the integrity of the weld as shown on Fig. 12.

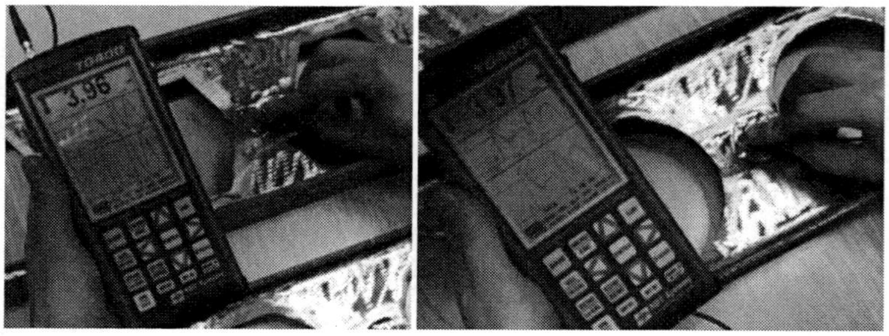

Fig. 12 Ultrasonic check the integrity of the weld

The analysis of the concentration of stress can be also advantageously used by optical methods, especially by the method reflection or transmission photoelasticimetry (Fig. 13) [5].

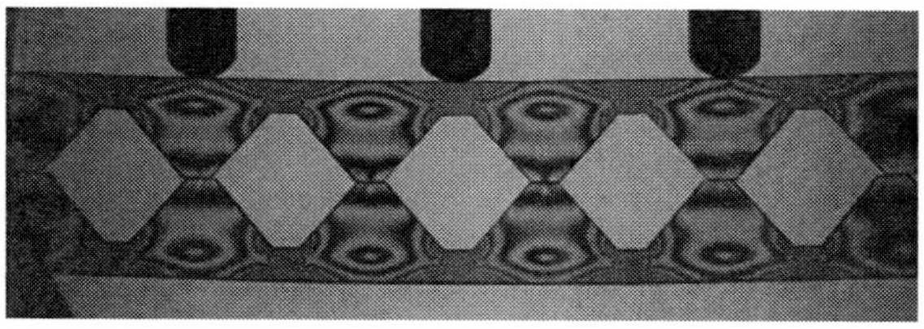

Fig. 13 Field of isochromatics in the CGH [5]

**Numerical solution**

Numerical solution for stress analysis by finite element method (FEM) was realized using SolidWorks software and Abaqus software. There were applied laws of model similarity in order to use the physical models for experimental modeling. The field of shear stresses $\tau_{zy}$ in the IPE 100, in the CGC and in the CGH shows Fig. 14, Fig. 15 and Fig. 16. The field of normal stresses $\sigma_z$ in the IPE 100, in the CGC and in the CGH shows Fig. 17, Fig. 18 and Fig. 19.

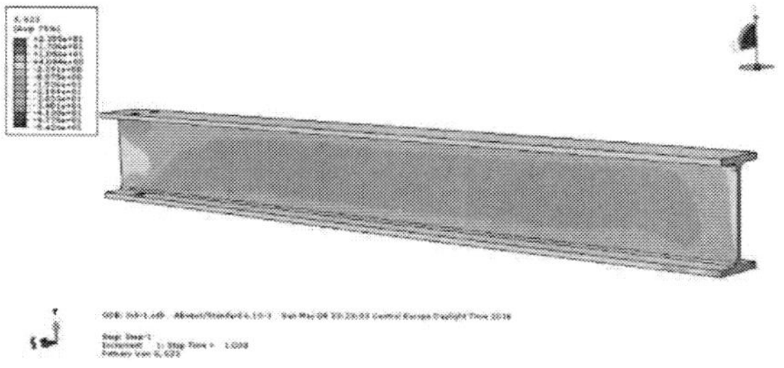

Fig. 14 Field of shear stresses $\tau_{zy}$ in the IPE 100

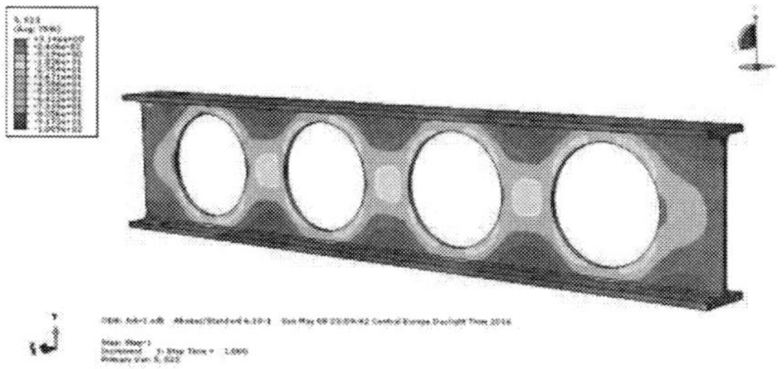

Fig. 15 Field of shear stresses $\tau_{zy}$ in the CGC

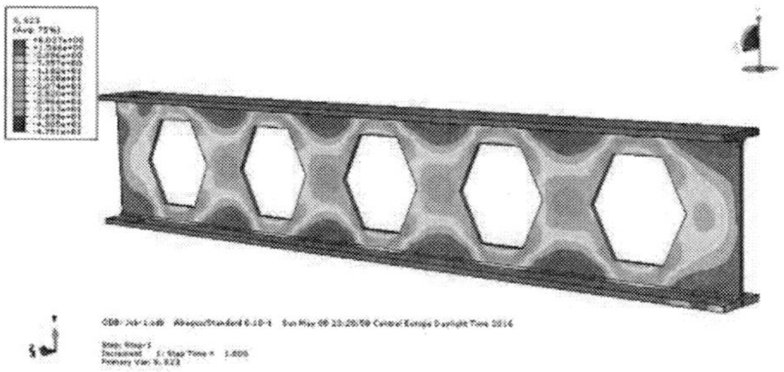

Fig. 16 Field of shear stresses $\tau_{zy}$ in the CGH

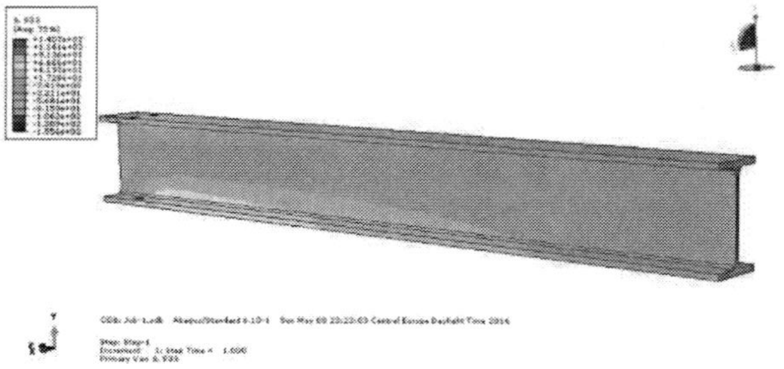

Fig. 17 Field of normal stresses $\sigma_z$ in the IPE 100

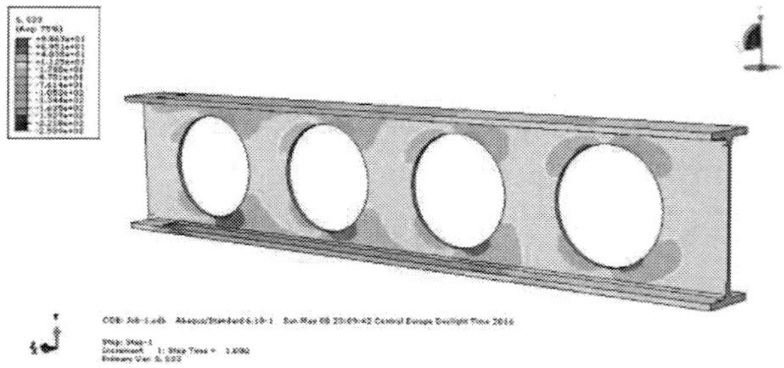

Fig. 18 Field of normal stresses $\sigma_z$ in the CGC

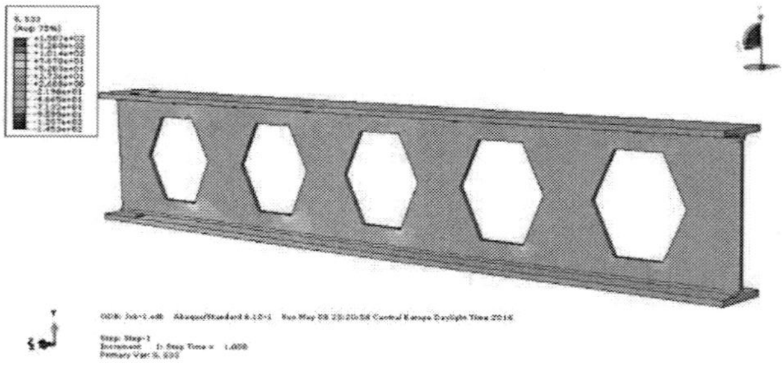

Fig. 19 Field of normal stresses $\sigma_z$ in the CGH

### Measured and calculated values of stresses

The data obtained by experimental and numerical analysis in selected critical points of the IPE, CGC and CGH are shown in Tab. 1 – Tab. 3 for input parameters: $F = 3\ 500$ N, $l_{IPE} = 720$ mm, $l_{CGC} = 650$ mm, $l_{CGH} = 720$ mm.

Tab. 1 Measured and calculated values of stresses in the IPE

| Position on the IPE | Stress (kind) | Value [MPa] Experimental analysis (strain gauges) | Value [MPa] Numerical analysis (FEM) |
|---|---|---|---|
| 5 | $\sigma_z$ | 55,70 | 52,00 |
| 6 | $\sigma_z$ | 55,50 | 51,70 |
| 8 | $\tau_{zy}$ | 15,10 | 10,80 |

Tab. 2 Measured and calculated values of stresses in the CGC

| Position on the CGC | Stress (kind) | Value [MPa] Experimental analysis (strain gauges) | Value [MPa] Numerical analysis (FEM) |
|---|---|---|---|
| 12 | $\tau_{zy}$ | 28,90 | 23,69 |
| 13 | $\sigma_z$ | 23,50 | 24,58 |
| 14 | $\sigma_z$ | 14,50 | 14,16 |
| 15 | $\sigma_z$ | 78,10 | 76,47 |
| 16 | $\sigma_z$ | 89,10 | 84,37 |
| 17 | $\tau_{zy}$ | 28,40 | 23,80 |

Tab. 3 Measured and calculated values of stresses in the CGH

| Position on the CGH | Stress (kind) | Value [MPa] Experimental analysis (strain gauges) | Value [MPa] Numerical analysis (FEM) |
|---|---|---|---|
| 7 | $\tau_{zy}$ | 38,90 | 19,77 |
| 9 | $\sigma_z$ | 74,50 | 67,20 |
| 10 | $\sigma_z$ | 70,20 | 75,19 |
| 11 | $\sigma_z$ | 73,60 | 72,24 |
| 18 | $\sigma_z$ | 86,60 | 88,35 |
| 19 | $\tau_{zy}$ | 29,80 | 19,19 |

## Conclusions

Based on a wide range of analytical, numerical and experimental strain and stress analyzes performed on castellated girders with circular and hexagonal holes and on beams without holes were obtained relevant data. These data serve to assess critical characteristics and behavior of standard beams and castellated girders.

## Acknowledgement

This work was supported by the Ministry of Education of Slovakia Foundation under Grant VEGA No. 1/0393/14 and VEGA No. 1/0751/16.

## References

[1] Information on https://www.youtube.com/watch?v=QVyskNPMXbk

[2] Information on http://scherersteel.com/joists-girders/image001/

[3] Information on https://www.emech.co.in/empower-9/contribution-in-building/

[4] Information on http://latteycivil.co.nz/newlook/services/bridge-work/transit-and-local-body-highway-bridges/

[5] M. Pástor, J. Bocko, T. Kula, C.S. Roszkos, Possibilities of Optimizing the Shape and Geometry of Castellated Beams Using Numerical and Experimental Modeling, in: American Journal of Mechanical Engineering, 2016, Vol. 4, No. 7, pp. 357-362.

[6] Information on http://www.noving.cz/en/ocelove-konstrukce/prolamovane-nosniky

[7] F. Procházka et al, Elements of Steel Structures (in Slovak), SNTL Praha, ALFA Bratislava, 1972, 352 p.

[8] A.S. Shaikh, P.B. Autade, Structural Analysis and Design of Castellated Beam in Cantilever Action, in: International Research Journal of Engineering and Technology (IRJET), Volume 03, Issue 08, 2016 163-170.

[9] S. Durif, A. Bouchaïr, O. Vassart, Validation of an analytical model for curved and tapered cellular beams at normal and fire conditions, in: Civil Engineering 57/1, 2013, pp. 83-95.

EAN 2017
55th conference on
experimental stress analysis

# Properties of Composites with Geopolymer Matrix Reinforced by Basalt Fabric

Iva Petríková[1,a], Bohdana Marvalová[1,b], Aleš Hrouda[1,c] & Lukáš Paur[1,d]

[1]Technical University of Liberec, Studentská 2, 46117 Liberec, Czech Republic

[a]iva.petrikova@tul.cz, [b]bohda.marvalova@tul.cz, [c]ales.hrouda@tul.cz, [d]lukas.paur@tul.cz,

**Keywords:** geopolymers, basalt fabric, composite, thermal treatment.

**Abstract.** The mechanical properties of composites made of basalt fabric reinforcement and geopolymer matrix have been investigated. Two different types of geopolymer matrix were investigated and two types of basalt fabric were used - namely weave and twill. The tensile and flexural properties of composite samples were investigated both without any thermal treatment and after heating to 500 ° C followed by slow cooling to room temperature. Tensile strength of basalt fabric alone was also investigated before and after the thermal treatment. The mechanical properties of composite and fabric have deteriorated as a result of heating but samples retained their integrity and partly also the strength. The basalt fibers and fabrics can be regarded as a good strengthening material for geopolymer composites used for fire resistant or temperature resistant panels.

## Introduction

Geopolymers are a class of amorphous aluminosilicate materials composed of cross-linked alumina ($AlO_4$) and silica ($SiO_2$) tetrahedra to form polysialates, with an alkali metal ion to balance the negative charge [1]. Geopolymers have shown potential as a low-cost, environmentally friendly structural material with the ability to maintain strength at elevated temperatures. Geopolymers can be used as a binder system for a broad variety of reinforcement materials. This property guarantees maximum load transfer from the geopolymer matrix to any reinforcement phase.

Basalt fibers can be considered environmentally friendly. Basalt is the most common reinforcing phase of the geopolymer that significantly improves the mechanical properties and high-temperature stability of composites. Young's modulus of basalt fiber varies between 78 and 90 GPa [2,3]. Compared to glass, basalt fiber has higher or comparable elastic modulus and tensile strength and higher chemical and thermal stability, as well as good thermal, electrical, and sound insulating properties.

The thermal insulating ability of basalt is three times that of asbestos, and due to such good insulating properties basalt is used in fire protection [4,5]. Basalt has electrically insulating properties which are 10 times better than glass. Basalt has much better chemical resistance than that of glass fiber, especially in the presence of strong alkalis [5,6,7].

Thermal stability of basalt fibers was investigated and reported in several papers. The strength of the basalt fibers after thermal treatment depends not only on the heating temperature but also on the time how long the fibers are exposed to high temperature. The comparison of the results of measurements [5,7,8] in Fig. 1 shows even the improvement of strength in the range of lower temperature but the results vary widely after heating to higher temperatures. The main factor determining the heat temperature stability of basalt fibers is their crystallization behaviour which depends on the chemical composition of basalt fiber [3].

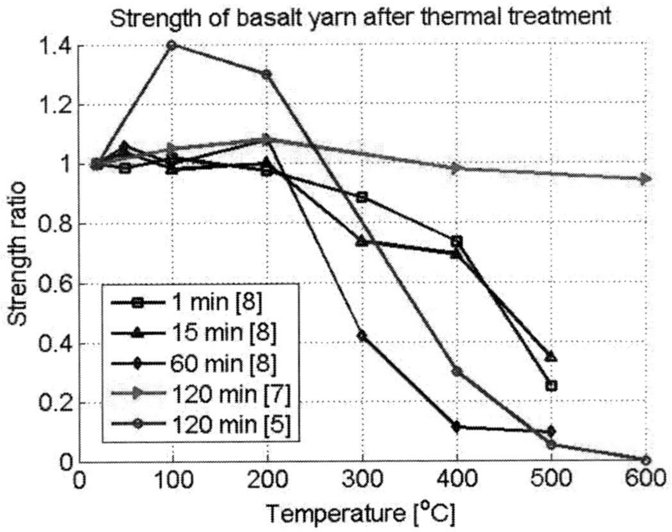


Fig. 1. Strength ratio of basalt yarns after thermal treatment.

The behaviour of the basalt fibres in geopolymer matrix during the heat treatment is quite different from the fibres alone due to chemical interaction [10]. The strength of the composite decreases as the result of increasing crystallisation and melting of the fibre occurred at higher temperatures. This process results in an increasingly brittle failure of the fiber and the composite. However, other studies [11,12] show that the use of basalt fibres as the reinforcement of geopolymers is justified for applications up to 600 °C.

## Experimental procedure

Two geopolymer resins with the working name B3P1 and FC4 were furnished by UniCRE Czech Republic. Geopolymer matrix FC4 consists of potassium water glass, potassium hydroxid (KOH), silica fume, metakaolinite and boric acid. Geopolymer matrix B3P1 consists of potassium water glass, metakaolinite and calcium. Composite panels 250x250 mm were prepared using the hand lay-up of six layers 0°/90° of 2-D plain weave basalt fabrics of weight 400 g/m2 furnished by Basaltex CZ. The chemical composition of the basalt fibers used in our composites is in Table 1.

Table 1. Chemical analysis of basalt fibers by spectrometer RX Bruker [9]

| Oxid | $SiO_2$ | $Al_2O_3$ | CaO | MgO | $TiO_2$ | $Na_2O$ | $K_2O$ | $Fe_2O_3$ | MnO | SrO | $ZrO_2$ | ZnO |
|---|---|---|---|---|---|---|---|---|---|---|---|---|
| % m | 54.39 | 17.6 | 8.92 | 3.62 | 1.12 | 2.66 | 1.66 | 9.77 | 0.14 | 0.4 | 0.04 | 0.02 |

The plates then were covered by separation and suction cloths. The process was followed by vacuuming in order to improve the infiltration of the matrix into the reinforcing fabric. Panels then were left in a compression press at ambient temperature for 4 weeks. The thickness of panels were about 2 mm. Samples of width 20 mm and length 200 mm were prepared from the cured panels by laser cutting. Tensile tests were performed at Instron 5967 with a video-extensometer and three point flexural tests were performed at TIRA 2850 testing machine.

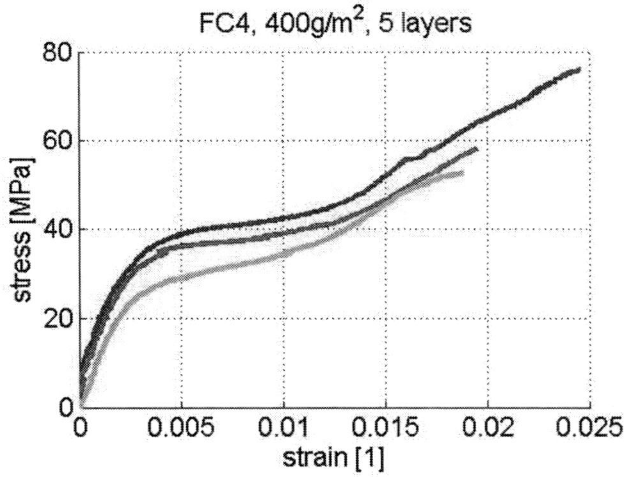


Fig. 2. Matrix FC4 with 5 layers of weave basalt fabric no heat treatment.

The strength in tension and in three point bending was measured after 30 min exposure to elevated temperature 500°C.The results of the tensile test of composite with FC4 matrix without the heat treatment are in the graphs in Fig.2. Both types of composite had the same strength without the heat treatment. Samples with B3P1 matrix maintained integrity at elevated temperatures but the strength decreased three times as can be seen in Fig. 3. The samples with FC4 matrix became brittle after the thermal treatment and the strength was unmeasurable. The results of three point bending are in Figs 4 - 6.

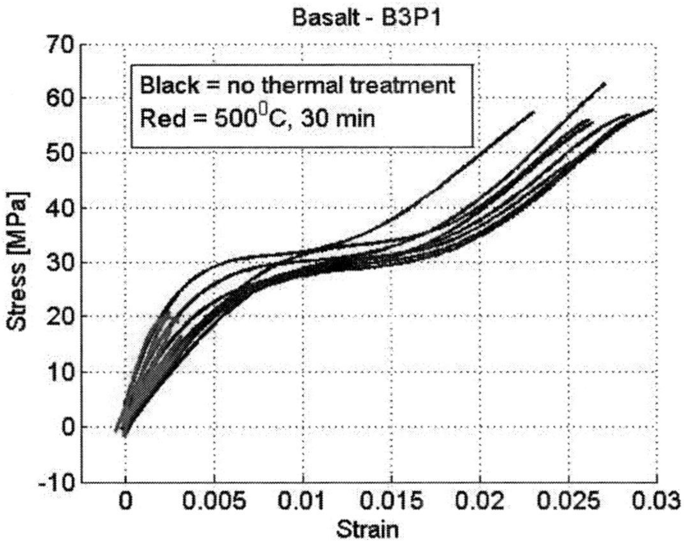


Fig. 3. Matrix B3P1 with 9 layers of twill basalt fabric.

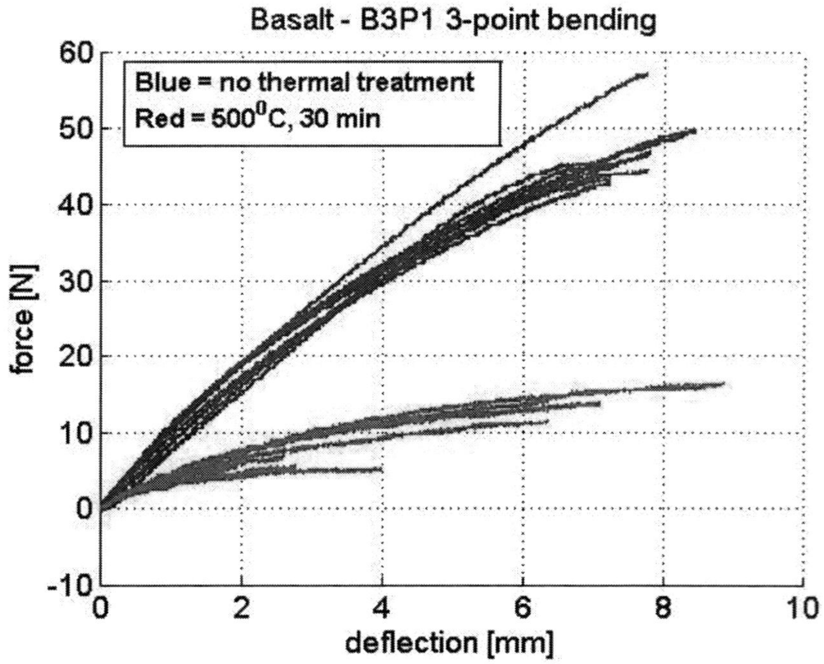


Fig. 4. Three point bending of the composite specimens with B3P1 matrix reinforced by 9 layers of twill basalt fabric.

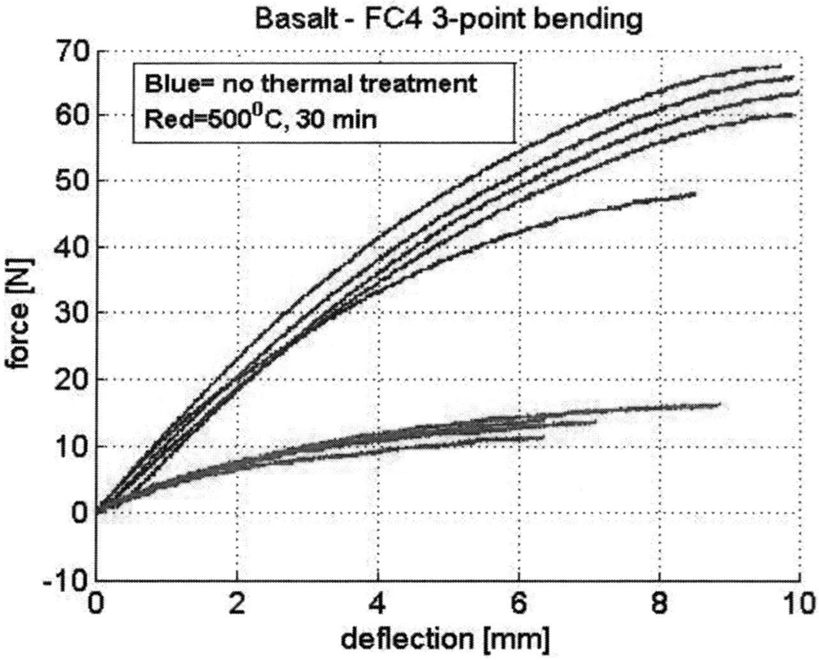


Fig. 5. Three point bending of the composite specimens with FC4 matrix reinforced by 9 layers of twill basalt fabric.

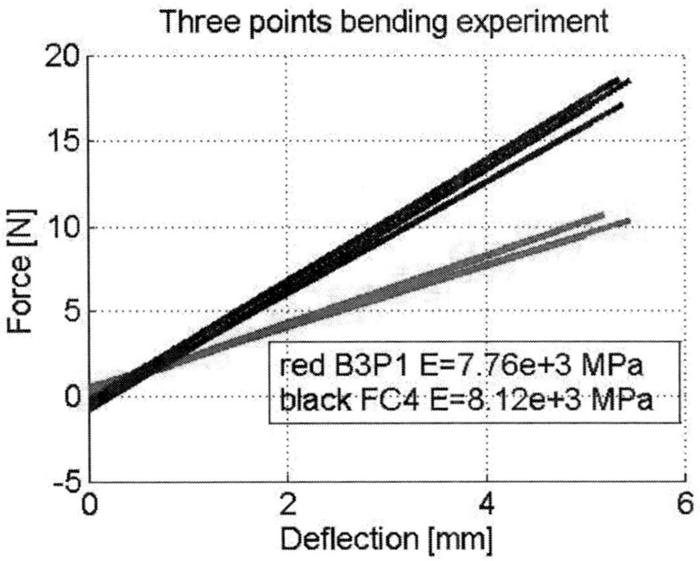


Fig. 6. Flexural modulus in three point bending of the composite specimens with reinforced by 5 layers of weave basalt fabric.

**Conclusion**

Thin composite panels were produced by hand lay-up followed by vacuuming to achieve the good penetration of geopolymer matrix within basalt fabric yarns. This process provided uniform mechanical properties throughout the entire panel. Tensile strength and three-point flexural strengths values were determined at ambient temperature for two groups of samples with different geopolymer matrix. Other specimens were exposed to elevated temperature at 500°C for 30 minutes in an oven without forced air circulation. The samples were subjected to the same tests. Both types of composite without thermal treatment had approximately the same strength values. The strength of samples dropped to approximately one third after the heat treatment but the material did not lose its cohesion. These properties can be further improved by a suitable composition of a geopolymer matrix which should not attack basalt fibers chemically. This problem is the subject of our further research.

Basalt fiber reinforced geopolymer composites are non-combustible materials which can be used for different applications where fire endurance and non-combustibility together with structural strength are required. Basalt reinforced geopolymer composites could be used in different industrial applications at elevated temperatures between 300°C-900°C. The composites are also suitable as panels in fire curtains or barriers inside buildings.

**Acknowledgement**

This work was supported by SGS 21120, Specific University Research Grant provided by the Ministry of Education, Youth and Sports of the Czech Republic. Special thanks to Pavlína Hájková, Ph.D. and Unipetrol Centre for Research and Education, Ústí n. L., Czech Republic, who had supplied geopolymer materials.

## References


[1] Ribero, D. and Kriven, W.M., 2016. Properties of geopolymer composites reinforced with basalt chopped strand mat or woven fabric. Journal of the American Ceramic Society,1-8, DOI: 10.1111/jace.14079

[2] Samal, S., Nhan, P.T., Petríková, I. and Marvalová, B., 2015. Improved Mechanical Properties of Various Fabric-Reinforced Geocomposite at Elevated Temperature. JOM, 67(7), pp.1478-1485.

[3] Fiore, V., Scalici, T., Di Bella, G. and Valenza, A., 2015. A review on basalt fibre and its composites. Composites Part B: Engineering, 74, pp.74-94.

[4] Samal, S., Nhan, P.T., Petríková, I., Marvalová, B., Vallons, K.A. and Lomov, S.V., 2015. Correlation of microstructure and mechanical properties of various fabric reinforced geo-polymer composites after exposure to elevated temperature. Ceramics International, 41(9), pp.12115-12129.

[5] Van de Velde, K., Kiekens, P. and Van Langenhove, L., 2003, July. Basalt fibres as reinforcement for composites. In Proceedings of 10th international conference on composites/nano engineering, University of New Orleans, New Orleans, LA, USA (Vol. 2026).

[6] Sim, J., Park, C. and Moon, D.Y., 2005. Characteristics of basalt fiber as a strengthening material for concrete structures. Composites Part B: Engineering, 36(6), pp.504-512.

[7] Wei, B., Cao, H. and Song, S., 2010. Tensile behavior contrast of basalt and glass fibers after chemical treatment. Materials & Design, 31(9), pp.4244-4250.

[8] Militký, J., Kovačič, V. and Rubnerová, J., 2002. Influence of thermal treatment on tensile failure of basalt fibers. Engineering Fracture Mechanics, 69(9), pp.1025-1033.

[9] Basaltex, Sumperk, basalt fabric producer, http://www.basaltex.cz/

[10] Welter, M., Schmuecker, M. and MacKenzie, K.J.D., 2015. Evolution of the Fibre-Matrix Interactions in Basalt-Fibre-Reinforced Geopolymer-Matrix Composites after Heating. J. Ceram. Sci. Tech, 6(1), pp.17-24.

[11]Samal, S., Marvalová, B., Petríková, I., Vallons, K.A., Lomov, S.V. and Rahier, H., 2016. Impact and post impact behavior of fabric reinforced geopolymer composite. Construction and Building Materials, 127, pp.111-124.

[12] Krystek, J., Laš, V., Pompe, V., and Hájková, P. Influence of Temperature on Selected Mechanical Properties of Geopolymer Composites, Proceedings of 55th Conference on Experimental Stress Analysis EAN2017, May 30th - June 1st, 2017, Nový Smokovec, Slovakia.

EAN 2017
55th conference on experimental stress analysis

# Reliability Assessment of Saint Joseph Church in Prague

Markova Jana[1, a], Sykora Miroslav [1,b] and Simunek Ivo[1,c]

[1]Klokner Institute, Czech Technical University in Prague, Czech Republic

[a]jana.markova@cvut.cz, [b]miroslav.sykora@cvut.cz, [c]ivo.simunek@cvut.cz

**Keywords:** Cracking; deformation; heritage; masonry; monitoring; movement; reliability assessment.

**Abstract.** Monitoring provides valuable information to be used for the decision making about the maintenance and rehabilitation of existing structures. The case study of a masonry church shows that numerous problems have to be commonly resolved during the optimally plan and perform monitoring; the major issues include the selection of technique, suitable position of monitoring devices on structures, data transmission and backup, consideration of uncertainties in measurements and defects of hardware. The study reveals how inspections and monitoring can support decisions related to optimal upgrade strategy and further use of the heritage structure.

## Introduction

Reliability assessment of heritage structures is a demanding task which commonly requires co-operation of different specialists [1]. Monitoring of structures is an important part of their reliability assessment. Monitoring during the preliminary assessment can contribute to planning of later monitoring during detailed assessment [2]. Presently Eurocodes are to be used for the verification of structures in all the CEN Member countries. However, a supplementary guidance is commonly needed for the verification of existing structures including heritage ones; this might be found in ISO 13822 [3]. For the operational applications in the Czech Republic, the new standard CSN 73 0038 [4] has been developed including information on properties of existing materials, on updating the basic variables, and also on experimental testing and monitoring.

The Klokner Institute was asked for the reliability assessment of the St. Joseph Church in Prague which is listed as a Czech national cultural heritage monument. The church was built within 1636-1653 as a part of the Capuchin monastery. The monastery was later rebuilt to neo-Romanesque barracks and recently transformed into a shopping centre Palladium. At present the church is thus situated closely to the shopping centre, hidden behind the wall which separates the church from the Republic Square and which creates a small courtyard decorated by two Baroque statues (Fig. 1). The interior of the church is quite simple. Lateral altars are decorated by two large paintings by the famous Czech painter Karel Skreta. The vertical walls of the church are made from stone masonry and ceilings include cylindrical vaults. The church is partly provided with cellars.

Fig. 1.

Fig. 2.

Various damages of the load bearing structures of the church appeared in the past and also during the construction of underground line B in 1990's of the last century. Partial reconstruction of some parts of the church was carried out from 2005 to 2007 when also the new shopping centre was built up. The appearance of various deformations and cracking in that time could be attributed to the construction of a new underground station and also the new shopping centre with underground parking garages.

After archaeological investigations, the Klokner Institute made the first survey of damages in 2004 in order to support decision about strengthening of the church foundations adjacent to the Palladium construction site by jet grouting. The findings made it possible to plan structural interventions. Selected walls were strengthened by Helifix ties and all rooms were renovated. However, it was shown that the strengthening was partly inefficient. damages across walls, paintings and various architectonic elements of the church

Several locations of the church were monitored for several years, focusing on structural movements, deformations and cracks in masonry walls and floors. The evaluation of monitoring results revealed that movements in some parts of the church are insignificant and in some cases even within the tolerances of a measurement technique. However, settlements of the whole structure in the direction to Palladium were identified.

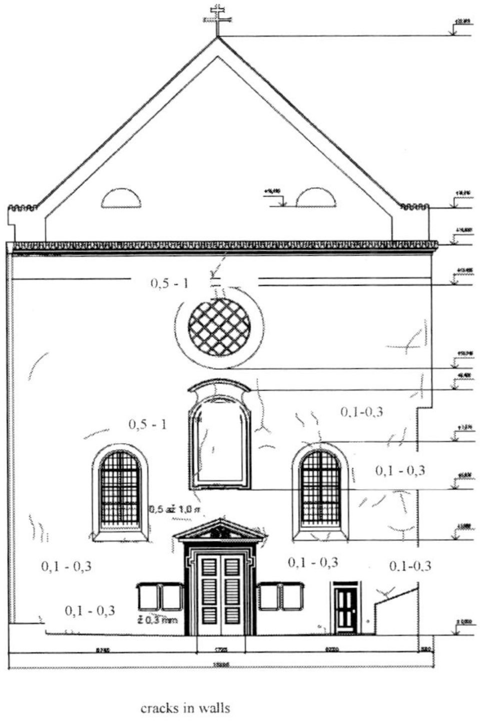


Fig. 3. Cracks in the front part of the church (in mm).

The results of the long-term monitoring carried out from 05/2005 to 06/2013 indicate that some monitored points settled about 5 mm, and the wall in the vicinity of Palladium up to 15 mm. Some damages and cracking resulted from the accidental horizontal movement of the sheet-pile wall during the construction of the Palladium centre which was also used for the stability assurance of the church foundations. The movement of the sheet-pile wall led to minor settlement and rotation of the church peripheral wall. Despite the fact that some church walls were partly strengthened by the Helifix bars, the cracking and other damages newly appeared and progressively developed.

## Survey of Damages and Monitoring

New survey of damages was undertaken (Fig. 3) and compared with the results of the previous one. Inspection revealed that several new damages appeared in the main church nave. In the wall neighbouring to the Palladium shopping centre, several new cracks of the width from 0.3 to 4 mm are visible. Main cracks extend to the arches and also to the ceilings, damaging the paintings. Cracks up to 4 mm in the masonry and considerable floor deformations appeared in the main nave.

The largest cracks of width of 5 mm were found in the east side of the church. In the west wall, the cracks in tops of the vaults width from 0.5 to 2 mm appeared. Considerable cracks exist between the west wall and the vault of the main nave. The cracks are present in most rooms of the church. Several wider cracks exist in the gable parts of the church. It might be assumed that some of them could be present before the construction of the new underground.

During the new survey, the cracks were documented in all rooms of the church including their shape, width and length. All the plaster targets which were placed on the church load bearing structures in the past (2005-2007) were checked. Inspection revealed that some of the monitored points suffered from no damage, while some others were affected by cracks.

The findings of the recent survey are critically compared to those from the 2004 survey. It is apparent that in some parts of the walls, ceilings and floors new cracks and deformations have occurred, e.g. cracking in tops of the vaults and bulging of the floor in the main church nave.

## Effects of Traffic

To estimate the influence of surrounding environment on the church, the traffic was monitored by a one channel vibrometer in May 2014. Passages of trams, buses, impacts of human activities and some other sources of vibrations were measured and evaluated as is reported in Table 1 (indicating maxima) accompanied by a floor layout with locations of the monitored points.

Table 1. Maximum monitored responses in mm/s.

| No. | Response $v_{ef}$ (mm/s) | vertical | laterally horizontal | longitudinally horizontal |
|---|---|---|---|---|
| 7 | lorry | 0.05 | 0.06 | 0.03 |
| 4 | tram | 0.08 | 0.045 | 0.03 |
| 7 | bus | 0.05 | 0.03 | 0.03 |
| 4 | pedestrians | 0.03 | 0.02 | 0.02 |
| 3 | minor impacts | 0.04 | 0.02 | 0.02 |
| 8 | bump | 0.02 | 0.01 | 0.15 |
| all | other | 0.01 | 0.01 | 0.01 |

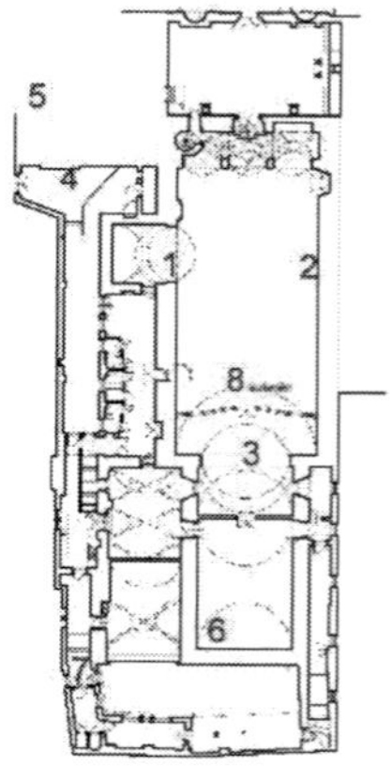


The results of monitoring are compared with the limiting values recommended in the national standard ČSN 73 0040 [5]. The load-bearing structure of the church might be classified as Resistance Class A (monument, deteriorated structure, built without prescriptive documents) and as Significance Class III (specific economic and social significance). For such a structure, the following limits are provided by the standard:

- $v_{ef} \leq 1.2$ mm/s – dynamic assessment is not required
- $v_{ef} \leq 4.5$ mm/s – no progressive damage is assumed to develop
- $v_{ef} \leq 7$ mm/s – minor damage including cracks of width up to 1 mm in joints of ceilings is expected to develop with likelihood 5%.

The measured maxima of the traffic effects are about two orders lower in comparison with the limiting values which might potentially have some influence on the cracks formation. It is thus concluded that the traffic in the vicinity of the St. Joseph church has insignificant influence on cracking of masonry structures.

## Destructive Test of Floor

One destructive test was made in the location of the floor deformation. The aim was to investigate the composition of the floor layers and reveal the cause of deformation. It was suspected that the floor might be influenced by the pressure of underground water. The composition of floor is illustrated in Fig. 4 where the cavity of the thickness from 0.03 m to 0.15 m under the layer of historic terrace.

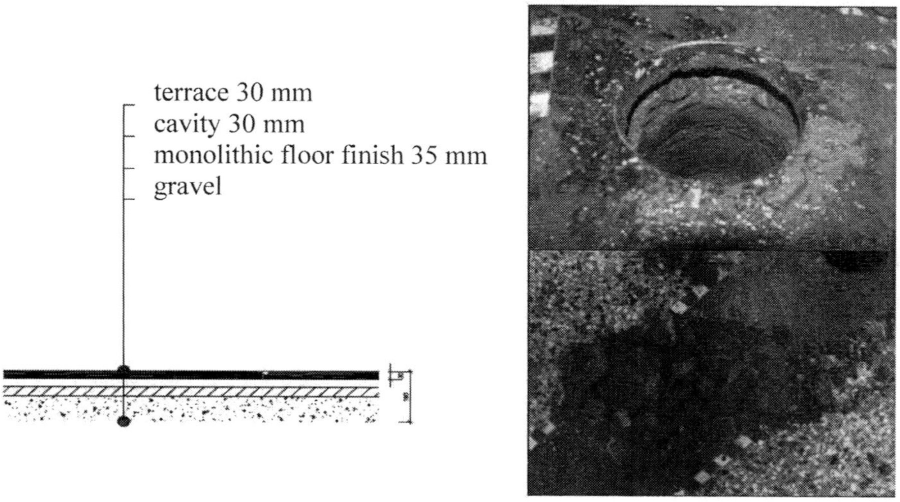


Fig. 4. Test pit in the location of the deformation of the floor.

No underground water nor injected mortar used for the soil strengthening were not detected by the test. It appears that the deformation of floor is caused by the horizontal movements of the neighbouring shopping centre.

**Conclusions**

The construction survey shows that the walls and vaults of the church, rehabilitated about ten years before the present investigations, are in many places newly damaged by cracks, mostly of about 0.4mm widths. In some cases, cracks exceeding 4 mm were found, mostly in the gable walls and in tops of the vaults. Some of the cracks were caused by the construction of the Prague underground line B, the others during the execution of the Palladium shopping centre. Recently, new significant cracks have appeared in the ceiling and tops of the vaults in the main church nave. In the lateral bays adjacent to the Palladium shopping centre, the floor is newly deformed, elevated up to 0.2 m.

The results of monitoring indicate the local settlement of foundations in some parts, deformations or movements of structures. An influence of a former embankment under the church foundations which could consolidate due to significant geotechnical activities during execution of Palladium shopping centre, might be assumed. Despite the injection of the soil and application of the sheet-pile walls for safeguarding the church and geotechnical works on Palladium building, their safeguarding effects might be locally reduced (also due to the technical problems with sheet-pile walls during execution of the shopping centre).

A destructive test of the deformed floor revealed a hollow without underground water or injection mortar due to former geotechnical protective works. It might be concluded that there might be an influence of local lateral pressures of structures, potentially in combination with uneven settlements of the foundations having impact on the deformation of floor in the main church nave.

It was recommended to further monitor movements of the structures, their possible effects on further progression of cracking and deformations. Results of such campaign would support enhanced reliability verification of the damaged masonry structure [6-8] On the basis of monitoring new effective measures can be accepted.

Informative monitoring of the effects of tram and road traffic leads to the conclusion that the public transport is insignificantly affecting reliability of the church. The current conditions of the load-bearing structural system of the church might be considered as acceptable and the church might be further used for its purpose by the Capuchin order and also by tourists.

Movements and crack development of selected structures of the church are being monitored. When no significant progress of cracks is observed, it is possible to repair the church. Otherwise relevant structural measures should be taken to increase stiffness and robustness of the load bearing structures of the historic monument.

## Acknowledgments

This study is an outcome of the research project NAKI DG16P02M050 "Optimisation of observations and assessment of heritage structures", supported by the Ministry of Culture of the Czech Republic.

## References

[1] M. Sýkora, M. Holický, J. Marková, T. Šenberger, PROBABILISTIC RELIABILITY ASSESSMENT OF EXISTING STRUCTURES focused on industrial heritage buildings, 1st ed., CTU Publishing House, Prague, 2016.

[2] Joint Committee on Structural Safety, Probabilistic Assessment of Existing Structures, RILEM Publications S.A.R.L., 2001.

[3] ISO 13822 Bases for design of structures - Assessment of existing structures, ISO, Geneva, 2010.

[4] CSN 73 0038 Assessment and verification of existing structures - Supplementary guidance, ÚNMZ, Prague, 2014.

[5] CSN 73 0040 Loads of technical structures by technical seismicity, ÚNMZ, Prague, 1996.

[6] J. Witzany, T. Čejka, M. Sýkora, M. Holický, Strength assessment of historic brick masonry, Journal of Civil Engineering and Management 22 (2016) 224-233.

[7] J. Witzany, T. Čejka, M. Sýkora, M. Holický, Assessment of compressive strength of historic mixed masonry, Journal of Civil Engineering and Management 22 (2016) 391-400.

[8] M. Sýkora, T. Čejka, M. Holický, J. Witzany, Probabilistic model for compressive strength of historic masonry, Safety, Reliability and Risk Analysis: Beyond the Horizon - Proceedings of the European Safety and Reliability Conference ESREL 2013, Amsterdam: Netherlands; 29 September - 2 October 2013, 2014, pp. 2645-2652.

EAN 2017
55th conference on experimental stress analysis

# Determination of the Cause of Failure Based on Fracture Marks on Parts of a Complicated Transmission Mechanism

G. Izrael[1,a], P. Élesztős[1,b], D. Kaločany[2,c]

[1]STU in Bratislava, Faculty of Mechanical Engineering, SK

[2]EDAG Engineering CZ, Prague 5, CZ

[a]gregor.izrael@stuba.sk, [b]pavel.elesztos@stuba.sk, [c]daniel.kalocany@gmail.com

**Keywords:** strain gauges, fatigue failure, burdening

**Abstract.** Device failure or a failure of its part can be caused by different factors. These include for example deficiency of assembly, operation conditions, or their closely related fatigue failure. In this paper, we deal with analyses of pivots fatigue failure and causes of the occurrence of this kind of failure in the transmission mechanism of a mixing device.

## Introduction

A breakdown of equipment can result in the inoperability of the machine, which will have a significant effect on production capacity and so on the economic indicators of the operator. Damage to a machine part may be caused by a variety of factors (working technology, the environment, material properties, operation period, assembly errors and the like) which influence the lifetime, not least the fatigue life of the machine. Such a case was the breakdown of the transmission mechanism of a piece of technological equipment for which it was necessary to analyze the breakdown and its possible reasons, and to propose the necessary measures to be taken.

## Breakdown, analysis of the reasons for the breakdown

The breakdown appeared on the gearbox of a mixing and dosing device whose output shaft, along with rotational movement, also performs a translational movement which is secured by means of a rather complex gearbox made up of a conical differential, excentres, a connecting rod and a so-called crosser. The damage, in the form of fatigue failure, was apparent in the "crosser" by the breaking off of both pins which are located on the sides (fig. 1).

Fig. 1 Damaged crosser with broken pins and caps

The operational fractures on the crosser pin were subjected to breakage analysis, on the basis of which the following facts were arrived at:

- On the crosser's left pin were clear marks of damage in the form of a fatigue fracture (fig. 2). From the figure it is clear that the fracture came from the side of the pin's rib stiffener and its initiation was in the place of the rounded transition from the crosser body to the pin (pin diameter 107 mm);
- the fracture area had conchoidal marks from the expansion of the fracture concentrated at the initiation point. The relatively rough area of the fatigue part of the fracture witnesses to intensive widening of the crack, leading to a sudden fracture. The opposite side of the fracture area showed characteristics of shearing (fig. 3);

Fig. 2 Fracture of the left crosser pin

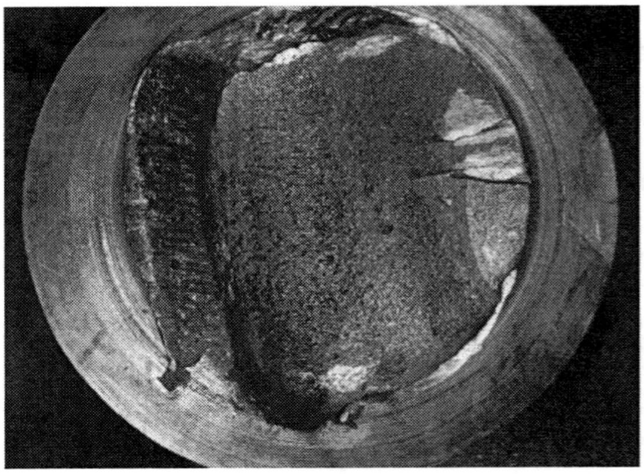

Fig. 3 Structure of the fracture area of the left crosser pin

- from the relation of the size of the area of the fatigue part of the fracture and of the sudden fracture it can be concluded that the amplitude of transition burdening exceeded the mean tension value. The direction of the expansion of the fatigue part and the position of the breakage, as well as the character of the surface of the fracture, suggest strain from unsymmetrical bending in the plane of the pin axis and in the movement of the crosser. The unsymmetrical bending stress results from the differentiated pin pressure as the crosser moved forwards – supplying the material and return movement;
- significant marks of expansion of the fatigue fracture were not found in the fracture area of the right pin (fig. 4). The surface remained rough, with a small facture area. The fracture here took place more quickly than on the left pin. The initiation of the fracture, the expansion direction, the fracture position and also the character of burdening were similar as in the case of the left pin.

From the inspection and analysis of the fractures of the pins on the individual sides of the crosser, there resulted the conclusion that the occurrence of the fatigue cracks were the effect of probably unbalanced burdening and to a large degree oversizing when placing the caps on the individual pins.

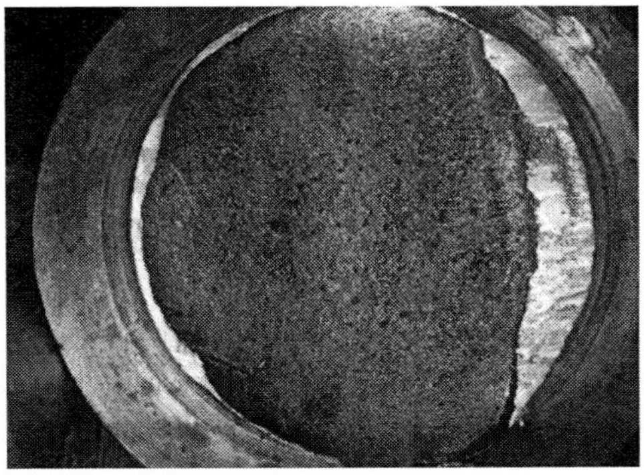

Fig. 4 Structure of the fracture area of the right crosser pin

When considering the complex tension state of the given part of the equipment (the crosser pins), oversizing in the fitting of the caps on the individual pins cannot be neglected. The crosser pins, to improve the frictional relations, are equipped with pin caps. These caps are heat-moulded onto the pins and after cooling provide a radial pressure on the pins which can produce a concentration of stress in the pins' translational zone. The distribution of these pressures along the diameter of the pin is axis-symmetrical. The goal of the analysis of the strain state was to judge the intensity of the effects of these strains which are superimposed on the other components of the internal forces (mainly arising from the bend stress of the crosser pins) and which can, under certain circumstances, have a large influence on the fatigue lifetime of these places.

On the equipment is used tolerance scale of pins and caps H7/r6:

pin: Ø 107 r6, which means => $\phi 107^{+0,076}_{+0,019}$

cap: Ø 107 H7, which means => $\phi 107^{+0,035}_{+0,00}$

maximal sizing: *0,076* mm

minimal sizing: *0,019* mm

For determining the normal pressure between the cap and the pin, the following simplifying conditions were introduced:

- The maximum possible sizing while observing the above-stated production tolerances was taken into consideration;
- It was assumed that the shaft is sufficiently thick and will not deform after cooling.

Under the stated assumptions the relative elongation in the cap can be expressed by the relation:

$$\varepsilon = \frac{\pi(D+\Delta D)-\pi D}{\pi D} = \frac{\Delta D}{D} = \frac{\Delta R}{R} \qquad [-] \qquad (1)$$

where:

$\varepsilon$ – is relative elongation [-]

$D$ – je the diameter of the shaft [mm]

$\Delta D$ – is the size of the tolerance [mm]

By means of Hook's law for uniaxial strain we then get:

$$\sigma = \frac{\Delta D}{D} E \qquad [\text{MPa}] \qquad (2)$$

where:

$E$ – is Young's modulus of compression [MPa]

After adjustment, the normal pressure $p$ (at maximum sizing) we can calculate with the relation:

$$p = \frac{\Delta D E t}{D^2} \qquad [\text{MPa}] \qquad (3)$$

After substitution of numerical values, the normal pressure derived from maximum sizing equalled $p = 41$ MPa. The concentration of stress in the transfer area from the pin to the thick part of the crosser was sought numerically by the finite elements method.

Fig. 5 Distribution of main stress on transition from pin to thick part of crosser

From the distribution of stress (fig. 5) it clearly follows that in the locality of maximum bending stress (where the fatigue crack initiated on the pins) there occurred significant stress from the moulded steel cap. Since these stresses are tensile (positive), with each rotation of the connecting rod they superimpose on the bending and increase the amplitude stress value.

Other significant factors affecting the fatigue lifetime of the equipment are assembly deviations of the excentres, looseness in the axial bearings, and similar.

As was stated above, the analyzed gearbox of the mixing equipment had to provide translational movement of the mixing worm in addition to its rotational motion. This property was achieved by the usage conception of two excentrically positioned connecting rods. If both excentres are synchronized completely accurately, they achieve closure at the very same moment, and so provide the necessary equal horizontal movement of both of the pins. Since the failures occurred just on the stated pins and after large-scale general maintenance (during the repairs the whole gearbox was disassembled and put together anew), it could be assumed that the excentres are not synchronized. If the extremes of the excentres are slewed against each other horizontal movement on the feeder pins will not be equal, and one side will be more stressed. For discovery of this fact, measuring was performed on the connecting rods wherein strain-gauge sensors were placed on the left and right sides of the upper flanges in the proximity of the crosser pins (fig. 6).

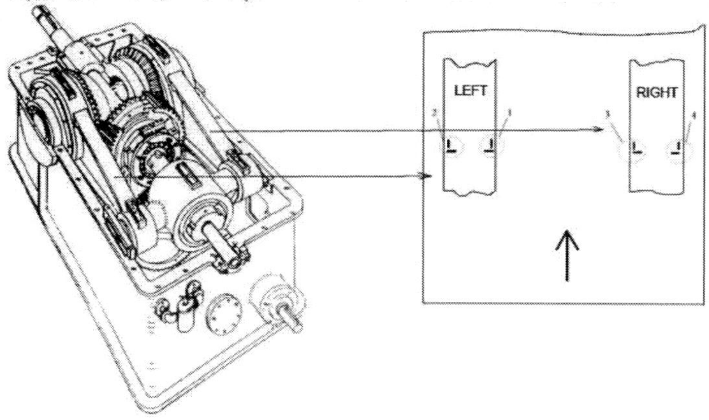


Fig. 6 Location of strain gauges on connecting rods

From the measured values it was clearly demonstrable that the assumption of inaccurate matching of the two excentres was confirmed. From the measured record (fig. 7) it was clear that in the movements of the mixing worm to the stroke (movement of the material) there was manifested on the left interior side about twice as much burdening as on the right exterior side, which clearly caused the bending stress on the crosser pins.

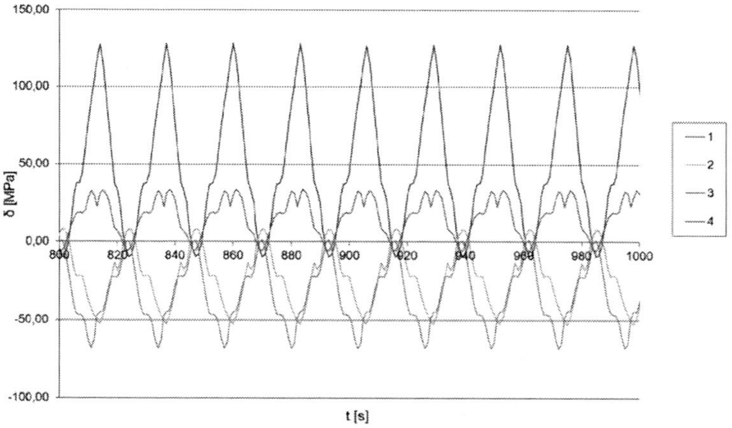


Fig. 7 Detail from the measured record

Maximum stress measured in individual places:

1. + 136,9 MPa
2. - 57,1 MPa
3. - 71,4 MPa
4. + 37,1 MPa

The stress values measured on the left connecting rod were transferred to the crosser pin and the stresses causing tensile and bending stress were separated:

$$\sigma_t = (\sigma_{max} + \sigma_{min})/2 = 136{,}9 - 57{,}1 = 39{,}9\ MPa \qquad [\text{MPa}] \qquad (4)$$

$$\sigma_o = (\sigma_{max} - \sigma_{min})/2 = 136{,}9 + 57{,}1 = 97\ MPa \qquad [\text{MPa}] \qquad (5)$$

The bending stress determined in this way caused the bending moment:

$$M_o = \sigma_o . W_{o,\,pin} = 13{,}178.10^6\ Nmm \qquad [N.mm] \qquad (6)$$

It is assumed that the bending across the sliding position of the junction of the connecting rod and the crosser is transferred onto the pin, where it causes the stress:

$$\sigma_{o,pin} = M_o/W_{o,pin} =$$

$$13{,}178.10^6/0{,}1.107^3 = 107{,}57\ MPa \qquad [MPa] \qquad (7)$$

With consideration of the concentration of the stress from the reduction of the diameter and rounding in the transition from the crosser body to the pin, which is expressed by the coefficient of the effect of the stress $\alpha$, the value in the critical place on the pin is:

$$\sigma_{max} = \alpha .\ \sigma_{o,pin} = 1{,}9.\ 107{,}57 = 204{,}38\ MPa \qquad [MPa] \qquad (8)$$

It is also necessary to add on the additional stress from the moulded cap, *41* MPa, from which it follows that the maximum stress will be $\sigma_{max} = 245{,}38$ MPa.
The value of this stress represents the lower border of the potential stress, and its value approaches the boundary fatigue of the material used. It should however be emphasized that these calculations are only approximate and cannot exclude inaccuracies that can significantly affect the results.

## Conclusion

The results of the measurements show that the premature breakdown on the crosser pins in the gearbox equipment was caused by added pressure, whose origin was not burdening from the working technology of mixing, but more probably due to geometric or assembly inaccuracies in the makeup of the gearbox equipment.

In the case of need it is possible to determine from the measured stress values on the connecting rods the load power of the movement of the mixer's shaft, which equals the differential of the stress on the tensile components in the measured places on the connecting rods.

## Acknowledgements

This publication is the result of the project implementation: Research of friction stir welding (FSW) application as an alternative to melting welding methods no. 26240220031 supported by the Research & Development Operational Programme funded by the ERDF. Funding from the VEGA grant 1/0740/16 is also greatly appreciated.

## References

[1] O. Ostertag, P. Sivák, Degradačné procesy a predikcia životnosti, Edícia vedeckej a odbornej literatúry, TU v Košiciach, (2010). In slovak.

[2] F. Trebuňa, F. Šimčák, J. Bocko: Failure analysis of storage tank. In: Engineering Failure Analysis. Vol. 16, no. 1 (2009), p. 26-38.

[3] F. Trebuňa, F. Šimčák, J. Bocko, M. Pástor: Analysis of causes of casting pedestal failures and the measures for increasing its residual lifetime, In: Engineering Failure Analysis. Vol. 18, no. 1(2011), p. 244-255.

[4] P. Slovák, L. Gulan: Strength verification of a boom loader module. Annals of Faculty of Engineering Hunedoara - Journal of Engineering, s. 37--40. ISSN 1584-2673.

EAN 2017
55th conference on experimental stress analysis

# Important Aspects of Testing Welds

Petr Brož [a], Daniel Dobiáš [b]

[1]University of West Bohemia, Faculty of Applied Sciences, Univerzitní 8, 306014 Pilsen, Czech Republic

[2]Czech Technical University in Prague, Klokner Institute, Šolínova 7, 166 08 Prague 6, Czech Republic

[a]broz.petr@tiscali.cz, [b]dobias@klok.cvut.cz

**Keywords:** Crack growth, driving force, fracture toughness, microstructure, residual stresses, weldment

**Abstract.** In the treatise presented, the fracture toughness testing of weldments is the matter. Consideration is primarily focused on the effect on testing specific microstructural zones, and elimination of residual stress influences on toughness measurement. Further, the effect of definite weld characteristics on interpretation of test outcomes is surveyed embracing statistical effects appearing from material in homogeneity.

## Introduction

Material discontinuities and defects affect weldments adversely. It can extremely influence their mechanical uprightness [1]. A listing of distinct kinds of weld imperfections is presented by the weld quality standards. These consist of rather different defects as arc strikes, cracks, inclusions, lack of fusion, misalignment, porosity, undercuts and additions that can be fixed to diverse ranges:

- Cracks and crack – like imperfections (ie hot and cold cracks, lack of fusion, incomplete penetration, et cetera) that must be prevented or – if they appear – are directly subject to fracture mechanics analysis.
- Geometric discontinuities that increase the local stresses (eg misalignment such as angular distortion or undercuts). This type of defects inclusive of welding residual stresses, may influence crack initiation and crack growth, too, and end failure.
- Material imperfections that function as crack initiation areas (ie slag inclusions), being eminent importance for fatigue life analyses and fatigue strength.
- Imperfections which evidently have no influence on fatigue life and fracture (eg porosity in low stress part of the constituent).

## Doing experiments with weld parts

Requisite considerations and parameters for an appropriate role of the test specimens are, as follows: welding process including filler material, base metal composition, joint thickness, preheat and interpass temperatures, heat input, detailed welding procedure, joint configuration, restraint, post-weld treatment, time between welding and testing, environment, test temperature.

Fatigue testing results in state of affairs that the fracture toughness can be crucially dependent on the foregoing factors. Some knowledge of the influences of the miscellaneous parameters is essential so as to achieve conservative limits in testing and constructional assessment. Sometimes hydrogen release heat handling must be realized prior to testing in order to assure same levels of hydrogen in the structure. That is made vital when the period between welding and service beginning is much longer than this between welding and testing.

For specimen preparing, the purpose is to stipulate the fracture toughness as regards of crack initiation or propagating (R – curve) for a precise target region, for example the weld centre line, the heat affected zone, at cetera (marked as weld positional, WP) and / or specific microstructure, eg, coarse grained heat affected zone (marked as SM). Occasionally, eg for narrow laser and electron beam welds, electro – discharge machining with thin diameter wire should be applied for inserting the notch. The example for specific microstructure (SM) notch placing is introduced in Table 1 [2].

| | Specimen orientation | Specimen geometry |
|---|---|---|
| 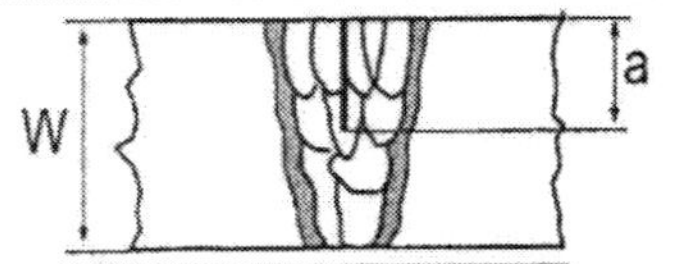  | NQ | $B \times B$ |
| 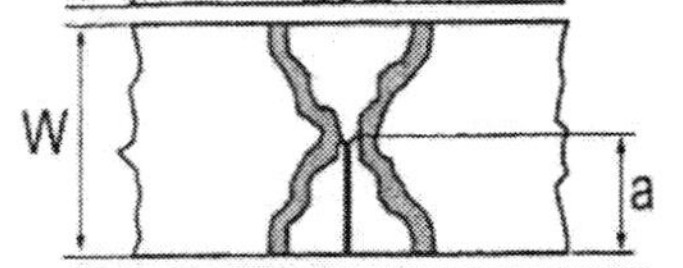  | NQ | $B \times B$ |
| 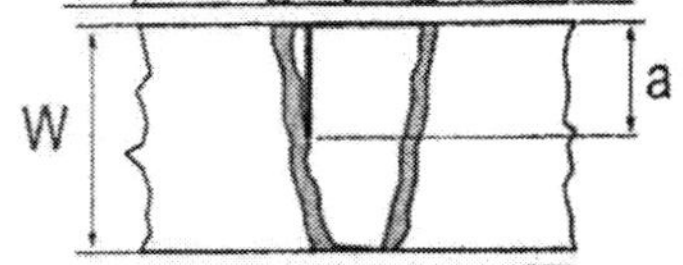  | NQ | $B \times B$ |
| 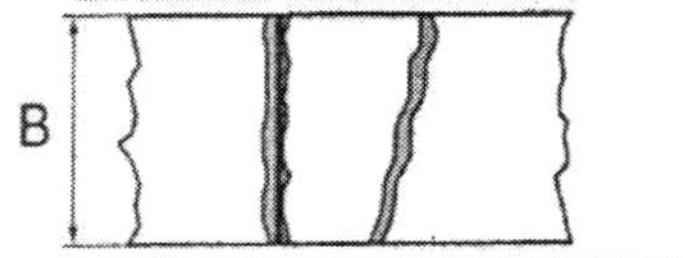  | NP | $B \times B$ or $B \times 2B$ |

Table 1. Specific microstructure notch location, according to [2]

## Pre – cracking

Prior to fatigue pre – cracking, it might be essential to adapt the welding residual stresses in specimen prior to testing. There are two incentives for that:

(i) In a fracture mechanics study, the residual stresses are taken into account when calculating the crack driving force in the constituent [1], [3]. From this standpoint, it would be perfect to totally take away the residual stresses in the test specimen.

(ii) As the welding residual stresses demonstrate inhomogeneous pattern through the section, they may affect fatigue crack growth like that the pre – crack front will develop an irregular form.

Even though the residual stresses are partially alleviated and redistributed owing to the extraction of the specimen and the insertion of the crack beginning notch, the remaining residual stresses could be large enough to affect the form of the pre – crack front and the issues of the fracture toughness so as to prevent this, diverse approaches are suggested for the handling of test specimens prior to or in course of pre – cracking (Fig. 1).

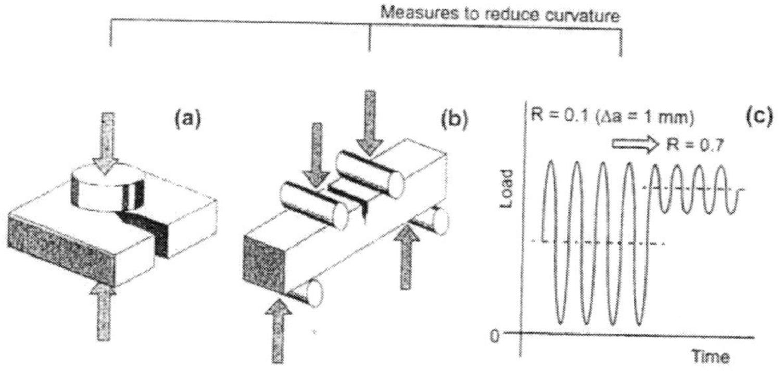


Fig. 1. Methods proposed for redistributing residual stresses prior to and during pre-cracking [1]

**Statistical profile of HAZ toughness**

Being a result of the scatter in the fracture toughness of weldments and in part of heat affected zone cracks, the definite statistical treatment becomes imperative. As a rule, that is based on the weakest link guiding principle and applies a three - parameter Weibull distribution fit, though even other types of mathematical distributions are used. The so – called VTT – Master Curve procedure included in the test standard ASTM E 1921 is often employed. This particularizes the aforementioned distribution of many replicate test results in the form

$$P = 1 - \exp\left\{-\left[\frac{(K_{mat} - K_{\min})}{K_0 - K_{\min}}\right]^m\right\} \tag{1}$$

where $P$ the failure probability of the test specimens, $K_{mat}$ the fracture toughness via the $K$ factor, $K_0$ the scale parameter, $K_{\min}$ the shift parameter and $m$ the shape parameter of the distribution. The coefficients $K_0$, $K_{\min}$ and $m$ are fit parameters, still in the Master Curve concept two these are fixed. For ferritic steels with yield strengths between $\sigma_Y = 275$ and 825 MPa, the shape parameter is given by $m = 4$ and the shift parameter by $K_{\min} = 20$ MPa $m^{1/2}$. The toughness $K_{mat}$, is formally specified from the critical $J$ integral, $J_{mat}$, or CTOD, $\Delta_{mat}$, by the relation

$$K_{mat} = \sqrt{J_{mat} \cdot E/(1 - \nu^2)} \tag{2}$$

or

$$K_{mat} = \sqrt{\beta \cdot \delta_{mat} \cdot \sigma_Y \cdot E/(1 - \nu^2)} \tag{3}$$

where $E$ means Young`s modulus, $v$ being Poison`s ratio and $\beta$ constraint factor that is conservatively chosen as $\beta = 1.5$ for steels with a strain hardening coefficient $N \geq 0.05$. The impact of the material inhomogeneity on statistical treatment is ilustrated in Fig. 2.

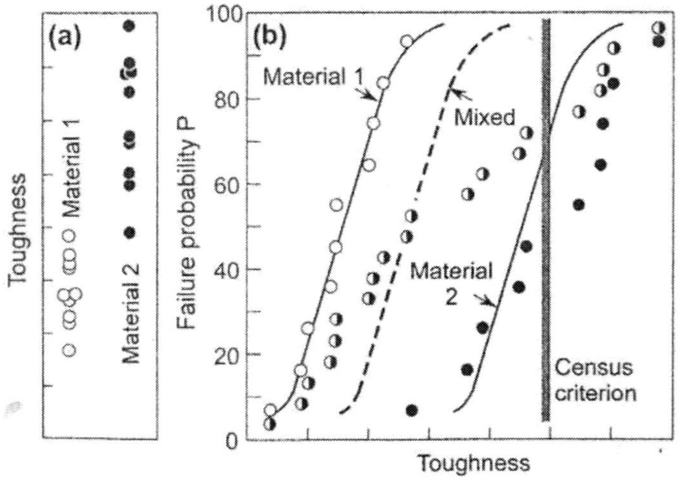


Fig. 2. Ilustration of the mixture effect of two materials with different toughness on the usual toughness distribution applying the censoring criterion [1]

**Fatigue crack growth qualities**

About the interrelation between fatigue crack, strength mismatch and welding residual stresses it is familiar the opinion that the welding residual stresses are important for fatigue crack propagation owing to their influence on the stress ratio $R$ (= $\sigma_{min}/\sigma_{max}$ or $K_{min}/K_{max}$). Compressive residual stresses decrease and tensile residual stresses increase the $R$ ratio. The varying residual stress profiles in another zones of the weldment such as the weld centre line or the fusion line are respected liable for crack propagation rates both faster and slower compared with the base metal while the different microstructures of base metal, weld metal and heat affected zone appear to play a minor part in the regime of the $da/dN - \Delta K$ curve. On the contrary, the near threshold range of the curves seems to be more influenced.

The $R$ – ratio is regarded at the crack driving force part of the component rather than at the material part. Residual stresses and strength mismatch have also an influence on the crack tip constraint, and it is known that such effects might play a part in the threshold and lower ranges of the $da/dN - \Delta K$ curves. In a recent investigation on a thermomechanically treated steel used in ship – building was founded the trend indicated in Fig. 3 [4].

Such as, Jones at al examined non – welded $C$ ($T$) and $M$ ($T$) samples of aircraft steel, austenitic ductile iron, cast steel and rail steel rigorously and discovered the crack growth rate not only influenced by the cyclic stress intensity factor range, $\Delta K$ and the $R$ – ratio still also by the crack length that is an obvious indication of constraint effects.

**Conclusion**

1. Characteristics in addition to conventional testing are the possibility of notch location choosing in connection with pre – and – post test metallography and the possible necessity for modifying the residual stress state before or during pre – cracking.
2. Usually, no pre – treatment of the specimens is required for welds that have been post weld heat treated for the aims of stress relief. The pre – treatment is also frequently unnecessary for surface notched ($B$ x $B$) specimens. The ISO test standards

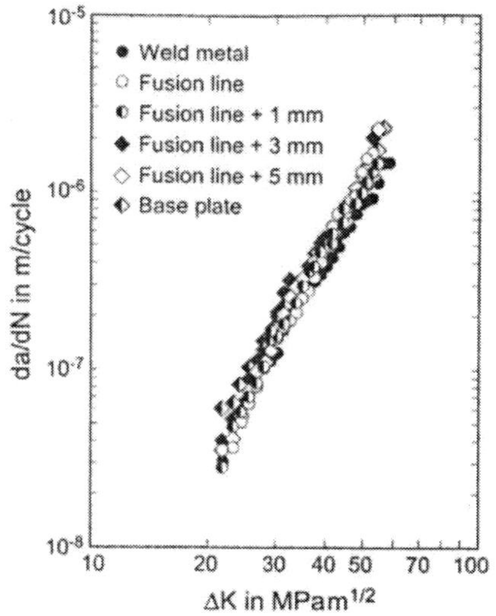


Fig. 3. Range $da/dN - \Delta K$ particulars for base metal, weld metal and HAZ, $R=0$ [4]

recommend the application of the shortest fatigue crack length authorized in the basic document ISO 12135 to minimize crack front curving and crack deviation, too, from the specified crack zone. If shallow notched specimens are used, ie $a_0/W$ < 0.45 ISO 15 653 affords solutions for $J$ and CTOD in that the plastic factor solution, $\eta_p$ , is based on crack mouth opening displacement gained from $J$ analyses. These solutions hold for $0.1 < a/W < 0.45$.

3. A minimum of 12 specimens essential for statistical describing the HAZ toughness of structural steels with yield limit up to 450 MPa in the ductile – brittle transition range is announced in BS 7910. Supplementary tests can be necessary to fulfil the metallographical validation criteria. These demands may be released prominently, if the HAZ shows upper shelf behaviour. After revision, the stated standard will incorporate the SINTAP/FITNET procedure as a choice when analyzing fracture toughness outcomes from welds. As the procedure is based on weakest link statistics, the fracture toughness inferred will be a function of crack front length in the structural element being appraised.
4. Some authors did not find any statistically substantial differences in the $da/dN - \Delta K$ despite of differences in materials strength, welding methods and $R$ ratio and even variable versus constant amplitude loading can be a result of the high tensile residual stresses in the weld centre. That would have maintained the crack open in course of the complete loading cycle such which no crack closure appeared that could have been influenced by aforementioned considerations.

## Acknoledgment


The authors gratefully acknowledge the financial support of the presented research by the University of West Bohemia in Pilsen. This article was prepared with support of the Grant Agency of the Czech Republic under the grant project GBP 105/12/G059.


## References


[1] U. Zerbst, M. Schodel, S. Webster, RA. Ainsworth, Fitness-for-service assessment of structures containing cracks, A workbook based on the European SINTAP/FITNET procedure, Elsevir, 2007.

[2] ISO 15653, Metallic materials-method for the determination of quasistatic fracture toughness of welds, International Organisation for Standardization (ISO), 2010
[3] P. Brož, Residual Stress Effects and Predicting Crack Growth in Weldments, Materials Structure & Micromechanics of Fracture VIII, MSMF8, 2016
[4] H.K. Lee, K.S. Kim, C.M. Kim, Fracture resistance of a steel weld joint under fatigue loading, Engng Fract Mech, 66, pp 403-19, 2000

EAN 2017
55th conference on experimental stress analysis

# Based reliability Assessment Method Applied to the Model of Structural Element

Petr Kesl[1, a] and František Plánička[2,b]

[1]University of West Bohemia in Pilsen, Faculty od Applied Sciences, Department of Mechanics, Univerzitní 8, 306 14 Plzeň, CZECH REPUBLIC

[2] University of West Bohemia in Pilsen, Faculty od Applied Sciences, Department of Mechanics, Univerzitní 8, 306 14 Plzeň, CZECH REPUBLIC

[a]pkesl@kme.zcu.cz, [b]planicka@kme.zcu.cz

**Keywords:** probabilistic SBRA Method, Monte Carlo simulation, serviceability of structure, strain, stress, modelling, steel beam, experimental results, results obtained by SBRA Method.

**Abstract.** The article deals with possibility of application of the simulation based reliability assessment SBRA Method in modelling of structures in the field of judgement of their serviceability. A steel simply supported beam of rectangular cross sectional area was taken into account. Strains of the model using electrical-resistance strain gage method were determined. Because of the same material of the model and real beam, strains and stresses are the same in both cases. At the first step a model of the beam was created and the necessary corresponding parameters were determined for real beam using modelling rules. It means that inverse procedure of modelling was used in that case. Stresses of the model and corresponding beam were determined using SBRA Method when following random variable quantities as section modulus of the beams W, Young´s modulus of elasticity E and acting force F were taken into account.

## Introduction

The article deals with an ability to assess the reliability of structural element on its model by means of the simulation based reliability assessment SBRA Method. The aim is to present a possibility of judgement of probability of failure of real component on its model. A simply supported steel beam of rectangular cross section area loaded by a simple concentrated force $F$ (Fig. 3) was taken into account as a model of real steel beam. That model was available. That´s why that opposite approach was chosen when for the existing model actual corresponding beam was searched using rules of modelling. Using modelling rules the bending stresses were determined in the real beam using experimentally determined strains and stresses in the model. Stresses of steel beam and its model were determined using SBRA Method too and that method was used for judgement of their probability failure. Obtained results were compared

The simulation Based Reliability Assessment Method is a probabilistic method using the Monte Carlo simulation [**1, 3**]. Substance of that method consist in repeated calculations of relatively simple equations, where variables as dimensions of the body, mechanical properties, loads, etc. can be insert. That variables can be constant or defined by histograms, respectively. Probability of failures of model and real beam were determined using Anthill software [**3**].

Modelling of engineering problems can be very often a way to solve them. It is generally based on the conditions

$$(\pi_i)_S = (\pi_i)_M, \quad i = 1, 2, ..., m \tag{1}$$

where $\pi_i$ are so called dimensionless parameters for structure (subscript S) and model (subscript *M*)

$$\pi = x_1^{e_1} \; x_2^{e_2} \; .....x_n^{e}, \tag{2}$$

if the solved problem depends on n variables $x_i$ (expressing physical, geometrical etc. quantities). Exponents $e_i$, $i = 1, 2, .., n$ has to fulfil the condition of dimensionless of $\pi$ terms, see [2].

$m = n - r$ represents number of independent $\pi$ terms, *r* is rank of so called dimensional matrix, see [2]. There is assumed that solved problem is described by *n* physical quantities $x_i$, $i = 1, 2, .., n$ containing $k$ so called primary quantities with primary units $\lfloor L_j \rfloor$, $j = 1, 2, .., k$. The necessary procedure is described in [2].

Variables incoming in the solved problem are: $l[mm]$ length , $W[mm^3]$ section modulus, $E[MPa]$ modulus of elasticity, $F[N]$ acting force , $\sigma[MPa]$ acting stress, $\varepsilon$ strain. There are two primary quantities length and force with two primary units in the solved problem $[mm]$ and $[N]$. It is possible in modelling of static tasks to consider Newton as basic dimension because time and mass are not in this type of tasks separate. Then number of independent terms will be $m = 6 - 2 = 4$m. Because the strain is dimensionless quantity there are only three remaining independent terms:

$$\pi_1 = l^3 / W \tag{3}$$

$$\pi_2 = F / El^2 \tag{4}$$

$$\pi_3 = \sigma / E \tag{5}$$

From Eq. 1 and Eq. 5 is obvious that stresses in the model and in the beam are for the equal material of the same magnitude. Scale for the model was chosen equal to 5. That is why the length of the beam was $l_B = 2500\,mm$.

From Eq. 1 and Eq.3 is corresponding section modulus of the beam

$$W_B = W_M \cdot l_B^3 / l_M^3 = 45.375 \cdot 10^3 \; mm^3$$

and from Eq.1 and Eq. 4 is corresponding force acting on the beam

$$F_B = F_M \cdot l_B^2 / l_M^2 = 25 F_M$$

Probability of failure is guided by so called safety function

$$P_{f(i)} = R_{(i)} = S_{(i)} \,, \tag{6}$$

where **S** is the load effect And **R** is the structural resistance. The probability of failure $P_f$ of a body can be then expressed as a ratio between the number $N_f$ of results that do not fulfil the defined before safety function and $N_t$ is total number of results [1], see Fig. 2.

$$P_f = \mathrm{N_f} / N_t \,. \tag{7}$$

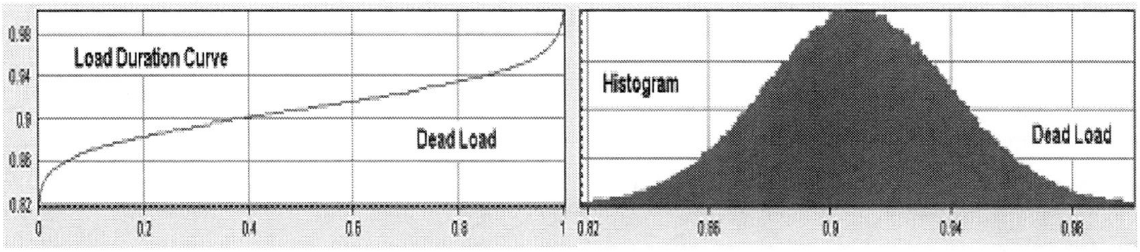


Fig. 1. Histogram of the Dead Load

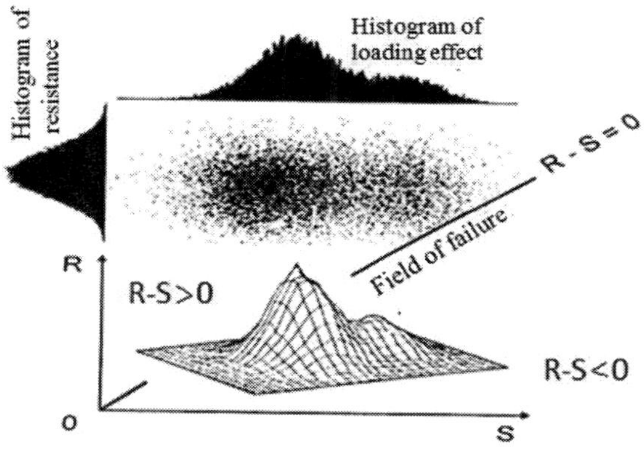


Fig. 2. Probability of failure

**Experimental results**

The used steel model is stated in Fig. 3. Loading forces were $F$ = 100, 200, 300, 400, 450 N. Strains were measured using electrical-resistance gages. Their positions are obvious in Fig. 3. Recorded values of strains and corresponding stresses are in the Tab. 1.

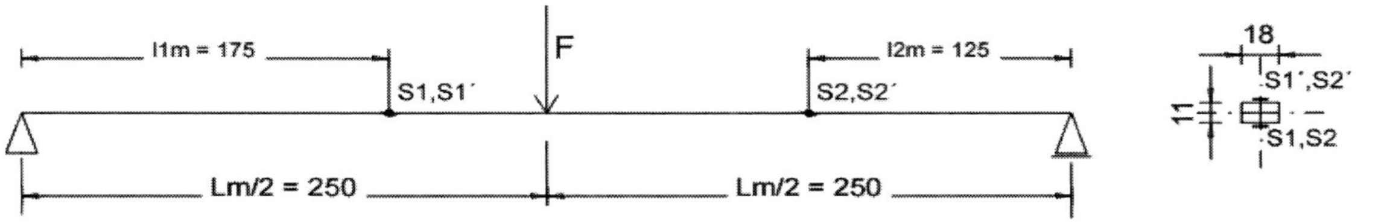


Fig. 3. Model of the beam

To determine corresponding state in real steel beam length of $l = 2500\,mm$ rules of modeling were used and loading forces $F_B$ and section modulus $W_B$ were determined. Material of model and of real beam was the same therefore stresses in the model and in the real beam are the same, steel S235 with Young´s modulus of elasticity $E = 2.1 \cdot 10^5$ $MPa$ was used, see the Tab. 1.

**Tab. 1** Results of experiment.

| | $F_m$[N] | 100 | 200 | 300 | 400 | 450 |
|---|---|---|---|---|---|---|
| Model experiment | $2\varepsilon_{1m} \cdot 10^{-3}$ | 0.230 | 0.460 | 0.685 | 0.920 | 1.030 |
| | $\sigma_{1m}$ [Mpa] | 24.15 | 48.30 | 72.45 | 96.60 | 108.15 |
| | $2\,\varepsilon_{2m} \cdot 10^{-3}$ | 0.160 | 0.320 | 0.474 | 0.636 | 0.710 |
| | $\sigma_{2m}$ [Mpa] | 16.80 | 33.60 | 49.77 | 67.78 | 74.55 |
| Beam | $F_B$[kN] | 2.5 | 5.0 | 7.5 | 10.0 | 11.25 |

**Tab. 2** Results obtained using Model and SBRA simulation.

| $F_m$[N] | 100 | | | 300 | | | 450 | | |
|---|---|---|---|---|---|---|---|---|---|
| $F_B$[kN] | 2,5 | | | 7,5 | | | 11,25 | | |
| | MODEL exp. | SBRA-m | SBRA-B | MODEL exp. | SBRA-m | SBRA-B | MODEL exp. | SBRA-m | SBRA-B |
| $2\varepsilon_{1m} \cdot 10^{-3}$ | 0.230 | - | - | 0.685 | - | - | 1.030 | - | - |
| $2\varepsilon_{2m} \cdot 10^{-3}$ | 0.160 | - | - | 0.474 | - | - | 0.710 | - | - |
| $\sigma_1$ [MPa] | 24.15 | 24.1089 | 24,1131 | 72.45 | 72.3371 | 72.3305 | 108.15 | 108.5044 | 108.5013 |
| $\sigma_2$ [MPa] | 16.80 | 16.8109 | 16.8094 | 49.77 | 50.4316 | 50.4297 | 74.55 | 75.6430 | 75.6484 |
| $\sigma_{1Det.}$[MPa] | 24.104 | | | 72.314 | | | 108.468 | | |
| $\sigma_{2Det.}$[MPa] | 16.804 | | | 50.413 | | | 75.618 | | |

m - Model, B – Beam, Det. – beam determined

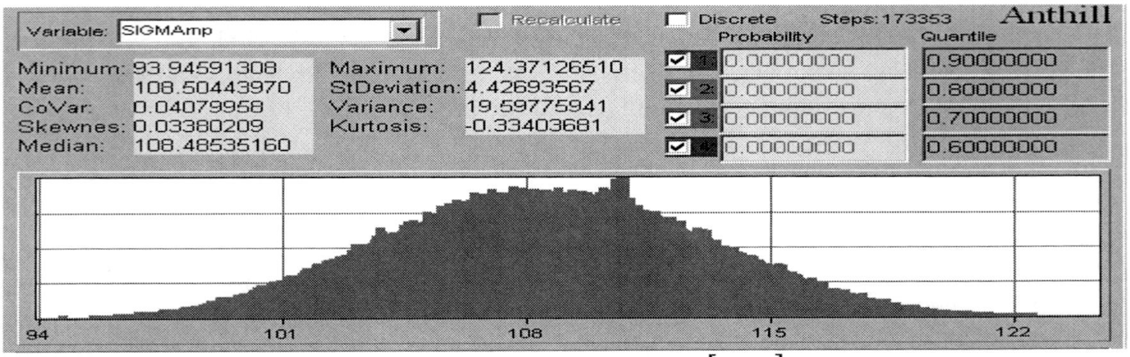


Fig. 4. The resulting stress $\sigma_{lm} = [MPa]\,\sigma_{1m}$.

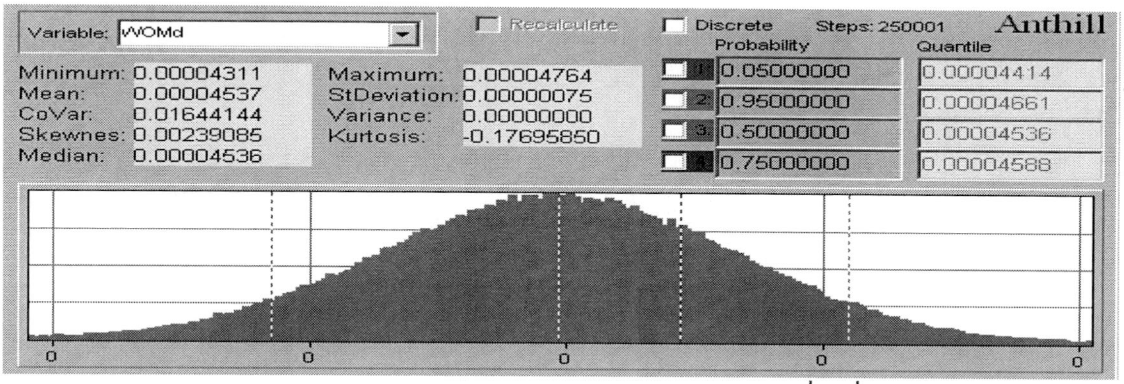


Fig 5. The histogram section modulus $\mathbf{W}_B = \lfloor m^3 \rfloor$.

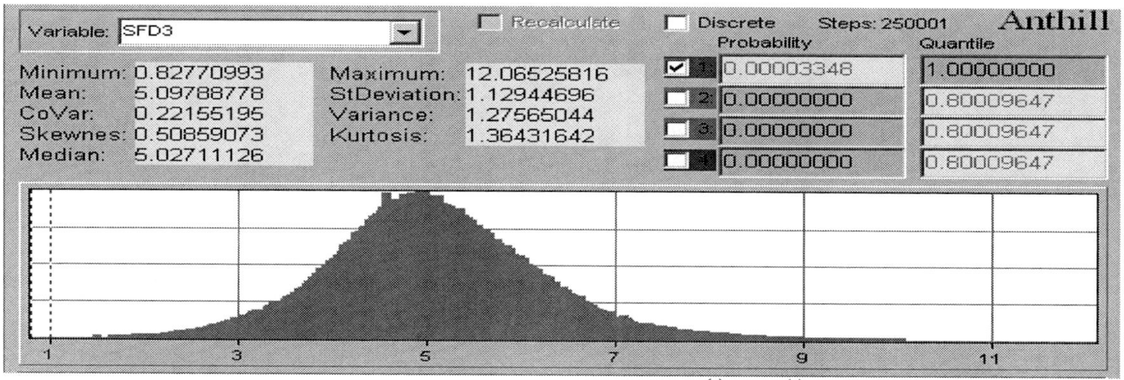


Fig. 6. Probability of failure $Pf(i)$, $SF(i)$.

---

In Fig.4. and Fig.5.there are for illustration presented histograms of the resultant stress and section modulus and in Fig. 6. histogram of probability of failure.

Probability of failures of the beam determined using SBRA method were probabilities of failure $\mathbf{P}_{f(i)B} = 3.476 \cdot 10^{-5}$ for beam and $\mathbf{P}_{f(i)m} = 3.348 \cdot 10^{-5}$ for the model

## Conclusions

Obtained results for the model and for the beam show very good correspondence. It gives a sure possibility for determination of failure probability of structure to determine it using corresponding model.


## Acknowledgement

The research work was supported by the Project SGS-2016-038.

## References

[1] Plánička, F., Šolc, M., Marek, P.: From Partial Factors Design to the SBRA Probabilitistic Method, Acta Mechanica Slovaca Vol.15, No 2, 2011, ISSN 1335-2393, pp. 22-27.

[2] Plánička, F., Krystek, J.: Dimensional Analysis and Modelling in Mechanics of Materials, Proceedings of 28th Danubia-Adria Symposium on Advances in Experimental Mechanics 2011, Siófok, Hungary, ISBN 978-963-9058-32-3, pp. 229-230.

[3] www.sbra-anthill.com/lit/publications.

EAN 2017
55th conference on experimental stress analysis

# Fabric Thickness Measurement under Tensile Forces

Diana Šimić Penava[1, a*], Željko Penava[2,b] and Željko Knezić[3,c]

[1]Faculty of Civil Engineering, University of Zagreb, Kačićeva 26, 10000 Zagreb, Croatia

[2, 3]Faculty of Textile Technology, University of Zagreb, Pr. b. Filipovića 28a, 10000 Zagreb, Croatia

[a]dianas@grad.hr, [b]zpenava@ttf.hr, [c]zknezic@ttf.hr

**Keywords:** thickness, woven fabrics, tensile force, extension.

**Abstract.** Woven fabric thickness (t) is its third dimension, which is considerably smaller than the fabric width and length, but at the same time it is very important because it affects the usage and properties of the fabric. During stretching, the fabric thickness is changing depending on the value of the tensile force. In an effort to be as precisely as possible to measure dynamic thickness changes of the fabric, and simultaneously collect all the required values about the change of the tensile force, extension and the area on which tensile force acts, new measuring apparatus are designed and manufactured. This measuring apparatus allows the simultaneous measurement of dynamic thickness of fabric, the fabric contraction and related tensile axial forces.

## Introduction

Under fabric thickness the greatest distance between the surface of the front side and the mutually parallel surface of the back side of the fabric is meant. Fabric thickness is measured in mm. Fabric thickness plays an important role in manual processing, design and end-use of textiles. Fabric thickness change, depending on the effect of various parameters [1]. In order to clarify the effects of some specific parameters on fabric thickness, such as yarn count change, change of warp and weft density, change of warp and weft thread tension in the weaving process, it is necessary to carry out experimental tests of fabric thickness change. To allow a real understanding of thickness behavior when a fabric is stretched, it has been necessary to develop a precise new device for measuring fabric thickness. A textile fabric is a very flexible material and it is easy to imagine that, during such a test, the sample moves a lot in the three spatial directions. Moreover, a textile fabric is made of interlaced yarns constituted of fibers twisted together [2]. Various apparatuses for measuring static thickness have been developed. Haven used a new fabric thickness-measurer [3], Shiefer used a compressometer for thickness and resilience measurements [4]. The non-contacting measurements [5, 6] (optical) were too heavy to be fixed on the fabric without deforming it. It came out that the fabric should carry the sensor during the test.

The aim of this work was to measure the dynamic change of fabric thickness using a newly constructed device when the fabric was in the axial state of stress in the weft and warp direction.

## Experiment

The experimental part deals with measurements of fabric thickness if tensile force acts on fabric samples until rupture. For this purpose a new device for dynamic measurements of fabric thickness changes, if tensile force acts on the fabric sample, was designed and constructed. The values of tensile forces, associated extensions and associated fabric thicknesses were found during testing.

Two cotton woven fabrics of different weft density and of the same warp density were available. In this work, only balanced orthogonal plain weave fabric was considered. Cotton yarn for the warp and weft had the same yarn count. Table 1 shows structural properties of the tested woven fabrics.

Table 1 Description of fabrics

| Fabric code | Yarn count (tex) | | Number of yarns ($cm^{-1}$) | | Mass ($g/m^2$) | Thickness $t_0$ (mm) |
|---|---|---|---|---|---|---|
| | warp | weft | warp | weft | | |
| F1 | 30 | 30 | 24 | 19 | 137 | 0.365 |
| F2 | 30 | 30 | 24 | 27 | 168 | 0.371 |

The yarn count was determined by the gravimetric method according to standard ISO 2060:1994. The number of threads per unit length was tested according to standard ISO 7211-2:1984. Initial woven fabric thickness $t_0$ is measured on newly-constructed apparatus at a pressure of 2 cN / $cm^2$. Before testing all specimens were conditioned under the conditions of standard atmosphere (relative air humidity 65 ± 2%, at a temperature of 20 ± 2°C).

For the purposes of this testing, standard samples with dimensions 300 x 50 mm were cut and clamped in the clamps of the tensile tester at a distance of $l_0$=200 mm and subjected to uniaxial tensile load till rupture. The pulling speed of the clamps was 100 mm/min. The samples were cut in the weft direction and in the warp direction. Five tests were carried out on the new designed device for the mentioned cutting directions of the specimen. For each cutting direction, the average values of axial force, associated extension and thickness, obtained from five measurements are shown in diagrams.

Using new constructed apparatus, a fabric test specimen is stretched and loaded at a constant pulling speed until the moment of specimen rupture. Figure 1 shows the schematic representation of the apparatus for measuring fabric thickness.

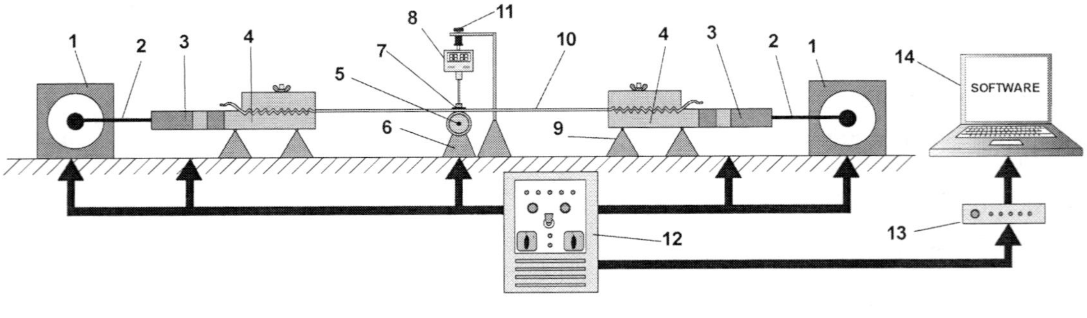


Fig. 1 Schematic representation of the device for measuring fabric thickness during extension

The constructed device is fully automated, computer-controlled and works at a constant pulling speed. Maximum measuring resolution of pulling (stretching) is 3.5 μm. The test specimen of fabric (10) is fastened between the movable fabric clamps. The clamps are moved by the electric drive (1) using the transmission system (2) at a constant speed, and they move by means of the screw rod (4) fastened to the bearing structure (1). The force transducer cell (3) is connected to the clamp on the one side, and on the other side the motor (1) moves the clamp by the transmission system (2). A force transducer cell for a range to 200

N is used. The electric motor work (1) is computer-controlled using the power and control unit with manual settings (12). In the middle of the test specimen a digital thickness gauge (8) with displacer (7) resting on the fabric specimen is placed. The displacer diameter is 2.54 mm, and the displacer area is 5.07 cm$^2$. The accuracy of thickness measurement is 0.001 mm. The full measurement range of the thickness gauge ranges from 0 to 25.4 mm. The displacer rests on the fabric under which the reference axle of the thickness gauge (5) is affixed to the support of the thickness measurement system (6). The applied digital thickness gauge is also furnished with displacement force setting (11).The whole system is connected to the computer (14) through the computer interface with A/D converter (13) and controlled by the serial port.

## Results and discussion

For each fabric (F1 and F2), results are shown on diagrams Figures 2-3. The measures have been made on 2 samples in each direction in warp and weft. The Microsoft Excel software was used for statistical analysis of data at $p<0.05$ for five measurements.

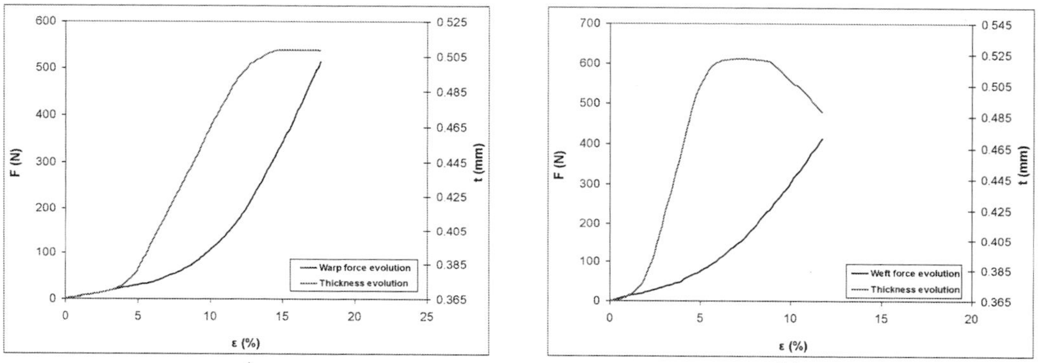


a) b)

Fig. 2 Diagram of tensile force- relative extension – thickness (F- ε- t) for fabric F1: a) force acts in the warp direction, b) force acts in the weft direction

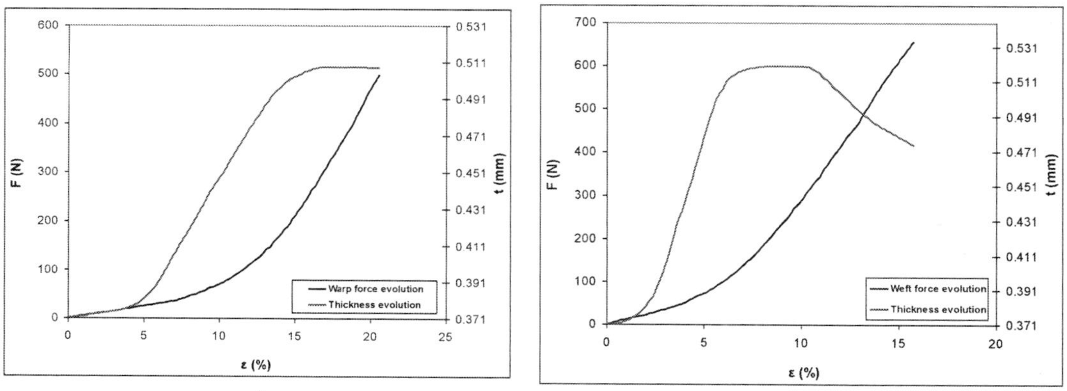


a) b)

Fig. 3 Diagram of tensile force- relative extension – thickness (F- ε- t) for fabric F2: a) force acts in the warp direction, b) force acts in the weft direction

The known mechanisms of the fabric longitudinal deformation during a tensile test can be described as follows: first, low decrimping of yarns, and fabric structure rearrangement due to internal stresses; these two mechanisms occur under low loads and provide low decrimping of stretched yarns; weave reaction and high de-crimping of the yarn in the direction of the test; and direct yarn strengthening when decrimping is complete. Before analyzing each fabric behavior, it is worth noticing the different initial thickness values, Table 1. These differences were expected because of the difference in fabric weft density. Fabric thickness measurement is highly dependant on the applied tensile force during the measure. Fabric thickness changes while the tensile force stretches the fabric. Thickness - axial force -extension curves in Figures 2 and 3 show typical non-linear relationships for fabric samples if tensile force acts in the weft or warp direction.

As is shown in Table 1, the number of yarns/cm is higher in warp than in weft. The woven fabric F1 has been tested because it is lightweight. Between 0 and 4% of strain in warp direction, and 0 and 2% of strain in weft direction, increasing thickness is very small. Between 4 and 12% of strain thickness is growing up from 0.375 to 0.505 mm within the warp. From 2% to 5% of strain in weft direction, thickness increases very fast from 0.370 to 0.525 mm.

For woven fabric F2, if the force acts in the warp direction in the extension range from $\varepsilon$=0 % to $\varepsilon$=4 % fabric thickness slightly increases to 0.381 mm. While the values of relative extension rise from $\varepsilon$=4 % to $\varepsilon$=15%, fabric thickness rapidly grows and reaches 0.511 mm. From 2% to 6% of strain in weft direction, thickness increases very fast from 0.379 to 0.529 mm.

Figures 2b and 3b show that in case if force acts in the weft direction with increasing weft density, the values of tensile forces and associated elongations at break of the fabric test specimen increase.

## Conclusion

Measurements of fabric dynamic thickness were performed on a newly constructed measuring apparatus which simultaneously records all necessary values of changing tensile force and associated extension. Fabrics of different weft densities reach approximately the same maximum values of relative fabric thickness with almost equal forces and relative extensions. It follows that weft density does not affect fabric thickness if it is subjected to the action of tensile force in the weft direction. Test results have shown that an increase in axial tensile forces on the fabric causes an increase in fabric thickness for all fabric densities.

## References

[1] B. Olofsson, A general model of a fabric as a geometric-mechanical structure, Journal of the Textile Institute Transactions. 55 (1964) 541-557.

[2] X. Xiao et al., Through-thickness permeability modelling of woven fabric under out-of-plane deformation, Journal of Materials Science. 49 (2014) 7563-7574.

[3] G. B. Haven, New Fabric Thickness-Measurer, Textile Research Journal. 3 (1933) 229-237.

[4] H. F. Schiefer, The compressometer: An instrument for evaluating the thickness, compressibility and compressional resilience of textiles and similar materials, Textile Research Journal. 10 (1933) 505-513.

[5] A. Majumdar, S. S. Saha, A novel fabric thickness and compression tester using magnetic inductance principle, Journal of the Textile Institute. 99 (2008) 359-362.

[6] J. Fan et al., A portable tester for nondestructively measuring fabric properties, Textile research Journal. 72 (2002) 21-26.

EAN 2017
55th conference on experimental stress analysis

# Interaction Enhancement between Recycled PET Macro-fiber Reinforcement and Cement Matrix via Plasma Treatment

TREJBAL Jan [1, a], PROŠEK Zdeněk [1,2,b], TESÁREK Pavel [1,c] and DEJDAR Lukáš [1,d]

[1] Faculty of Civil Engineering, Czech Technical University in Prague, Thákurova 7, 166 29 Prague, Czech Republic

[2] CTU in Prague, UCEEB, Třinecká 1024, 273 43 Buštěhrad, Czech Republic

[a] jan.trejbal@fsv.cvut.cz, [b] zdenek.prosek@fsv.cvut.cz, [c] pavel.tesarek@fsv.cvut.cz, [d] lukas.dejdar@fsv.cvut.cz

**Keywords:** Fiber Reinforcement; Polymeric Macro-Fibers; Interfacial Properties; Surface Modification; Plasma Treatment.

## Abstract

Presented work deals with an interaction enhancement between recycled PET macro-fibers reinforcement and cement matrix via plasma treatment (oxygen low-pressure inductively coupled plasma). Wettability between demineralized water / treated fibers and adhesion between fibers / cement matrix were studied in order to assess the treatment effect. The wettability was measured using an optical goniometer, while the adhesion was examined by means of indirect mechanical testing – displacement controlled four-point bending tests of cement samples reinforced with reference and 30 s treated fibers. Consequently, an impact of the treatment on fiber tensile strength was tested. It was shown, that the plasma modification increased the fiber wettability by the water almost three-times already after 5 s treatment, while fiber tensile strength stayed almost unaffected. Bending tests revealed that the prismatic cement samples reinforced with treated fibers reached to higher toughness in elastic response, to higher residual-strength in post-cracking phase and to lower maximal limit of elasticity in comparison with reference samples.

## Introduction

Steel fibers used as reinforcement in concrete mixtures can be replaced with polymeric-based macro fibers, if certain circumstances respected [1, 2]. The usage of the polymeric fibers is regulated by ČSN EN 14889-2 and EN 14845-2. According to the second cited standard, the residual strength of fiber reinforced concrete (amount of fibers 4 $kg/m^3$, measured if CMOD is equal to 3.5 mm) obtained during bending test has to be higher than 1.4 MPa. Unfortunately, in civil engineering praxis, a compliance of mentioned minimal residual strength is difficult to achieve because of adverse polymeric fiber surface properties. These properties can be described as smooth and chemically inert in relation to other concrete mixture phases [3-5]. As a consequent of these drawbacks, both strong chemical as well as physical interaction between fiber surfaces and surrounded matrix is not guaranteed [3, 5]. In order to eliminate this phenomenon and to achieve stronger interaction between the two materials (bond), some modifications should be applied. In contrast with mechanical (surface roughening) and chemical (wet oxidation, polymeric chain opening) modification methods, a plasma treatment seems to be a good alternative combining both the physical (ion bombardment) and the chemical (surface activation by functional groups implementation) modification. Moreover, compared to the other mentioned methods, plasma treatment modifies the fiber surfaces gently – fiber mechanical properties stay with no decrease [3, 6]. For these purposes, we used oxygen plasma treatment known as dry oxidation modification in this study.

## Materials and Methods

**Fibers.** Recycled PET (polyethylene terephthalate) macro-fibers having dimeter equal to 260 μm and length 20 mm made by Spokar (Czech Republic manufacturer) were treated by plasma treatment in order to change their surface properties such as wettability by water and surface free energy.

**Cement samples.** Fiber reinforced cement prismatic samples having dimension equal to 40 × 40 × 160 mm was made from CEM 42.5 II R. The water to cement ratio was equal to 0.4. The curing and hardening of mixtures conducted for 28 days after water admixing. An amount of fibers was 2 % wt.

**Plasma treatment.** Oxygen plasma treatment in inductively coupled low-pressure cool plasma system was done (power 100 W, $O_2$ flow 50 sccm). The treatment process differed only in exposition time (from 5 to 60 seconds).

**Wettability measurements.** Wettability of both, the reference and plasma treated fibers were examined through contact angle measurement using a direct optical method, according to [7]. Fibers were wetted by demineralized water.

**Mechanical tests.** Both the fiber tensile strength (force needed to fiber break) of single fiber and bending strength tests (four-point bending test) of fibers-reinforced cement samples were conducted using Web Tiv Ravestain FP100 loading frame. The experiment was displacement controlled at constant rate of 0.6 mm/min.

## Result and Discussion

The contact angles between plasma treated fibers and demineralized water observed using direct optical method decreased more than three-times after 5 s plasma treatment in comparison with reference ones, as captured in Fig. 1. Based on these findings, it can be clearly said that the plasma treatment duration lasting longer than 5 s is no effective in terms of wettability. On the other hand, to attain desirable physical changes (roughening) onto fiber surfaces by ion bombardment, the treatment time should be longer lasting. We estimated that the most effective treatment time is approximately 30 s. This finding is consistent with our earlier published researches, for example with [8].

Tensile strength tests of single fibers showed that their tensile strength (more precisely said force needed to fiber break) stayed almost unaffected regardless to the treatment time. A slight decrease revealed (max 15 %) obtained during testing can be attributed to surface damage caused by ion bombardment. The results are summarized in Fig. 2.

The four-point bending tests showed that the samples reinforced with 30 s plasma treated fibers were characterized by steeper strength increase in the liner-elastic response compared to reference samples containing no treated fibers. This phenomenon can be probably explained by an elimination of shrinkage cracks caused by stronger interaction between the two materials. Significant changes were observed in the post-cracking phase of samples response. Residual bending strength of samples reinforced with treated fibers was considerably higher in comparison with reference ones. Force needed to attain mid-span deflection of 1 mm oscillated about 4 N and 3 N in the case of samples containing modified and reference fibers, respectively. This behavior clearly pointed out to the desirable interphase interaction increase between fiber surfaces and cement matrix due to plasma treatment. Force-displacement diagram captured during bending test is shown in Fig. 3.

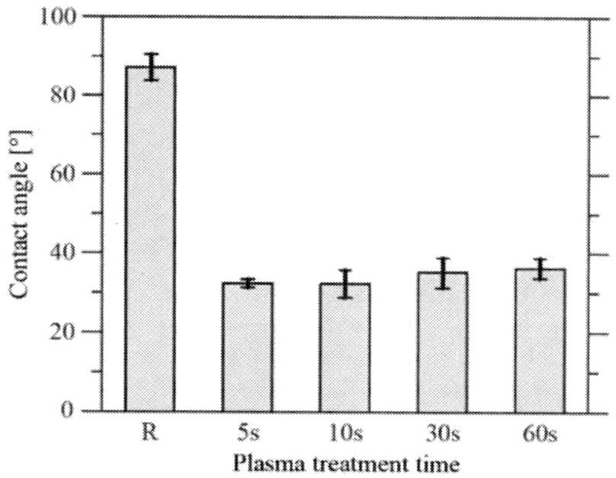


Fig. 1: Contact angle sizes depending on plasma treatment time obtained during contact angles measurements.

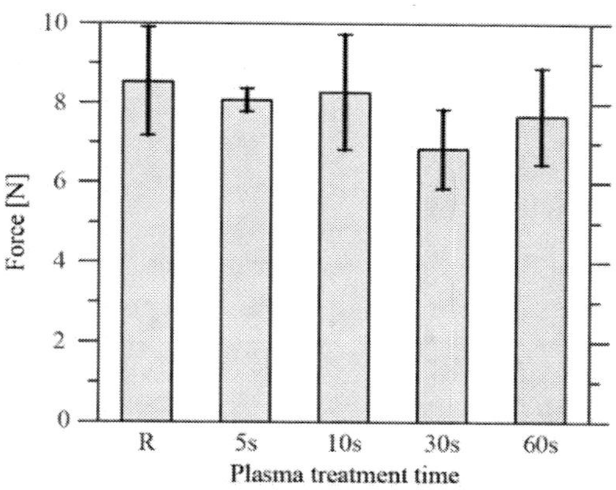


Fig. 2: Force needed to break the fibers depending on treatment time.

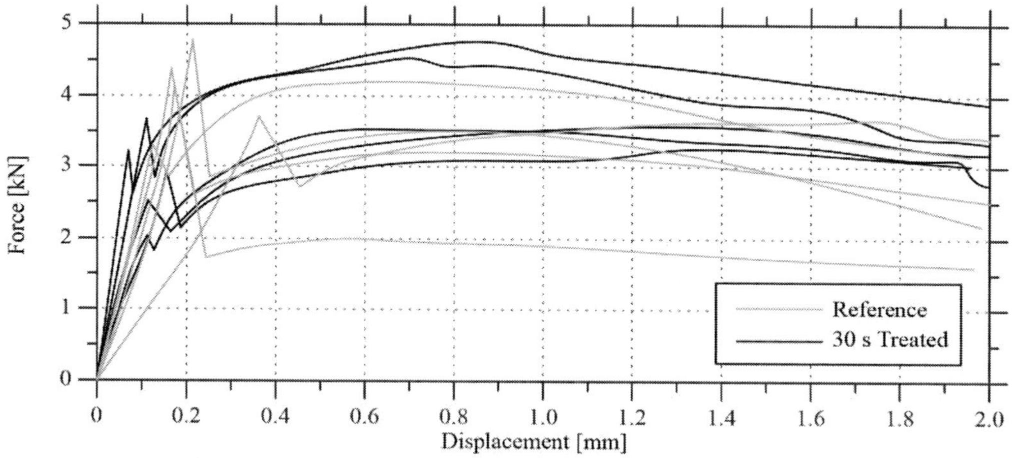


Fig. 3: Loading force depending on specimen mid-span displacement obtained during four-point bending tests.

## Conclusion

The aim of this work was to eliminate general polymeric fiber drawbacks such as smooth and inert surface and to attain their bond increase with cement matrix. Polymeric macro-fibers (polyethylene terephthalate) having 260 μm in diameter and 20 mm length were surface modified in order to increase their wettability and roughness. To achieve desirable changes, low-pressure cool inductively coupled oxygen plasma treatment was done. The wettability of modified fibers by distilled water was examined using a direct horizontal goniometer. The impact of ion bombardment on fiber mechanical properties was investigated by means of tensile strength testing. The adhesion increase between the two materials was examined indirectly by bending tests of cement samples reinforced with reference and modified fibers.
It was shown that the plasma treatment increased fiber wettability by water almost three times already after 5 s of treatment duration, while fiber tensile strength styed nearly unchanged. Significant changes of bending response were found out during four-point bending test of prismatic cement samples reinforced with reference and modified fibers. Samples containing

modified fibers exhibited a higher residual strength in the post-cracking phase compared with reference ones. This bending response plainly pointed out to an increase of mechanical cohesion between modified fibers and cement matrix due to ion bombardment causing fiber surface roughness increase.


## Acknowledgements

This was financially supported by CTU in Prague – project SGS16/201/OHK1/3T/11 and by the Grant Agency of the Czech Republic research project 15-12420S.


## References


[1] M. Luňáček, P. Suchánek, R. Bader, Aplication of biocomponent syntetic macro fibres concrix in tunneling, Tunnel. 8 (2012) 54-56.

[2] M. di Prisco, G. Plizzari, L. Vandewalle, Fibre reinforced concrete: new design perspectives, Material and Structures. 42 (2009) 1261-1281.

[3] J. Trejbal et al., Impact of surface plasma treatment on the performance of PET fiber reinforcement in cementitious composites, Cement & Concrete Research. 89 (2016) 276-278.

[4] B. Felekoglu, K. Tosun, B. Baradan, A comparative study on the flexural performance of plasma treated polypropylene fiber reinforced cementitious composites, Journal of Materials Processing Technology. 209 (2009) 5133-5144.

[5] V. Machovič et al., Microstructure of interfacial transition zone between pet fibres and cement paste, Acta Geodyn. Geomater. 10 (2013) 121-127.

[6] S.E. Elsaka et al., Influence of chemical surface treatments on adhesion of fiber posts to composite resin core materials, Dental Materials. 29 (2013) 550-558.

[7] J. Trejbal el al., A direct optical method for contact angle measurement on micro-fibers, in: NANS 2015, P. Tesárek (Eds), Nanomaterials and nanotechnology in civil engineering, CTU in Prague, Prague, 2015, pp. 90-96.

[8] J. Trejbal, P. Tesárek, Š. Potocký, Time depend wettability deterioration of plasma treated polymeric macro-fibers, Key Engineering Materials. 731 (2017) 43-48.

# SECTION 5

## Monitoring of operational loadings and operational states of structures and devices

EAN 2017
55th conference on experimental stress analysis

# Experimental analysis of shearing power during chain production

BERKA Ondřej[1, a], LOPOT František[1,b] ,DUB Martin[1,c] and HOŠTIČKA Vladimír[1,d]

[1]CTU in Prague, Faculty of mechanical Engineering, Deparment of Design and Machine Components, Technická 4, 166 07, Prague 6, Czech Republic

[a]ondrej.berka@fs.cvut.cz, [b]frantisek.lopot@fs.cvut.cz, [c]martin.dub@fs.cvut.cz, [d]vladimir.hosticka@fs.cvut.cz

**Keywords:** Roller chain; Measurement, Punching tools

## Introduction

This paper is focused on measuring of shearing power during chain production. One of the most important factors for chain quality is accuracy of manufacturing. Default accuracy of manufacturing is given by standards (e.g. DIN 8187, ČSN 02 3311) [2]. Experimental analysis is dealing with roller chains for motorcycle chain gears. These chains are very dynamically stressed. Therefore that is very important for accuracy of manufacturing. Production cutting tools are made with prescribe accuracy and that is subject to subsequent review. Next influence is action force on cutting tools during the shearing. Shearing power deforms cutting tools, this leads to an increasing of the accuracy of manufacturing. For this purpose it decides to determine the deformation of these components (shearing power). Shearing power and deformation are analyzed on cutting tools for roller chains.

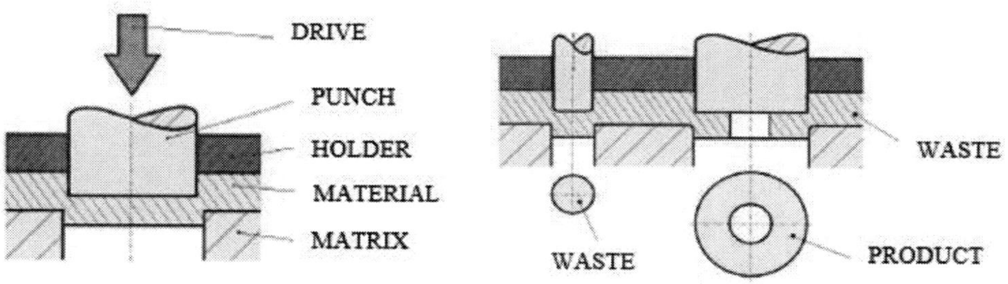


Fig. 1 Cutting tools and roller chain production [2]

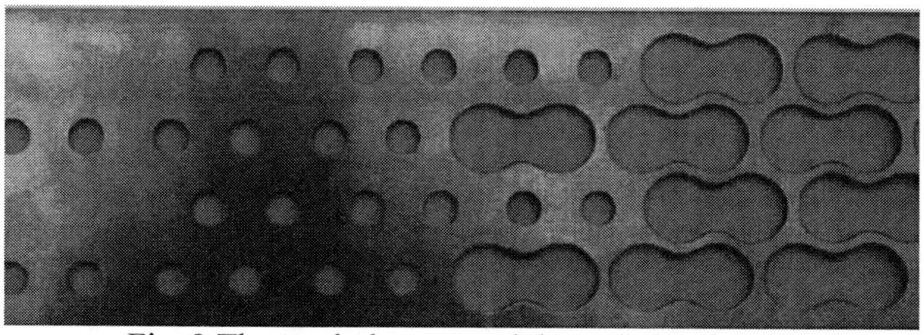

Fig. 2 The steel plate passed through the tool [2]

## Measurement

The measurement was made by strain gauges on cutting tools (Fig. 3). Tensile/compressive stress and bending stress were measured for deformation analysis on punches and four strain gauges were installed on each of the punch (Fig. 4). Each strain gauge was connected separately to the quarter bridge.

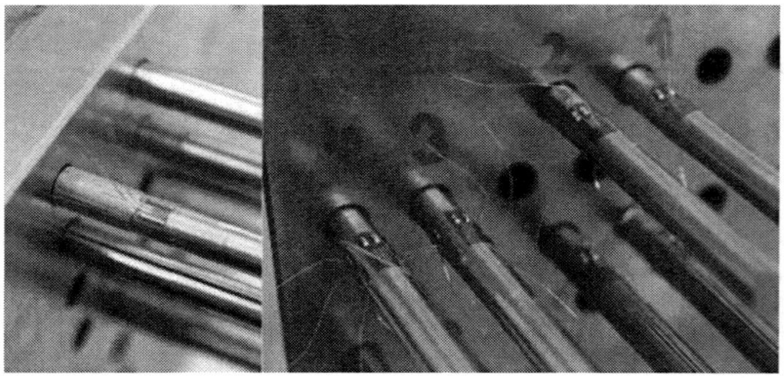

Fig. 3 Punches with strain gauges [2]

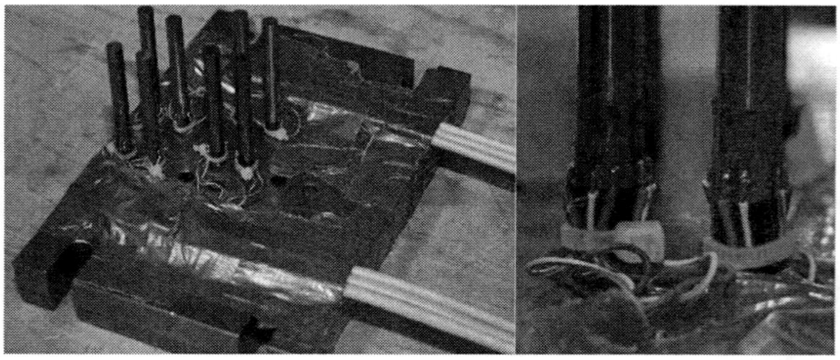

Fig. 4 Connected strain gauges on punches [2]

Tension is measured by all four strain gauges SG1, SG2, SG3 and SG4 (Eq. 1). Following punches loading are bend. That is determined in two planes. The first plane is given by strain gauges SG1 and SG3, the other plane is given by strain gauges SG2 and SG4. Therefore bending stress is calculated in this two planes by Eq. 2 and Eq. 3. [1]

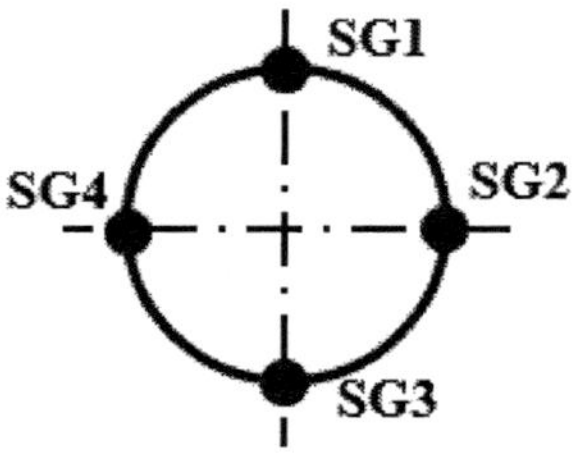


Fig. 5 Arrangement of the strain gauges on punches

$$\sigma_t = \frac{E \cdot (\varepsilon_1 + \varepsilon_2 + \varepsilon_3 + \varepsilon_4)}{4} \qquad (1)$$

$$\sigma_{b1} = \frac{E \cdot (\varepsilon_1 - \varepsilon_3)}{2} \tag{2}$$

$$\sigma_{b2} = \frac{E \cdot (\varepsilon_2 - \varepsilon_4)}{2} \tag{3}$$

**Results**

The tensile force is given by Eq.4 and the resultant bend by Eq. 6.

$$F_t = \sigma_t \cdot A = \frac{E \cdot (\varepsilon_1 + \varepsilon_2 + \varepsilon_3 + \varepsilon_4)}{4} \cdot A \tag{4}$$

$$M_b = \sqrt{M_{b1}^2 + M_{b2}^2} = \sqrt{(\sigma_{b1} \cdot W_o)^2 + (\sigma_{b2} \cdot W_o)^2} \tag{5}$$

$$M_b = \sqrt{\left(\frac{E \cdot (\varepsilon_1 - \varepsilon_3)}{2} \cdot W_o\right)^2 + \left(\frac{E \cdot (\varepsilon_2 - \varepsilon_4)}{2} \cdot W_o\right)^2} \tag{6}$$

Time behavior of the punches tensile force and punches bend are presented on Fig.6 a Fig.7. Cycle punches of operation were divided into the several parts for the tensile force measurement – getting into the contact with material, penetrating through material, getting out of material, broaching in the final hole, finishing in bottom dead center and pulling out.

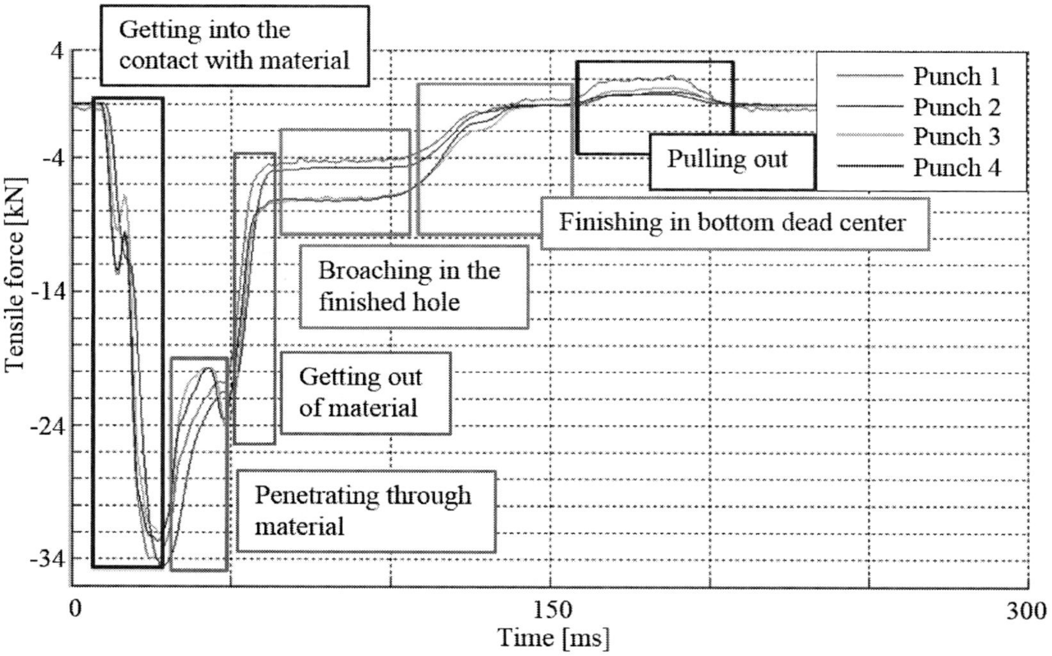


Fig. 6 Time behavior of the punches tensile force

Bend measurement cycle of operation was divided into three parts - penetrating through material, getting out of material and broaching in the final hole.

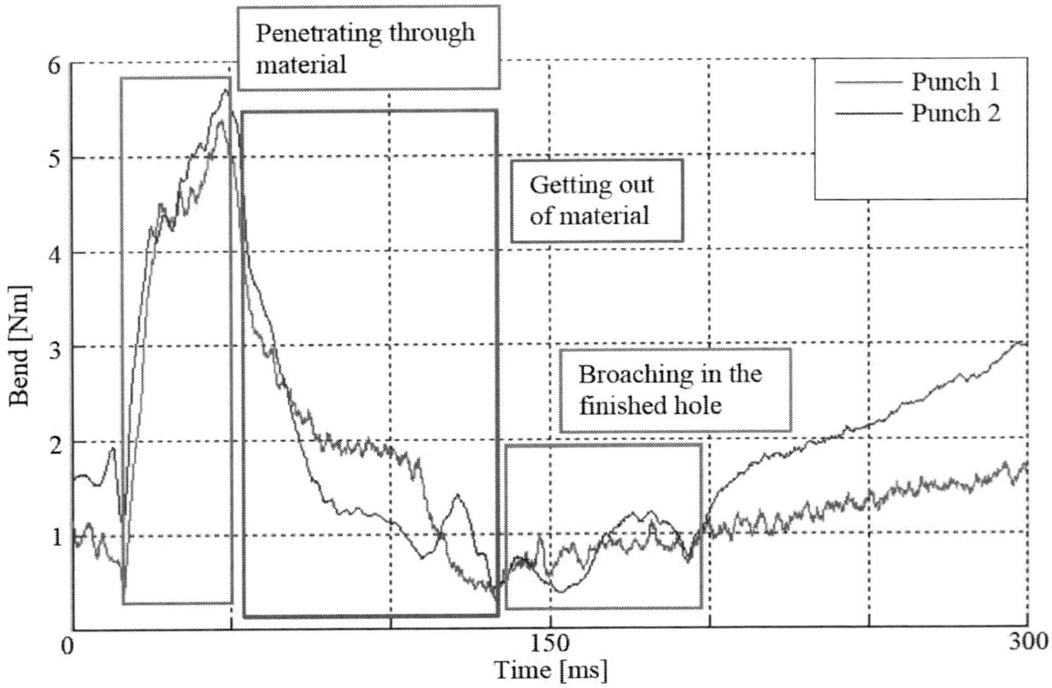


Fig. 7 Time behaviour of the punches bend

**Conclusions**

These analysis enables to determine shearing power during roller chain production, when tension/pressure and bend are measured (Fig. 8). Parasitic bend has a negative impact on accuracy manufactured roller chain. Bend was found based on the measurement and bend is caused accuracy of manufacturing, therefor adjustment was designed for the punching tools.

Fig. 8 Measurement on the punching tools

**References**


[1] M. Dub, J. Kolář, F. Lopot, V. Dynybyl, O. Berka, Experimental analysis of stress state and motion of tram gearbox hinge, EAN 2014 - 52nd International Conference on Experimental Stress Analysis (2014).

[2] V. Hoštička, Analysis of construction of a cutting tool and experimental defection of shearing forces, CTU in Prague (2015).

EAN 2017
55th conference on experimental stress analysis

# Modelling of the flow of streams of cohesionless and cohesive bulk materials in a conveyor discharge point with a flat conveyor belt

Joanna Cyganiuk[1, a], Piotr Kuryło[2, b]

[1, 2] University of Zielona Gora, Licealna 9 Street, 65-417 Zielona Gora, Poland

[a] j.cyganiuk@ibem.uz.zgora.pl, [b] p.kurylo@ibem.uz.zgora.pl

**Keywords:** pouring points, discharge point, bulk materials, adhesion, modelling of the flow, material velocity

**Abstract.** The paper presents the analysis of flow conditions of cohesive and cohesionless bulk materials in a conveyor discharge point of a flat conveyor belt. The analysis was carried out for stationary flows at high velocities. It presents mathematical methods of description of velocity of material leaving a throwing point of a flat conveyor belt as well as final equations which enable the determination of velocity of the material after it has left the throwing point (with the accuracy sufficient for practical use). Next, the velocity calculated for the proposed mathematical description (for selected material groups) has been compared with the velocity obtained from mathematical relations commonly used by engineers. The proposed equations for determining the velocity of material beyond the point have been proved useful, since they enable excluding the indirect equations. Finally, the difference between the values of velocity obtained with the proposed and indirect equations has been determined and the relative error for the proposed method has been calculated.

## 1. Introduction

In transport conveyor systems built on the basis of belt conveyors, pouring points are important parts. They are crucial elements allowing transporting bulk materials on different levels of this type of system. An important issue in a conveyor transport is to ensure proper flow of material through pouring points. This involves ensuring continuity of the transported material stream and ensuring the required transport capacity of the system. Pouring points used in transport systems often take advantage of gravity to transport materials. These types of pouring points include e.g. chutes [1], [20]. Another pouring points that cooperate with belt conveyors are impact plates [7], [9], [10], [11], [16], [18], [19]. They are to slow down the velocity of material to the value corresponding with the velocity of the receiving conveyor in the feed point.

A throwing point with a flat conveyor belt, discussed further, uses a driven belt for moving and throwing the material. However, the point is not inclined as described in [5], [6], [8], [14]. This means that the angle of inclination of the conveyor belt is 0° [16], [17].

In transport systems, slow belt conveyors are most commonly used. They operate with velocities below 3 m/s [1], [20], and equations determining the velocity of cohesive and cohesionless material beyond the conveyor pulley are described in [14]. However, these equations should not be used for fast conveyors. Equations for fast conveyors are presented in [14]. They do not take adhesion into consideration, but include the influence of air resistance on the velocity of the transported material beyond the head pulley. In the work [17], the influence of adhesion on the velocity of material beyond the point was also omitted. However the work included material, kinematic and dynamic parameters. In the work [4] the adhesion phenomenon also was omitted.

In the case of discharge belt conveyors [15], the construction of this type of conveyors and the discharge method of the material are different from the ones used for a belt conveyor feeding the next pouring point. Thus, equations proposed in [15] designed for estimation the velocity of material beyond the pulley, should not be used for fast feeding belt conveyors.

In [12], only kinematic and geometric parameters for estimation of the velocity of material falling down onto a belt conveyor were taken into consideration, whereas in [2], authors discuss simple engineering equations describing the velocity of material flowing from a belt conveyor. They take into consideration the coefficient of friction of the material against the belt but do not take into consideration forces and adhesion affecting onto the material stream.

Works [5], [6], [8] take into account kinematic, dynamic and material parameters including adhesion, but equations presented there are designed for inclined conveyors with ascending and/or descending belts.

Author in [7] analyse the trajectory of material beyond a throwing point depending on the velocity of a belt conveyor, but they do not analyse the velocity of the material beyond the point where it reaches the next pouring point. In addition, work [3] refers to the prediction of the trajectory of the material beyond the throwing point, but in this case, both fast and slow conveyors were taken into consideration.

Therefore, it was important to propose equations describing the velocity of material beyond the pulley of a not inclined belt conveyor, equipped with the belt without cleats for both cohesive and cohesionless materials.

## 2. Analysis of flow conditions of cohesionless and cohesive materials in the throwing point with a flat belt

In conveyor transport systems, a throwing point refers to a head pulley of the belt conveyor. From the pulley, the material is transferred further, as a feed, onto another conveyor, which, most often, is located lower and the feed is delivered with the use of a chute [1], [20] or an impact plate [1], [6], [7], [9-11], [16], [20]. The material from the throwing point may also be thrown directly onto hillocks.

The analysis of flow conditions of cohesive and cohesionless bulk materials in a throwing point with a flat belt is shown in Figure 1 and 2. The analysis was carried out both for fast conveyors, where the belt transports the material with the velocity above 3 [m/s], and for stationary flow. A belt without cleats was analysed and the influence of air resistance was not considered.

Figures 1 and 2 illustrate magnitudes taken into consideration in the analysis of the flow conditions that affect the process of discharging the material from the conveyor's pulley. Geometric, kinematic and dynamic conditions of the motion of elementary mass dm in a throwing point, which is not inclined, were taken into consideration.

In a throwing point discharging cohesionless materials, the elementary mass dm is subject to the following forces describing dynamic conditions of its flow (Fig. 1): gravitational force dG [N], centrifugal force dF [N], normal force $dN_{xs}$ [N], tangential force $dT_{xs}$ [N]. While determining the relations describing the discharge of the martial from the pulley, the inertial force dJ [N] formed as a result of action of gravitational force dG, centrifugal force dF and forces $dN_{xs}$, $dT_{xs}$ was also considered.

In the case of cohesive materials, the dynamic conditions of the flow of the material in the throwing point, besides the above mentioned forces, also adhesion force $dF_a$ (Fig. 2) was considered. The adhesion force forms between the belt and the transported material.

Kinematic conditions of the flow of material in a throwing point with a flat belt (Fig. 1 and 2) include: the velocity of the conveyor belt $v_t$ [m/s], the velocity of bulk material stream flowing into the head pulley $v_p$ [m/s], which is equal to the velocity of the belt, and the velocity of material leaving the head pulley $v_w$ [m/s].

Geometric conditions of material flowing out from the throwing point include the following parameters [8]: angle of conveyor descent $\alpha$ [°] (in this case $\alpha$=0°), angle of the material stream flowing out of the head pulley $\beta_t$ [°] (in this case $\beta_t$ is the angle of material flowing into another point, e.g. impact point), angle coordinate which describes the position of infinitesimal mass of material on the head pulley $\varphi$ [°] [8], radius of the head pulley $r_b$ [m], the average radius of curvature of material stream $R_{śr}$ [m], thickness of the layer of material stream at the moment it flows onto the head pulley $h_m$ [m] [8], thickness of the layer of material stream leaving the head pulley $h_w$ [m] [8].

Kinematic and geometric conditions of material flowing out of a head pulley with a flat belt for cohesive and cohesionless materials are the same. whereas in the case of dynamic conditions – the difference is the adhesion force appears in case of cohesive materials.

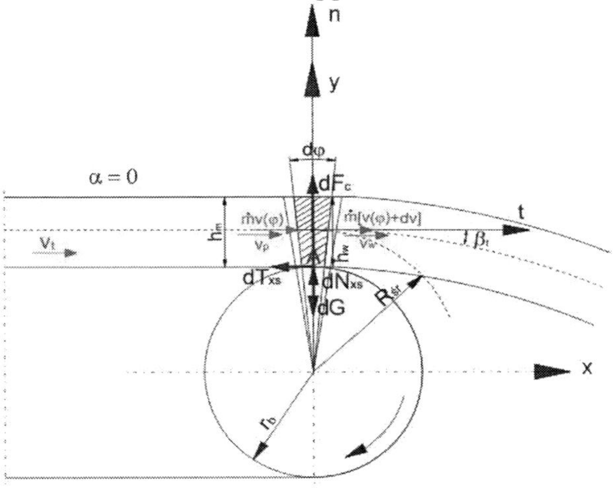


Fig. 1. Geometric, kinematic and dynamic conditions of the flow of a stream of cohesionless bulk material on a head pulley of a conveyor with a flat belt [own elaboration]

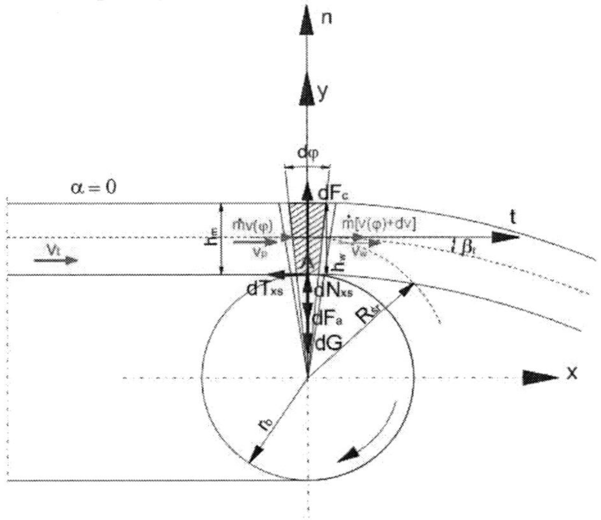


Fig. 2. Geometric, kinematic and dynamic conditions of the flow of a stream of cohesive bulk material on a head pulley of a conveyor with a flat belt [7]

## 3. Mathematical model of the flow of a stream of cohesionless and cohesive bulk material in a throwing point with a flat belt

In the analysis of the discharge of material from a throwing point with a flat belt, it was assumed that the material is separated from the pulley in point A (Fig. 1 and 2). This is the point in which the belt overlaps the head pulley. Regardless of the type of transported material, Equation 1 must be satisfied to let the material leave the belt [1]:

$$\frac{v^2}{R} > g \cdot \cos\alpha \tag{1}$$

where: v – belt velocity [m/s], R - radius of the pulley [m], g – gravitational acceleration [$m/s^2$], α - the angle of the conveyor descend [°] [1], [8].

For cohesive materials, also Equation 2, taking into consideration adhesion and high velocities of the conveyor, must be satisfied [14]:

$$\frac{F_r - A_d}{\cos\alpha} \geq 1 \tag{2}$$

where:

$$A_d = \frac{\sigma_a}{\gamma \cdot h_m} \tag{3}$$

$$F_r = \frac{v^2}{g \cdot R} \tag{4}$$

where: v – belt velocity [m/s], γ - specific gravity of bulk material [N/m3], $\sigma_a$ – adhesion [$N/m^2$], $h_m$ - thickness of the material stream [m], g – gravitational acceleration [$m/s^2$], R - radius of the pulley [m] [8].

To obtain mathematical relations allowing determination of velocity v(φ) [m/s] of cohesive and cohesionless bulk material beyond the throwing points with a flat belt, a system of equations was used. The system differs depending on the type of material, and for cohesionless material it includes:

- equation of continuity [8]:

$$\dot{m} = \rho \cdot v(\varphi) \cdot A(\varphi) \tag{5}$$

- equation of equilibrium [8]:

$$\dot{m}(\vec{v} + d\vec{v}) - \dot{m}\vec{v} = d\vec{G} + d\vec{F}_c + d\vec{N}_{xs} + d\vec{T}_{xs} \tag{6}$$

- equation describing the contact friction condition on surfaces of walls restricting the path of the flow of a bulk material stream for static conditions and cohesionless materials (according to [14]):

$$\tau_{ws} = \sigma_n \mu_{xs} \tag{7}$$

For cohesive materials, the system of equations includes:

- equation of continuity [8]:

$$\dot{m} = \rho \cdot v(\varphi) \cdot A(\varphi) \tag{8}$$

- equation of equilibrium [8]:

$$\dot{m}(\vec{v} + d\vec{v}) - \dot{m}\vec{v} = d\vec{G} + d\vec{F}_c + d\vec{F}_a + d\vec{N}_{xs} + d\vec{T}_{xs} \tag{9}$$

- equation describing the contact friction condition on the surfaces of walls restricting the path of the flow of a bulk material stream for static conditions and cohesive materials (according to [14]):

$$\tau_{ws} = (\sigma_n + \sigma_a)\mu_{xs} \tag{10}$$

In Equations 7 and 9, $\mu_{xs}$ is a contact friction coefficient for static conditions. The general form of the system of equations for cohesionless material takes the form 11 [8]:

$$\begin{cases} \dot{m} = \rho \cdot v(\varphi) \cdot A(\varphi) \\ \dot{m}(\vec{v} + d\vec{v}) - \dot{m}\vec{v} = d\vec{G} + d\vec{F}_c + d\vec{N}_{xs} + d\vec{T}_{xs} \\ \tau_{ws} = \sigma_n \mu_{xs} \end{cases} \tag{11}$$

whereas for cohesive materials, the form of the system of equations is as follows [8]:

$$\begin{cases} \dot{m} = \rho \cdot v(\varphi) \cdot A(\varphi) \\ \dot{m}(\vec{v} + d\vec{v}) - \dot{m}\vec{v} = d\vec{G} + d\vec{F}_c + \vec{F}_a + \left|d\vec{N}_{xs}\right| + \left|d\vec{T}_{xs}\right| \\ \tau_{ws} = (\sigma_n + \sigma_a)\mu_{xs} \end{cases} \tag{12}$$

A detailed description of parameters included in the systems of Equations 11 and 12 is the same as presented in publication [8], where $A(\varphi)$ – cross-section of the stream [$m^2$], $\rho$ - bulk material density [$kg/m^3$], $\dot{m}$ - mass flow [kg/s].

Projection of Equation 6 onto directions of the assumed coordinate system <n,t> (Fig. 1) will generate a system of equation as a function of angular parameter $\varphi$, which after simplification, for cohesionless materials, takes the form:

$$\begin{cases} \mu_{xs}\left|d\vec{F}_c\right| - \mu_{xs}\left|d\vec{G}\right| = 0 \\ \dot{m}\, dv = 0 \end{cases} \tag{13}$$

For cohesive materials projection of Equation 9 onto direction of the assumed coordinate system <n,t> (Fig. 2) will also generate a system of equations as a function of angular parameter $\varphi$, a simplified form of which is as follow:

$$\begin{cases} \mu_{xs}\left|d\vec{F}_c\right| - \mu_{xs}\left|d\vec{F}_a\right| - \mu_{xs}\left|d\vec{G}\right| = 0 \\ \dot{m}\, dv + \mu_{xs}\left|d\vec{F}_a\right| = 0 \end{cases} \tag{14}$$

The solutions to Equations 13 and 14 are respectively differential equations, where for cohesionless materials the solution takes the following form 15:

$$\frac{dv^2(\varphi)}{d\varphi} - 2\mu_{xs} v^2(\varphi) = -2\, g\, \mu_{xs} R(\varphi) \tag{15}$$

and for cohesive materials – the solution is presented by 16:

$$\frac{dv^2(\varphi)}{d\varphi} - 2\mu_{xs} v^2(\varphi) = -2\mu_{xs} g\, R(\varphi)\left[1 + \frac{2\sigma_a}{\gamma h(\varphi)}\right] \tag{16}$$

Equations 15 and 16 are a special case of the Bernoulli equation, which can be written in the form (according to [16]):

$$y' + P(\varphi)y = Q(\varphi) \tag{17}$$

By integrating Equation 17, equations that allow determining the velocity of the material beyond the throwing point for cohesive and cohesionless materials were obtained. For cohesionless materials the velocity is described by Equation 18:

$$v_w = \sqrt{C_w\, e^{2\mu_{xs}\varphi} + g\, R(\varphi)} \tag{18}$$

and for cohesive materials by Equation 19:

$$v_w = \sqrt{C_w\, e^{2\mu_{xs}\varphi} + g\, R(\varphi)\left[1 + \frac{2\sigma_a}{\gamma h(\varphi)}\right]} \tag{19}$$

The description of parameters included in Equations 11 and 12 is the same as presented in publication [8], where $C_w$ - integration constant, $R(\varphi)$ - an average radius of curvature of the stream [m], h $\varphi$) - thickness of the material stream (according to Fig. 2, it is the height of the material stream lying on the conveyor belt).

Equations 18 and 19 contain integration constants, which describe Equation 20 for cohesionless materials, and Equation 21 for cohesive materials:

$$C_W = e^{-2\mu_{xs}\varphi}\left\{v_p^2 - g\,R(\varphi)\right\} \tag{20}$$

$$C_W = e^{-2\mu_{xs}\varphi}\left\{v_p^2 - g\,R(\varphi)\left[1 + \frac{2\sigma_a}{\gamma h(\varphi)}\right]\right\} \tag{21}$$

Equations 18 and 19 as well as 20 and 21 differ one from another by an element consisting adhesion.

To obtain a solution, boundary conditions, which are the same for both cohesive and cohesionless materials, were taken. Thus, the following boundary conditions, allowing determining the integration constant, were assumed [5], [6]:

$$\varphi = \alpha \tag{22}$$

$$R(\varphi) = R(\alpha) = R_0 = r_b + 0{,}5h_m \tag{23}$$

$$h_m = \frac{\dot{m}}{\rho v_t B} \tag{24}$$

$$v(\varphi) = v(\alpha) = v_0 = v_t \tag{25}$$

$$A(\varphi) = A(\alpha) = A_0 = \frac{\dot{m}}{v(\alpha)\rho} = \frac{\dot{m}}{v_t\,\rho} = h_m\,B \tag{26}$$

Velocity $v_w$ allows assuming the following boundary condition [5], [6]:

$$\varphi = \varphi_W = \beta_t \tag{27}$$

The value of the velocity of material beyond the throwing point is obtained by applying a simple iteration method, assuming initial conditions determined from Equations 22-27.
To obtain the exact approximation of velocity $v_w$ of material leaving the point, the following relation should be satisfied:

$$\left|\frac{v_{wn} - v_{w(n-1)}}{v_{wn}}\right| \le \delta_{v_W}\ [\%] \tag{28}$$

where $\delta_{vw}$ is an acceptable relative deviation of the estimated velocity $v_w$. It is assumed that velocity $v_w$ was correctly determined if the value of the deviation ranges 1%-2%.

Parameter α is constant because the conveyor is not inclined, unlike in solutions presented in [5], [6], and [8].

Equations 18 and 19 were tested to verify if they comply with the assumption that the velocity of material directly leaving the head pulley of the conveyor is equal to the velocity of the belt. Both equations proved to be compliant with the assumption.

Equations 18 and 19 can also be used to determine the velocity of material beyond the throwing point, in the function of depending on the assumed angle of material inflow to the next point, e.g. an impact plate (Fig. 3). In this case, it may be assumed in the proposed equations that angle $\beta_t$ is equal to the angle at which the material falls onto a surface (e.g. impact plate). Such an approach is a simplification, but returns results sufficient for practical use.

## 4. Indirect equations

Indirect equations allow determining the velocity of material leaving the throwing point which cooperates with an impact point presented in [16], where they were applied for a conveyor with an ascending and descending belt. These equations can also be used for a flat conveyor. They are suitable mainly for a conveyor supplied with an impact plate. They return best results for such application.

Figure 3 presents the flow conditions of material on a conveyor with a flat belt (the conveyor is not inclined), whereas Equations 29-32 allow determining the numerical values of parameters that can be seen in Figure 3. These equations are usable regardless the material.

Indirect equations (Fig. 3) include [16]:

- equation describing an angle of material flowing onto impact point $\xi_0$:

$$\xi_0 = \mathrm{arctg}\left( \mathrm{tg}\beta_t - \frac{g \cdot x_0}{2 \cdot v_{w1}^2 \cdot \cos^2 \beta_t} \right) \tag{29}$$

- equation describing the velocity of material flowing onto impact point $v_0$:

$$v_0 = v_{w1} \cdot \cos\beta_t \cdot \sqrt{\mathrm{tg}^2 \xi_0 + 1} \tag{30}$$

- equations describing distance parameters, according to Fig. 3:

$$x_0 = s_0 + (r_b + 0{,}5 \cdot h_0) \cdot \sin\beta_t \tag{31}$$

$$y_0 = \mathrm{tg}\beta_t \cdot x_0 - \frac{g \cdot x_0^2}{2 \cdot v_{w1} \cdot \cos^2 \beta_t} \tag{32}$$

where $\beta_t$ is an inclination angle of the material stream leaving the conveyor. In this case $\beta_t$ is equal 0 because the conveyor is not inclined.

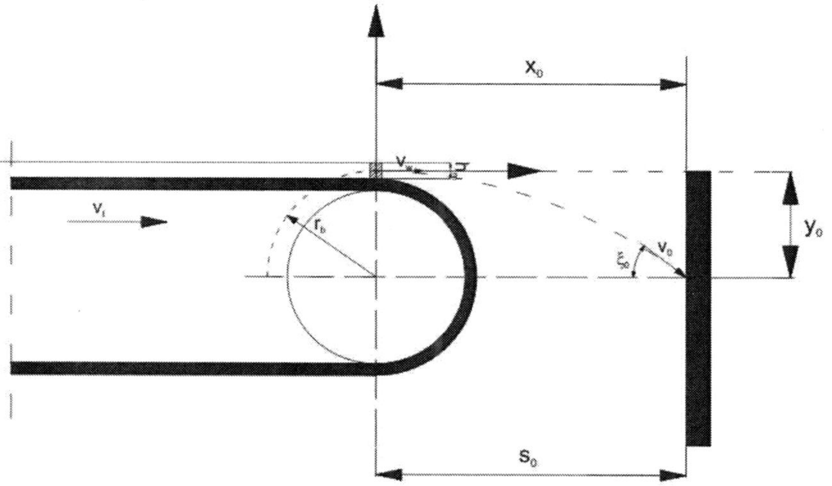


Fig.3. Cooperation of a throwing point with a flat belt and an impact point with a not inclined plate [elaborated based on [16]]

## 5. Comparison of the presented calculation methods

Table 1 was presents an example demonstrating the usefulness of the proposed equations. It also includes a comparison between the results obtained with the use of the proposed equations and the results obtained with the currently used indirect equations. In the proposed equations, the indirect equations are omitted, assuming that the angle of the outflow of material from a head pulley is equal to the angle of the inflow of material to the impact plate ($\beta_t = \xi_0$). This is a considerable simplification in relation to the currently used indirect equations, where not only the velocity of material flowing into the impact plate, but also the angle at which it flows into the plate are determined.

Table 1 presents the comparison of the results obtained with indirect equations and with the proposed equations for both cohesive and cohesionless materials. Table 1 also contains estimation of the percentage error for results obtained from the proposed equations in relation to the results obtained with the use of indirect equations. It also presents the difference between the obtained velocities of material at the moment when it falls onto the plate. Initial conditions of the material outflow are the same for both calculation cases.

Some of the values describing the parameters of the outflow of material from the throwing point were taken from [8] and [20].

Table 1. Determination of the velocity of material beyond the discharge point in relation of the analyzed angle at which material flows into impact point [own elaboration]

| Cohesionless material | Cohesive material |
|---|---|
| - required mass flow $\dot{m} = 350$ [kg/s],<br>- belt velocity $v_p = 3{,}15$ [m/s],<br>- belt width $B_t = 0{,}8$ [m],<br>- pulley radius $r_b = 0{,}25$ [m],<br>- angle of inclination of the conveyor belt $\alpha = 0$ [°],<br>- gravitational acceleration $g = 9{,}81$ [m/s$^2$],<br>- material bulk density $\rho = 850$ [kg/m$^3$],<br>- specific gravity $\gamma = 8340$ [N/m$^3$],<br>- friction coefficient for static conditions $\mu_{xs} = 0{,}50$,<br>- angle at which bulk material flows into subsequent point $\beta_t = \xi 0 = 41$ [°],<br>- distance between the head pulley and the point at which material is discharged $s_0 = 0{,}9$ [m], | - required mass flow $\dot{m} = 350$ [kg/s],<br>- belt velocity $v_p = 3{,}15$ [m/s],<br>- belt width $B_t = 0{,}8$ [m],<br>- pulley radius $r_b = 0{,}25$ [m],<br>- angle of inclination of the conveyor belt $\alpha = 0$ [°],<br>- gravitational acceleration $g = 9{,}81$ [m/s$^2$],<br>- material bulk density $\rho = 1380$ [kg/m$^3$],<br>- specific gravity $\gamma = 13540$ [N/m$^3$],<br>- friction coefficient for static conditions $\mu_{xs} = 0{,}51$,<br>- angle at which bulk material flows into subsequent point $\beta_t = \xi 0 = 41$ [°],<br>- distance between the head pulley and the point at which material is discharged $s_0 = 0{,}9$ [m],<br>- adhesion $\sigma_a = 350$ [N/m$^2$], |
| CALCULATIONS<br>- *proposed equations*<br>the first step of approximation<br>boundary conditions determining the integration constant<br>$\varphi = \alpha = 0$<br>$v(\varphi) = v(\alpha) = v_0 = v_t = 3{,}1$ [illegible]<br>$h_m = \frac{\dot{m}}{B_t \rho v_t} = \frac{350}{[illegible]} = 0{,}$ [illegible]<br>[illegible]<br>[illegible]<br>integration constant<br>$C_{w1} = e^{-2\mu_{xs}\alpha}\left\{v_t^2 - g\,R_{w1}\right\}$<br>$C_{w1} = 6{,}671$<br>boundary condition determining the velocity<br>$\varphi = \varphi_W = \beta_t = 4$ [illegible]<br>velocity | CALCULATIONS<br>- *proposed equations*<br>the first step of approximation<br>boundary conditions determining the integration constant<br>$\varphi = \alpha = 0$<br>$v(\varphi) = v(\alpha) = v_0 = v_t = 3{,}1$ [illegible]<br>$h_m = \frac{\dot{m}}{B_t \rho v_t} = \frac{350}{[illegible]} = 0{,}$ [illegible]<br>[illegible]<br>[illegible]<br>integration constant<br>[illegible]<br>$C_{w1} = 6{,}976$<br>boundary condition determining the velocity<br>$\varphi = \varphi_W = \beta_t = 4$ [illegible]<br>velocity |

| | |
|---|---|
| [illegible]<br>$v_{w1} = 4{,}110$<br>deviation value<br>[illegible]<br>[illegible]<br>approximation is insufficient<br>the second step of approximation<br>[illegible]<br>[illegible]<br>[illegible]<br>integration constant<br>[illegible]<br>$C_{w2} = 6{,}855$<br>velocity<br>[illegible]<br>$v_{w2} = 4{,}133$<br>deviation value<br>[illegible]<br>- *indirect equations*<br>$\beta_t = \alpha = 0$<br>[illegible]<br>$x_0 = 0{,}9$<br>[illegible]<br>$\xi_0 = 41$<br>[illegible]<br>$v_0 = 4{,}216$<br>- relative error of the determination of the velocity with both methods<br>$\delta_{v_r} = 2\%$ | [illegible]<br>$v_{w1} = 4{,}351$<br>deviation value<br>[illegible]<br>[illegible]<br>approximation is insufficient<br>the second step of approximation<br>[illegible]<br>[illegible]<br>[illegible]<br>integration constant<br>[illegible]<br>$C_{w2} = 7{,}112$<br>velocity<br>[illegible]<br>$v_{w2} = 4{,}422$<br>deviation value<br>[illegible]<br>- *indirect equations*<br>$\beta_t = \alpha = 0$<br>[illegible]<br>$x_0 = 0{,}9$<br>[illegible]<br>$\xi_0 = 41$<br>[illegible]<br>$v_0 = 4{,}216$<br>- relative error of the determination of the velocity with both methods<br>$\delta_{v_r} = 5\%$ |

The relative error $\delta_{vr}$ for results achieved with the proposed equations and with the indirect equations amounts respectively 2% for cohesionless materials and 5% for cohesive materials, which is acceptable for practical use. But the value of the error increases with the growing distance between the impact plate and head pulley. The difference between the velocity determined with the use of the indirect equations and the proposed equations for cohesionless materials amounts 0,08 [m/s], and for cohesive materials amounts 0,21 [m/s]. The minimal permissible distance between the throwing point and the impact plate was taken into consideration.

Differences obtained in calculations for values of velocity allow statement that the proposed equations can be used for estimation of the velocity of material beyond the throwing point. The indirect equations take into consideration kinematic and geometric conditions, whereas the proposed equations take also into consideration dynamic parameters, adhesion, the coefficient of friction of material against the belt surface, as well as material parameters such as material bulk density and specific gravity.

## 6. Conclusions

The presented analysis of the conditions of the flow of cohesionless and cohesive materials in a throwing point with a flat belt (non inclined conveyor) as well as the proposed equations obtained from the analysis are suitable for engineering calculations and for estimation of the velocity of material beyond the head pulley with the accuracy sufficient for practical use. It is especially important if the value of the velocity beyond the point must be known, which enables the correct selection of parameters of pouring points that cooperate with conveyors, and for maintaining the constant capacity of the transport system. The proposed equations consider not only kinematics and dynamics parameters of the flow but also material parameters.

The proposed equations, comparing to the indirect equations, give good compliance of 2% for cohesionless materials and 5% for cohesive materials. Such value of the relative error was obtained for an impact plate operating in angular pouring points, whereas for parallel pouring points, where distance between the head pulley and impact plate increases to 1.2 m [20], the relative error can achieves almost 10%.

The proposed equations provide a tool for engineers allowing estimating the value of velocity of material beyond the throwing point, which assures the correct capacity of the transport system consisting of belt conveyors. The proposed equations prove to be useful in cases where a throwing point follows or precedes an impact point, especially in properly selected distance between cooperating points, but they don't consider the air resistance.

The proposed equations assume that the angle at which material falls onto the subsequent point is known. However, the equations may turn out to be useless for the assumed angle at which material flows out of the head pulley if the distance is larger than 2 m.

## References

[1]. J. Antoniak, Przenośniki taśmowe w górnictwie podziemnym i odkrywkowym, Wydawnictwo Politechniki Śląskiej, Gliwice, 2007,

[2]. J. Antoniak, Obliczenia przenośników stosowanych w górnictwie, Wydawnictwo Śląsk, Katowice, 1964,

[3]. P.C. Arnold, G.L. Hill, Predicting the discharge trajectory from belt conveyors; Bulk solids handling 10/4 (1990) 379-382,

[4]. H. Colijn, Mechanical conveyors for bulk solids, Elsevier Science Publisher B. V. , New York, 1985,

[5]. J. Cyganiuk, F. Przystupa, Modelowanie przepływów strumieni kohezyjnych materiałów ziarnistych w węzłach zrzutowych z taśmą wznoszącą, Górnictwo Odkrywkowe, 3-4 (2011) 50-54,

[6]. J. Cyganiuk, Modelowanie przepływu strumieni materiałów ziarnistych we współpracujących węzłach przesypowych systemów transportu przenośnikowego, Górnictwo Odkrywkowe, 3-4 (2007) 28-32,

[7]. J. Cyganiuk, Safety of granular materials in conveyor transport systems, Machinery and equipment safety in industry , Sekcja Wydawnicza Wydziału Zarządzania Politechniki Częstochowskiej, Częstochowa 2013 72-92,

[8]. J. Cyganiuk, P. Kuryło, E. Tertel, Modelling of the flow of streams of cohesionless and cohesive bulk materials in a conveyor discharge point of a descending belt conveyor, Procedia Engineering, 96 (2014) 28-38,

[9]. J.Cyganiuk, Modelling of the flow of bulk material of group III in an impact pouring point with a flat impact plate, Systems : Journal of Transdisciplinary Systems Science, 12/ 3 (2007) .67-76,

[10]. J. Cyganiuk, F.W. Przystupa, Wpływ kąta pochylenia płyty odbojowej na prędkość materiału opuszczającego węzeł odbojowy, Górnictwo Odkrywkowe, 4-5 (2009) 30-34,

[11]. J. Cyganiuk, Modelowanie przepływu niekohezyjnych materiałów ziarnistych na płaskich płytach odbojowych, Górnictwo Odkrywkowe, 3-4 (2013) 39-45,

[12]. W. Czuba, K. Furmanik, Analysis of a grain motion in the transfer area of the belt conveyor, Maintenance and Reliability, 15/ 4 (2013) 390-396,

[13]. D. Hastie, P. Wypych, Conveyor Belt trajectories – comparing predicted to experimental results, Bulk solids handling, 30/8 (2010) 438-445,

[14]. Z. Korzen, Mechanics of Belt conveyor discharge process as affected by air drag, Bulk solids handling, 9/3 (1989) 289-297,

[15]. Z. Korzen, Mechanics Mechanics of bulk solid stream flow in throwing belt conveyors, Bulk solids handling, 10/1, (1990) 55-64,

[16]. Z. Korzen, The dymanics of bulk solids flow on impact plates of belt conveyor systems, Bulk solid handling, 8/6 (1988) 689-697,

[17]. Z. Korzen, Zum abwurfvorgang bei gurtforderen, Fordern und heben 1894, 34/5 380-388,

[18]. F.W. Przystupa, Cyganiuk J.,: Modelowanie przepływu materiału ziarnistego typu cement w węźle odbojowym z płaską płytą odbojową, Górnictwo Odkrywkowe 3-4 (2007) 103-108,

[19]. В. Л. Сакович, В.П. Кукса, Движение груза в пересыпных устройствах ленточных конвейров, Горные страительные дорожные машины, Техника, (1967) 78-86,

[20]. T. Żur, Przenośniki taśmowe w górnictwie, Wydawnictwo Śląsk, Katowice, 1979,

EAN 2017
55th conference on experimental stress analysis

# Comparison of Methods used for the residual Stress Analysis in a Pipe made from Polypropylene

František Fojtík[1, 2, a], Paška Zbyněk[2,b], Kolář Petr[2,c]

[1]IT4Innovations, VŠB – Technical University of Ostrava, 17. listopadu 15/2172 Ostrava-Poruba, Czech Republic

[2]Faculty of Mechanical Engineering, VŠB – Technical University of Ostrava, Department of Applied Mechanics, 17. listopadu, 2172/15 Ostrava, Czech Republic

[a]frantisek.fojtik@vsb.cz, [b]zbynek.paska@vsb.cz, [c]petr.kolar@vsb.cz

**Keywords:** Slitting method; hole drilling method; residual stresses.

**Abstract.** This contribution is focused on the problematics of measuring residual stresses in a pipe, which is made from polypropylene. The pipe is usually used for supply. Two experimental approaches were used to analyse the residual stresses. The slitting method was applied to find out the hoop residual stresses. The results were then confronted with those gained from the hole drilling methodology. The distribution of the residual stresses along the thickness of the tube was investigated by these two experimental methods. The FEM analysis was performed and the results were compared with the results from the measurements.

## Introduction

It is well known that inside every fabricated part there is some level of the residual stresses. The producers are trying to reduce the amount of the residual stresses by applying many techniques. Lifespan of component is then usually decreased, which is caused by influence of the residual stresses. Therefore it is desirable to know their amount and distribution.

The residual stresses in materials are caused by external forces, deformation of the component, or by the temperature acting on the component (in the whole component or only its pieces), which must cause elasticity plasticity state inside the component. Two experimental approaches were used to analyse the residual stresses in the polypropylene pipe. The first method is the slitting method, which was in publication [1, 2, 3] applied for analysis of the pipe made from different material. In the above-mentioned literature, apart from the dependence of lifespan on residual stresses, changes of stresses and other mechanical parameters were also investigated in time with interval of several weeks. In this contribution, the changes of mechanical parameters in time are not observed, even if it takes place in polypropylene material. The results obtained by the slitting method and by the hole drilling method are also compared. If these experimental methodologies are applied to plastic materials, then some difficulties may occur [4, 5, 6]. The experimental results are then confronted with those obtained from the FEM analysis.

## Characteristics of the pipe

The outer diameter of the pipe, which was used for residual stresses measurement, is 63mm. The thickness of the wall is 10.5mm. Pressure category is denoted PN 20. The pipe is made from polypropylene – PPR. This type of the pipe is used for cold water distribution to the maximum pressure of 1 MPa. Some parameters of the pipe are listed in Table 1. The parameters were adopted from the catalogue lists. The elastic modulus was determined by three point bend tests.

| Property | Value | Unit |
|---|---|---|
| Density | 905 | $[kg \cdot m^{-3}]$ |
| Elastic modulus | 850 | $[MPa]$ |
| Poisson's ratio | 0.40 | $[-]$ |
| Yield strength | 25 | $[MPa]$ |
| Ductility at yield point | 10 | $[\%]$ |
| Coefficient of thermal expansion | 0.00012 | $[K^{-1}]$ |
| Coefficient of thermal conduction ($20°C$) | 0.24 | $[W \cdot m^{-1} \cdot K^{-1}]$ |
| Specific thermal capacity ($20°C$) | 2000 | $[J \cdot kg^{-1} \cdot K^{-1}]$ |

Tab. 1. Physical and thermal parameters of the material.

**Determination of the hoop stress by the slitting methodology**

Procedure of the slitting method is based on the methodologies described in literature [1, 2, 3]. The wall of the pipe was imaginary divided into 10 layers and then sliced up into rings, which were 10mm wide. In total, 13 specimens were obtained. Specimen number one was without machining. Turning was used to remove first six layers from the inner side of the specimens denoted 2 to 6. In the same way, the first six layers were removed from the outer side of the specimens denoted 8 to 13 (Fig. 1). A CNC controlled turning machine was used for this task.

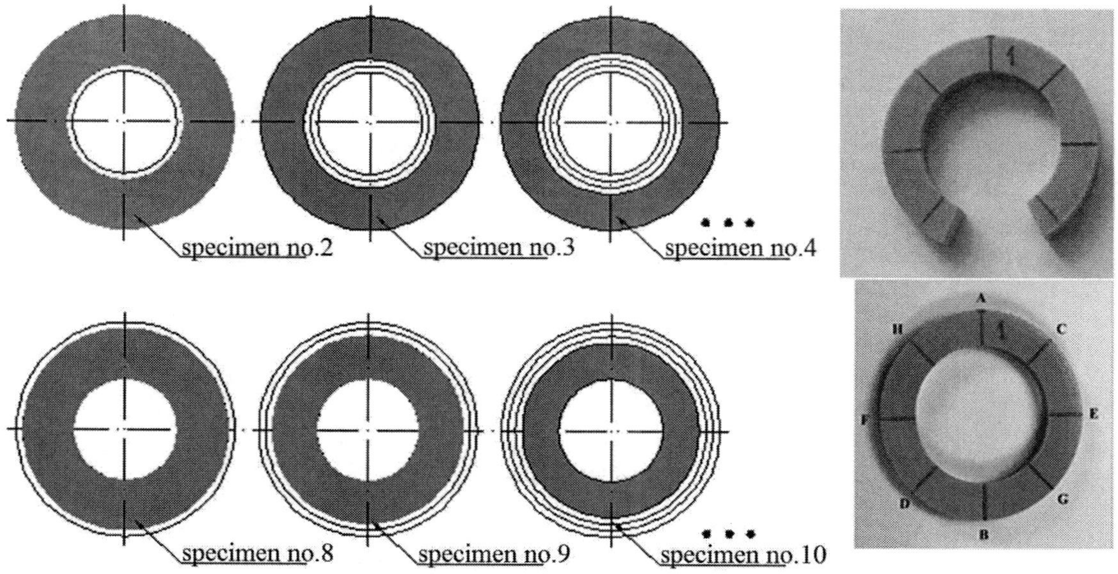


Fig. 1 Scheme of the turning operation on the left side. On the right side, specimen number one before and after the cutting can be seen. The points of measurement are marked.

Outer and inner diameter of the specimens were measured in several places and several cuts. (Figure 1). The measurements were performed by applying Wenzel LH65 X3M with the touch probe SP25M with modulus SM25-1. For each ring, the arithmetic mean value was determined from the measured values in different places. The change of the outer diameter was, after the segment was cut, measured in the same way and is shown in Figure 2.

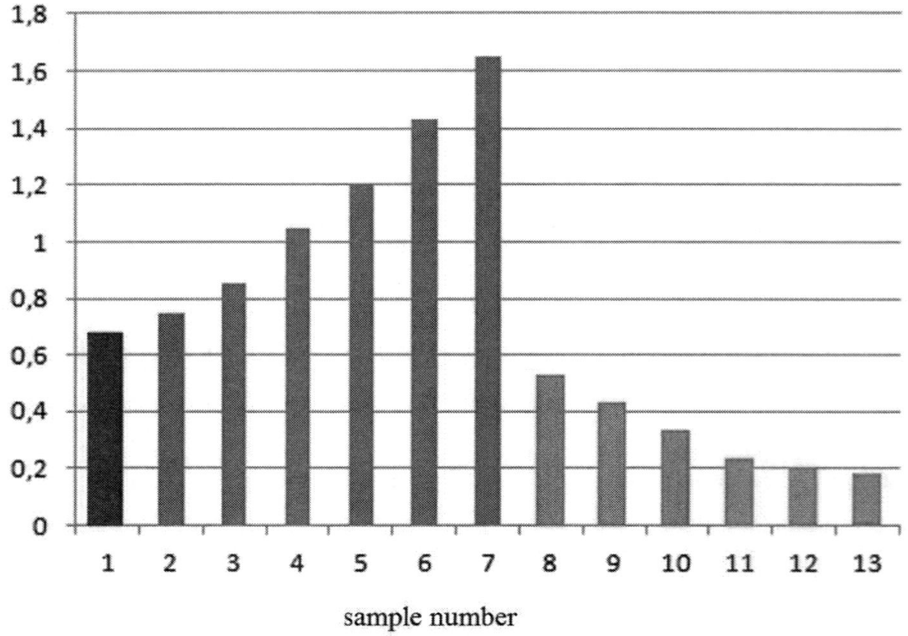


Fig. 2. Change of the outer radius.

The change in the curvature of the curved beam neutral surface in Figure. 2 plays the key role for the stress distribution determination in the wall of the pipe:

$$\frac{1}{R^I} - \frac{1}{R} = \frac{M}{E * A_s * e * R}, \tag{1}$$

where $R$ is the radius of neutral surface in original state, $R^I$ is the radius of the neutral surface in the deformed state, $M$ is the bending force, $E$ is elastic modulus, $A_s$ denotes cross section area, and $e$ is the eccentricity.

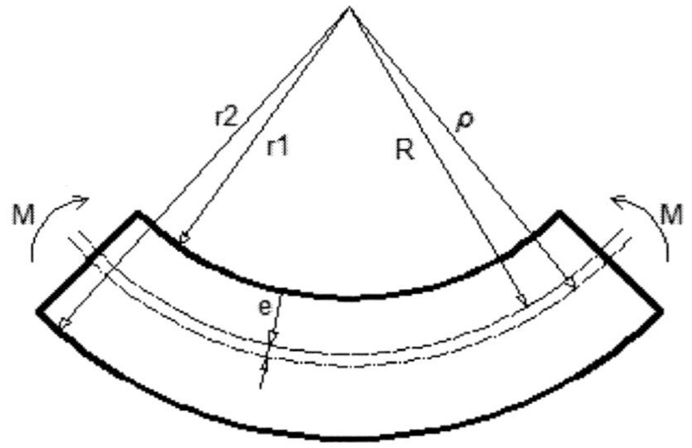


Fig. 3. Scheme of curved beam element.

If the wall of the pipe is divided into $n$ layers with the same width, then each layer $i$ has radius $r_i$, stress components $\sigma_i$ and moment $M_i$. The bending force depends on the position $R$ of the neutral surface and can be determined by equation:

$$M_i = F_i(R - r_i) = \sigma_i \cdot \Delta r \cdot W * (R - r_i)\,, \qquad (2)$$

where $\Delta ri = r_{i+1} - r_i$ a $W$ is axial dimension of the ring. Overall bending force corresponds to the change of the curvature of the non machined ring:

$$M = \sum_{i=1}^{n} Mi\,. \qquad (3)$$

The condition of equilibrium must be satisfied:

$$\sum_{i=1}^{n} \sigma i = 0\,. \qquad (4)$$

The distribution of the residual stress along the wall of the tube is plotted in Fig. 6 in the manner from the inner to outer diameter. On the inner diameter, there is the tensile stress, but on the outer diameter, there is the compressive stress. The stress in the absolute value is bigger on the outer diameter. This is due to manufacturing process, when cooling is only on the outer surface. The fluctuations in the upper half of the graph may be the consequence of the inaccurate measurements.

### Determination of the residual stress by the hole drilling methodology

The procedure of measuring residual stresses by the hole drilling method can be found in literature [5]. The procedure according to standard ASTM E 837 – 13a [7] is also widely used. The integral methodology or the methodology of power series are suitable for determination of the stress gradient under the surface [5]. The strain gauge rosette RY 1.5/120M was glued on the pipe. Another rosette was added to eliminate influence of temperature. A special milling-cutter for plastic materials was used to drilling hole 1.8mm of diameter to the depth of 2mm.

The procedure was performed in accordance with the ASTM standard. The calibration constants were determined from the experimental setup on beam with equivalent strength, and these constants are valid for this material, strain gauge rosette, and chosen procedure. The measurement was carried out on inner and outer surface of the pipe. The results of the residual stress gradient in the circumferential direction, obtained by the hole drilling method, across the depth of the hole can be also seen in Figure 6. On the outer surface (0.05 mm under the original surface), the hoop stress value of -5.15 MPa was determined by the numerical approach. On the inner surface, the value of 1.41 MPa was established. The main advantage of this method is determination of the principal stresses and their direction, so stresses in any direction can be calculated. Furthermore, we obtain idea about the residual stress in the axial direction of the pipe.

### Residual stresses from the FEM analysis

Analysis was performed in the MSC Patran software with the MSC Marc solver [8]. Simulation was, due to symmetry, carried out on one quarter of the model. Material parameters are listed in Table 1. Isoparametric quadratic elements QUAD 4 were used for representation of the finite element model.

The inverse approach was used with regard to the manufacturing process - heating up of the model. The initial temperature condition of 283 K (the temperature of cooling bath) was applied on the cooled surfaces of the model. In the next step, another temperature condition of 403 K (the temperature of the pipe when it is coming to the cooling bath) is applied on the outer surface of the pipe. The pipe is cooled only on the side of the outer diameter. The time of the simulation was 144 s, which is exactly the cooling time in the manufacturing process.

The boundary conditions are illustrated in Figure 4. The values of the normal stress in the circumferential directions are shown in Figure 5. The calculated residual hoop stress on the outer surface takes the value of -6.12 MPa and on the inner surface 4.07 MPa.

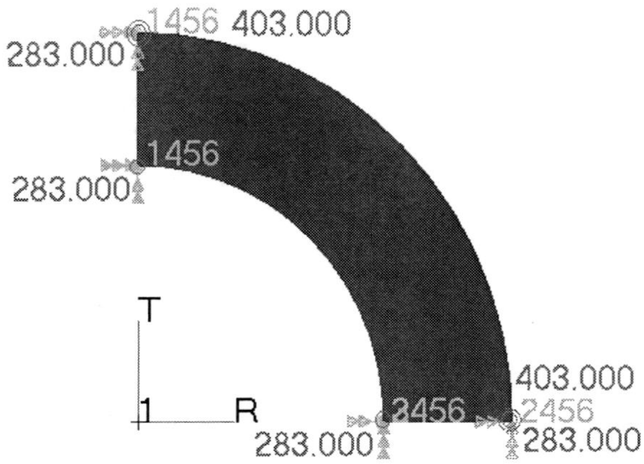


Fig. 4. Boundary conditions and thermal load.

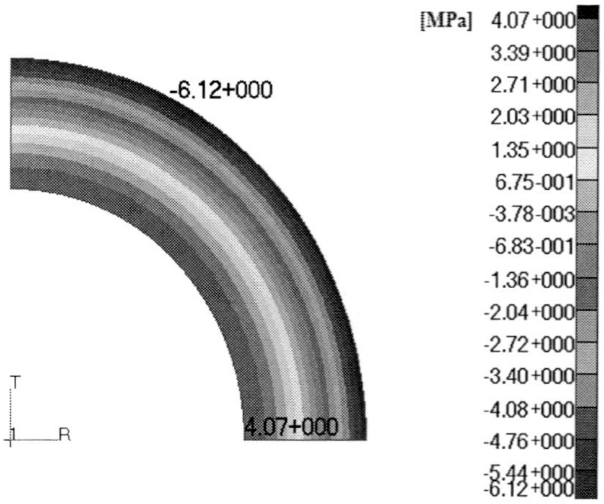


Fig. 5. Normal stress distribution in circumferential direction.

**Comparison of results**

The results obtained by the experimental measurements and FEM analysis are shown in Figure 6. It also shows the circumferential stresses calculated according to the slitting method. On the outer surface, the hoop stress has the value of -6.84 MPa. On the inner surface, which corresponds to the first removed layer, the hoop stress has the value of 3.37 MPa.

The residual stresses were also investigated by the hole drilling method. This methodology enables principal stresses and their direction to be determined, and thus the stresses in any direction can be calculated. Furthermore, we obtain idea about the residual stress in the axial direction of the pipe in one measurement. The results of the residual stress gradient in the circumferential direction, obtained by the hole drilling method, across the depth of the hole can also be seen in Figure 6. On the outer surface (0.05 mm under the original surface), the hoop stress value of -5.15 MPa was determined by the numerical approach. On the inner surface, the value of 1.41 MPa was established. From the measurement, it is apparent that on the outer surface the residual stresses correspond with the results obtained by the slitting method. On the side of the inner diameter, some difference can be seen. The difference may be caused by the manufacturing process, when the pipe is extruded over the mandrel. This process caused deformation in the surface layer and new distribution of the residual stresses so the slitting method cannot catch them. The results gained from the slitting method are shown in Figure 6, when the thick curved beam theory was implemented. More information can be found in literature [1].

The distribution of the residual stresses across the thickness of the pipe, which was determined by the FEM approach, is shown in Figure 6. The values of the normal stresses in the circumferential direction are shown in Figure 5. The calculated residual hoop stress on the outer surface takes the value of -6.12 MPa and on the inner surface 4.07 MPa.

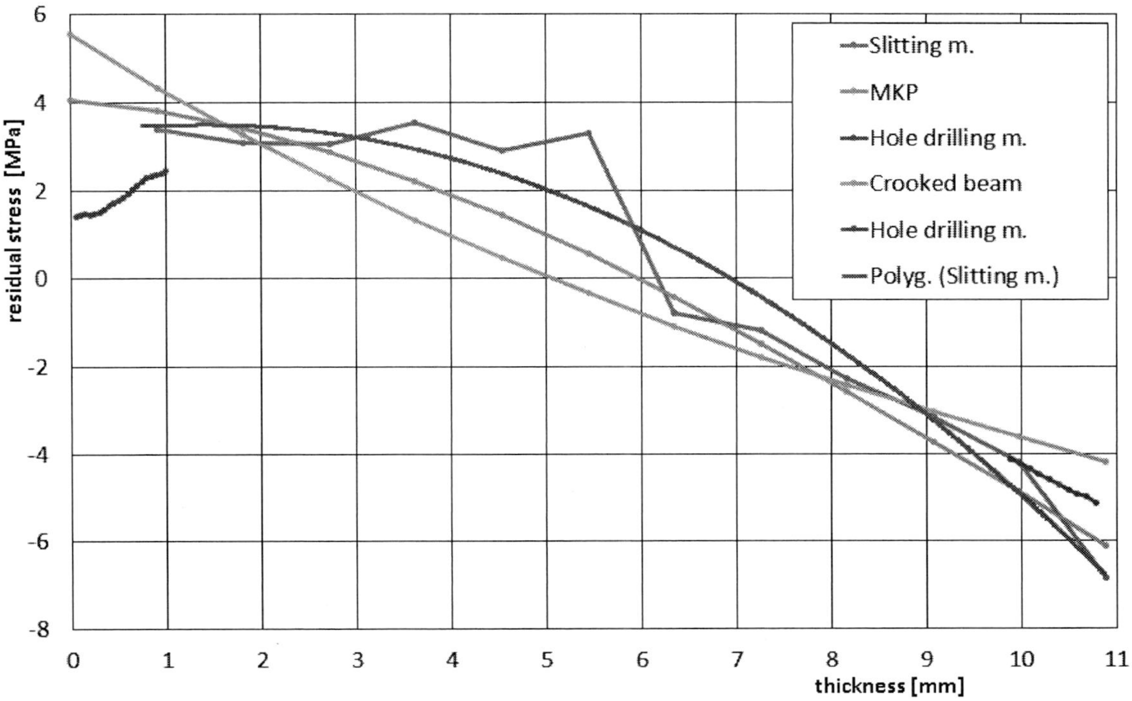


Fig. 6. The distribution of the residual stresses across the thickness of the pipe and approximation by the polynomial function.

## Conclusion

This contribution is focused, in detail manner, on the comparison of the two experimental methodologies, the slitting method and the hole drilling method, carried out on the pipe, which is made from polypropylene. The experimental results have been verified with the results of the residual stresses obtained by the FEM approach. The distribution of the residual stresses across the thickness of the pipe can be seen in Figure 6. Quite good correspondence between the experimental methodologies and the FEM approach is observed on the outer surface. On the inner surface the situation is different. The evaluation of the residual stresses by the slitting method is carried out after the first layer is removed, then the results obtained from the slitting method agree quite well with the results from the FEM analysis. The results from the hole drilling method are different from those obtained from the FEM analysis. This is probably caused by manufacturing, when the pipe is extruded through the thorn. This processing caused the deformation of the outer surface due to cooling of the pipe. The residual stresses are then redistributed and cannot be caught by the slitting method.


## Acknowledgement

This work was supported by The Ministry of Education, Youth and Sports from the National Programme of Sustainability (NPU II) project "IT4Innovations excellence in science - LQ1602", by the Czech Science Foundation (GA15-18274S) and partially supported by the Specific Research (SP2017/136). The support is gratefully acknowledged.

## References


[1] P. Kolar, Measurement of Residual Stresses in Plastics Separation Methods, Thesis, VŠB – TU Ostrava, 2014.

[2] J. Kucera, J. Krivanek, Morphology and internal stress distribution in the PP pipe wall. Deformation und Bruchverhalten von Kunstoffen. 2007, Merseburg.

[3] J. Z.W. Guan, J.C. Boot, A method to predict triaxial residual stresses in plastic pipes, Polymer Engineering and Science, October 1, 2004

[4] N. S. Rossini, M. Dassisti, K. Y. Benyounis, A. G. Olabi, Methods of measuring residual stresses in components. 2012, vol. 35, s. 572- 588

[5] G. S. Schajer, Practical Residual Stress Measurement Methods, Wiley, 2013.

[6] ASTM E 837 – 13a: Determining Residual Stresses by the Hole – Drilling Strain-Gage Method.

[7] K. Kolarik, Z. Pala, N. Ganev, F. Fojtik, Combining xrd with hole-drilling method in residual stress gradient analysis of laser hardened c45 steel. Adv Mater Res 996:277–282, 2014. doi:10.4028/www.scientific.net/AMR.996.277.

[8] Patran Element Library. http://web.mscsoftware.com/patran/current/ html_patran/wwhelp/
wwhimpl/js/html/wwhelp.htm#href=Marc/marc05_ examples 2.7.6.html

EAN 2017
55th conference on experimental stress analysis

# Experimental and Numerical Determination of Riffle Barrel Properties

DOUBRAVA Karel[1, a], SHÁNĚL Vít[1, b], FUMFERA Jaromír[1, c], RŮŽIČKA Milan[1, d] and STEINBAUER Pavel[1, e]

[1]Czech Technical University in Prague, Faculty of Mechanical Engineering,

Department of Mechanics, Biomechanics and Mechatronics,

Technická 4, Praha 6, Czech

[a]Karel.Doubrava@fs.cvut.cz, [b]Vit.Shanel@fs.cvut.cz, [c]Jaromir.Fumfera@fs.cvut.cz, [d]Milan.Ruzicka@fs.cvut.cz, [e] Pavel Steinbauer@fs.cvut.cz

**Keywords:** Riffle Barrel; Strain Gauge; FEM, Hybrid Composite Barrel, EMA.

**Abstract.** The paper describes an experimental procedure for determining the properties of a conventional steel barrel. The deformation behavior during shooting was monitored by strain gauges and the modal properties determined by experimental modal analysis. The results obtained were used to debug the numerical model of rifle barrel shooting. Modal analysis of hybrid barrel samples were made to assess the consistency of the modal properties of produced prototypes.

## Introduction

Accuracy and precision of shooting is one of the requirements for the development of rifle. In the framework of research project is the means to achieve the goal of using a hybrid barrel with steel core and composite wrap. The actual behavior of a conventional barrel during firing was monitored by strain gauges and accelerometer. The obtained data helped us to verify boundary conditions of the numerical model of shot from the barrel.

## Experimental Investigation

Strain gauges were used to assess the behavior of the gun barrel during firing. Dominant bending stress has been assumed in the preparation phase of the experiment, therefore, strain gauges were installed into half bridge wiring. A limited number of available channels with sufficient sampling frequency (96 kHz) was another reason for selecting half bridge wiring with the exclusion of the influence of temperature and purely tensile strain component. Bending strain were monitored in two sections. Four strain gauges were evenly installed in each section in the longitudinal direction.

Flexural properties of the test barrel were evident after the first series of experiments, the measured bending waveforms strain had a frequency corresponding to the first natural frequency of the barrel. The size of the bending deformation in the vertical direction was approximately three times bigger than the size of the bending strain in the horizontal direction.

Making of operational modal analysis was planning, but the accelerometer was damaged during the measurement, high jerk values were probably the reason. Experimental modal analysis was used for determination of dynamic properties of the barrel and the whole gun.

The question of the size of other strain components during shot was the reason another series of experiments [1]. Torsional component of strain being monitored by the shear strain gages, (V-shaped strain gauge with two grids and half bridge connection) installed between sections with longitudinal strain gauges.

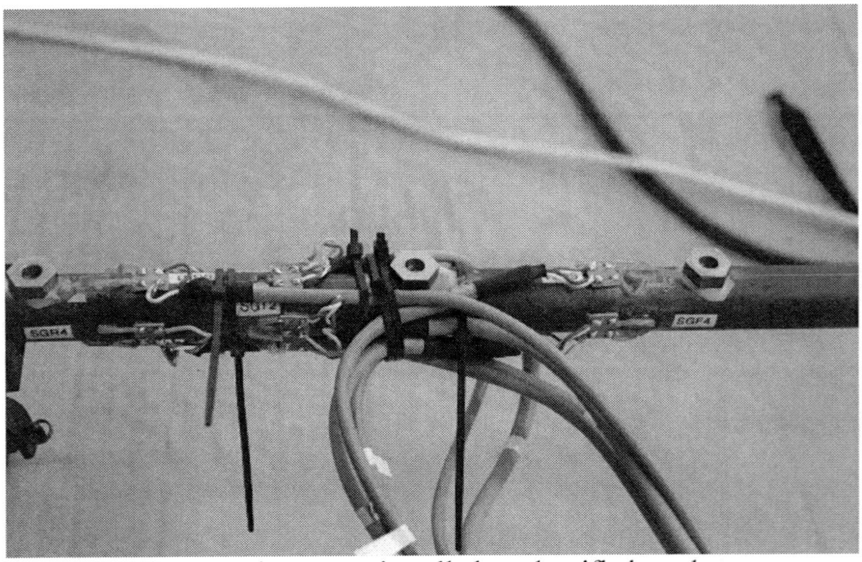

Fig. 1 Strain gauges installed on the rifle barrel

Whole set of strain gauges can be seen in Fig. 1. Gauges used in previous measurements for monitoring vertical bending, were connected via a three wire quarter bridge circuit. With regard to possible parasitic effect of temperature on the measured strain, a temperature of barrel surface was monitored by platinum resistance thermometer. Four channels were measured with high sampling frequency (96 kHz), the remaining channels were monitored with a lower frequency (19 kHz) by means other strain gauge units. Both units were synchronized by means fireware connection.

Muzzle velocity was measured by means photocell, when small scatter of velocities was recorded.

Purely tensile deformations were measured immediately after firing, these were quickly replaced by the bending deformations of a larger size. Shear strain measured during firing are much smaller than the bending deformation. The temperature of the barrel after the shot has risen by a few degrees of Celsius.

## Numerical Analysis

Numerical model of same situation was made in Finite Element software, when elements with eight nodes and reduced integration and explicit solution were used [2]. The rifle barrel was modeled using planar symmetry and together with bolt assembly. The shot was modeled as a pressure impulse in the chamber of the barrel with propagation of pressure in barrel by means user subroutine. The simulation was computed until the bullet left the rifle barrel. The strains were recorded in places of strain gauges. Computed deformation of the rifle barrel during the shooting can be seen in Fig. 2.

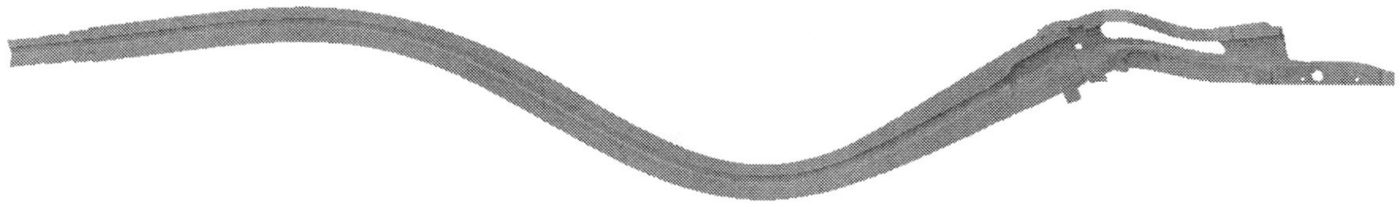

Fig. 2 Magnified deformation of the rifle barrel during the shooting computed by FEM.

Example of comparing measured and computed bending strain in one section can be seen in Fig. 3 Strain data were normalized to biggest value of measured strain.

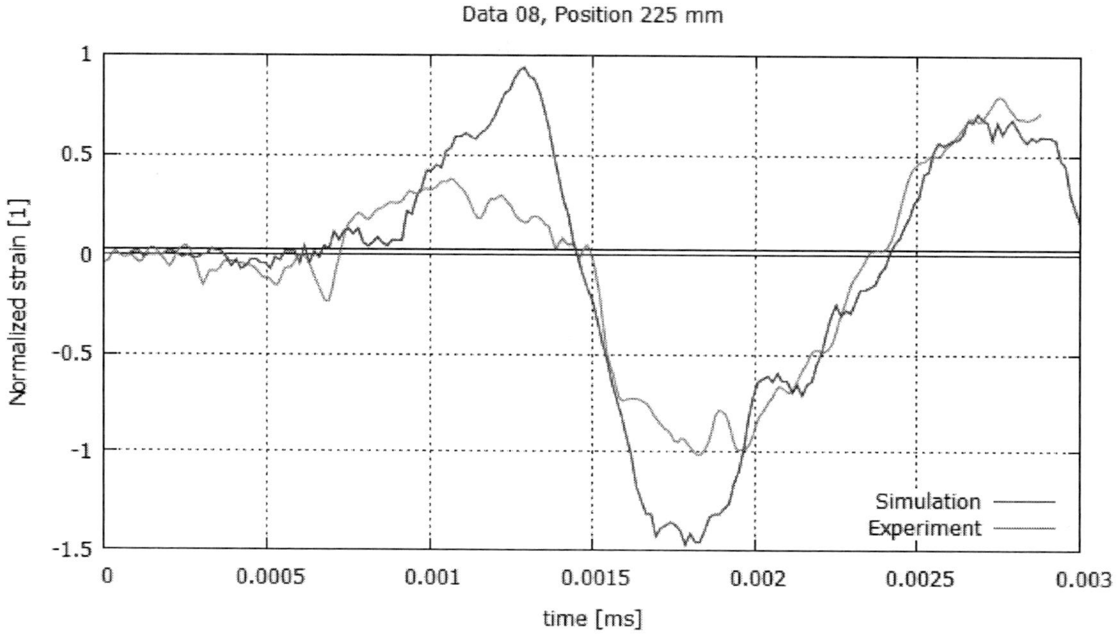


Fig. 3 Comparing of normalized measured and computed bending strain.

Based on the comparison of calculated and measured data, it is possible to state the suitability of the created numerical model for further research on hybrid composite barrel.

## Modal Properties of Hybrid Barrel Samples

The question of the effect of the composition used for the manufacture of hybrids barrel was examined with regard to the modal properties of the final product.

Modal properties characterize the transverse stiffness of the barrel and its attenuation properties. Each measured sample was attached to rubber hinges with great compliance (see Fig. 4). For the excitation, the impact hammer was used to measure the excitation time in the plastic tip of the tool. The results of the measurements were processed in the ModalView software environment by the Polyfit algorithm to obtain their own frequencies, the respective modal damping and the oscillation shapes.

A total number of 20 different test specimens of composite prototypes of barrel were measured. The measurement showed good repeatability for the samples produced by the same technology and the same design. The results are suitable for verification of computational and optimization design tools.

Fig. 4 Barrel sample fixation during EMA measurement.

## Conclusions

Characteristics of a conventional barrel were identified during shooting. A numerical model, with a good agreement with experiment, was created. Measurement of the modal properties of the prototypes specimens is used to identify the composition that was used to produce a functional porotype on the hybrid barrel. This barrel will be subjected to measurements to compare properties with the original conventional product.


## Acknowledgement

This work was supported by the Technological Agency of the Czech Republic in the project TH01010772 -The development of accurate rifle with a composite hybrid barrel


## References


[1] K. Doubrava, V. Shánĕl, M. Růžička, Measuring Load the Gun Barrel during Firing [Report] 2016, 12105/16/52, CTU in Prague (in Czech)

[2] V. Shánĕl, M. Španiel, K. Doubrava, Application and Development of Numerical Models in Ballistic, Engineering Mechanics 2016 - Book of full texts. Prague: Institute of Thermomechanics, AS CR, v.v.i., 2016. pp. 506-509. ISSN 1805-8248. ISBN 978-80-87012-59-8.

# Comparison of Fans' Behavior and Grandstand Dynamic Response during Two Hockey Matches Between the Same Teams

Martin Verner[1, a], Michal Polák[1, b] and Tomáš Plachý[1, c]

[1]Czech Technical University in Prague, Faculty of Civil Engineering, Department of Mechanics, Thákurova 7, Praha 6, Czech Republic

[a]martin.verner@fsv.cvut.cz, [b]polak@fsv.cvut.cz, [c]plachy@fsv.cvut.cz

**Keywords:** Grandstand, Vibration, Fans' Behavior, Hockey Match.

**Abstract.** The main aim of two experiments described in the paper was to study ice hockey fans behavior in a selected grandstand sector and simultaneously grandstand vibrations induced by them as the experimental basis for specification of new spectator's crowd load models. These experiments followed on the two series of similar experiments focused on spectators induced grandstand vibrations and fans behavior during some football matches that were performed at the Letná football stadium in Prague in the 2013-14 and 2014-15 seasons. The paper summarizes and compares the basic results of two experiments focused on the fans behavior and its influence on the level of induced grandstand vibrations during two ice hockey games between the same teams. The experiments were carried out at the ice hockey stadium in Pilsen that is the home stadium of the HC Škoda Plzeň ice hockey club.

## Introduction

Substantial dynamic forces can be induced by a group of sport spectators moving synchronously. Consequently, grandstand vibration amplitudes which can be seen by naked eye could be occasionally caused. In history, there are known cases when the large vibration of a grandstand caused panic of spectators or even its breakdown in a rare situation [1].

At the present time, the satisfactory dynamic load of grandstands is not described in any standard with acceptable precision. Nevertheless, new dynamic models of spectator's crowd were developed based on stochastic approaches [2, 3]. There is also a lack of real experiments which could be used for verification or improvement of the models [1, 4, 5].

The main objective of the experiments introduced in this paper was to study spectators in a selected grandstand sector during two ice hockey games between the same teams. The experiments were focused on the types of fans behavior, their activity and its influence on the level of induced grandstand vibrations as the experimental basis for specification, improvement and verification of new spectator's crowd load models [2, 3]. These experiments followed two series of similar experiments focused on fans behavior and induced grandstand vibrations during some football matches that were performed at the Letná football stadium in Prague in the 2013-14 and 2014-15 seasons [6, 7].

The experiments were realized at the ice hockey stadium in Pilsen that is the home stadium of the HC Škoda Pilsen ice hockey club (see Fig. 1) during two ice hockey games against HC Mountfield Hradec Králové on November 20th 2015, which HC Škoda won 6:2, and on February 2nd 2016, that HC Škoda lost 3:4.

## Description of the Stadium

The basic load bearing members of the stadium roof are cables which are anchored to the perimeter reinforced concrete columns. The reinforced concrete elements that form the surface of the roof above the grandstands are supported by the cables. The roof section above the ice rink is transparent. The grandstand of the HC Škoda Pilsen stadium is the reinforced concrete frame structure consisting of T- shaped horizontal beams, sloping main beams and columns mainly (see Fig. 2). We investigated the sector where the most active fans of the home team were gathered (see Fig. 1). This structure is naturally less sensitive to dynamic excitation than the steel cantilever grandstand observed in the course of the experiments performed on the Letná football stadium in Prague [6, 7].

Fig. 1: The observed grandstand sector, where the most active home fans are usually gathered, without fans (left) and with the most active fans (right)

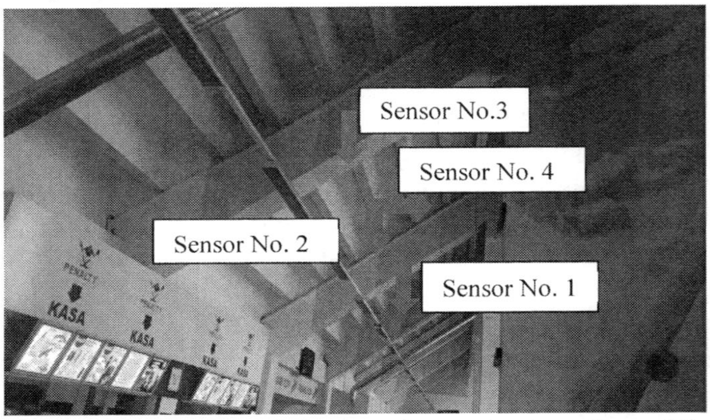


Fig. 2: The bottom view on the sector and the location and orientation of the accelerometers used by carried out experiments.

## The Experiment Arrangement

Four piezoelectric acceleration transducers Brüel&Kjær of type 4507 B0 were placed under the investigated sector of the grandstand to measure the response caused by the most active fans (Fig. 2). The same transducers were used for both experiments. Two transducers (Sensors No. 1 and 3 in Fig. 2) were placed in the middle of two sloping main girders of the grandstand to measure vertical accelerations. The remaining two transducers were placed in the middle of the T-shaped beam to measure vertical (Sensor No. 2 – Fig.2) and horizontal (Sensor No. 4 – Fig. 2.) accelerations. The behavior of fans (Fig. 1 – right) was monitored by camera placed on the opposite side of the stadium.

## Results

The fans' behavior and the induced response of the grandstand were monitored during two ice hockey matches between the same teams HC Škoda Plzeň and HC Mountfield Hradec Králové. Very similar types of spectators' behavior were observed during two studied hockey games as the types, which were monitored during the football matches [6, 7]. It means these ones - static state, walking and running, jumping, swaying, hand clapping, goal scoring and the Mexican wave. However, some differences could be found. During the ice hockey matches, the fans did not use the jumping type of cheering at all. The other difference is that the ice hockey spectators used more often the cheering type with hand clapping. The comparison of the vertical and horizontal accelerations in the middle of the monitored beam induced by the cheering of type hand clapping observed during two matches between the same teams are shown in Fig. 3 and Fig. 4.

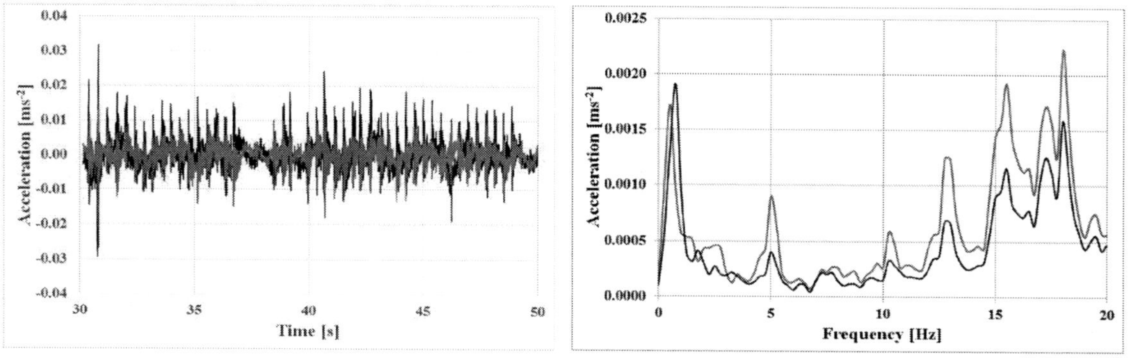


Fig. 3: The vertical (red) and horizontal (black) accelerations in the middle of T-beam caused by the hand clapping type of cheering - November 20th 2015.

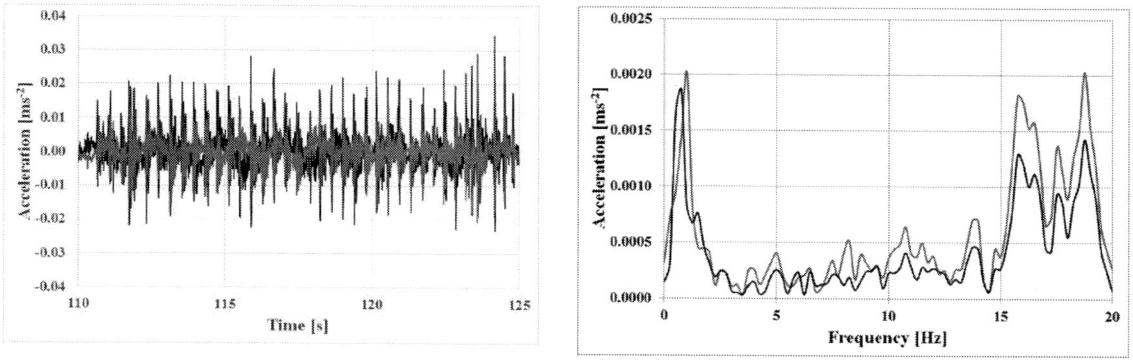


Fig. 4: The vertical (red) and horizontal (black) accelerations in the middle of T-beam caused by the hand clapping type of cheering - February 2nd 2016.

## Conclusions

The paper compares the results of two experiments focused on the fans behavior and its influence on the level of induced grandstand vibrations during two ice hockey games between the same teams HC Škoda Plzeň and HC Mountfield Hradec Králové. During both experiments, the same types of cheering were observed and the character of the dynamic response induced by spectators was very similar too (Fig.3 and Fig. 4), only the amplitudes depend on the number of fans presented in the observed stadium sector. The character of cheering was categorized to different types as for football matches [6, 7]. During the ice hockey matches, the fans did not use the jumping type of cheering at all in contrary to football matches. The Table 1 summarizes the fans involvement in main types of cheering during the last third of both monitored games when the fans are usually most active. There are no big differences between both games.

The paper has been supported by the Czech Science Foundation project No. GA15-15728S.

Table 1: The time duration and fans involvement in the particular types of cheering evaluated during the last third of two studied ice hockey games.

| | 1st match | | | 2nd match | | |
|---|---|---|---|---|---|---|
| | Time [mm:ss] | Fans involved in cheering [%] | Standard deviation [%] | Time [mm:ss] | Fans involved in cheering [%] | Standard deviation [%] |
| Static state | 09:22 | 96 | 5 | 09:30 | 98 | 2 |
| Hand clapping | 18:15 | 63 | 21 | 17:41 | 68 | 19 |
| Swaying | 02:04 | 68 | 25 | 04:23 | 69 | 22 |
| Goal scoring | 00:19 | 88 | 11 | 01:36 | 97 | 4 |
| Mexican wave | 00:40 | 13 | 4 | 00:00 | 0 | 0 |

**References**

[1] C.A. Jones, P. Reynolds, A. Pavic, Vibration serviceability of stadia structures subjected to dynamic crowd loads: A literature review, Journal of Sound and Vibration 330 (2011) 1531–1566.

[2] O. Rokos, J. Maca, Stochastic approach in the human-induced vibration serviceability assessment of grandstands, in proc.: Proceedings of the IX International Conference on Structural Dynamics EURODYN2014 (2014), eds. A. Cunha, E. Caetano, P. Ribeiro, G. Müller, Faculdade de Engenharia da Universidade do Porto, Porto Portugal, pp. 2591-2598.

[3] O. Rokos, J. Maca, The response of grandstands driven by filtered Gaussian white noise processes, Advances in engineering software 72 (2014) 85 – 94.

[4] P. Reynolds, A. Pavic, Vibration performance of a large cantilever grandstand during an international football match, Journal of Performance of Constructed Facilities 20 (2006) 202 – 212.

[5] A. Caprioli, M. Vanali, A. Cigada, One year of structural health monitoring of the Meazza Stadium in Milan: Analysis of the collected data, in proc.: Proceedings of the 27th Conference and Exposition on Structural Dynamics 2009 IMAC-XXVII (2009), Society for Experimental Mechanics Inc., Orlando Florida USA, 9p.

[6] M. Verner, M. Polak, T. Plachy, An Experimental Study Focused to Spectators-induced Vibrations of a Cantilever Grandstand during two Sport Matches, in proc.: Proceedings of 53rd International Conference on Experimental Stress Analysis 2015 (2015), eds. P. Padevet, P. Bittnar, Czech Technical University in Prague Faculty of Civil Engineering, Český Krumlov, pp. 460–465.

[7] M. Verner, T. Plachy, M. Polak, An Experiment Focused on Fans Behaviour and Induced Grandstand Vibrations during a Football Match, Applied Mechanics and Materials 837 (2016) 75–78.

EAN 2017
55th conference on experimental stress analysis

# Calibration of Material Parameters during Billet Straightening

Martin Fusek[1, 2, a], Radim Halama[1, 2, b] Zdeněk Poruba[1,c]

[1]Faculty of Mechanical Engineering, VŠB – Technical University of Ostrava, 17. listopadu 15/2172, 708 33 Ostrava-Poruba, Czech Republic

[2]IT4Innovations, VŠB – Technical University of Ostrava, 17. listopadu 15/2172, 708 33 Ostrava-Poruba, Czech Republic

[a]martin.fusek@vsb.cz, [b]radim.halama@vsb.cz, [c]zdenek.poruba@vsb.cz

**Keywords:** Digital image correlation method, automatization, billet straightening, mechanical parameters, FEM

**Abstract.** This contribution deals with the use of image digital correlation method for the determination of displacements, strains and stresses field on the 3-D mechanical part – a billet resulting by division from continuous casting process. Results of the measurement are used for the verification of identification method for the online determination of mechanical parameters of the material from billet bending. The motivation for this task is the preparation of the experimental algorithm for the automated evaluation of the part curvature and the proposal of the intervention for curvature adjustment so that the resulting shape suits to tolerance requirements for the product given.

## Introduction

Straightening of semi-finished products – billets resulting by division from continuous casting process, is required by further subscribers. The as straight billet as possible within tolerances given by the size of semi-finisher products is required. The straightening process is made manually at present. That means that the billet is manipulated by manually controlled machines. The straightening process is controlled by the machinist and is based on his experiences. It is obvious that the productivity and accuracy of the straightening process is in hands of the operator. The development of the automated straightener is currently in progress.

However, such device needs to identify automatically material properties of straightened semi-finished product so that the controlling program could carry out adequate and effective straightening interventions. The approach can be twofold. The first one loads parameters of the straightened material from the database. The second approach identifies material properties online during the straightening process. The first approach seems to be easier however it is prone to errors – wrong material, inaccurate values of material parameters, etc. For this reason, the second approach of automated parameters identification is being developed.

The straightening process is further predicted online. Based on the optimization algorithm, interventions of the straightening machine using action members – straightening rollers are proposed.

In following chapters, the experiment and related activities leading to the proposal of identification algorithm for obtaining material properties and further development of optimization SW is described.

**Experiment**

The experiment itself was carried out in premises of the company KOMA – Industry Ltd., Ostrava – Vítkovice, Czech Republic. The straightening machine for billet adjustment was used. The straightening machine was fitted with sensors for measurement of displacements and forces. The measurement of forces is based on the record of pressure rates in working rolls of the straightening machine. For evaluation of the cylinder displacement the incremental encoder is used. The accuracy of sensors is given by requirements of the praxis and similar displacement and pressure sensors will be used for the automated straightening machine designed.

The image digital correlation method was used for the measurement of displacements of the straightened billet. The digital image correlation method is the optical method which uses the sequence of images in which the changes are identified [1]. On the base of changes identified, the field of displacements, strains and stresses can be evaluated. The digital image correlation method is not tied only to problems of mechanics of solids, but it can also be used in other areas of science and technology. The device Mercury RT 2x2.3Mpx@40Hz was applied in the measurement. The software Mercury RT made by the company Sobriety Ltd. was used for the evaluation of images obtained.

The mechanical part measured is depicted in Fig. 1 (left). Transverse dimensions of the billet used in the experiment were 200 x 200 mm. The material of the billet was steel 34CrMo4. The straightening process itself is performed so that the billet measured is pressed against rests and is formed using the ram. The straightening ram itself is placed symmetrically in relation to both rests. The chosen results obtained from the measurement and subsequent evaluation are depicted in Fig. 2.

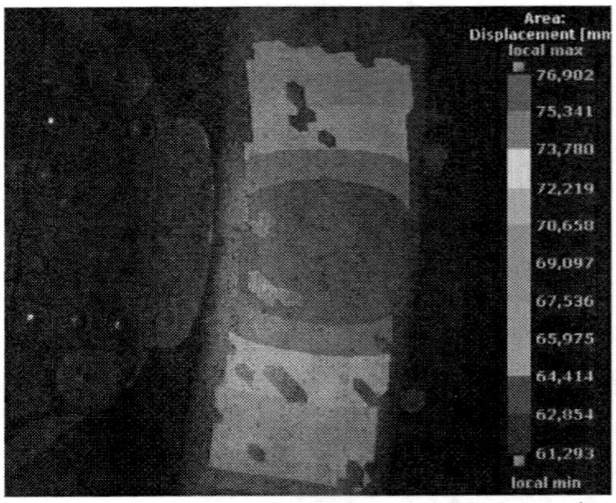


Fig. 1 – The part measured placed in straightening machine (left). DIC measurement results – total displacement [mm] (right).

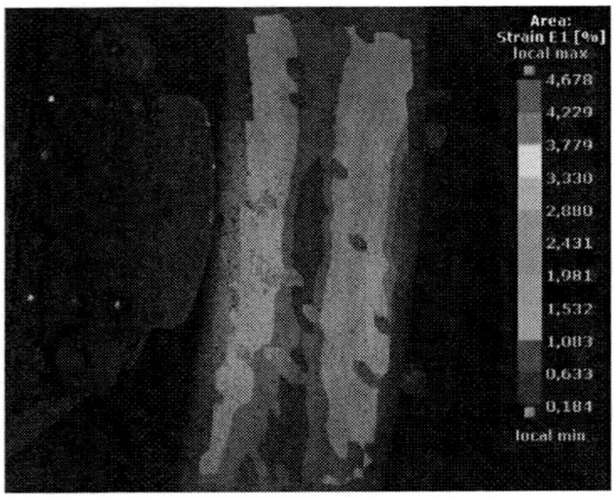


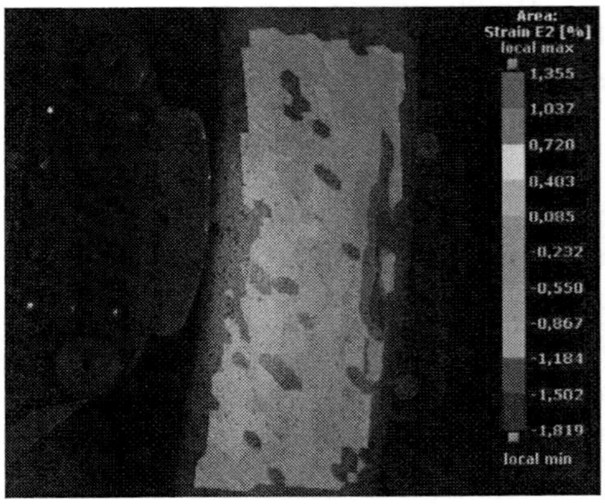

Fig.2 – DIC contours of first principal strain (left), DIC contours of second principal strain (right).

Further, specimens determined for standard material tests in the laboratory were taken from the testing billet. The sampling was performed in the location which was not influenced by previous straightening attempts. The sampling was realized both in transverse and longitudinal direction of the testing billet. The record from the conventional tension test is depicted on Fig. 3.

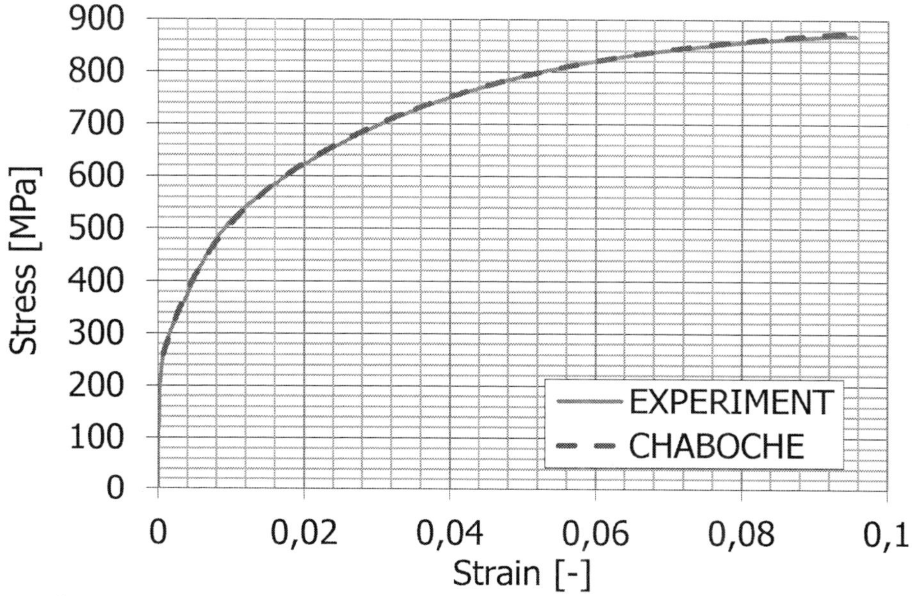


Fig. 3 – Stress-strain curve of 34CrMo4 steel from tensile test and its description by Chaboche model

**Simulation**

The automated straightening process should be realized in several phases. Firstly, the initial shape of the straightened billet is detected. Secondly, material properties are identified. Followingly, the optimization simulation of the straightening process is realized. On the base of its results the straightening operations themselves are performed. This procedure will be done in closed loop until the billet is straightened to the tolerance required. This straightening process will be also carried out in the second plane.

The identification of parameters will be performed at the beginning of the straightening and will be repeated on the base of need – for instance in case of long straightening time and "cooling down" of the billet.

The identification is carried out on the base of reverse approach using own solver. This approach is being tested at present. Since the development of own solver is not finished yet the commercial FEM solver ANSYS was used.

For the description of the stress-strain behavior, the Chaboche model with two backstress parts was used.

$$\alpha \quad = \alpha_1 + \alpha_2 \tag{1}$$

For the simplicity, this constitutive model is described for uniaxial loading, i.e. in scalar form. The evolution rule of Chaboche for the backstress [2] is given by differential equation

$$d\alpha_i = C_i d\varepsilon_p - \gamma_i \alpha_i dp \tag{2}$$

where $C_i$ and $\gamma_i$ are material parameters, $d\varepsilon_p$ is the increment of plastic strain and $dp$ is an increment of accumulated plastic strain. All material parameters fitted on data for examined steel are stated in the Table 1.

Table 1 Cyclic plasticity model parameters

| Material Parameters |
|---|
| $E$=197GPa; $\nu$=0.3<br>$\sigma_Y = 226MPa$ ; $C_{1-2} = 36907, 15180\ MPa$<br>$\gamma_{1-2} = 214, 30$ |

The FEM model was created using BEAM188 finite elements. The appropriate macro was used so that parameters of the model could be easily changed. The refinement of the finite element mesh was realized in the surrounding of the loading force so that the development of the plastic deformation could be correctly described. The span of rests was assumed the same as in real case - 1560 mm. The maximal deflection specified in the middle corresponded to the experimentally determined value, approx. 77 mm.

Results in the form of contours of axial deformation and axial stress in the surrounding of loading applied is depicted in Fig. 4. The good agreement with the experiment is obvious from Fig. 2.

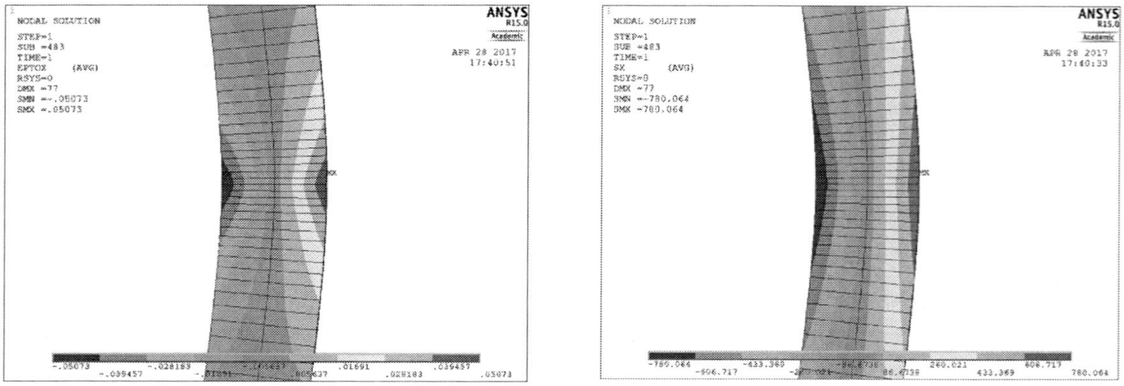


Fig.4 – FEM contours of axial strain (left), contours of axial stress (right).

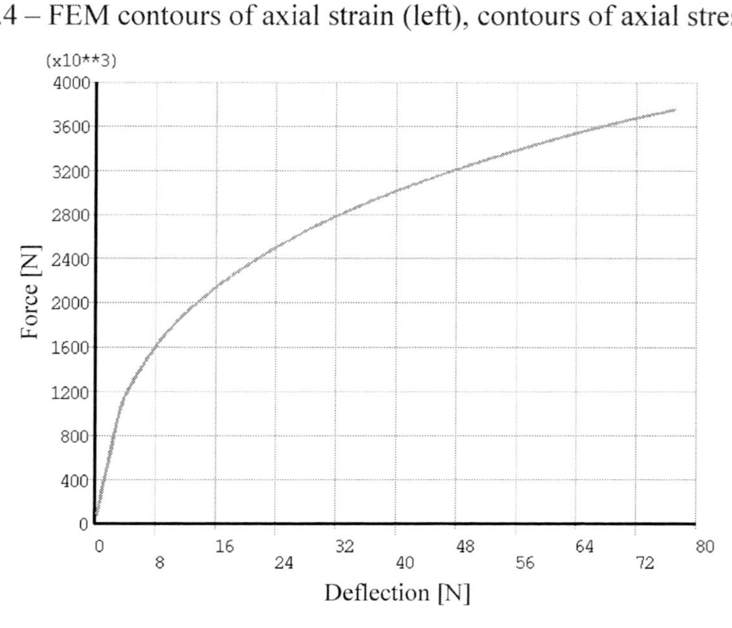

Fig.5 – Dependency of force on deflection obtained by FEM.

Based on the FEM simulation, the dependence of the force on the maximal deflection of the beam was evaluated as well – see Fig. 5. Only the maximal force of 4.17e6 N was known from the experiment. Its value was obtained from the working pressure.

## Conclusions

The measurement of displacements and strains on the testing billet using the image digital correlation method was performed. Results are used for the verification and adjustment of the identification algorithm of material properties. The algorithm has been developed based on the billet material tension tests performed. The billet straightening process using the three-point bending is significantly different from the straightening of metal sheets and belts [3]. The straightening optimization can be realized through the modification of design, e.g. for circular rods [4], or through the adjustment of technology using sensors. In our case, the last mentioned case was used. The development of the subsequent straightening algorithm for the expert system is temporary in progress.

## Acknowledgement

This work was supported by The Ministry of Education, Youth and Sports from the National Programme of Sustainability (NPU II) project „IT4Innovations excellence in science - LQ1602", from the Specific Research Project (SP2017/136), and also by the Czech Science Foundation (GA15-18274S).

## References

[1] F. Trebuňa, R. Huňady, M. Hagara, Experimental methods of mechanics - Digital image correlation. Košice: TU in Košice, 2016. ISBN 978-80-5532-346-6. (In Slovak)

[2] J.L. Chaboche, J. Lemaitre, Mechanics of Solid Materials, first ed., Cambridge University Press, Cambridge, 1990.

[3] J. Petruška, T. Navrat, F. Sebek, A new model for fast analysis of leveling process, Advanced Materials Research 586, 2012, pp. 389-393.

[4] J. Petruška, T. Návrat, F. Šebek, Novel approach to computational simulation of cross roll straightening of bars, Journal of Materials Processing Technology 233, 2016, pp. 53-67.

EAN 2017
55th conference on experimental stress analysis

# Identification of the torque strut force load

LUFINKA Aleš[1,a], NOVÁK Ondřej[1,b] and PETRŮ Michal[1,c]

[1]Technical university of Liberec, Studentská 2, 46117 Liberec 1, Czech Republic

[a]ales.lufinka@tul.cz, [b]novak.ondra1@seznam.cz, [c]michal.petru@tul.cz

**Keywords:** Torque strut, force measurement, strain gage sensor.

**Abstract.** A torque strut is member used as a fixing element of the car drive unit. Because a breaking of its metal part during the tests occurred, the measurement of forces applied to the strut during real usage in the car was required. The paper describes the load measurement methodology and it also explains the causes of a strut failure.

## Introduction

A torque strut is member used as a fixing element of the car drive unit. It is placed between a gearbox and suspension subframe. It provides a flexible transmission of the forces that are generated by a torque of the engine. It consists of metal and rubber parts (Fig. 1 left).

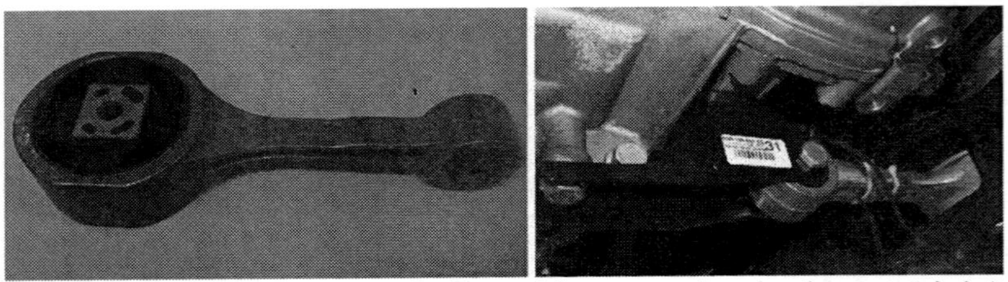

Fig. 1 The torque strut (left) and its mounting in the car (right)

Each new torque strut type has to be tested and its final tests are carried out in a real car (Fig. 1 right). Simulation of inappropriate launch of the car is one of these tests. It consist in a fast releasing of a clutch under revolutions that limiter allows. This process is repeated several times for the first and reverse gear. The car "jumps" and torque strut is loaded by high force peaks. The first gear generates tensile stress and the reverse one generates compressive stress.

One of the torque strut types was very often broken during this test. Thus it was necessary to determine the real loading force acting in this strut and to compare it with expected value of this strut design. The torque strut real loading during its real usage in the car had to be measured.

## The real force measurement methodology

If the torque strut is assembled in a car, there is no space for standard force sensor (Fig. 1 right). Therefore, it was decided to measure the force by the strut relative elongation ε because both values are proportional according to Young's modulus E (Eq. 1) (of course only before the material yield point attainment) [1].

$$E = \frac{\sigma}{\varepsilon} = \frac{F}{A \cdot \varepsilon} \tag{1}$$

where $\sigma$ is the normal tension, $F$ is the loading force and $A$ is the cross-sectional area

Because the Young's modulus and the cross-sectional area may be considered constant, the force can be determined from the relative elongation (Eq. 2) which can be measured by strain gauge sensors.

$$F = E \cdot A \cdot \varepsilon \quad (2)$$

Two strain gauges sensors 1-LY41-6/350 [2] were glued to the opposite sides of the strut at its smallest cross-section place (Fig. 2 left). Then the strut was placed into the calibration device, loaded by the axial force 10 kN and the strain gauge signals were calibrated to the force.

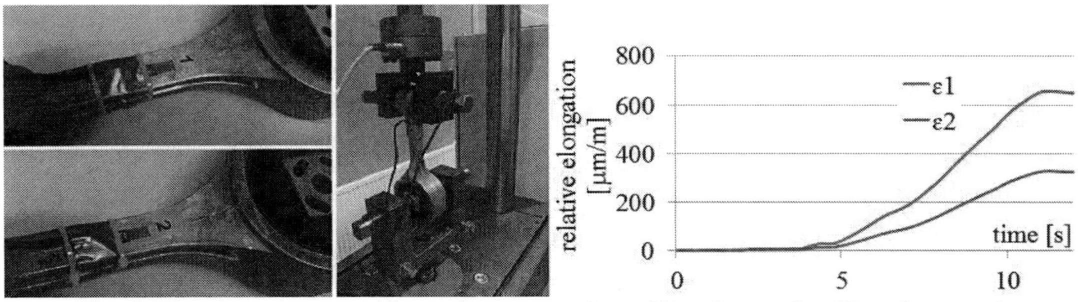


Fig. 2 The torque strut with strain gauges, its calibration and calibration result

This calibration showed first unexpected results (Fig. 2 right). Signal from one strain gauge was greater than from the second one. If the symmetry strut is expected, both signals should be practically identical during the strut axial loading.

**The real force measurement result**

The torque strut with the strain gauges was assembled to the car and test described above was performed. Six launches for first and reverse gear were carried out. Test results are summarized in the Fig. 3.

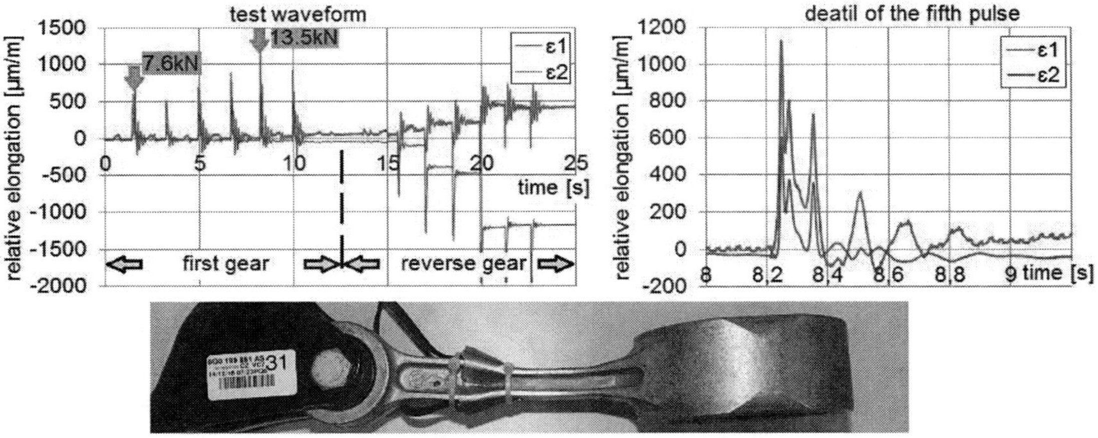


Fig. 3 The torque strut test results

Strain gauge signals showed same trends as the calibration, one was higher than the other one. Already after the fifth launch (the force peak was about 13.5 kN and the relative elongation 1117 µm/m) the strain gauge signals did not return to the zero value, the strut was permanently deformed and its material yield point was exceeded (see the waveform detail in the Fig. 3 right). One signal remained positive and the other one negative value, the strut was bent. It was unexpected result because the strut is loaded by tensile force in the first gear. The strut bending was gradually magnified during the next launches in reverse gear that create compressive loadings. Based on this result it was decided to check the strut cross section.

### The strut cross section checking

The torque strut exhibited during calibration different values of voltage and therefore different deformations although strain gauges were placed symmetrically. The causes may be various, for example a defect of material, the load applied out of symmetry axis or asymmetric cross-section. For this purpose, the cut-outs from strut applicable for image analysis were created. The strut was divided by cuts perpendicular to the separation plane into two parts and finished by the milling to ensure the flatness of the cut with sufficient roughness. Then, the images was prepared using a microscope Nikon Eclipse LV100 and subjected to measurement using image analysis system NIS-Elements. One sample from a tested (T) and a new (N) strut were prepared. The wall thickness and the size of the area left and right rib were measured at four points (Fig. 4). The results can be found in the Tab.1. There are also shown the percentage difference between measured parameters of the thickness and the area of the left and right ribs.

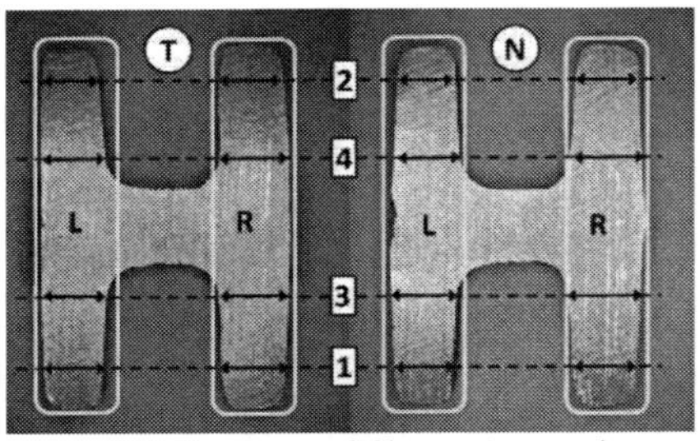


Fig. 4 The strut cross sections, check areas and lines (tested strut left, new strut right)

Tab. 1 The strut dimension check results

| dist. [mm] | tested strut (T) | | | | | new unused strut (N) | | | | |
|---|---|---|---|---|---|---|---|---|---|---|
| area [mm$^2$] | dist.1 | dist.2 | dist.3 | dist.4 | area | dist.1 | dist.2 | dist.3 | dist.4 | area |
| rib L | 3.120 | 3.027 | 3.520 | 3.467 | 65.34 | 3.093 | 3.160 | 3.613 | 3.587 | 67.01 |
| rib R | 3.360 | 3.413 | 3.867 | 3.773 | 72.65 | 3.480 | 3.414 | 3.613 | 3.933 | 75.03 |
| difference | 0.240 | 0.386 | 0.347 | 0.306 | 7.32 | 0.387 | 0.254 | 0.000 | 0.346 | 8.02 |
| [%] | **7.7** | **12.8** | **9.9** | **8.8** | **11.2** | **12.5** | **8.0** | **0.0** | **9.6** | **12.0** |

The cause of unusual result was revealed - the strut casting is poorly manufactured and the cross-section is asymmetrical. Left area is about 12 % smaller than the right one. The wall thickness differences are significant and probably it is a reason of different unexpected behavior that leads to the bending strut. The torque strut was then redesigned in accordance with these findings and the new type of the strut was identically tested.

### The redesigned strut real force measurement

The new strut type was tested identically as the first one. Two strain gauge sensors were glued to the opposite sides of the strut and calibrated by the 10 kN force. No unexpected result was achieved during the calibration, both strain gauge signals were almost identical (Fig. 5 right). It indicated the symmetrical strut casting.

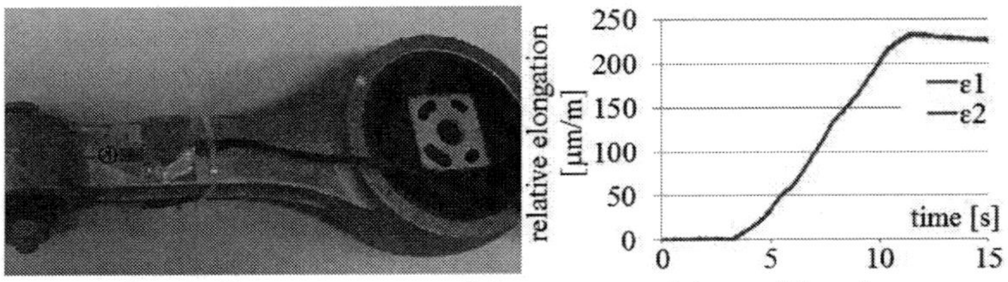


Fig. 5 The new type of the strut and its calibration

Than the strut was assembled to the car and the same test was repeated five times so that approximately 25 force peaks for first and the same for reverse gears were generated. Both strain gauge signals returned to the zero value after all tests, no permanent deformation was measured. Test pulses were different because their size and shape were influenced by the car driver (Fig. 6).

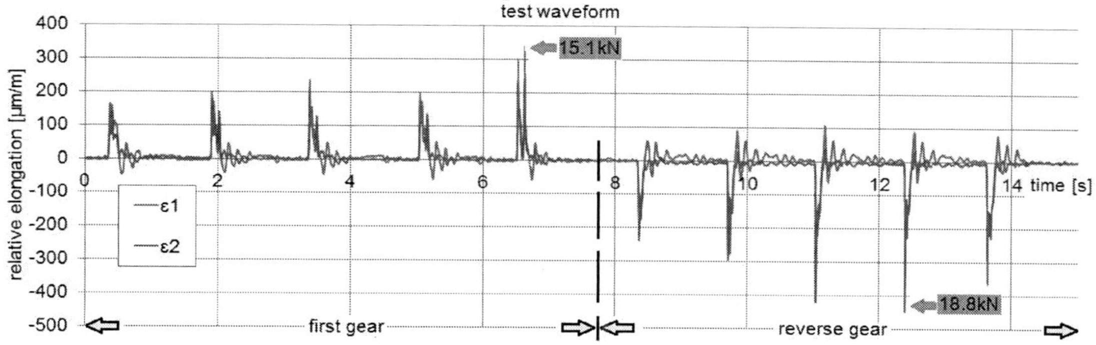


Fig. 6 New redesigned strut measurement example

Therefore the force peaks average value was calculated and the maximal values of the relative elongation and force were determined for both load direction. The absolute maximum of force peak was 18.8 kN and it caused relative elongation 439 µm/m (Tab. 2). The strut material yield point was exceeded by the relative elongation about 1117 µm/m during the first strut type testing. If we assume that the material of both the struts is the same and the strut new type maximum relative elongation is only 439 µm/m, the normal stress acting in the new strut is at safe level and critical material yield point will not be reached.

Tab.2 Test peaks summarization

| | force peaks average value | maximum force | maximum rel. elongation |
|---|---|---|---|
| first gear | 9.83 kN | **15.1** kN | **352** µm/m |
| reverse gear | 14.64 kN | **18.8** kN | **439** µm/m |

## Conclusions

Test results showed that the first strut type cross section was undersized, its deformation was permanent after the force pulse of only 13.5 kN. In addition, its casting was poorly manufactured, it is asymmetrical and it causes the strut bending deformation. All these errors were eliminated during the new strut type design. The cross section area was increased and both halves of its casting are symmetric. No permanent deformation was observed during its testing, although the force peaks ranged up to 20 kN, the redesigned strut is safe for use.

## Acknowledgment

The results of this project LO1201 were obtained with co-funding from the Ministry of Education, Youth and Sports as part of the targeted support from the "Národní program udržitelnosti I" programme.

## References

[1] E. A. Avallone, T. Baumeister, Ali M. Sadegh Marks' standard handbook for mechanical engineers, 11th ed. New York: McGraw-Hill, 2006, ISBN 978-007-1508-155.

[2] Y Series Universal Strain Gauges, information on http://www.hbm.com/en/0364/strain-gauges-for-stress-analysis.

# Experimental analysis of welding tooling for ITk ATLAS

M. Rachač[1, a], M. Dub[1, b], M. Janda[1], P. Malý[1], J. Kamenický[1], O. Berka[1], V. Dynybyl[1], P. Kubový[2], F. Lopot[1,2,c]

[1]CTU in Prague, Faculty of mechanical engineering, Department of designing and machine components, Technická 4, 166 07, Prague, Czech Republic

[2]Charles University, Faculty of sports and physical education, Dep. of anatomy and biomechanics, Lab. of extrem loading, José Martího 31, 162 52, Prague, Czech Republic

[a]marek.rachac@fs.cvut.cz, [b]martin.dub@fs.cvut.cz, [c]frantisek.lopot@fs.cvut.cz

**Keywords:** Welding tooling, ITk ATLAS

## Introduction

The paper deals with the testing of a recently developed welding tooling which will be used for welding of a new type of rails inside a mock-up of ATLAS internal detector in CERN, readiness for its upgrade.

The detector is used for detection of radiation at collisions of particles in the accelerator by means of CCDs which underwent a rapid development in the last decade in terms of performance, speed and sensitivity, and that is why the overall upgrade of the detector is being undertaken.

On the detection carriers (staves) exchange, the complete reconstruction of related systems, such as wiring, cooling circuits design and system for creating and maintaining a vacuum in XHV is bound. The detector as an assembly creates a highly complex structure where all errors resulting from installation of individual components are very difficult to detect. This is one of main reasons for mock-up construction.

The preparation of the mock-up corresponding in detail with the final design of the inner detector is crucial because the mock-up enables training and check of all assembly procedures in real conditions but without radiation. And thus the health risks for staff are minimalized in fact.

For the detector upgrade, with an effort to reduce the total costs, the mock-up constructed for present ATLAS version is reused and in detail adapted to the intended engineering design.

## Tooling design

Firstly, the force limit, which the tooling after the installation must withstand without any change of its geometry and position. This demand is justified by the fact that technicians will be moving in the surrounding space during the installation of the rails and their contact with the tooling cannot be avoided (*Fig.* 1).

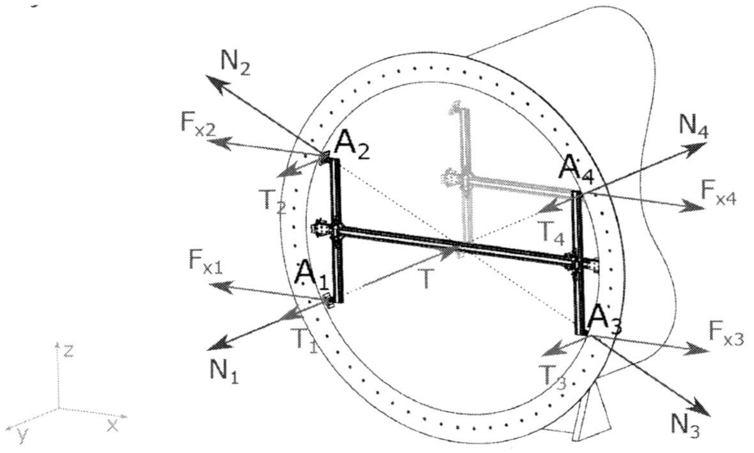


Fig. 1 - Forces diagram

On the other hand, there was a requirement for the maximal allowable deformation of the detector tube caused by the tooling installation (*Fig.* 2).

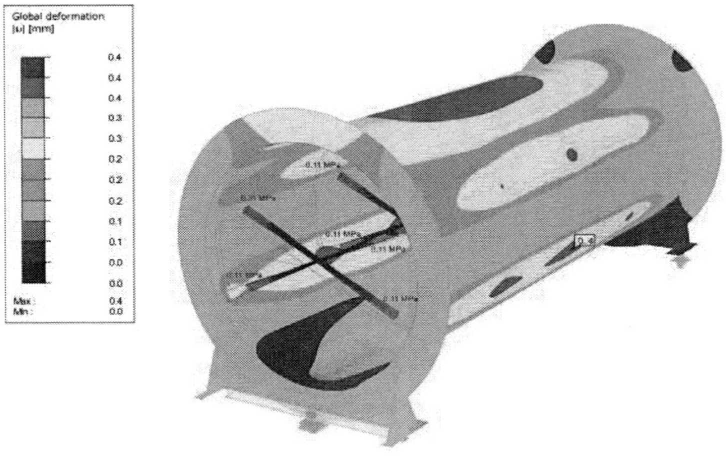


Fig. 2 - FEM analysis

The tooling is designed according to the request from CERN from standardized Bosch al-profiles which are also used in their other workshops (*Fig.* 3).

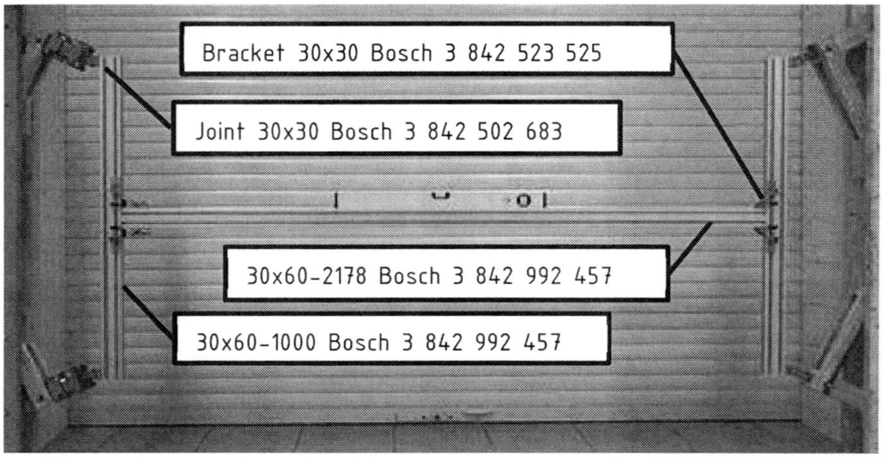


Fig. 3 - Parts of welding tooling

The tooling consist of T-part, vertical beam and an independently adjustable horizontal beam with terminals for fastening and adjustment of rails according to instructions from CERN people. In the tube, the tooling is secured by strutting of T-part and vertical beam using two strutting bolts. The contact area of endplates of the tooling is shaped according to the tube diameter. The end plates are made from a hard wood to prevent damages of the inner surface of the tube. Fig. *3* provides an illustrative picture of the design solution of the tooling.

## Measurement and instrumentation

Needed experiment was carried out in Laboratory of Extreme Loading because of appropriate space and instrumentation. Sensor application is obviously seen in *Fig.* 4**Chyba! Nenašiel sa žiaden zdroj odkazov.**.

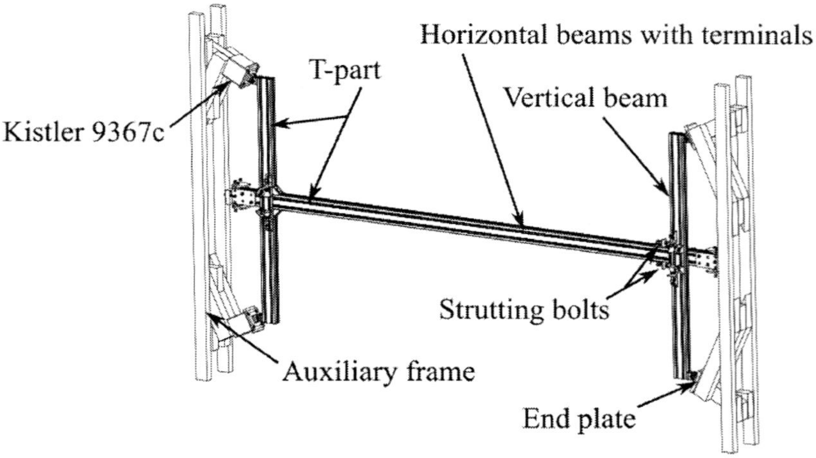


Fig. 4 - The tooling installed into the auxiliary frame

The tooling was installed into the auxiliary frame simulating the inner cylindrical surface of the detector tube. Contact forces in endplates of the tooling are detected by peizoelectric sensors Kirstler 9-series with sensitivity of 0.001 N. The torque on strutting bolts was monitored by tenzometric spanner Kistler with sensitivity 0.01 Nm in range from 0 to 10 Nm.

All data were acquired by Dewetron Sirius station using Kistler amplifiers with 1 kHz sampling frequency.

In order to avoid both damage of the tooling and exceeding of the deformation limit of the detector tube, the tightening torque was restricted on 4.5 Nm by a slip clutch application in the spanner. The operating stuff was instructed to tighten firstly the upper and then the bottom strutting bolt always either about one half of the turn allowed by surrounding space or up to the limit value of the torque. A short breaks of about 20s between particular steps of the tightening proces were added in order to identify each step in recording definitely (*Fig.* 5).

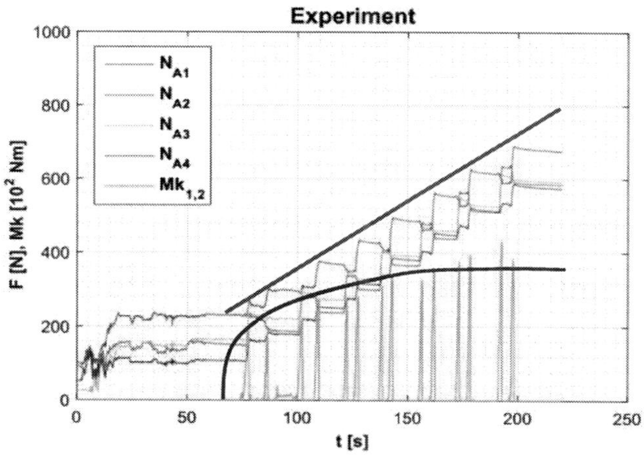


Fig. 5 - Graph Contact force, Torgue - Time

The data acquisition was started before inserting the tooling into the auxiliary frame and stopped when the contact force on endplates reached the limit.

## Results

It is important to assess the measured data were both in the coordinates "contact force – torque" and torque, contact force – time" [1].

From acquired time courses, the force magnitudes related to the peaks of the torque (see *Fig.* 5) are read and separated. By this way, the graphs with force-torque coordinates could be installed (*Fig.* 6).

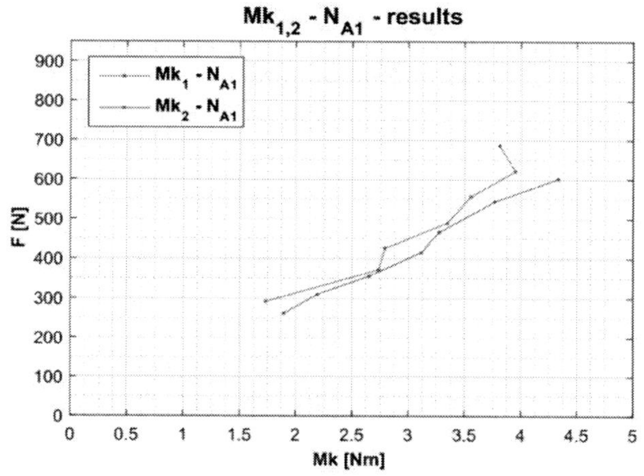


Fig. 6 - Graph Contact force - Torgue

But besides the above mentioned dependency, the time aspect seen from the graph in Fig. 5 must be considered in. The detailed analysis of the course of the experiment implies the knowledge essential for creation of a simple assembly procedure.

Here is not enough information about tightening of bolts by defined torque, but it is necessary to monitor the number of revolutions of the bolts after the moment of their contact with a vertical beam.

The courses of mentioned graphs show that for correct installation of the tooling in the detector tube, its geometrical change as the result of turns of strutting bolts is crucial. Due to this careful approach, the Relatively uniform distribution of the force from Strutting bolts to the contact forces on endplates of the tooling with a minimal risk of its damage or of exceeding the limit of deformation in the detector tube during installation.

## Conclusion

The acquired data brought important findings to set up a simple assembly instructions for actual conditions of the tooling installation.

Because there is impossible to count with lengthy installation of measuring equipment in the real assembling conditions, the number od halfturns of strutting bolts calculated always from initial contact between the bolt and the vertical beam (tightened by hand) was assigned as the appropriate and effective quantificator for the immediate state of the tooling instalation. This parameter shall be monitored and controlled by the operating stuff directly without need of any special equipment. The installation of the tooling is finished reaching of prescribed number of halfturns for both strutting bolts.

During writing this paper, the tooling was installed in CERN and succesfully used to the intended purpose (*Fig*. 7).

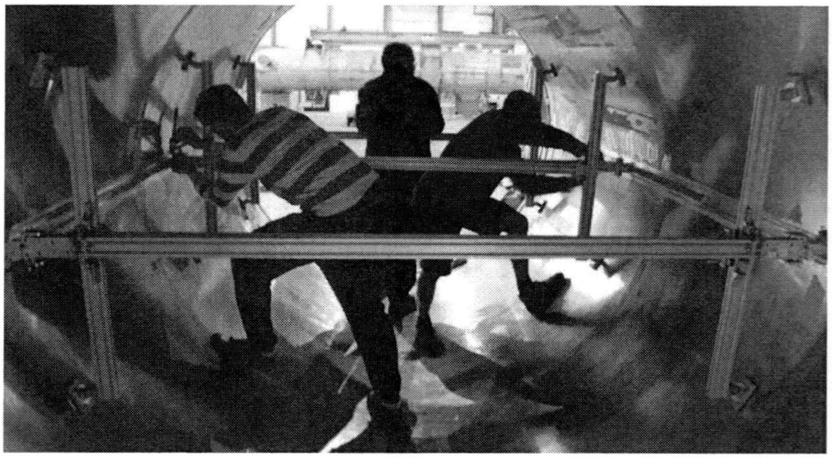

Fig. 7 - Installation in CERN


## Acknowledgement

The project and the paper were supported by grant no. SGS16/069/OHK2/1T/12.


## References


[1] Janda, M., Starý, F., Dub, M., Berka, O., Lopot, F., Malý, P., . . . Dynybyl, V. Experimental evaluation of a prestressed bolt connection mounting prestress. EAN 2015 proceedings, 149-152. (2015).

EAN 2017
55th conference on experimental stress analysis

# Strain Gauge Measurement on the Oilseed Screw Press

Pavel Malý[1,a], František Starý[1,b], Vojtěch Dynybyl[1,c] and Jan Hoidekr[1,d]

[1]CTU in Prague, Department of Designing and Machine Components, Technická 4, 166 07 Prague 6, Czech Republic

[a]pavel.maly@fs.cvut.cz, [b]frantisek.stary@fs.cvut.cz, [c]vojtech.dynybyl@fs.cvut.cz, [d]jan.hoidekr@fs.cvut.cz

**Keywords:** Experimental Testing, Measurement, Strain Gauge, Oilseed Screw Press.

**Abstract.** This paper gives the description of the preparation, implementation and execution of the experimental measurement on the oilseed screw press. Set of nine strain gauges placed on the basket ribs of the machine were used to obtain the distribution of the pressure along the basket during the operation. Moreover, special pressure gauge was used to obtain the absolute value of the pressure in the chosen places of the basket.

## Introduction

Measurements and experimental testing on machines during real conditions become important in these days. The results from experiments are primarily used for analysis of real operational parameters that can be further used for the subsequent optimization of the specific parts or structural units. The measurements are also used for validation of different computational simulations and analysis so they can be compared with the reality, [1], [2].

Various data can be measured on this specific oilseed screw press. In addition to the described measurement on the basket ribs with help of strain gauges and measurement of the pressure using special gauge, temperature of the processing material in the machine or forces on the shaft were also measured.

## Method

The strain gauge measurement on the basket ribs of the oilseed screw press was carried out to obtain the relative values used for definition of the pressure distribution along the basket of the machine. The absolute value of the pressure in the basket is not possible to obtain using strain gauge in this method, however special calibrated pressure sensor can be used. Another possibility for pressure measurement in the basket is to use special measuring element, [3].

**Investigated machine.** The measurement described in this text was carried out on the oilseed screw press (figure 1) in the premises of the machine producer during the operation of pressing the oil [4]. So the main aim of the experiment was to obtain the parameters of the real working process.

Fig. 1: Oilseed screw press FS 1010, [4]

**Instrumentation and measurement protocol.** The scheme of the machine is shown in figure 2, where the shaft with screws and inserts, or the position and numbering of strain gauges on the ribs of the basket can be seen. The material input to the oil press is in radial direction to the axis of the shaft in the place of rib number 17, the processed material flows in the direction of the decreasing strain gauges numbers to the axial output at the end of the shaft. During the flow of the processed material through the working chamber the oil is extracted.

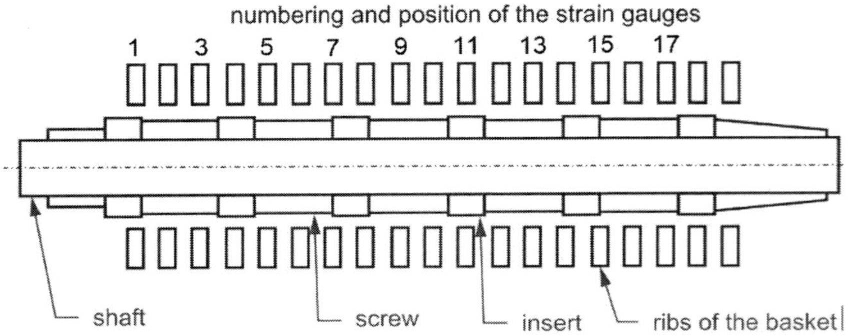


Fig. 2: Scheme of the ribs and shaft of the tested machine

The set of nine T rosette strain with two measured grids gauges (1-XY91, [5]) on the ribs of the basket were used. The configuration of the strain gauges was half-bridge type for temperature compensation, the position of strain gauge resistors is in figure 3

The special calibrated miniature pressure transducer Burster Model 81530 connected as full-bridge was used to measure absolute value of the pressure along the basket of the press in the specified places [6].

Fig. 3: Strain gauge on the rib of the basket

The operational conditions in the machine are unfavourable (high temperature, dirt, abrasion, oil etc.) so special procedures for the measurement preparation had to be used (special glue, strain gauge protection, cabling, temperature compensation).

The special two components epoxy adhesive X280 from HBM was used for the installation of strain gauges, [5], because this adhesive offers high temperature resistance (up to 200°C). The curing time is eight hours at room temperature with applied pressure of 10 to 50 N/cm $^2$, therefore special fixture was used during the adhesive curing (see figure 4).

Fig. 4: Applying of the pressure during the adhesive curing

The protected set of the strain gauges on the ribs with wiring are shown in figure 5. Special tough putty AK22 from HBM was used to protect the strain gauges and PTFE wiring with the operating temperature up to 200°C had to be used.

Fig. 5: Final measured ribs of the basket with protected strain gauges and wiring

**Data acquisition and processing.** The process of pressing was controlled by the PLC system with possibility of measurement and archiving of some operational parameters (motor load and revolutions, seed mass flow etc.). The National Instruments apparatus and software (PXI, LabView) were used for the strain gauge measurement with the sampling frequency of 1000 Hz. The data processing was performed in LabView software. The example of obtained data from strain gauges in graphical form for a long term measurement is shown in figure 6.

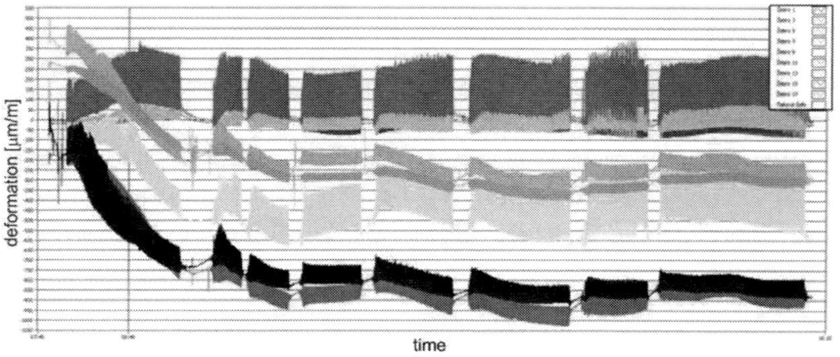


Fig. 6: Measured and processed data

**Experiment outcomes**. The required outcome from the measurement is the pressure distribution along the oil press basket for different operating states and processed material. The pressure of the processed material in the machine is supposed to be increased on the each insert, whereas the function of screws is to transport the material so the pressure is lower there.

## Results and conclusions

The described experimental method can be used to obtain the real operational parameters of the oilseed screw press. The processed results correspond to the expectation so the measured data provide information for the design of the machine for specific oilseed and optimization of the pressing process. The obtained pressure distribution along the basket can be used as a boundary condition for the process simulations.

## Acknowledgment

This research has been realized using the support of Technological Agency, Czech Republic, project TA04010992, Research and development of technologies for obtaining vegetable oils and cakes with an emphasis on quality of cakes as feed and the appropriateness of used of the construction materials.. This support is gratefully acknowledged.

## References

[1] M. Dub, J. Kolář, F. Lopot, V. Dynybyl, O. Berka, Experimental analysis of stress state and motion of tram gearbox hinge, EAN 2014 - 52nd International Conference on Experimental Stress Analysis (2014).

[2] O. Berka, M. Dub, F. Lopot, V. Dynybyl, Experimental analysis of stress state and motion of tram gearbox hinge, EAN 2015 - 53rd Conference on Experimental Stress Analysis (2015).

[3] J. Hoidekr, P. Malý, F. Starý, P. Mossóczy, V Dynybyl, Complement of calibration of measuring element using FEA, EAN 2016 - 54th International Conference on Experimental Stress Analysis (2016)

[4] Information on http:// www.farmet.cz

[5] Information on http://www.hbm.cz

[6] Information on www.burster.com

# Long-term cyclic test - analysis and monitoring

Ivan Jeřábek[a], Lukaš Horký[b] and Karel Weigel[c]

Doosan Bobcat Engineering s.r.o., Czech Republic

[a]ivan.jerabek@doosan.com, [b]lukas.horky@doosan.com, [c]karel.weigel@doosan.com

**Keywords:** Long-term cyclic tests, analysis, algorithm

Long-term cyclic tests are considered as the standard verification methods for new design of mini excavators and loaders at Doosan Bobcat Engineering. For the implementation of long term validation tests hydraulic loading system from Inova Company is used (Fig. 1).

Fig. 1 Cyclist test of hydraulic cylinders

There is a rig for each test which allows a long term load of a tested specimen and monitoring of its operating parameters. Load cycle is defined (load spectrum on the tested specimen) and a number of cycles for each test. It is followed by setup of test control system and definition of all other monitored parameters. Monitoring of operating parameters generate a huge volume of data (Fig. 2) which is difficult to process and analyse. Simple statement that the tested sample complies with the required load is often not sufficient for the evaluation of the test.

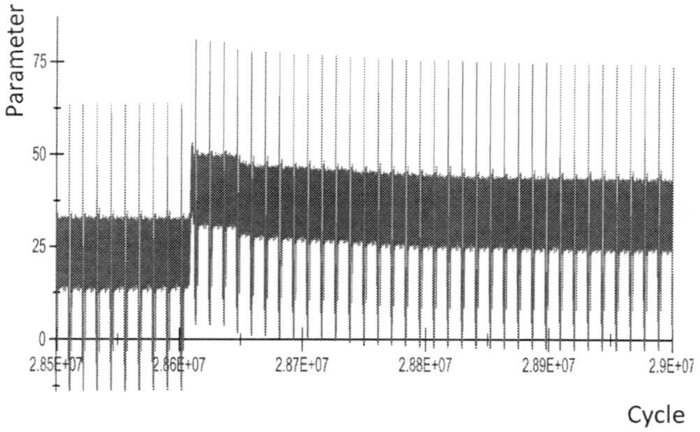


Fig. 2 Details of monitoring parameter

For monitoring of long term test an algorithm has been proposed that compares and evaluates deviation of monitored parameter values from a reference values. Algorithm divides all control and monitoring channels into separate channels. Each channel is analysed independently.

The algorithm assumes:

- data file describing cyclic process for each channel
- all channels are synchronized – all channels are using common time axis
- one channel is "reference value" that is used to define load of the test. Reference value is considered as correct "theoretical value".
- First cycle of each data file is "correct".

Analysis algorithm is divided into three procedures

- "etalon" definition: "etalons" are defined for each channel. "Etalon" is defined as first cycle of the test (Fig. 3 Time 0-100 min.). Cycle length is defined in configuration input file or calculated from reference channel.

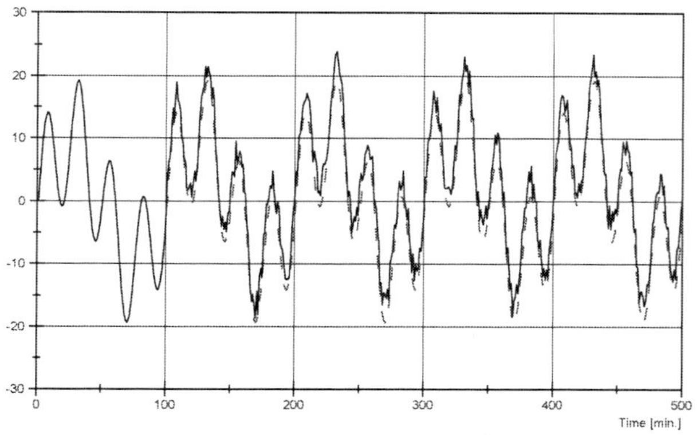


Fig. 3 Detail of monitored parameter (red line referent value, black line real measurement value)

- calculation deviations of cycles values. Deviation is defined as difference between cycle value and etalon. For the comparison is supposed that the first cycle in each channel is correct and can be used as an etalon for the repeated cycles that will follow.

It is possible to compare all following cycles in each channel with the etalon and calculate their deviation. The result for each cycle of all channels creates a data files that is statistically analysed.

- statistical data analysis: for each deviation data file average value, standard deviation, minimum and maximum are calculated.

Statistical analysis result is a parameter which represents the cycle and allows deciding if the cycle is or is not within required limits.

- Average value which represents the offset from the reference value
- Standard deviation represents variance values from the reference value
- Maximum and minimum value for the time line deviation determination

The parameters can be displayed in a chart to get a better view (Fig. 4.)

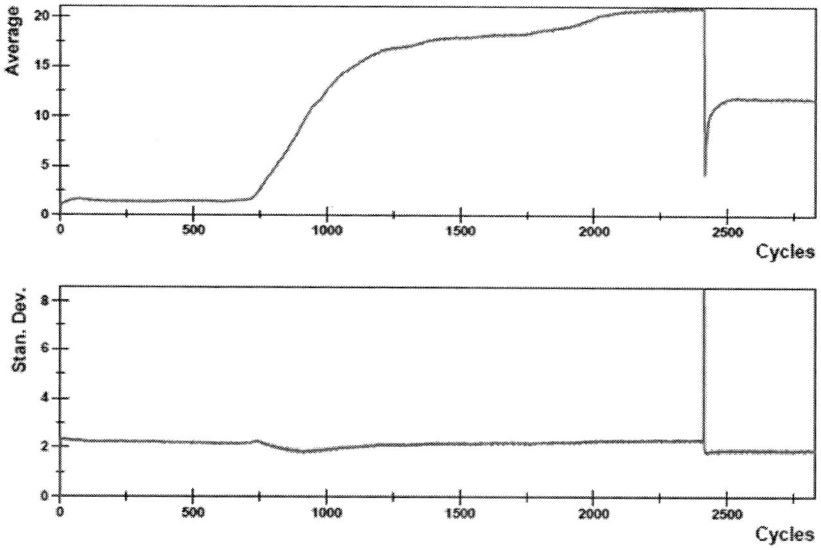


Fig. 4 Analysis Resuls

## References

[1] NI Diadem reference manual

[2] Information on http://www.ni.com

[3] Jiří Neubauer; Marek Sedlačík; Oldřich Kříž, Základy statistiky, Grada e-book

[4] http://test.bobcatdobris.cz/en

EAN 2017
55th conference on experimental stress analysis

# Synergy Related Issues In Multi-Criterial Designing Of Technological Processes

Piotr Kuryło[1, a], Joanna Cyganiuk[2,b]

[1, 2] University of Zielona Gora, Licealna 9 Street, 65-417 Zielona Gora, Poland

[a] p.kurylo@ibem.uz.zgora.pl, [b] j.cyganiuk@ibem.uz.zgora.pl

**Keywords:** synergy, coherence, technological process, concurrent synergy

**Abstract.** A technological process may be realised according to various criteria, with the use of various methods and with the help of various technical measures, i.e. multi-alternatively. The objective of the paper is to determine the relations between the number of the assumed criteria and the optimal number of the assumed alternatives at defined synergy and coherency coefficients.

## 1. Introduction

The contemporary designing and planning of the production process should take into account all elements of the planned technology that affect the quality of the finished product, the speed of realization of the production process, and consequently, the delivery time. Evaluation of the effectiveness of a technological process is a complex task, involving analyses of criteria according to which it is possible to realise any arbitrary technological processes. Basically, two criteria, i.e. costs and time, are taken into account, and sometimes also the criterion of energy efficiency. Designing and planning of the production process should also take into account all the problems concerning the design of machine components, also in terms of multi-criterial and multi-variant designing.

Technological preparation of production should be preceded by a full analysis of the technological documentation from the point of view of standardization, unification, rational selection and material savings as well as optimal development of semi-finished products with the use of various types of treatment technologies [1]. The analysis should take into account the impact of the applied technology on the environment where the production process is to be implemented. Optimizing the designing process, in turn, should be based on the analysis of the criteria according to which the planned production will be realized.

It is necessary, therefore, to optimize and systematize the production process yet at its conceptual stage playing a significant role in reducing the quantity of media involved in the production process. Thus, such an analysis should take into account three fundamental criteria, i.e. costs, time and energy efficiency.

The analysis should also take into account the relationship between the number of adopted criteria and the optimal number of accepted technological alternatives at defined ratios of synergy and coherency between the parameters for the implemented technology and its optimization.

## 2. The analysis of basic concepts [2,3,4]

*Synergy* – means a collaboration of various factors, which add up more to the whole than the sum of individual separate activities. The term is a general concept and applied in various fields.

*Technical synergy*, on the other hand, is a mutual interaction between more than two technical factors (parameters).

In terms of organization, synergy means that separate departments of an organization cooperate, and through the mutual influence they become more efficient than if they worked separately. For instance, if each department of a small company collaborates with one financial department, their efficiency would be greater than if each department had their own financial section.

In technology, synergy is advantageous due to the positive difference between the result obtained thanks to cooperation on the one hand and a total result of occurrence of separated and isolated parameters on the other hand. When the difference is negative, then dis-synergy occurs.

*Synergism* – means a collaboration of two or more factors e.g. components, substances, manufacturing processes [2,3,4].

Prof. Burakowski singled out three basic types of synergies, namely [1,5]:

- *Total synergism*, when :

$$a > (a_1+a_2) \text{ (i.e. when „2+2=>4")} \tag{1}$$

where: a- a factor of a process (e.g. process parameter) – global (general) effect
$a_1$ – process parameter (factor No. 1)
$a_2$ – process parameter (factor No. 2)

- *Additive synergism*, when:

$$a = (a_1+a_2) \text{ (i.e. when „2+2=4")} \tag{2}$$

- *Subtractive synergism*, when:

$$a < (a_1+a_2) \text{ (i.e. when „2+2=<4")} \tag{3}$$

Thus, total synergism occurs when at least two factors interact positively with a result greater than the sum of their separate interactions. Subtractive synergism occurs when at least two factors interact positively with a result lesser than the sum of their separate interactions but greater than a separate interaction of each of the factors apart.

A situation when the factors (parameters) do not affect each other's performance may be referred to as a neutralism, i.e. when [5]:

$$\text{„2+2=4"} \tag{4}$$

whilst the situation when the factors reduce the effect of each other's performance may be referred to as antagonism, i.e. when [5]:

$$\text{„2+2<2"} \tag{5}$$

Obviously, antagonism is the opposite of synergism.

It should be noted, however, that synergism involves the occurrence of at least two factors (at least two constraints, two parameters). In the case when two factors occur, synergism is referred to as double-factor synergism [2÷5]. Assuming one factor (parameter) $a_1$, and the other $a_2$, and their synergistic effect a, it is possible to determine the synergy ratio as follows [2÷5]:

$$k_{a_1} = \frac{a}{a_1} \tag{6}$$

$$k_{a_2} = \frac{a}{a_2} \tag{7}$$

when synergism is the effect of interaction of more than two factors, then the respective synergy ratios may take the form :

$$k_{a_1} = \frac{a}{a_1},\ k_{a_2} = \frac{a}{a_2},\ \ldots,\ k_{a_x} = \frac{a}{a_x} \tag{8}$$

In technology, the effect of, at least, two factors on a given system may alter one or more features of this system, in other words may alter its characteristic.

## 2.1. Types and measurements of synergism [2]

*Synergism* results in a better efficiency (increased mechanical properties) than the sum of their separate performance. It is equal to the sum of their separate performances and sometimes may be less efficient but always greater than separate performance of single factors. In all the cases, new properties may be obtained (properties which do not occur at operation of a single factor). Thus, it is possible to differentiate (e.g. for a determined property a) Hence it stands out (for example, for a specific property a) [2]:

- *synergism total (full), the so-called „effect 2+2 = 5" for a > (a1 + a2),*
- *synergistic additive (additive), the so-called „effect 2+2 = 4" for a =(a1 + a2),*
- *synergism incomplete (incomplete), the so-called "effect 2+2 = 3" for a < (a1 + a2).*

Fig. 1 presents a graphical interpretation of synergism in engineering.

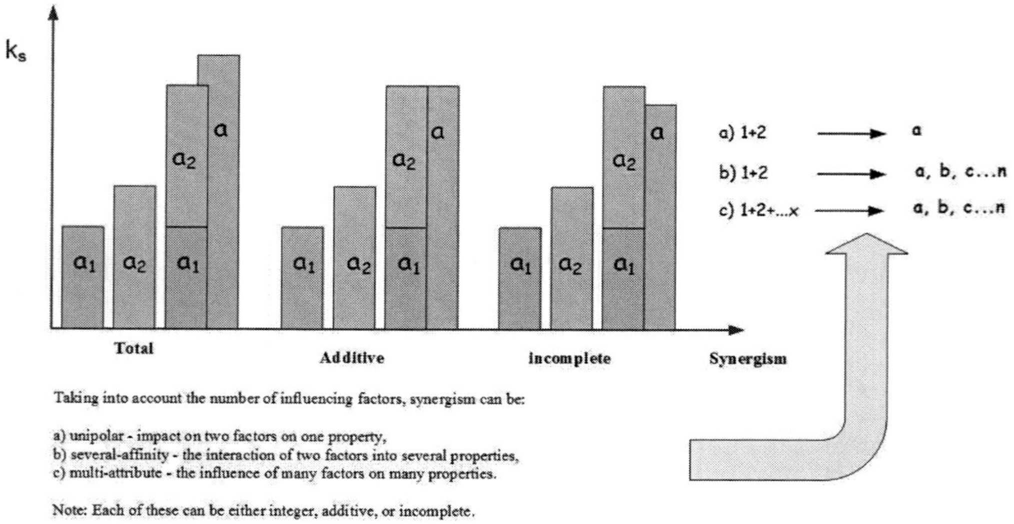


Fig. 1. Graphical representation of synergism in engineering [2]

## 2.2. Synergism in engineering

Considerations of technical synergism may be presented on the example of the determination of properties of a surface layer (WW).

The properties of non-modified surface layers may turn out to be insufficient for their exploitation and so they are subject to modification with the use of various technologies [2, 4, 6÷7].

A technological process of manufacturing a product includes a range of technologies (operations, treatments and transitions), applied in a certain order which results in alterations in a whole object or in a part of its volume. Thus, a given technological system describes synergism of a manufacturing technology (Fig. 2) [2].

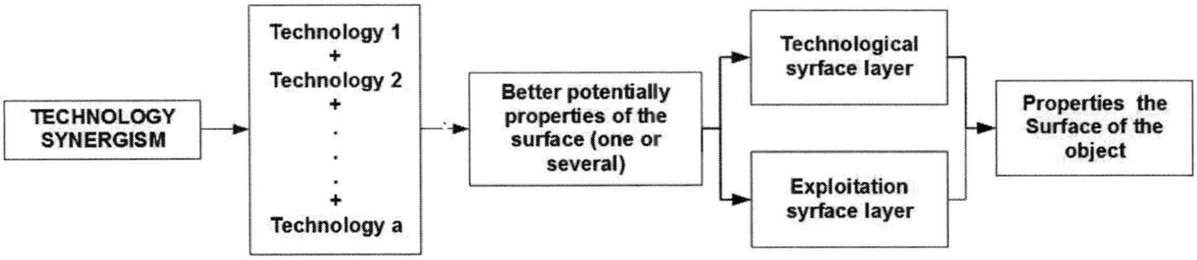


Fig. 2. A structure of synergism of manufacturing [2]

A required state of surface subjected to exploitation (at a given external constraints) requires proper selection of the number of technologies and the order of their application (1,2.3. ...n) to obtain the effect of synergy.

*Synergism in engineering* is, thus, the effect of interactions of the subsequent technologies of the manufacturing process on the properties of the resultant object.

All methods of technological manufacturing of a surface layer, depending on the phenomena prevailing or the type of physical or chemical effect on the core or surface, can be divided into 6 groups [2,3,4]:

1. *mechanical* — uses the pressure of a tool (machining) or the pressure and kinetic energy (shot peening) in order to cold strengthen the metal surface layer,
2. *thermal* — uses phenomena related to thermal operations such as: laser, electron, plasma or arc heating,
3. thermal-*chemical* — uses joined interaction of heat and chemical medium active for the processed material, e.g. nitriding, galvanising,
4. *electrochemical* — it uses electrochemical reduction (electrolytic coating),
5. *chemical* — uses chemical reactions (paint coatings),
6. *physical* — uses various physical phenomena taking place under atmospheric pressure or reduced contribution of ions or metallic or non-metallic element.

It is impossible to apply a single method to manufacture an object of required properties. It is always necessary to apply a number of technologies, and the order of the applied technologies (operations, treatments, and transitions) can not be arbitrary (tab. 2.1.) [3,4].

Table 1 presents values of ratios of synergy of hardness for a selected surface of an object made of 30HGSA alloy, whose surface layer was modified.

It should be emphasized that the correctly elaborated production process, starting with the correctly designed structure of a product, through manufacturing, and finally – correctly elaborated exploitation process – determines the type of synergism (total, additive or subtractive) that occurs between the considered technologies.

Further example of the considered task – synergism related to resistance to wear of friction surfaces (Tab. 2).

Tab. 1. Examples of synergy ratios obtained through the application of various treatment technologies [2,4]

| Type of steel | Method of modification of the surface layer | Surface hardness [HV] | Ratio of synergy of hardness [$k_H$] |
|---|---|---|---|
| 30HGSA | normalizing | 236 | 1,00 |
| | normalizing+shot peening | 394 | 1,67 |
| | normalizing+induction hardening | 620 | 2,63 |
| | normalizing+induction hardening+shot peening | 723 | 3,10 |
| | Thermal improvement | 342 | 1,00 |
| | Thermal improvement+shot peening | 442 | 1,29 |
| | Thermal improvement+induction hardening+shot peening | 810 | 2,37 |

Tab. 2. Synergy ratios in a technical application for various treatment technologies [2,4]

| Treatment | | Synergy ratio $k_{zm}$ |
|---|---|---|
| Mechanical treatment | Rolling | 1,1÷1,4 |
| | Shot peening | 1,1÷1,3 |
| Thermal – chemical treatment | Carburizing and hardening | 1,1÷2,2 |
| | Nitriding | 1,1÷1,25 |
| | Nitrocarburizing | 1,3 |
| | Cyaniding | 1,8 |
| Thermal treatment | Induction hardening | 1,2÷1,6 |
| Galvanic treatment | Chroming | 0,66÷0,9 (-) |
| | Nickeling | 1÷0,66 (-) |

To analyse the complexity of the process of defining the values of the synergism ratio for a given manufacturing technology, the notion of *technical synergism* should be defined.

Synergism of a technological process can be divided into:

- *synergism of a design*:
  e.g. synergism resulting from technical preparation of production,
- *synergism of a process:*
  e.g. synergism of metallurgical processes,
  e.g. synergism of removal machining,
  e.g. synergism of forming machining,
  e.g. synergism of finishing machining,
  synergism of removal-free machining,
  e.g. synergism of thermal treatment,
  e.g. synergism of thermal-chemical treatment.

- *synergism of technological assembly*:
  - e.g. synergism of a technological process of assembling mounting units,
  - e.g. synergism of a technological process of assembling mounting systems.

The division depends on the technologies applied and the number of operations used in the manufacturing process.

*Coherence* - cohesion, connectivity, compactness, is an attempt to formulate opinions and to make decisions which comply with the former ones. In management, it is important to maintain coherence between the company's objectives, objectives of individual employers, company's economical needs on the one hand and the environmental needs, marketing needs and the needs of research and development – on the other hand [15].

Thus, considering the issue only in terms of synergism determination seems to be insufficient. Without the evaluation of the coherence of processes carried, it is difficult to state univocally whether the selected technology is the best and optimal option.

## 3. Research

The carried out research included the analysis of achievable structures of technical design. On the basis of the review of bibliography [10÷25], a new complex structure of technical designing of technological processes was elaborated. The proposed structure of technical design introduces, inter alia, the analysis of mutual influence of the used technologies yet at the design stage as an indispensable component in the structure of technical design (Fig. 3.). The proposed structure became a starting point for the subsequent analysis of synergy ratios as well as for the determination of the coherency coefficients.

Experience has shown that "it is not advisable to apply all the criteria simultaneously". For instance, for high-quality products, irrespective application of the time criterion may not lead to satisfactory results. Faster does not always mean better. The same applies to products which are costly to manufacture.

Thus, designing technological processes should be based on "healthy" multi-criterial and multi-alternative designing which takes into consideration full synergism and coherency between all the technologies applied in the production process. Multi-criterial approach should be critical in the search for optimal solutions.

Optimization of manufacturing processes leads to the formation of mathematical models whose components, i.e. sets of possible solutions and optimization criteria, should reflect leading ideas of the company's policy. Optimization depends on the existence of alternative solutions and may be related to all the components of the production process (Fig. 4.).

The implemented optimization should also consider synergism and coherency at each step of their implementation.

The need for the multi-alternative and multi-criterial approach to process design is presented in Manufacturing Interfaces Annals of the CIRP [10]. The paper emphasises the necessity to present the sets of solutions for possible manufacturing processes in a comprehensive manner. The set of evaluation criteria (optimization criteria) constitutes one element of a mathematical model which enables taking into account factors dominating in the company's policy.

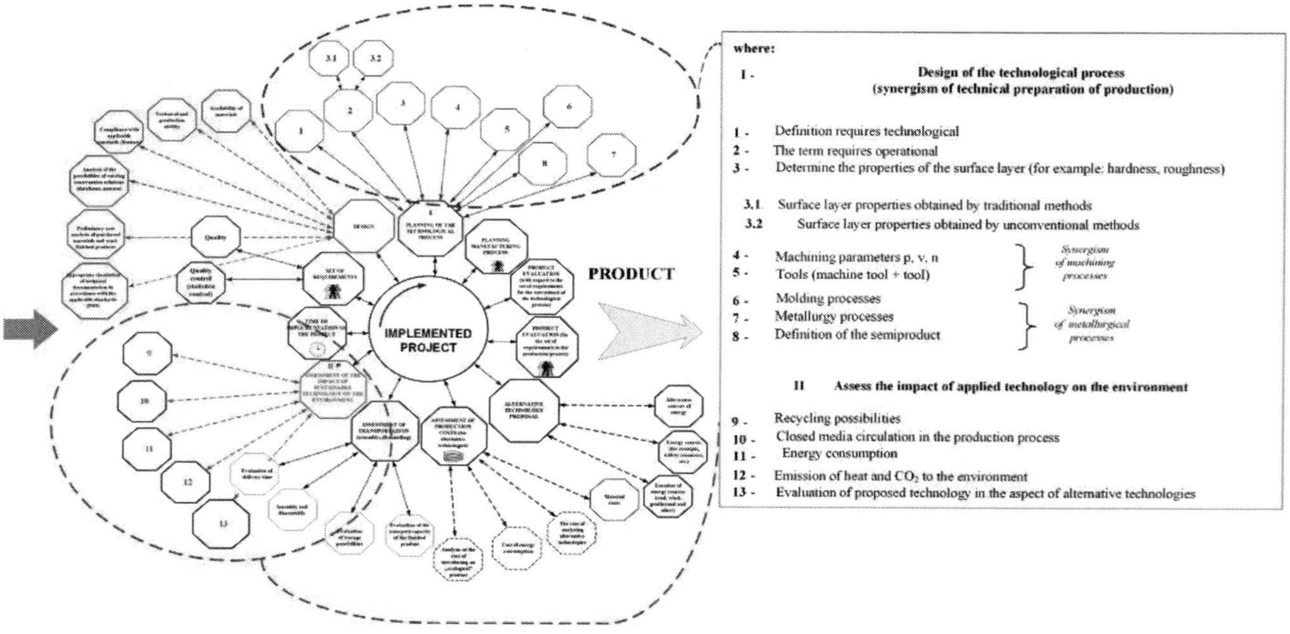


Fig. 3. Multi-criterial structure of technical design [8÷9]

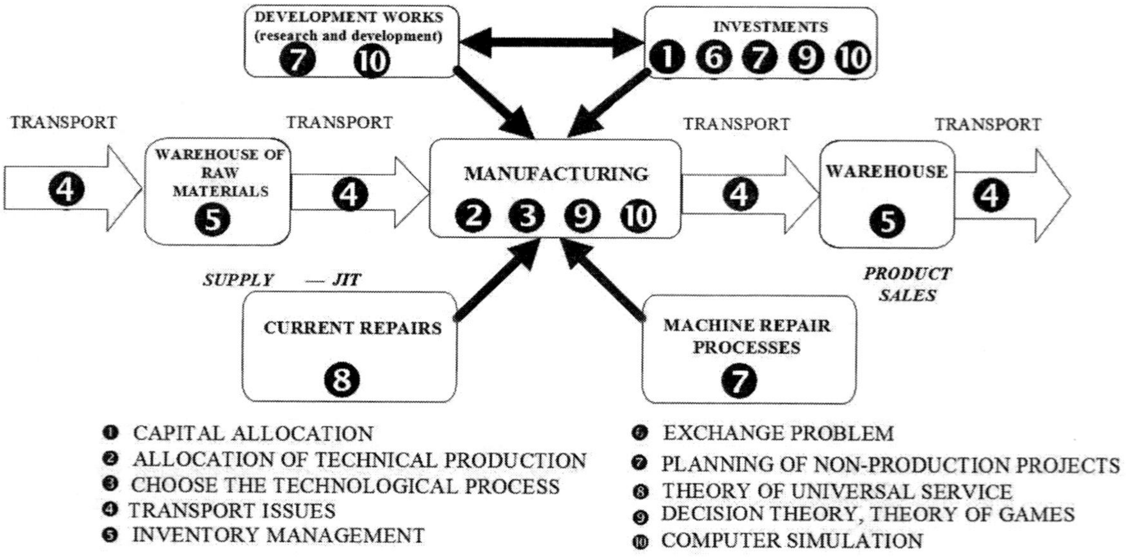


Fig. 4. Optimization applications in a production process [11]

While optimizing the company's activities, it is especially important to optimize manufacturing processes. The idea refers to both optimization of machining processes (referred to as parametric optimization taking into account the synergy of technical preparation of the production process) and the structure of the processes (structural optimization). The optimization of the conditions (parameters) of machining is complementary to the optimization of the structure. These two issues are closely related. The optimization of the structure requires a prior determination of process parameters (values of which should be as close as possible to the optimal ones), whereas the optimization of parameters requires prior optimization of the structure. To solve the arising contradiction, iterative procedure should be followed. First, the structure of the process is selected (the structure should be as close as possible to the optimal one) with typical values of the parameters. Next, the optimization of the parameters should be carried out which may be followed by the subsequent optimization of the structure, etc. [12÷13].

The analysis of the possibilities of optimization to be carried out in manufacturing processes requires the knowledge of a number of sciences such as mathematics, chemistry or strength of materials (Fig. 5).

The objective of parametric optimization (taking into account the process synergy) is the adoption of such values of parameters (out of the available ones in given conditions – within the range of solutions permissible by the determined restrictions [14]) which ensure extreme value of the assumed optimization criterion.

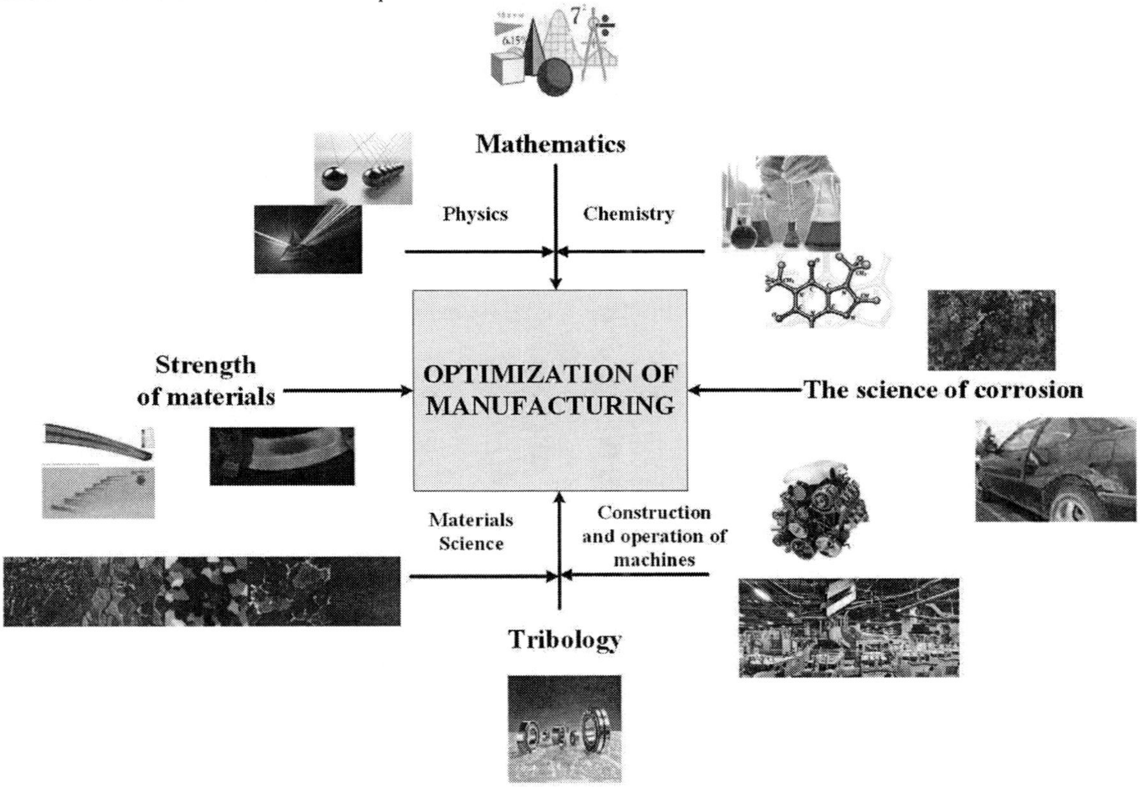


Fig. 5. The importance of basic sciences in the optimization of manufacturing [43]

### 3.1. Criteria for multi-criterial optimization

Each manufacturing process should compromise between the product quality (increasing its reliability) and the strive to reduce its manufacturing and exploitation costs (cost of raw materials including). Finding the best solution for the manufacturing process involves a properly prepared optimization analysis and the selection of criteria according to which the optimization is possible (Fig. 6).

The result of optimization largely depends on the task formulation, and particularly on the selection of criteria against which the alternative solutions of the process are evaluated, taking into account both the process synergy and coherency. In the optimization of manufacturing processes, the following criteria are most frequently adopted [41]:

- performance criteria (functional, aesthetic);
- technical criteria (general technical criteria, material and manufacturing criteria);
- economic criteria (production costs, exploitation costs).

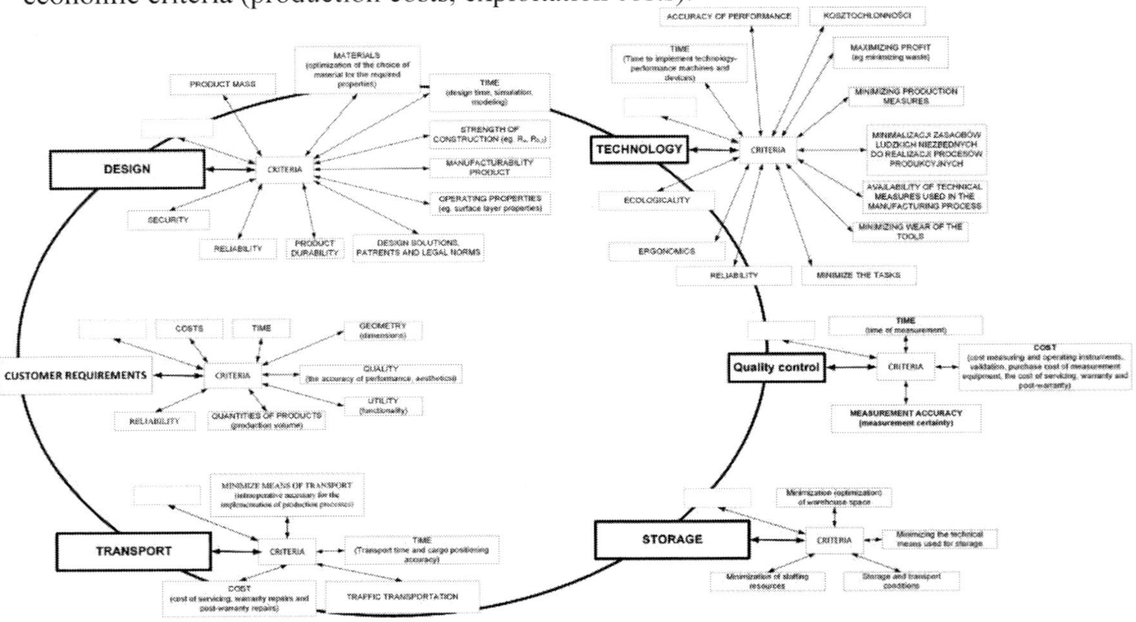


Fig. 6. Diagram of an optimization analysis and the optimization criteria [43]

The selected technology determines essentially the functionality and aesthetic of the product. It influences the quality of product properties such as: the accuracy of the shape and dimensions, stereometric structure of the surface, friction coefficient, mechanical properties, thermal conductivity, electrical conductivity, external appearance, mass. Providing the required quality of the product is the basic condition to be met while deciding on an option of the manufacturing process. *General-technical criteria* should take into account the production volume, i.e. the total number of products to be manufactured throughout the whole period of the product production, the production season (i.e. the expected number of years the product is to be manufactured without its sound modernization, the date for the production completion, etc.). *Manufacturing criteria* include the company's equipment (manufacturing devices and appliances), accessories, research and development backgrounds and staff. *Material criterion* should include the selection of the type of a semi-finished product, the accuracy of its dimensions and shape, its mechanical properties and material deficiency. *Economical criteria* include, first of all, production costs which in majority of cases are sufficient for the comparison and selection of the alternatives of the manufacturing process. When the considered options of the manufacturing process vary by just a single cost component, e.g. material consumption or labour costs, then in order to select the suitable option it is sufficient to determine and compare this particular cost component, irrespectively to it being a direct or indirect cost.

A number of technical-economic indicators may be used to compare and select the best (optimal) option of a manufacturing process, such as: an indicator of cost per unit, workload, the use of material, machines and their exploitation efficiency. Theoretically, the selection of a manufacturing option is not difficult. It is sufficient to determine the cost of the product, being the sum of manufacturing costs and exploitation costs, and the optimization problem is reduced to a single criterion task. However, a question arises how to put it in practice, at the stage of planning the production process, particularly while launching the production of a new product, i.e. when a precise determination of manufacturing costs, and particularly exploitation costs, is very difficult, and sometimes even impossible, due to the fact that it might require information available only after a trial sample has been manufactured and time consuming exploitation analyses have been carried out. Therefore, the evaluation of the options was based not on a single criterion, i.e. the product manufacturing costs, but on two or more criteria which are available yet at the planning stage, and the optimisation problem becomes thus a multi-criterial task. The evaluation criteria should be mutually independent, to allow the determination of their influences irrespectively of each other. However, this condition is difficult to be met in practice, therefore the literature describes it as a condition of "required independence" [28].

In the most general case, the evaluation of manufacturing processes may be based not only on criteria of deterministic (precise and sharp) and probabilistic nature (which in [39] were referred to as probabilistic-statistic criteria), but also criteria of a fuzzy nature (blurred) [28÷34], and additionally the effectiveness of mutual interaction of the assumed technical parameters (synergy and coherency) should be measured. At the stage of planning a process of manufacturing machine parts, the precise determination of the values of the deterministic criteria are impossible to be assessed. In such cases, subjective, score-based criteria should be used for the evaluation of the options of manufacturing processes [35÷39].

Generally, optimisation may be divided into [26]:

- a single-criterion based evaluation, when only a single criterion is required to achieve the ideal state,
- a multi-criterial based evaluation (vector, poly-optimisation), when a number of criteria condition achieving the ideal state.

*Optimisation* is a field of science which deals with methods of selecting optimal actions connected with human activities in the field of technology, economy, etc. [27].

*Structure optimisation* deals with the issues referring the selection of shape parameters and physical parameters of widely understood structures [27].

The shape of a structure is determined on the bases of the structure parameters. The structure parameters include: structure topology (the number and type of the structure elements), cross-section shapes, cross-section dimensions, the kind of material used, weight, physical, chemical and mechanical parameters of the material, structure vibrations, power supply, energy consumption, efficiency, functionality, exploitation parameters, colour, ergonomic parameters and many more [27].

The structure parameters include all the features characterising its shape, dimensions, power, rotations, etc., i.e. measurable parameters which can be presented with mathematical notions.

Apart from that, a structure is described with the help of non-measurable parameters, such as: aesthetics, appearance, etc., described with the blurred notions. Parameters may be determined before the planning stage, or may be determined with the use of optimisation procedures, and then they are referred to as *decision-making factors* (project factors).

*Optimisation criterion* is a basic optimisation notion. It is used for comparing particular solutions. In mathematical terms, a criterion is called an *objective function*. The optimisation criterion is selected at the initial planning stage, and has to meet the requirements for optimal planning and designing, may be selected out of a number of structure parameters or may be a combination of a number of parameters. A universal yard stick for measuring the quality of structural solutions is its cost. The decision-making problem is to express costs with the use of parameters [27].

An exemplary structure of criteria for the selection of a technological process is presented in Fig. 7.

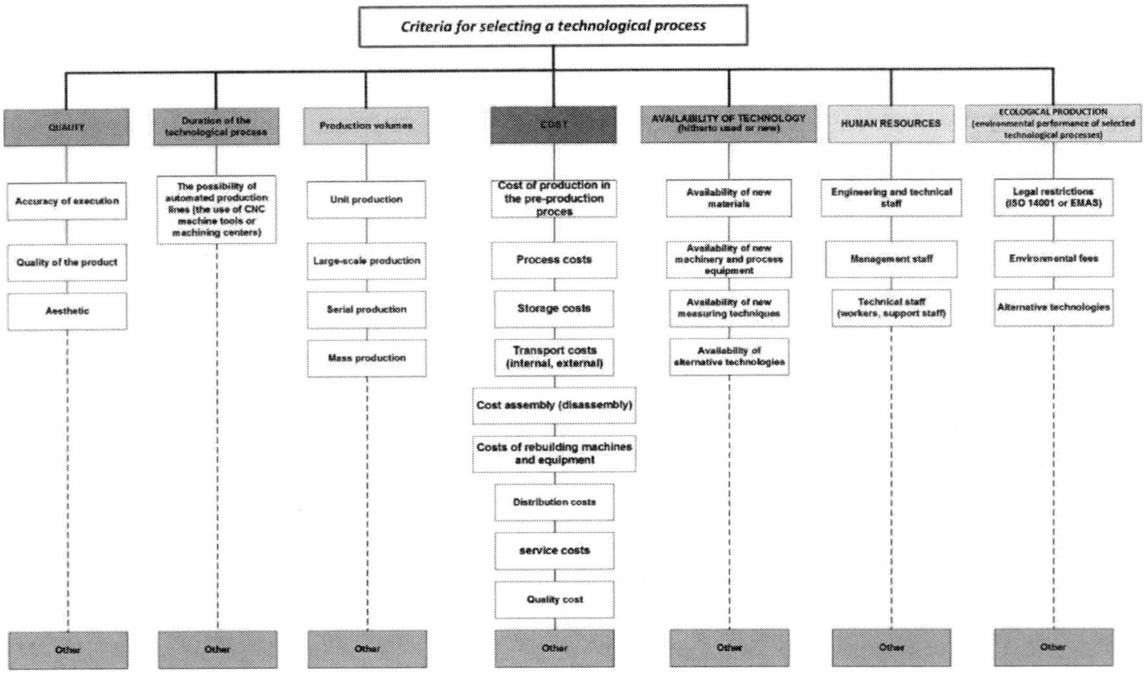


Fig. 7. A structure of criteria for the selection of a technological process [8÷9, 43, 122]

One of the most important stages of optimal designing of each manufacturing process is the selection of an optimisation procedure as a tool for solving an engineering problem. The selection is not an easy issue since there is no single universal method, equally efficient for all engineering problems. There is a kind of a psychological barrier connected with the abundance of procedures to choose from, which hinders the decision making process of solving the optimisation task. It is thus an example of optimisation in an optimal designing.

The concept of technical designing which takes into account design and process synergism consists in development and implementation activities in the fields of planning, designing and the elaboration of technology and manufacturing.

In the technical design, each new project is elaborated by an interdisciplinary team e.g. an SE team (simultaneous engineering) consisting in: designers-constructors, technologists salespeople, and representatives of procurement and finance departments as well as suppliers of manufacturing equipment and subsuppliers of basic parts of the planned product. The operation of the interdisciplinary SE team allows introducing any necessary changes in the product design and the manufacturing technology yet at the early stages of the product formation and the production planning. Thanks to it, the costs of the alterations are definitely lower and the new product may be launched faster [26].

Therefore, there is a need to combine the available methods of designing technological processes (taking into account synergy and coherency coefficients) with concurrent engineering as the most efficient and universal method of the realisation of a product which takes into account all stages of the manufacturing enterprise implementation. Tight collaboration of constructors, assembly workers, managers, planners service people and other specialists engaged in the product manufacturing process will allow to develop algorithms for methods of structure designing and planning of technological processes, and thus - typisation of elements of machine parts in the whole designing process.

It needs to be also stated that all works connected with the new product designing should be carried out concurrently taking into account design and process synergism. Also the environmental impact of the planned technology should be determined. The paper systematises the notion of the design of technological processes through the formation of a unified structure which takes into account the most essential aspects of the design and implementation of a new product to be launched onto the modern market. The proposed structure, as mentioned before, includes most essential issues from the point of view of the manufacturing technology in terms of technical design. It should be stressed that all "components" of designing of a technological process may be developed or simplified depending on the complexity of the designed product. It is also possible to determine the environmental impact for each component, with taking into account the synergy and coherency between the components of the manufacturing process. The final effect of the research implementation will be typisation of elements of the parts of machines in the whole process of technical design and the elaboration of a computer program solving the problem of a complete design of elements of machines from the stage of concept formation to the stage of a ready-made product (taking into account the basic criteria).

### 3.2. The selected methods of multi-criterial optimisation taking into account deterministic criteria

Among the multiplied methods of multi-criterial optimisation which exploit criteria of a deterministic nature, the most useful ones for solving engineering problems are the following [40; 42]:

- method of weights,
- method of weights with a normalized criterion,
- method of a global criterion,
- minimax method,
- minimax method with weights,
- utility function method,
- method of limited criteria,
- lexicographic method,
- method of objective planning.

*Method of weights* – the problem of multi-criterial optimisation is reduced to single-criterial optimisation thanks to the formation of a substitutive criterion. The method allows obtaining a certain representation of a set of alterations optimal in Pareto sense. The values of weights are almost always determined arbitrarily, and most frequently according to the decisive body, with taking into account subjective factors which is a drawback of this method. A method useful for solving engineering tasks is also a *method of a global criterion* which assumes finding the ideal solution which satisfies the condition of the minimum of each criterion being considered irrespectively of the others. Next, a global criterion is formulated by determining "distances" between the points: optimal and ideal (illustrating the optimal variant and most frequently a fictitious ideal variant) [41].

In the global criterion method [40], the best variant (optimal) is determined by vector, which minimizes the assumed form of the local global criterion.

Another method is the *minimax method.* It is close to the global criterion method as it assumes the minimisation of maximal deviations from ideal solutions. The deviations are assigned weights [40]. The weights represent directly the "importance" of particular criteria. The values of the weight coefficients are determined in a similar way to the one used in the weight method or in the normalised criterion weight method. The next method utilised in the multi-criterial optimisation is the *utility function method.* The method requires finding the extreme value of the utility function [41]. The utility function must be determined on the basis of the analysis of objectives which the solution is to be suited for. However, the determination of the function may be difficult in certain instances. The solution to the problem is the intersection of the set of compromises and the set of the curves of the utility function. The weight coefficients are determined as for the weight method, the weight with a normalised criterion method as well as in the minimax method completed with weights, i.e. arbitrarily in accordance to the decision-making person's knowledge and other subjective factors. The subsequent method is the *limited criteria method* [41]. The method is applied when it is possible to determine the maximum values for the particular criteria. The basic disadvantage of the application of the method is the necessity of determining restrictions which would allow satisfying particular criteria as well as the existence of a non-empty set of solutions. Moreover, it is necessary to decide which criterion will be minimised while solving the task. The next method is the *lexicographic method* [41] which consists in ordering the components of the criterion according their importance. The best variant is the one which minimises the criterion starting with the most important component to the least important one. The procedure finishes after an unambiguous variant has been obtained. The next method is the *objective planning method* [41]. The application of this method involves the determination of values of criteria which should be obtained. Detailed characteristics for the above mentioned methods are discussed in [40; 42].

## 4. Conclusions

Multi-criterial planning of manufacturing processes, particularly for new products, is implemented with incomplete information. Lack of information and limited measures necessary to acquire it, make the optimal option of a process structure easier to be determined basing on roughly assessed criteria. In a number of cases, they may be determined by experts in selected fields. Subjective score-based criteria may be very helpful in these cases. The selection of the optimal variant based on the mentioned above criteria may be realised on the ground of Yager's method. The method enables ranking of the evaluated options from the best to the worst in terms of the assumed weights. The weights of the criteria are evaluated by specialists by comparing subsequent pairs of criteria with the use of matrices.

In a multi-criterial planning of manufacturing processes for products similar to the ones that have already been manufactured, it is generally possible to determine (with sufficient accuracy) the values of criteria assumed for assessment. In these cases, a method with the optimum in the sense of Pareto is most frequently used. The weights of the assessment criteria are equal. In most cases, the method involves two stages. The first stage involves the determination of the set of non-dominated variants (the set of Pareto-optimal variants). The second stage involves the selection of the best variant out of this set. It usually entails introducing an additional (most frequently - new) criterion. The elimination of the least preferable variants, which is performed with the use of the method "optimum in the Pareto sense", is inefficient when two or more criteria are considered. Therefore Yager method has been modified in terms of its practical application for the assessment of variants of criteria obtained from calculations or measurements (the so called deterministic criteria).

For example, for machining purposes, at the stage of planning the processes of manufacturing of machine parts, the parameters of machining are selected from a catalogue of cutting tools, which is usually guided by the criterion of a unit cost of manufacturing. The selection of optimal parameters of machining, in terms of two (or more) criteria and also in terms of the value of coefficients of synergy and coherency between the used technologies is generally rare and used only in specific cases, e.g. for finishing of high quality surfaces.

Applying the multi-criterial optimisation of the structure of manufacturing processes, parameters for manufacturing of selected parts of machines as well as synergy of technological processes enable formulation general, informative conclusions:

- multi-criterial approach to the evaluation of the variants of manufacturing processes has become a basic means of considering the complex diversity of requirements that are to be met by machining processes,

- computer support for designing of optimal manufacturing processes must enable multi-criterial optimisation which takes into account various criteria, most often subjective score-based criteria or relative – used in the case of new products, as well as deterministic criteria for products which are similar to the ones that have been manufactured before,
- optimisation criteria, treated as deterministic criteria in conventional models (e.g. as costs) must be repeatedly treated as non-deterministic (e.g. subjective score-based criteria or subjective relative assessments). This is caused by lack of information at the early stages of planning a process of manufacturing of new machines parts. manufacturing design,
- optimisation criteria applied to the evaluation should be independent, and their number, for the determination of the set of non-dominated solutions (Pareto-optimal), should not exceed three, which is beneficial due to the possibility of obtaining an inconsiderable size of Pareto set,
- multi-crterial optimisation of the structure of a manufacturing process is the best when it is a single-stage procedure, e.g Yager's method or its modifications,
- mutli-criterial optimisation of parameters of operations and treatments usually brings good results if it consists of two stages: the determination of a set of non-dominated solutions (Pareto-optimal), and then (for a Pareto set with two or more elements) the selection of the best solution out of this set in terms of an additional (most often new) criterion,
- practical effects resulting from the application of multi-criterial optimisation depend mainly on skilful selection of the evaluation criteria (optimisation criteria) at particular stages of optimisation procedure,
- synergy and coherency between particular technologies (process parameters) need to be taken into account at each stage of implementation of a technological process (from planning to the manufacturing of a final product),
- taking into account synergy and coherency in multi-criterial and multi-variant planning of technological processes brings much better results in terms of efficiency and effectiveness of each manufacturing process.

**References**


[1]. Kuryło P., own materials,

[2]. T. Burakowski, Rozważania o synergizmie w inżynierii powierzchni, Wyd. Politechniki Radomskiej, Radom 2004,

[3]. T. Burakowski, Synergizm w eksploatacji. Problemy Eksploatacji 4, 43, (2001) 89-102,

[4]. T. Burakowski, Współczynnik synergizmu w technologii. Inżynieria Materiałowa no 6, (2000) 245-247,

[5]. M. Kupczyk, Ostrza skrawające z twardymi i supertwardymi powłokami. Wytwarzanie – Badanie – Eksploatacja, Wyd. Politechniki Poznańskiej Poznań 2005r,

[6]. I. Kamińska-Szmaj, Słownik wyrazów obcych Wyd. Europa, Wrocław 2001r,

[7]. I.Kamińska-Szmaj, M. Jarosz i inni, Słownik Wyrazów Obcych, Wydawnictwa Europa. ISBN 83-87977-08-X, Wrocław 2001r,

[8]. P. Kuryło, R. Kielec S. Janik Wpływ optymalizacji procesów projektowo-konstrukcyjnych na ochronę środowiska naturalnego. Studia i Materiały: Referaty i Ekologia pogranicza. VI Polsko-Niemiecka Konferencja Naukowa. IBN, Gorzów Wielkopolski 2001r.

[9]. P. Kuryło Wielokryterialne projektowanie procesów technologicznych w aspekcie synergii. Proizvodstvo. Tehnologiâ. Ekologiâ red. L. E. Švarcburg-Moskva: Moskovskij Gosudarstvennyj Tehnologičeskij Universitet "Stankin", Moskva (2011) pp. 189÷199,

[10]. F.J.A.M. Houten: Manufacturing Interfaces. Annals of the CIRP 41, 2, (1992) 699÷710.

[11]. J. Szadkowski: Optymalizacja strukturalna i parametryczna w projektowaniu procesów technologicznych obróbki skrawaniem. Sympozjum Inauguracyjne Szkoły Podstaw

Optymalizacji Procesów Technologicznych i Systemów Produkcyjnych. PAN, Sekcja Podstaw Technologii Komitetu Budowy Maszyn, TNOiK, Bielsko-Biała 1986,

[12]. J. Szadkowski: Artificial Inteligence Approach to Structural and Parametrical Optimatization of Multi-Tool-Machining Processes, Gepgyartastechnologia (Feitigungstechnik) Budapest (l992) 9÷10,

[13]. S. Szulc: Struktura technologicznych procesów budowy maszyn, Archiwum Budowy Maszyn 9, 1, (1962) 87÷112,

[14]. A. Samek, J. Duda, Budowa struktur procesów technologicznych obróbki z zastosowaniem transformatorów, operatorów i identyfikatorów dla potrzeb projektowania wspomaganego komputerowo. Konferencja nt. Automatyzacja Napędów i Procesów Technologicznych. Częstochowa 1986,

[15]. A. Samek, J. Duda,Elementy struktury sieciowego modelu dla zautomatyzowanego generowania procesów technologicznych. Prace Naukowe ITBM Politechniki Wrocławskiej 36. Seria: Konferencja 13 nt. Komputerowe Wspomaganie Konstruowania Obrabiarek i Procesów Obróbki Skrawaniem CAD-CAM-CIM `88. Wrocław 1988,

[16]. A. Samek, J. Duda, Opis i systematyka zdarzeń w procesie technologicznym dla celów wspomaganego komputerowo projektowania. Materiały na Konferencję Naukową nt. Podstawy Projektowania Procesów i Systemów Produkcyjnych. Monografia 150. Kraków, (1993) 185÷200,

[17]. J. Kaczmarek, Podstawy obróbki wiórowej, ściernej i erozyjnej, WNT, Warszawa. 1970,

[18]. J. Dmochowski, Podstawy obróbki skrawaniem, PWN, Warszawa 1983,

[19]. W. Grzesik, Podstawy skrawania materiałów metalowych, WNT, Warszawa 1998,

[20]. R. Saravanan, P. Asokan, Sachidanandam M. Comparative Analysis of Conventional and Non-Conventional Optimization Techniques for CNC Turning Process. Intemational Joumal of Advanced Manufacturing Technology, 17, 7, (2001) 471÷476,

[21]. W. Przybylski, Technologia obróbki nagniataniem. WNT, Warszawa 1987,

[22]. Z. Galas, I. Nykowski, Żółkiewski Z., Programowanie wielokryterialne, Państwowe Wydawnictwo Ekonomiczne Warszawa 1987,

[23]. M. Ostwald, Optymalne projektowanie trójwarstwowych konstrukcji powłokowych, Poznań, Wydawnictwo Politechniki Poznańskiej, Rozprawy nr 290, Poznań 1993,

[24]. M. Ostwald, Współczesne problemy optymalizacji. Od lekcji natury do zastosowań W technice, Zielona Góra, Wyd. Uniwersytetu Zielonogórskiego Współczesne Problemy Techniki, Technika 3, Zielona Góra 2003 pp. 99-135,

[25]. A. Borkowski, S. Jendo, Mathematical Programming, wz M. Save, Prager W. (ed.), Structural Optimization, vol. 2, New York and London, Plenum Press, 199,

[26]. A. Brandt (red.), Kryteria i metody optymalizacji konstrukcji, Państwowe Wydawnictwo Naukowe ,Warszawa 1977,

[27]. R. Knosala, W. Pedrycz, Komputerowy system wspomagający proces oceny rozwiązań konstrukcyjnych. Zeszyty Naukowe Politechniki Śląskiej 1012, Mechanika 86, pp. 143÷163, Gliwice l987,

[28]. J. Kacprzyk, Zbiory rozmyte w analizie systemowej. PWN, Warszawa 1986,

[29]. R. Knosala, Methoden zur Bewertung von Bauelementen als Voraussetzung fur die Entwiccklung von Baukastensystemem. Technische Universitat Dresden. Dissertation B, Dresden 1989,

[30]. R. Knosala, Objektivierung des Bewertungsprozesses beim Konstrurieren. Konstruction 43, (1991) 344÷352,

[31]. R. Knosala, S. Płonka: Objektivierung der Bewertung bei technischer Produktionsvorbereitung. Proceedings of Intemational Conference on Computer Integrated Manufacturing, Zakopane (1994) 51÷77,

[32]. A. Breiing, R. Knosala, Bewerten technischen Systeme (T heoretische und methodische grundlagen bewertungstechnischer Entscheidungshilfen). Springer-Verlag, Berlin-Heidelberg 1997,

[33]. R.R. Yager, Fuzzy Decision Making including Unequal Objectives. Fuzzy Sets and Systems 1, (1978) 87÷95,

[34]. R.R Yager, Multiple Objective Decision - Machining Using Fuzzy Sets. Intemational Journal on Man - Machines Studies 9, (1977) 375÷382,

[35]. S. Płonka Model optymalizacji wielokryterialnej struktury procesów technologicznych. Postępy Technologii Maszyn i Urządzeń 18, 1, (l994) 25-60,

[36]. S. Płonka: Metody oceny i wyboru optymalnej struktury procesu technologicznego. Rozprawy Naukowe 48. Budowa i Eksploatacja Maszyn 31. Wydawnictwo Politechniki Łódzkiej, Filia w Bielsku-Białej, Bielsko-Biała l998,

[37]. S. Płonka, Metoda oceny i wyboru optymalnej struktury procesu technologicznego. Materiały Konferencji Naukowo-Technicznej nt. Projektowanie Procesów Technologicznych TPP`98. Poznań-Czemiejewo, (199) 153÷162,

[38]. J. Szwabowski Polioptymalizacja parametryczna wielozabiegowych operacji toczenia zarysów złożonych. Prace Naukowe 504, Politechnika Szczecińska, Szczecin 1993,

[39]. J. Szwabowski, Zastosowanie kryterium minimalnej ceny jednostkowej w polioptymalizacji parametrycznej wielozabiegowych operacji tokarskich. Postępy Technologii Maszyn i Urządzeń 18, 3, (1994) 57-77,

[40]. A. Osyczka, Optymalizacja wielokryterialna. Sympozjum Inauguracyjne Szkoły Podstaw M Optymalizacji Procesów Technologicznych i Systemów Produkcyjnych. PAN, Sekcja Podstaw Technologii KBM. TNOiK, Bielsko-Biała, 1986, pp. 37-48,

[41]. S. Jendo, W. Marks, O optymalizacji wielokryterialnej konstrukcji. Archiwum Inżynierii Lądowej 30, 1, l984 l÷21,

[42]. W. Stadler, Preference Optimality in Multicriteria Control and Programming Problems.Joumal of Nonlinear Analysis Theory, Methods and Applications, 4, 1 (1980),

[43]. J. Kostka, P. Frankovský, M. Pástor, F. Trebuňa, F. Šimčák, Application of PhotoStress method in stress analysis of structural elements under consideration of centrifugal force effect. Procedia Engineering, 96, (2014), 235-241.

[44]. P. Frankovský, O. Ostertag, E. Ostertagová, F. Trebuňa, J. Kostka and M. Výrostek, Experimental analysis of stress fields of rotating structural elements by means of reflection photoelasticity. Applied Optics, 56(11), (2017), 3064-3070.

[45]. P. Kuryło Multikriterálna optimalizácia výrobných procesov tézy habilitačnej prednášky k habilitačnému pokračovaniu za docenta v študijnom odbore: 5.2.51 výrobne technológie. Technická Univerzita v Košiciach, Košice 2016,

# Monitoring of damage of the composite leaf spring using methods of acoustic emission

Władysław Papacz [1,a], Peter Frankovský [2,b], Ján Kostka [2,c], Daniel Kottfer [2,d]

[1] University of Zielona Gora, ul. prof. Z. Szafrana 4, 65-516 Zielona Gora, Poland

[2] Technical University of Košice, Faculty of Mechanical Engineering, Letná 9, 042 00 Košice, Slovakia

[a] W.Papacz@ibem.uz.zgora.pl, [b] peter.frankovsky@tuke.sk, [c] jan.kostka@tuke.sk, [d] daniel.kottfer@tuke.sk

**Keywords:** composite leaf spring, acoustic emission, fatigue strength, AE sensor.

**Abstract.** The paper presents the application of acoustic emission (AE) for the operational testing of the composite leaf springs. The method applied is characterized by a high sensitivity of detection and can be used as a way of monitoring the condition of structures in real time. The characteristics of the acoustic emission enable the evaluation of damage accumulation and intensity of their formation and propagation. The state of a composite structure of the leaf springs can be, perfect monitored using the acoustic emission. Any change in the stiffness of the leaf spring was accompanied by a change in density of the acoustic emission. There is a noticeable regularity between the degree of degradation of the structure of the leaf spring and the level of emissions. The increase in the rate of degradation was accompanied always by the increase in the level of acoustic emission. Immediately prior to the decrease of stiffness occurred the marked increase in the level of the AE. Knowing the characteristics of the AE for the leaf spring, the status of its structure can be assessed.

## Introduction

Use of polymer composite materials for the construction of vehicle suspension elements is not only a simple replacement of steel with a different material. Composite materials allow any formation of transverse sections of elements being produced, at the same time affecting the stiffness of such elements. Thus, it is possible to construct suspensions with the prescribed characteristics as well as integrated elements of suspensions which work as springy and leading elements. Due to the above it has been possible to simplify the construction of vehicle suspension and to lower considerably its weight and costs of production. Moreover, a greater capacity of composite materials to accumulate springy energy may allow the use of smaller and simpler dampers. Numerous researchers [1,2,3] prove that springy elements constructed from composites materials in comparison with steel springs are characterized by higher fatigue durability, they are resistant to corrosion and they are many times lighter. A great advantage of composite materials is the way they get damaged under fatigue loading. Elements made of these materials, as opposed to steel elements, do not break or get damaged rapidly. This in turn increases considerably the safety of vehicle operation. The above mentioned properties of composite materials will decide about their wider application in the construction of vehicles, including also the construction of springy elements of suspensions. Increasing competition and innovation in automobile sector tends to modify the existing products by usage of the new and advanced materials – polymer composites.

During the operation the composite leaf springs are subject to heavy loads and the environment with high humidity, chemical and physical aggressiveness. Under these conditions, the composite structure may change due to a nucleation and growth and accumulation of damages [5]. The formation and accumulation of defects in the material during operation lead to changes in the technical state of construction. However, the "local" individual damages do not result in the destruction of the entire leaf spring. This justifies the need to monitor its state for safe operation. For this purpose you can use the method of non-destructive examination (non-destructive examination—NDE), which allows the examination of the material state periodically or continuously in real time [4,6].

Damage of the composite leaf springs, due to operation, may arise as a result of rupture of the boundary layer fiber-matrix (called debonding), matrix cracking, fiber breakage and delamination. The local damage (the order of several microns) do not cause significant deterioration of its performance characteristics. Damages occurring in the individual layers of the composite have greater impact on the condition of the leaf spring These can be individual layers transverse cracks, splits of roving bands, longitudinal and intra-layer cracks (local delamination) and breaking or shearing of roving bands [5]. These damages usually lead to destruction of the leaf spring. In order to study the composite leaf springs, it is possible to use the method of measuring the acoustic emission. It is characterized by high sensitivity of detection and can be used as a method of monitoring the condition of structures in real time. The characteristics of the acoustic emission enable the evaluation of damage accumulation, the intensity of their formation and propagation. In next part of article, the study of fatigue strength and AE will become proven.

AE is a result of disturbance over test. During knowledge resources [11], a sound source is being if under the influence of vibrations of sufficient energy to make the human reaction. For AE test this value is changed: in frequency domain to max. 10 MHz (in this cas to 1MHz) and in magnitude value (confine in discriminator). This disorder was first described by Isaac Newton, as rapid changes in pressure and vibration of air. This means that identical air density at each point in space is the same time is a silence. Changes in air density, causing changes in sound pressure, that is atmospheric pressure disturbances are called acoustic density [12]. In addition to atmospheric disturbances, the acoustic wave is accompanied by disorder temperature of the medium. A full description of the acoustic wave should contain all the parameters characterizing it (density acoustic, acoustic pressure, temperature, acoustic, acoustic deflection of particles, particle velocity of acoustic, acoustic particle acceleration). AE test uses accelerometers sensors, with high sensitivity and flat response (which is important for valuable tests) in the bandwidth of operation.

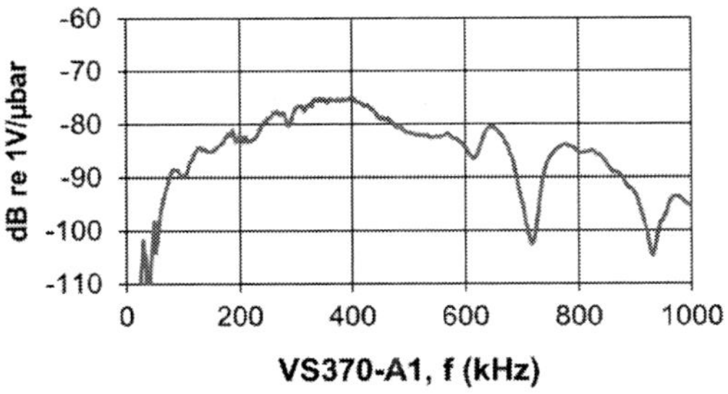


*Fig. 1. The characteristic of AE sensor VS370-A1 [13]*

Because it is complicated (Figure 1 presents example characteristic of single AE sensor) to obtain sensor with all mentioned necessary parameters, often in one measurement head are a few sensors, for different frequency range.

Piezoelectric sensors built from semiconductor have a large internal resistance. Therefore, it is necessary signal conditioning as possible to the sensor in order to avoid additional inductance and capacitance and thus increase the impedance of the entire measurement path. Conditioning is to join an operational amplifier integrator circuit with a capacitor in the feedback block [14].

In the case of piezoelectric materials working conditions (e.g. temperature) are very important. In this case pyroelectric effect is also very important (induction of surface load during temperature) [14].

## The test program, test equipment

The term acoustic emission [AE] applies to both physical phenomena and methods of measurement. In the first case, the term AE is understood as generation of elastic waves in a given environment as a result of the local and dynamic release of elastic energy. EA method relies on the detection, analysis and measuring the emitted wave parameters, which can be sourced by motion of dislocation, phase transitions associated with changes in the microstructure of the material, formation and propagation of macro-and micro-cracks, breakage of fibers and matrix in composites. The metering of AE creates, therefore, the possibility of a direct linkage of the information contained in the AE signals with the cause of their excitation.

The correlation between the acoustic emission and degradation of the composite leaf spring structure undergoing fatigue testing on the testing stand was observed. The force bending leaf spring was measured and the acoustic emission was registered. Long-term fatigue tests on testing stand were carried out according to the author's method, which involved conducting tests for the four levels of load of the leaf spring. The loads were determined using the equation

$$P_K = k_z \cdot Q_i \quad (1)$$

where:

$i = 1$ or 2, $K$=1, 2, 3, 4,

$Q_1$ - static load of the leaf spring with vehicle mass and car cargo,

$Q_2$ - static load of the leaf spring with vehicle mass without cargo,

$K_z$ - dynamic load coefficient.

Full characterization of the leaf spring fatigue loads are shown in Table 1.

*Table 1. Characteristics of the leaf spring fatigue loads*

| | Level I $P_1$=1.6$Q_1$ [N] | Level II $P_2$=1.3$Q_1$ [N] | Level III $P_3$=$Q_1$ [N] | Level IV $P_4$=$Q_2$ [N] |
|---|---|---|---|---|
| $P_{max.}$ | 13000 | 10562.5 | 8125 | 6400 |
| $P_{min.}$ | 6200 | 3762.5 | 1425 | 0 |
| $P_{śr.}$ | 9600 | 7162.5 | 4775 | 3200 |
| $P_a$ | 3400 | 3400 | 3350 | 3200 |

Each leaf spring was loaded with a single value of the force until its failure. The deflection of leaf spring was set before test, and maintained constant until destruction of the spring. With increase of the number of load cycles, decrease in the stiffness of leaf spring was observed; at constant deflection the decrease manifested itself by a decline of extorting force. Extortions were carried out in the sinusoidal cycle unilaterally positive. The amplitude of extortions was 0.085m, which represented 70% of the maximum of deflection. The study was conducted in the cycle of fixed-kinematic extortions whose frequency was 1.0 Hz. If the leaf spring has not been destroyed and reached 6 million cycles of fatigue, the test was stopped.

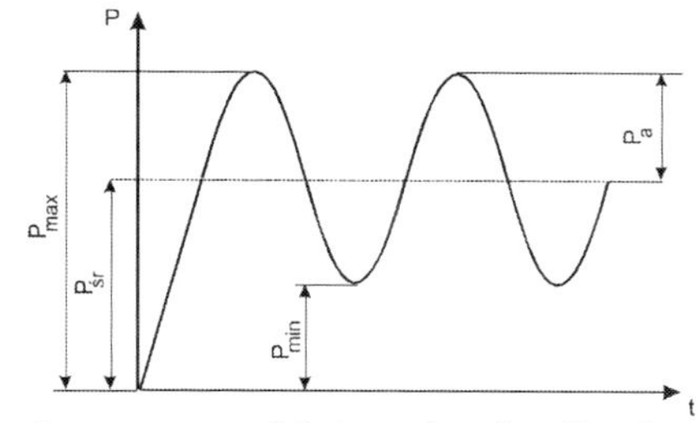


*Fig. 2. The time course of fatigue loads of leaf spring*

Schematic of the composite leaf springs testing stand is shown in Figure 3. The leaf spring tested - 1 was deformed cyclically by a hydraulic servomotor - 3. The servomotor stroke was established at the beginning of the testing. The leaf spring loading force was measured indirectly by the strain gauge system - 2 and the strain gauge bridge -7. The acoustic emission was measured using a head - 4. The measured values were stored in the recorder - 9, and hence transmitted to PC and collected in the data sets. The acoustic emission was measured at every 100 thousands of fatigue cycles. The average density of the emission for fifty consecutive cycles close to the assumed number of fatigue cycles was determined.

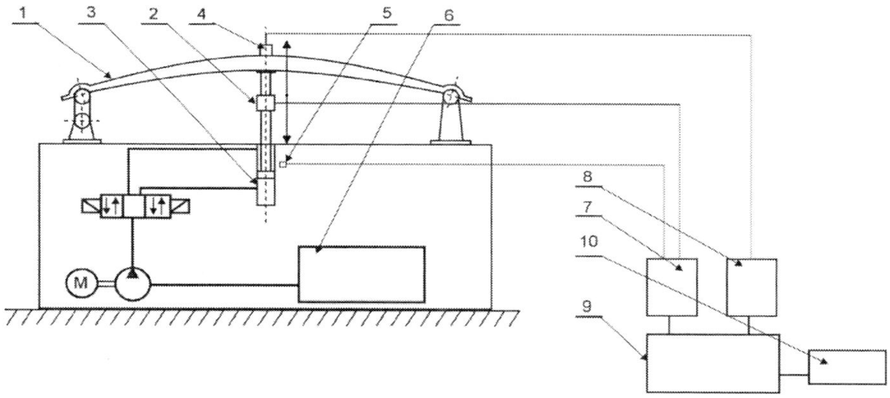


*Fig.3. Scheme of the leaf spring fatigue test stand using AE:1-tested spring, 2- dynamometer, 3-hydraulic servomotor, 4-piezoelectric head for receiving signals AE, 5-position sensors, 6- hydraulic system, 7- tensometric bridge, 8- analyzer AE-3, 9- recorder, 10- computer PC*

An electromechanical transducer (piezoelectric) was the receiver of the induced stress waves, which was fixed to the surface of the material being tested. The acoustic signals received by the transducer were converted into electrical impulses (discrete in time domain), which were then amplified and recorded by the AE analyzer (Fig. 4). AE is saved by system in digital form, thus – under digital signal processing (DSP).

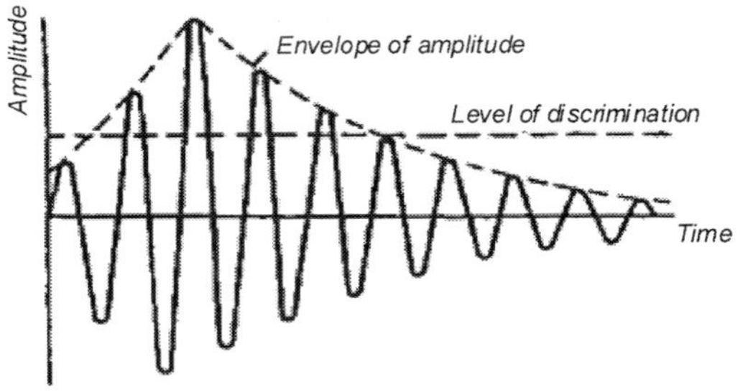


*Fig. 4.The AE counts (elastic wave amplitudes above the level of discrimination) and the AE event (the wave envelope) received by the transducer [7]*

Form of the recorded AE signal depends not only on its source of origin, but also on the type of converter that changes the elastic wave energy into an electric signal from the environment in which the signal propagates and the location and method of mounting the sensor. Taking as a criterion the method of attachment and signal transduction in the leaf spring testing, an acoustic emission sensor SEA (Schall-Emissions-Aufnehmer) was used that is characterized by low noise and which transmits frequencies within the band from 0 Hz to 1 MHz. The measuring apparatus, whose diagram is shown in Figure 5, can record many descriptors characterizing the source signal, allows their registration in real time and analyze the impulse shapes, their distribution and the frequency characteristics. Signal, after process of discrimination (which is made in comparator) is filtered by low – pass filter (0 – 1MHz) in next step signal is sampling and quantizing. In this case, according to Shannon – Kotielnikov – Whitaker – Niquist law [8 - 10], minimum sampling frequency is 2MHz.

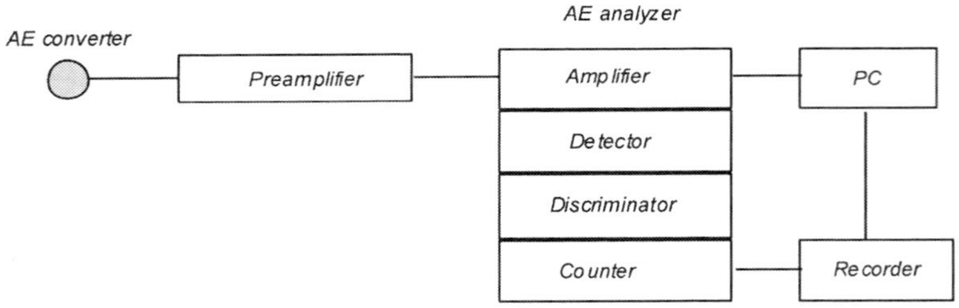


*Fig. 5. Block diagram of the AE metering system*

Analysis of changes in the structure of the test material on the basis of descriptors EA is difficult and requires a lot of preliminary tests in order to correctly choose the parameters of the measurement channel and properly interpret the results. In this study, it was decided, after the pilot tests, to apply the easiest measurable signals:

- count rate – the number of counts in a given time, and
- the sum of AE counts – the sum of amplitudes of the stress waves exceeding the established threshold of discrimination.

The rate of AE counts and energy of the emitted signals have proven to be the most sensitive indicators of the leaf spring structure destructive changes. These descriptors are associated with the amplitude and energy of the original signal, and thus indirectly determine the energy characteristics of the AE sources. This fact makes these parameters reflect well the changes in the structure of the material being tested.

During DSP operation is possible to pre-diagnosis (failures predict), not only for whole leaf spring, but also particular parts of system. Moreover it is possible to detect of failure, under one frequency test. In this case necessary is provide signal processing methodology.

The aim of all the signal digital processing operations performed during the investigations was to change the shape of the signal spectrum in a replicable way using standard methods. The first operation is signal windowing. In this case, windowing in the time domain since it is limited to the multiplication of the discrete vibration signal and the discrete window spectrum. Naturally, one could use a window in the frequency domain, but this would require the convolution of the two discrete signals.

A rectangular spectrum would have an ideal windowing sequence for damping uncharacteristic (from the investigation point of view) parts of the spectrum and simultaneously amplifying its characteristic parts. The ideal window would not distort the signal and prevent spectral leakage (an effect in which a part of the signal component, not situated by the frequencies for which the analysis is made, appears in all the output discrete signal values after transformation to the frequency domain [8, 9]). This case od processing is a base for comparison analysis – thus it must be done on proper working leaf and this value is the reference value.

**Conclusions from the study of composite springs**

The process of fatigue destruction of the leaf spring resulted due to the formation, near the neutral axis, of micro-cracks which connected with each other and gradually led to the formation of local macro-cracks. The continuation of destruction process proceeded similarly. Macro-cracks united and weaken the entire structure. As a result, the composite leaf spring underwent dissection near its neutral axis (Fig. 6). Then the weakening of the structure occurred which was expressed by the significant decrease in stiffness. The gustily running damage threatening the security have never been observed.

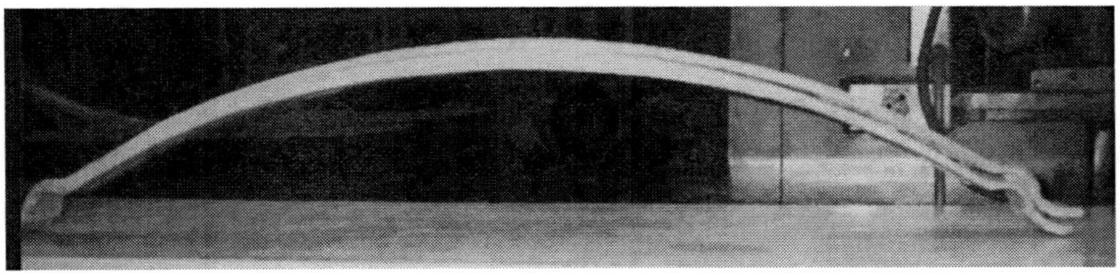

*Fig.6. Example of damaged composite spring*

In fatigue tests, the 18% decrease of bending force compared to its initial value was assumed as the criterion of leaf spring failure. Then the first micro- cracks of spring occurred. Thus, the leaf spring for which the bending force decreased by 18% compared to its initial value was regarded as damaged. It should be emphasized that the leaf spring was still able to carry the load reduced.

The leaf springs loaded with the bending force of about 60% greater than the allowable static load (13kN) became destroyed very quickly. In accordance with the accepted criterion, the failure occurred after 700 thousand fatigue cycles. Typical course of changes of the extorting force at the above load is shown in Figure 6. The fall of extorting force and thus the reduction of composite leaf spring stiffness were accompanied by an increase in the density of the acoustic emission. In the leaf spring, a rapid process of its structure degradation in the form of splitting of bands of roving, longitudinal and intra-layer cracks, occurred. This process was accompanied by a gradual increase of the acoustic emission.

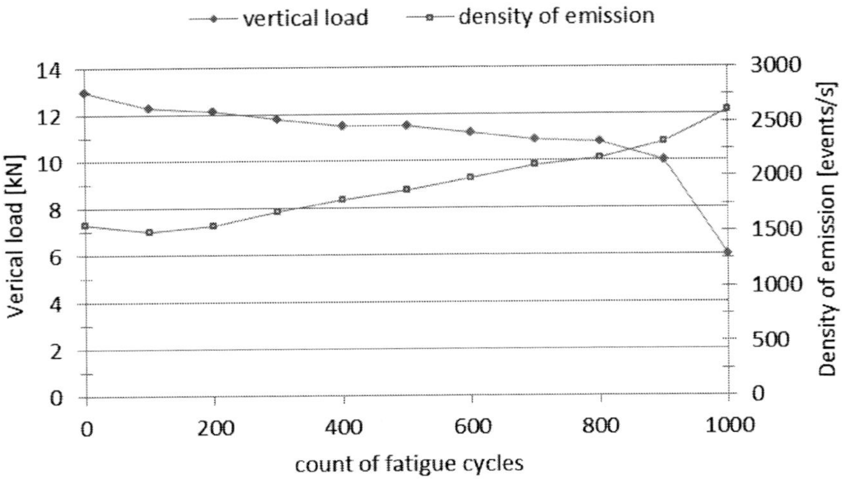


*Fig.7. Changes in the leaf spring bending force, initial load 13 kN*

Figure 8 shows the typical course of the force extorting the bending of springs and the EA density as the function of the number of fatigue cycles. In this case, the initial load was 130% of the static load resulting from the weight of the vehicle and the permissible cargo mass. The leaf spring kept operational efficiency up to 1.6 million fatigue cycles. The total lifetime of the leaf spring can be clearly divided into three periods. The first lasted about 200 thousand of the fatigue cycles characterized by a rapid decrease in stiffness accompanied by the acoustic emission. The second period included the stabilization of both rigidity and acoustic emission. It lasted up to 2.8 million cycles. The stiffness of the leaf spring and the acoustic emission level stabilized. The third period was the period of stiffness decline accompanied by the EA increase. The period lasting about 500 thousand cycles between 2.2 million and about 2.7 million cycles can also be distinguished. In this range, a small increase in the EA compared to the immediately preceding period is noticeable.

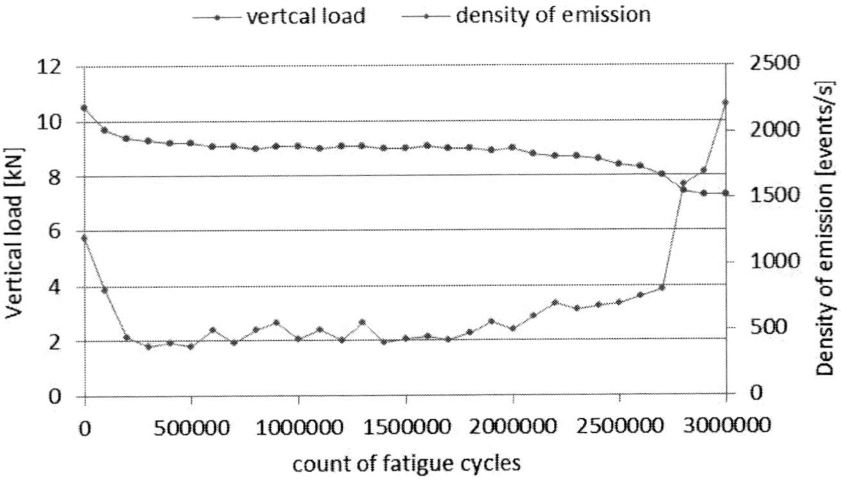


*Fig.9. Changes in the leaf spring bending force, initial load 10.6 kN*

Figure 9 shows the course of the leaf spring bending force and corresponding EA as the function of the amount of fatigue cycles for the initial load of 100% of the permissible static load (8.12 kN). Figure 10 presents the bending force course resulting from the burden of their leaf spring with the vehicle mass without cargo (6.4 kN). Loading of the leaf spring with the force corresponding to the maximum static load led to the destruction of the leaf spring after reaching approximately 5.7 million cycles. In contrast, the loading with the forces corresponding to the static load of vehicle without cargo did not lead to the destruction of the leaf springs. For the 8.12kN load the course of stiffness and EA are similar to the load of 10.6 kN force. In turn, in case of the lower load with the force of 6.4 kN, the leaf spring destruction process and the accompanying EA were not observed. The leaf spring reached 6 million cycles and the test was stopped.

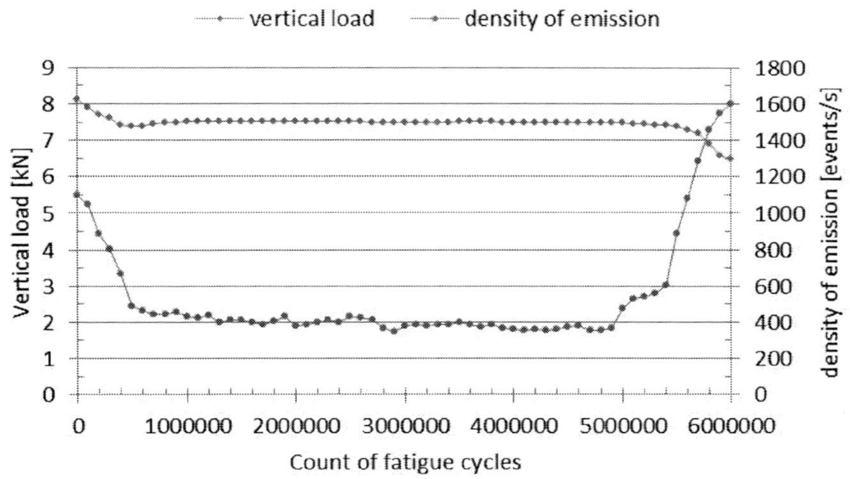


*Fig.10. Changes in the leaf spring bending force, initial load 8.12 kN*

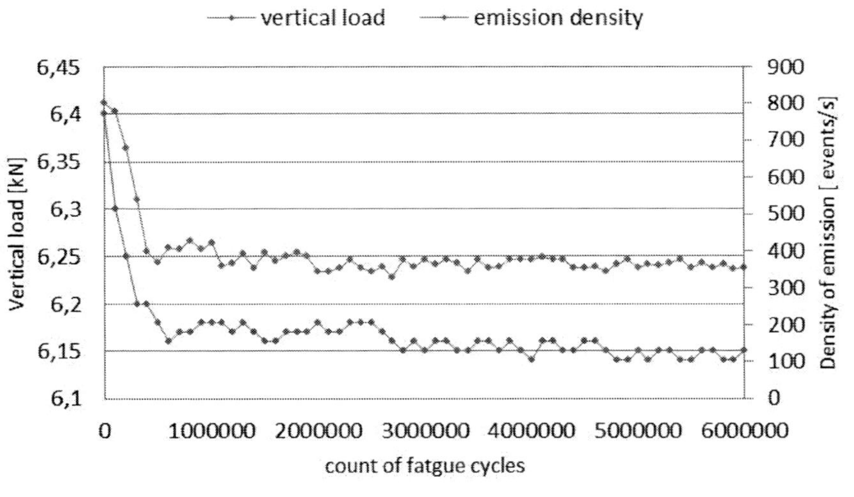

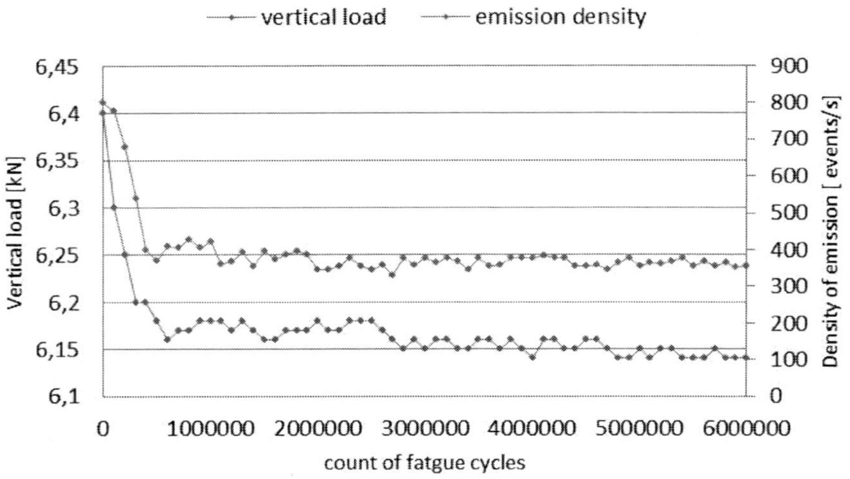


*Fig.11. Changes in the leaf spring bending force, initial load 6.40 kN*

Regardless of the load, during the initial period a decline of the spring stiffness and decrease of the acoustic emission density occurred. During this period the phenomena of accumulation and disclosure of the composite technology failures occurred. With the increase in the number of cycles, the stiffness of the leaf spring and the acoustic emission stabilized and remained at a relatively constant level until the leaf spring destruction.

## Conclusions

The leaf spring destruction process was due to the spring stratification in the neutral axis. Nonetheless the leaf spring is still able to carry the loads smaller by about 20%. The overall durability of the leaf spring can be divided into three distinct periods: the initial decrease in stiffness of the leaf spring, the period of stiffness stabilization, and the final interval where an accelerated decline in the stiffness occurred. The decrease in stiffness of the leaf spring, characteristic for the last period of durability, can be taken as the criterion for destruction of the leaf spring. The lengths of subsequent periods are dependent on the value of the leaf spring bending loads. In the case of the maximum load used in tests, the lack of period of rigidity stabilization was evident. The construction was subject to the explosive growth of micro-cracks within the first load cycles. This resulted in the destruction the leaf spring in a short time.

When the springs were loaded with smaller forces, the decrease of stiffness ended after 300 thousand cycles followed by a stabilization period of stiffness. This phenomenon was caused by the accumulation of existing micro-cracks and discontinuities, introduced into the structure of the leaf spring as the result of technological errors. Simultaneously, the level of internal stresses did not generated the new damages.

The state of structure of the composite from which the leaf springs were made can be monitored using the acoustic emission. Any change in the stiffness of the leaf spring was accompanied by the change in density of the acoustic emission. There is a noticeable regularity between the degree of degradation of the leaf spring structure and the level of emissions. The increase of the degradation level was accompanied always by the increase in acoustic emission. The increase in the level of AE directly before the stiffness decrease is also interesting. Thus, knowing the AE characteristics for the leaf spring, it could be used to diagnose the state of the leaf spring structure.

The use of DSP in the research, based on AE, gives the opportunity to pre – diagnostic research and of leaf springs. Using the signal conditioning is possible not only to test a variety of materials (e.g. polymer), but also the individual parts of the system.

## References


[1] Pozhilarasu V., Parameshwaran, T.Pillai: Performance comparison of conventional and composite leaf spring, International Journal of Engineering Science and Technology (IJEST), Vol. 4 No.12 December 2012, Pages 4827-4831

[2] Abdul Rahim et al :Developing a composite based elliptic spring for automotive applications. Journal of materials and design. 2010, Pages:475-484

[3] H.A. Al–Qureshi: Automobile leaf springs from composite materials. Journal of Materials Processing Technology. 2001, Pages 58 –61

[4] Trebuňa F., Masláková K., Frankovský, P.: Residual stress measurements / - 2011. In: MMaMS 2011 : Modeling of Mechanical and Mechatronical Systems: proceedings of the 4th international conference : Herlany, Slovakia, 20. - 22. 9. 2011. - Košice : TU, 2011 S. 487-491. - ISBN 978-80-553-0731-2

[5] Bełzowski A.: Mechanical degradation of polymer composites. Methods for assessment of the long-term strength and the extent of damage, Publishing House of the Wrocław Technical University, Wroclaw, 2002.

[6] Ochelski S.: Experimental methods of the mechanics of structural composites, WNT, Warsaw, 2004.

[7] Moliński W.: Detection of the formation and development of cracks in the wood using the method of acoustic emission (AE). Annals of the Poznań Agricultural University. Scientific dissertations. Issue 288/1998

[8] Cempel C.: Wibroakustyka stosowana. Warsaw, PWN 1989.

[9] Doppler Laser Vibrometry, http://www.dipmec.unian.it/misure/strumenti/LDV.

[10] Lyons R.G.: "Understanding Digital Signal Processing". Pearson Education 2011.

[11] Singiresu S. Rao: Mechanical Vibrations. Singapore, Prentice Hall 2005.

[12] Markiewicz R.: Dźwięki i fale. Poznań, Wyd. Naukowe Uniwersytetu im. Adama Mickiewicza 2004.

[13] http://www.vallen.de/sites/default/files/sov1212.pdf

[14] Marciniak W.: Przyrządy półprzewodnikowe i układy scalone. Warszawa, WNT 1979.

[15] J. Vavro, M. Kopecký, P. Koštial, H. Hajská, M. Fandáková, The Experimental Measurement of the Tyre Impurities Circulation by the Dynamic Loading and Application of the Experiment Evaluation by the Method of the Active Factor Experiment, Machine Dynamics Problems, Warsaw University of Technology, 2004.

[16] P. Kopas, M. Sága, M. Uhríčik, Three-dimensional S-N curve method to estimate fatigue life of aluminium alloy under combined loading and comparing with fatigue criteria, Technológ 5 (2013) 231-234.

[17] Jančo, R., Écsi, L., Élesztös, P.: FSW Simulation of Aluminium plates by SYSWeld – Part I. The International Conference on Numerical and Experimental Solution of Welding, 15 – 19, (2014)

# SECTION 6

## Modelling and design of mechatronic systems

EAN 2017
55th conference on experimental stress analysis

# HMI of teleoperated robots

Ladislav Jurišica[1, a], František Duchoň[1, b], Martina Szabová[1, c] and Zuzana Mikulová[1, d]

[1] Slovak university of technology in Bratislava, Ilkovičova 3, 812 19 Bratislava, Slovakia

[a]ladislav.jurisica@stuba.sk, [b]frantisek.duchon@stuba.sk, [c]martina.szabova@stuba.sk, [d]zuzana.mikulova@stuba.sk

**Keywords:** HMI, robot

**Abstract.** Automation of discrete manufacturing and service processes requires application of robots. In the article, we discuss the telerobot control, especially focused on applications in technical environment. The main task of HMI is to create the technology functionality clear, readable and comprehensible. HMI has to be designed with regards to human operator. HMI attributes can have an influence to: quality of performed tasks, time necessary to train the operators to perform particular work, create less stressful work environment with minimization of operators' exhaustion, increase of reliability and quality of performed tasks and thus reduction of operation costs.

## Introduction

Robots are becoming more accessible to ordinary users who do not have technical education. This applies not only to household applications, but this trend is well seen in industrial production. Sophisticated robots become now easier to use especially particularly in terms of interaction with them. In the article, we discuss the telerobot control, especially focused on applications in technical environment. The main task of HMI is to create the technology functionality clear, readable and comprehensible. HMI has to be designed with regards to human operator. HMI attributes can have an influence to: quality of performed tasks, time necessary to train the operators to perform particular work, create less stressful work environment with minimization of operators' exhaustion, increase of reliability and quality of performed tasks and thus reduction of operation costs. All these issues are discussed, because it seems that this part of robotics will be very important in the future and not everybody pay the attention to HMI.

## HMI and telecontrol

Easier tasks might be performed by industrial robots with program control, more difficult tasks require so-called adaptive robots and the most difficult tasks are for robots with artificial intelligence. Robots can perform the activity based on created program or program with a possibility of correction according to changes in environment or autonomously [7]. From the point of view of operator, the autonomous systems are advantageous. However, not all of the tasks can be performed by autonomous robots due to the high demands on system control (decision making) hence its complexity and high price. Semi-autonomous systems or systems with various forms of operator intervention to robot's activity are applied. In these systems, it is important for operator's exhaustion reduction to increase level of robot control, that operator interferes to control only if necessary and in highest possible control level, so the robot can perform all routines automatically. It is important for the system of security inspection to be working during interactive control of robot, so the operator cannot input to perform a dangerous operation.

The complexity of autonomous action control by the robots is defined by the necessity of an external environment description – the working area of the robot. That is difficult during programming robot's actions in not determined conditions of the working area. Therefore it is necessary to use telecontrol of the robot [4,10,11,12,13]. It might be so-called mobile manipulator or stationary robot with regards to specific conditions in the working area. While teleoperating, operator has a chance to evaluate conditions in the working area and based on that to control robot's activity. Use of telecontrol is suitable when robot works in specific conditions (e.g. unknown and changing environment, cooperation with a human in the operation, cooperation with other robots, etc.). Telecontrol requires a feedback from working environment, for example inform of visual image, sound, position, orientation, motion dynamics, etc. In number of cases it is needed to use force-torque or tactile sensors.

It is necessary during telecontrol to:

- Define operation stages, which ones should be controlled by operator and which automatically.
- Minimize any risks of incorrect sequence during operation and ensuing dangerous consequences.
- Decide which sensor information is needed for operator to perform operation and recognize his reaction times, which is related to risk of instability caused by presence of human in control loop. It is necessary to deal with the mnemonic quality of the system design and an operator's "transfer function".

**Control architecture**

Control architecture is determined based on general requirements. It also depends on control objectives (position and its derivation, force, impedance) and chosen control (standard, adaptive, with UI elements).

The level of operator's participation in robot control depends not only on the complexity of operations, but significantly on system structure and telecontrol principles. During command, replicating and semi-autonomous manipulator operation, operator continuously controls motion of executive system. Supervised and dialog control allows operator to radically relieve of direct control of the simple and repetitive robot movements [7].

In supervised control, all information about system state is shown on control panel for the operator. Based on this information the operator makes decisions about further actions and sends appropriate control commands. The control system decides on the current control program of the activity with regards to control commands and autonomously performs preprogrammed required actions.

Dialog – interactive systems are another stage of telecontrol. Adaptive control is applied in interactive systems and control program can involve algorithms of artificial intelligence. Control system itself creates solutions, which are presented to operator for approval or asks him for decisions in situations, for which it cannot find solutions alone. Consequently, it checks operator's decisions so it can improve system security (it does not perform an unsecure operation).

Up-to-date use of telecontrol is for example in medicine, for drone control, for control robot in the water, underground, in space, etc. In technical practice, it is using in not determined conditions in extreme environments, where it is necessary to perform complex movements, difficult operations, installation work with use of universal tools and equipment, etc. [5,6,14,15,16,17,18]. Telecontrol is suitable if: tasks are unstructured and not repeated, key parts of task require difficult manipulation (especially eye – hand coordination), object recognition or understanding of situation. The condition is that requirements of information security for operator and transmission of control commands cannot limit the communication line restrictions (bandwidth, time delay, interference).

The necessary condition for reliable operation during telecontrol is an appropriate information and communication system of the robot – Man Machine Interface MMI or Human Machine Interface HMI (Interface consists of hardware and software, that allows user inputs. Those are submitted as signals for the machines, which return desired result of action to operator) [1,8,9]. It is also used the name HCI - Human Computer Interaction [3]. It is a discipline dealing with the design, assessment and implementation of interactive computing systems in terms of users. It also studies main effects in connection with this interaction. Motivation and goals of HCI are to develop and improve: safety, utility, efficiency, effectiveness, usability and attractiveness of the system.

Information and action exchanges in the robot operation between the operator and the robot through a user interface are covered by HRI (Human Robot Interface) [2]. When designing the interface, the limited human sensory, motor and cognitive abilities should be kept in mind. Ergonomics deals with such kind of issues in order to create consistency among the technical solution, product function and adaptation to human abilities and needs. Use of ergonomics during system design leads to: a reduction in incapacity for work and occupational diseases, increase of work performance, reducing errors and spoilage, improve mental state of workers, minimizing signs of mental and physical fatigue. The main areas of ergonomics interest in terms of HMI design are focused on arrangement of control elements and indicators and appropriate use of color and sound. That results to good understanding of information about controlled process by the operator. HMI interface enables operator to take full advantage of his ability to control the process and reduces operator's workload.

Realization of HMI is connected with a sufficient amount of robot sensors and working environment sensors as well as HW and SW tools for information exchange to operator and for entering control commands for the robot and other equipment in working area. One of the options to create HMI is to use virtual or augmented reality. The general block diagram of telerobot use is shown in Fig. 1.

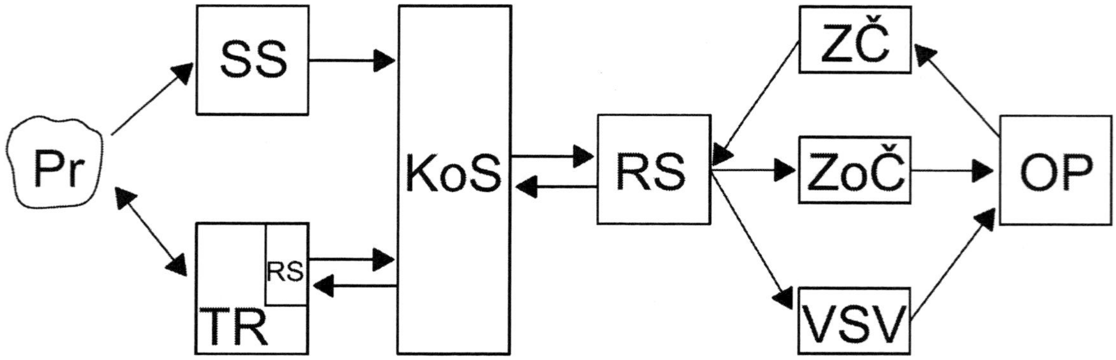


Fig. 1: Scheme of telecontrolled robot – Pr is environment, SS – sensor system, TR – telerobot, RS – control system, KoS – communication system, ZČ – input elements, ZoČ – display elements, VSV – sensory feedback, OP – operator.

If there is a sufficient amount of information from sensors (visual, sound, force-torque, tactile, position, speed, acceleration, orientation... [7,19,20,21]) presented to operator, it can lead to his feeling of physical presence at the same place as robot and sensors are. It is called telepresence. To ensure the perfect telepresence, the information transferred from the robot working area to operator should be for all human senses (in robotics in technical environment it is unlikely to utilize all 5 operator's senses).

The issues during telecontrol are caused usually by the possibility of communication failure, bandwidth limit of the communication channel, the communication delay, the possibility of noise and disturbances in information transmission. The main problem is also the high psychological stress of operator during telecontrol. Problems are caused by that operator is not moving in control workplace. Although robot is moving, the operator often has only a partial view of working area (has no stereopsis). He has to react to high number of various outputs. That highlights the importance of quality of HMI designated for this kind of control.

People obtain 90% of information from the surroundings by sight. Acquisition of visual information by technical resources is associated also to camera movement, focusing, resize the view angle or image spectrum, type of lightning, etc. [24]. For more precise scene recognition, there might be added another data to visual scene for operator from other sensors, e.g. about distance or temperature of different objects in scene, alternatively about forces and torques. An important role during system design belongs to optimal sensor deployment on the robot. It depends on complexity of addressed tasks, level of system intelligence, solo or group deployment of robots, etc. Visual systems are used with stationary cameras, cameras placed on special manipulators, cameras placed on robot's manipulators or the combination of above. Additional features for effective usage of visual system may be provided by special scene lightning. The information for operator is presented on screens or by video glasses. 3D view can be achieved also by augmented reality.

The use of sound is technically quite simple, but on the other hand the record of sound can be affected by other audio sources, which are unrelated to telecontrol activity. When the information transfer is bi-directional, it is the natural way of communication for the operator. The complexity of sound recognition is in accurate and reliable recognition of content and meaning of the communication – input of commands for performing activity. For complete sound and voice communication it is necessary to resolve the question of secure and reliable decode of commands from operator and also the speech synthesizer for exchange data from sensors to operator.

Other essential sensors for telecontrol are touch sensors, force sensors, torque sensors, position sensors, speed sensors and acceleration sensors [22,23]. This information can be provided to operator on display panels or some other data (force, torque, touch) directly by actuators of appropriate kinematic scheme at the operator's workplace. Feedback force-torque effect on operator needs a suitable kinematic solution of control elements at his workplace too. Within this kind of control, transformation of forces and torques size can be performed from the working area to operator. Such control increases the quality of telepresence, operation performance by the operator has high accuracy and decreases the risk of operator's exhaustion.

For mobile telerobots it is essential to be equipped in addition to above by sensors for localization and navigation. For technical environment, the essentials are mainly visual systems, optical systems, laser systems, ultrasonic sensor systems, or even GPS.

Depending on the complexity of the transferred information, the requirements apply to communication system. In complex applications, it is desired to have multi-channel system with high-speed information transfer with high resistance to interference. There are applications where cable connection between operator and the robotic system might be used, but there are also ones where the wireless data transfer is required. Examples of telerobotic systems are shown in Fig. 2-4.

Fig. 2: Telerobot Scorpio - Ability of underpass the car, ability of explosive device disposal without the need of manipulation with it, possibility of remote control including the video transmission by radio signal, possibility of remote control including the video transmission through cable connection, simple control, obstacle avoidance and overpass, possibility of control and reconnaissance by visual systems, measuring of distance between obstacles and telerobotic system, detection of explosive device and measuring of distance between this device and muzzle of water gun and consequent disposal, possibility of simple implementation of additional extensions.

Fig. 3: Telerobot Teodor – main parameters: length 300 mm, width 685 mm, height 1 240 mm, weight 375 kg, speed max. 3 km/h, payload 350 kg, operation time 3-4 hours. Control station: weight 86 kg. Control panel: weight 9 kg.

Fig. 4: Mobile manipulator MT-15KM - all movements are provided by electric motors. Main parameters: rotation + 45°, main arm lift - 30° to + 105°, auxiliary arm lift + 105°, wrist rotation ± n x 360°, jaws opening max. 260 mm, weight 86 kg, payload - unloading 1,5 m - 15 kg, payload - unloading 0,5 m - 40 kg, installed power 475 W.

**Conclusions**

Telecontrol is still in focus of research institutions. Decision-making possibilities of the operator, processing the amount of information about the complex and changing environment, speed of evaluation the system state and the solution acceptance by a technical system are the things that still prefer operator's participation on control instead of fully-autonomous system. We can conclude the rapid development in all telerobotics subsystems. There is a signifficant increase of computing power, progress in sensor parameters, which improves HMI features, simplifies system control, reduces demands on operator training and exhaustion of operator during his work.

**Acknowledgments**

This work was supported by grants APVV-14-0894, VEGA 1/0065/16, and Req-00347-0001.

**References**

[1] Gérer,A.: HMI – nové poznatky a najlepšie skúsenosti pri tvorbe operátorského rozhrania. seriál, ATP Journal, 2013,

[2] Human-Robot Interaction. A Research Portal for the HRI Community. http://humanrobotinteraction.org/

[3] Šperka,M.: Interakcia človek - počítač. Podporný materiál k predmetu Interakcia človeka s počítačom, prednášanom v zimnom semestri 2009 na FI MU v Brne.
[4] Aracil,R., Buss,M., Cobos,S., Ferre,M., Hirche,S., Kuschel,M., Peer,A.: The Human Role in Telerobotics. In.: Advances in Telerobotics, pp.11-24, 2007.
[5] Sotiris Avgousti, Eftychios G. Christoforou, Andreas S. Panayides, Sotos Voskarides, Cyril Novales, Laurence Nouaille, Constantinos S. Pattichis, Pierre Vieyres: Medical telerobotic systems: current status and future trends. BioMedical Engineering OnLine2016
[6] Garrec,P., Perrot,Y.: Force feedback nuclear telerobotics in France: R&D results and industrial achievements. ceatech Interactive robotics, CERN, 2015.
[7] Jurišica,L., Hubinský,P., Kardoš,J.: Robotika, FEI STU Bratislava, 2005. https://www.researchgate.net/profile/Peter_Hubinsky/publication/267783300_Robotika/links/54b2357e0cf2318f0f93f383.pdf
[8] https://en.wikipedia.org/wiki/HMI
[9] https://en.wikipedia.org/wiki/Human%E2%80%93robot_interaction
[10] https://en.wikipedia.org/wiki/Telerobotics
[11] http://www.learnaboutrobots.com/telerobotics.htm
[12] http://www.ieee-ras.org/telerobotics
[13] https://www.nap.edu/read/4761/chapter/13
[14] http://www.krafttelerobotics.com/
[15] http://www.euroc-project.eu/index.php?id=559
[16] http://telerobotics.weebly.com/uses-and-types.html
[17] http://bionics.seas.ucla.edu/publications/BC_07.pdf
[18] http://www.techradar.com/news/world-of-tech/future-tech/telepresence-rise-of-the-telerobots-1184036
[19] https://www.intorobotics.com/types-sensors-target-detection-tracking/
[20] Intelligent Robots: A Feast for the Senses . RIA. http://www.robotics.org/content-detail.cfm/Industrial-Robotics-Industry-Insights/Intelligent-Robots-A-Feast-for-the-Senses/content_id/5530
[21] Muthukkumar Kadavasal Sivaraman, James H. Oliver: Sensor Augmented Virtual Reality BasedTeleoperation Using Mixed Autonomy. Iowa State University.2009. http://lib.dr.iastate.edu/cgi/viewcontent.cgi?article=1047&context=me_pubs
[22] Jánoš Rudolf, Hajduk Mikuláš, Semjon Ján, Šidlovská Ľuboslava: Design of Hybrid Mobile Service Robot. In: Applied Mechanics and Materials, Vol. 245, 2013, pp. 255-260.
[23] Židek Kamil, Líška Ondrej: Accelerometers usability for danger tilt off-highway vehicles and signal filtration with kalman filter. In: Journal of Applied Science in the Thermodynamics and Fluid Mechanics, Vol. 4, no. 2, 2010, pp. 1-6.
[24] Hargaš Libor, Hrianka Miroslav, Koniar Dušan: Image processing and analysis : a practical approach - text book. Žilina : University of Žilina, 2008.

EAN 2017
55th conference on experimental stress analysis

# Improvements of VFH* Method

Andrej Babinec[1,a], Marian Kľúčik[1,b] and Zuzana Mikulová[1,c]
[1]FEI STU, Ilkovičiva 3, 812 19 Bratislava, Slovakia
[a]andrej.babinec@stuba.sk, [b]marian.klucik@stuba.sk, [c]zuzana.mikulova@stuba.sk

**Keywords:** Mobile robot, navigation, moving obstacles.

## Introduction

A mobile robot operating in unknown environment should have got a reliable control system [1] with implementation of a reactive navigation as well as a path planning. The reactive navigation is necessary because it allows the robot to avoid unknown obstacles and therefore it protects the robot and environment against any mechanical damages [2,3]. Forasmuch as current robots have modern sensors on board, e.g. laser rangefinders or visual systems, they have wider knowledge about the larger surrounding. This is why the local forms of path planning have to be employed. It makes the robot motion more fluent and produces shorter trajectories.

There are many algorithms focused on reactive navigation and path planning. In particular, Vector Field Histogram methods [4,5,6] are unique because they are very effective in obstacle avoidance and they are easily customizable. Even more, VFH* method yet combines reactive navigation with local path planning into one complex algorithm. Unfortunately, these algorithms are natively designed for static environment so moving objects are in most cases impassable obstacles.

Further in the article, a new reactive navigation of mobile robot based on VFH* method is presented. The algorithm is modified so that the robot is able to plan a local avoidance maneuver even in the presence of moving obstacles. The modification consists of several improvements, which are described in following section.

## Modifications in the Method

Considering the non-static environment it was necessary to design how to include the information about mobile obstacles into the algorithm's computation. Recognition of moving objects is discussed in many papers [7,8,9]. However, for the purpose of reactive navigation it is necessary to know only a final representation of moving objects, because they enter the decision-making process. So we need to create some form to store information about the obstacles suitable for further processing during local path planning. Details are presented in first subsection.

The original VFH* method includes VFH+ method, which is its essential component. In VFH+ method, polar histograms are created from local grid map. The histograms consist of $n$ bins. Each bin corresponds to particular angle sector coming out from the robot position. The near space around the robot (active window) is therefore divided into $n$ angle sectors. Each histogram bin carries the value proportional to the amount of occupied cells in that direction. The potential passages are identified in valleys of the primary histogram. Then the single one passage from the set of identified passages is chosen that leads the robot to the goal along the smooth and safe trajectory. For the purpose of taking a robot size into account and for simplifying it to a point, the occupied cells in a grid map are enlarged by the circle with radius $r_{r+s}$ and the enlargement angle $\gamma_{i,j}$ is defined by Eq. 1 [5]. For each obstacle cell in the map, the angle $\gamma_{i,j}$ designates the membership of the sector $k$ by specific rules. These rules had to be modified in order to adapt the method to dynamic environment. Second subsection of this section is focused on this modification in geometrical constrains and its consequences.

$$\gamma_{i,j} = \arcsin \frac{r_{r+s}}{d_{i,j}} \tag{1}$$

The VFH* method produces look-ahead tree through the free space in near robot surrounding based on potential passages given by VFH+. The optimal branch in the look-ahead tree predicts the future change of the robot's direction. As this change is known at the end of the tree development stage, it is convenient to process it and properly deflect the robot's next direction. These improvements are contained in third subsection.

**Representation of Mobile Obstacle.** In histogramic algorithms, static objects are stored in a grid map that is given as an input to the navigation algorithm. Besides, moving objects have to be included. While the laser rangefinder is utilized in many robotic systems, it is convenient to use the laser data directly in moving objects representation. The output of the rangefinder is an array of couples ($\varphi_i$, $d_i$). It describes the sensed distance to an obstacle in a particular direction. Moving obstacle can be detected after an appropriate filtration and segmentation. Then the set of distances measured towards the moving obstacle from the point where the obstacle was recognised for the first time is included to a data structure in Eq. 2 along with the robot position and the recognized velocity vector of the obstacle. The symbols are explained by the Fig. 1, where $x^R$, $y^R$, $\varphi^R$ denote the robot position where the obstacle was detected for the first time, $k$ is the index of the first laser ray that belongs to the obstacle and $n$ is the number of the laser rays that belong to the obstacle.

$$O^m = \{x^R, y^R, \varphi^R, \{d_k,...,d_{k+n}\}, k, n, v_D, \varphi_D\}. \tag{2}$$

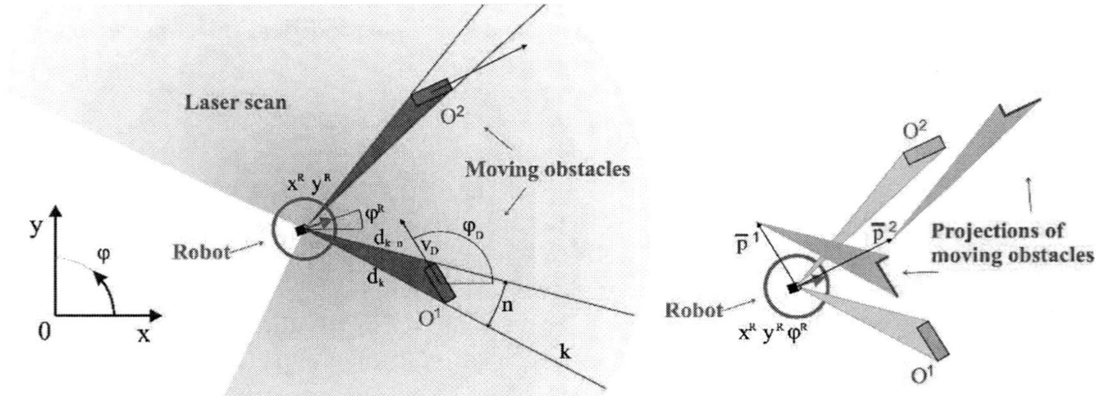


Fig. 1 The laser scanner segments which serve as the representations of moving obstacles, and their example of projections.

This representation allows preserving the original object shape. The objects are inserted to the grid map just when they have to be presented during a computation. After the processing, the objects are removed from the grid.

This approach allows to simulate and predict a motion of the moving obstacle $O^m$ by shifting the reference point ($x^R$, $y^R$) along the projection vector $\bar{p}^m$, which goes out from the reference point in the direction of obstacle motion $\varphi_D$. Therefore, it is necessary to take into consideration also a time difference between current state and predicted state. Inclusion of this time difference into the computation of the look-ahead tree is crucial improvement of the method VFH*, which allows employment of the method into the non-static environment.

**Improvement in Geometrical Constraints.** A definition of so called enlargement angle in original VFH* does not count with the possibility that the robot and obstacle could intersect and it is valid only for $d_{i,j} \geq r_{r+s}$. In the virtual scene, when the algorithm predicts motion of mobile obstacles it is possible that predicted projection of the obstacle hit the projection of the robot. Such a node in look-ahead tree that leads to the collision has to be closed so the optimal branch of the tree should be allowed to grow through other passages in free space. It is necessary that these virtual collisions are mathematically possible in order to be recognised. Therefore, the definition of the enlargement angle was extended, which can be seen in Eq. 3.

$$\gamma_{i,j} = \begin{cases} \arcsin \dfrac{r_{r+s}}{d_{i,j}} & if \quad d_{i,j} \geq r_{r+s} \\ \dfrac{\pi}{2} & if \quad r_r < d_{i,j} < r_{r+s} \\ \pi & if \quad d_{i,j} \leq r_r \end{cases} \tag{3}$$

**Improvement of Final Decision Step.** During the stage of look-ahead tree development, a direction $\theta_i$, in which the robot projection came from the node $i$-1 to node $i$, is stored. After determination of optimal branch, directions $\theta_i$ of the branch are compared with each other. The direction that is significantly different from current robot orientation $\varphi^R$, affects a future robot's motion mostly. The node $i$ with maximal $\delta_i$ according Eq. 4 is called a deflecting node. Then the direction from the root node in position $(x^R,y^R)$ to the deflecting node is called a deflecting direction $\theta_d$ (see Fig. 2).

$$\delta_i = \left|\varphi^R - \theta_i\right| \tag{4}$$

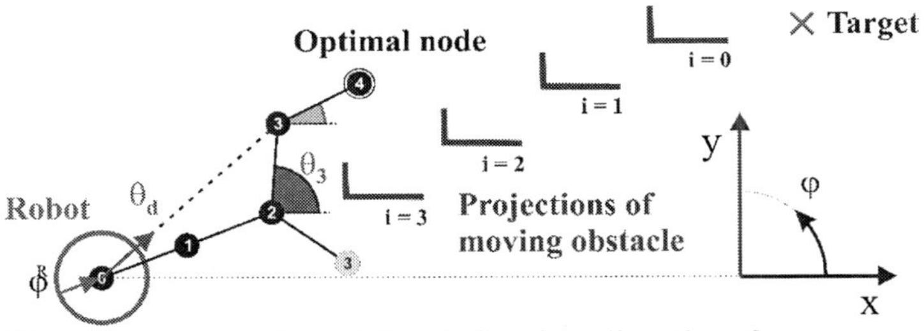

Fig. 2 Determination of the deflecting direction $\theta_d$.

If the deflecting direction respects the restrictions in the primary node (root) given by the histograms of VFH+ method, it becomes a desired direction for the next robot step. If $\theta_d$ lies outside of optimal opening, another direction is used according the Fig. 3.

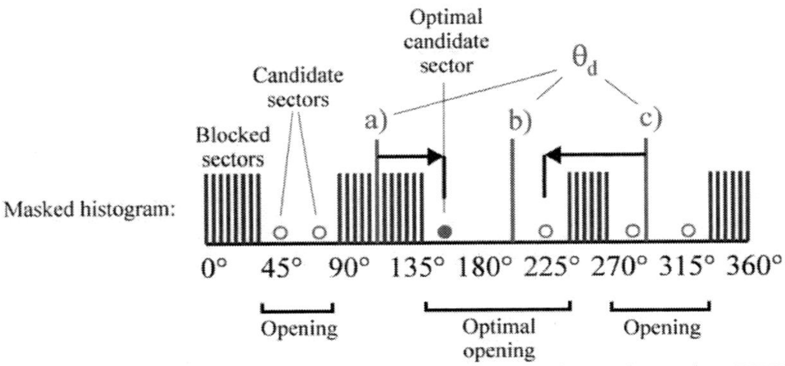


Fig. 3 Restrictions for the deflecting direction given by VFH+ method. $\theta_d$ lies: a) in blocked sector – the nearest candidate sector of the optimal opening is used; b) in optimal opening between its marginal candidate sectors – $\theta_d$ is used; c) in another opening – the nearest candidate sector of the optimal opening is used.

## Verification

The proposed improved method was implemented to our indoor differentially driven mobile robot, which was equipped with laser scanner. During experiments all process data had been recorded in order to be visualized. In following figures, light blue large circle represents the robot and blue curve is its overall recorded path from start (asterisk) to goal (cross). Moving obstacles are depicted by colored square cells of the grid map. Green cells numbered by zero represent the actual position of the obstacles in time $t$ along with actual robot position in the plane. Other colored squared cells represent projected positions of the obstacles into future using respective time differences. These projections are utilized during look-ahead tree development. In each node of the tree (small circles), there are only corresponding projections of the obstacles (with the same number and color) taken into computation of next node. Blue node is the optimal node on the optimal branch. Deep-violet arrow means actual robot direction $\varphi^R$. Light-violet arrow is the deflected direction $\theta_d$ after consideration of all restrictions.

In the experiment, three moving obstacles were sent in different directions so that they should intersect expected path of the robot. In Fig. 4, there are shown three instants of time with visualized data that had been recorded during the experiment. The corresponding look-ahead tree is shown for each instant of time. The development of the tree was affected by projections of the moving obstacles. It is clear that optimal branch is bending in such way that the algorithm leads the robot to avoid the obstacles in their predicted positions.

A performance of the proposed method is compared with original VFH*. It is obvious, that VFH* in Fig. 5 failed because it led the robot into the collision with the mobile obstacle in time 11.5s. This unmodified method does not create projections of obstacles so it reacts only on green ones with zero labels which are in this case considered to be static obstacles.

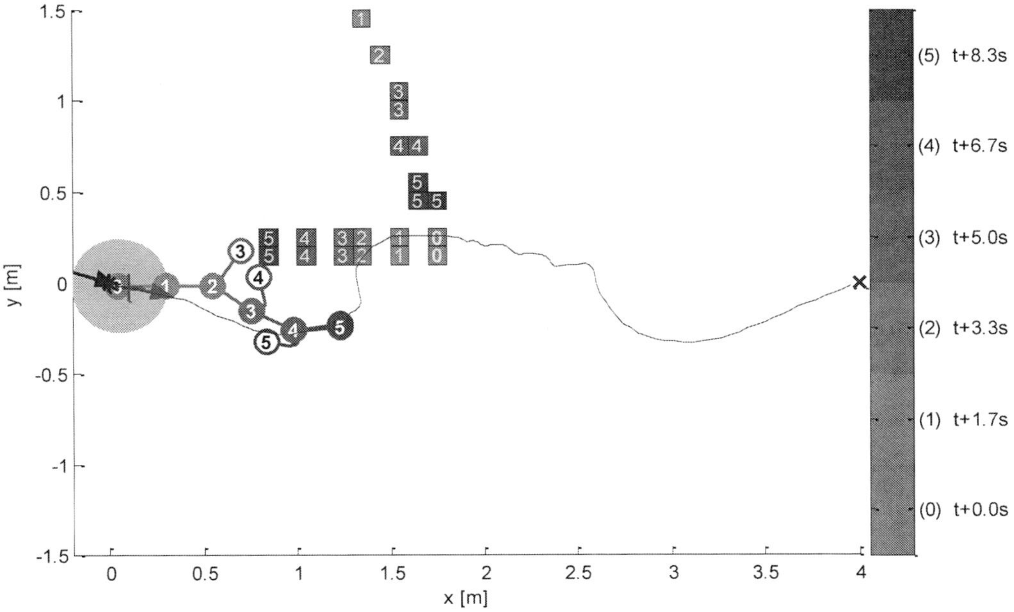
y [m]
x [m]
1.5
1
0.5
0
-0.5
-1
-1.5
0
0.5
1
1.5
2
2.5
3
3.5
4
(5) t+8.3s
(4) t+6.7s
(3) t+5.0s
(2) t+3.3s
(1) t+1.7s
(0) t+0.0s

(a)

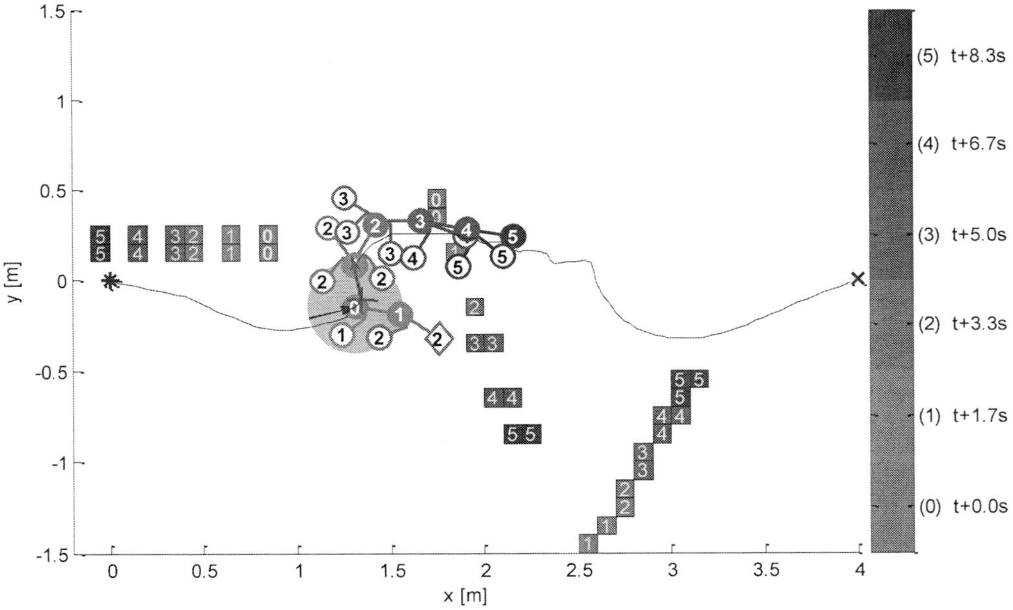
y [m]
x [m]
1.5
1
0.5
0
-0.5
-1
-1.5
0
0.5
1
1.5
2
2.5
3
3.5
4
(5) t+8.3s
(4) t+6.7s
(3) t+5.0s
(2) t+3.3s
(1) t+1.7s
(0) t+0.0s

(b)

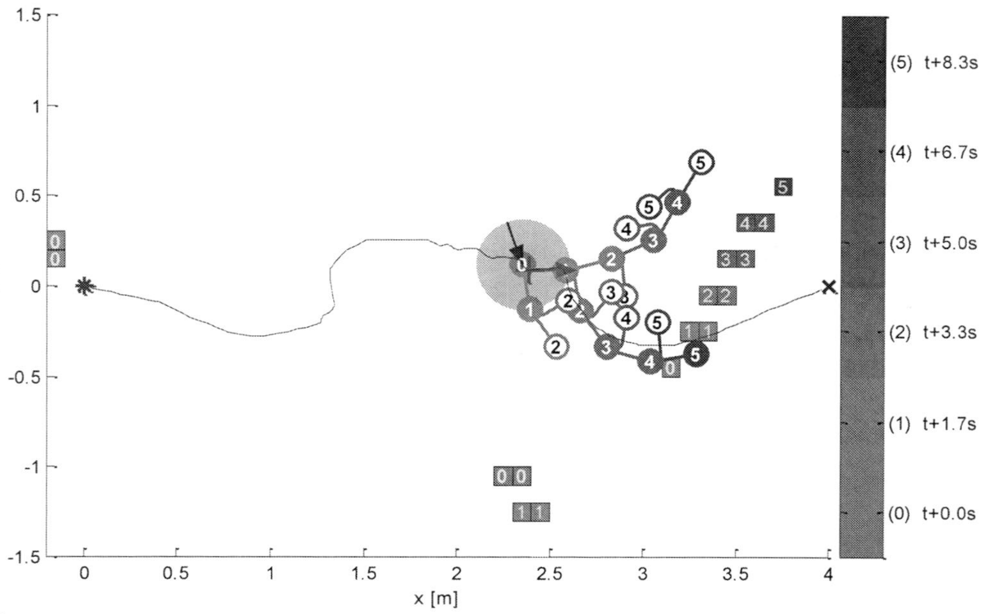


(c)

Fig. 4 Performance of improved VFH* in the presence of three moving obstacles. Three moments are displayed: a) $t$=0.3s , b) $t$=9.3s , c) $t$=18.7s

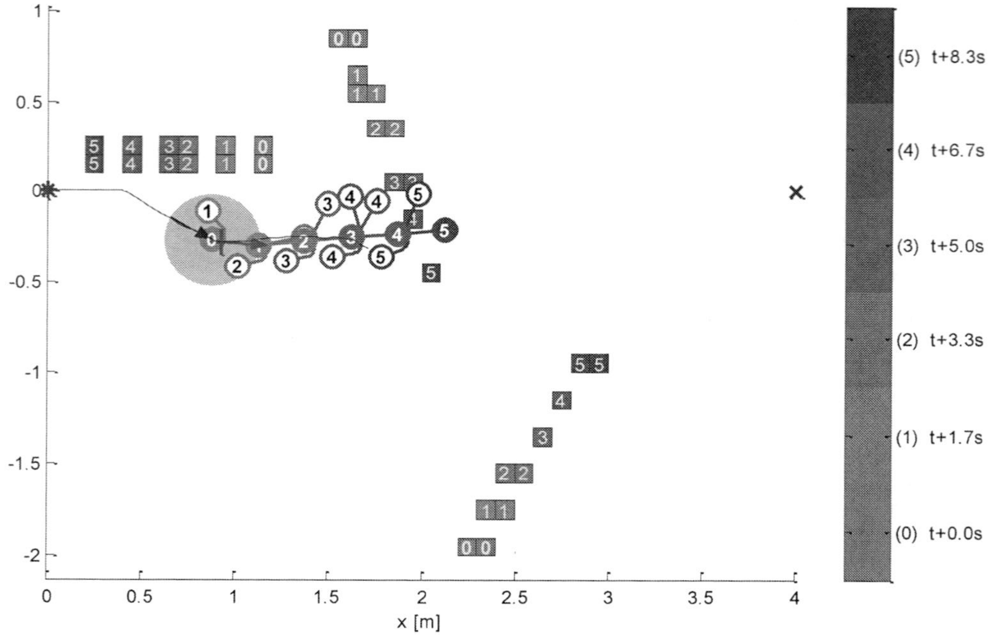


Fig. 5 Performance of original VFH* in the presence of three moving obstacles. A time stamp of the developed look-ahead tree is $t$=6.3s, the collision occurred in the time 11.5s.

**Conclusions**

The experiments in the article show that the modified VFH* is capable of collision-free navigation of the robot to the target also in environment with multiple obstacles. On the other hand, original VFH* failed because it does not take into account the time-varying environment during tree development stages. Nevertheless, mainly the navigation part of the problem is studied in this paper. Although the uniform rectilinear obstacle motion is assumed, the method is not dependent on this type of motion in general. In real world, obstacles may move along arbitrary trajectories and their movement may be predicted using some extrapolation method or using an algorithm that regards a probability of changing the obstacle direction. Therefore, further research will be focused on reliable detection of dynamic objects, their classification and prediction of their movement.

**Acknowledgement**

This article was created with the support of the Ministry of Education, Science, Research and Sport of the Slovak Republic by grants VEGA 1/0752/17 and VEGA 1/0065/16.

**References**

[1] A. Simonová, L. Hargaš, D. Koniar, Z. Loncová and M. Hrianka, Mathematical analysis of complete operation cycle of a system with two-position controller, *ELEKTRO*, Strbske Pleso, 2016, pp. 213-215, doi: 10.1109/ELEKTRO.2016.7512067

[2] Ľ. Miková and M. Kelemen, Concept of Mobile Robots for Movement in an Unknown Environment, *American Journal of Mechanical Engineering*, Vol. 4, No. 7, 2016, pp 262-265.

[3] M. Kelemen, I. Virgala, T. Kelemenová, Ľ. Miková, P. Frankovský, T. Lipták and M. Lörinc, Distance Measurement via Using of Ultrasonic Sensor, *Journal of Automation and Control*, 2015, Vol. 3, No. 3, pp. 71-74, DOI:10.12691/automation-3-3-6

[4] J. Borenstein and Y. Koren, The Vector Field Histogram – Fast Obstacle Avoidance for Mobile Robots, in *IEEE Journal of Robotics and Automation*, 7/3, 1991, pp. 278-288.

[5] I. Ulrich and J. Borenstein, VFH+: Reliable Obstacle Avoidance for Fast Mobile Robots, in *Proceedings of the IEEE International Conference of Robotics and Automation*, Leuven, Belgium, 1998.

[6] I. Ulrich and J. Borenstein, VFH*: Local Obstacle Avoidance with Look-Ahead Verification, in *Proceedings of the IEEE International Conference of Robotics and Automation*, San Francisco, CA, 2000.

[7] R. Labayrade, C. Royere, D. Gruyer and D. Aubert, Cooperative Fusion for Multi-Obstacles Detection With Use of Stereovision and Laser Scanner, in *Autonomous Robots*, Vol. 19, 2005, pp. 117-140.

[8] D. Ferguson, M. Darms, C. Urmson and S. Kolski, Detection, Prediction, and Avoidance of Dynamic Obstacles in Urban Environments, in *IEEE Intelligent Vehicles Symposium*, 2008, pp. 1149-1154.

[9] A. Ess, B. Leibe, K. Schindler and L. van Gool, Moving Obstacle Detection in Highly Dynamic Scenes, in *IEEE International Conference on Robotics and Automation*, Kobe, 2009, pp. 4451-4458.

EAN 2017
55th conference on experimental stress analysis

# Virtual Instrumentation in the Diagnostic and Control Process for Industrial Applications

Dušan Koniar[1a], Libor Hargaš[1b], Matúš Danko[1c], Michal Taraba[1d], Tomáš Uriča[1e], Anna Simonová[1f] and Branislav Hanko[2g]

[1]Dpt. of Mechatronics and Electronics, University of Žilina, Univerzitná 1, 010 26 Žilina, SK

[2] BH Motorsport, Družstevná 6, 038 53 Turany, SK

[a]dusan.koniar@fel.uniza.sk, [b]libor.hargas@fel.uniza.sk, [c]matus.danko@fel.uniza.sk, [d]michal.taraba@fel.uniza.sk, [e]tomas.urica@fel.uniza.sk, [f]anna.simonova@fel.uniza.sk, [g]info@bhmotorsport.sk

**Keywords:** virtual instrumentation, process control, diagnostics, industry applications

**Abstract.** This paper brings a look on integration of virtual instrumentation to the industrial (mechatronic) applications. Virtual instrumentation conception enables higher level of user´s flexibility, system manageability and also cost saving. We designed two mechatronic systems with integrated virtual instrumentation and these systems are primarily intended for educational purposes (for specialization automation and mechatronics), but also they can be implemented immediately in industry. Replacement of human labour with automated approach follows the actual trends in many branches of industry.

## Introduction

Nowadays, virtual instrumentation (VI) serves in many applications across several fields in science and industry. The simplest definition of VI says, that VI is replacement of huge amount of conventional measuring devices with software (virtual) one using universal I/O hardware (measuring cards) and suitable sensors [1]. VI also can take control over the processes in manufacturing, automation and industry with driving actuators (based on the processing of acquired information and signals from process). In addition, implementation of VI reduces the costs (usually the one universal instrument against several specialized device); work area is also more manageable in sense of VI's flexibility, complexity and modular solution.

Very important part of process in manufacturing or industry is diagnostics or automated measurement of critical quantities. Automated approach using VI can significant decrease the human faults in this task.

In this paper we want to introduce two models, where the integration of VI in mechatronic/industry solution can be present. First model is used for diagnostic purposes in car industry and also could be used for demonstrating functionality and communication of various parts in car (sensors, electric devices, mechanical devices) with control unit. The networking is based on CAN (Control Area Network) bus [2]. This model was developed also for educational purposes with support of BH Motorsport, Turany (Slovakia). Second model is based on machine vision algorithms for quality control in industry (contactless measurement, inspection and sorting). The key element is an inspection camera sensor and designing of robust algorithm for image processing and understanding.

In the next section, models and their functionality will be presented.

**Car diagnostic model based on CAN networking and VI**

As mentioned in the Introduction, this model based on VI and CAN (Fig. 1) is dedicated for car diagnostic and also for demonstrating car communication between parts and control unit (suitable for education). This model can operate in two modes:

- linked with real car control unit via CAN;
- link with gateway (via CAN) simulating car control unit (simulation mode).

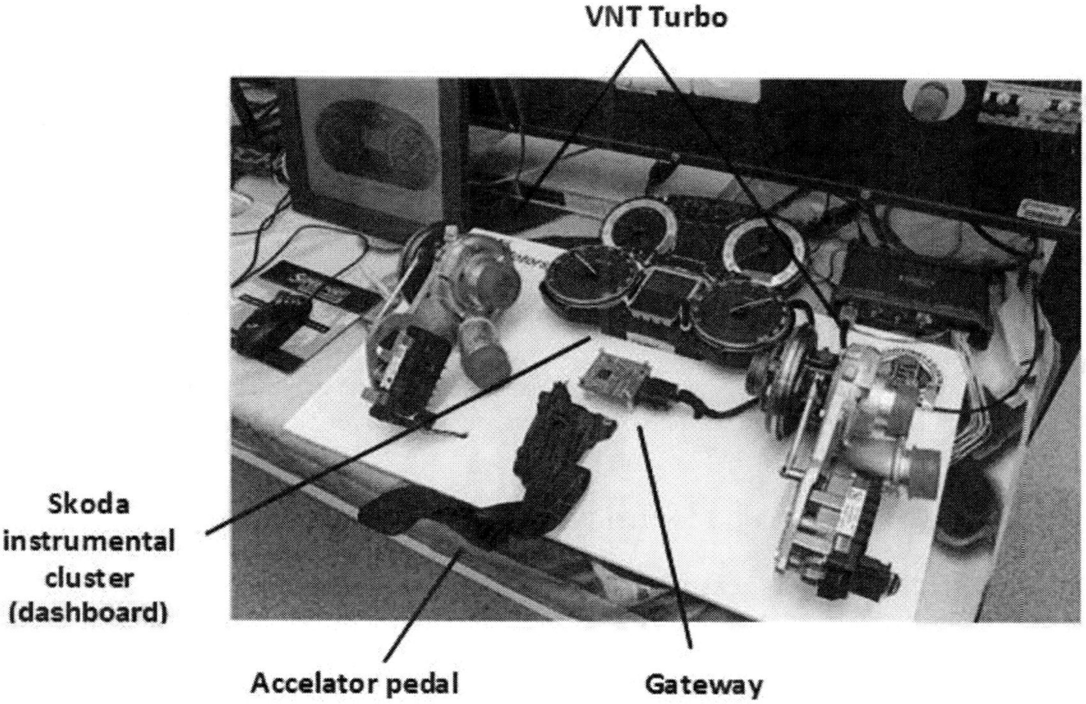


*Fig. 1 Car diagnostic model*

In both modes, National Instruments cRIO (FPGA controller) with high-speed duplex CAN channel and touch panel is a display portal for visualizing the processes.

The main part of model in simulation mode is **gateway** which simulates CAN message transmitted from engine control unit. Simulated CAN messages contain data for processing engine revolutions, speed of vehicle, coolant temperature and accelerator pedal position.

Second element is **accelerator pedal**. The accelerator pedal position sensor detects the amount of travel of the accelerator pedal. This sensor outputs a voltage signal, which corresponds to the amount of pedal travel, to the A/D converter of the gateway. The accelerator-pedal module (APM) comprises an accelerator pedal and a potentiometer or a non-contacting Hall sensor as angular-position sensor. This sensor registers the movement and the position of the accelerator pedal. From this information, the engine management calculates the required torque and accordingly addresses the throttle device and the injection system. The accelerator-pedal module can output analog or digital signals [3].

Next section is two **turbochargers with variable geometry**. In a VNT turbo, the exhaust flow through the turbine wheel is controlled by a row of vanes that move to match the exact boost requirements of the engine. At low engine speed, the variable nozzle turbine vanes close to restrict flow of exhaust gases through the turbine, thereby increasing boost pressure. At high engine speeds, the vanes open to maximize the exhaust gas flow, thereby avoiding turbo over-speed and maintaining the boost pressure required by the engine. Position of the vanes is controlling by servo engines directly through CAN bus.

The last segment of model is **instrument cluster of Škoda Octavia**. Instrument cluster (dashboard) shows only speed, rpm and coolant temperature. Messages of other indicators are not simulated and fuel gauge is analog.

**CAN communication**

The Controller Area Network (CAN) is a serial communications protocol that efficiently supports distributed real-time control with a very high level of data integrity. Though conceived and defined by BOSCH in Germany for automotive applications, CAN is not restricted to that industry. CAN fulfils the communication needs of a wide range of applications, from high-speed networks to low-cost multiplex wiring. For example, in automotive electronics, engine control units, sensors and anti-skid systems may be connected using CAN, with bit-rates up to 1 Mbit/s. At the same time, it is cost effective to build CAN into vehicle body electronics, such as lamp clusters and electric windows, to replace the wiring harness otherwise required [2].

The content of a message is described by an identifier. The identifier does not indicate the destination of the message, but describes the meaning of the data, so that all nodes in the network are able to decide by message filtering whether the data is to be acted upon by them or not. Within a CAN network, it is guaranteed that a message is accepted simultaneously either by all nodes or by no node. Thus, data consistency is a property of the system achieved by the concepts of multicast and by error handling.

Message transfer between nodes (sensors, parts) is manifested and controlled by four different frame types:

- A **Data frame** carries data from a transmitter to the receivers.
- A **Remote frame** is transmitted by a bus node to request the transmission of the Data frame with the same identifier.
- An **Error frame** is transmitted by any node on detecting a bus error.
- An **Overload frame** is used to provide for an extra delay between the preceding and the succeeding Data or Remote frames [4].

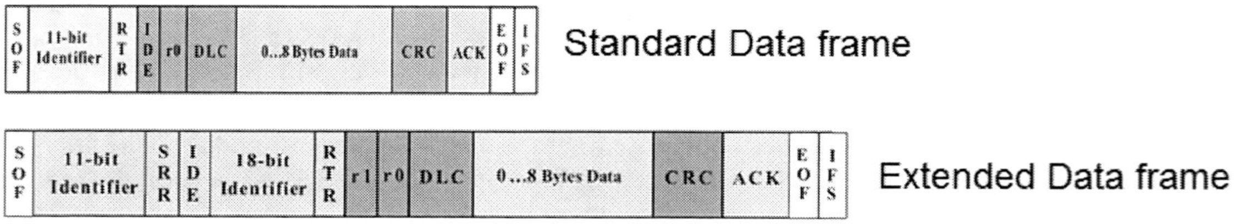


*Fig. 2 Vision inspection model for industrial application*

Two types of Data frames (Fig. 2) are defined in CAN protocol. Standard Data frames can contain 8-byte data and are used for simple CAN structure, on the other side, Extended Data frame is used for communication where the nodes are divided into groups and classes.

Whenever the bus is free, any node may start to transmit a message. If two or more nodes start transmitting messages at the same time, the bus access conflict is resolved by bit-wise arbitration using the identifier. The mechanism of arbitration (Fig. 3) guarantees that neither information nor time is lost. If a Data frame and a Remote frame with the same identifier are initiated at the same time, the Data frame prevails over the Remote frame. During arbitration, every transmitter compares the level of the bit transmitted with the level that is monitored on the bus.

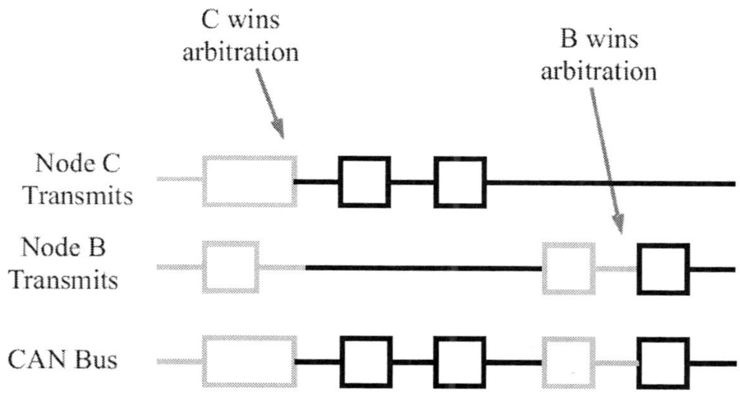


*Fig. 3 Arbitration of CAN bus*

If these levels are equal the node may continue to send. When a recessive level is sent, but a dominant level is monitored, the node has lost arbitration and must withdraw without sending any further bits [2].

**Visualizing the data from CAN bus**

Visualization of the data acquired from CAN bus of a model is done in LabVIEW application as a virtual instrument (virtual Škoda dashboard) (Fig. 4) and also physically on real instrumentation cluster of Škoda Octavia. Virtual instrument runs on NI TouchPanel.

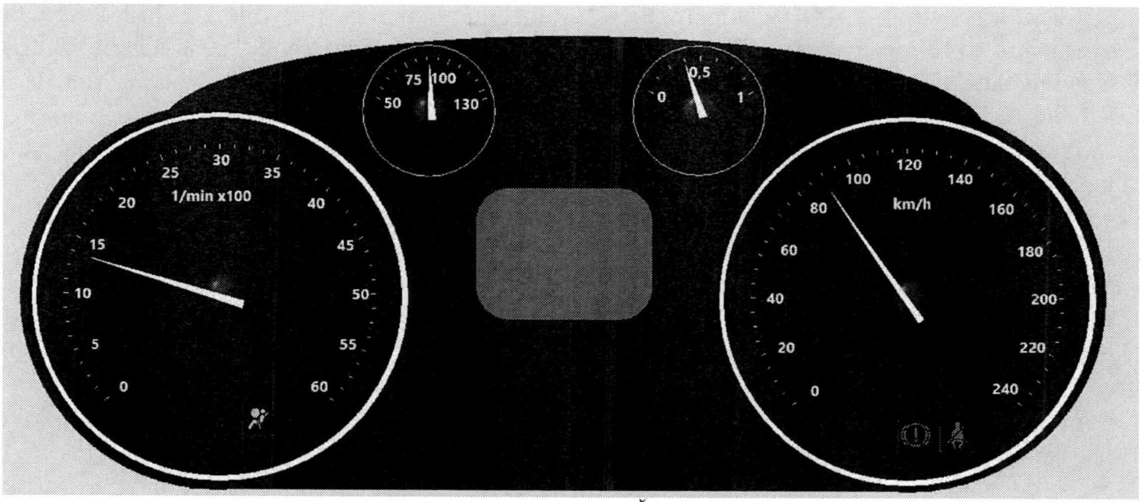


*Fig. 4 Virtual instrument – Škoda dashboard*

The I/O part of virtual instrument consists of NI CompactRIO 9082 FPGA controller with NI 9853 high-speed duplex 1 Mb/s CAN bus module (Fig. 5) [5] – [8].
In the LabVIEW virtual instrument, we selected relevant Standard Data frames acquisition with relevant identifiers and data content. The main part of application was frame filtering for sorting the data for individual indicators of the screen.

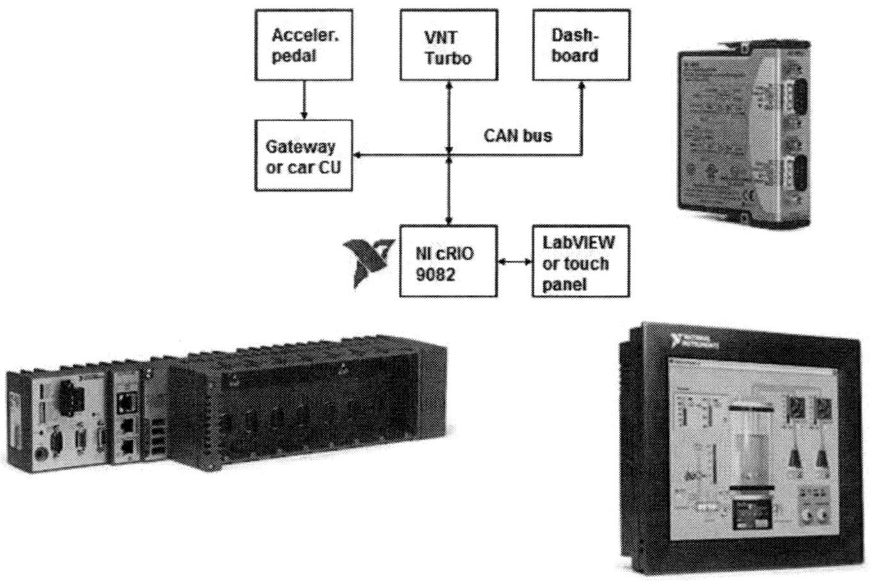


*Fig. 5 Interconnection of parts in model*

In some indicators we had to normalize or interpolate the values (e.g. coolant temperature) from raw sensor data.

**Visual inspection system for industry applications based on VI**

Visual inspection can be defined as a common method of quality control, data acquisition, and data analysis. Visual inspection provides quality and quantity checking and control of components.

Visual inspection is simple as compared to other methods. Despite this, it still has several advantages over more high-tech methods. Compared to other methods, it is far more cost effective. For similar reasons also one of the easiest techniques to perform inspection. It is also one of the most reliable techniques. The main tasks of visual inspection are control product type, quantity, dimensions, completeness, obvious defects, colour, material and inspection speed.

Before a system is selected the all requirements for the machine vision system should be clearly defined. Not all criteria can be met by all system types so that the choice is quickly narrowed down to the appropriate test system. The starting point for this consideration is the object to be inspected and the process environment in which the system must be installed. Detailed knowledge of this is essential to establish a stable plant process and to implement an economically viable application. [9] – [11]

A variety of basic conditions and parameters must be weighed for the selection of the image processing system (Fig. 6):

The description of created video inspection system is presented in the following paragraph. The hardware and software platform for machine vision is virtual instrumentation. The main part of visual system is a camera. The camera must be suitable for required task. The IEEE 1394 camera MARLIN F-046B was used for this visual inspection system. This camera based specification is: grayscale, 60 fps (frames per second) and resolution 780x582. The goal of video inspection system is sort and quantified components such as screws, screw pads and nuts. In application the quantified element can be selected and algorithm then search the captured scene. Search algorithm is based on matching techniques. The most important technique is pattern matching technique. This algorithm is based on cross correlation algorithm (1).

$$R(i,j)=\frac{\sum_{x=0}^{L-1}\sum_{y=0}^{K.1}\left(w(x,y)-\overline{w}\right)\left(f(x+i,y+j)-\overline{f}(i,j)\right)}{\sqrt{\sum_{x=0}^{L-1}\sum_{y=0}^{K.1}\left(w(x,y)-\overline{w}\right)^{2}}\sqrt{\sum_{x=0}^{L-1}\sum_{y=0}^{K.1}\left(f(x+i,y+j)-f(i,j)\right)^{2}}} \qquad (1)$$

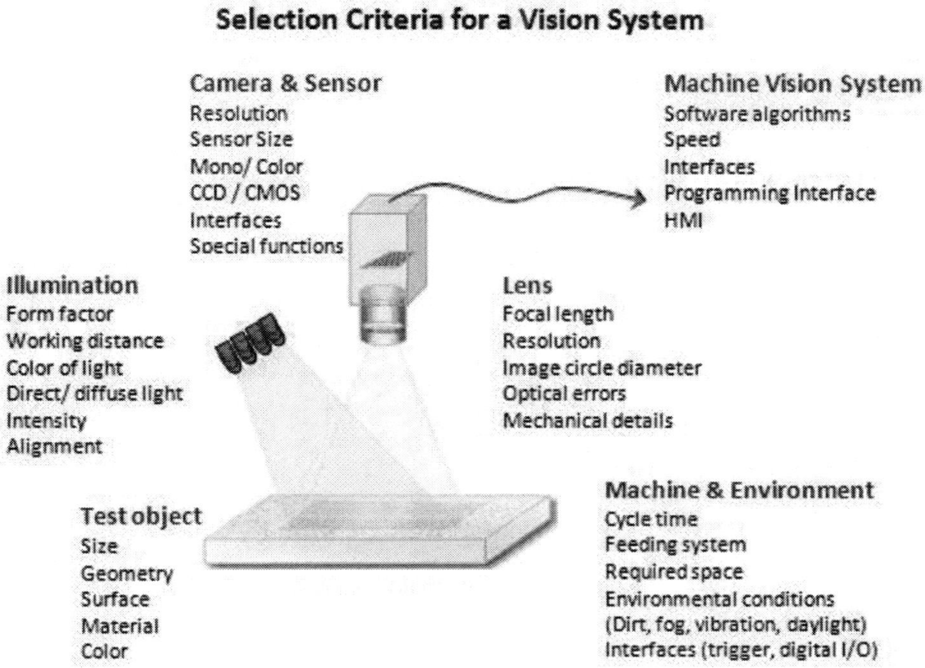


*Fig. 6 Selection criteria for a vision system*

In first step of algorithm the searched component is selected and then the shape of element is described by mathematical functions. In the next step of algorithm the whole captured image is scanned to detect the similarities with template. The areas in image is classified by score value, which determine the similarity with template. Based user setting the most significant areas with similarity is then sign as an area with coincidence. The higher value of score means the higher similarity with template.

The concept of workplace is shown at Fig. 7. The workplace consist from inspection camera with lens, stand for conveyor belt simulation and PC with evaluation software. The connection between inspection camera and PC is realized through measuring card. This card support communication with virtual instrumentation LabVIEW. The application software then receive data from camera and evaluate and displayed images ten times per second. The inspection can perform faster if the visualisation will be realized only in case of fault detection. The condition of evaluation is set by user (which components are wanted for evaluation, how fast the inspection is performed).

The results of described algorithms is shown at Fig. 8. The robustness of algorithm can be seen at Fig. 8 where is shown the result of inspection for screw pad. In presented result is shown that the screw pads can be detect when are alone in simulated conveyor belt or when are overlapped with another components. The algorithm searches screw pads on simulated conveyor belt according to user's criteria. The user can select component which will be searched and maximal numbers of matches for each component.

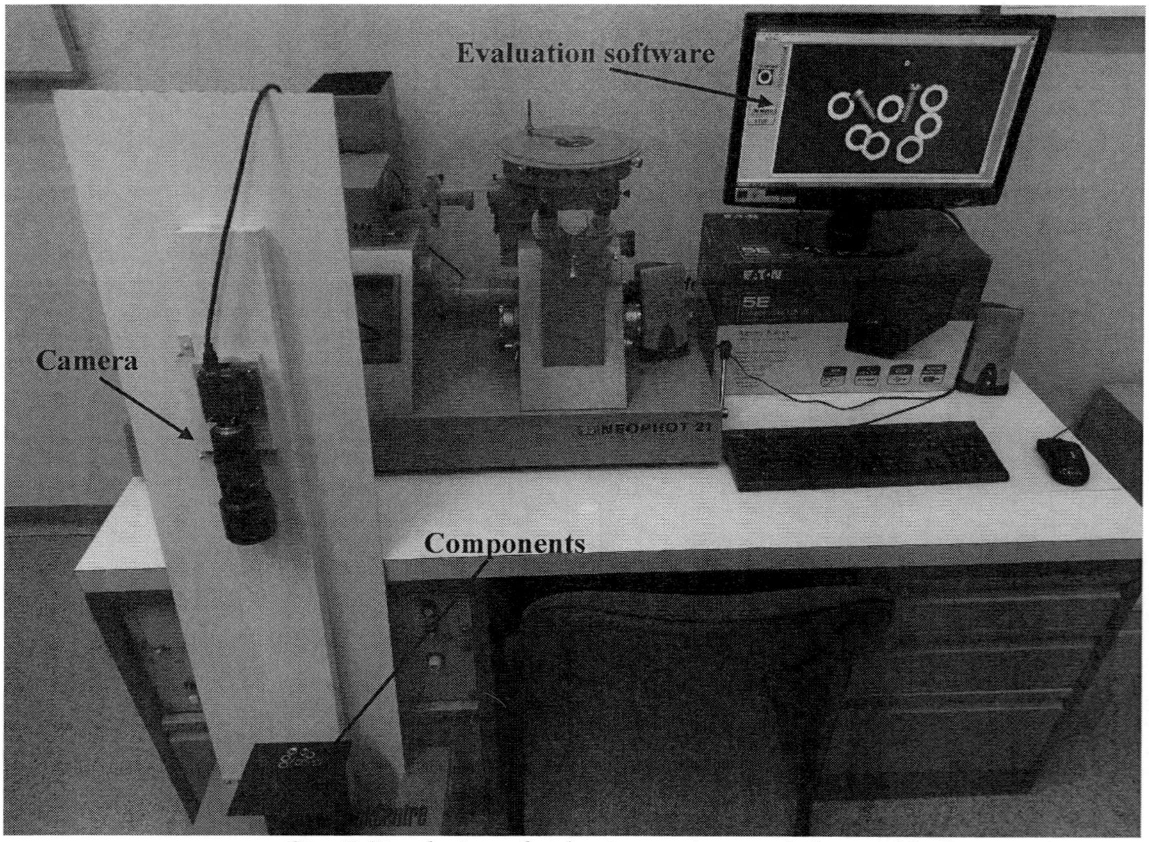

*Fig. 7 Simulation of video inspection workplace*

*Fig. 8 Result of video inspection for screw pad*

## Conclusions

In this paper were presented two models, which can be used in mechatronic applications or on the other hand the models can be used in teaching process. The first model present possibility of virtual instrumentation in real time applications. The industry PC implemented in cRIO and CAN interface was described in case of visualization of processes during control of vehicle. The simulation panel contains components which is used in real cars and the software created through virtual instrumentation can control these components. Another mode of software is visualization of the components state based on CAN protocol.

In the second model is shown possibility of virtual instrumentation exploitation in video inspection applications. The presented system was designed for detecting of selected components and determination of numbers of these components.

**Acknowledgments**

The results were supported by following grants: *KEGA 071ŽU-4/2017*: Key Competences Formation and Effective Support of Students Mobility at Technology Faculties: Modelling, Design and Assessment of Flexible Education Concept; *APVV-15-0462:* Research on sophisticated methods for analysing the dynamic properties of respiratory epithelium's microscopic elements and *APVV-15-0571*: Research on optimal control of energy flow in the system of electric vehicles.

**References**

[1] C. R. Dorf, The Electrical Engineering Handbook (Third Edition) Systems, Controls, Embedded Systems, Energy and Machines, CRC Press 2006, ISBN 0-8493-7346-6

[2] Freescale Semiconductor, Inc., Bosch Controller Area Network (CAN) Version 2.0

[3] Robert Bosch GmbH, Gasoline systems, Robert Bosh GmbH, Germany, 2013

[4] Texas Instruments, Introduction to the Controller Area Network (CAN), 2008

[5] National instruments, Getting started guide NI 9853, 2015

[6] National instruments, Operating instruction and specification NI 9802, 2009

[7] National instruments, Getting started guide NI cRIO-9082, 2016

[8] National instruments, User manual NI cRIO-9082, 2016

[9] http://www.vision-doctor.com/en/specification-of-mv-system.html

[10]Ľ Miková, M Kelemen, Concept of Mobile Robots for Movement in an Unknown Environment, American Journal of Mechanical Engineering, ISSN: 2328-4102, 2016

[11]T Kelemenová, P Frankovský, I Virgala, Ľ Miková, M. Kelemen, Machines for inspection of pipes, Acta Mechatronica, 2016

EAN 2017
55th conference on experimental stress analysis

# Quality observation of micro electro mechanic systems with the use of virtual instrumentation

Libor Hargaš, Anna Simonová, Michal Taraba, Tomáš Uriča, Matúš Danko, Dušan Koniar[1]

[1]Department of Mechatronics and Electronics

Faculty of Electrical Engineering, University of Žilina

Žilina, Slovakia

libor.hargas@fel.uniza.sk, anna.simonova@fel.uniza.sk, michal.taraba@fel.uniza.sk, tomas.urica@fel.uniza.sk, matus.danko@fel.uniza.sk, dusan.koniar@fel.uniza.sk

**Keywords:** LabVIEW, Arduino, lighting control of the light-emitting diode, LED driver.

**Abstract.** The article deals with the control of the light level of four-colour RGBA LEDs for illumination of micro electro mechanic samples of a microscope while the methods of control of the light level are introduced and discussed first. In the next part of the article, four LED drivers for four-colour LEDs are selected. Furthermore, this article deals with the creation of the program for the colour mixing and the brightness control of the light emitted by LEDs, while the program itself is implemented in LabVIEW graphical programming software. System is dedicated for inspection tasks in mechatronic or industry applications, especially for microscopic items. Cooperation between virtual instrumentation and industrial application is also shown.

## Introduction

The impact of new technologies produced and developed even more powerful and fuel-efficient LEDs that surpass the properties of filament lamps. They may be found in a wide range of applications from street lighting, through architectural lighting to light microscopy. The main purpose of this paper is light microscopy of micro electro mechanic systems (MEMS) such as microelectronic components, devices, machine components and micro switches. This microscope can also be used for detection of fast moving MEMS components in computers (Hard disc) and optical systems. An important feature of these light sources are small size and high power density. With the use of multicolour power LED, colour mixing of emitted light is possible to achieve the best contrast of image and surfaces details. LED diodes can be switched at high frequencies, because LED is electronic component with a small inertia. This feature is useful for detection of fast moving MEMS where the stroboscopic effect is used. For control of the LED diode suitable LED driver has to be used, so the constant current trough diode is ensured. For the scanning of the microscopic sample by video camera switching frequency is important parameter of LED drivers, by which the brightness of the LED diode is controlled.

## Brightness control of a LED diode, electrical source

The construction of the PN junctions of a LED comprises two semiconductor materials, characterized by high purity. These materials are alloyed with the trace amounts of adequate materials. By their presence, either an excess of electrons (as the N-type materials) or on the contrary, the lack of them (P-type material) [1]. By attaching a DC voltage to the PN junction, recombination of each electron and hole occurs.

TABLE 1. Wavelenght, material and forward voltage of LEDs

| | Red | Green | Blue |
|---|---|---|---|
| **Wavelength [mm]** | 600-630 | 500-520 | 440-460 |
| **Material** | ALGaInP | InGaN | InGaN |
| **Forward voltage [V]** | 1.7-2 | 3.2 | 3.4-3.5 |

During the recombination of each electron-hole pair, a specific quantum of energy is released, which can be radiated out of the crystal. Electric power is converted directly to the light of a certain colour. Thanks to the latest technological processes, which ensure high product purity, the flexibility of production processes is increased. Therefore, the yellow, green, red and orange LEDs are produced by the same technology and their colour may be controlled by changing the size of the band gap [1].

## LED electric model

As it is shown in Figure 2, LEDs are non-linear passive electronic components. In case of low voltage to a PN junction, diode is not conductive and does not emit light. If the voltage increases and reaches the threshold voltage, the LED current starts to rise sharply and causes light emitting. If the increase of voltage continues, LED is quickly overheated and can be destroyed. Therefore, it is necessary to connect a source of constant current [2]. The LED can be replaced with an equivalent circuit of the Zener diode and series resistance ESR. This model can be used as a load, which would replace the expensive LEDs in the development of a new power supply for LEDs. LEDs have besides ta series resistance also a dynamic resistance. Some manufacturers of LEDs indicate dynamic resistance in the data sheet, but in the most cases it is necessary to determine it from the volt-ampere characteristics. As it is shown in Figure 2, LEDs are non-linear passive electronic components. In case of low voltage to a PN junction, diode is not conductive and does not emit light. If the voltage increases and reaches the threshold voltage, the LED current starts to rise sharply and causes light emitting. If the increase of voltage continues, LED is quickly overheated and can be destroyed. Therefore, it is necessary to connect a source of constant current [2]. The LED can be replaced with an equivalent circuit of the Zener diode and series resistance ESR. This model can be used as a load, which would replace the expensive LEDs in the development of a new power supply for LEDs. LEDs have besides ta series resistance also a dynamic resistance. Some manufacturers of LEDs indicate dynamic resistance in the data sheet, but in the most cases it is necessary to determine it from the volt-ampere characteristics.

## LED thermal model

This section is focused on a thermal model of the temperature control of a LED (Fig. 1). Thermal properties and operating temperature significantly influence the light output of a LED. For the optimal operation of LEDs, it is necessary to regulate the temperature of PN junction and remove the excess heat from its case to the surroundings through the cooler. Unless there is a sufficient heat dissipation from the LED, the chip temperature may increase to values that cause a drop-in efficiency and a significant reduction in its life time. Optimal operation of LEDs can be achieved by a good thermal design and suitable installation of LEDs' lighting. Thermal properties are analogous to electrical parameters of resistance, capacitance, voltage and current. The thermal resistance of a material is the ability to resist the transfer of a heat flow.

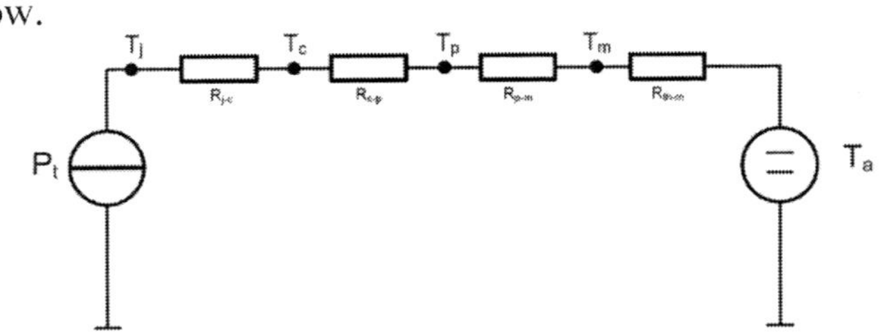


*Fig. 1 Equivalent thermal circuit of the cooling [3]*

If the material has a high thermal resistance, it transmits heat very slowly. Heat hardness is characterized by a thermal resistance, which is similar to the electrical resistance. The thermal calculation is made from a thermal model of the circuit elements. In the following figure (Fig. 1) there is shown the equivalent thermal circuit cooling, which is composed of the multiple thermal resistances [3], [4]. $T_j$ is the temperature of the PN junction, $T_c$ is the temperature of the case of the LED (PCB), $T_p$ is the temperature of insulation pad, $T_m$ is the temperature of the cooler and $T_a$ is the temperature of the ambient. The thermal resistances are: $R_{j\text{-}c}$ - the temperature resistance between PN junction and case, $R_{c\text{-}p}$ - the temperature resistance of the circuit board, $R_{p\text{-}m}$ - the temperature insulation pad and $R_{th\text{-}m}$ - the temperature resistance of the cooler [3]. Thermal resistance represents the ability of a material to conduct heat. Under the influence of the thermal resistance, the one side of the material, by which the heat flow is entered, is heated to the temperature υ1 whereas the side of the material from the heat flow exits is taken to υ2 [5]. Thermal resistance is calculated as follows:

$$R_{\upsilon_{12}} = \frac{\upsilon_1 - \upsilon_2}{P}. \tag{1}$$

$R_{tot}$ - total thermal resistance is calculated by adding the thermal resistance of the materials, that are in the direction of the heat transfer from the chip to the cooler. This temperature is emitted to ambient. LEDs convert the part of the received energy to heat and part to portion of the light. The following equation determines the power dissipation of a LED:

$$P_t = V_f \cdot I_f, \tag{2}$$

where $U_f$ and $I_f$ are RMS values of voltage and current flowing through the diode. The temperature difference is calculated as follows:

$$\Delta T = T_j - T_a = P_t \cdot R_{tot}. \tag{3}$$

## Digital control of the brightness of a LED

Digital control of a LED's brightness is based on the PWM modulation. LED is continuously switched on and switched off. This process must be fast enough so that the human eye d not register the blinking of a LED. In practice, PWM dimming frequencies in range of 400-1200 Hz are used. The control of brightness is realized by changing the PWM duty cycle. This method of dimming a LED does not fit into the machine industry or anywhere where it might occur a strobe effect. From Fig. 2 b.) it is clear, that the current which flows through a LED during the light emitting is always the same, only the mean value of the current is changed. This results in a constant colour temperature of the emitted light.

## Design of the system for controlling the brightness of LEDs

LED lighting has been realized as a light for a desktop microscope Olympus, which is powered by the grid. Light control is implemented in a graphical programming software LabVIEW. Lighting control system for the microscope consists of the following blocks (Fig. 2).

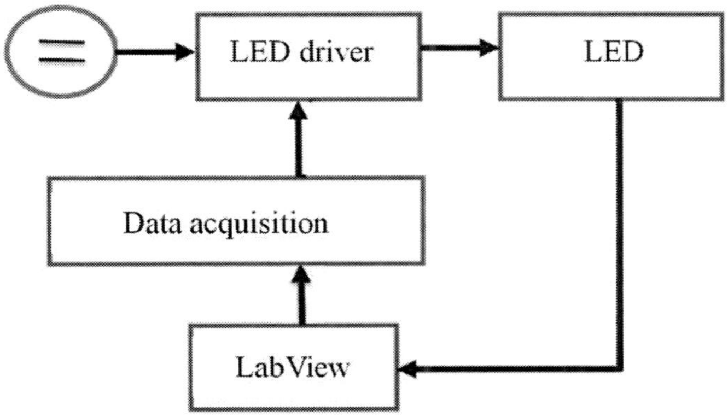


*Fig. 2 Block diagram of the system*

## The light source for the microscope

As the light source the LED LZ4-00MA00 has been selected. It is a four-colour RGBA LED so that in addition to the three basic colours red, green and blue contains a fourth amber colour. This colour is used to obtain a lower colour temperature of white colour. The patented design has unique thermal and optical power. LED achieves high light output 10W.

TABLE 2. Luminous flux of all colours

| luminous flux [lm]/colour | Red | Green | Blue | Amber |
|---|---|---|---|---|
| If=700mA | 115 | 155 | 30 | 75 |
| If=1000mA | 160 | 200 | 40 | 95 |

## RGBA LED Driver for LED diode LZ4-00MA00

Available RGBA driver is intended for low-power applications and therefore it is able to deliver a small amount of current to LEDs. For power applications there are available RGB LED drivers. The disadvantage of such types of the drivers is that they are capable of driving only three LEDs. That's why the RGBA monochrome LED driver was chosen a for each colour of the emitted light. For powering one LED, the topology type step-down converter was chosen. The best driver that meets the criteria of PWM brightness is LED driver LED 5000. The calculation of the circuit elements for step-down converter for LED 5000 is described in [9].

## Software control of the LED brightness in graphical software LabVIEW

The aim of the program was to achieve control of LEDs' brightness by PWM adjustment. As the LED contains 4 colours (RGBA), the brightness control has been done for all colours consequently and thus the brightness control of the white light is acquired. Another requirement was to independently control the brightness of each LED and colour mixing based on the additive model. The platform Arduino Nano is used as the control card. Arduino Nano is a platform with ATmega328 microcontroller. Arduino is used as the measuring card communicating with LabVIEW via USB interface. Arduino Nano has fourteen digital input-output pins and eight analogue inputs. Arduino is able to communicate with the following interfaces: SPI, I2C, USB. Arduino is able to deliver up to 40 mA current to each pin. The communication between LabVIEW and Arduino can be done using two interfaces LIFA and LINX. For interfacing Arduino with LabVIEW, a newer version of LINX is chosen as LIFA is unsupported. To establish communication, the add-on LabVIEW VI package manager, which is an open-source, needs to be downloaded and installed. VI package manager itself has implemented more installation interfaces. One of them is referred to the LIFA, which is still used even though it is unsupported [8].

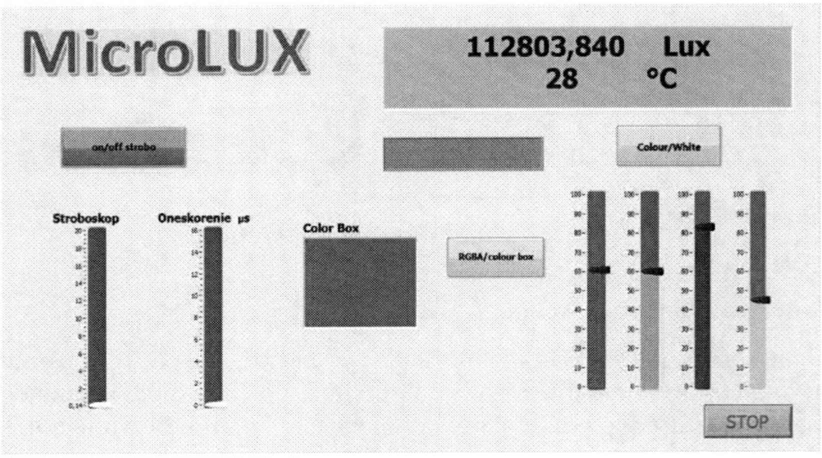


*Fig. 3 Control panel*

**The main program loop**

Prior to the main loop of the program the COM port Arduino is conducted. The main loop of the program consists of an infinite while loop where the *Case* structure is integrated, which can acquire two states: true and false. If the case structure acquires true state, it performs part of the program which controls the intensity of light emitted white light. If the case structure acquires false state, it performs part of the program which controls the light intensity of each colour separately. In either case, the adjusted PWM is in range of 0-255.

**Front panel in the LabVIEW**

The front panel of GUI (Fig. 3) consists of the sliders that adjust the brightness and also the indicator window. Switching between the mixture of colours and the light intensity control of white colour is realized by a switch *colours/white*. By clicking this button, the colour mixing panel is replaced by a panel dimmable white colour and vice versa. By pressing the button *RGBA /Colour box*, the specific colour from colour palette is chosen or the colour can be manually set by the RGBA sliders.

**Experimental observation of microelectronic components**

In Fig. 4, there is shown a sample of miniature ribbon cable of CD drive. Quality of ribbon cable can be reviewed from Fig. 4. An important parameter for evaluating the quality of ribbon cable is the resistance of conductive paths. Another parameter is the dielectric strength of two neighbouring conducting paths. Both cases of considering the quality of ribbon cable are based on the edge contrast of the observed sample. The best contrast is achieved by using of a colour exposition. In Fig. 4, there are shown samples with different colour exposition.

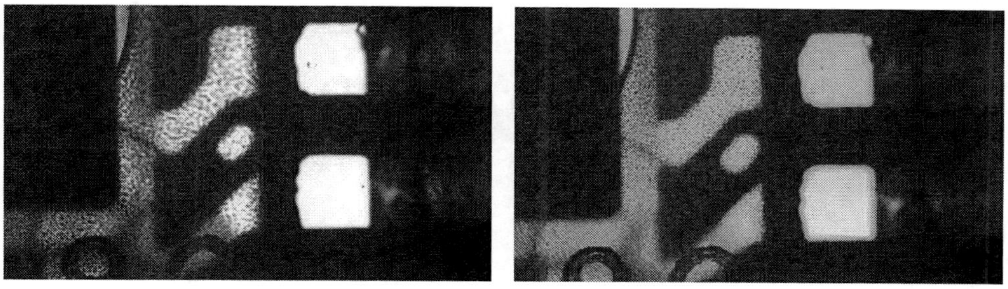

*Fig. 4 Ribbon cable of CD ROM drive*

Detail accuracy visibility is also demonstrated under the various wavelengths and lighting conditions.

Quality of micro-gears is crucial for many micro-mechanic applications. With our system we took some experimental inspection of gears under variable lighting conditions. Gears were ca. 3 mm in diameter. In Fig. 5 we can see the damage of gear tooth.

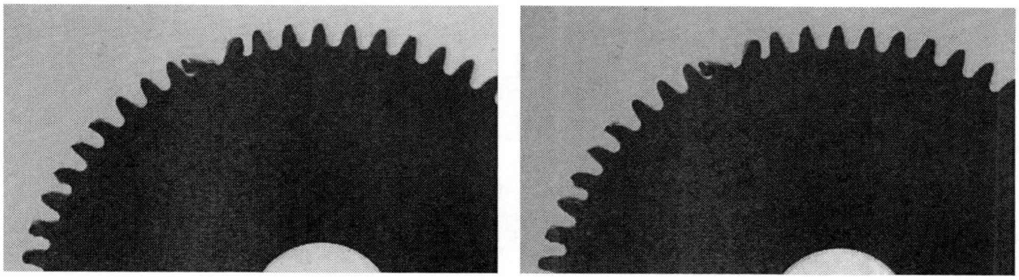

*Fig. 5 Micro-gear with damaged tooth*

Entire system was implemented to the light microscopy Olympus CX21 and images were acquired via high-speed CMOS colour camera Basler A504kc (Fig. 6).

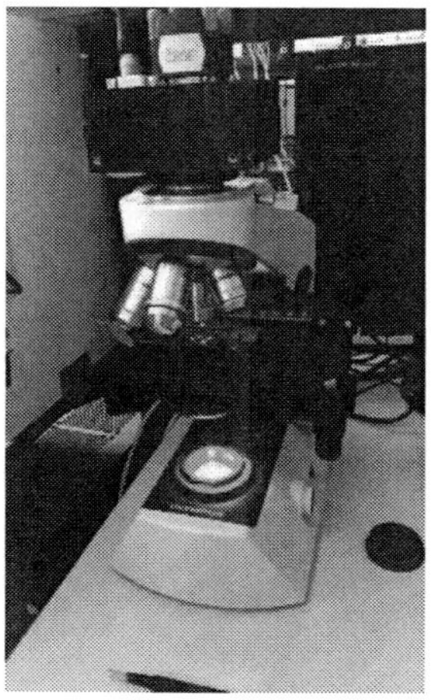

*Fig. 6 Workstation with microscope and camera*

## Conclusions

In the paper we proposed a solution for inspection of a micro mechanical parts. During the work we declared a cooperation between virtual instrumentation and mechatronic system. The system is modular with open architecture and user can easily change anyone of the integrated parts (e.g. replacement of camera, replacement of microscope or LED module based on requirements of industrial task). The system can be immediately implemented to industry or it can be also used for educational purposes. Usage of virtual instrumentation and resulting automated approach to industry solutions decreases human faults in many jobs and follows the world trends.


## Acknowledgments

The results were supported by following grants: *KEGA 071ŽU-4/2017*: Key Competences Formation and Effective Support of Students Mobility at Technology Faculties: Modelling, Design and Assessment of Flexible Education Concept and *APVV-15-0462:* Research on sophisticated methods for analysing the dynamic properties of respiratory epithelium's microscopic elements.


## References


[1] Jiří Vrábel a kol. 2013. Světlo a osvětlování. Praha: FCC Public, 2013. 624 s. ISBN 978-80-86534-21-3

[2] Recom-lighting.: HIGH POWER LED AND DC/DC DRIVER APPLICATIONS. [online]. [cit. 2017.5.4]. Available at: http://www.recomlighting.com/uploads/tx_recomdownloads/WP_Lighting_DC-DC_Driver_1210.4.pdf

[3] HASAN YILMAZ.: Design, application and comparsion of single stage flyback and sepic pfc ac/dc converters for power led lighting application. [online].[cit. 2017.5.4]. Available at: https://etd.lib.metu.edu.tr/upload/12615097/index.pdf

[4] Čuntala, J., Kondelová, A., Hock, O., Pridala, M.: Electro-Thermal Modeling of Power LED Using COMSOL Environment, In: 2016 ELEKTRO 11TH INTERNATIONAL CONFERENCE, pp. 127-130, ISBN:978-1-4673-8698-2

[5] MICROCHIP.: Controlling High Brightness LEDs Using the dsPIC® "GS" Series of Switch Mode Power Supply Digital Signal Controllers. [online].[cit. 2017.5.4]. Available at: http://www.microchip.com/stellent/groups/SiteComm_sg/documents/DeviceDoc/en550157.pdf

[6] Dobrucký, B.- Špánik, P.: Modelovanie a simulácia výkonových polovodičových štruktúr. Žilina: EDIS,1999.203s. ISBN 80-710-575-4

[7] Špánik, p., Drgoňa, p., Frivaldský, M., Kuchta, j.: Elimination of Current Sensing in Digital Control System for Resonant Converter, In: International Conference on Applied Electronics 2013, 10-12 September, Pilsen, IEEE CFP1369A-PRT, pp. 269-273, ISSN 1803-7232

[8] The Official Site of Arduino. Available at: https://www.arduino.cc/

[9] STMicroeletronics.: STMicroeletronics.: LED5000.[online].[cit. 2017.5.4]. Available at:http://www2.st.com/content/ccc/resource/technical/document/datasheet/d3/89/c0/98/72/a9/43/3b/DM00070687.pdf/files/DM00070687.pdf/jcr:content/translations/en.DM00070687.pdf

# Bond Graph Methodology for Modeling of Dynamical Systems and Derive the Equations of Motion

Peter Frankovský[a], Alexander Gmiterko[b], Darina Hroncová[c], Michal Kelemen[d], Oskar Ostertag[e], Lukáš Dominik[f]

Department of Mechatronics, Faculty of Mechanical Engineering, Technical University of Košice, Park Komenského 8, 042 00 Košice, Slovakia

[a]peter.frankovsky@tuke.sk,[b]alexander.gmiterko@tuke.sk, [c]darina.hroncova@tuke.sk, [d]michal.kelemen@tuke.sk, [e]oskar.ostertag@tuke.sk, [f]lukas.dominik@tuke.sk

**Keywords:** Bond Graph, Mechanism, Simulation, Equation of State, Dynamical Systems

**Abstract.** The aim of the thesis is the application of Bond Graph methodology for modeling dynamic systems. In the first section of the article the bond graph notation is defined and its underlying concept is explained. In the following parts is solved an example electromechanical model of elevator. Electromechanical model of elevator is solved by this approach at the level of its physical behavior. Technique of the bond graph theory is demonstrated and its place in the process of modeling of dynamic system of elevator behavior is discussed. From a bond graph diagram of the electromechanical system of an elevator, using a step-by-step procedure, system equations may be generated. As a starting point a model of an elevator system is taken.

## Introduction

Generally, mechatronics symbolizes the synergetic integration of mechanical engineering with electronics, informatics and control theory. On the one hand, this multidisciplinarity has many positive features, but on the other hand, the design process is complex and requires the cooperation of several engineering specialties. Each expert for the given area is however prone to see the problem from his perspective. These differences in the views often lead to misconceptions and misunderstandings between different specialists. Ultimately, this may decrease the market position and business value of the product.

In order to achieve the desired properties of the mechatronic system its inherent complexity must be underpinned already during its design. The way how to do this has been known since ancient times: divide et impera (divide and rule). The design process is usually decomposed into several parts which perform some activity. One example of a possible design process is shown in Fig. 1 [1].

We can notice that the design process is divided into two parts: abstract and concrete. In the abstract part the mechatronic system is modelled on an abstract level, which helps to understand and define the problem area, logical structure and logical behaviour of the mechatronic systems. For the description of the logical structure and logical behaviour are used formal languages such as UML (Unified Modelling Language).

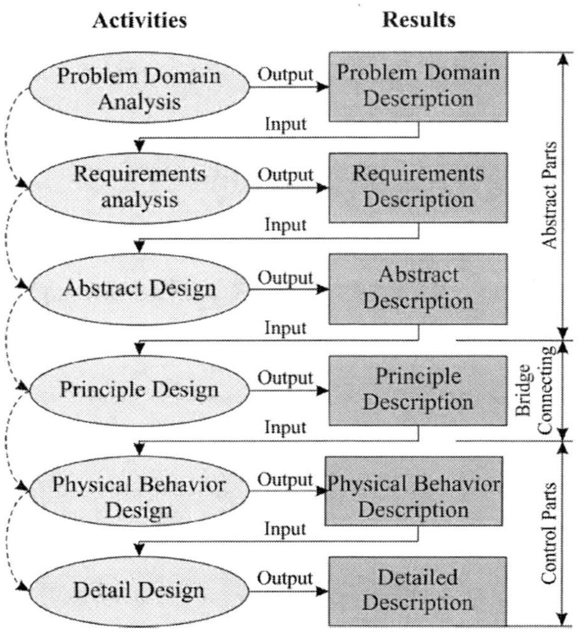


**Fig. 1 Mechatronics design process activity**

In the concrete part an important role is played by realisation factors such as the used technologies (mechanics, electronics or software), size, power etc. Here, the physical behaviour of the bodies (motors, sensors, actuators, regulators etc.) and their networking is important. By the term "physical behaviour" we mean the behaviour of the system which depends on physical parameters such as weight, spring stiffness, inductance etc. and is mathematically described by differential respectively difference equations [3, 4].

In this article, we focused on the area of modelling of the physical behaviour of lumped-parameter mechatronic systems using performance charts (bond graphs). In traditional approaches to physical modelling multidisciplinary system, the governing equations are derived from a combination of Newton's laws, Kirchhoff's laws, Bernoulli's equations, and other fundamental governing equations in different domains of knowledge [2]. The method of bond graphs uses the opposite approach. First, the modelled dynamic system is distributed to subsystems and their mutual bonds are shown by a directed graph. At first arises the graphical simulation scheme - oriented graph, from which is then derived the mathematical model of physical behaviour [3-5] in the form of state space equations.

## Bond Graphs Characteristic

The bond graph technique is a graphical language of modelling. Component energy ports are connected by bonds that specify the transfer of energy between system components. Components regardless of any energy domains are connected through a lossless line called power bond. The direction in which the power flow is assigned, a positive value is indicated by a half-arrow on one end of the bond as it is shown in Fig. 2. The language of bond graphs aspires to express general class physical systems through power interactions. In the bond graph method, power consists of two variables which are known as generalized effort generalized flow denoted by e and f respectively. The power flowing in the bond is defined as the product of an effort and a flow variable. The factors of power i.e., effort and flow, have different interpretations in different physical domains. Yet, power can always be used as a generalized coordinate to model coupled systems residing in several energy domains. A subsystem is represented by a closed line with a name. This line represents the frontiers of the subsystem [2, 5].

For each energy interchange of the system with its environment we associate to it an energetic port of a defined type (mechanical energy, electrical energy, etc.). A unidirectional semi headed arrow shows the energy interchange throw this port and carries the data relative to the power transported (e: effort and f: flow). These two variables are necessary and sufficient to describe the energetic transfers inside the system. They correspond to a couple of variables in each energetic domain.

**Table 1 Effort and flow variables in some physical domains**

| Energetic domain | Effort e | Flow f |
|---|---|---|
| Translational mechanics | Force | Velocity |
| Rotational mechanics | Torque | Angular velocity |
| Electricity | Voltage | Current |
| Hydraulic | Pressure | Volume flow rate |
| Thermal | Temperature | Entropy change rate |
| | Pressure | Volume change rate |
| Magnetic | Magneto-motive Force | Magnetic flow |

The elementary components/subsystems are classified by their energetic behavior (energy dissipation, energy storage, etc.), by their function inside the system (flow sensor, etc.).

**Table 2 Effort and flow variables in some physical domains**

| | |
|---|---|
| **Junction 0:**<br>all efforts are equal Ex: Parallel connection in electrics. | f-f1-f2-f3=0<br>e=e1=e2=e3 |
| **Junction 1:**<br>all flows are equal Ex: Series connection in electrics. | f=f1=f2=f3<br>e-e1-e2-e3=0 |

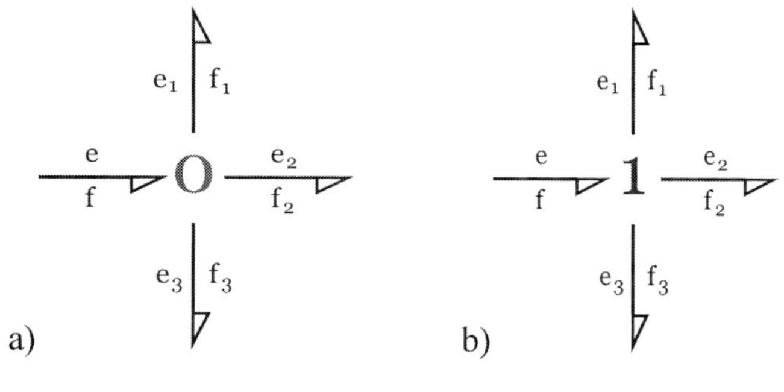


**Fig. 2 a) Example of 0-junction; b) Example of 1-junction**

Junctions can connect two or more bonds. There are only two kinds of junctions, the 1 and the 0-junction. They conserve power and are reversible. They simply represent system topology and hence the underlying layer of junctions and two-port elements in a complete model (also termed the Junction Structure) is power conserving.

1-junctions have equality of flows and the efforts sum up to zero with the same power orientation. Such a junction represents a common mass point in a mechanical system, a series connection (with same current flowing in all elements) in an electrical network and a hydraulic pipeline representing flow continuity, etc. So, a 1-junction is governed by the following rules: The flows on the bonds attached to a 1-junction are equal and algebraic sum of the efforts is zero. The signs in the algebraic sum are determined by the half-arrow directions in a bond graph [2, 6].

0-junctions have equality of efforts while the flows sum up to zero, if power orientations are taken positive toward the junction. This junction represents a mechanical series, electrical node point and hydraulic pressure distribution point or pascalian point. So, a 0-junction is governed by the following rules: The efforts on the bonds attached to a 0-junction are equal and the algebraic sum of the flows is zero. The signs in the algebraic sum are determined by the half-arrow directions in a bond graph [5, 7].

Modeling is started by indicating the physical structure of the system. The bonds are first interpreted as interactions of energy, and then the bonds are endowed with the computational direction, interpreting the bonds as bilateral signal flows. During modeling, it need not be decided yet what the computational direction of the bond variables is. Not that, determining the computational direction during modeling restricts submodel reuse. It is however necessary to derive the mathematical model (set of differential equations) from the graph. The process of determining the computational direction of the bond variables is called causal analysis. The result is indicated in the graph by the so-called causal stroke, indicating the direction of the effort, and is called the causality of the bond (Fig. 3) [4, 7].

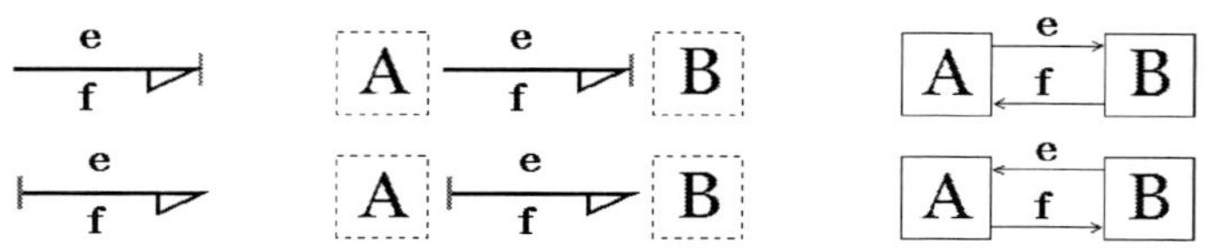


**Fig. 3 Determine the signal direction of the effort and flow**

The last information that is necessary for simulation purposes indicated in the chart is causality. Bond graphs have a notion of causality, indicating which side of a bond determines the instantaneous effort and which determines the instantaneous flow. It is necessary to determine which of the two power signal is for that multi-port independent variable and thus to the multi-port through a performance bond entering and which is dependent variable, and thus to the multi-port through a performance bond go out. Causality is characterized with perpendicular line to binding at the site of the ports, where is input values of effort (e) as an independent variable [8-10].

Because the interaction energy is a function of two variables, so for each element of the bond graph there are two possible options for input and output (Fig. 4):

a) Effort e – input, flow f – output,

b) Effort e – output, flow f – input.

In describing by energy variables on the edge must respect the following principles illustrated in Fig. 4.

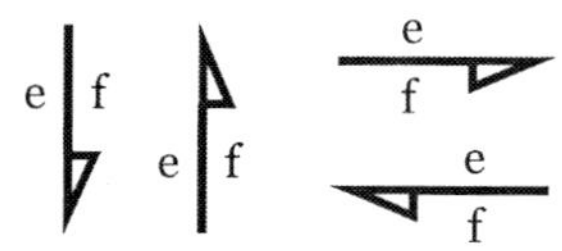


**Fig. 4 Principles of describing oriented edges with bond variables**

## Model of the Elevator Mechanism

The aim of the simulation using Bond Graph is for us to build the model of elevator mechanism. The article describes how to use graphical formulation of system-modeling techniques for engineering systems involving power interactions which in English literature to named Bond Graphs to solve simple model of lifting equipment - lifts. In various parts of the contributions is given formalism making procedure for establishing performance graph electrical power and mechanical parts for lifts and subsequently derived an equation of state of the system [10-12]. Solving simple model of the lift consists of a DC motor, gear, drum and load Fig. 5 is shown.

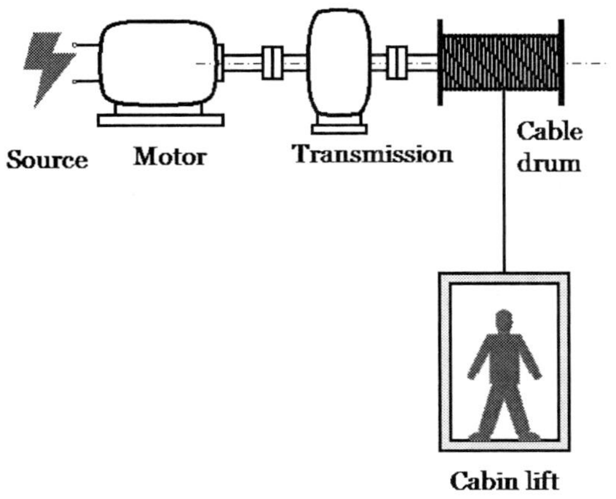


**Fig. 5 Model of the elevator mechanism**

The process of construction of the bond graph and consequently the equation of state is described in the steps below.

Modeling is a part of the first electric motor and the mechanical part of the engine with a mechanical lift parts. The mechanical part of the lift is composed of mechanical parts acting rotary motion and mechanical parts acting translational movement [13]. Plans of components elevator model is illustrated in the following figure (Fig. 6) where: $u_z$ - voltage source $u_i$ - induced voltage, $\omega_1$ - angular velocity in the motor output, $M_h$ - drive torque at the output of the electric motor, $I_M$ - inertia rotor engine, $R_{l1}$ - mechanical resistance of the motor, $R_{l2}$ - gearbox mechanical resistance, $\omega_2$ - angular velocity of the gearbox output, $I_B$ - inertia drum, $m_K$ - cabin weight, $m_C$ - the weight of the passenger, $k_C$, $b_C$ - stiffness and damping feet of the passenger.

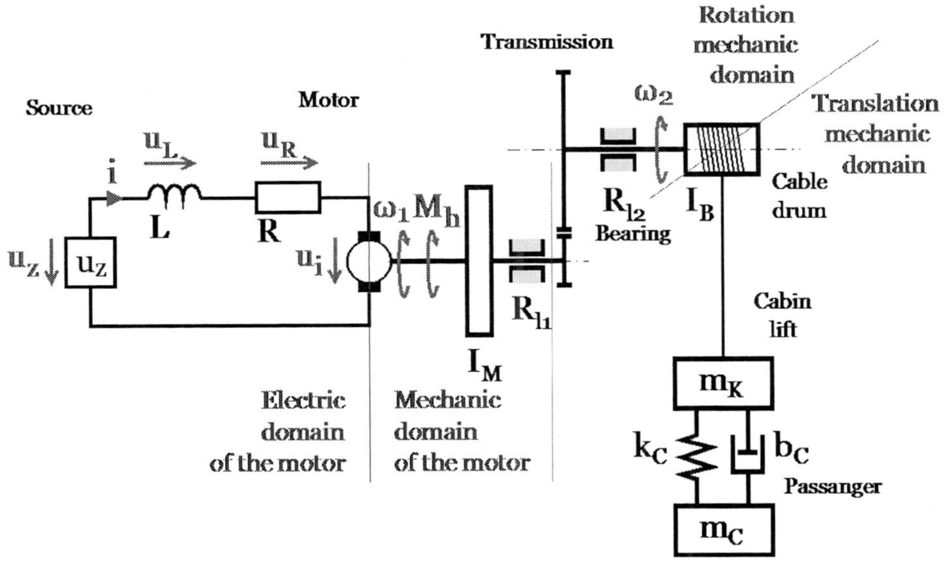


**Fig. 6 Parameters of the elevator mechanism with domain information**

Description of the flow of current and voltage in the scheme of DC motor is shown in Fig. 7 and schematic diagram of electric and mechanic domain of the motor of the elevator model is shown in Fig. 8.

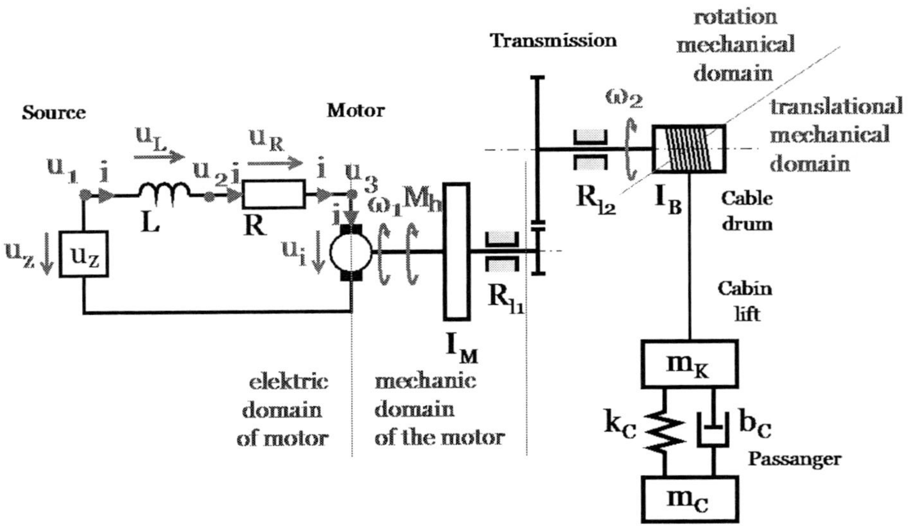


**Fig. 7 Parameters of the elevator mechanism with domain information and flow of the voltage and current**

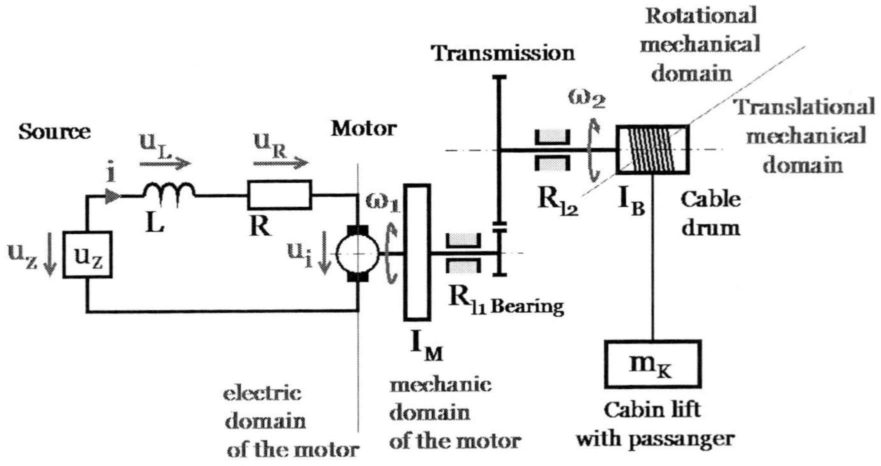


**Fig. 8 Schematic diagram of electric and mechanic domain of the motor of the elevator model**

The electrical and mechanical part of the engine and the description of voltage and current flow are shown in Fig. 9.

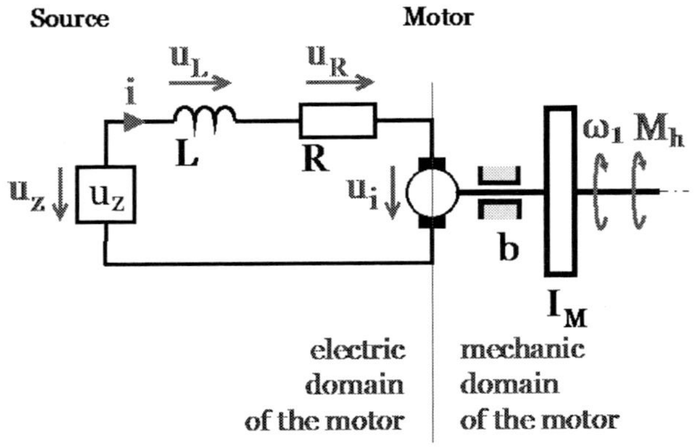


**Fig. 9 Schematic diagram of electric and mechanic domain of the motor**

## Mechanic Domain of the Motor

In this part is shown the procedure for establishing bond graphs step-by-step. We identify the individual components of the system according to the first step of construction of the bond graph [4, 11]:

Source effort, e – $u_Z$ SE: $u_Z$
Inductor, L (Inductance) I:L
Resistor, R R:R

Gyrator – DC motor as the inverter electrical power to mechanical power of acting rotary motion. The result is the following bond graph of the electric motor (Fig. 10).

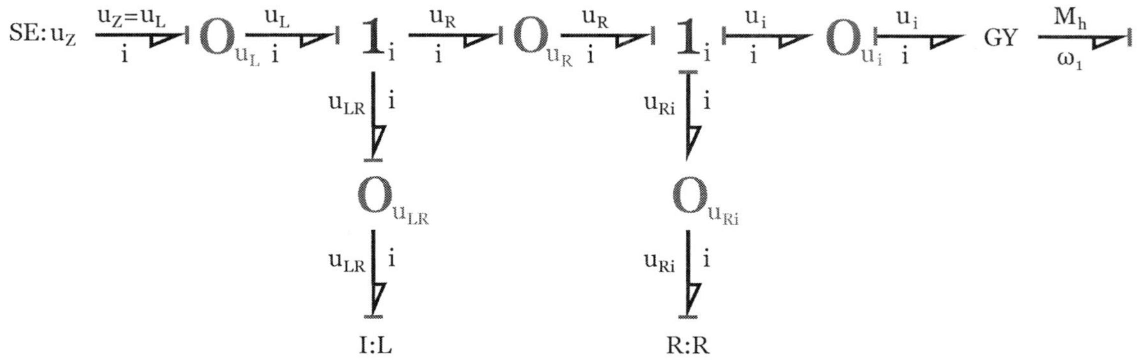


**Fig. 10 Bond graph of electric domain of the motor**

We simplify bond graph by the rules of simplifying. In the simplified model we marked the integral causality with vertical lines in the Fig. 10. The parts of bond graphs within the ellipse are going to be replaced by single edge found above the ellipse in the Fig. 11.

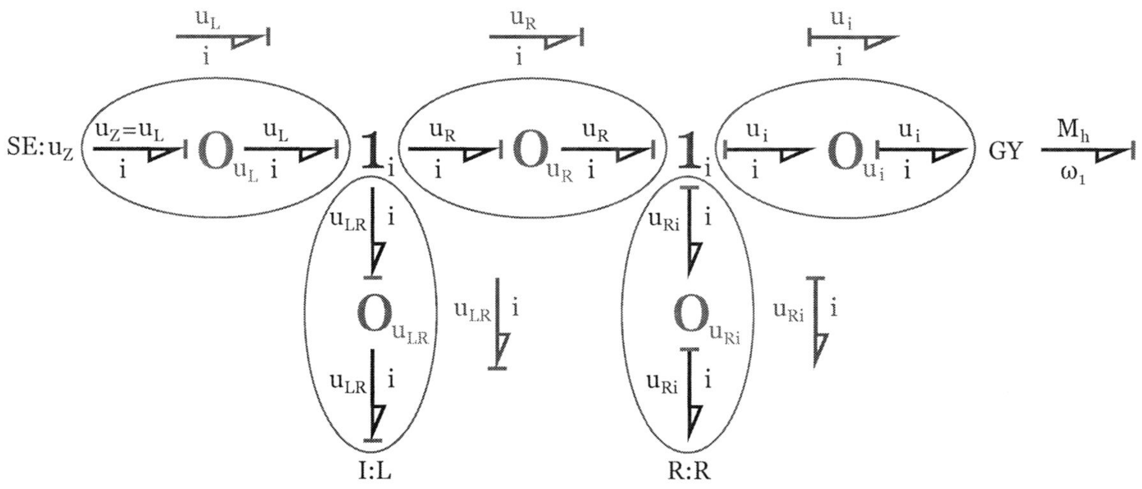


**Fig. 11 The parts of bond graphs within the ellipse are going to be replaced by single edge**

While $u_i=u_R$, we use reduction in 1-junction (Fig. 12).

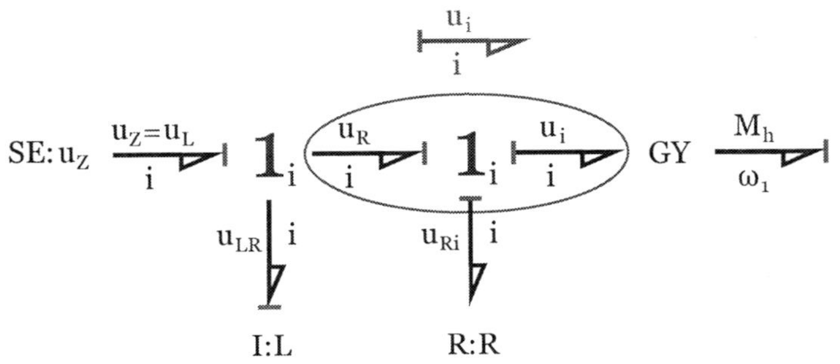


**Fig. 12 Mark of reduction in 1-junction**

The result is the following bond-graph of the electric domain of the DC electric motor with the suggested connecting the mechanical parts (Fig. 13).

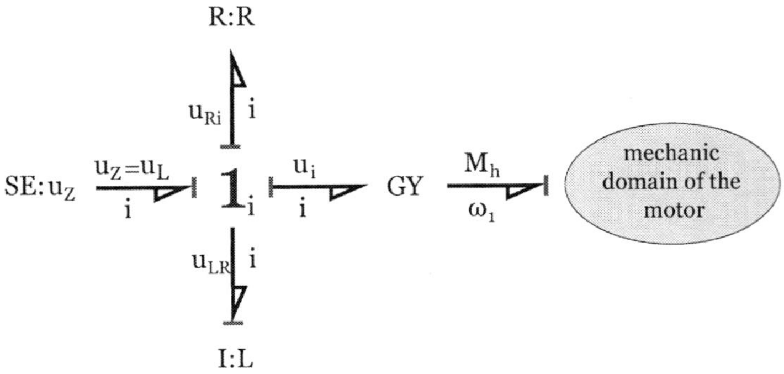


**Fig. 13 The resulting bond graph of the electric part of the motor**

## The Mechanic Domain of the Motor and Elevator

In the first case we consider a model of a mechanical part of the lift, which consists of rotation and translation part of lift mechanism. The contemplated is the mechanical part of the electric motor ($I_M$, $R_{l1}$) [11].

According to the rules for the compilation of bond graph of mechanical systems is a model for said rotational and translational mechanical parts of the lift (Fig. 14) constructed bond graph in the next steps. Rotating part will contain axis of the winding drum and the gate transformer (gateway) and the translational part of the cord will actually exit gate [14].

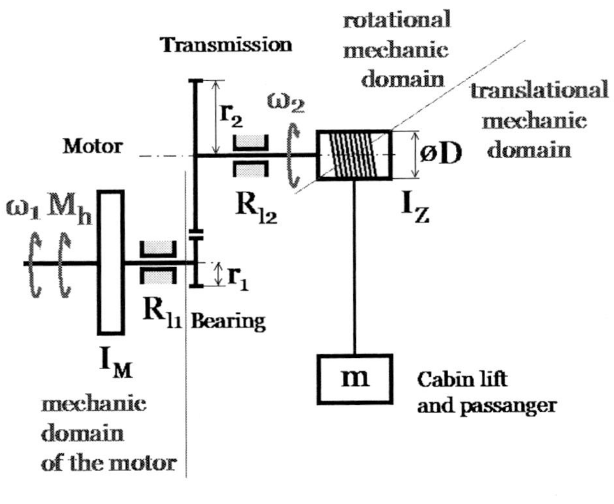


**Fig. 14 Schematic diagram of the mechanical elevator model**

The load will affect the force of gravity there will be a source effort with SE: mg.

**Step 1:** Identify all elements of the system:

Source is effort: e - $M_h$, SE: $M_h$,

Source is effort: e - mg, SE: mg,

Inductor: I - L , m, $I_M$ , $I_Z$, I:L, I:m, I:$I_M$, I:$I_Z$,

Resistor: R - $R_{l1}$, $R_{l2}$, R:$R_{l1}$, R:$R_{l2}$.

Transformer - in the system shown in Fig. 14 it is a transmission (TF:$r_2/r_1$) and drum (TF:D/2).

**Step 2:** In the system we mark the reference angular velocity ωref in rotational movement and the reference velocity vref in translational motion (Fig. 15).

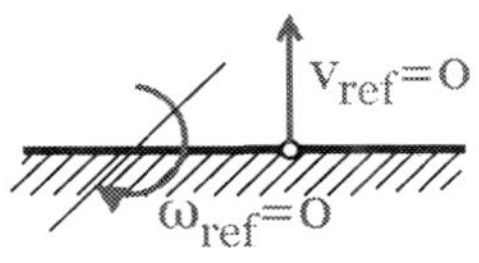


**Fig. 15 Reference angular velocity and translational velocity**

**Step 3:** In the model of the mechanical system to identify and select all junctions with different velocity (flows) and give them a unique name in Fig. 16. In the solution model to name and mark velocity: $\omega_1$, $\omega_2$, $v_1$.

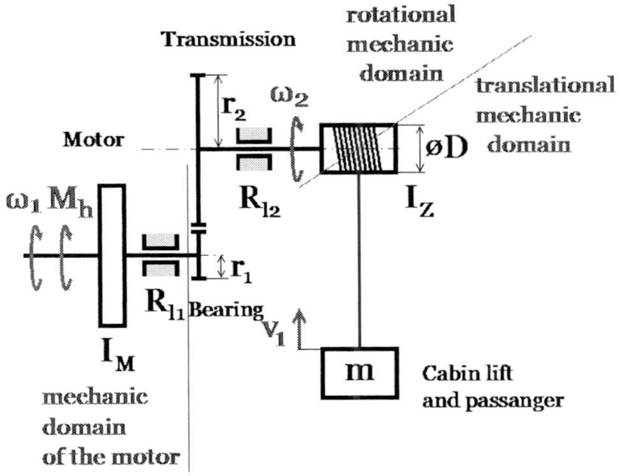


**Fig. 16 Mark different velocity**

**Step 4:** Mark significant points with common speed (the velocity reference is not because it is zero) by the type of junctions. Velocity detected in step 3 draw using 1-junction as in the mechanical system. The reference velocity is not rendered because it has zero velocity. Velocity marked using 1-junction (Fig. 17).

$$1_{\omega_1} \qquad 1_{\omega_2} \qquad 1_{v_1}$$

**Fig. 17 First skeleton of bond graph with velocity shown as 1-junctions**

**Step 5:** Identified angular velocity differences and linear velocity differences. No differences velocities in the system we have.

**Step 6:** Mark junction 0 and construct difference velocities using the junction 0. No differences velocities in the system we have and not connect 0-junction. Junction structure is now ready and can be connected to individual elements. Connect the transformer to the junction 1 connect and source effort which is drive torque Mh and source effort the load, the gravitational force mg (Fig.18).

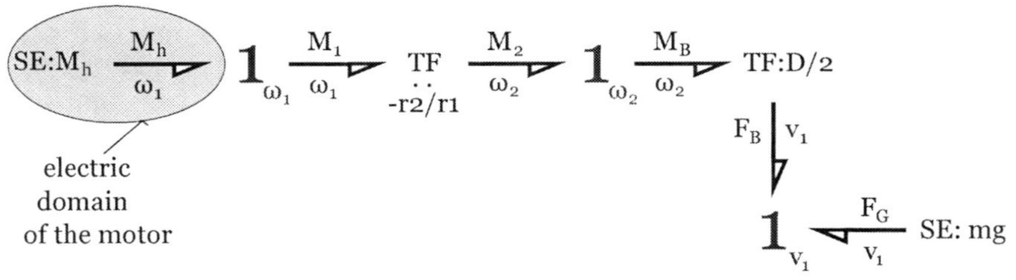


**Fig. 18 Mark velocity with junction 1 and source efforts**

**Step 7:** All elements are connected to the appropriate junctions, as shown in Fig. 19.

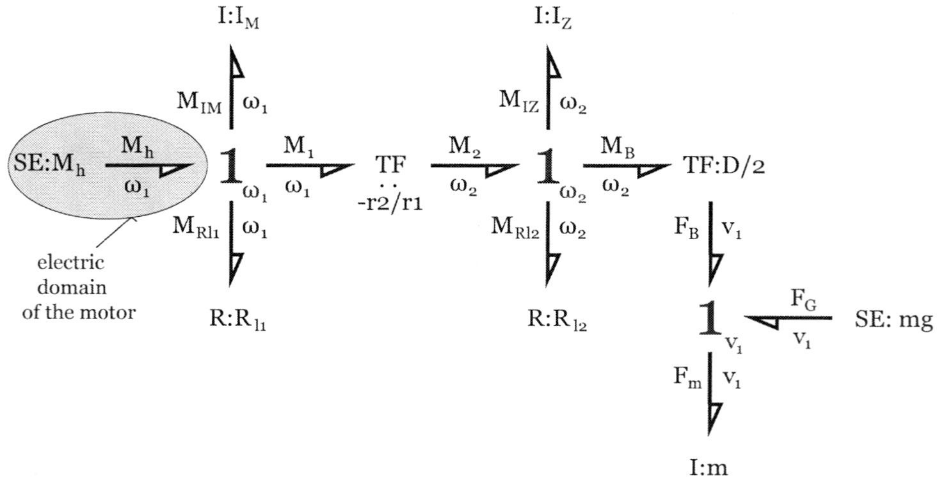


**Fig. 19 Bond graph of the mechanic domain of elevator - connecting the individual elements to the 1-junction**

**Step 8:** Simplified bond graph.
**Step 9:** Marking integral causality (Fig. 20).

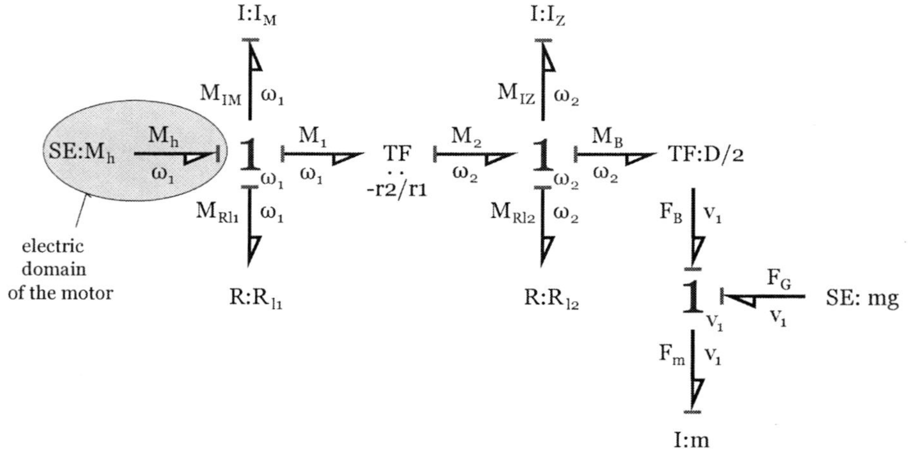


**Fig. 20 The resulting bond graph of the mechanical part**

The following section then simplify the model so that we to reduce sliding mechanical part and get a model with only rotating mechanical parts (Fig. 21). The inertia characteristics of the slider are contained in the moment of inertia $I_Z$ (Fig. 22). Bond graph in this section is different from the resolution model of reduced further by changing the transformer comprises a rotational motion to linear, and takes into account the source efforts sliding materials (Fig. 20). This section derives the bond graph of reduced mechanical parts, which in the next parts, we convert to the signal diagram and then the state diagram.

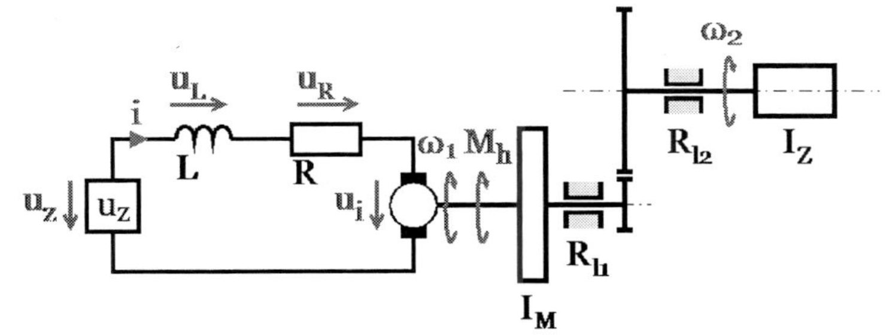


**Fig. 21 Adjusted schematic diagram of the model of lift**

The results of bond graphs applications are shown in the bond graph diagram of the mechanical part of motor and in the bond graph diagram of the rotational mechanic part of elevator (Fig. 22). In formulating the dynamic equations that describe the system, causality defines, for each modeling element, which variable is dependent and which is independent. By propagating the causation graphically from one modeling element to the other, analysis of large-scale models becomes easier.

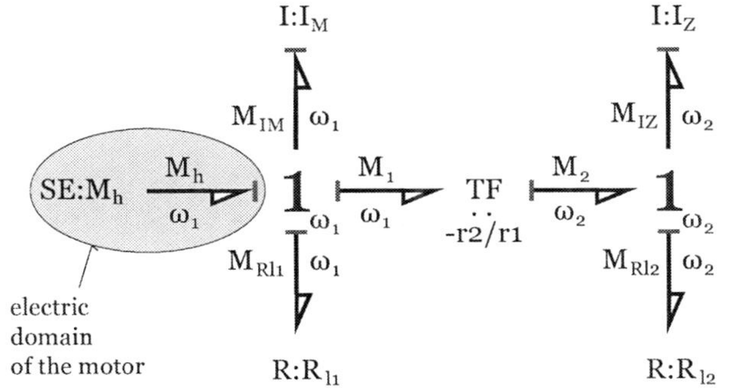


**Fig. 22 The resulting bond graph of the mechanical part**

## The Conversion from Causal Bond Graph to the State Block Diagram

Once a fully augmented bond graph is available, the equations for the system can be developed in a very orderly fashion. Frequently, when the system is small or uncomplicated in structure, state-space equations can be written down directly. However, as system size and complexity grow, the need for an organized procedure for equation generation becomes apparent.

Very general and powerful procedures are available for producing sets of system equations. There are three simple steps to be followed:

1. select input and energy state variables;
2. formulate the initial set of system equations;
3. reduce the initial equations to state-space form.

Selection of inputs is straightforward. For each source element write on the graph the input variable to the system. These variables will appear in the final state equations if they have any effect on system behavior. The list of input variables will be called **U**. The state variables can be placed in a vector called the state vector **X**. The derivative of the state variables will always be either an effort or a flow variable.

State diagram is a special case of block diagrams. In contrast to the bond graph, which shows the flow of power, in the State diagram illustrates the transmission of signals (physical quantities) represented power and energy values, collectively known as *e* (effort), *f* (flow), *p* (generalized momentum), *q* (generalized coordinates). State diagram peculiarity lies in the fact that the whole dynamics is modeled elementary energy accumulators integrators (Fig. 23a). Other blocks are block diagrams of state for constant multiplication and summation member.

Appliances energy - resistors are elements (Fig. 23b) which convert the energy into heat. Examples are electrical resistors and dampers in the mechanics. The natural direction of flow is from the source to the resistor [4].

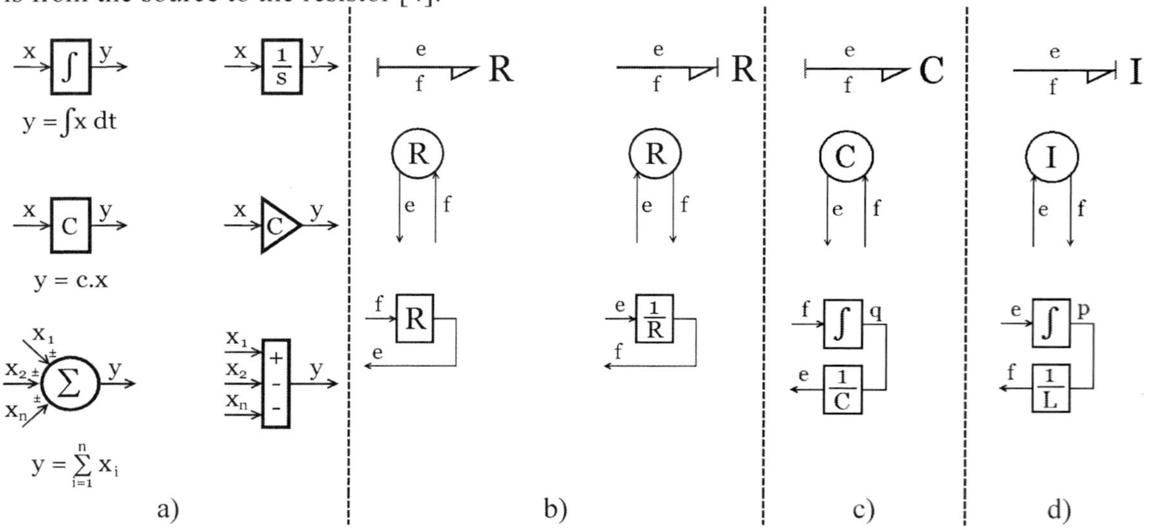


**Fig. 23 Bond graph element: a) blocks of state diagram, b) the 1-port resistor, c) the 1-port Capacitor, d) the 1-port Inductor**

The capacitor accumulates and releases energy (Fig. 23c). The physical element is such a spring, flexible shaft, capacitor, open tanks, pressure vessels or electrical battery. The capacitor is usually selected flow from the source to the capacitor. Inductor accumulates and releases energy (Fig. 23d). Physically, such a factor is the material body, flywheel, or long pipe coil. For integral causality we need to enter the inductors effort (force).

Transformers (Fig. 24a) and gyrators (Fig. 24b) to performed the regulatory function of the power converter circuits that transform the performance of a single physical nature to another.

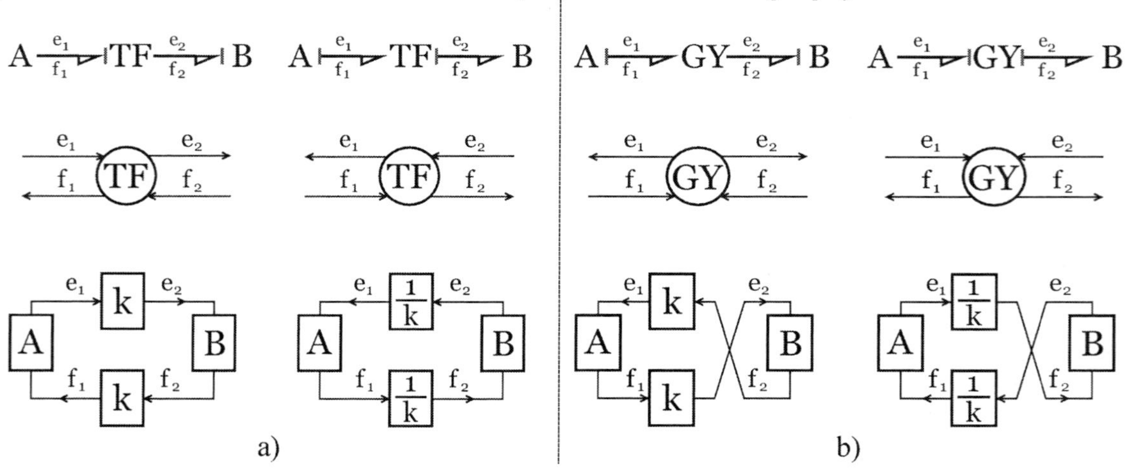


**Fig. 24 Type of a) transformers and b) gyrators**

**Schematic Diagram of the Electric and Mechanic System of Elevator**

In the first case we consider the model of the mechanism of the lift (Fig. 25), which consists of the mechanics domain performing the rotational motion (transmission), and the mechanics domain performing the translational movement (translational motion of the elevator cabin). We consider the mechanical part of the electric motor ($I_M$ - moment of inertia of the engine, $R_{l1}$ - viscous friction in the motor bearing). The weight of the gearbox is neglected [15, 16].

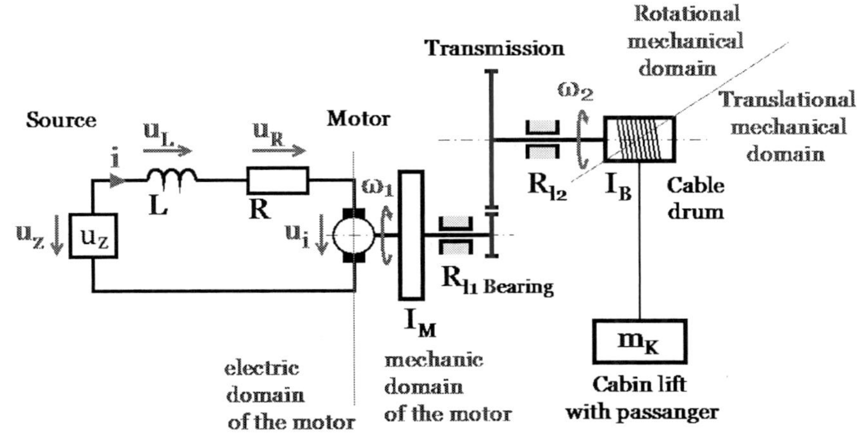


**Fig. 25 Schematic diagram of the elevator**

This model is simplified in that we reduce the mechanical portion acting linear motion and get a model with only mechanical parts acting rotary motion (Fig. 26). Inertial properties of the acting sliding movement are contained in the moment of inertia $I_Z$.

According to the rules for the compilation of performance graph of mechanical systems it is for that model of the lift (Fig. 26) designed bond graph marked with integral causality.

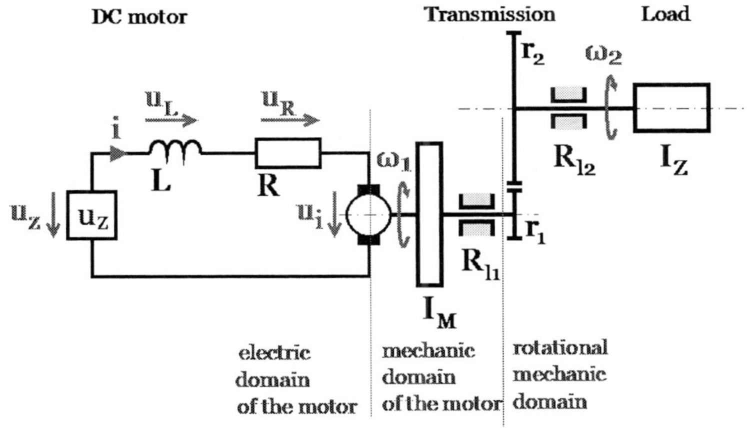


**Fig. 26 Adjusted schematic diagram model of the lift**

The resulting graph of electrical power and a reduced mechanical part is after editing in the form (Fig. 27).

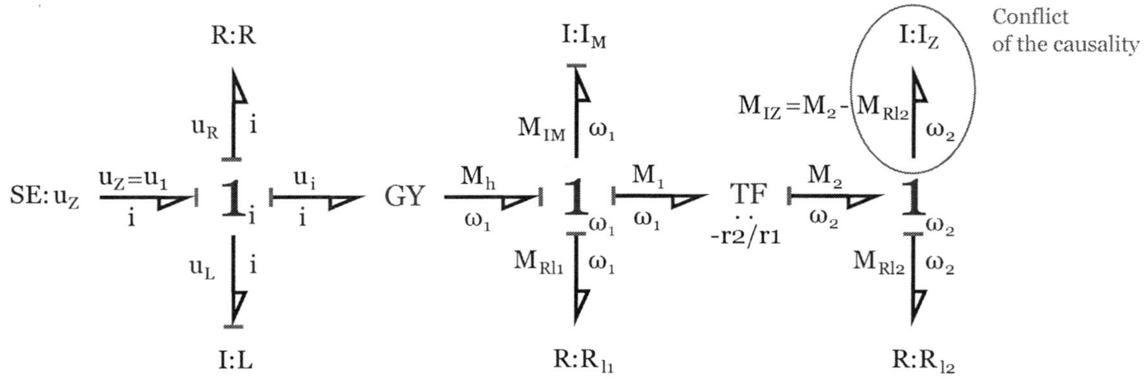


**Fig. 27 The resulting bond graph of the DC motor and mechanical parts of the elevator acting rotational motion with designating the conflict of the causality**

Conflict of the causality in bond graph of the DC motor and rotational mechanic part of the lift is shown in Fig. 28.

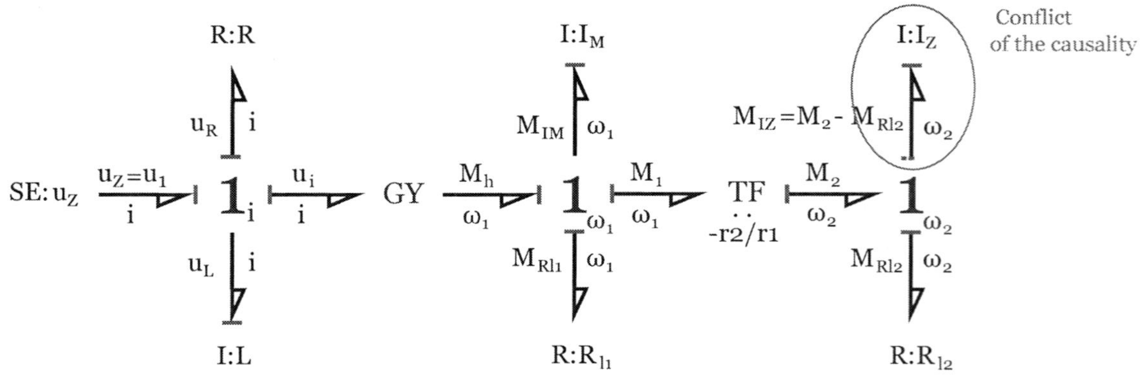


**Fig. 28 Conflict of the causality in the bond graph of the DC motor and rotational mechanic part of the elevator**

There are two ways to resolve this problem of conflict of causality. The first method assumes a certain torsion flexibility of a long shaft to the gearbox output. This is physically justified because there is a perfectly rigid body. The second way we use if we want to have a model with fewer equation of state. In this case we accept a derivative causality (Fig. 27) by drawing it in the state diagram.

## Expansion to Block Diagrams

The next section describes a three-step transition from the bond graph to the block diagram for the model of the lift in which the mechanical part consists only of the rotating part (Fig. 26). Block diagram will be prepared for the bond graph in Fig. 28. We consider the ideal source effort SE:$u_z$ and derivative causality in integrator I: $I_M$.

**Step 1:** All symbols of nodes and elements of the bond graph shall be marked circle.

**Step 2:** Individual edge bond graph will be replaced by a pair of signal edges and circled symbols nodes shall be connected by these pairs of signal edges (Fig. 29). The orientation signal edge is done in accordance with the marked causality (Fig. 28). Where it is intended that the ideal source of effort is the voltage already is a bond graph with the pair of signal edges in the form of Fig. 29.

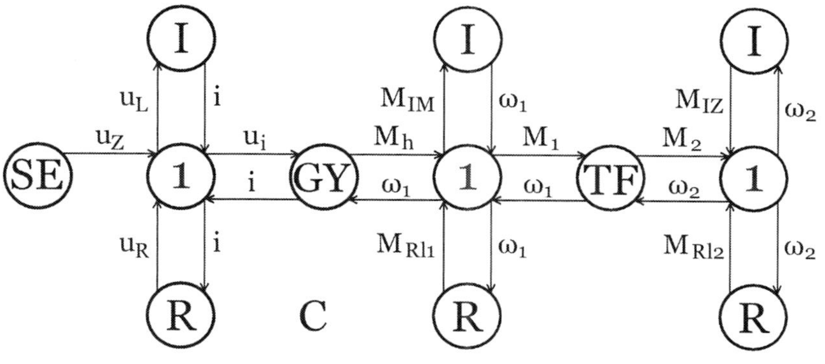


**Fig. 29 Expansion of bonds to bilateral signal flows of the elevator model**

**Step 3:** All nodes are replaced circled by the block structure of Fig. 23 and Fig. 24. Using the rules set out in the literature [5, 6] will be set up state diagram in Fig. 30, which serves as a basis for drawing up the equation of state.

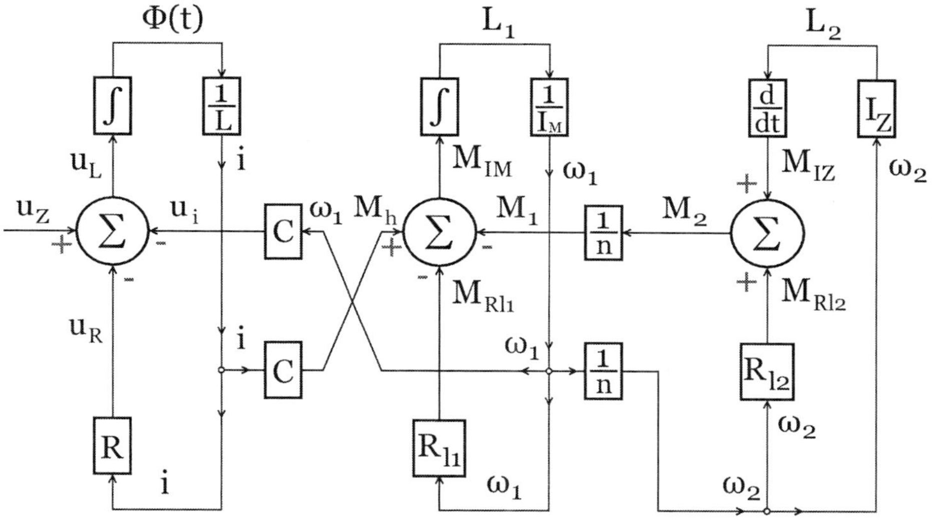


**Fig. 30 The resulting block diagram of the model in standard form**

When drawing up models of dynamic systems they are always based on certain assumptions. Any real system can be complicated. In engineering practice is an effort to assemble as possible the simplest models. But simplification must be to the extent that the model gave us the answer to the problem studied. Therefore, when compiling models it is necessary to decide whether some of inductors, capacitors and resistors can be neglected. Such a simplification, if bad, may lead to such a fit of the model to its use for the next job is difficult. In the bond graph will be arise algebraic loops or we have to use a derivation causality.

## Derivation of Equation from Causal Bond Graphs

In the block diagram of Fig. 30 has the inductor $I_Z$ differential causality. Energy variable momentum $L_2$ is not a state variable. State variable for this dynamic system are: inductance $\Phi$ and momentum $L_1$. Energy variable $L_2$ is with state variable in algebraic relationship [17]. Therefore, before we write the equation of state, we see this relationship.

$$L_2 = I_Z.\omega_2 = I_Z.\frac{1}{n}.\omega_1 = I_Z.\frac{1}{n}.\frac{1}{I_M}.L_1, \tag{1}$$

$$\frac{dL_2(t)}{dt} = \frac{I_Z}{n.I_M}.\frac{dL_1(t)}{dt}. \tag{2}$$

From block diagram (Fig. 29) for state variables we write:

$$\frac{d\Phi(t)}{dt} = u_L, \tag{3}$$

$$\frac{dL_1(t)}{dt} = M_{IM}, \qquad IM \equiv I_M, \tag{4}$$

we obtain

$$\begin{aligned} u_L(t) &= u_Z(t) - u_R(t) - u_i(t) = \\ &= u_Z(t) - R.i(t) - C.\omega_1(t) = \\ &= u_Z(t) - R.\frac{1}{L}.\Phi(t) - C.\frac{1}{I_M}.L_1(t). \end{aligned} \tag{5}$$

First state equation is in form:

$$d\Phi(t) = -\frac{R}{L}.\Phi(t) - \frac{C}{I_M}.L_1(t) + u_Z(t) \tag{6}$$

The state equation from block diagram (Fig. 30) could be expressed as:

$$\begin{aligned} M_{IM}(t) &= M_h(t) - M_{RI1}(t) - M_1(t) = \\ &= C.i(t) - R_{I1}.\omega_1(t) - \frac{1}{n}.M_2(t) = \\ &= \frac{C}{L}.\Phi(t) - \left(\frac{R_{I1}}{I_M} + \frac{R_{I2}}{nI_M}\right).L_1(t) - \frac{I_Z}{n^2 I_M}.\frac{dL_1(t)}{dt}, \end{aligned} \tag{7}$$

where

$$\frac{dL_1(t)}{dt} = \frac{C}{L}.\Phi(t) - \left(\frac{R_{I1}}{I_M} + \frac{R_{I2}}{nI_M}\right).L_1(t) - \frac{I_Z}{n^2 I_M}.\frac{dL_1(t)}{dt}, \tag{8}$$

$$\left(1 + \frac{I_Z}{n^2 I_M}\right)\frac{dL_1(t)}{dt} = \frac{C}{L}.\Phi(t) - \left(\frac{R_{I1}}{I_M} + \frac{R_{I2}}{nI_M}\right).L_1(t). \tag{9}$$

Second state equation is in form:

$$\frac{dL_1(t)}{dt}=\frac{C}{LQ}.\Phi(t)-\frac{\left(\frac{R_{I1}}{I_M}+\frac{R_{I2}}{nI_M}\right)}{Q}.L_1(t), \tag{10}$$

where

$$Q=\left(1+\frac{I_Z}{n^2 I_M}\right). \tag{11}$$

The equation of state of the elevator are in the matrix form:

$$\begin{bmatrix}\frac{d\Phi(t)}{dt}\\ \frac{dL_1(t)}{dt}\end{bmatrix}=\begin{bmatrix}\frac{R}{L} & -\frac{C}{I_M}\\ \frac{C}{L.Q} & -\frac{\frac{R_{I1}}{I_M}+\frac{R_{I2}}{nI_M}}{Q}\end{bmatrix}.\begin{bmatrix}\Phi(t)\\ L_1(t)\end{bmatrix}+\begin{bmatrix}1\\0\end{bmatrix}.u(t). \tag{12}$$

The state variables for this dynamic system are the inductance Φ(t) and the momentum L1(t). Their another solution may be for example in Matlab/Simulink [18, 19].

## Conclusions

Most commonly used software systems for simulation of dynamic systems requires as input data in the form of model equations in block diagram form (Matlab/Simulink) [15]. Create equations of motion or draw the block diagrams it is not simple in more complex systems. Therefore, it is preferred to use a bond graphs methodology, it can create the bond graphs and described the equations of motion less demanding.

Modeling of dynamic systems decomposed into subsystems and their interactions shows a direct graph. This will create a first graphical simulation scheme, which subsequently derived a mathematical model that is assembled from oriented graph then put together a mathematical model. Bond graph is usually compiled easily and be accused of it better dynamic characteristics of the system than the equation of state or other graphical model specially designed for electrical, mechanical and other systems. Construction of bond graph of complex systems often helps to understand the physical nature of their activities.

Bond graphs represent a convenient tool for physical systems analysis and a precise way to represent a mathematical model. Often schematic diagrams are not entirely clear about whether certain effects are to be included or neglected in the model.

This advantage will appreciate in relation to the signal block diagram. Construction of the power graph of complex systems often helps to understand the physical nature of their activities. Formalism performance graphs and algorithmic approach to generating describing differential equations is useful for the analysis of dynamic systems with the transformation of various forms of energy (mixed systems - mechanical, electrical, hydraulic and others) occurring in mechatronic systems and there is also the largest application which is shown in the analysis of electric and mechanic domain of the elevator. Many systems involving two or more forms of energy, such as mechanical, electrical and hydraulic, there are no standard schematic diagrams that clearly indicate assumptions made in the modeling process.

This method allows a deeper insight into a dynamic system and his internal interaction and thus to better understand the behavior of the system. The inner workings of mechanical systems are better understood, visible feedback, reverse influence. A practical example of an elevator systems model is presented as the application of this methodology [20- 25].

The motivation in writing this article is help the users, get familiar with the technique and develop confidence in using it. In that form it can work to serve the educational purpose of finding out about the Bond Graphs theory.

**Acknowledgements**

This work was supported in part by the Ministry of Education of the Slovakia Foundation under grant projects VEGA No. 1/0872/16 and grant projects VEGA No. 1/0393/14.

**Nomenclature**

| | |
|---|---|
| f | : Flow |
| e | : Effort |
| $u_z$ | : Voltage source |
| $u_i$ | : Induced voltage |
| $\omega_1$ | : Angular velocity in the motor output |
| $\omega_2$ | : Angular velocity of the gearbox output |
| $M_h$ | : Drive torque at the output of the electric motor |
| $I_M$ | : Inertia rotor engine |
| $I_B$ | : Inertia drum |
| $R_{l1}$ | : Mechanical resistance of the motor |
| $R_{l2}$ | : Gearbox mechanical resistance |
| $m_K$ | : Cabin weight |
| $m_C$ | : Weight of the passenger |
| $k_C$ | : Stiffness |
| $b_C$ | : Damping feet of the passenger |
| $\Phi(t)$ | : Inductance |
| $L_1(t)$ | : Momentum |

**References**

[1] A. Shakeri, A methodology for development of mechatronics systems. *Ph.D. thesis*, NTNU Trondheim, (1998).

[2] K. Ogata, System dynamics. New Jersey: Prentice Hall, (1998).

[3] A. Gmiterko, D. Hroncová, Modeling of Mechatronic Systems by the Bond graph Methodology, (in Slovak). 1. vyd. Košice: Strojnícka fakulta TU, (2013).

[4] J. U. Thoma, Introduction to bond graphs and their applications. Elsevier, (2013).

[5] D. C. Karnopp, D. L. Margolis, R. C. Rosenberg, System dynamics: modeling, simulation, and control of mechatronic systems. John Wiley & Sons (2012).

[6] D. Shuvra, Mechatronic Modeling and Simulation Using Bond Graphs. CRC Press, Taylor and Francis Group, United States of America, (2009).

[7] P. Horáček, Systems and Models, (in Czech). Vydavatelství ČVUT, Praha, (1999).

[8] W. Zhongshuang, L. Nianli, A Review of Bond Graph Theory and Its Application. *Mechanical science and technology*, 27 (1) (2008), 72.

[9] W. Borutzky, Bond Graph Modelling and Simulation of Mechatronics Systems, An Introduction into the Methodology. Bonn-Rhein-Sieg University of Applied Sciences, Germany, (2006).

[10] R. Merzouki, A. K. Samantaray, P. M. Pathak, B. O. Bouamama, Intelligent mechatronic systems: modeling, control and diagnosis. Springer Science & Business Media, (2012).

[11] P. Frankovský, F. Trebuňa, D. Hroncová, A. Gmiterko, J. Kostka, Modelling of mechatronic systems using bond graphs. Dynamical systems - Theory and Applications: Mechatronics and Life Sciences. Łódź: Department of Automation, Biomechanics and Mechatronics, (2015), pp. 163-174.

[12] E. R. Reberson, R. Schwertassek, Dynamics of multibody systems. Springer Science & Business Media, (2012).

[13] M. T. Islam, C. Yin, S. Jian, L. Rolland, Dynamic analysis of Scissor Lift mechanism through bond graph modeling. In Advanced Intelligent Mechatronics (AIM), 2014 IEEE/ASME International Conference on IEEE, (2014), pp. 1393-1399.

[14] M. Vacková, Machine safety and risk assessment methods, (in Slovak). *Košická bezpečnostná revue*. 4 (1) (2014), pp.197-203.

[15] S. E. Lyshevski, Control systems theory with engineering applications. Springer Science & Business Media, 2012.

[16] B. Friedland, Control system design: an introduction to state-space methods. Courier Corporation, 2012.

[17] R. Margetts, R. F. Ngwompo, M. F. da Cruz, Construction and analysis of causally dynamic hybrid bond graphs. Proceedings of the Institution of Mechanical Engineers, Part I: *Journal of Systems and Control Engineering*, 227 (3) (2013), pp. 329-346.

[18] V Pirč, E. Ostertagová, Mathematic with MATLAB, TU Košice, Košice, (2007), ISBN 978-80-8073-820-4.

[19] V. Pirč, E. Ostertagová, On the relatioship between the initial and the boundary value problems for second order linear differential equations with delay. Applied & computing matematics, 2 (1997), pp. 93-97.

[20] C. W. de Silva, Mechatronic Engineering. Mechatronics: Fundamentals and Applications, (2015).

[21] A. A. Mahfouz, M. K. Mohammed, F. A. Salem, Modeling, Simulation and Dynamics Analysis Issues of Electric Motor, for Mechatronics Applications, Using Different Approaches and Verification by MATLAB/Simulink. International Journal of Intelligent Systems and Applications, 5 (5) (2013), 39.

[22] A. J. Calvo, C. Álvarez-Caldas, L. J. San Román, Analysis of Dynamic Systems Using Bond Graph and SIMULINK. In New Trends in Educational Activity in the Field of Mechanism and Machine Theory, Springer International Publishing, (2014), pp. 155-162.

[23] I. V.Abramov, Y. R. Nikitin, A. I. Abramov, E. V. Sosnovich, P. Božek, (2014). Control and diagnostic model of brushless DC motor. Journal of Electrical Engineering, 65(5), 277-282.

[24] W. Papacz, P. Kuryło, E. Tertel, (2013). The Composite Control Arm–Analysis of the Applicability in Conventional Suspension. American Journal of Mechanical Engineering, 1(7), 161-164.

[25] G. Domek, A. Kołodziej, (2016). Design of the tendon structure in timing belts. Procedia Engineering, 136, 365-369.

EAN 2017
55th conference on experimental stress analysis

# The problem of cooperation of a flat belts with elements of mechatronic systems

DOMEK Grzegorz[1, a], KOŁODZIEJ Andrzej[1,b] WILCZYŃSKI Michał[1,c] KRAWIEC Piotr[2,d]

[1]Kazimierz Wielki Uniwersity, Chodkiewicza 30, 85-064 Bydgoszcz, Poland

[2]Poznan University of Technology, Marii Skłodowskiej Curie 5, 60-965 Poznan, Poland

[a]gdomek@ukw.edu.pl, [b]a.kolodziej@pwsz.kalisz.pl, [c]mwilczynski@whm.pl, [d]p.krawiec@put.poznan.pl

**Keywords:** flat belts, gear with flat belts, conveyor systems.

**Abstract.** A flat belt in power transmission and conveying systems cooperates with several elements consisting of timing pulleys, tensioners or guiding rails. Expectations concering coeficient of friction vary according to application. They depend strongly on characteristics of the process as well as the type of friction. In recent constructions, producers of belts are very much concerned about achieving as much slippery surface as possible. The work describes the problem of friction on different surfaces as well as its influence on gear lifetime. Research results confirm that on many surfaces bigger coefficient of friction is expected.

## Introduction

A part of the system fulfilling at the same time the role of a transmission and a coupling – is a flat belt transmission. Mark Twain once said: "the reports of my death have been greatly exaggerated". Similarly, the following phrase can be reformulated: use of flat belts in advanced structures of machines is unjustified and firmly obsolete. Transmissions with flat belts were among the first flexible link transmissions and after introduction of other types of belts for the torque transfer, their role has been limited to a belt conveyor belts. They were called conveyor belts and process belts, which was a bad translation from the German language and it seemed that their function is very simple. Transfer of a product from point A to point B, or being a component of the manufacturing process (Fig. 1). Over the last years their functions and quality have started to significantly change as structures which may fulfill the strictest requirements of modern machines have been developed. In some machine drives they reach highest speeds which are unavailable for other belt types. They are also used in presses, where wooden or plastic plates are pressed under high pressure. Flat belts are also used in drives with the highest torques [1].

Fig. 1 Conveyor belt system in PU foam plant

**Flat belt operation problems**

A fundamental feature of belt transmissions is the ability to transfer torque to significant distances. This task can also be accomplished by other flexible link transmissions and drive shafts. Belts are used because they are lightweight and the transmission does not emit excessive noise. "Heavy" solutions cause loss of energy, but users consider them to be more durable. This is incorrect thinking, because the durability problem is solved by selection of an appropriate material also in the case of other transmissions. The conveyor belt and drive belt life depends on both volumetric and energetic wear. Belt transmissions are usually made of artificial and natural polymers, whose consumption of energy of bonds between chains of polymers causes material structure weakening and, as a consequence, cracking. Belts are also made from crystalline materials, like for example steel belts, and in this case material fatigue may occur after some time [2,3]. Volumetric wear is mainly caused by the abrasion, although there is also crushing of the material. Crushing of polymer belts occurs in the first cycles of belt operation, and of course after a long period of time due to material creeping. Abrasion occurs in all drive belts and conveyor belts and it results from the friction or combined form-fitting and friction coupling between the belt and pulleys. In a real belt transmission, the belt slip over the pulley is associated with the compensation of the driving strand extension in relation to the slack strandextension. Thus, since there is slipping in each transmission then also abrasion occurs there. Belts and pulleys are chosen so as the belt-pulley friction couple ensures maximum efficiency of the transmission, and the wear resistance is the second most important design factor. Pulleys highly resistant to abrasion are selected first, then the belt made of material resistant to abrasion, but the priority remains the friction factor. It often seems to designers that the friction factor is directly proportional to the abrasion resistance, but unfortunately it is not. Similarly, the hardness of the belt and the pulley does not directly translate into abrasion resistance as this is related to the internal structure of the material (Fig.2).

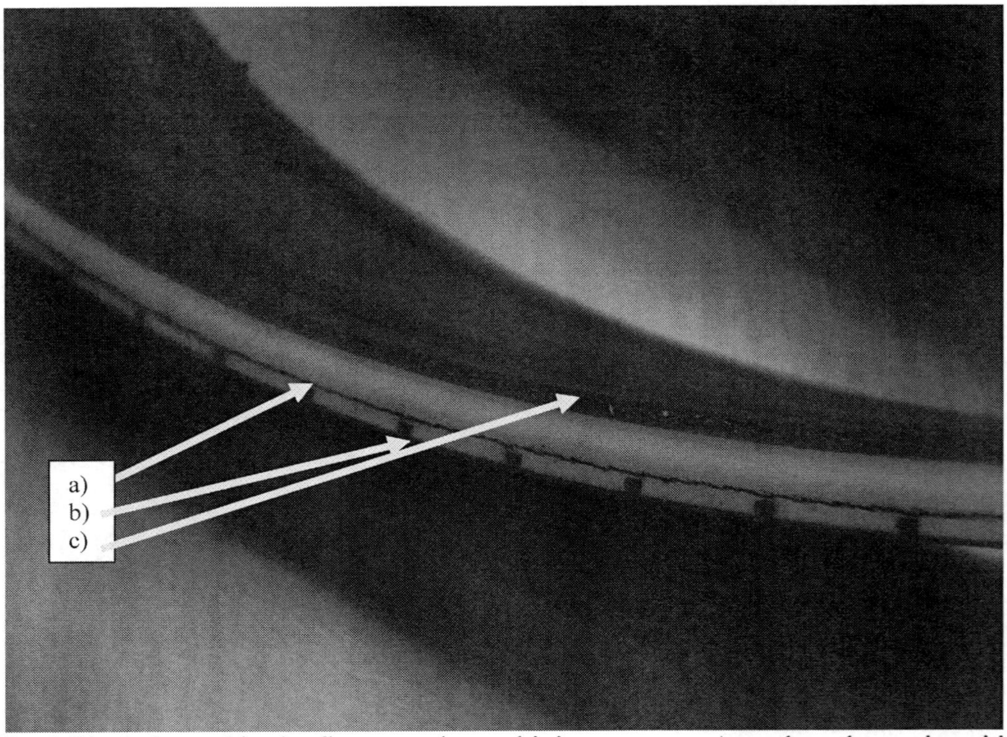


Fig. 2 Flat belt used in the floor panels machining process: a) steel cord, running sid, c) back side with abrasion resistant covering

The first flat belts were made of leader, and then this material was abandoned as flimsy one. Currently a number of belts which are built on modern matrix are covered with leader as this material is very resistant to abrasion. Presence of permanent slip on pulleys does not damage the belt and, upon reduction of the peripheral force, the belt can be further operated. Belts of such structure are used in the heaviest chippers in plants manufacturing wooden plates. Material abrasion is a phenomenon to which special attention is paid in the case of tires on the wheels of motor vehicles - rubber, industrial vehicles - polyurethane and railway rolling stock - steel. Similarly, in the case of belts working in extreme operating conditions different types of rubber are selected, while polymers are selected for indoor facilities and steel in the case of enormous pressures and very difficult operating conditions. Belt operation circumstances are often very complex. For example, belts dispensing banknotes in atms must not fail, although the ambient conditions between the winter and summer change diametrically [4]. The same applies to belts working in explosion hazard environments or chemically aggressive environments. In cigarettes production systems, the belt material penetrates the tobacco and is combusted together with it, therefore the range of materials for such belts is very limited. Conveyor belts in sugar production industry are subject to very intense abrasion, i.e. half of the belt thickness is abraded after two years of operation. There is abrasion product in the manufactured goods and its quantity amounts to hundreds of kilograms. The material of the belt must hold a food grade (be certified for contact with food) and already for several years the European Norms regulate the amount of material that can migrate into food from the belt.

## Coupling between a belt and a pulley

The friction coupling of a weightless thread with a pulley was described by Euler it his papers and although this issue has been later addressed by many authors, the extension of the Euler's formula allows to understand phenomena occurring in the flat belt transmission. In the actual transmission one should indicate the phenomena occurring between the belt and the pulley and between belt material and the cord [6,7]. Modern types of fibers used in the carrier layer of the belt are characterized by better and better mechanical properties. The driving strand to slack strand extension ratio gets closer and closer to unity. The coupling quality depends on the belt on pulley friction factor $\mu_z$ and internal belt friction factor $\mu_w$. The coupling of the transmission can be considered as follows:

$$\frac{dS_1}{dS_2} = e^{\beta\mu}, \qquad (1)$$

where:

$$\mu = \mu_z + \mu_w \qquad (2)$$

while $-\beta$ is the contact angle between the belt and the pulley in radians. Higher tension resistance of the belt load-carrying layer improves the running surface operating conditions. Its elongation is smaller and fatigue damages are avoided. The problems of the modern belts include rheological phenomena and cohesion of the cord and belt material as well as the belt linear velocity. If the driving strand extension $\Delta L_C$ to the slack strand extension $\Delta L_b$ ratio approaches unity, one should consider material strain below the load-carrying layer consisting in its compression and horizontal displacement in relation to the cord. Thus, if:

$$\frac{\Delta L\,c}{\Delta L b} \gg 1 \qquad (3)$$

then a coupling in the flat belt transmission gear:

$$\frac{dS_1}{dS_2} = f(\sigma, h, D, F_r, F_o) \qquad (4)$$

is the function of the belt material $-\sigma$, contact surface with the pulley $-D$, peripheral force $F_o$and the pre-stress force $F_r$ as well as the belt layer height $h$ below the load-carrying layer.

## Guiding the belt run on pulleys

The major problem seems to be flat belt slipping over the pulley as this caused abandoning of that kind of transmissions (Fig. 3). However, that problem was solved many years ago. A belt pulley, at least the driving one, should be of a barrel shape. It is possible to provide the running track side of the belt with wedge-shaped profiles cooperating with an appropriately shaped pulley. Mechanical systems correcting the belt running direction are also used in many transmission systems [5].

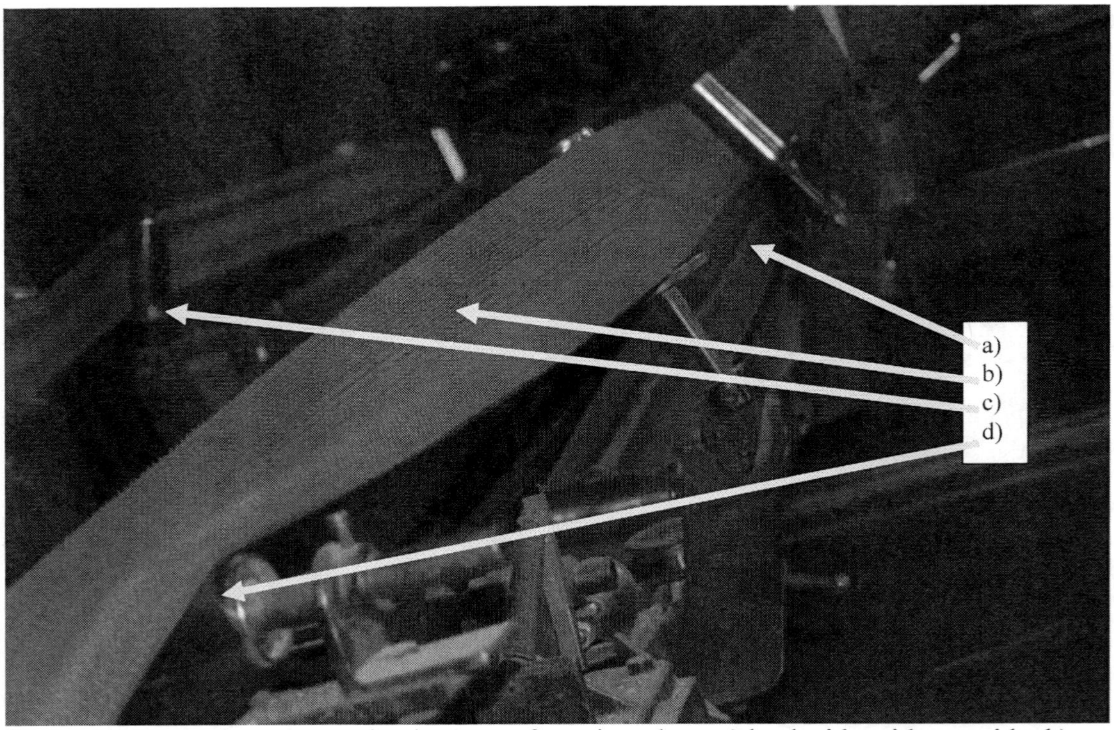

Fig. 3 Belt system in paper packaging manufacturing plant: a) back side with v- guide, b) top side with special structure, c) guide roll for top side, d) guide roll for bottom side, with gap.

Those systems compensate the effect of changes in the force acting over the belt width as well as installation and structural errors of mechanical systems. The manufacturing accuracy of machines where belt transmissions gears are mounted is also a cause for damage of other belt types.

**Selection of the drive drum diameter and the belt bending problem**

Designers have been using calculation software provided by belt manufacturers to select belt transmission components for many years. Such software includes a number of implemented limits and constant values. Obtained results do not coincide with those achieved from own calculations. The software excludes too extreme calculation results, too narrow belts, too small pulley diameters, and additionally provides a number of parameters defining properties of a belt. By putting those parameters in a complex software, a manufacturer prevents this knowledge from dissemination. Sometimes the best way to select the belt to pulleys is belt selection according to the minimum pulley diameter over which the belt can run. Thus the "strongest" belt that can cooperate with a pulley of a given diameter is selected. The belt is probably often over-dimensioned, but it certainly fulfills its function and shall not break over the pulleys.

**Summary**

In modern flat belts the load-carrying layer is made of materials ensuring the lowest possible elongation while maintaining flexibility over small diameter pulleys. A cord made of carbon fiber or highest grade steel provide high stability of flat belt load-carrying layer. Thus the coupling with pulleys depends to a large extent on the material used on the running side. Like in Euler's formula, it was possible to approach an unstretchable cord, but the problem of the strand weight and thickness remains unsolved. That is why materials of the belt running side ensuring high friction factor over a pulley material are used, but also those materials that are unstretchable and light and for which this layer is thinnest one. Like in glued joints, the thinner the running layer the higher the efficiency of the transmission. Many new studies on flat belt transmissions were prepared in recent years and transmissions of that type still have many successful applications ahead.

## References


[ 1] G. Domek, Motion analysis of timing belt used in control systems, American Journal of Mechanical Engineering, vol 1, No.7 2013.

[ 2] M. Dudziak, G. Domek, A. Kołodziej, Contact problems between the hub and the shaft with a three-angular shape of cross-section for different angular positions, Procedia Engineering 96(2014) 50-58.

[ 3] G. Domek, M. Dudziak, A. Kołodziej, Description of timing belt coupling using a pendulum with variable coupling point, Advanced Technologies in Mechanics, Vol 2, No 1(2) 2015 p. 2-8, ISSN 2392-0327.

[ 4] M. Dudziak, G. Domek, A. Kołodziej, Variation of static parameters of cooperation in axisymmetric connection,Procedia Engineering Vol. 136, 2016, p.56-62.

[ 5] G. Domek, A. Kołodziej, Modification of Joints in Polymer Timing Belts, Machine Dynamics Recherche, Vol 39, No 1 (2015), p. 15-20.

[ 6] R. Dyja, E. Gawronska, N Sczygiol, The effect of mechanical interactions between the casting and the mold on the conditions of heat dissipation: A numerical model, Archives of Metallurgy and Materials, vol. 60, no.3, p.1901-1909, 2015.

[ 7] H. Dressing, F. Holzweissig, Dynamics of Machinery, Theory and Applications, Springer Verlag, Berlin, Heidelberg 2010.

EAN 2017
55th conference on experimental stress analysis

# Methodology for probabilistic model creation of lidar measurement

Dekan Martin[1,a], Chovanec Luboš[1,b], Szabová Martina[1,c] and Babinec Andrej[1,d]

[1]Faculty of Electrical Engineering and Information Technology STU in Bratislava, Ilkovičova 3, Bratislava, Slovakia

[a]martin.dekan@stuba.sk, [b]lubos.chovanec@stuba.sk, [c]martina.szabova@stuba.sk, [d]andrej.babinec@stuba.sk

**Keywords:** Probabilistic model, laser rangefinder, mixed pixel.

**Abstract.** The core of this article comprises several approaches to probabilistic modeling of Hokuyo type LRF. Motivation for creating these models was the failure to faithfully represent LRF features in existing models. Moreover, the use of mixed pixel error as a mean of laser rangefinder identification is described.

## Introduction

Nowadays, mobile robots preform their operations in dynamically changing environment. That is why they need to sense this environment with powerful and reliable sensors, which can provide sufficient amount of data. Historically there are many sensors used for environment sensing. However, for reliable and effective navigation of the robot the laser rangefinders (LRF) are mostly used. Due to continual decline in price they are already used in hobby robotics or in service robots for domestic tasks [1]. When working with a sensor it is necessary to know the principle of sensor's measurement and its properties[9,11]. There are several measurement principles used in LRFs [2,3] precisely triangulation, time of flight (TOF), frequency modulation continuous wave (FMCW) and phase shift measurement (PSM). These measurement principles are well known and the details about them can be found in [1] or in [2]. Properties of LRF can be defined as product details, but they can be extended by properties, which are not defined by producer. Such not well defined properties may be represented by measurements to other materials than those mentioned in catalogue or by measurements when LRF fails [3-5]. In [4] authors identified properties of LRF Hokuyo URG-04LX in detail. Measuring principle of this rangefinder is similar to UTM-30LX type.

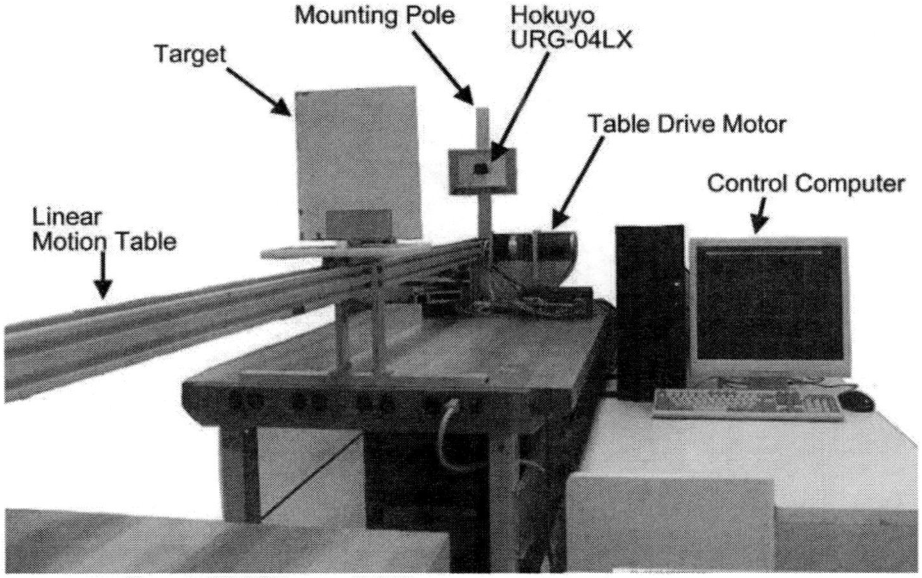


Fig.1. Setup for laser parameter measurement [4]

According to their results there is a Warm-up time of 80 minutes before the measurements stabilize as well as an error dependence on the type of the surface measuring (Fig 2.).

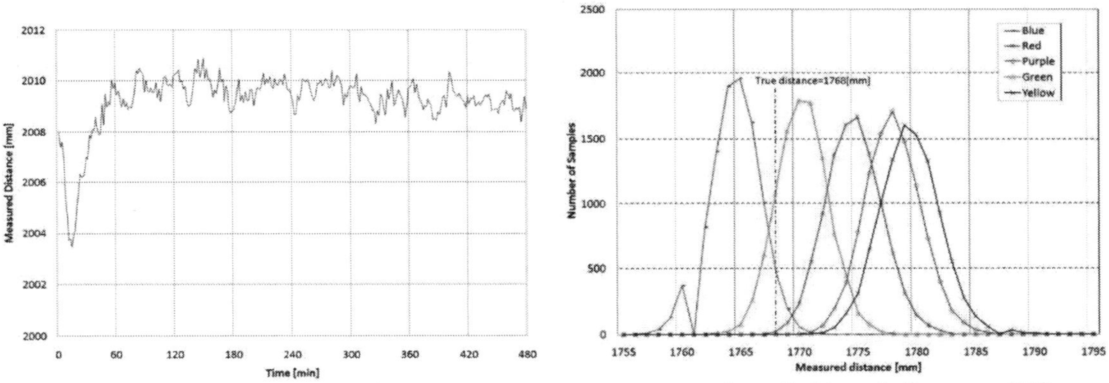


Fig 2. Parameters of LRF Hokuyo URG-04LX, warm up time (left) and distance difference on different surfaces (right) [4]

In [6] a picture of a wall measured with LRF Hokuyo URG-04LX was taken using a camera without an infrared filter. A ruler fixed on the wall was used to determine the beam diameter as shown in Fig. 3.

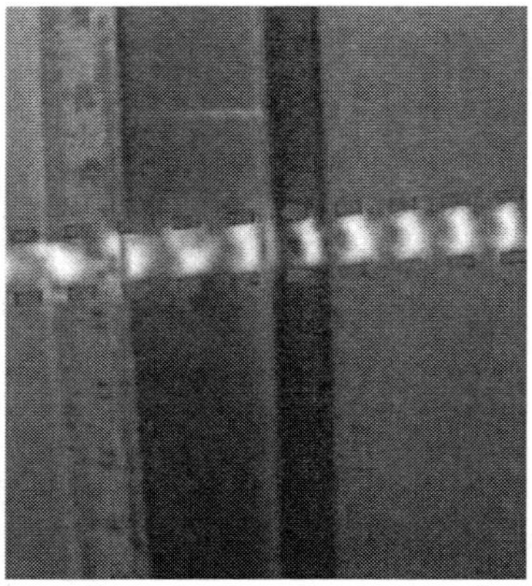

Fig 3. Laser beam pattern measured on a wall [6]

**Probabilistic model creation**

The Errors in measurement may significantly affect data interpretation and subsequent procedures. In mobile robotics, this can lead to destruction of robot, damage of objects or human injuries[10]. That is why it is needed to determine the type and size of errors in measurement as accurately as possible. In determining of laser rangefinder's errors, it is suitable to utilize the parameters that were specified by the manufacturer. However, these parameters are defined for specific conditions. In the case of mobile robotics such conditions may not been always respected. Hence it is needed to perform the measurements, which will determine rangefinder's parameters[8]. While the researchers in [4] and [6] focused mainly on distance error, in this work, we focused on the accuracy of measurement in respect of laser beam's rotation in space.
The result of these parameters is a probabilistic model. The concept of proposed model is derived from the model of Hokuyo URG-04LX used in [4]. This concept is focused on two basic parameters of rangefinder. The first parameter is derived from measurement's accuracy, e.g. stability of measured distances within the repeated measurement without the change in measurements conditions. The results of this parameter were already published in [7] and were very similar to [4]. Second parameter is an accuracy of measurement in respect of laser beam's rotation in space.

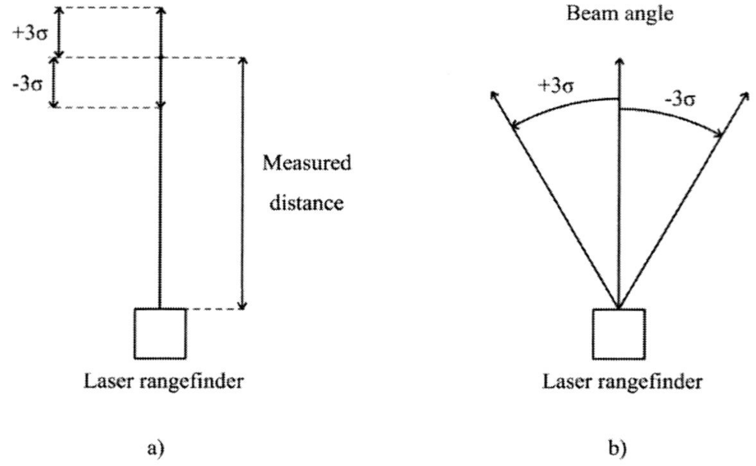


Fig 4. Parameters of LRF used in probabilistic model

The value of this parameter could be measured directly by taking picture as in [6]. But this method requires costly hardware and its result are questionable, because it can't determine whether the nearest beams found in consecutive measurements belongs to the same beam index in measurements. In our work, we used different approach. The LRF suffers from an error on the boundary of two objects named mixed pixel. This error arises when the reflected energy is distributed between these two objects, and the resulting distance is somewhere between these objects. Thus, when the heading of the laser beams between two consecutive measurements differ the location of mixed pixel should differ too. This error was used to determine the second parameter of proposed model.

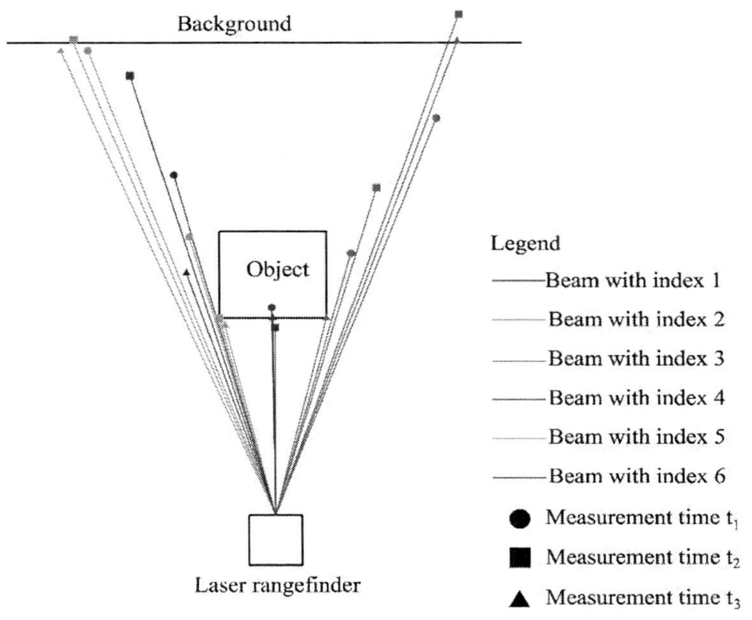


Fig 5. Principle of mixed pixel occurrence

Identification of mixed pixels can be used for determination of laser beam angle distribution. At the beginning, the boundary points of measured object are determined. At first, average standard deviation in measured distances for individual beams is computed (Fig. 6b). Adjacent beams without the sudden step in standard deviation correspond to the objects. The last beams of which standard deviation is under the threshold are boundary beams, i.e. boundary points (red dots at Fig. 6b). Beams where the standard deviation is different from the average standard deviations of the corresponding object are characterized by mixed pixels. These beams are consequently analyzed.

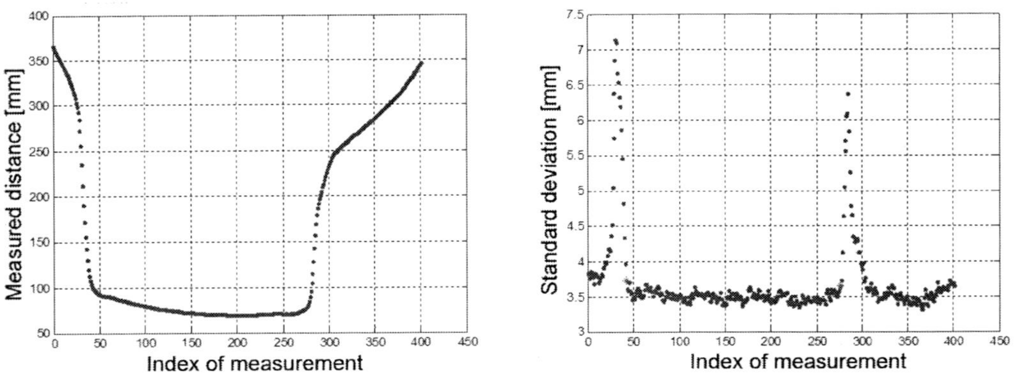


a) b)

Fig 6. Average distance for individual laser beams measured on an object (a) and its corresponding standard deviation (b)

Analysis consists of the determination of ratio between the measurements on object and mixed pixels occurrence. The results are shown in Fig. 7. It is clear that seven adjacent beams to boundary beam securely cover the area of mixed pixels occurrence.

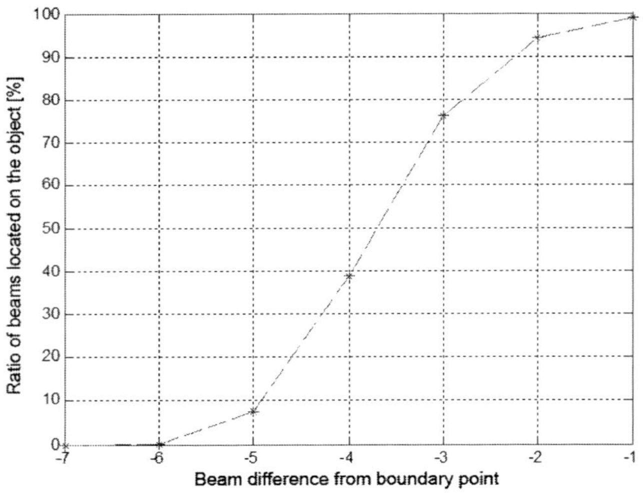


Fig 7. Ratio between the measurement located on object and as a mixed pixel.

Based on the frequencies of mixed pixel occurrence derived from this analysis variance of laser beam angle is determined.

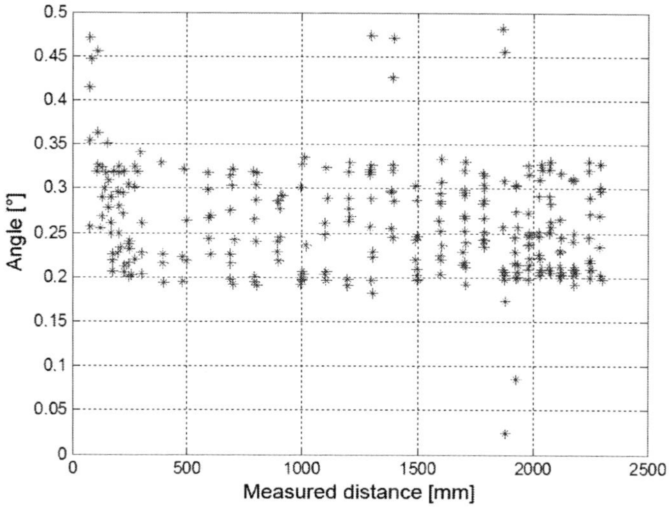


Fig 8. Variance of laser beam angle calculated for different measured distance and relative position between laser and object.

Resultant standard deviation was calculated as an average from the obtained values in Fig 8. and it is equal to $\sigma_r$=0.3616°. The model can be expressed as:

$$f(x,y)=\frac{1}{2\pi\sqrt{\sigma_x^2\sigma_y^2}}e^{-\left(\frac{(x-\mu_x)}{2\sigma_x^2}+\frac{(y-\mu_y)}{2\sigma_y^2}\right)}$$

$$\sigma_y=\tan(\sigma_r)x \qquad (1)$$

This model is dependent on measured distance (Fig. 9), which better corresponds to the characteristics of used LRF. However, as it can be seen in Fig 8, standard deviation of laser beam angle increases with decreasing measured distance. Therefore, Hokuyo LRF was submitted to further analysis and the model was extended.

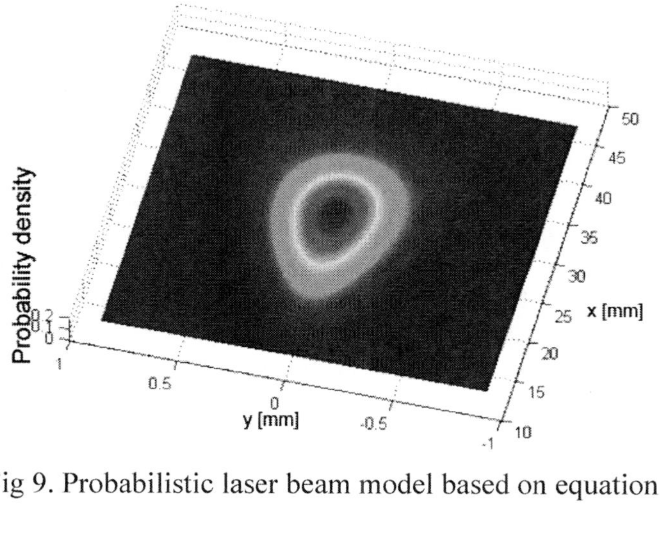


Fig 9. Probabilistic laser beam model based on equation (1).

The next concept of Hokuyo LRF model is focused on three parameters. The first two parameters are derived from the previous concept. However, in this case the second parameter was determined on the basis of new experiment. This parameter is defined as the value of accuracy concerning the distribution of laser beam angle and in this case, it also takes into account the width of the laser beam. The real width of the laser beam is constant at the same distance measurement, but it is distributed by some probability in the space due to inaccuracies in the sensor hardware. In the case when the width of the laser beam is neglected, the distribution of the laser beam angle is illustrated in Fig. 10. Angles $\alpha_{e1}$, $\alpha_{e2}$, $\alpha_{e3}$, and $\alpha_{e4}$ represent the variance in laser beam angle, which is dependent on the measured distance. It is evident that influence of the laser beam width increases with decreasing measured distance.

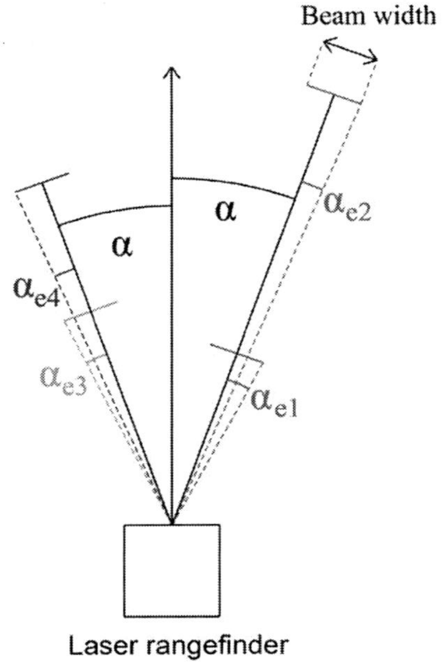


Fig 10. Variance of laser beam with marked beam width

It is clear that the variance of laser beam angle is greater at small distances. For this reason, such part of the measurements will be taken into account, where the influence of the laser beam width on the distribution of laser beam angle is significant. Influence of the laser beam width $a$ on the variance of laser beam angle $\alpha$ is illustrated in Fig. 11. The situation in Fig. 11 describes the measurement, when the laser beam $l'_2$ seemingly measures distance to the object. In fact, it should not fall on this object and the measurement in this angle should include measurement to the background object. The whole issue is caused due to the laser beam width $a$ and due to the variance of laser beam angle $\alpha$. As it can be seen, influence of the laser beam width on the variance of the laser beam angle is not only function of the measured distance, but it is also the function of the relative position between LRF and measured object. This is expressed by the angle $\beta$, which represents the angle between the beam perpendicular to the object $l_0$ and the beam, which represents the boundary of the object $l_1$. However, due to the inaccuracies of the measurement (caused by the above-described effects), the boundary of the object can be represented by the beam $l_2$ – virtual boundary of the object. The difference between $\alpha$ and $\alpha'$ is that $\alpha'$ is not taking into account the width of the laser beam. Easier said, if the model uses only $\alpha'$ it is the second concept of LRF model. If the model uses $\alpha$, it is the third concept of LRF model, because $\alpha$ is derived from $\alpha'$ and $a$.

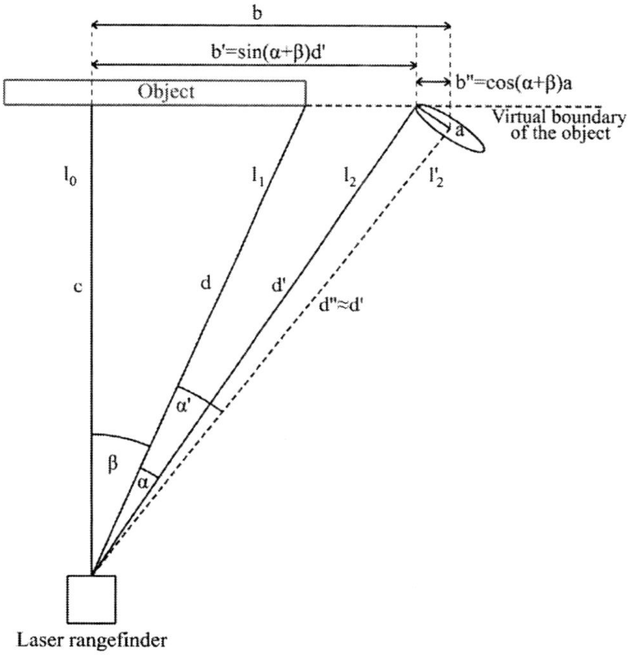

Fig 11. Principle of laser beam variance calculation with beam with taken into account.

It is clear that:

$$b = b' + b'' \tag{2}$$

By the expressions of individual sinuses in the triangles:

$$\sin(\alpha' + \beta) = \sin(\alpha + \beta) + \cos(\alpha + \beta)\frac{a}{d'} \tag{3}$$

There are two unknown variables in the equation (6) – width of the angle $a$ and angle $\alpha$. Seeing that the influence of the laser beam width decreases with increasing measured distance, it can be determined from the measuring of such distance, where this influence is negligible. For this reason, distributions in laser beam angle in Fig. 8 have been divided into two sets. First set contained measurements over the distance of 500 mm and the second set contained measurements under the distance of 500 mm. The angle $\alpha$ was determined on the basis of the first set (assumption that there is no influence of the laser beam width) and it is equal to 0.2473°. When compared to the angular resolution of the Hokuyo LRF, it may seem that two adjacent beams may be replaced in some cases by each other. However, this has not been confirmed by any experiment. And therefore, it appears that the variance of the laser beam angle consists of two parts. The first part is represented by constant variance of angle in whole measurement and second part is represented by variance of angle in individual laser beams. According to this assumption, variance of angle in individual laser beams is not one that would cause the exchange of adjacent beams. However, to confirm this assumption it is necessary to realize further experiments, which are not included in this article.

Regression curve defined by the method of least squares was used for determining the unknown laser beam width $a$:

$$min = \left[ sin(\alpha + \beta) + cos(\alpha + \beta)\frac{a}{d'} - sin(\alpha' + \beta) \right]^2 \tag{4}$$

The minimum of the regression curve for the data shown in Fig. 12 is at $a$ equal to 0.135 mm.

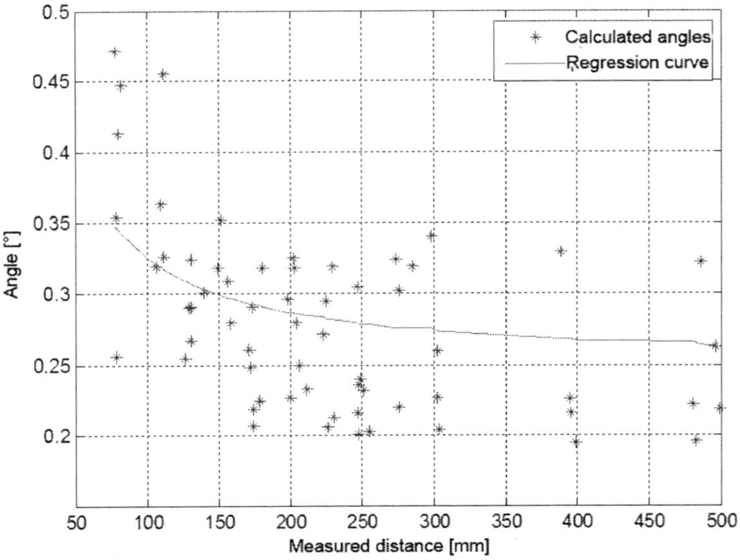


Fig 12. Variance of laser beam angle at short distances with regression curve that characterize the variance with beam width taken into account.

Resulting probabilistic model of Hokuyo LRF:

$$f(x,y) = \frac{1}{2\pi\sqrt{\sigma_x^2\sigma_y^2}} e^{-\left(\frac{(x-\mu_x)}{2\sigma_x^2}+\frac{(y-\mu_y)}{2\sigma_y^2}\right)}$$

$$\sigma_y = a + tan(\sigma_r)x \qquad (5)$$

Where $a$ is laser beam width equal to 0.135 mm, $\sigma_r$ is the standard deviation in laser beam angle equal to 0.2473°, and $\sigma_r$ is standard deviation in the direction of measured distance and it is equal to 5.6894 mm [7]. Resulting model is shown in Fig. 13.

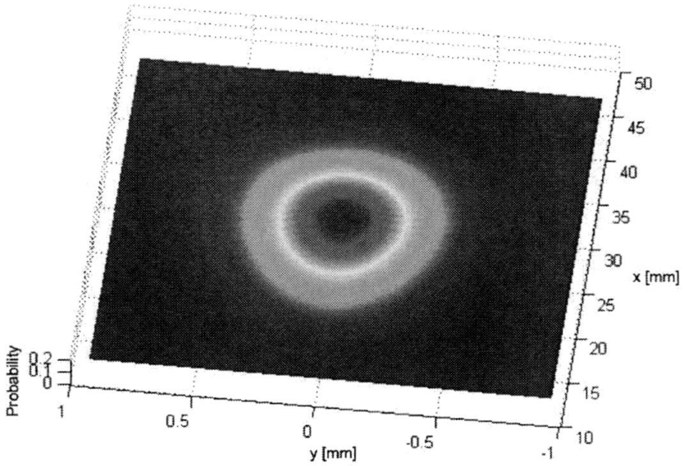


Fig 13. Final probabilictic model of LRF Hokuyou UTM30-LX

## Conclusions

Before using the data from the sensors in mobile robotics, it is very important to understand their interpretation. Therefore, different models of sensors are usually used. We were not satisfied with existing models of Hokuyo type LRF. They were analyzed and the conclusion was that neither of them reflects the real nature of the measurement or expensive calibration equipment is needed for the estimation. Therefore, a new model of such sensor was created. It considers the width of the beam and it can be created without any other equipment.

## Acknowledgments

This work was supported by national grants VEGA 1/0065/16 and VEGA 1/0752/17

## References

[1] K. Konolige, J. Augenbraun, N. Donaldson, C. Fiebig and P. Shah, "A low-cost laser distance sensor", in Proc. IEEE International Conference on Robotics and Automation, 2008, pp. 3002-3008.

[2] S. M. Nejad and S. Olyaee, "Comparison of TOF, FMCW and Phase-Shift Laser Range-Finding Methods by Simulation and Measurement", Quartarly Journal of Technology & Education, vol. 1, no. 1, pp. 11-18, Autumn 2006.

[3] P. Labecki, M. Nowicki and P. Skrzypczynski, "Characterization of a compact laser scanner as a sensor for legged mobile robots", Management and Production Engineering Review, vol. 3, no. 3, pp. 45-52, October 2012.

[4] Y. Okubo, C. Ye and J. Borenstein, "Characterization of the Hokuyo URG-04LX laser rangefinder for mobile robot obstacle negotiation", in Proc. Unmanned Systems Technology XI, 2009, pp. 1-10.

[5] J. Tuley, N. Vandapel and M. Hebert, "Analysis of Artifacts in 3-D LADAR Data", in Proc. IEEE International Conference on Robotics and Automation, 2005, pp. 2203-2210.

[6] F. Pomerleau, A. Breitenmoser, M. Liu, F. Colas, R. Siegwart. "Noise characterization of depth sensors for surface inspections." in 2nd International Conference on Applied Robotics for the Power Industry (CARPI), 2012, Zurich, Switzerland. 2012, <10.1109/CARPI.2012.6473358>. <hal-01142707>

[7] M. Dekan, F. Duchoň, L. Jurišica and A. Vitko, "Probabilistic Model of Laser Rangefinder," AD ALTA: Journal of Interdisciplinary Research, vol. 1, no. 2, pp. 151-155, 2011.

[8] Vagaš, M., M. Sukop, and J. Semjon. "The calibration issues of 3D vision system by using two 2D camera sensors." International Scientific Herald 3.2 (2012): 264-237.

[9] Virgala, Ivan, and Peter Frankovský, Mária Kenderová. "Friction Effect Analysis of a DC Motor." American Journal of Mechanical Engineering 1.1 (2013): 1-5.

[10] I. Virgala, M. Kelemen, E. Prada and R. Surovec, "Motion analysis of snake robot segment," 2013 IEEE 11th International Symposium on Applied Machine Intelligence and Informatics (SAMI), Herl'any, 2013, pp. 145-148. doi: 10.1109/SAMI.2013.6480963

[11] Loncová, Z., Hargaš, L., Koniar, D., Hrianka, M., & Simonová, A. (2015). Parameters measurement of the object in a video sequence. IFAC-PapersOnLine, 48(4), 211-214.

EAN 2017
55th conference on experimental stress analysis

# Probabilistic Localization of Robot in Outdoor Environment Using GNSS

Szabová Martina [1, a], Duchoň František [1,b], Dekan Martin [1,c], Chovanec Ľuboš [1,d]

[1] Faculty of Electrical Engineering and Information Technology STU in Bratislava, Ilkovičova 3, Bratislava, Slovakia

[a] martina.szabova@stuba.sk, [b] frantisek.duchon@stuba.sk, [c] martin.dekan@stuba.sk, [d] lubos.chovanec@stuba.sk

**Keywords:** GNSS, Kalman filter, Particle Filter, Robot localization

**Abstract.** This paper describes a method to determine the two-dimensional position of a mobile robot in known map in an outdoor environment from GNSS measurements. The map is composed of roads from Open Street Map. The algorithm is divided in three parts. First part is preprocessing of measured data by Kalman filter. Second part is localization in the known map – road identification, mobile robot is localized on one specific road. The last part is precise localization on this specific road by particle filter. Algorithm is working with different methods (Kalman filter, fuzzy logic, particle filter) together and obtains good results. The algorithm's performance is demonstrated by using GNSS receiver GTPA010. Verification tests are described at the end of this paper.

## Introduction

In mobile robotics, localization is a problem that can be solved by many different approaches. The choice for a best localization method vastly depends on the type of the environment where the robot should move as well as on the task that it should solve [1-4]. For example, while the GNSS localization will give good results in outdoor environment it would be useless in localization of robot inspecting sewers [5]. Another parameter that affects the choice of localization method is whether the environment in which robot moves is known or unknown and whether the environment is static or dynamic [6-7]. For known or partially known environments many absolute localization methods exist [8]. The most widely used absolute localization methods in outdoor environment are methods based on GNSS [9-12]. This is mostly thanks to its simplicity for end-user and cost of the receivers [10]. For indoor environments, there exist GNSS-like systems, but these require additional cost and effort for transmitter installation [13]. Another approach for robot localization is landmark detection. These systems require either costly sensor (e.g. laser rangefinder) [14] or they are computationally demanding (visual systems) [15, 2]. In the current research, many of these methods are integrated into a SLAM system [16, 6]. Some of these methods require sensors capable of distance measurement or environmental feature detection [17]. Every approach has its advantages as well as disadvantages. Some of the disadvantages, such as multipath problem or sky coverage in GNSS systems, arise from the used sensor, or processing methods used for localization, for example local minimum in ICP algorithm. The aim of this paper is to propose a post processing method that can be used as good initial guess for SLAM algorithms as well as a basic stand-alone method for mobile robot localization on a road. For this purpose

series of existing algorithms were used. These algorithms were integrated into a complex platform. According to our results, using multiple types of processing eliminates some of the disadvantages of respective algorithms and yields to better results than the results from using these algorithms on their own.
For measurements GNSS receiver GTPA010 was used. This GNSS receiver is sending data in NMEA sentences. Due to low sampling resolution of the sensor, steps in measurements are occurring. These steps need to be eliminated e.g. by preprocessing. Preprocessing methods are described in the first part. Requirements for preprocessing methods are characterized as low pass filter structure, which eliminates false measurements and evaluates trend in robot's movement. In this part Kalman filter is applied.
The second part of this paper describes proposed algorithm. It is based on localization of the mobile robot in a map. The map consists of roads that are obtained from Open Street Map framework (OSM). Thus the first step of mobile robot localization is identification of the road where it is possible for the robot to be present. This type of localization does not solve exact robot position, only the possible road (or part of the road). Therefore, particle filter is applied on these results. Particle filter is creating possible positions - particles of the mobile robot on the road. Density of the particles shows where the mobile robot could be. Verification tests of the complex algorithm are in the section 4 in this paper.

## 2 Theoretical backgrounds

In the following section theory of all methods that were used in the paper is described. First step of the proposed algorithm is preprocessing of the GNSS data. Good preprocessing method is essential for the next parts of the algorithm. In our research Kalman filter was applied. Kalman filter is a recursive filter which means that all previous measurements are incorporated in the last filtered measurement. Kalman filter in general solves the problem of state **x** estimation.
Kalman filter theory, which was used in this paper, is based on [11, 14, 16, 18-19].
The mobile robot is usually moving with slow velocity. In [19] is described how to set initial value of the covariance matrix **Q** for slow movement, which will be applied later.

$$\begin{aligned}\mathrm{diag}(\mathbf{A}) &= \begin{pmatrix}1 & 1 & 1 & 1\end{pmatrix}\\ \mathrm{diag}(\mathbf{Q}) &= \begin{pmatrix}Q'_{dx}\Delta t & Q'_{dy}\Delta t & Q'_{dz}\Delta t & \sigma^2_{\delta t}\end{pmatrix}\end{aligned} \tag{1}$$

The Kalman filter output is minimum mean-square error estimate of the real state. Kalman filter covariance matrixes were set according to parameters of our sensor that were identified through experiment. State vector **x** and its covariance matrix **P** were initialized to zero values.

When the mobile robot is moving through crossroads it is hard to decide on which of the roads it is located. In this kind of situation the algorithm needs to work with a hypothesis that the robot is located on any of these roads. Fuzzy logic allows to work with a set of information where sum of the probabilities is bigger than 1. Truth of sentence is described by membership function, which evaluates the sentence by the value of probability. These sets are called fuzzy sets. Membership function are typically trapezoidal, triangular or in Z or S form [20, 18].
In this paper fuzzy logic is used for probability calculation, on which road in the map is mobile robot located on. Each of the roads can have probability between 0 and 1, with 0.5 being a threshold value where the algorithm presumes that the robot is located on that road.

Thus fuzzy logic application for the mobile robot localization allows the algorithm to work with more than one road with a probability higher than 0.5.
After this localization one or more roads have higher probability than 0.5 with no exact position on these roads. Due to this, it is necessary to apply next method for more precise localization. For this problem, particle filter is applied.
Particle filter is Bayes filter. In general, particle filter algorithm consists of 4 steps (figure 1). In the first step particles are created randomly. For best results particles should have uniform distribution. In the next step the state of each particle is changed according to a control signal. The third step occurs when a new measurement is obtained. According to this measurement, probability is calculated for each particle. The probability is higher for particles that are located near the measurement and lower for more distant particles. In the last step of the algorithm particles are resampled from previous particles according to their probability [8, 15, 18, 21].

**Algorithm Particle_filter($\mathcal{X}_{t-1}, u_t, z_t$):**
  $\bar{\mathcal{X}}_t = \mathcal{X}_t = \emptyset$
  for $m = 1$ to $M$ do
    sample $x_t^{[m]} \sim p(x_t \mid u_t, x_{t-1}^{[m]})$
    $w_t^{[m]} = p(z_t \mid x_t^{[m]})$
    $\bar{\mathcal{X}}_t = \bar{\mathcal{X}}_t + \langle x_t^{[m]}, w_t^{[m]} \rangle$
  endfor
  for $m = 1$ to $M$ do
    draw $i$ with probability $\propto w_t^{[i]}$
    add $x_t^{[i]}$ to $\mathcal{X}_t$
  endfor
  return $\mathcal{X}_t$

**Fig. 1**: Algorithm of Particle filter [23]

### 3 Algorithm

In the following text, the basic principles of proposed localization method are described. The map where the mobile robot is localized is based on roads obtained from Open Street Map. Probability for each road that is near the robot is determined. This value describes odds that the mobile robot is located on the road. The probability is determined as a fuzzy function of the distance between GNSS position and its projection on the road. The function is determined on the basis of experimental measurements in different conditions. Fuzzy membership function is based on:

- precision of the sensor,
- shortest distance between filtered position,
- and the road and HDOP information from GGA sentence.

For better clarity, the road probability in following figures is plotted in a color scale (fig. 2). Thanks to the fuzzy logic it is possible to set higher probability than 0.5 to more than one road (e.g. situation on a crossroad).

**Fig. 2:** Map of identified roads

This method of mobile robot localization does not calculate the exact robot position. The output of this algorithm is set of roads, with probability that is higher than 0.5 e.g. there is a chance that the robot is on one of these roads. To determine the robot position on the road particle filter is used. Particle filter generates 500 particles - positions, that are distributed on all roads with probability higher than 0.5. The count of the particles on a particular road is determined by its probability and sum of the probabilities of all roads with probability higher than 0.5.

Probability of the particles that is used in 5$^{th}$ step of the algorithm is defined as fuzzy function which is based on distance. Control signal was obtained from mobile robot. Particle filter algorithm is shown in figure 4. New particles have uniform probability within a limited distance from crossroad location. When the road is identified again the new particles are generated from the previous ones that were associated with this road with respect to their probability - resampling. Result of this algorithm is possible mobile robot location on the road.

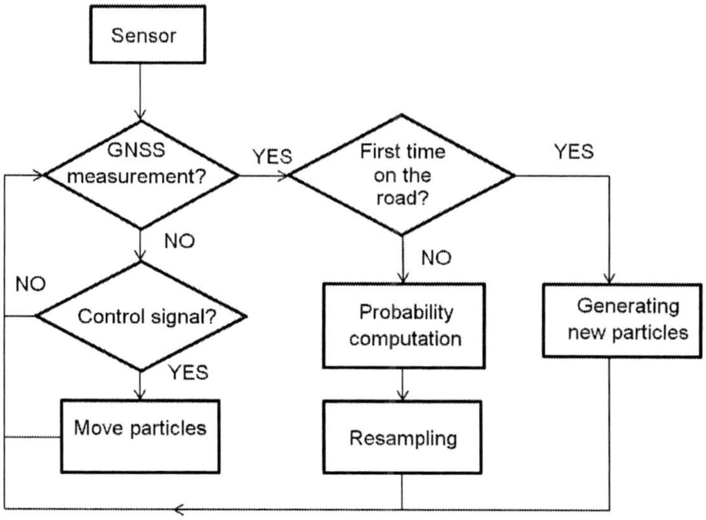


**Fig. 3**: Algorithm of particle filter

## 4 Verification

Following section deals with verification of proposed algorithm in real life experiments. The aim of these experiments is to determine whether the algorithm works correctly. Four measurements were made in park with huge road net for verification of the proposed algorithm. Every measurement contains more than 3 000 samples. Information about the roads from OSM was used as reference. All movements passed through crossroads and part of the movement was under trees or close to buildings. Measurements were made in varied time periods with different satellite constellations. Situation in measurement 3 is shown in figure 4 and detail is in figure 5.

**Fig. 4:** Measurement 3 in Google Map. Red - measured data, blue - filtered data, green – road OSM data

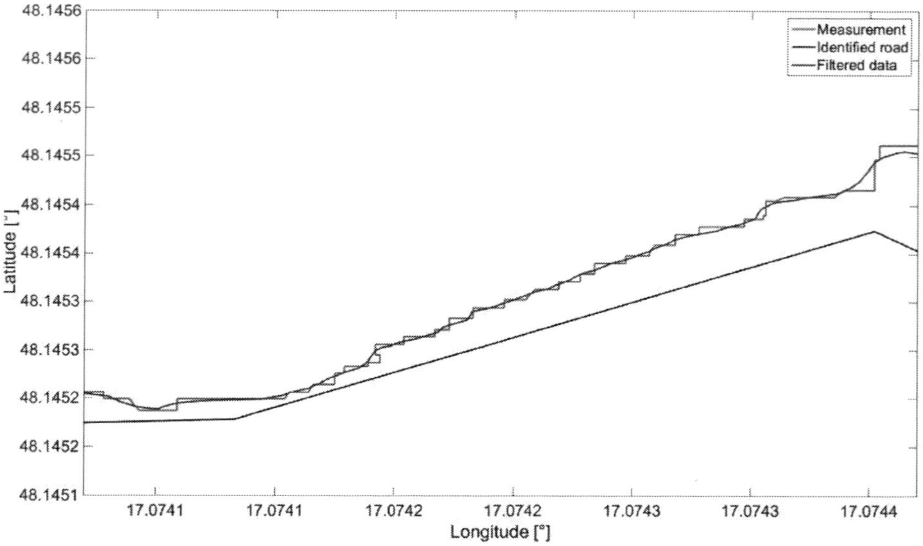


**Fig. 5:** Measurement 3 detail: Red - measured data, blue - filtered data, black - identified road in OSM

First test was performed on measured and filtered data. The aim of it was to validate Kalman filter application. Table 1 contains data from this test and describes RMS of filtered and measured data. The first column contains RMS for filtered data; the second column contains RMS for measured data. RMS is used to describe which data is numerically closer to the road.

**Table 1** RMS of filtered and measured data

| | RMS Filtered data [m] | RMS Measured data [m] |
|---|---|---|
| Measurement 1 | 2.90 | 2.85 |
| Measurement 2 | 3.57 | 3.71 |
| Measurement 3 | 2.49 | 2.56 |
| Measurement 4 | 3.09 | 5.68 |

In this part of the first test filtered data had better results. Measurement 2 has better results for measured data. This may be caused due to different reasons. E.g. OSM allows to edit road information by users, that is why it is not known how was the road measured (wide road may have been measured in the middle or on the right or left side of the road). Other measurements are filtered closer to reference road. Generally filtered data are better than measured data.

Mobile robot movements are linear and continuous. Measured data are stepped and that is not similar to real robot movement. Kalman filter application suppresses these steps and therefore filtered movement looks more like a real robot motion. In this respect filtered data exceeds the measured data significantly. Choice of Kalman filter and its setting may be considered as sufficient (figure 5).

The second test was based on the same measurements as the previous one. The aim of this test is to verify the capability of the algorithm to correctly determine on which road the mobile robot is moving. This test is divided into two parts. First part consists of the mobile robot movement on a single road. Table 2 shows data from this first part and it is split into three columns. The first column (F - false) describes when no road was identified, the second column (FP – false positive) describes when a road was identified, but the mobile robot was moving on another road, the third column (P-positive) describes when a road was identified and the mobile robot was moving on this road.

**Table 2** Identification on single road. F - false, FP - false positive and P - positive identification

| Type of movement | Single road | | |
|---|---|---|---|
| Identification | F [%] | FP [%] | P [%] |
| Measurement 1 | 11.9 | 13.1 | 75.0 |
| Measurement 2 | 4.2 | 0.0 | 95.8 |
| Measurement 3 | 5.5 | 0.0 | 94.6 |
| Measurement 4 | 0.0 | 0.0 | 100.0 |

When mobile robot was moving on one road, this road was correctly identified in more than 75% of all measurements. Only in one measurement the road was not correctly identified and robot had identified its movement on other road. This could be caused by real road width, which is in contrast with map representation in OSM, which is represented as a line without width. Shift between the real position and GNSS position can be caused by bad visibility e.g. due to trees. The wrong identification happened in less than 5% in our experiments.

The second part is devoted to a situation where the robot is moving through crossroads of two or three roads. The crossroad is defined as a small area around a point where two or more roads are merging. These line segments are approximately 4-8 meters long. Table 3 describes data from second part of this test and it is split into two main columns. First main column is for two roads crossroads, second column is for three and more roads crossroads. Each of main columns is split to three subcolumns (F, FP, P). The column F describes when at least one of the roads from the crossroad had smaller probability than 0.5, the column FP describes when all roads in crossroad are identified but the mobile robot is moving through another crossroad or outside the crossroad area, the column P describes when all roads in crossroad are identified and the mobile robot is moving through this crossroad. In the fourth measurement robot was moving only on one road. For this reason the fourth measurement is not listed in the table 3.

**Table 3** Identification on crossroad. F - false, FP - false positive and P - positive identification

| Type of movement | Crossroads | | | | | |
|---|---|---|---|---|---|---|
| | 2 roads | | | 3 roads | | |
| Identification | F [%] | FP [%] | P [%] | F [%] | FP [%] | P [%] |
| Measurement 1 | 0.0 | 0.0 | 21.0 | 0.0 | 0.0 | 79.0 |
| Measurement 2 | 0.0 | 0.0 | 100.0 | 0.0 | 0.0 | 0.0 |
| Measurement 3 | 0.0 | 0.0 | 23.9 | 0.0 | 0.0 | 76.1 |

When robot was moving through a crossroad all roads were identified correctly. The algorithm was working correctly and it also performed well when robot was moving close to the trees.

The output from the previous part of the algorithm is particular road or crossroad. To determine exact position on the road particle filter is applied for each identified road. The test for verification of the particle filter was based on real mobile robot position and the position gained from particle filter. The output of the particle filter is not only one possible position, but set of possible positions with its probability. Thus for this verification test one particle position had to be chosen from this set. The position was generated for each road that had particles as a center of gravity (CoG) of these particles. Results of this test are shown in table 4. Some results of particle filter are shown in figure 6.

**Table 4** RMS of distance between CoG of particles and real position of the mobile robot

| | RMS of distance - CoG and real | |
|---|---|---|
| | Single road [m] | Crossroad [m] |
| Measurement 1 | 1.68 | 1.41 |
| Measurement 2 | 1.89 | 1.93 |
| Measurement 3 | 1.23 | 1.15 |
| Measurement 4 | 1.10 | - |

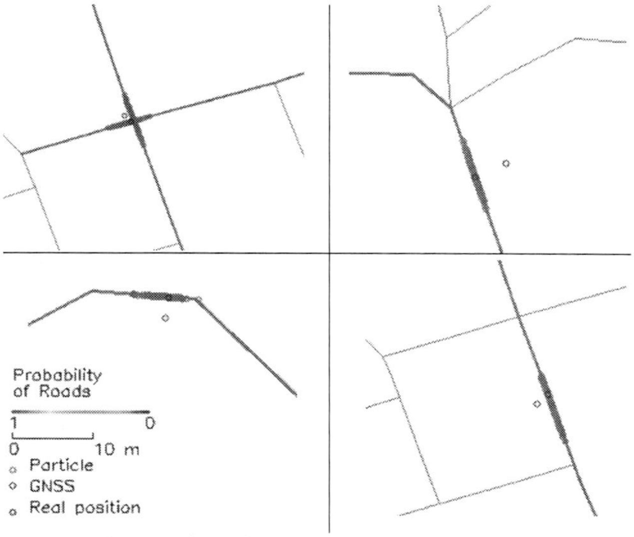


**Fig. 6**: Results of algorithm with particle filter

Average distance between CoG of particles and real position was between 1 and 2 meters. Accuracy of used GNSS receiver is 3 meters. Thus our algorithm can improve GNSS measured position by 33%.

**7 Conclusion**

In this paper results of GNSS positioning with filtering and road identification algorithms have been presented. In the introduction GNSS parameters and methods was described. Afterwards, the basics and parametrization of the filters were explained. Next section described the whole algorithm, which also uses particle filter for exact position identification. Results of several experiments, which support proposed algorithm, are also presented. Verification was made in several different environments and it shows that the algorithm is well designed. Every method improves results in some part. Kalman filter eliminates Gaussian noise. Implementation of fuzzy logic in road identification allows working with crossroads and not only with one road, and particle filter localizes robot on these roads. Combination of the methods used in our algorithms leads to better results than using these methods separately. Accuracy was improved to approximately 2 meters compared to 3 meters accuracy of GNSS receiver. Algorithm is proposed so that system can be expanded or changed. It can be expanded to by new sensor, which can improve data preprocessing or localization on the road. Moreover, proposed algorithm is able to work in real time. This is necessary condition for usage in mobile robot localization. The future work is to expand the system by INS sensor and implement data fusion of INS and GNSS, which will improve the algorithm. Other improvement of the algorithm is self-learning. Its aim is to make fuzzy function and covariance matrixes to learn from previous measurements.

**Acknowledgments** This work was supported by national grants VEGA 1/0065/16 and VEGA 1/0752/17, and authors would like to thank program Aktion Slowakei-Österreich and for support of prof. Bernhard Hofmann-Wellenhof.

**References**

[1] Ouyang, R.W., Wong, A.K.-S. Kam Tim Woo. GPS Localization Accuracy Improvement by Fusing Terrestrial TOA Measurements. In: IEEE International Conference on Communications (ICC); 23-27 May 2010; Cape Town

[2] Debiao Lu, Schnieder, E. Performance Evaluation of GNSS for Train Localization. Intelligent Transportation Systems, IEEE Transactions. 2014; 1054-10510. DOI: 10.1109/TITS.2014.2349353
[3] Matias, B., et al.. High-accuracy low-cost RTK-GPS for an unmannned surface vehicle. In: OCEANS 2015 - Genova; 18-21 May 2015; Genova
[4] Hofmann-Wellenhof B., Legat K., Wieser M. Navigation Principles of Positioning and Guidance. Springer: Wien; 2003
[5] Aidan F. Browne, Vutetakis D. Localization of Autonomous Mobile Ground Vehicles in a Sterile Environment: A survey. In: ASEE Southeast Section Conference; 14-17. April 2015
[6] Jae-Hong Shim, Young-Im Cho A Mobile Robot Localization using External Surveillance Cameras at Indoor. Procedia Computer Science. 2015, DOI 10.1016/j.procs.2015.07.242
[7] Burian, F., Florian, T., Zalud, L. The Identification of Drivers Behaviour Through the Use of GPS and Odometry. In: 17th International Conference on Soft Computing MENDEL 2011, p. 492-496
[8] Linhai Xie, Xiaohong Xu Mobile Robot Self-localization System Based on Multi-sensor Information Fusion in Indoor Environment. In Proceedings of the 2015 Chinese Intelligent Automation Conference, pp. 61-69. Springer, Berlin; 2015
[9] Ioannis Kostavelis, Antonios Gasteratos Semantic mapping for mobile robotics tasks: A survey. In: Robotics and Autonomous Systems. 2015, 86-103
[10] Yezhang Tu, Zhiwu Huang, Xiaoyong Zhang, Wentao Yu, Yang Xu, Bin Chen The Mobile Robot SLAM Based on Depth and Visual Sensing in Structured Environment. Robot Intelligence Technology and Applications 3.2015; pp. 343-357
[11] Yang Tian, Ruipeng Gao, Kaigui Bian, Fan Ye, Tao Wang, Yizhou Wang, Xiaoming Li Towards ubiquitous indoor localization service leveraging environmental physical features. In: INFOCOM. 2014 Proceedings IEEE.2014 DOI 10.1109/INFOCOM.2014.6847924
[12] Nejdl Lukas et al. Remote-controlled robotic platform ORPHEUS as a new tool for detection of bacteria in the environment. ELECTROPHORESIS. 2014;
[13] Xian-lun TANG, La-mei LI, Bo-jie JIANG Mobile robot SLAM method based on multi-agent particle swarm optimized particle filter. The Journal of China Universities of Posts and Telecommunications. 2014; 78-86
[14] Chen G., Pham T. T. Introduction to Fuzzy Sets, Fuzzy Logic and Fuzzy Control Systems. CRC Press. 2001
[15] Ivan Virgala, Michal Kelemen, Tatiana Kelemenová, Tomáš Lipták, Matej Poláček Design of Mobile Inspection Robot. American Journal of Mechanical Engineering. 2014;
[16] Jorge Fuentes-Pacheco, José Ruiz-Ascencio, Juan Manuel Rendón-Mancha Visual simultaneous localization and mapping: a survey. Artificial Intelligence Review. 2015; 55-81
[17] Quoc-Hung Nguyen, et al. A Visual SLAM System on Mobile Robot Supporting Localization Services to Visually Impaired People. In: Computer Vision - ECCV, 2015; 716-729
[18] Thrun S., Burgard W., Fox D.: Probabilistic Robotics. MIT Press, Massachusetts; 2005
[19] Hung-Yuan Chung, Chun-Cheng Hou, Yu-Shan Chen: Indoor Intelligent Mobile Robot Localization Using Fuzzy Compensation and Kalman Filter to Fuse the Data of Gyroscope and Magnetometer. Industrial Electronics, IEEE Transactions. 2015;6436 – 6447
[20] Subirana J.S., Zornoza J.M.J., Hernándes-Pajares M. GNSS Data Processing, Vol I: Fundamentals and Algorithms. ESA Communications. Noordwijk; 2013
[21] Dieter Fox, Sebastian Thrun, Wolfram Burgard, Frank Dellaert Particle Filters for Mobile Robot Localization. In: Sequential Monte Carlo Methods in Practice, Springer. 2015;401-428

EAN 2017
55th conference on experimental stress analysis

# Optimization of Hexacopter Stability Control with Disturbances in its Stochastic State Space Dynamics Model

I. N. Ibrahim, M. A. Al Akkad and E. V. Sosnovich
Department of Mechatronics and Robotics
Kalashnikov Izhevsk State Technical University
Izhevsk, Russia

ibrncfe@gmail.com, aimanakkad@yandex.ru, ellasosnovich@gmail.com

**Keywords:** Linear Quadratic Regulator, Stochastic State Space, Hexacopter Dynamics, Disturbances, Nonlinear Control, Coupled and Underactuated Models.

**Abstract.** In this paper, the behavior of a hexacopter with the existence of disturbances inserted into the equations of motion was explored. These disturbances emulate the motion of a robotic arm attached to the body of the aircraft model. Initially, kinematics and dynamics are described. Then, equations of motion are derived for modeling and disturbances analysis. The derived dynamic model reflects the real motion of the hexacopter with respect to the earth, which is also characterized by nonlinearity, time variance, underactuation and coupling among the equations' variables. The state space method was used to write the equations in addition to decoupling and linearization techniques in order to design the controllers. Finally, PID and LQR controllers and simulation results are presented. Both PID and LQR controllers have been investigated to control the aircraft model. The results of comparison between them indicate that the LQR controller is kind of an optimization technique and used to stabilize the attitude of the hexacopter. It consumes low energy with respect to the PID controller in addition to rejecting the perturbation and noise in the stochastic state space of the aircraft model.

## 1. Introduction

Nowadays, Drones invade several application domains [1,2,3,20,24,26]. The control of an aerial robot such as a hexacopter requires studying its dynamics in order to account for gravity effects, aerodynamic forces [1,2,21,22,23] and disturbances. This work will focus on the modeling, simulation, and control of a hexacopter type UAV. The reason for choosing a hexacopter is challenging in the control field because it's a highly nonlinear, multivariable, coupled and underactuated system, in addition to its advantages such as high maneuverability and stationary flight [4,5]. Underactuated systems, defined as a mechanical system in which the dimension of the configuration space exceeds that of the control input space, that is, with fewer control inputs than degrees of freedom [2]. Modeling of such a system is not a trivial problem due to the coupled dynamics of the aerial vehicle [3]. The contributions of this work are deriving an accurate and detailed mathematical model of a hexacopter UAV in regard to the earth frame. The equation of motion of the whole system was designed using the Newton-Euler formulation for translational and rotational dynamics of a rigid body [4,5,6,7,8,25,26]. The disturbances are presented as an outdoor environment for simulation and it was considered as a so complicated problem in addition to the stochastic idea in the state space model [9, 10], which is omitted in most of the literature. This paper includes the disturbances, which represent the robotic arm movement of a robotic arm attached to the hexacopter in the equations of force and moment. Lucia [3] and Hasan [4] work in comparison to this research relied on a complex mechanism through the mathematical modeling of certain arms of fixed design, and adding it to the flying object equations. This is considered limited, complex, and does not cover the changes that may occur in the air, and weather conditions to which the aircraft is exposed in the air far from laboratory conditions. The disturbances have been measured using LabVIEW software and used to build a stochastic state space model [11], which is a simple type of representation with the existence of noise. Two types of control methods were investigated, using proportional integral deferential PID controller and linear quadratic regulator LQR controller in order to get a good response in addition to low energy consumption. Bouabdallah implemented such a type of controller in the closed loop system to stabilize the angular attitude of a UAV [12] [13] [14]. Castillo implemented an LQR controller too, to stabilize a quadcopter in hover situation [15]. Hoffmann et al. used the LQR as a solution to the energy problem in order to apply lower costs on attitude deviations by varying the Q matrix, but this degraded the tracking performance. A good compromise is found in [16]. This paper is structured as follows: initially, kinematics and dynamics are described. Then, equations of motion are derived for modeling and disturbances analysis. Finally, PID and LQR controllers and simulation results are presented. The future work will focus on developing attitude, position, and altitude robust controllers to obtain proper strategies for autopilot stabilization and trajectory control.

## 2. Reference Systems for the UAV Hexacopter

To describe the hexacopter motion only two reference systems are necessary: earth inertial frame (E-frame) and body-fixed frame (B-frame). The inertial frame is the system that uses the North, East, and Down (NED) coordinates and the origin of this reference system is fixed in one point located on the earth surface as shown in Fig. 1, and the (X, Y, Z) axes are directed to the North, East, and Down, respectively. The mobile frame $(X_B, Y_B, Z_B)$ is the body fixed frame that is centered in the hexacopter center of gravity (CG) and oriented as shown in Fig. 1. The angular position of the body frame with respect to the inertial one is usually defined by means of the Euler angles: roll $\phi$, pitch $\theta$, and yaw $\psi$, where vector: $\sigma = [\phi\ \theta\ \psi]^T, \phi\ and\ \theta \in ]-\frac{\pi}{2}, \frac{\pi}{2}[;\ \psi \in ]-\pi, \pi[$. The inertial frame position of the vehicle is given by vector $\xi = [x\ y\ z]^T$ [4, 16, 17]. The transformation from the body frame to the inertial frame is realized by using the well-known rotation matrix $C_b^n$ [16, 17]. while the transformation matrix for angular velocities from the body frame to the inertial one is $S$ [3, 18]. Where $\dot{\sigma} = S\ .\ \Omega$ , $\dot{\xi} = C_b^n\ .\ V$ , the angular velocity is defined by vector $\Omega = [p\ q\ r]^T$ , and the linear velocity is defined by vector $V = [u\ v\ w]^T$ in the body frame.

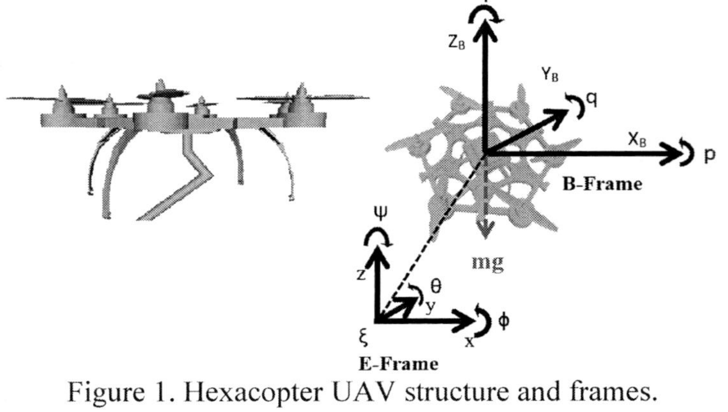


Figure 1. Hexacopter UAV structure and frames.

## 3. Dynamical Model of Hexacopter

To describe the hexacopter dynamics, that is assumed to be a rigid body and has a symmetrical structure, Newton-Euler equations [3, 18] that govern linear and angular motion are used. This model has the The thrust and torque as in [15] are: $|T_i| = \rho C_T A R^2 \omega_i^2$ , and $|Q_i| = \rho C_Q A R^3 \omega_i^2$, Where blade rotation is with angular velocity $\omega$, the blade radius is $R$, $C_T$ and $C_Q$ are respectively thrust and torque coefficients, $\rho$ is the air density and A is the disc area, while the opposing force to the travelling of the hexacopter in air is the drag force and can be expressed by the following equation at the earth's frame: $F_{AI} = K_{TI}.\dot{\xi}$, where $K_{TI}$ is a diagonal matrix related to the aerodynamic friction constant $k_t$ [15, 17]. the gravity force in the earth frame according to [17] is: $F_{GI} = m[0 \quad 0 \quad g]_E^T$. Other forces like the Coriolis force from the earth, the wind, and Euler forces are considered as a disturbance, summarized as $F_{DI}$ in the earth frame: $F_{DI} = [F_{dIx} \quad F_{dIy} \quad F_{dIz}]_E^T$. the inertia matrix of the aircraft is J so according to [3, 4, 17] is defined as the following form: $J = [J_{xx}\ 0\ 0\ ;0\ J_{yy}\ 0\ ;0\ 0\ J_{zz}]^T$; $J \in R_{3\times3}$. $m$ is the mass of hexacopter and $l$ is the distance from CG to the center of the propeller. The attitude of the vehicle in the air change, by controlling the angular velocity of motors, then the thrust moment vector is defined as $M_T = [M_p \quad M_q \quad M_r]^T$, where $M_p, M_q, M_r$ are the moments about the axes $X_B, Y_B, Z_B$ in the body frame [3, 4, 17]. The aerodynamic moment is expressed by the following equation: $M_{AI} = K_{RI}.\Omega^2 = K_{RI}[\dot{\phi}^2 \quad \dot{\theta}^2 \quad \dot{\psi}^2]^T$ , where $K_{RI}$ is a diagonal matrix related to the rotational aerodynamic friction constant by the parameter $K_r$ [15, 17]. Disturbance moment is the total of the disturbances affecting the torque around the aircraft axes expressed as $M_{DI} = [M_{dI\phi} \quad M_{dI\theta} \quad M_{dI\psi}]^T$. Propeller Gyroscopic effect $M_{gyro} = [-J_r\dot{\theta}\omega_r \quad J_r\dot{\phi}\omega_r \quad 0]^T$ [3], Where $J_r$ is the rotational inertia of the propeller $[NmS^2]$ and the $\omega_r$ is the overall propeller speed $[rad/s]$, as $\omega_r = -\omega_1 + \omega_2 - \omega_3 + \omega_4 - \omega_5 + \omega_6$. Yaw counter moment is the differences in rotational acceleration of the propellers defined as $M_{counter} = [0 \quad 0 \quad J_r\dot{\omega}_r]^T$ [3]. When the transformation matrix for angular velocities $S \rightarrow I$ the hexacopter tends to the stable point, therefore the equations of the angular rate will be related to the earth frame, in addition to considering some assumptions as: $\frac{Jr}{Jx} = \frac{Jr}{Jy} = \frac{Jr}{Jz} \rightarrow 0$ is a small effect around zero, $a = \frac{k_t}{m} \rightarrow 0$ and $J_x = J_y$. Therefore, the equations that govern the translational and rotational motion with respect to the earth frame are shown in Eq. 1, The control input to the system is vector $U = [U_x \quad U_y \quad U_z \quad U_p \quad U_q \quad U_r]^T$.

$$\begin{pmatrix} \ddot{x} = U_x - \frac{k_t}{m}\dot{x} + \frac{F_{dIx}}{m} \\ \ddot{y} = U_y - \frac{k_t}{m}\dot{y} + \frac{F_{dIy}}{m} \\ \ddot{z} = U_z - \frac{k_t}{m}\dot{z} - g + \frac{F_{dIz}}{m} \end{pmatrix} \text{and} \begin{pmatrix} \ddot{\phi} = U_p + b_1\dot{\theta}\dot{\psi} + c_1\,\dot{\phi}^2 + \frac{M_{dI\phi}}{Jx} \\ \ddot{\theta} = U_q + b_2\dot{\phi}\dot{\psi} + c_2\,\dot{\theta}^2 + \frac{M_{dI\theta}}{Jy} \\ \ddot{\psi} = U_r + c_3\,\dot{\psi}^2 + \frac{M_{dI\psi}}{Jz} \end{pmatrix} \tag{1}$$

where

$$\begin{cases} U_p = \frac{M_p}{Jx} = \frac{\sqrt{3}\rho l c_T A R^2}{2Jx}(\omega_3^2 + \omega_6^2 - \omega_4^2 - \omega_5^2) = a_\phi(\omega_3^2 + \omega_6^2 - \omega_4^2 - \omega_5^2) \\ U_q = \frac{M_q}{Jy} = \frac{\rho l c_T A R^2}{2Jy}(\omega_3^2 + \omega_5^2 + 2\omega_1^2 - \omega_4^2 - \omega_6^2 - 2\omega_2^2) = a_\theta(\omega_3^2 + \omega_5^2 + 2\omega_1^2 - \omega_4^2 - \omega_6^2 - 2\omega_2^2) \\ U_r = \frac{M_r}{Jz} = \frac{\rho c_Q A R^3}{2Jz}(\omega_1^2 + \omega_4^2 + \omega_6^2 - \omega_2^2 - \omega_3^2 - \omega_5^2) = a_\psi(\omega_1^2 + \omega_4^2 + \omega_6^2 - \omega_2^2 - \omega_3^2 - \omega_5^2) \\ b_1 = \frac{(Jy - Jz)}{Jx}, b_2 = \frac{(Jz - Jx)}{Jy}, c_1 = -\frac{k_r}{Jx}, c_2 = -\frac{k_r}{Jy}, c_3 = -\frac{k_r}{Jz} \end{cases} \tag{2}$$

$$\begin{cases} U_x = (\cos\phi\cos\psi\sin\theta + \sin\phi\sin\psi)\, u_T/m = a_x(\phi,\theta,\psi) \sum_{i=1}^{6} \omega_i^2 \\ U_y = (\cos\phi\sin\theta\sin\psi - \sin\phi\cos\psi)\, u_T/m = a_y(\phi,\theta,\psi) \sum_{i=1}^{6} \omega_i^2 \\ U_z = (\cos\phi\cos\theta)\, u_T/m = a_z(\phi,\theta) \sum_{i=1}^{6} \omega_i^2 \\ u_T = \sum_{i=1}^{6} |T_i| = \rho C_T A R^2 \sum_{i=1}^{6} \omega_i^2 \end{cases} \tag{3}$$

## 4. State Space Model

The dynamic model presented in translational and rotational equation set can be rewritten in the state-space form: $\dot{X} = f(X) + g(X,U) + \delta$, where δ is the disturbances, and $X \in \mathcal{R}^{12}$ is the vector of state variables given as follows: $X = [x \ \ \dot{x} \ \ y \ \ \dot{y} \ \ z \ \ \dot{z} \ \ \phi \ \ \dot{\phi} \ \ \theta \ \ \dot{\theta} \ \ \psi \ \ \dot{\psi}]^T$, Then we can derive the final equations of the system, which governs the transitional and rotational of hexacopter, with respect to the earth frame in state space as follows:

$$\begin{pmatrix} \dot{x}_2 \\ \dot{x}_4 \\ \dot{x}_6 \end{pmatrix} = \begin{pmatrix} -a & 0 & 0 \\ 0 & -a & 0 \\ 0 & 0 & -a \end{pmatrix} \begin{pmatrix} x_2 \\ x_4 \\ x_6 \end{pmatrix} + \begin{pmatrix} U_x \\ U_y \\ U_z \end{pmatrix} + \begin{pmatrix} F_{dIx}/m \\ F_{dIy}/m \\ \frac{F_{dIz}}{m} - g \end{pmatrix} = \begin{pmatrix} U_x \\ U_y \\ U_z \end{pmatrix} + \begin{pmatrix} F_{dIx}/m \\ F_{dIy}/m \\ \frac{F_{dIz}}{m} - g \end{pmatrix} \tag{4}$$

$$\begin{pmatrix} \dot{x}_8 \\ \dot{x}_{10} \\ \dot{x}_{12} \end{pmatrix} = \begin{pmatrix} b_1 x_{10} x_{12} + c_1 x_8^2 \\ b_2 x_8 x_{12} + c_2 x_{10}^2 \\ b_3 x_8 x_{10} + c_3 x_{12}^2 \end{pmatrix} + \begin{pmatrix} 1 & 0 & 0 \\ 0 & 1 & 0 \\ 0 & 0 & 1 \end{pmatrix} \begin{pmatrix} U_p \\ U_q \\ U_r \end{pmatrix} + \begin{pmatrix} \frac{M_{dI\phi}}{Jx} \\ \frac{M_{dI\theta}}{Jy} \\ \frac{M_{dI\psi}}{Jz} \end{pmatrix} . \tag{5}$$

The control input to the system is vector $U = [U_x \ \ U_y \ \ U_z \ \ U_p \ \ U_q \ \ U_r]^T$, where the following equation Eq. 21 gives the relation between the angular speed of the propellers and the control inputs. The model of disturbances, which simulates the robotic arm, was presented as shown in Fig. 2. The final concluded equations represent nonlinearity, coupling and time-variance in the variables that are not considered in other papers like [3, 4], which use simplification techniques as well as linearization tools. Any change in the input variables leads to changes in most of the output variables. With respect to the control input vector U, it is clear that the rotational subsystem is fully-actuated, while the translational subsystem is underactuated as it is dependent on both the translational state variables $x_1$ to $x_6$ and the rotational ones $x_7$ to $x_{12}$.

$$\begin{bmatrix} U_x \\ U_y \\ U_z \\ U_p \\ U_q \\ U_r \end{bmatrix} = \begin{bmatrix} +a_x & +a_x & +a_x & +a_x & +a_x & +a_x \\ +a_y & +a_y & +a_y & +a_y & +a_y & +a_y \\ +a_z & +a_z & +a_z & +a_z & +a_z & +a_z \\ 0 & 0 & +a_\phi & -a_\phi & -a_\phi & +a_\phi \\ 2a_\theta & -2a_\theta & +a_\theta & -a_\theta & +a_\theta & -a_\theta \\ +a_\psi & -a_\psi & -a_\psi & +a_\psi & -a_\psi & +a_\psi \end{bmatrix} \cdot \begin{bmatrix} \omega_1^2 \\ \omega_2^2 \\ \omega_3^2 \\ \omega_4^2 \\ \omega_5^2 \\ \omega_6^2 \end{bmatrix} \tag{6}$$

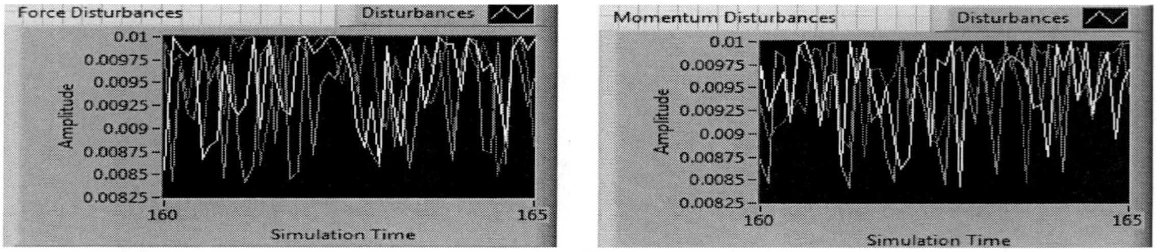

Figure 2. Force and momentum disturbances resulted from the robotic arm.

## 5. Proportion, Integral, And Derivative Controller Design

Four linear and classical PID controllers are presented, and their parameters were tuned using Ziegler Nichols algorithm [19]. The controllers speed here is considered suitable in a way that enables tracking the vibrations and changes in order to avoid the vibrations in the output of the aircraft as much as possible. The aim of these controllers is to control attitude and altitude of the vehicle at desired trajectory in space. The PID mathematical model is described as [19]:

$$u(t) = k_C[e(t) + \frac{1}{T_i}\int e(t).d(t) + T_d\frac{d}{dt}e(t)] \quad (7)$$

Parameter $k_c$ is the proportional gain, $T_i$ is the integral time, and $T_d$ is the derivative time. These parameters are defined, in order to get the best performance by decreasing vibrations, steady-state errors, and response time. Fig. 3 shows a block diagram of PIDs controlling hexacopter. The parameters of each PID are defined as:

$$u_T(t) = 24.19\left[e(t) + \frac{1}{0.08299}\int e(t).d(t) + 0.01328\frac{d}{dt}e(t)\right] \quad (8)$$

$$u_p(t) = 0.001\left[e(t) + \frac{1}{6.1}\int e(t).d(t) + 0.1\frac{d}{dt}e(t)\right] \quad (9)$$

$$u_r(t) = -0.00103\left[e(t) + \frac{1}{6.3}\int e(t).d(t) + 0.04073\frac{d}{dt}e(t)\right] \quad (10)$$

$$u_y(t) = 0.006\left[e(t) + \frac{1}{20}\int e(t).d(t) + 0.4\frac{d}{dt}e(t)\right] \quad (11)$$

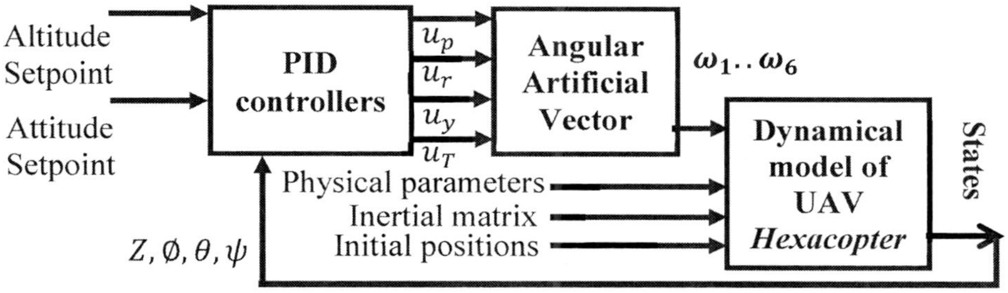


Figure 3. A block diagram of PIDs connected to Hexacopter model.

To control a hexacopter an application was conducted by LabVIEW using Runge-Kutta 2 method with a fixed step of 0.05 [sec]. The PID controllers were implemented to stabilize the altitude and attitude as shown from Fig. 4 to Fig. 8. The curves show the control signal and the fast response of the controllers with the existence of disturbances that emulate the existence of an arm in a non-laboratory environment. These controllers avoid the occurrence of vibrations in the output variables of the flying object as possible. The tuning process has been achieved after many experimental trials. Our scenario is the hovering flight at the altitude of multi-levels in the air according to table 1. The system's parameters, which has been used in the simulation model, are listed in table 2. Correlations were analyzed for all parameters of motion equations.

Table 1. Stability Response of PID Controllers.

| PID Controller | Response Time [S] | Steady-State Error |
|---|---|---|
| Altitude | 22.5 | -0.0911 [m] |
| Pitch | 8.15 | 0.01955 [deg] |
| Roll | 5.72 | 0.0941 [deg] |
| Yaw | 5.58 | 0.00173 [deg] |

Table 2. Parameters used in the simulation.

| | | |
|---|---|---|
| m=4 [kg] | g=9.806 [m/s$^2$] | l=0.5 [m] |
| $J_x$,$J_y$=3.8e-4 [kg.m.s$^2$/rad] | R =0.15 [m] | $k_t$=4.8e-3 [kg.s/m] |
| $J_z$=7.1e-4 [kg.m.s$^2$/rad] | A =0.071 [m$^2$] | $k_r$=6.4e-5 [kg.m.s/rad] |
| $C_T$=0.01458 | $C_Q$=1.037e-3 | ρ=1.293 [kg/m$^3$] |

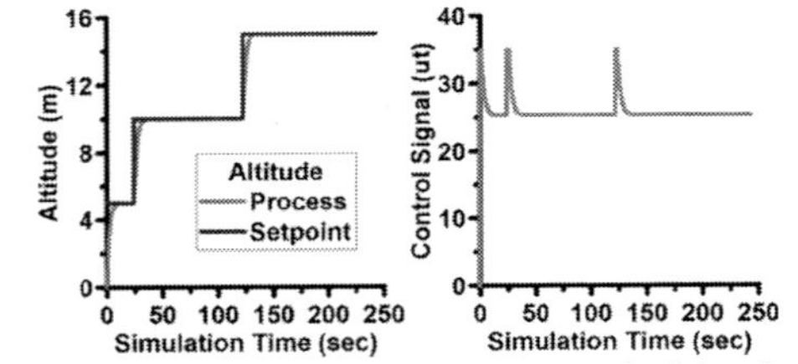


Figure 4. Stability response of altitude and control signal.

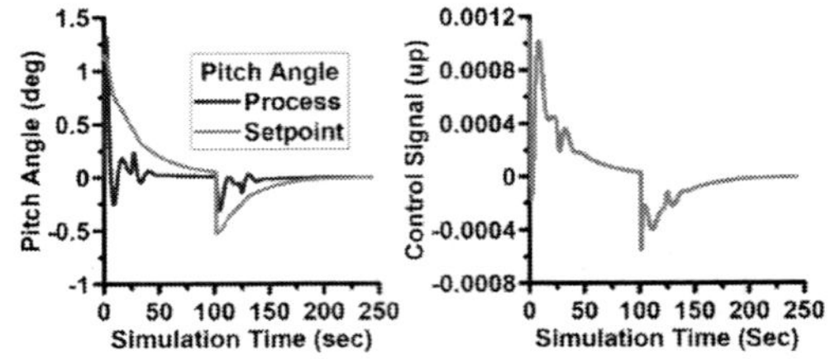


Figure 5. Stability response of pitch angle and control signal.

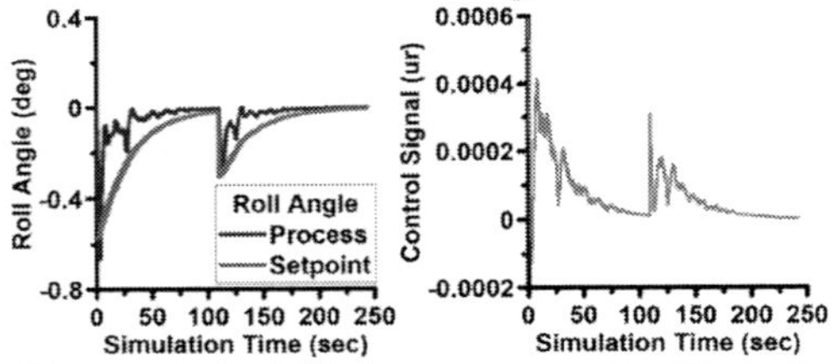


Figure 6. Stability response of roll angle and control signal.

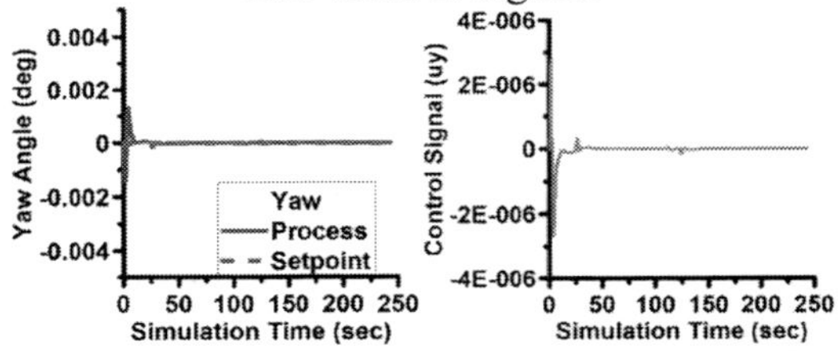


Figure 7. Stability response of yaw angle and control signal.

Figure 8. Stability response of X and Y coordinate systems.

## 6. Linear Quadratic Regulator

This kind of controller aims to minimize the following quadratic cost function [15, 16]:

$$J = \int_0^{\infty}[X^T(t).Q.X(t) + U^T(t).R.U(t)]dt$$

Using a feedback controller gain $k$ such that $U(t) = -kX(t)$, where $Q, R$ are weighting matrices. The task in LQR design is to choose appropriate weighting matrices. Where $Q, R$ diagonal matrices, $Q$ limit the amplitude of the state variables while $R$ limit the amplitude of the inputs and these coefficients treat the optimization and energy terms in the model. The main goal is to make the vehicle reach its desired position as fast as possible. The state space representation of the dynamic model is $X = AX + BU + \delta_M$, Where: $A$ is a state matrix, $B$ is a input matrix, $X = [\emptyset \quad \theta \quad \psi \quad \dot{\emptyset} \quad \dot{\theta} \quad \dot{\psi}]^T$ is a state vector, $U = [M_p \quad M_q \quad M_r]^T$ is a control signal $\delta_M = [M_{dI\phi} \quad M_{dI\theta} \quad M_{dI\psi}]^T$ is disturbance vector of moments, and finally reference vector is $X_r = [\emptyset_r \quad \theta_r \quad \psi_r \quad \dot{\emptyset}_r \quad \dot{\theta}_r \quad \dot{\psi}_r]^T$. The LQR controller was implemented on a state vector to follow the reference vector. The model was linearized around the hover situation after decoupling the equations of motion in Eq. 1, and writing the stochastic state space model in addition to add disturbances to emulate the real motion in the sky. It is considered weak to make an approximation in describing the whole aircraft system in the state space, because the translational movement equations in the earth frame are dependent on the output of the rotational movement equations, therefore the system is divided into two state spaces. A linear–quadratic regulator (LQR) controller was designed only for the rotational movement. The random state space allows adding the disturbances to the model therefore it is considered closer to the real aircraft model, where its information is taken based on the covariance and mean calculations of previously studied disturbances' measurements. Fig. 9 illustrates the application of the LQR controller to the aircraft. The closed loop simulation results using LabVIEW are shown on table (3). Fig. 10 shows the model's states after applying the LQR controller. Figures 11 and 12 show the results of applying the LQR controller and the control signals. The amplitude values of the control signals are very small in comparison to the PID controller, leading to reduction in the consumed power.

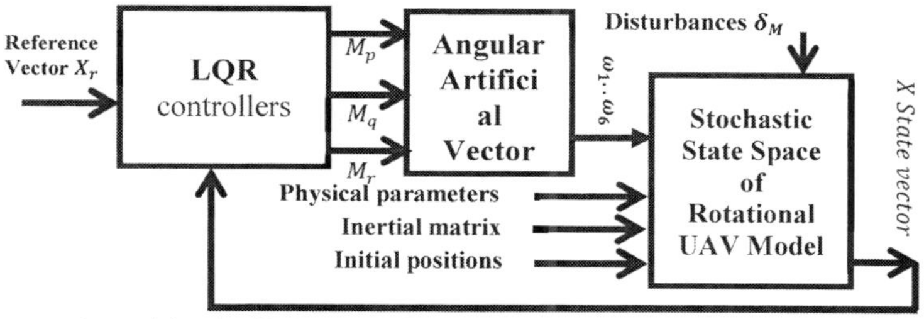


Figure 9. A block diagram of PIDs connected to Hexacopter model.

Table (3) Stability Response of LQR Controller.

| Time Response Parametric Data | $\phi$ | $\theta$ | $\psi$ |
|---|---|---|---|
| **Settling Time [s]** is the time required for the response to reach 1% of its final value. | 0.05 | 0.05 | 0.05 |
| **Rise Time [s]** is the time required for the dynamic system response to rise from 10% of its final value to 90% of its final value. | 0.05 | 0.05 | 0 |
| **Peak Time [s]** is the time required for the dynamic system response to reach the peak value of its first overshoot. | 0.05 | 0.05 | 0.05 |
| **Peak value** returns the value at which the maximum absolute value of the time response occurs. | 1.16842 | 1.16842 | 1.09014 |
| **Overshoot** is the dynamic system response value that most exceeds unity, expressed as a percent. | 1.75277E-5 | 1.75277E-5 | 1.87856E-5 |
| **Steady-State Gain** is the final value of the signal after transient responses decay. | 1.16842 | 1.16842 | 1.09014 |

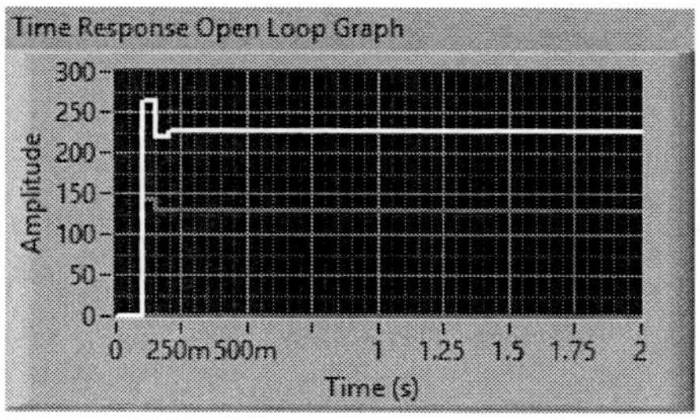


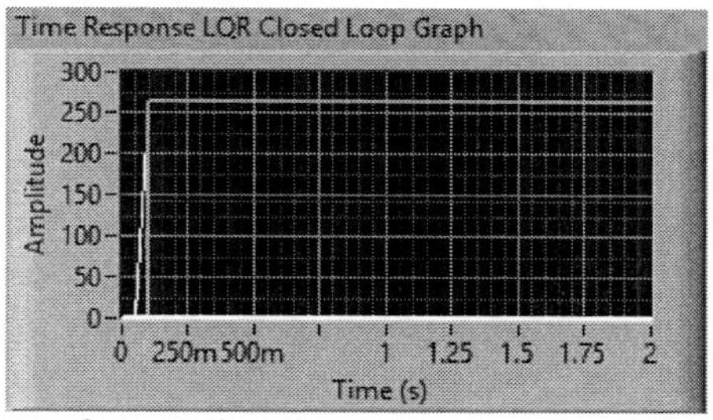


Figure (10). Time response of open and closed loop for stochastic state space model.

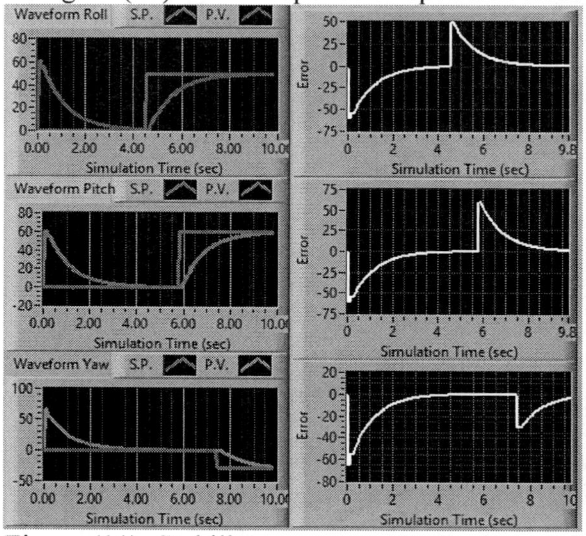


Figure (11). Stability response of roll, pitch and yaw angles and their error signals.

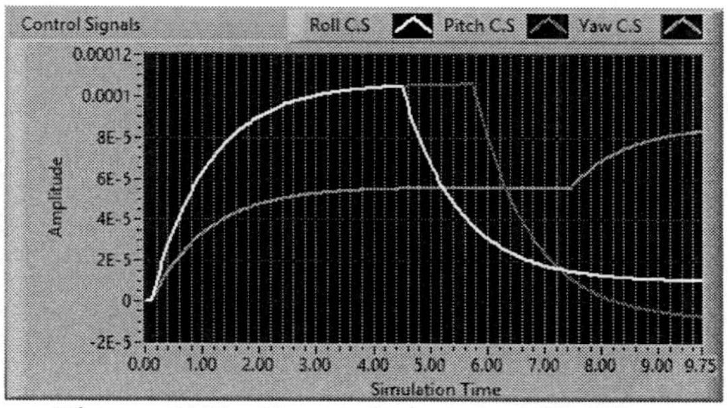


Figure (12). Control signals of LQR controller.

## Conclusions

In this paper, the behavior of a hexacopter with the existence of disturbances inserted into the equations of motion was explored. These disturbances emulate the motion of a robotic arm attached to the body of the aircraft model. The derived dynamic model reflects the real motion of the hexacopter with respect to the earth, which is also characterized by nonlinearity, time variance, underactuation and coupling among the equations' variables. The state space method was used to write the equations in addition to decoupling and linearization techniques in order to design the controllers. Both PID and LQR controllers have been investigated to control the aircraft model. The results of comparison between them indicate that LQR controller is kind of an optimization technique and used to stabilize the attitude of the hexacopter. It consumes low energy with respect to the PID controller in addition to rejecting the perturbation and noise in the stochastic state space of the aircraft model.

## References


[1] Bouadi, H., and M. Tadjine. "Nonlinear observer design and sliding mode control of four rotors helicopter." *Vorld Academy of Science, Engineering and Technology* 25 (2007): 225-229.

[2] Abas, Norafizah, Ari Legowo, and Rini Akmeliawati. "Parameter identification of an autonomous quadrotor." *Mechatronics (ICOM), 2011 4th International Conference On*. IEEE, 2011.

[3] Fogelberg, Johan. "Navigation and Autonomous Control of a Minicopter in Indoor Environments." *MSc Thesis, Department of Automatic Control, Lund University*, ISSN 0280-5316 (2013).

[4] Moussid, Mostafa, Adil Sayouti, and Hicham Medromi. "Dynamic Modeling and Control of a HexaRotor using Linear and Nonlinear Methods."*International Journal of Applied Information Systems* 9.5 (2015).

[5] Sanca, Armando S., P. J. Alsina, and F. Cerqueira Jés de Jesus. "Dynamic modeling with nonlinear inputs and backstepping control for a hexarotor Multi-aerial vehicle." *Robotics Symposium and Intelligent Robotic Meeting (LARS), 2010 Latin American*. IEEE, 2010.

[6] Di Lucia, Stefano, Gian Diego Tipaldi, and Wolfram Burgard. "Attitude stabilization control of an aerial manipulator using a quaternion-based backstepping approach." Mobile Robots (ECMR), 2015 European Conference on. IEEE, 2015.

[7] Ibrahim N. I., Oumran B., A Coupled and Underactuated Dynamic Model of Microcopter Using LabVIEW, IV Всероссийская научно-техническая конференция аспирантов, магистрантов и молодых ученых с международным участием (Young Scientists To Accelerate Scientific And Technological Progress In The Xxi Century), Ижевск 20 – 21 апреля 2016, PP. 981-990.

[8] Ibrahim N. I., Attitude and Altitude Stabilization of Microcopter using PID Controllers with Disturbances in LabVIEW, IV Всероссийская научно-техническая конференция аспирантов, магистрантов и молодых ученых с международным участием (Young Scientists To Accelerate Scientific And Technological Progress In The Xxi Century), Ижевск 20 – 21 апреля 2016, PP. 989-996.

[9] Bouabdallah, Samir, and Roland Siegwart. "Backstepping and sliding-mode techniques applied to an indoor micro quadrotor." Proceedings of the 2005 IEEE international conference on robotics and automation. IEEE, 2005.

[10] Grzonka S, Grisetti G, Burgard W. Towards a navigation system for autonomous indoor flying, In Robotics and Automation, 2009. ICRA'09. IEEE International Conference on 2009 May 12, pp. 2878-2883

[11] Bouabdallah, S., Murrieri, P. and Siegwart, R. (2005), "Towards autonomous indoor micro VTOL", Autonomous Robots, [Online], vol. 18, no. March, 2005, pp. 14/03/2007 available at: http://www.springerlink.com/content/llug3r141ll85726/fulltext.pdf.
[12] Bouabdallah, S., Murrieri, P. and Siegwart, R. (2004), "Design and control of an indoor micro quadrotor", 2004 IEEE International Conference on Robotics and Automation, April 2004, New Orleans, pp. 4393.
[13] Bouabdallah, S., Noth, A. and Siegwart, R. (2004), PID vs LQ control techniques applied to an indoor micro quadrotor, Swiss Federal Institute of Technology.
[14] Castillo, P., Lozano, R. and Dzul, A., (2005), Stabilization of a Mini Rotorcraft with Four Rotors, IEEE Control Systems Magazine.
[15] Hoffmann, G. M., Rajnarayan, D. G., Waslander, S. L., Dostal, D., Jang, J. S. and Tomlin, C. J. (2004), "The stanford testbed of autonomous rotorcraft for multi agent control", Digital Avionics Systems Conference, Vol. 2, 2004, pp. 12.E.4- 121-10.
[16] D. Julián, M. Colorado, *Towards Miniature MAV Autonomous Flight: A Modeling & Control Approach*, (Master thesis of Science in Robotics and Automation, Technical University of Madrid, April 2009).
[17] O. Magnussen, K. E. Skjonhaug, *Modeling, Design and Experimental Study for a Quadcopter System Construction* (Master Thesis in Faculty of technology and science, University of Agder-Kristiansand S, Norway, 2011).
[18] D. Sjoholm, M. Biel, *Automatic Control of a Quadrotor in the Smart Building* (KTH, School of Engineering Sciences (SCI), Master Thesis of Science in Engineering, KTH Royal Institute of Technology, 2014. Stockholm-Sweden, 2014).
[19] K. J. Astrom, R. M. Murray, *Feedback Systems* (Princeton University Press, ISBN-13: 978-0691135762, Vol.2, 11b, 2012).
[20] M. FABIÁN, R. BOSLAI, P. IŽOL, J. JANEKOVÁ, J. FABIANOVÁ, G. FEDORKO, P. BOŽEK. *Use of parametric 3D modelling* - tying parameter values to spreadsheets at designing molds for plastic injection. In Manufacturing technology. Vol. 15, No. 1 (2015), online, s. 24-31. ISSN 1213-2489.
[21] V. GOGA, P. BOŽEK. Application of INS to mechatronic systems control and regulation using virtual dynamic models. In *Proceedings of the 2013 International Conference on Process Control [elektronický zdroj] : Štrbské Pleso, Slovakia, June 18-21, 2013*. 1st. ed. Piscataway : IEEE, 2013, s.CD-ROM, p.185-189. ISBN 978-80-227-3951-1.
[22] P. BOŽEK. Control of a robotic arm on the principle of separate decision of an inertial navigation system. In *Applied Mechanics and Materials*. Vol. 611 (2014), s. 60-66. ISSN 1660-9336.
[23] P. BOŽEK, P. POKORNÝ, J. SVETLÍK, A. LOZHKIN, I. ARKHIPOV. The` calculations of Jordan curves trajectory of the robot movement. In *International Journal of Advanced Robotic Systems*. Vol. 13, no. 5 (2016), online, [7]s. ISSN 1729-8806.
[24] M. Kelemen, I. Virgala, P. Frankovský, T. Kelemenová, L. Miková. Amplifying System for Actuator Displacement. International Journal of Applied Engineering Research, 11(15), (2016).8402-8407.
[25] M. Kelemen, I. Virgala, T. Kelemenová, Ľ. Miková, , P. Frankovský, T. Lipták, M. Lörinc, M. (2015). Distance Measurement via Using of Ultrasonic Sensor. Journal of Automation and Control, 3(3), 71-74.
[26] P. Frankovský, O. Ostertag, E. Ostertagová, F. Trebuňa, J. Kostka and M. Výrostek. (2017). Experimental analysis of stress fields of rotating structural elements by means of reflection photoelasticity. Applied Optics, 56(11), , 3064-3070.

EAN 2017
55th conference on experimental stress analysis

# The proposal of estimation method of mating between pulleys and cogbelt

KRAWIEC Piotr[1, a], DOMEK Grzegorz[2, b], ADAMIEC Jarosław[1, c], WALUŚ Konrad[1, d], Warguła Łukasz[1, e]

[1]Poznan University of Technology, Marii Skłodowskiej Curie 5, 60-965 Poznan, Poland

[2]Kazimierz Wielki Uniwersity, Chodkiewicza 30, 85-064 Bydgoszcz, Poland

[a]piotr.krawiec@put.poznan.pl, [b]gdomek@ukw.edu.pl, [c]jarosław.adamiec@put.poznan.pl, [d]konrad.walus.put.poznan.pl[1], [e]lukasz.warguła@put.poznan.pl

**Keywords:** CAD, FEA, high speed camera, wrap and contact

## Abstract

Two estimation methods of mating (meshing) between pulleys and cogbelt of uneven-running strand transmission are presented in this work. The first one is based on the application of high speed camera, and the second one is based on CAD model and FEA software in order to determine the teeth number of pulley and cogbelt during meshing. The performed estimation allows to indicate the reasons of incorrect mating due to e.g. modelling and manufacturing of cogbelt pulleys.

## Introduction

In belt transmissions with synchronous timing belts, load is transmitted as a shaped and frictional contact. The turning moment is trasmitted by the appropriate shape of the belt cooperating with the timing belt pulleys.
A transmission with a synchronous belt requires little initial tension and therefore loads shafts and bearings less than flat belts or V-belts.

To increase the drive of a transmission, the wrapping angle of the small wheel is increased by means of the tension roll. The synchronous belt is equipped with the carrying input consisting of glass fibres, steel fibres or aramid fibres. Carrying belts exibit such properties as: high resistance to stretching, little extension and high elasticity. Furthermore, fibres shall have appropriate fatigue strength. Decreasing the cross-section of a single fibre results in an increased elasticity of the belt and decreased stress in a single fibre.

The so-called 'polygonal effect' is induced in a belt transmission with a timing belts. This effect influences a non-uniform transmission of rotary motion.The phenomenon is induced by the change in the diameters of the arc of contact in the wheel. The real diameter of the wheel changes within dimentions $d_{max}$ and $d_{min}$ and depends on the number of cogs and the wheel graduation.

A timing belt pulley wheel is, in fact, a polygon that consists of sections of chords formed by the heads of cogs. At the moments the belt meshing and demeshing, the foot of a cog contacts the polygon and the point of contact between the cog outline and the polygon is a kinematic centre of the rotating wheel as long as it is increasing.

At meshing and demeshing of one cog in a loaded belt, the pivoting point O of the belt is situated opposite the loaded surface of the side of the belt pulley cog - points a) and b) (Fig. 1).

The lower part of Figure 1 - points c) and d) show the shift of the pivoting points O at meshing and demeshing in the strand of an unloaded belt. The pivoting point O is situated on the extended by the load surface of the belt wheel cog .

As a result of the polygonal effect, despite the constant angular velocity ω of the driving wheel, the velocities of the belt strand as well as of the driven wheel change. The ratio of velocities of wheels depends on the number of cogs. Therefore, the greater number of cogs in the wheel, the more the non-uniformity of the drive transfer is diminished and the lesser influence of the polygonal effect.

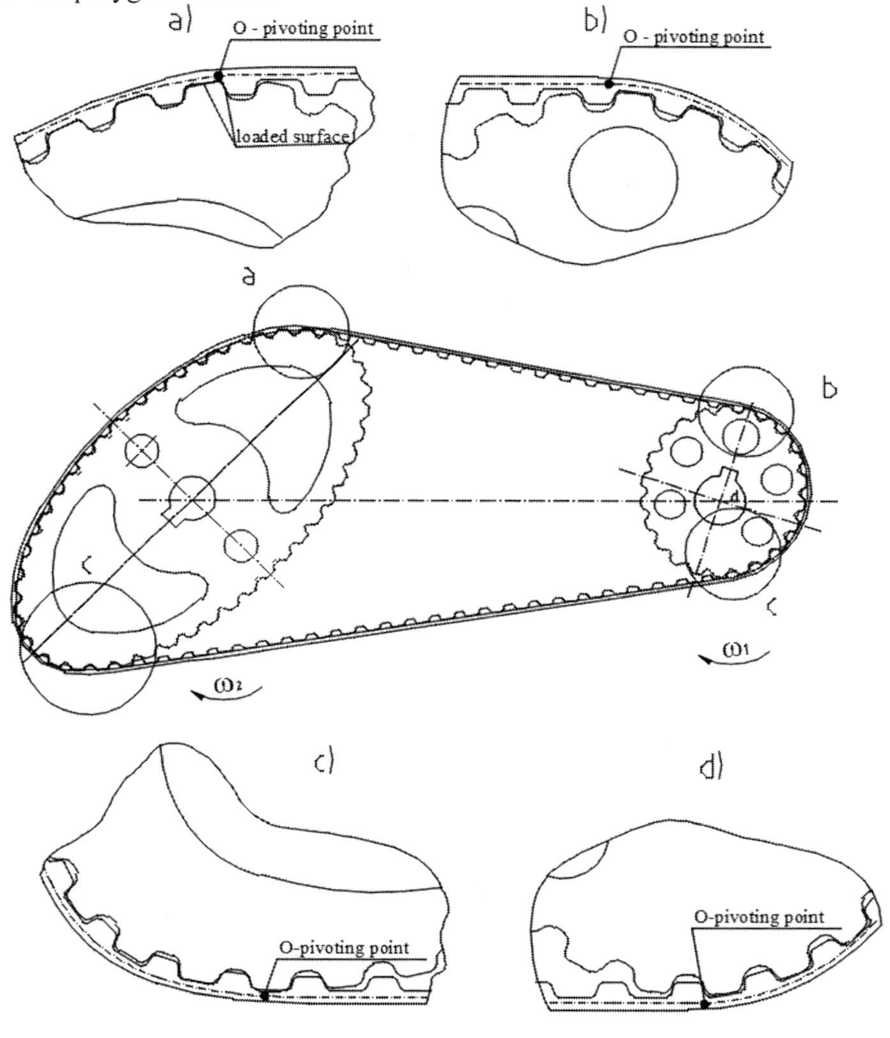


Fig. 1. The scheme of the synchronous belt meshing: a) demeshing out of the driving wheel, b) meshing with the driven wheel, c) meshing with the driving wheel, d) demeshing out of the driven wheel

This phenomenon is particularly important during mating between pulleys and cogbelt of uneven-running transmission. The characteristic feature of this type of transmission is a periodical variation of the velocity ratio during one rotation cycle due to the application of pulleys with noncircular envelope. Design and modelling problems of envelope of gears are presented in studies [1-5]. The present state of knowledge is the reason for investigation and explanation of phenomena during the transmission work – this will be done with the application of FEM.

**Numerical model**

The examined numerical model of a strand transmission was elaborated in the Abaqus program. The model was described as a two dimensional state of stresses. Geometrical models (solid models) of pulleys and cogbelt were prepared in AutoCAD and Autodesk Inventor programs. The transmission elements were transferred to the Abaqus program in DXF files. The initial configuration of the transmission – before assembly – is presented in figure 2. The pulleys and the cogbelt were discretizated by 4-node and 3-node finite elements for a plane state of stresses with reduced integration and linear shape functions (CPS4R and CPS3 according to Abaqus denotation).

The cogbelt reinforcement was defined as 2-node bar element with linear shape functions (T2D2 according to Abaqus denotation). The bar elements were defined on the nodes of elements of plane state of stresses. The FEM model of the transmission consisted of 18 084 elements and 18 710 nodes and had 37 418 degrees of freedom.

Aluminium was taken as a material of wheels. The material is isotropic and elastic with the following properties:

- Young's modulus – 73 GPa,
- Poisson ratio – 0,33,
- density – 2800 kg/m$^3$.

Hooke's classical model of elastic material was taken for material of wheels [6].

The FEM model of the cogbelt had a composite structure (two materials) – fig. 2. The rubber, which had been used as a body of the cogbelt, was described as isotropic and nonlinear-elastic material [7-10].

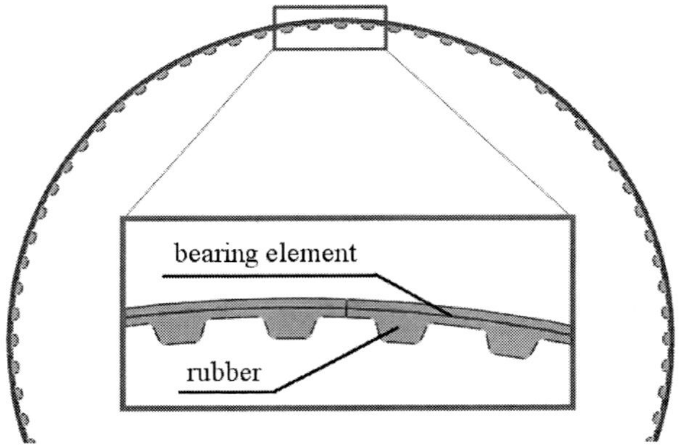


Fig. 2. Composite structure (two materials) of the cogbelt

The exemplification of the transmission work was done in two steps:

1) cogbelt stretching,
2) a simulation of the transmission work. This process has a quasi-static character.

In first step, a kinematic excitation (horizontal displacement) was applied to the reference nodes of the rigid body constraints of both pulleys – this excitation caused a displacement of the pulleys in order to get the specified force in the cogbelt, i.e. 150 kN.

The speed of a vertical displacement and a rotation on axis were excited as zero. The simulation of this process lasted 2 seconds (fig. 3). The diagram of the energy balance for the first step confirms the quasi-static character of the process (fig. 4). This figure shows that after selection of a backlash (ca 1.5 s), most of the work of external forces is transferred to the system in a form of the internal energy (initial stresses in the cogbelt) and is partially dissipated because of friction between the pulleys and the cogbelt. The kinematic energy gets a small value (< 20 mJ) mainly due to the cogbelt vibrations which are excited by a given excitation.

Fig. 3. Kinematic excitation (horizontal displacement – brown arrows) which stretches the cogbelt

The second step is a simulation of the transmission work. The following two analysis were conducted:

- Z1 – excited rotation of the pulley K1 during a free rotation of the pulley K2,
- Z2 – excited rotation of the pulley K1 during the pulley K2 braking with a torque of 10 N·m.

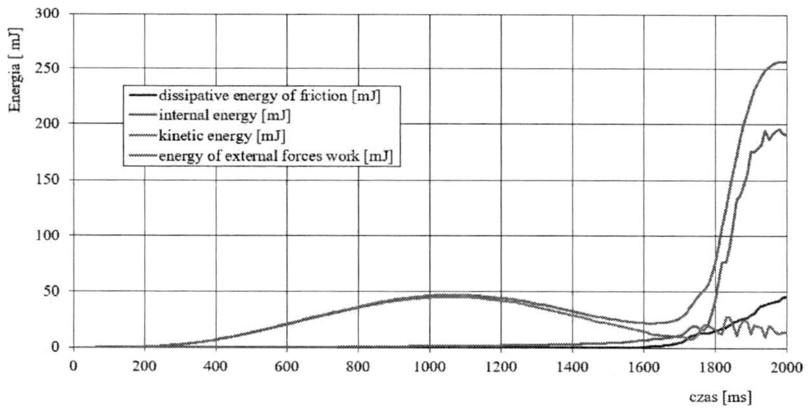


Fig. 4. Energy balance for the first step (cogbelt stretching)

## Contact modelling with FEA

Contact was defined as a system of two pairs of contact surfaces between:

- tooth surfaces of wheel K1 and cogbelt,
- tooth surfaces of wheel K2 and cogbelt.

In Abaqus contact constraints are automatically defined for nodes which are placed on surfaces of contact pair. Contact conditions are forced with the application of penalty method. The friction is defined with the use of classical Coulomb's friction law – the value of 0.3 was taken for the friction coefficient. In Abaqus/Explicit contact conditions (penetrations and contact reactions) are checked in every increment. For each successive form of deformation the contact constraints are taken as a result of constraints which are imposed to calculations of velocity and acceleration. Contact reactions are calculated during determining the vector of nodal, external forces $P$. The procedure of explicit integration of equation of motion does not demand the application of iterative algorithm – this allows to solve effectively strong discrete and complex contact algorithms such as dynamic simulation of transmission work. This criterion was decisive during selection of calculation method for dynamic simulation.

Figures 5 and 6 present number of teeth in wrap and contact in function of angle of rotation for driving and driven wheel.

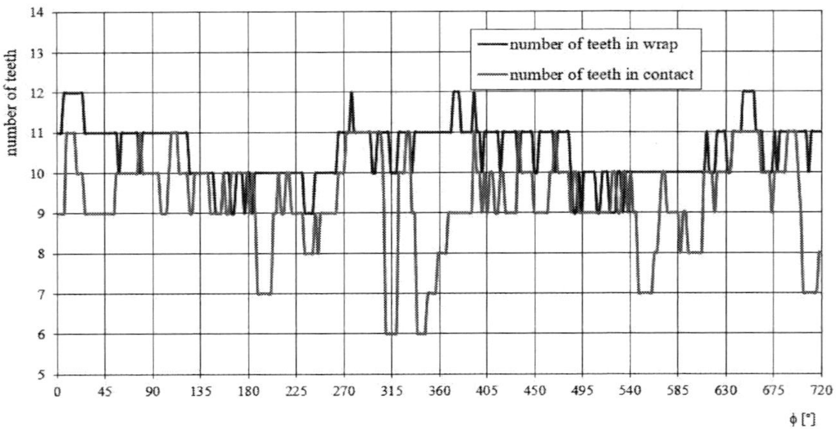


Fig. 5. Number of teeth of small wheel in wrap and contact (FEA method)

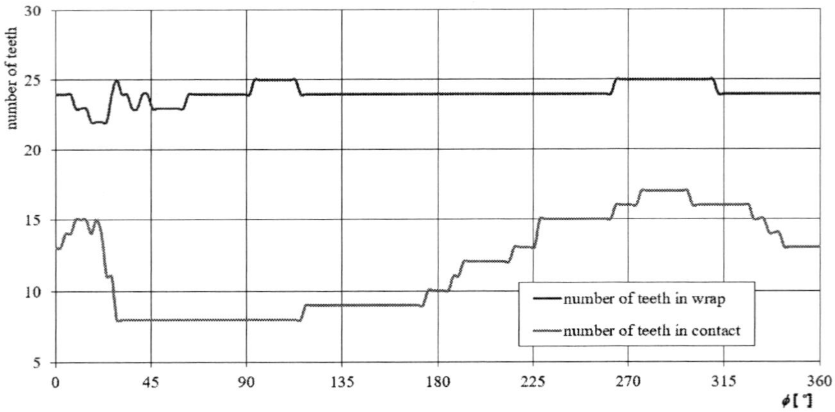


Fig. 6. Number of teeth of elliptical wheel in wrap and contact (FEA method)

## Test stand and measurements

A measuring set with CMOS camera High Speed Star 5 series (LaVision manufacturer) was applied to register a coupling process between cogbelt and noncircular cogbelt pulley. This device allows to register directly high speed processes such as fuel injection and load motion (Fig. 7). The photos can be done with the speed 250 000 frames per second (fps), but only for reduced resolution of matrix CMOS 128×16 pixels. The spatial resolution was the main priority for the research measurements, so it was decided to set the speed to the value of

3000 fps and it was sufficient value for the research tests. Maximal resolution of camera is equal to 1024×1024 points for the speed 3000 fps (it responds to physical resolution of matrix for detector size of 17×17 micrometers). The chosen camera allows to register monochrome images and works in spectral range from 380 to 800 nm. For applied configuration the camera was equipped with lens with focal distance 50 mm. In order to get optimal (maximal) filling of view field of camera, it was necessary to set the camera in a distance from several to ca. 30 cm from the investigated objects. This causes observable distortions of view which can make difficulties in the analysis of image and can introduce some errors into measured quantities.

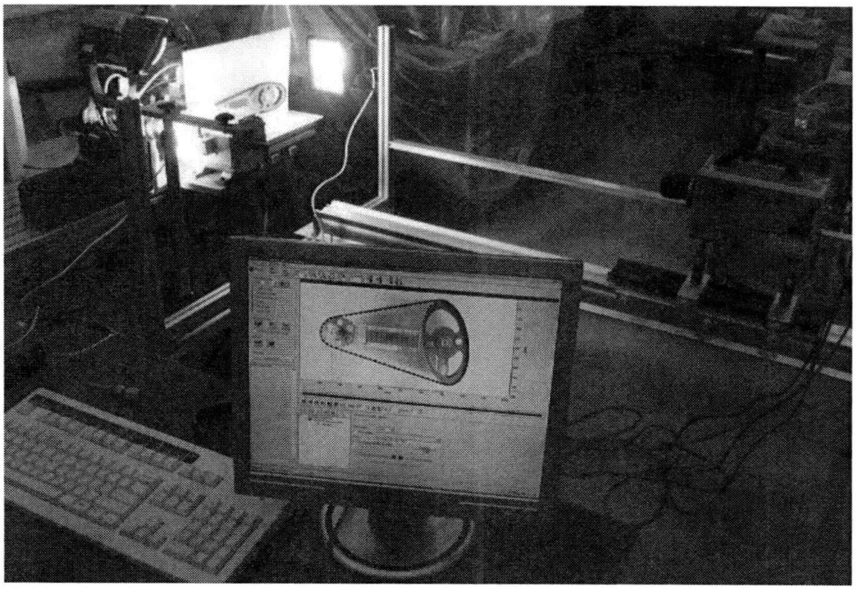

Fig. 7. Test stand for image acquisition of work of uneven-running strand transmission with cogbelt

The investigated transmission was illuminated with two halogen lamps – each one with the power of 500 W. The lamps were set together in order to minimalize the presence of shadow in the field of view.

## Methodology of image processing and analysis

From the point of view of image processing and analysis the proposed approach should be defined as model based recognition, because the essential element of elaborated algorithms is the assumption of model accessibility of recognized object. This model is directly imported by software from AutoCAD files.

In elaborated process of image analysis the following assumption was made: the processing of each frame of video stream is independently done.

This process consists of two main stages:

- CAD model and image fitting,
- estimation of chosen variables for state of investigated object (cogbelt pulley): angle of rotation, rate of misfit.

The elaborated software was implemented with the application of known library of image analysis and computer vision OpenCV [11, 12]. This is very universal and the most applied environment for developing the image applications – it is accessible within a framework of open license software.

Figures 8 and 9 show quantitative list of teeth of noncircular cogbelt pulleys which are in wrap and contact.

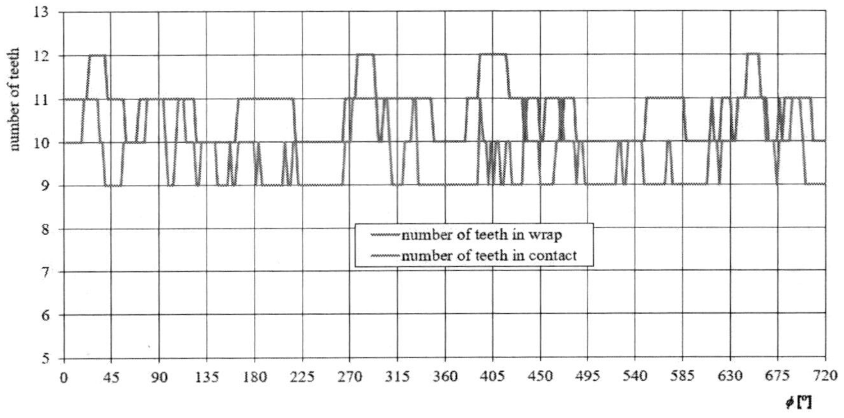


Fig. 8. Number of teeth of small wheel in wrap and contact (optical method)

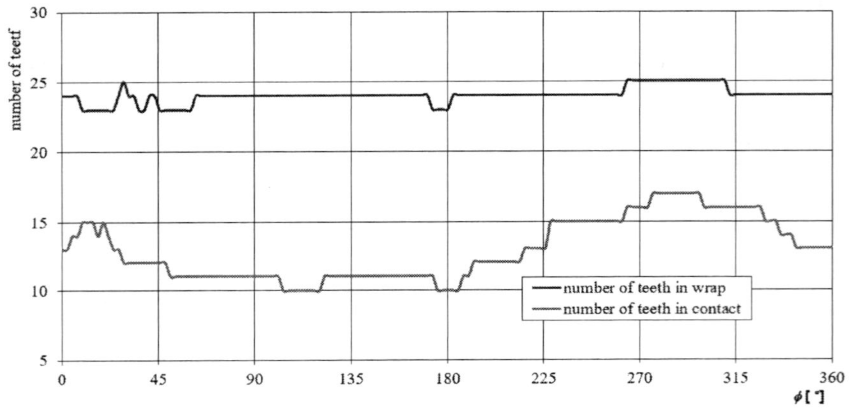


Fig. 9. Number of teeth of elliptical wheel in wrap and contact (optical method)

On the basis of diagrams one can show the angles of rotation of wheels for which the number of teeth is excessively reduced and this can results in very high load of cogbelt and uneven-running motion of transmission.

## Conclusions

A proper mating between pulleys and cogbelt in strand transmission is a base to apply full loading of such type of drive.

On the basis of performed analyses of images one can conduct the quantitative estimation of the number of teeth in wrapping angle and meshed teeth i.e. the number of loaded teeth. The accepted method of estimation requires improvements, but after some corrections it can be applied to estimate strand transmissions with noncircular cogbelt pulleys and classical strand transmissions. The aim of research with the application of finite element analysis was the quantitative and qualitative analysis of operation of strand transmission with noncircular cogbelt pulleys. The operation of cogbelt during transmission work was analysed. On the basis of this analysis the following parameters were evaluated: the number of cogbelt teeth in wrapping angle and number of meshed teeth of pulley and cogbelt.

## References


[1] G. Domek, Motion analysis of timing belt used in control systems, American Journal of Mechanical Engineering, vol 1, No.7 (2013) M.

[2] P. Krawiec, A. Marlewski , Designing irregular gears via splines, 9th International Conference on Modeling and Applied Simulation MAS, Fes Marocco, (2010), pp. 125-130.

[3] M. Kujawski, P. Krawiec, Analysis of generation capabilities of noncircular cogbelt pulleys on the example of a gear with an elliptical pitch line Journal of Manufacturing Science and Engineering - Transactions of the ASME, 2011, Vol. 133, Iss. 5, pp. 0510061–0510067.

[4] P. Krawiec, A. Marlewski, Profile design of noncircular belt pulleys, Journal of Theoretical and Applied Mechanics, 54(2) (2016) pp. 561–570.

[5] Nagel T., Zahnriemengetriebe. München, Hanser Verlag, 2008.

[6] Zienkiewicz O.C., Taylor R.L., The Finite Element Method, Elsevier, 2006.

[7] ABAQUS Manuals, ABAQUS, Inc. Providence, 2005.

[8] R.L.Burden, J.D.Faires, Numerical Analysis. Boston, PWS Publishers, 1985.

[9] Chapra S.C., Canale R.P., Numerical Methods for Engineers. Boston, McGraw Hill, 1998.

[10] M.A.Crisfield, Non-linear Finite Element Analysis of Solid and Structures, Wiley, New York, 1997.

[11] Open Source Computer Vision Library: Reference Manual, 2001 http://software.intel.com.

[12] J.C.Russ, The Image Processing Handbook, Thrid Edition, CRC Press, Springer, IDEE Press,1999.

EAN 2017
55th conference on experimental stress analysis

# The Use of Geometric Mechanics Concept to Modeling of Ackerman Steered Car-like Vehicle

Peter Frankovský[a], Alexander Gmiterko[b], Lukáš Dominik[c], Darina Hroncová[d], Ivan Virgala[e], Ľubica Miková[f]

Department of Mechatronics, Faculty of Mechanical Engineering, Technical University of Košice, Park Komenského 8, 042 00 Košice, Slovakia

[a]peter.frankovsky@tuke.sk,[b]alexander.gmiterko@tuke.sk, [c]lukas.dominik@tuke.sk, [d]darina.hroncova@tuke.sk, [e]ivan.virgala@tuke.sk, [f]lubica.mikova@tuke.sk

**Keywords:** Ackermann steered vehicle, kinematic model, nonholonomic constraints, geometric mechanics

**Abstract.** This paper provides an approach of differential geometry for modelling of the four-wheeled Ackermann-steered vehicle. Nonholonomic constraints which represent no-slip and no-slide conditions are modelled in the local coordinate systems by iterative Jacobian method. In this paper are derived kinematic equations of motion of the vehicle moving in plane.

## Introduction

The motion characteristics of a vehicle play general role in planning its path. The vehicle moving in plane has generally three degrees of freedom - translation along two axes and rotation about axis perpendicular to the plane. But not every one of these movements is possible. It is necessary to consider nonholonomic constraints resulting from velocity conditions imposed on wheels. This means that when the vehicle is in motion there must be no-slip and no-slide on wheels.

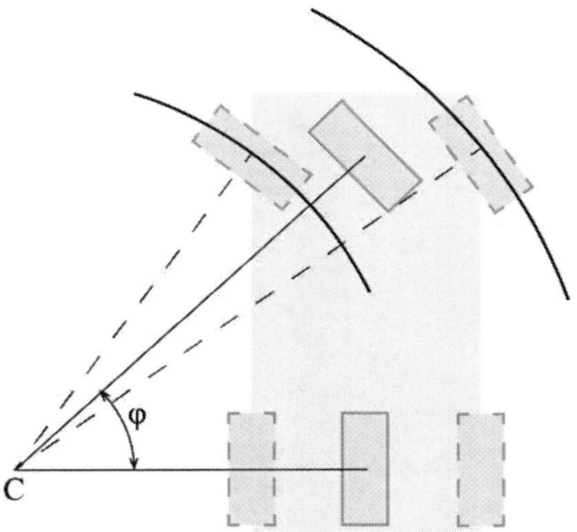


**Fig. 1 Geometry of the Ackermann steered vehicle**

The importance of Ackermann steering geometry (Fig. 1) is in prevention of wheels to slip sideways while vehicle is moving along a curved path. Geometrical solution is such that all wheels have their axes arranged as the radii of circles with common center.

## System modelling

Based on this idea we can simplify the four wheeled vehicle model into two wheeled bicycle model as is shown in (Fig. 2). The motion characteristics of this simplified model are the same as characteristics of four wheeled vehicle.

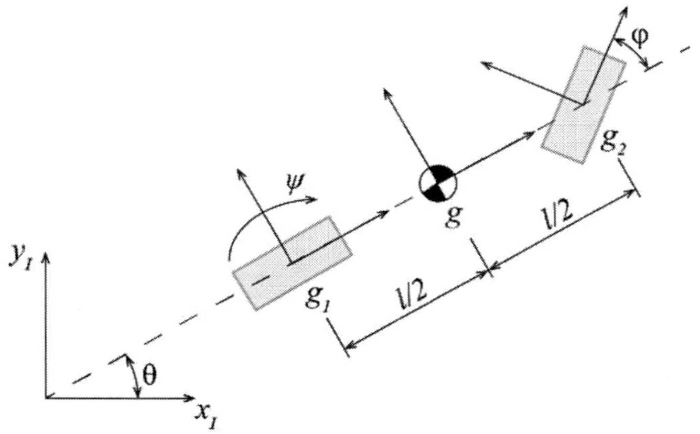


**Fig. 2 Simplified model of Ackermann steered four wheeled vehicle**

We can define nonholonomic constraints of the moving vehicle in inertial coordinate system or in vehicle's local coordinate systems which is easier. In local coordinate systems $g_1$ and $g_2$ these constraints are expressed as no-slip and no-slide conditions

$$\xi_{g_1}^{x} - r\dot{\psi} = 0, \tag{1}$$

$$\xi_{g_1}^{y} = 0, \tag{2}$$

$$\xi_{g_2}^{y} = 0, \tag{3}$$

where $\dot{\psi}$ is angular velocity of drive wheel, $r$ is radius of wheel and $\xi$ are velocities in local coordinate systems.

Position and orientation of local coordinate systems $g_1$ and $g_2$ relative to coordinate system $g$ located at the vehicle's center of gravity is given by

$$\begin{aligned} g_{1,g} &= \left(-\tfrac{l}{2}, 0, 0\right), \\ g_{2,g} &= \left(\tfrac{l}{2}, 0, \varphi\right). \end{aligned} \tag{4}$$

Velocities expressed in local coordinate systems $\xi_{g_1}$ and $\xi_{g_2}$ can be mapped with velocities expressed in the vehicle's center of gravity $\xi_g$. The map from vehicle's center of gravity to local velocities is the adjoint inverse action

$$\xi_{g_1} = Ad^{-1}_{g_1,g}.\xi_g = \begin{bmatrix} 1 & 0 & 0 \\ 0 & 1 & -\frac{l}{2} \\ 0 & 0 & 1 \end{bmatrix} \begin{bmatrix} \xi^x \\ \xi^y \\ \xi^\theta \end{bmatrix} = \begin{bmatrix} \xi^x \\ \xi^y - \frac{l}{2}\xi^\theta \\ \xi^\theta \end{bmatrix},$$

$$\xi_{g_2} = Ad^{-1}_{g_2,g}.\xi_g = \begin{bmatrix} \cos\varphi & \sin\varphi & \frac{l}{2}\sin\varphi \\ -\sin\varphi & \cos\varphi & \frac{l}{2}\cos\varphi \\ 0 & 0 & 1 \end{bmatrix} \begin{bmatrix} \xi^x \\ \xi^y \\ \xi^\theta \end{bmatrix} = \begin{bmatrix} \xi^x\cos\varphi + \xi^y\sin\varphi + \xi^\theta\frac{l}{2}\sin\varphi \\ -\xi^x\sin\varphi + \xi^y\cos\varphi + \xi^\theta\frac{l}{2}\cos\varphi \\ \xi^\theta \end{bmatrix}. \tag{5}$$

Substituting velocity components $\xi^x_{g_1}$, $\xi^y_{g_1}$ and $\xi^y_{g_2}$ from equations (5) into nonholonomic constraints (1, 2, 3) we obtain differential equations in a Pfaffian form

$$\begin{aligned} &\xi^x - r\dot{\psi} = 0, \\ &\xi^y - \tfrac{l}{2}\xi^\theta = 0, \\ &-\xi^x\sin\varphi + \xi^y\cos\varphi + \xi^\theta\tfrac{l}{2}\cos\varphi = 0. \end{aligned} \tag{6}$$

The above equations we can rewrite into matrix form

$$\begin{bmatrix} 1 & 0 & 0 \\ 0 & 1 & -\frac{l}{2} \\ -\sin\varphi & \cos\varphi & \frac{l}{2}\cos\varphi \end{bmatrix} \begin{bmatrix} \xi^x \\ \xi^y \\ \xi^\theta \end{bmatrix} + \begin{bmatrix} -r & 0 \\ 0 & 0 \\ 0 & 0 \end{bmatrix} \begin{bmatrix} \dot{\psi} \\ \dot{\varphi} \end{bmatrix} = \begin{bmatrix} 0 \\ 0 \\ 0 \end{bmatrix}. \tag{7}$$

By editing the above equation we obtain relation representing the influence of shape speeds acting on local velocity of the vehicle

$$\begin{bmatrix} \xi^x \\ \xi^y \\ \xi^\theta \end{bmatrix} = \begin{bmatrix} r & 0 \\ \frac{r}{2}\tan\varphi & 0 \\ \frac{r}{l}\tan\varphi & 0 \end{bmatrix} \begin{bmatrix} \dot{\psi} \\ \dot{\varphi} \end{bmatrix}. \tag{8}$$

In geometric mechanics, the equation of motion for principally kinematic system are generally expressed as kinematic reconstruction equation

$$\xi = -\mathbf{A}(r)\dot{\mathbf{r}}, \tag{9}$$

where $\xi$ is the velocity vector of the vehicle's center of gravity, $\mathbf{A}$ is the local connection associated with system constraints and $\dot{\mathbf{r}}$ is vector of the shape velocities.

Finally the mathematical model of the Ackermann steered vehicle expressed in inertial coordinate system is given by equation

$$\begin{bmatrix} \dot{x} \\ \dot{y} \\ \dot{\theta} \end{bmatrix} = \begin{bmatrix} \cos\theta & -\sin\theta & 0 \\ \sin\theta & \cos\theta & 0 \\ 0 & 0 & 1 \end{bmatrix} \begin{bmatrix} r & 0 \\ \frac{r}{2}\tan\varphi & 0 \\ \frac{r}{l}\tan\varphi & 0 \end{bmatrix} \begin{bmatrix} \dot{\psi} \\ \dot{\varphi} \end{bmatrix} = \begin{bmatrix} r\cos\theta - \frac{r}{2}\sin\theta\tan\varphi & 0 \\ r\sin\theta + \frac{r}{2}\cos\theta\tan\varphi & 0 \\ \frac{r}{l}\tan\varphi & 0 \end{bmatrix} \begin{bmatrix} \dot{\psi} \\ \dot{\varphi} \end{bmatrix}. \tag{10}$$

## Conclusions

The aim of this paper was targeted to mathematical modelling of the four wheeled vehicle with nonholonomic constraints. Nonholonomic constraints were derived in coordinate system set in center of gravity of the vehicle by iterative Jacobian method. Based on these constraints was formed kinematic model of the vehicle.

## Acknowledgements

The authors would like to thank to Slovak Grant Agency – project VEGA 1/0872/16.

## References

[1] H.Choset, R. Hatton, An introduction to geometric mechanics and differential geometry, 2011.

[2] M. Egerstedt, X. Hu, H. Rehbinder, A. Stotsky, Path planning and robust tracking for a car-like robot.

[3] E. Kardoss, B. Kiss, Path planning and tracking control for an automatic parking assist system.

[4] A. Gupta, R. Divekar, M. Agrawal, Autonomous parallel parking system for ackerman steering four wheelers, 2010.

[5] S. Jagannathan, S. Q. Zhu, F. L. Lewis, Path planning and control of a mobile base with nonholonomic constraints, 2009.

[6] A. M. Bloch, Nonholonomic mechanics and control, 2015.

[7] P. Božek, Y. Turygin, Measurement of the operating parameters and numerical analysis of the mechanical subsystem. Measurement Science Review, 14(4), pp. 198-203, 2014.

EAN 2017
55th conference on experimental stress analysis

# 3D Vision Methods for Enhanced Human-robot Interaction

Tölgyessy Michal[1,a], Dekan Martin[1,b] Rodina Jozef[1,c]

[1]Faculty of Electrical Engineering and Information Technology STU in Bratislava, Slovakia

[a]michal.tolgyessy@stuba.sk, [b]martin.dekan@stuba.sk, [c]jozef.rodina@stuba.sk

**Keywords:** HRI, robot motion, gesture recognition

**Abstract.** This article presents an overview of several original human-robot interaction methods. All of them are gesture based. Human motion and hand position is captured by a 3D sensor. Presented methods have been experimentally verified on real mobile robot platforms.

## Introduction

Robots are becoming a more and more visible part of human lives. UAV drones with onboard visual systems make real-time videos at music festivals, or capture weddings and other social events from above. Many people own autonomous vacuum cleaners. This trend stresses more and more the need for natural human-machine interaction. In the field of robotics the term HRI (Human-Robot Interaction) became a standard name. Since it is a vast field different researchers concentrate on different topics. Some focus on gesture control of UAV's [1][2]. Others on pointing to physical objects and evaluating robot perception [3][4]. Gesture based control has even been implemented in communication with robotic swarms [5][6].
Research work presented in this paper focuses primarily on hand-gesture based human-robot communication. Perception is based on 3D visual sensing. Sensor used for experimental evaluation is the Kinect [7] [8] [9] [10] [11] [12]. Two of the presented methods rely on 3D joint positions. To extract 3D data from depth stream mainly the work of Shotton et. al. was used [13].

## 1. A plane-based gesture algorithm for mobile robot control

Algorithm is based on position of human operator's hands in relation to his or her torso. These positions define discrete movement commands for a mobile robot that are being continuously sent over a wireless network. Firstly, a set of points to define the *torso plane* are chosen. In the example in Fig. 1 they are right shoulder (A), left shoulder (B) and right hip (C). Position of hands ($H_1$ and $H_2$) is then analyzed in relation to the ABC plane.

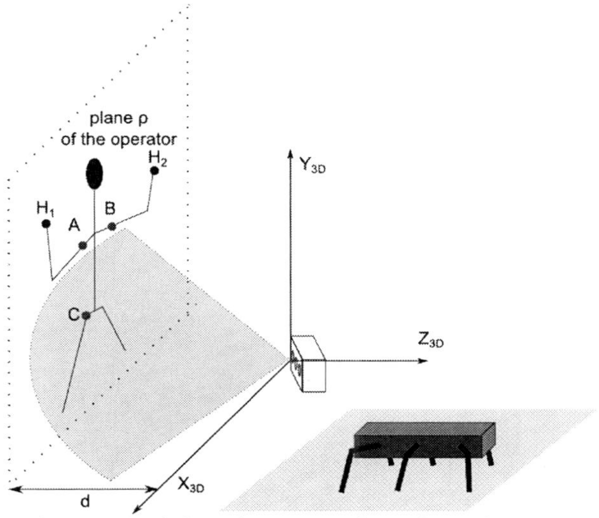

Fig 1. Obtaining the torso plane for robot control

To ensure the ability to work with different gestures 3 zones are introduced. Each hand can be present at any given time in only one of these zones (Fig. 2).

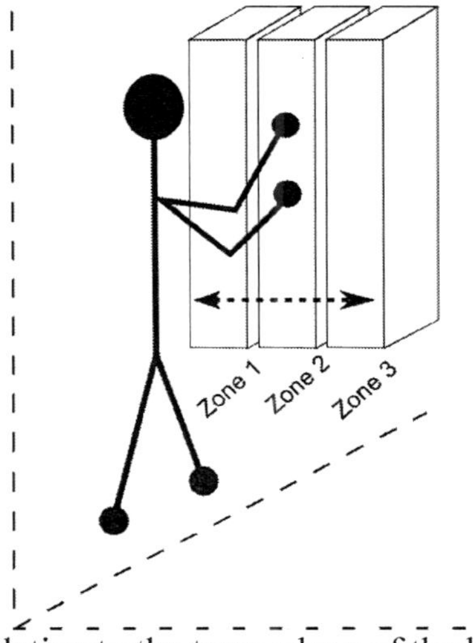

Fig. 2 Zones relative to the torso plane of the human operator

To determine the zone in which a hand is currently present it is necessary to execute these steps:

1. Compute vectors **AB** and **AC** from operator's body joints
2. Compute the normal vector **n** of the torso **ABC** plane from cross product **ABxAC**
3. Obtain the parameters of $ax + by + cz + d = 0$ by applying **n**
4. Distance between hand and torso plane is computed thusly:

$$D(H,\rho) = \frac{ax_H+by_H+cz_H+d}{\sqrt{a^2+b^2+c^2}} \qquad (1)$$

Commands for the robot are defined by presence of hands in particular zones:

- STOP – both right and left hand in Zone 2

- ROTATION RIGHT – right hand in Zone 1, left hand in Zone 3
- ROTATION LEFT – right hand in Zone 3, left hand in Zone 1
- MOVE FORWARDS – both right and left hand in Zone 3
- MOVE BACKWARDS – both right and left hand in Zone 1

**Experiment**

To evaluate this gesture recognition method a candy carrier robot was controlled in a laboratory as well as crowded environment (Fig. exp1).

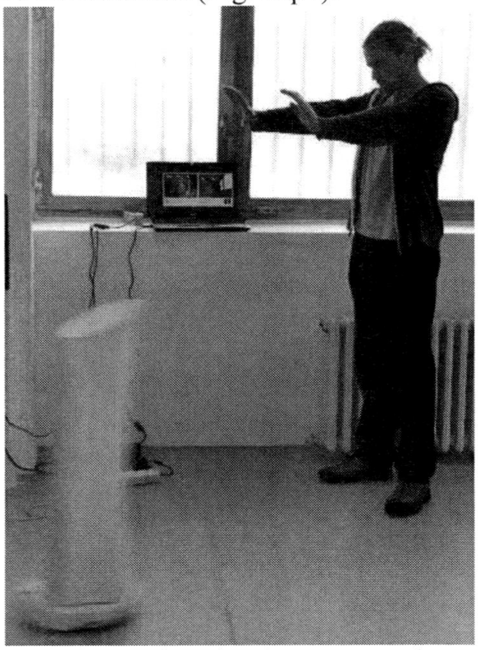

Fig. exp1 Experimental evaluation on a candy carrier robot

## 2. Controlling a mobile robot via pointing gesture

Second of the presented algorithms allows a human operator to point to a spot in the ground and guide the robot to this spot. To achieve this, 3D sensor is mounted directly on the robot. It detects position of elbow (A) and wrist (B) of the operator and computes a line defined by these points. Subsequently, robot finds the intersection of this line with the plane (ground) where both robot and human are present. An illustration of this scenario is in Fig. 3.

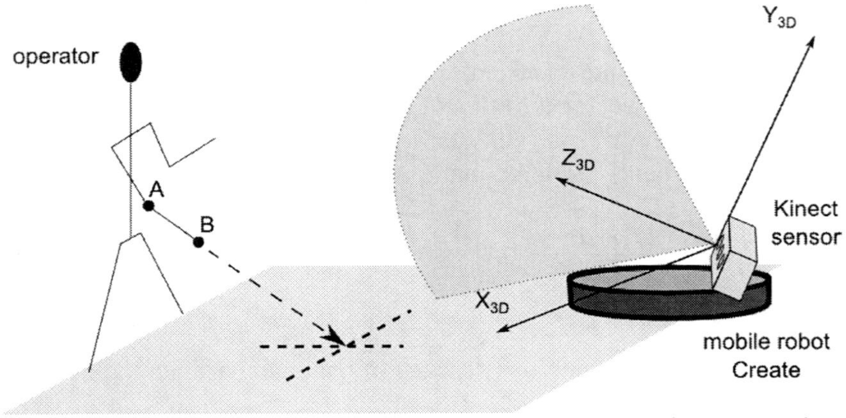


Fig. 3 Human operator points to the spot, where robot Create is supposed to move

Movement is begun when operator's hand is raised. A flowchart of the whole process is in Fig. 4.

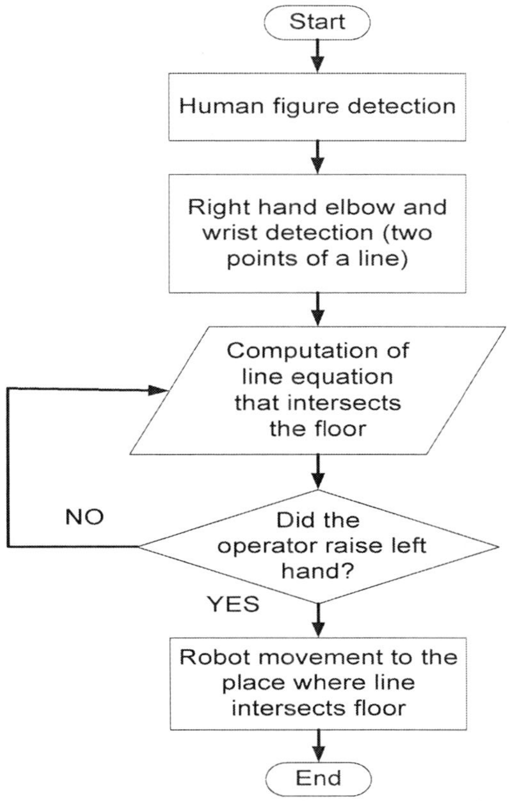


Fig. 4 State machine flowchart of the method

To compute necessary vectors standard methods of analytical geometry are used. For this, fundamental geometrical features need to be extracted (Fig. 5).

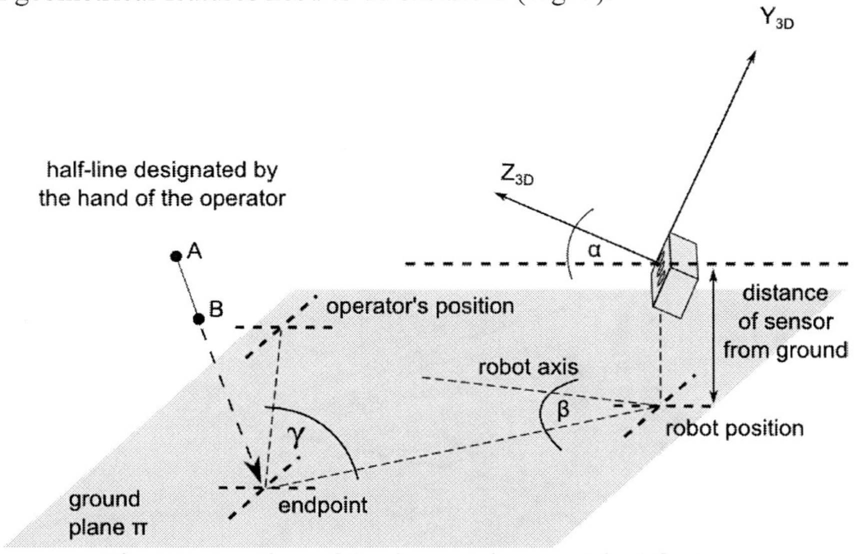


Fig. 5 Extraction of fundamental geometrical features

3D sensor mounted on robot is not centered and it is also rotated, therefore transformation between coordinate system of the robot (R) and Kinect (K) is needed (Fig. 6a, Fig 6b).

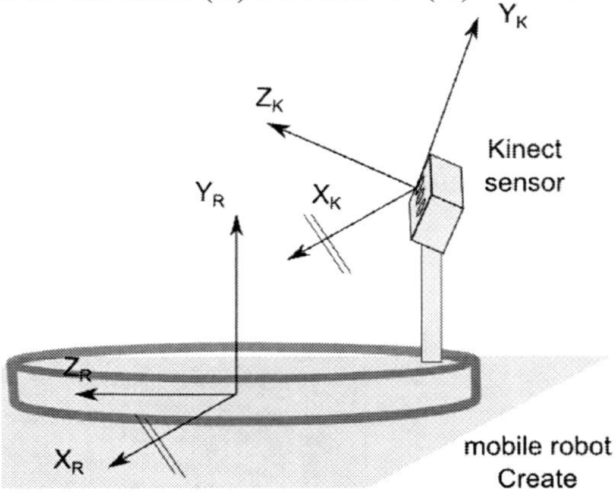

Fig. 6a Illustration of the two coordinate systems

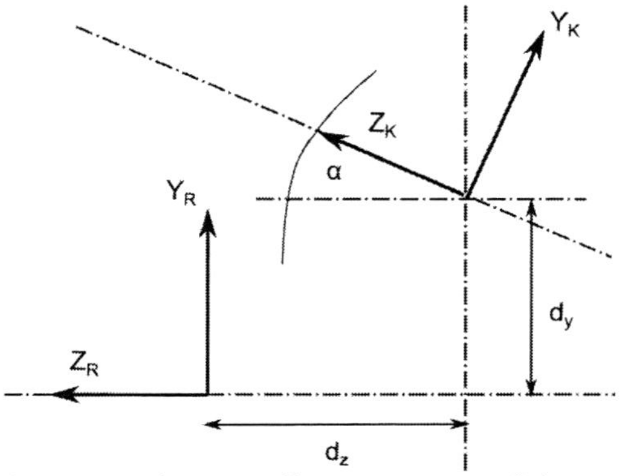

Fig. 6b Transformation between the coordinate system of the sensor and the coordinate system of the robot

Transformation between the two coordinate systems is defined by:

$$\begin{pmatrix} X_R \\ Y_R \\ Z_R \\ 1 \end{pmatrix} = \begin{pmatrix} 1 & 0 & 0 & 0 \\ 0 & \cos\alpha & \sin\alpha & d_y \\ 0 & -\sin\alpha & \cos\alpha & d_z \\ 0 & 0 & 0 & 1 \end{pmatrix} \cdot \begin{pmatrix} X_K \\ Y_K \\ Z_K \\ 1 \end{pmatrix}$$

Final vector features are illustrated in Fig. 7. Vector **v** is the basis for distance and rotations of robot movement.

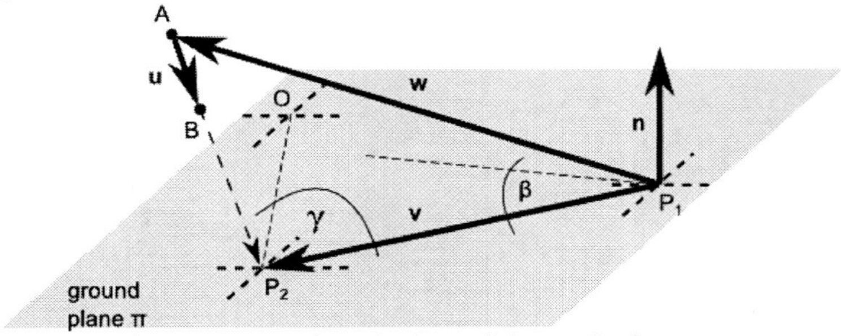

Fig. 7 Vector description of the main features

**Experiment**

To test this method experimental evaluation was done in laboratory environment. Several participants were stood in front of the robot and pointed to a desired spot in the ground. From experiment data probability density models were computed and evaluated. Illustration of experiment scenario is in Fig. exp2.

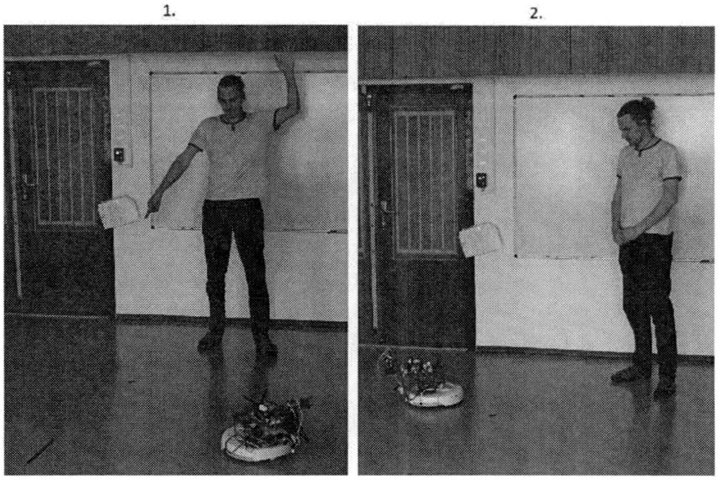


Fig. exp2 Laboratory experiment

## 3. Controlling a group of robots to perform a common task by gestures only

Basic idea of third algorithm is to use two hands of a human operator for every action issued to a group of robots. One hand of the operator is used to select a robot from the group. The other is used to control motion of the selected robot. Each robot is assigned a number – index that is used for its identification (ID). Number of raised fingers on operator's right hand matches the ID of the currently controlled robot. Therefore, if the operator wishes to control a hexapod robot with ID 3, he needs to raise 3 arbitrary fingers on his right hand. A healthy human has 5 fingers, so it is possible to pick from a group of 2-5 robots, theoretically 2-6 robots if a fist with no fingers raised was used, too.

Left hand is left to manage the motion control process itself. To define several motion types we compare the position of both hands in 3D space. The workspace of left hand is divided into 3 zones (Fig. 8). If both hands are close enough (overhead viewpoint) the hand controlling the robot is considered to be present in Zone 2. If it crosses a defined threshold in front of the right hand it's in Zone 3, if it's put behind right hand it's in Zone 1. This way 3 motion types for the robot can be selected, however this can be insufficient. That is why position of the thumb of left hand is also considered to increase the number of motion types.

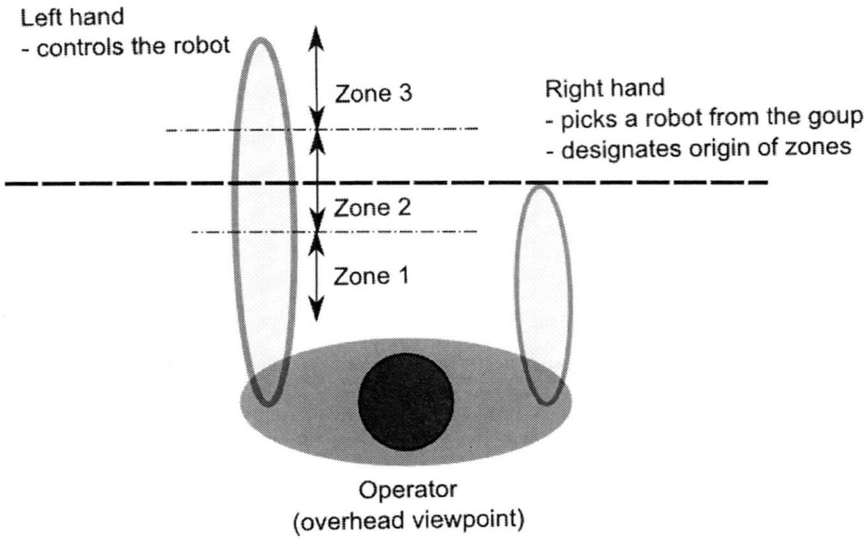

Fig. 8 Overhead viewpoint of the human operator

Fig. 9 shows an example of possible gestures of the two operator's hands. First case shows a robot with ID 3 rotating left (thumb position), second case shows a robot with ID 5 rotating right and last case a robot with ID 2 being in the state of halt.

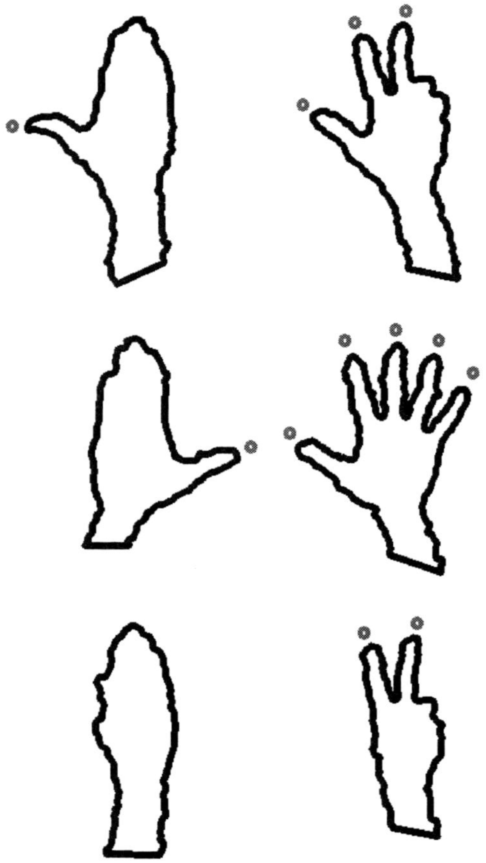
Fig. 9 Gesture examples

Adding the zone concept to examples from Fig. 9 leads to 9 motion types. All of them are listed in the following table.

| Left hand thumb position | Left hand zone | Robot index | Motion type |
|---|---|---|---|
| Left | 3 | 1-5 | Left forward circular motion |
| Right | 3 | 1-5 | Right forward circular motion |
| Middle | 3 | 1-5 | Forward motion |
| Left | 2 | 1-5 | Rotation left |
| Right | 2 | 1-5 | Rotation right |
| Middle | 2 | 1-5 | Stop |
| Left | 1 | 1-5 | Left backward circular motion |
| Right | 1 | 1-5 | Right backward circular motion |
| Middle | 1 | 1-5 | Backward motion |

To segment fingertips, contours of operator's hands are found. Within these contours all sharp vertexes are searched for and subsequently, fingers are classified. Fingertip vertexes are the sharpest. Therefore, to test the sharpness vectors **w** and **v** from nearby contour points are formed and their angle is computed based on simple dot product (Fig. 10).

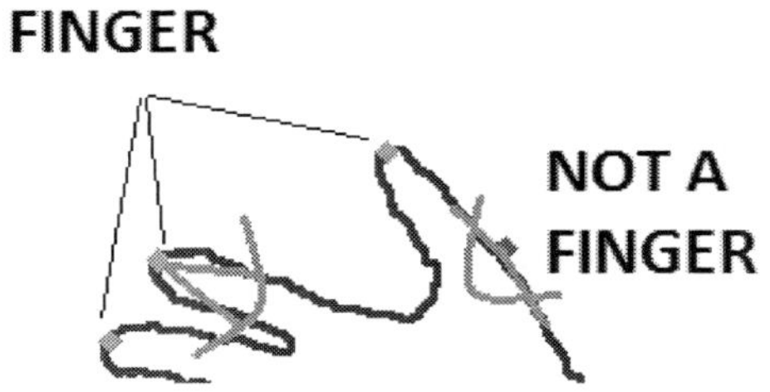


Fig. 10 Finger detection

**Experiment**

To test this method 3 different robots were used in laboratory environment. Hexapod, iRobot Create and a ball robot were controlled by one operator using wireless technology as shown in Fig. exp3.

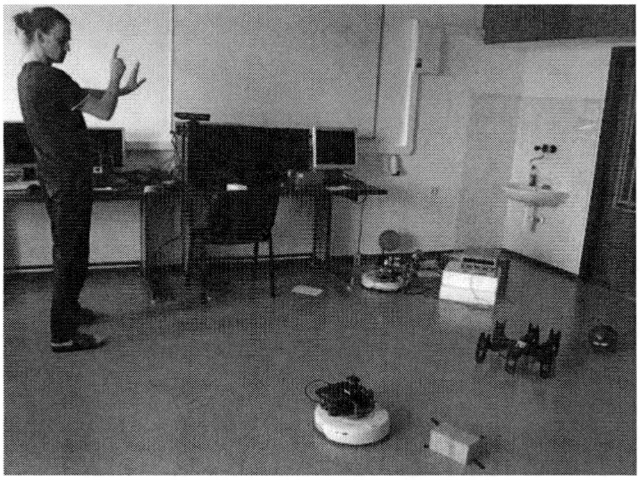

Fig. exp3 Experiment in laboratory environment

## Conclusions

We have successfully implemented three human-robot interaction methods. All of them are based on 3D vision. Methods were tested by human operators on real mobile robotic systems. We believe our research contributes to the more and more spreading field of HRI.

**Acknowledgments**
This work was supported by national grants Stimuly Micsrostep 0691/2015 and VEGA 1/0752/17.

## References

[1] G. Costante, E. Bellocchio, P. Valigi, E. Ricci, "Personalizing vision-based gestural interfaces for HRI with UAVs: A transfer learning approach," in Proc. IEEE International Conference on Intelligent Robots and Systems, 2014, pp. 3319 – 3326.
[2] V.M. Monajjemi, J. Wawerla, R. Vaughan, G. Mori, "HRI in the sky: Creating and commanding teams of UAVs with a vision-mediated gestural interface," in Proc. IEEE/RSJ International Conference on Intelligent Robots and Systems, 2013, pp. 617-623.
[3] C. P. Quintero, R. T. Fomena, A. Shademan, N. Wolleb, T. Dick, and M. Jagersand, "SEPO: Selecting by pointing as an intuitive human-robot command interface," in Proc. IEEE International Conference on Robotics and Automation, 2013, pp. 1166-1171.
[4] C. Høilund, V. Krüger, and T. B. Moeslund, "Evaluation of human body tracking system for gesture-based programming of industrial robots," in Proc. IEEE Conference on Industrial Electronics and Applications, 2012, pp. 477-480.
[5] A. Giusti, J. Nagi, L.M. Gambardella, S. Bonardi and G.A. Di Caro, "Human-swarm interaction through distributed cooperative gesture recognition," in Proc. 7th ACM/IEEE International Conference on Human-Robot Interaction (HRI), 2012, 401–401.
[6] J. Alonso-Mora, R. Siegwart, B. Perun and P. Beardsley, "Human-Robot Swarm Interaction for Entertainment," in Proc. 9th ACM/IEEE International Conference on Human-Robot Interaction (HRI), 2014, 98–98.
[7] Freedman et. al., "Depth Mapping Using Projected Patterns," U.S. Patent 2010/0118123 A1, May 13, 2010.
[8] A. Reichinger. (2011, April). Kinect Pattern Uncovered. [Online]. Available: http://azttm.wordpress.com/2011/04/03/kinect-pattern-uncovered/
[9] B. Klug. (2010, Dec.). Microsoft Kinect: The AnandTech Review. [Online]. Available: http://www.anandtech.com/show/4057/microsoft-kinect-the-anandtech-review/2
[10] K. Konolige, P. Mihelich. (2012, Dec.). Kinect Operation. [Online]. Available: http://wiki.ros.org/kinect_calibration/technical
[11] M. R. Andersen, T. Jensen, P. Lisouski, A. K. Mortensen, M. K. Hansen, T. Gregersen, and P. Ahrendt, "Kinect Depth Sensor Evaluation for Computer Vision Applications," Department of Engineering, Aarhus University. Denmark. 37 pp. - Technical report ECE-TR-6, Feb. 2012.
[12] J. Kramer, N. Burrus, F. Echtler, D. C. Herrera, and M. Parker, "Hardware," in Hacking the Kinect, Apress, 2012, pp. 11–13.
[13] J. Shotton, R. Girshick, A. Fitzgibbon, T. Sharp, M. Cook, M. Finocchio, R. Moore, P. Kohli, A. Criminisi, A. Kipman, A. Blake, "Efficient human pose estimation from single depth images," in Proc. IEEE Transactions on Pattern Analysis and Machine Intelligence, 2013, pp. 2821-2840.